STROUSE

Fundamentals of Building Construction

BAY CRANE
BAY CRANE

FUNDAMENTALS OF BUILDING CONSTRUCTION

MATERIALS AND METHODS

FIFTH EDITION

Edward Allen and **Joseph Iano**

JOHN WILEY & SONS, INC.

Frontispiece: World Trade Construction site, 2008. Photo by Andrew Watts.

This book is printed on acid-free paper. ♾

Published by John Wiley & Sons, Inc., Hoboken, New Jersey

Published simultaneously in Canada

Library of Congress Cataloging-in-Publication Data:
Allen, Edward, 1938-
Fundamentals of building construction : materials and methods /
Edward Allen, Joseph Iano. – 5th ed.
p. cm.
ISBN 978-0-470-07468-8 (cloth)
1. Building. 2. Building materials. I. Iano, Joseph. II. Title.
TH145.A417 2008
629–dc22

2008036198

Printed in the United States of America

10 9 8 7 6 5 4 3 2 1

CONTENTS

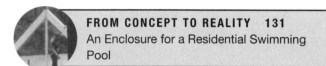

8 Brick Masonry 297

9 Stone and Concrete Masonry 337

10 Masonry Wall Construction 377

11 Steel Frame Construction 411

12 Light Gauge Steel Frame Construction 489

16 Roofing 651

17 Glass and Glazing 707

18 Windows and Doors 747

19 Designing Exterior Wall Systems 783

20 Cladding with Masonry and Concrete 809

PREFACE TO THE *FIFTH EDITION*

First published almost a quarter century ago, *Fundamentals of Building Construction: Materials and Methods*, now in its fifth edition, has wrought a revolution in construction education. It has been instrumental in making a previously unpopular area of study not merely palatable but vibrant and well liked. It has taken a body of knowledge once characterized as antithetical to design excellence and has made it widely recognized as being centrally relevant to good building design. It has replaced dry, unattractive books with a well designed, readable volume that students value and keep as a reference work. It was the first book in its field to be even-handed in its coverage and profusely and effectively illustrated throughout. It was the first to release the teacher from the burden of explaining everything in the subject, thereby freeing class time for discussions, case studies, field trips, and other enrichments.

Gaining a useful knowledge of the materials and methods of building construction is crucial and a necessity for the student of architecture, engineering, or construction, but it can be a daunting task. The field is huge, diverse, and complex, and it changes at such a rate that it seems impossible to ever master. This book has gained its preeminent status as an academic text in this field because of its logical organization, outstanding illustrations, clear writing, pleasing page layouts, and distinctive philosophy:

It is **integrative**, presenting a single narrative that interweaves issues of building science, materials science, legal constraints, and building craft so that the reader does not have to refer to separate parts of the book to make the connections between these issues. Building techniques are presented as whole working systems rather than component parts.

It is **selective** rather than comprehensive. This makes it easy and pleasant for the reader to gain a basic working knowledge that can later be expanded, without piling on so many facts and figures that the reader becomes confused or frightened away from wanting to learn about construction. Reading other texts was once like trying to drink from a fire hose; reading this one is like enjoying a carefully prepared meal.

It is **empowering** because it is structured around the process of designing and constructing buildings. The student of architecture will find that it features the design possibilities of the various materials and systems. Students interested in building or managing the construction process will find its organization around construction sequences to be invaluable.

The book is necessarily **complex** without being complicated. It avoids the dilemma of having to expand ad infinitum over time by presenting the basic construction systems, each in sufficient detail that the student is brought to an operational level of knowledge. It deals, as its subtitle indicates, with fundamentals.

We have made many changes in the book between the fourth and fifth editions. Chapter 1 includes new coverage of the role of the construction contractor in the process of making buildings. Included in this section are a discussion of different models and contractual arrangements for the delivery of construction services, the scheduling and management of construction, and evolving trends in the delivery of both design and construction services.

A new series of sidebars, *Building Enclosure Essentials*, is introduced. These treat topics critical to the performance of the building envelope, such as the flow of heat, air, and moisture through the exterior walls and roof.

Coverage of sustainable construction and green building rating systems has been updated and expanded, both within the body of the text and within the sidebars dedicated to considerations of sustainability for different materials which appear in almost every chapter.

The book tracks a number of trends that are discernable in the building industry: Sustainable design is becoming increasingly mainstream and growing in sophistication. New forms of contractual relationships between the owner, architect, and builder encourage more streamlined and cooperative design and building processes. The materials of construction themselves continue to evolve, with higher strength steel and concrete, high-performance glazing, waste reduction and materials recycling.

We continue to expand the book's use of photorealistic renderings. These figures play an important part in achieving the authors' goal of making complex building details readily understandable, while also appealing to the eye.

We continue to take maximum advantage of the ever-expanding World Wide Web. The text's encyclopedic details, along with an array of additional resources for both students and teachers, are readily available via its dedicated web site (www.wiley.com/constructioneducation). A test bank, PowerPoint slide show, review questions, Instructor's Manual, and more can be found there. Coauthor Joseph Iano's personal web site (www.ianosbackfill.com) provides an outlet for additional content and up-to-date coverage of new developments in the field. The selected list of web site addresses included in the reference section at the end of each chapter provides links to the other most relevant resources available on the Web, providing starting points for students' further explorations.

With this edition we have thoroughly updated references to contemporary building standards, codes, and practices, to ensure that the text remains the most current and accurate source of information on construction fundamentals available. Every chapter has been revised to reflect the latest versions of MasterFormat, the International Building Code, LEED, ANSI, and ASTM standards. Industry-specific standards, such as those from the American Concrete Institute (ACI), American Institute of Steel Construction (AISC), and American Architectural Manufacturers Association (AAMA), to name just a few, have been thoroughly updated in the appropriate sections of the text as well.

The updated and expanded companion *Exercises in Building Construction* and its answer key continue to provide a unique and invaluable tool for helping students to understand the real-world application of building construction knowledge to the design and construction of buildings.

Despite the extensive scope of this latest revision, every change has been carefully reviewed by the authors and an independent board to be sure that the text remains up-to-date, accurate, and consistent with its original principles. In this way, as the book continues to change over time, the essential qualities that make it an educational success are preserved and strengthened.

The authors' special thanks go to the talented Lon R. Grohs, producer of the text's stunning photorealistic illustrations, and in this latest edition, of the cover art as well. We are also grateful to the many photographers and organizations who have furnished new information and illustrations.

The people of John Wiley & Sons, Inc. continue, as always, to demonstrate great professionalism. Amanda L. Miller, Vice President and Publisher, has for many years been a source of wisdom and support. Paul Drougas, Editor, has been invaluable for his industry knowledge, patience, and sense of humor. He is a true friend. Lauren Olesky, Assistant Developmental Editor, was reliable and helpful through all stages of this revision. Donna Conte, Senior Production Editor, continues, as in previous revisions, to oversee the most difficult task of managing production and schedules with grace and perseverance.

We especially offer our thanks to the many teachers, students, and professionals who have purchased and used this work. Your satisfaction is our greatest reward, your loyalty is greatly appreciated, and your comments are always welcome!

Joseph Iano dedicates this *Fifth Edition* to Lesley, Allen, Paul, and Ethan.

E.A., South Natick, Massachusetts
J.I., Seattle, Washington

Fundamentals of Building Construction

1

MAKING BUILDINGS

An ironworker connects a steel wide-flange beam to a column.
(Courtesy of Bethlehem Steel Company)

We build because most human activities cannot take place outdoors. We need shelter from sun, wind, rain, and snow. We need dry, level platforms for our activities. Often we need to stack these platforms to multiply available space. On these platforms, and within our shelter, we need air that is warmer or cooler, more or less humid, than outdoors. We need less light by day, and more by night, than is offered by the natural world. We need services that provide energy, communications, and water and dispose of wastes. So, we gather materials and assemble them into the constructions we call buildings in an attempt to satisfy these needs.

Learning to Build

Throughout this book many alternative ways of building are described: different structural systems, different systems of enclosure, and different systems of interior finish. Each system has characteristics that distinguish it from the alternatives. Sometimes a system is distinguished chiefly by its visual qualities, as one might acknowledge in choosing one type of granite over another, one color of paint over another, or one tile pattern over another. However, visual distinctions can extend beyond surface qualities; a designer may prefer the massive appearance of a masonry bearing wall building to the slender look of an exposed steel frame on one project, yet would choose the steel for another building whose situation is different. Again, one may choose for purely functional reasons, as in selecting terrazzo flooring that is highly durable and resistant to water instead of more vulnerable carpet or wood in a restaurant kitchen. One could choose on purely technical grounds, as, for example, in electing to posttension a long concrete beam for greater stiffness rather than rely on conventional steel reinforcing. A designer is often forced into a particular choice by some of the legal constraints identified later in this chapter. A choice is often influenced by considerations of environmental sustainability. And frequently the selection is made on purely economic grounds. The economic criterion can mean any of several things: Sometimes one system is chosen over another because its first cost is less; sometimes the entire life-cycle costs of competing systems are compared by means of formulas that include first cost, maintenance cost, energy consumption cost, the useful lifetime and replacement cost of the system, and interest rates on invested money; and, finally, a system may be chosen because there is keen competition among local suppliers and/or installers that keeps the cost of that system at the lowest possible level. This is often a reason to specify a very standard type of roofing material, for example, that can be furnished and installed by any of a number of companies, instead of a newer system that is theoretically better from a functional standpoint but can only be furnished by a single company that has the special equipment and skills required to install it.

One cannot gain all the knowledge needed to make such decisions from a textbook. It is incumbent upon the reader to go far beyond what can be presented here—to other books, to catalogs, to trade publications, to professional periodicals, and especially to the design office, the workshop, and the building site. There is no other way to gain much of the required information and experience than to get involved in the art and business of building. One must learn how materials feel in the hand; how they look in a building; how they are manufactured, worked, and put in place; how they perform in service; how they deteriorate with time. One must become familiar with the people and organizations that produce buildings—the architects, engineers, materials suppliers, contractors, subcontractors, workers, inspectors, managers, and building owners—and learn to understand their respective methods, problems, and points of view. In the meantime, this long and hopefully enjoyable process of education in the materials and methods of building construction can begin with the information presented in this textbook.

Go into the field where you can see the machines and methods at work that make the modern buildings, or stay in construction direct and simple until you can work naturally into building-design from the nature of construction.

Frank Lloyd Wright, *To the Young Man in Architecture*, 1931

Sustainability

In constructing and occupying buildings, we expend vast quantities of the earth's resources and generate a significant portion of the earth's environmental pollution: The U.S. Green Building Council reported in 2008 that buildings account for 30 to 40 percent of the world's energy use and associated greenhouse gas emissions. Construction and operation of buildings in the United States accounted for more than one-third of this country's total energy use and the consumption of more than two-thirds of its electricity, 30 percent of its raw materials, a quarter of its harvested wood, and 12 percent of its fresh water. Building construction and operation is responsible for nearly half of this country's total greenhouse gas emissions and close to a third of its solid waste stream. Buildings are also significant emitters of particulates and other air pollutants. In short, building construction and operation cause many forms of environmental degradation that place an increasing burden on the earth's resources and jeopardize the future of the building industry and societal health and welfare.

Sustainability may be defined as **meeting the needs of the present generation without compromising the ability of future generations to meet their needs.** By consuming irreplaceable fossil fuels and other nonrenewable resources, by building in sprawling urban patterns that cover extensive areas of prime agricultural land, by using destructive forestry practices that degrade natural ecosystems, by allowing topsoil to be eroded by wind and water, and by generating substances that pollute water, soil, and air, we have been building in a manner that will make it increasingly difficult for our children and grandchildren to meet their needs for communities, buildings, and healthy lives.

On the other hand, if we reduce building energy usage and utilize sunlight and wind as energy sources for our buildings, we reduce depletion of fossil fuels. If we reuse existing buildings imaginatively and arrange our new buildings in compact patterns on land of marginal value, we minimize the waste of valuable, productive land. If we harvest wood from forests that are managed in such a way that they can supply wood at a sustained level for the foreseeable future, we maintain wood construction as a viable option for centuries to come and protect the ecosystems that these forests support. If we protect soil and water through sound design and construction practices, we retain these resources for our successors. If we systematically reduce the various forms of pollution emitted in the processes of constructing and operating buildings, we keep the future environment cleaner. And as the industry becomes more experienced and committed to designing and building sustainably, it becomes increasingly possible to do these things with little or no increase in construction cost while creating buildings that are less expensive to operate and more healthful for their occupants for decades to come.

Realization of these goals is dependent on our awareness of the environmental problems created by building activities, knowledge of how to overcome these problems, and skill in designing and constructing buildings that harness this knowledge. While the practice of sustainable design and construction, also called *green building*, remains a relatively recent development in the design and construction industry, its acceptance and support continue to broaden among public agencies, private developers, building operators and users, architectural and engineering firms, contractors, and materials producers. With each passing year, green building techniques are becoming less a design specialty and more a part of mainstream practice.

The Building Life Cycle

Sustainability must be addressed on a life-cycle basis, from the origins of the materials for a building, through the manufacture and installation of these materials and their useful lifetime in the building, to their eventual disposal when the building's life is ended. Each step in this so-called *cradle-to-grave* cycle raises questions of sustainability.

Origin and Manufacture of Materials for a Building

Are the raw materials for a building plentiful or rare? Are they renewable or nonrenewable? How much of the content of a material is recycled from other uses? How much *embodied energy* is expended in obtaining and manufacturing the material, and how much water? What pollutants are discharged into air, water, and soil as a result of these acts? What wastes are created? Can these wastes be converted to useful products?

Construction of the Building

How much energy is expended in transporting a material from its origins to the building site, and what pollutants are generated? How much energy and water are consumed on the building site to put the material in place? What pollutants are associated with the installation of this material in the building? What wastes are generated, and how much of them can be recycled?

Use and Maintenance of the Building

How much energy and water does the building use over its lifetime as a consequence of the materials used in its construction and finishes? What problems of indoor air quality are caused by these materials? How much maintenance do these materials require, and how long will they last? Can they be recycled? How much energy and time are consumed in maintaining these materials? Does this maintenance involve use of toxic chemicals?

Demolition of the Building

What planning and design strategies can be used to extend the useful life of buildings, thereby forestalling resource-intensive demolition and construction of new buildings? When demolition is inevitable, how will the building be demolished and disposed of, and will any part of this process cause pollution of air, water, or soil? Can demolished materials be recycled into new construction or diverted for other uses rather than disposed of as wastes?

One model for sustainable design is nature itself. Nature works in cyclical processes that are self-sustaining and waste nothing. More and more building professionals are learning to create buildings that work more nearly as nature does, helping to leave to our descendants a stock of healthful buildings, a sustainable supply of natural resources, and a clean environment that will enable them to live comfortably and responsibly and to

FIGURE 1.1
The LEED-NC 2009 Project Scorecard. The document shown here was in draft status at the time of this publication. See the U.S. Green Building Council web site for the most current version of this document. ***(Courtesy of U.S. Green Building Council)***

LEED for New Construction and Major Renovation 2009 Project Scorecard

Project Name:
Project Address:

Yes	?	No		Sustainable Sites	26 Points
Y			Prereq 1	**Construction Activity Pollution Prevention**	Required
			Credit 1	**Site Selection**	1
			Credit 2	**Development Density & Community Connectivity**	5
			Credit 3	**Brownfield Redevelopment**	1
			Credit 4.1	**Alternative Transportation,** Public Transportation Access	6
			Credit 4.2	**Alternative Transportation,** Bicycle Storage & Changing Rooms	1
			Credit 4.3	**Alternative Transportation,** Low-Emitting & Fuel-Efficient Vehicles	3
			Credit 4.4	**Alternative Transportation,** Parking Capacity	2
			Credit 5.1	**Site Development,** Protect or Restore Habitat	1
			Credit 5.2	**Site Development,** Maximize Open Space	1
			Credit 6.1	**Stormwater Design,** Quantity Control	1
			Credit 6.2	**Stormwater Design,** Quality Control	1
			Credit 7.1	**Heat Island Effect,** Non-Roof	1
			Credit 7.2	**Heat Island Effect,** Roof	1
			Credit 8	**Light Pollution Reduction**	1

Yes	?	No		Water Efficiency	10 Points
			Prereq 1	**Water Use Reduction,** 20% Reduction	Required
			Credit 1.1	**Water Efficient Landscaping,** Reduce by 50%	2
			Credit 1.2	**Water Efficient Landscaping,** No Potable Use or No Irrigation	2
			Credit 2	**Innovative Wastewater Technologies**	2
			Credit 3.1	**Water Use Reduction,** 30% Reduction	2
			Credit 3.2	**Water Use Reduction,** 40% Reduction	2

Yes	?	No		Energy & Atmosphere	35 Points
Y			Prereq 1	**Fundamental Commissioning of the Building Energy Systems**	Required
Y			Prereq 2	**Minimum Energy Performance:** 10% New Bldgs or 5% Existing Bldg Renovations	Required
Y			Prereq 3	**Fundamental Refrigerant Management**	Required
			Credit 1	**Optimize Energy Performance**	1 to 19
				12% New Buildings or 8% Existing Building Renovations	1
				16% New Buildings or 12% Existing Building Renovations	3
				20% New Buildings or 16% Existing Building Renovations	5
				24% New Buildings or 20% Existing Building Renovations	7
				28% New Buildings or 24% Existing Building Renovations	9
				32% New Buildings or 28% Existing Building Renovations	11
				36% New Buildings or 32% Existing Building Renovations	13
				40% New Buildings or 36% Existing Building Renovations	15
				44% New Buildings or 40% Existing Building Renovations	17
				48% New Buildings or 44% Existing Building Renovations	19
			Credit 2	**On-Site Renewable Energy**	1 to 7
				1% Renewable Energy	1
				5% Renewable Energy	3
				9% Renewable Energy	5
				13% Renewable Energy	7
			Credit 3	**Enhanced Commissioning**	2
			Credit 4	**Enhanced Refrigerant Management**	2
			Credit 5	**Measurement & Verification**	3
			Credit 6	**Green Power**	2

Yes ? No

pass these riches on to their descendants in a never-ending succession.

Assessing Green Buildings

In the United States, the most widely adopted method for rating the environmental sustainability of a building's design and construction is the U.S. Green Building Council's Leadership in Energy and Environmental Design, or *LEED*™, rating system. LEED for New Construction and Major Renovation projects, termed LEED-NC, groups sustainability goals into categories including site selection and development, efficiency in water use, reductions in energy consumption and in the production of atmospheric ozone-depleting gases, minimizing construction waste and the depletion of nonrenewable resources, improving the quality of the indoor environment, and encouraging innovation in sustainable design and construction practices (Figure 1.1). Within each category are specific *credits* that contribute points toward a building's overall assessment of sustainability. Depending on the total number of points accumulated, four levels of sustainable design are recognized, including, in order of increasing performance, Certified, Silver, Gold, and Platinum.

The process of achieving LEED certification for a proposed new building begins at the earliest stages of project conception, continues throughout the design and construction of the project, and involves the combined efforts of the owner, designer, and builder. During this process, the successful achievement of individual credits is documented and submitted to the

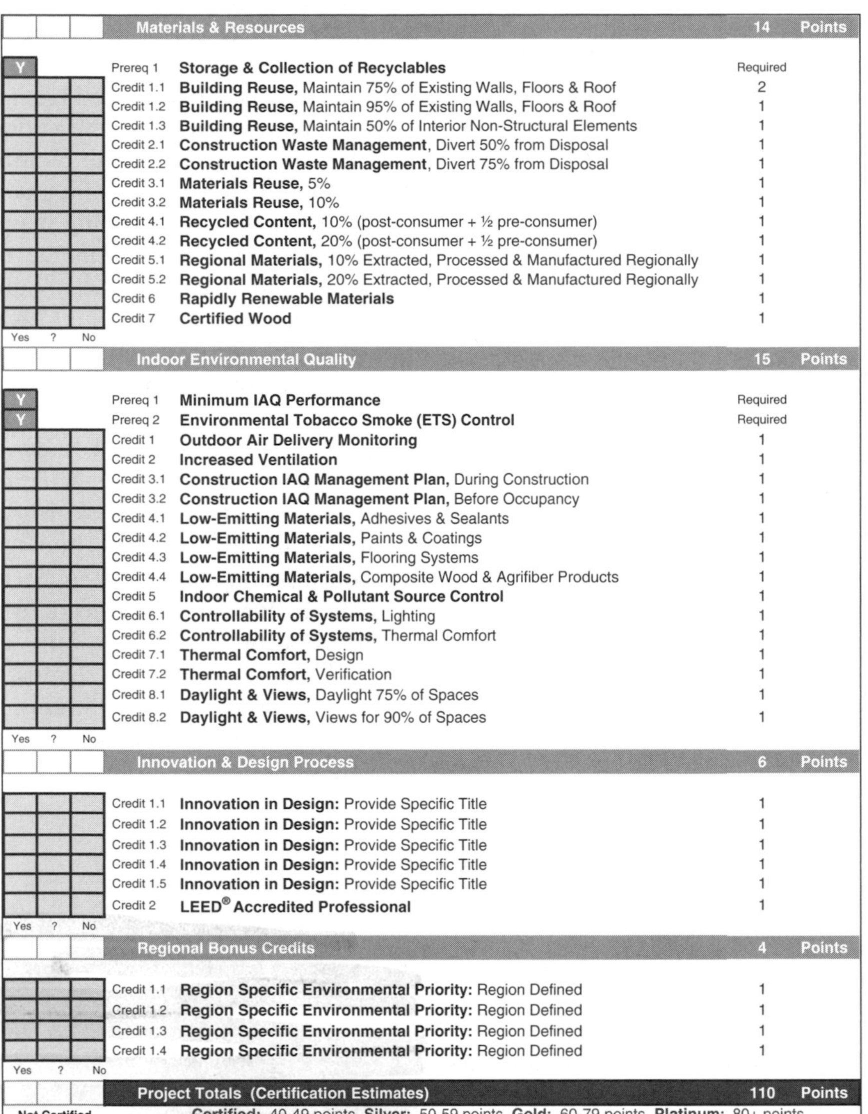

Yes	?	No	Materials & Resources		14 Points
Y			Prereq 1	**Storage & Collection of Recyclables**	Required
			Credit 1.1	**Building Reuse,** Maintain 75% of Existing Walls, Floors & Roof	2
			Credit 1.2	**Building Reuse,** Maintain 95% of Existing Walls, Floors & Roof	1
			Credit 1.3	**Building Reuse,** Maintain 50% of Interior Non-Structural Elements	1
			Credit 2.1	**Construction Waste Management,** Divert 50% from Disposal	1
			Credit 2.2	**Construction Waste Management,** Divert 75% from Disposal	1
			Credit 3.1	**Materials Reuse,** 5%	1
			Credit 3.2	**Materials Reuse,** 10%	1
			Credit 4.1	**Recycled Content,** 10% (post-consumer + ½ pre-consumer)	1
			Credit 4.2	**Recycled Content,** 20% (post-consumer + ½ pre-consumer)	1
			Credit 5.1	**Regional Materials,** 10% Extracted, Processed & Manufactured Regionally	1
			Credit 5.2	**Regional Materials,** 20% Extracted, Processed & Manufactured Regionally	1
			Credit 6	**Rapidly Renewable Materials**	1
			Credit 7	**Certified Wood**	1
Yes	?	No	Indoor Environmental Quality		15 Points
Y			Prereq 1	**Minimum IAQ Performance**	Required
Y			Prereq 2	**Environmental Tobacco Smoke (ETS) Control**	Required
			Credit 1	**Outdoor Air Delivery Monitoring**	1
			Credit 2	**Increased Ventilation**	1
			Credit 3.1	**Construction IAQ Management Plan,** During Construction	1
			Credit 3.2	**Construction IAQ Management Plan,** Before Occupancy	1
			Credit 4.1	**Low-Emitting Materials,** Adhesives & Sealants	1
			Credit 4.2	**Low-Emitting Materials,** Paints & Coatings	1
			Credit 4.3	**Low-Emitting Materials,** Flooring Systems	1
			Credit 4.4	**Low-Emitting Materials,** Composite Wood & Agrifiber Products	1
			Credit 5	**Indoor Chemical & Pollutant Source Control**	1
			Credit 6.1	**Controllability of Systems,** Lighting	1
			Credit 6.2	**Controllability of Systems,** Thermal Comfort	1
			Credit 7.1	**Thermal Comfort,** Design	1
			Credit 7.2	**Thermal Comfort,** Verification	1
			Credit 8.1	**Daylight & Views,** Daylight 75% of Spaces	1
			Credit 8.2	**Daylight & Views,** Views for 90% of Spaces	1
Yes	?	No	Innovation & Design Process		6 Points
			Credit 1.1	**Innovation in Design:** Provide Specific Title	1
			Credit 1.2	**Innovation in Design:** Provide Specific Title	1
			Credit 1.3	**Innovation in Design:** Provide Specific Title	1
			Credit 1.4	**Innovation in Design:** Provide Specific Title	1
			Credit 1.5	**Innovation in Design:** Provide Specific Title	1
			Credit 2	**LEED® Accredited Professional**	1
Yes	?	No	Regional Bonus Credits		4 Points
			Credit 1.1	**Region Specific Environmental Priority:** Region Defined	1
			Credit 1.2	**Region Specific Environmental Priority:** Region Defined	1
			Credit 1.3	**Region Specific Environmental Priority:** Region Defined	1
			Credit 1.4	**Region Specific Environmental Priority:** Region Defined	1
			Project Totals (Certification Estimates)		110 Points
Not Certified				**Certified:** 40-49 points **Silver:** 50-59 points **Gold:** 60-79 points **Platinum:** 80+ points	

Green Building Council, which then makes the final certification of the project's LEED compliance.

The U.S. Green Building Council continues to refine and improve upon LEED-NC and is expanding its family of rating systems to include existing buildings (LEED-EB), commercial interiors (LEED-CI), building core and shell construction (LEED-CS), homes (LEED-H), and other categories of construction and development. Through international sister organizations, LEED is being implemented in Canada and other countries. Other green building programs, such as the Green Building Initiative's *Green Globes,* the National Association of Home Builders' *Green Home Building Guidelines,* and the International Code Council and National Association of Home Builders' jointly developed *National Green Building Standard,* offer alternative assessment schemes.

Some green building efforts focus more narrowly on reducing building energy consumption, a measure of building performance that frequently correlates closely with the generation of greenhouse gas emissions and global warming trends. The American Society of Heating, Refrigerating and Air-Conditioning Engineers' *Advanced Energy Design Guides* and the U.S. Environmental Protection Agency's *Energy Star* program both set goals for reductions in energy consumption in new buildings that exceed current national standards. These standards can be applied either as stand-alone programs or as part of a more comprehensive effort to achieve certification through LEED or some other green building assessment program.

Buildings can also be designed with the goal of zero energy use or carbon neutrality. A *net zero energy* building is one that consumes no more energy than it produces, usually when measured on an annual basis to account for seasonal differences in building energy consumption and on-site energy production. Net zero energy use can be achieved using current technology combining on-site renewable energy generation (such as wind or solar power), passive heating and cooling strategies, a thermally efficient building enclosure, and highly efficient mechanical systems and appliances.

A *carbon-neutral* building is one that causes no net increase in the emission of carbon dioxide, the most prevalent atmospheric greenhouse gas. If emissions due only to building operation are considered, the calculation is similar to that for a net zero energy building. If, however, the embodied carbon in the building's full life cycle—from materials extraction and manufacturing, through building construction and operations, to demolition, disposal, and recycling—is considered, the calculation becomes more complex. Carbon-neutral calculations may also consider the site on which the building resides. For example, what is the carbon footprint of a fully developed building site, including both its buildings and unbuilt areas, in comparison to that of the site prior to construction or in comparison to its natural state prior to human development of any kind? Another possible consideration is, what role, if any, should *carbon offsetting* (funding of off-site activities that reduce global carbon emissions, such as planting of trees), play in such calculations? Questions such as these and the concepts of sustainability and how they relate to building construction will continue to evolve for the foreseeable future.

Considerations of sustainability are included throughout this book. In addition, a sidebar in nearly every chapter describes the major issues of sustainability related to the materials and methods discussed in that chapter. These will be helpful in weighing the environmental costs of one material against those of another, and in learning how to build in such a way that we preserve for future generations the ability to meet their building needs in a reasonable and economical manner. For more information on organizations whose mission is to raise our awareness and provide the knowledge that we need to build sustainably, see the references listed at the end of this chapter.

The Work of the Design Professional: Choosing Building Systems

A building begins as an idea in someone's mind, a desire for new and ample accommodations for a family, many families, an organization, or an enterprise. For any but the smallest buildings, the next step for the owner of the prospective building is to engage, either directly or through a hired construction manager, the services of building design professionals. An architect helps to organize the owner's ideas about the new building, develops the form of the building, and assembles a group of engineering specialists to help work out concepts and details of foundations, structural support, and mechanical, electrical, and communications services.

. . . the architect should have construction at least as much at his fingers' ends as a thinker his grammar.

Le Corbusier, *Towards a New Architecture*, 1927

This team of designers, working with the owner, then develops the scheme for the building in progressively finer degrees of detail. Drawings and written specifications are produced by the architect–engineer team to document how the building is to be made and of what. The drawings and specifications are submitted to the local government building authorities, where they are checked for conformance with zoning ordinances and building codes before a permit is issued to build. A general contractor is selected, either by negotiation or by competitive bidding, who then hires subcontractors to carry out many specialized portions of the work. Once construction begins, the general contractor oversees the construction process while the building inspector, architect, and engineering consultants observe the work at frequent intervals to be sure that it is carried out according to plan. Finally, construction is finished, the building is made ready for occupancy, and that original idea, often initiated years earlier, is realized.

Although a building begins as an abstraction, it is built in a world of material realities. The designers of a building—the architects and engineers—work constantly from a knowledge of what is possible and what is not. They are able, on the one hand, to employ a seemingly limitless palette of building materials and any of a number of structural systems to produce a building of almost any desired form and texture. On the other hand, they are inescapably bound by certain physical limitations: how much land there is with which to work; how heavy a building the soil can support; how long a structural span is feasible; what sorts of materials will perform well in the given environment. They are also constrained by a construction budget and by a complex web of legal restrictions.

Those who work in the building professions need a broad understanding of many things, including people and culture, the environment, the physical principles by which buildings work, the technologies available for utilization in buildings, the legal restrictions on building design and use, the economics of building, and the contractual and practical arrangements under which buildings are constructed. This book is concerned primarily with the technologies of construction—what the materials are, how they are produced, what their properties are, and how they are crafted into buildings. These must be studied, however, with reference to many other factors that bear on the design of buildings, some of which require explanation here.

Zoning Ordinances

The legal restrictions on buildings begin with local *zoning ordinances*, which govern the types of activities that may take place on a given piece of land, how much of the land may be covered by buildings, how far buildings must be set back from adjacent property lines, how many parking spaces must be provided, how large a total floor area may be constructed, and how tall the buildings may be. In larger cities, zoning ordinances may include fire zones with special fire-protection requirements, neighborhood enterprise districts with economic incentives for new construction or revitalization of existing buildings, or other special conditions.

Building Codes

In addition to its zoning ordinances, local governments regulate building activity by means of *building codes*. Building codes protect public health and safety by setting minimum standards for construction quality, structural integrity, durability, livability, accessibility, and especially fire safety.

Most building codes in North America are based on one of several *model building codes*, standardized codes that local jurisdictions may adopt for their own use as an alternative to writing their own. In Canada, the *National Building Code of Canada* is published by the Canadian Commission on Building and Fire Codes. It is the basis for most of that country's provincial and municipal building codes. In the United States, the *International Building Code*® is the predominant model code. This code is published by the International Code Council, a private, nonprofit organization whose membership consists of local code officials from throughout the country. It is the basis for most U.S. building codes enacted at the state, county, and municipal levels. The International Building Code (IBC) is the first unified model building code in U.S. history. First published in March 2000, it was a welcome consolidation of a number of previous competing regional model codes.

Building-code-related information in this book is based on the IBC. The IBC begins by defining *occupancy groups* for buildings as follows:

- Groups A-1 through A-5 are public Assembly occupancies: theaters, auditoriums, lecture halls, nightclubs, restaurants, houses of worship, libraries, museums, sports arenas, and so on.
- Group B is Business occupancies: banks, administrative offices, higher-education facilities, post offices, banks, professional offices, and the like.
- Group E is Educational occupancies: schools for grades K through 12 and day-care facilities.
- Groups F-1 and F-2 comprise industrial processes using moderate-flammability or noncombustible materials, respectively.
- Groups H-1 through H-5 include various types of High Hazard occupancies in which toxic, corrosive, highly flammable, or explosive materials are present.
- Groups I-1 through I-4 are Institutional occupancies in which occupants under the care of others may not be able to save themselves during a fire or other building emergency, such as health care facilities, custodial care facilities, and prisons.
- Group M is Mercantile occupancies: stores, markets, service stations, and salesrooms.
- Groups R-1 through R-4 are Residential occupancies, including apartment buildings, dormitories, fraternity and sorority houses, hotels, one- and two-family dwellings, and assisted-living facilities.
- Groups S-1 and S-2 include buildings for Storage of moderate- and low-hazard materials, respectively.
- Group U is Utility buildings. It comprises agricultural buildings, carports, greenhouses, sheds, stables, fences, tanks, towers, and other secondary buildings.

The IBC's purpose in establishing occupancy groups is to distinguish various degrees of need for safety in buildings. A hospital, in which many patients are bedridden and cannot escape a fire without assistance from others, must be built to a higher standard of safety than a hotel or motel. A warehouse storing noncombustible masonry materials, which is likely to be occupied by only a few people, all of them able-bodied, can be constructed to a lower standard than a large retail mall building, which will house large quantities of combustible materials and will be occupied by many users varying in age and physical capability. An elementary school requires more protection for its occupants than a university building. A theater needs special egress provisions to allow its many patrons to escape quickly, without stampeding, in an emergency.

These definitions of occupancy groups are followed by a set of definitions of *construction types*. At the head of this list is Type I construction, made with highly fire-resistant, noncombustible materials. At the foot of it is Type V construction, which is built from combustible wood framing—the least fire-resistant of all construction types. In between are Types II, III, and IV, with levels of resistance to fire falling between these two extremes.

With occupancy groups and construction types defined, the IBC proceeds to match the two, stating which occupancy groups may be housed in which types of construction, and under what limitations of building height and area. Figure 1.2 is reproduced from the IBC. This table gives values for the maximum building height, in both feet and number of stories above grade, and the maximum area per floor for every possible combination of occupancy group and construction type. Once these base values are adjusted according to other provisions of the code, the maximum permitted size for a building of any particular use and type of construction can be determined.

This table concentrates a great deal of important information into a very small space. A designer may refer to it with a particular building type in mind and find out what types of construction will be permitted and what shape the building may take. Consider, for example, an office building. Under the IBC, a building of this type belongs to Occupancy Group B, Business. Reading across the table from left to right, we find immediately that this building may be built to any desired size, without limit, using Type I-A construction.

Type I-A construction is defined in the IBC as consisting of only noncombustible materials—masonry, concrete, or steel, for example, but not wood—and meeting minimum requirements for resistance to the heat of fire. Looking at the upper table in Figure 1.3, also reproduced from the IBC, we find under Type I-A construction a listing of the required *fire resistance ratings*, measured in hours, for various parts of our proposed office

TABLE 503
ALLOWABLE HEIGHT AND BUILDING AREAS[a]
Height limitations shown as stories and feet above grade plane.
Area limitations as determined by the definition of "Area, building," per story

		TYPE OF CONSTRUCTION								
		TYPE I		TYPE II		TYPE III		TYPE IV	TYPE V	
		A	B	A	B	A	B	HT	A	B
	HGT(feet)									
GROUP	HGT(S)	UL	160	65	55	65	55	65	50	40
A-1	S	UL	5	3	2	3	2	3	2	1
	A	UL	UL	15,500	8,500	14,000	8,500	15,000	11,500	5,500
A-2	S	UL	11	3	2	3	2	3	2	1
	A	UL	UL	15,500	9,500	14,000	9,500	15,000	11,500	6,000
A-3	S	UL	11	3	2	3	2	3	2	1
	A	UL	UL	15,500	9,500	14,000	9,500	15,000	11,500	6,000
A-4	S	UL	11	3	2	3	2	3	2	1
	A	UL	UL	15,500	9,500	14,000	9,500	15,000	11,500	6,000
A-5	S	UL	UL	UL	UL	UL	UL	UL	UL	UL
	A	UL	UL	UL	UL	UL	UL	UL	UL	UL
B	S	UL	11	5	4	5	4	5	3	2
	A	UL	UL	37,500	23,000	28,500	19,000	36,000	18,000	9,000
E	S	UL	5	3	2	3	2	3	1	1
	A	UL	UL	26,500	14,500	23,500	14,500	25,500	18,500	9,500
F-1	S	UL	11	4	2	3	2	4	2	1
	A	UL	UL	25,000	15,500	19,000	12,000	33,500	14,000	8,500
F-2	S	UL	11	5	3	4	3	5	3	2
	A	UL	UL	37,500	23,000	28,500	18,000	50,500	21,000	13,000
H-1	S	1	1	1	1	1	1	1	1	NP
	A	21,000	16,500	11,000	7,000	9,500	7,000	10,500	7,500	NP
H-2[d]	S	UL	3	2	1	2	1	2	1	1
	A	21,000	16,500	11,000	7,000	9,500	7,000	10,500	7,500	3,000
H-3[d]	S	UL	6	4	2	4	2	4	2	1
	A	UL	60,000	26,500	14,000	17,500	13,000	25,500	10,000	5,000
H-4	S	UL	7	5	3	5	3	5	3	2
	A	UL	UL	37,500	17,500	28,500	17,500	36,000	18,000	6,500
H-5	S	4	4	3	3	3	3	3	3	2
	A	UL	UL	37,500	23,000	28,500	19,000	36,000	18,000	9,000
I-1	S	UL	9	4	3	4	3	4	3	2
	A	UL	55,000	19,000	10,000	16,500	10,000	18,000	10,500	4,500
I-2	S	UL	4	2	1	1	NP	1	1	NP
	A	UL	UL	15,000	11,000	12,000	NP	12,000	9,500	NP
I-3	S	UL	4	2	1	2	1	2	2	1
	A	UL	UL	15,000	10,000	10,500	7,500	12,000	7,500	5,000
I-4	S	UL	5	3	2	3	2	3	1	1
	A	UL	60,500	26,500	13,000	23,500	13,000	25,500	18,500	9,000
M	S	UL	11	4	4	4	4	4	3	1
	A	UL	UL	21,500	12,500	18,500	12,500	20,500	14,000	9,000
R-1	S	UL	11	4	4	4	4	4	3	2
	A	UL	UL	24,000	16,000	24,000	16,000	20,500	12,000	7,000
R-2	S	UL	11	4	4	4	4	4	3	2
	A	UL	UL	24,000	16,000	24,000	16,000	20,500	12,000	7,000
R-3	S	UL	11	4	4	4	4	4	3	3
	A	UL	UL	UL	UL	UL	UL	UL	UL	UL
R-4	S	UL	11	4	4	4	4	4	3	2
	A	UL	UL	24,000	16,000	24,000	16,000	20,500	12,000	7,000
S-1	S	UL	11	4	3	3	3	4	3	1
	A	UL	48,000	26,000	17,500	26,000	17,500	25,500	14,000	9,000
S-2[b, c]	S	UL	11	5	4	4	4	5	4	2
	A	UL	79,000	39,000	26,000	39,000	26,000	38,500	21,000	13,500
U[c]	S	UL	5	4	2	3	2	4	2	1
	A	UL	35,500	19,000	8,500	14,000	8,500	18,000	9,000	5,500

For SI: 1 foot = 304.8 mm, 1 square foot = 0.0929 m^2.

UL = Unlimited, NP = Not permitted.

a. See the following sections for general exceptions to Table 503:

1. Section 504.2, Allowable height increase due to automatic sprinkler system installation.
2. Section 506.2, Allowable area increase due to street frontage.
3. Section 506.3, Allowable area increase due to automatic sprinkler system installation.
4. Section 507, Unlimited area buildings.

b. For open parking structures, see Section 406.3.

c. For private garages, see Section 406.1.

d. See Section 415.5 for limitations.

FIGURE 1.2
Height and area limitations of buildings of various types of construction, as defined in the 2006 IBC. *(Portions of this publication reproduce tables from the* 2006 International Building Code, *International Code Council, Inc., Washington, D.C. Reproduced with Permission. All rights reserved.)*

FIGURE 1.3
Fire resistance of building elements as required by the 2006 IBC. *(Portions of this publication reproduce tables from the* 2006 International Building Code, *International Code Council, Inc., Washington, D.C. Reproduced with Permission. All rights reserved.)*

building. For example, the first line states that the structural frame, including such elements as columns, beams, and trusses, must be rated at 3 hours. The second line also mandates a 3-hour resistance for *bearing walls,* which serve to carry floors or

TABLE 601
FIRE-RESISTANCE RATING REQUIREMENTS FOR BUILDING ELEMENTS (hours)

BUILDING ELEMENT	TYPE I		TYPE II		TYPE III		TYPE IV	TYPE V	
	A	B	A[e]	B	A[e]	B	HT	A[e]	B
Structural frame[a]	3[b]	2[b]	1	0	1	0	HT	1	0
Bearing walls Exterior[g] Interior	 3 3[b]	 2 2[b]	 1 1	 0 0	 2 1	 2 0	 2 1/HT	 1 1	 0 0
Nonbearing walls and partitions Exterior	See Table 602								
Nonbearing walls and partitions Interior[f]	0	0	0	0	0	0	See Section 602.4.6	0	0
Floor construction Including supporting beams and joists	2	2	1	0	1	0	HT	1	0
Roof construction Including supporting beams and joists	$1^1/_2$[c]	1[c, d]	1[c, d]	0[d]	1[d]	0[d]	HT	1[c, d]	0

For SI: 1 foot = 304.8 mm.

a. The structural frame shall be considered to be the columns and the girders, beams, trusses and spandrels having direct connections to the columns and bracing members designed to carry gravity loads. The members of floor or roof panels which have no connection to the columns shall be considered secondary members and not a part of the structural frame.
b. Roof supports: Fire-resistance ratings of structural frame and bearing walls are permitted to be reduced by 1 hour where supporting a roof only.
c. Except in Group F-1, H, M and S-1 occupancies, fire protection of structural members shall not be required, including protection of roof framing and decking where every part of the roof construction is 20 feet or more above any floor immediately below. Fire-retardant-treated wood members shall be allowed to be used for such unprotected members.
d. In all occupancies, heavy timber shall be allowed where a 1-hour or less fire-resistance rating is required.
e. An approved automatic sprinkler system in accordance with Section 903.3.1.1 shall be allowed to be substituted for 1-hour fire-resistance-rated construction, provided such system is not otherwise required by other provisions of the code or used for an allowable area increase in accordance with Section 506.3 or an allowable height increase in accordance with Section 504.2. The 1-hour substitution for the fire resistance of exterior walls shall not be permitted.
f. Not less than the fire-resistance rating required by other sections of this code.
g. Not less than the fire-resistance rating based on fire separation distance (see Table 602).

TABLE 602
FIRE-RESISTANCE RATING REQUIREMENTS FOR EXTERIOR WALLS BASED ON FIRE SEPARATION DISTANCE[a, e]

FIRE SEPARATION DISTANCE = X (feet)	TYPE OF CONSTRUCTION	OCCUPANCY GROUP H	OCCUPANCY GROUP F-1, M, S-1	OCCUPANCY GROUP A, B, E, F-2, I, R, S-2, U[b]
$X < 5$[c]	All	3	2	1
$5 \le X < 10$	IA Others	3 2	2 1	1 1
$10 \le X < 30$	IA, IB IIB, VB Others	2 1 1	1 0 1	1[d] 0 1[d]
$X \ge 30$	All	0	0	0

For SI: 1 foot = 304.8 mm.

a. Load-bearing exterior walls shall also comply with the fire-resistance rating requirements of Table 601.
b. For special requirements for Group U occupancies see Section 406.1.2
c. See Section 705.1.1 for party walls.
d. Open parking garages complying with Section 406 shall not be required to have a fire-resistance rating.
e. The fire-resistance rating of an exterior wall is determined based upon the fire separation distance of the exterior wall and the story in which the wall is located.

roofs above. *Nonbearing walls or partitions*, which carry no load from above, are listed in the third line, referring to Table 602, which gives fire resistance rating requirements for exterior walls of a building based on their proximity to adjacent buildings. (Table 602 is included in the lower portion of Figure 1.3.) Requirements for floor and roof construction are defined in the last two lines of Table 601.

Looking across Table 601 in Figure 1.3, we can see that fire resistance rating requirements are highest for Type I-A construction, decrease to 1 hour for various intermediate types, and fall to zero for Type V-B construction. In general, the lower the construction type numeral, the more fire-resistant the construction system is. (Type IV construction is somewhat of an anomaly, referring to *Heavy Timber* construction consisting of large wooden members that are relatively slow to catch fire and burn.)

Once fire resistance rating requirements for the major parts of a building have been determined the design of these parts can proceed, using building assemblies meeting these requirements. Tabulated fire resistance ratings for common building materials and assemblies may come from a variety of sources, including the IBC itself, as well as a from catalogs and handbooks issued by building material manufacturers, construction *trade associations*, and organizations concerned with fire protection of buildings. In each case, the ratings are derived from full-scale laboratory tests of building components carried out in accordance with an accepted standard fire test protocol to ensure uniformity of results. (This test, ASTM E119, is described more fully in Chapter 22 of this book.) Figures 1.4 to 1.6 show sections of tables from catalogs and handbooks to illustrate how such fire resistance ratings are commonly presented.

In general, when determining the level of fire resistance required for a building, the greater the degree of fire resistance, the higher

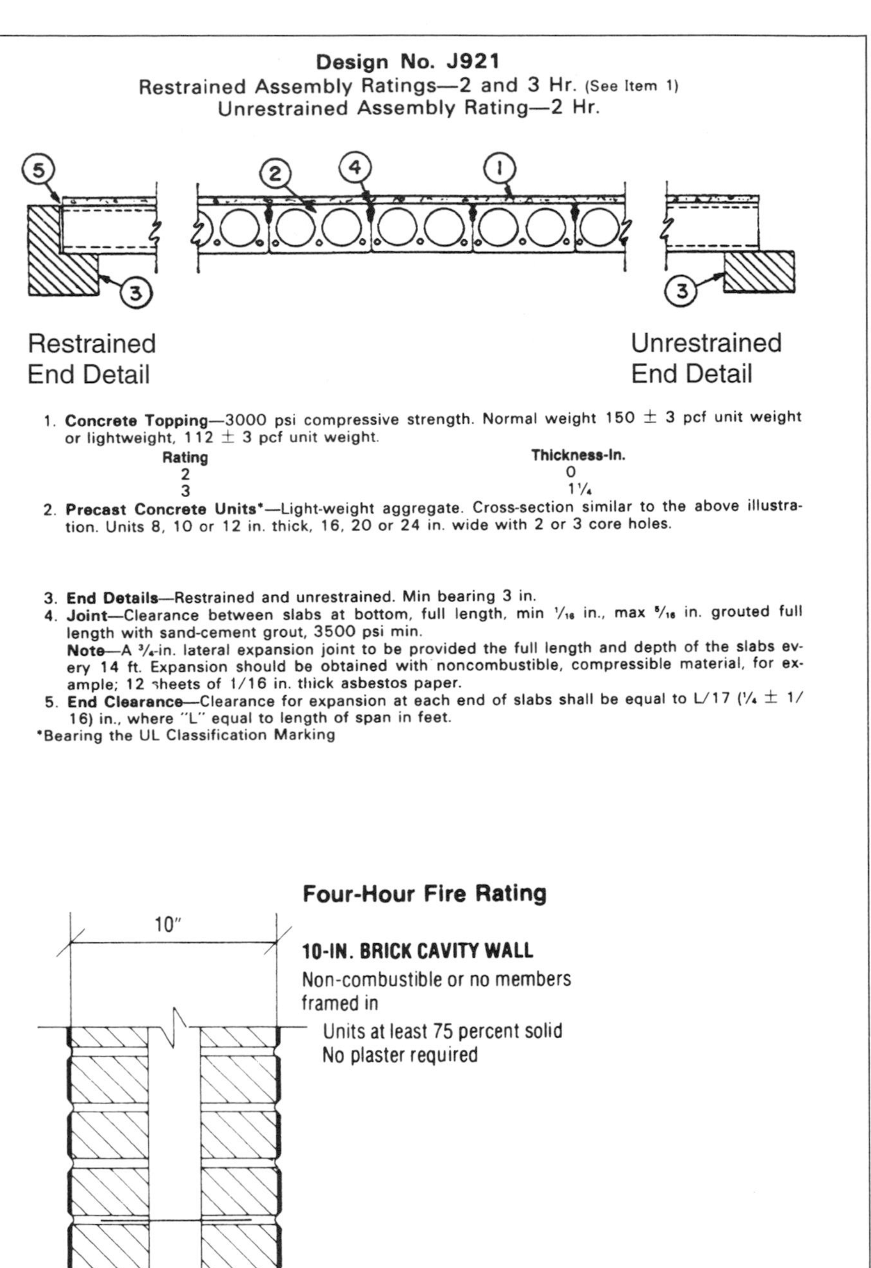

Design No. J921
Restrained Assembly Ratings—2 and 3 Hr. (See Item 1)
Unrestrained Assembly Rating—2 Hr.
5
2
4
1
3
3
Restrained End Detail
Unrestrained End Detail
1. Concrete Topping—3000 psi compressive strength. Normal weight 150 ± 3 pcf unit weight or lightweight, 112 ± 3 pcf unit weight.
Rating / Thickness-In.
2 / 0
3 / 1 1/4
2. Precast Concrete Units*—Light-weight aggregate. Cross-section similar to the above illustration. Units 8, 10 or 12 in. thick, 16, 20 or 24 in. wide with 2 or 3 core holes.
3. End Details—Restrained and unrestrained. Min bearing 3 in.
4. Joint—Clearance between slabs at bottom, full length, min 1/16 in., max 5/16 in. grouted full length with sand-cement grout, 3500 psi min.
Note—A 3/4-in. lateral expansion joint to be provided the full length and depth of the slabs every 14 ft. Expansion should be obtained with noncombustible, compressible material, for example; 12 sheets of 1/16 in. thick asbestos paper.
5. End Clearance—Clearance for expansion at each end of slabs shall be equal to L/17 (1/4 ± 1/16) in., where "L" equal to length of span in feet.
*Bearing the UL Classification Marking
Four-Hour Fire Rating
10"
10-IN. BRICK CAVITY WALL
Non-combustible or no members framed in
Units at least 75 percent solid
No plaster required
References 1, 7 and 9

FIGURE 1.4
Examples of fire resistance ratings for concrete and masonry structural elements. The upper detail, taken from the Underwriters Laboratories *Fire Resistance Directory*, is for a precast concrete hollow-core plank floor with a poured concrete topping. *Restrained* and *unrestrained* refer to whether or not the floor is prevented from expanding longitudinally when exposed to the heat of a fire. The lower detail is from literature published by the Brick Institute of America. *(Reprinted with permission of Underwriters Laboratories Inc. and the Brick Institute of America, respectively.)*

Design No. A814
Restrained Assembly Rating—3 Hr.
Unrestrained Assembly Rating—3 Hr.
Unrestrained Beam Rating—3 Hr.

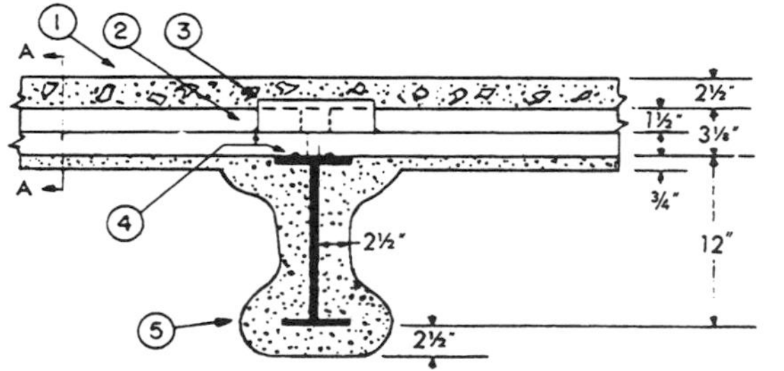

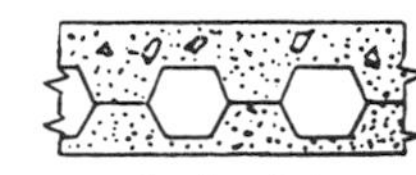

Beam—W12×27, min size.

1. **Sand-Gravel Concrete**—150 pcf unit weight 4000 pcf compressive strength.
2. **Steel Floor and Form Units***—Non-composite 3 in. deep galv units. All 24 in. wide, 18/18 MSG min cellular units. Welded to supports 12 in. O.C. Adjacent units button-punched or welded 36 in. O.C. at joints.
3. **Cover Plate**—No. 16 MSG galv steel.
4. **Welds**—12 in. O.C.
5. **Fiber Sprayed***—Applied to wetted steel surfaces which are free of dirt, oil or loose scale by spraying with water to the final thickness shown above. The use of adhesive and sealer and the tamping of fiber are optional. The min and density of the finished fiber should be 11 pcf and the specified fiber thicknesses require a min fiber density of 11 pcf. For areas where the fiber density is between 8 and 11 pcf, the fiber thickness shall be increased in accordance with the following formula:

$$\text{Thickness, in.} = \frac{(11)\ (\text{Design Thickness, in.})}{\text{Actual Fiber Density, pcf.}}$$

Fiber density shall not be less than 8 pcf. For method of density determination refer to General Information Section.

*Bearing the UL Classification Marking.

Design No. X511
Rating—3 Hr.

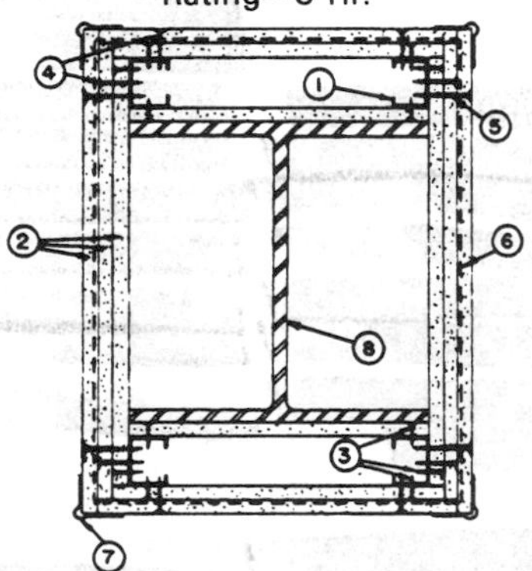

1. **Steel Studs**—1⅝ in. wide with leg dimensions of 1-5/16 and 1-7/16 in. with a ¼-in. folded flange in legs, fabricated from 25 MSG galv steel, ¾- by 1¾-in. rectangular cutouts punched 8 and 16 in. from the ends. Steel stud cut ½ in. less in length than assembly height.
2. **Wallboard, Gypsum***—½ in. thick, three layers.
3. **Screws**—1 in. long, self-drilling, self-tapping steel screws, spaced vertically 24 in. O.C.
4. **Screws**—1⅝ in. long, self-drilling, self-tapping steel screws, spaced vertically 24 in. O.C., except on the outer layer of wallboard on the flange, which are spaced 12 in. O.C.
5. **Screws**—2¼ in. long, self-drilling, self-tapping steel screws, spaced vertically 12 in. O.C.
6. **Tie Wire**—One strand of 18 SWG soft steel wire placed at the upper one-third point, used to secure the second layers of wallboard only.
7. **Corner Beads**—No. 28 MSG galv steel, 1¼ in. legs or 27 MSG uncoated steel, 1⅜ in. legs,

FIGURE 1.5
Fire resistance ratings for a steel floor structure and column, respectively, taken from the Underwriters Laboratories *Fire Resistance Directory*. *(Reprinted with permission of Underwriters Laboratories Inc.)*

FIGURE 1.6
A sample of fire resistance ratings published by the Gypsum Association, in this case for an interior partition. *(Courtesy of the Gypsum Association)*

Fire Rating	Sound Rating STC	GA File No.	DETAILED DESCRIPTION	SKETCH AND DESIGN DATA: Fire	SKETCH AND DESIGN DATA: Sound
1 HR	**30 to 34**	**WP 3620**	**Construction Type: Gypsum-Veneer Base, Veneer Plaster, Wood Studs** One layer ½" type X gypsum veneer base applied at right angles to each side of 2 x 4 wood studs 16" o.c. with 5d etched nails, 1¾" long, 0.099" shank, ¼" heads, 8" o.c. Minimum 1/16" gypsum-veneer plaster over each face. Stagger vertical joints 16" and horizontal joints each side 12". Sound tested without veneer plaster. (LB)	Thickness: 4⅞" Approx. Weight: 7 psf Fire Test: UC, 1-12-66 Sound Test: G & H IBI-35FT. 5-26-64	

the cost. Most frequently, therefore, buildings are designed with the lowest level of fire resistance permitted by the building code. Our hypothetical office building could be built using Type IA construction, but does it really need to be constructed to this high standard?

Let us suppose that the owner desires a five-story building with 30,000 square feet per floor. Reading across the table in Figure 1.2, we can see that in addition to Type I-A construction, the building can be of Type I-B construction, which permits a building of eleven stories and unlimited floor area, or of Type II-A construction, which permits a building of five stories and 37,500 square feet per floor. But it cannot be of Type II-B construction, which allows a building of only four stories and 23,000 square feet per floor. It can also be built of Type IV construction but not of Type III or Type V.

Other factors also come into play in these determinations. If a building is protected throughout by an approved, fully automatic sprinkler system for suppression of fire, the IBC provides that the tabulated area per floor may be quadrupled for a single-story building or as much as tripled for a multistory building (depending on additional considerations omitted here for the sake of simplicity). A one-story increase in allowable height is also granted under most circumstances if such a sprinkler system is installed. If the five-story, 30,000-square-foot office building that we have been considering is provided with such a sprinkler system, a bit of arithmetic will show that it can be built of any construction type shown in Figure 1.2 except Type V.

If more than a quarter of the building's perimeter walls face public ways or open spaces accessible to firefighting equipment, an additional increase of up to 75 percent in allowable area is granted in accordance with another formula. Furthermore, if a building is divided by fire walls having the fire resistance ratings specified in another table (Figure 1.7), each divided portion may be considered a separate building for purposes of computing its allowable area, which effectively permits the creation of a building many times larger than Figure 1.2 would, at first glance, indicate.

TABLE 705.4
FIRE WALL FIRE-RESISTANCE RATINGS

GROUP	FIRE-RESISTANCE RATING (hours)
A, B, E, H-4, I, R-1, R-2, U	3[a]
F-1, H-3[b], H-5, M, S-1	3
H-1, H-2	4[b]
F-2, S-2, R-3, R-4	2

a. Walls shall be not less than 2-hour fire-resistance rated where separating buildings of Type II or V construction.

b. For Group H-1, H-2 or H-3 buildings, also see Sections 415.4 and 415.5.

FIGURE 1.7
Fire resistance requirements for fire walls, according to the 2006 IBC. *(Portions of this publication reproduce tables from the* 2006 International Building Code, *International Code Council, Inc., Washington, D.C. Reproduced with Permission. All rights reserved.)*

The IBC also establishes standards for natural light, ventilation, means of emergency egress, structural design, construction of floors, walls, and ceilings, chimney construction, fire protection systems, accessibility for disabled persons, and many other important factors. The International Code Council also publishes the *International Residential Code (IRC)*, a simplified model code specifically addressing the construction of detached one- and two-family homes and townhouses of limited size. Within any particular building agency, buildings of these types may fall under the requirements of either the IBC or the IRC, depending on the code adoption policies of that jurisdiction.

The building code is not the only code with which a new building must comply. Health codes regulate aspects of design and operation related to sanitation in public facilities such as swimming pools, food-service operations, schools, or health care facilities. Energy codes establish standards of energy efficiency for buildings affecting a designer's choices of windows, heating and cooling systems, and many aspects of the construction of a building's enclosing walls and roofs. Fire codes regulate the operation and maintenance of buildings to ensure that egress pathways, fire protection systems, emergency power, and other life-safety systems are properly maintained. Electrical and mechanical codes regulate the design and installation of building electrical, plumbing, and heating and cooling systems. Some of these codes may be locally written. But like the building codes discussed above, most are based on national models. In fact, an important task in the early design of any major building is determining what agencies have jurisdiction over the project and what codes and regulations apply.

Other Constraints

Other types of legal restrictions must also be observed in the design and construction of buildings. The *Americans with Disabilities Act (ADA)* makes accessibility to public buildings a civil right of all Americans, and the *Fair Housing Act* does the same for much multifamily housing. These *access standards* regulate the design of entrances, stairs, doorways, elevators, toilet facilities, public areas, living spaces, and other parts of affected buildings to ensure that they are accessible and usable by physically handicapped members of the population.

The U.S. *Occupational Safety and Health Administration (OSHA)* controls the design of workplaces to minimize hazards to the health and safety of workers. OSHA sets safety standards under which a building must be constructed and also has an important effect on the design of industrial and commercial buildings.

An increasing number of states have limitations on the amount of *volatile organic compounds (VOCs)* that building products can release into the atmosphere. VOCs are organic chemical compounds that evaporate readily. They can act as irritants to building occupants, they contribute to air pollution, and some are greenhouse gases. Typical sources of VOCs are paints, stains, adhesives, and binders used in the manufacture of wood panel products.

States and localities have conservation laws that protect wetlands and other environmentally sensitive areas from encroachment by buildings. Fire insurance companies exert a major influence on construction standards through their testing and certification organizations (Underwriters Laboratories and Factory Mutual, for example) and through their rate structures for building insurance coverage, which offer strong financial incentives to building owners for more hazard-resistant construction. Building contractors and construction labor unions have standards, both formal and informal, that affect the ways in which buildings are built. Unions have work rules and safety rules that must be observed; contractors have particular types of equipment, certain kinds of skills, and customary ways of going about things. All of these vary significantly from one place to another.

Construction Standards and Information Resources

The tasks of the architect and the engineer would be impossible to carry out without the support of dozens of standards-setting agencies, trade associations, professional organizations, and other groups that produce and disseminate information on materials and methods of construction, some of the most important of which are discussed in the sections that follow.

Standards-Setting Agencies

ASTM International (formerly the American Society for Testing and Materials) is a private organization that establishes specifications for materials and methods of construction accepted as standards throughout the United States. Numerical references to ASTM standards—for example, ASTM C150 for portland cement, used in making concrete—are found throughout building codes and construction specifications, where they are used as a precise shorthand for describing the quality of materials or the requirements of their installation. Throughout this book, references to ASTM standards are provided for the major building materials presented. Should you wish to examine the contents of the standards themselves, they can be found in the ASTM references listed at the end of this chapter. In Canada, corresponding standards are set by the Canadian Standards Association (CSA).

The *American National Standards Institute (ANSI)* is another private organization that develops and certifies North American standards for a broad range of products, such as exterior windows, mechanical components of buildings, and even the accessibility requirements referenced within the IBC itself (ICC/ANSI A117.1). Government agencies, most notably the U.S. Department of Commerce's *National Institute of Science and Technology (NIST)* and the National Research Council Canada's *Institute for Research in Construction (NRC-IRC)*, also sponsor research and establish standards for building products and systems.

Construction Trade and Professional Associations

Design professionals, building materials manufacturers, and construction trade groups have formed a large number of organizations that work to develop technical standards and disseminate information related to their respective fields of interest. The Construction Specifications Institute, whose MasterFormat™ standard is described in the following section, is one example. This organization is composed both of independent building professionals, such as architects and engineers, and of industry members. The Western Wood Products Association, to choose an example from among hundreds of *trade associations*, is made up of producers of lumber and wood products. It carries out research programs on wood products, establishes uniform standards of product quality, certifies mills and products that conform to its standards, and publishes authoritative technical literature concerning the use of lumber and related products. Associations with a similar range of activities exist for virtually every material and product used in building. All of them publish technical data relating to their fields of interest, and many of these publications are indispensable references for the architect or engineer. A considerable number of the standards published by these organizations are incorporated by reference into the building codes. Selected publications from professional and trade associations are identified in the references listed at the end of each chapter in this book. The reader is encouraged to obtain and explore these publications and others available from these various organizations.

MasterFormat

The *Construction Specifications Institute (CSI)* of the United States and its Canadian counterpart, *Construction Specifications Canada (CSC)*, have evolved over a period of many years a comprehensive outline called *MasterFormat* for organizing information about construction materials and systems. MasterFormat is used as the outline for construction specifications for the vast majority of large construction projects in these two countries, it is frequently used to organize construction cost data, and it forms the basis on which most trade associations' and manufacturers' technical literature is cataloged. In some cases, MasterFormat is used to cross-reference materials information on construction drawings as well.

MasterFormat is organized into 50 primary *divisions* intended to cover the broadest possible range of construction materials and buildings systems. The portions of MasterFormat relevant to the types of construction discussed in this book are as follows:

Procurement and Contracting Requirements Group
- Division 00 —Procurement and Contracting Requirements

Specifications Group
- General Requirements Subgroup
  - Division 01 —General Requirements
- Facility Construction Subgroup
  - Division 02 —Existing Conditions
  - Division 03 —Concrete
  - Division 04 —Masonry
  - Division 05 —Metals
  - Division 06 —Wood, Plastics, and Composites
  - Division 07 —Thermal and Moisture Protection
  - Division 08 —Openings
  - Division 09 —Finishes
  - Division 10 —Specialties
  - Division 11 —Equipment
  - Division 12 —Furnishings
  - Division 13 —Special Construction
  - Division 14 —Conveying Equipment
- Facilities Services Subgroup
  - Division 21 —Fire Suppression
  - Division 22 —Plumbing
  - Division 23 —Heating, Ventilating, and Air Conditioning
  - Division 25 —Integrated Automation
  - Division 26 —Electrical
  - Division 27 —Communications
  - Division 28 —Electronic Safety and Security
- Site and Infrastructure Subgroup
  - Division 31 —Earthwork
  - Division 32 —Exterior Improvements
  - Division 33 —Utilities

These broadly defined divisions are further subdivided into *sections*, each describing a discrete scope of work usually provided by a single construction trade or subcontractor. Individual sections are identified by six-digit codes, in which the first two digits correspond to the division numbers above and the remaining four digits identify subcategories and individual units within the division. Within Division 05 – Metals, for example, some commonly referenced sections are:

- Section 05 10 00 — Structural Steel Framing
- Section 05 21 00 — Steel Joist Framing
- Section 05 31 00 — Steel Decking
- Section 05 40 00 — Cold-Formed Metal Framing
- Section 05 50 00 — Metal Fabrications

Almost every chapter in this book gives MasterFormat designations for the information it presents to help the reader know where to look in construction specifications and other technical resources for further information. The full MasterFormat system is contained in the volume referenced at the end of this chapter.

MasterFormat organizes building systems information primarily according to work product, that is, the work of discrete building trades, making it especially well suited for use during the construction phase of building. Other organizational systems, such as *Uniformat*™ and *OmmiClass*™, offer a range of alternative organizational schemes suitable to other phases of the building life cycle and other aspects of building functionality. See the references at the end of this chapter for more information about these systems.

The Work of the Construction Professional: Constructing Buildings

Providing Construction Services

An owner wishing to construct a building hopes to achieve a finished project that meets its functional requirements and its expectations for design and quality, at the lowest possible cost, and on a predictable schedule. A contractor offering its construction services hopes to produce quality work, earn a profit, and complete the project in a timely fashion. Yet, the process of building itself is fraught with uncertainty: It is subject to the vagaries of the labor market, commodity prices, and the weather; despite the best planning efforts unanticipated conditions arise, delays occur, and mistakes are made; and the pressures of schedule and cost inevitably minimize the margin for miscalculation. In this high-stakes environment, the

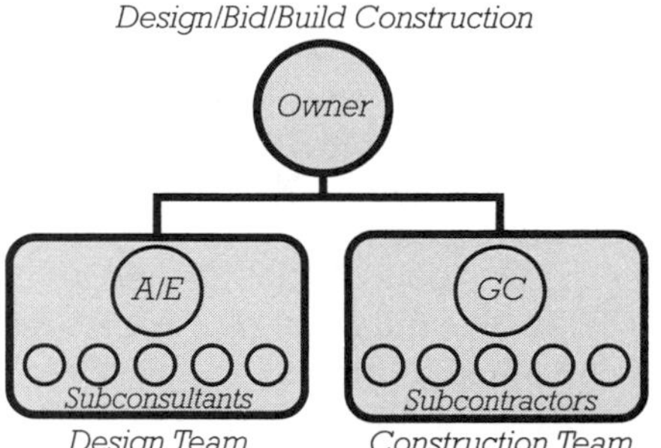

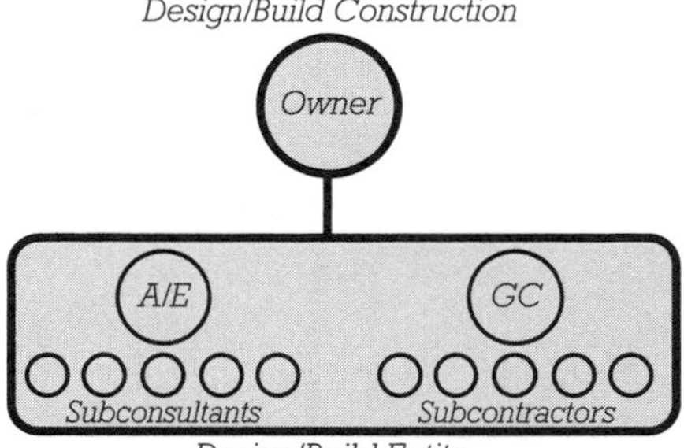

Figure 1.8
In design/bid/build project delivery, the owner contracts separately with the architect/engineer (A/E) design team and the construction general contractor (GC). In a design/build project, the owner contracts with a single organizational entity that provides both design and construction services.

relationship between the owner and contractor must be structured to share reasonably between them the potential rewards and risks.

Construction Project Delivery Methods

In conventional *design/bid/build* project delivery (Figure 1.8), the owner first hires a team of architects and engineers to perform design services, leading to the creation of drawings and technical specifications, referred to collectively as the *construction documents*, that comprehensively describe the facility to be built. Next, construction firms are invited to bid on the project. Each bidding firm reviews the construction documents and proposes a cost to construct the facility. The owner evaluates the submitted proposals and awards the construction contract to the bidder deemed most suitable. This selection may be based on bid price alone, or other factors related to bidders' qualifications may also be considered. The construction documents then become part of the construction contract, and the selected firm proceeds with the work. On all but small projects, this firm acts as the *general contractor*, coordinating and overseeing the overall construction process but frequently relying on smaller, more specialized *subcontractors* to perform significant portions or even all of the construction work. During construction, the design team continues to provide services to the owner, helping to ensure that the facility is built according to the requirements of the documents as well as answering questions related to the design, changes to the work, payments to the contractor, and similar matters. Among the advantages of design/bid/build project delivery are its easy-to-understand organizational structure, well-established legal precedents, and ease of management. The direct relationship between the owner and the design team ensures that the owner retains control over the design and provides a healthy set of checks and balances during the construction process. Also, with design work completed before the project is bid, the owner starts construction with a fixed construction cost and a high degree of confidence regarding the final costs of the project.

In design/bid/build project delivery, the owner contracts with two entities, and design and construction responsibilities remain divided between these two throughout the project. In *design/build* project delivery, one entity ultimately assumes responsibility for both design and construction (Figure 1.8). A design/build project begins with the owner developing a conceptual design or program that describes the functional or performance requirements of the proposed facility but does not detail its form or how it is to be constructed. Next, using this conceptual information, a design/build organization is selected to complete all remaining aspects of the project. Selection of the designer/builder may be based on a competitive bid process similar to that described above for design/bid/build projects, on negotiation and evaluation of an organization's qualifications for the proposed work, or on some combination of both. Design/build organizations themselves can take a variety of forms: a single firm encompassing both design and construction expertise, a construction management firm that subcontracts with a separate design firm to provide those services, or a joint venture between two firms, one specializing in construction and the other in design. Regardless of the internal structure of the design/build organization, the owner contracts with this single entity throughout the remainder of the project, which assumes responsibility for all remaining design and construction services. Design/build project delivery gives the owner a single source of accountability for all aspects of the project. It also places the designers and constructors in a collaborative relationship, introducing construction expertise into the design phases of a project and allowing the earliest possible consideration of constructability, cost control, construction scheduling, and similar matters. This delivery method also readily accommodates fast track construction, a scheduling technique for reducing construction time that is described below.

Other delivery methods are possible: An owner may contract separately with a design team and a *construction manager*. As in design/build construction, the construction manager participates in the project prior to the onset of construction, introducing construction expertise during the design stage. Construction management project delivery can take a variety of forms and is frequently associated with especially large or complex

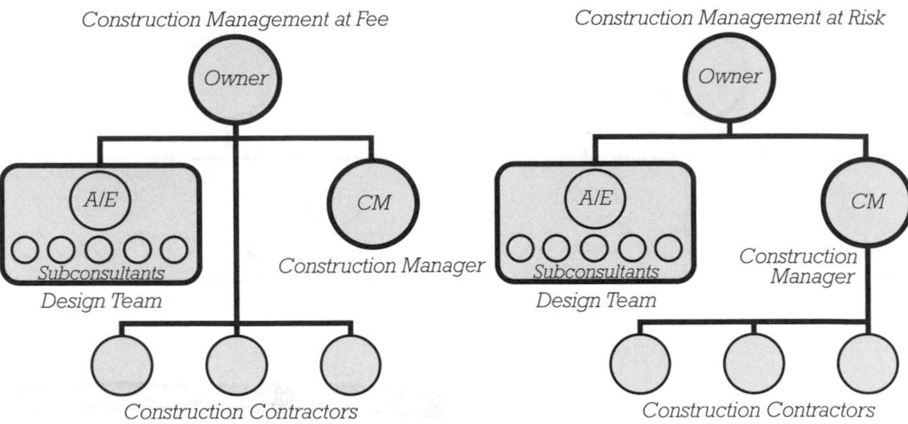

FIGURE 1.9
In its traditional role, a construction manager (CM) *at fee* provides project management services to the owner and assists the owner in contracting directly for construction services with one or more construction entities. A CM at fee is not directly responsible for the construction work itself. A CM *at risk* acts more like a general contractor and takes on greater responsibility for construction quality, schedule, and costs. In either case, the A/E design team also contracts separately with the owner.

projects (Figure 1.9). In *turnkey* construction, an owner contracts with a single entity that provides not only design and construction services, but financing for the project as well. Or design and construction can be undertaken by a *single-purpose entity,* of which the owner, architect, and contractor are all joint members. Aspects of these and other project delivery methods can also be intermixed, allowing many possible organizational schemes for the delivery of design and construction services that are suitable to a variety of owner requirements and project circumstances.

Paying for Construction Services

With *fixed fee* or *lump sum* compensation, the contractor or other construction entity is paid a fixed dollar amount to complete the construction of a project regardless of that entity's actual costs. With this compensation method the owner begins construction with a known, fixed construction cost and assumes minimal risk for unanticipated increases in cost. On the other hand, the construction contractor assumes most of the risk of unforeseen costs but also stands to gain from potential savings. Fixed fee compensation is most suitable to projects where the scope of the construction work is well defined at the time that the construction fee is set, as is the case, for example, with conventional design/bid/build construction.

As an alternative, compensation may be set on a *cost plus a fee* basis, where the owner agrees to pay the construction entity for the actual costs of construction—whatever they may turn out to be—plus an additional fee. In this case, the construction contractor is shielded from most cost uncertainty, and it is the owner who assumes most of the risk of added costs and stands to gain the most from potential savings. Cost plus a fee compensation is most often used with projects where the scope of construction work is not fully known at the time that compensation is established, a circumstance most frequently associated with construction management or design/build contracts.

With fixed fee compensation, the builder assumes most of the risk related to unanticipated construction costs; with cost plus a fee compensation, the owner assumes most of this risk. Between these two extremes, many other fee-structuring arrangements can be used to allocate varying degrees of risk between the two parties.

Sequential versus Fast Track Construction

In *sequential construction* (Figure 1.10), each major phase in the design and construction of a building is completed before the next phase begins. Sequential construction can take place under any of the project delivery methods described previously. It is frequently associated with design/bid/build construction, where the separation of design and construction phases fits naturally with the contractual separation between design and construction service providers.

Phased construction, also called *fast track construction,* aims to reduce the time required to complete a project by overlapping the design and construction of various project parts (Figure 1.10). By allowing construction to start sooner and by overlapping the work of design and construction, phased construction can reduce the time required to complete a project. However, phased construction also introduces its own risks. Since construction on some parts of the project begins before all design is complete, an overall cost for the project cannot be established until a significant portion of construction is underway. Phased construction also introduces more complexity into the design process and increases the potential for costly design errors (for example, if foundation design does not adequately anticipate the requirements of the not yet fully engineered structure above). Phased construction can be applied to any construction delivery method discussed above. It is frequently associated with design/build and construction management project delivery methods, where the early participation of the construction entity provides resources that are helpful in managing the complex coordination of overlapping design and construction activities.

Construction Scheduling

Constructing a building of any significant size is a complex and costly endeavor, requiring the combined efforts of countless participants and the coordination of myriad tasks. Managing this process requires an in-depth understanding of the work required, of the ways in which different aspects of the work depend upon each other, and of the constraints on the sequence in which the work must be performed.

Figure 1.11 captures one moment in the erection of a tall building. The process is led by the construction of the building's central, stabilizing core structures (in the photograph, the pair of concrete towerlike structures extending above the highest floor levels). This work is followed by the construction of the surrounding floors, which rely, in part, on the previously completed cores for support. Attachment of the exterior skin can follow only after the floor plates are in place and structurally secure. And as the building skin is installed and floor areas become enclosed and protected from the weather, further operations, such as the roughing in of mechanical and electrical systems, and eventually, the installation of finishes and other elements, can proceed in turn. This simple example illustrates considerations that apply to virtually every aspect of building construction and at every scale from a building's largest systems to its smallest details: Successful construction requires a detailed understanding of the tasks required and their interdependencies.

The construction project schedule is used to analyze and represent construction tasks, their relationships, and the sequence in which they must be performed. Development of the schedule is a fundamental part of construction project planning, and regular updating of the schedule throughout the life of the project is essential to its successful management. In a *Gantt chart,* a series of horizontal bars represent the duration of various tasks or groups of tasks that make up the project. Gantt charts provide an easy-to-understand representation of construction tasks and their relationships in time. They can be used to provide an overall picture of a project schedule, with only a project's major phases represented (Figure 1.10), or they can be expanded to represent a larger number of more narrowly defined tasks at greater levels of project detail (Figure 1.12).

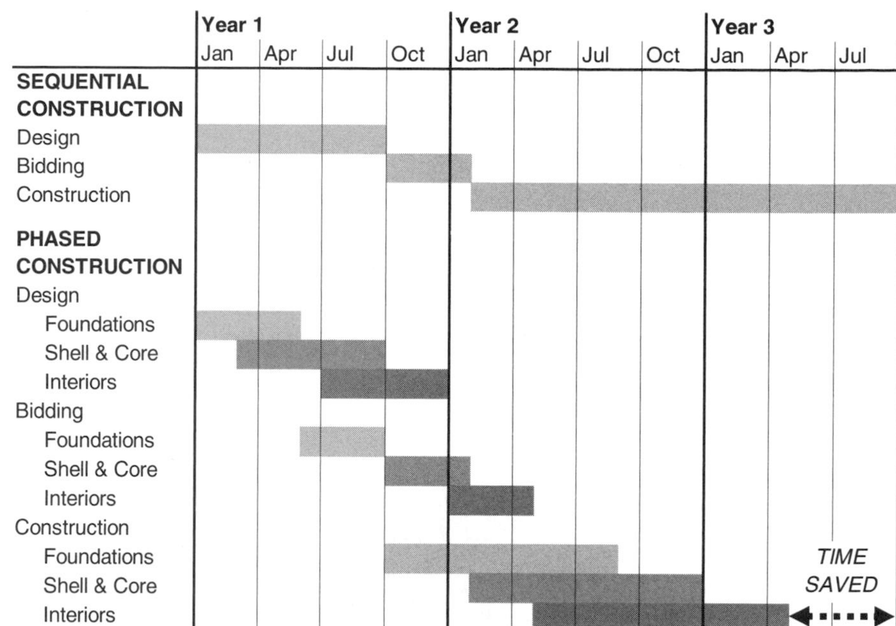

FIGURE 1.10
In sequential construction, construction does not begin until design is complete. In phased construction, design and construction activities overlap, with the goal of reducing the overall time required to complete a project.

The *critical path* of a project is the sequence of tasks that determines the least amount of time in which a project can be completed. For example, the construction of a building's primary structural system is commonly on the critical path of a project schedule. If any of the tasks on which the completion of this system depends—such as design, shop drawing production and review, component fabrication, materials delivery, or erection on site—are delayed, then the final completion date of the project will be extended. On the other hand, other systems not on the critical path have more flexibility in their scheduling, and delays (within limits) in their execution will not necessarily impact the overall project schedule.

The *critical path method* is a technique for analyzing collections of tasks and optimizing the project schedule to minimize the duration and cost of a project. This requires a detailed breakdown of the work involved in a project and the identification of dependencies between the parts (Figure 1.13). This information is combined with considerations of cost and resources available to perform the work, and then analyzed, usually with the assistance of computer software, to identify optimal scheduling scenarios. Once the critical path of a project has been established, the elements on this path are likely to receive a high degree of scrutiny during the life of the project, since delays in any of these steps will directly impact the overall project schedule.

Managing Construction

Once a construction project is underway, the general contractor or

FIGURE 1.11
In this photo, the construction sequence of a tall building is readily apparent: A pair of concrete core structures lead the construction, followed by concrete columns and floor plates and, finally, the enclosing curtain wall.

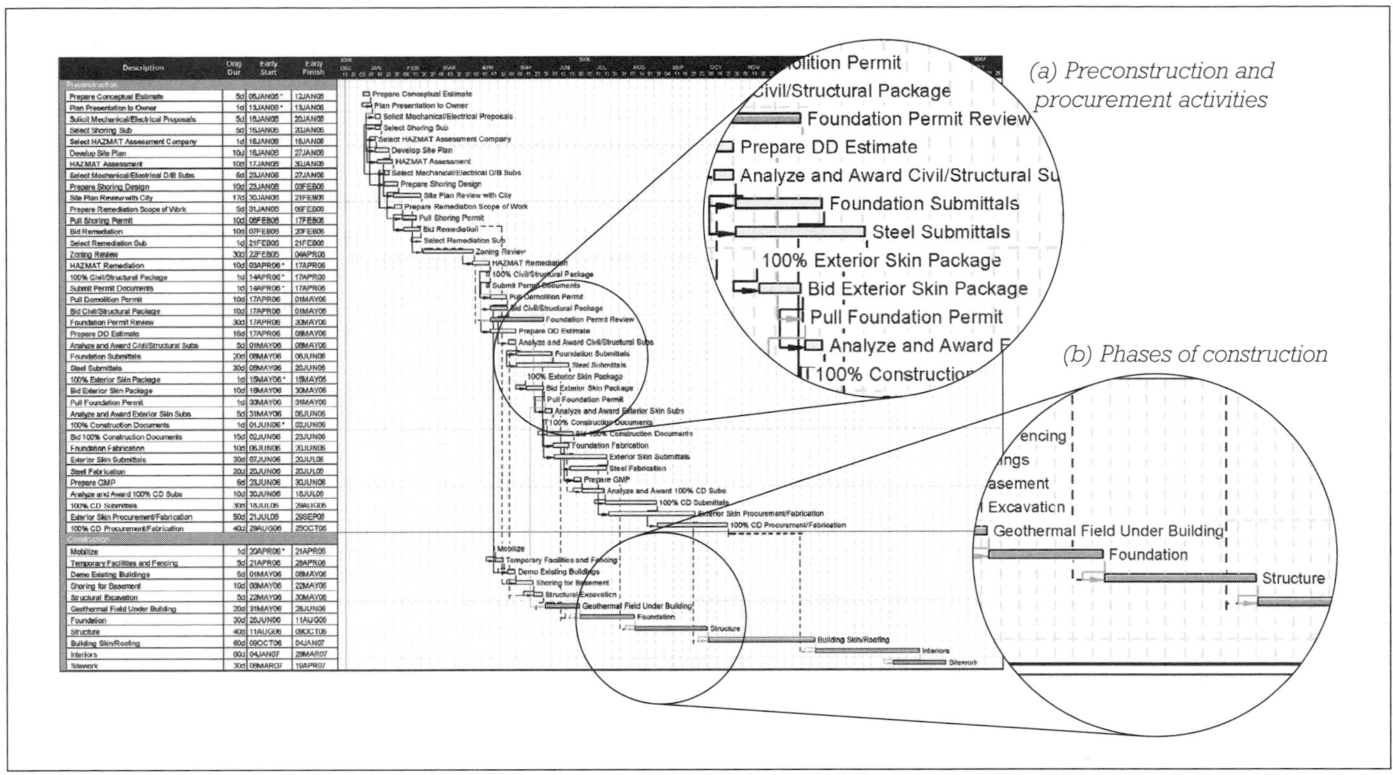

FIGURE 1.12
In a Gantt chart, varying levels of detail can be represented. In this example, roughly the top three-quarters of the chart is devoted to a breakdown of preconstruction and procurement activities such as bidding portions of the work to subtrades, preparing cost estimates, and making submittals to the architect (*a*). Construction activities, represented more broadly, appear in the bottom portion (*b*).

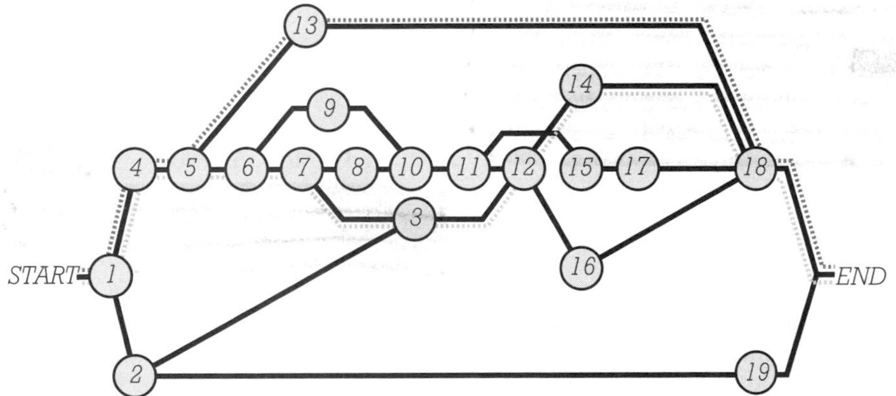

FIGURE 1.13
The critical path method depends on the detailed analysis of work tasks and their relationships to generate an optimal construction schedule. Shown here is a schematic network diagram representing task dependencies. For example, task 6 cannot begin until tasks 1, 4, and 5 are completed, and tasks 7 and 9 cannot begin until task 6 is finished. The dashed lines on the diagram trace two of many possible paths from the start to the end of the diagram. To determine the critical path for this collection of tasks, all such paths must be identified and the time required to complete each one calculated. The path requiring the most time to complete is the critical path, that is, the sequence of activities that determines the least time in which the collection of tasks as a whole can be completed.

construction manager assumes responsibility for day-to-day oversight of the construction site, management of trades and suppliers, and communications between the construction team and other major parties, such as the owner and the designer. On projects of any significant size, this may include responsibility for filing construction permits, securing the project site, providing temporary power and water, setting up office trailers and other support facilities, providing insurance coverage for the work in progress, managing personnel on site, maintaining a safe work environment, stockpiling materials, performing testing and quality control, providing site surveying and engineering, arranging for cranes and other construction machinery, providing temporary structures and weather protection, disposing or recycling of construction waste, soliciting the work of subtrades and coordinating their efforts, submitting product samples and technical information to the design team for review, maintaining accurate records of the construction as it proceeds, monitoring costs and schedules, managing changes to the work, protecting completed work, and more.

Trends in the Delivery of Design and Construction Services

Improving Collaboration Among Team Members

The design and construction industry continues to evolve, testing innovative organizational structures and project delivery methods in which designers, builders, and owners assume less adversarial and less compartmentalized roles. Such approaches share characteristics such as:

- Contractual relationships and working arrangements that foster collaboration between project members
- Participation of the construction contractor during the design phases of a project
- Overlapping of design and construction activities to reduce the "time to market"
- Expanded definitions of project services to encompass the full life cycle of a project—from its original conception, through planning, design and construction, to postconstruction occupancy—to better serve the needs of the building owner

The growth of design/build in the construction marketplace is one example of this trend: Between 1980 and 2005 the share of private, nonresidential construction work performed as design/build construction increased from roughly 5 percent of the total market to an estimated 30 to 40 percent. Alternatives to traditional design/bid/build project delivery have gained increased acceptance in the public construction sector as well. Other new practice models, with such names as *teaming*, *concurrent design*, *integrated practice*, or *alliancing*, combine efficient project delivery methods with innovations in team member relationships in a variety of ways, with the aim of aligning all parties' efforts with the shared goal of a finished product of the highest possible quality and value to the owner.

Improving Efficiency in Production

Other efforts within the construction industry focus on improvements in the efficiency of construction methods themselves. Unlike factory production, much building construction takes place outdoors, is performed within constrained and often physically challenging work areas, and is executed by a highly fragmented workforce. Despite the differences in these production environments, the construction industry is looking to lessons learned in factory production for approaches to improving the quality and efficiency of its own processes. Such so-called *lean construction* methods attempt to:

- Eliminate wasteful activities
- Structure the methods of production and the supply chain of materials and products to achieve the quickest and most reliable workflow
- Decentralize information and decision making so as to put control of construction processes into the hands of those most familiar with the work and most capable of improving it

Current estimates of labor inefficiency in building construction run as high as 35 to 40 percent, and estimates of materials wastage are 20 percent or more. The challenge of lean construction is to restructure the way in which construction materials and building components are manufactured, delivered, and assembled so as to reduce these inefficiencies and improve the quality of the delivered product.

Improving Information Management

Developments in information technology also are influencing the way buildings are designed and constructed. Most notable is *building information modeling (BIM)*, the computerized, three-dimensional modeling of building systems. Unlike the two-dimensional representation of building systems characteristic of conventional *computer-aided design (CAD)*, BIM involves an intelligent model. Components are not only represented geometrically, but are also linked to data describing their intrinsic properties and their relationships to other components. Originally developed for use in highly capital-intensive industries such as aerospace and automobile manufacturing, this modeling technology is now finding increased application in the design and construction of buildings.

BIM has the potential to impact many aspects of the building life cycle. It can aid the design team with the visualization and realization of complex geometries. It can improve coordination between design disciplines—for example, searching out "collisions" between mechanical system ductwork and structural framing or other such physical interferences between systems—and it can facilitate the modeling of building energy use and the performance of other building systems. For the builder, BIM can be used to improve coordination of trades, to drive the automated fabrication or preassembly of building components, and to integrate cost and schedule data more closely with building design. For the building owner, information accumulated in the model during design and construction can be carried forward for use with postconstruction building operations and facilities planning.

As the implementation of BIM matures, it is expected to have a profound impact as a communication tool used to improve the coordination and sharing of information among all of the parties to a project. As the integrated building model is shared across the traditional boundaries of disciplines and project phases, the boundaries of responsibility between the designers, constructors, and owners will also blur, and new, more integrated relationships between these parties will likely be required to fully enable the potential of this technology.

Recurring Concerns

Certain themes are woven throughout this book and surface again and again in different, often widely varying contexts. These represent a set of concerns that fall into two broad categories: building performance and building construction.

The performance concerns relate to the inescapable problems that must be confronted in every building: fire; building movement of every kind, including foundation settlement, structural deflections, and expansion and contraction due to changes in temperature and humidity; the flow of heat and air through building assemblies; water vapor migration and condensation; water leakage; acoustical performance; aging and deterioration of materials; cleanliness; building maintenance; and others.

The construction concerns are associated with the practical problems of getting a building built safely, on time, within budget, and to the required standard of quality: division of work between the shop and the building site; optimum use of the various building trades; sequencing of construction operations for maximum productivity; convenient and safe worker access to construction operations; dealing with inclement weather; making building components fit together; and quality assurance in construction materials and components through grading, testing, and inspection.

To the novice, these matters may seem of minor consequence when compared to the larger and often more interesting themes of building form and function. To the experienced building professional, who has seen buildings fail both aesthetically and physically for want of attention to one or more of these concerns, they are issues that must be resolved as a matter of course before the work of a building project can be allowed to proceed.

CSI/CSC

MasterFormat Sections for Procurement of Construction and General Project Requirements

00 10 00	SOLICITATION
00 11 00	Advertisements and Invitations
00 30 00	AVAILABLE INFORMATION
00 40 00	PROCUREMENT FORMS AND SUPPLEMENTS
00 41 00	Bid Forms
00 50 00	CONTRACTING FORMS AND SUPPLEMENTS
00 52 00	Agreement Forms
00 70 00	CONDITIONS OF THE CONTRACT
01 10 00	SUMMARY
01 11 00	Summary of Work
01 30 00	ADMINISTRATIVE REQUIREMENTS
01 31 00	Project Management and Coordination
01 32 00	Construction Progress Documentation Construction Progress Schedule
01 40 00	QUALITY REQUIREMENTS
01 41 00	Regulatory Requirements
01 50 00	TEMPORARY FACILITIES AND CONTROLS
01 70 00	EXECUTION AND CLOSEOUT REQUIREMENTS
01 80 00	PERFORMANCE REQUIREMENTS
01 81 00	Facility Performance Requirements Sustainable Design Requirements

Selected References

1. Allen, Edward. *How Buildings Work* (3rd ed.). New York, Oxford University Press, 2005.

What do buildings do, and how do they do it? This book sets forth in easily understandable terms the physical principles by which buildings stand up, enclose a piece of the world, and modify it for human use.

2. U.S. Green Building Council. New Construction & Major Renovation, Version 2.2, Reference Guide. Washington, DC, 2006.

This guide is an essential reference for building designers wishing to comply with the Green Building Council's LEED for New Construction rating system. As the standard continues to be revised, look for updated versions.

3. Williams, Daniel E. *Sustainable Design: Ecology, Architecture, and Planning.* Hoboken, NJ, John Wiley & Sons, Inc., 2007.

This book provides a comprehensive treatment of the ecological, social, and economic basis for sustainable design in architecture.

4. ASTM International. *ASTM Standards in Building Codes.* Philadelphia, updated regularly.

This volume contains most of the ASTM standards referenced in standard building construction practice.

5. The Construction Specifications Institute and Construction Specifications Canada. *MasterFormat™ 2004 Edition.* Alexandria, VA, and Toronto, 2004.

This handbook lists the full set of MasterFormat numbers and titles under which construction information is most commonly organized.

6. International Code Council, Inc. *International Building Code®.* Falls Church, VA, updated regularly.

This is the predominant U.S. model building code, used as the basis for the majority of U.S. state, county, and municipal building codes.

7. Canadian Commission on Building and Fires Codes. *National Building Code of Canada.* Ottawa, National Research Council of Canada, updated regularly.

This is the model building code used as the basis for most Canadian provincial and municipal building codes.

8. Allen, Edward, and Joseph Iano. *The Architect's Studio Companion* (4th ed.). Hoboken, NJ, John Wiley & Sons, Inc., 2007.

This designer's reference tabulates building code requirements, simplifying determination of the allowable heights and areas for any building under the IBC or the Canadian Building Code. It explains clearly what each construction type means, relating it to actual construction materials and structural systems, and it gives extensive rules of thumb for structural systems, mechanical systems, and egress planning.

9. Halpin, Daniel W. *Construction Management* (3rd ed.). Hoboken, NJ, John Wiley & Sons, Inc., 2005.

This book covers the full range of contemporary construction management topics.

10. Elvin, George. *Integrated Practice in Architecture.* Hoboken, NJ, John Wiley & Sons, Inc., 2007.

How is the design industry responding to the evolution of design practice models and building delivery methods? This book provides case studies and insights into these recent trends.

11. Allen, Edward and Patrick Rand. *Architectural Detailing* (2nd ed). Hoboken, NJ, John Wiley & Sons, Inc., 2007.

How are the many functional, constructional, and aesthetic requirements of building resolved in the detailing of building systems? This book presents a systematic treatment of the principles and practice of design and the detailing of building assemblies.

Web Sites

Making Buildings

Author's supplementary web site: **www.ianosbackfill.com/01_making_buildings**

Whole Building Design Guide: **www.wbdg.org**

Sustainability

AIA Sustainability Resource Center: **http://www.aia.org/susn_rc_default**

American Society of Heating, Refrigerating and Air-Conditioning Engineers, Engineering for Sustainability: **www.engineeringforsustainability.org**

Architects/Designers/Planners for Social Responsibility: **www.adpsr.org**
Canada Green Building Council: **www.cagbc.org**
Climate Change, Global Warming, and the Built Environment—Architecture 2030: **www.architecture2030.org**
Green Building Initiative (GBI): **www.thegbi.com**
Green Globes: **www.greenglobes.com**
NAHB, Green Home Building Guidelines: **www.nahbrc.org/greenguidelines**
NAHB Research Center, National Green Building Standard: **www.nahbgreen.org**
National Renewable Energy Laboratory (NREL), Buildings Research: **www.nrel.gov/buildings**
Sustainable Buildings Industry Council (SBIC): **www.sbicouncil.org**
U.S. Environmental Protection Agency Energy Star Program: **www.energystar.org**
U.S. Green Building Council (USBGC): **www.usgbc.org**

The Work of the Design Professional: Choosing Building Systems

American Institute of Architects (AIA): **www.aia.org**
American Planning Association (APA): **www.planning.org**
Canadian Codes Centre: **irc.nrc-cnrc.gc.ca/codes**
International Code Council (ICC): **www.iccsafe.org**
U.S. Department of Justice, Americans with Disabilities Act (ADA): **www.ada.gov**

Construction Standards and Information Resources

American National Standards Institute (ANSI): **www.ansi.org**
ASTM International: **www.astm.org**
Canadian Standards Association (CSA): **www.csa.ca**
Construction Specifications Canada (CSC): **www.csc-dcc.ca**
Construction Specifications Institute (CSI): **www.csinet.org**
National Institute of Building Sciences (NIBS): **www.nibs.org**
National Research Council Canada, Institute for Research in Construction (NRC-IRC): **irc.nrc-cnrc.gc.ca**
Underwriters Laboratories, Inc. (UL): **www.ul.com**
U.S. Department of Commerce, National Institute of Standards and Technology (NIST): **www.nist.gov**
U.S. Department of Energy, Building Energy Codes: **www.energycodes.gov**

The Work of the Construction Professional: Constructing Buildings

Associated General Contractors of America (AGC): **www.agc.org**
Building Owners and Managers Association (BOMA): **www.boma.org**
Construction Management Association of America (CMAA): **cmaanet.org**
The Construction Users Roundtable (CURT): **www.curt.org**
Design-Build Institute of America (DBIA): **www.dbia.org**

Key Terms

sustainability
green building
cradle-to-grave
embodied energy
LEED
LEED credit
Green Globes
Green Home Building Guidelines
National Green Building Standard
Advanced Energy Design Guides
Energy Star program
net zero energy
carbon-neutral
carbon offsetting
zoning ordinance
building code
model building code
National Building Code of Canada
International Building Code (IBC)
occupancy group
construction type
fire resistance rating
bearing wall
nonbearing wall, partition
Heavy Timber construction
trade association
International Residential Code (IRC)
Americans with Disabilities Act (ADA)
Fair Housing Act
access standard
Occupational Health and Safety Administrations (OSHA)
volatile organic compound (VOC)
ASTM International
American National Standards Institute (ANSI)
National Institute of Science and Technology (NIST)
Institute for Research in Construction (NRC-IRC)
trade association
Construction Specifications Institute (CSI)
Construction Specifications Canada (CSC)
MasterFormat
specification division
specification section
Uniformat
OmmiClass
design/bid/build
construction document
general contractor
subcontractor
design/build
construction manager
turnkey
single-purpose entity
fixed fee compensation, lump sum compensation
cost plus a fee compensation
sequential construction
phased construction, fast track construction
Gantt chart
critical path
critical path method
teaming
concurrent design
integrated practice
alliancing
lean construction
building information modeling (BIM)
computer-aided design (CAD)

Review Questions

1. Who are the members of the typical team that designs a major building? What are their respective roles?

2. What are the major constraints under which the designers of a building must work?

3. What types of subjects are covered by zoning ordinances? By building codes?

4. In what units is fire resistance measured? How is the fire resistance of a building assembly determined?

5. Compare and contrast design/bid/build and design/build construction. What is the difference between a construction manager and a general contractor? What is the difference between lump sum and cost plus a fee compensation? What is fast track construction, and what types of contracts and fee compensation is it mostly commonly associated with?

6. If you are designing a five-story office building (Occupancy Group B) with 35,000 square feet per floor, what types of construction will you be permitted to use under the IBC if you do not install sprinklers? How does the situation change if you install sprinklers? In each case, what level of fire protection is required for the structural frame of the building?

Exercises

1. Have each class member write to two or three trade associations at the beginning of the term to request their lists of publications, and then have each send for some of the publications. Display and discuss the publications.

2. Repeat the above exercise for manufacturers' catalogs of building materials and components.

3. Obtain copies of your local zoning ordinance and building code (they may be in your library). Look up the applicable provisions of these documents for a specific site and building. What setbacks are required? How large a building is permitted? What construction types may be employed? What types of roofing materials are permitted? What are the restrictions on interior finish materials? Outline in complete detail the requirements for emergency egress from the building.

4. Arrange permission to "shadow" an architect or CM during visits to a construction site or during project meetings related to a construction project. Take notes. Interview the architect or CM about their role and the challenges they have encountered. Report back to the class what you have learned

FRANKI

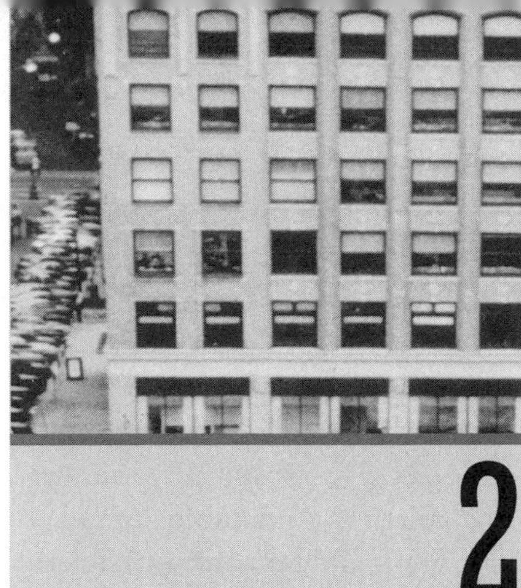

2

FOUNDATIONS

Foundation work in progress for a midrise hotel and apartment building in Boston. The earth surrounding the excavation is retained with steel sheet piling supported by steel walers and tiebacks. Equipment enters and leaves the site via the earth ramp at the bottom of the picture. Although a large backhoe at the right continues to dig around old piles from a previous building on the site, the installation of pressure-injected concrete pile footings is already well underway, with two piledrivers at work in the near and far corners and clusters of completed piles visible in the center of the picture. Concrete pile caps and column reinforcing are under construction in the center of the excavation. *(Courtesy of Franki Foundation Company)*

The function of a *foundation* is to transfer the structural loads reliably from a building into the ground. Every building needs a foundation of some kind: A backyard toolshed will not be damaged by slight shifting of its foundation and may need only wooden skids to spread its load across an area of the ground surface sufficient to support its weight. A wood-framed house needs greater stability than a toolshed, so its foundation reaches through the unstable surface to underlying soil that is free of organic matter and unreachable by winter frost. A larger building of masonry, steel, or concrete weighs many times more than a house, and its foundations pierce the earth until they reach soil or rock that is competent to carry its massive loads; on some sites, this means going 100 feet (30 m) or more below the surface. Because of the variety of soil, rock, and water conditions that are encountered below the surface of the ground and the unique demands that buildings make upon their foundations, foundation design is a highly specialized field combining aspects of geotechnical and civil engineering that can be sketched here only in its broad outlines.

Foundation Requirements

A building foundation must support different kinds of loads:

- *Dead load,* the combined weight of all the permanent components of the building, including its own structural frame, floors, roofs, and walls, major permanent electrical and mechanical equipment, and the foundation itself
- *Live loads,* nonpermanent loads caused by the weights of the building's occupants, furnishings, and movable equipment
- *Rain and snow loads,* which act primarily downward on building roofs
- *Wind loads,* which can act laterally (sideways), downward, or upward on a building
- *Seismic loads,* horizontal and vertical forces caused by the motion of the ground relative to the building during an earthquake
- Loads caused by soil and hydrostatic pressure, including *lateral soil pressure loads,* horizontal pressures of earth and groundwater against basement walls; in some instances, *buoyant uplift* forces from underground water, identical to the forces that cause a boat to float; and in others, lateral force *flood loads* that can occur in areas prone to flooding
- In some buildings, *horizontal thrusts* from long-span structural systems such as arches, rigid frames, domes, vaults, or tensile structures

A satisfactory foundation for a building must meet three general requirements:

1. The foundation, including the underlying soil and rock, must be safe against a structural failure that could result in collapse. For example, the foundation for a skyscraper must support the great weight of the building above on a relatively narrow base without danger of overturning.

2. During the life of the building, the foundation must not settle in such a way as to damage the structure or impair its function. (Foundation settlement is discussed more fully in the next section.)

3. The foundation must be feasible, both technically and economically, and practical to build without adverse effects on surrounding property. For example, New York City's tallest buildings tend to cluster on the central and southern portions of Manhattan Island, where the underlying bedrock is closest to the surface and foundations for such buildings are easiest and least expensive to construct.

Foundation Settlement

All foundations settle to some extent as the earth materials around and beneath them adjust to the loads of the building. Foundations on bedrock settle a negligible amount. Foundations in other types of soil may settle much more. As an extreme example, Mexico City's Palace of Fine Arts has settled more than 15 feet (4.5 m) into the clay soil on which it is founded since it was constructed in the early 1930s. However, building foundation settlement is normally limited to amounts measured in millimeters or fractions of an inch.

> **We must never trust too hastily to any ground. . . . I have seen a tower at Mestre, a place belonging to the Venetians, which, in a few years after it was built, made its way through the ground it stood upon . . . and buried itself in earth up to the very battlements.**
>
> **Leon Battista Alberti, *Ten Books on Architecture,* 1452.**

Where *foundation settlement* occurs at roughly the same rate throughout all portions of a building, it is

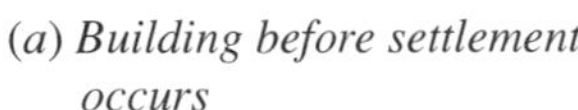

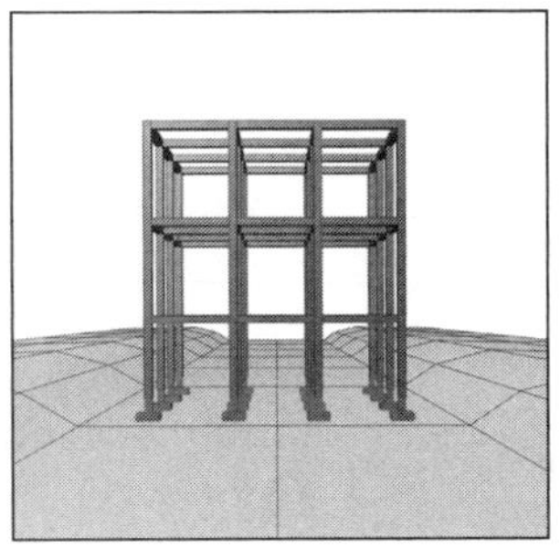

(*a*) *Building before settlement occurs* (*b*) *Uniform settlement* (*c*) *Differential settlement*

FIGURE 2.1
Uniform settlement (*b*) is usually of little consequence in a building, but differential settlement (*c*) can cause severe structural damage.

termed *uniform settlement.* Settlement that occurs at differing rates between different portions of a building is termed *differential settlement.* When all parts of a building rest on the same kind of soil, and the loads on the building and the design of its structural system are uniform throughout, differential settlement is normally not a concern. However, where soils, loads, or structural systems differ between parts of a building, different parts of the building structure may settle by substantially different amounts, the frame of the building may become distorted, floors may slope, walls and glass may crack, and doors and windows may not work properly (Figure 2.1). Most foundation failures are attributable to excessive differential settlement. Gross failure of a foundation, in which the soil fails completely to support the building, is extremely rare.

EARTH MATERIALS

Classifying Earth Materials

For the purposes of foundation design, *earth materials* are classified according to particle size, the presence of organic content, and, in the case of finer grained soils, sensitivity to moisture content:

- *Rock* is a continuous mass of solid mineral material, such as granite or limestone, which can only be removed by drilling and blasting. Rock is never completely monolithic, but is crossed by a system of joints (cracks) that vary in quantity and extent and that divide the rock into irregular blocks. Despite these joints, rock is generally the strongest and most stable material on which a building can be founded.
- *Soil* is a general term referring to any earth material that is particulate.
- If an individual particle of soil is too large to lift by hand or requires two hands to lift, it is a *boulder.*
- If it takes the whole hand to lift a particle, it is called a *cobble.*
- If a particle can be lifted easily between thumb and forefinger, the soil is *gravel.* In the Unified Soil Classification System (Figure 2.2), gravels are classified as having more than half their particles larger than 0.19 inch (4.75 mm) in diameter but none larger than 3 inches (76 mm).
- If individual soil particles can be seen but are too small to be picked up individually, the soil is *sand.* Sand particles range in size from about 0.19 to 0.003 inch (4.75–0.075 mm). Both sand and gravel are referred to as *coarse-grained soils.*
- Individual *silt* particles are too small to be seen with the unaided eye and range in size from 0.003 to 0.0002 inch (0.075–0.005 mm). Like coarse-grained soil particles, silt particles are roughly spherical, or equidimensional, in shape.
- *Clay* particles are plate-shaped rather than spherical (Figure 2.3) and smaller than silt particles, less than 0.0002 inch (0.005 mm) in size. Both sands and silts are also referred to as *fine-grained soils.*
- Peat, topsoil, and other *organic soils* are not suitable for the support of building foundations. Because of their high organic matter content, they are spongy, they compress easily, and their properties can change over time due to changing water content or biological activity within the soil.

Properties of Soils

The ability of a coarse-grained soil (gravel or sand) to support the weight of a building depends primarily on the strength of the individual soil particles and the friction between them. Imagine holding a handful of spherical, smooth ball bearings: If you squeeze the bearings, they easily slide past one another in your hand. There is little friction between them. However, if you squeeze a handful of crushed stone, whose particles have rough, angular facets, the frictional forces between the particles are large, and there will be little movement between them. This resistance to sliding, or *shear resistance,* of the crushed stone is also directly proportional to the confining force pushing the particles together. Thus, sand confined by surrounding soil within the earth can support a heavy building, whereas a conical pile of sand deposited loosely on the surface of the ground can support very little, because there is little or no shear resistance between the unconfined particles. Soils that behave in this manner are termed *frictional* or *cohesionless.*

			Group Symbol	Descriptive Names of Soil within This Group
Coarse-Grained Soils	Gravels	Clean Gravels	GW	Well-graded gravel or well-graded gravel with sand, little or no fines
			GP	Poorly graded gravel or poorly graded gravel with sand, little or no fines
		Gravels with Fines	GM	Silty gravel, silty gravel with sand
			GC	Clayey gravel, clayey gravel with sand
	Sands	Clean Sands	SW	Well-graded sand or well-graded sand with gravel, little or no fines
			SP	Poorly graded sand or poorly graded sand with gravel, little or no fines
		Sands with Fines	SM	Silty sand, silty sand with gravel
			SC	Clayey sand, clayey sand with gravel
Fine-Grained Soils	Silts and Clays	Liquid Limit < 50	ML	Silt or silt-sand-gravel mixtures, low plasticity
			CL	Lean clay or clay-sand-gravel mixtures, low plasticity
			OL	Organic clay or silt (clay or silt with significant organic content), or organic clay- or silt-sand-gravel mixtures, low plasticity
		Liquid Limit = 50	MH	Elastic silt, silt-sand-gravel mixtures
			CH	Fat clay or clay-sand-gravel mixtures, high plasticity
			OH	Organic clay or silt (clay or silt with significant organic content), or organic clay- or silt-sand-gravel mixtures, high plasticity
	Highly Orgzanic Soils		PT	Peat, muck, and other highly organic soils

FIGURE 2.2
The Unified Soil Classification System, from ASTM D 2487. The group symbols are a universal set of abbreviations for soil types, as seen for example, in Figure 2.8.

In fine-grained soils, particles are smaller, particle surface area is larger in relation to size and weight, and the spaces between particles, or *soil pores,* are smaller. As a consequence, surface forces also affect the properties of these soils. The properties of silts are more sensitive to the amount of water in the soil than are those of coarse-grained soils. With sufficient moisture content, capillary forces can reduce friction between particles and change the state of silt from solid to liquid.

Clay particles, being extremely small and flatter in shape, have surface-area-to-volume ratios hundreds or thousands of times greater than even those of silt. Electrostatic repulsive and attractive forces play an important role in clay soil's properites, as do variations in the arrangement, or *fabric,* of the particles in sheets or other structures that are more

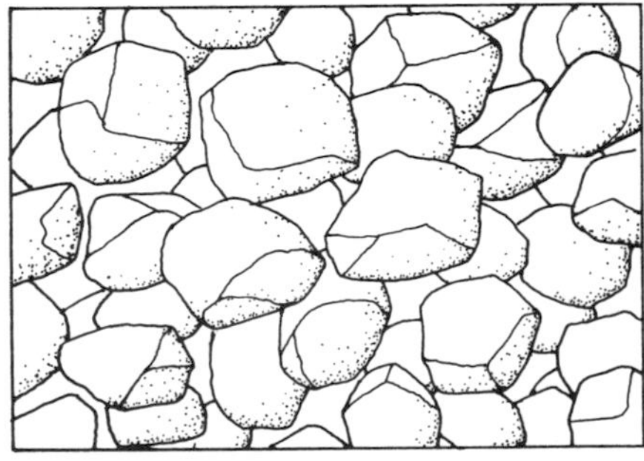

FIGURE 2.3
Silt particles (*top*) are approximately equidimensional granules, while clay particles (*bottom*) are platelike and much smaller than silt. (A circular area of clay particles has been magnified to make the structure easier to see.)

EXCAVATION IN FRICTIONAL SOIL

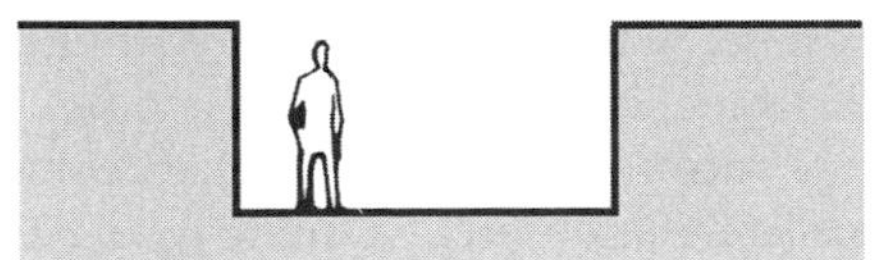

EXCAVATION IN HIGHLY COHESIVE SOIL

FIGURE 2.4
Excavations in frictional and highly cohesive soils.

complex than the simple close-packing typical of spherical particles in coarse-grained soils and silts. As a result, clays are generally *cohesive*; that is, even in the absence of confining force, they retain measurable shear strength. Put simply, cohesive soils tend to stick together. It is often possible to dig vertical-walled excavations in clay soil (Figure 2.4). There is sufficient shear strength in the unconfined soil to prevent the soil wall from sliding into the excavation. In contrast, a cohesionless soil such as sand must be excavated at a much more shallow angle to avoid the collapse of the excavation wall. Cohesive soils also tend to be hard when dry and moldable, or *plastic*, when moist. Some silts also exhibit cohesive properties, though generally to a lesser extent than clays.

Soils for Building Foundations

Generally, soil groups listed toward the top of Figure 2.2 are more desirable for supporting building foundations than those listed further down. The higher-listed soils tend to have better *soil engineering properties*, that is, they tend to have greater loadbearing capacity, to be more stable, and to react less to changes in moisture content. Rock is generally the best material on which to found a building. When rock is too deep to be reached economically, the designer must choose from the strata of different soils that lie closer to the surface and design a foundation to perform satisfactorily in the selected soil.

Figure 2.5 gives some conservative values of *loadbearing capacity* for various types of soil. These values give only an approximate idea of the relative strengths of different soils; the strength of any particular soil is also dependent on factors such as the presence or absence of water, the depth at which the soil lies beneath the surface, and, to some extent, the manner in which the foundation acts upon it. In practice, the designer may also choose to reduce the pressure of the foundations on the soil to well below these values in order to reduce the potential for building settlement.

The *stability* of a soil is its ability to retain its structural properties under the varying conditions that may occur during the lifetime of the building. In general, rock, gravels, and sands tend to be the most stable soils, clays the least stable, and silts somewhere in between. Many clays change size under changing subsurface moisture conditions, swelling considerably as they absorb water and shrinking as they dry. In the presence of highly expansive clay soils, a foundation may need to be designed

TABLE 1804.2
ALLOWABLE FOUNDATION AND LATERAL PRESSURE

CLASS OF MATERIALS	ALLOWABLE FOUNDATION PRESSURE (psf)[d]	LATERAL BEARING (psf/f below natural grade)[d]	LATERAL SLIDING	
			Coefficient of friction[a]	Resistance (psf)[b]
1. Crystalline bedrock	12,000	1,200	0.70	—
2. Sedimentary and foliated rock	4,000	400	0.35	—
3. Sandy gravel and/or gravel (GW and GP)	3,000	200	0.35	—
4. Sand, silty sand, clayey sand, silty gravel and clayey gravel (SW, SP, SM, SC, GM and GC)	2,000	150	0.25	—
5. Clay, sandy clay, silty clay, clayey silt, silt and sandy silt (CL, ML, MH and CH)	1,500[c]	100	—	130

For SI: 1 pound per square foot = 0.0479 kPa, 1 pound per square foot per foot = 0.157 kPa/m.
a. Coefficient to be multiplied by the dead load.
b. Lateral sliding resistance value to be multiplied by the contact area, as limited by Section 1804.3.
c. Where the building official determines that in-place soils with an allowable bearing capacity of less than 1,500 psf are likely to be present at the site, the allowable bearing capacity shall be determined by a soils investigation.
d. An increase of one-third is permitted when using the alternate load combinations in Section 1605.3.2 that include wind or earthquake loads.

FIGURE 2.5
Presumptive surface bearing values of various soil types, from the 2006 IBC. Classes 3, 4, and 5 refer to the soil group symbols in Figure 2.2. *(Portions of this publication reproduce tables from the* 2006 International Building Code, *International Code Council, Inc., Washington, D.C. Reproduced with Permission. All rights reserved)*

with underlying void spaces into which the clay can expand to prevent structural damage to the foundation itself. When wet clay is put under pressure, water can be slowly squeezed out of it, with a corresponding gradual reduction in volume. In this circumstance, long-term settlement of a foundation bearing on such soil is a risk that must be considered. Taken together, these properties make many clays the least predictable soils for supporting buildings. (In Figure 2.2, the fine-grained soil groups indicated as having a liquid limit greater than 50 are generally the ones most affected by water content, exhibiting higher plasticity (moldability) and greater expansion when wet and lower strength when dry.

In regions of significant earthquake risk, stability of soils during seismic events is also a concern. Sands and silts with high water content are particularly susceptible to *liquefaction*, that is, a temporary change from solid to liquid state during cyclic shaking. Soil liquefaction can lead to loss of support for a building foundation or excessive pressure on foundation walls.

The drainage characteristics of a soil are important in predicting how water will flow on and under building sites and around building substructures. Where a coarse-grained soil is composed of particles mostly of the same size, it has the greatest possible volume of void space between particles, and water will pass through it most readily. Where coarse-grained soils are composed of particles with a diverse range of sizes, the volume of void space between particles is reduced, and such soils drain water less efficiently. Coarse-grained soils consisting of particles of all sizes are termed *well graded* or *poorly sorted*, those with a smaller range of particle sizes are termed *poorly graded* or *well sorted*, and those with particles mostly of one size are termed *uniformly graded* (Figure 2.6).

Because of their smaller particle size, fine-grained soils also tend to drain water less efficiently: Water passes slowly through very fine sands and silts and almost not at all through many clays. A building site with clayey or silty soils near the surface drains poorly and is likely to be muddy and covered with puddles during rainy periods, whereas a gravelly site is likely to remain dry. Underground, water passes quickly through strata of gravel and sand but tends to accumulate above layers of clay and fine silt. An excellent way to keep a basement dry is to surround it with a layer of uniformly graded gravel or crushed stone. Water passing through the soil toward the building cannot reach the basement without first falling to the bottom of this porous layer, from where it can be drawn off in perforated pipes before it accumulates (Figures 2.60–2.62). It does little good to place perforated drainage pipes directly in clay or silt because water cannot flow through the impervious soil toward the pipes.

Rarely is the soil beneath a building site composed of a single type. Beneath most buildings, soils of various types are arranged in superimposed layers (*strata*) that were formed by various past geologic processes. Frequently, soils in any one layer are themselves also mixtures of different soil groups, bearing descriptions such as *well-graded gravel with silty clay and sand, poorly graded sand with clay, lean clay with gravel*, and so on. Determining the suitability of any particular site's soils for support of a building foundation, then, depends on the behaviors of the various soils types and how they interact with each other and with the building foundation.

Subsurface Exploration and Soils Testing

Prior to designing a foundation for any building larger than a single-family house (and even for some single-family

FIGURE 2.6
Two gravel samples, illustrating differences in range of particle sizes or *grading*. The left-hand sample, with a diverse range of particle sizes, comes from a well-graded sandy gravel. On the right is a sample of a uniformly graded gravel in which there is little variation in size among particles. *(Photos by Joseph Iano)*

houses), it is necessary to determine the soil and water conditions beneath the site. This can be done by digging *test pits* or by making *test borings*. Test pits are useful when the foundation is not expected to extend deeper than roughly about 16 feet (3 m), the maximum practical reach of small excavating machines. The strata of the soil can be observed in the pit, and soil samples can be taken for laboratory testing. The level of the *water table* (the elevation at which the soil is saturated and the pressure of the groundwater is atmospheric) will be readily apparent in coarse-grained soils if it falls within the depth of the pit because water will seep through the walls of the pit up to the level of the water table. On the other hand, if a test pit is excavated below the water table in clay, free water will not seep into the pit because the clay is relatively impermeable. In this case, the level of the water table must be determined by means of an observation well or special devices that are installed to measure water pressure. If desired, a *load test* can be performed on the soil in the bottom of a test pit to determine the stress the soil can safely carry and the amount of settlement that should be anticipated under load.

If a pit is not dug, borings with standard penetration tests can give an indication of the bearing capacity of the soil by the number of blows of a standard driving hammer required to advance a sampling tube into the soil by a fixed amount. Laboratory-quality soil samples can also be recovered for testing. Test boring (Figure 2.7) extends the possible range of exploration much deeper into the earth than test pits and returns information on the thickness and locations of the soil strata and the depth of the water table. Usually, a number of holes are drilled across the site; the information from the holes is coordinated and interpolated in the preparation of drawings that document the subsurface conditions for the use of the engineer who will design the foundation (Figure 2.8).

Laboratory testing of soil samples is important for foundation design. By passing a dried sample of coarse-grained soil through a set of sieves with graduated mesh sizes, the particle size distribution in the soil can be determined. Further tests on fine-grained soils assist in their identification and provide information on their engineering properties. Important among

FIGURE 2.7
A truck-mounted drilling rig capable of core drilling to 1500 feet (450 m).
(Courtesy of Acker Drilling Company, Inc.)

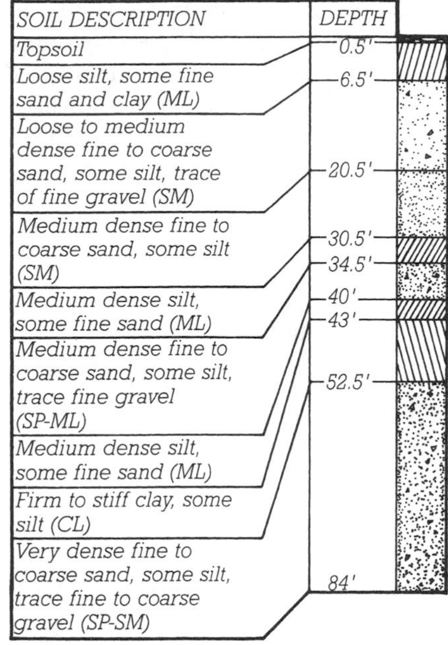

SOIL DESCRIPTION
DEPTH
Topsoil
Loose silt, some fine sand and clay (ML)
Loose to medium dense fine to coarse sand, some silt, trace of fine gravel (SM)
Medium dense fine to coarse sand, some silt (SM)
Medium dense silt, some fine sand (ML)
Medium dense fine to coarse sand, some silt, trace fine gravel (SP-ML)
Medium dense silt, some fine sand (ML)
Firm to stiff clay, some silt (CL)
Very dense fine to coarse sand, some silt, trace fine to coarse gravel (SP-SM)
0.5'
6.5'
20.5'
30.5'
34.5'
40'
43'
52.5'
84'

FIGURE 2.8
A typical log from a soil test boring indicating the type of soil in each stratum and the depth in feet at which it was found. The abbreviations in parentheses refer to the Unified Soil Classification System, and are explained in Figure 2.2.

these are tests that establish the *liquid limit*, the water content at which the soil passes from a plastic state to a liquid state, and the *plastic limit*, the water content at which the soil loses its plasticity and begins to behave as a solid. Additional tests can determine the water content of the soil, its permeability to water, its shrinkage when dried, its shear and compressive strengths, the amount by which the soil can be expected to consolidate under the load, and the rate at which consolidation will take place (Figure 2.9). The last two qualities are helpful in predicting the rate and magnitude of foundation settlement in a building.

The information gained through subsurface exploration and soil testing is summarized in a written *geotechnical report*. This report includes the results of both the field tests and the laboratory tests, recommended types of foundations for the site, recommended depths and bearing stresses for the foundations, and an estimate of the expected rate of foundation settlement. This information can be used directly by foundation and structural engineers in the design of the excavations, dewatering and slope support systems, foundations, and substructure. It is also used by contractors in the planning and execution of sitework.

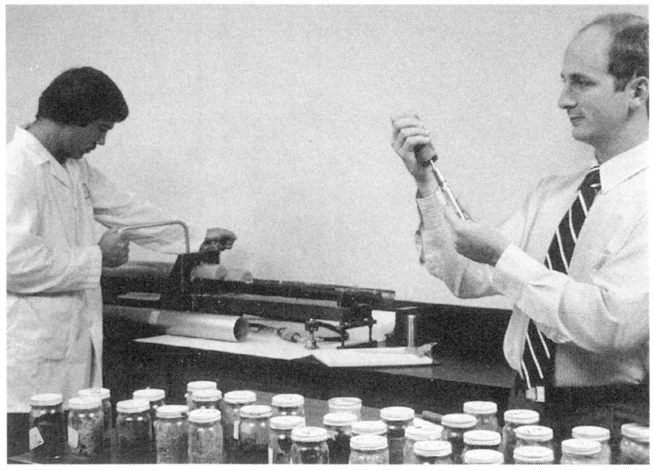
(*a*)

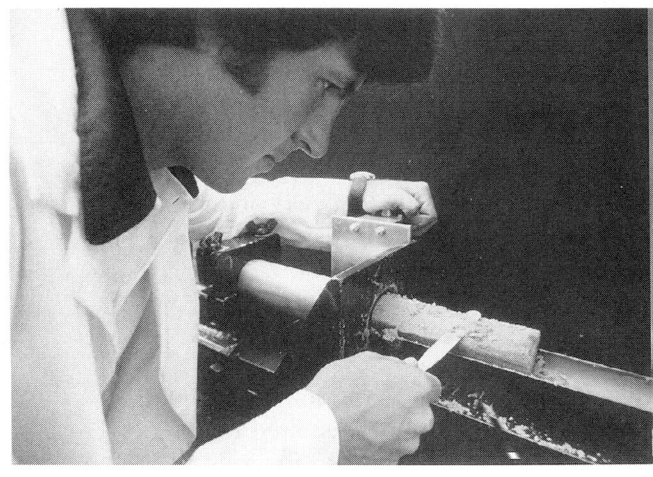
(*b*)

FIGURE 2.9
Some laboratory soil testing procedures. (*a*) To the right, the hardness of split spoon samples is checked with a penetrometer to be sure that they come from the same stratum of soil. To the left, a soil sample from a Shelby sampling tube is cut into sections for testing. (*b*) A section of undisturbed soil from a Shelby tube is trimmed to examine the stratification of silt and clay layers. (*c*) A cylindrical sample of soil is set up for a triaxial load test, the principal method for determining the shear strength of soil. The sample will be loaded axially by the piston in the top of the apparatus and also circumferentially by water pressure in the transparent cylinder. (*d*) A direct shear test, used to measure the shear strength of cohesionless soils. A rectangular prism of soil is placed in a split box and sheared by applying pressure in opposite directions to the two halves of the box. (*e*) One-dimensional consolidation tests in progress on fine-grained soils to determine their compressibility and expected rate of settlement. Each sample is compressed over an extended period of time to allow water to flow out of the sample. (*f*) A panel for running 30 simultaneous constant-head permeability tests to determine the rate at which a fluid, usually water, moves through a soil. (*g*) A Proctor compaction test, in which successive layers of soil are compacted with a specified tamping force. The test is repeated for the same soil with varying moisture content, and a curve is plotted of density achieved versus moisture content of the soil to identify the optimum moisture content for compacting the soil in the field. Not shown here are some common testing procedures for grain size analysis, liquid limit, plastic limit, specific gravity, and unconfined compression. *(Courtesy of Ardaman and Associates, Inc., Orlando, Florida)*

(*c*)

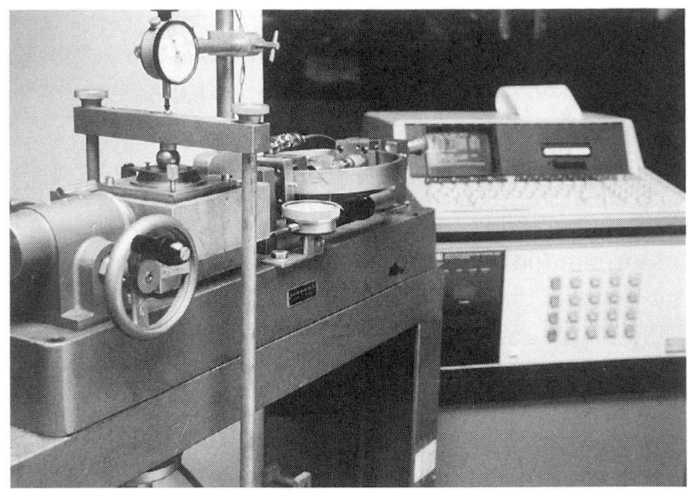

(*d*)

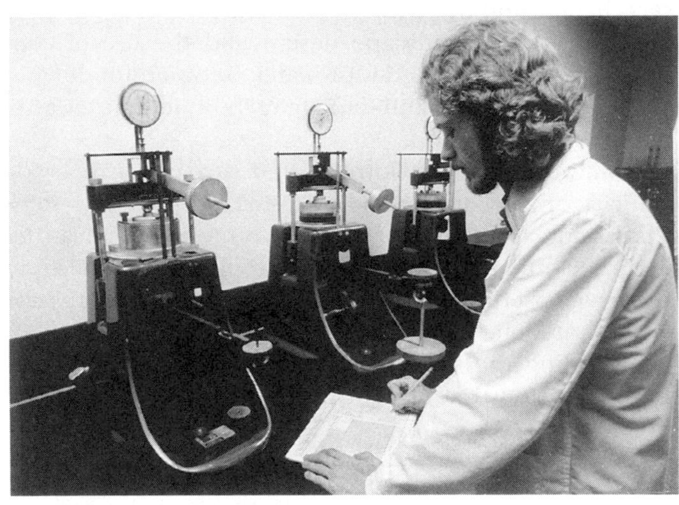

(*e*)

(*f*)

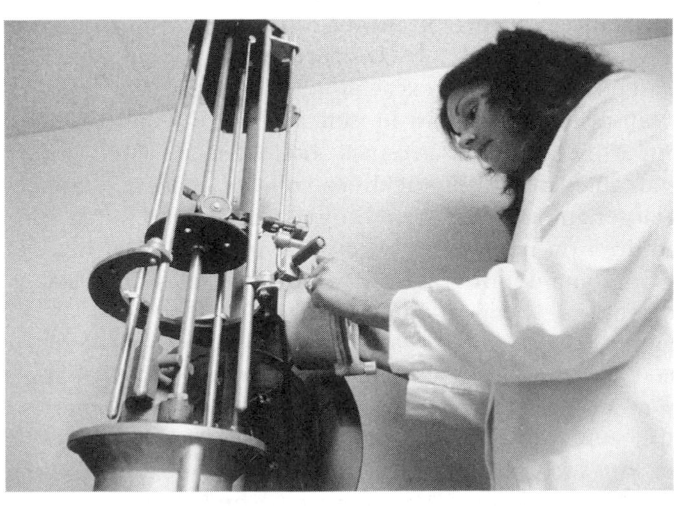

(*g*)

Considerations of Sustainability in Site Work, Excavations, and Foundations

Building sites should be selected and developed so as to protect and conserve natural habitats and resources, to promote biodiversity, to preserve quality open space, and to minimize pollution and unnecessary energy consumption.

Site Selection

Buying and renovating an existing building rather than building a new one saves a great deal of building material and energy. If the existing building has been scheduled for demolition, its renovation also avoids the disposal of an enormous quantity of material in a landfill.

Building in an urban area with existing infrastructure, rather than in undeveloped land unconnected to other community resources, protects open space, natural habitats, and natural resources.

Building on a damaged or polluted site, and designing the building so that it helps to restore the site, benefits the environment rather than degrades it.

Avoiding construction on prime agricultural land prevents the permanent loss of this productive use of the land.

Avoiding construction on undeveloped land that is environmentally sensitive protects the wildlife and natural habitats such land supports. This includes floodplains, land that provides habitat for endangered or threatened species, wetlands, mature forest lands and prairies, and land adjacent to natural bodies of water.

Avoiding construction on public parkland or land adjacent to bodies of water that support recreational use prevents the permanent loss of public resources.

Selecting a building that is well connected to existing networks of public transportation, and to pedestrian and bicycle paths, pays environmental dividends every day for the life of the building by saving fuel, reducing air pollution from automobiles, and minimizing commute times.

Site Design

Minimizing the building footprint and protecting and enhancing portions of the site with natural vegetation protects habitat and helps maintain biodiversity.

Appropriate landscape design and the use of captured rainwater, recycled wastewater, or other nonpotable sources of water for landscape irrigation minimize wasteful water consumption.

Minimizing impervious ground surface (such as for vehicle parking) and providing a surface drainage system that conducts water to areas on the site where it can be absorbed into the ground works to replenish natural aquifers, avoid overloading of storm sewer systems, and reduce water pollution.

Grading the site to appropriate slopes and planting vegetation that holds the soil in place will prevent erosion.

Excavation

At least some *excavation* is required for every building. Organic topsoil is subject to decomposition and to shrinking and swelling with changes in moisture content. It is excellent for growing lawns and landscape plants but unsuitable for supporting buildings. Often, it is scraped away from the building area and stockpiled to one side for redistribution over the site after construction of the building is complete. After the topsoil has been removed, further digging is necessary to place the footings out of reach of water and wind erosion. In colder climates, foundations must be placed below the level to which the ground freezes in winter, the *frost line*, or they must be insulated in such a way that the soil beneath them cannot freeze. Otherwise, a foundation can be lifted and damaged by soil that expands slightly as it freezes. Or, under certain soil and temperature conditions, upward migration of water vapor from the pores in the soil can result in the formation of *ice lenses*, thick layers of frozen water crystals than can lift foundations by even larger amounts.

Excavation is required on many sites to place the footings at a depth where soil of the appropriate bearing capacity is available. Excavation is frequently undertaken so that one or more levels of basement space can be added to a building, whether for additional habitable rooms, for parking, or for mechanical equipment and storage. Where footings must be placed deep to get below the frost line or reach competent soil, a basement is often bargain-rate space, adding little to the overall cost of the building.

In particulate soils, a variety of excavating machines can be used to loosen and lift the soil from the ground: bulldozers, shovel dozers, backhoes, bucket loaders, scrapers, trenching machines, and power shovels of every type. If the soil must be moved more than a short distance, dump trucks come into use.

In rock, excavation is slower and many times more costly. Weak or highly fractured rock can sometimes be broken up with power shovels, tractor-mounted rippers, pneumatic hammers, or drop balls such as those used in building demolition. Blasting, in which explosives are placed

Large existing trees cannot be replaced except by growing new ones over a period of many decades. Planting of new trees is good, but preservation of existing trees is even better.

Distinctive site features such as rock formations, forests, grasslands, streams, marshes, and recreational paths and facilities, if destroyed for new construction, can never be replaced.

Providing shade, vegetated or reflective roof surfaces, reflective paving materials, and open pavement systems reduces heat island effects and creates an improved microclimate for both humans and wildlife.

Minimizing nighttime light pollution is a benefit to humans and nocturnal wildlife.

Avoiding unnecessary shading of adjacent buildings protects those buildings' sources of natural illumination and useful solar heat, minimizing their unnecessary consumption of electricity and heating fuel.

Siting a building for best exposure to sun and wind maximizes solar heat gain in winter and minimizes it in summer to save heating and air conditioning fuel. It also allows utilization of daylight to replace electric lighting.

In general, a building should be designed in such a manner that the site does the heaviest work of environmental modification through good orientation to sun and wind, trees that are placed so as to provide shade and windbreaks, and use of below-grade portions of the building for thermal mass. The passive shell of the building can do most of the rest of the work through orientation of windows with respect to sunlight, good thermal insulation and airtightness, utilization of thermal mass, and energy-efficient windows. The active heating, cooling, and lighting equipment should serve only to fine-tune the interior environment, using as little fossil fuel and electricity as possible.

Construction Process

It is essential to protect trees and sensitive areas of the site from damage during construction.

A building should comply with all local conservation laws relating to soils, wetlands, and stormwater.

Topsoil should be stockpiled carefully during construction and reused on the site.

It is important to guard against soil erosion by water and wind during construction, as well as the sedimentation of streams and sewers, or the polluting of air with dust or particulates that can result.

Vehicle tires compact soil so that it cannot absorb water or support vegetation. Thus, it is important to develop minimal, well-marked access routes for trucks and construction machinery that minimize soil compaction, as well as minimizing noise, dust, air pollution, and inconvenience to neighboring buildings and sites.

Construction machinery should be selected and maintained so that it pollutes the air as little as possible.

Surplus excavated soils should be reused either on the site or on another site nearby.

Construction wastes should be recycled as much as possible.

and detonated in lines of closely spaced holes drilled deep into the rock, is often necessary. In developed areas where blasting is impractical, rock can be broken up with hydraulic splitters.

Excavation Support

If the site is sufficiently larger than the area to be covered by the building, the edges of the excavation can be sloped back or *benched* at an angle such that the soil will not slide back into the hole. This angle, called the *angle of repose*, can be steep for cohesive soils such as the stiffer clays, but it must be shallow for frictional soils such as sand and gravel. On constricted sites, the soil surrounding an excavation must be held back by some kind of *slope support* or *excavation support* capable of resisting the pressures of earth and groundwater (Figure 2.10). Such construction can take many forms, depending on the qualities of the soil, depth of excavation, equipment and preferences of the contractor, proximity of surrounding buildings, and level of the water table.

Shoring

The most common types of slope support, or *shoring*, are soldier beams and lagging, and sheet piling. With

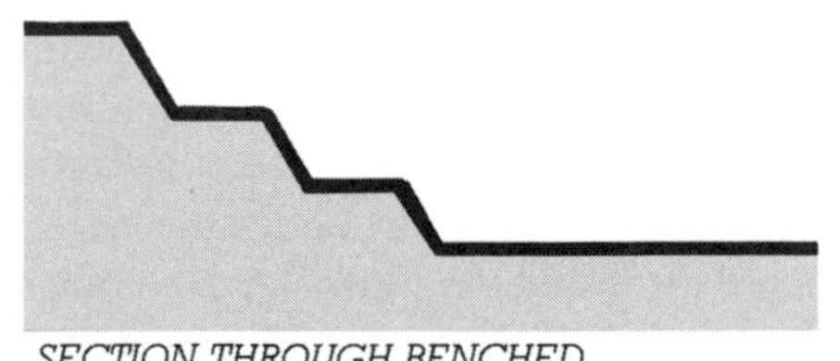

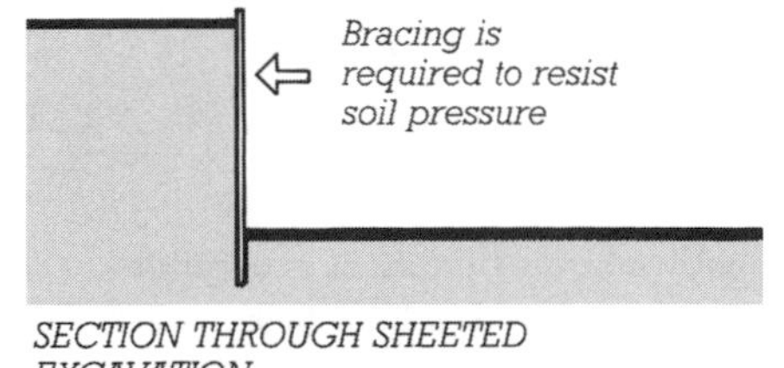

FIGURE 2.10
On a spacious site, an excavation can be benched. When excavating close to property lines or nearby buildings, some form of slope support, such as sheeting, is used to retain the soil around the excavation.

soldier beams and lagging, steel columns called *H-piles* or *soldier beams* are driven vertically into the earth at small intervals around an excavation site before digging begins. As earth is removed, the *lagging*, usually consisting of heavy wood planks, is placed against the flanges of the columns to retain the soil outside the excavation (Figures 2.11 and 2.12). *Sheet piling* or *sheeting* consists of vertical planks of wood, steel, or precast concrete that are placed tightly against one another and driven into the earth to form a solid wall before excavation begins (Figures 2.13 and 2.14). Most often, shoring is temporary, and it is removed as soil is replaced in the excavation. However, it may also be left in place to become a permanent part of the building's substructure. This may be necessary, for example, where shoring is located extremely close to a property line and there is no practical way to remove it after completion of construction without disturbing adjacent property or structures.

Slope support may also take the form of *pneumatically applied concrete*, also called *shotcrete*, in which excavation proceeds first and then the sloped sides are reinforced with a relatively stiff concrete mixture sprayed directly from a hose onto the soil. This method works well where the soil is sufficiently cohesive to hold a steep slope at least temporarily. The hardened concrete reinforces the slope and protects against soil erosion. (Figure 2.15).

Slurry Walls

A *slurry wall* is a more complicated and expensive form of excavation support that is usually economical only if it becomes part of the permanent foundation of the building. The first steps in creating a slurry wall are to lay out the wall's location on the surface of the ground with surveying instruments and to define the location and thickness of the wall with shallow poured concrete *guide walls* (Figures 2.16 and 2.17). When the formwork has been removed from the guide walls, a special narrow *clamshell bucket*, mounted on a crane, is used to excavate the soil from between the guide walls. As the narrow trench deepens, the tendency of its earth walls to collapse is counteracted by filling the trench with a viscous mixture of water and bentonite clay, called a *slurry*, which exerts pressure against the earth walls, holding them in place. The clamshell bucket is lowered and raised through the slurry to continue excavating the soil from the bottom of the trench until the desired depth has been reached, often a number of stories below the ground. Slurry is added as required to keep the trench full at all times.

Meanwhile, workers have welded together cages of steel bars designed to reinforce the concrete wall that will replace the slurry in the trench. Steel tubes whose diameter corresponds to the width of the trench are driven vertically into the trench at predetermined intervals to divide it into sections of a size that can be reinforced and concreted conveniently. The concreting of each section begins with the lowering of a cage of reinforcing bars into the slurry. Then concrete is poured into the trench from the bottom up, using a funnel-and-tube arrangement called a *tremie*. As the concrete rises in the trench, it displaces the slurry, which is pumped out into holding tanks, where it is stored for reuse. After the concrete reaches the top of the trench and has hardened sufficiently, the vertical pipes on either side of the recently poured section are withdrawn from the trench, and the adjoining sections are poured. This process is repeated for each section of the wall. When the concrete in all the trenches has cured to its intended strength, earth removal begins inside the wall, which serves as sheeting for the excavation.

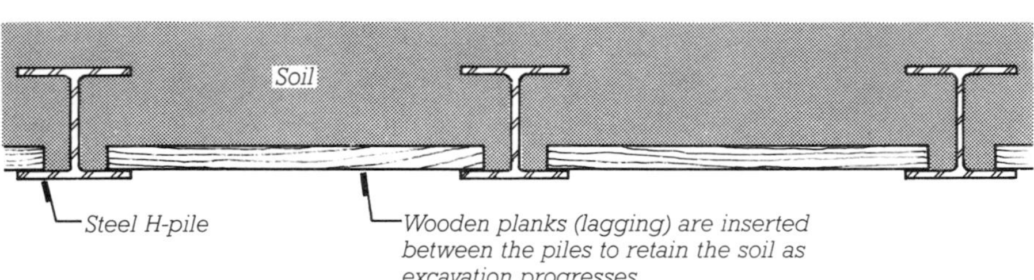

FIGURE 2.11
Soldier beams and lagging, seen in horizontal section.

FIGURE 2.12
Soldier beams and lagging. Lagging planks are added at the bottom as excavation proceeds. The drill rig is boring a hole for a tieback to brace a soldier beam. *(Courtesy of Fr anki Foundation Company)*

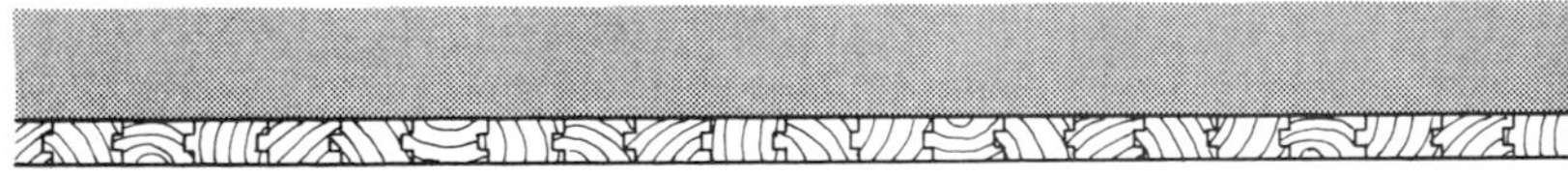

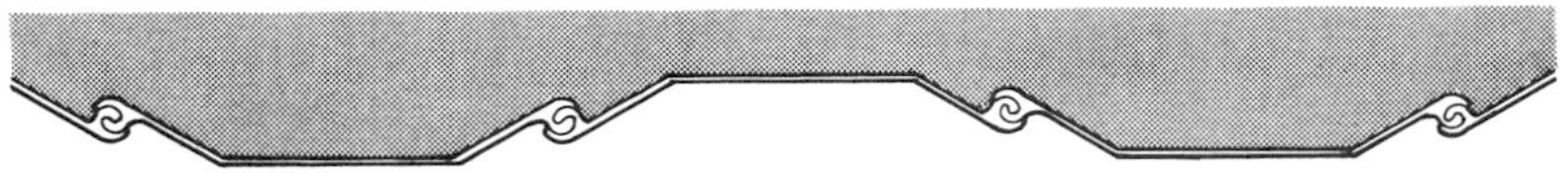

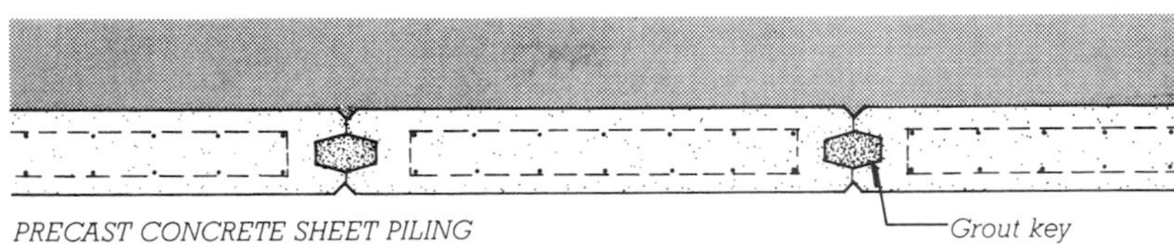

FIGURE 2.13
Horizontal sections through three types of sheet piling. The shading represents the retained earth.

FIGURE 2.14
Drilling tieback holes for a wall of steel sheet piling. Notice the completed tieback connections to the horizontal waler in the foreground. The hole in the top of each piece of sheet piling allows it to be lifted by a crane. *(Courtesy of Franki Foundation Company)*

FIGURE 2.15
Where slope support turns the corner in this excavation and the soil can be sloped at a lesser angle, less expensive shotcrete takes the place of soldier beams and lagging. *(Photo by Joseph Iano)*

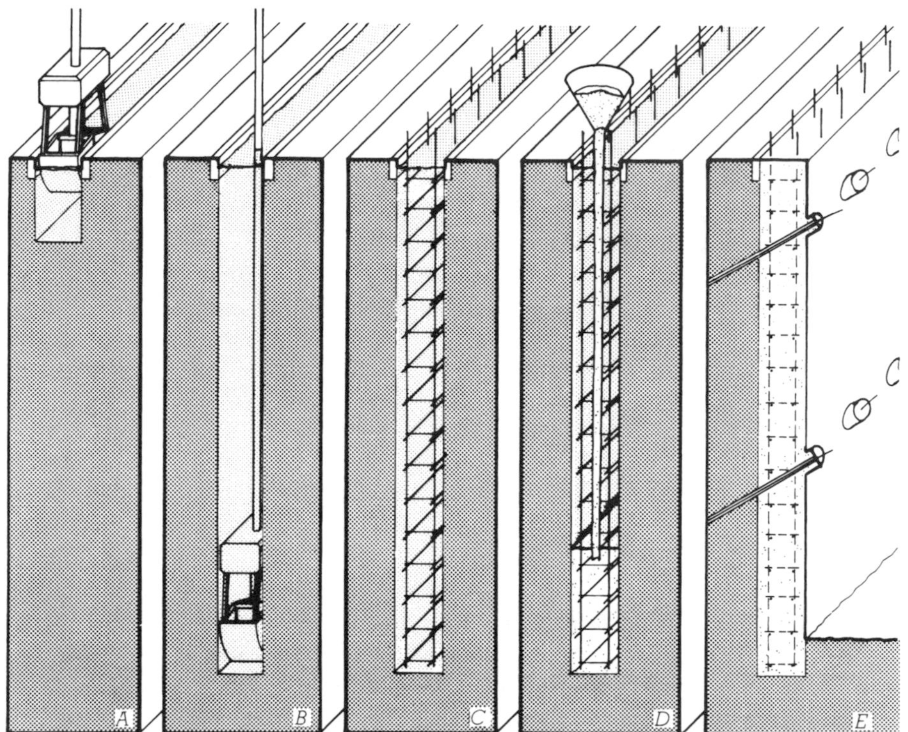

Figure 2.16
Steps in constructing a slurry wall. (*a*) The concrete guide walls have been installed, and the clamshell bucket begins excavating the trench through a bentonite clay slurry. (*b*) The trench is dug to the desired depth, with the slurry serving to prevent collapse of the walls of the trench. (*c*) A welded cage of steel reinforcing bars is lowered into the slurry. (*d*) The trench is concreted from the bottom up with the aid of a tremie. The displaced slurry is pumped from the trench, filtered, and stored for reuse. (*e*) The reinforced concrete wall is tied back as excavation progresses.

(*a*)

(*b*)

Figure 2.17
Constructing a slurry wall. (*a*) The guide walls are formed and poured in a shallow trench. (*b*) The narrow clamshell bucket discharges a load of soil into a waiting dump truck. Most of the trench is covered with wood pallets for safety.

(*c*)

(*d*)

(*e*)

FIGURE 2.17 (*continued*)
(*c*) A detail of the narrow clamshell bucket used for slurry wall excavation. (*d*) Hoisting a reinforcing cage from the area where it was assembled, getting ready to lower it into the trench. The depth of the trench and height of the slurry wall are equal to the height of the cage, which is about four stories for this project. (*e*) Concreting the slurry wall with a tremie. The pump just to the left of the tremie removes slurry from the trench as concrete is added. *(Photos b, c, and d courtesy of Franki Foundation Company. Photos a and e courtesy of Soletanche)*

In addition to the sitecast concrete slurry wall described in the preceding paragraphs, *precast concrete slurry walls* are built. The wall is prestressed and is produced in sections in a precasting plant (see Chapter 15), then trucked to the construction site. The slurry for precast walls is a mixture of water, bentonite clay, and portland cement. Before a section is lowered by a crane into the slurry, its face is coated with a compound that prevents the clay–cement slurry from adhering to it (Figure 2.18). The sections are installed side by side in the trench, joined by tongue-and-groove edges or synthetic rubber gaskets. After the portland cement has caused the slurry to harden to a soillike consistency, excavation can begin, with the hardened slurry on the inside face of the wall dropping away from the coated surface as soil is removed. The primary advantages of a precast slurry wall over a sitecast one are better surface quality, more accurate wall alignment, a thinner wall (due to the structural efficiency of prestressing), and improved watertightness of the wall because of the continuous layer of solidified clay outside.

Soil Mixing

Soil mixing is a technique of adding a modifying substance to soil and blending it in place by means of augers or paddles rotating on the end of a vertical shaft (Figure 2.19). This technique has several applications, one of which is to remediate soil contaminated with a chemical or

FIGURE 2.18
Workers apply a nonstick coating to a section of precast concrete slurry wall as it is lowered into the trench.

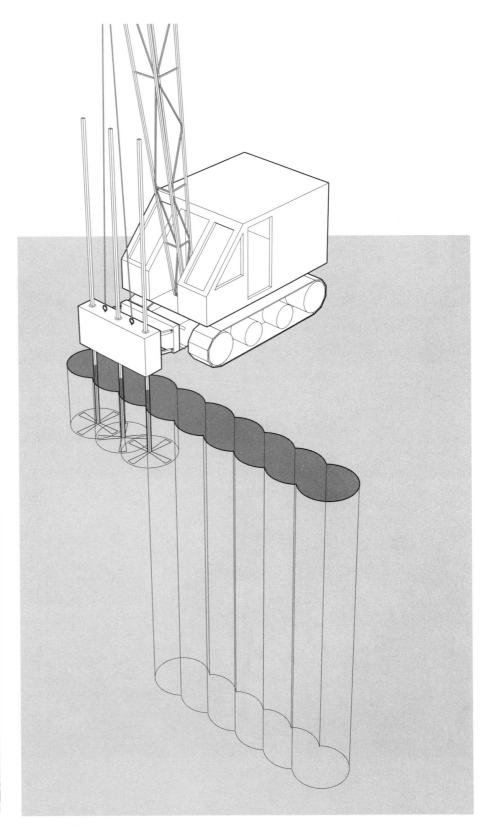

FIGURE 2.19
Soil mixing.

biological substance by blending it with a chemical that renders it harmless. Another is to mix portland cement and water with a soil to create a cylinder of low-strength concrete in the ground. A linear series of these cylinders can serve as a cutoff wall against water penetration or as excavation support (Figure 2.20). Soil mixing can also serve to stabilize and strengthen areas of weak soil.

Bracing

All forms of slope support and excavation support must be braced against soil and water pressures as the excavation deepens (Figure 2.21). *Crosslot bracing* utilizes temporary steel wide-flange columns that are driven into the earth by a piledriver at points where braces will cross. As the earth is excavated down around the sheeting and the columns, tiers of horizontal bracing struts, usually

(*a*)

(*b*)

FIGURE 2.20
Soil-mixed excavation support. (*a*) Excavation proceeds after completion of soil mixing. Bracing consists of soldier piles, walers, and tiebacks. The soldier piles are inserted during soil mixing, before the soil/cement mixture hardens. The walers and tiebacks are installed later, as excavation progresses. (*b*) Fully excavated soil-mixed sheeting. This excavation support system must be strong enough to resist the soil pressures caused by the adjacent buildings. *(Photographs courtesy of Schnable Foundation Company)*

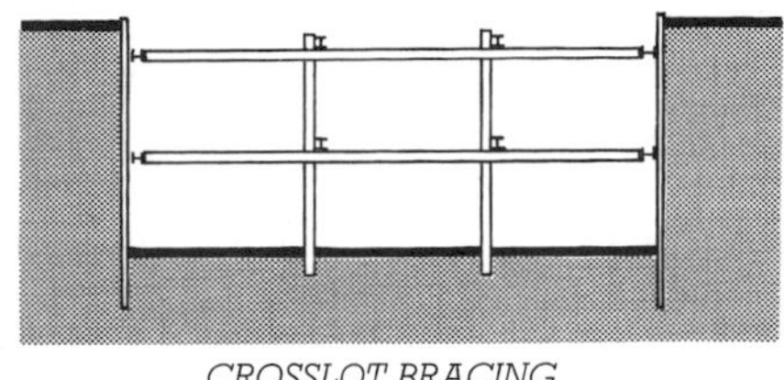
CROSSLOT BRACING

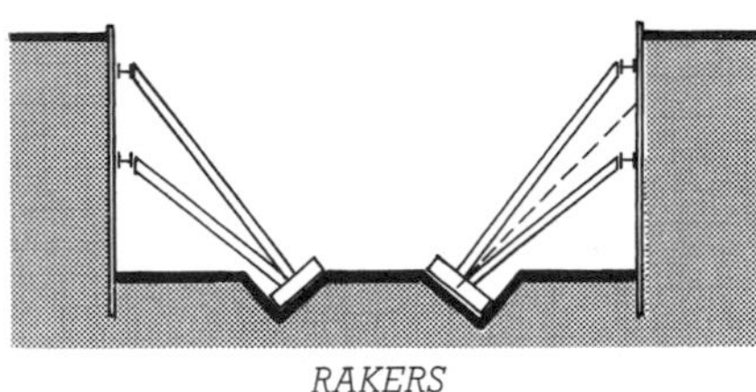
RAKERS

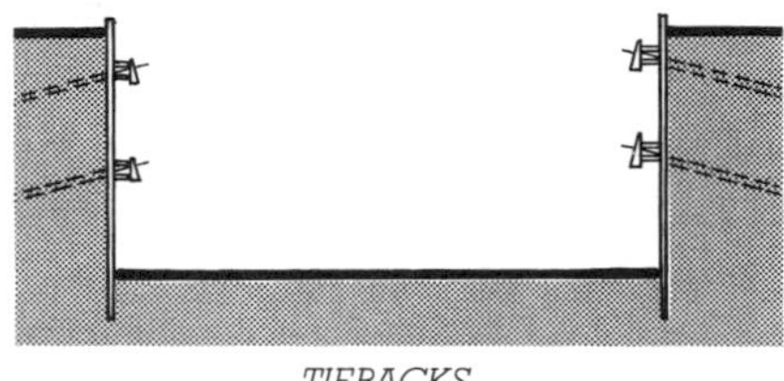
TIEBACKS

FIGURE 2.21
Three methods of bracing shoring drawn in cross section. The connection between the waler and the brace, raker, or tieback needs careful structural design. The broken line between rakers indicates the mode of excavation: The center of the hole is excavated first with sloping sides, as indicated by the broken line. The heel blocks and uppermost tier of rakers are installed. As the sloping sides are excavated deeper, more tiers of rakers are installed. Notice how the tiebacks leave the excavation totally free of obstructions.

of steel, are added to support *walers*, which are beams that span across the face of the sheeting. Where the excavation is too wide for crosslot bracing, sloping *rakers* are used instead, bearing against *heel blocks* or other temporary footings.

Both rakers and crosslot bracing, especially the latter, are a hindrance to the excavation process. A clamshell bucket on a crane must be used to remove the earth between the braces, which is much less efficient and more costly than removing soil with a shovel dozer or backhoe in an open excavation.

Where subsoil conditions permit, *tiebacks* can be used instead of braces to support the sheeting while maintaining an open excavation. At each level of walers, holes are drilled at intervals through the sheeting and the surrounding soil into rock or a stratum of stable soil. Steel cables or tendons are then inserted into the holes, grouted to anchor them to the rock or soil, and stretched tight with hydraulic jacks (posttensioned) before they are fastened to the walers (Figures 2.22–2.24).

Excavations in fractured rock can often avoid sheeting altogether, either by injecting grout into the joints of the rock to stabilize it or by drilling into the rock and inserting *rock anchors* that fasten the blocks together (Figure 2.25).

In some cases, vertical walls of particulate soils can be stabilized by *soil nailing*. A soil nail is similar to a rock anchor: It is a length of steel reinforcing bar that is inserted into a nearly horizontal hole drilled deep into the soil. Grout is injected into the hole to bind the soil nail to the surrounding soil. Large numbers of closely spaced nails are used to knit a large block of soil together so that it behaves more like weak rock than particulate soil.

Bracing and tiebacks in excavations are usually temporary. Their function is taken over permanently by the floor structure of the basement

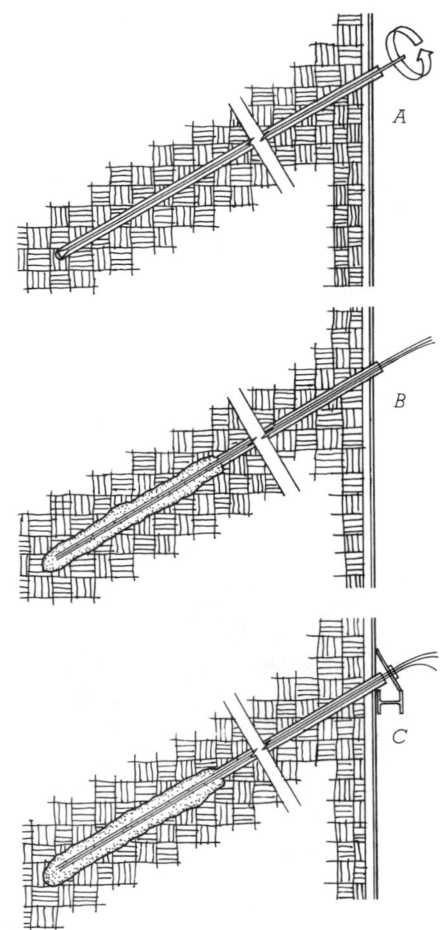

FIGURE 2.22
Three steps in the installation of a tieback to a soil anchor. (*a*) A rotary drill bores a hole through the sheeting and into stable soil or rock. A steel pipe casing keeps the hole from caving in where it passes through noncohesive soils. (*b*) Steel prestressing tendons are inserted into the hole and grouted under pressure to anchor them to the soil. (*c*) After the grout has hardened, the tendons are tensioned with a hydraulic jack and anchored to a waler.

(*a*)

FIGURE 2.23
Installing tiebacks. (*a*) Drilling through a slurry wall for a tieback. The ends of hundreds of completed tiebacks protrude from the wall.

(*continued*)

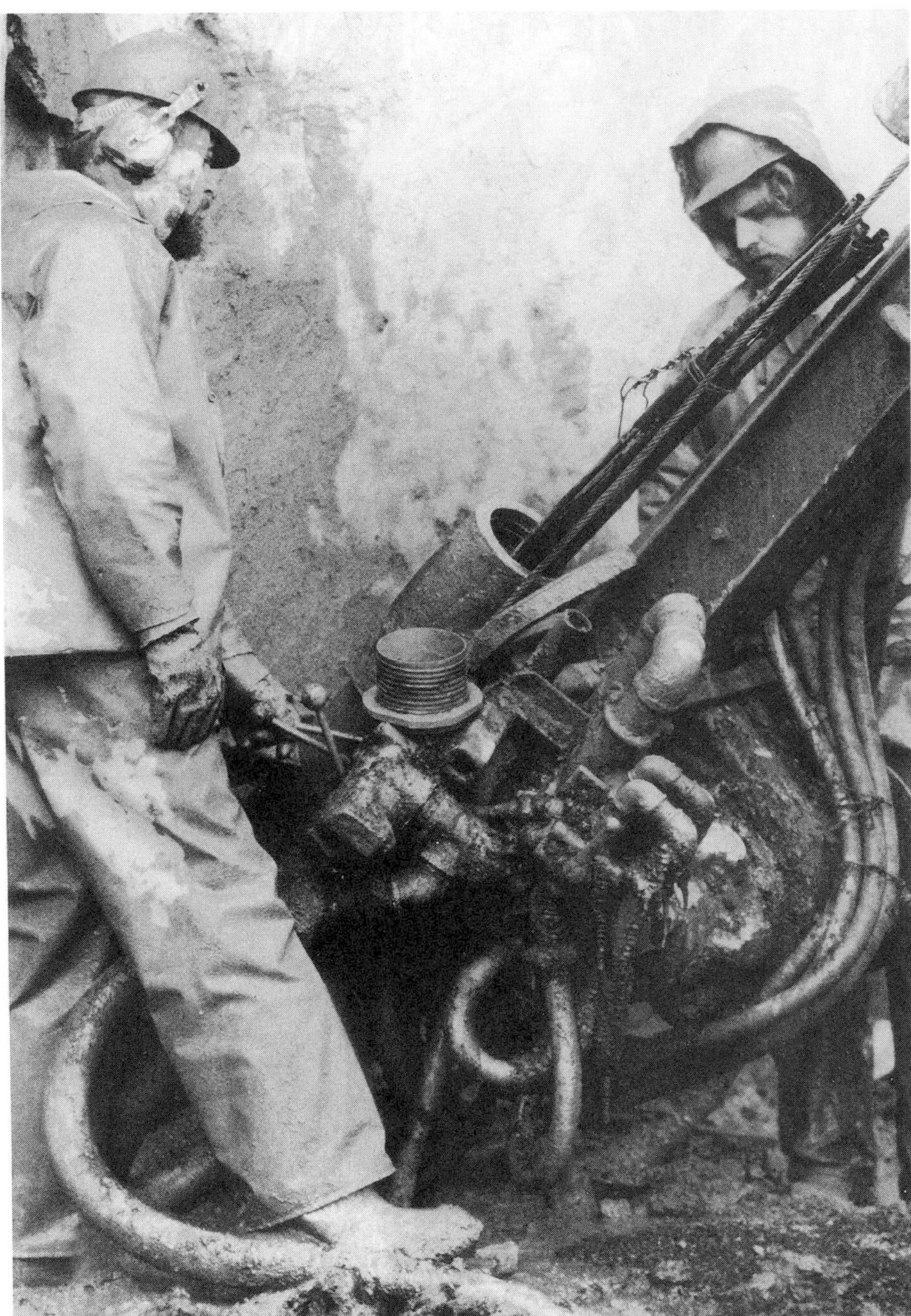

(*b*)

FIGURE 2.23 (*continued*)
(*b*) Inserting prestressing tendons into the steel casing for a tieback. The apparatus in the center of the picture is for pressure injecting grout around the tendons.

(c)

FIGURE 2.23 (*continued*)
(c) After the tendons have been tensioned, they are anchored to a steel chuck that holds them under stress, and the cylindrical hydraulic jack is moved to the next tieback. *(Photos courtesy of Franki Foundation Company)*

FIGURE 2.24
Slurry wall and tieback construction used to support historic buildings around a deep excavation for a station of the Paris Metro. *(Photo courtesy of Soletanche)*

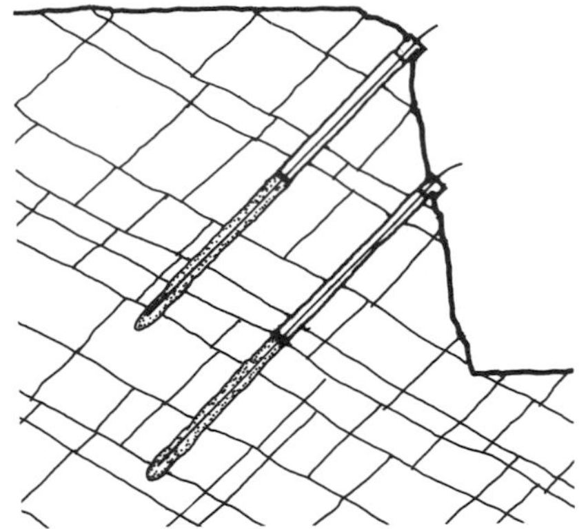

FIGURE 2.25
Rock anchors are similar to tiebacks but are used to hold jointed rock formations in place around an excavation.

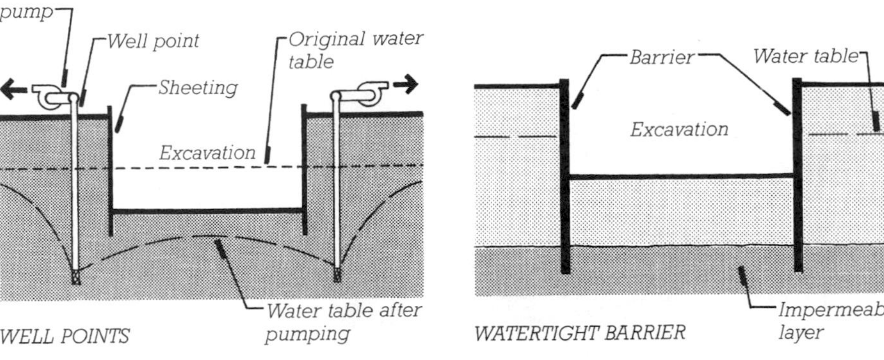

FIGURE 2.26
Two methods of keeping an excavation dry, viewed in cross section. The water sucked from well points depresses the water table in the immediate vicinity to a level below the bottom of the excavation. Watertight barrier walls work only if their bottom edges are inserted into an impermeable stratum that prevents water from working its way under the walls.

levels of the building, which is designed specifically to resist lateral loads from the surrounding earth as well as ordinary floor loads.

Dewatering

During construction, excavations must be kept free of standing water. Such water may come from precipitation or it may come from groundwater seepage originating from any of a number of sources, such as surface water percolating through the soil, underground streams, perched water moving over impervious soil strata, or adjacent permanently saturated soil areas where the excavation extends below the water table. Some shallow excavations in relatively dry soil conditions may remain free of standing water without any intervention. But most excavations require some form of *dewatering*, or extraction of water from the excavation or surrounding soil. The most common method of dewatering is to remove water by pumping as it accumulates in pits, called *sumps*, created at low points in the excavation. Where the volume of groundwater flowing into the excavation is great, or with certain types of soils, particularly sands and silt, that may be softened by constant seepage, it may be necessary to keep groundwater from entering the excavation at all. This can be done either by pumping water from the surrounding soil to depress the water table below the level of the bottom of the excavation or by erecting a watertight barrier, such as a slurry wall, around the excavation (Figure 2.26).

Well points are commonly used to depress the water table. These are vertical sections of pipe with screened openings at the bottom that keep out soil particles while allowing water to enter. Closely spaced well points are driven into the soil around the entire perimeter of the excavation. These are connected to horizontal header pipes leading to pumps that continually draw water from the system and discharge it away from the building site. Once pumping has drawn down the water table in the area of the excavation, work can continue "in the dry" (Figures 2.27 and 2.28). For excavations deeper than the 20 feet (6 m) or so that cannot be drained by a suction pump stationed at ground level, two rings of well points may be required, the inner ring being driven to a deeper level than the outer ring, or a single ring of deep wells with submersible force pumps may have to be installed.

In some cases, well points may not be practical: they may have insufficient capacity to ensure that an excavation remains dry; restrictions on the disposal of groundwater may limit their use; reliability due to power outages may be a concern; or lowering of the water table may have serious adverse effects on neighboring buildings by causing consolidation and settling of soil under their foundations or by exposing untreated wood foundation piles, previously protected by total immersion in water, to decay. In these cases, a *watertight barrier wall* may be used as an alternative (Figure 2.26). Slurry walls and soil mixed walls (pages 40–45) can make excellent watertight barriers. Sheet piling can also work, but it tends to leak at the joints. *Soil freezing* is also possible. In this method, an array of vertical pipes similar to well points is used to continuously circulate coolant at temperatures low enough to freeze the soil around an excavation area, resulting in a temporary but reliable barrier to groundwater. Watertight barriers must resist the hydrostatic pressure of the surrounding water, which increases with depth, so for deeper excavations, a system of bracing or tiebacks is required. A watertight barrier also works only if it reaches into a stratum of impermeable soil such as clay. Otherwise, water can flow beneath the barrier and rise up into the excavation.

Figure 2.27
Well point dewatering. Closely spaced vertical well points connect to the larger-diameter header pipe. *(Courtesy of Griffin Dewatering Corporation)*

Figure 2.28
An excavation is kept dry despite the close proximity of a large body of water. Two dewatering pumps are visible in the foreground. A pair of header pipes and numerous well points can be seen surrounding the excavation. (*Courtesy of Griffin Dewatering Corporation)*

FOUNDATIONS

It is convenient to think of a building as consisting of three major parts: the *superstructure*, which is the aboveground portion of the building; the *substructure*, which is the habitable below-ground portion; and the foundations, which are the components of the building that transfer its loads into the soil (Figure 2.29).

There are two basic types of foundations: shallow and deep. *Shallow foundations* are those that transfer the load to the earth at the base of the column or wall of the substructure. *Deep foundations*, either piles or caissons, penetrate through upper layers of incompetent soil in order to transfer the load to competent bearing soil or rock deeper within the earth. Shallow foundations are generally less expensive than deep ones and can be used where suitable soil is found at the level of the bottom of the substructure, whether this be several feet or several stories below the ground surface.

The primary factors that affect the choice of a foundation type for a building are:

- Subsurface soil and groundwater conditions
- Structural requirements, including foundation loads, building configurations, and depth

Secondary factors that may be important include:

- Construction methods, including access and working space
- Environmental factors, including noise, traffic, and disposal of earth and water
- Building codes and regulations
- Proximity of adjacent property and potential impacts on that property
- Time available for construction
- Construction risks

The foundation engineer is responsible for assessing these factors and, working together with other members of the design and construction team, selecting the most suitable foundation system.

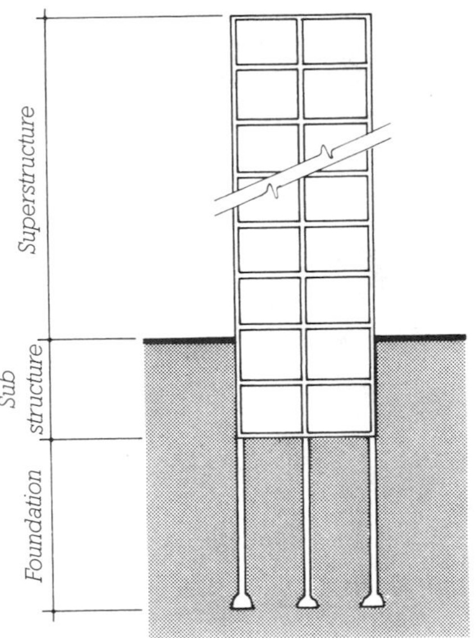

FIGURE 2.29
Superstructure, substructure, and foundation. The substructure in this example contains two levels of basements, and the foundation consists of bell caissons. (In some buildings, the substructure and foundation may be partly or wholly the same.)

Shallow Foundations

Most shallow foundations are simple concrete *footings*. A *column footing* is a square block of concrete, with or without steel reinforcing, that accepts the concentrated load placed on it from above by a building column and spreads this load across an area of soil large enough that the allowable bearing stress of the soil is not exceeded. A *wall footing* or *strip footing* is a continuous strip of concrete that serves the same function for a loadbearing wall (Figures 2.30 and 2.31).

To minimize settlement, footings are usually placed on undisturbed soil. Under some circumstances, footings may be constructed over *engineered fill*, which is earth that has been deposited under the supervision of a soils engineer. The engineer, working from the results of laboratory compaction tests on samples taken from the soil used for filling, makes sure that the soil is deposited in thin layers at a controlled moisture content and compacted in accordance with detailed procedures that ensure a known loadbearing capacity and long-term stability.

Footings appear in many forms in different foundation systems. In climates with little or no ground frost, a concrete *slab on grade* with thickened edges is the least expensive foundation and floor system that one can use and is applicable to one- and two-story buildings of any type of construction (see Chapter 14 for further information on slabs on grade). Or, in colder regions the edges of a slab on grade may be supported with deeper wall footings that bear on soil below the frost line. For floors raised above the ground, either over a *crawlspace* or a *basement*, support is provided by concrete or masonry foundation walls supported on concrete strip footings (Figure 2.32). When building on slopes, it is often necessary to step the footings to maintain the required depth of footing at all points around the building (Figure 2.33). If soil conditions or earthquake precautions require it, column footings on steep slopes may be linked together with reinforced concrete *tie beams* to avoid possible differential slippage between footings.

Footings cannot legally extend beyond a property line, even for a building built tightly against it. If the outer toe of the footing were simply cut off at the property line, the footing would not be symmetrically loaded by the column or wall and would tend to rotate and fail. *Combined footings* and *cantilever*

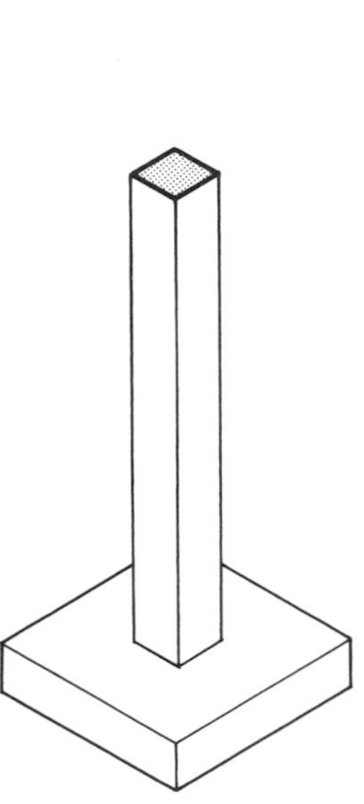

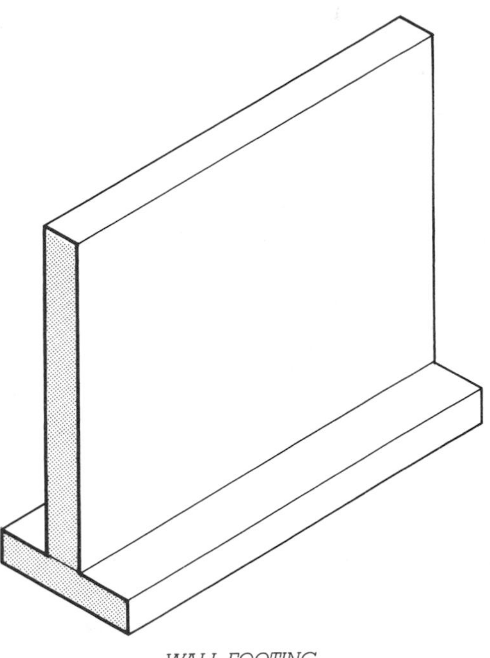

FIGURE 2.30
A column footing and a wall footing of concrete. The steel reinforcing bars have been omitted from this illustration for clarity. The role of steel reinforcing in the structural performance of concrete elements is explained in Chapter 13.

FIGURE 2.31
These concrete foundation walls for an apartment building, with their steel formwork not yet stripped, rest on wall footings. For more extensive illustrations of wall and column footings, see Figures 14.5, 14.7, and 14.11, and 14.13 *(Reprinted with permission of the Portland Cement Association from Design and Control of Concrete Mixtures; Photos: Portland Cement Association, Skokie, IL)*

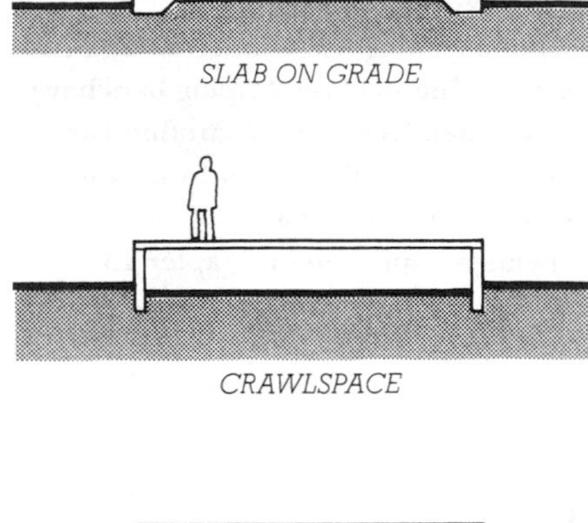

FIGURE 2.32
Three types of substructures with shallow foundations. The slab on grade is the most economical under many circumstances, especially where the water table lies near the surface of the ground. A crawlspace is often used under a floor structure of wood or steel, and gives much better access to underfloor piping and wiring than a slab on grade. Basements provide usable space for building occupants.

footings solve this problem by tying the footings for the outside row of columns to those of the next row in such a way that any rotational tendency is neutralized (Figure 2.34).

In situations where the allowable bearing capacity of the soil is low in relation to the weight of the building, column footings may become large enough that it is more economical to merge them into a single *mat* or *raft foundation* that supports the entire building. Mats for very tall buildings may be 6 feet (1.8 m) thick or more and are heavily reinforced (Figure 2.35).

Where the bearing capacity of the soil is low and settlement must be carefully controlled, a *floating foundation* is sometimes used. A floating foundation is similar to a mat foundation, but is placed beneath a building at a depth such that the weight of the soil removed from the excavation is equal to the weight of the building above. One story of excavated soil weighs about the same as five to eight stories of superstructure, depending on the density of the soil and the construction of the building (Figure 2.36).

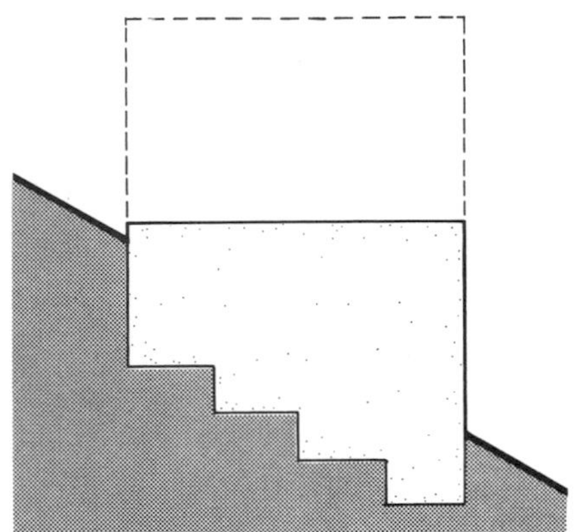

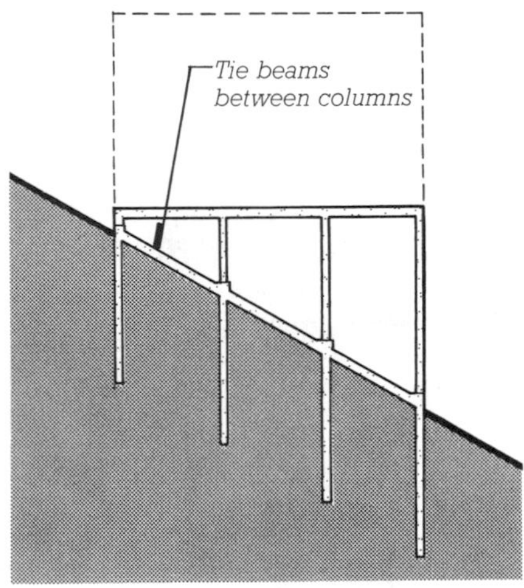

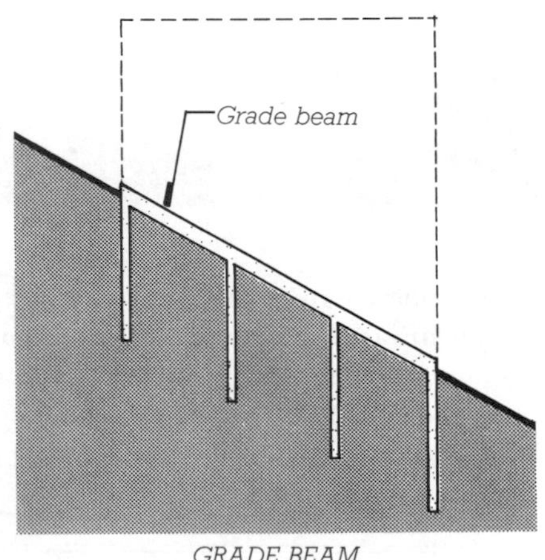

FIGURE 2.33
Foundations on sloping sites, viewed in a cross section through the building. The broken line indicates the outline of the superstructure. Wall footings are stepped to maintain the necessary distance between the bottom of the footing and the surface of the ground. Separate column foundations, whether caissons, as shown here, or column footings, are often connected with reinforced concrete tie beams to reduce differential movement between the columns. A grade beam differs from a tie beam by being reinforced to distribute the continuous load from a bearing wall to separate foundations.

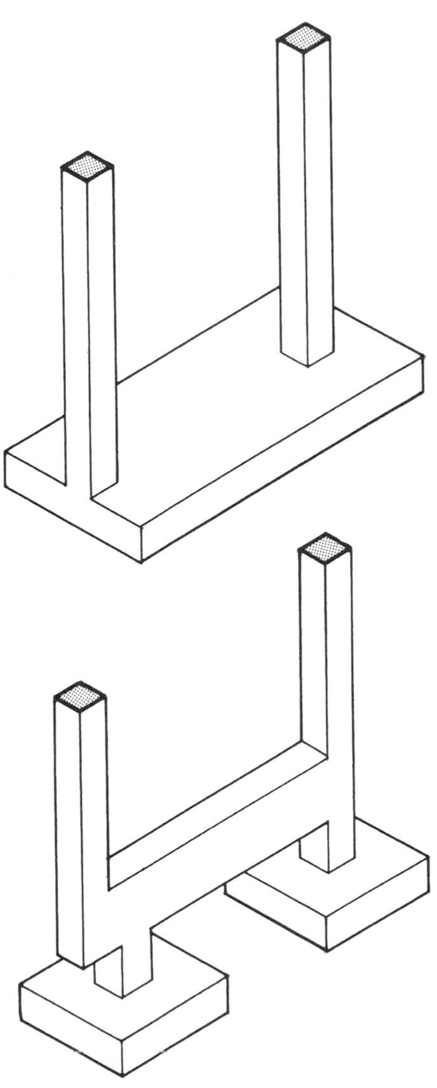

FIGURE 2.34
Either a combined footing (top) or a cantilevered footing (bottom) is used when columns must abut a property line. By combining the foundation for the column against the property line, at the left, with the foundation for the next interior column to the right in a single structural unit, a balanced footing design can be achieved. The concrete reinforcing steel has been omitted from these drawings for the sake of clarity.

FIGURE 2.35
Pouring a large foundation mat. Six truck-mounted pumps receive concrete from a continuous procession of transit-mix concrete trucks and deliver this concrete to the heavily reinforced mat. Concrete placement continues nonstop around the clock until the mat is finished to avoid "cold joints," weakened planes between hardened concrete and fresh concrete. The soil around this excavation is supported with a sitecast concrete slurry wall. Most of the slurry wall is tied back, but a set of rakers is visible at the lower right. *(Courtesy of Schwing America, Inc.)*

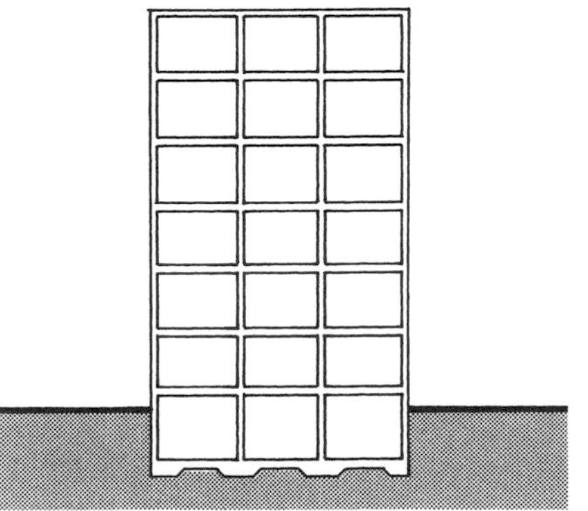

FIGURE 2.36
A cross section through a building with a floating foundation. The building weighs approximately the same as the soil excavated for the substructure, so the stress in the soil beneath the building is the same after construction as it was before.

Deep Foundations

Caissons

A *caisson*, or *drilled pier* (Figure 2.37), is similar to a column footing in that it spreads the load from a column over a large enough area of soil that the allowable stress in the soil is not exceeded. It differs from a column footing in that it extends through strata of unsatisfactory soil beneath the substructure of a building until it reaches a more suitable stratum. A caisson is constructed by drilling or hand-digging a hole, belling (flaring) the hole out at the bottom as necessary to achieve the required bearing area, and filling the hole with concrete. Large *auger drills* (Figures 2.38 and 2.39) are used for drilling caissons; hand excavation is used only if the soil is too full of boulders for the drill. A temporary cylindrical steel casing is usually lowered around the drill as it progresses to support the soil around the hole. When a firm bearing stratum is reached, the bell, if required, is created at the bottom of the shaft either by hand excavation or by a special *belling bucket* on the drill (Figure 2.40). The bearing surface of the soil at the bottom of the hole is then inspected to be sure it is of the anticipated quality, and the hole is filled with concrete, withdrawing the casing as the concrete rises. Reinforcing is seldom used in the concrete except near the top of the caisson, where it joins the columns of the superstructure.

Caissons are large, heavy-duty foundation components. Their shaft

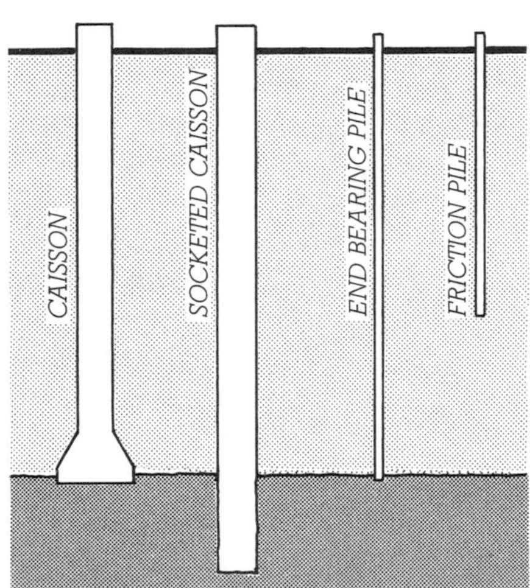

FIGURE 2.37
Deep foundations. Caissons are concrete cylinders poured into drilled holes. They reach through weaker soil (light shading) to bear on competent soil beneath. The end bearing caisson at the left is belled as shown when additional bearing capacity is required. The socketed caisson is drilled into a hard stratum and transfers its load primarily by friction between the soil or rock and the sides of the caisson. Piles are driven into the earth. End bearing piles act in the same way as caissons. The friction pile derives its load-carrying capacity from friction between the soil and the sides of the pile.

FIGURE 2.38
A 6-foot- (1828-mm)-diameter auger on a telescoping 70-foot (21-m) bar brings up a load of soil from a caisson hole. The auger will be rotated rapidly to spin off the soil before being reinserted in the hole. *(Courtesy of Calweld Inc.)*

FIGURE 2.39
For cutting through hard material, the caisson drill is equipped with a carbide-toothed coring bucket. *(Courtesy of Calweld Inc.)*

diameters range from 18 inches (460 mm) up to 8 feet (2.4 m) or more. *Belled caissons* are practical only where the bell can be excavated in a cohesive soil (such as clay) that can retain its shape until concrete is poured. Where groundwater is present, the temporary steel casing can prevent flooding of the caisson hole during its construction. But where the bearing stratum is permeable, water may be able to fill the hole from below and caisson construction may not be practical.

A *socketed caisson* (Figure 2.37) is drilled into rock at the bottom rather than belled. Its bearing capacity comes not only from its end bearing, but from the frictional forces between the sides of the caisson and the rock as well. Figure 2.41 shows the installation of a rock caisson or drilled-in caisson, a special type of socketed caisson with a steel H-section core.

FIGURE 2.40
The bell is formed at the bottom of the caisson shaft by a belling bucket with retractable cutters. The example shown here is for an 8-foot- (2.44-m)-diameter shaft and makes a bell 21 feet (6.40 m) in diameter. *(Courtesy of Calweld Inc.)*

(*a*) (*b*)

FIGURE 2.41

Installing a rock caisson. (*a*) The shaft of the caisson has been drilled through softer soil to the rock beneath and cased with a steel pipe. A churn drill is being lowered into the casing to begin advancing the hole into the rock. (*b*) When the hole has penetrated the required distance into the rock stratum, a heavy steel H-section is lowered into the hole and suspended on steel channels across the mouth of the casing. The space between the casing and the H-section is then filled with concrete, producing a caisson with a very high load-carrying capacity because of the composite structural action of the steel and the concrete. *(Courtesy of Franki Foundation Company)*

Piles

A *pile* (Figure 2.37) is distinguished from a caisson by being forcibly driven into the earth rather than drilled and poured. It may be used where noncohesive soils, subsurface water conditions, or excessive depth of bearing strata make caissons impractical. The simplest kind of pile is a *timber pile*, a tree trunk with its branches and bark removed; it is held small end down in a *piledriver* and beaten into the earth with repeated blows of a heavy mechanical hammer. If a pile is driven until its tip encounters firm resistance from a suitable bearing stratum such as rock, dense sands, or gravels, it is an *end bearing pile*. If it is driven only into softer material, without encountering a firm bearing layer, it may still develop a considerable load-carrying capacity through frictional resistance between the sides of the pile and the soil through which it is driven; in this case, it is known as a *friction pile*. (Some piles rely on a combination of end bearing and friction for their strength.) Piles are usually driven closely together in clusters that contain 2 to 25 piles each. The piles in each cluster are later joined at the top by a reinforced concrete *pile cap*, which distributes the load of the column or wall above among the piles (Figures 2.42 and 2.43).

If . . . solid ground cannot be found, but the place proves to be nothing but a heap of loose earth to the very bottom, or a marsh, then it must be dug up and cleared out and set with piles made of charred alder or olive wood or oak, and these must be driven down by machinery, very closely together. . . .

Marcus Vitruvius Pollio, Roman Architect, *The Ten Books on Architecture,* 1st century B.C.

End bearing piles work essentially the same as caissons and are used on sites where a firm bearing stratum can be reached by the piles, sometimes at depths of 150 feet (45 m) or more. Each pile is driven "to refusal," the point at which little additional penetration is made with continuing blows of the hammer, indicating that the pile is firmly embedded in the bearing layer. Friction piles work best in silty, clayey, and sandy soils. They are driven either to a predetermined depth or until a certain level of resistance to hammer blows is encountered, rather than to refusal as with end bearing piles. Clusters of friction piles have the effect of distributing a concentrated load from the structure above into a large volume of soil around and below the cluster, at stresses that lie safely within the capability of the soil (Figure 2.44).

The loadbearing capacities of piles are calculated in advance based on soil test results and the properties of the piles and piledriver. To verify

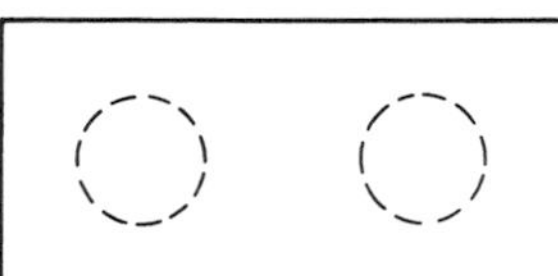

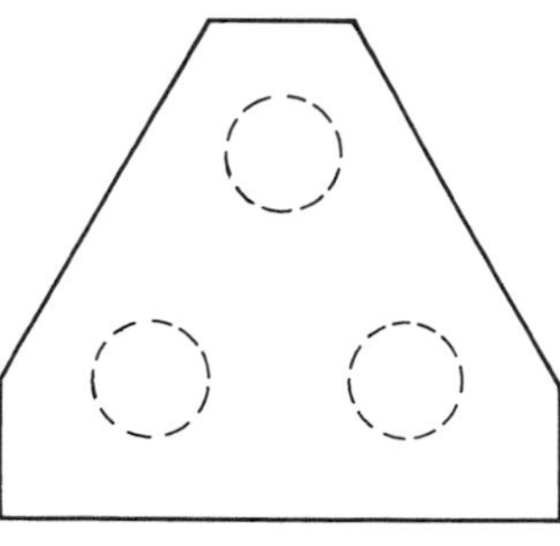

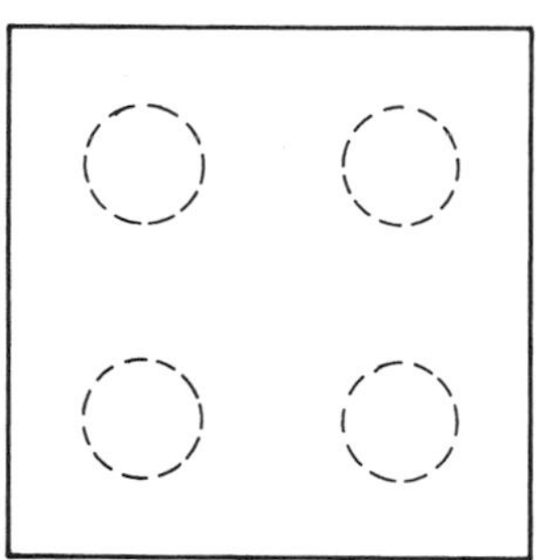

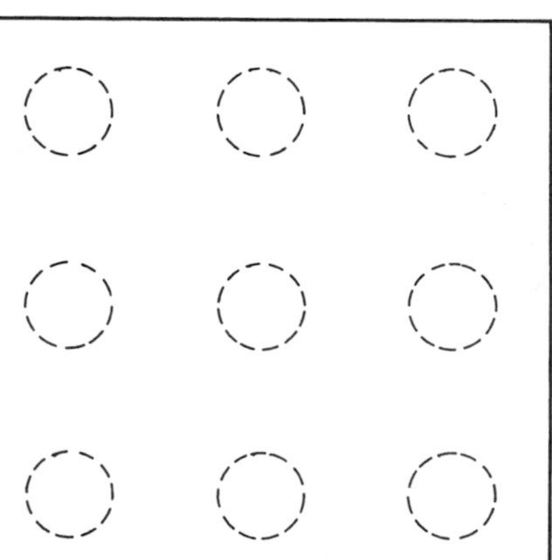

FIGURE 2.42
Clusters of two, three, four, and nine piles with their concrete caps, viewed from above. The caps are reinforced to transmit column loads equally into all the piles in the cluster, but the reinforcing steel has been omitted here for the sake of clarity.

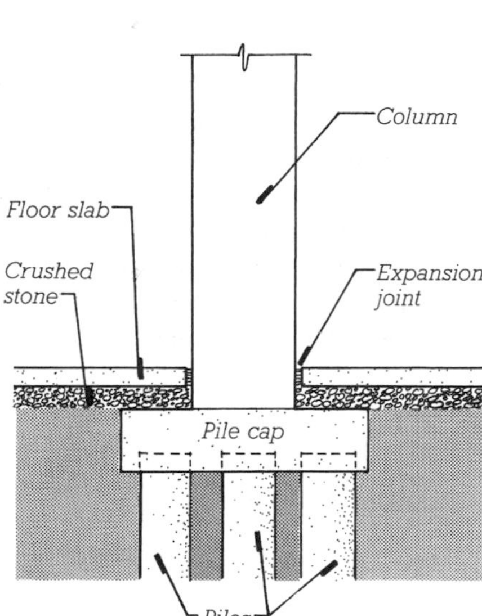

FIGURE 2.43
An elevation view of a pile cap, column, and floor slab.

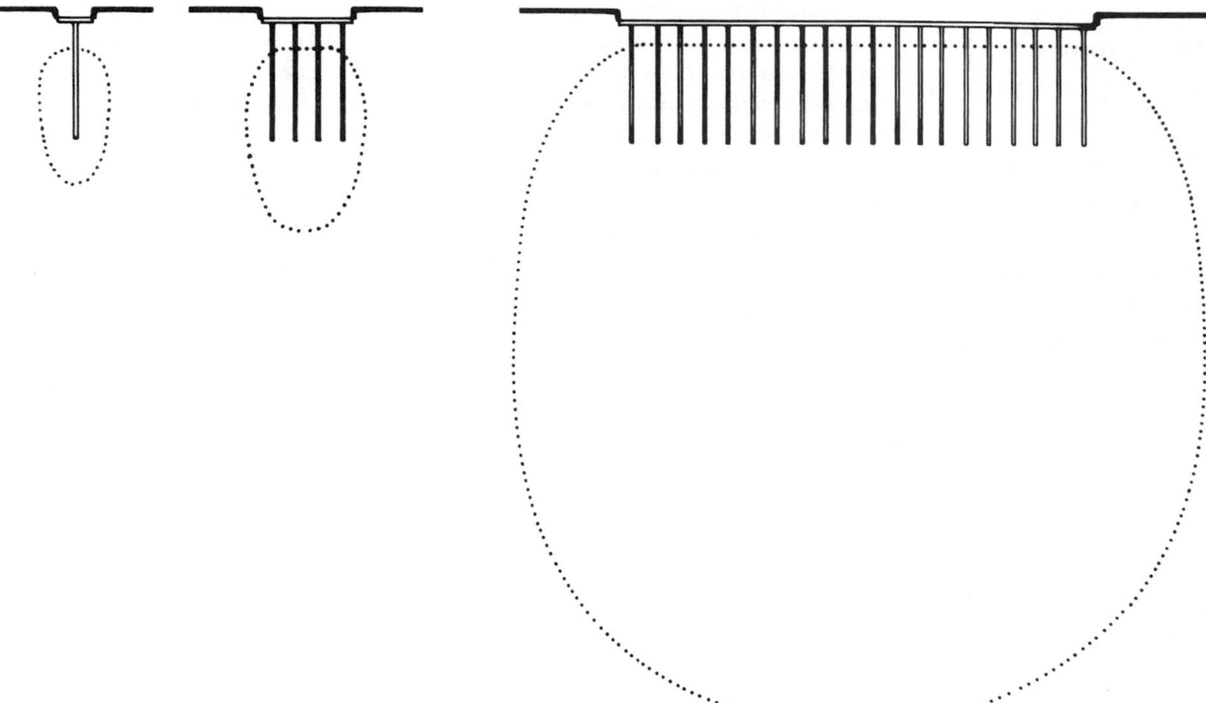

FIGURE 2.44
A single friction pile (left) transmits its load into the earth as an equal shear pressure along the bulb profile indicated by the dotted line. As the size of the pile cluster increases, the piles act together to create a single larger bulb of higher pressure that reaches deeper into the ground. A building with many closely spaced clusters of piles (right) creates a very large, deep bulb. Care must be taken to ensure that large-pressure bulbs do not overstress the soil or cause excessive settlement of the foundation. The settlement of a large group of friction piles in clay, for example, will be considerably greater than that of a single isolated pile.

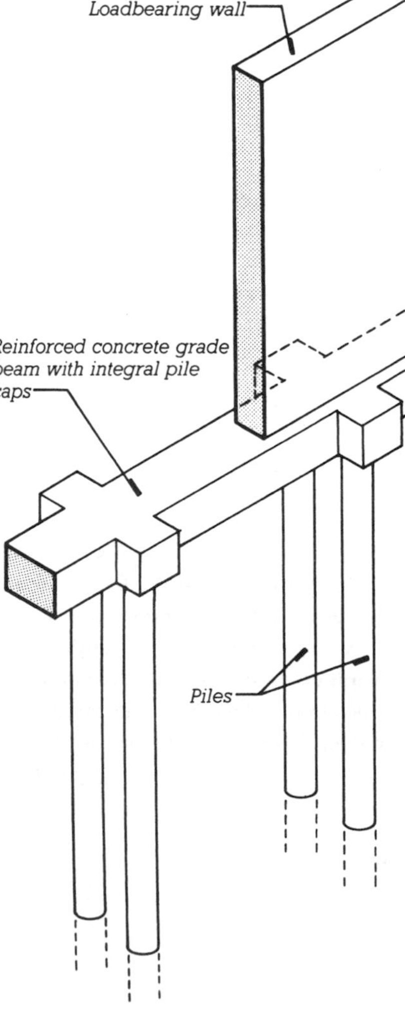

FIGURE 2.45
In order to support a loadbearing wall, pile caps are joined by a grade beam. The reinforcing in the grade beam is similar to that in any ordinary continuous concrete beam and has been omitted for clarity. In some cases, a concrete loadbearing wall can be reinforced to act as its own grade beam.

the correctness of the calculation, test piles are often driven and loaded on the building site before foundation work begins.

Where piles are used to support loadbearing walls, reinforced concrete *grade beams* are constructed between the pile caps to transmit the wall loads to the piles (Figure 2.45). Grade beams are also used with caisson foundations for the same purpose.

Pile Driving

Pile hammers are massive weights lifted by the energy of steam, compressed air, compressed hydraulic fluid, or a diesel explosion, then dropped against a block that is in firm contact with the top of the pile. Single-acting hammers fall by gravity alone, while double-acting hammers are forced downward by reverse application of the energy source that lifts the hammer. The hammer travels on tall vertical rails called *leads* (pronounced "leeds") at the front of a piledriver (Figure 2.46). It is first hoisted up the leads to the top of each pile as driving commences, then follows the pile down as it penetrates the earth. The piledriver mechanism includes lifting machinery to raise each pile into position before driving.

FIGURE 2.46
A piledriver hammers a precast concrete pile into the ground. The pile is supported by the vertical structure (leads) of the piledriver and driven by a heavy piston mechanism that follows it down the leads as it penetrates deeper into the soil. (© *David van Mill, Netherlands*)

In certain types of soil, piles can be driven more efficiently by vibration than by hammer blows alone, using a vibratory hammer mechanism. Where vibrations from hammering could be a risk to nearby existing structures, some lightweight pile systems can also be installed by rotary drilling or hydraulic pressing.

Pile Materials

Piles may be made of timber, steel, concrete, and various combinations of these materials (Figure 2.47). Timber piles have been used since Roman times, when they were driven by large mechanical hammers hoisted by muscle power. Their main advantage is that they are economical for lightly loaded foundations. On the minus side, they cannot be spliced during driving and are, therefore, limited to the length of available tree trunks, approximately 65 feet (20 m). Unless pressure treated with a wood preservative or completely submerged below the water table, they will decay (the lack of free oxygen in the water prohibits organic growth). Relatively small hammers must be used in driving timber piles to avoid splitting them. Capacities of individual timber piles lie in the range of 10 to 55 tons (9000 to 50,000 kg).

Two forms of steel piles are used, H-piles and pipe piles. *H-piles* are special hot-rolled, wide-flange sections, 8 to 14 inches (200 to 355 mm) deep, which are approximately square in cross section. They are used mostly in end bearing applications. H-piles displace relatively little soil during driving. This minimizes the upward displacement of adjacent soil, called *heaving*, that sometimes occurs when many piles are driven close together. Heaving can be a particular problem on urban sites, where it can lift adjacent buildings.

H-piles can be brought to the site in any convenient lengths, welded together as driving progresses to form any necessary length of pile, and cut off with an oxyacetylene torch when the required depth is reached. The cutoff ends can then be welded onto other piles to avoid waste. Corrosion can be a problem in some soils, however, and unlike closed pipe piles and hollow precast concrete piles, H-piles cannot be inspected after driving to be sure they are straight and undamaged. Allowable loads on H-piles run from 30 to 225 tons (27,000 to 204,000 kg).

Steel *pipe piles* have diameters of 8 to 16 inches (200 to 400 mm). They may be driven with the lower end either open or closed with a heavy steel plate. An open pile is easier to drive than a closed one, but its interior must be cleaned of soil and inspected before being filled with concrete, whereas a closed pile can be inspected and concreted immediately after driving. Pipe piles are stiff and can carry loads from 40 to 300 tons (36,000 to 270,000 kg). They displace relatively large amounts of soil during driving, which can lead to upward heaving of nearby soil and buildings. The larger sizes of pipe piles require a very heavy hammer for driving.

Minipiles, also called *pin piles* or *micropiles*, are a lightweight form of steel piles made from steel bar or pipe 2 to 12 inches (50 to 300 mm) in diameter. Minipiles are inserted into holes drilled in the soil and grouted in place. When installed within existing buildings, they may also be forced into the soil by hydraulic jacks pushing downward on the pile and upward on the building structure. Since no hammering is required, they are a good choice for repair or improvement of existing foundations where vibrations

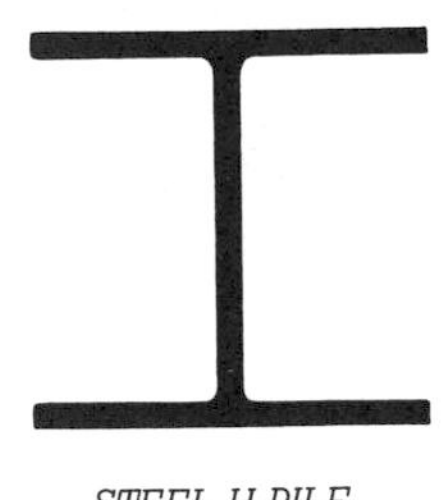

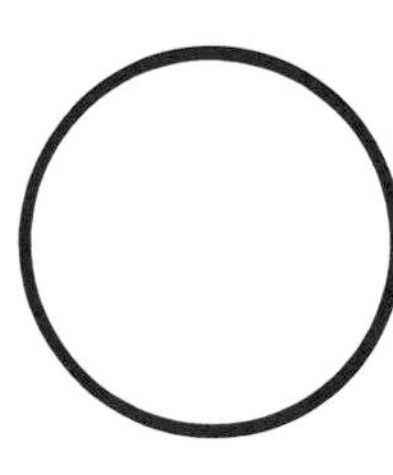

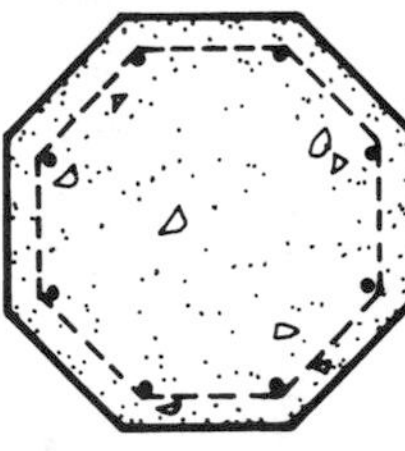

FIGURE 2.47
Cross sections of common types of piles. Precast concrete piles may be square or round instead of the octagonal section shown here and may be hollow in the larger sizes.

from the hammering of conventional piles could damage the existing structure or disrupt ongoing activities within the building (Figure 2.54). Where vertical space is limited, such as when working in the basement of an existing building, minipiles can be installed in individual sections as short as 3 feet (1 m) that are threaded end-to-end as driving progresses. Minipiles can reach depths as great as 200 feet (60 m) and have working capacities as great as 200 to 300 tons (180,000 to 270,000 kg).

Precast concrete piles are square, octagonal, or round in section, and in large sizes often have open cores to allow inspection (Figures 2.47–2.49). Most are prestressed, but some for smaller buildings are merely reinforced (for an explanation of prestressing, see pages 544–548). Typical cross-sectional dimensions range from 10 to 16 inches (250 to 400 mm) and bearing capacities from 45 to 500 tons (40,000 to 450,000 kg). Advantages of precast piles include high load capacity, an absence of corrosion or decay problems, and, in most situations, a relative economy of cost. Precast piles must be handled carefully to avoid bending and cracking before installation. Splices between lengths of precast piling can be made effectively with mechanical fastening devices that are cast into the ends of the sections.

A *sitecast concrete pile* is made by driving a hollow steel shell into the ground and filling it with concrete. The shell is sometimes corrugated to increase its stiffness; if the

FIGURE 2.48
Precast, prestressed concrete piles. Lifting loops are cast into the sides of the piles as crane attachments for hoisting them into a vertical position.
(Courtesy of Lone Star/San-Vel Concrete)

FIGURE 2.49
A driven cluster of six precast concrete piles, ready for cutting off and capping.
(Photo by Alvin Ericson)

corrugations are circumferential, a heavy steel *mandrel* (a stiff, tight-fitting liner) is inserted in the shell during driving to protect the shell from collapse, then withdrawn before concreting. Some shells with longitudinal corrugations are stiff enough that they do not require mandrels. Some types of mandrel-driven piles are limited in length, and the larger diameters of sitecast piles (up to 16 inches, or 400 mm) can cause ground heaving. Load capacities range from 45 to 150 tons (40,000 to 136,000 kg). The primary reason to use sitecast concrete piles is their economy.

There is a variety of proprietary sitecast concrete pile systems, each with various advantages and disadvantages (Figure 2.50). Concrete *pressure-injected footings* (Figure 2.51) share characteristics of piles, piers, and footings. They are highly resistant to uplift forces, a property that is useful for tall, slender buildings in which there is a potential for overturning of the building, and for tensile anchors for tent and pneumatic structures. *Rammed aggregate piers* and *stone columns* are similar to pressure-injected footings, but are constructed of crushed rock that has been densely compacted into holes created by drilling or the action of proprietary vibrating probes.

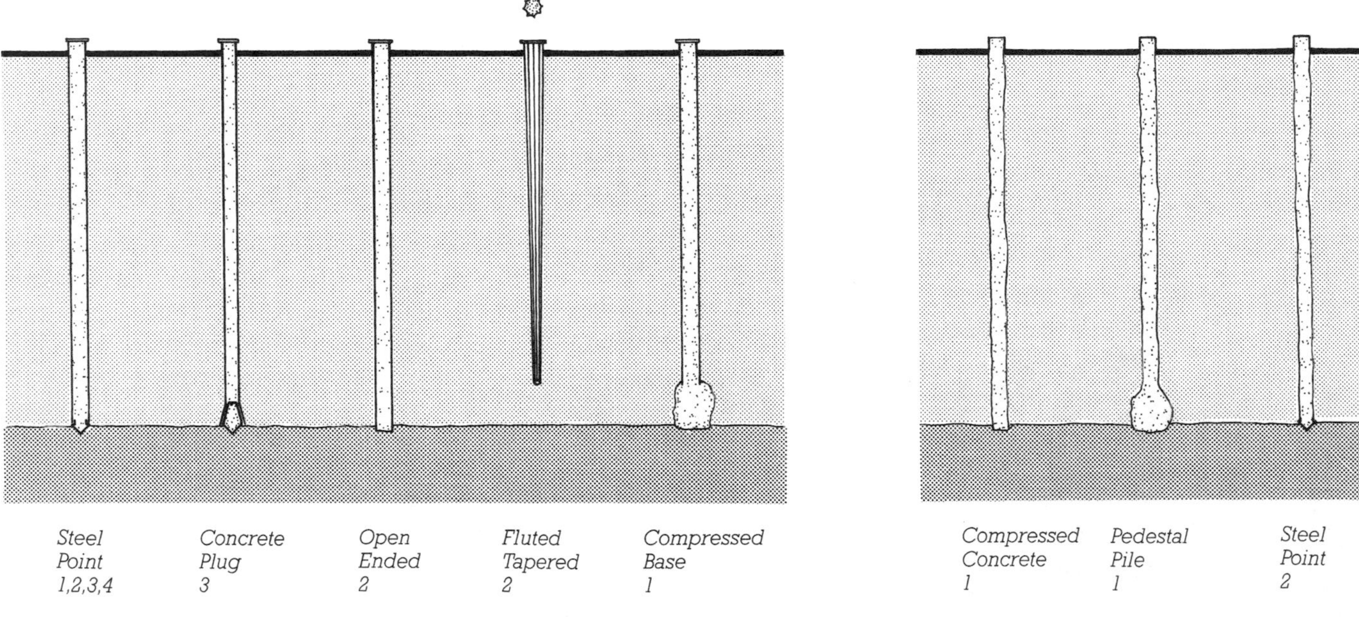

FIGURE 2.50

Some proprietary types of sitecast concrete piles. All are cast into steel casings that have been driven into the ground; the uncased piles are made by withdrawing the casing as the concrete is poured and saving it for subsequent reuse. The numbers refer to the methods of driving that may be used with each: 1. Mandrel driven. 2. Driven from the top of the tube. 3. Driven from the bottom of the tube to avoid buckling it. 4. Jetted. Jetting is accomplished by advancing a high-pressure water nozzle ahead of the pile to wash the soil back alongside the pile to the surface. Jetting has a tendency to disrupt the soil around the pile, so it is not a favored method of driving under most circumstances.

(*a*) (*b*) (*c*)

(*d*) (*e*) (*f*)

FIGURE 2.51

Steps in the construction of a proprietary pressure-injected, bottom-driven concrete pile footing. (*a*) A charge of a very low-moisture concrete mix is inserted into the bottom of the steel drive tube at the surface of the ground and compacted into a sealing plug with repeated blows of a drop hammer. (*b*) As the drop hammer drives the sealing plug into the ground, the drive tube is pulled along by the friction between the plug and the tube. (*c*) When the desired depth is reached, the tube is held and a bulb of concrete is formed by adding small charges of concrete and driving the concrete out into the soil with the drop hammer. The bulb provides an increased bearing area for the pile and strengthens the bearing stratum by compaction. (*d, e*) The shaft is formed of additional compacted concrete as the tube is withdrawn. (*f*) Charges of concrete are dropped into the tube from a special bucket supported on the leads of the driving equipment. *(Courtesy of Franki Foundation Company)*

Seismic Base Isolation

In areas where strong earthquakes are common, buildings are sometimes placed on *base isolators*. When significant ground movement occurs, the base isolators flex or yield to absorb a significant portion of this movement; as a result, the building and its substructure move significantly less than they would otherwise, reducing the forces acting on the structure and lessening the potential for damage. A frequently used type of base isolator is a multilayered sandwich of rubber and steel plates (Figure 2.52). The rubber layers deform in shear to allow the rectangular isolator to become a parallelogram in cross section in response to relative motion between the ground and the building. A lead core deforms enough to allow this motion to occur, provides damping action, and keeps the layers of the sandwich aligned.

Underpinning

Underpinning is the process of strengthening and stabilizing the foundations of an existing building. It may be required for any of several reasons: The existing foundations may never have been adequate to carry their loads, leading to excessive settlement of the building over time. A change in building use or additions to the building may overload the existing foundations. New construction near a building may disturb the soil around its foundations or require that its foundations be carried deeper. Whatever the cause, underpinning is a highly specialized task that is seldom the same for any two buildings. Three different alternatives are available when foundation

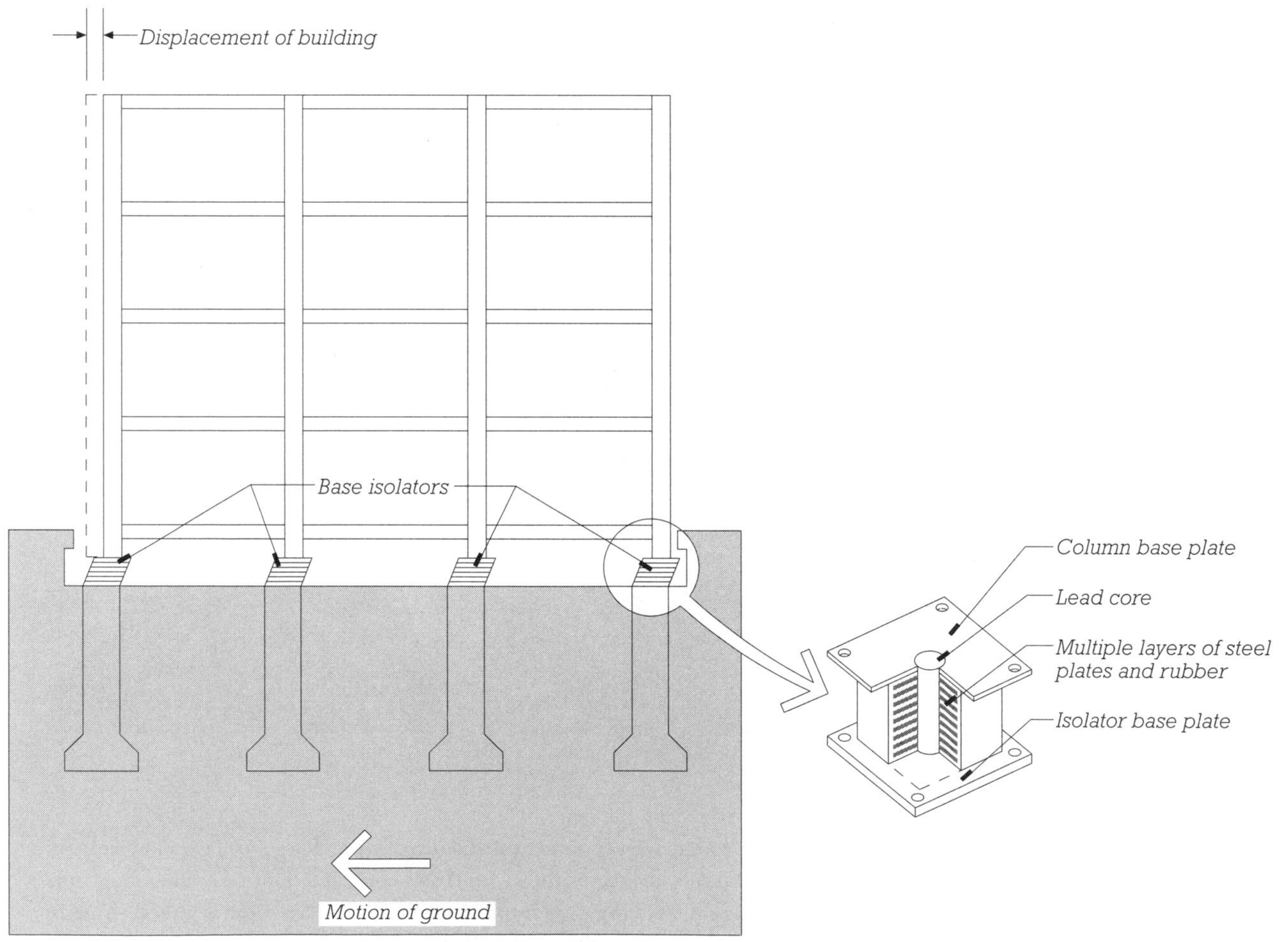

Figure 2.52
Base isolation.

capacity needs to be increased: The foundations may be enlarged; new, deep foundations can be inserted under shallow ones to carry the load to a deeper, stronger stratum of soil; or the soil itself can be strengthened by grouting or by chemical treatment. Figures 2.53 and 2.54 illustrate in diagrammatic form some selected concepts of underpinning.

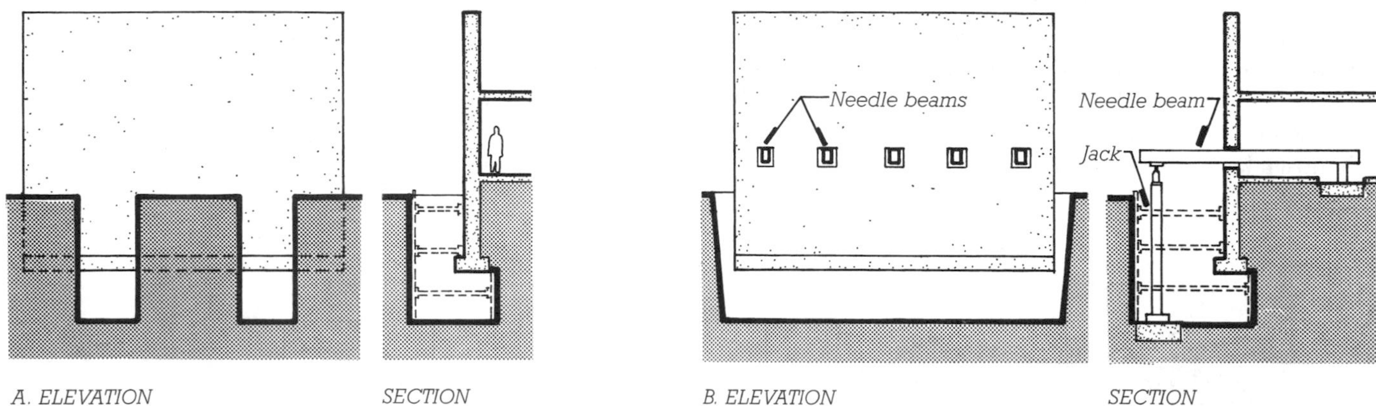

FIGURE 2.53
Two methods of supporting a building while carrying out underpinning work beneath its foundation, each shown in both elevation and section. (*a*) Trenches are dug beneath the existing foundation at intervals, leaving the majority of the foundation supported by the soil. When portions of the new foundations have been completed in the trenches, using one of the types of underpinning shown in Figure 2.54, another set of trenches is dug between them and the remainder of the foundations is completed. (*b*) The foundations of an entire wall can be exposed at once by needling, in which the wall is supported temporarily on needle beams threaded through holes cut in the wall. After underpinning has been accomplished, the jacks and needle beams are removed and the trench is backfilled.

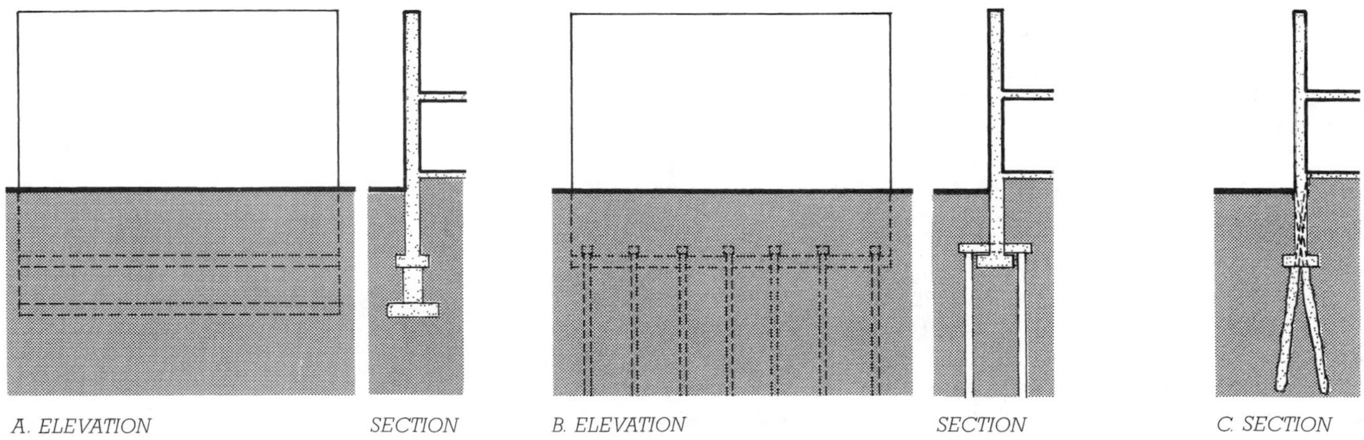

FIGURE 2.54
Three types of underpinning. (*a*) A new foundation wall and footing are constructed beneath the existing foundation. (*b*) New piles or caissons are constructed on either side of the existing foundation. (*c*) Minipiles are inserted through the existing foundation. Minipiles do not generally require excavation or temporary support of the building.

Retaining Walls

A *retaining wall* holds soil back to create an abrupt change in the elevation of the ground. A retaining wall must resist the pressure of the earth that bears against it on the uphill side. Retaining walls may be made of masonry, preservative-treated wood, coated or galvanized steel, precast concrete, or, most commonly, sitecast concrete.

The structural design of a retaining wall must take into account such factors as the height of the wall, the character of the soil behind the wall, the presence or absence of groundwater behind the wall, any structures whose foundations apply pressure to the soil behind the wall, and the character of the soil beneath the base of the wall, which must support the footing that keeps the wall in place. The rate of structural failure in retaining walls is high relative to the rate of failure in other types of structures. Failure may occur through fracture of the wall, overturning of the wall due to soil failure, lateral sliding of the wall, or undermining of the wall by flowing groundwater (Figure 2.55). Careful engineering design and site supervision are crucial to the success of a retaining wall.

There are many ways of building retaining walls. For walls less than 3 feet (900 mm) in height, simple, unreinforced walls of various types are often appropriate (Figure 2.56). For taller walls, and ones subjected to unusual loadings or groundwater, the type most frequently employed today is the cantilevered concrete

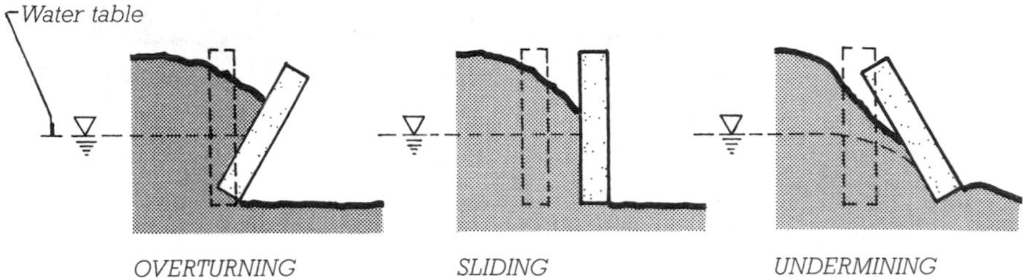

Figure 2.55
Three failure mechanisms in retaining walls. The high water table shown in these illustrations creates pressure against the walls that contributes to their failure. The undermining failure is directly attributable to groundwater running beneath the base of the wall, carrying soil with it.

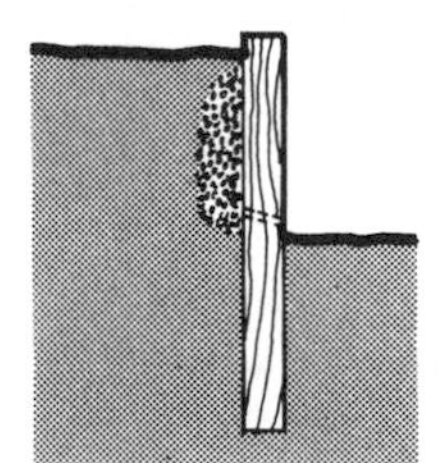

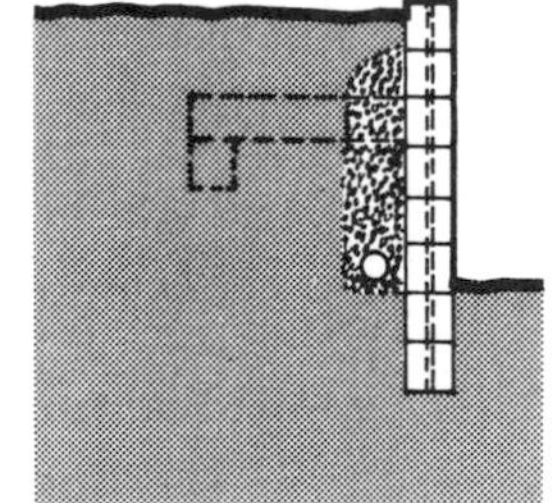

Figure 2.56
Three types of simple retaining walls, usually used for heights not exceeding 3 feet (900 mm). The deadmen in the horizontal timber wall are timbers embedded in the soil behind the wall and connected to it with timbers inserted into the wall at right angles. The timbers, which should be pressure treated with a wood preservative, are held together with very large spikes or with steel reinforcing bars driven into drilled holes. The crushed-stone drainage trench behind each wall is important as a means of relieving water pressure against the wall to prevent wall failure. With proper engineering design, any of these types of construction can also be used for taller retaining walls.

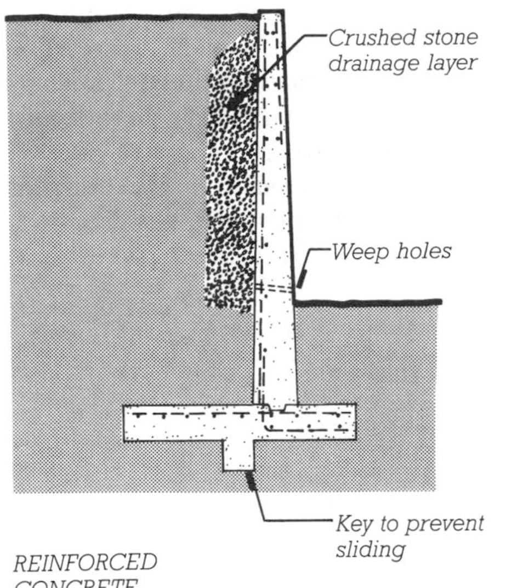

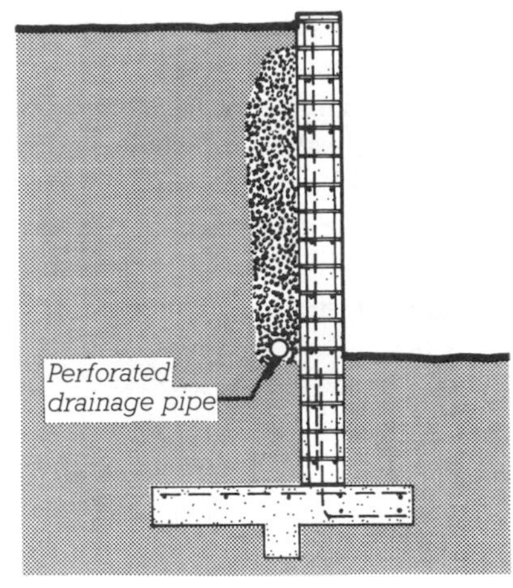

FIGURE 2.57
Cantilevered retaining walls of concrete and concrete masonry. The footing is shaped to resist sliding and overturning, and drainage behind the wall reduces the likelihood of undermining. The pattern of steel reinforcing (broken lines) is designed to resist the tensile forces in the wall.

FIGURE 2.58
A segmental retaining wall consisting of specially made concrete blocks designed to interlock and prevent sliding. The wall leans back against the soil it retains; this reduces the amount of soil the wall must retain and makes it more stable against the lateral push of the soil. *(Courtesy of VERSA-LOK Retaining Wall Systems)*

retaining wall, two examples of which are shown in Figure 2.57. The shape and reinforcing of a cantilevered wall can be custom designed to suit almost any situation. Proprietary systems of interlocking concrete blocks are also used to construct sloping *segmental retaining walls* that need no steel reinforcing (Figure 2.58).

Earth reinforcing (Figure 2.59) is an economical alternative to conventional retaining walls in many situations. Soil is compacted in thin layers, each containing strips or meshes of galvanized steel, polymer fibers, or glass fibers, which stabilize the soil in much the same manner as the roots of plants. *Gabions* are another form of earth retention in which corrosion-resistant wire baskets are filled with cobble- or boulder-sized rocks and then stacked to form retaining walls, slope protection, and similar structures.

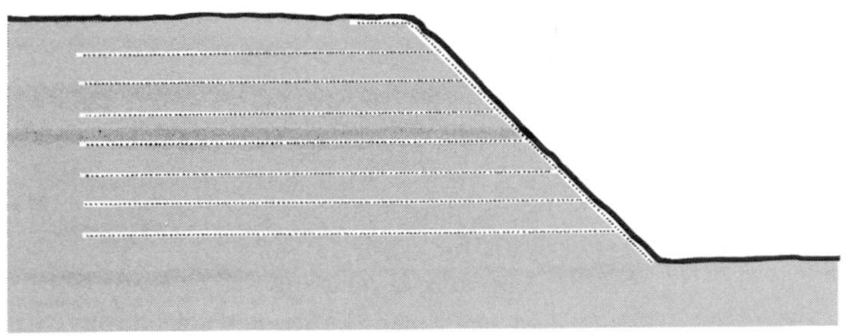

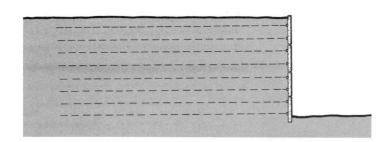

FIGURE 2.59
Two examples of earth reinforcing. The embankment in the top section was placed by alternating thin layers of earth with layers of synthetic mesh fabric. The retaining wall in the lower section is made of precast concrete panels fastened to long galvanized steel straps that run back into the soil.

Geotextiles

Geotextiles are flexible fabrics made of chemically inert plastics that are highly resistant to deterioration in the soil. They are used for a variety of purposes relating to site development and the foundations of buildings. As described in the accompanying text, in *earth reinforcement* or *soil reinforcement,* a plastic mesh fabric or grid is used to support an earth embankment or retaining wall. The same type of mesh may be utilized in layers to stabilize engineered fill beneath a shallow footing (Figure A), or to stabilize marginal soils under driveways, roads, and airport runways, acting very much as the roots of plants do in preventing the movement of soil particles.

Another geotextile that is introduced later in this chapter is drainage matting, an open matrix of plastic filaments with a feltlike filter fabric laminated onto one side to keep soil particles from entering the matrix. In addition to providing free drainage around foundation walls, drainage matting is often used beneath the soil in the bottoms of planter boxes and under heavy paving tiles on rooftop terraces, where it maintains free passage for the drainage of water above a waterproof membrane.

Synthetic filter fabrics are wrapped over and around subterranean crushed stone drainage layers such as the one frequently used around a foundation drain. In this position, they keep the stone and pipe from gradually becoming clogged with fine soil particles carried by the groundwater. Similar fabrics are used at grade during construction, acting as temporary barriers to filter soil out of water that runs off a construction site, thus preventing contamination of lakes, streams, or stormwater systems.

Special geotextiles are manufactured to stake down on freshly cut slopes to prevent soil erosion and encourage revegetation; some of these are designed to decay and disappear into the soil as plants take over the function of slope stabilization. Another type of geotextile is used for weed control in landscaped beds, where it allows rainwater to penetrate the soil but blocks sunlight, preventing weeds from sprouting.

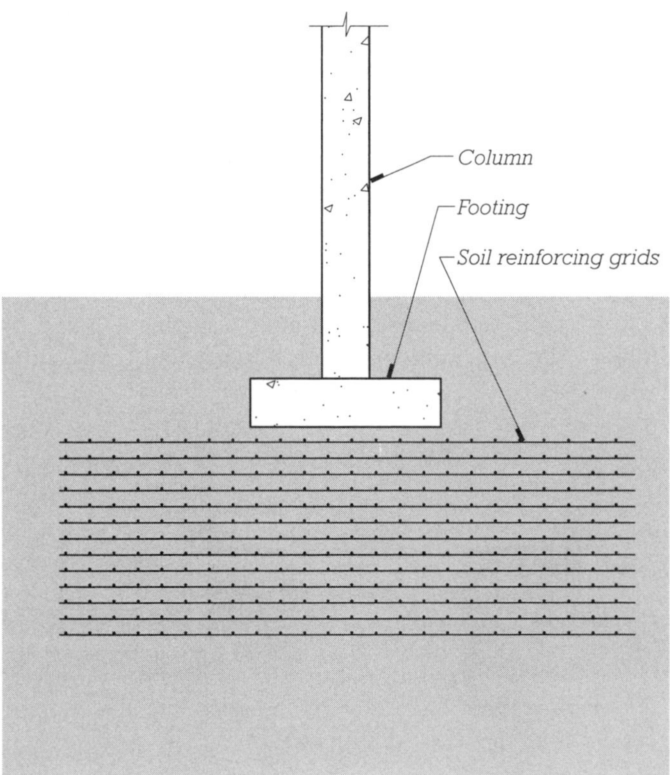

Waterproofing and Drainage

Where building substructures enclose basements, parking garages, or other usable space, groundwater must be kept out. Concrete alone is rarely adequate for this purpose. Moisture can migrate through its microscopic pores, or through other pathways created by shrinkage cracks, form tie holes (form ties, used in sitecast concrete construction, are explained in Chapter 14), utility penetrations, and the joints between concrete pours. To ensure a substructure's resistance to water entry, two approaches are used: drainage and waterproofing. *Drainage* draws groundwater away from a foundation, reducing the volume and pressure of water acting on the foundation's walls and slabs. Waterproofing acts as a barrier, stopping water that reaches the foundation from passing though to the interior.

Drainage, consisting of some combination of *drainage backfill* (well-sorted crushed stone or gravel), drainage mat, and perforated drain piping, is used with almost every building substructure (Figure 2.60). *Drainage mat* is a manufactured component that may be made of a loose mat of stiff, inert fibers, a plastic egg-crate structure, or some other very open, porous material. It is faced on the outside with a filter fabric that prevents fine soil particles from entering and clogging the drainage passages in the mat. Any subterranean water that approaches the wall descends through the porous material of the mat to the drain pipe at the footing. *Perforated drain piping* is frequently laid around the outside perimeter of a building foundation. The pipes are 4 or 6 inches (100 or 150 mm) in diameter and provide an open channel in the crushed stone bed through which water can flow by gravity either "to daylight" at a lower elevation on a sloping site, to a municipal storm sewer system, or to a sump pit that can be automatically pumped dry whenever it fills. The pipes are laid at least 6 inches (150 mm) below the top of the basement floor slab to maintain the groundwater level safely below that of the slab. Perforations in the pipes face downward so that water is drained from the lowest possible level. Where groundwater conditions are severe, rows of perforated pipe may be installed under the basement slab as well (Figure 2.61).

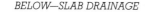

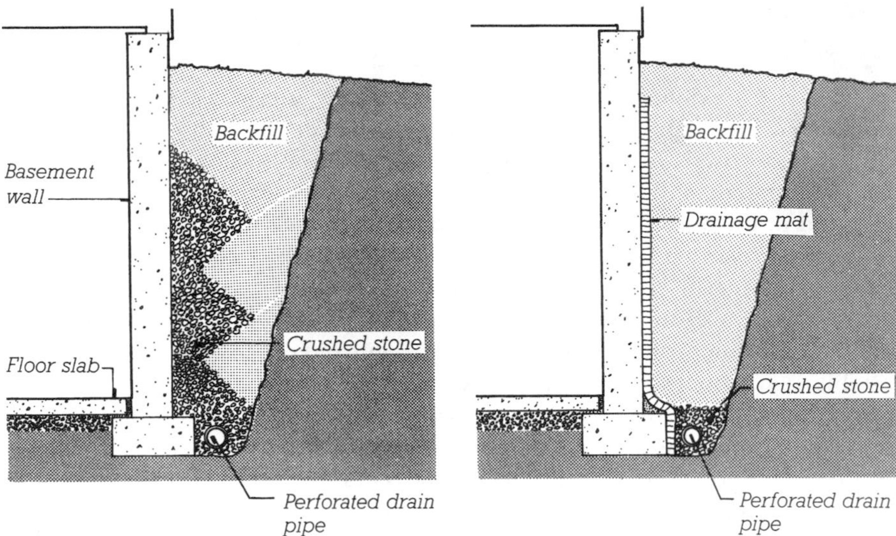

Figure 2.60
Two methods of relieving water pressure around a building substructure by drainage. The gravel drain (left) is hard to do well because of the difficulty of depositing the crushed stone and backfill soil in neatly separated, alternating layers. The drainage mat (right) is easier and often more economical to install.

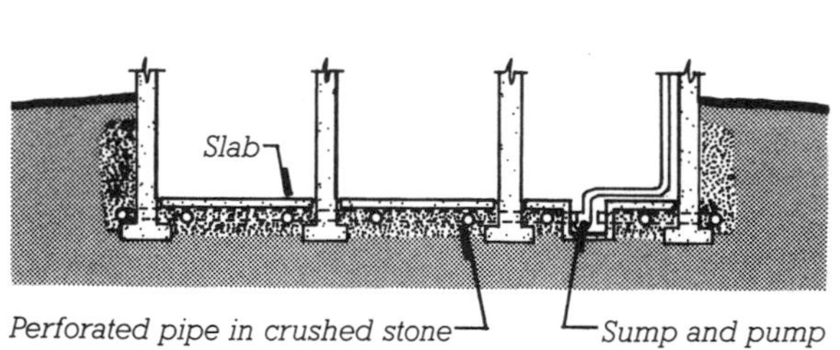

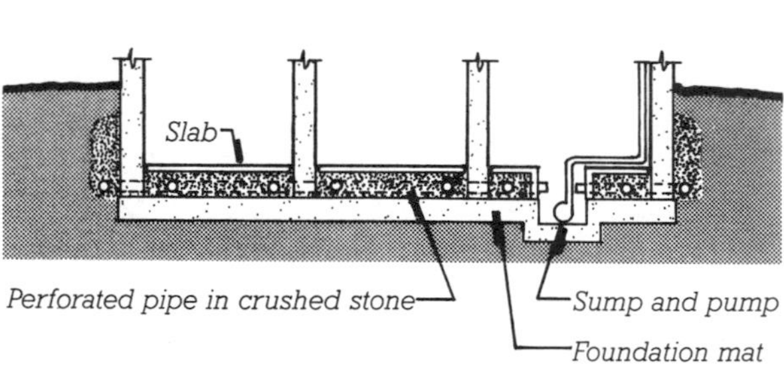

Figure 2.61
For a high degree of security against substructure flooding, drainage both around and under the basement is required, as seen here in a section view. Above-slab drainage is used in buildings with mat foundations.

On most foundations, some form of water-repelling barrier is also used to protect against the passage of groundwater. *Dampproofing* is a moisture-resistant cement plaster or asphalt compound commonly applied to residential basement walls and to other substructures where groundwater conditions are mild or waterproofing requirements are not critical. *Cement plaster dampproofing*, or *parge coating*, is light gray in color and troweled on. *Asphalt* or *bituminous dampproofing* is dark in color and is applied in liquid form by spray, roller, or trowel. Dampproofing is less expensive and less resistant to water passage than true waterproofing.

Waterproofing, unlike dampproofing, can prevent the passage of water even under conditions of hydrostatic pressure. It is used where groundwater conditions are severe or the need to protect subgrade space from moisture is critical. Waterproof membranes are most commonly formulated from plastics, asphalt compounds, or synthetic rubbers and come in a great variety of forms.

Liquid waterproofing is applied by spray gun, roller, or squeegee and then allowed to cure in place. It is easy to install and easy to form around complex shapes. When fully cured, the finished membrane is seamless and fully bonded to the underlying substrate. However, because liquid membranes are formed in the field, they are subject to uneven application, and the surfaces to which they are applied must be clean, smooth, and dry to ensure reliable adhesion of the membrane.

Preformed *sheet membrane waterproofing* may be adhered or mechanically fastened to substructure walls or laid loosely over horizontal surfaces (Figure 2.62). Fabricated under controlled factory conditions, sheet membranes are reliably uniform in material quality and thickness. However, they can be more difficult to form around complex shapes, and the seams between sheets, which are sealed in the field, may be subject to

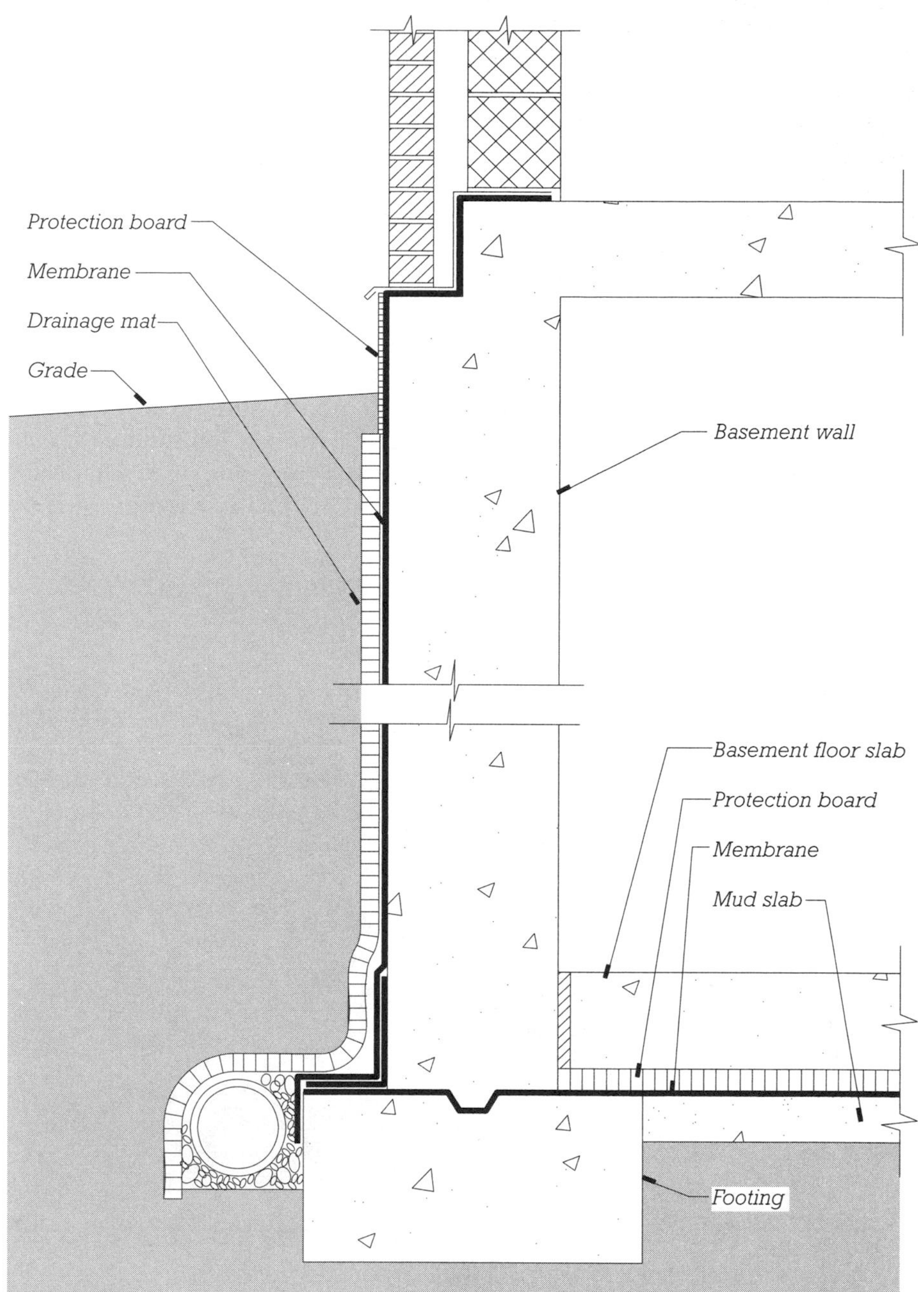

FIGURE 2.62
A diagrammatic representation of the placement of sheet membrane waterproofing around a basement. A mud slab of low-strength concrete was poured to serve as a base for placement of the horizontal membrane. Notice that the vertical and horizontal membranes join to wrap the basement completely in a waterproof enclosure.

lapses in quality. Sheet membranes that are loosely laid or mechanically fastened can be used over substrates that will not bond with liquid-applied or adhered sheet membranes. They are also a good choice where substrate cracking or movement may be expected, because such movement is less likely to stress or damage the membrane. An advantage of adhered membranes (both sheet and liquid) is that in the case of a defect, water cannot travel far under the membrane, limiting the extent of water damage that may occur and simplifying the tracing of leaks.

Bentonite waterproofing is made from sodium bentonite, a naturally occurring, highly expansive clay. It is most often applied as preformed sheets consisting of dry clay sandwiched within corrugated cardboard, geotextile fabric, or plastic sheets (Figure 2.63). When bentonite comes in contact with moisture, it swells to several times its dry volume and forms an impervious barrier to the further passage of water. Bentonite sheets can be placed directly on the soil under a concrete slab on grade or mechanically attached to uncured, damp concrete walls. In slurry form, bentonite can be sprayed even onto highly irregular, rough stone walls. The swelling behavior of bentonite clay also allows it to adjust to cracking and movement in the substrate.

Integral waterproofing includes cementitious plaster or crystalline admixtures for concrete or mortar that react chemically to stop up the pores of these materials and render them watertight. It may be applied to the surface of existing concrete or masonry or used as an admixture in new concrete. Unlike most other waterproofing materials, many

FIGURE 2.63
Waterproofing in progress on a concrete foundation. Leftmost: the bare foundation wall remains exposed. Middle: bentonite waterproofing panels are fastened in place. These panels are lined on the outer face with a black-colored, high-density plastic that adds to the waterproofing qualities of the panel. Right: drainage mat has been installed over the waterproofing. The mat's outer face of filter fabric is lightly dimpled, telegraphing the egg crate structure of the underlying molded plastic panel. The top edge of the mat is secured in place with an aluminum *termination bar* that holds the panel in place and keeps dirt and debris from falling behind the panel. Lower right: white perforated drain piping can be seen, temporarily supported on wood blocking and running alongside the footing. *(Photo by Joseph Iano)*

integral waterproofing materials can be applied as *negative side waterproofing*, that is, applied to the inner side of a concrete wall acting to resist water passage from the opposite side.

Blind-side waterproofing is installed prior to the pouring of concrete walls. This occurs most commonly when a substructure wall is built close to a property's edge, and excavation cannot be enlarged beyond the property line to permit workers access to the outer face of the wall after its construction. Drainage matting is first applied directly to the excavation sheeting, and then any of a number of possible waterproofing membranes are applied over the drainage mat. Later, the concrete wall is poured against the membrane. The sheeting remains permanently in place (Figure 2.64).

Joints in construction require special attention to ensure watertightness. Preformed *waterstops* made of plastic, synthetic rubber, or metal can be cast into the mating concrete edges of both moving and nonmoving joints to block the passage of water (Figures 2.65 and 2.66). Waterstops for nonmoving joints such as between concrete pours of a wall or slab can also be made of strips of bentonite or mastic that are temporarily adhered to the edge of one pour. After the adjacent pour is complete, these stops remain embedded in the joint, where they form a watertight barrier (Figure 2.67).

Most waterproofing systems are inaccessible once building construction is complete; they are expected to perform for the life of the building, and even small defects in installation can allow the passage of large volumes of water. For these reasons, waterproofing membranes are inspected carefully

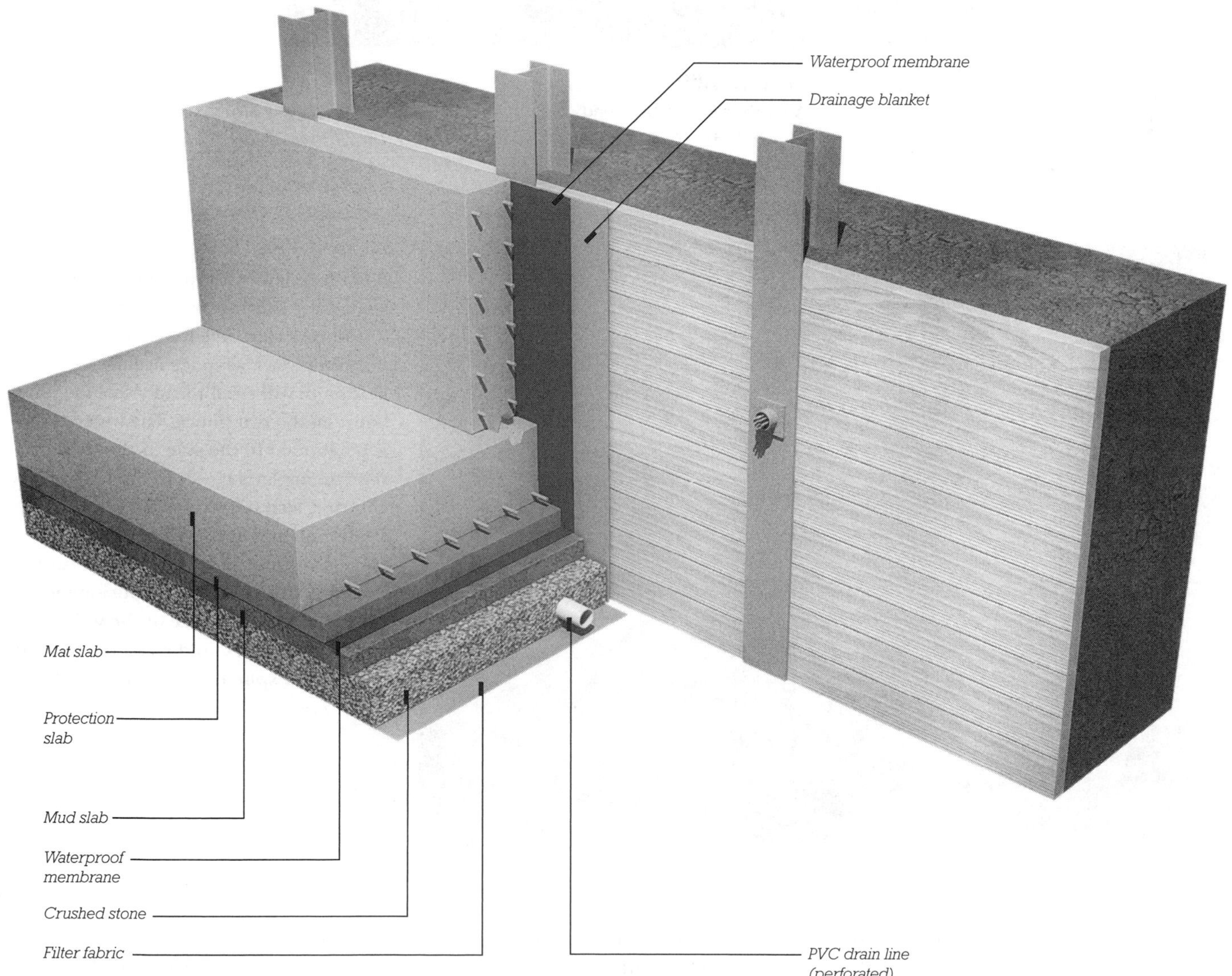

FIGURE 2.64
Blind-side waterproofing is used where there is no working space between a sheeted excavation and the outside of the foundation wall. The drainage mat and waterproof membrane are applied to the sheeting; then the basement wall is poured against them.

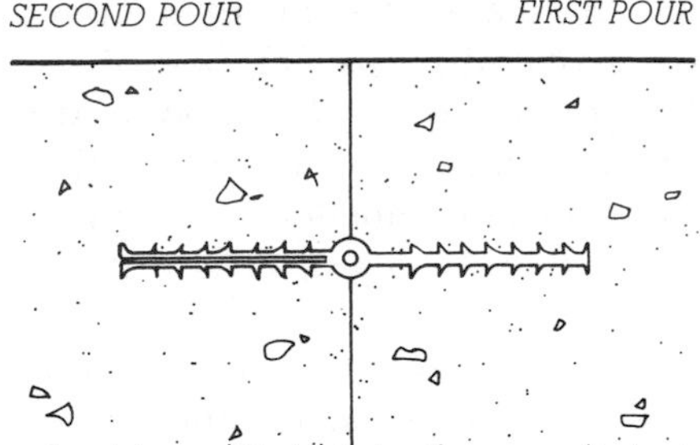

FIGURE 2.65
A synthetic rubber waterstop is used to seal against water penetration at movement joints and at joints between pours of concrete in a foundation. The type shown here is split on one side so that its halves can be placed flat against the formwork where another wall will join the one being poured. After the concrete has been poured and cured and the formwork has been removed from the first wall, the split halves are folded back together before the next wall is poured.

FIGURE 2.66
A rubber waterstop ready for the next pour of a concrete wall, as diagrammed in Figure 2.65. *(Courtesy of Vulcan Metal Products, Inc., Birmingham, Alabama)*

FIGURE 2.67
A bentonite waterstop is adhered to a concrete footing prior to casting of the concrete wall above. Later, if groundwater seepage occurs, the bentonite will swell to fully seal the joint between the two pours. The waterstop is positioned to the side of the steel reinforcing bars closer to the wall's exterior, also protecting the reinforcing from moisture and corrosion. However, because of bentonite's expansive force, the waterstop must not be positioned too close to the surface of the wall, or when it swells, it could cause portions of concrete to split away or spall. *(Photo by Joseph Iano)*

during construction and horizontal membranes are often *flood tested* (submerged for an extended period time while leak-checking is performed) to detect the presence of defects while repairs can still be easily made. Once inspection and testing are complete, membranes are covered with a *protection board,* insulation board, or drainage matting to shield the membrane from prolonged exposure to sunlight and to prevent physical damage during soil backfilling or subsequent construction operations.

Basement Insulation

Comfort requirements, heating fuel efficiency, and building codes often require that basement walls be thermally insulated to limit the loss of heat from basements to the soil outside. Thermal insulation may be applied either inside or outside the basement wall. Inside the wall, mineral batt or plastic foam insulation may be installed between wood or steel furring strips, as shown in Figure 23.4. Alternatively, polystyrene foam or glass fiber insulation boards, typically 2 to 4 inches (50 to 100 mm) thick may be placed on the outside of the wall, held by either adhesive, fasteners, or the pressure of the soil. Proprietary products are available that combine insulation board and drainage mat in a single assembly.

Shallow Frost-Protected Foundations

In *shallow frost-protected foundations*, extruded polystyrene foam insulation boards can be used in cold climates to construct footings that lie above the normal frost line in the soil, resulting in lower excavation costs. Continuous layers of insulation board are placed around the perimeter of the building in such a way that heat flowing into the soil in winter from the interior of the building maintains the soil beneath the footings at a temperature above freezing (Figure 2.68). Even beneath unheated buildings, properly installed thermal insulation can trap enough geothermal heat around shallow foundations to prevent freezing.

Backfilling

After the basement walls have been waterproofed or dampproofed, insulating boards or protection boards have been applied, drainage features have been installed, and internal constructions that support the basement walls, such as interior walls and floors, have been completed, the area around a substructure is *backfilled* to restore the level of the ground. (A substructure built tightly against sheeted walls of an excavation needs little or no backfilling.) The backfilling operation involves placing soil back against the outside of the basement walls and compacting it there in layers, taking care not to damage drainage or waterproofing components or to exert excessive soil pressure against the walls. An open, fast-draining soil such as gravel or sand is preferred for backfilling because it allows the perimeter drainage system around the basement to do its work. Compaction must be sufficient to minimize subsequent settling of the backfilled area.

In some situations, especially in backfilling utility trenches under roadways and floor slabs, settling can be virtually eliminated by backfilling with *controlled low-strength material* (*CLSM*), which is made from portland cement and/or fly ash (a byproduct of coal-burning power plants), sand, and water. CLSM, sometimes called "flowable fill," is brought in concrete mixer trucks and poured into the excavation, where it compacts and levels itself, then hardens into soil-like material. The strength of CLSM is matched to the situation: For a utility trench, CLSM is formulated so that it is weak enough to be excavated easily by ordinary digging equipment when the pipe needs servicing, yet as strong as a good-quality

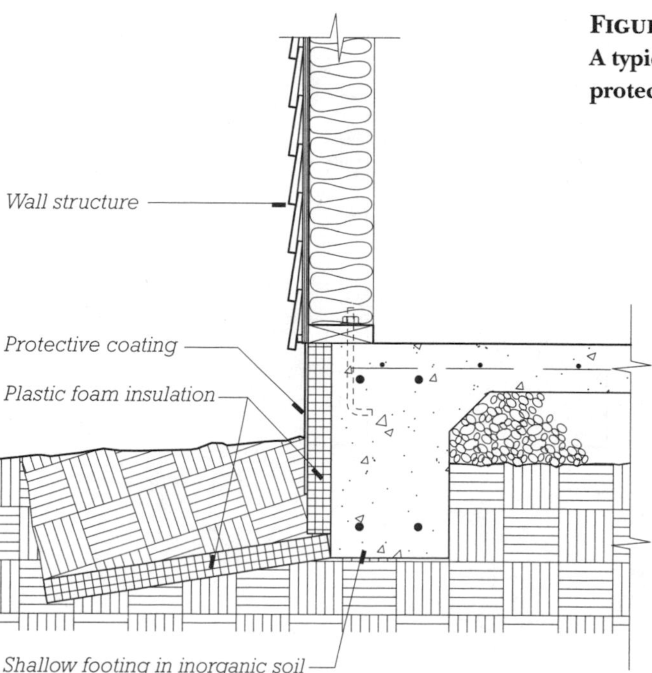

Figure 2.68
A typical detail for a shallow frost-protected footing.

compacted backfill. CLSM has many other uses in and around foundations. It is often used to pour *mud slabs*, which are weak concrete slabs used to create a level, dry base in an irregular, often wet excavation. The mud slab serves as a working surface for the reinforcing and pouring of a foundation mat or basement floor slab and is often the surface to which a waterproofing membrane is applied. CLSM is also used to replace pockets of unstable soil that may be encountered beneath a substructure or to create a stable volume of backfill around a basement wall.

Up–Down Construction

Normally, the substructure of a building is completed before work begins on its superstructure. If the building has several levels of basements, however, substructure work can take many months or even years. In such a case, *up–down construction* is sometimes an economical option, even if its first cost is somewhat more than that of the normal procedure, because it can save considerable construction time.

As diagrammed in Figure 2.69, up–down construction begins with installation of a perimeter slurry wall. Internal steel columns for the substructure are lowered into drilled,

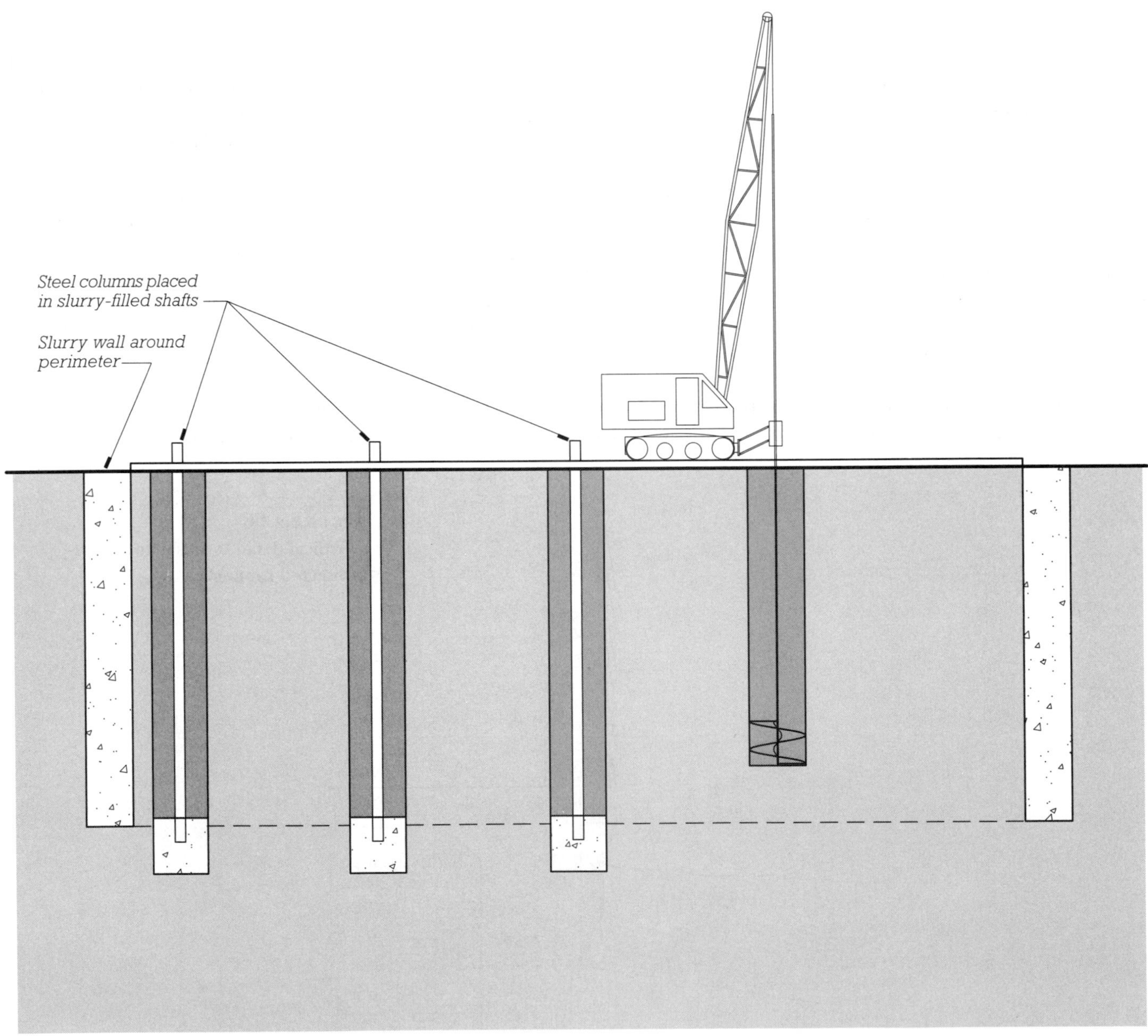

(*a*)

Figure 2.69
Up–down construction.

slurry-filled holes, and concrete footings are tremied beneath them. After the ground floor slab is in place and connected to the substructure columns, erection of the superstructure may begin. Construction continues simultaneously on the substructure, largely by means of mining machinery: A story of soil is excavated from beneath the ground floor slab and a level mud slab of CLSM is poured. Working on the mud slab, workers reinforce and pour a concrete structural slab for the floor of the topmost basement level and connect this floor to the columns. When the slab is sufficiently strong, another story of soil is removed from beneath it, along with the mud slab. The process is repeated until the substructure is complete, by which time the superstructure has been built many stories into the air.

Designing Foundations

It is a good idea to begin the design of the foundations of a building at the same time as architectural design work commences. Subsurface conditions beneath a site can strongly influence fundamental decisions about a building—its location on the site, its size and shape, its weight, and the required degree of flexibility of its construction. On a large building project, at least three designers are involved in these decisions: the architect, who has primary responsibility for the location and form of the building; the structural engineer, who has primary responsibility for its physical integrity; and the foundation engineer, who must decide, on the basis of site exploration and laboratory reports, how best to support it in the earth. More often than not, it is possible for the foundation engineer to design foundations for a building design dictated entirely by the architect. In some cases, however, the cost of the foundations may consume a much larger share of the construction budget than the architect has anticipated, unless certain

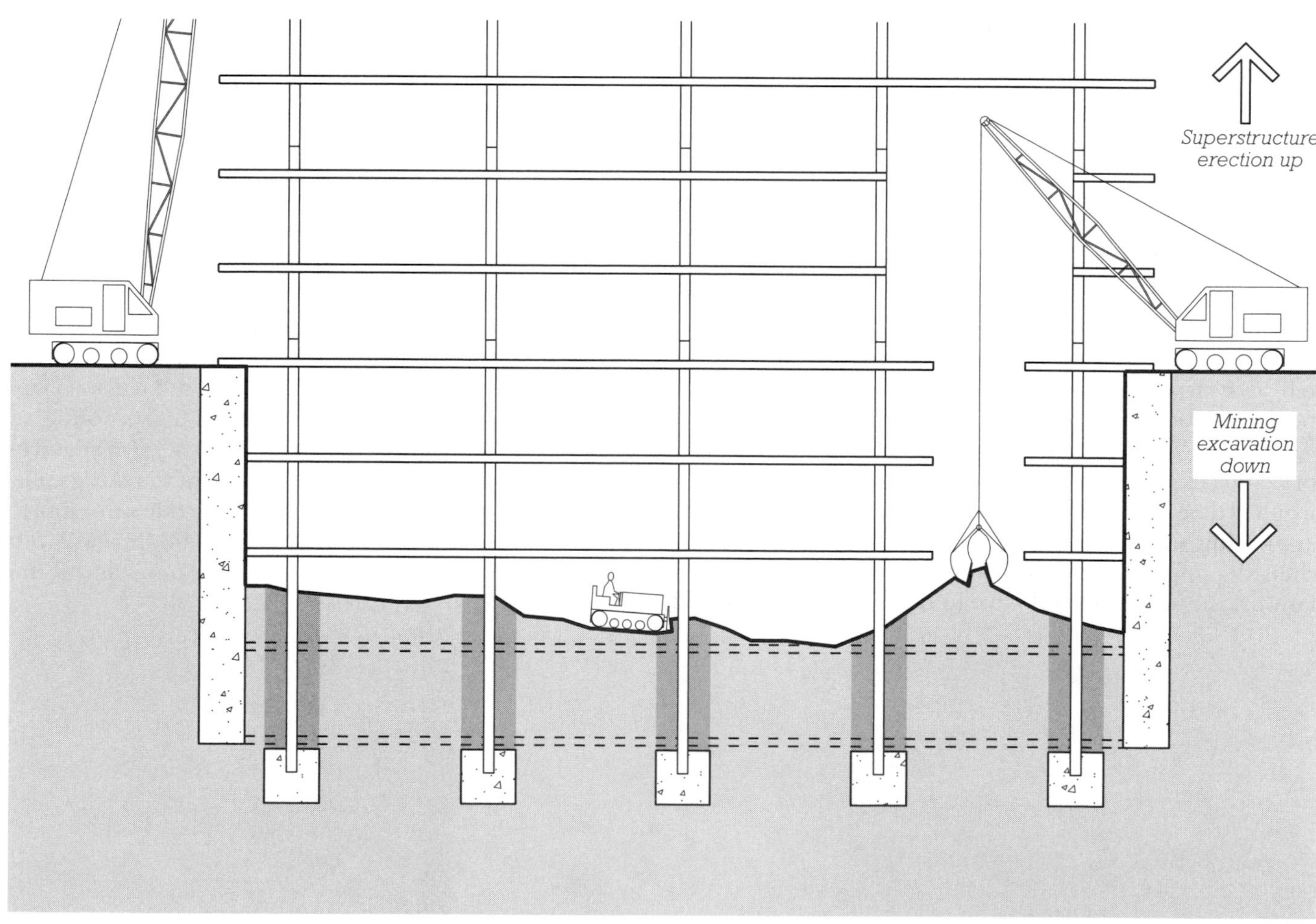

(*b*)

compromises can be reached on the form and location of the building. It is safer and more productive for the architect to work with the foundation engineer from the outset, seeking alternative site locations and building configurations that will result in the fewest foundation problems and the lowest foundation cost.

In designing a foundation, a number of different design thresholds need to be kept in mind. If the designer crosses any of these thresholds, foundation costs take a sudden jump. Some of these thresholds are:

- *Building below the water table.* If the substructure and foundations of a building are above the water table, minimal effort will be required to keep the excavation dry during construction. If the water table is penetrated, even by an inch, expensive steps will have to be taken to dewater the site, strengthen the slope support system, waterproof the foundation, and either strengthen the basement floor slab against hydrostatic uplift pressure or provide for adequate drainage to relieve this pressure. For an extra inch or foot of depth, the expense would probably not be justified; for another story or two of useful building space, it might be.
- *Building close to an existing structure.* If the excavation can be kept well away from adjacent structures, the foundations of these structures can remain undisturbed and no effort and expense are required to protect them. When digging close to an existing structure, and especially when digging deeper than its foundations, the structure will have to be temporarily braced and may require permanent underpinning with new foundations. Furthermore, an excavation at a distance from an existing structure may not require sheeting, while one immediately adjacent almost certainly will.
- *Increasing the column or wall load from a building beyond what can be supported by a shallow foundation.* Shallow foundations are far less expensive than piles or caissons under most conditions. If the building grows too tall, however, a shallow foundation may no longer be able to carry the load, and a threshold must be crossed into the realm of deep foundations. If this has happened for the sake of an extra story or two of height, the designer should consider reducing the height by broadening the building. If individual column loadings are too high for shallow foundations, perhaps they can be reduced by increasing the number of columns in the building and decreasing their spacing.

For buildings at the scale of one- and two-family dwellings, foundation design is usually much simpler than for large buildings because foundation loadings are low. The uncertainties in foundation design can be reduced with reasonable economy by adopting a large factor of safety in calculating the bearing capacity of the soil. Unless the designer has reason to suspect poor soil conditions, the footings are usually designed using rule-of-thumb allowable soil stresses and standardized footing dimensions. The designer then examines the actual soil when the excavations have been made. If it is not of the quality that was expected, the footings can be hastily redesigned using a revised estimate of soil-bearing capacity before construction continues. If unexpected groundwater is encountered, better drainage may have to be provided around the foundation or the depth of the basement decreased.

Foundation Design and the Building Codes

Because of the public safety considerations that are involved, building codes contain numerous provisions relating to the design and construction of excavations and foundations. The IBC defines which soil types are considered satisfactory for bearing the weight of buildings and establishes a set of requirements for subsurface exploration, soil testing, and submission of soil reports to the local building inspector. It goes on to specify the methods of engineering design that may be used for the foundations. It sets forth maximum loadbearing values for soils that may be assumed in the absence of detailed test procedures (Figure 2.5). It establishes minimum dimensions for footings, caissons, piles, and foundation walls and contains lengthy discussions relating to the installation of piles and caissons and the drainage and waterproofing of substructures. The IBC also requires engineering design of retaining walls. In all, the building code attempts to ensure that every building will rest upon secure foundations and a dry substructure.

CSI/CSC

MasterFormat Sections for Earthwork, Foundations, and Below-Grade Waterproofing

07 10 00	DAMPPROOFING AND WATERPROOFING
07 11 00	Dampproofing
07 13 00	Sheet Waterproofing
07 14 00	Fluid-Applied Waterproofing
07 16 00	Cementitious and Reactive Waterproofing
07 17 00	Bentonite Waterproofing
31 10 00	SITE CLEARING
31 20 00	EARTH MOVING
31 22 00	Grading
31 23 00	Excavation and Fill
	Dewatering
31 30 00	EARTHWORK METHODS
31 32 00	Soil Stabilization
31 34 00	Soil Reinforcement
31 40 00	SHORING AND UNDERPINNING
31 50 00	EXCAVATION SUPPORT AND PROTECTION
31 60 00	SPECIAL FOUNDATION AND LOAD-BEARING ELEMENTS
31 62 00	Driven Piles
31 63 00	Bored Piles
	Drilled Concrete Piers and Shafts
	Drilled Minipiles
32 32 00	RETAINING WALLS

Selected References

1. Ambrose, James E. *Simplified Design of Building Foundations* (2nd ed.). Hoboken, NJ, John Wiley & Sons, Inc., 1998.

After an initial summary of soil properties, this small book covers simplified foundation computation procedures for both shallow and deep foundations.

2. Liu Cheng and Jack Evett. *Soils and Foundations* (7th ed.). Upper Saddle River, NJ, Prentice Hall, Inc., 2008.

This textbook offers a detailed discussion of the engineering properties of soils, subsurface exploration techniques, soil mechanics, and shallow and deep foundations but is well suited to the beginner.

3. Schroeder, W. L., et al. *Soils in Construction* (5th ed.). Upper Saddle River, NJ, Prentice Hall, Inc., 2003.

A well-illustrated, clearly written, moderately detailed survey of soils, soil testing, subsurface construction, and foundations.

4. Henshell, Justin, and C. W. Griffin. *Manual of Below-Grade Waterproofing Systems.* Hoboken, NJ, John Wiley & Sons, Inc.,1999.

A comprehensive presentation of means for keeping water out of building substructures.

5. National Roofing Contractors Association. *The NRCA Waterproofing Manual.* Rosemont, IL, updated regularly.

This manual provides detailed guidelines for the application of waterproofing and dampproofing to building substructures.

Web Sites

Foundations

Author's supplementary web site: **www.ianosbackfill.com/02_foundations/**
Whole Building Design Guide, Foundation Walls: **www.wbdg.org/design/env_bg_wall.php**

Subsurface Exploration and Soils Testing

Ardaman & Associates (geotechnical consulting): **www.ardaman.com**

Dewatering

Griffin Dewatering: **www.griffindewatering.com**

Deep Foundations

Case Foundation Company: **www.casefoundation.com**
Geopier Foundation: **www.geopier.com**
Layne Christensen Company: **www.laynegeo.com**
Nicholson Construction Company: **www.nicholsonconstruction.com**
Schnabel Foundation Company: **www.schnabel.com**

Waterproofing

CETCO Building Materials Group: **www.cetco.com/BMG**
Grace Construction Products: **www.na.graceconstruction.com**

Key Terms and Concepts

foundation
dead load
live load
rain load
snow load
wind load
seismic load
lateral soil pressure load
buoyant uplift
flood load
horizontal thrust
foundation settlement
uniform settlement
differential settlement
earth material
rock
soil
boulder
cobble
gravel
sand
coarse-grained soil
silt
clay
fine-grained soil
organic soil
shear resistance
frictional soil, cohesionless soil
soil pore
soil particle fabric
cohesive soil
plastic
soil engineering properties
soil loadbearing capacity
stability
liquefaction
well-graded soil, poorly sorted soil
poorly graded soil, well-sorted soil
uniformly graded soil
soil grading
soil strata
test pit
test boring
water table
load test
liquid limit
plastic limit
geotechnical report
excavation
frost line
ice lens
benched excavation
angle of repose
slope support
excavation support
shoring
H-pile
soldier beam
lagging
sheet piling
sheeting
pneumatically applied concrete, shotcrete
slurry wall
guide wall
clamshell bucket
slurry
tremie
precast concrete slurry wall
soil mixing
bracing
crosslot bracing
waler
raker
heel block
tieback
rock anchor
soil nailing
dewatering
sump
well point
watertight barrier wall
soil freezing
superstructure
substructure
shallow foundation
deep foundation
footing
column footing
wall footing, strip footing
engineered fill
slab on grade
crawlspace
basement
tie beam
combined footing
cantilever footing
mat foundation, raft foundation
floating foundation
caisson, drilled pier

auger drill
belling bucket
belled caisson
socketed caisson
pile
timber pile
piledriver
end bearing pile
friction pile
pile cap
grade beam
pile hammers
piledriver lead
H-pile
heaving
pipe pile
minipile, pin pile, micropile
precast concrete pile
sitecast concrete pile
mandrel
pressure-injected footing
rammed aggregate pier
stone columns
base isolator
underpinning
retaining wall
segmental retaining wall
earth reinforcing
gabions
geotextile
earth reinforcement, soil reinforcement
drainage
drainage backfill
drainage mat
perforated drain piping
dampproofing
cement plaster dampproofing, parge coating
asphalt dampproofing, bituminous dampproofing
waterproofing
liquid waterproofing
sheet membrane waterproofing
bentonite waterproofing
integral waterproofing
negative side waterproofing
blind-side waterproofing
waterstop
flood test
protection board
shallow frost-protected foundations
backfill
controlled low-strength material (CLSM)
mud slab
up–down construction

Review Questions

1. What is the nature of the most common type of foundation failure? What measures can be taken to prevent its occurrence?

2. Explain in detail the differences among sand, silt, and clay, especially as they relate to the foundations of buildings.

3. List three different ways of sheeting an excavation. Under what circumstances would sheeting not be required?

4. Under what conditions would you use a watertight barrier instead of well points when digging below the water table?

5. If shallow foundations are substantially less costly than deep foundations, why do we use deep foundations?

6. What soil conditions favor the use of belled caissons? What soil conditions favor piles over caissons? What type of piles are especially well suited to repair or improvement of existing foundations and why?

8. List and explain some cost thresholds frequently encountered in foundation design.

9. For each of the following, suggest a type of waterproofing: (a) A foundation with many complex shapes, (b) a foundation subject to significant movement or cracking after the waterproofing is applied, (c) an existing foundation wall that can only be accessed and waterproofed from the interior side.

10. Explain the difference between waterproofing and dampproofing. When is either an appropriate choice for protecting a foundation from moisture?

12. List the components of a typical foundation drainage system and their functions.

Exercises

1. Obtain the foundation drawings and soils report for a nearby building. Look first at the log of test borings. What sorts of soils are found beneath the site? How deep is the water table? What type of foundation do you think should be used in this situation (keeping in mind the relative weight of the building)? Now look at the foundation drawings. What type is actually used? Why?

2. What type of foundation and substructure is normally used for houses in your area? Why?

3. Look at several excavations for major buildings under construction. Note carefully the arrangements made for slope support and dewatering. How is the soil being loosened and carried away? What is being done with the excavated soil? What type of foundation is being installed? What provisions are being made for keeping the substructure permanently dry?

3

Wood

A logger fells a large Douglas fir. *(Weyerhaeuser Company Photo)*

Wood is perhaps the best loved of all the materials that we use for building. It delights the eye with its endlessly varied colors and grain patterns. It invites the hand to feel its subtle warmth and varied textures. When it is fresh from the saw, its fragrance enchants. We treasure its natural, organic qualities and take pleasure in its genuineness. Even as it ages, bleached by the sun, eroded by rain, worn by the passage of feet and the rubbing of hands, we find beauty in its transformations of color and texture.

Wood earns our respect as well as our love. It is strong and stiff, yet by far the least dense of the materials used for the beams and columns of buildings. It is worked and fastened easily with small, simple, relatively inexpensive tools. It is readily recycled from demolished buildings for use in new ones, and when finally discarded, it biodegrades rapidly to become natural soil. It is our only renewable building material, one that will be available to us for as long as we manage our forests with an eye to the perpetual production of wood.

But wood, like a valued friend, has its idiosyncrasies. A piece of lumber is never perfectly straight or true, and its size and shape can change significantly with changes in the weather. Wood is peppered with defects that are relics of its growth and processing. Wood can split; wood can warp; wood can give splinters. If ignited, wood burns. If left in a damp location, it decays and harbors destructive insects. The skillful designer and the seasoned carpenter, however, know all these things and understand how to build with wood to bring out its best qualities, while neutralizing or minimizing its problems.

Trees

Wood comes from trees and is produced through natural growth processes. Understanding tree physiology is essential to knowing how to build with wood.

Tree Growth

The trunk of a tree is covered with a protective layer of dead *bark* (Figure 3.1). Inside the dead bark is a layer of living bark composed of hollow longitudinal cells that conduct nutrients from the leaves to the roots and other living parts of the tree. Inside this layer of living bark lies a very thin layer, the *cambium,* which creates new bark cells toward the outside of the trunk and new wood cells toward the inside. The thick layer of living wood cells inside the cambium is the *sapwood.* In this zone of the tree, nutrients are stored and sap is pumped upward from the roots to the leaves and distributed laterally in the trunk. At the inner edge of this zone, sapwood dies progressively and becomes *heartwood.* In many species of trees, heartwood is easily distinguished from sapwood by its darker color. Heartwood no longer participates in the life processes of the tree but continues to contribute to its structural strength. At the very center of the trunk, surrounded by heartwood, is the *pith* of the tree, a small zone of weak wood cells that were the first year's growth.

An examination of a small section of wood under a low-powered microscope shows that it consists primarily of tubular cells whose long axes are parallel to the long axis of the trunk. The cells are structured of tough *cellulose* and are bound together by a softer cementing substance called *lignin.* The direction of the long axes of the cells is referred to as the *grain* of the wood. Grain direction is important to the designer of wooden buildings because the appearance and physical properties of wood parallel to grain and perpendicular to grain are very different.

In temperate climates, the cambium begins to manufacture new sapwood cells in the spring, when the air is cool and groundwater is plentiful, conditions that favor rapid growth. Growth is slower during the heat of the summer, when water is scarce. *Springwood* or *earlywood* cells are therefore larger and less dense in substance than *summerwood* or *latewood* cells. Concentric bands of springwood and summerwood make up the annual growth rings in a trunk that can be counted to determine the age of a tree. The relative proportions of springwood and summerwood also have a direct bearing on the structural properties of the wood a given tree will yield because summerwood is stronger and stiffer than springwood. A tree grown under continuously moist, cool conditions grows faster than another tree of the same species grown under warmer, drier conditions, but its wood is not as dense or as strong.

Softwoods and Hardwoods

Softwoods come from coniferous trees and *hardwoods* from broadleafed trees. The names can be deceptive because woods from some coniferous trees may actually be harder than woods from some broadleafed trees. Nevertheless, the distinction between these two types of woods is a useful one. Softwood trees have a relatively simple microstructure, consisting mainly of large longitudinal cells (*tracheids*) together with a small percentage of radial cells (*rays*), whose function is the storage and radial transfer of nutrients (Figure 3.2). Hardwood trees

FIGURE 3.1
Summerwood rings are prominent and a few rays are faintly visible in this cross section of an evergreen tree. But the cambium, which lies just beneath the thick layer of bark, is too thin to be seen, and heartwood cannot be distinguished visually from sapwood in this species. *(Courtesy of Forest Products Laboratory, Forest Service, USDA)*

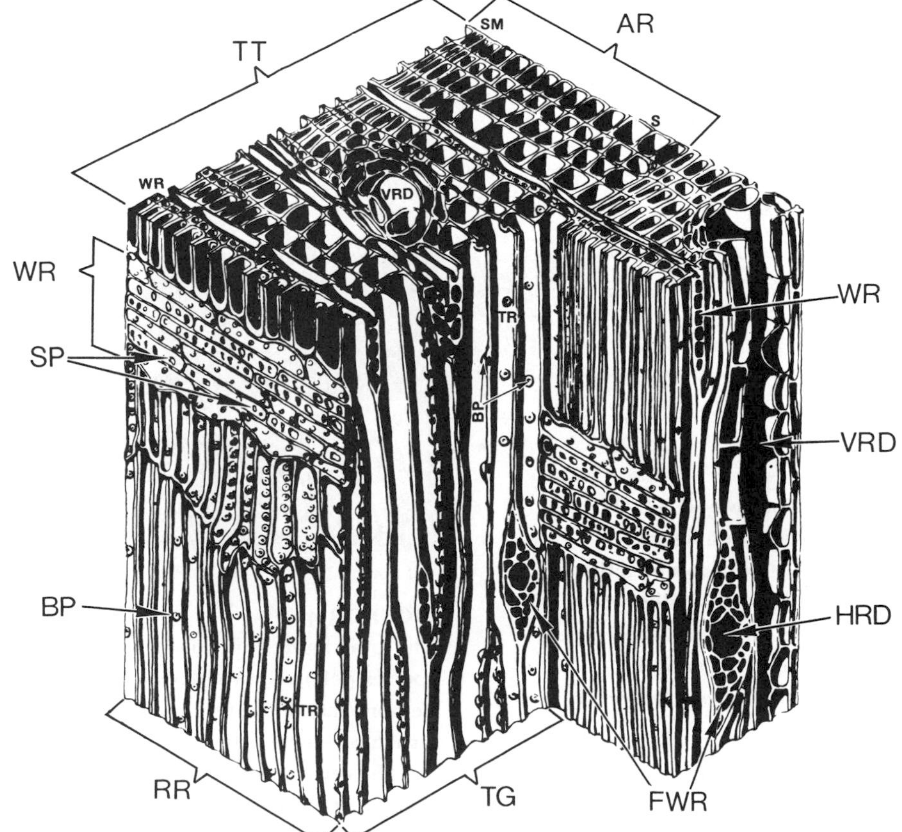

FIGURE 3.2
Vertical cells (tracheids, labeled TR) dominate the structure of a softwood, seen here greatly enlarged. But rays (WR), which are cells that run radially from the center of the tree to the outside, are clearly in evidence. An annual ring (labeled AR) consists of a layer of smaller summerwood cells (SM) and layer of larger springwood cells (S). Simple pits (SP) allow sap to pass from ray cells to longitudinal cells and vice versa. Resin is stored in vertical and horizontal resin ducts (VRD and HRD), with the horizontal ducts centered in fusiform wood rays (FWR). Border pits (BP) allow the transfer of sap between longitudinal cells. The face of the sample labeled RR represents a radial cut through the tree, and TG represents a tangential cut. *(Courtesy of Forest Products Laboratory, Forest Service, USDA)*

are more complex in structure, with a much larger percentage of rays and two different types of longitudinal cells: small-diameter *fibers* and large-diameter *vessels* or *pores*, which transport the sap of the tree (Figure 3.3).

When cut into lumber, softwoods generally have a coarse and relatively uninteresting grain structure, while many hardwoods show beautiful patterns of rays and vessels (Figure 3.4). Most of the lumber used today for building framing comes from softwoods, which are comparatively plentiful and inexpensive. For fine furniture and interior finish details, hardwoods are often chosen. A few softwood and hardwood species widely used in North America are listed in Figure 3.5, along with the principal uses of each; however, it should be borne in mind that literally thousands of species of wood are used in construction around the world and that

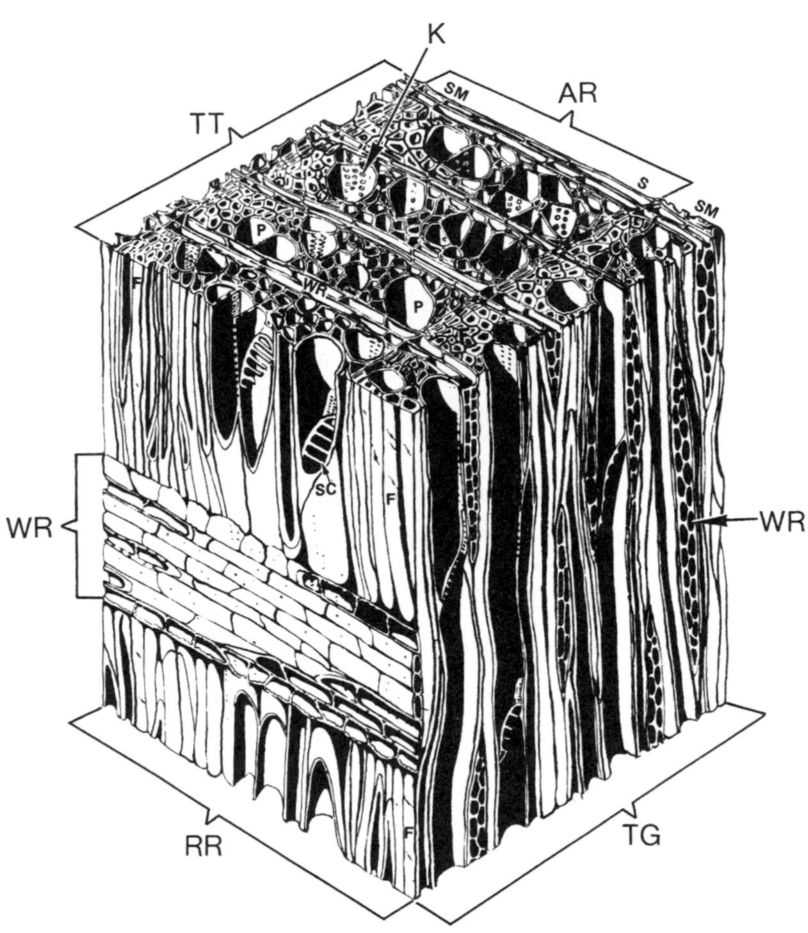

FIGURE 3.3
Rays (WR) constitute a large percentage of the mass of a hardwood, as seen in this sample, and are largely responsible for the beautiful grain figures associated with many species. The vertical cell structure is more complex than that of a softwood, with large pores (P) to transport the sap and smaller wood fibers (F) to give the tree structural strength. Pore cells in some hardwood species end with crossbars (SC), while those of other species are entirely open. Pits (K) pass sap from one cavity to another. *(Courtesy of Forest Products Laboratory, Forest Service, USDA)*

(*a*)

(*b*)

FIGURE 3.4
The grain figures of two softwood species (left) and two hardwoods (right) demonstrate the difference in cellular structure between the classes of woods: (*a*) The cells in sugar pine are so uniform that the grain structure is almost invisible except for scattered resin ducts. (*b*) Vertical-grain Douglas fir shows very pronounced dark bands of summerwood.

the available species vary considerably with geographic location. In North America, the major lumber-producing forests are in the western and eastern mountains of both the United States and Canada. Other regions, chiefly in the southeastern United States, also produce significant quantities.

Certified Wood

Certified wood comes from forests that are managed for their long-term ecological sustainability and economic viability. The most widely recognized wood certifying body is the *Forest Stewardship Council (FSC)*. FSC certification encompasses a broad range of social, economic, and ecological principals. FSC criteria address respect for laws and international treaties, long-term tenure and land ownership rights and responsibilities, rights of indigenous peoples, community relations, and workers' rights. They encourage

(*c*)

(*d*)

(*c*) Red oak exhibits large open pores amid its fibers. (*d*) This quartersliced mahogany veneer has a pronounced "ribbon" figure caused by varying light reflections off its fibers. *(Photos by the Edward Allen)*

SOFTWOODS	HARDWOODS
Used for Framing, Sheathing, Paneling	***Used for Moldings, Paneling, Furniture***
Alpine fir	Ash
Balsam fir	Beech
Douglas fir	Birch
Eastern hemlock	Black walnut
Eastern spruce	Butternut
Eastern white pine	Cherry
Englemann spruce	Lauan
Idaho white pine	Mahogany
Larch	Pecan
Loblolly pine	Red oak
Lodgepole pine	Rosewood
Longleaf pine	Teak
Mountain hemlock	Tupelo gum
Ponderosa pine	White oak
Red spruce	Yellow poplar
Shortleaf pine	
Sitka spruce	
Southern yellow pine	
Western hemlock	
White spruce	
Used for Moldings, Window and Door Frames	***Used for Finish Flooring***
Ponderosa pine	Pecan
Sugar pine	Red oak
White pine	Sugar maple
Used for Finish Flooring	Walnut
Douglas fir	White oak
Longleaf pine	
Decay-Resistant Woods, Used for Shingles, Siding, Outdoor Structures	
California redwood	
Southern cypress	
Western red cedar	
White cedar	

FIGURE 3.5
Some species of woods commonly used in construction in North America, listed alphabetically in groups according to end use. All are domestic except Lauan (Asia), Mahogany (Central America), Rosewood (South America and Africa), and Teak (Asia).

efficient use of forest products and services to ensure the economic viability and social benefits of forestry operations. They ensure forest management practices that prevent overharvesting, conserve biological diversity and natural resources, and protect the environment. The FSC accredits local certifying bodies around the globe.

Considerations of Sustainability in Wood Construction

Wood: A Renewable Resource

• Wood is the only major structural material that is renewable.

• In the United States and Canada, tree growth each year greatly exceeds the volume of harvested trees, though many timberlands are not managed in a sustainable manner.

• On other continents, many countries long ago felled the last of their forests, and many forests in other countries are being depleted by poor management practices and slash-and-burn agriculture. Particularly in the case of tropical hardwoods, it is wise to investigate sources and to ensure that the trees were grown in a sustainable manner.

• Some panel products can be manufactured from rapidly renewable vegetable fibers, recoverable and recycled wood fibers, or recycled cellulose fibers.

• Bamboo, a rapidly renewable grass, can replace wood in the manufacture of flooring, interior paneling, and other finish carpentry applications. In other parts of the world, bamboo is used for the construction of scaffolding, concrete formwork, and even as the source of fibrous material for structural panels analogous to wood-based oriented strand board (OSB), particleboard, and fiberboard.

Forestry Practices

• Two basic forms of forest management are practiced in North America: sustainable forestry, and clearcutting and replanting. The clearcutting forest manager attains sustainable production by cutting all the trees in an area, leaving the stumps, tops, and limbs to decay and become compost, setting out new trees, and tending them until they are ready for harvest. In sustainable forestry, trees are harvested more selectively from a forest in such a way as to minimize damage to the forest environment and maintain the biodiversity of its natural ecosystem.

• Environmental problems often associated with logging of forests include loss of wildlife habitat, soil erosion, pollution of waterways, and air pollution from machinery exhausts and burning of tree wastes. A recently clearcut forest is a shockingly ugly tangle of stumps, branches, tops, and substandard logs left to decay. It is crisscrossed by deeply rutted, muddy haul roads. Within a few years, decay of the waste wood and new tree growth largely heal the scars. Loss of forest area may raise levels of carbon dioxide, a greenhouse gas, in the atmosphere, because trees take up carbon dioxide from the air, utilize the carbon for growth, and give back pure oxygen to the atmosphere.

• The buyer of wood products can support sustainable forestry practices by specifying products certified as originating from sustainable forests, those that are managed in a socially responsible and environmentally sound manner. FSC-certified wood products, for example, satisfy the requirements of LEED and all other major green building assessment programs.

Mill Practices

• Skilled sawyers working with modern computerized systems can convert a high percentage of each log into marketable wood products. A measure of sawmill performance is the lumber recovery factor (LRF), which is the net volume of wood products produced from a cubic meter of log.

• Manufactured wood products such as oriented strand board, particleboard, I-joists, and laminated strand lumber efficiently utilize most of the wood fiber in a tree and can be produced from recycled or younger-growth, rapidly renewable materials; finger-jointed lumber is made by gluing end to end short pieces of lumber that might otherwise be treated as waste. The manufacturer of large, solid timbers generates more unused waste and yields fewer products from each log.

• Kiln drying uses large amounts of fuel but produces more stable, uniform lumber than air drying, which uses no fuel other than sunlight and wind.

• Mill wastes are voluminous: Bark may be shredded to sell as a landscape mulch, composted, burned, or buried in a landfill. Sawdust, chips, and wood scraps may be burned to generate steam to power the mill, used as livestock bedding, composted, burned, or buried in a landfill.

• Many wood products can be manufactured with significant percentages of recoverable or recycled wood, plant fiber, or paper materials.

Transportation

• Because the major commercial forests are located in concentrated regions of the United States and Canada, most lumber must be shipped considerable distances. Fuel consumption is minimized by planing and drying the

A second wood certification organization is the *Sustainable Forestry Initiative (SFI)*. Originally formed under the auspices of the American Forest & Paper Association, the SFI is now a fully independent body that accredits forestry practices in the United States and Canada.

lumber before it is shipped, which reduces both weight and volume.

• Some wood products can be harvested or manufactured locally or regionally.

Energy Content

• Solid lumber has an embodied energy of roughly 1000 to 3000 BTU per pound (2.3 to 7.0 MJ/kg). An average 8-foot-long 2 × 4 (2.4-m-long 38 × 89 mm) has an embodied energy of about 17,000 BTU (40 MJ). This includes the energy expended to fell the tree, transport the log, saw and surface the lumber, dry it in a kiln, and transport it to a building site.

• Manufactured wood products have higher embodied energy content than solid lumber, due to the glue and resin ingredients and the added energy required in their manufacture. The embodied energy of such products ranges from about 3000 to 7500 BTU per pound (7.0 to 17 MJ/kg).

• Wood construction involves large numbers of steel fasteners of various kinds. Because steel is produced by relatively energy-intensive processes, fasteners add considerably to the total energy embodied in a wood frame building.

• Wood does not have the lowest embodied energy of the major structural materials when measured on a pound-for-pound basis. However, when buildings of comparable size, but structured with either wood, light gauge steel studs, or concrete, are compared, most studies indicate that those of wood have the lowest total embodied energy of the three. This is due to wood's lighter weight (or, more precisely, its lesser density) in comparison to these other materials, as well as the relative efficiency of the wood light frame construction system.

Construction Process

• A significant fraction of the lumber delivered to a construction site is wasted: It is cut off when each piece is sawed to size and shape and ends up on the scrap heap, which is usually burned or taken to a landfill. On-site cutting of lumber also generates considerable quantities of sawdust. Construction site waste can be reduced by designing buildings that utilize full standard lengths of lumber and full sheets of wood panel materials.

• Wood construction lends itself to various types of prefabrication that can reduce waste and improve the efficiency of material usage in comparison to on-site building methods.

Indoor Air Quality (IAQ)

• Wood itself seldom causes IAQ problems. Very few people are sensitive to the odor of wood.

• Some of the adhesives and binders used in glue-laminated lumber, structural composite lumber, and wood panel products can cause serious IAQ problems by giving off volatile organic compounds such as formaldehyde. Alternative products with low-emitting binders and adhesives are also available.

• Some paints, varnishes, stains, and lacquers for wood also emit fumes that are unpleasant and/or unhealthful.

• In damp locations, molds and fungi may grow on wood members, creating unpleasant odors and releasing spores to which many people are allergic.

Building Life Cycle

• If the wood frame of a building is kept dry and away from fire, it will last indefinitely. However, if the building is poorly maintained and wood elements are frequently wet, wood components may decay and require replacement.

• Wood is combustible and gives off toxic gases when it burns. It is important to keep sources of ignition away from wood and to provide smoke alarms and easy escape routes to assist building occupants in escaping from burning buildings. Where justified by building size or type of occupancy, building codes require sprinkler systems to protect against the rapid spread of fire.

• When a building is demolished, wood framing members can be recycled directly into the frame of another building, sawn into new boards or timbers, or shredded as raw material for oriented-strand materials. There is a growing industry whose business is purchasing and demolishing old barns, mills, and factories and selling their timbers as *reclaimed lumber.*

A study commissioned by the Canadian Wood Council compares the full life cycle of three similar office buildings, one each framed with wood, steel, or concrete and all three operated in a typical Canadian climate. In this study, total embodied energy for the wood building is about half of that for the steel building and two-thirds of that for the concrete building. The wood building also outperforms the others in measures of greenhouse gas emissions, air pollution, solid waste generation, and ecological impact.

Recognition of wood certification organizations and specific credit requirements vary among the major green building assessment programs. When choosing certified woods for a particular project, the specifier should verify the particular requirements of the green building program selected for that project.

Lumber

Sawing

The production of *lumber*, lengths of squared wood for use in construction, begins with the felling of trees and the transportation of the logs to a sawmill (Figure 3.6). Sawmills range in size from tiny family operations to giant semiautomated factories, but the process of lumber production is much the same regardless of scale. Each log is stripped of its bark, then passed repeatedly through a large *headsaw*, which may be either a circular saw or a bandsaw, to reduce the log to untrimmed slabs of lumber

Figure 3.6
Loading logs onto a truck for their trip to the sawmill. *(Photo by Donald K. O'Brien)*

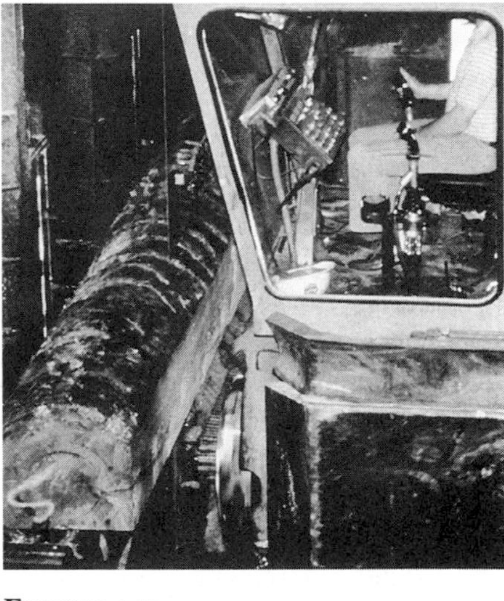

Figure 3.7
In a large mechanized mill, the operator controls the high-speed bandsaw from an overhead booth. *(Courtesy of Western Wood Products Association)*

Figure 3.8
Sawn lumber is sorted into stacks according to its cross-sectional dimensions and length. *(Courtesy of Western Wood Products Association)*

(Figure 3.7). The *sawyer* judges (with the aid of a computer in the larger mills) how to obtain the maximum marketable wood from each log and uses hydraulic machinery to rotate and advance the log in order to achieve the required succession of cuts. As the slabs fall from the log at each pass, a conveyor belt carries them away to smaller saws that reduce them to square-edged pieces of the desired widths (Figure 3.8). The sawn pieces at this stage of production have rough-textured surfaces and may vary slightly in dimension from one end to the other.

The sequence and pattern in which a log is sawn affect the orientation of grain within the finished lumber pieces. Lumber for structural uses is typically *plainsawn* (also called *flatsawn*), a method of dividing the log that produces the maximum yield of useful pieces and therefore the greatest economy (Figure 3.9). In plainsawn lumber, the orientation of annual growth rings within the board varies along its width, with large portions of the board's wider face dominated by a grain pattern in which annual rings are oriented close to parallel with the face. Such *flat-grain* lumber is characterized by a tendency to warp or distort during seasoning, to vary in surface appearance, and to erode relatively quickly and unevenly when used in applications such as flooring and exterior siding. Where such characteristics are undesirable, lumber may be *quartersawn, riftsawn,* or *edge-sawn.* These sawing methods produce *edge-grain* or *vertical gain* lumber, in which annual growth rings run more consistently perpendicular to the pieces' wider faces. Edge-grain boards tend to remain flat despite changes in moisture content, their faces have tighter and more pleasing grain figures, and their wearing qualities are improved because there are no broad areas of soft springwood exposed in the face, as there are in flatsawn lumber. In some wood species, edge-grain lumber also holds paint better than when plainsawn. Because edge-grain sawing methods produce smaller boards and more waste than plainsawing, they increase the cost of the finished lumber and are typically reserved for hardwoods and uses such as finish flooring, interior trim, and furniture, where added expense is justified.

Seasoning

Growing wood contains a quantity of water that can vary from about 30 percent to as much as 300 percent of the

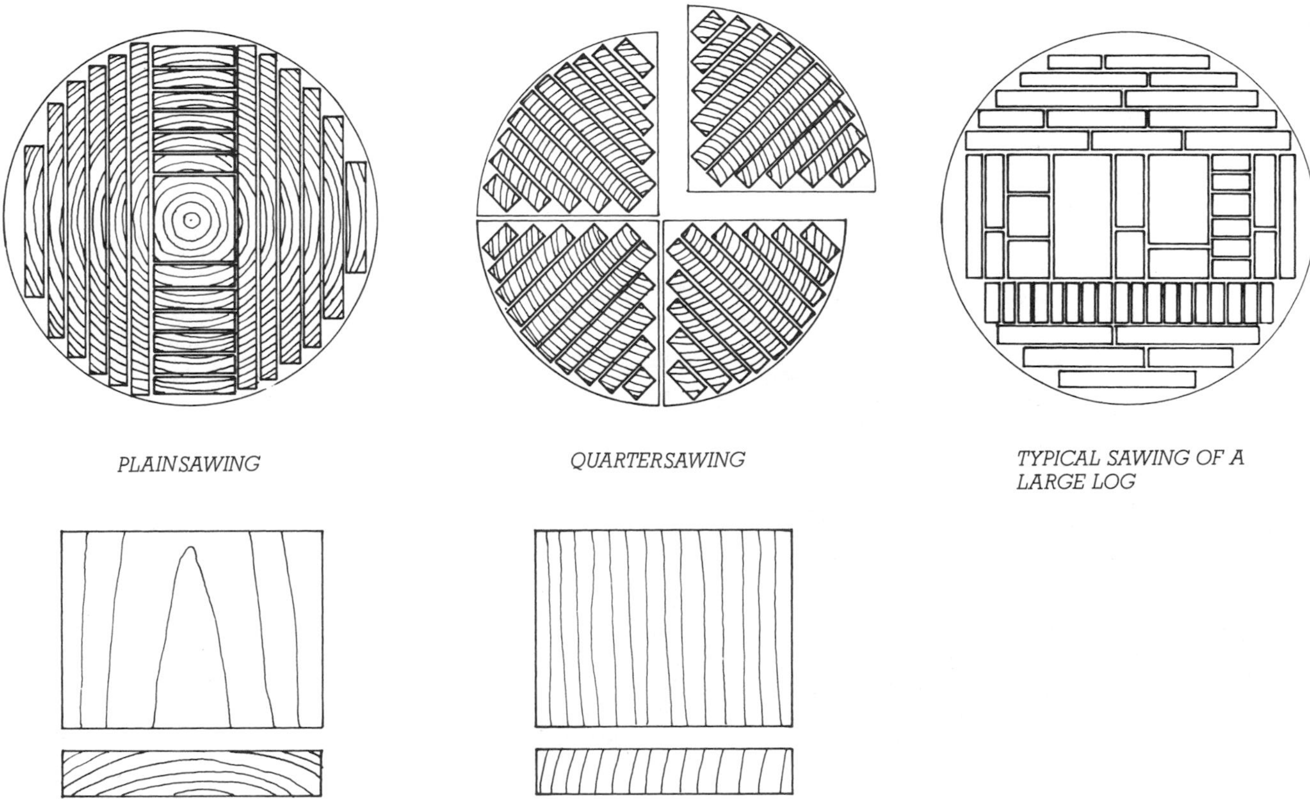

FIGURE 3.9
Plainsawing produces boards with a broad grain figure, as seen in the end and top views below the plainsawn log. Quartersawing produces a vertical grain structure, which is seen on the face of the board as tightly spaced parallel summerwood lines. A large log of softwood is typically sawn to produce some large timbers, some plainsawn dimension lumber, and, in the horizontal row of small pieces seen just below the heavy timbers, some pieces of vertical-grain decking.

oven-dry weight of the wood. After a tree is cut, this water starts to evaporate. First to leave the wood is the *free water* that is held in the cavities of the cells. When the free water is gone, the wood contains about 26 to 32 percent moisture as *bound water* held within the cellulose of the cell walls. As the bound water begins to evaporate, the wood starts to shrink, and the strength and stiffness of the wood begin to increase. The shrinkage, stiffness, and strength increase steadily as the moisture content continues to decrease. Wood can be dried to any desired moisture content, but framing lumber is considered to be *seasoned* when its moisture content is 19 percent or less. For framing applications that require closer control of wood shrinkage, lumber seasoned to a moisture content of 15 percent, labeled "MC 15," is produced. It is of little use to season ordinary framing lumber to a moisture content below about 13 percent, because wood is hygroscopic and will take on or give off moisture, swelling or shrinking as it does so, in order to stay in equilibrium with the moisture in the surrounding air. Lumber for interior finish carpentry and architectural woodwork is normally seasoned to a moisture content in the range of 5 to 11 percent, bringing it as close as possible to its final *equilibrium moisture content (EMC)*, that is, the moisture content at which it is expected to arrive in the completed, conditioned building.

Most lumber is seasoned at the sawmill, either by air drying in loose stacks for a period of months or, more commonly, by drying within a kiln under closely controlled conditions of temperature and humidity for a period of days (Figures 3.10 and 3.11). Seasoned lumber is stronger and stiffer than unseasoned (*green*) lumber and more dimensionally stable. It is also lighter in weight, which makes it more economical to ship. *Kiln drying* is generally preferred to *air drying* because it can be done faster, and it produces lumber with fewer distortions and more uniform quality.

Wood does not shrink and swell uniformly with changes in moisture content. Moisture shrinkage along the length of the log (*longitudinal shrinkage*) is negligible for practical purposes. Shrinkage in the radial direction (*radial shrinkage*) is very large by comparison, and shrinkage around the circumference of the log (*tangential shrinkage*) is about half again greater than radial shrinkage (Figure 3.12). If an entire log is seasoned before sawing, it will shrink very little along its length, but it will

Nor do I ever come to a lumber yard with its citylike, graduated masses of fresh shingles, boards and timbers, without taking a deep breath of its fragrance, seeing the forest laid low in it by processes that cut and shaped it to the architect's scale of feet and inches. . . .

Frank Lloyd Wright, *Architecture Record,* December, 1928.

FIGURE 3.10
For proper air drying, lumber is supported well off the ground. The *stickers,* which keep the boards separated for ventilation, are carefully placed above one another to avoid bending the lumber, and a watertight roof protects each stack from rain and snow. *(Courtesy of Forest Products Laboratory, Forest Service, USDA)*

FIGURE 3.11
Measuring moisture content in boards in a drying kiln. *(Courtesy of Western Wood Products Association)*

grow noticeably smaller in diameter, and the difference between the tangential and radial shrinkage will cause it to check, that is, to split open all along its length (Figure 3.13).

These differences in shrinkage rates are so large that they cannot be ignored in building design. In constructing building frames of plainsawn lumber, a simple distinction is made between parallel-to-grain shrinkage, which is negligible, and perpendicular-to-grain shrinkage, which is considerable. The difference between radial and tangential shrinkage is not considered because the orientation of the annual rings in plainsawn lumber is random and unpredictable. As we will see in Chapter 5, wood building frames are carefully designed to equalize the amount of wood loaded perpendicular to grain from one side of the structure to the other in order to avoid the noticeable tilting of floors and tearing of wall finish materials that would otherwise occur.

The position in a log from which a piece of lumber is sawn determines in large part how it will distort as it dries. Figure 3.14 shows how the differences between tangential and radial shrinkage cause this to happen.

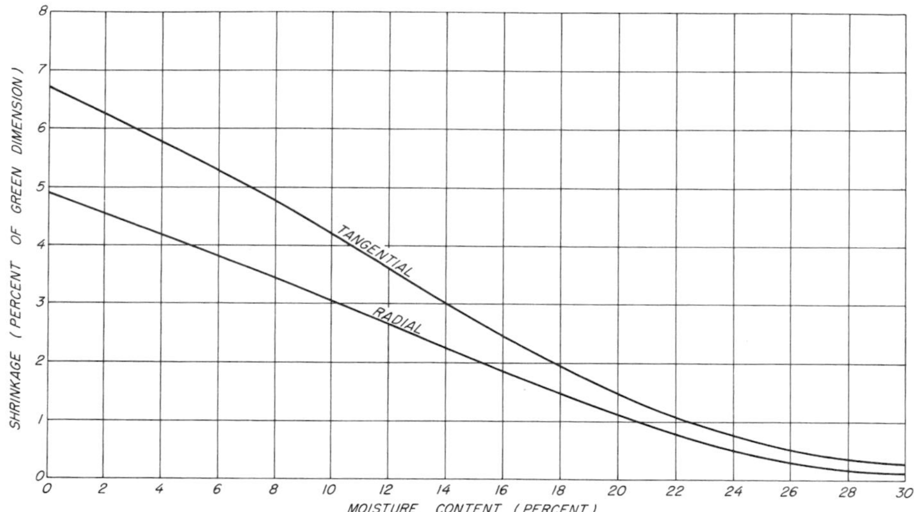

Figure 3.12
Shrinkage of a typical softwood with decreasing moisture content. Longitudinal shrinkage, not shown on this graph, is so small by comparison to tangential and radial shrinkage that it is of no practical consequence in wood buildings. *(Courtesy of Forest Products Laboratory, Forest Service, USDA)*

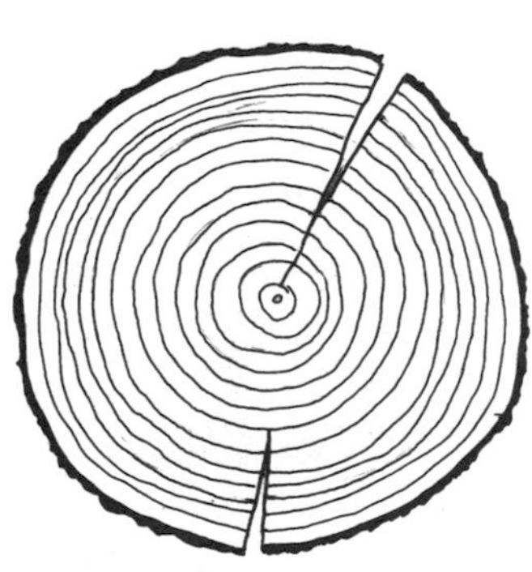

Figure 3.13
Because tangential shrinkage is so much greater than radial shrinkage, high internal stresses are created in a log as it dries, resulting inevitably in the formation of radial cracks called *checks*.

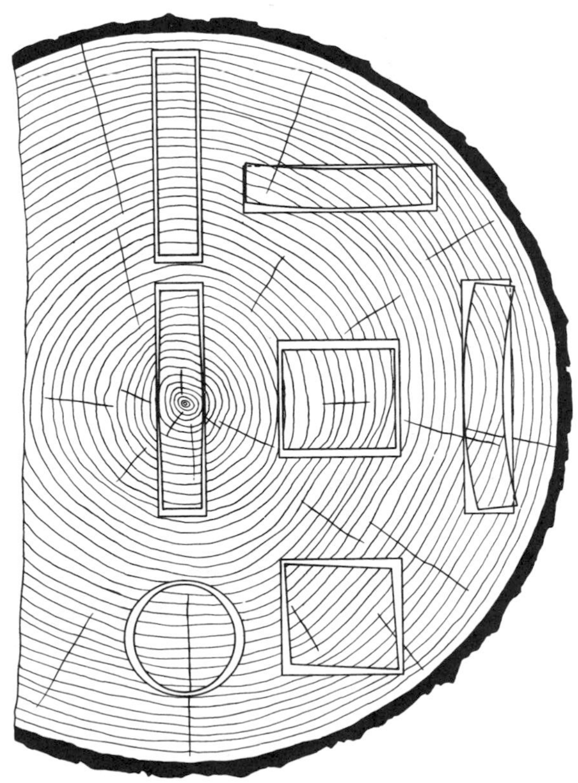

Figure 3.14
The difference between tangential and radial shrinkage also produces seasoning distortions in lumber. The nature of the distortion depends on the position the piece of lumber occupied in the tree. The distortions are most pronounced in plainsawn lumber (upper right, extreme right, lower right). *(Courtesy of Forest Products Laboratory, Forest Service, USDA)*

These effects are pronounced, and are readily predicted and observed in everyday practice.

Surfacing

Lumber is *surfaced* to make it smooth and more dimensionally precise. Rough (*unsurfaced*) lumber is often available commercially and is used for many purposes, but surfaced lumber is easier to work with because it is more square and uniform in dimension and less damaging to the hands of the carpenter. Surfacing is done by high-speed automatic machines, or planes, whose rotating blades smooth the surfaces of the piece and round the edges slightly. Most lumber is surfaced on all four sides, *surfaced four sides (S4S)*, but hardwoods are often *surfaced two sides (S2S)*, leaving the two edges to be finished by the craftsman.

Lumber is usually seasoned before it is surfaced, which allows the planing process to remove some of the distortions that occur during seasoning, but for some framing lumber this order of operations is reversed. The designation *S-DRY* in a lumber gradestamp indicates that the piece was surfaced (planed) when in a seasoned (dry) condition, and *S-GRN* indicates that it was planed when green.

Lumber Defects

Almost every piece of lumber contains one or more discontinuities in

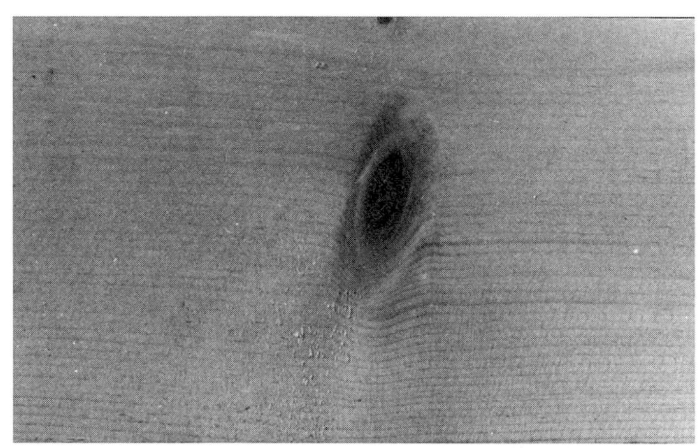

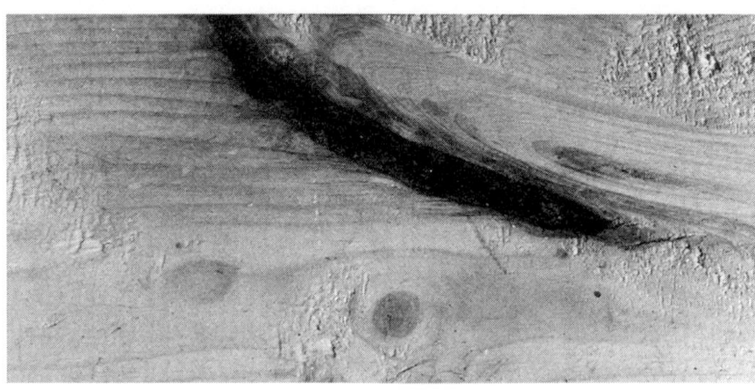

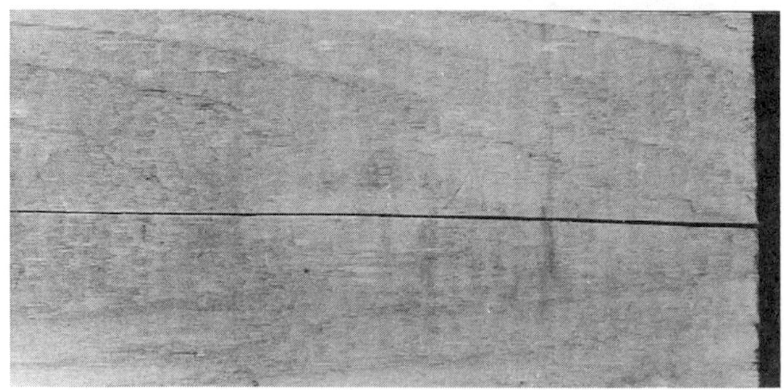

its structure caused by *growth characteristics* of the tree from which it came or *manufacturing characteristics* that were created at the mill (Figures 3.15 and 3.16). Among the most common growth characteristics are *knots,* which are places where branches joined the trunk of the tree; *knotholes,* which are holes left by loose knots dropping out of the wood; *decay;* and *insect damage.* Knots and knotholes reduce the strength of a piece of lumber, make it more difficult to cut and shape, and are often considered detrimental to its appearance. Decay and insect damage that occurred during the life of the tree may or may not affect the useful properties of the piece of lumber, depending on whether the organisms are still alive in the wood and the extent of the damage.

Manufacturing characteristics arise largely from changes that take place during the seasoning process because of the differences in rates of shrinkage with varying orientations to the grain. *Splits* and *checks* are usually caused by shrinkage stresses. *Crooking, bowing, twisting,* and *cupping* all occur because of nonuniform shrinkage. *Wane* is an irregular rounding of edges or faces that is caused by sawing pieces too close to the perimeter of the log. Experienced carpenters judge the extent of these defects and distortions in each piece of lumber and decide accordingly where and how to use the piece in the building. Checks are of little consequence in framing lumber, but

FIGURE 3.15 (opposite page)
Surface features often observed in lumber include, in the left-hand column from top to bottom, a knot cut crosswise, a knot cut longitudinally, and a bark pocket. To the right are a gradestamp, wane on two edges of the same piece, and a small check. The gradestamp indicates that the piece was graded according to the rules of the American Forest Products Association, that it is #2 grade Spruce–Pine–Fir, and that it was surfaced after drying. The 27 is a code number for the mill that produced the lumber. *(Photos by Edward Allen)*

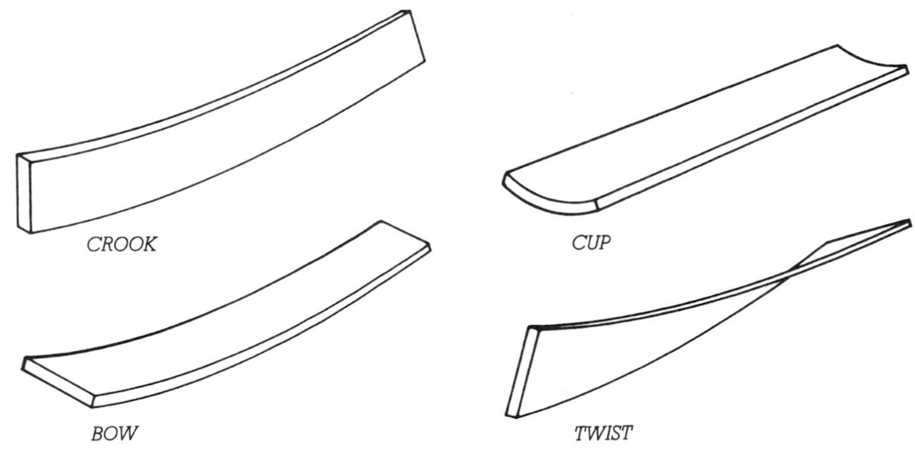

FIGURE 3.16
Four types of seasoning distortions in dimension lumber.

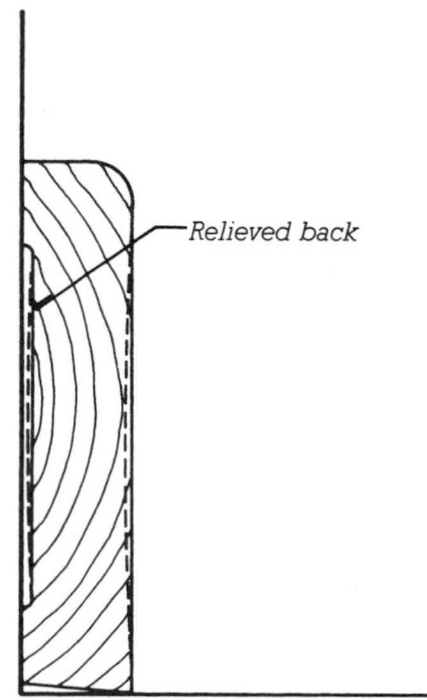

FIGURE 3.17
The effects of seasoning distortions can often be minimized through knowledgeable detailing practices. As an example, this wood baseboard, seen in cross section, has been formed with a relieved back, a broad, shallow groove that allows the piece to lie flat against the wall even if it cups (broken lines). The sloping bottom on the baseboard ensures that it can be installed tightly against the floor despite the cup. The grain orientation in this piece is the worst possible with respect to cupping. If quartersawn lumber were not available, the next best choice would have been to mill the baseboard with the center of the tree toward the room rather than toward the wall.

a joist or rafter with a crook in it is usually placed with the convex edge (the "crown") facing up to allow the floor or roof loads to straighten the piece. Badly bowed wall studs, floor joists, or roof rafters may be straightened by sawing or planing away the crown before being covered by wallboard, subflooring, or sheathing. Badly twisted pieces are put aside to be cut up for blocking. The effects of cupping in flooring and interior baseboards and trim are usually minimized by using well-seasoned, quartersawn stock and by shaping the pieces so as to reduce the likelihood of distortion (Figure 3.17).

Lumber Grading

Each piece of lumber is graded either for appearance or for structural strength and stiffness, depending on its intended use, before it leaves the mill. Lumber is sold by species and grade; the higher the grade, the higher the price. Grading offers the architect and the engineer the opportunity to build as economically as possible by using only as high a grade as is required for a particular use. In a specific building, the main beams or columns may require a high structural grade of lumber, while the remainder of the framing members will perform adequately in an intermediate, less expensive grade. For blocking, the lowest grade is perfectly adequate. For finish trim that will be coated with a clear finish, a high appearance grade is desirable; for painted trim, a lower grade will suffice.

Structural grading of lumber may be done either visually or by machine. In *visual grading*, trained inspectors examine each piece for ring density and for growth and manufacturing characteristics, then judge it and stamp it with a grade in accordance with industry-wide grading rules (Figure 3.18). In *machine grading*, an automatic device assesses the structural properties of the wood and stamps a grade automatically on the piece. This assessment is made either by flexing each piece between rollers and measuring its resistance to bending (*machine stress-rated*) or by scanning the wood electronically to determine its density. *Appearance grading*, naturally enough, is done visually. Figure 3.19 outlines a typical grading scheme for framing lumber, and Figure 3.20 outlines the appearance grades for nonstructural lumber. Light framing lumber for houses and other small buildings is usually ordered as "#2 and better" (a mixture of #1 and #2 grades) for floor joists and roof rafters and as "Stud" grade for wall framing. "Economy" grade, not shown in the table, is reserved for lumber not intended for structural purposes.

FIGURE 3.18
A grader marks the grade on a piece with a lumber crayon preparatory to applying a gradestamp. *(Courtesy of Western Wood Products Association)*

FIGURE 3.19
Standard structural grades for western softwood lumber. For each species of wood, the allowable structural stresses for each of these grades are tabulated in the structural engineering literature. *(Courtesy of Western Wood Products Association)*

Structural Properties of Wood

The strength of a piece of wood depends chiefly on its species, its grade, and the direction in which the load acts with respect to its grain. For example, in both tension and compression, wood is typically several times stronger parallel to grain than perpendicular to grain. With its usual assortment of defects, it is stronger in compression than in tension. Allowable strengths (structural stresses that include factors of safety) vary tremendously with species and grade. For example, allowable compressive strength parallel to grain for commercially available grades and species of framing lumber varies from 325 to 1700 psi (2.24 to 11.71 MPa), a difference of more than five times. Figure 3.21 compares the average structural properties of framing lumber to those of some other common structural materials—brick masonry, steel, and concrete. Of the four materials, only wood and steel have useful tensile strength. Defect-free wood is comparable to steel on a strength-per-unit-weight basis, but with the ordinary run of defects, an average piece of lumber is somewhat inferior to steel by this yardstick.

When designing a wooden structure, the architect or engineer determines the maximum stresses that are likely to occur in each of the structural members and selects an appropriate species and grade of lumber for each. In a given locale, a limited number of species and grades are usually available in retail lumberyards, and it is from these that the selection is made. It is common practice to use a stronger but more expensive species (Douglas fir or Southern pine, for example) for highly stressed major members and to use a weaker, less expensive species (such as Eastern hemlock) or species group (Hemlock–Fir, Spruce–Pine–Fir) for the remainder of the structure. Within each species, the designer selects grades based on published tables

DIMENSION LUMBER GRADES Table **2.1**

Product	Grades	WWPA Western Lumber Grading Rules Section Reference	Uses
Structural Light Framing (SLF) 2″ to 4″ thick 2″ to 4″ wide	SELECT STRUCTURAL NO.1 NO.2 NO.3	(42.10) (42.11) (42.12) (42.13)	Structural applications where highest design values are needed in light framing sizes.
Light Framing (LF) 2″ to 4″ thick 2″ to 4″ wide	CONSTRUCTION STANDARD UTILITY	(40.11) (40.12) (40.13)	Where high-strength values are not required, such as wall framing, plates, sills, cripples, blocking, etc.
Stud 2″ to 4″ thick 2″ and wider	STUD	(41.13)	An optional all-purpose grade designed primarily for stud uses, including bearing walls.
Structural Joists and Planks (SJ&P)	SELECT STRUCTURAL NO.1 NO.2 NO.3	(62.10) (62.11) (62.12) (62.13)	Intended to fit engineering applications for lumber 5″ and wider, such as joists, rafters, headers, beams, trusses, and general framing.

STRUCTURAL DECKING GRADES Table **2.2**

Product	Grades	WWPA Western Lumber Grading Rules Section Reference	Uses
Structural Decking 2″ to 4″ thick 4″ to 12″ wide	SELECTED DECKING	(55.11)	Used where the appearance of the best face is of primary importance.
	COMMERCIAL DECKING	(55.12)	Customarily used when appearance is not of primary importance.

TIMBER GRADES Table **2.3**

Product	Grades	WWPA Western Lumber Grading Rules Section Reference	End Uses
Beams and Stringers 5″ and thicker, width more than 2″ greater than thickness	DENSE SELECT STRUCTURAL* DENSE NO. 1* DENSE NO. 2* SELECT STRUCTURAL NO.1 NO.2	(53.00 & 170.00) (53.00 & 170.00) (53.00 & 170.00) (70.10) (70.11) (70.12)	Grades are designed for beam and stringer type uses when sizes larger than 4″ nominal thickness are required.
Post and Timbers 5″ × 5″ and larger, width not more than 2″ greater than thickness	DENSE SELECT STRUCTURAL* DENSE NO. 1* DENSE NO. 2* SELECT STRUCTURAL NO.1 NO.2	(53.00 & 170.00) (53.00 & 170.00) (53.00 & 170.00) (80.10) (80.11) (80.12)	Grades are designed for vertically loaded applications where sizes larger than 4″ nominal thickness are required.

*Douglas Fir or Douglas Fir-Larch only.

APPEARANCE LUMBER GRADES					Table 2.5
	Product	Grades[1]	Equivalent Grades in Idaho White Pine	WWPA Grading Rules Section Number
Highest Quality Appearance Grades	Selects *(all species)*	B & BTR SELECT C SELECT D SELECT	SUPREME CHOICE QUALITY	10.11 10.12 10.13
	Finish *(usually available only in Doug Fir and Hem-Fir)*	SUPERIOR PRIME E		10.51 10.52 10.53
	Special Western Red Cedar Pattern Grades	CLEAR HEART A GRADE B GRADE		20.11 20.12 20.13
General Purpose Grades	Common Boards (WWPA Rules) *(primarily in pines, spruces, and cedars)*	1 COMMON 2 COMMON 3 COMMON 4 COMMON 5 COMMON	COLONIAL STERLING STANDARD UTILITY INDUSTRIAL	30.11 30.12 30.13 30.14 30.15
	Alternate Boards (WCLIB Rules) *(primarily in Doug Fir and Hem-Fir)*	SELECT MERCHANTABLE CONSTRUCTION STANDARD UTILITY ECONOMY		**WCLIB[3]** 118-a 118-b 118-c 118-d 118-e
	Special Western Red Cedar Pattern[2] Grades	SELECT KNOTTY QUALITY KNOTTY		**WCLIB[3]** 111-e 111-f

[1]Refer to WWPA's *Vol. 2, Western Wood Species* book for full-color photography and to WWPA's *Natural Wood Siding* for complete information on siding grades, specification, and installation.

FIGURE 3.20
Nonstructural boards are graded according to appearance. *(Courtesy of Western Wood Products Association)*

Material	Working Strength in Tension[1]	Working Strength in Compression[1]	Density	Modulus of Elasticity
Wood (framing lumber)	300–1000 psi 2.1–6.9 MPa	600–1700 psi 4.1–12 MPa	30 pcf 480 kg/m^3	1,000,000–1,900,000 psi 6900–13,000 MPa
Brick masonry (including mortar, unreinforced)	0	250–1300 psi 1.7–9.0 MPa	120 pcf 1900 kg/m^3	700,000–3,700,000 psi 4800–25,000 MPa
Structural steel	24,000–43,000 psi 170–300 MPa	24,000–43,000 psi 170–300 MPa	490 pcf 7800 kg/m^3	29,000,000 psi 200,000 MPa
Concrete (unreinforced)	0	1000–4000 psi 6.9–28 MPa	145 pcf 2300 kg/m^3	3,000,000–4,500,000 psi 21,000–31,000 MPa

[1]Allowable stress or approximate maximum stress under normal loading conditions.

FIGURE 3.21
Comparative physical properties of four common structural materials: wood (shaded row), brick masonry, steel, and concrete. Wood has significant strength in both tension and compression. The ranges of values of wood strength and stiffness reflect differences among species and grades of lumber.

of allowable stresses. The higher the structural grade, the higher the allowable stress. But the lower the structural grade, the less costly is the lumber.

There are many factors other than species and grade that influence the useful strength of wood. These include the length of time the wood will be subjected to its maximum load, the temperature and moisture conditions under which it serves, and the size and shape of the piece. Certain fire-retardant treatments also reduce the strength of wood slightly. All these factors are taken into account when engineering a building structure of wood.

Lumber Dimensions

Lumber sizes in the United States are given as *nominal dimensions* in inches, such as 1 × 2 ("one by two"), 2 × 10,

Nominal Dimension	Actual Dimension
1″	¾″ (19 mm)
5/4″	1″ (25 mm)
2″	1½″ (38 mm)
3″	2½″ (64 mm)
4″	3½″ (89 mm)
5″	4½″ (114 mm)
6″	5½″ (140 mm)
8″	7¼″ (184 mm)
10″	9¼″ (235 mm)
12″	11¼″ (286 mm)
Over 12″	¾″ (19 mm) less

Figure 3.22
The relationship between nominal and actual dimensions for the most common sizes of kiln-dried lumber is given in this simplified chart, which is extracted from the complete chart in Figure 3.23.

STANDARD SIZES—FRAMING LUMBER Table **2.4**
Nominal & Dressed (Based on *Western Lumber Grading Rules*)

		Nominal Size		Dressed Dimensions: Thicknesses & Widths (inches)		
Product	**Description**	**Thickness (inches)**	**Width (inches)**	**Surfaced Dry**	**Surfaced Unseasoned**	**Length (feet)**
DIMENSION	S4S	2	2	1½	1 9/16	6′ and longer, generally shipped in multiples of 2′
		3	3	2½	2 9/16	
		4	4	3½	3 9/16	
			5	4½	4 5/8	
			6	5½	5 5/8	
			8	7¼	7½	
			10	9¼	9½	
			12	11¼	11½	
			over 12	off ¾	off ½	
				Thickness (unseasoned)	**Width (unseasoned)**	
TIMBERS	Rough or S4S (shipped unseasoned)	5 and larger		1/2 off nominal (S4S). See 3.20 of WWPA Grading Rules for Rough.		6′ and longer, generally shipped in multiples of 2′
		Thickness	**Width**	**Thickness (dry)**	**Width (dry)**	
DECKING	2″ (Single T&G)	2	5	1½	4	6′ and longer, generally shipped in multiples of 2′
			6		5	
			8		6¾	
			10		8¾	
			12		10¾	
	3″ and 4″ (Double T&G)	3	6	2½	5¼	
		4		3½		

Abbreviations: FOHC—Free of Heart Center; S4S—Surfaced four sides; T&G—Tongued and grooved; Rough Full Sawn—Unsurfaced lumber cut to full specified size

FIGURE 3.23
A complete chart of nominal and actual dimensions for both framing lumber and finish lumber. *(Courtesy of Western Wood Products Association)*

and so on. At one time, sawn lumber may have approached these actual dimensions. Today however, subsequent to sawing, seasoning, and surfacing, true sizes are less. By the time a kiln-dried 2 × 10 reaches the lumberyard, its *actual dimensions* are close to 1½ by 9¼ inches (38 by 235 mm). The relationship between nominal lumber dimensions (which are always written without inch marks) and actual dimensions (which are written with inch marks) is given in simplified form in Figure 3.22 and in more complete form in Figure 3.23. Anyone who designs or constructs wooden buildings soon commits the simpler of these relationships to memory. Because of changing moisture content and manufacturing tolerances, however, it is never wise to assume that a piece of lumber will conform precisely to its expected dimensions. Wood members vary in size seasonally with changes in humidity and temperature. In hot, dry locations such as attics, wood framing may shrink to dimensions substantially below its original measurements. Members in older buildings may have been manufactured to full nominal dimensions or to earlier standards of actual dimensions such as 1⅝ inches or 1¾ inches (41 or 44 mm) for a nominal 2-inch member.

Pieces of lumber less than 2 inches in nominal thickness (38 mm actual thickness) are called *boards.* Pieces ranging from 2 to 4 inches in nominal thickness (38 to 89 mm actual thickness) are referred to collectively as *dimension lumber.* Pieces nominally 5 inches (actual size 114 mm) and more in thickness are termed *timbers.*

Dimension lumber is usually supplied in 2-foot (610-mm) increments of length. The most commonly used lengths are 8, 10, 12, 14, and 16 feet (2.44, 3.05, 3.66, 4.27, and 4.88 m), but retailers frequently stock rafter material in lengths to 24 feet (7.32 m). Actual lengths are usually a fraction of an inch longer than nominal lengths.

Lumber in the United States is priced by the *board foot.* Board foot measurement is based on nominal dimensions, not actual dimensions. A board foot of lumber is defined as a solid volume 12 square inches in nominal cross-sectional area and 1 foot long. A 1 × 12 or 2 × 6 10 feet long contains 10 board feet. A 2 × 4 10 feet long contains $[(2 \times 4)/12] \times 10 = 6.67$ board feet, and so on. Prices of dimension lumber and timbers in the United States are usually quoted on the basis of dollars per thousand board feet. In other parts of the world, lumber is sold by the cubic meter. Because the board foot is based on lumber nominal dimensions and metric sizes use actual dimensions, there is no direct conversion between board feet and cubic meters.

The architect and engineer specify lumber for a particular construction use by designating its species, grade, seasoning, surfacing, nominal size, and chemical treatment, if any. When ordering lumber, the contractor must additionally give the required lengths of pieces and the required number of pieces of each length.

Wood Products

Much of the wood used in construction is processed into manufactured products such as laminated wood, wood panels, or composites of various types. These products were originally designed to overcome various shortcomings of solid wood structural members. With diminishing forest quality and a new awareness of sustainability, however, these products have assumed new importance. Emphasis in the forest products industry is steadily shifting away from dimension lumber and focusing on maximum utilization of wood fiber from each tree. Year by year, a larger and larger percentage of the wood fiber used in buildings is in the form of manufactured wood products.

Glue-Laminated Wood

Large structural members are often produced by joining many smaller strips of wood together with glue to form *glue-laminated wood* (called *glulam* for short). There are three major reasons to laminate: size, shape, and quality. Any desired size of structural member can be laminated, up to the capacities of the hoisting and transportation machinery needed to deliver and erect it, without having to search for a tree of sufficient girth and height. Wood can be laminated into shapes that cannot be obtained in nature: curves, angles, and varying cross sections (Figure 3.24). Quality can be specified and closely controlled in laminated members because defects can be cut out of the wood before laminating. Seasoning is carried out before the wood is laminated (largely eliminating the checks and distortions that characterize solid timbers), and the strongest, highest-quality wood can be placed in the parts of the member that will be subjected to the highest structural stresses. The fabrication of laminated members adds to the cost per board foot but provides structural members that are smaller in size than solid timbers of equal load-carrying capacity. In many cases, solid timbers are simply not available at any price in the required size, shape, or quality.

Individual laminations are most commonly 1½ inches (38 mm) thick except in curved members with small bending radii, where ¾-inch (19-mm) stock is used. End joints between individual pieces are either *finger jointed* or *scarf jointed.* These types of joints allow the glue to transmit tensile and compressive forces longitudinally from one piece to the next within a lamination (Figure 3.25). Adhesives are chosen according to the moisture conditions under which the member will serve. Any size member can be laminated. For use in residential construction, standard sizes range from 3⅛ to 6¾ inches (79 to 171 mm) in width and from 9 to 36 inches (229 to 914 mm) in depth. For larger buildings, standard sizes up to 14¼ inches (362 mm) wide and 75 inches (1905 mm) deep are not uncommon.

"Your chessboard, sire, is inlaid with two woods: ebony and maple. The square on which your enlightened gaze is fixed was cut from the ring of a trunk that grew in a year of drought: you see how its fibers are arranged? Here a barely hinted knot can be made out: a bud tried to burgeon on a premature spring day, but the night's frost forced it to desist. . . . Here is a thicker pore: perhaps it was a larvum's nest. . . ." The quantity of things that could be read in a little piece of smooth and empty wood overwhelmed Kublai; Polo was already talking about ebony forests, about rafts laden with logs that come down the rivers, of docks, of women at the windows. . . .

Italo Calvino *Invisible Cities,* 1978.

FIGURE 3.24
Glue laminating a U-shaped timber for a ship. Several smaller members have also been glued and clamped and are drying alongside the larger timber. *(Courtesy of Forest Products Laboratory, Forest Service, USDA)*

Where glue-laminated beams will be exposed to the weather or to high levels of moisture in the completed construction, the laminations may be preservative-treated to protect against decay.

Hybrid glulam beams substitute composite laminated veneer lumber (see the next section) for the usual solid wood top and bottom laminations in the beam. Since the highest stresses in the beam occur at its top and bottom edges, substituting a stronger material in just these locations results in a beam that is 20 percent stronger and 15 percent stiffer overall. In *FRP reinforced glulam* beams, structural capacity is increased by gluing a thin strip of high-strength fiber-reinforced plastic (FRP) between the

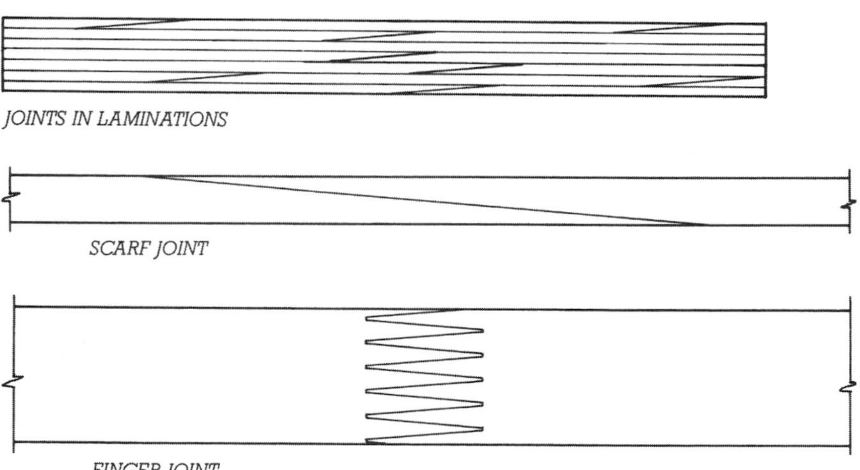

FIGURE 3.25
Joints within a lamination of a glue-laminated beam, seen in the upper drawing in a small-scale elevation view, must be scarf jointed or finger jointed to transmit the tensile and compressive forces from one piece of wood to the next. The individual pieces of wood are prepared for jointing by high-speed machines that mill the scarf or fingers with rotating cutters of the appropriate shape.

first and second laminations nearest the edge of the beam (usually the lower one) that acts in tension. The fibers used—aramid, glass, carbon, or high-performance polyethylene—are much stiffer and stronger than wood, and are oriented longitudinally and embedded in a plastic matrix prior to being fabricated into the beam. The result is a savings of 25 to 40 percent in the volume of wood in comparison to a conventional glulam.

Structural Composite Lumber

Structural composite lumber, also called *engineered lumber,* products are substitutes for solid lumber made from wood veneers or wood fiber strands and glue. *Laminated strand lumber* (*LSL*) and *oriented strand lumber (OSL)* are made from shredded wood strands, coated with adhesive, pressed into a rectangular cross section, and cured under heat and pressure (the wood strands used in the manufacture of LSL are longer than those used in OSL). LSL and OSL are the least strong and least expensive of the composite lumber products. They are used mainly for rim boards and short-span headers. *Laminated veneer lumber (LVL)* is made from thin wood veneer sheets, as wide as the member is deep, that are glued and laminated into thicker members. LVL is similar in appearance to plywood except without crossbands (Figure 3.26). *Parallel strand lumber* (*PSL*) is made from long, thin strips of wood veneer glued and pressed in a process similar to that for LSL and OSL, but with the veneer strips arranged more uniformly parallel than the strands in those other products. PSL is the heaviest, strongest, and most expensive of the composite lumber products (Figure 3.27). LVL and PSL are most commonly used for longer-span headers and floor beams.

Studs and posts may also be made of any of the composite lumber types described in the previous paragraph, as well as from *structural finger-jointed lumber* in which short lengths of solid lumber scrap are finger-jointed and glued end-to-end into longer lengths. Composite studs and posts are well suited for use in the framing of walls that are especially tall or that have very large openings, or wherever else long, straight, and especially strong, stiff members are needed.

Structural composite lumber products make productive use of wood materials that are rapidly renewable or that might otherwise be treated as waste, and they offer many of the benefits of glue-laminated wood: dimensional stability, structural strength up to three times that of conventional solid material, availability of large sizes and long lengths, and consistent quality. Particularly where these products are used in exposed interior applications, however, attention should also be paid to the types of adhesives and binders used in the products and the potential for offgassing of volatile organic compounds or formaldehyde.

Wood-Plastic Composite Decking and Nonstructural Lumber

Wood-plastic composite (WPCs) products are made from wood fibers and plastics of various types, mixed with other ingredients, such as ultraviolet stabilizers, pigments, lubricants, and biocides, which are then heated and pressed, extruded, or injection-molded into final form. In comparison to their solid wood counterparts, WPC materials offer more consistent material quality, freedom from defects and distortion, and, depending on their formulation, superior resistance to moisture. They are used most notably for exterior decking, as well as for exterior railing systems and finish trim, both interior and exterior. Like structural composites, WPC products make productive use of rapidly renewable or waste materials. Some WPCs also have high recycled materials content.

WPC decking, most commonly made from blends of polyethylene or polypropylene and wood fiber, is available in sizes matching conventional solid wood decking in lengths up to 20 feet (6.1 m). It may be fastened with corrosion-resistant nails or screws, or with concealed hardware that engages the edges of the boards. A variety of maintenance-free colors and textured finishes are available, some remarkably similar in appearance to genuine hardwoods.

Ingredients and manufacturing processes used in the making of *composite wood trim* vary widely, as do the workability, surface qualities, and durability of the finished products. Blends of plastic and wood similar to those used for the manufacture of composite decking may be used. Alternatively, formulations with a higher proportion of wood fiber—more similar in composition to traditional wood panel products, such as plywood, OSB, and fiberboard, discussed later in this chapter—may be used. Products may be prefinished, factory primed, encapsulated within a dense plastic outer shell, or coated with a resin-impregnated paper that improves the quality of field-applied finishes.

In comparison to solid lumber, nonstructural composite lumber expands and contracts more with changes in temperature, so greater allowance for thermal movement must be made during installation. And in the case of spanning members such as decking, the lesser stiffness of composites necessitates closer spacing of joists or other members on which the members are supported.

Wood trim made up from shorter lengths of finger-jointed and glued solid wood material is also available as an alternative to conventional finish lumber. In comparison to solid pieces, finger-joint stock is more stable and more consistently free of defects. It also makes use of short-length scraps that might otherwise be treated as waste.

FIGURE 3.26
An LVL beam, made of veneers similar to those used in the manufacture of plywood, resting on its side in preparation for installation on a concrete foundation wall. When placed in its final position, the beam will be rotated so that the laminations are oriented vertically in the beam. Also visible in the photograph are a preservative-treated wood sill (left) and an LSL rim board (right), both of which are discussed later in this chapter. *(Photo by Joseph Iano)*

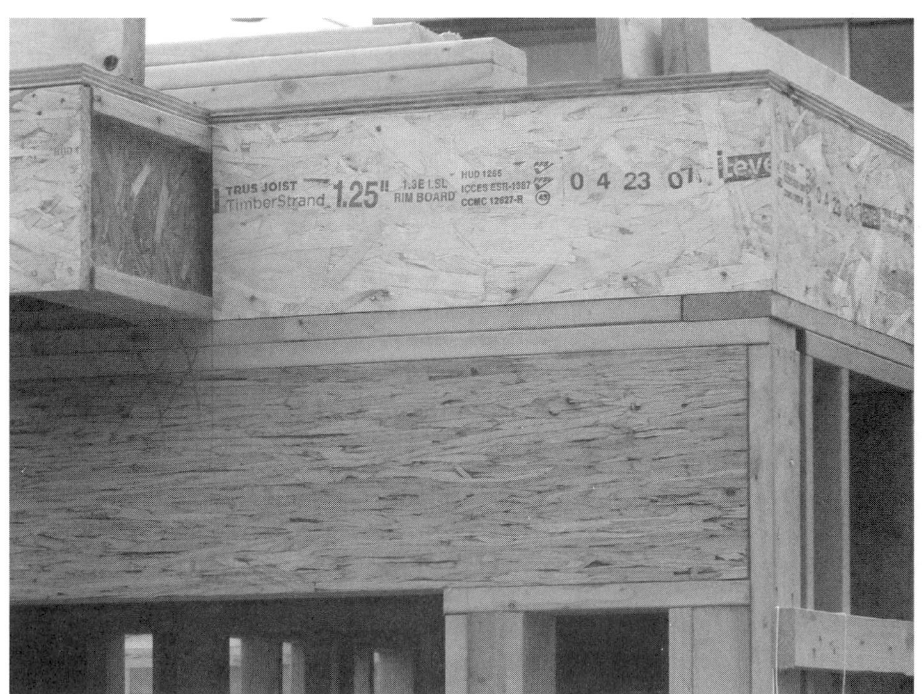

FIGURE 3.27
A view of corner framing for a residential garage. The lower beam is a PSL. A high-strength member such as this is needed to span the roughly 9-foot (2.7-m)-wide garage opening while supporting the floor load above. There are two solid wood top plates above the PSL. Above the top plates is an LSL rim board that encloses and steadies the floor framing behind it. This lower-strength member is adequate for this nonstructural function. Several manufactured wood products discussed later in this chapter are also in view: Above the rim board, the laminated edge of the plywood subfloor is visible, and to the left of the rim board, an I-joist is exposed where the floor framing projects beyond the plane of the wall below. *(Photo by Joseph Iano)*

A Naturally Grown Building Material

Of the major building materials, wood is the only one that is organic in origin. This accounts for much of its uniqueness as a material of construction. Trees grow naturally in the forest, and most of the work of "manufacturing" wood is done for us by the solar-powered processes of growth within the tree itself. This makes wood a renewable resource and inexpensive to produce: Trees need only be harvested, cut to size, and dried before they are ready for use in construction. On the other hand, we have only limited control of the quality and properties of wood. Unlike steel, concrete, or masonry, we can do little to refine solid wood to suit our needs. Rather, we must accept its natural strengths and limitations.

Wood is well suited to our building structures. The tree itself is a wooden structure, a tower, erected for the purpose of displaying leaves to the sun.* The tree is subject to the same forces as the buildings we erect: It supports itself against the pull of gravity; it withstands the pressures of wind and the accumulation of snow and ice on its limbs; and it resists the natural stresses of the environment, such as changes in temperature and moisture, and attack by other organisms. Wood has evolved to perform these jobs well in the tree. Not surprisingly, it also performs them well in our buildings.

In nature, the greatest forces acting on a tree, such as the pressure of the wind, tend to act in bursts of relatively short duration. In response, wood has evolved to be strongest when resisting forces that act briefly rather than those that act over longer periods of time. When we build with wood, we recognize this unique property. In the engineering design of a wooden structure, the maximum loads that the structure may carry are increased as the length of time these loads are expected to act decreases. This increase, called the *load duration factor*, ranges from 15 to 100 percent over the basic design strength of the material and is applied to wooden structures when considering the effects of wind, snow, earthquake, and impact—the same types of loads the tree has evolved to withstand most effectively.

The structural form of the tree is also well suited to its environment. Tree branches are supported at one end only—an arrangement called a *cantilever*. The joint supporting a cantilevered branch must be strong and stiff, efficiently utilizing the material's structural capabilities to the utmost. Yet despite the stiffness of the joint, the branch it supports can deflect relatively large amounts since it is fastened at one end only. Under heavy loads, such as wet snow, the branch can droop and shed its load. Likewise, under strong winds, the tree and its branches can sway and shed much of the force acting upon them. The tree's capacity to bend and sway helps it to survive forces that might otherwise break it.

When we build with wood, we cannot directly exploit the efficient structural form of the tree. The large deflections characteristic of this form are unsuitable for buildings. They would cause discomfort to building occupants and place undue stress on other building components. Nor can we exploit the naturally strong joints that join a

*Brayton F. Wilson and Robert R. Archer, "Tree Design: Some Biological Solutions to Mechanical Problems," *BioScience*, Vol. 29, No. 5, p. 293, 1979.

Plastic Lumber

Lumberlike products with a plastic content of 50 percent or more are referred to as *plastic lumber*. The term *recycled plastic lumber (RPL)* may also be applied to this product category, as the most common material used in the manufacture of plastic lumber, high-density polyethylene (HDPE), is obtained from the recycling of post-consumer waste such as milk jugs and detergent bottles. The use of plastic lumber reduces the demand for harvested wood and, where recycled ingredients are used, diverts solid waste from landfills. However, not all plastic lumber originates from recycled materials, and the RPL moniker alone should not be relied upon where recycled content is an important criterion for product selection.

In the manufacturing of plastic lumber, one or more plastic resins may be used alone, or may be mixed with rubber, wood waste, glass fiber, or other materials, and then molded or extruded into solid or hollow shapes mimicking those of conventional solid lumber. In its finished state, plastic lumber is resistant to sun, water, and insects, does not require protective coatings or finishes, and is nontoxic, durable, and maintenance free, making it an especially attractive alternative to preservative-treated lumber for exterior applications. However, like WPCs, plastic lumber is more flexible than solid wood and must be supported at more closely spaced intervals. In addition, plastic lumber expands and contracts more with changes in temperature and so must be installed with greater allowances for thermal movement. Exterior decking, made with HDPE, polystyrene, or polyvinyl chloride (PVC), and finish trim products claim the largest share of the plastic wood market in building construction.

Structural-grade plastic lumber (SGPL), most commonly made from HDPE reinforced with glass fibers, can be formulated to be at least as

tree's limbs, despite their benefits in stiffness and economy of material. In preparing wood for our uses, we cut the tree into pieces of convenient size and shape, destroying the continuity between its parts. When we reassemble these pieces, we must devise new ways to join them. Despite the many methods developed for fabricating wood connections, joints with strength and stiffness comparable to the tree's are rarely achievable. Rather, we rely on much simpler types of connections that suffice when a beam is supported at both ends.

Wood in a tree performs many functions that wood in a building does not, because wood is involved with all aspects of the tree's life processes and growth. In order to meet numerous and specialized tasks, wood has evolved a complex internal structure that is highly directional. The fact that wood is anisotropic, that is, it has different properties in different directions, affects virtually all of its properties. The direction of the grain in a piece of wood influences its production, shaping, fastening, strength, durability, aging, and beauty. No other major building material, perhaps with the exception of some specialized reinforced concrete systems, has such directional characteristics.

Consider two wood building systems, the log cabin and the heavy timber frame. Log cabin construction is appropriate to the most primitive construction technology. Logs found on or near the site are prepared for use with a minimum of labor and simple tools. Once assembled, the logs provide structural support and act as interior finish, exterior cladding, and insulation. Though a practical solution to providing shelter, log cabins use wood inefficiently and result in a structure that is dimensionally unstable due to the large stacks of cross-grain lumber. Where more sophisticated tools and methods are available, heavy timber takes the place of log cabin construction. In this system, wood members are assembled in ways that more closely resemble the tree form. The resulting structure is stronger, requires less material, and is more stable. A simple change in the orientation of the wood members produces a completely different building system, one that is more efficient, durable, and comfortable.

Modern developments in the wood construction materials industry reflect efforts to overcome the limitations of wood in its natural state and to make the most efficient use of the material. Wood panels made of plywood are larger and more stable than those that can be produced from solid boards while minimizing the adverse effects of wood grain and natural defects. Panel products such as OSB and hardboard make use of wood waste that might otherwise be discarded. Glue-laminated and composite lumber beams can be produced in sizes and of a quality exceeding those available in solid timbers. Chemical treatments can protect wood from fire or decay. Thus, the trend is toward using wood less in its natural state and more as a raw material in sophisticated manufacturing processes to produce more refined and higher-quality construction components. Yet despite such advances, the tree itself remains an inspiration in its grace and strength, as well as in the lessons it offers for our understanding of building materials and methods.

strong as conventional solid wood, though less stiff and more prone to long-term creep under permanent loads. SGPL planks, joists, beams, posts, piles, and other manufactured elements are used in the construction of decks, docks, piers, other types of exterior and marine structures, and even vehicular-capacity bridges.

Wood Panel Products

Wood in panel form is advantageous for many building applications (Figure 3.28). The panel dimensions are usually 4 by 8 feet (1220 by 2440 mm). Panels require less labor for installation than individual boards because fewer pieces must be handled, and wood panel products are fabricated in such a way that they minimize many of the limitations of boards and dimension lumber: Panels are more nearly equal in strength in their two principal directions than solid wood; shrinking, swelling, checking, and splitting are greatly reduced. Additionally, panel products make more efficient use of forest resources than solid wood products through less wasteful ways of reducing logs to building products and through utilization in some types of panels of material that would otherwise be thrown away—branches, undersized trees, and mill wastes. Many wood panel products are made largely from recovered or recycled wood waste, and panels made from rapidly renewable vegetable fibers are also available.

Structural Wood Panel Types

Structural wood panel products fall into three general categories (Figure 3.29): *Plywood* panels are made up of thin layers of wood *veneer* glued together. The grain on the front and back face veneers runs in the long direction of the sheet, whereas the grain in one or more interior crossbands runs perpendicular, in the shorter

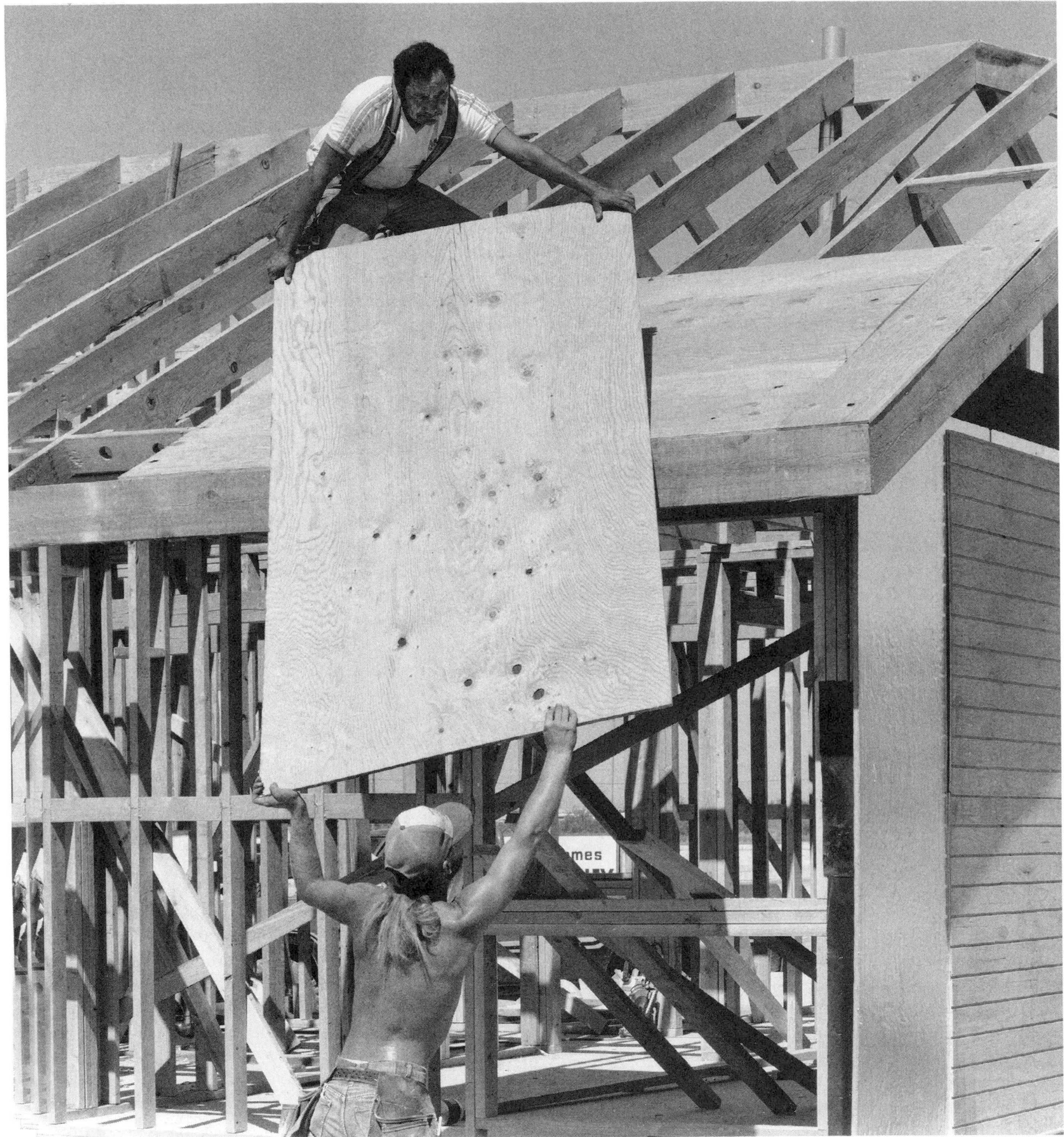

FIGURE 3.28
Plywood is made of veneers selected to give the optimum combination of economy and performance for each application. This sheet of roof sheathing plywood is faced with a D veneer on the underside and a C veneer on the top side. These veneers, though unattractive, perform well structurally and are much less costly than the higher grades incorporated into plywood made for uses where appearance is important. *(Courtesy of APA–The Engineered Wood Association)*

FIGURE 3.29
Five different wood panel products, from top to bottom: plywood, composite panel, waferboard, OSB, and particleboard. *(Courtesy of APA–The Engineered Wood Association)*

direction. There is always an odd number of layers in plywood, which equalizes the effects of moisture movement, but an interior layer may be made up of a single veneer or of two veneers with their grains running in the same direction. *Composite panels* have two parallel face veneers bonded to a core of reconstituted wood fibers. *Nonveneered panels* are of several different types:

- *Oriented strand board (OSB)* is made of long shreds (strands) of wood compressed and glued into three to five layers. The strands are oriented in the same manner in each layer as the grains of the veneer layers in plywood. Because of the length and controlled orientation of the strands, OSB is generally stronger and stiffer than the other types of nonveneered panels. Because it can be produced from small trees and even branches, OSB is generally more economical than plywood. It is the material most commonly used for sheathing and subflooring of light frame wood buildings. OSB is also sometimes called *waferboard*, a name that applies more precisely to a similar type of panel made of large flakes of wood that has been largely replaced by OSB.
- *Particleboard* is manufactured in different density ranges, and is made up of smaller wood particles than OSB or waferboard that are compressed and bonded into panels. It finds use in buildings mainly as a base material for wood veneer and plastic laminate. It is also used commonly as an *underlayment panel* to create an especially smooth substrate for the application of resilient flooring.
- *Fiberboard* is a very fine-grained board made of wood fibers and synthetic resin binders intended for interior uses only. The processing of the raw wood products in fiberboard manufacturing is more intensive than that in the manufacture of particleboard, resulting in a panel that is dimensionally more stable, stiffer, better able to hold fasteners, and superior in its working and finishing characteristics. The most commonly used form of fiberboard is *medium-density fiberboard* (*MDF*). It is used in the production of cabinets, furniture, moldings, paneling, and many other manufactured products. Though the names are similar, care should be taken not to confuse MDF panels with medium-density overlay (MDO) plywood panels, discussed below.

Plywood Production

Veneers for plywood and composite panels are *rotary sliced*: Logs are soaked in hot water to soften the wood. Then each log is rotated in a large lathe against a stationary knife that peels off a continuous strip of veneer, much as paper is unwound from a roll (Figures 3.30 and 3.31). The strip of veneer is clipped into sheets that pass through a drying kiln where, in a few minutes, their moisture content is reduced to roughly 5 percent. The sheets are then assembled into larger sheets, repaired as necessary with patches glued into the sheet to fill open defects, and graded and sorted according to quality (Figure 3.32). A machine spreads glue on the veneers as they are laid atop one another in the required sequence and grain orientations. The glued veneers are transformed into plywood in presses that apply elevated temperatures and pressures to create dense, flat panels. The panels are trimmed to size, sanded as required, and graded and gradestamped before shipping. Veneers of Grade B and higher are always sanded smooth, but panels intended for sheathing are left unsanded because sanding slightly reduces their thickness, which diminishes the structural strength of the panel. Panels intended for subfloors and floor underlayment are lightly *touch sanded* to produce a flatter, smoother surface without significantly reducing their structural performance.

Standard plywood panels are 4 by 8 feet (1220 by 2440 mm) in surface area and range in thickness from 1/4 to 1 1/8 inch (6.4 to 28.4 mm). Longer panels are manufactured for siding and industrial uses. Actual surface dimensions of structural plywood are approximately 1/8 inch (3 mm) less than nominal, permitting panels to be installed with gaps between them to allow for moisture expansion. Plywood panels intended for use as subflooring can be manufactured with tongue-and-groove edges that eliminate unevenness in the subfloor that could telegraph through the finish flooring.

Where an especially smooth and durable surface is required, plywood may be finished with a resin-treated overlay on one or both sides to make *medium-density overlay (MDO)* or *high-density overlay (HDO)* plywood.

(*a*)

(*b*)

(*c*)

(*d*)

FIGURE 3.30
Plywood manufacture. (*a*, *b*) A 250-horsepower lathe spins a softwood log as a knife peels off a continuous sheet of veneer for plywood manufacture. (*c*) An automatic clipper removes unusable areas of veneer and trims the rest into sheets of the proper size for plywood panels. (*d*) The clipped sheets are fed into a continuous forced-air dryer, along whose 150-foot (45-m) path they will lose about half their weight in moisture. (*e*) Leaving the dryer, the sheets have a moisture content of about 5 percent. They are graded and sorted at this point in the process. (*f*) The higher grades of veneer are patched on this machine, which punches out defects and replaces them with tightly fitted wood plugs. (*g*) In the layup line, automatic machinery applies glue to one side of each sheet of veneer and alternates the grain direction of the sheets to produce loose plywood panels. (*h*) After layup, the loose panels are prepressed with a force of 300 tons per panel to consolidate them for easier handling. (*i*) Following prepressing, panels are squeezed individually between platens heated to 300 degrees Fahrenheit (150°C) to cure the glue. (*j*) After trimming, sanding, or grooving as specified for each batch, the finished plywood panels are sorted into bins by grade, ready for shipment. *(Photos b and i courtesy of Georgia-Pacific; others courtesy of APA–The Engineered Wood Association)*

(*e*)

(*f*)

(*g*)

(*h*)

(*i*)

(*j*)

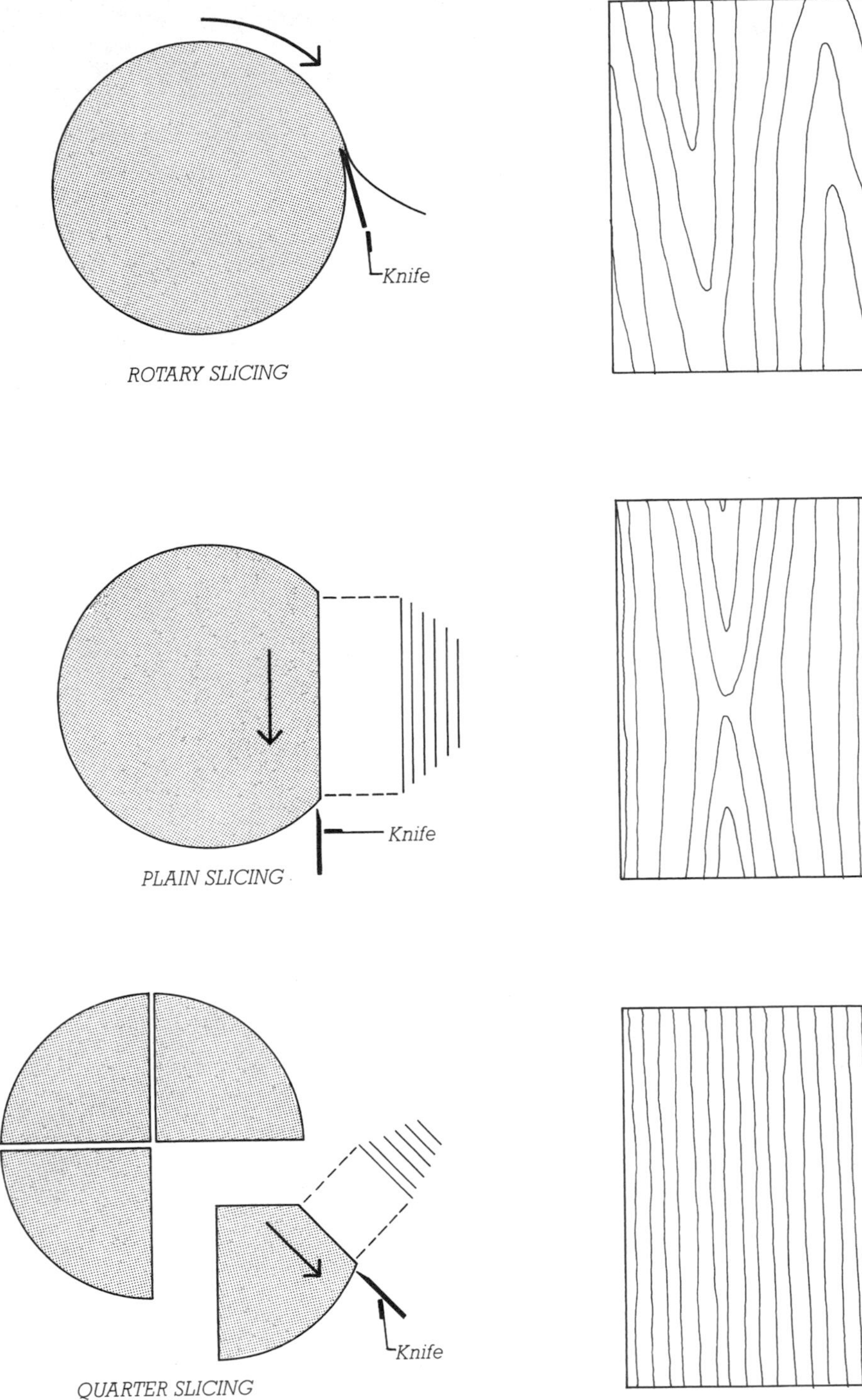

FIGURE 3.31
Veneers for structural plywood are rotary sliced, which is the most economical method. For better control of grain figure in face veneers of hardwood plywood, flitches are plainsliced or quartersliced. The grain figure produced by rotary slicing, as seen in the detail to the right, is extremely broad and uneven. The finest figures are produced by quarterslicing, which results in a very close grain pattern with prominent rays.

Overlaid panels are used in the construction of concrete forms, cabinetwork, furniture, exterior siding, signage, and other applications where the greatest durability and highest surface finish quality are required.

Unlike veneers for structural panels, veneers for hardwood plywoods intended for interior paneling and cabinetwork are usually sliced from square blocks of wood called *flitches* in a machine that moves the flitch vertically against a stationary knife (Figure 3.31). *Flitch-sliced veneers* are analogous to quartersawn lumber: They exhibit a much tighter and more interesting grain figure than rotary-sliced veneers. They can also be arranged on the plywood face in such a way as to produce symmetrical grain patterns.

Composite panels and nonveneered panels are manufactured by analogous processes to the same set of standard sizes as plywood and to some larger sizes as well. Panels of all kinds used for sheathing and subflooring can also be manufactured with water-resistant coatings and edge sealers, and in the case of OSB, special resins, to provide better resistance to extended periods of wetting during construction.

Specifying Structural Wood Panels

For structural uses such as subflooring and sheathing, wood panels may be specified either by thickness or by *span rating*. The span rating is determined by laboratory load testing and is given on the gradestamp on the back of the panel, as shown in Figures 3.33 and 3.34. The purpose of the span rating system is to permit the use of many different species of woods and types of panels while achieving the same structural objectives. Under normal loading conditions, any panel with a span rating of 32/16 may be used as roof sheathing over rafters spaced 32 inches (813 mm) apart or as subflooring over joists spaced 16 inches (406 mm) apart. The long dimension

FIGURE 3.32
Veneer grades for softwood plywood. Plywood with A grade face veneer is costly and limited in availability. It is used only in applications demanding the highest quality appearance. Plywood with B face veneer is used for concrete formwork construction and for less demanding appearance applications. Plywood with C-plugged face veneer is used for underlayment and combination subflooring/underlayment, where an especially smooth substrate is required to avoid telegraphing irregularities through the finish flooring material. Plywood for sheathing and subflooring is most commonly specified as "CDX," consisting of a C grade face, a D grade face, and constructed with exterior glue (the panel is installed with the C face oriented toward the weather). C grade veneer is also the lowest grade permitted in any ply of an Exterior rated panel, that is, a siding panel intended for permanent exposure to the weather. *(Courtesy of APA–The Engineered Wood Association)*

TABLE 1

VENEER GRADES

Grade	Description
A	Smooth, paintable. Not more than 18 neatly made repairs, boat, sled, or router type, and parallel to grain, permitted. Wood or synthetic repairs permitted. May be used for natural finish in less demanding applications.
B	Solid surface. Shims, sled or router repairs, and tight knots to 1 inch across grain permitted. Wood or synthetic repairs permitted. Some minor splits permitted.
C Plugged	Improved C veneer with splits limited to 1/8-inch width and knotholes or other open defects limited to 1/4 x 1/2 inch. Wood or synthetic repairs permitted. Admits some broken grain.
C	Tight knots to 1-1/2 inch. Knotholes to 1 inch across grain and some to 1-1/2 inch if total width of knots and knotholes is within specified limits. Synthetic or wood repairs. Discoloration and sanding defects that do not impair strength permitted. Limited splits allowed. Stitching permitted.
D 	Knots and knotholes to 2-1/2-inch width across grain and 1/2 inch larger within specified limits. Limited splits are permitted. Stitching permitted. Limited to Exposure 1 or Interior panels.

of the sheet must be placed perpendicular to the length of the supporting members. A 32/16 panel may be plywood, composite, or OSB, may be composed of any accepted wood species, and may be any of several thicknesses, so long as it passes the structural tests for a 32/16 rating.

The designer must also select from three *exposure durability classifications* for structural wood panels: Exterior, Exposure 1, and Exposure 2. Panels marked "Exterior" are suitable for use as siding or in other applications permanently exposed to the weather. "Exposure 1" panels have fully waterproof glue but do not have veneers of as high a quality as those of Exterior

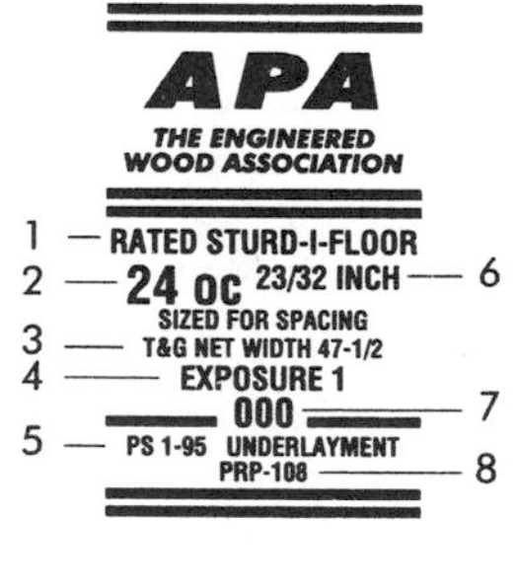

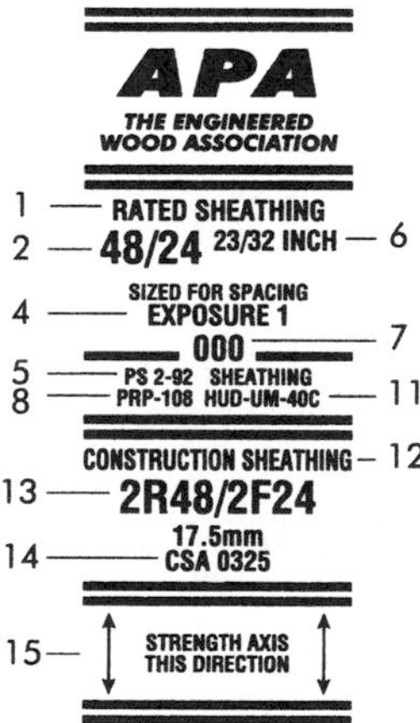

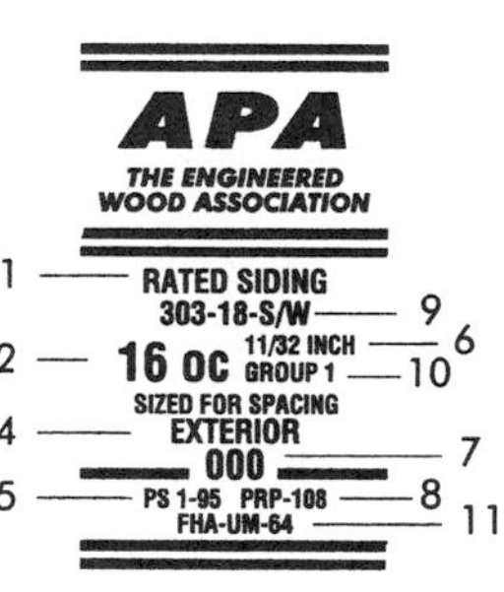

FIGURE 3.33
Typical gradestamps for structural wood panels. Gradestamps are found on the back of each panel. *(Courtesy of APA–The Engineered Wood Association)*

1 Panel grade
2 Span Rating
3 Tongue-and-groove
4 Exposure durability classification
5 Product Standard
6 Thickness
7 Mill number
8 APA's Performance Rated Panel Standard
9 Siding face grade
10 Species group number
11 HUD/FHA recognition
12 Panel grade, Canadian standard
13 Panel mark – Rating and end-use designation, Canadian standard.
14 Canadian performance-rated panel standard
15 Panel face orientation indicator

TABLE 2

GUIDE TO APA PERFORMANCE RATED PANELS[a][b]
FOR APPLICATION RECOMMENDATIONS, SEE FOLLOWING PAGES.

Panel	Typical Trademark	Typical Trademark	Description
APA RATED SHEATHING Typical Trademark	APA THE ENGINEERED WOOD ASSOCIATION RATED SHEATHING 40/20 19/32 INCH SIZED FOR SPACING EXPOSURE 1 000 PS 2-92 SHEATHING PRP-108 HUD-UM-40C	APA THE ENGINEERED WOOD ASSOCIATION RATED SHEATHING 24/16 7/16 INCH SIZED FOR SPACING EXPOSURE 1 000 PRP-108 HUD-UM-40C	Specially designed for subflooring and wall and roof sheathing. Also good for a broad range of other construction and industrial applications. Can be manufactured as plywood, as a composite, or as OSB. EXPOSURE DURABILITY CLASSIFICATIONS: Exterior, Exposure 1, Exposure 2. COMMON THICKNESSES: 5/16, 3/8, 7/16, 15/32, 1/2, 19/32, 5/8, 23/32, 3/4.
APA STRUCTURAL I RATED SHEATHING[c] Typical Trademark	APA THE ENGINEERED WOOD ASSOCIATION RATED SHEATHING STRUCTURAL I 32/16 15/32 INCH SIZED FOR SPACING EXPOSURE 1 000 PS 1-95 C-D PRP-108	APA THE ENGINEERED WOOD ASSOCIATION RATED SHEATHING 32/16 15/32 INCH SIZED FOR SPACING EXPOSURE 1 000 STRUCTURAL I RATED DIAPHRAGMS-SHEAR WALLS PANELIZED ROOFS PRP-108 HUD-UM-40C	Unsanded grade for use where shear and cross-panel strength properties are of maximum importance, such as panelized roofs and diaphragms. Can be manufactured as plywood, as a composite, or as OSB. EXPOSURE DURABILITY CLASSIFICATIONS: Exterior, Exposure 1. COMMON THICKNESSES: 5/16, 3/8, 7/16, 15/32, 1/2, 19/32, 5/8, 23/32, 3/4.
APA RATED STURD-I-FLOOR Typical Trademark	APA THE ENGINEERED WOOD ASSOCIATION RATED STURD-I-FLOOR 24 oc 23/32 INCH SIZED FOR SPACING T&G NET WIDTH 47-1/2 EXPOSURE 1 000 PS 2-92 SINGLE FLOOR PRP-108 HUD-UM-40C	APA THE ENGINEERED WOOD ASSOCIATION RATED STURD-I-FLOOR 20 oc 19/32 INCH SIZED FOR SPACING T&G NET WIDTH 47-1/2 EXPOSURE 1 000 PRP-108 HUD-UM-40C	Specially designed as combination subfloor-underlayment. Provides smooth surface for application of carpet and pad and possesses high concentrated and impact load resistance. Can be manufactured as plywood, as a composite, or as OSB. Available square edge or tongue-and-groove. EXPOSURE DURABILITY CLASSIFICATIONS: Exterior, Exposure 1, Exposure 2. COMMON THICKNESSES: 19/32, 5/8, 23/32, 3/4, 1, 1-1/8.
APA RATED SIDING Typical Trademark	APA THE ENGINEERED WOOD ASSOCIATION RATED SIDING 24 oc 19/32 INCH SIZED FOR SPACING EXPOSURE 1 000 PRP-108 HUD-UM-40C	APA THE ENGINEERED WOOD ASSOCIATION RATED SIDING 303-18-S/W 16 oc 11/32 INCH GROUP 1 SIZED FOR SPACING EXTERIOR 000 PS 1-95 PRP-108 FHA-UM-64	For exterior siding, fencing, etc. Can be manufactured as plywood, as a composite or as an overlaid OSB. Both panel and lap siding available. Special surface treatment such as V-groove, channel groove, deep groove (such as APA Texture 1-11), brushed, rough sawn and overlaid (MDO) with smooth- or texture-embossed face. Span Rating (stud spacing for siding qualified for APA Sturd-I-Wall applications) and face grade classification (for veneer-faced siding) indicated in trademark. EXPOSURE DURABILITY CLASSIFICATION: Exterior. COMMON THICKNESSES: 11/32, 3/8, 7/16, 15/32, 1/2, 19/32, 5/8.

(a) Specific grades, thicknesses and exposure durability classifications may be in limited supply in some areas. Check with your supplier before specifying.

(b) Specify Performance Rated Panels by thickness and Span Rating. Span Ratings are based on panel strength and stiffness. Since these properties are a function of panel composition and configuration as well as thickness, the same Span Rating may appear on panels of different thickness. Conversely, panels of the same thickness may be marked with different Span Ratings.

(c) All plies in Structural I plywood panels are special improved grades and panels marked PS 1 are limited to Group 1 species. Other panels marked Structural I Rated qualify through special performance testing. Structural II plywood panels are also provided for, but rarely manufactured. Application recommendations for Structural II plywood are identical to those for APA RATED SHEATHING plywood.

FIGURE 3.34
A guide to specifying structural panels. Plywood panels for use in wall paneling, furniture, and other applications where appearance is important are graded by the visual quality of their face veneers rather than their structural properties.
(Courtesy of APA–The Engineered Wood Association)

panels; they are suitable for structural sheathing and subflooring, which will be protected from the weather once the building is finished, but must often endure long periods of wetting during construction. "Exposure 2" is suitable for panels that will be fully protected from weather and will be subjected to only a minimum of wetting during construction. About 95 percent of structural panel products are classified as Exposure 1.

For plywood panels intended as finish surfaces, the quality of the face veneers is of obvious concern and should be specified by the designer (Figure 3.32). For example, A-B plywood (with an A grade veneer on one face and a B grade veneer on the other) might be specified for cabinet construction, while less expensive C-D plywood is adequate for exterior sheathing that will be concealed in the finished construction. For higher-quality architectural woodwork, fine flitch-sliced hardwood face veneers may be selected rather than rotary-sliced softwood veneers and the matching pattern of the veneers specified.

Other Panel Products

Hardboard is a thin, dense panel made of highly compressed wood fibers. It is available in various thicknesses and surface finishes, and in some formulations is durable against weather exposure. Hardboard is produced in configurations for residential siding and roofing as well as in general-purpose and pegboard panels of standard dimension.

Insulating fiberboard sheathing is a low-density panel, usually ½ or ¾ inch (13 or 19 mm) thick, made of wood or vegetable fibers and binders, and coated with asphalt for water resistance. It has some thermal insulating value and is used in wood construction chiefly as nonstructural wall sheathing, although some panels of this type are strong enough to also contribute lateral force resistance to a building frame. Other *cellulosic fiberboard* panels are made from finely processed recycled paper waste and are used for wall sheathing, acoustical isolation, carpet underlayment, and even structural roof decking. Panels of these types are low in cost, make productive use of waste or recycled materials, and help to conserve forest resources.

Hardwood plywood panels, made from birch, maple, poplar, or alder veneers, are popular for use in cabinetry and other interior finish carpentry.

I shall always remember how as a child I played on the wooden floor. The wide boards were warm and friendly, and in their texture I discovered a rich and enchanting world of veins and knots. I also remember the comfort and security experienced when falling asleep next to the round logs of an old timber wall; a wall which was not just a plain surface but had a plastic presence like everything alive. Thus sight, touch, and even smell were satisfied, which is as it should be when a child meets the world.

Christian Norberg-Schulz, *Wooden Houses*, 1979.

Wood Chemical Treatments

Chemical treatments are used to counteract two major weaknesses of wood: its combustibility and its susceptibility to attack by decay and insects. *Fire-retardant treatment (FRT)* is accomplished by placing lumber in a vessel and impregnating it under pressure with certain chemical salts that greatly reduce its combustibility. Fire-retardant-treated wood is expensive, so it is little used in single-family residential construction. Its major uses are roof sheathing in attached houses and framing for nonstructural partitions and other interior components in buildings of fire-resistant construction.

Preservative-treated wood is used where decay or insect resistance is required, such as with wood that is used in or near the ground, that is exposed to moisture in outdoor structures such as marine docks, fences, and decks, or that is used in areas of high termite risk. *Creosote* is an oily derivative of coal that is widely used to treat wood in engineering structures. But the odor, toxicity, and unpaintability of creosote-treated wood make it unsuitable for most purposes in building construction. *Pentachlorophenol* is also impregnated as an oil solution, and as with other oily preservatives, wood treated with it cannot be painted. Preservative-treated wood is frequently referred to as *pressure-treated wood*, though, more accurately, this latter term refers to fire-retardant and preservative treatments, since both are typically applied using pressure impregnation processes.

The most widely used wood preservatives in building construction are waterborne salts, which permit subsequent painting or staining. For many decades, the most common of these was *chromated copper arsenate (CCA)*, which imparts a greenish color to the wood. However, due to concerns over its toxicity, especially for children, CCA-treated lumber has been phased out of most residential and commercial building construction use in favor of lumber treated with salts that do not contain arsenic. *Alkaline copper quat (ACQ)* and *copper boron azoles (CBA* and *CA)* are compounds that rely primarily on high concentrations of copper for their preservative properties. Though arsenic-free, copper-based preservatives are also considered potentially hazardous and must be handled with appropriate precautions. Borate compounds, such as *sodium borate (SBX)*, are the least toxic to humans. But borate-treated wood can only be used above ground in applications protected from the weather.

Preservatives can be brushed or sprayed onto wood, but long-lasting protection (30 years and more) can only be accomplished by *pressure impregnation*, which drives the preservative chemicals deeply into the fibers of the wood. To improve absorption, some wood species are punctured with an array of small cuts in the wood's surface, called *incising*, prior to the preservative treatment. Incising improves retention of the chemical preservative but also somewhat lowers the structural capacity of the wood member (Figure 3.35). Wood treated with waterborne salts may be sold without drying, which is appropriate for use in the ground or for rough framing. But wood that is *kiln-dried after treatment* is lighter, more stable, and a better choice for finish work or where appearance is important.

The concentration of preservative necessary to protect any given lumber product varies, depending on the particular treatment chemical, the species of wood, and the severity of the environment in which the product will be used. To simplify the specification and selection of treated lumber, the American Wood Preservers Association (AWPA) *Use Category System* is used, in which higher Use Category numbers correspond to more intensive preservative treatment and suitability for more severe exposures (Figure 3.36). For example, wood deck posts, with ends in direct contact with the earth, should be treated to a higher Use Category (UC4A) than wood decking intended for use only above ground (UC3B).

The heartwood of some species of wood is naturally resistant to decay and insects and can be used instead of preservative-treated wood. The IBC recognizes redwood, cedar, black locust, and black walnut as decay-resistant species and redwood and Eastern red cedar as termite-resistant. Because the sapwood of these species is no more resistant to attack than the wood of any other tree, "All-Heart" grade should be specified. A more comprehensive listing of wood species and their relative decay-resistance can be found in the USDA Forest Products Laboratory's *Wood Handbook*, listed in the references at the end of this chapter. WPCs and plastic lumber, which are immune to decay or insect attack, are also becoming increasingly popular alternatives to preservative-treated lumber.

Most wood-attacking organisms need both air and moisture to live. Accordingly, most can be kept out of wood by constructing and maintaining a building so that its wood components are kept dry. This includes keeping all wood well clear of the soil, ventilating attics and crawlspaces to remove moisture, using good construction detailing to shed water and keep it out of building assemblies, using air and vapor retarders properly in conjunction with insulation to prevent the accumulation of condensation within exterior walls and roofs, and repairing roof and plumbing leaks as soon as they occur. Somewhat counterintuitively, wood that is fully and continuously submerged in fresh water is immune to decay because the water does not provide sufficient oxygen for decay-causing organisms to survive. On the other hand, where wood is only periodically or partially submerged and both moisture and oxygen are plentiful, it is highly vulnerable to decay. Submersion in salt water does not prevent deterioration because

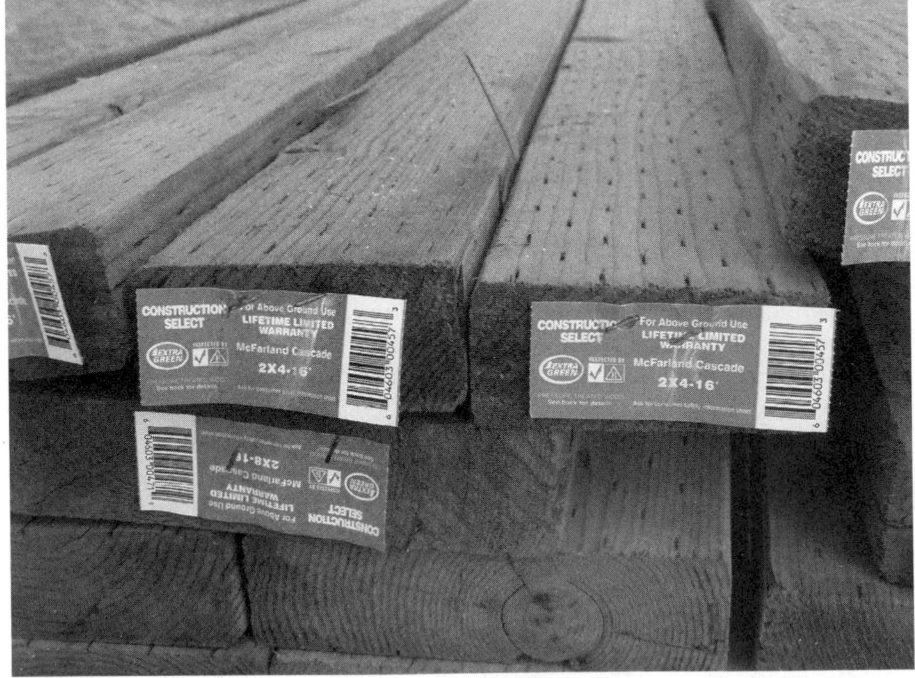

FIGURE 3.35
Pressure preservative-treated lumber. Incising marks are visible on the board surfaces. Each piece has a label stapled to one end indicating the degree of preservative treatment, in this case sufficient for above-ground use. This means that these boards are suitable, for example, for exterior decking or railings, or for use as foundation sill plates, but they are not appropriate for use in direct contact with soil. As part of the treatment process, this lumber has also been stained a light brown color to provide a more attractive finish appearance. *(Photo by Joseph Iano)*

Use Category	Service Conditions	Typical Uses
UC1	Interior construction, above ground, dry	Interior framing, woodwork, and furnishings; for resistance to insect attack only
UC2	Interior construction, above ground, damp	Interior construction; for resistance to insect attack and/or moisture
UC3	Exterior construction, above ground	
UC3A	Not exposed to prolonged wetting, finish coated, readily sheds water	Exterior painted or stained siding, millwork, and trim
UC3B	Exposed to prolonged wetting, unfinished or poorly drained	Exterior decking, deck framing, railings, and uncoated millwork
UC4	Ground contact or fresh water	
UC4A	Normal exposure conditions; non-critical, replaceable components	Posts for fences and decks
UC4B	High decay potential; critical or difficult-to-replace components	Permanent wood foundations
UC4C	Extreme decay potential; critical components	Pilings, utility poles in severe decay environments

Figure 3.36
Part of the AWPA Use Category System, listing the categories most commonly specified for pressure preservative–treated wood used in building construction. The lumber in Figure 3.35, indicated as suitable for above-ground use, has been treated to meet the requirements of Use Category UC3B. *(Copyrighted material reproduced with permission from the American Wood Preservers Association, www.awpa.com)*

marine organisms can attack wood under these conditions.

The heavy concentrations of copper used in the most common wood preservatives make wood treated with these chemicals corrosive to many metals. Plain steel or aluminum fasteners (see below) and hardware should not be used in contact with such wood. Rather, nails, screws, bolts, joist hangers, hold-downs, column bases, and other such components must be of corrosion-resistant metals such as stainless steel or heavily galvanized steel, or of some other metal not adversely affected by the copper content of these chemicals. Chemical preservative treatments continue to evolve. New, improved copper formulations, alternative nonmetallic treatments, and heat-treating processes are appearing on the market that can match the levels of protection of currently established products but that are also environmentally less hazardous, less energy-intensive to produce, and less corrosive to metal fasteners and hardware.

Wood Fasteners

Fasteners have always been the weak link in wood construction. The interlocking timber connections of the past, laboriously mortised and pegged, were weak because much of the wood in a joint had to be removed to make the connection. In today's wood connections, which are generally based on metal devices, it is usually impossible to insert enough nails, screws, or bolts in a connection to develop the full strength of the members being joined. Adhesives and toothed plates are often capable of achieving this strength but are largely limited to factory installation. Fortunately, most connections in wood structures depend primarily on direct bearing of one member on another for their strength, and a variety of simple fasteners are adequate for the majority of purposes.

Nails

Nails are sharp-pointed metal pins that are driven into wood with a hammer or a mechanical nail gun. *Common nails* and the slightly more slender *box nails* have flat heads and are used for most structural fastening in light frame construction. *Finish nails* are even smaller in diameter and are used to fasten finish woodwork, where they are less obtrusive than common nails (Figure 3.37).

In the United States, the size of a nail is measured in *pennies*, abbreviated "d." One explanation for this strange unit is that it originated long ago as the price of 100 nails of a given size. It persists today despite the loss of its original meaning. Figure 3.38 shows the dimensions of the various sizes of common nails. Box nails and finish nails are thinner in diameter but have the same length as common nails of equal penny size.

Nails are ordinarily furnished *bright*, meaning that they are made of

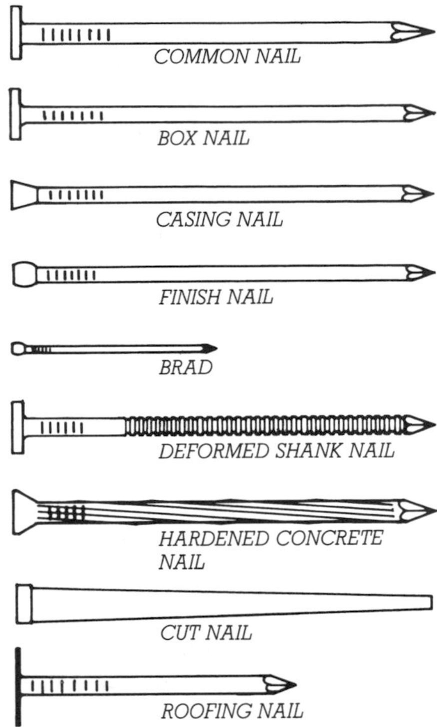

FIGURE 3.37
Most nailed framing connections are made with common nails, box nails, or their machine-driven equivalent. Box nails are also used for fastening wood shingles and other types of siding. Casing nails, finish nails, and brads are used for attaching finish components; their heads are set below the surface of the wood with a steel punch, and the holes are filled before painting. Deformed shank nails, which are more resistant to withdrawal from the wood than smooth shank nails, are used for attaching gypsum wallboard, sheathing, subflooring, and floor underlayment, materials that cannot be allowed to work loose in service. The most common deformation pattern is the ring-shank pattern shown here. Concrete nails can be driven short distances into masonry or concrete for attaching furring strips and sleepers. Cut nails, once used for framing connections, now serve mostly for attaching finish flooring because their square tips punch through the wood rather than wedge through, minimizing splitting of brittle woods. Roofing nails have large heads to prevent tearing of soft asphalt shingles.

plain, uncoated steel. Nails that will be exposed to the weather should be of a corrosion-resistant type, such as *hot-dip galvanized*, aluminum, or stainless steel. (The thinner zinc coating on electrogalvanized nails is not suitable for exterior exposure.) Corrosion resistance is particularly important for nails in exterior siding, trim, and decks, which would be stained by rust leaching from bright nails. For more information on the corrosion resistance of metals, see pages 505–508.

The three methods of fastening with nails, *face nailing, end nailing,* and *toe nailing,* are shown in Figure 3.39. Each of these methods has its uses in building construction, as illustrated in Chapters 5 and 6. Nails are the favored means of fastening wood because they are inexpensive, require no predrilling of holes under most conditions, and can be installed extremely rapidly.

Specialty Nails

In an effort to make nailing easier for the carpenter, nails made of more slender wire with names like *sinkers* and *coolers* have been introduced. A 10d sinker is a bit shorter than a 10d common nail, has a smaller diameter, and is coated with a resin that acts as a lubricant when the nail is driven—all of these differences being intended to reduce the driving effort. Sinkers and coolers may be used to fasten nonstructural wallboard or panels, or they may be used in structural connections where the role of the nail is merely to hold members in alignment while the loads are transferred from one member to the other by direct bearing of wood on wood. Where force must be transmitted through the nails themselves, however, the strength of the connection depends on both the length and the diameter of the nails, and sinkers and coolers should not be substituted for common nails except where structural calculations or manufacturer test results have demonstrated the suitability of these nails types.

Joist hanger nails are the same diameter as the equivalent common nail, but shorter. For example, a nail the same diameter as a 10d common nail but only 1½ inches (38 mm) long (sometimes referred to as an "N10" nail) may be used when nailing a joist hanger directly into the side of a single nominal 2-inch (actual size 38 mm) member, where a longer nail would protrude from the wood member's opposite side. Other specialty nails may have oversized heads, deformed shafts that provide increased resistance to shear and withdrawal compared to conventional nails, or other special features. They should be used only in locations permitted by the building code or as recommended by the fastener manufacturer.

FIGURE 3.38
Standard sizes of common nails, reproduced full size. The abbreviation "d" stands for "penny." The length of each nail is given below its size designation. The three sizes of nail most often used in light frame construction, 16d, 10d, and 8d, are shaded.

Machine-Driven Nails

Although driving a nail by hand is fast and easy, driving several nails per minute for an entire working day is fatiguing. Most carpenters now do the majority of their nailing of both framing and finish work with powered *nail guns.* These use *collated nails* that are joined in linear arrays for swift loading into the magazine of the gun. Guns may be powered pneumatically,

FIGURE 3.39
Face nailing is the strongest of the three methods of nailing. End nailing is relatively weak and is useful primarily for holding framing members in alignment until gravity forces and applied sheathing make a stronger connection. Toe nailing is used in situations where access for end nailing is not available. Toe nails are surprisingly strong; load tests show them to carry about five-sixths as much load as face nails of the same size.

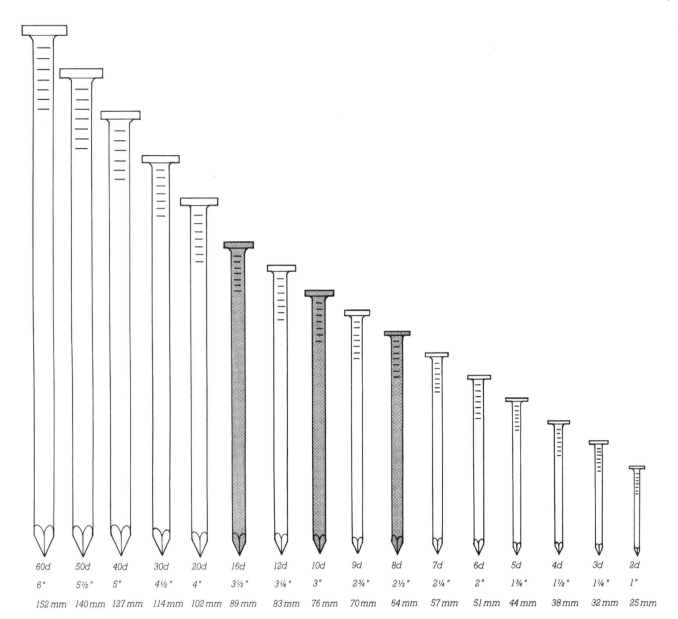

60d	50d	40d	30d	20d	16d	12d	10d	9d	8d	7d	6d	5d	4d	3d	2d
6"	5½"	5"	4½"	4"	3½"	3¼"	3"	2¾"	2½"	2¼"	2"	1¾"	1½"	1¼"	1"
152 mm	140 mm	127 mm	114 mm	102 mm	89 mm	83 mm	76 mm	70 mm	64 mm	57 mm	51 mm	44 mm	38 mm	32 mm	25 mm

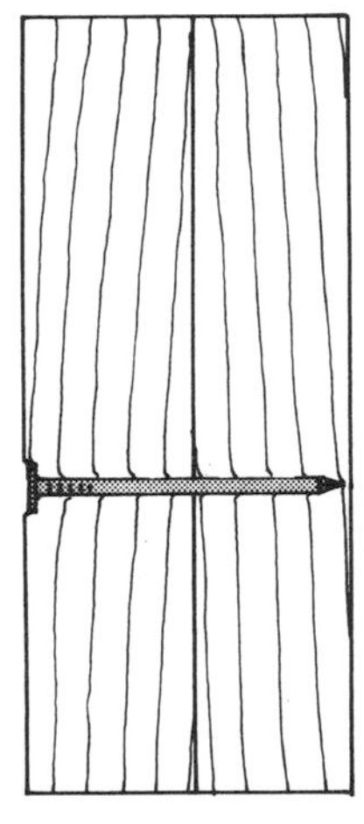
FACE NAIL

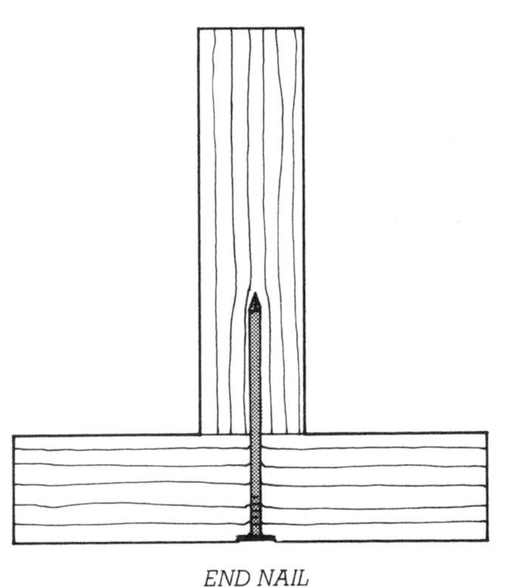
END NAIL

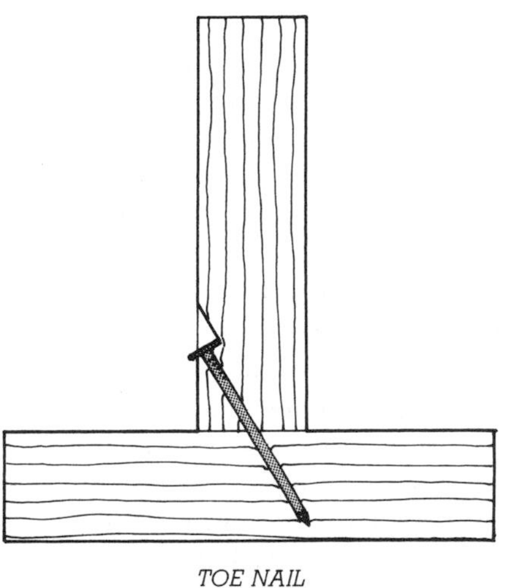
TOE NAIL

with air delivered to the gun via hose from an air compressor, or by internal combustion, with flammable gas stored in a cartridge within the gun. When the nose of the nail gun is pressed against a solid surface and the trigger is pulled, a piston drives a nail with a single instantaneous blow. Various sizes of nail guns drive everything from 16d common nails to inch-long finish nails, with significant productivity gains over hand nailing. For example, floor sheathing can be fastened at the rate of one or more nails per second per carpenter using nail guns. Installation of finish components such as window casings and baseboards is much speedier because the nail gun sets each nail below the surface of the wood, ready for filling and painting, and bent nails, a constant problem with small-diameter finish nails driven by hand, are rare. Increasingly, common and finish nails that are not collated are marketed primarily to homeowners, because professional builders use them only in situations where a bulky nail gun cannot reach a tight location.

Various types of U-shaped, gun-driven *staples* find uses in construction. The smallest staples are used to attach sheathing paper and thermal insulation. Heavier-duty staples serve as fasteners for cabinet components, wood flooring, shingles, sheathing, and underlayment panels. Staples are not always an equal substitute for nailing. Care must be taken to ensure that a staple meets building code requirements and the recommendations of the manufacturer of the material that is being fastened.

Wood Screws and Lag Screws

Screws are spiral-threaded fasteners installed by turning action whereby the threads draw the screw tightly into the material being fastened (Figure 3.40). In comparison to nails, screws cost more and take longer to install, but they can be inserted with greater precision, can exert greater clamping force between joined pieces, have greater holding power, and can be backed out and reinserted if a component needs to be adjusted or remounted. Traditional *wood screws* require predrilled *pilot holes* into which the screw is inserted and then tightened with a screwdriver. Common uses include joining of cabinetry parts, installation of wide-plank flooring, mounting of hardware such as hinges, and other finish woodworking applications.

Self-drilling wood screws do not require predrilled pilot holes and can be installed more quickly with power screwdrivers. They are used for attaching subflooring to floor framing (to reduce floor squeaking), mounting gypsum wallboard to wall studs (to avoid nail popping), attaching exterior decking to deck framing (to resist loosening of decking boards caused by moisture-related expansion and contraction), and wherever else the greater precision and holding power of screws are benefits. Like nails, self-drilling screws are also available in collated strips for use in self-feeding power guns.

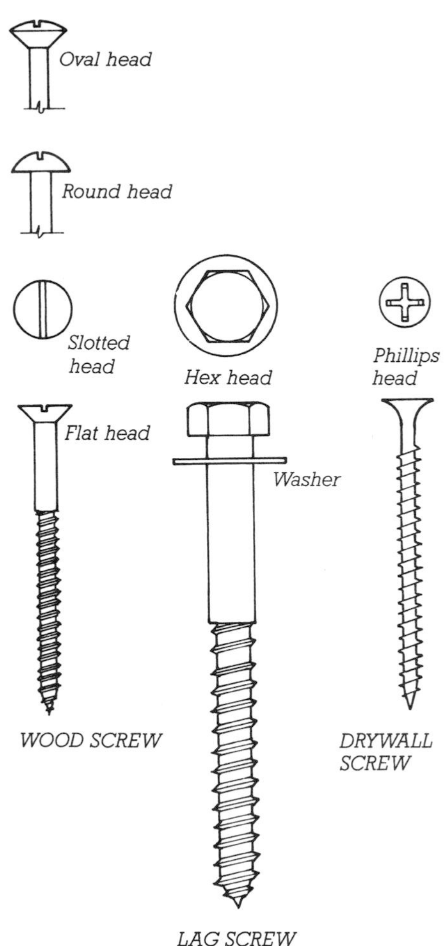

FIGURE 3.40
Some common screw types. Flat-head screws are used without washers and are driven flush with the surface of the wood. Round-head screws are used with flat washers and oval-head screws with countersunk washers. The drywall screw does not use a washer and is the only screw shown here that does not require a predrilled hole. Slotted and Phillips heads are most common. Screw heads with square and star-shaped recesses to mate with special driver bits of the same shape are becoming popular because of the very positive engagement that they create between the driver bit and the screw.

Large screws for heavy structural connections are called *lag screws*. They have square or hexagonal heads and are driven with a wrench rather than a screwdriver. In all cases, screw types should be matched to the fastening application with care. For example, *drywall screws*, which are relatively slender and brittle, are not suitable for structural applications.

Specialty Screws

Screws seem to come in an endless variety of styles (Figure 3.41). Alternate driver shapes, such as square or star-shaped, engage and hold screws more reliably and can transmit greater torque than traditional slotted or Phillips drivers. Wider, steeper thread patterns improve screw-holding strength and

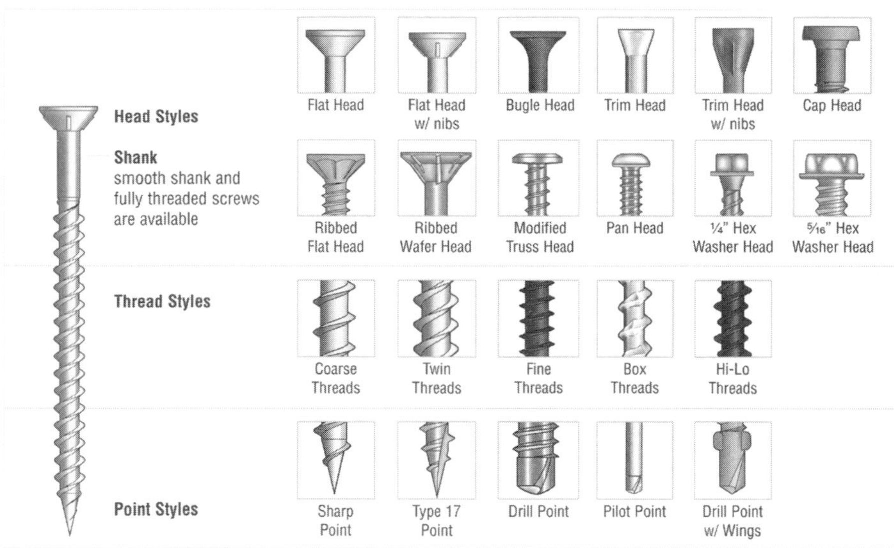

FIGURE 3.41
A manufacturer's chart illustrating variations in styles of power-driven screw heads, shanks, threads, and points. *(Courtesy of Simpson Strong-Tie Company, Inc.)*

allow faster driving. Specially contoured screw heads recess themselves neatly below the surface of the wood. Multipitch thread patterns improve a screw's ability to draw fastened pieces tightly together. Screws with organic coatings can be matched to the finish color of the fastened material and provide corrosion resistance at less cost than fasteners made of galvanized or stainless steel. Slender, small-headed screws can be installed almost as unobtrusively as finish nails. In structural applications, longer, larger-diameter self-drilling screws can be installed more quickly and easily than conventional lag screws or bolts.

Bolts

Bolts are used mainly for structural connections in heavy timber framing and, less frequently, in wood light framing for fastening ledgers, beams, or other heavy applications. Commonly used bolts range in diameter from 3/8 to 1 inch (10 to 25 mm) in almost any desired length. Flat steel disks called *washers* are inserted under the heads and nuts of bolts to distribute the compressive force from the bolt across a greater area of wood (Figure 3.42).

Timber Connectors

Various types of specialized timber connectors provide increased load-carrying capacity over bolts. The *split-ring connector* (Figure 3.43) is used in conjunction with a bolt and is inserted in matching circular grooves in the mating pieces of wood. It provides greater capacity by spreading the load across a much greater area of wood than can be done with a bolt alone. The split permits the ring to adjust to wood shrinkage. Split rings are useful primarily in heavy timber construction.

Toothed Plates

Toothed plates (Figure 3.44) are used in factory-produced lightweight roof and floor trusses (Figure 3.47). They are inserted into the wood with hydraulic presses, pneumatic presses, or mechanical rollers and act as metal splice plates, each with a very large number of built-in nails. They are extremely effective connectors because no drilling or gluing is required, they can be installed rapidly, and their multiple closely spaced points interlock tightly with the fibers of the wood.

Sheet Metal and Metal Plate Framing Devices

Dozens of ingenious sheet metal and metal plate devices are manufactured for strengthening common connections in wood framing. The most frequently used is the *joist hanger*, but all of the devices shown in Figures 3.45 and 3.46 find extensive use. There are two parallel series of this type of device, one made of sheet metal for use in light framing and one made of thicker metal plate for heavy timber and laminated wood framing. The devices for light framing are attached with nails and the heavier devices with bolts or lag screws. Where such

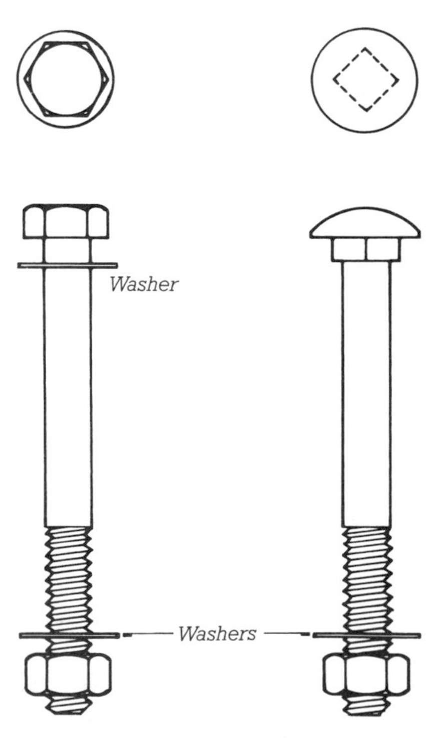

FIGURE 3.42
Both machine bolts and carriage bolts are used in wood construction. Carriage bolts have a broad button head that needs no washer and a square shoulder under the head that is forced into the drilled hole in the wood to prevent the bolt from turning as the nut on the opposite end is tightened.

framing devices will be exposed to the weather or to lumber treated with corrosive chemical treatments, they should be made from corrosion-resistant metal such as heavily galvanized steel or stainless steel. To prevent galvanic corrosion between dissimilar metals, fasteners for metal framing devices should always be made of the same material as the framing device: Galvanized framing devices should be fastened with galvanized fasteners, and stainless steel devices should be fastened with stainless steel fasteners.

Adhesives

Adhesives are widely employed in the factory production of plywood and panel products, laminated wood, wood structural components, and cabinetwork. Some such adhesives, most notably urea-formaldehyde, can release formaldehyde or other unhealthful gases long after the manufacturing process is complete and the products have been placed into service in buildings. When used for interior carpentry, such products can be the source of indoor air quality problems.

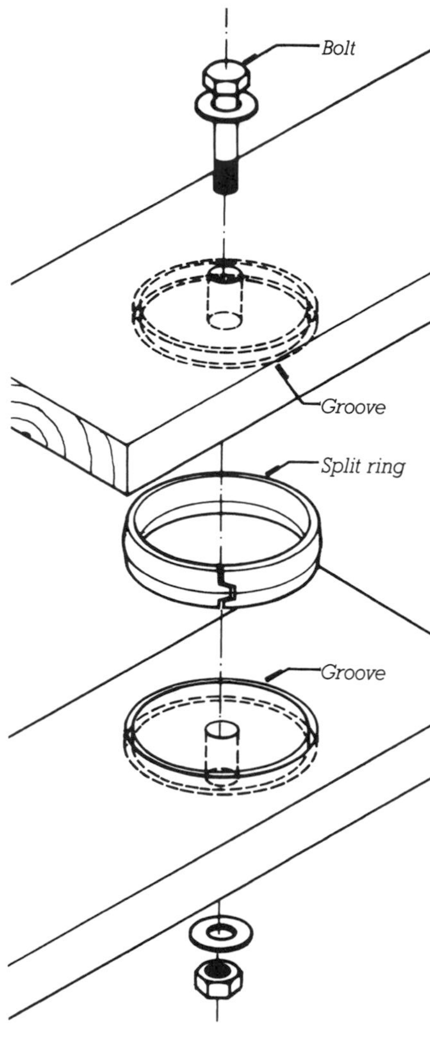

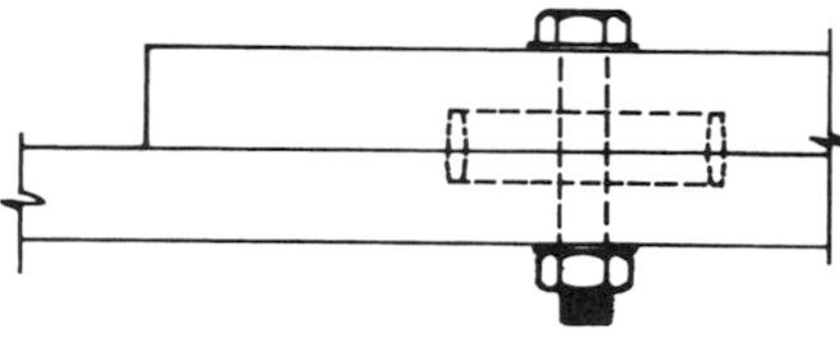

FIGURE 3.43
Split rings are high-capacity connectors used in heavily loaded joints of timber frames and trusses. After the center hole has been drilled through the two pieces, they are separated and the matching grooves are cut with a special rotary cutter driven by a large power drill. The joint is then reassembled with the ring in place.

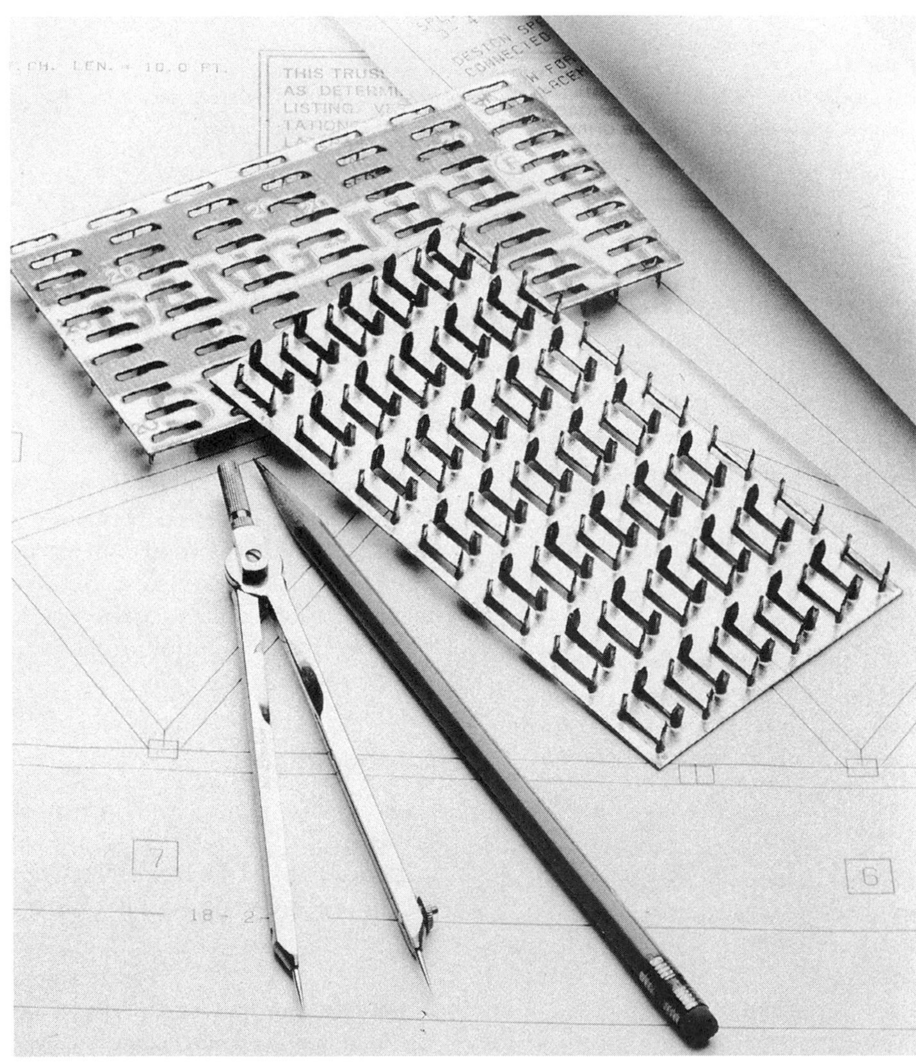

FIGURE 3.44
Manufacturers of toothed-plate connectors also manufacture the machinery to install them and provide computer programs to aid truss fabricators in designing and detailing trusses for specific buildings. The truss drawing in this photograph was generated by such a program. The small rectangles on the drawing indicate the positions of toothed-plate connectors at the joints of the truss. *(Courtesy of Gang-Nail Systems, Inc.)*

Where products manufactured with adhesives or binders will be used indoors, those made with a more chemically stable exterior adhesive called *phenol-formaldehyde*, or with alternative formaldehyde-free adhesives, should be considered. Finishes or veneers that encapsulate composites also reduce emissions from the glues and binders used in their manufacture.

Adhesives are used less on the construction site, where it is more

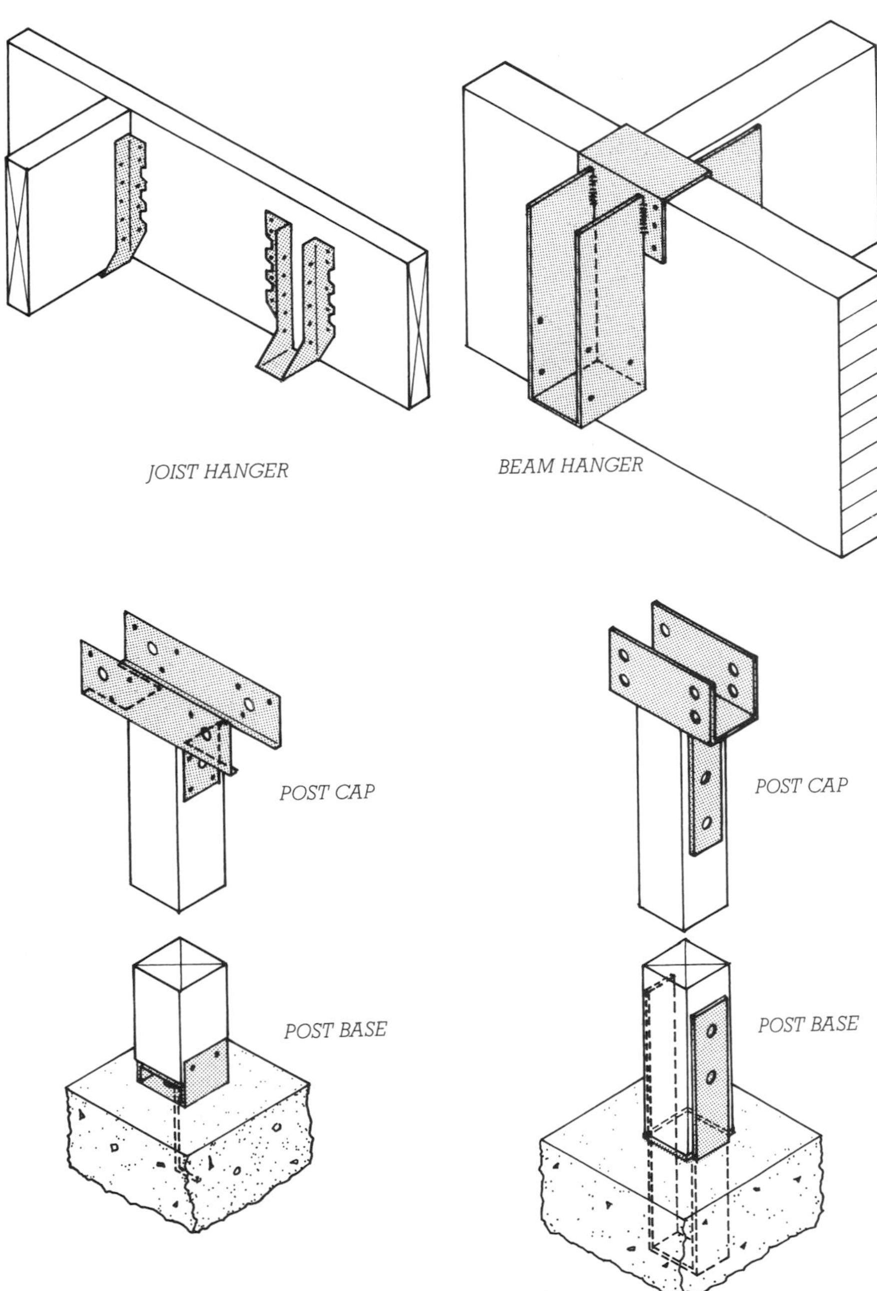

FIGURE 3.45
Joist hangers are used to make strong connections in floor framing wherever wood joists bear on one another at right angles. They are attached with special short nails driven through the holes punched in the hangers. The heavier steel beam hangers are used primarily in laminated wood construction. Post bases serve the twofold function of preventing water from entering the end of the post and anchoring the post to the foundation. The bolts and lag screws used to connect the wood members to the heavier connectors are omitted from this drawing.

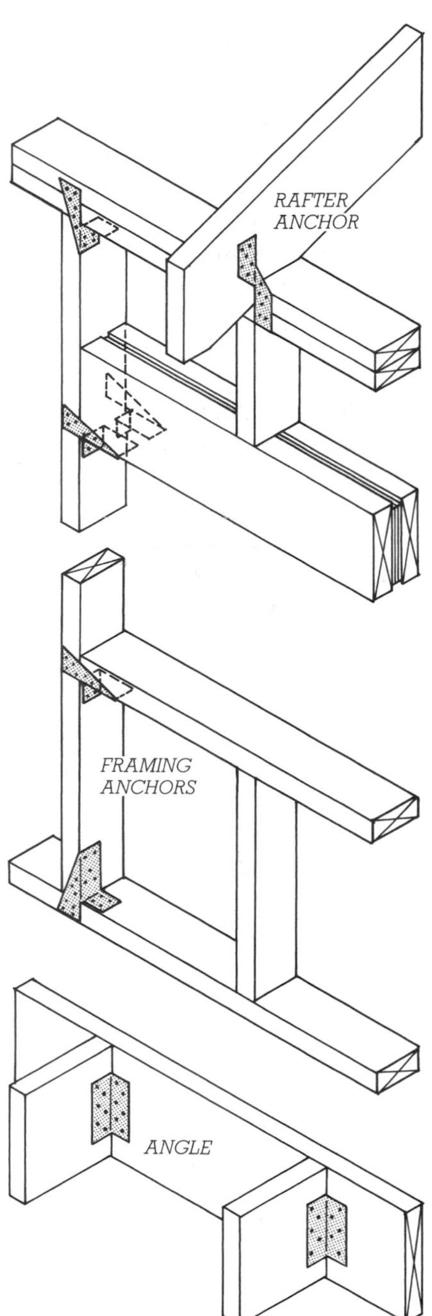

FIGURE 3.46
The sheet metal connectors shown in this diagram are less commonly used than those in Figure 3.45, but are invaluable in solving special framing problems and in reinforcing frames against wind uplift and earthquake forces.

difficult to clamp and hold glued joints and to maintain controlled environmental conditions until the adhesive has cured. In rough carpentry work, adhesive is most commonly used for securing subflooring panels to wood framing (Figure 5.27). In this application, mastic sealant is applied using a sealant gun. The joint between the subfloor panel and the joist framing is then clamped by nailing or screwing the panels to the framing, the fasteners serving as the primary structural connection and the adhesive acting primarily to reduce squeaks in the floor. Adhesives of various types are also used in finish carpentry, usually in combination with nails or screws, to improve the strength and stability of joints in finish trim, site-built cabinets, and other architectural woodwork.

Manufactured Wood Components

Dimension lumber, structural panel products, mechanical fasteners, and adhesives may be used in combination to manufacture a number of highly efficient structural components that offer advantages to the designer of wooden buildings.

Trusses

Trusses for both roof and floor construction are manufactured in small, highly efficient plants in every part of North America. Most are based on 2 × 4s and 2 × 6s joined with toothed-plate connectors. The designer or builder need only specify the span, roof pitch, and desired overhang detail. The truss manufacturer then uses a preengineered design for the specified truss or custom engineers a truss design and develops the necessary cutting patterns for its constituent parts. The manufacture and transportation of trusses are shown in Figure 3.47, and several uses of trusses are depicted in Chapter 5.

Roof trusses use less wood than a comparable frame of conventional rafters and ceiling joists. Like floor trusses, they span the entire width of the building in most applications, allowing the designer greater freedom to locate interior partitions anywhere they are needed. The chief disadvantages of roof trusses as they are most commonly used are that they make the attic space unusable and generally restrict the designer to the spatial monotony of a flat ceiling throughout the building. Truss shapes can be designed and manufactured to overcome both of these limitations, however (Figure 3.47*c*).

Wood I-Joists

Manufactured wood I-shaped members, called *I-joists*, are used for framing of both roofs and floors (Figure 3.48). The flanges of the members may be made from solid lumber, laminated veneer lumber, or laminated strand lumber. The webs may be plywood or OSB, though in this particular application, OSB's greater shear strength makes it a superior choice over plywood. Like trusses, these components use wood more efficiently than conventional dimension lumber rafters and joists, and they can span farther between supports. They are also lighter in weight than corresponding solid members, lack crooks and bows, are more dimensionally stabile, and are available in lengths up to 40 feet (12.2 m). Because I-joists span farther than conventional framing lumber, floors framed with these components are sometimes more prone to uncomfortable vibration when subjected to normal occupant live loads. Extra stiffening elements or oversized joists may be used to counteract this tendency. Chapter 5 shows the use of I-joists in wood framing.

Panel Components

Dimension lumber and wood structural panels lend themselves readily to many forms of *prefabrication*. In comparison to conventional on-site construction methods, factory prefabrication allows smaller on-site work crews, faster on-site construction, more consistent quality, protection from weather delays, efficient use of materials, and reduced waste.

At its simplest, prefabrication may consist of factory-assembled *framed panels*, sections of framing, usually 4 feet (1220 mm) wide, sheathed with a sheet of plywood or OSB, trucked to the construction site, and rapidly nailed together into a complete frame, sheathed and ready for finishing. For greater structural efficiency, loadbearing panels for walls, floors, and roofs may be prefabricated utilizing top and bottom sheets of plywood or OSB as primary load-carrying members. These sheets are joined firmly together by either a stiff plastic foam core to make a *structural insulated panel (SIP)*, or by dimension lumber framing to make a *stressed-skin panel (SSP)*. SIPs (pronounced "sips"), in particular, are a popular choice for the construction of highly energy-efficient homes and small buildings (Figure 3.49).

In *panelized construction*, whole sections of walls or floors are framed and sheathed in the factory, and then trucked to the construction site and installed in rapid succession. While prefabrication of more complex assemblies—including, for example, insulation, wiring, windows, doors, and exterior and interior finishes—is technically feasible, it is less frequently done due to the added difficulties of joining up all such systems in the field and obtaining the needed building code inspections in a factory setting.

Factory-Built Housing

Houses may be built entirely in a factory, at times even complete with furnishings, and then transported to prepared foundations, where they are set in place and made ready for occupancy in a matter of hours or days. A *manufactured home* (referred

(*a*)

(*b*)

(*c*)

FIGURE 3.47
Manufacturing and transportation of wood trusses. (*a*) Factory workers align the wood members of a roof truss and position toothed-plate connectors over the joints, tapping them with a hammer to embed them slightly and keep them in place. The roller marked "Gantry" then passes rapidly over the assembly table and presses the plates firmly into the wood. (*b*) The trusses are transported to the construction site on a special trailer. (*c*) Trusses can be designed and produced in almost any configuration.
(Photos courtesy of Wood Truss Council of America)

FIGURE 3.48
A bundle of I-joists delivered to the construction site. The webs are made from OSB and the flanges from solid lumber. *(Photo by Joseph Iano)*

to in the past as a *mobile home*) is built on its own permanent, towable chassis and is designed to comply with a federal building code administered by the U.S. Department of Housing and Urban Development. A factory-built *modular home* is designed to comply with the building code

FIGURE 3.49
Three types of prefabricated wood panels. The framed panel is identical to a segment of a conventionally framed wall, floor, or roof. The facings on the stressed-skin panel are bonded by adhesive to thin wood spacers to form a structural unit in which the facings carry the major stresses. A SIP, also sometimes called a *sandwich panel*, functions structurally in the same way as an SSP, but its facings are bonded to a core of insulating foam instead of wood spacers.

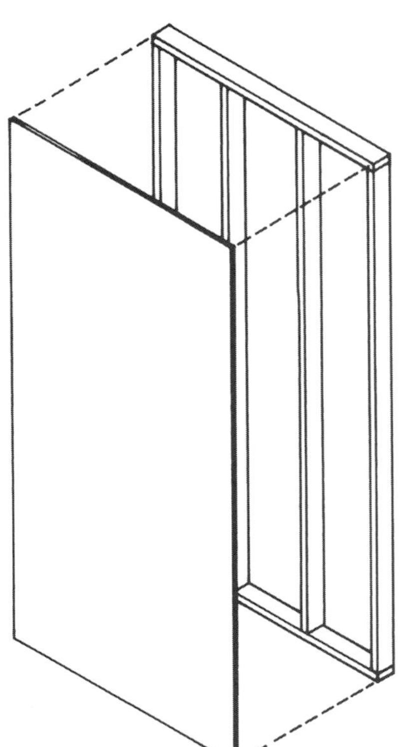

FRAMED PANEL

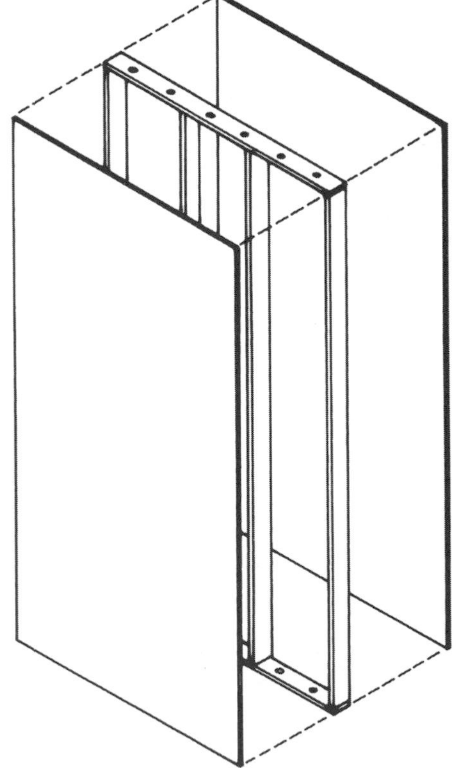

STRESSED-SKIN PANEL

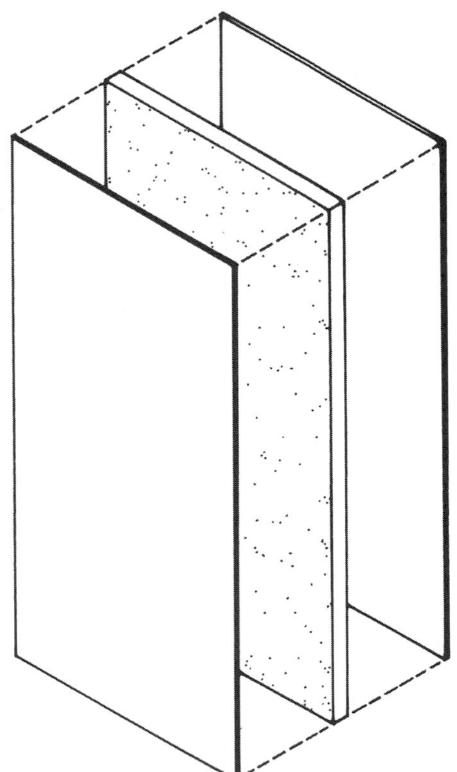

STRUCTURAL INSULATED PANEL

in effect where the home will be located and is transported on a conventional flat-bed tractor trailer. Both manufactured and modular homes are constructed in units not wider than 14 to 16 feet (4.27 to 4.88 m) to allow for transportation on public roads. Multiple units may be combined side-by-side or even stacked vertically to make the complete home.

At their least expensive, manufactured homes are sold at a fraction of the price of conventionally constructed houses of the same floor area. This is due in part to the economies of factory production and mass marketing and in part to the use of components that are lighter and less costly, and therefore shorter in life expectancy. However, both manufactured and modular homes may also be constructed to levels of cost and quality equal to those of conventionally constructed houses, but with the potential to significantly reduce the duration of construction on site.

Though most frequently associated with residential construction, modular techniques may also be applied to the construction of larger commercial or institutional building types.

Types of Wood Construction

Wood construction has evolved into two major systems of on-site construction: The hand-hewn frames of centuries past have become the heavy timber frames of today, used both for single-family houses and for larger buildings. And from the heavy timber frame has sprung wood light frame construction, which is the dominant system for houses, small commercial structures, and apartment buildings even as tall as four or five stories. These two systems of framing are detailed in Chapters 4 and 5.

CSI/CSC

MasterFormat Sections for Wood, Wood-Plastic Composites, and Plastic Lumber Materials

06 0500	COMMON WORK RESULTS FOR WOOD, PLASTICS, AND COMPOSITES
	Wood, Plastic, and Composite Fastening
	Wood Treatment
06 1000	ROUGH CARPENTRY
06 11 00	Wood Framing
06 12 00	Structural Panels
	Stressed Skin Panels
	Structural Insulated Panels
06 16 00	Sheathing
06 17 00	Shop-Fabricated Structural Wood
	Laminated Veneer Lumber
	Parallel Strand Lumber
	Wood I-Joists
	Rim Boards
	Shop-Fabricated Wood Trusses
06 18 00	Glue-Laminated Construction
06 2000	FINISH CARPENTRY
06 5000	STRUCTURAL PLASTICS
06 6000	PLASTIC FABRICATIONS
06 6500	Plastic Simulated Wood Trim
06 7000	STRUCTURAL COMPOSITES
06 7300	Composite Decking

Selected References

1. Hoadley, R. Bruce. *Understanding Wood: A Craftsman's Guide to Wood Technology* (2nd ed.). Newtown, CT, Taunton Press, 2000.

Beautifully illustrated and produced, this volume vividly explains and demonstrates the properties of wood as a material of construction.

2. USDA Forest Products Laboratory. *Wood Handbook: Wood as an Engineering Material.* Ottawa, Canada, Algrove Publishing Limited, 2002.

This Forest Products Laboratory reference has been in publication since the 1930s. It provides comprehensive coverage of wood properties and the use of wood in construction. This book can also be viewed for free on the Forest Product Laboratory's web site, www.fpl.fs.fed.us.

2. Western Wood Products Association. *Western Woods Use Book.* Portland, OR, updated regularly.

This looseleaf binder houses a complete reference library on the most common species of dimension lumber, timbers, and finish lumber. It includes the National Design Specification for Wood Construction, the standard for engineering design of wood structures.

3. APA–The Engineered Wood Association. *Binder: Architects' Reference Materials.* Tacoma, WA, update regularly.

This is a complete looseleaf guide to manufactured wood products, including plywood, OSB, waferboard, SIPs, and glue-laminated wood. This organization's web site is also an excellent source of technical information on structural panels, glue-laminated wood, and structural composite lumber.

4. Canadian Wood Council. *Wood Reference Handbook.* Ottawa, 1995.

A superbly illustrated encyclopedic reference on wood materials.

Web Sites

Wood

Author's supplementary web site: **www.ianosbackfill.com/03_wood**
Forest Stewardship Council: **www.fsc.org**
Sustainable Forestry Initiative: **www.aboutsfi.org**

Lumber

American Wood Council: **www.awc.org**
Canadian Wood Council: **www.cwc.ca**
Hardwood Council: **www.hardwoodcouncil.com**
U.S. Forest Products Laboratory: **www.fpl.fs.fed.us**
Western Wood Products Association: **www.wwpa.org**

Wood Products

American Institute of Timber Construction: **www.aitc-glulam.org**
APA–The Engineered Wood Association: **www.apawood.org**

Wood Panel Products

APA–The Engineered Wood Association: **www.apawood.org**
Composite Panel Association: **www.pbmdf.com**
Hardwood Plywood & Veneer Association: **www.hpva.org**

Wood Chemical Treatments

AWPA–The American Wood Preservers Association: **www.awpa.com**

Manufactured Wood Components

Manufactured Housing Institute: **www.manufacturedhousing.org**
Structural Insulated Panel Association: **www.sips.org**
WTCA (formerly Wood Truss Council of America): **www.sbcindustry.com**

Key Terms and Concepts

bark
cambium
sapwood
heartwood
pith
cellulose
lignin
grain
springwood, earlywood
summerwood, latewood
softwood
hardwood
tracheid
ray
fiber
vessel, pore
certified wood
Forest Stewardship Council (FSC)
Sustainable Forestry Initiative (SFI)
reclaimed lumber
lumber
headsaw
sawyer
plainsawn, flatsawn
flat-grain
quartersawn, riftsawn, edge-sawn
edge-grain, vertical grain
free water
bound water
seasoned
equilibrium moisture content (EMC)
green lumber
kiln drying
air drying
sticker
longitudinal shrinkage
radial shrinkage
tangential shrinkage
surfaced
unsurfaced
surfaced four sides (S4S)
surfaced two sides (S2S)
S-DRY
S-GRN
growth characteristics
manufacturing characteristics
knot
knothole
decay
insect damage
split
check
crook
bow
twist
cup
wane
structural grading
visual grading
machine grading
machine-stress rated
appearance grading
nominal dimension
actual dimension
board
dimension lumber
timber
board foot
glue-laminated wood, glulam
finger jointed
scarf jointed
hybrid glulam beam
FRP reinforced glulam
structural composite lumber, engineered lumber
laminated strand lumber (LSL)
oriented strand lumber (OSL)
laminated veneer lumber (LVL)
parallel strand lumber (PSL)
structural finger-jointed lumber
wood-plastic composite (WPC)
composite wood trim
plastic lumber
recycled plastic lumber (RPL)
structural-grade plastic lumber (SGPL)
load duration factor
cantilever
structural wood panel
plywood
veneer
composite panel
nonveneered panel
oriented strand board (OSB)
waferboard
particleboard
underlayment panel
fiberboard
medium-density fiberboard (MDF)
rotary sliced
touch sanded
medium-density overlay (MDO)
high-density overlay (HDO)
flitch
flitch-sliced veneer
span rating
exposure durability classification
hardboard
insulating fiberboard sheathing
cellulosic fiberboard
hardwood plywood
fire-retardant treatment (FRT)
preservative-treated wood
creosote
pentachlorophenol
pressure-treated wood
chromated copper arsenate (CCA)
alkaline copper quat (ACQ)
copper boron azoles (CBA and CA)
sodium borate (SBX)
pressure impregnation
incising
kiln-dried after treatment
Use Category System
nail
common nail
box nail
finish nail
penny (d)
bright nail
hot-dip galvanized nail
face nailing
end nailing
toe nailing
sinker, cooler
joist hanger nail
nail gun
collated nail
staple
screw
wood screw
pilot hole
self-drilling wood screw
lag screw
drywall screw
bolt
washer
split-ring connector
toothed plate
joist hanger
adhesive
phenol-formaldehyde
truss
I-joist
prefabrication
framed panel
structural insulated panel (SIP)
stressed-skin panel (SSP)
panelized construction
manufactured home
modular home

Review Questions

1. Discuss the changes in moisture content and the effects of these changes on a piece of dimension lumber from the time the tree is cut, through its processing, and until it has been in service in a building for an entire year.

2. What are the differences between plainsawn and quartersawn lumber? What applications are appropriate for each?

3. Give the actual cross-sectional dimensions of the following pieces of kiln-dried lumber: 1 × 4, 2 × 4, 2 × 6, 2 × 8, 4 × 4, 4 × 12.

4. Why is wood laminated? What types of wood products make use of lamination?

5. What are the advantages of using of structural composite lumber in comparison to solid wood?

6. What is meant by a span rating of 32/16? What types of wood products are rated this way?

7. Name the most appropriate plywood veneer grade for each of the following applications: (a) the outer face of a wall or roof sheathing panel, (b) subflooring to receive a wood finish flooring, (c) underlayment to receive a thin, resilient vinyl floor covering, (d) plywood to receive a paint finish in an application demanding the highest possible finish quality.

8. For what reasons might you specify preservative-treated wood? What alternative materials may be used where preservative-treated wood is normally required?

9. Which common species of wood have naturally decay- and insect-resistant heartwood?

10. Why are nails the fasteners of choice in wood construction?

Exercises

1. Visit a nearby lumberyard or building materials supply center. Examine and list the species, grades, and sizes of lumber carried in stock. For what uses are each of these intended? While at the yard, look also at the available range of fasteners.

2. Pick up a number of scraps of dimension lumber from a shop or construction site. Examine each to see where it was located in the log before sawing. Note any drying distortions in each piece: How well do these correspond to the distortions you would have predicted? Measure the width and thickness of each scrap and compare your measurements to the specified actual dimensions for each.

3. Assemble samples of as many different species of wood as you can find. Learn how to tell the different species apart by color, odor, grain figure, ray structure, relative hardness, and so on. What are the most common uses for each species?

4. Visit a construction site and list the various types of lumber and wood products being used. Look for a gradestamp on each and determine why the given grade is being used for each use. If possible, look at the architect's written specifications for the project and see how the lumber and wood products were specified.

PROJECT: An Enclosure for a Residential Swimming Pool

ARCHITECT: Edward Allen

The budget for this project, an enclosure for a small swimming pool attached to a single-family residence, was tight. The span of the roof was a bit too long to frame with standard rafters, so the architect decided to make his first use of custom-made wood roof trusses assembled with bolts. These were to be spaced 4 feet (1.2 m) apart and exposed on the interior of the building as a major feature of its architecture. To stay within the budget, the trusses would be made of ordinary 2-inch framing lumber. Standard factory-made roof trusses were considered, but were felt to be unsuitable because their toothed-plate connectors were judged to be rather unattractive.

The roof pitch was fixed at 12/12 (45 degrees) very early in the design process. To give a soaring feeling to the interior space, the bottom chords of the trusses were to be sloped at 6/12 to create a scissors configuration (Figure A).

The architect calculated the assumed loads on the roof and the maximum forces in the members of the truss. Then he determined the required sizes of the truss members, double 2 × 10s for the top chords and single 2 × 4s for the other members. Next, he calculated the required number of bolts in each joint of the truss. Finally, he began to draw large-scale details of the joints, which revealed a nasty surprise: Using the bolt spacings and edge clearances required by code, it was impossible to fit the required number of bolts into the outermost joints of the truss. No matter what size bolts were tried, there was not nearly enough space available in the wood to contain them. Split rings did not improve the situation because there was not enough space for them, either.

In further investigating this problem, the architect discovered that bolts are expensive to buy and labor intensive to install. When the required number of bolts for all the trusses were multiplied by the cost per bolt, it was evident that bolted trusses would cost far too much to be possible within the project's modest budget.

Further study of wood engineering literature revealed that nails might be used to make the truss connections rather than bolts. Ten penny hot-dip galvanized nails would resist rusting in the damp air above the pool and would cost just pennies each to install, compared to installed bolt costs of several dollars each. A nail would not transfer nearly as many pounds of force as a bolt, but many more nails than bolts could be fitted into a joint, because there were no rules in the code about nail spacings, except a provision that the nails should not split the wood.

So far, so good—but back at the drawing board, constructing the truss joints at full scale on tracing paper, it became apparent that so many nails were needed in the outermost joints that it was almost physically impossible to fit them all in, and splitting of the wood would be inevitable.

The only solution was to change the configuration of the truss so as to reduce the forces in the truss joints to the point that nailed connections would be possible and reasonable. Examination of the member force calculations showed that this could best be done by decreasing the slope of the bottom chords of the truss so as to increase the angle between the top and bottom chords where they meet over the supports. By trial and error, a slope was found that would reduce the member forces enough to allow the joints to be nailed (Figure B).

There were still so many nails required in each joint that they would have to be very closely spaced, so the architect made full-scale cardboard templates to guide the builder in placing the nails in regular patterns that would be visually attractive. These patterns minimized splitting forces in the wood by locating the nails along skew lines so that they did not line up along the grain pattern (Figure C). As an additional precaution against splitting, holes would be drilled for the nails in accordance with building code requirements. The architect also visited local lumberyards to find wood for the trusses that would look good and not have an excessive tendency to split. Two kinds of framing lumber were readily available, Hem–Fir and Spruce–Pine–Fir (SPF), both cut from stands of mixed tree species. The Hem–Fir already was splitting in the piles at the lumberyard from drying stresses. The SPF was the more attractive of the two. It was also somewhat softer and more even-grained, which would make it less likely to split. Happily, it was no more expensive than the Hem–Fir.

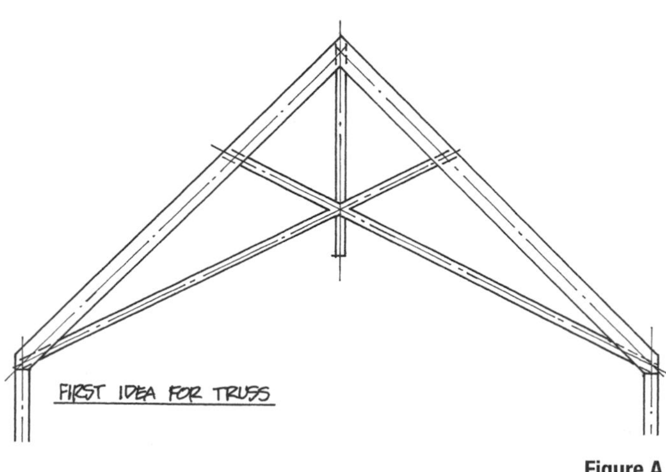

Figure A

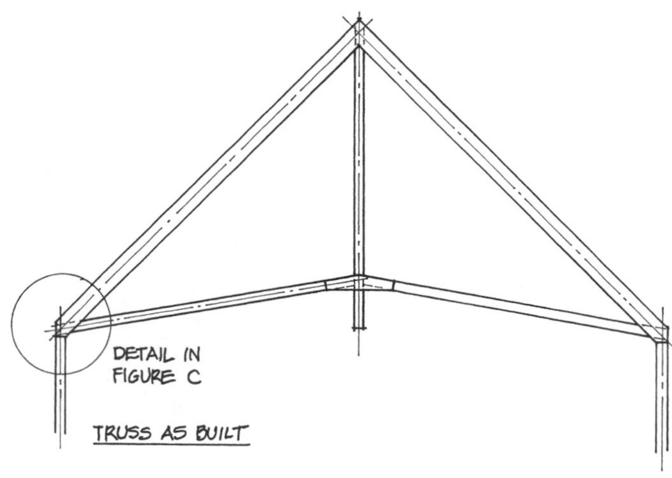

Figure B

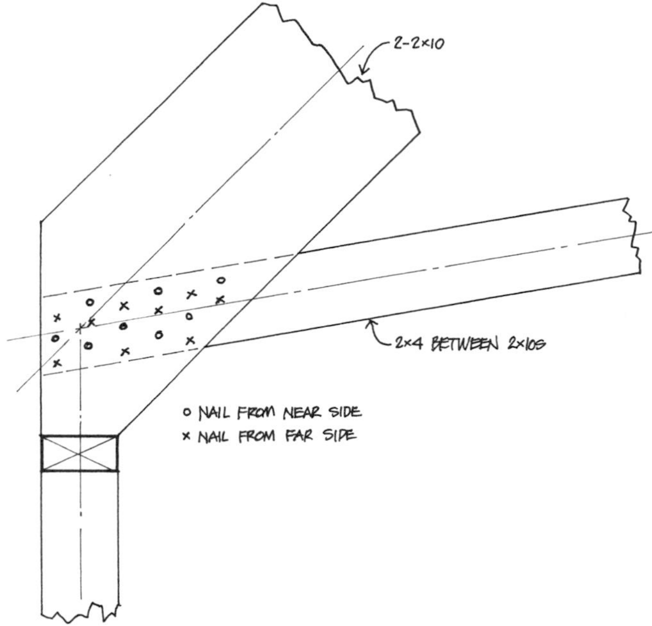

Figure C

When it came time to fabricate and erect the trusses, the architect spent a couple of days working with the builder on the building site. A first truss was cut, drilled, and assembled on a patio adjacent to the pool. The assembly process proved to be easy and quick, and the wood showed no tendency to split. Two carpenters needed less than a day to fabricate the remainder of the trusses in a stack on top of the first one (Figure D).

Figure D

Figure E

Figure F

A small truck-mounted crane was rented from a local neon sing company to install the trusses. It took only an hour to lift and place all of them (Figure E). The interior space was now easy to visualize: It had both the soaring quality that had originally been sought and a rich visual complexity that was created by the repetition of the many members of the trusses (Figure F, G) Many lessons had been learned:

- The design of a truss is not a linear process. It requires give-and-take to arrive at a solution that is visually satisfactory, sufficiently strong, and efficient to build.
- One cannot simply draw a form for a truss and assume that it can be made to work. The truss must have sufficient depth, and its more highly stressed members must meet at angles that are not too acute.
- Working closely with the builder makes it possible to do things that would otherwise be impossible.
- Time spent investigating the qualities and properties of materials is time well spent, both for the immediate project and for future ones. Innovation on one project is a stepping stone to further innovation on the next.

Figure G

4

HEAVY TIMBER FRAME CONSTRUCTION

Architect Nils Finne combines heavy timber framing with robust stonework and finely detailed exterior finish carpentry in this well-crafted residence. *(Architect: FINNE Architects www.FINNE.com; photograph by Art Grice)*

Wood beams have been used to span roofs and floors of buildings since the beginning of civilization. The first timber-framed buildings were crude pit houses, lean-tos, teepees, and basketlike assemblages of bent saplings. In earliest historic times, roof and floor timbers were combined with masonry loadbearing walls to build houses and public buildings. In the Middle Ages, braced wall frames of timber were built for the first time (Figures 4.1 and 4.2). The British carpenters who emigrated to North America in the 17th and 18th centuries brought with them a fully developed knowledge of how to build efficiently braced frames, and for two centuries North Americans lived and worked almost exclusively in buildings framed with hand-hewn wooden timbers joined by interlocking wood-to-wood connections (Figures 4.3 and 4.4). Nails were rare and expensive, so they were used only in door and window construction and, sometimes, for fastening siding boards to the frame.

Until two centuries ago, logs could be converted to boards and timbers only by human muscle power. To make timbers, axemen skillfully scored and hewed logs to reduce them to a rectangular profile. Boards were produced slowly and laboriously with a long, two-man pit saw, one man standing in a pit beneath the log, pulling the saw down, and the other standing above, pulling it back up. At the beginning of the 19th century, water-powered sawmills began to take over the work of transforming tree trunks into lumber, squaring timbers and slicing boards in a fraction of the time that it took to do the same work by hand.

Most of the great industrial mills of 19th-century North America, which manufactured textiles, shoes, machinery, and all the goods of civilization, consisted of heavy sawn timber floors and roofs supported by masonry exterior walls (Figures 4.5 and 4.6). The house builders and barn raisers of the early 19th century switched from hand-hewn to sawn timbers as soon as they became available. Many of these mills, barns, and houses still survive, and with them survives a rich tradition of heavy timber building that continues to the present day.

Figure 4.1
Braced wall framing was not developed until the late Middle Ages and early Renaissance, when it was often exposed on the face of the building in the style of construction known as *halftimbering*. The space between the timbers was filled with brickwork or with *wattle and daub*, a crude plaster of sticks and mud, as seen here in Wythenshawe Hall, a 16th-century house near Manchester, England. *(Photo by James Austin, Cambridge, England)*

FIGURE 4.2
European timber house forms generally followed a progression of development from crude pit dwellings, made of earth and tree trunks, to cruck frames, to braced frames. The *crucks* (curved timbers), hewn by hand from appropriately shaped trees, were precursors to the laminated wood arches and rigid frames that are widely used today.

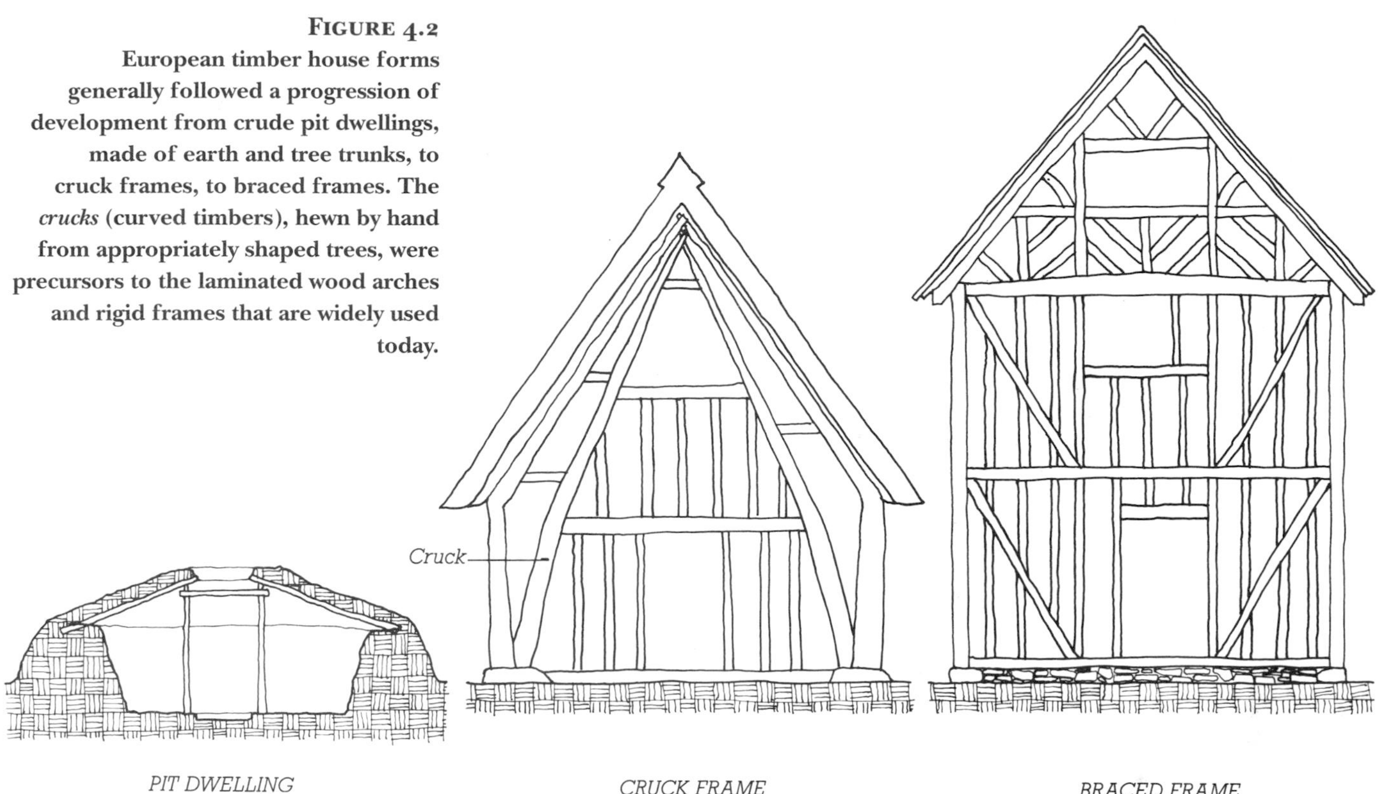

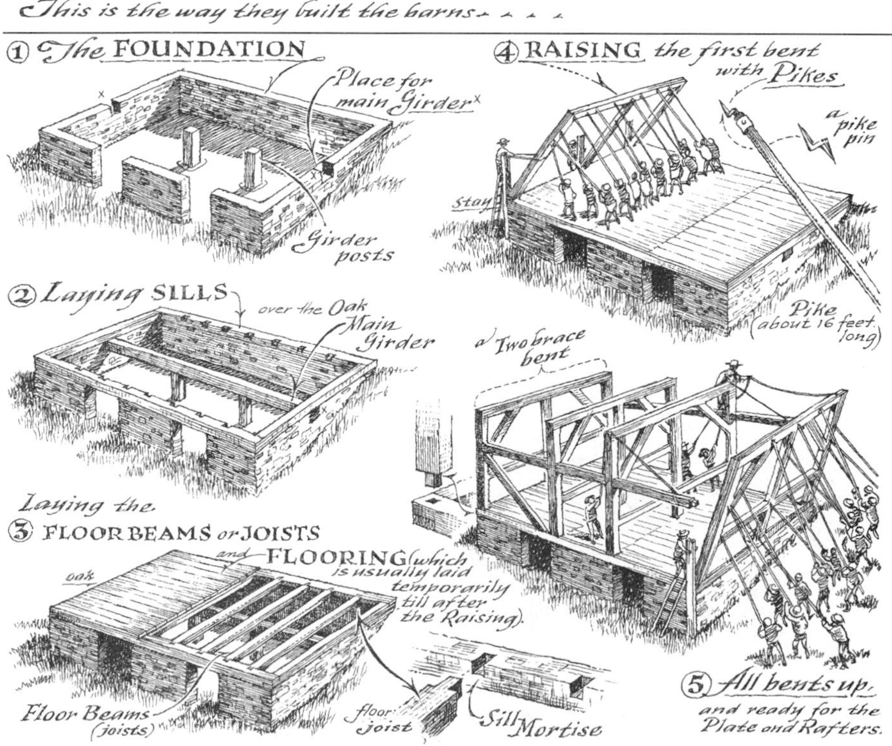

FIGURE 4.3
The European tradition of heavy timber framing was brought to North America by the earliest settlers and was used for houses and barns until well into the 19th century. *(Drawing by Eric Sloane; courtesy of the artist)*

(*a*)

(*b*)

(*c*)

FIGURE 4.4
Traditional timber framing has been revived in recent years by a number of builders who have learned the old methods of joinery and updated them with the use of modern power tools and equipment. (*a*) Assembling a *bent* (a plane of columns, beams, rafters, and braces). (*b*) The completed bents are laid out on the floor, ready for raising. (*c*) Raising the bents, using a truck-mounted crane, and installing floor framing and roof *purlins* (smaller, secondary framing members that span across the primary beams or rafters). (*d*) The completed frame. (*e*) Enclosing the timber-framed house with sandwich panels consisting of waferboard faces bonded to an insulating foam core. *(Courtesy of Benson Woodworking, Inc., Alstead, New Hampshire)*

(d)

(e)

The old country builder, when he has to get out a cambered beam or a curved brace, goes round his yard and looks out the log that grew in the actual shape, and taking off two outer slabs by handwork in the sawpit, chops it roughly to shape with his side-axe and works it to the finished face with the adze, so that the completed work shall for ever bear the evidence of his skill . . .

Gertrude Jekyll, English garden designer, writing in 1900.

FIGURE 4.5
Most 19th-century industrial buildings in North America were constructed of heavy timber roofs and floors supported at the perimeter on masonry loadbearing walls, a construction method that came to be known to as Mill construction. This impressive group of textile mills stretches for a distance of 2 miles (3 km) along the Merrimac River in Manchester, New Hampshire. *(Photo by Randolph Langenbach)*

FIGURE 4.6
The generously sized windows in the mills provided plenty of daylight to work by. Columns were of wood or cast iron. Most New England mills, like this one, were framed very simply: Their floor decking is carried by beams running at right angles to the exterior walls, supported at the interior on two lines of columns. Notice that the finish flooring runs perpendicular to the structural decking. Overhead sprinklers provide additional fire safety to a construction method that already has inherent fire-resistive qualities. *(Photo by Randolph Langenbach)*

FIRE-RESISTIVE HEAVY TIMBER CONSTRUCTION

Large timbers, because of their greater capacity to absorb heat, are much slower to catch fire and burn than smaller pieces of wood. When exposed to fire, a heavy timber beam, though deeply charred by gradual burning, will continue to support its load long after an unprotected steel beam exposed to the same conditions has collapsed. If the fire is not prolonged, a fire-damaged heavy timber beam or column can often be sandblasted afterward to remove the surface char and continue in service. For these reasons, building codes recognize heavy timber framing that meets certain specific requirements as having fire-resistive properties.

In the International Building Code (IBC), for a building to be classified as *Type IV Heavy Timber (HT) construction,* its wooden structural members must meet certain minimum size requirements and its exterior walls must be constructed of noncombustible materials. Minimum permitted sizes for solid wood timbers are summarized in Figure 4.7. Glue-laminated members used in this type

Considerations of Sustainability in Heavy Timber Construction

In addition to the issues of sustainability of wood production and use that were raised in the previous chapter, there are issues that pertain especially to heavy timber frame construction:

It is wasteful to saw large, solid timbers from logs: In most instances, only one or two timbers can be obtained from a log, and it is often difficult to saw smaller boards from the leftover slabs.

Glue-laminated timbers and composite timbers utilize wood fiber much more efficiently than solid timbers.

Recycled timbers from demolished mills, factories, and barns are often available. Most of these are from old-growth forests in which trees grew slowly, producing fine-grained, dense wood. As a result, many have structural properties that are superior to those of new-growth timbers. Recycled timbers may be used as is, resurfaced to give them a new appearance, or resawn into smaller members. However, they often contain old metal fasteners. Unless these are meticulously found and removed, they can damage saw blades and planer knives, causing expensive mill shutdowns while repairs are made.

Continuous bending action of beams may be created by splicing beams at points of inflection rather than over supports, as shown in Figures 4.15, 4.20 and 4.21. This reduces maximum bending moments, allowing timber sizes to be reduced substantially.

Timbers do not lose strength with age, although they do sag progressively if they are overloaded. When a heavy timber building is demolished at some time in the future, its timbers can be recycled, even if they were obtained as recycled material for the building that is being demolished.

A heavy timber frame enclosed with foam core sandwich or stressed-skin panels is relatively airtight and well insulated, with few thermal bridges. Heating and cooling of the building will consume relatively little energy.

The glues and finish coatings used with glue-laminated timbers may give off gases such as formaldehyde that can cause indoor air quality (IAQ) problems. It is wise to determine in advance what glues and coatings are to be used, and to avoid ones that may cause IAQ problems.

of construction must meet similar requirements. Exterior walls may be constructed of concrete, masonry, or metal cladding. Historically, this combination of fire-resistive wood framing and noncombustible exterior was referred to as *Mill construction,* reflecting its origins in 19th-century brick masonry mill structures, or as *Slow-burning construction* (Figures 4.8–4.11). Traditionally, the edges of the timbers were also *chamfered* (beveled at 45 degrees) to eliminate the thin edges of wood that catch fire most easily, but this is no longer a code requirement.

Wood Shrinkage in Heavy Timber Construction

Where the edges of floors and roofs of a heavy timber frame are supported on concrete or masonry, special attention must be given to the potential for differential shrinkage between the outer walls and the interior wood column supports. Wood, compared to masonry or concrete, expands and contracts more with changes in moisture content, particularly in the direction perpendicular to its grain. These changes occur over a period of years as large timbers gradually dry and seasonally with changes in ambient conditions. A heavy timber frame building is detailed to minimize the effects of this differential shrinkage by eliminating cross-grain wood from the interior lines of support. In traditional Mill construction, cast iron

	Supporting Floor Loads	Supporting Roof and Ceiling Loads Only
Columns	8 × 8 (184 × 184 mm)	6 × 8 (140 × 184 mm)
Beams and Girders	6 × 10 (140 × 235 mm)	4 × 6 (89 × 140 mm)
Trusses	8 × 8 (184 × 184 mm)	4 × 6 (89 × 140 mm)
Decking	3″ decking plus 1″ finish (64-mm decking plus 19-mm finish)	2″ decking, or 1⅛″ plywood (38-mm decking, or 29-mm plywood)

Figure 4.7
Minimum sizes for solid wood members used in Type IV Heavy Timber construction, as specified in the IBC.

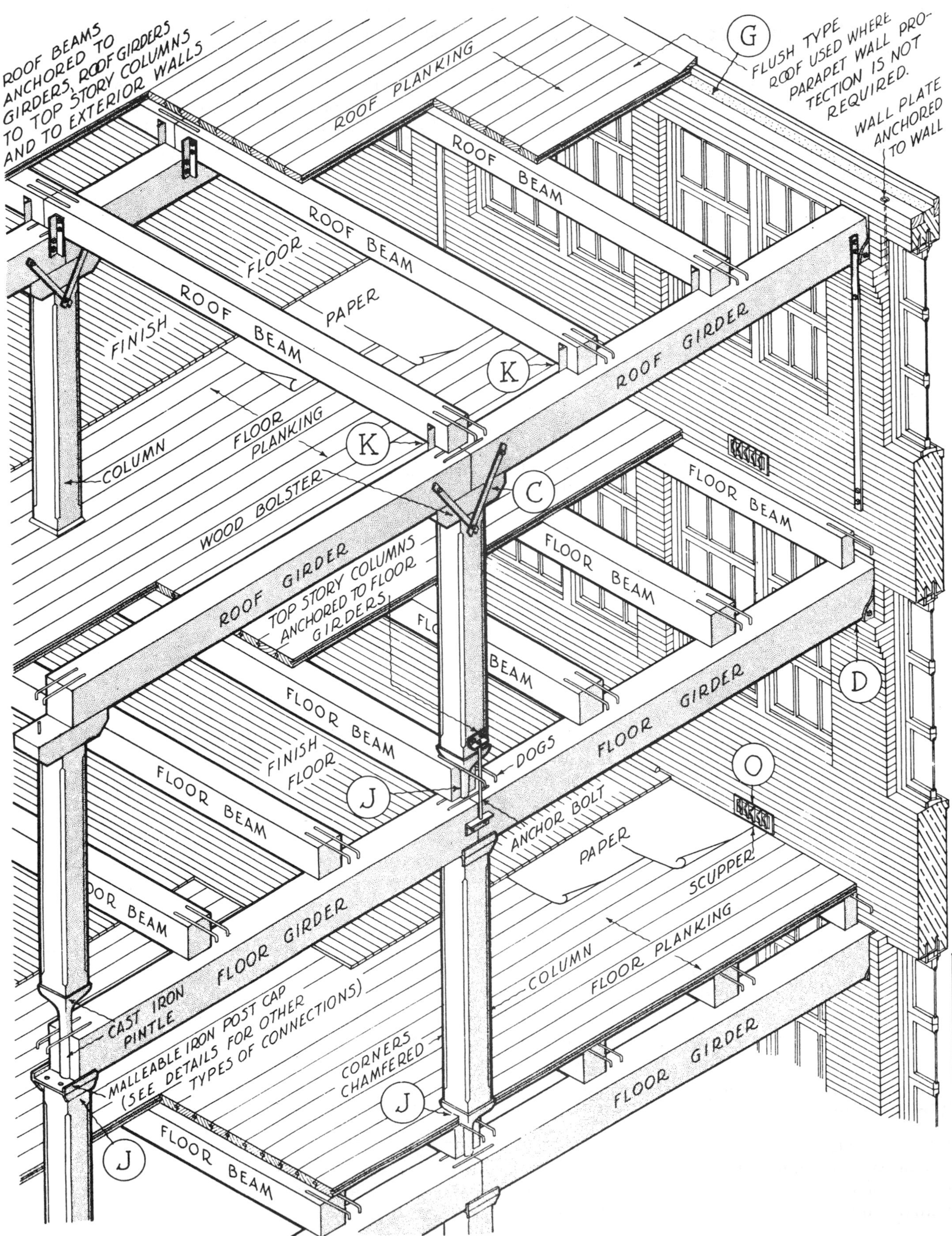

ROOF BEAMS ANCHORED TO GIRDERS, ROOF GIRDERS TO TOP STORY COLUMNS AND TO EXTERIOR WALLS
ROOF PLANKING
G
FLUSH TYPE ROOF USED WHERE PARAPET WALL PROTECTION IS NOT REQUIRED.
WALL PLATE ANCHORED TO WALL
ROOF BEAM
ROOF BEAM
FLOOR
PAPER
ROOF BEAM
FINISH
K
ROOF GIRDER
FLOOR PLANKING
K
COLUMN
WOOD BOLSTER
C
FLOOR BEAM
FLOOR BEAM
ROOF GIRDER
TOP STORY COLUMNS ANCHORED TO FLOOR GIRDERS
FLOOR BEAM
D
FLOOR BEAM
FLOOR GIRDER
DOGS
FINISH FLOOR
FLOOR BEAM
J
O
ANCHOR BOLT
PAPER
SCUPPER
FLOOR BEAM
FLOOR GIRDER
COLUMN
FLOOR PLANKING
CAST IRON PINTLE
MALLEABLE IRON POST CAP (SEE DETAILS FOR OTHER TYPES OF CONNECTIONS)
CORNERS CHAMFERED
J
FLOOR GIRDER
J
FLOOR BEAM

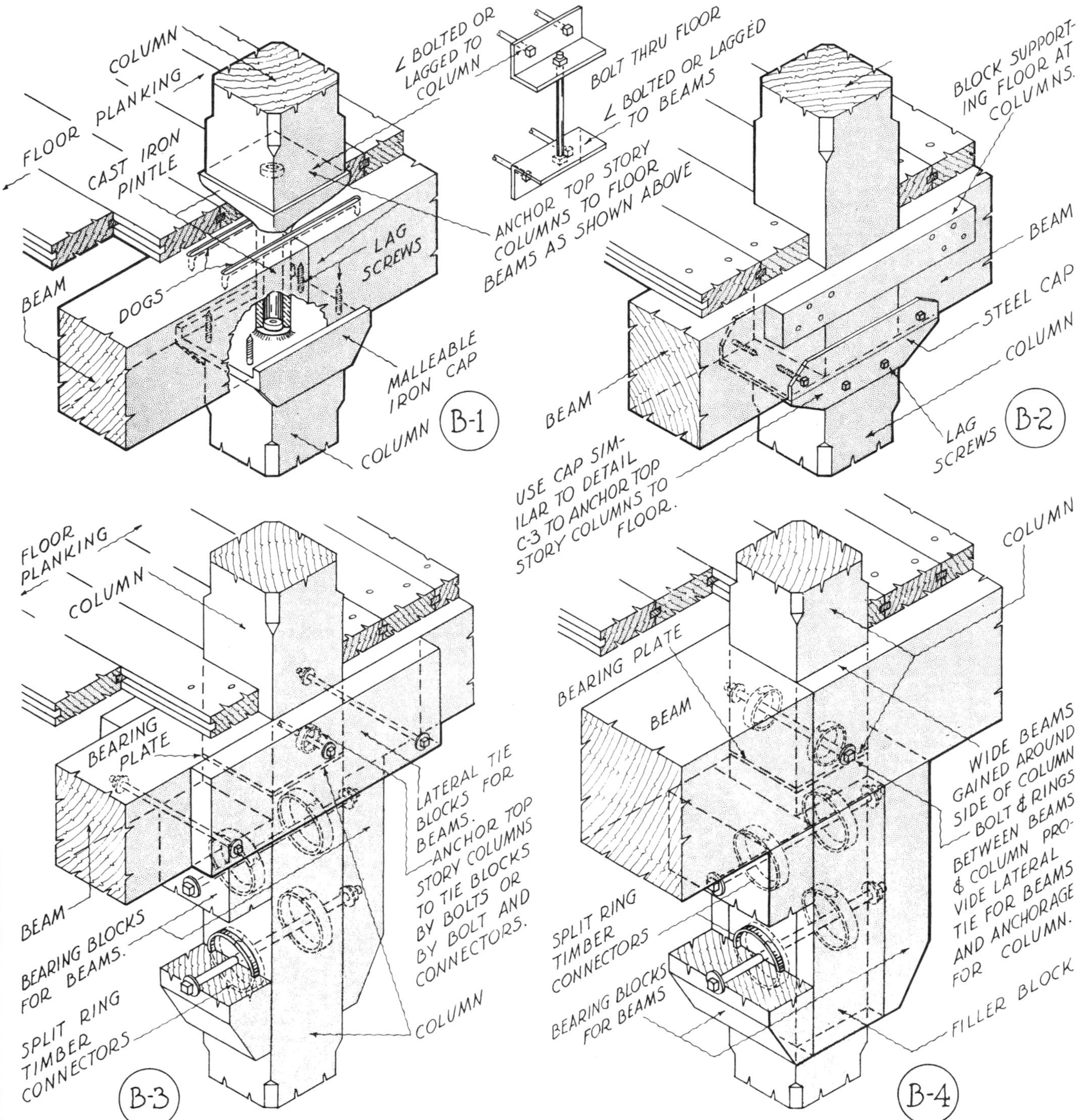

FIGURE 4.8 (opposite page)
Traditional Mill construction bypasses problems of wood shrinkage at the interior lines of support by using cast iron pintles (see the lower connection detail in the left-most column in the illustration) to transmit the column loads through the beams and girders. *Iron dogs* tie the beams together over the girders. A long steel strap anchors the roof girder to a point sufficiently low in the outside wall that the weight of the masonry above the anchorage point is enough to resist wind uplift on the roof. *(Heavy timber construction details courtesy of the National Forest Products Association, Washington, DC)*

FIGURE 4.9
Four alternative details for interior girder–column intersections. Detail B-1 avoids wood shrinkage problems with an iron pintle, while the other three details bring the columns through the beams with only a steel bearing plate between the ends of the column sections. Split-ring connectors are used in the lower two details to form a strong enough connection between the bearing blocks and the columns to support the loads from the beams; it would take a much larger number of bolts to do the same job. *(Heavy timber construction details courtesy of the National Forest Products Association, Washington, DC)*

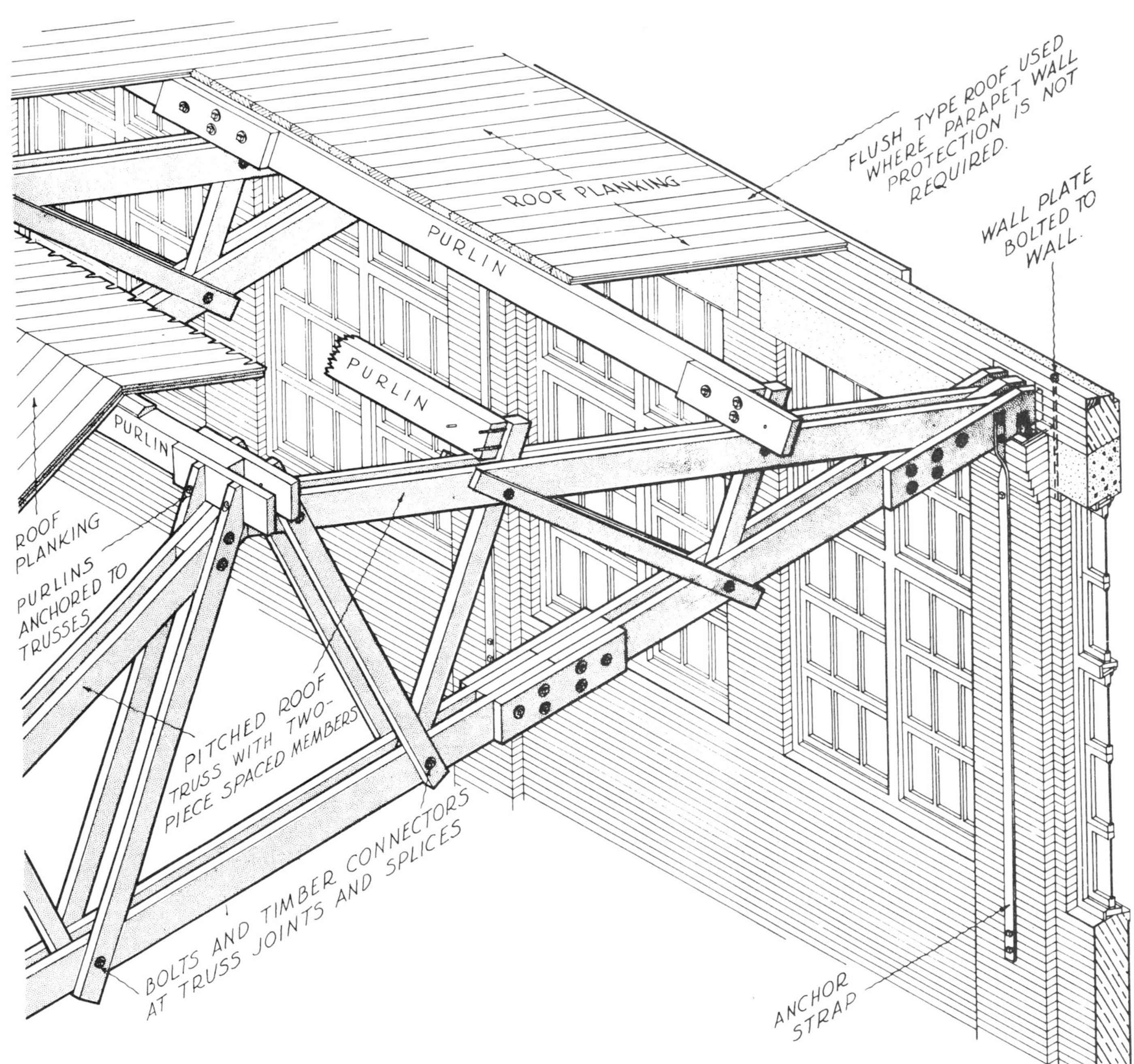

FIGURE 4.10
Heavy timber roof trusses for Mill construction. Split-ring connectors are used to transmit the large forces between the overlapping members of the truss. A long anchor strap is again used at the outside wall, as explained in the legend to Figure 4.8. *(Heavy Timber construction details courtesy of the National Forest Products Association, Washington, DC)*

pintles or steel caps carry the column loads past the cross-grain of the beams at each floor, so that the beams and girders can shrink without causing the floors and roof to sag (Figures 4.8 and 4.9). In contemporary practice, a glue-laminated column may be fabricated as a single piece running the entire height of the building, with the beams connected to the column by wood *bearing blocks* or welded metal connectors; or columns may be butted directly to one another at each floor with the aid of metal connectors (Figures 4.13, 14.14, and 4.17).

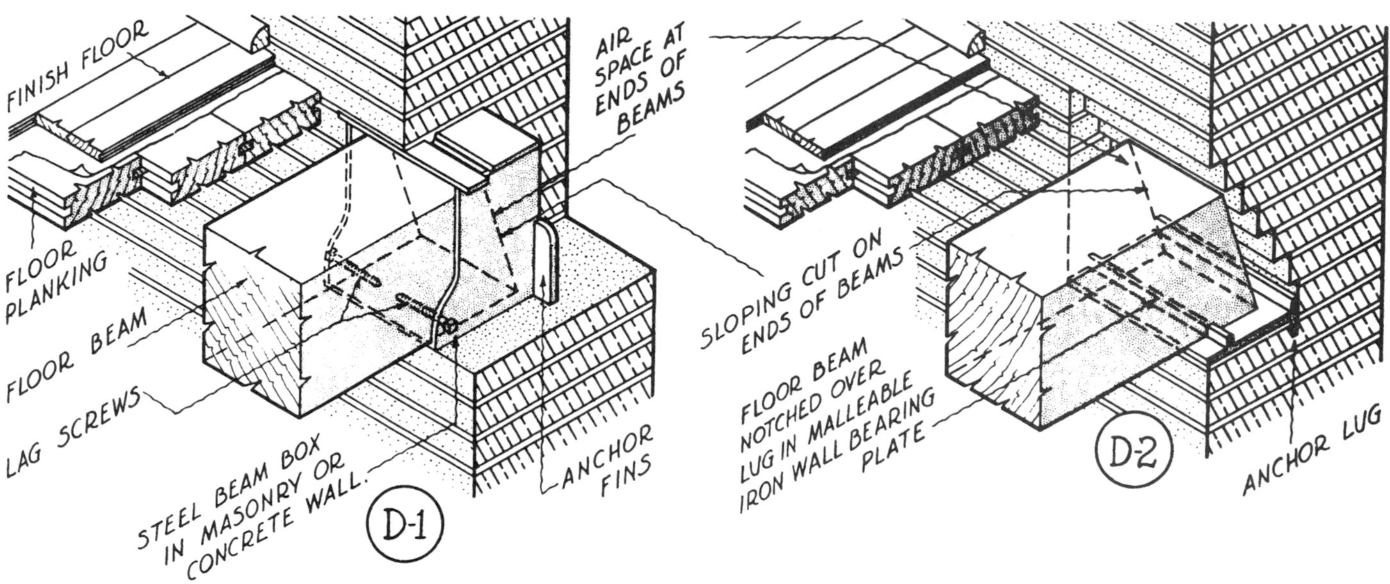

Figure 4.11
Two alternative details for the bearing of a beam on masonry in traditional Mill construction. In each case, the beam end is firecut to allow it to rotate out of the wall if it burns through (Figure 4.12), but it is also anchored against pulling away from the wall by means of either lag screws or a lug on the iron bearing plate. *(Heavy timber construction details courtesy of the National Forest Products Association, Washington, DC)*

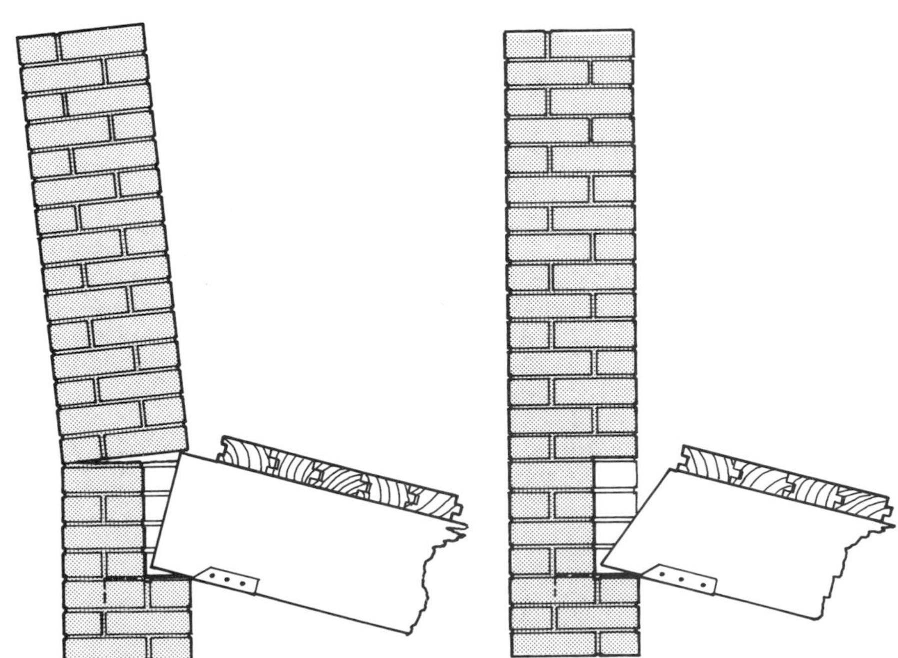

FIGURE 4.12
A timber beam that burns through in a prolonged fire is likely to topple its supporting masonry wall (left) unless its end is *firecut* (right). In this illustration, the beam is anchored to the wall in each case by a steel strap anchor.

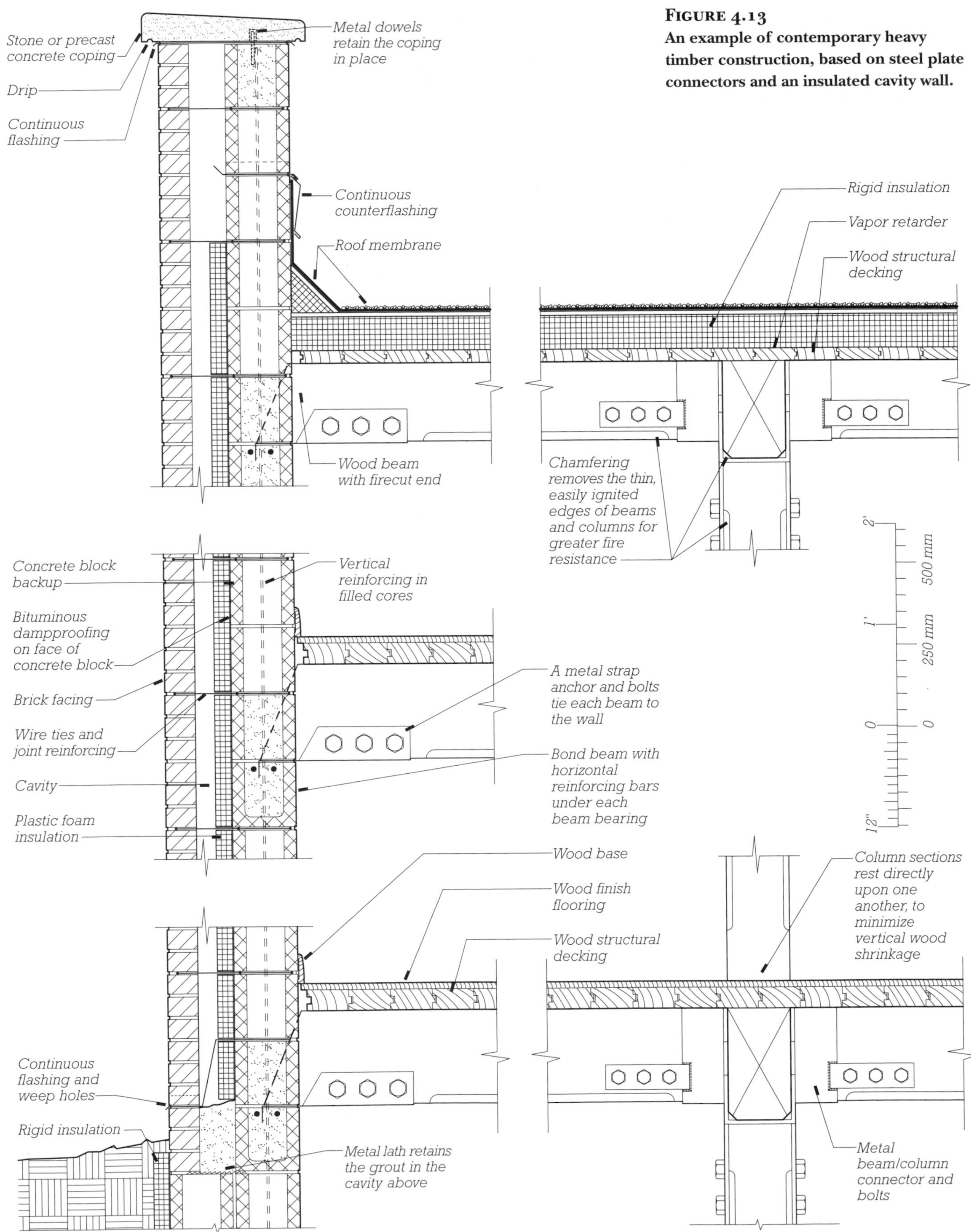

FIGURE 4.13
An example of contemporary heavy timber construction, based on steel plate connectors and an insulated cavity wall.

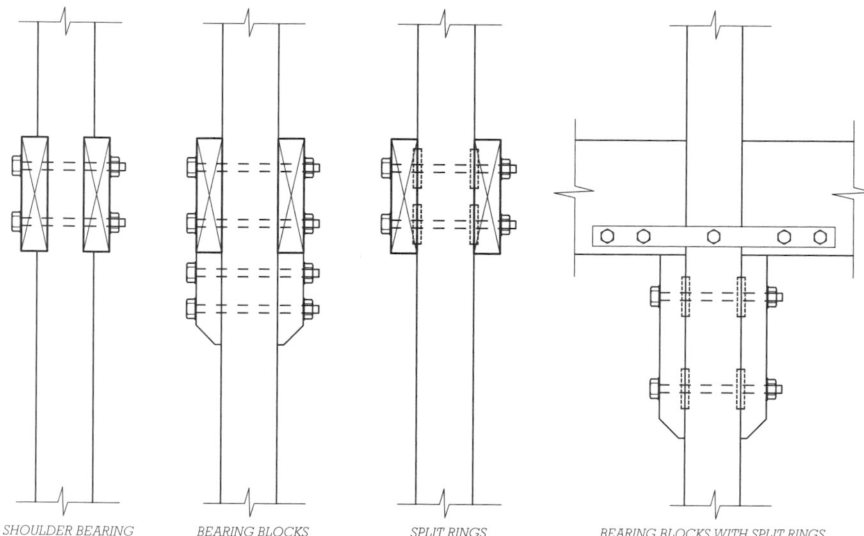

FIGURE 4.14
Some typical beam–column connections for contemporary heavy timber construction. The first three examples are for double beams sandwiched on either side of the column, and the fourth shows single beams in the same plane as the column. In the shoulder bearing connection, the beams are recessed into the column by an amount that allows the load to be transferred safely by wood-to-wood bearing; the bolts serve only to keep the beams in the recesses. Bearing blocks allow more bolts to be inserted in a connection than can fit through the beams, and each bolt in the bearing blocks can hold several times as much load as one through the beam because it acts parallel to the grain of the wood rather than perpendicular. It is generally impossible to place enough bolts in a beam–column joint to transfer the load successfully without bearing blocks unless split rings are used, as shown in the third example. The steel straps and bolts in the fourth example hold the beams on the bearing blocks.

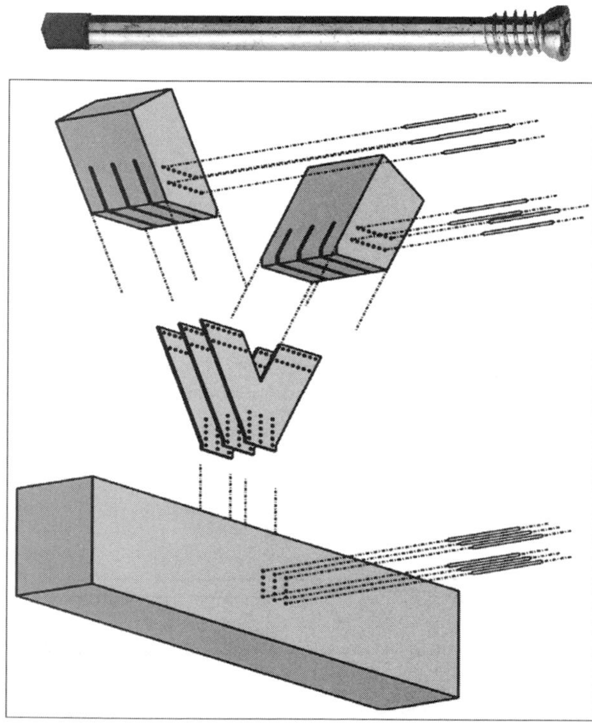

FIGURE 4.15
This contemporary heavy timber fastening system uses proprietary self-drilling steel dowels (top) in combination with embedded steel plates spanning the joint between members to create high-strength connections. The type of joint diagrammed in the lower portion of this figure can be used as a substitute for the lapped members and split-ring connectors illustrated in Figure 4.10. *(Courtesy of SFS intec, Inc., Wyomissing, PA, www.sfsintecusa.com)*

FIGURE 4.16
A completed heavy timber truss joint using the fastening system illustrated in the previous figure. *(Courtesy of SFS intec, Inc., Wyomissing, PA, www.sfsintecusa.com)*

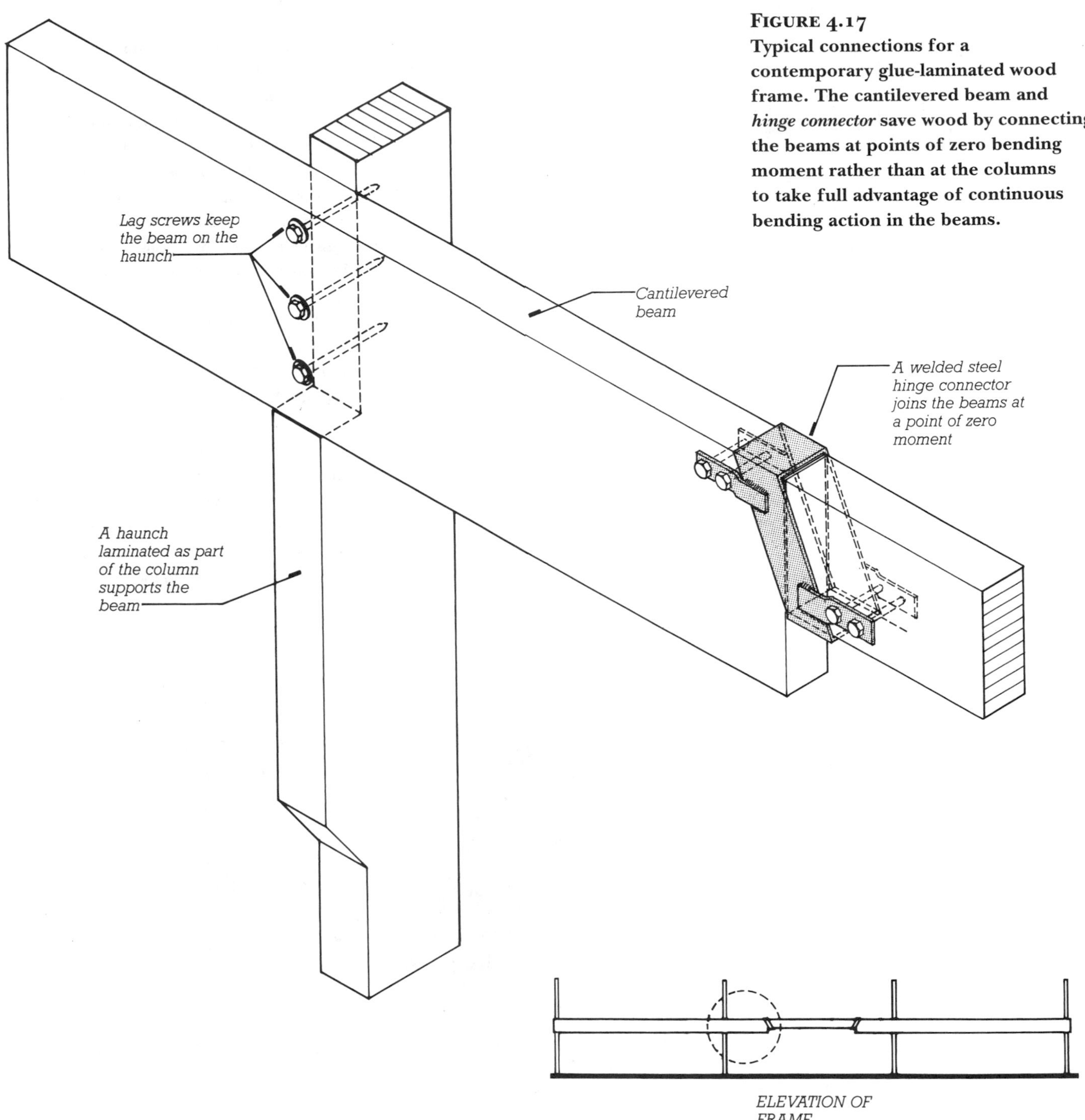
Lag screws keep the beam on the haunch
Cantilevered beam
A welded steel hinge connector joins the beams at a point of zero moment
A haunch laminated as part of the column supports the beam
ELEVATION OF FRAME

FIGURE 4.17
Typical connections for a contemporary glue-laminated wood frame. The cantilevered beam and *hinge connector* save wood by connecting the beams at points of zero bending moment rather than at the columns to take full advantage of continuous bending action in the beams.

Anchorage of Timber Beams and Masonry Walls

Where heavy timber beams join masonry or concrete walls, three problems must be solved: First, the beam must be protected from decay caused by moisture seeping through the wall. This is done by leaving a ventilating airspace of at least ½ inch (13 mm) between the masonry and all sides of the beam except the bottom, unless the beam is chemically treated to resist decay. The second and third problems have to be solved together: The beam must be securely anchored to the wall so that it cannot pull away under normal service, yet it must be able to rotate freely so that it does not pry the wall apart if it burns through and collapses during a severe fire (Figure 4.12). Traditional and more contemporary methods of accomplishing these dual needs are shown in Figures 4.11 and 4.13, respectively.

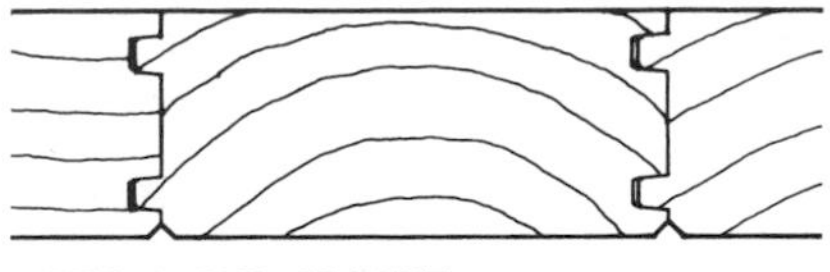

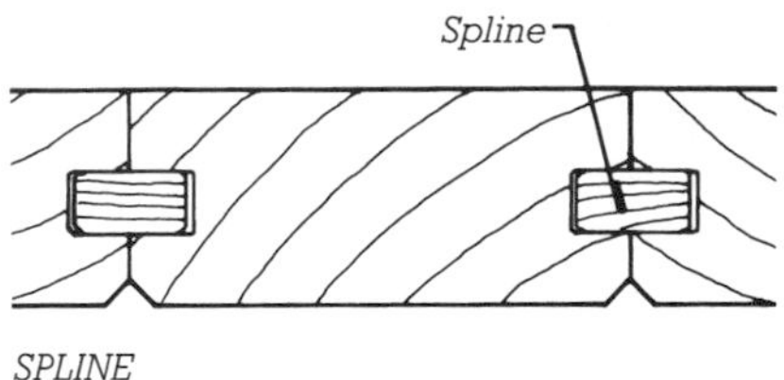

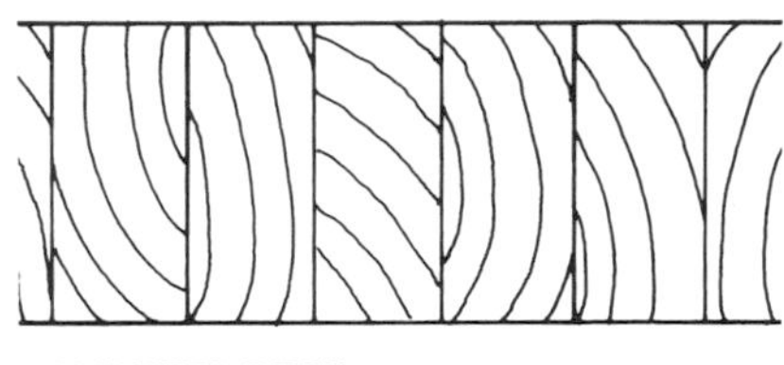

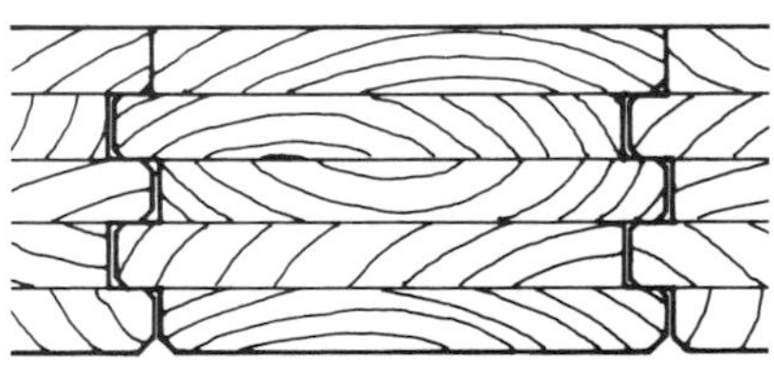

Figure 4.18
Large-scale cross sections of four types of Heavy Timber decking. *Tongue-and-groove decking* is the most common, but the other three types are slightly more economical of lumber because wood is not wasted in the milling of the tongues. *Laminated decking* is a traditional type for longer spans and heavier loads; it consists of ordinary dimension lumber laid on edge and spiked together. *Glue-laminated decking* is a modern type. In the example shown here, five separate boards are glued together to make each piece of decking. Decking of any type is usually furnished and installed in random lengths. The end joints do not necessarily line up over beams; rather, they are staggered to avoid creating zones of structural weakness. The *splines*, tongues, or nails allow the narrow strips of decking to share concentrated structural loads as if they constituted a continuous sheet of solid wood.

Floor and Roof Decks for Heavy Timber Buildings

Building codes require that Type IV Heavy Timber buildings have floors and roofs of solid wood construction without concealed cavities. Figure 4.18 shows different types of decking used for these purposes. Minimum permissible thicknesses of decking are given in Figure 4.7. To meet code requirements, floor decking must also be covered with a finish floor consisting of nominal 1-inch (19-mm) tongue-and-groove boards laid at right angles or diagonally to the structural decking. In some circumstances, ½-inch (13-mm) plywood or other composite wood panels are also permitted as the finish layer.

Combustible Buildings Framed with Heavy Timber

Heavy timber members may also be used in buildings that do not meet the building code definition of Type IV Heavy Timber construction. They may be used in combination with smaller wood framing members or in buildings with exterior walls of combustible construction. These buildings, classified under the IBC as Type V (Wood Light Frame) construction, bring the architectural qualities and structural performance of beam-and-decking framing to smaller, residential, commercial, religious, and institutional buildings. In these applications, there are no restrictions (other than structural) on the minimum size of timbers, minimum thickness of decking, or exterior wall material, and light framing of nominal 2-inch (38-mm) lumber can be incorporated as desired.

Lateral Bracing of Heavy Timber Buildings

A heavy timber frame building with an exterior masonry or concrete bearing wall is normally braced against wind and seismic forces by the shear resistance of its exterior walls, working together with the diaphragm action of its roof and floor decks. In areas of high seismic risk, the walls must be heavily reinforced both vertically and horizontally, and the decks may have to be specially nailed or overlaid with plywood to increase their shear resistance, as well as specially anchored to the perimeter walls. In buildings with framed exterior walls, diagonal bracing or shear panels must be provided. Seismic upgrades to historic heavy timber and masonry buildings often require the insertion of new steel-braced frames or reinforced concrete shear walls in order to meet contemporary lateral force resistance requirements.

Building Services in Heavy Timber Buildings

The heavy timber structure poses some special problems for the designer because the exposed framing and decking do not provide the concealed cavities that are present in light frame structures and other

FIGURE 4.19
When floors and roofs are heavy timber framed, lighting, communications wiring, ductwork, fire sprinklers, and other services are exposed to view. *(Photo by Joseph Iano)*

conventional building systems. Roof thermal insulation cannot be hidden in spaces between ceiling joists or roof rafters and, instead, must be placed on top of the roof deck. If the roof is nearly flat, it can be insulated and roofed in the manner shown for low-slope roofs in Chapter 16, or, if the roof is steeply pitched, a nailing surface for the shingles or other roofing must be added on the outside of the insulation. Electrical wiring for lighting fixtures that are mounted on the underside of the decking must either run through exposed metal conduits below the deck, which may be visually unsatisfactory, or be channeled through the insulation above the deck. Ductwork for heating and cooling must remain exposed. If the walls and partitions are made of masonry or stressed-skin panels, special arrangements are necessary for installing wiring, plumbing, and mechanical system components in the walls as well (Figure 4.19).

Longer Spans in Heavy Timber

For buildings that require spans longer than 20 to 30 feet (6 to 9 m), the maximum usually associated with framing of sawn timbers, the designer may select from among several alternative types of timber structural devices.

FIGURE 4.20
Installing tongue-and-groove roof decking over laminated beams and girders. *(Courtesy of American Institute of Timber Construction)*

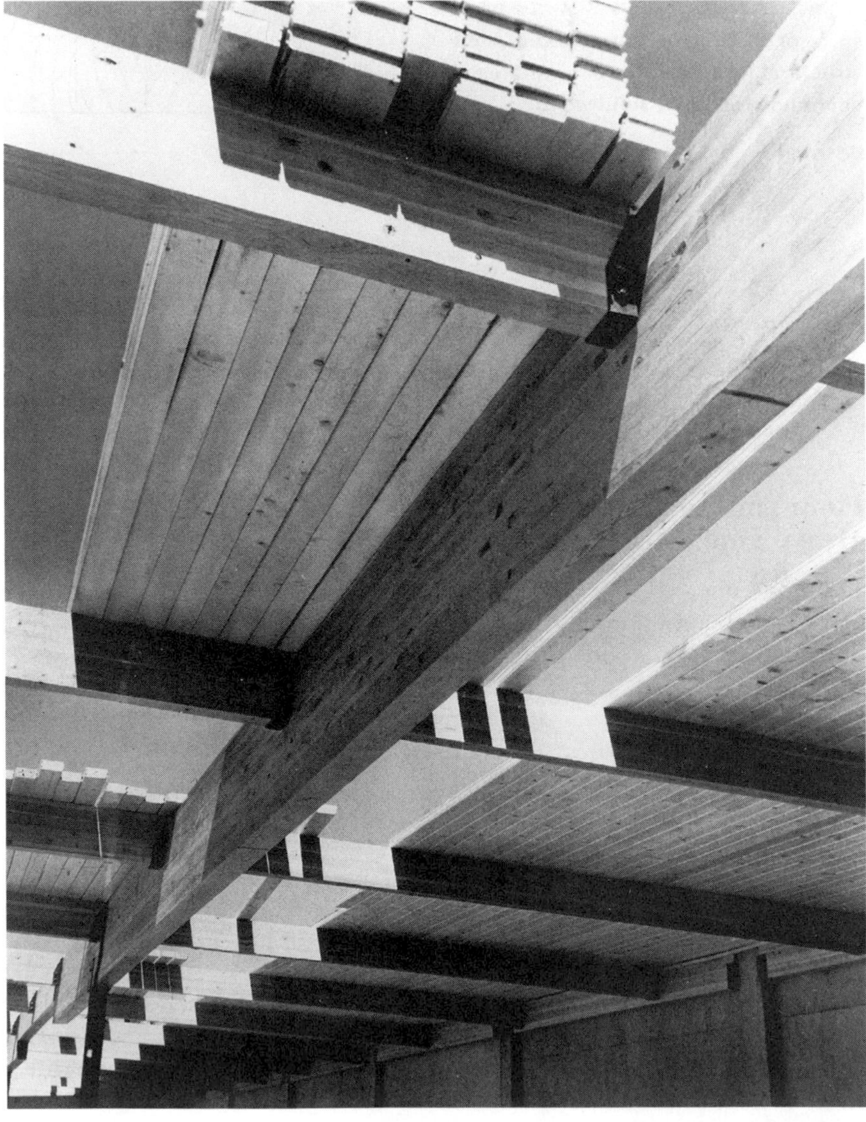

Large Beams

With large, old-growth trees no longer readily available, very large timber beams are usually built up of laminations or strands rather than sawn. Such beams are stronger and more dimensionally stable than sawn wood beams and can be made in the exact size and shape desired (Figures 4.20 and 4.21). Hybrid or FRP reinforced glulam beams, as described in Chapter 3, can be useful when especially longer span are needed or heavy loads must be supported.

Rigid Frames

The cruck (Figure 4.2), cut from a bent tree, was a form of *rigid frame* or *portal frame*. Today's rigid frames are glue-laminated to shape and find wide use in longer-span buildings. Standard configurations are readily available (Figures 4.22–4.24), or the designer may order a custom shape. Rigid frames exert a horizontal thrust, so they must be tied together at the base with *steel tension rods*, also called *tie rods*. In laminated wood construction, rigid frames are often called *arches*, acknowledging that the two structural forms act in very nearly the same manner.

Trusses

The majority of wood trusses built each year are light roof trusses of nominal 2-inch (38-mm) lumber joined by toothed plates (see Figures 3.44, 3.47, 5.65, and 5.66). For larger buildings, however, *heavy timber trusses* may be used. Their joints are made with steel bolts and

FIGURE 4.21
This roof uses glue-laminated beams to support proprietary long-span trusses made of wood with steel tube diagonals. Notice that the beams are hinged in the manner shown in Figure 4.17. The trusses, roof joists, and plywood roof deck are prefabricated into panels to reduce installation time. *(Courtesy of Trus Joist MacMillan)*

FIGURE 4.22
Three-hinged arches of glue-laminated wood carry laminated wood roof purlins. The short crosspieces of wood between the purlins are temporary ladders for workers. *(Courtesy of American Institute of Timber Construction)*

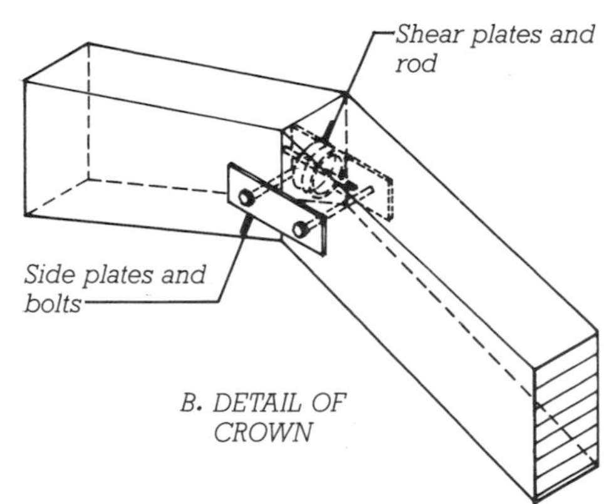

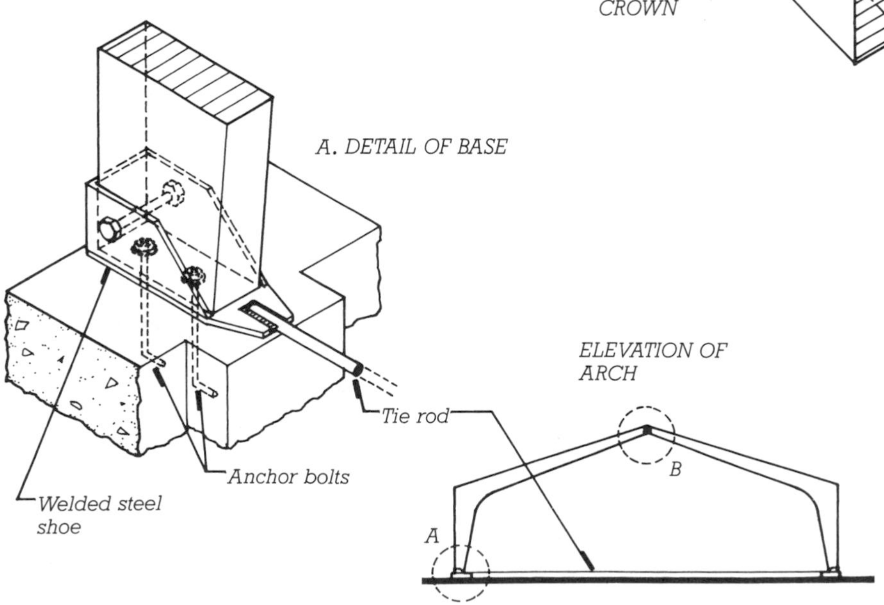

FIGURE 4.23
Typical details for three-hinged arches of laminated wood. The tie rod is later covered by the floor slab.

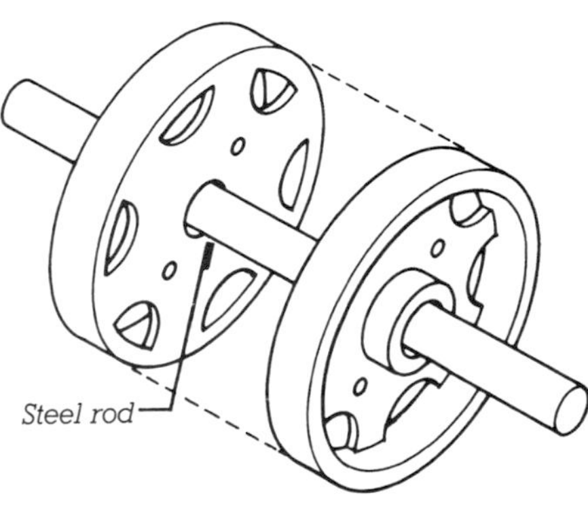

FIGURE 4.24
The *shear plates* in the crown connection of the arch, shown here in a larger-scale detail, are recessed into grooves in the wood and serve to spread any force from the steel rod across a much wider surface area of wood to avoid crushing and splitting.

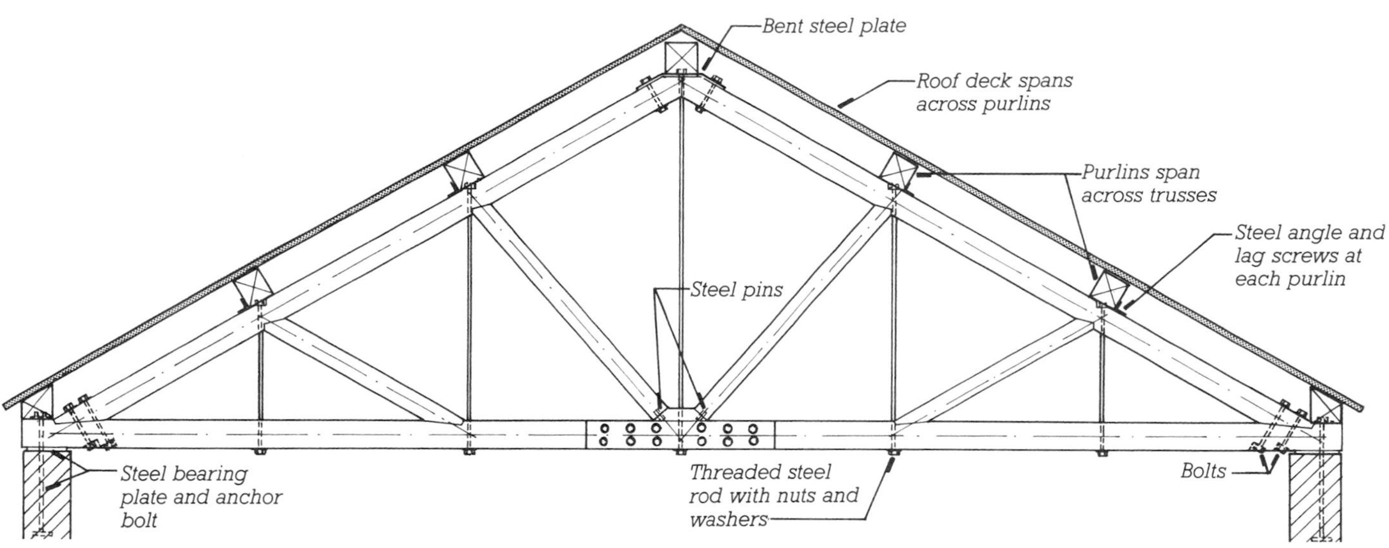

FIGURE 4.25
A heavy timber roof truss with steel rod tension members. This type of truss is easy to construct but cannot be used if predicted wind uplift forces are sufficiently strong to cause the forces in the tension members to reverse (such slender steel rods cannot resist compressive forces). The center splice in the lower chord of the truss is required only if it is impossible to obtain single pieces of lumber long enough to reach from one end of the truss to the other. Compare this mode of truss construction with that shown in Figures 4.10, 4.15, and 4.16; still another common mode is to form the truss of a single layer of heavy members connected by steel side plates and bolts, as shown in Figures 4.27 and 4.28.

FIGURE 4.26
Laminated wood roof trusses with steel connector plates. The steel rods are for lateral stability, to keep the bottom chords of the trusses from moving sideways. *(Woo & Williams, Architects. Photographer: Richard Bonarrigo)*

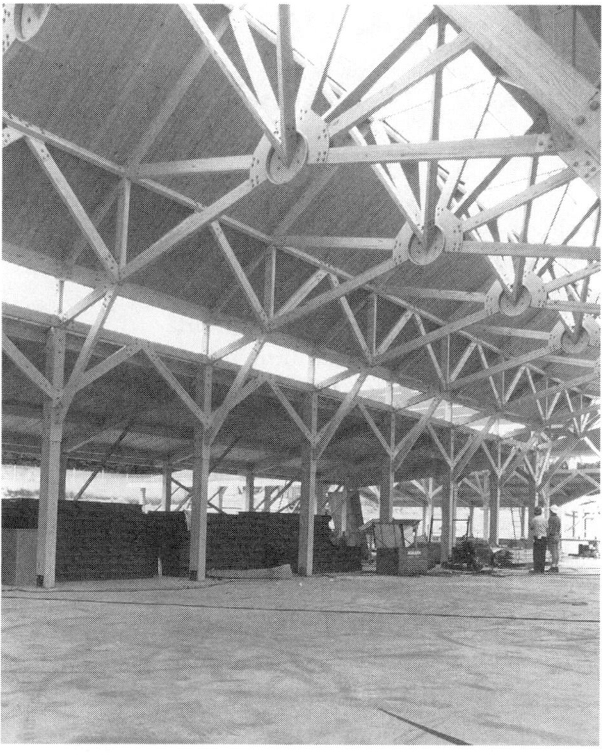

FIGURE 4.27
Roof trusses for the Montville, New Jersey, Public Library consist of a single plane of glue-laminated members joined by custom-designed steel side plates and bolts. An extensive scheme of diagonal bracing resists wind and seismic forces in both axes of the building.

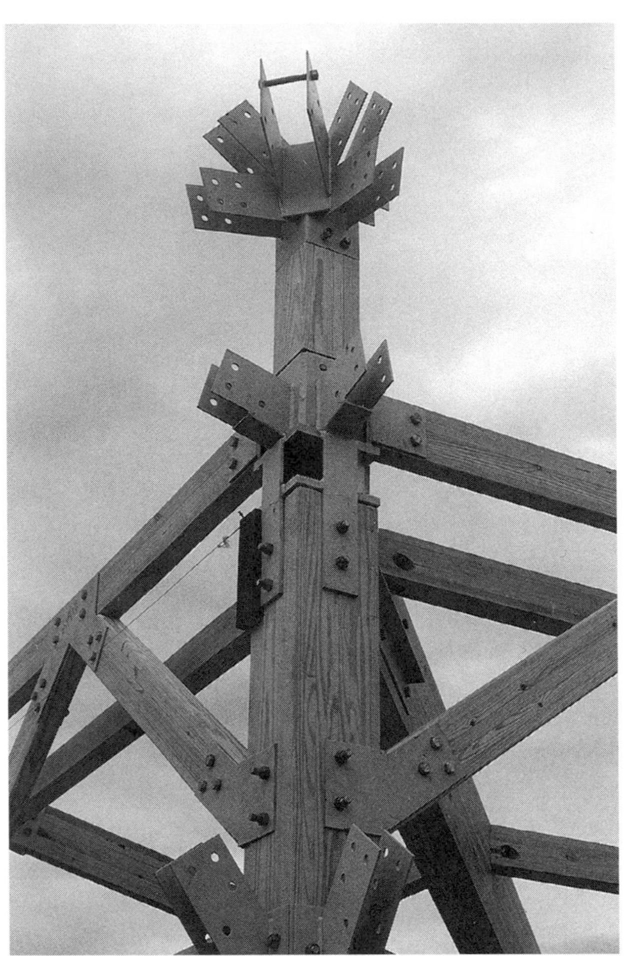

Figure 4.28
The Montville Public Library timbers are joined with custom-designed connectors that are welded together from steel plates. *(Design and photographs by Eliot Goldstein, The Goldstein Partnership, Architects)*

welded steel plate connectors or split-ring connectors. Sawn, laminated, and structural composite timbers are used, sometimes in combination with steel rod tension members. Many shapes of heavy timber truss are possible, and spans of over 100 feet (30 m) are common (Figures 4.10 and 4.25–4.28).

Arches and Domes

Long curved timbers for making *heavy timber vaults* and *heavy timber domes* are easily fabricated in laminated wood and are widely used in athletic arenas, auditoriums, suburban retail stores, warehouses, and factories (Figures 4.29–4.31). Arched structures, like rigid frames, exert lateral thrusts that must be countered by tie rods or suitably designed foundations.

Figure 4.29
Semicircular laminated wood arches support the timber roof of the Back Bay Station in Boston. Notice the use of horizontal steel tie rods to resist the thrust of the arches at the base. *(Architects: Kallmann, McKinnell & Wood. Photo © Steve Rosenthal)*

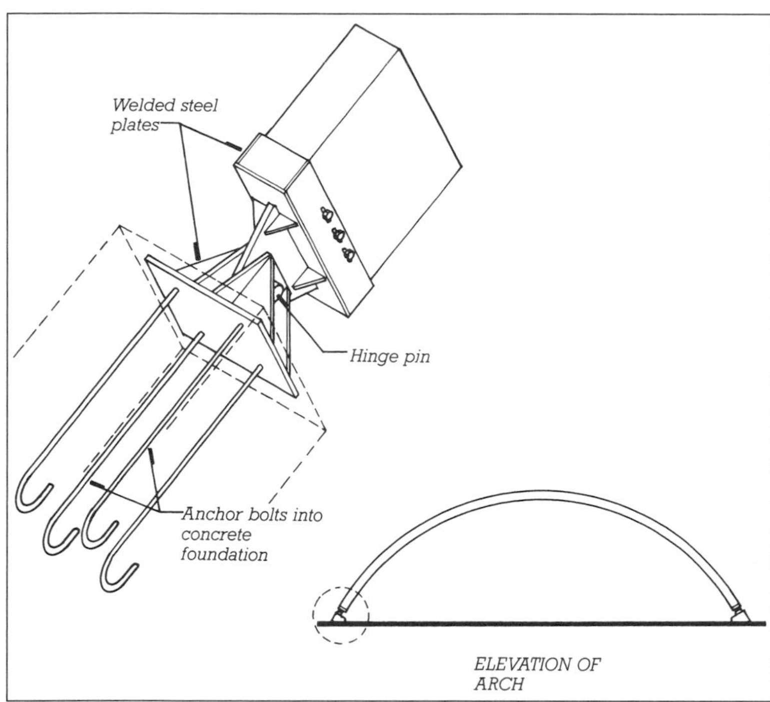

FIGURE 4.30
A typical hinged foundation connection for an arch or dome, made of welded steel plates. The hinge pin allows for rotation between the arch and the foundations, which avoids many kinds of forces that might otherwise be placed on the structure by wood shrinkage and foundation movement.

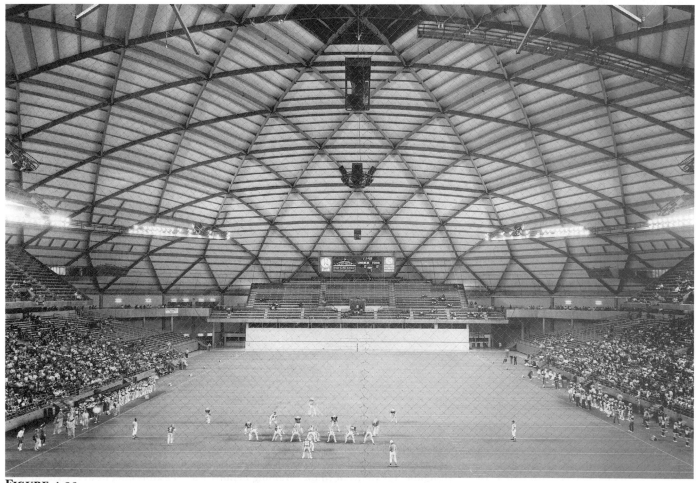

FIGURE 4.31
This laminated wood dome spans 530 feet (161.5 m) to cover a 25,000-seat stadium and convention center in Tacoma, Washington.
(Architects: McGranahan, Messenger Associates. Structural engineers: Chalker Engineers, Inc. Photo by Gary Vannest. Courtesy of American Wood Council)

For Preliminary Design of a Heavy Timber Structure

- Estimate the nominal depth of **wood roof decking** at 1/45 of its span. Estimate the depth of wood floor decking at 1/35 of its span. Standard nominal depths of wood decking are 2, 3, 4, 6, and 8 inches (actual size 38, 64, 89, 140, and 184 mm).
- Estimate the depth of **solid wood beams** at 1/15 of their span and the depth of **glue-laminated beams** at 1/20 of their span. Add a nominal 6 inches (150 mm) to these depths for girders. The width of a solid wood beam or girder is usually 1/4 to 1/2 of its depth. The width of a glue-laminated beam typically ranges from 1/7 to 1/4 of its depth.
- Estimate the depth of timber **triangular roof trusses** at 1/5 to 1/2 of their span and the depth of **bowstring trusses** at 1/2 to 2/3 of their span.
- To estimate the size of a **wood column**, add up the total roof and floor area supported by the column. A nominal 6-inch (actual size 140 mm) column can support up to about 400 square feet (37 m^2) of area, an 8-inch (actual size 184 mm) column 1000 square feet (93 m^2), a 10-inch (actual size 235 mm) column 1500 square feet (140 m^2), a 12-inch (actual size 286 mm) column 2500 square feet (230 m^2), and a 14-inch (actual size 337 mm) column 3500 square feet (325 m^2). Wood columns are usually square or nearly square in proportion.

For actual sizes of solid timbers in conventional units, see Figures 3.22 and 3.23. Standard sizes of glue-laminated timbers are given in Chapter 3. For a building that must qualify as Type IV Heavy Timber construction under the IBC, minimum timber sizes are given in Figure 4.7.

These approximations are valid only for purposes of preliminary building layout, and must not be used to select final member sizes. They apply to the normal range of building occupancies such as residential, office, commercial, and institutional buildings. For manufacturing and storage buildings, use somewhat larger members.

For more comprehensive information on preliminary selection and layout of a structural system and sizing of structural members, see Edward Allen and Joseph Iano, *The Architect's Studio Companion* (4th ed.), New York, John Wiley & Sons, Inc., 2007.

Heavy Timber and the Building Codes

The table in Figure 1.2 shows the range of height and area limits for buildings of different occupancies built of Heavy Timber (Type IV) construction and of Light Frame (Type V) construction (applicable when heavy timber framing and light frame construction are mixed). Notice that when considering any particular occupancy, the allowable heights and areas for Heavy Timber construction are comparable to those for 1-Hour-Protected Noncombustible (Type IIA) construction (steel, concrete, or masonry) and are greater than those for protected and unprotected Type V construction or unprotected steel (Type IIB). These limits are an explicit recognition of the inherent fire-resistive qualities of heavy timber framing. (Recall also that allowable floor areas in this table can be increased by installing an automatic fire suppression sprinkler system in the building, as outlined in Chapter 1 of this book.)

Uniqueness of Heavy Timber Framing

Heavy timber cannot span as far or with such delicacy as steel, and it

Figure 4.32
Architects Greene and Greene of Pasadena, California, were known for their carefully wrought expressions of timber framing in houses such as this one, built for David B. Gamble in 1909. *(Photo by Wayne Andrews)*

cannot mimic the structural continuity or smooth shell forms of concrete, yet many people respond more positively to the idea of a timber building than they do to one of steel or concrete. To some degree, this response may stem from the color, grain figure, and warmer feel of wood. In part, it may also come from the pleasant associations that people have with the sturdy, satisfying houses our ancestors erected from handhewn timbers only a few generations ago. Most people today live in dwellings where none of the framing is exposed. Exposed ceiling beams in one's house or apartment are considered an amenity, and shopping or restaurant dining in a converted mill building of heavy timber and masonry construction is generally thought to be a pleasant experience. There is something in all of us that derives satisfaction from seeing wood beams at work.

In economic reality, heavy timber must compete successfully on the basis of price with other construction materials, and as our forests have diminished in quality and shipping costs have risen, timber is no longer an automatic choice for building a mill or any other type of structure. For many buildings, however, heavy timber is an economical alternative to steel and concrete, particularly in situations where the appearance and feel of its large wooden members will be highly valued by those who use it, in regions close to commercial forests, or where code provisions or fire insurance premiums create a financial incentive.

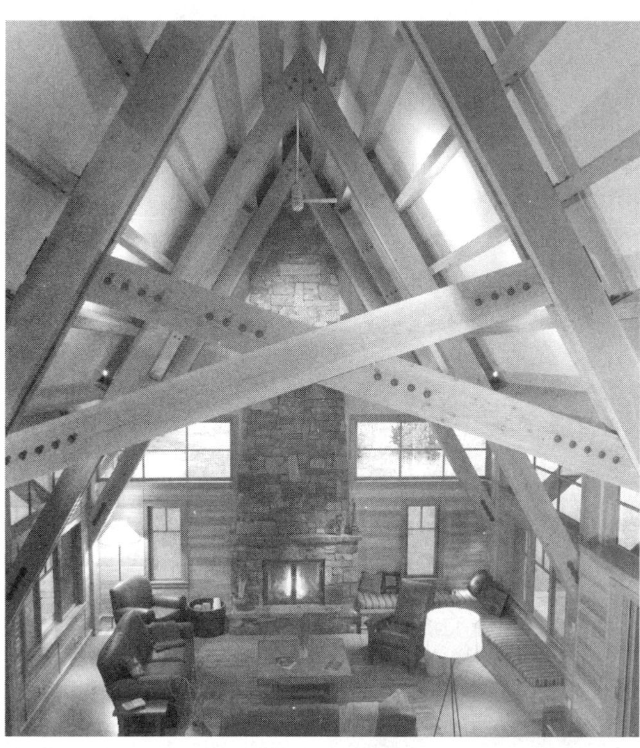

Figure 4.33
Dramatic exposed timber scissor trusses support a steeply sloping roof structure in this vacation cabin designed by Nils Finne in the North Cascade Mountains of Washington State. *(Architect: FINNE Architects www.FINNE.com; photograph by Art Grice)*

Figure 4.34
Each of the attached dwellings in Sea Ranch Condominium #1 at Sea Ranch, in northern California, built in 1965, is framed with a simple cage of unplaned timbers sawn from trees taken from another portion of the site. The diagonal members are wind braces. *(Architects: Moore, Lyndon, Turnbull, and Whitaker. Photo by Edward Allen)*

(*a*)

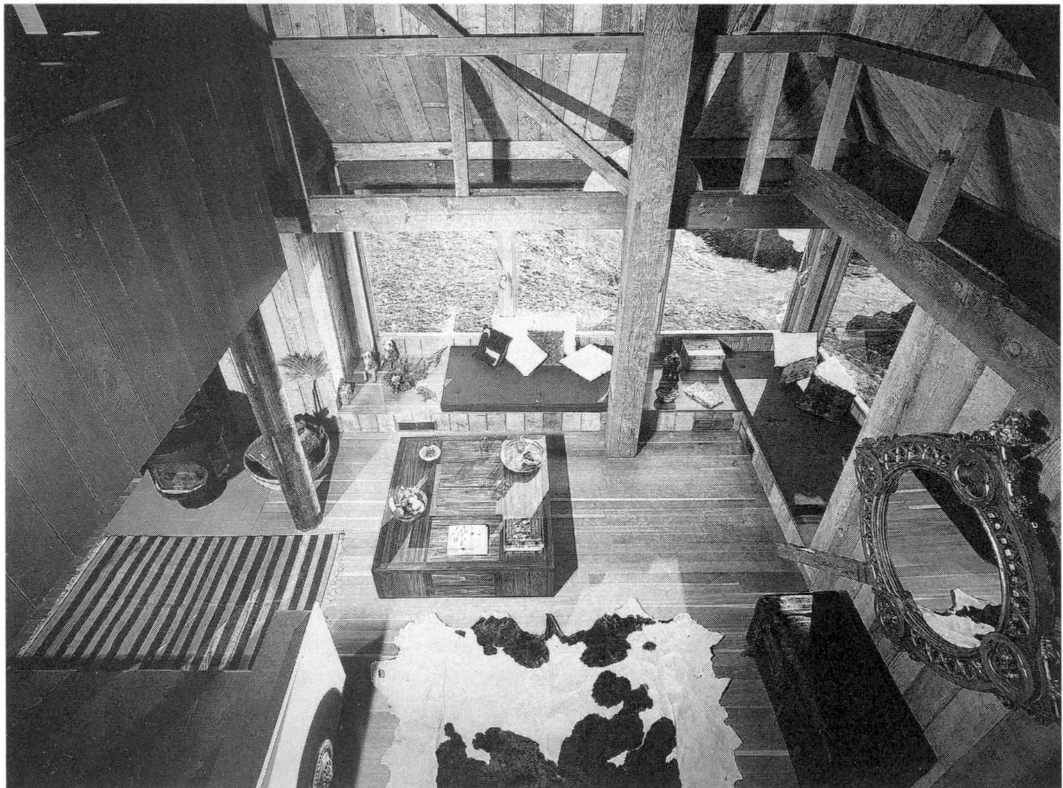
(*b*)

FIGURE 4.35
(*a*) The exterior of Sea Ranch Condominium #1, an interior view of which is shown in the opening photograph of this chapter, is sheathed with vertical 2-inch (38-mm) unplaned tongue-and-groove decking and clad with ¾-inch (19-mm) tongue-and-groove redwood siding. (*b*) Another view of the exposed timbers and connectors inside a dwelling in this building. The flooring is vertical-grain Douglas fir. *(Photos © Morley Baer)*

CSI/CSC

MasterFormat Sections for Heavy Timber Frame Construction

06 10 00	ROUGH CARPENTRY
06 13 00	Heavy Timber Heavy Timber Construction
06 15 00	Wood Decking Timber Decking Laminated Wood Decking
06 18 00	Glue-Laminated Construction

Selected References

1. American Institute of Timber Construction. *Timber Construction Manual* (5th ed.). New York, John Wiley & Sons, Inc., 2004.

This is a comprehensive design handbook for timber structures, including detailed engineering procedures as well as general information on wood and its fasteners.

2. American Institute of Timber Construction. *Typical Construction Details.* AITC 104–2003. Englewood, CA, 2003.

This 32-page reference contains dozens of examples of how to detail connections in heavy timber buildings. Especially instructive are the bad examples that are presented as lessons in what to avoid. This reference can also be viewed for free on the American Institute of Timber Construction's web site, www.aitc-glulam.org.

3. Benson, Tedd. *Timber-Frame Home: Design, Construction and Finishing.* Middletown, CT, Taunton Press, 1997.

The author is one of the most experienced traditional timber framers in the world. This is a comprehensive, richly illustrated, authoritative guide to timber house construction.

4. Goldstein, Eliot E. *Timber Construction for Architects and Builders.* New York, McGraw-Hill, 1998.

This thoroughly practical book is based on the author's firsthand experience in designing timber structures, as well as recognized timber engineering practice.

Web Sites

Heavy Timber Frame Construction

Author's supplementary website: **www.ianosbackfill.com/04_heavy_timber_frame_construction**

American Institute of Timber Construction: **www.aitc-glulam.org**

American Wood Council: **www.awc.org**

Canadian Wood Council: **www.cwc.ca**

Timber Framers Guild: **www.tfguild.org**

Key Terms and Concepts

halftimbering
wattle and daub
cruck
bent
purlin
Type IV Heavy Timber (HT) construction
Mill construction, Slow-burning construction
chamfered
iron dog
pintle
bearing block
hinge connector
firecut
tongue-and-groove decking
laminated decking
glue-laminated decking
spline
rigid frame, portal frame
steel tension rod, tie rod
arch
shear plate
heavy timber truss
heavy timber vault
heavy timber dome

Review Questions

1. Why does Heavy Timber construction receive relatively favorable treatment from building codes and insurance companies?

2. What are the important factors in detailing the junction of a wood beam with a masonry loadbearing wall? Draw several ways of making this joint.

3. Draw from memory one or two typical details for the intersection of a wood column with a floor of a building of Heavy Timber construction.

Exercises

1. Determine from the code table in Figure 1.2 whether a building you are currently designing could be built of Heavy Timber construction and what modifications you might make in your design so that it will conform to the requirements of Heavy Timber construction.

2. Find a barn or mill that was constructed in the 18th or 19th century and sketch some typical connection details. How is the structure stabilized against horizontal wind forces?

3. Obtain a book on traditional Japanese construction from the library and compare Japanese timber joint details with 18th- or 19th-century American practice.

5

Wood Light Frame Construction

A New England school of arts and crafts is housed in a cluster of small buildings of wood light frame construction that cling to a dramatic mountainside site overlooking the ocean. *(Architect: Edward Larrabee Barnes. Photo by Joseph W. Molitor)*

Wood light frame construction is the most versatile of all building systems. There is scarcely a shape it cannot be used to construct, from a plain rectilinear box to cylindrical towers to complex foldings of sloping roofs with dormers of every description. During the century and a half since it first came into use, wood light framing has served to construct buildings in styles ranging from reinterpretations of nearly all the historical fashions to uncompromising expressions of every contemporary architectural philosophy. It has assimilated without difficulty during this same period a bewildering and unforeseen succession of technical improvements in building: central heating, air conditioning, gas lighting, electricity, thermal insulation, indoor plumbing, prefabricated components, and electronic communications cabling.

Light frame buildings are easily and swiftly constructed with a minimal investment in tools. Many observers of the building industry have criticized the supposed inefficiency of light frame construction, which is carried out largely by hand methods on the building site, yet it has successfully fought off competition from industrialized building systems of every sort, partly by incorporating their best features, to remain the least expensive form of durable construction. It is the common currency of small residential and commercial buildings in North America today.

Wood light frame construction has its deficiencies: If ignited, it burns rapidly; if exposed to dampness, it decays. It expands and contracts by significant amounts in response to changes in humidity, sometimes causing chronic difficulties with cracking plaster, sticking doors, and buckling floors. The framing itself is so unattractive to the eye that it is seldom left exposed in a building. These problems can be controlled, however, by clever design and careful workmanship, and there is no arguing with success: Frames made by the monotonous repetition of wooden joists, studs, and rafters are likely to remain the number one system of building in North America for a long time to come.

FIGURE 5.1
Carpenters apply plywood roof sheathing to a platform-framed apartment building. The ground floor is a concrete slab on grade. The edge of the wooden platform of the upper floor is clearly visible between the stud walls of the ground and upper floors. Most of the diagonal bracing is temporary, but permanent let-in diagonal braces occur between the two openings at the lower left and immediately above in the rear building. The openings have been framed incorrectly, without supporting studs for the headers. *(Courtesy of Southern Forest Products Association)*

History

Wood light frame construction was the first uniquely American building system. It was developed in the first half of the 19th century when builders recognized that the closely spaced vertical members used to infill the walls of a heavy timber building frame were themselves sufficiently strong that the heavy posts of the frame could be eliminated. Its development was accelerated by two technological breakthroughs of the period: Boards and small framing members of wood had recently become inexpensive for the first time in history because of the advent of the water-powered sawmill; additionally, machine-made nails had become remarkably cheap compared to the hand-forged nails that preceded them.

The *balloon frame* was the earliest wood framing system to be constructed exclusively of slender, closely spaced wooden members: *joists* for the floors, *studs* for the walls, *rafters* for the sloping roofs. Heavy posts and beams were completely eliminated and, with them, the difficult, expensive mortise-and-tenon joinery they required. There was no structural member in a frame that could not be handled easily by a single carpenter, and each of the hundreds of joints was made with lightning rapidity with two or three nails. The impact of this new building system was revolutionary: In 1865, G. E. Woodward could write that "A man and a boy can now attain the same results, with ease, that twenty men could on an old-fashioned frame . . . the Balloon Frame can be put up for forty percent less money than the mortise and tenon frame."

The balloon frame (Figure 5.2) used full-length studs that ran continuously for two stories from foundation to roof. In time, it became apparent that these were too long to erect efficiently. Furthermore, the tall, hollow spaces between studs acted as

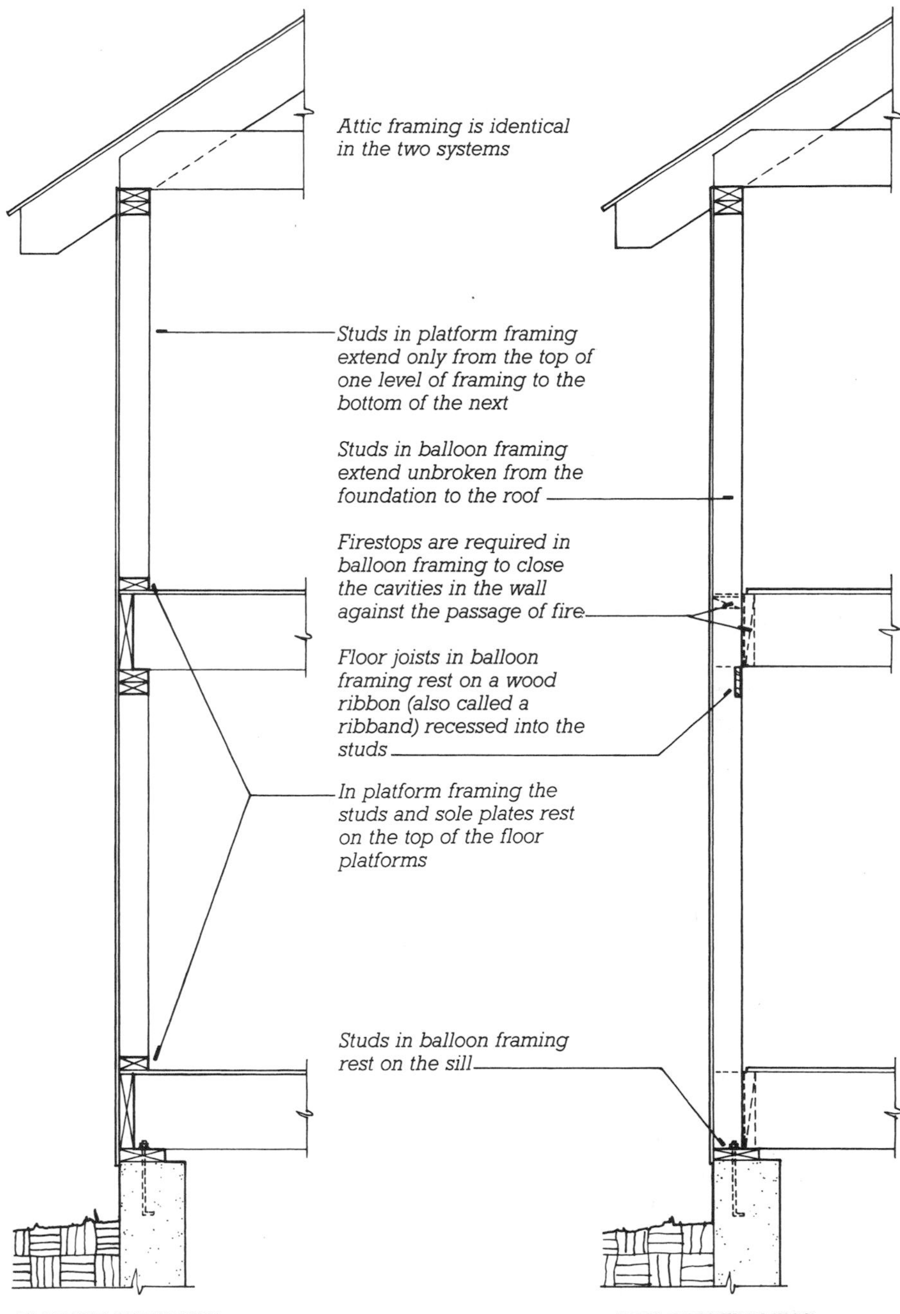
Attic framing is identical in the two systems
Studs in platform framing extend only from the top of one level of framing to the bottom of the next
Studs in balloon framing extend unbroken from the foundation to the roof
Firestops are required in balloon framing to close the cavities in the wall against the passage of fire
Floor joists in balloon framing rest on a wood ribbon (also called a ribband) recessed into the studs
In platform framing the studs and sole plates rest on the top of the floor platforms
Studs in balloon framing rest on the sill
PLATFORM FRAMING
BALLOON FRAMING

Figure 5.2
Comparative framing details for platform framing (*left*) and balloon framing (*right*). Platform framing is much easier to erect and is the only light framing system used today. However, a platform frame settles considerably as the wood dries and shrinks. If nominal 12-inch (300-mm) joists are used to frame the floors in these examples, the total amount of loadbearing cross-grain wood between the foundation and the attic joists is 33 inches (838 mm) for the platform frame and only 4½ inches (114 mm) for the balloon frame.

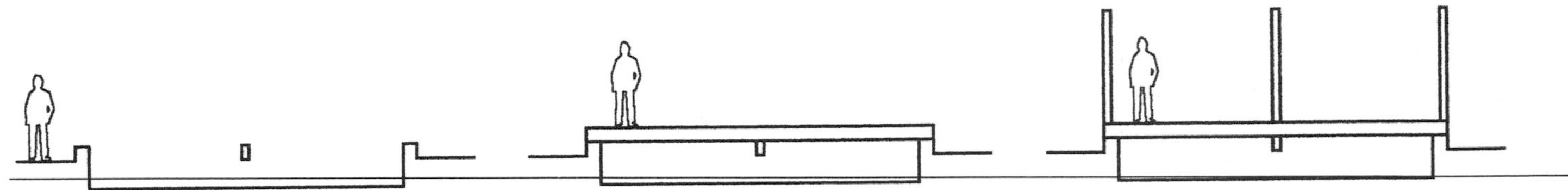

multiple chimneys in a fire, spreading the blaze rapidly to the upper floors, unless they were closed off with wood or brick *firestops* at each floor line. Several modified versions of the balloon frame were subsequently developed in an attempt to overcome these difficulties. The most recent of these, the *platform frame,* is now the universal standard.

Platform Frame

Although complex in its details, the platform frame is simple in concept (Figure 5.3). A floor platform is built. Loadbearing walls are erected upon it. A second-floor platform is built upon these walls and a second set of walls upon this platform. The attic and roof are then built upon the second set of walls. There are, of course, many variations: A concrete slab that lies directly on the ground is sometimes substituted for the ground-floor platform; a building may be one or three stories tall instead of two; and several types of roofs are frequently built that do not incorporate attics. The essentials, however, remain: A floor platform is completed at each level, and the walls bear upon the platform rather than directly upon the walls of the story below.

The advantages of the platform frame over the balloon frame are several: It uses short, easily handled lengths of lumber for the wall framing. Its vertical hollow spaces are automatically firestopped by the platform framing at each floor. Its platforms are convenient working surfaces for the carpenters who build the frame. The major disadvantage of the platform frame is that each platform constitutes a thick layer of wood whose grain runs horizontally. This leads inevitably to a relatively large amount of vertical shrinkage in the frame as excess moisture dries from the wood, which can cause distress in the exterior and interior finish surfaces.

A conventional platform frame is made entirely of nominal 2-inch members, which are actually 1½ inches (38 mm) in thickness. These are ordered and delivered cut to the nearest 2-foot (600-mm) length, then measured and sawn to exact length on the building site. All connections are made with nails, using either face nailing, end nailing, or toe nailing (Figure 3.39) as required by the characteristics of each joint. Nails are driven either by hammer or nail gun. In either case, the connection is quickly made because the nails are installed without drilling holes or otherwise preparing the joint.

Each plane of structure in a platform frame is made by aligning a number of pieces of framing lumber parallel to one another at specified intervals, nailing these to crosspieces at either end to maintain their spacing and flatness, and then covering the plane of framing with *sheathing,* a facing layer of boards or panels that join and stabilize the pieces into a single structural unit, ready for the application of finish materials inside and out (Figure 5.4). In a floor structure, the parallel pieces are the floor joists, and the crosspieces at the ends of the joists are called *headers, rim joists,* or *band joists.* The sheathing on a floor is known as the *subfloor.* In a wall structure, the parallel pieces are the *studs,* the crosspiece at the bottom of the wall is the *sole plate* or *bottom plate,* and the crosspiece at the top (which is doubled for strength if the wall bears a load from above) is called the *top plate.* In a sloping roof, the rafters are headed off by the top plates at the lower edge of the roof and by the *ridge board* at the peak.

Openings are required in all these planes of structure: for windows and doors in the walls; for stairs and chimneys in the floors; and for chimneys, skylights, and dormers in the roofs. In each case, these are made by heading off the opening: Openings in floors are framed with headers and *trimmers* (Figure 5.17), which usually must be doubled to support the higher loads placed on them by the presence of the opening. In walls, *sills* head off the bottoms of window and door openings, while trimmer studs on the sides provide support to

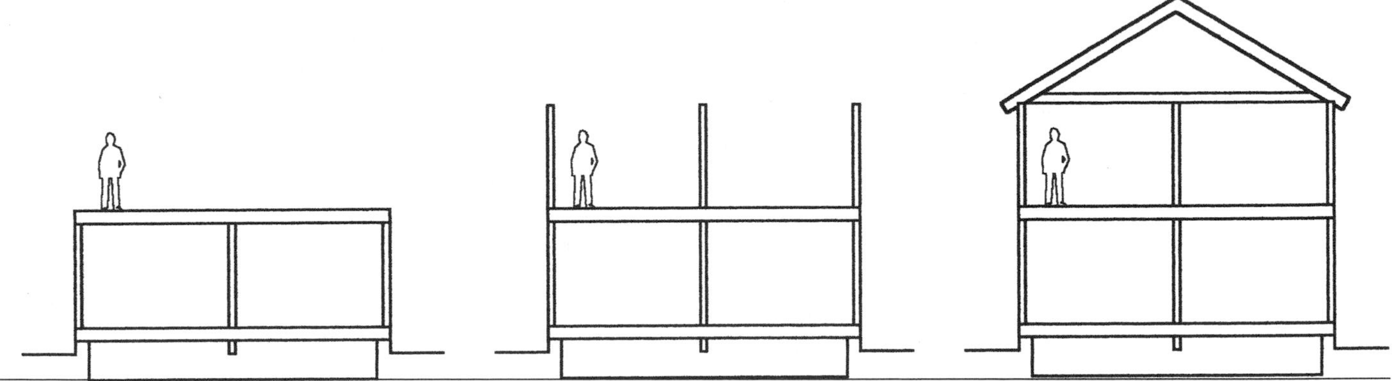

FIGURE 5.3
The concept of platform framing, shown in cross section, reading from left to right: A foundation wall is constructed. A ground-floor platform is framed and sheathed. Ground-floor wall frames are assembled horizontally on the platform, then tilted into their final positions. A second-floor platform is constructed on top of the ground-floor walls, and the process of wall construction is repeated. The attic floor and roof are added.

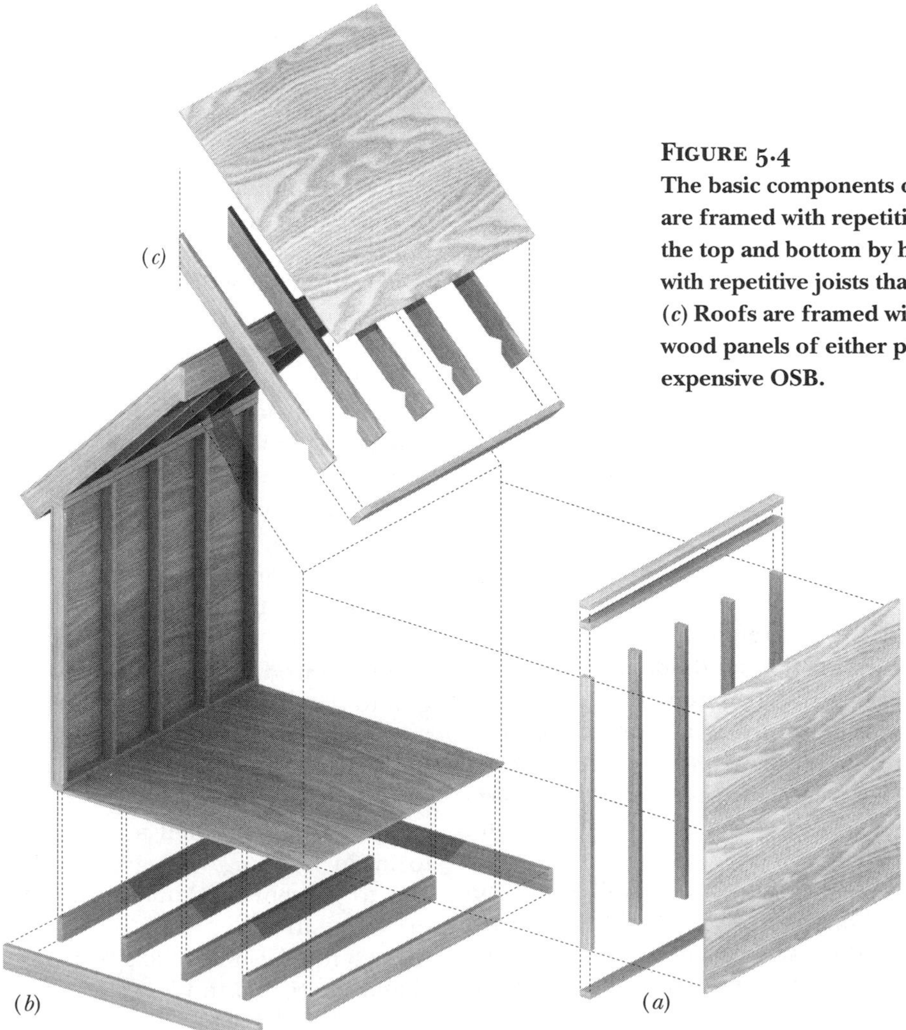

FIGURE 5.4
The basic components of platform frame construction. (*a*) Walls are framed with repetitive vertical studs that are connected at the top and bottom by horizontal plates. (*b*) Floors are framed with repetitive joists that are connected at their ends by headers. (*c*) Roofs are framed with rafters. Every surface is sheathed with wood panels of either plywood (as shown) or, more commonly, less expensive OSB.

Considerations of Sustainability in Wood Light Frame Construction

In addition to the issues of sustainability of wood production and use that were raised in Chapter 3, there are issues that pertain especially to wood light frame construction:

- A wood light frame building can be designed to minimize waste in several ways. It can be dimensioned to utilize full sheets and lengths of wood products. Most small buildings can be framed with studs 24 inches (610 mm) o.c. rather than 16 inches (406 mm). A stud can be eliminated at each corner by using small, inexpensive metal clips to support the interior wall finish materials. If joists and rafters are aligned directly over studs, the top plate can be a single member rather than a double one. Floor joists can be spliced at points of inflection rather than over girders; this reduces bending moments and allows use of smaller joists. Roof trusses often use less wood than conventional rafters and ceiling joists.
- Laminated strand lumber and rim joists, wood I-joists, laminated veneer lumber beams and headers, glue-laminated girders, parallel strand lumber girders, and OSB sheathing are all materials that utilize trees more efficiently than solid lumber. Finger-jointed studs made up of short lengths of scrap lumber glued together may replace solid full-length wood studs.
- Framing carpenters can waste less lumber by saving cutoffs and reusing them rather than throwing them automatically on the scrap heap. In some localities, scrap lumber can be recycled by shredding it for use in OSB production. The burning of construction scrap should be discouraged because of the air pollution it generates.
- Although the thermal efficiency of wood light frame construction is inherently high, it can be improved substantially by various means, as shown in Figures 7.17–7.21. Wood framing is much less conductive of heat than light-gauge steel framing. Steel framing of exterior walls is not a satisfactory substitute for wood framing unless the heat flow path through the steel framing members can be broken with a substantial thickness of insulating foam.

loadbearing headers across the tops (Figure 5.32).

Sheathing, a layer of wood panels nailed over the outside face of the framing, is a key component of platform framing. The end nails that connect the plates to the studs have little holding power against uplift of the roof by wind, but the sheathing connects the frame into a single strong unit from foundation to roof. The rectilinear geometry of the parallel framing members has no useful resistance to *wracking* by lateral forces such as wind, but rigid sheathing panels brace the building effectively against these forces. Sheathing also furnishes a surface to which shingles, boards, and flooring are nailed for finish surfaces. In buildings constructed without sheathing, or with sheathing materials that are too weak to brace the frame, such as insulating plastic foam, diagonal bracing must be applied to the wall framing to impart lateral stability.

Foundations for Light Frame Structures

Foundations for light framing, originally made of stone or brick, are now made in most cases of sitecast concrete or concrete block masonry (Figures 5.5–5.11). These materials are highly conductive of heat and usually must be insulated to meet code requirements for energy conservation (Figures 5.8, 5.9, 5.12, and 5.13). Where concrete and masonry construction methods are not practical, such as in extremely cold regions, foundations may be constructed entirely of preservative-treated wood (Figures 5.14 and 5.15). Such *permanent wood foundations* can be constructed in any weather by the same crew of carpenters that will frame the building; they are readily insulated in the same manner as the frame of the house they support; and they easily accommodate the installation of electrical wiring, plumbing, and interior finish materials. Insulating concrete formwork (ICF) foundation systems using permanent insulating forms are easy to construct, eliminate the need for formwork removal, and provide integral insulation (Figure 14.12). Proprietary precast concrete foundation systems relying on factory-fabricated reinforced concrete panels are rapidly erected on site and can be used to construct basement structures of consistent strength and quality. They may be manufactured with insulation integral to the precast panel or designed to readily accept insulation added in the field.

All basements also need to be carefully dampproofed and drained to avoid flooding with groundwater and to prevent the buildup of water pressure in the surrounding soil that could cause the walls to cave in (Figures 5.6, 5.7, and 5.11).

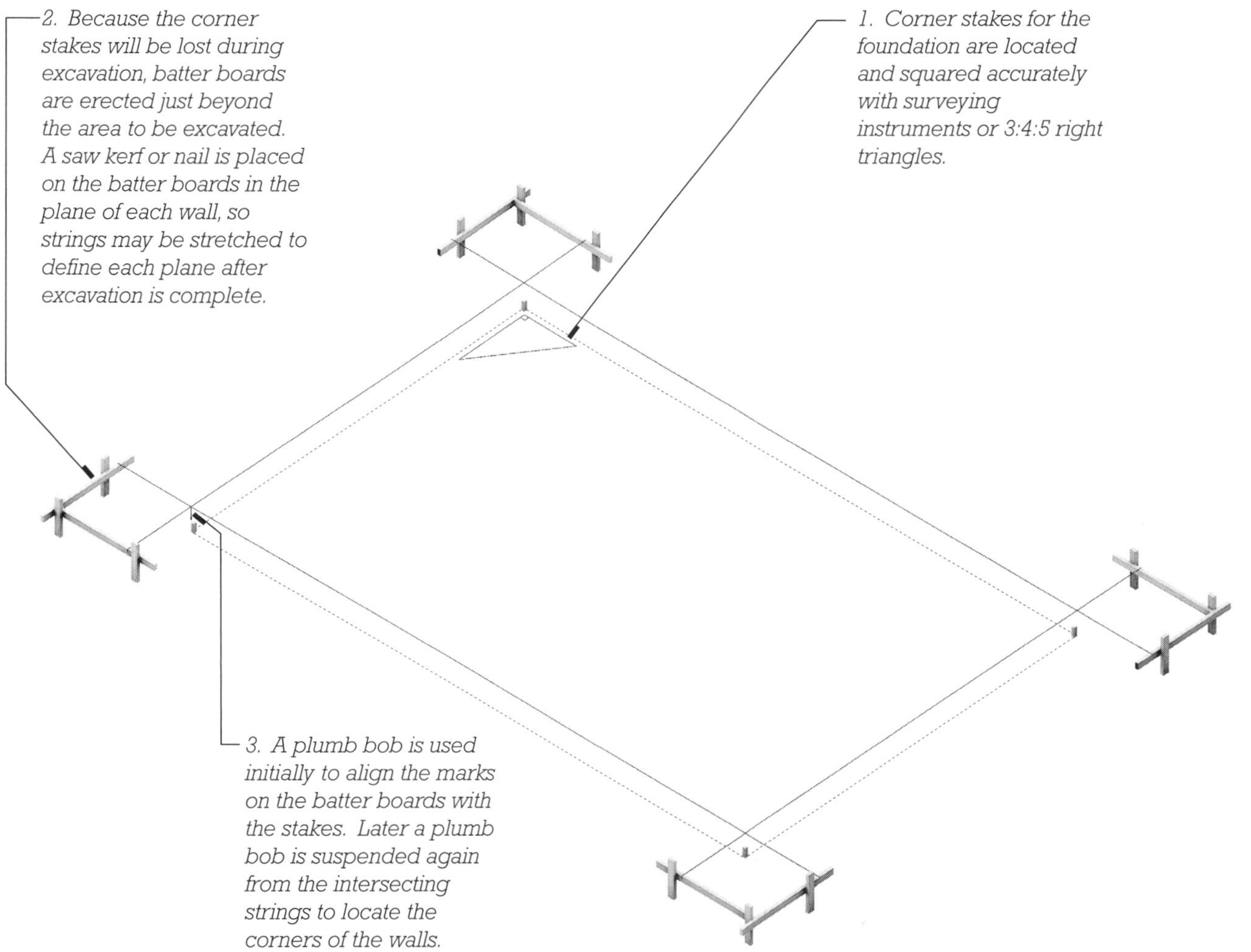

The construction of a platform frame building begins with the driving of stakes to fix its position on the site, and the placing of batter boards as reference marks for the builder.
2. Because the corner stakes will be lost during excavation, batter boards are erected just beyond the area to be excavated. A saw kerf or nail is placed on the batter boards in the plane of each wall, so strings may be stretched to define each plane after excavation is complete.
1. Corner stakes for the foundation are located and squared accurately with surveying instruments or 3:4:5 right triangles.
3. A plumb bob is used initially to align the marks on the batter boards with the stakes. Later a plumb bob is suspended again from the intersecting strings to locate the corners of the walls.
1

FIGURE 5.5
Step One in the construction of a simple platform frame building: establishing the position, shape, and size of the building on the site. After the corners of the foundation have been staked, *batter boards* are erected safely beyond the limits of the excavation area. String lines crossing over the corner stakes are stretched from the batter boards, and their positions on the boards are preserved with notches or nails in the boards. Later, after excavation, the strings are stretched again, and the foundation corners can be relocated at the bottom of the excavation. This drawing begins a series of isometric drawings that will follow the erection of a wood light frame building step by step throughout the course of this chapter.

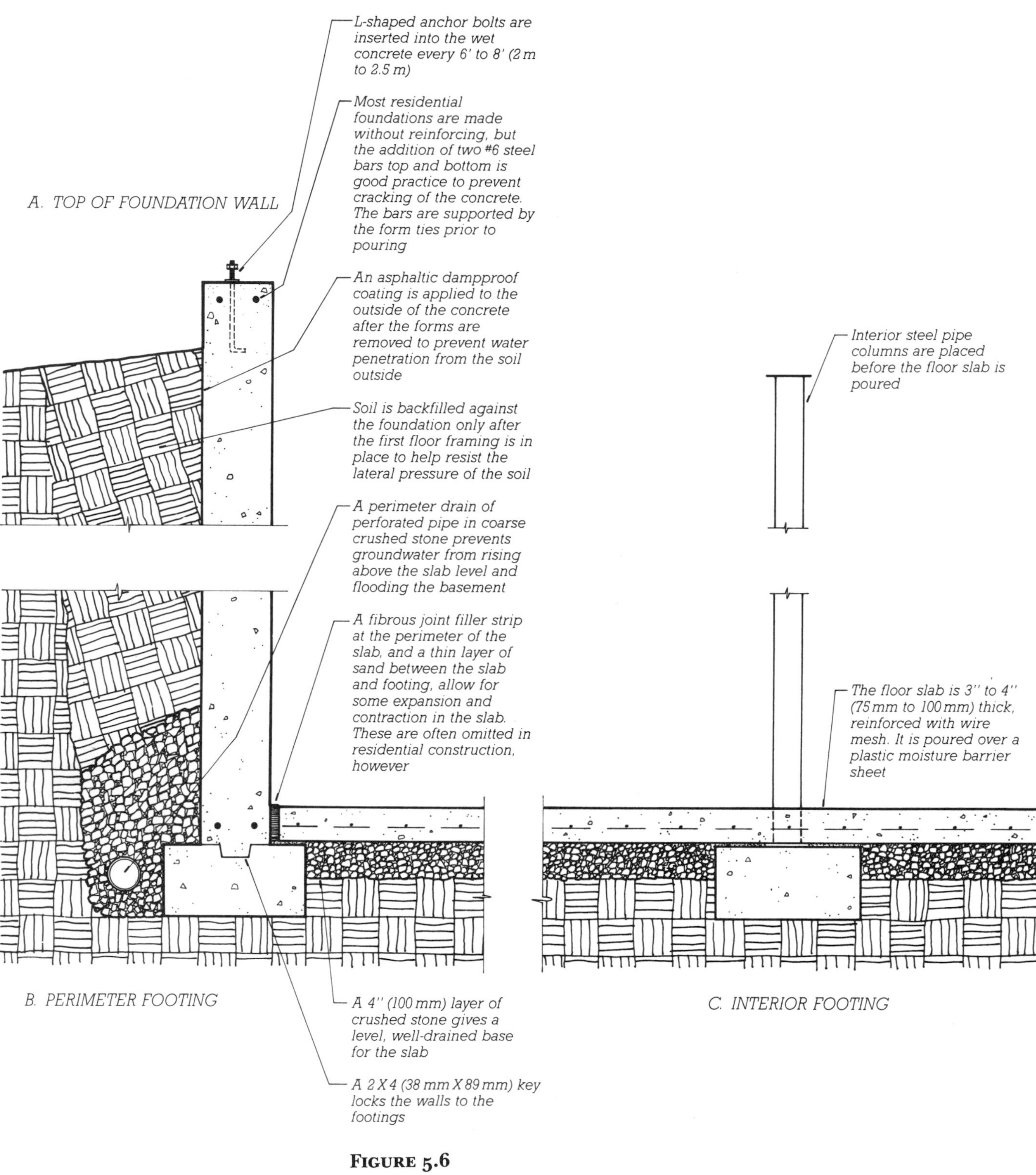

L-shaped anchor bolts are inserted into the wet concrete every 6' to 8' (2 m to 2.5 m)
Most residential foundations are made without reinforcing, but the addition of two #6 steel bars top and bottom is good practice to prevent cracking of the concrete. The bars are supported by the form ties prior to pouring
A. TOP OF FOUNDATION WALL
An asphaltic dampproof coating is applied to the outside of the concrete after the forms are removed to prevent water penetration from the soil outside
Interior steel pipe columns are placed before the floor slab is poured
Soil is backfilled against the foundation only after the first floor framing is in place to help resist the lateral pressure of the soil
A perimeter drain of perforated pipe in coarse crushed stone prevents groundwater from rising above the slab level and flooding the basement
A fibrous joint filler strip at the perimeter of the slab, and a thin layer of sand between the slab and footing, allow for some expansion and contraction in the slab. These are often omitted in residential construction, however
The floor slab is 3" to 4" (75 mm to 100 mm) thick, reinforced with wire mesh. It is poured over a plastic moisture barrier sheet
B. PERIMETER FOOTING
C. INTERIOR FOOTING
A 4" (100 mm) layer of crushed stone gives a level, well-drained base for the slab
A 2 X 4 (38 mm X 89 mm) key locks the walls to the footings

FIGURE 5.6
Typical details for a sitecast concrete foundation and basement for a platform frame building. Details *A*, *B*, and *C* are keyed to the circled portions of the drawing in Figure 5.7.

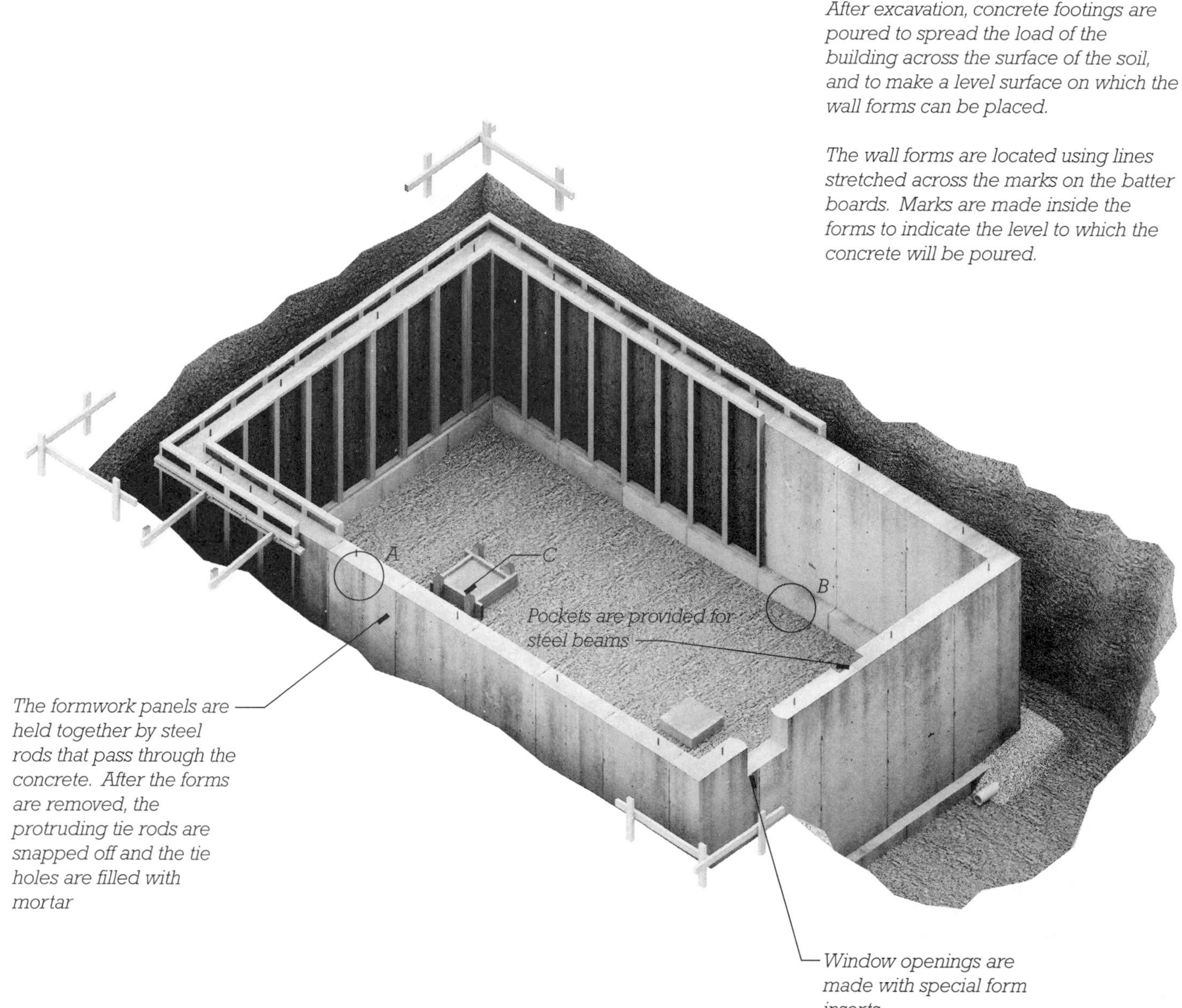

After excavation, concrete footings are poured to spread the load of the building across the surface of the soil, and to make a level surface on which the wall forms can be placed.
The wall forms are located using lines stretched across the marks on the batter boards. Marks are made inside the forms to indicate the level to which the concrete will be poured.
A
C
B
Pockets are provided for steel beams
The formwork panels are held together by steel rods that pass through the concrete. After the forms are removed, the protruding tie rods are snapped off and the tie holes are filled with mortar
Window openings are made with special form inserts

2

FIGURE 5.7
Step Two in the construction of a typical platform frame building: excavation and foundations. The letters *A*, *B*, and *C* indicate portions of the foundation that are detailed in Figure 5.6.

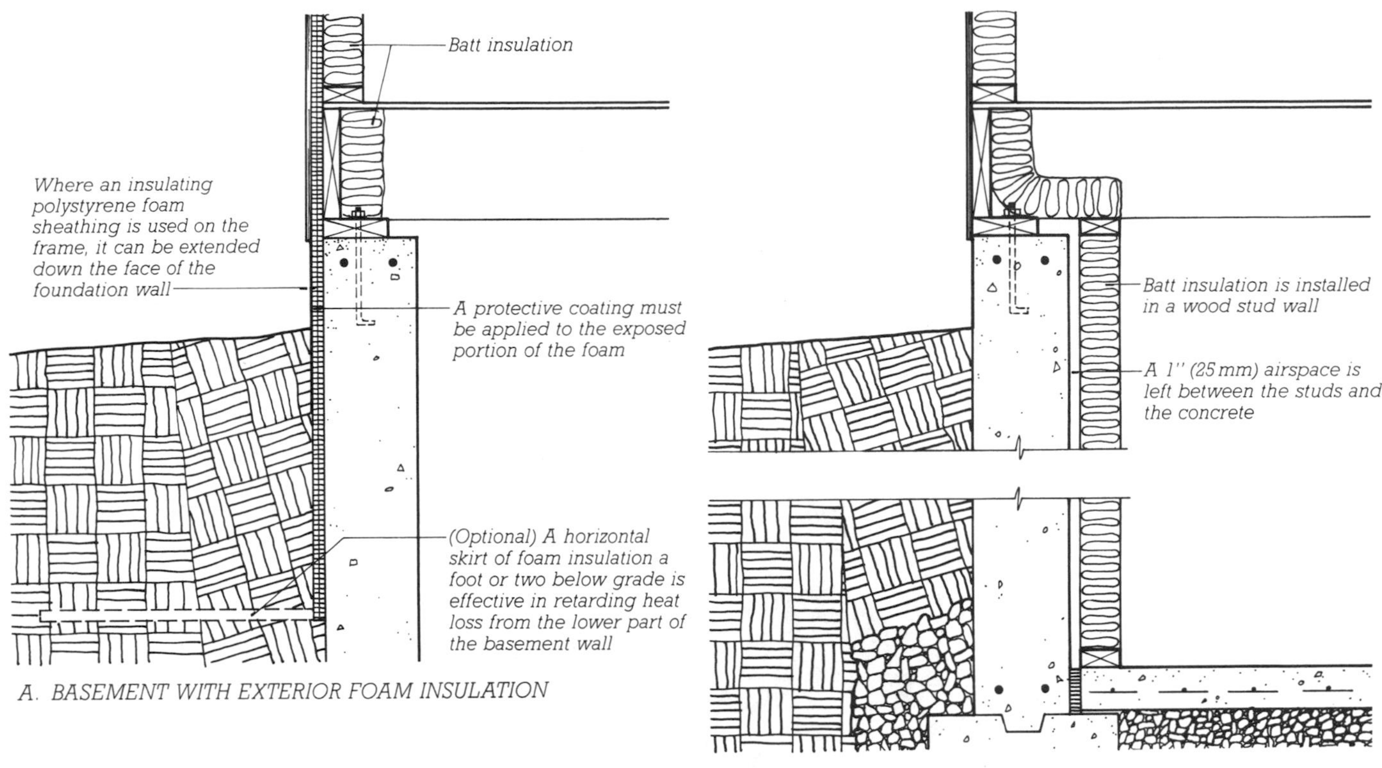
Batt insulation
Where an insulating polystyrene foam sheathing is used on the frame, it can be extended down the face of the foundation wall
A protective coating must be applied to the exposed portion of the foam
(Optional) A horizontal skirt of foam insulation a foot or two below grade is effective in retarding heat loss from the lower part of the basement wall
A. BASEMENT WITH EXTERIOR FOAM INSULATION
Batt insulation is installed in a wood stud wall
A 1" (25 mm) airspace is left between the studs and the concrete
B. BASEMENT WITH INTERIOR BATT INSULATION

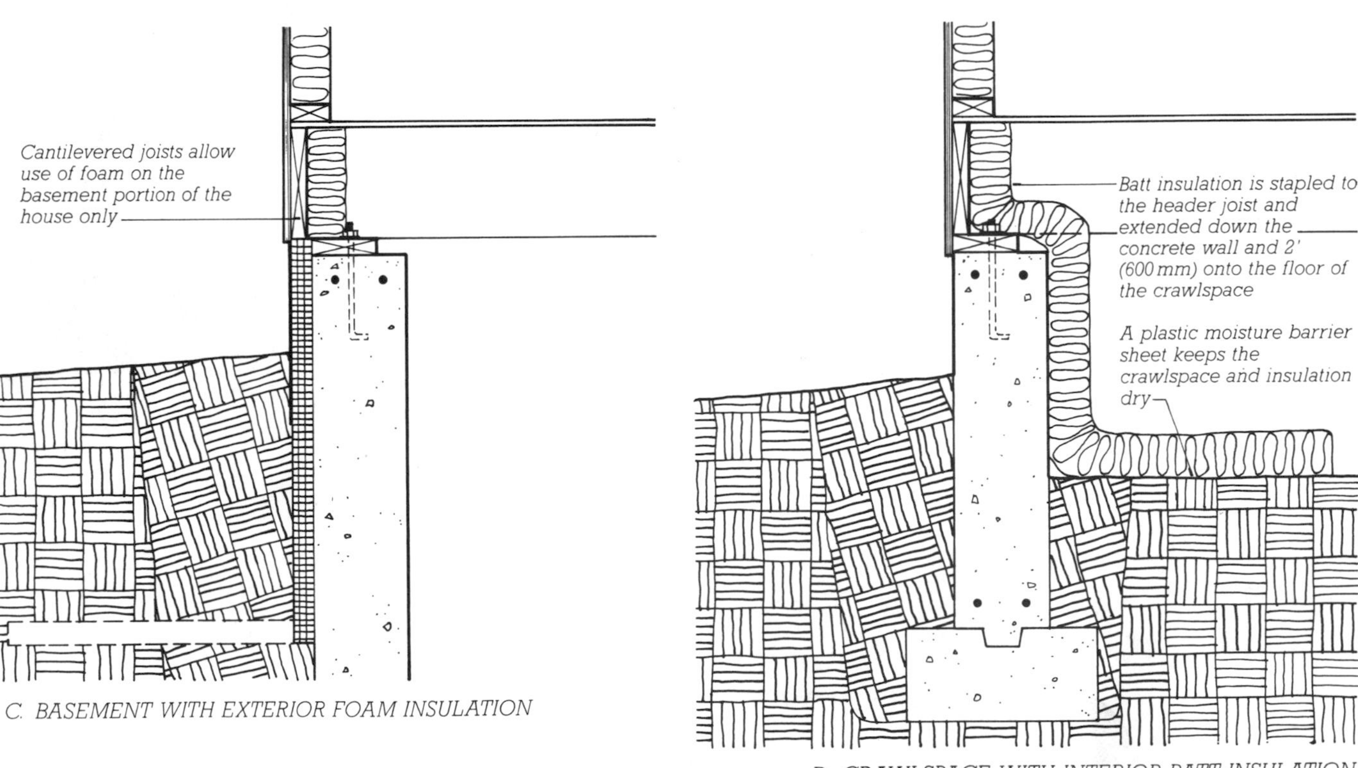
Cantilevered joists allow use of foam on the basement portion of the house only
C. BASEMENT WITH EXTERIOR FOAM INSULATION
Batt insulation is stapled to the header joist and extended down the concrete wall and 2' (600 mm) onto the floor of the crawlspace
A plastic moisture barrier sheet keeps the crawlspace and insulation dry
D. CRAWLSPACE WITH INTERIOR BATT INSULATION

FIGURE 5.8
Most building codes require thermal insulation of the foundation. These are three different ways of adding insulation to a cast-in-place concrete or concrete masonry basement and one way of insulating a crawlspace foundation. The crawlspace may alternatively be insulated on the outside with panels of plastic foam. The interior batt insulation shown in *B* is commonly used but raises questions about how to avoid possible problems arising from moisture accumulating between the insulation and the wall.

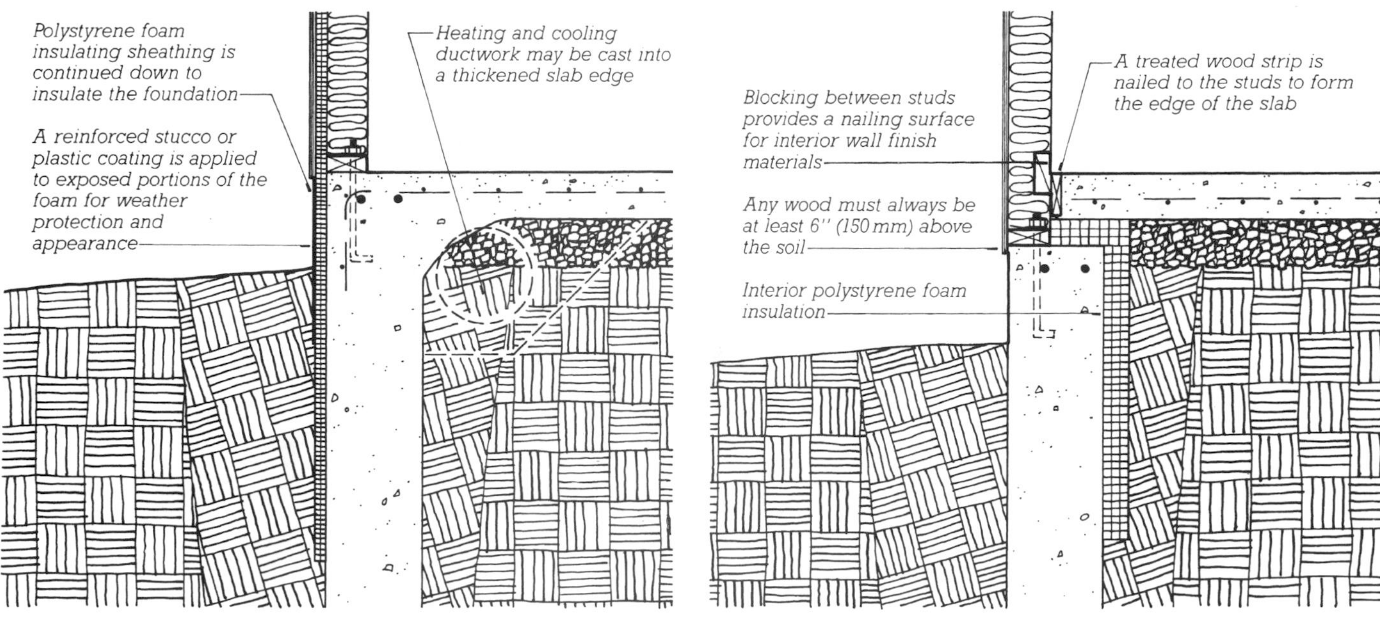
Polystyrene foam insulating sheathing is continued down to insulate the foundation
A reinforced stucco or plastic coating is applied to exposed portions of the foam for weather protection and appearance
Heating and cooling ductwork may be cast into a thickened slab edge
Blocking between studs provides a nailing surface for interior wall finish materials
Any wood must always be at least 6" (150 mm) above the soil
Interior polystyrene foam insulation
A treated wood strip is nailed to the studs to form the edge of the slab

A. SLAB ON GRADE WITH EXTERIOR INSULATION

B. SLAB ON GRADE WITH INTERIOR INSULATION

2 X 6 (38 mm X 140 mm) studs give a 3½" (89 mm) bearing on the slab plus a 2" (51 mm) projection to cover the edge of the foam insulation

2" (51 mm) foam insulation with protective coating

The slab is thickened to at least 12" (300 mm) under bearing partitions and posts

The width of the thickened slab is determined by the loadbearing capacity of the soil and the magnitude of the load

C. SLAB ON GRADE WITH EXTERIOR INSULATION

D. INTERIOR FOOTING FOR SLAB ON GRADE

FIGURE 5.9

Some typical concrete slab-on-grade details with thermal insulation. In areas where termites are common, there must be a metal flashing (*termite shield*) similar to that shown in Figure 5.18 *C* or other interruption that passes through the foam insulation above ground level to prevent termites from tunneling undetected through the insulation to reach the wood structure. This applies to the foam-insulated basement details in Figure 5.8 as well.

Figure 5.10
Erecting formwork for a sitecast concrete foundation wall. The footing has already been cast and its formwork removed. It is visible in front of the worker at the left. *(Photo by Joseph Iano)*

Figure 5.11
Masons construct a foundation of concrete masonry. The first coat of parging, which is portland cement plaster applied to help dampproof the foundation, has already been applied to the outside of the wall, and the drainage layer of crushed stone has been backfilled in place. The projecting pilaster in the center of the wall will support a beam under the center of the main floor. After a second coat of parging, the outside of the foundation will be coated with an asphaltic dampproofing compound. *(Reprinted with permission of the Portland Cement Association from* Design and Control of Concrete Mixtures, *12th edition; Photos: Portland Cement Association, Skokie, IL)*

FIGURE 5.12
A concrete masonry foundation with polystyrene foam insulation applied to its outer surface. *(Photograph provided by The Dow Chemical Company)*

FIGURE 5.13
Following completion of the exterior siding, a worker staples a glass fiber reinforcing mesh to the foam insulation on the exposed portions of the basement wall. Next, he will trowel onto the mesh two thin coats of a cementitious, stucco-like material that will form a durable finish coating over the foam. *(Photograph provided by The Dow Chemical Company)*

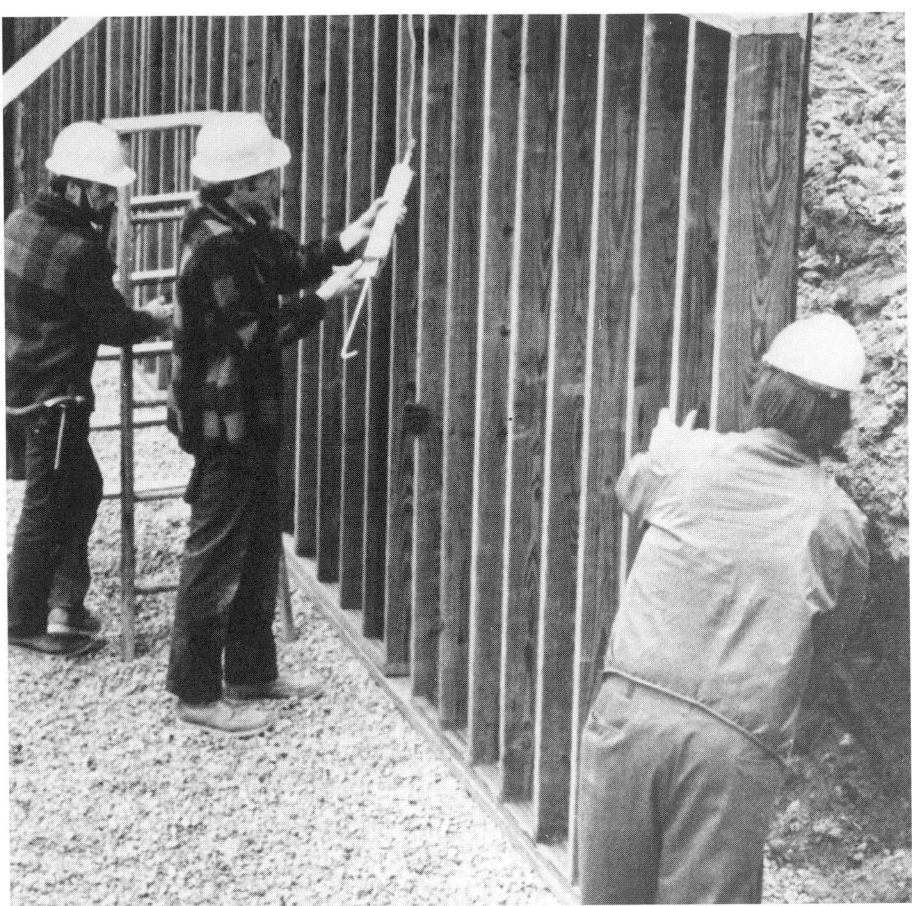

FIGURE 5.14
Erecting a permanent wood foundation. One worker applies a bead of sealant to the edge of a panel of preservative-treated wood components as another prepares to push the next panel into position against the sealant. The panels rest on a horizontal preservative-treated plank, which, in turn, rests on a drainage layer of crushed stone. Wood foundations can be constructed in any weather and can be insulated in the same way as the superstructure of the building.
(Courtesy of APA–The Engineered Wood Association)

FIGURE 5.15
The difference in color makes it clear where the preservative-treated wood foundation leaves off and the untreated superstructure of the building begins.
(Courtesy of APA–The Engineered Wood Association)

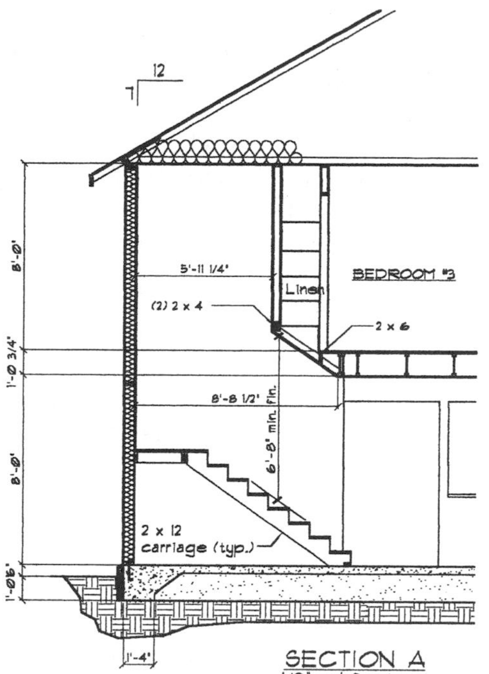

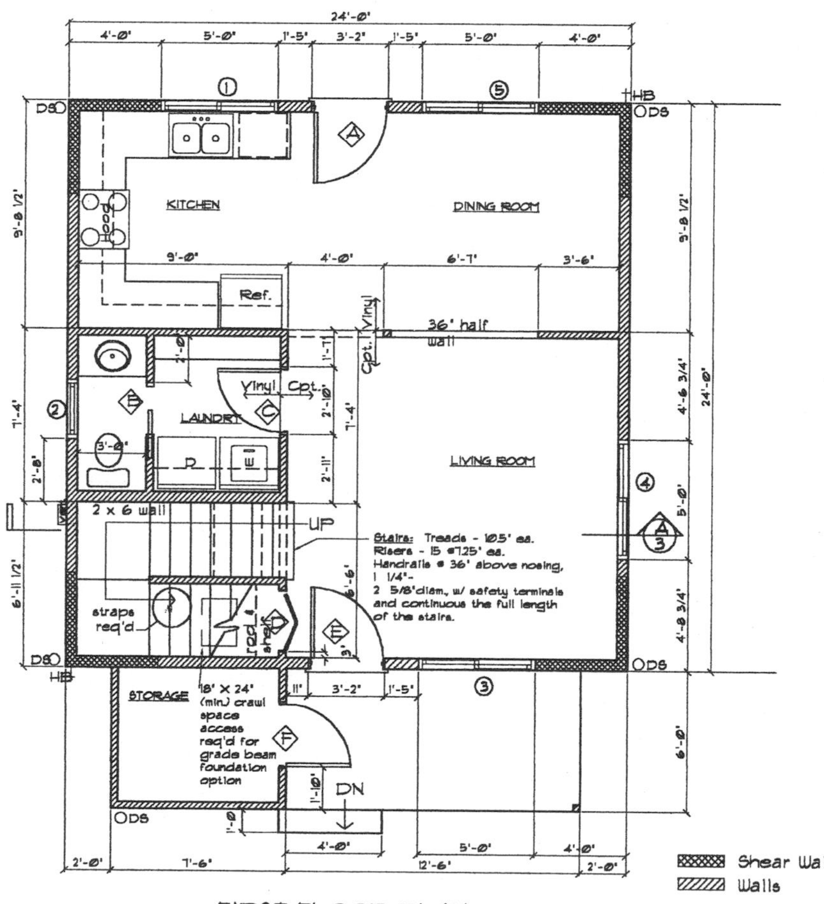

Figure 5.16
A floor plan and building section are two important components of the construction drawings for a simple house with wood light framing. The ground floor is a concrete slab on grade. Notice that portions of the walls have been designated on the floor plan as shear walls; shear walls are discussed beginning on page 185.

Building the Frame

Planning the Frame

An experienced carpenter can frame a simple building from the most minimal drawings, but the framing for a larger or custom-designed structure may need to be planned as carefully as for a steel- or concrete-framed building (Figure 5.16). The architect or engineer determines an efficient layout and the appropriate sizes for joists and rafters, and communicates this information to the carpenters by means of *framing plans* (Figures 5.17 and 5.51). For most purposes, member sizes can be determined using standardized structural tables that are part of residential building codes, or, for more complex framing or special conditions, custom engineering may be required. Larger-scale *section details*, similar to those seen throughout this chapter, are prepared for major connections in the building system. The *architectural floor plans* serve to indicate the locations and dimensions of walls, partitions, and openings, and the *exterior elevations* show the outside faces of the building, with vertical dimensions or elevations indicated as required. For most buildings, *building sections* are also drawn that cut completely through the building, showing the dimensional relationships of the various floor levels and roof planes and the slopes of the roof surfaces. *Interior elevations* are often prepared for kitchens, bathrooms, and other rooms with elaborate interior features.

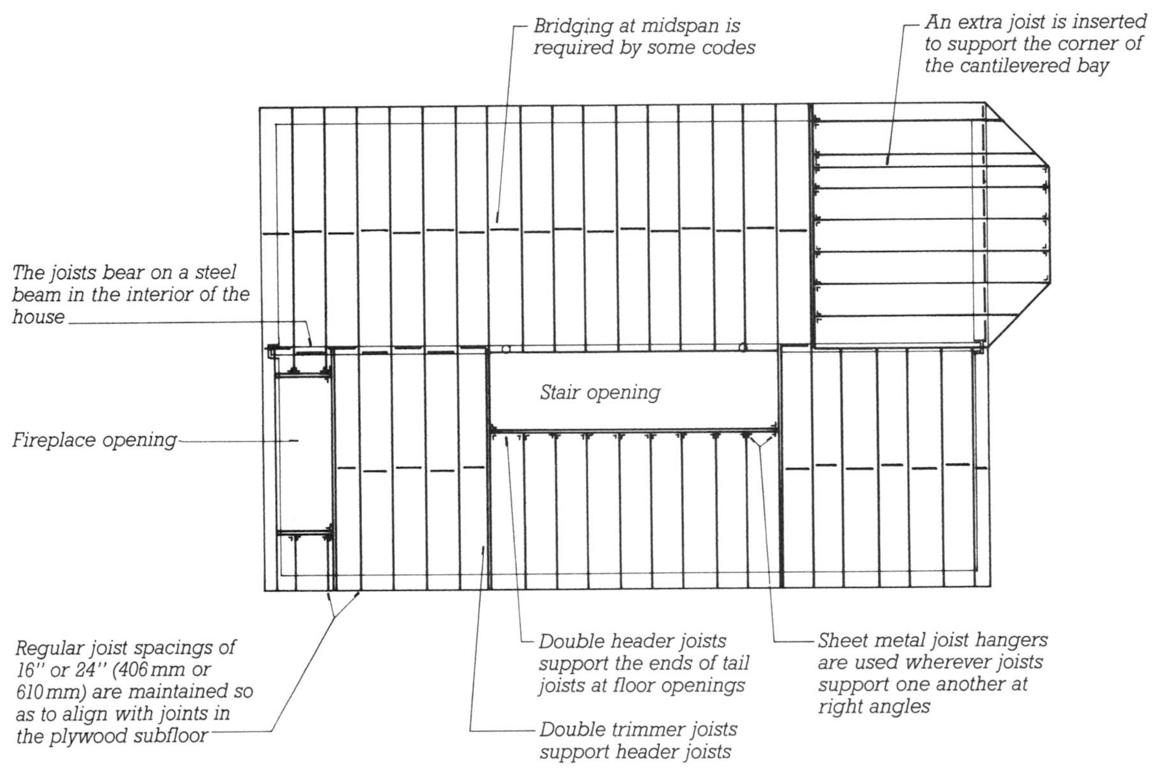
Bridging at midspan is required by some codes
An extra joist is inserted to support the corner of the cantilevered bay
The joists bear on a steel beam in the interior of the house
Stair opening
Fireplace opening
Regular joist spacings of 16" or 24" (406 mm or 610 mm) are maintained so as to align with joints in the plywood subfloor
Double header joists support the ends of tail joists at floor openings
Double trimmer joists support header joists
Sheet metal joist hangers are used wherever joists support one another at right angles
FLOOR FRAMING PLAN

FIGURE 5.17
A framing plan for the ground-floor platform of the building shown in Figure 5.19.

Erecting the Frame

The erection of a typical platform frame (referred to, imprecisely, as *rough carpentry* in the architect's specifications) can best be understood by following this chapter's sequential isometric diagrams, beginning with Figure 5.19. Notice the basic simplicity of the building process: A platform is built on top of the foundation. Walls are assembled horizontally on

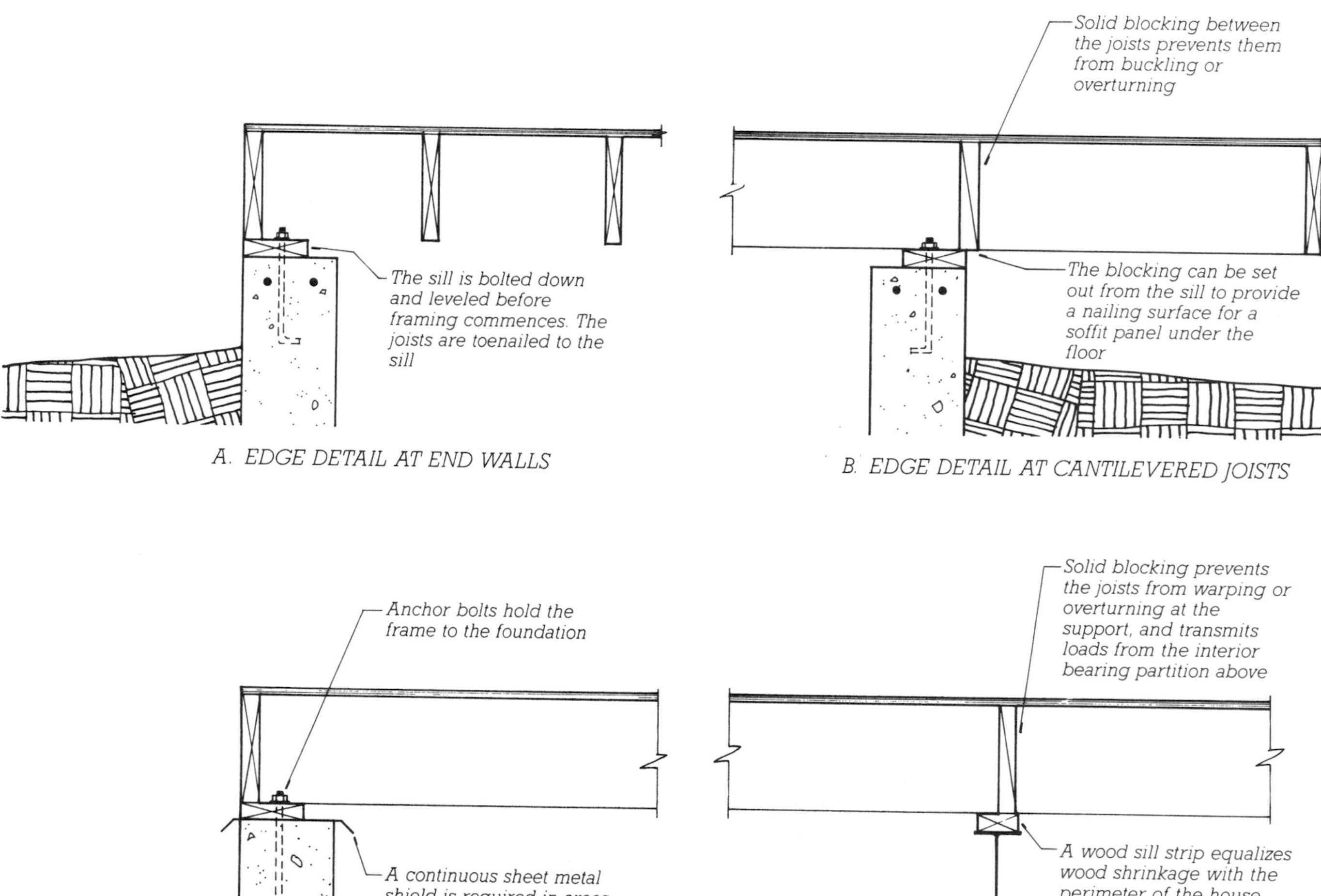
The sill is bolted down and leveled before framing commences. The joists are toenailed to the sill
A. EDGE DETAIL AT END WALLS
Solid blocking between the joists prevents them from buckling or overturning
The blocking can be set out from the sill to provide a nailing surface for a soffit panel under the floor
B. EDGE DETAIL AT CANTILEVERED JOISTS
Anchor bolts hold the frame to the foundation
A continuous sheet metal shield is required in areas with a high risk of termite infestation
C. EDGE DETAIL AT SIDE WALLS
Solid blocking prevents the joists from warping or overturning at the support, and transmits loads from the interior bearing partition above
A wood sill strip equalizes wood shrinkage with the perimeter of the house
D. DETAIL OF INTERIOR JOIST BEARING

FIGURE 5.18
Ground-floor framing details, keyed to the lettered circles in Figure 5.19. A compressible or resilient sill sealer strip should be installed between the sill and the top of the foundation to reduce air leakage but is not shown on these diagrams. Using accepted architectural drafting conventions, continuous pieces of lumber are drawn with an X inside and intermittent blocking with a single diagonal. The metal termite shield in detail *C* is used in areas where the risk of termite infestation is high. It prevents subterranean termites from traveling undetected from cracks in the concrete into the wood framing above.

the platform and tilted up into place. Another platform or a roof is built on top of the walls. Most of the work is accomplished without the use of ladders or scaffolding, and temporary bracing is needed only to support the walls until the next level of framing is installed and sheathed.

The details of a platform frame are not left to chance. While there are countless regional and personal variations in framing details and techniques, the sizes, spacings, and connections of the members in a platform frame are standardized and closely regulated by building codes, even down to the size and number of nails for each connection

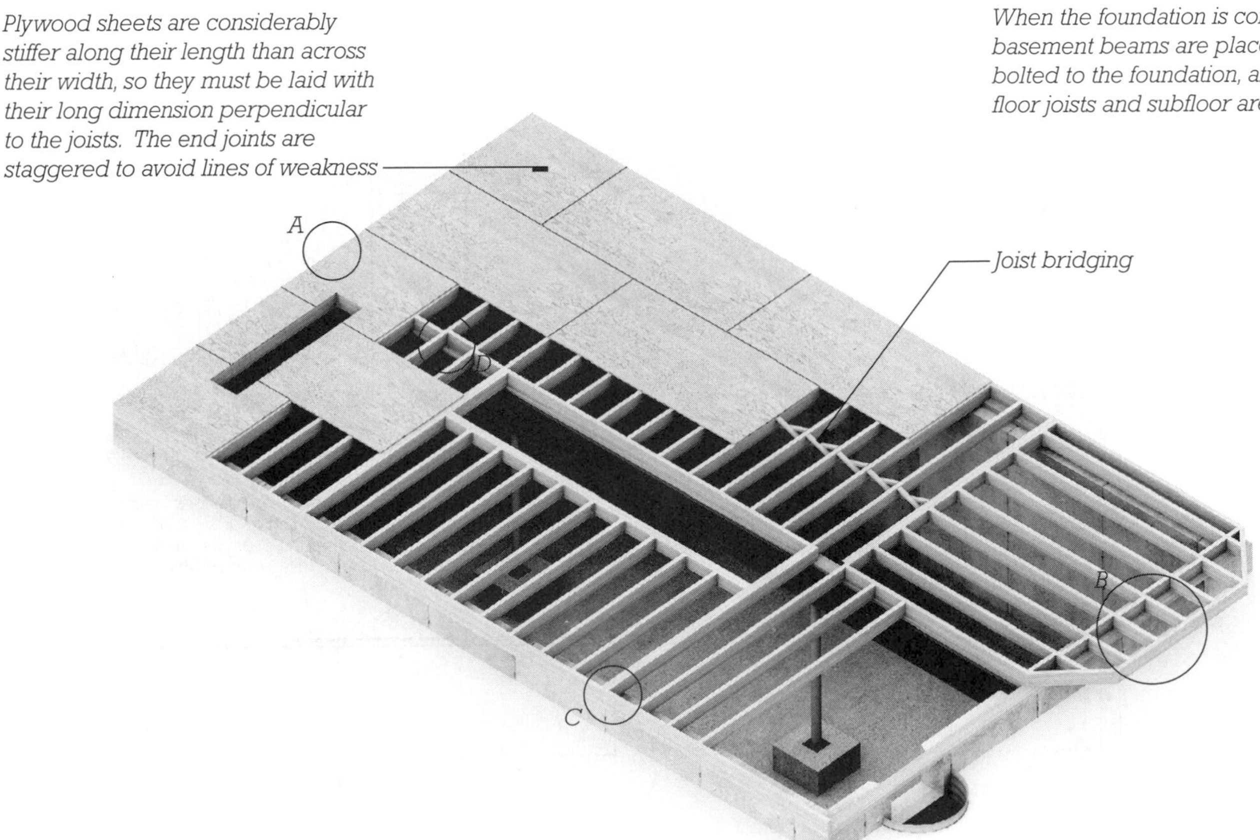

Plywood sheets are considerably stiffer along their length than across their width, so they must be laid with their long dimension perpendicular to the joists. The end joints are staggered to avoid lines of weakness
When the foundation is complete, basement beams are placed, sills are bolted to the foundation, and the first floor joists and subfloor are installed.
Joist bridging
A
B
C
D
3

FIGURE 5.19
Step Three in erecting a typical platform frame building: the ground- floor platform. Compare this drawing with the framing plan in Figure 5.17. Notice that the direction of the joists must be changed to construct the cantilevered bay on the end of the building. A cantilevered bay on a long side of the building could be framed by merely extending the floor joists over the foundation. The letters *A*, *B*, and *C* represent portions of the framing detailed in Figure 5.18.

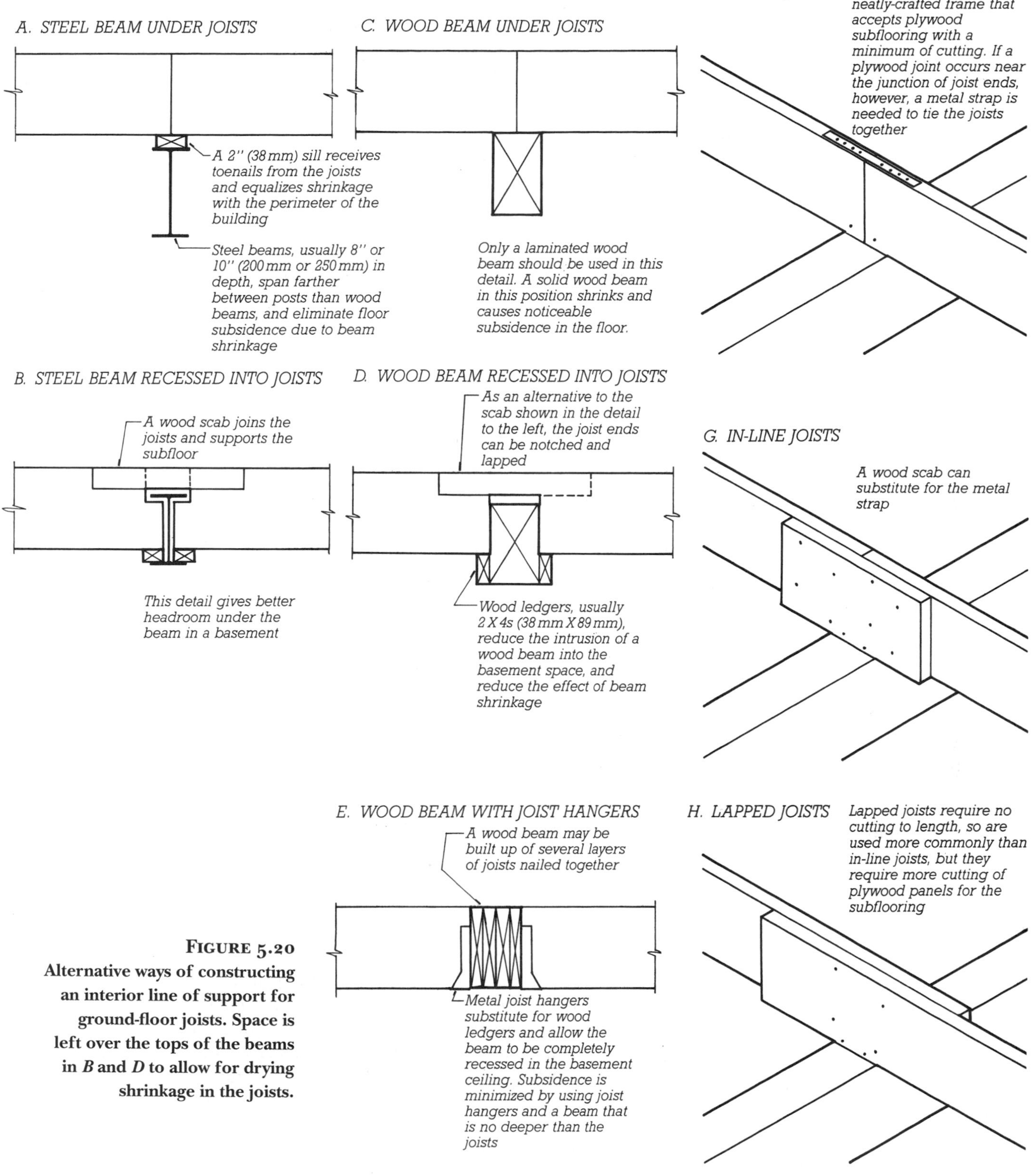

A. STEEL BEAM UNDER JOISTS
A 2" (38 mm) sill receives toenails from the joists and equalizes shrinkage with the perimeter of the building
Steel beams, usually 8" or 10" (200 mm or 250 mm) in depth, span farther between posts than wood beams, and eliminate floor subsidence due to beam shrinkage
C. WOOD BEAM UNDER JOISTS
Only a laminated wood beam should be used in this detail. A solid wood beam in this position shrinks and causes noticeable subsidence in the floor.
F. IN-LINE JOISTS
In-line joists make a neatly-crafted frame that accepts plywood subflooring with a minimum of cutting. If a plywood joint occurs near the junction of joist ends, however, a metal strap is needed to tie the joists together
B. STEEL BEAM RECESSED INTO JOISTS
A wood scab joins the joists and supports the subfloor
This detail gives better headroom under the beam in a basement
D. WOOD BEAM RECESSED INTO JOISTS
As an alternative to the scab shown in the detail to the left, the joist ends can be notched and lapped
Wood ledgers, usually 2 X 4s (38 mm X 89 mm), reduce the intrusion of a wood beam into the basement space, and reduce the effect of beam shrinkage
G. IN-LINE JOISTS
A wood scab can substitute for the metal strap
E. WOOD BEAM WITH JOIST HANGERS
A wood beam may be built up of several layers of joists nailed together
Metal joist hangers substitute for wood ledgers and allow the beam to be completely recessed in the basement ceiling. Subsidence is minimized by using joist hangers and a beam that is no deeper than the joists
H. LAPPED JOISTS
Lapped joists require no cutting to length, so are used more commonly than in-line joists, but they require more cutting of plywood panels for the subflooring

FIGURE 5.20
Alternative ways of constructing an interior line of support for ground-floor joists. Space is left over the tops of the beams in *B* and *D* to allow for drying shrinkage in the joists.

Connection	Common or Box Nail Size
Built-up girders and beams	10d face nails, spaced 32″ apart, each edge staggered top and bottom; plus 2–10d nails at each end or splice
Floor joists sill or beam (below)	3–8d toe nails
Tail joists to headers, headers to trimmers	Use joist hangers
Blocking between joists or rafters, to beam, sill or top plate	3–8d toe nails
Sole plate to joist or blocking (below)	16d face nails 16″ apart
Stud to sole plate	2–16d end nails or 3-8d toe nails
Stud to top plate	2–16d end nails
Double studs, built-up corner studs	10d face nails 24″ apart
Built-up header with ½″ spacer	16d face nails 16″ apart along each edge
Let-in bracing	2–8d face nails each stud or plate
Exterior corner studs to abutting corner studs	16d face nails 24″ apart on each edge
Double top plate	10d face nails 24″ apart
End laps in top plate	8–16d face nails
Corner and intersection laps in top plate	2–10d face nails
Ceiling joists to top plate	3–8d toe nails
Ceiling joists, laps over partitions	3–10d face nails
Rim joist to top plate	8d toe nails 6″ apart
Rafter to top plate	2–16d toe nails
Rafter to ridge board, valley, or hip rafter	4–16d toe nails or 3–16d face nails
Rafter to ceiling to joist	3–10d face nails
Collar tie to rafter	3–10d face nails
Structural sheathing up to ½″ thick, subfloors and walls	6d common nails only, 6″ apart at edges and 12″ apart at intermediate framing
Structural sheathing up to 1″ thick, subfloors walls, and roofs	8d common nails only, 6″ apart at edges and 12″ apart at intermediate framing

FIGURE 5.21
Platform framing members are fastened according to this nailing schedule, which framing carpenters know by memory and which is incorporated into most building codes and construction standards.

(Figure 5.21) and the thicknesses and nailing patterns of the sheathing panels.

Attaching the Frame to the Foundation

The *foundation sill plate*, usually made of preservative-treated wood for added resistance to insects and moisture, is bolted to the foundation as a base for the wood framing (Figure 5.18, *A*, *B*, and *C*). A single sill, as shown in the details here, is all that is required by code, but in better-quality work, the sill may be doubled or made from thicker stock for greater stiffness. A thicker sill plate may also be required in areas of high seismic risk, to provide a stronger connection between the wood framing and the foundation anchor bolts. Because the top of a foundation wall is usually somewhat uneven, the sill may also be shimmed up at low spots with wood shingle wedges or plastic shims to provide a more level, even base for subsequent framing. A *sill sealer*, made of compressible or resilient material, should be inserted between the sill and the foundation to reduce air infiltration through the gap. Some sill sealer materials also discourage moisture wicking up from the foundation into the wood framing (Figure 5.22). Conventional anchor bolts are sufficient to hold most buildings on their foundations, but tall frames or those subject to high winds

FIGURE 5.22
Carpenters apply a preservative-treated wood sill to a sitecast concrete foundation. A strip of compressible glass fiber sill sealer has been placed on the top of the concrete wall, and the sill has been drilled to fit over the projecting anchor bolts. Before each section of sill is bolted tightly, it is leveled as necessary with wood shingle or plastic shims between the concrete and the wood. As the anchor bolts are tightened, the sill sealer squeezes down to a negligible thickness. A length of bolted-down sill is visible at the upper right. The basement windows were clamped into reusable steel form inserts and placed in the formwork before the concrete was cast. After the concrete was cast and the formwork was stripped, the steel inserts were removed, leaving a neatly formed concrete frame around each window. A pocket in the top of the wall for a steel beam can be seen at the upper left. *(Photo by Edward Allen)*

or strong earthquakes may require more elaborate attachments (Figures 5.39–5.41).

Floor Framing and Bridging

Floor framing (Figure 5.19) is laid out in such a way that the ends of uncut 4-foot by 8-foot (1.2-m by 2.4-m) subflooring panels will fall directly over joists; otherwise, many panels will have to be cut, wasting both materials and time. Standard joist spacings are 16 or 24 inches o.c. (406 or 610 mm o.c.; "o.c." stands for "on center," meaning that the spacing is measured from center to center of the joists). Though less common, a joist spacing of 19.2 inches (486 mm) can also be used. Any of these spacings automatically provides a joist to support each end of every 8-foot (2.4-m) panel (Figure 5.23).

Wood composite I-joists or manufactured trusses are frequently used in place of solid wood joists. Their

FIGURE 5.23
Installing floor joists. Blocking will be inserted between the joists over the two interior beams to prevent overturning of the joists. *(Photo by Edward Allen)*

FIGURE 5.24
Various types of manufactured joists and floor trusses are often used instead of dimension lumber. This I-joist has laminated veneer lumber (LVL) flanges and a plywood web. I-joists are manufactured in very long pieces and tend to be straighter, stronger, stiffer, and lighter in weight than sawn joists. *(Courtesy of Trus Joist MacMillan)*

FIGURE 5.25
These floor trusses (shown here being set up for a demonstration house in a parking lot) are made of sawn lumber members joined by toothed-plate connectors. The OSB web at each end provides a section of truss that can be sawn by workers to easily adjust the length of the truss if necessary. Trusses are deeper than sawn joists or I-joists but can span farther. And even though they are deeper than other joist products, they may result in a thinner floor ceiling assembly, as pipes and ductwork can run through the passages within the truss rather than below it. *(Courtesy of Wood Truss Council of America)*

FIGURE 5.26
Applying OSB subflooring. Note that the longer dimension of the panel runs perpendicular to the joist framing.
(Courtesy of APA–The Engineered Wood Association)

advantages include the ability to span greater distances, straightness and uniformity, and reduced drying shrinkage of the floor platform. These joist types are most frequently spaced at 19.2 or 24 inches o.c. (488 or 610 mm o.c.) I-joists, with their thin webs, often require reinforcing where concentrated loads occur, such as at joist ends, at cantilevers, or where interior loadbearing partitions bear on the joists (Figures 5.44 and 5.45).

After floor framing is complete, subflooring is fastened in place. To reduce squeaking in the finished floor and to increase floor stiffness, adhesive may be used in conjunction with mechanical fasteners (Figures 5.26 and 5.27). Deformed shank nails or screws, either of which hold subflooring more securely than common nails, also contribute to the prevention of squeaks. Plywood and OSB panels must be laid with the grain of their face layers perpendicular to the direction of the joists because these panels are considerably stiffer in this orientation (that is, a standard 4-foot by 8-foot (1.2-m by 2.4-m) panel should be laid with its longer dimension perpendicular to the joists). Sheathing and subflooring panels are normally manufactured ⅛ inch (3 mm) short in each face dimension so that they may be spaced slightly apart at all their edges to prevent floor buckling during construction from the expansion of storm-wetted panels. *Bridging*, consisting of solid blocking or wood or metal crossbracing inserted between joists at midspan or at intervals not exceeding 8 feet (2.4 m), is a traditional feature of floor framing (Figures 5.19, 5.28). (The term *blocking* refers, in general, to short lengths of lumber inserted between framing members for any of a variety of purposes, such as bracing, reinforcing, or providing solid backing where fasteners are required for attachment of finish materials or equipment.) Its function is to hold the joists straight and to help them share concentrated loads. The International Residential Code requires bridging only for joists deeper than 2 × 12 feet (38 × 286 mm). In better-quality construction, however, bridging is frequently used in all floors, regardless of depth, to improve stiffness and reduce vibration. Solid blocking is also required wherever floor joists span continuously across supporting beams to resist overturning of the joists (Figure 5.18*B, D*).

Where the ends of joists butt into supporting headers, as around stair openings and at changes of joist direction for projecting bays, end nails and toe nails cannot transfer the full load between members, and sheet metal joist hangers must be used. Each hanger provides a secure pocket for the end of the joist and punched holes into which a number of special short nails are driven to make a safe connection.

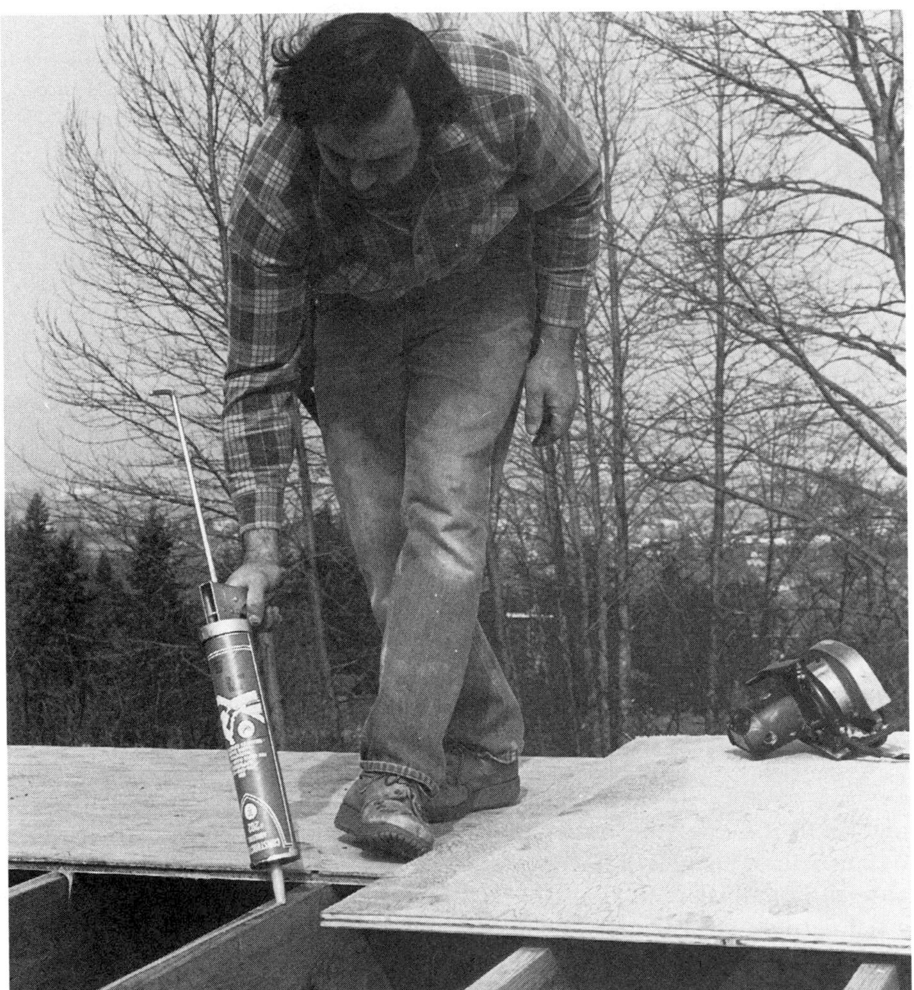

FIGURE 5.27
For a stiffer, quieter floor, subflooring should be glued to the joists. The adhesive is a thick mastic that is squeezed from a sealant gun on the tops of the joists just before the subflooring is laid in place. *(Courtesy of APA–The Engineered Wood Association)*

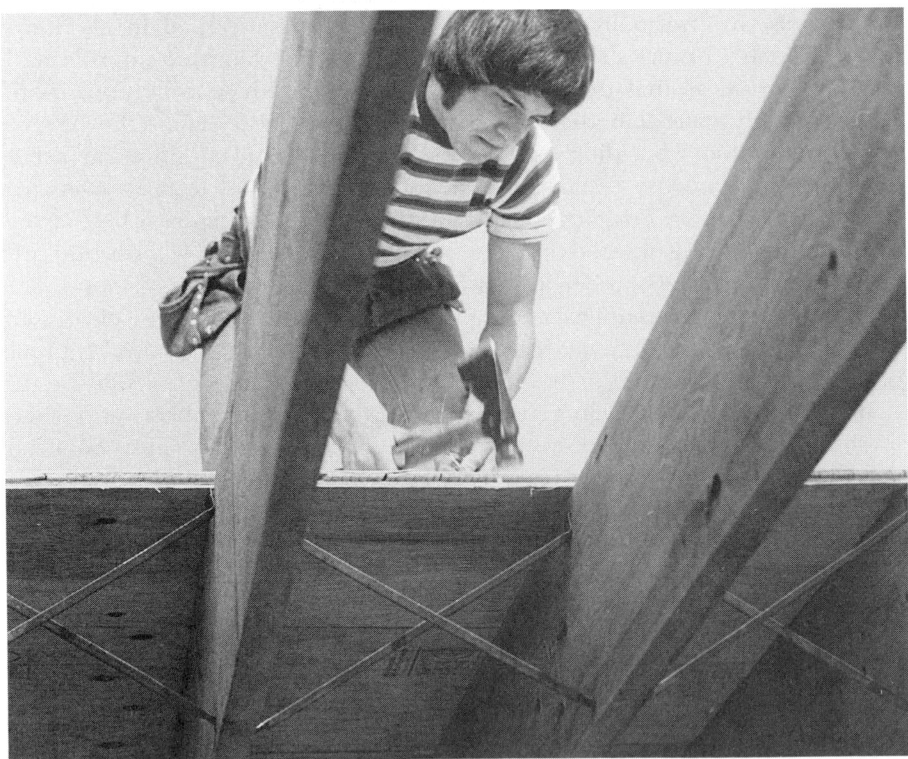

FIGURE 5.28
Bridging between joists may be solid blocks of joist lumber, wood crossbridging, or, as seen here, steel crossbridging. During manufacture, the thick steel strip is folded across its width into a V- shape to increase its stiffness. Steel crossbridging with one end toothed requires only one nail per piece and no cutting to size, so it is the fastest to install. *(Courtesy of APA–The Engineered Wood Association)*

Wall Framing, Sheathing, and Bracing

Wall framing, like floor framing, is laid out in such a way that a framing member, in this case a stud, occurs under each vertical joint between sheathing panels (Figures 5.29–5.32). The lead carpenter initiates wall framing by laying out the stud locations on the top plate and sole plate of each wall. Other carpenters follow behind to cut the studs and headers and assemble the walls in a horizontal position on the subfloor. As each wall frame is completed, it is tilted up and nailed into position, with temporary bracing as needed (Figures 5.34–5.37).

Long studs for tall walls often must be larger than a 2 × 4 in order to resist wind forces. Because long pieces of sawn dimension lumber tend not to be straight and may be less readily available, studs manufactured of structural composite lumber may be used instead. The International Residential Code requires horizontal wood blocking to be installed between studs at midheight in wall frames that are taller than 10 feet (3 m). The purpose of this blocking is to stop off the cavities between studs to restrict the spread of fire.

Headers over window and door openings must be sized in accordance with building code criteria. Typically, a wall opening header consists of two nominal 2-inch members standing on edge, separated by a plywood spacer that serves to make the header as thick as the depth of the wall studs. Headers for long spans and/or heavy loads are often made of laminated veneer lumber or parallel strand lumber, either of which is stronger and stiffer than dimension lumber. Several manufacturers market prefabricated headers that include thermal insulation as a way of reducing the heat loss that occurs through these difficult-to-insulate assemblies. At either end, headers rest on shortened studs called *trimmer or jack studs*, which themselves are nailed to full-height *king studs*. At the bottom of a window opening, the *rough sill* is supported on *cripple studs* (Figure 5.32).

Each corner and partition intersection must furnish nailing surfaces for the edge of each plane of exterior and interior finish materials. This requires a minimum of three studs at each intersection, unless special metal clips are used to reduce the number to two (Figure 5.32).

Sheathing was originally made of solid boards, usually 6 to 10 inches (150 to 250 mm) wide. If applied horizontally, these boards did little to brace the building against wracking caused by the forces of wind or earthquake. If applied on a diagonal, however, they produced a rigid frame. Today, walls are sheathed with either plywood or OSB, which provide permanent, very stiff bracing (Figure 5.63), or with let-in bracing as described below. Panels are applied as soon as possible after the wall is framed, often while it is still lying on the floor platform.

In regions of strong winds or earthquakes, wall sheathing plays a very important part in the lateral stability the structure. A properly

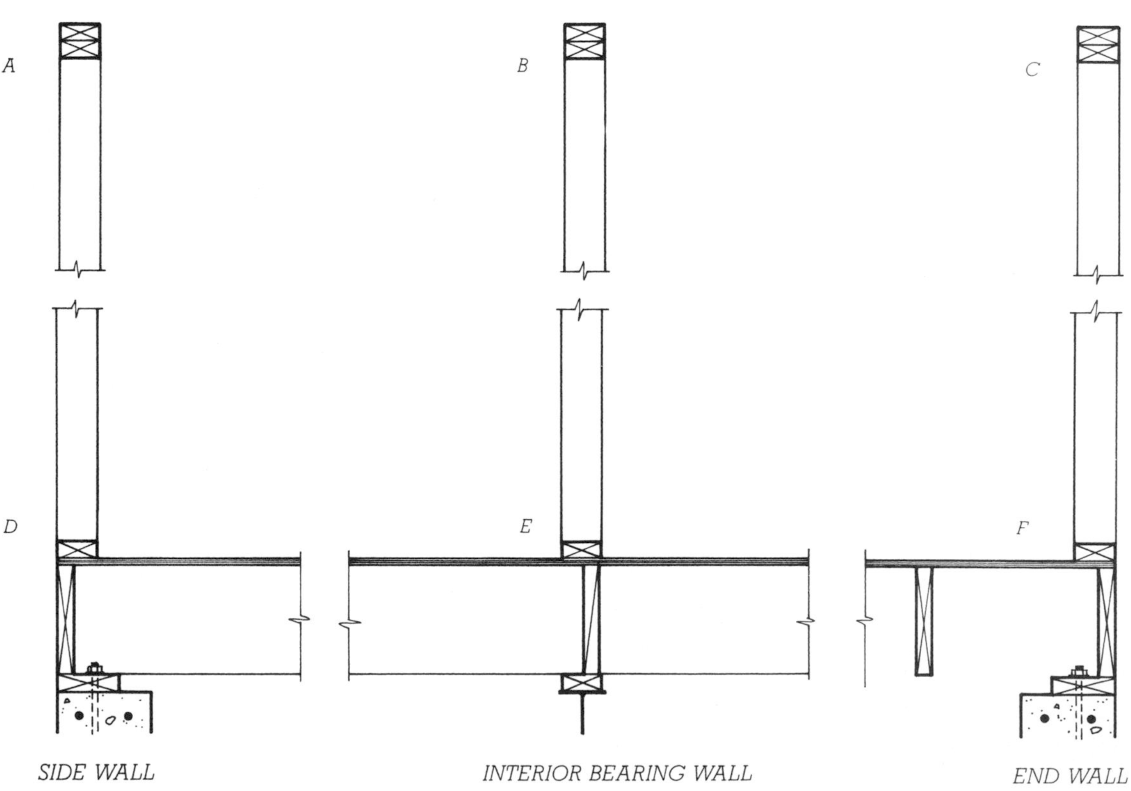

FIGURE 5.29 Typical ground-floor wall framing details, keyed by letter to Figure 5.30.

sheathed wall acts as a *shear wall* to resist lateral forces. Both interior and exterior walls can act as shear walls, which must be provided in both east–west and north–south orientations and must be distributed more or less symmetrically in the floor plan. Stresses within shear walls are proportional to their length, with shorter walls being subjected to higher stresses than those that are longer. Where walls have large openings for windows or doors, the remaining solid portions that can contribute to lateral force resistance may become relatively short and therefore exposed to very high stresses. In such cases, sheathing may have to be attached with larger nails at very close spacings, the horizontal edges of sheathing panels may have to be supported by wood blocking to keep them from buckling, and studs over which sheathing panels join may need to be larger in size, such as 3 × or 4 × (64 mm or 89 mm) members, in order to hold the required nails without splitting. Where the needed strength cannot be reliably achieved using site-construction methods, factory-fabricated panels made of wood or steel components may be used (Figure 5.38). Shear walls subject to high forces may

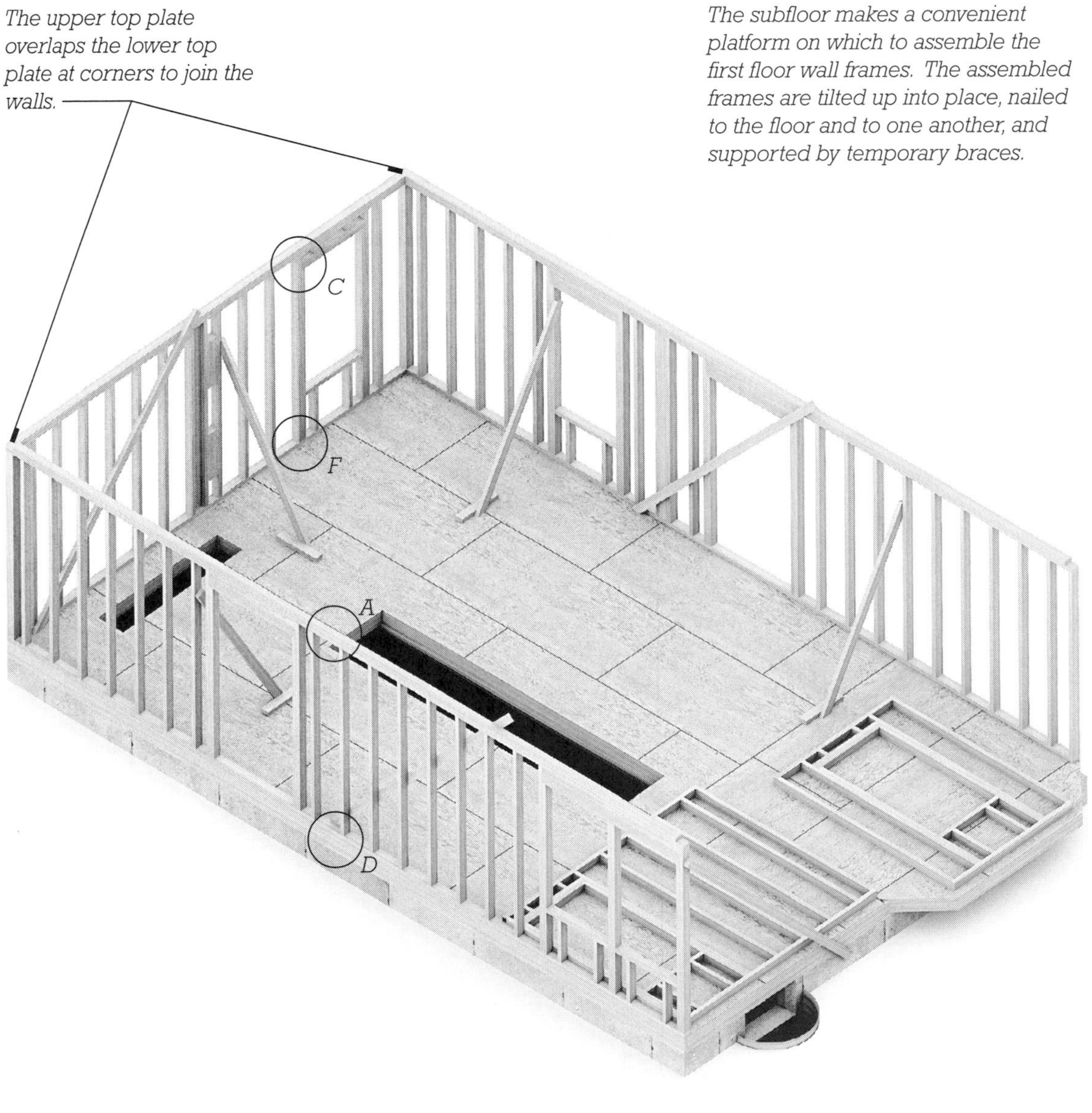

The upper top plate overlaps the lower top plate at corners to join the walls.

The subfloor makes a convenient platform on which to assemble the first floor wall frames. The assembled frames are tilted up into place, nailed to the floor and to one another, and supported by temporary braces.

C

F

A

D

FIGURE 5.30
Step Four in erecting a platform frame building: The ground-floor walls are framed. The letters *A*, *B*, and *C* indicate portions of the framing that are detailed in Figure 5.29.

also require *hold-downs* to prevent the walls from pulling up off the foundation or floor platform (Figures 5.39–5.41). Consultation with a structural engineer is recommended (and often legally required) when building in areas that are earthquake-prone or subject to very high wind forces.

Sheathing panels made from wood or paper fiber, plastic foam, and glass fiber are intended principally as thermal insulation and to provide a base for water-resistant building paper or house wrap. Most panels of these types are nonstructural, so walls sheathed with such panels rely on *let-in diagonal* bracing or strategically located structural panels for lateral force resistance. Let-in bracing may be made of wood members, such as 1 × 4 (19 × 89 mm) boards, or light steel members that are recessed into the outer face of the studs of the wall before it is sheathed (Figures 5.32 and 5.42).

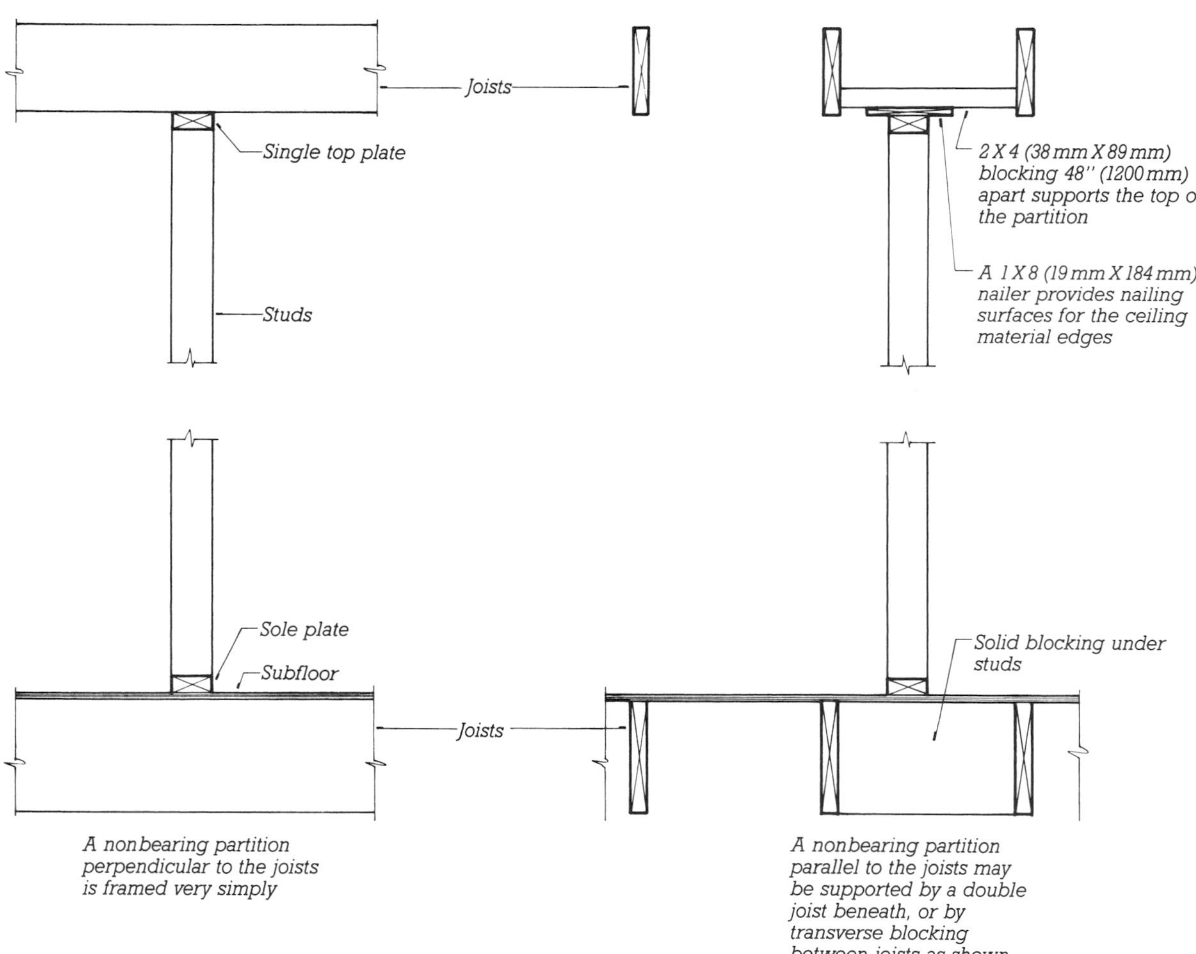
Joists
Single top plate
Studs
2 X 4 (38 mm X 89 mm) blocking 48" (1200 mm) apart supports the top of the partition
A 1 X 8 (19 mm X 184 mm) nailer provides nailing surfaces for the ceiling material edges
Sole plate
Subfloor
Joists
Solid blocking under studs
A nonbearing partition perpendicular to the joists is framed very simply
A nonbearing partition parallel to the joists may be supported by a double joist beneath, or by transverse blocking between joists as shown

FIGURE 5.31
Framing details for nonloadbearing interior partitions.

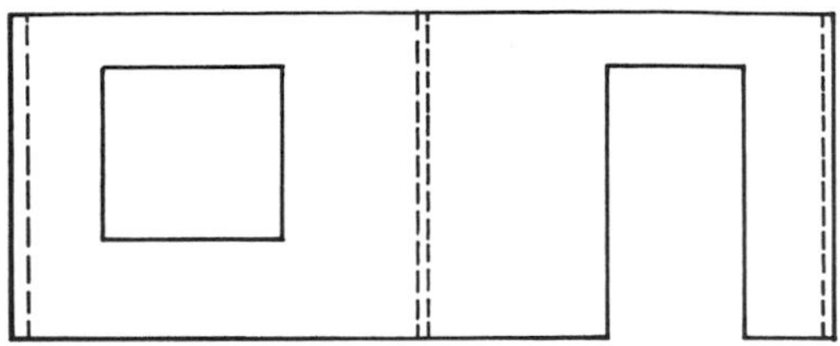

1. This is the layout of a typical exterior wall. It meets two other exterior walls at the corners, and a partition in the middle. It has two rough openings, one for a window and one for a door.

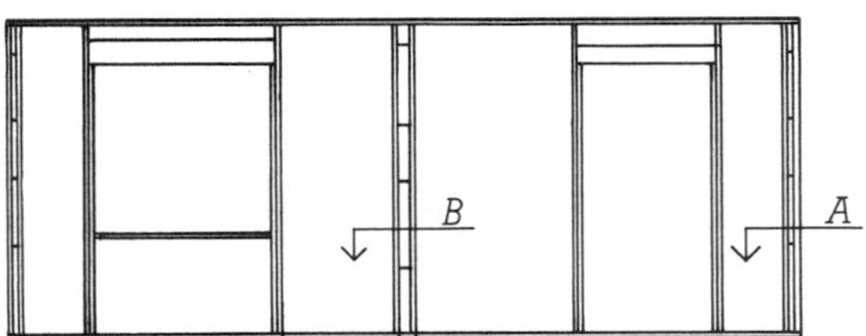
B
A

2. The framer begins by marking all the stud and opening locations on the sole plate and top plate. The "special" studs are cut and assembled first: two corner posts, a partition intersection, and full-length studs and supporting studs for the headers over the openings.

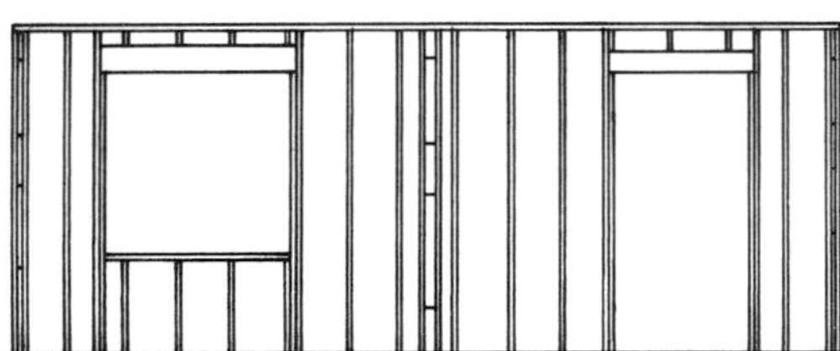

3. The wall is next filled with studs on a regular 16" (400 mm) or 24" (600 mm) spacing, to provide support for edges of sheathing panels.

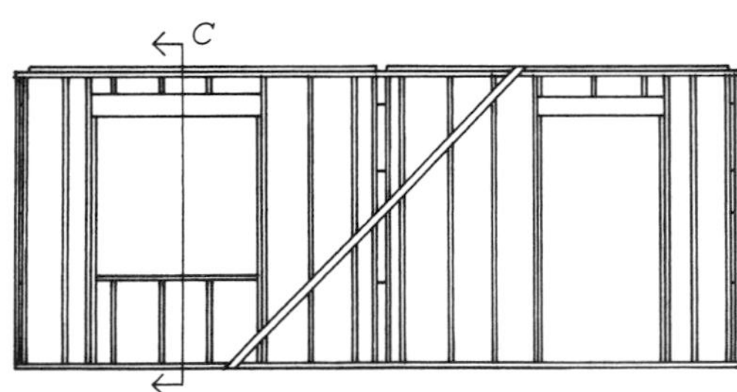
C

4. Diagonal bracing, usually 1X4 (19 mm X 89 mm), is let into the face of the frame if the building will not have rigid sheathing. The second top plate may be added before the wall is tilted up, or after.

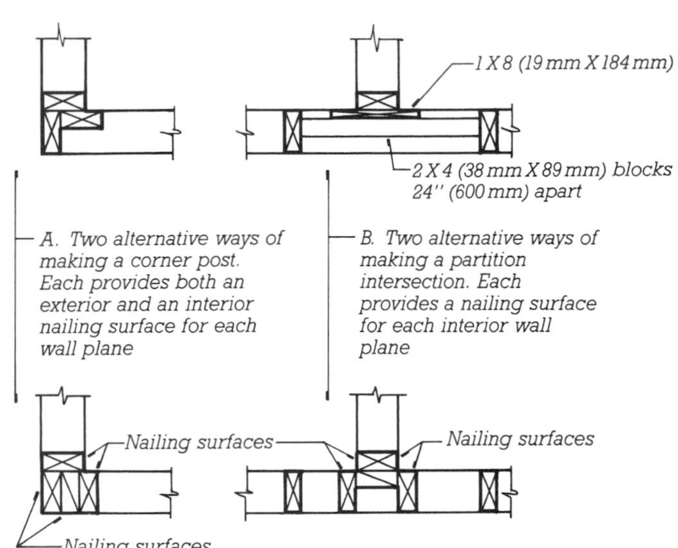
1X8 (19 mm X 184 mm)
2X4 (38 mm X 89 mm) blocks 24" (600 mm) apart
A. Two alternative ways of making a corner post. Each provides both an exterior and an interior nailing surface for each wall plane
B. Two alternative ways of making a partition intersection. Each provides a nailing surface for each interior wall plane
Nailing surfaces
Nailing surfaces
Nailing surfaces

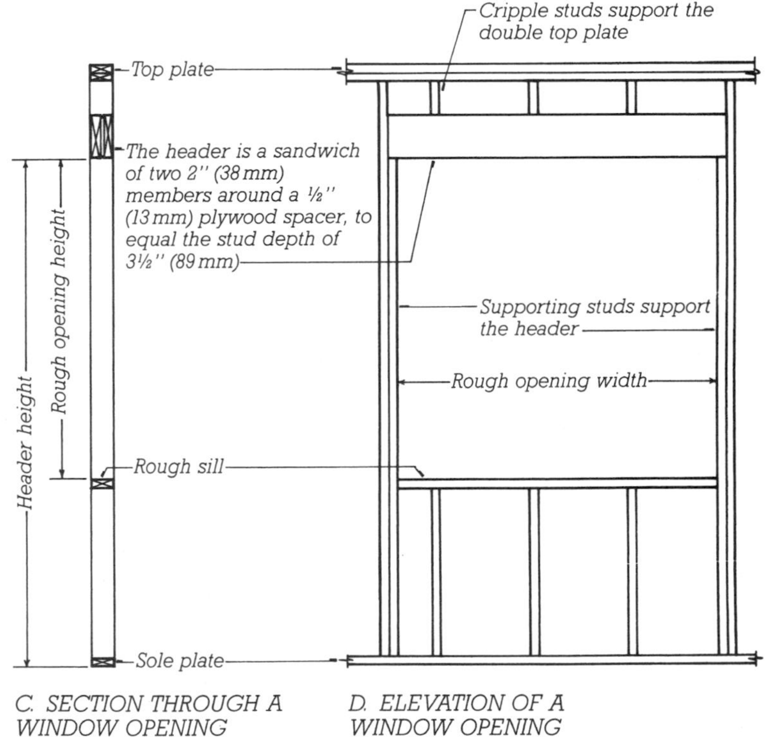
Cripple studs support the double top plate
Top plate
The header is a sandwich of two 2" (38 mm) members around a ½" (13 mm) plywood spacer, to equal the stud depth of 3½" (89 mm)
Supporting studs support the header
Rough opening width
Rough opening height
Header height
Rough sill
Sole plate
C. SECTION THROUGH A WINDOW OPENING
D. ELEVATION OF A WINDOW OPENING

FIGURE 5.32

Steps in the framing of a typical wall and details at wall intersections and a window opening. In *D*, the short studs above the header and below the rough sill are called *trimmer* or *cripple studs*.

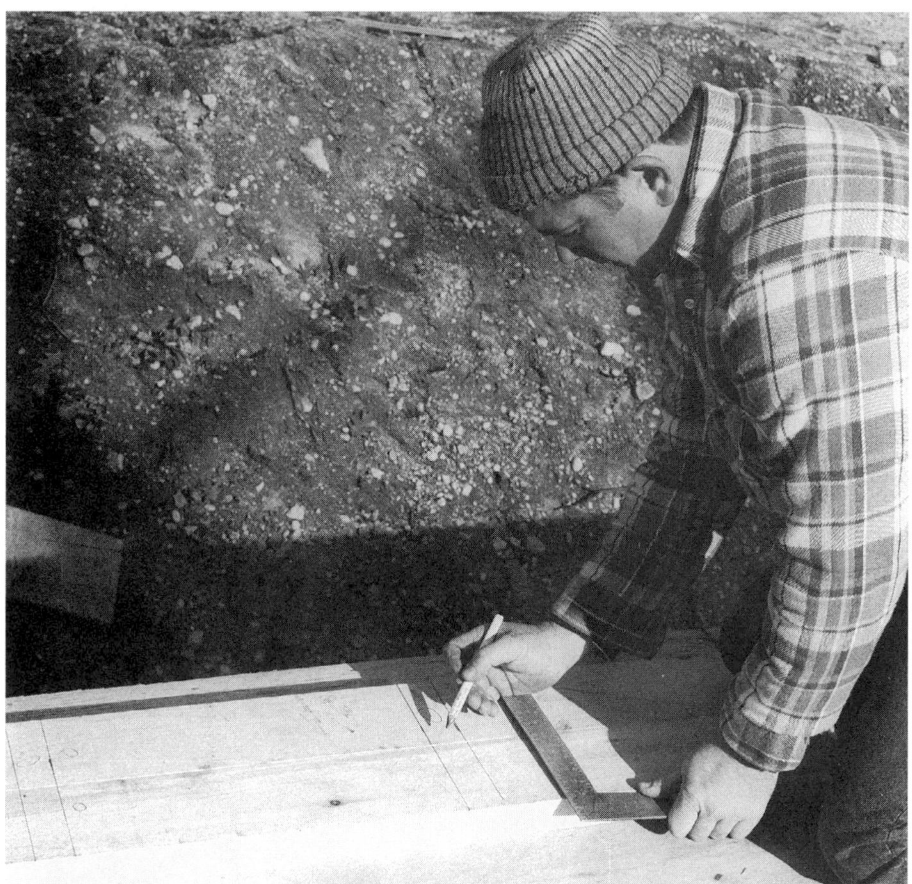

FIGURE 5.33
As the first step in constructing a wall frame, the chief carpenter aligns the top plate and sole plate side by side on the floor platform and marks each stud location on both of them simultaneously with a pencil and framing square. *(Photo by Edward Allen)*

FIGURE 5.34
Nailing studs to a plate, using a pneumatic nail gun. The triple studs are for a partition intersection. *(Courtesy of Senco Products, Inc.)*

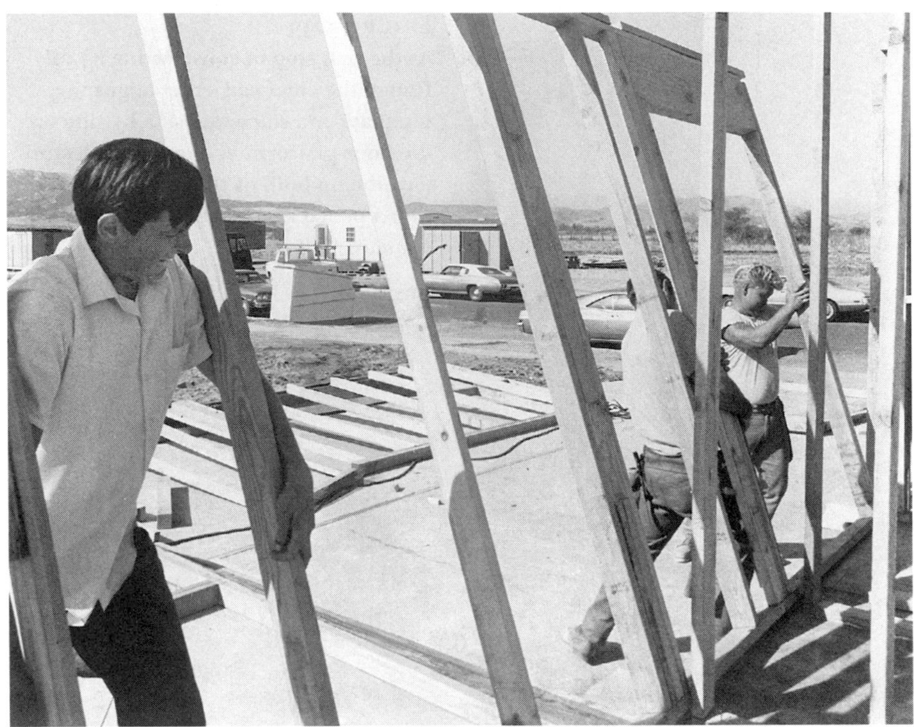

Figure 5.35
Tilting an interior partition into position. The gap in the upper top plate will receive the projecting end of the upper top plate from another partition that intersects at this point. *(Courtesy of APA–The Engineered Wood Association)*

Figure 5.36
Fastening a wall to the floor platform. Horizontal blocking between studs as seen toward the left in this photo is typically installed to provide a solid substrate for the later attachment of bathroom hardware, exterior panels, or any number of possible other items. *(Courtesy of Senco Products, Inc.)*

FIGURE 5.37
Lower-level wall framing is held up by temporary diagonal bracing until the floor joist framing is in place above and wall bracing or sheathing is complete, after which the frame becomes completely self-bracing. The outer walls of this building are framed with 2 × 6 (38 × 140 mm) studs to allow for a greater thickness of thermal insulation, whereas the interior partitions are made of 2 × 4 (38 × 89 mm) stock. *(Photo by Joseph Iano)*

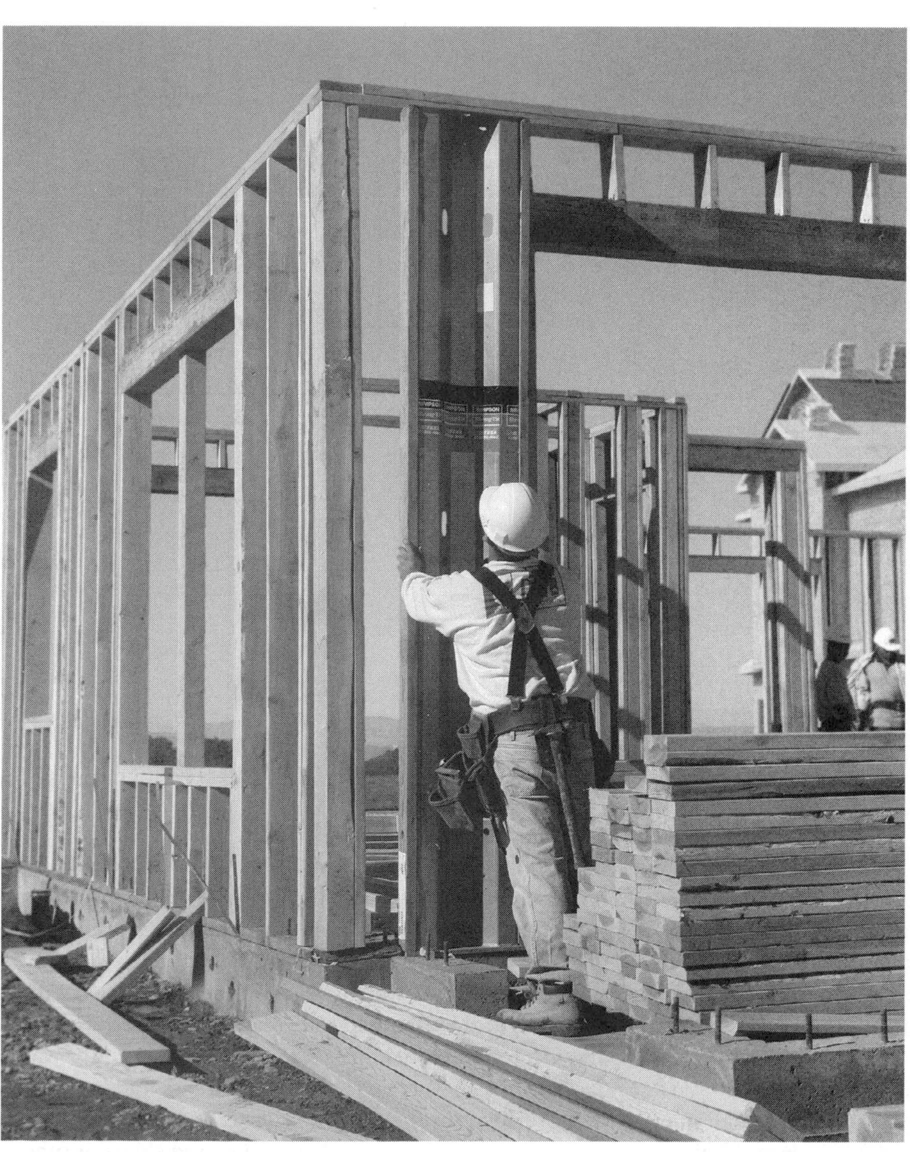

FIGURE 5.38
Prefabricated shear wall panels are especially useful in garage wall framing or other walls with large openings where only narrow, solid sections of wall remain to resist lateral forces. They may be made of all wood components or, where even greater strength is required, of wood and metal, as shown here. The corrugated galvanized steel in this panel is over 1/8 inch (3.5 mm) thick, and the panel's capacity to resist lateral forces is several times greater than that of comparable all-wood prefabricated panels. The bottom of the panel will be anchored with bolts embedded 21 inches (533 mm) or more into the concrete foundation, and the sides and top will be screw-fastened to the surrounding wood framing. Holes in the panel can accommodate wiring runs within the wall. *(Courtesy of Simpson Strong-Tie Company Inc.)*

FIGURE 5.39
Strap tie hold-downs are made of galvanized steel straps with hooked or deformed ends cast into the concrete foundation wall (*top*). After the wall is framed, the exposed length of strap may be nailed directly to framing or, as seen here, nailed through the sheathing into the studs or posts behind (*bottom*). The number, size, and spacing of nails used to fasten the strap depend on the magnitude of the loads that must be resisted and the capacity of the wood member to hold nails without splitting. In the top image, anchor bolts cast into the top of the foundation wall that will be used to secure the sill plate in place are also visible. *(Photos by Joseph Iano)*

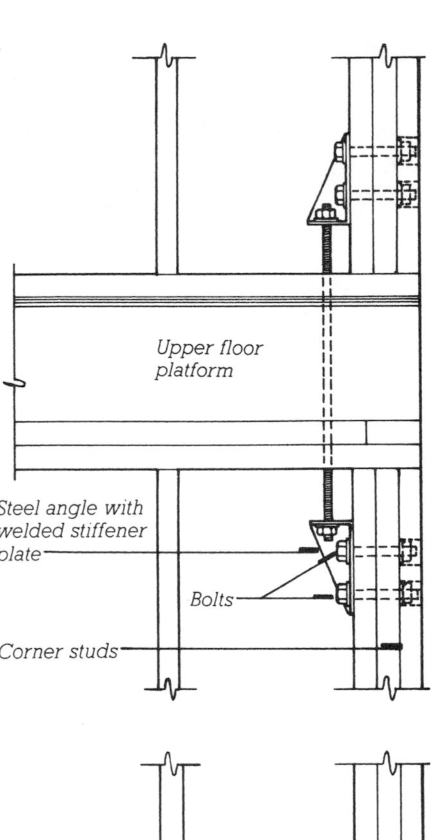

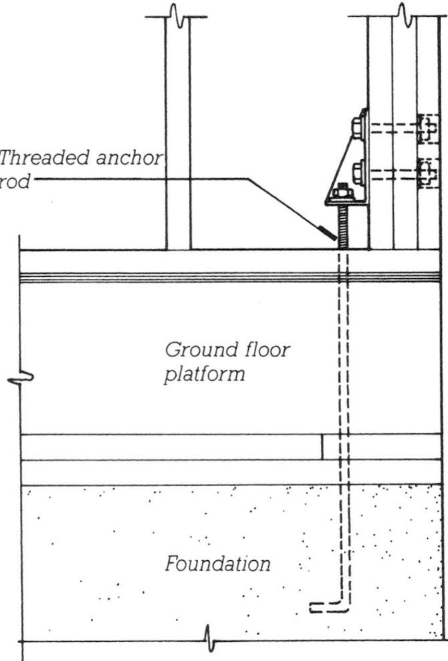

FIGURE 5.40
Hold-downs made from threaded rod and steel plate anchors can resist much greater forces than strap tie hold-downs. They may be used at each floor level to securely tie the full height of the building frame to its foundation. For the type shown here, the nuts at the ends of the threaded rods may require retightening after the first heating season to compensate for wood shrinkage, which can mean that access holes must be provided through the interior wall surfaces. Other models rely on spring-loaded tapered shims or other mechanical compensation so as to be self-adjusting.

FIGURE 5.41
A heavy-duty seismic hold-down similar to that illustrated in Figure 5.40. The anchor rod with one end cast into the concrete foundation wall protrudes through the preservative-treated wood sill where its threaded end is bolted to the anchor. The anchor in turn is bolted to a 4 × 4 (89 × 89 mm) post with five bolts of substantial diameter. A conventional foundation anchor bolt with an oversized square washer is also partially visible to the right of the hold-down. Also note the thicker-than-normal, 3-inch nominal (64-mm) sill plate, a common feature in wood light framing designed for high seismic forces. *(Photo by Joseph Iano)*

FIGURE 5.42
Applying a panel of insulating foam sheathing. Because this type of sheathing is too soft and weak to brace the frame, diagonal bracing is inserted into the outside faces of the studs at the corners of the building. Steel bracing, nailed at each stud, is used in this frame and is visible just to the right of the carpenter's leg. *(Courtesy of The Celotex Corporation)*

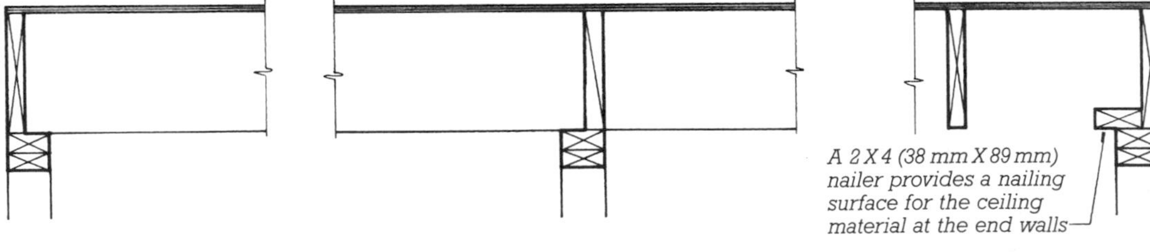
A 2X4 (38 mm X 89 mm) nailer provides a nailing surface for the ceiling material at the end walls
A. SIDE WALL
B. INTERIOR BEARING WALL
C. END WALL

FIGURE 5.43
Details of the second-floor platform, keyed to the letters on Figure 5.46. The extra piece of lumber on top of the top plate in *C*, End Wall, is continuous blocking whose function is to provide a nailing surface for the edge of the finish ceiling material, which is usually either gypsum board or veneer plaster base.

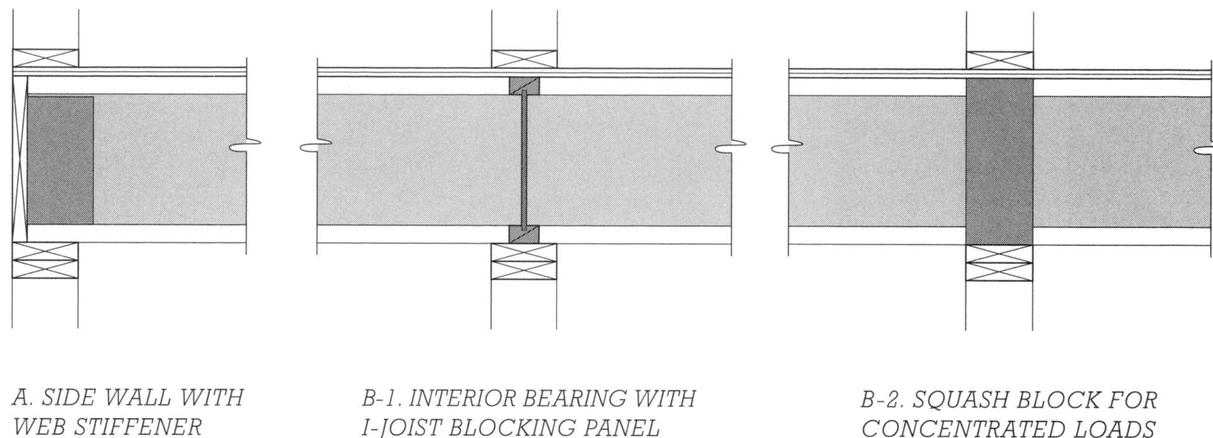

FIGURE 5.44
Alternative details to those shown in Figure 5.43 for a floor framed with I-joists rather than solid lumber. For clarity, the web of the I-joist is rendered light gray and blocking with a darker shade. The web stiffener shown in *A* is cut slightly shorter than the height of the web of the joist, and is installed so that a small space remains between the top of the blocking and the underside of the top flange. This prevents the blocking from prying the flanges apart if the I-joist itself shrinks or is compressed slightly (see also Figure 5.45). The I-joist blocking panel in *B-1* is exactly analogous to the solid blocking shown in Figure 5.45 *B*. A squash block is shown in *B-2*. This is a short section of 2x (38-mm) framing, installed vertically like a very short stud on either side of the I-joist. Squash blocks are used under points of concentrated load, such as under loadbearing posts or studs on either side of a large opening in a wall above. They are cut slightly longer than the full height of the joist to ensure that the loads are transmitted through the blocks and not the joist.

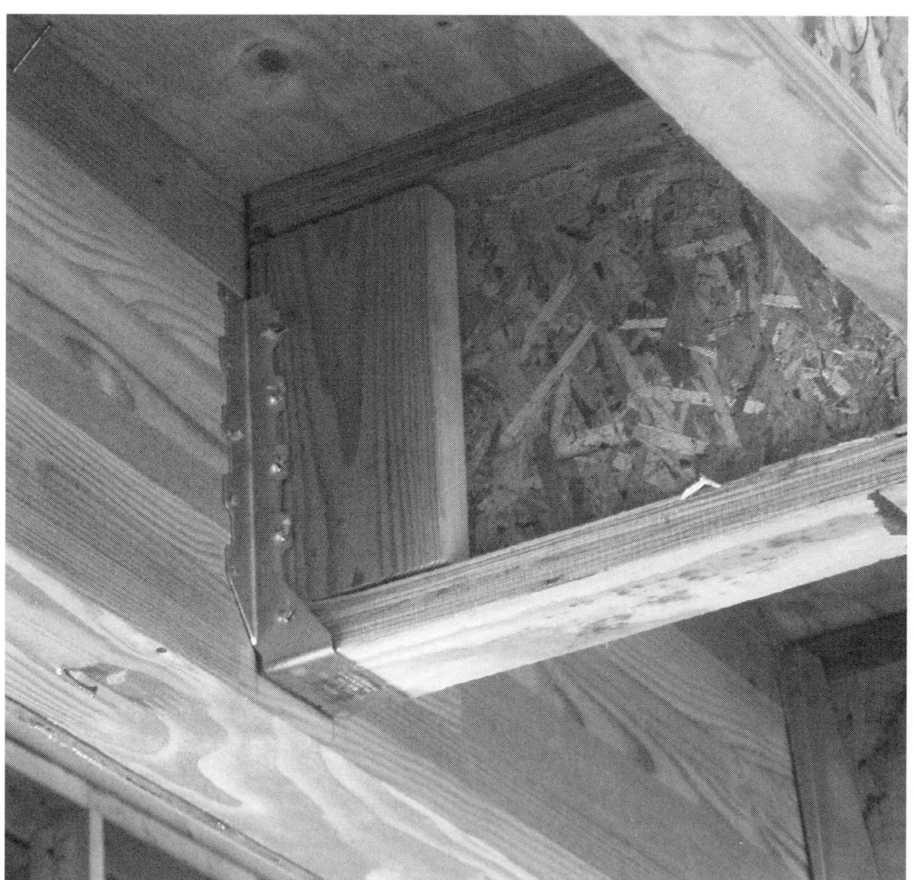

FIGURE 5.45
Blocking where an I-joist frames into a glue-laminated wood beam, as illustrated in Figure 5.44*A*. This blocking acts as a web stiffener and also prevents the I-joist from rotating within the metal hanger, which is not tall enough to restrain the upper flange of the joist. The need for blocking depends on the size of the I-joist, the magnitude of the loads, and the type of hanger used. *(Photo by Joseph Iano)*

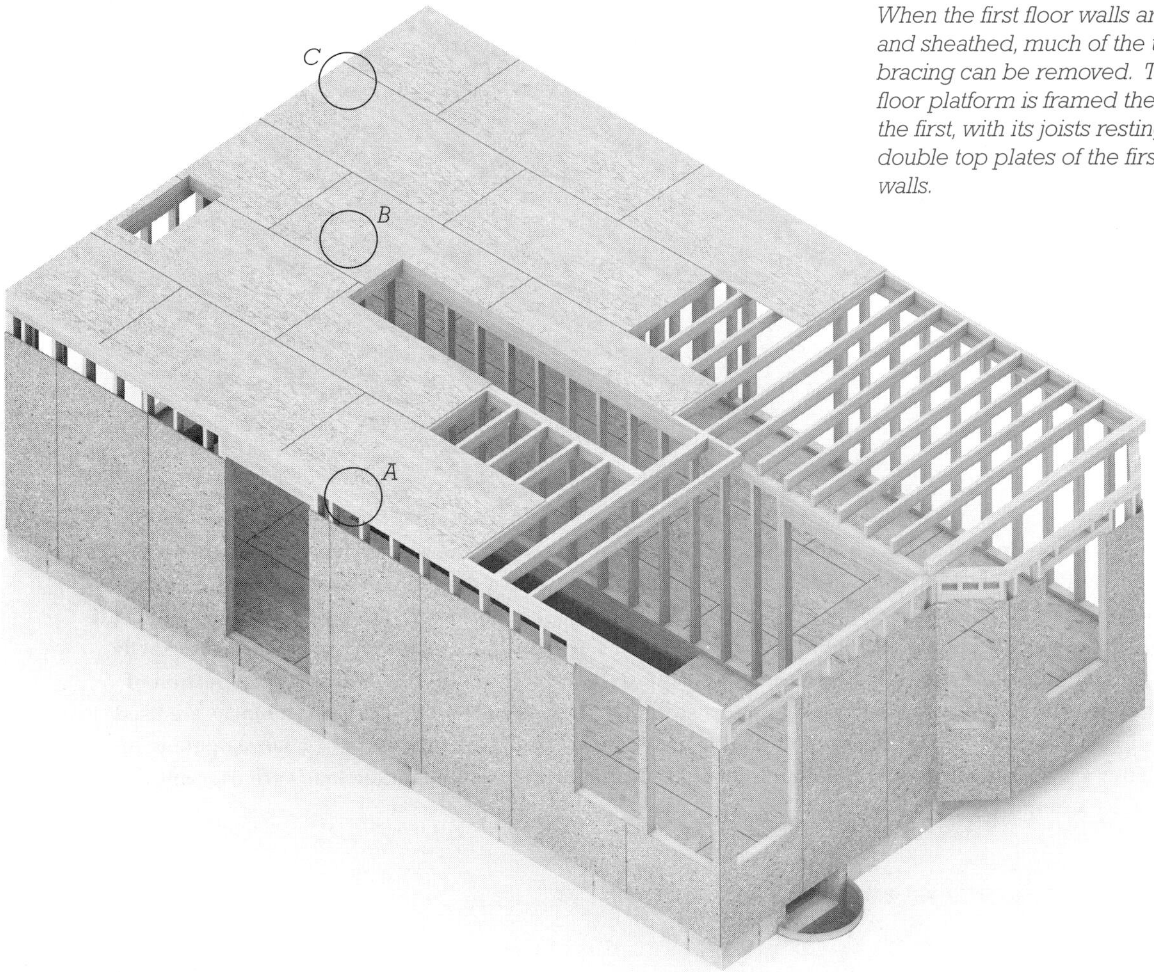

When the first floor walls are complete and sheathed, much of the temporary bracing can be removed. The second floor platform is framed the same as the first, with its joists resting on the double top plates of the first floor walls.

5

FIGURE 5.46
Step Five in erecting a two-story platform frame building: building the upper floor platform. The letters *A*, *B*, and *C* indicate portions of the framing that are detailed in Figure 5.44.

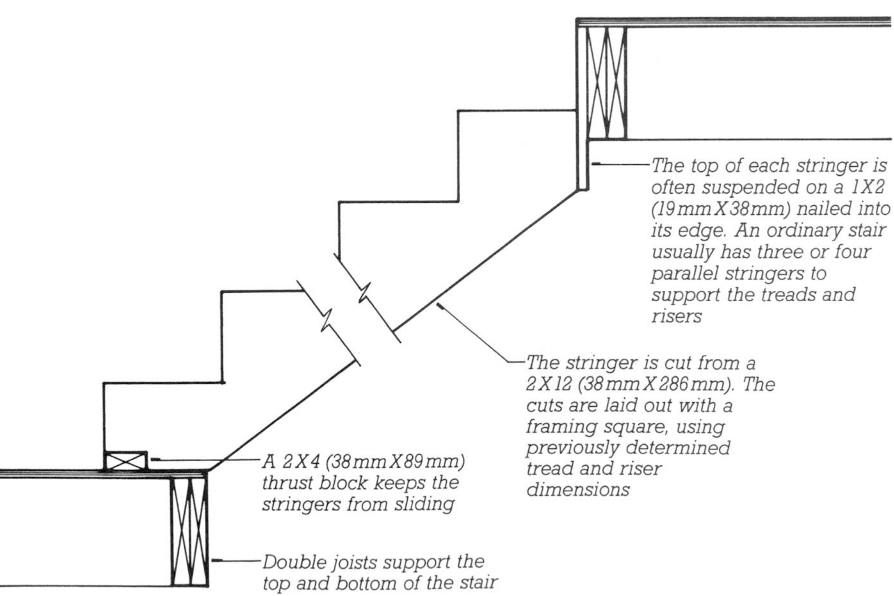

The top of each stringer is often suspended on a 1X2 (19mm X 38mm) nailed into its edge. An ordinary stair usually has three or four parallel stringers to support the treads and risers
The stringer is cut from a 2X12 (38mm X 286mm). The cuts are laid out with a framing square, using previously determined tread and riser dimensions
A 2X4 (38mm X 89mm) thrust block keeps the stringers from sliding
Double joists support the top and bottom of the stair

FIGURE 5.47
Interior stairways are usually framed as soon as the upper-floor platform is completed. This gives the carpenters easy up-and-down access during the remainder of the work. Temporary treads of joist scrap or plywood are nailed to the stringers. These will be replaced by finish treads after the wear and tear of construction are finished.

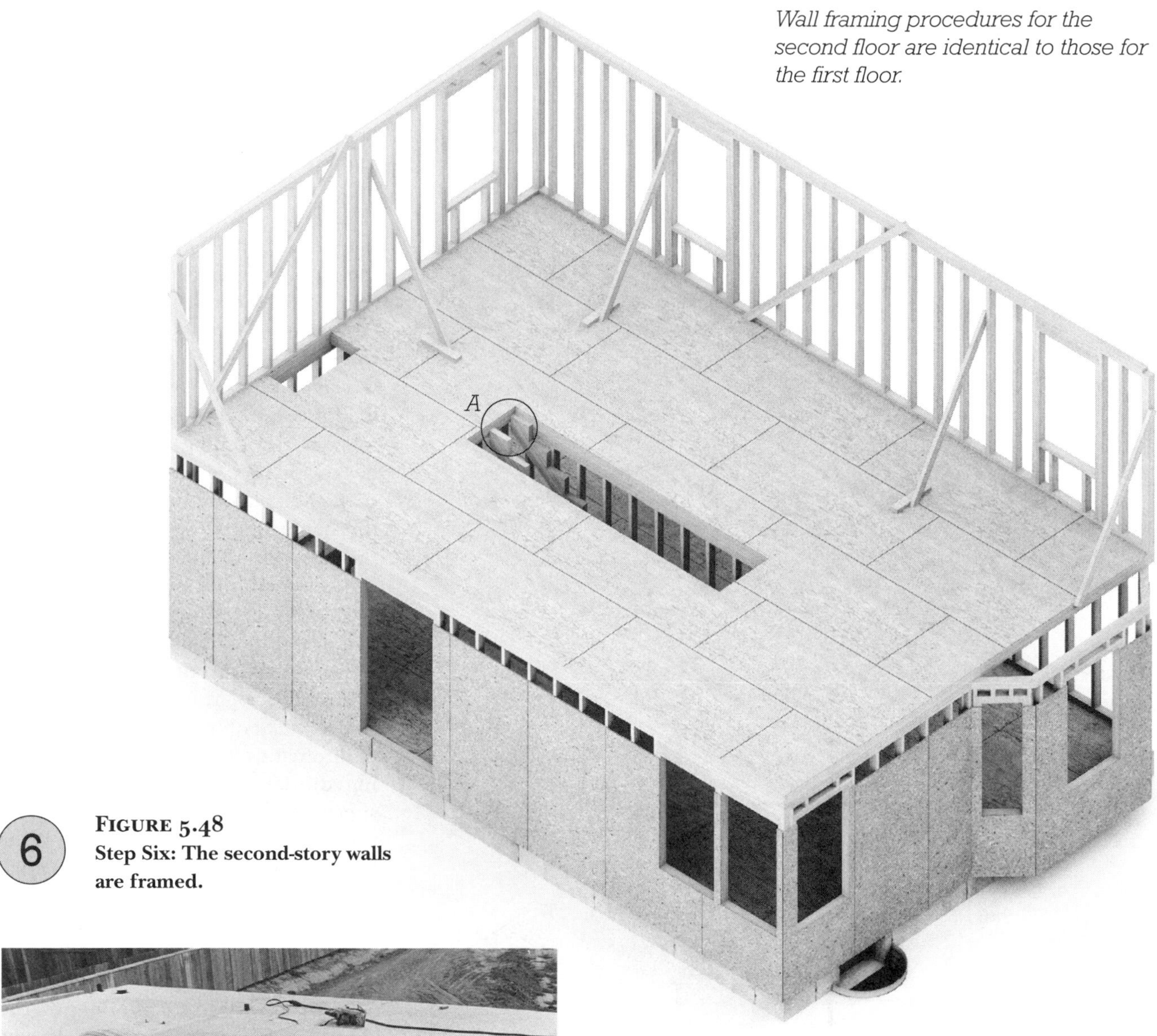

Wall framing procedures for the second floor are identical to those for the first floor.
A

6

FIGURE 5.48
Step Six: The second-story walls are framed.

FIGURE 5.49
Installing upper-floor combination subflooring/underlayment. The grade of the plywood panels used is C–C Plugged. Its top (plugged) face has all surface voids filled and is lightly sanded, making it sufficiently smooth to allow carpeting to be installed directly over the subfloor without additional underlayment. The long edges of the panels have interlocking tongue-and-groove joints to prevent deflection of panel edges under a heavy concentrated load such as a standing person or the leg of a piano. *(Courtesy of APA–The Engineered Wood Association)*

Roof Framing

The generic roof shapes for wood light frame buildings are shown in Figure 5.50. These are often combined to make roofs that are suited to the covering of more complex plan shapes and building volumes.

For structural stability, the rafters in gable and hip roofs must be securely tied together at the top of the supporting walls by well-nailed ceiling joists to make what is, in effect, a series of triangular trusses. If the designer wishes to eliminate the ceiling joists to expose the sloping underside of the roof as the finished ceiling surface, a beam or bearing wall must be inserted at the ridge or a system of exposed horizontal ties must be used in place of the ceiling joists. Sometimes a designer wishes to raise the ceiling joists or exposed ties to a higher elevation than the tops of the wall plates. This greatly increases the bending forces in the rafters and should be done only after consultation with a structural engineer.

Although a college graduate architect or engineer would find it difficult to lay out all the cuts of the rafters for a sloping roof using trigonometry, a skilled carpenter, without resorting to mathematics, has little problem making the layout if the *pitch* (slope) is specified as a ratio of *rise* to *run*. Rise is the vertical dimension and run is the horizontal. In the United States, pitch is usually given on the architect's drawings as inches of rise per foot (12 inches) of run. A skilled carpenter uses these two figures on the two edges of a *framing square* to lay out the rafter as shown in Figures 5.52 and 5.58. The actual length of the rafter is never figured, nor does it need to be, because

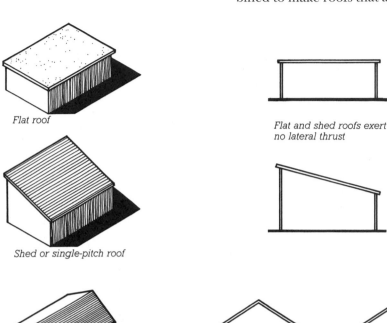

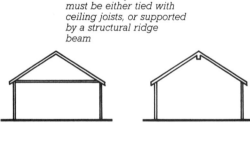

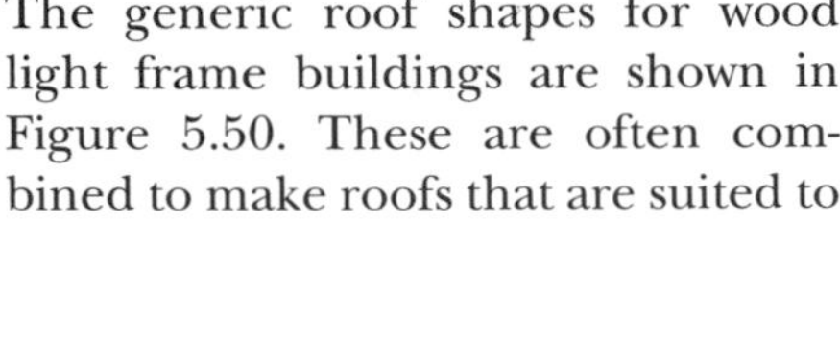

Flat roof
Flat and shed roofs exert no lateral thrust

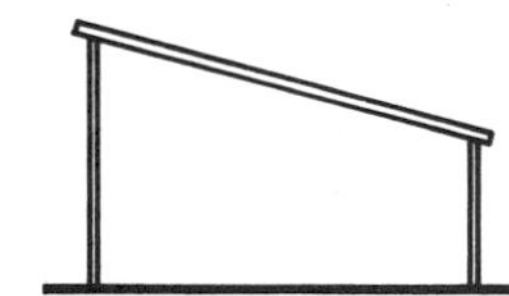

Shed or single-pitch roof

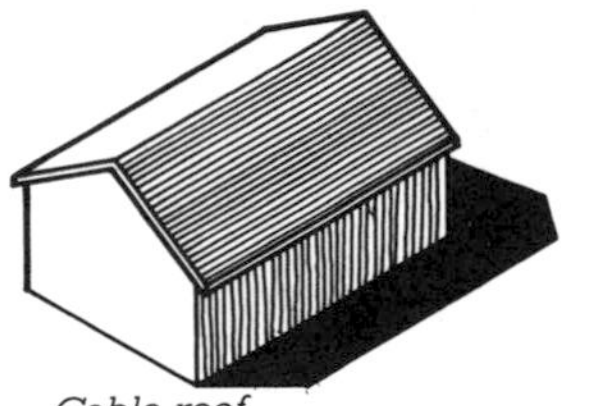
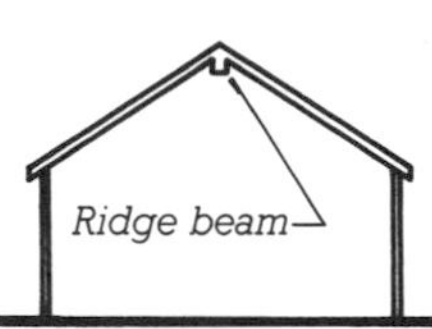
Gable roof
Ceiling joist
Ridge beam
Gable and hip rafters must be either tied with ceiling joists, or supported by a structural ridge beam

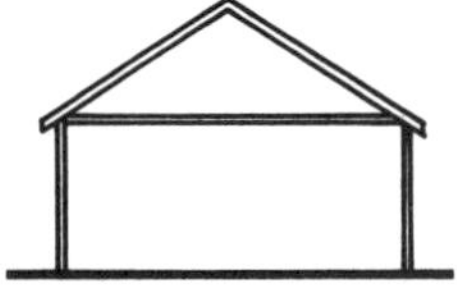

Hip roof

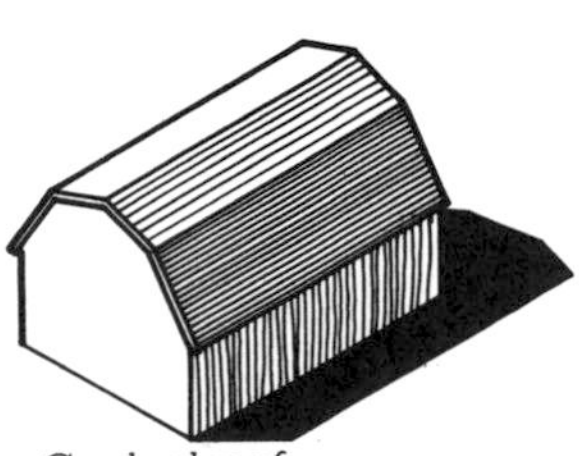
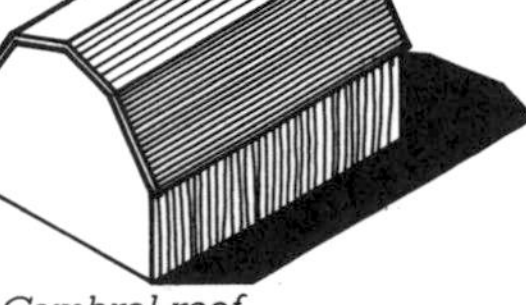
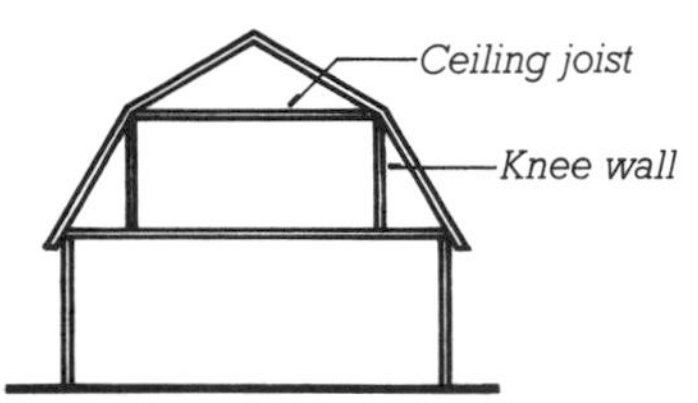
Gambrel roof
Ceiling joist
Knee wall
Gambrel and mansard roofs require both knee walls and ceiling joists for structural stability

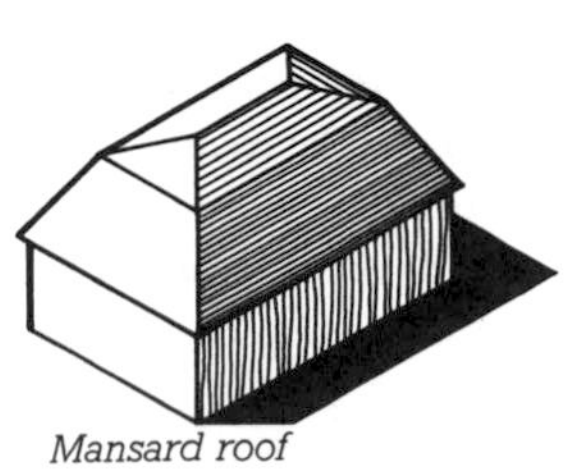
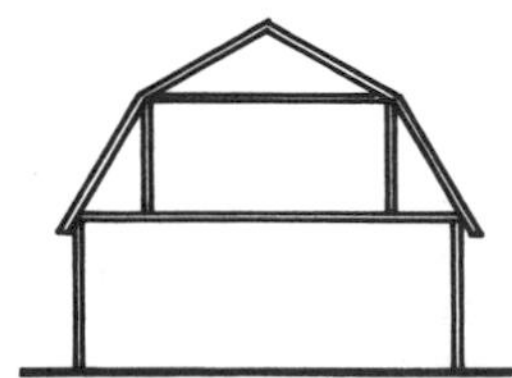
Mansard roof

FIGURE 5.50
Basic roof shapes for wood light frame buildings.

all the measurements are made as horizontal and vertical distances with the aid of the square. Today, many carpenters prefer to do rafter layout with the aid of tables that give actual rafter lengths for various pitches and horizontal distances; these tables are stamped on the framing square itself or printed in pocket-size booklets. Also available are hand-held calculators that are specially programmed to find dimensions of rafters.

Hips and *valleys* introduce another level of trigonometric complexity in rafter layout, but the experienced carpenter has little difficulty even here: Again, he or she can use published tables for hip and valley rafters or do the layout the traditional way, as illustrated in Figure 5.56. The head carpenter lays out only one rafter of each type by these procedures. This then becomes the *pattern rafter* from which other carpenters trace and cut the remainder of the rafters (Figure 5.59).

The balloon frame is closely connected with the level of industrialization which had been reached in America [in the early 19th century]. Its invention practically converted building in wood from a complicated craft, practiced by skilled labor, into an industry. . . . This simple and efficient construction is thoroughly adapted to the requirements of contemporary architects . . . elegance and lightness [are] innate qualities of the balloon-frame skeleton.

Sigfried Giedion, *Space, Time and Architecture: The Growth of a New Tradition,* 1967.

In areas that are subject to hurricanes, the building codes may require that rafters be attached to their supporting walls with sheet metal rafter anchors such as the one shown in Figure 3.46. The type, size, and spacing of the nails that attach the roof sheathing to the rafters are also closely controlled. The intent of both of these measures is to reduce the likelihood that the roof will be blown off in high winds.

Similar to requirements for bridging in floor joists, the International Residential Code requires bridging for ceiling joists and rafters where the depth of these members exceeds their thickness by a ratio of 6 to 1 based on nominal dimensions (for conventional solid wood framing, members deeper than 2 × 12 or 38 × 286 mm). Where the undersides of the joists or rafters remain unfinished, wood strapping nailed perpendicular to the bottoms of the joists or rafters may be used instead of solid blocking or crossbracing installed between framing members.

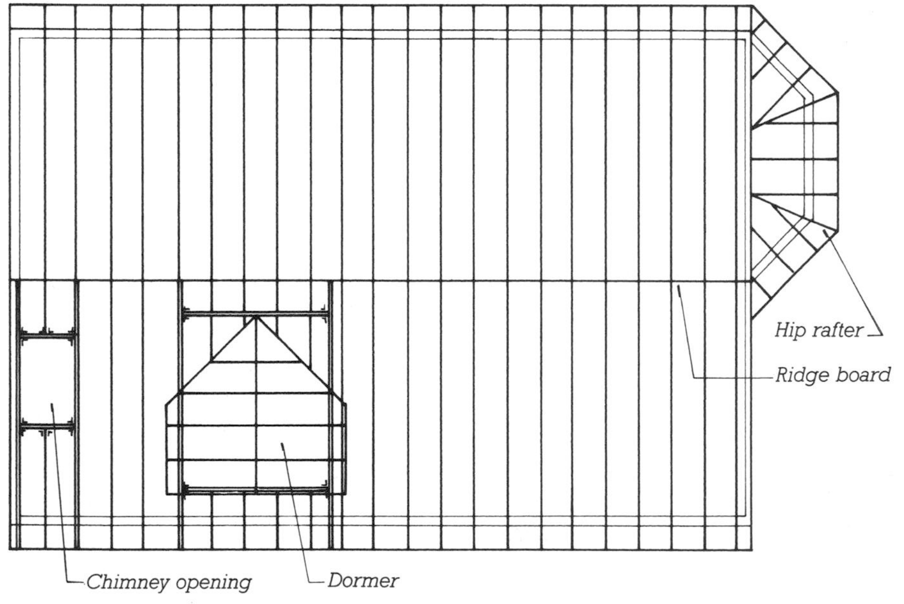

FIGURE 5.51
A roof framing plan for the building illustrated in Figure 5.53. The dormer and chimney openings are framed with doubled header and trimmer rafters. The dormer is then built as a separate structure that is nailed to the slope of the main roof.

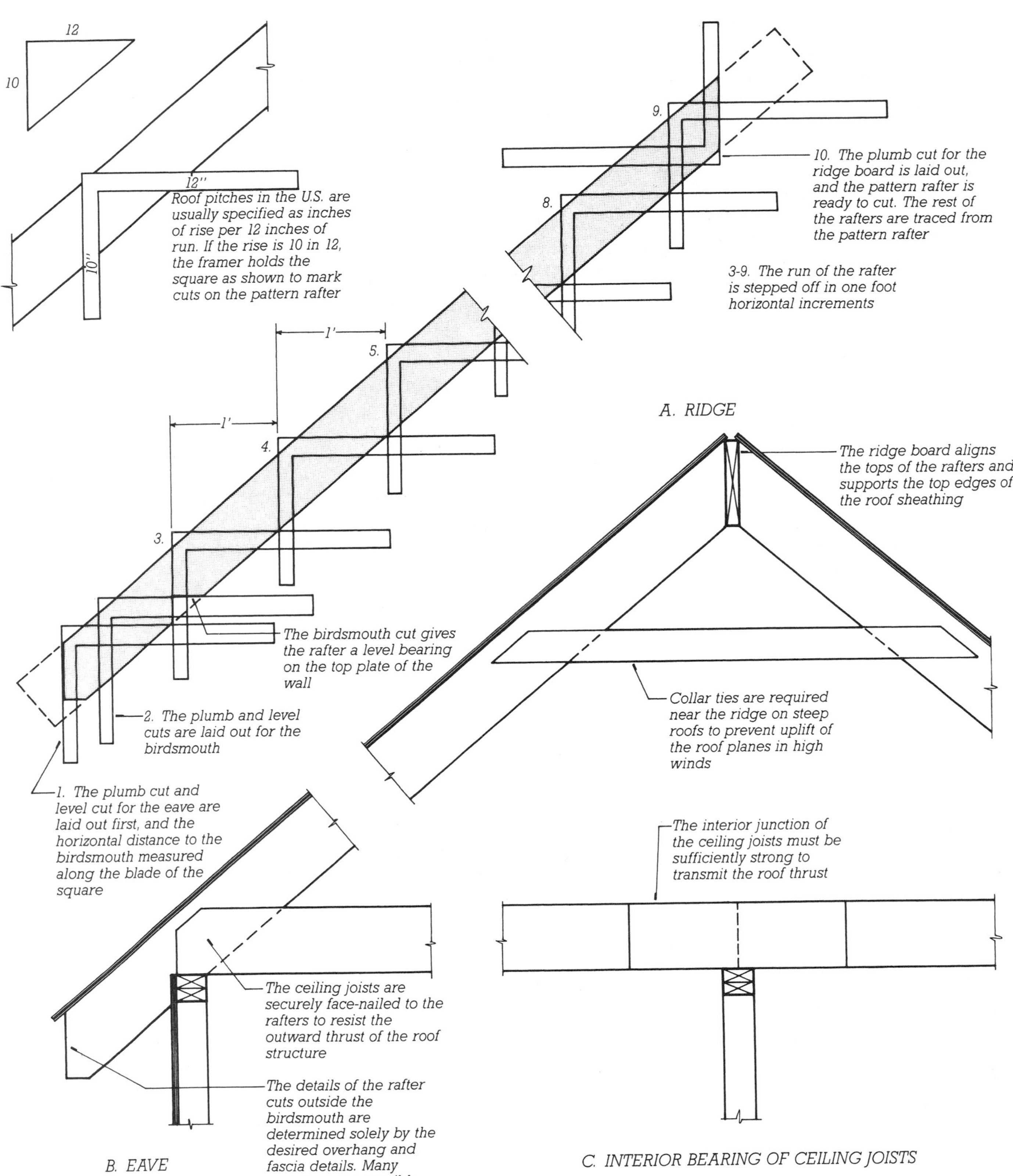

12
10
12''
10''
Roof pitches in the U.S. are usually specified as inches of rise per 12 inches of run. If the rise is 10 in 12, the framer holds the square as shown to mark cuts on the pattern rafter
9.
8.
10. The plumb cut for the ridge board is laid out, and the pattern rafter is ready to cut. The rest of the rafters are traced from the pattern rafter
3-9. The run of the rafter is stepped off in one foot horizontal increments
1'
5.
4.
3.
A. RIDGE
The ridge board aligns the tops of the rafters and supports the top edges of the roof sheathing
The birdsmouth cut gives the rafter a level bearing on the top plate of the wall
2. The plumb and level cuts are laid out for the birdsmouth
Collar ties are required near the ridge on steep roofs to prevent uplift of the roof planes in high winds
1. The plumb cut and level cut for the eave are laid out first, and the horizontal distance to the birdsmouth measured along the blade of the square
The interior junction of the ceiling joists must be sufficiently strong to transmit the roof thrust
The ceiling joists are securely face-nailed to the rafters to resist the outward thrust of the roof structure
The details of the rafter cuts outside the birdsmouth are determined solely by the desired overhang and fascia details. Many variations are possible
B. EAVE
C. INTERIOR BEARING OF CEILING JOISTS

FIGURE 5.52
Roof framing details and procedures: The lettered details are keyed to Figure 5.53. The remainder of the page shows how a framing square is used to lay out a pattern rafter, reading from the first step at the lower end of the rafter to the last step at the top.

The ceiling joists above the second floor (which also serve as the attic floor joists) are toenailed to the tops of the second floor walls. A few rafters are then erected to support the ridge board, and the remainder of the rafters are put up. Double headers and trimmers are used around openings in the roof.

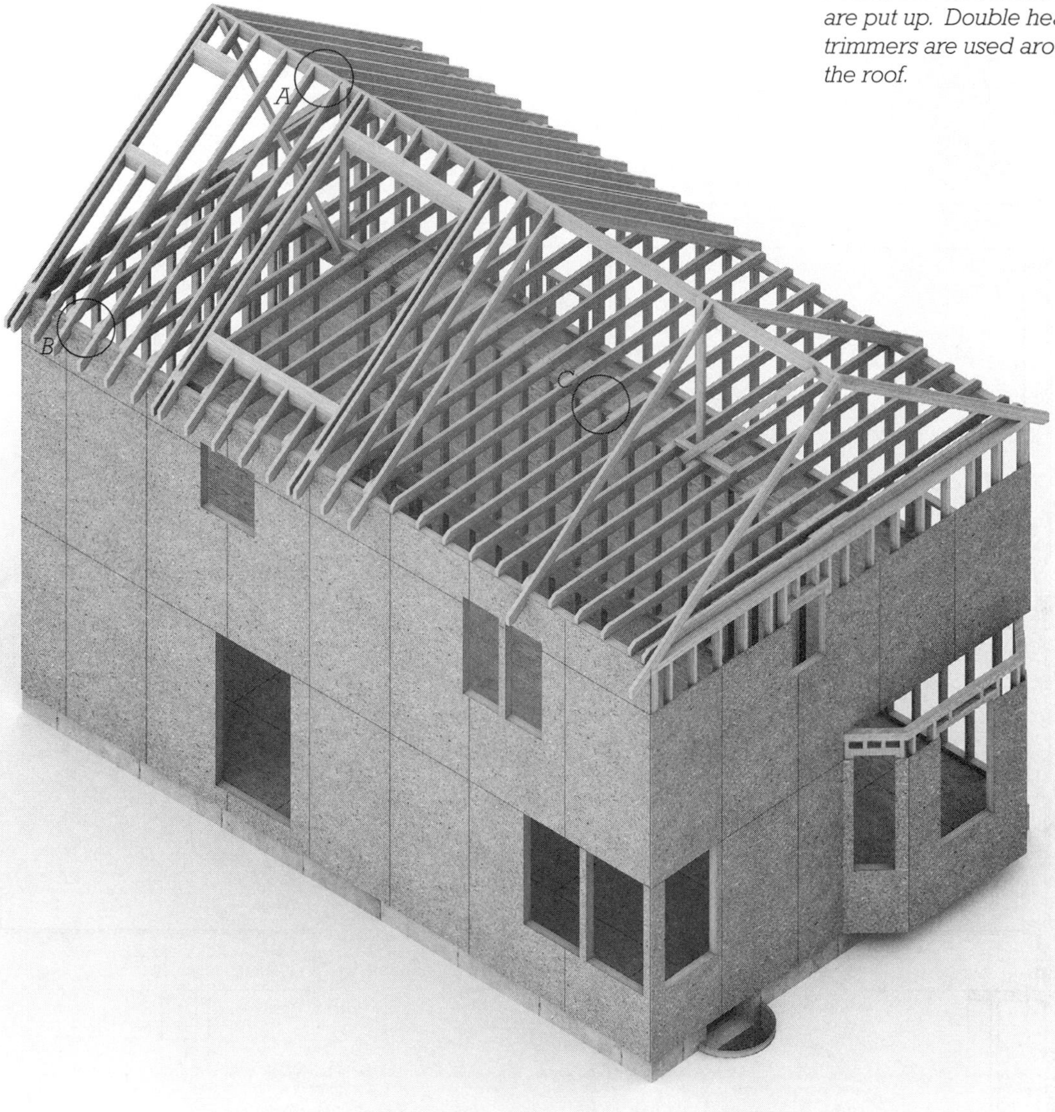

FIGURE 5.53
Step Seven: framing the attic floor and roof. The outer ends of attic floor joists are not usually headed off, but instead are face nailed to the rafter pairs that overlap them.

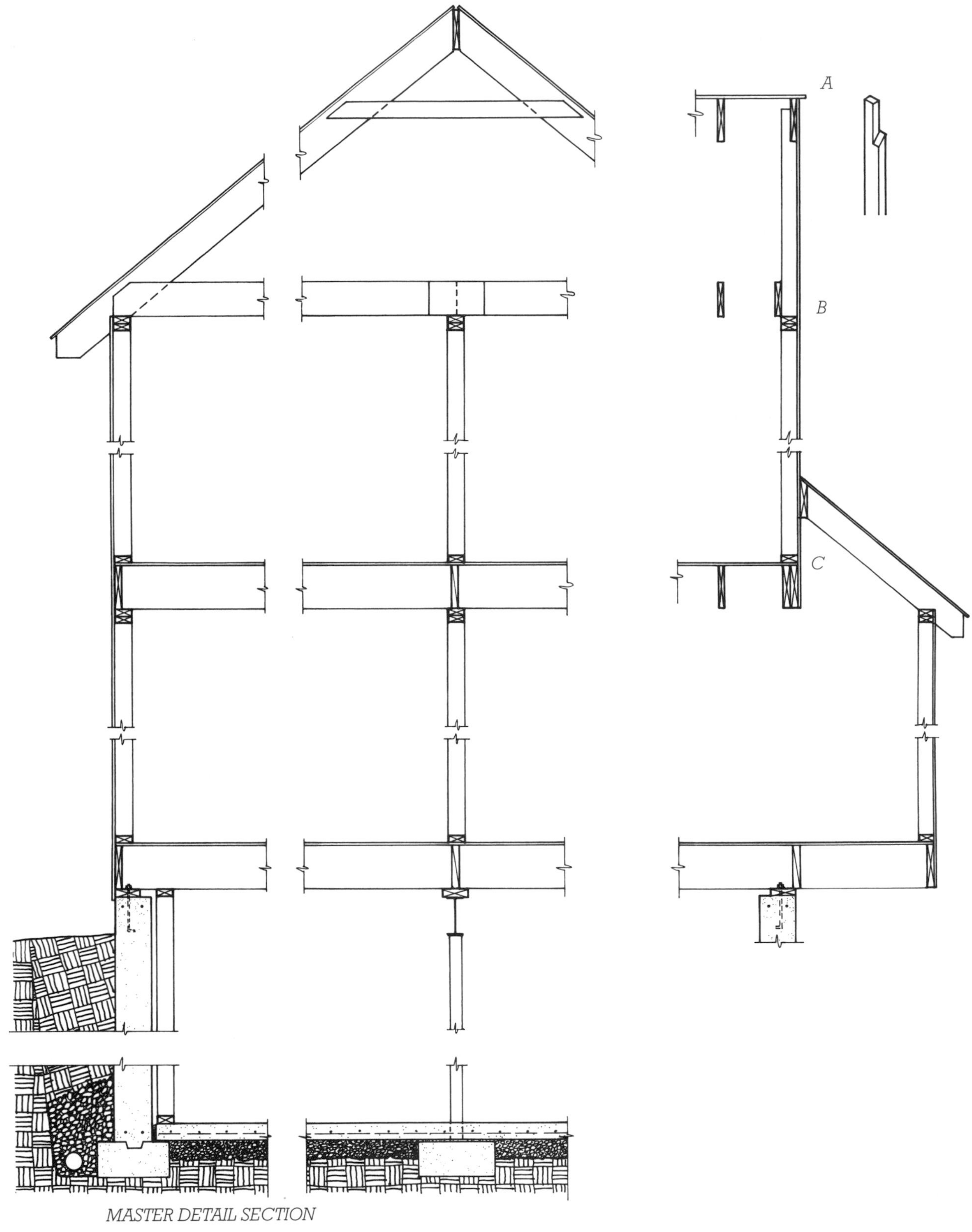

FIGURE 5.54
A summary of the major details for the structure shown in Figure 5.55, aligned in relationship to one another. The gable end studs are cut as shown in *A* and face nailed to the end rafter.

The framing of the building is completed with installation of the roof sheathing, the gable end walls, and the dormer.

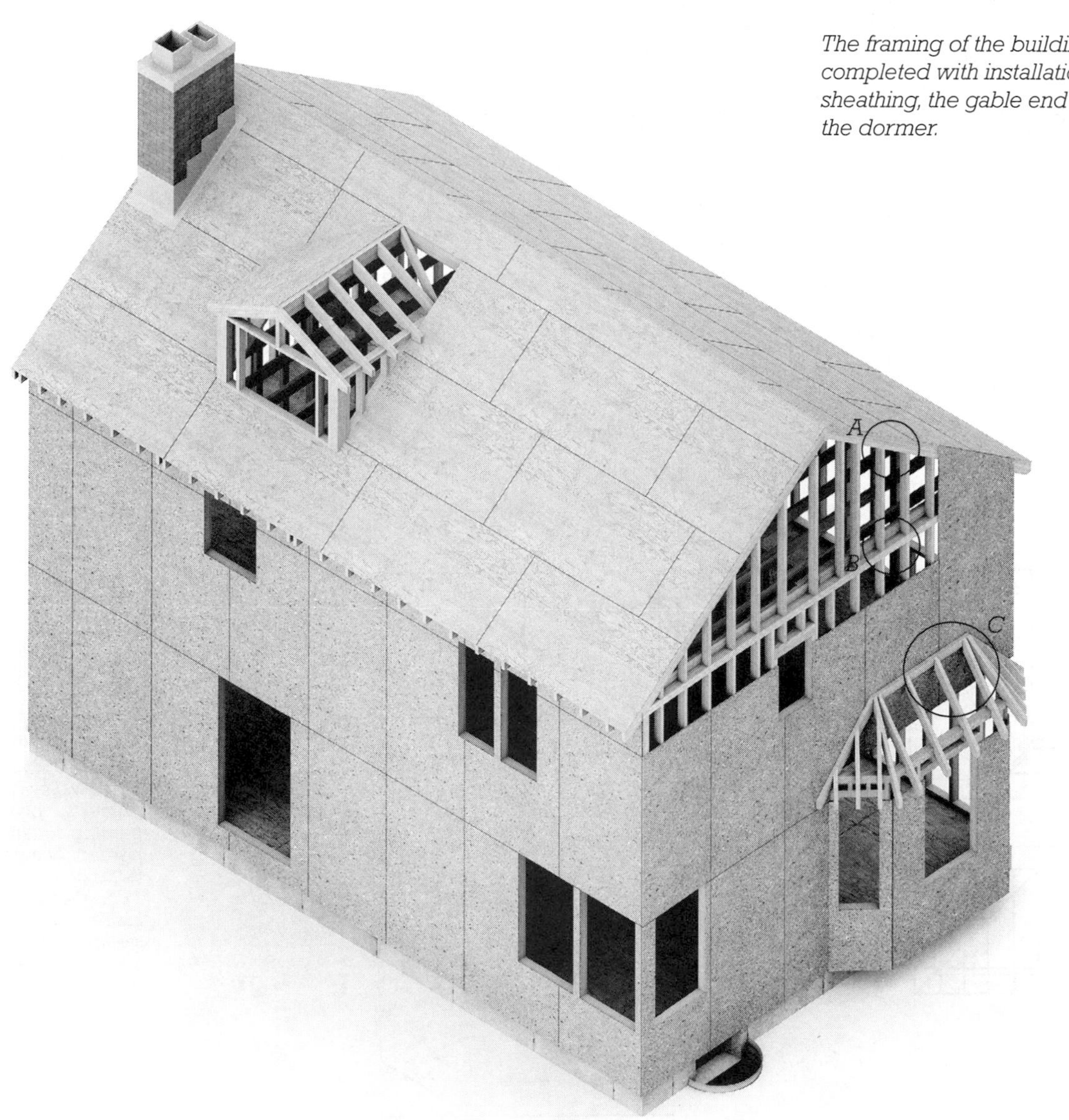

8

FIGURE 5.55
Step Eight: The sheathed frame is completed. The letters *A*, *B*, and *C* indicate portions of the framing that are detailed in Figure 5.54.

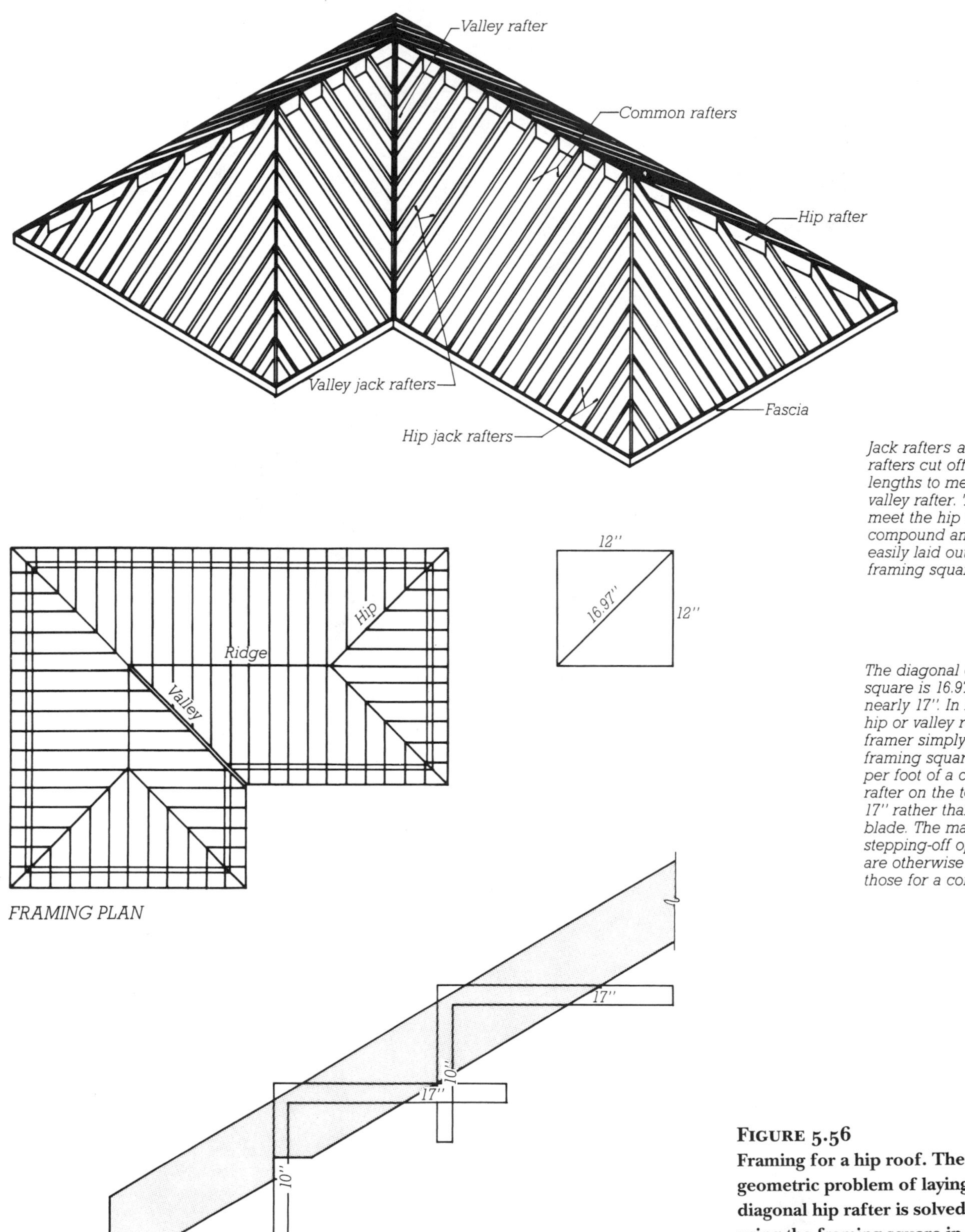

Jack rafters are common rafters cut off at varying lengths to meet a hip or valley rafter. The jacks meet the hip or valley at a compound angle that is easily laid out with a framing square

The diagonal of a 12" square is 16.97", or very nearly 17". In laying out a hip or valley rafter, the framer simply aligns the framing square to the rise per foot of a common rafter on the tongue, and 17" rather than 12" on the blade. The marking and stepping-off operations are otherwise identical to those for a common rafter

FIGURE 5.56
Framing for a hip roof. The difficult geometric problem of laying out the diagonal hip rafter is solved easily by using the framing square in the manner shown. Toward the bottom of the rafter shown in the lower part of this figure is a *birdsmouth cut*, an angled notch in the rafter that allows the rafter to seat securely on the top plate of the wall.

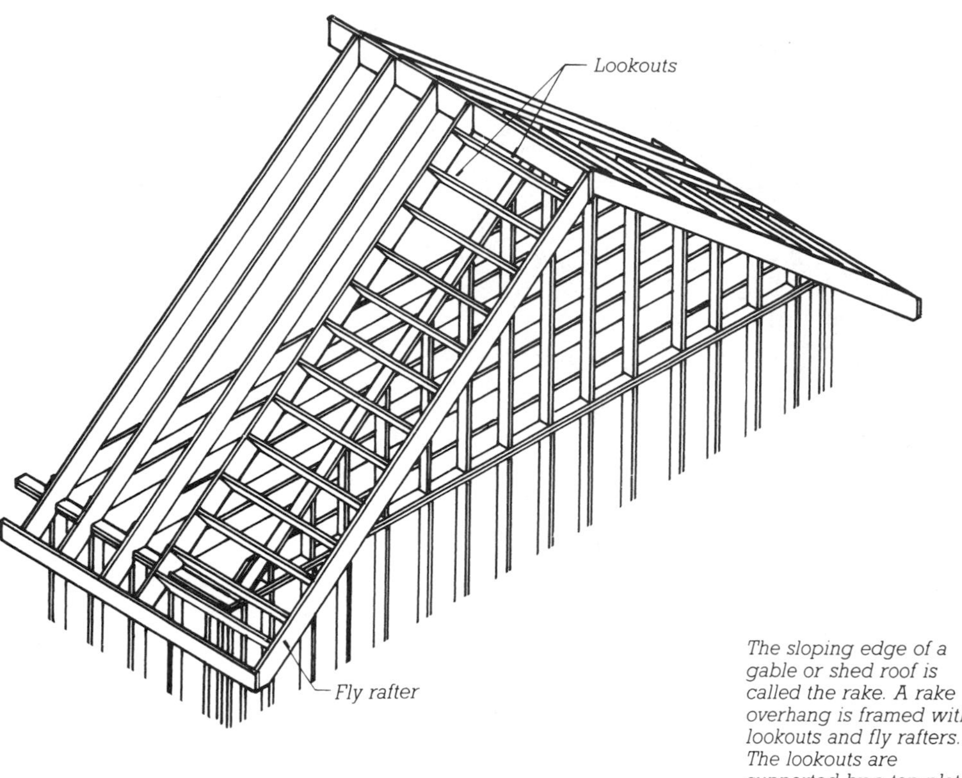

Lookouts
Fly rafter
The sloping edge of a gable or shed roof is called the rake. A rake overhang is framed with lookouts and fly rafters. The lookouts are supported by a top plate over the gable end studs

FIGURE 5.57
Framing for an overhanging *rake*.

FIGURE 5.58
A framing square being used to mark rafter cuts. The run of the roof, 12 inches, is aligned with the edge of the rafter on the blade (the wider leg) of a square, and the rise, 7 inches in this case, is aligned on the tongue (the narrower leg) of the square. A pencil line along the tongue will be perfectly vertical (*a plumb cut*) when the rafter is installed in the roof, and one along the blade will be horizontal (*a level cut*). True horizontal and vertical distances can be measured on the blade and tongue, respectively. The layout of these types of cuts can also be seen in Figures 5.52 and 5.56. *(Photo by Edward Allen)*

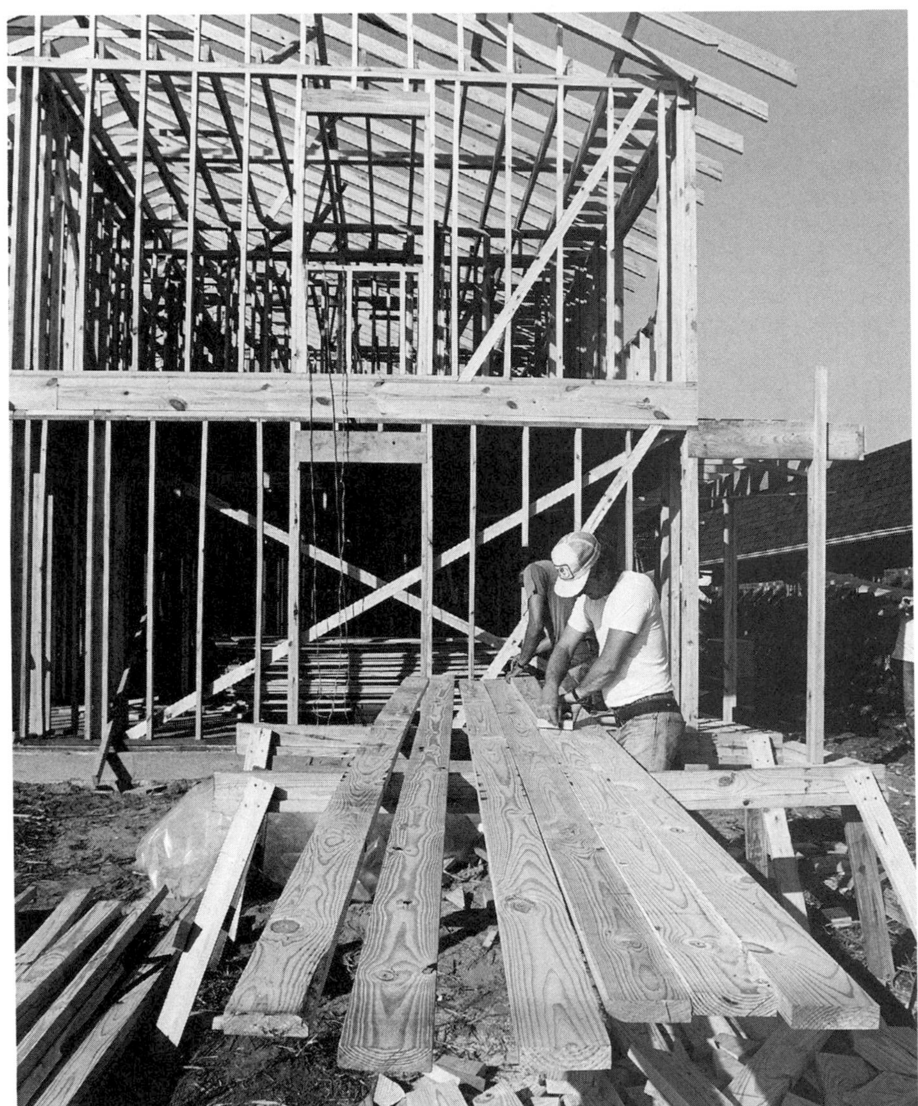

FIGURE 5.59
Tracing a pattern rafter to mark cuts for the rest of the rafters. The corner of the building behind the carpenters has let-in corner braces on both floors, and most of the rafters are already installed. *(Courtesy of Southern Forest Products Association)*

FIGURE 5.60
I-joists may be used as rafter material instead of solid lumber. *(Courtesy of Trus Joist MacMillan)*

FIGURE 5.61
Applying plywood roof sheathing to a half-hipped roof. Blocking between rafters at the wall line (far left in the photo) has been drilled with large holes for attic ventilation. The line of horizontal blocking between studs is to support the edges of plywood siding panels applied in a horizontal orientation. *(Courtesy of APA–The Engineered Wood Association)*

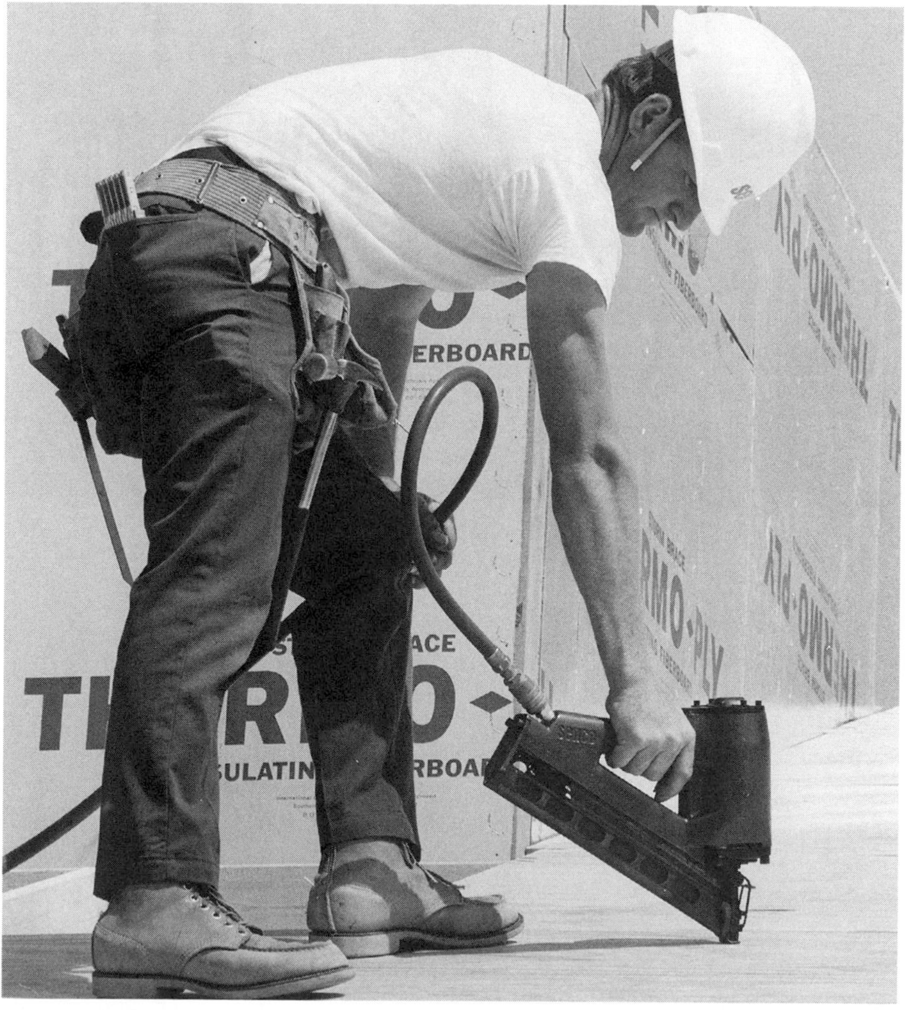

FIGURE 5.62
Fastening roof sheathing with a pneumatic nail gun. *(Courtesy of Senco Products, Inc.)*

FIGURE 5.63
A light wood frame structure fully sheathed with OSB structural panels.

Variations on Wood Light Frame Construction

Framing for Increased Thermal Efficiency

The 2 × 4 (38 × 89 mm) has been the standard-size wall stud since light framing was invented. In recent years, however, pressures for heating fuel conservation have led to energy code requirements for more thermal insulation than can be inserted in the cavities of a wall framed with members only 3½ inches (89 mm) deep. One solution is to frame walls with 2 × 6 (38 × 140 mm) studs, usually at a spacing of 24 inches (610 mm), creating an insulation cavity 5½ inches (140 mm) deep. Alternatively, 2 × 4-framed walls may be covered either inside or out with insulating plastic foam sheathing, thus reaching an insulation value about the same as that of a conventionally insulated 2 × 6-framed wall. For even greater insulation performance, 2 × 6 studs and insulating sheathing may be used in tandem. Alternatively, in very cold climates, even deeper wall assemblies that can achieve greater insulation values may be constructed of two separate layers of wall studs or of vertical truss studs made up of pairs of ordinary studs joined at intervals by plywood plates. Some of these constructions methods are illustrated in Figures 7.17 through 7.21.

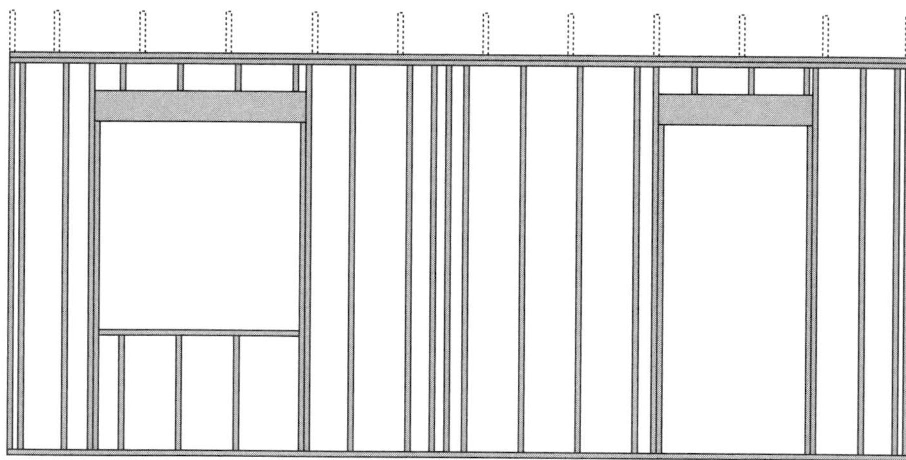

A. Conventional wall framing with studs @ 16 o.c.

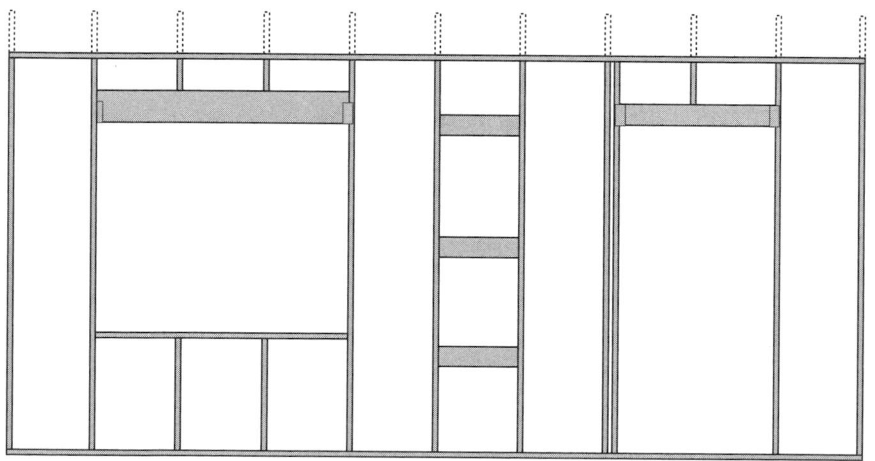

B. Wall framed with advanced framing techniques

Figure 5.64
Comparison of walls framed with conventional and advanced techniques. Wall *A* is framed as explained in Figure 5.32. Studs are spaced at 16 inches (406 mm) o.c., the layout of the wall and its openings is not coordinated with the framing module, and standard details are used for corners, openings, and other features. In wall *B*, studs are spaced at 24 inches (610 mm) o.c., the length of the wall and the location and size of its openings have been coordinated to the greatest extent possible with the 24-inch module, and redundant framing members have been eliminated. Note that with the single top plate, floor joists or roof rafters (shown dashed in these figures) bearing on wall *B* must fall directly over studs below. The total length of framing lumber used in wall *B* is half of that required for wall *A*. Though only a foot and a half shorter in length, wall *B* also can be sheathed with five standard-sized sheathing panels, whereas wall *A* requires six. Even if wall *A* is constructed of 2 × 4 studs and wall *B* is constructed of 2 × 6 studs, the overall savings in materials and reduction in waste in wall *B* are substantial.

Framing for Optimal Lumber Usage

In wood light frame structures constructed with *advanced framing techniques* (also called *optimum value engineering*), special attention is given to minimizing the use of redundant or structurally unnecessary wood members, thereby reducing the amount of lumber required to construct the frame and, once the frame is insulated, increasing its thermal efficiency (Figure 5.64). A variety of techniques may be used, including:

- Spacing framing members at 24 inches (610 mm) rather than 16 inches (406 mm) o.c.: Wider spacing of framing members reduces the amount of lumber required. In exterior walls, thermal efficiency is also improved by the reduction in thermal bridging area in comparison to walls framed with more closely spaced members.
- Designing to a 24-inch (610-mm) module: When the outside dimensions of a framed structure conform to a 24-inch module, sheathing panel waste is minimized. Planning rough opening sizes and locations in floors, walls, and roofs to conform, where possible, to modular dimensions can reduce waste even further. Designing to modular dimensions also reduces wastage of interior wallboard.
- Using single top plates in all walls, both bearing and nonbearing: In the case of bearing walls, this requires floor or roof framing members to align directly over studs in the walls that support them.
- Minimizing other unneeded framing members: Don't use headers

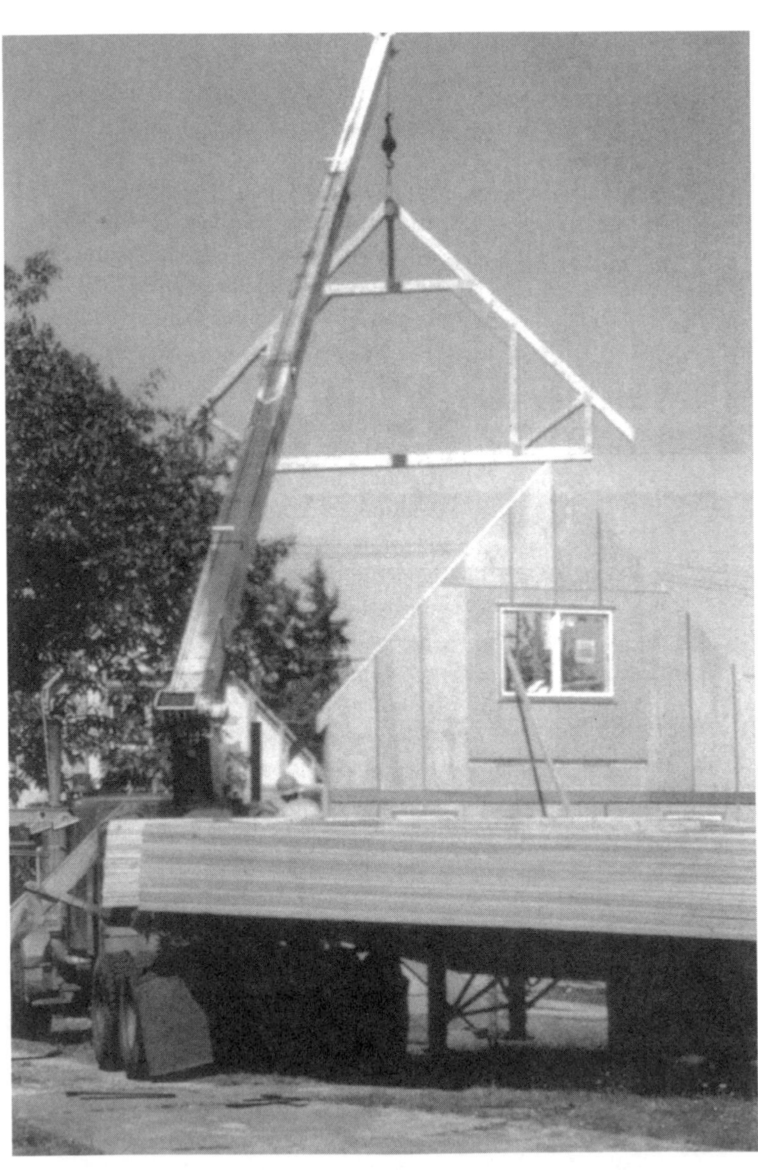

FIGURE 5.65
Roof trusses are typically lifted to the roof by a boom mounted on the delivery truck. This is one of a series of identical attic trusses, which will frame a habitable space under the roof. *(Photo by Rob Thallon)*

over openings in nonbearing walls, since they are not needed; in bearing walls, use headers only as deep as required for the loads and span. Where corner studs serve only to provide nailing surfaces or support for wallboard, use other, less wasteful blocking techniques or metal clips designed for this purpose instead. Replace jack studs supporting headers at window and door openings with metal hangers; eliminate unneeded cripple studs under rough sills. All of these techniques save lumber and, in exterior walls, increase energy efficiency by reducing thermal bridging through solid framing members.

• Eliminating unneeded plywood and OSB wall sheathing: Where let-in bracing is structurally adequate for lateral force resistance, eliminate structural panel wall sheathing entirely and cover walls with insulating sheathing for better thermal efficiency. Where structural panels are required, use the minimum extent of panels necessary.

Advanced framing techniques rely on unconventional framing methods and significantly reduce redundancy in the building frame. For these reasons, they should not be used without guidance from a structural engineer or other qualified design professional, and special review and approval from local building authorities may be required. Nevertheless, where these techniques are used, significant benefits can be realized. According to the National Association of Home Builders Partnership for Advancing Technology in Housing, advanced framing techniques can reduce the amount of lumber used in a wood light frame structure by up to 19 percent and improve the energy efficiency of the insulated structure by as much as 30 percent.

Prefabricated Framing Assemblies

Roof trusses, and to a lesser extent *floor trusses*, are used in platform frame buildings because of their speed of erection, economy of material usage, and long spans. Though many floor trusses are light enough to be lifted and installed by two carpenters, most truss assemblies are erected with the aid of a small crane that often is attached to the truck on which the trusses are delivered (Figure 5.65). Roof trusses are particularly slender in proportion, usually only 1½ inches (38 mm) thick and capable of spanning 24 to 32 feet (7.5 to 10 m). They must be temporarily braced during construction to prevent buckling or the dominolike collapse of all the trusses until they are adequately secured permanently by the application of roof sheathing panels and interior finishes (Figure 5.66).

Manufactured wall panels have been adopted more slowly than roof and floor trusses, and are used mostly by large builders who build hundreds or thousands of houses per year. For the smaller builder, wall framing can be done on site with the same amount of material as with panels and with

FIGURE 5.66
A roof framed with prefabricated trusses. Approximately midway up the upper chords of the trusses, temporary strapping ties the trusses to one another for bracing. Other diagonal bracing, not visible in this photograph, ties the trusses to floor or ceiling framing to prevent the entire row from collectively tipping sideways. *(Photo by Joseph Iano)*

For Preliminary Design of a Wood Light Frame Structure

Estimate the depth of **wood rafters** on the basis of the horizontal (not slope) distance from the outside wall of the building to the ridge board in a gable or hip roof and the horizontal distance between supports in a shed roof. A 2 × 4 rafter spans approximately 7 feet (2.1 m), a 2 × 6 10 feet (3.0 m), a 2 × 8 14 feet (4.3 m), and a 2 × 10 17 feet (5.2 m).

The depth of **wood light roof trusses** is usually based on the desired roof pitch. A typical depth is one-quarter of the width of the building, which corresponds to a 6⁄12 pitch in a gable truss. Trusses are generally spaced 24 inches (600 mm) o.c. and can span up to approximately 65 feet (20 m).

Estimate the depth of **wood floor joists** as follows: 2 × 6 joists span up to 9 feet (2.7 m), 2 × 8 joists 11 feet (3.4 m), 2 × 10 joists 14 feet (4.3 m), and 2 × 12 joists 17 feet (5.2 m).

Estimate the depth of manufactured **wood I-joists** as follows: 9½-inch (240-mm) joists span 16 feet (4.9 m), 11 7⁄8 -inch (300-mm) joists 19 feet (5.8 m), 14-inch (360-mm) joists span 23 feet (7.0 m), and 16-inch (400-mm) joists span 25 feet (7.6 m).

Estimate the depth of **wood floor trusses** as 1⁄18 of their span. Typical depths of floor trusses range from 12 to 28 inches (305–710 mm) in 2-inch (51-mm) increments.

2 × 4 **wood studs** 24 inches (600 mm) o.c. can support attic and roof loads only. Either 2 × 4 studs 16 inches (400 mm) o.c. or 2 × 6 studs 24 inches o.c. can support one floor plus attic and roof. 2 × 6 studs 16 inches o.c. can support two floors plus attic and roof.

Framing members in light frame buildings are usually spaced either 16 or 24 inches (400 or 600 mm) o.c. For actual sizes of dimension lumber in both conventional and metric units, see Figure 3.22.

These approximations are valid only for purposes of preliminary building layout and must not be used to select final member sizes. They apply to the normal range of building occupancies such as residential, office, commercial, and institutional buildings. For manufacturing and storage buildings, use somewhat larger members.

For more comprehensive information on preliminary selection and layout of a structural system and sizing of structural members, see Edward Allen and Joseph Iano, *The Architect's Studio Companion* (4th ed.), New York, John Wiley & Sons, Inc., 2007.

little or no additional overall expenditure of labor, especially when the building requires walls of varying heights and shapes.

Wood Light Frame Construction and the Building Codes

As shown in the table in Figure 1.2, the International Building Code (IBC) allows buildings of almost every occupancy group to be constructed with wood platform framing, classified as Type V construction, but with relatively severe restrictions on height and floor area. For example, a commercial office building, Occupancy Group B, built of Type V-B (unprotected) construction may be two stories in height and 9000 square feet (835 m^2) in area per floor. In comparison, if built of Type IV heavy timber construction, the same building may be five stories tall and

Figure 5.67
This proprietary fire wall consists of light-gauge metal framing, noncombustible insulation, and gypsum board. The metal framing is attached to the wood structure on either side with special clips (not shown here) that break off if the wood structure burns through and collapses, leaving the wall supported by the undamaged clips on the adjacent building. *(Photographs of Area Separation Wall supplied by Gold Bond Building Products, Charlotte, North Carolina)*

FIGURE 5.68
This townhouse has collapsed completely as the result of an intense fire, but the houses on either side, protected by fire walls of the type illustrated in Figure 5.67, are essentially undamaged.
(Photographs of Area Separation Wall supplied by Gold Bond Building Products, Charlotte, North Carolina)

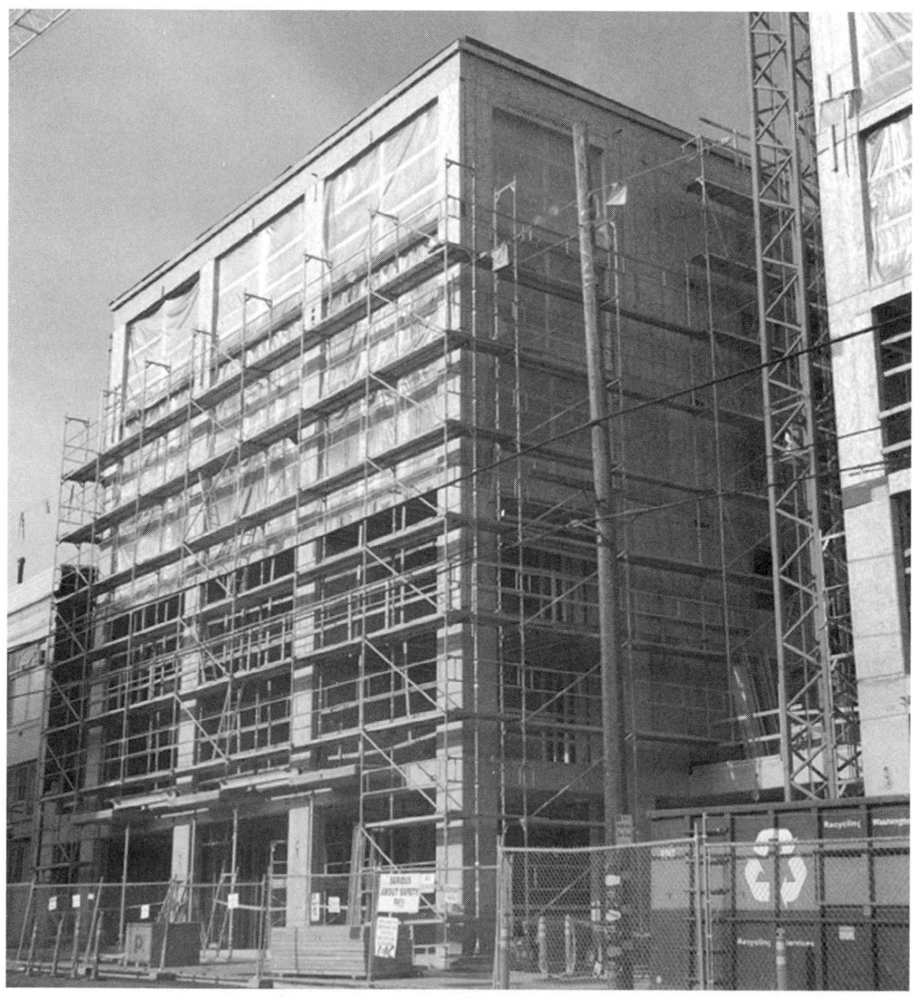

FIGURE 5.69
This mixed-use urban building is being constructed of cast-in-place concrete up to the second-floor level and of light wood framing above. The building code recognizes a variety of such structural system combinations, allowing light wood frame construction to be used in buildings larger in size than otherwise permitted for those constructed exclusively of this system.

up to 36,000 square feet (3345 m^2) per floor and, if built of more fire-resistive noncombustible construction types, even larger. (Recall also that allowable floor areas in this table can in many circumstances be increased by installing an automatic fire suppression sprinkler system in the building, as outlined in Chapter 1 of this book.)

Building size may also be increased by subdividing a structure with fire walls, an approach commonly used in the construction of attached dwellings and row townhouses. A fire wall separates a single structure into separate portions, each of which may have a floor area as large as that normally permitted for an entire building. The required fire resistance of the fire wall is given in a separate table in the code, reproduced in this book as Figure 1.7. Additionally, the fire wall must extend from the foundation through the roof and must be constructed so that it remains standing even if the construction on one side burns completely away. The traditional fire wall is made of brick or concrete masonry, but lighter, less expensive systems using metal framing and gypsum board may also be used (Figures 5.67 and 5.68).

Though platform frame construction is the least fire-resistive of all construction types, building code limits placed on it are sufficiently flexible to allow its use for a diverse range of building uses (Figure 5.69). Furthermore, its economies are such that most building owners will choose platform frame construction over more fire-resistant types of construction if given the opportunity. And despite their vulnerability to fire, the comprehensive life-safety requirements of modern building codes ensure that buildings of wood light frame construction are safe places for their occupants. A study published by the Canadian Wood Council of residential fire deaths in Canada for the years 1993–1995 shows that the rate of fire deaths in Type VA dwellings is about the same as the death rate in dwellings made of noncombustible types of construction. Another Canadian study finds that the rate of death in dwellings fell steadily from about 3.6 per 100,000 residents in 1975 to 1.4 per 100,000 in 1995, with much of the decline attributable to the installation of smoke detectors.

Uniqueness of Wood Light Frame Construction

Wood light framing is popular because it is an extremely flexible and economical way of constructing small buildings. Its flexibility stems from the ease with which carpenters with ordinary tools can create buildings of astonishing complexity in a variety of geometries. Its economy can

Figure 5.70
The W. G. Low house, built in Bristol, Rhode Island, in 1887 to the design of architects McKim, Mead, and White, illustrates both the essential simplicity of wood light framing and the complexity of which it is capable. *(Photo by Wayne Andrews)*

be attributed in part to the relatively unprocessed nature of the materials from which it is made, and in part to mass-market competition among suppliers of components and materials and local competition among small builders.

Platform framing is the one truly complete and open system of construction that we have. It incorporates structure, enclosure, thermal insulation, mechanical installations, and finishes into a single constructional concept. Thousands of products are made to fit it: competing brands of windows and doors; interior and exterior finish materials; electrical, plumbing; and heating products. For better or worse, it can be dressed up to look like a building of wood or of masonry in any architectural style from any era of history. Architects have failed to exhaust its formal possibilities, and engineers have failed to invent a new environmental control system that it cannot assimilate. Wood light frame

FIGURE 5.71
In the late 19th century, the sticklike qualities of wood light framing often found expression in the exterior ornamentation of houses. *(Photo by Edward Allen)*

construction can be used to construct the cheapest and most mundane buildings. Yet, one can look to the best examples of the Carpenter Gothic, Queen Anne, and Shingle-style buildings of the 19th century, or the Bay Region and Modern styles of more recent times, to realize that wood light framing also gives the designer the freedom to make a finely crafted building that nurtures life and elevates the spirit.

FIGURE 5.72
Light wood frame construction is used to build the vast majority of housing in North America. It remains to this day the most affordable and flexible construction system for buildings of this type. *(Project: New Holly Phase I, by Weinstein A|U Architects & Planners. Photo by Joseph Iano)*

FIGURE 5.73
In this contemporary New England cottage, designer Dennis Wedlick has exploited the sculptural possibilities of platform framing. *(Photo: © Michael Moran)*

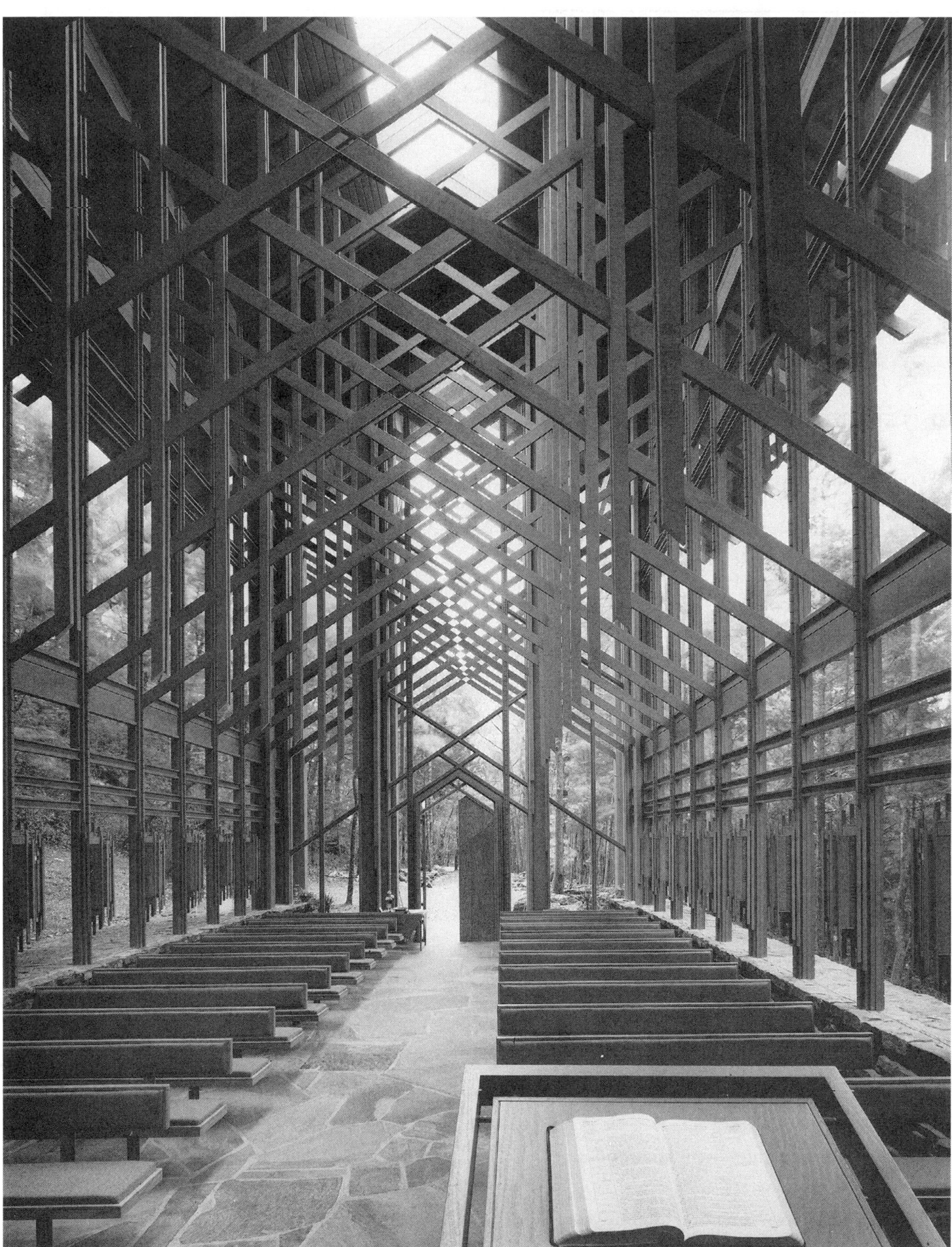

FIGURE 5.74
The Thorncrown Chapel in Eureka Springs, Arkansas, designed by Fay Jones and Associates, Architects, combines large areas of glass with special framing details to create a richly inspiring space. To avoid damage to the sylvan site, all materials were carried in by hand rather than on trucks. Thus, all framing was done with nominal 2-inch (38-mm) lumber rather than heavy timbers. *(Photo by Christopher Lark. Courtesy of American Wood Council)*

CSI/CSC

MasterFormat Sections for Wood Light Framing

06 10 00	**ROUGH CARPENTRY**
06 11 00	Wood Framing
06 12 00	Structural Panels
06 14 00	Treated Wood Foundations
06 16 00	Sheathing

Selected References

1. American Forest & Paper Association. *Details for Conventional Wood Frame Construction.* Washington, DC, 2001.

This 55-page publication is available as a free download from the American Wood Council at www.awc.org. It provides an excellent introduction to wood light framing methods and their use in residential construction, and includes design and construction guidelines and extensive illustrated details.

2. APA – The Engineered Wood Association. *Performance Rated I-Joists, Construction Details for Floor and Roof Framing.* Tacoma, WA, 2004.

This 55-page publication is available as a free download from the APA – Engineered Wood Association at www.apawood.org. It provides extensive guidelines and illustrated details for wood light framing with engineered I-joists.

3. Thallon, Rob. *Graphic Guide to Frame Construction.* Newtown, CT, Taunton Press, 2000.

Unsurpassed for clarity and usefulness, this is an encyclopedic collection of details for wood platform frame construction.

4. Allen, Edward, and Rob Thallon. *Fundamentals of Residential Construction* (2nd ed.). New York, John Wiley & Sons, Inc., 2002.

The 628 pages of this book expand upon the chapters on residential-scale construction in the book you are now reading, giving full details of every aspect, including plumbing, mechanical and electrical systems, and landscaping.

5. International Code Council, Inc. *International Residential Code.* Falls Church, VA, updated regularly.

This is the definitive legal guide for platform frame residential construction throughout most of the United States. It includes details of every aspect of construction in both wood light frame and light-gauge steel.

Web Sites

Wood Light Frame Construction

Author's supplementary web site: **www.ianosbackfill.com/05_wood_light_frame_construction**

American Wood Council: **www.awc.org**

APA – Engineered Wood Association: **www.apawood.org**

Canadian Wood Council: **www.cwc.ca**

Fine Homebuilding magazine: **www.finehomebuilding.com**

Journal of Light Construction magazine: **www.jlconline.com**

NAHB Toolbase Services: **www.toolbase.org**

Wood Design & Building magazine: **www.wood.ca**

Key Terms and Concepts

wood light frame construction
balloon frame
joist
stud
rafter
firestop
platform frame
sheathing
header
rim joist
band joist
subfloor
stud
sole plate, bottom plate
top plate
ridge board
trimmer
sill
wracking
permanent wood foundation
batter board
termite shield
framing plan
section detail
architectural floor plan
exterior elevation
building section
interior elevation
rough carpentry
foundation sill plate
sill sealer
bridging
blocking
trimmer, jack stud
king stud
rough sill
cripple stud
shear wall
hold-down
let-in diagonal bracing
flat roof
shed roof, single-pitch roof
gable roof
hip roof
gambrel roof
mansard roof
ceiling joist
ridge beam
knee wallpitch
rise
run
framing square
hip
valley
pattern rafter
collar tie
dormer
birdsmouth cut
common rafter
valley rafter
valley jack rafter
hip rafter
hip jack rafter
fascia
lookout
fly rafter
rake
plumb cut
level cut
advanced framing techniques, optimum value engineering
roof truss
floor truss

Review Questions

1. Draw a series of very simple section drawings to illustrate the procedure for erecting a platform frame building, starting with the foundation and continuing with the ground floor, the ground-floor walls, the second floor, the second-floor walls, and the roof. Do not show details of connections but simply represent each plane of framing as a heavy line in your section drawing.

2. Draw from memory the standard detail sections for a two-story platform frame dwelling. *Hint:* The easiest way to draw a detail section is to draw the pieces in the order in which they are put in place during construction. If your simple drawings from Question 1 are correct, and if you follow this procedure, you will not find this question too difficult.

3. What are the differences between balloon framing and platform framing? What are the advantages and disadvantages of each? Why has platform framing become the method of choice?

4. Why is little added firestopping required in platform framing?

5. Why is a steel beam or glue-laminated wood beam preferred to a solid wood beam at the foundation level?

6. How is a platform frame building braced against wind and earthquake forces?

7. Light framing of wood is highly combustible. In what different ways does a typical building code take this fact into account?

Exercises

1. Visit a building site where a wood platform frame is being constructed. Compare the details that you see on the site with the ones shown in this chapter. Ask the carpenters about the procedures you see them using. Where their details differ from the ones illustrated, make up your own mind about which is better and why.

2. Develop floor framing and roof framing plans for a building you are designing. Estimate the approximate sizes of the joists and rafters using the rules of thumb provided in this chapter.

3. Make thumbnail sketches of 20 or more different ways of covering an L-shaped building with combinations of sloping roofs. Start with the simple ones (a single shed, two intersecting sheds, two intersecting gables) and work up to the more elaborate ones. Note how the varying roof heights of some schemes could provide room for a partial second-story loft or for high spaces with clerestory windows. How many ways do you think there are of covering an L-shaped building with sloping roofs? Look around as you travel through areas with wood frame buildings, especially older areas, and see how many ways designers and framers have roofed simple buildings in the past. Build up a collection of sketches of ingenious combinations of sloping roof forms.

4. Build a scale model of a platform frame from basswood or pine, reproducing accurately all its details, as a means of becoming thoroughly familiar with them. Better yet, build a small frame building for someone at full scale, perhaps a toolshed, playhouse, or garage.

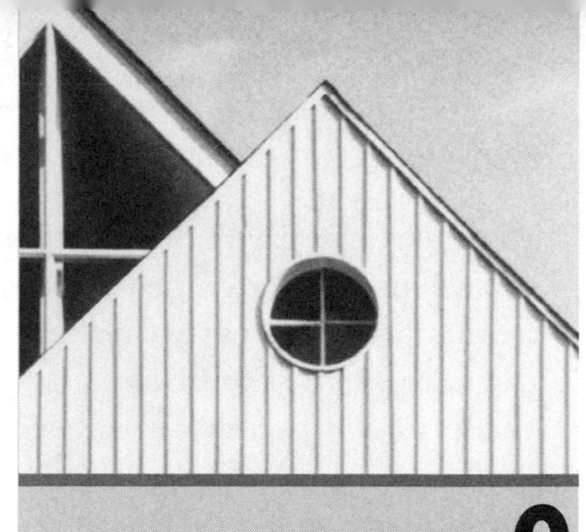

6

Exterior Finishes for Wood Light Frame Construction

Architects Hartman–Cox clad this small church with an economical but attractive siding of plywood and vertical wood battens, all painted white. The roofs are of asphalt shingles and the windows are of wood. The rectangular windows are double hung. *(Photo by Robert Lautman. Courtesy of American Wood Council)*

As the rough carpentry of a platform frame building nears completion, a logical sequence of exterior finishing operations begins. First, the eaves and rakes of the roof are finished, which permits the roof to be shingled so as to offer as much weather protection as possible to subsequent operations. When the roof has been completed, the windows and doors are installed. Then the siding is applied. The building is now "tight to the weather," allowing interior finishing work to take place safely sheltered from sun, rain, snow, and wind. The outside of the building is now ready for exterior painting or staining. Finish grading, landscaping, and paving work may commence as electricians, plumbers, sheet metal workers, drywallers, finish carpenters, and flooring installers swarm about the interior of the building.

Protection from the Weather

Roofing Underlayment

In the interest of protecting the building structure from weather as early as possible, *roofing underlayment* is installed soon after the roof framing is completed and sheathed. The most common material used for this purpose is one or two layers of *building felt*, made of matted cellulose fibers saturated with asphalt, often referred to simply as "tarpaper." During this stage of construction, the roofing underlayment serves as a temporary barrier to rain, allowing the structure to start drying out in preparation for interior work. Once the underlayment is covered by the final roofing, it continues to serve as a permanent backup to the roofing, reducing the chance that wind-driven rain or minor leakage can penetrate into the roof structure. Traditionally, building felt was designated as either 15-lb or 30-lb, referring to the weight of material used to cover 100 square feet (9.2 m^2) of roof area. Contemporary standards specify roofing felt as either No. 15 or No. 30. Though still frequently referred to as 15-lb or 30-lb felt, today's felts actually weigh slightly less than these names suggest. Proprietary roofing underlayment membranes, made from polypropylene and polyethylene fabrics, may also be substituted for traditional building felt. These *synthetic roofing underlayments* tend to be lighter, more tear-resistant, less sensitive to prolonged exposure to sunlight, and more costly than building felt.

Wall Moisture Barrier

Later during construction, the sheathed walls of the structure are also covered with a protective layer intended to repel water and restrict the leakage of air. This layer, variously referred to as a *moisture barrier, water-resistive barrier, weather-resistive barrier*, or *air barrier*, must be applied before siding is installed. But whether it is applied before or after the installation of doors and windows is usually up to the builder. The traditional materials used are the same No.15 felt used for roofing underlayment or an asphalt-saturated paper with similar water-resistive properties called *Grade D sheathing paper*. More recently, concern for building energy efficiency has led to the development of airtight papers made of synthetic fibers as substitutes for traditional building felts or papers. These *housewraps* are manufactured in sheets as wide as 10 feet (3 m) in order to minimize seams (Figure 6.1). To further reduce air leakage, the seams that remain may be sealed with self-sticking tape provided by the housewrap manufacturer. (For a more in-depth discussion of the control of air leakage through the building envelope, see pages 800–803.)

Like roofing underlayment, the wall moisture barrier provides temporary protection from rain and wind during construction and, once finish siding is installed, also acts as a permanent secondary line of defense against moisture penetration. While it is important that the wall moisture barrier resist penetration of liquid water, it must nevertheless also allow water vapor to pass relatively easily. Especially during cold months, interior humidity that tends to move through the wall assembly must not become trapped and allowed to accumulate, but rather, must be allowed to continue to pass freely through the wall to the exterior.

Roofing

Finishing the Eaves and Rakes

Before a roof can be shingled, the *eaves* (horizontal roof edges) and *rakes* (sloping roof edges) must be completed. Several typical ways of doing this are shown in Figures 6.2–6.4. When designing eave and rake details, the designer should keep several objectives in mind: The edges of the roof shingles should be positioned and supported in such a way that water flowing over them will drip free of the trim and siding below. Provision must be made to drain rainwater and snowmelt from the roof without damaging the structure below. In most roof assemblies, the eaves must be ventilated to allow free circulation of air beneath the roof sheathing. And siding, which is not installed until later, must fit easily against or into the eave and rake trim.

FIGURE 6.1
Building felt covers most of the roof and housewrap completely covers the walls of this house under construction in preparation for the application of roofing and siding. The bottom few feet of the roof sheathing remain exposed, awaiting the application of a waterproof ice barrier material as described later in this chapter. Lightweight housewraps such as seen in this photograph have to a large extent replaced building felt or paper for protecting walls. *(Photo by Rob Thallon)*

FIGURE 6.2
Two typical details for the rakes (sloping edges) of sloping roofs. The upper detail has no rake overhang, and the lower has an overhang supported on lookouts.

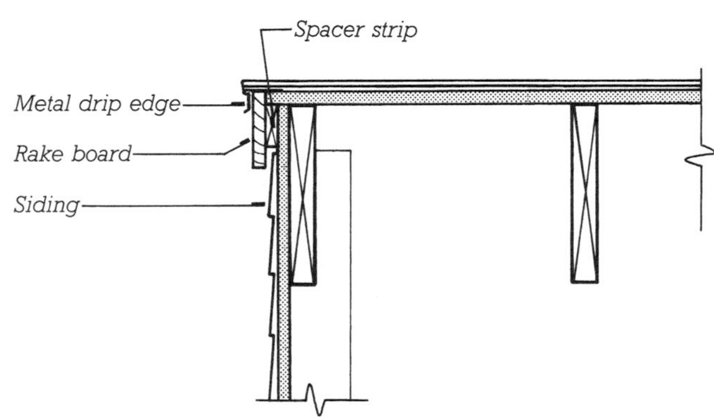

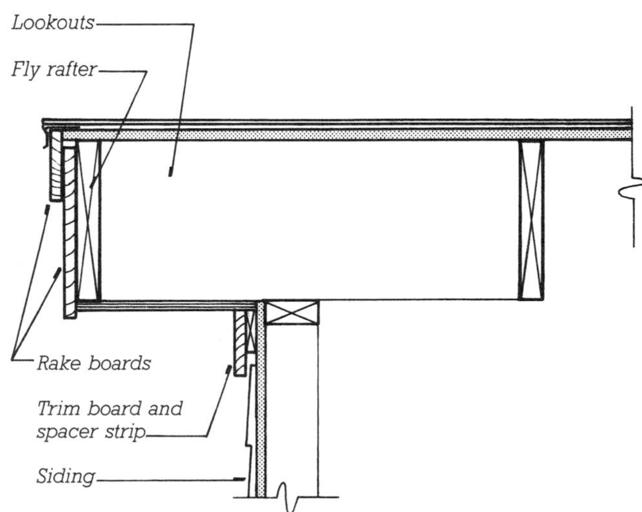

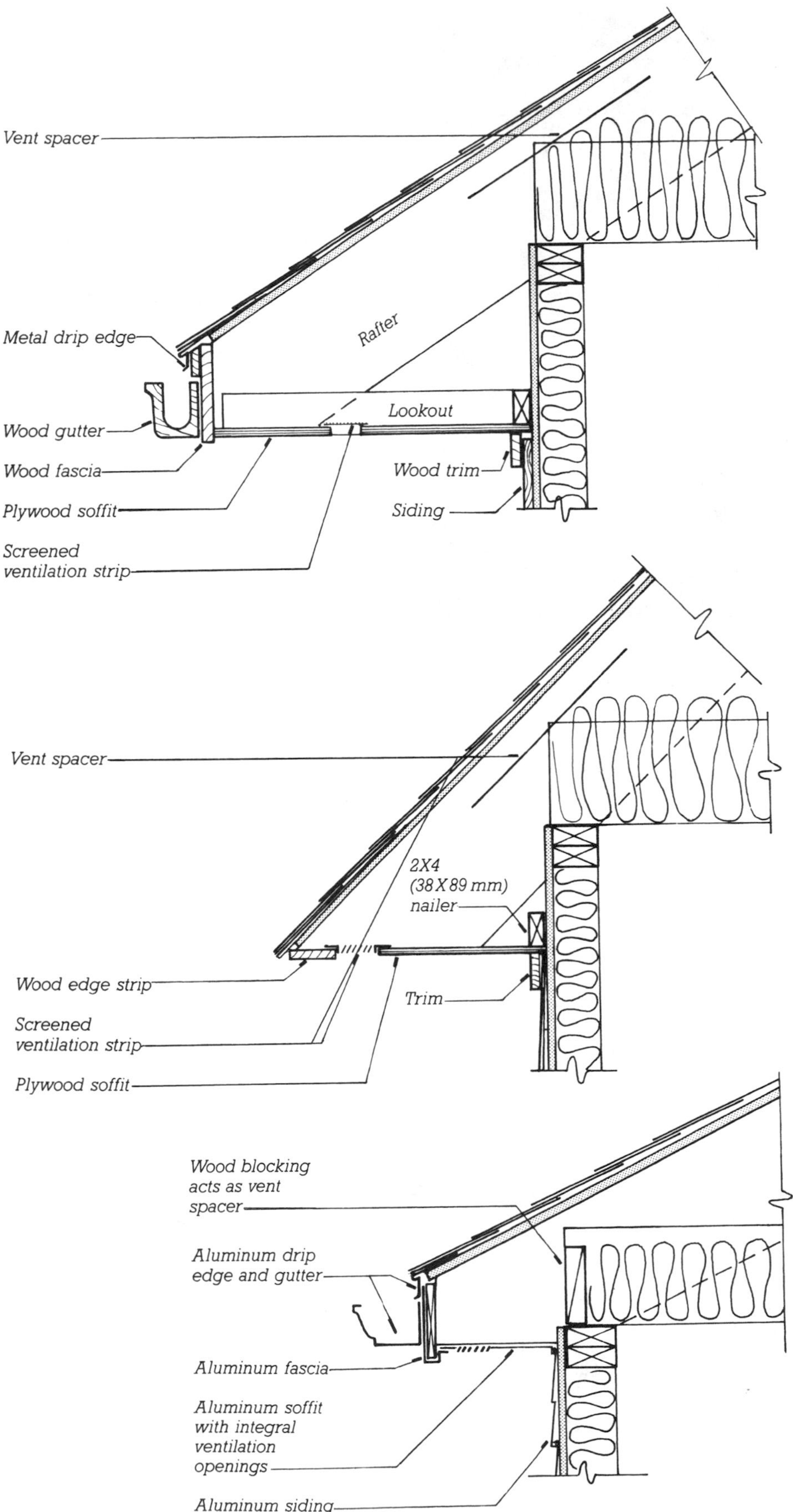

Figure 6.3
Three ways, from among many, of finishing the eaves of a light frame wood building. The top detail has a fascia and gutter, both of wood. The gutter is spaced slightly away from the fascia on blocks or shims to prevent moisture from being trapped between them, where it could lead to decay. The width of the overhang may be varied by the designer, and a metal or plastic gutter may be substituted for the wood one. The sloping line at the edge of the ceiling insulation indicates a vent spacer, as shown in Figure 6.6. The middle detail has no fascia or gutter; it works best for a steep roof with a sufficient overhang to drain water well away from the walls below. The bottom detail is finished entirely in aluminum. It shows wood blocking as an alternative to vent spacers for maintaining free ventilation through the attic.

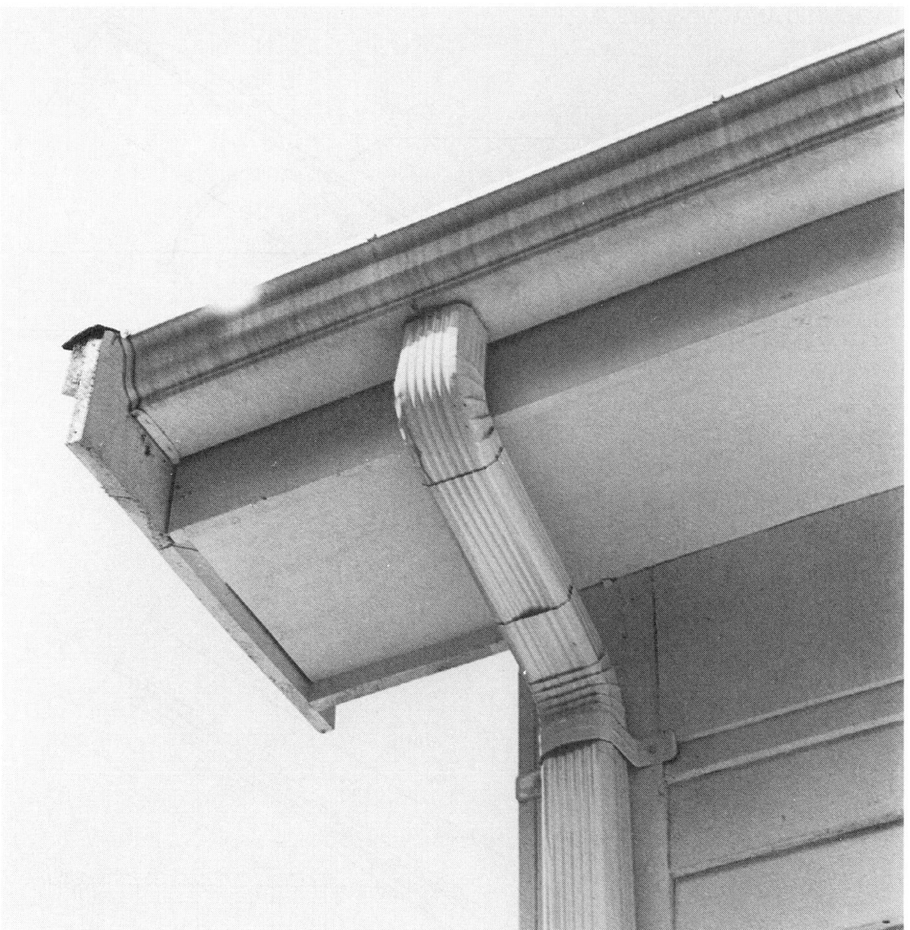

FIGURE 6.4
A poor example of a wood eave detail with an aluminum gutter and downspout. The carpentry at the intersection of the eave and the rake shows some poorly fitted joints and badly finished ends, and the soffit is not ventilated at all, which may lead to the formation of ice dams on the roof (see Figure 6.5) unless constructed as an unventilated roof, as explained in the text. *(Photo by Edward Allen)*

Roof Drainage

Gutters and *downspouts* (downspouts are also called *leaders*) are installed on the eaves of a sloping roof to remove rainwater and snowmelt without wetting the walls or causing splashing or erosion on the ground below. *External gutters* may be made of wood, plastic, aluminum, or other sheet metal and are fastened to the outer edge of the roof eave (Figures 6.3 and 6.4). *Internal* or *concealed gutters*, which are recessed into the surface of the roof, are rare. They are difficult to waterproof, and if they do leak, they can cause significantly more damage to the building structure and its finishes than can a leaky external gutter. Gutters are sloped toward the points at which downspouts drain away the collected water. On larger buildings, gutters and downspouts are sized using rainfall intensity data for the region in which the building is located and flow capacity formulas found in building and plumbing codes. At the bottom of each downspout, the water must be conducted away from the building in order to prevent soil erosion or basement flooding. A simple precast concrete *splash block* can spread the water and direct it away from the foundation, or a system of underground piping can collect the water from the downspouts and conduct it to a storm sewer, a *dry well* (an underground seepage pit filled with coarse crushed stone), or a drainage ditch.

Where collection of rainwater from roofs is not required by the building code or other regulations, gutters may be omitted and their associated problems of clogging and ice buildup avoided. Such roofs should be designed with eaves of sufficient depth to protect the wall below from excessive wetting by water falling from the roof edge. To prevent soil erosion and mud spatter from the falling water, the drip line at ground level below should be provided with a bed of crushed stone or other suitable surface material.

Roof Overhangs and Rain Protection

Roof eaves and other overhangs play an important role in protecting buildings from the weather. Where generous overhangs are provided, the volume of rain that reaches the building wall is greatly reduced in comparison to walls without such protection, and the chance for water to penetrate into the walls is much less. As a rule of thumb, a roof overhang can effectively protect a vertical portion of wall approximately

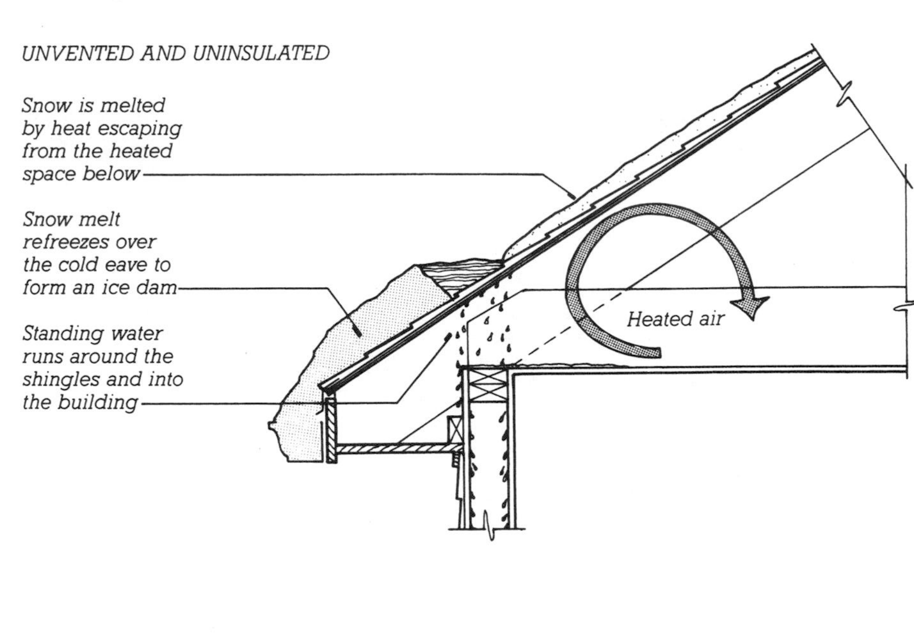
UNVENTED AND UNINSULATED
Snow is melted by heat escaping from the heated space below
Snow melt refreezes over the cold eave to form an ice dam
Standing water runs around the shingles and into the building
Heated air

FIGURE 6.5
Ice dams form because of inadequate insulation combined with a lack of ventilation, as shown in the top diagram. The lower two diagrams show how insulation, attic ventilation, and vent spacers are used to minimize the melting of snow on the roof. In areas prone to ice damming, building codes also require the installation of a strip of water-resistant ice barrier material under the shingles in the area of the roof where ice dams are likely to form (see Figure 6.9).

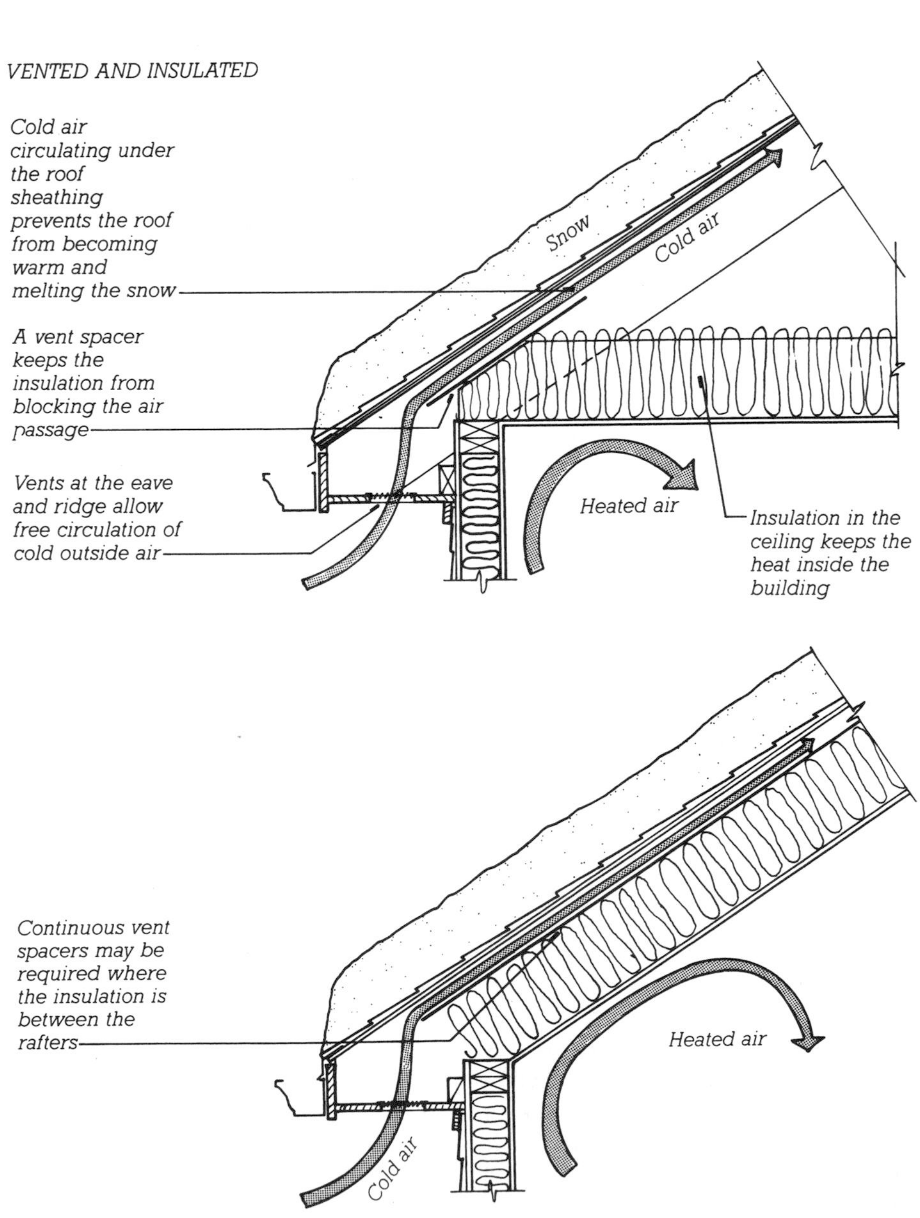
VENTED AND INSULATED
Cold air circulating under the roof sheathing prevents the roof from becoming warm and melting the snow
Snow
Cold air
A vent spacer keeps the insulation from blocking the air passage
Vents at the eave and ridge allow free circulation of cold outside air
Heated air
Insulation in the ceiling keeps the heat inside the building
Continuous vent spacers may be required where the insulation is between the rafters
Heated air
Cold air

twice as tall as the depth of the overhang. (For example, a 2-foot, or 600-mm, overhang can protect approximately 4 feet, or 1200 mm, of wall below). And even overhangs only 1 to 2 feet (600 to 1200 mm) deep can provide significant protection to the structure below in comparison to a wall with no overhangs at all. The design of walls without overhangs should be approached with caution: These walls are prone to staining, leaking, decay, and premature deterioration of the windows, doors, and siding.

Ventilated Roofs

In cold climates, snow on a sloping roof tends to be melted by heat escaping through the roof from the heated space below. At the eave overhang, however, the shingles and gutter, which are not above heated space, can be much colder, and the snowmelt can refreeze and begin to build up layers of ice. Soon an *ice dam* forms and meltwater accumulates above it, seeping between the shingles, through the roof sheathing, and into the building, causing damage to walls and ceilings (Figure 6.5). Ice dams can be largely eliminated by keeping the entire roof cold enough so that snow will not melt. In a *ventilated roof* this is accomplished by continuously passing outside air under the roof sheathing through vents at the eave and ridge. In buildings with an attic, the attic itself is ventilated and kept as cold as possible. Insulation is installed

FIGURE 6.6
To maintain clear ventilation passages where thermal insulation materials come between the rafters, vent spacers of plastic foam or wood fiber are installed as shown here. The positioning of the blocking or vent spacers is shown in Figures 6.3 and 6.5. Vent spacers can be installed along the full slope of a roof to maintain free ventilation between the sheathing and the insulation where the interior ceiling finish material is applied to the bottom of the rafters. *(Proper Vent;™ is registered trademark of Poly-Foam Inc.)*

in the ceilings below to retain the heat in the building. Where there is no attic and the insulation is installed between the rafters, the spaces between the rafters are ventilated by means of air passages between the insulation and the roof sheathing (Figure 6.6).

Soffit vents create the required ventilation openings at the eave; these usually take the form of a continuous slot covered either with screening or with a perforated aluminum strip made especially for the purpose (Figure 6.7). As an alternative to soffit vents, proprietary panels with hollow ventilation channels can be installed under the lowest several courses of roof shingles. A continuous slot running across the roof sheathing allows exterior air to circulate under the shingles, through the panel, and into the ventilation space under the roof sheathing. Ventilation openings at the ridge can be either *gable vents* just below the roof line in each of the end walls of the attic or a continuous *ridge vent,* a screened cap that covers the ridge of the roof and draws air through gaps in the roof sheathing on either side of the ridge board (Figure 6.8). Building codes establish minimum area requirements for roof ventilation openings based on the floor area beneath the roof being ventilated.

In regions prone to ice damming, building codes may also require that an *ice barrier* be installed along the eaves. This is usually a self-adhered sheet of polymer-modified bitumen, referred to as *rubberized underlayment* or by one product of this type's proprietary trade name that has fallen into colloquial usage, *ice and water shield.* The sheet must extend from the lowest edge of the roof to a line that is 2 feet (600 mm) inside the insulated space of the building (Figure 6.9). The soft, sticky bitumen seals tightly around the shingle nails that are driven through it and prevents penetration of water that may accumulate behind an ice dam. Alternatively, the lower portion of the sloped roof may be covered with sheet metal or other roofing material that is sufficiently waterproof to protect the roof from damage if ice damming occurs.

Roof ventilation serves other purposes as well. It keeps a roof cooler in summer by dissipating solar heat absorbed by the shingles. A lower roof temperature extends the life of the roofing materials and reduces the heat load on the building interior. For this reason, many asphalt shingle manufacturers require that their shingles be installed on ventilated roofs as part of their product warranty conditions. Roof ventilation also helps dissipate moisture that collects under the roof sheathing that may originate from leaks in the roofing or, in colder months, that may be generated by the condensation of interior humidity that comes in contact with the cold underside of the roof sheathing.

Unventilated Roofs

Unventilated roofs can be constructed with insulated foam panels applied over the roof sheathing or with spray foam insulation applied below the sheathing between the roof rafters (Figure 6.10). When unventilated roofs are used in cold climates, protection from ice damming depends on ice barriers and the installation of sufficient insulation to reduce heat flow through the roof assembly, thereby slowing the melting

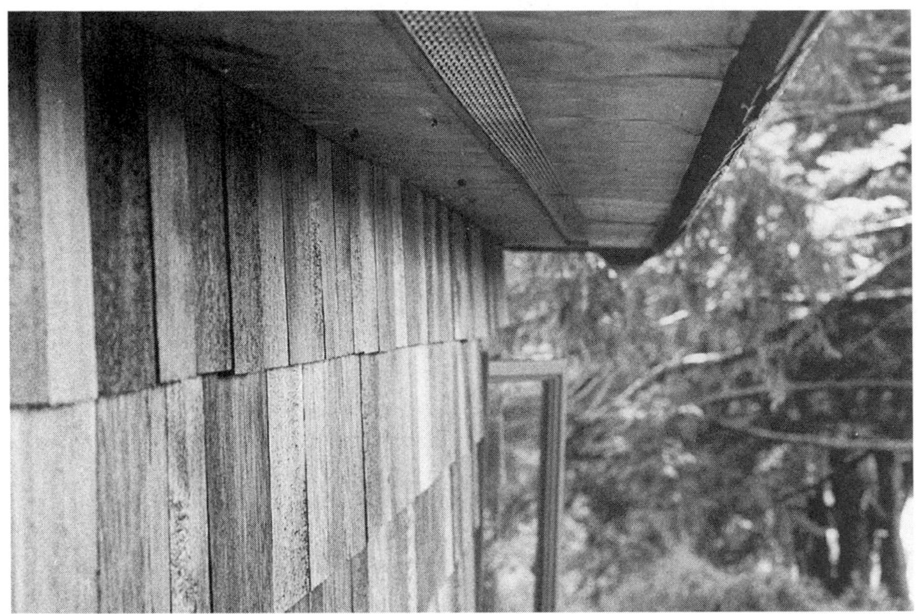

FIGURE 6.7
A continuous soffit vent made of perforated sheet aluminum permits generous airflow to all the rafter spaces but keeps out insects. The walls of this building are covered with cedar shingles. *(Photo by Edward Allen)*

FIGURE 6.8
This building has both a louvered gable vent and a continuous ridge vent to discharge air that enters through the soffit vents. The gable vent is of wood and has an insect screen on the inside. The aluminum ridge vent is internally baffled to prevent snow or rain from entering even if blown by the wind. Vertical square-edged shiplap siding is used on this building. The roofing is a two-layered asphalt shingle designed to mimic the rough texture of a wood shake roof. *(Photo by Edward Allen)*

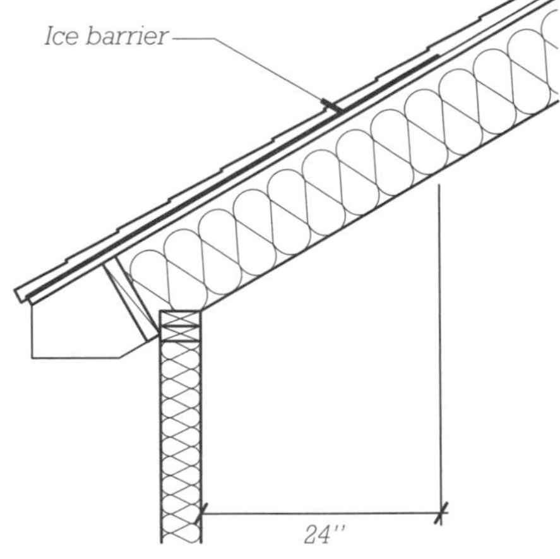

FIGURE 6.9
In cold climates, building codes typically require that an ice barrier be installed beneath the lowest courses of shingles. This must extend at least 24 inches (600 mm) over the insulated portion of the inhabited space.

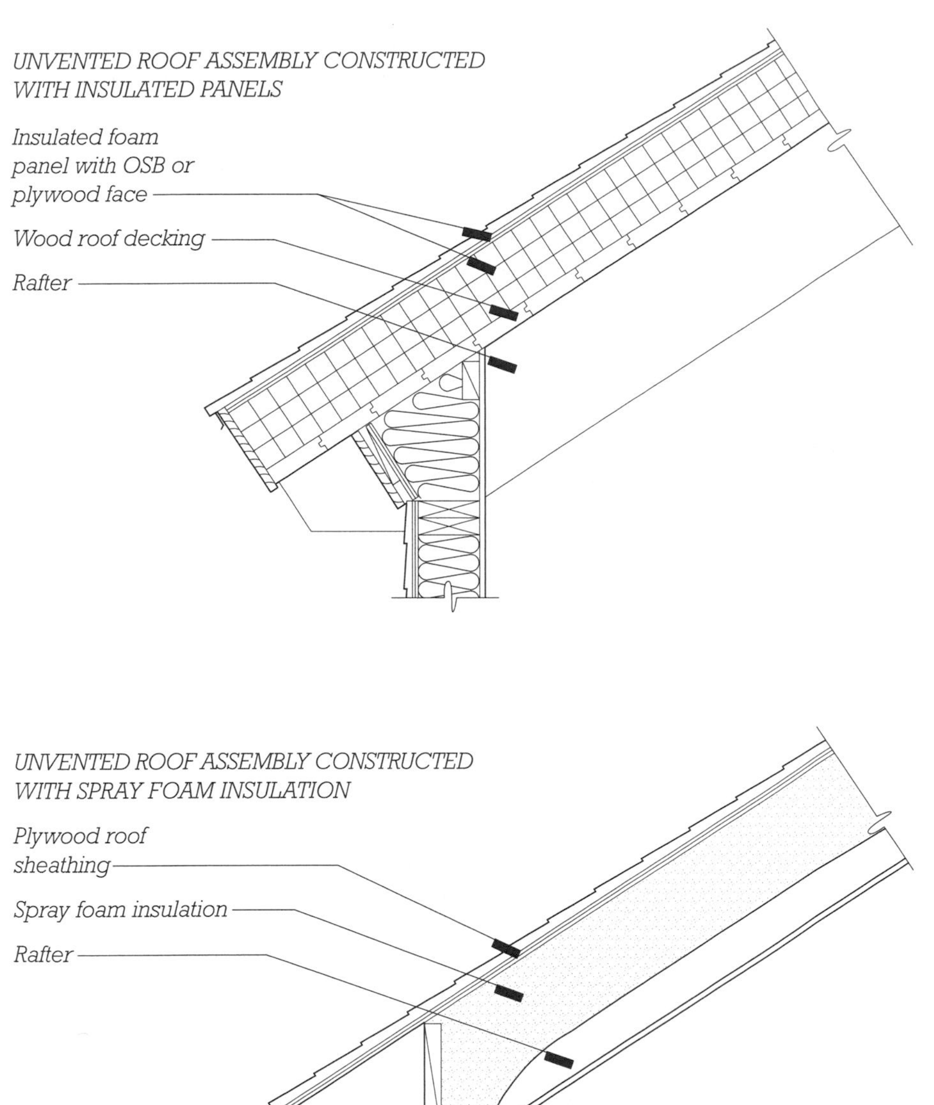

FIGURE 6.10
Two ways to build an unventilated roof assembly. In the top diagram, insulated foam panels are installed on top of wood decking. The insulated panels have a plywood or OSB face that provides a surface that can hold the nails or staples with which the roof underlayment and shingles are attached. The panels themselves are secured to the wood roof decking below with screws that are long enough to penetrate through the thickness of the panel into the roof decking or rafters below. In the lower diagram, foam insulation is sprayed beneath the roof sheathing, and the framing and insulation are concealed by a gypsum board ceiling. Both of these assemblies depend on adequate levels of insulation to keep the roof surface cold enough to avoid ice damming, and on the restriction of air leakage and the diffusion of humidity to prevent moisture from reaching the outer layers of the roofing assembly where condensation could occur.

of snow on the roof. Condensation resistance in unventilated roofs depends on the inability of significant quantities of air or water vapor to pass through the foam insulation to the colder layers of the roof assembly where condensation could occur. With insulated panels, special care must be taken to seal the joints between panels so that air leakage around panel edges does not occur. With spray foam insulation, the International Building Code (IBC) requires that the foam have very low permeability to airflow. Where high roof temperatures are a concern (due to the lack of ventilation below the roof sheathing), shingles or other roof coverings with high solar reflectance may be considered, or insulated panels with integral ventilation channels may be used to create a ventilated assembly.

Unventilated roof assemblies run a higher risk of entrapment of condensed moisture within the roof construction than ventilated assemblies. For this reason, unventilated roof assemblies are not necessarily appropriate for all climates and building codes place various restrictions on their use. Before using such assemblies, the designer should verify their suitability to their particular project circumstances and locale.

Advantages of unventilated roofing systems include simplicity of detailing, a thinner roof profile, and no risk of windblown water entry into the roof assembly through poorly baffled ridge vents. However, where the potential for condensation is high, for example in very cold climates or in roofs over high-humidity spaces, such assemblies should be studied and detailed carefully to ensure that moisture conditions are adequately controlled.

Roof Shingling

Wood light frame buildings can be roofed with any of the systems described in Chapter 16. Because *asphalt shingles* are much less expensive than any other type of roofing, and because they are highly resistant to fire, they are used in the majority of cases. They are applied either by the carpenters who have built the frame or by a roofing subcontractor. Figures 6.11 and 16.40–16.43 show how an asphalt shingle roof is applied. Wood shingle and shake roofs (Figures 16.37–16.39), clay and concrete tile roofs (Figure 16.46), and architectural sheet metal roofs (Figures 16.48–16.52) are also fairly common. Low-slope roofing, either a built-up or a single-ply membrane, may also be applied to a light frame building (Figures 16.4–16.33).

Windows and Doors

Flashings

Prior to the installation of windows and doors, *flashings* are installed to prevent water from seeping through gaps around the edges of these components. Flashings may be made of any of a number of corrosion-resistant metals, plastic, synthetic fabrics similar to housewraps, or modified bitumen sheets similar to the rubberized underlayment used for ice barriers on roofs. The choice of materials and the complexity of the detailing vary with the type of door or window frame, the severity of the exposure to wind and rain, and the detailing of the wall trim (Figures 6.12 and 6.13).

Installing Windows and Doors

Windows and doors are covered in detail in Chapter 18. *Nail-fin* or

Figure 6.11
Applying asphalt shingles to a roof. Shingles and other roof coverings are covered in detail in Chapter 16. *(Photo by Edward Allen)*

FIGURE 6.12

Steps in installing a vinyl-clad wood window in a wall that has been covered with housewrap: *1.* With a single strip of self-sticking, flexible, waterproof sheet material, a U-shaped flashing is formed along the sill and up onto the side jambs. *2.* A bead of sealant is placed around the opening. *3.* The window unit is pushed into the opening from the outside. Its exterior flange is bedded in the sealant. *4.* One corner of the window unit is anchored with a nail through its flange and into the framing of the wall. With the aid of a level, a measuring tape, and thin wooden wedges, the unit is carefully squared as further nails are driven through the flanges. *5, 6.* Strips of self-sticking waterproof sheet are adhered to the window flanges and housewrap over the side jambs and head for airtightness and watertightness.

FIGURE 6.13
In the installation on the left, the flashings and window are installed before the housewrap. Synthetic fabric flashings are installed into the opening, followed by the window unit itself. Later, the housewrap will be overlapped with the flashings and taped to make a continuous barrier to air and water. In the example on the right, housewrap is installed first, and then adhesive-backed flashings are lapped over the housewrap and into the window opening. The window will be installed later. The bottom of this opening is also flashed with a copper sheet metal *sill pan*. The sill pan provides extra protection from leakage, capturing water that may penetrate into the space between the window frame and the rough framing and directing it back to the exterior. *(Photos by Edward Allen and Joseph Iano)*

flanged windows are simple to install because they include a continuous flange around the perimeter through which nails are driven into the sheathing and studs (perpendicular to the wall plane) to fasten the units securely in place. The flange also provides a convenient surface for the application of self-sticking flashings. Most plastic-clad, aluminum, and solid plastic windows are manufactured as nail-fin types, as are some, but not all, wood windows. *Finless* or *flangeless windows* and most doors do not have nailing flanges. They are fastened through the sides of the window or door frame into the supporting studs (parallel to the wall plane) with long finish nails or screws. The fasteners are either covered by additional trim or they are countersunk, with the resulting holes later filled with putty and painted over. A typical entrance door and details for its installation are shown in Figures 6.14 and 6.15.

Everybody loves window seats, bay windows, and big windows with low sills and comfortable chairs drawn up to them. . . . A room where you feel truly comfortable will always contain some kind of window place.

Christopher Alexander et al., *A Pattern Language*, 1977

Figure 6.14
A six-panel wood entrance door with flanking sidelights and a fanlight above. A number of elaborate traditional entrance designs such as this are available from stock for use in light frame buildings. *(Courtesy of Morgan Products Ltd.)*

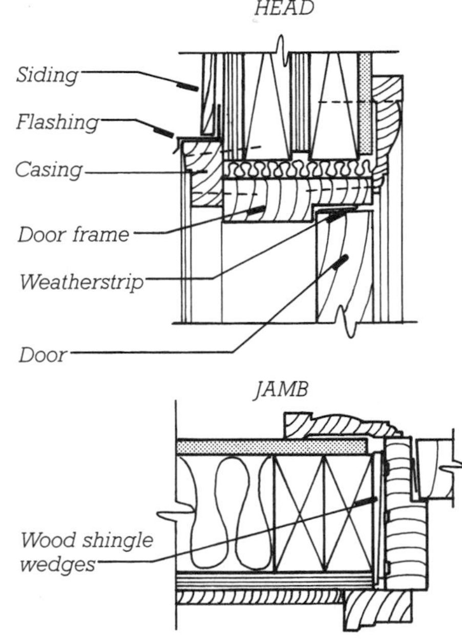

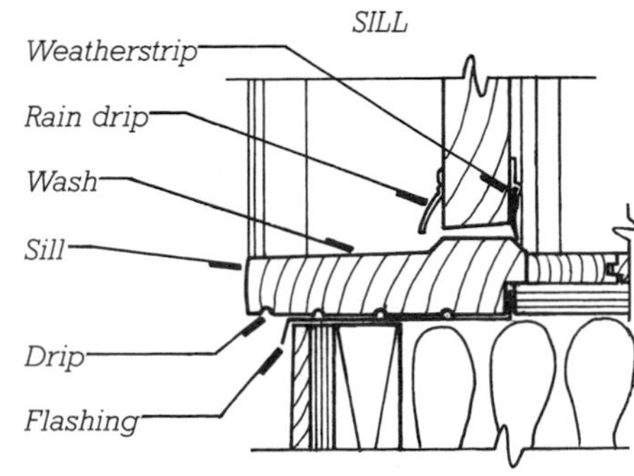

Figure 6.15
Details of an exterior wood door installation. The door opens toward the inside of the building. The flashing in the head intercepts water running down the wall and directs it away from the door and frame. The sill flashing prevents water that seeps past the sill from leaking into the floor structure.

Paints and Coatings

Paints and other architectural coatings (stains, varnishes, lacquers, sealers) protect and beautify the surfaces of buildings. A good coating job begins with thorough surface preparation to make the surface (called the *substrate*) ready to receive the coating. The coating materials must be carefully chosen and skillfully applied using the proper tools and techniques. And to finish the job, environmental conditions must be right for the drying or curing of the coating.

Material Ingredients

Most architectural coating materials are formulated with four basic types of ingredients: vehicles, solvents, pigments, and additives.

- The *vehicle* or *binder* provides adhesion to the substrate and forms a film over it; it is often referred to as a *film-former.*
- *Solvents* are volatile liquids used to improve the working properties of the paint or coating. The most common solvents used in coating materials are water and hydrocarbons, but turpentine, alcohols, ketones, esters, and ethers are also used.
- *Pigments* are finely divided solids that add color, opacity, and gloss control to the coating material. They also impart hardness, abrasion resistance, and weatherability to the coating.
- Additives modify various properties of the coating material. Driers, for example, are additives that hasten the curing of the coating. Other additives may relate to ease of application, resistance to fading, and other functions.

Solvent-Based Coatings and Water-Based Coatings

Coatings fall into two major groups, *solvent-based* and *water-based.* Solvent-based coatings use solvents other than water, most commonly alkyd resins or sometimes natural oils or polyurethane resins. These coatings cure by solvent evaporation, oxidation of the vehicle, or moisture curing from reaction of the vehicle with humidity in the air. Cleanup after painting is usually done with mineral spirits. Water-based coatings use water as the solvent. Most vehicles in water-based coatings are vinyl or acrylic resins. Cleanup is done with soap and water. In everyday use, solvent-based paints are most often referred to as "oil paints" or "alkyd paints" and the water-based paints as "latex paints."

Water-based and solvent-based coatings each have uses for which they are preferred and other uses for which they are more or less interchangeable. Most latex paints generate fewer odors during their application than comparable alkyd paints, an important consideration for renovation work where building occupants may be exposed to paint fumes. Where used for exterior finishing, latex paints produce a more flexible coating able to stretch without breaking when subjected to expansion and contraction of the substrate due to changes in temperature and humidity. Latex coatings are also more breathable, lessening the chance of moisture being trapped behind painted siding or trim.

Historically, alkyd paints gave a smoother, harder surface and were the preferred type of coating for interior painting, especially where a durable, high-gloss finish was desired. Alkyd primers also offered superior covering ability and reliable adhesion to even problematic substrates. However, solvent-based coatings emit significantly more *volatile organic compounds (VOCs)* than water-based coatings. As pollution regulations and sustainability-influenced indoor air quality standards have placed increasingly strict limits on VOC emissions, the use of solvent-based coatings has decreased markedly. In response, manufacturers have worked to improve the formulation of water-based coatings to the extent that, today, water-based coatings are available that can equal the performance of solvent-based coatings in most applications.

Types of Paints and Coatings

The various types of architectural coatings can be defined by the relative proportions of vehicle, solvent, pigment, and additives in each.

Paints contain relatively high amounts of pigments. The highest pigment content is in flat paints, those that dry to a completely matte surface texture. Flat paints contain a relatively low proportion of film-forming vehicle.

Paints that produce more glossy surfaces are referred to as *enamels.* A high-gloss enamel contains a very high proportion of vehicle and a relatively low proportion of pigment. The vehicle cures to form a hard, shiny film in which the pigment is fully submerged. A semigloss enamel has a somewhat lower proportion of vehicle, though still more than a flat paint.

Stains range from transparent stains to semitransparent and solid stains. Transparent stains are intended only to change the color of the substrate, usually wood and sometimes concrete. They contain little or no vehicle or pigment, a very high proportion of solvent, and a dye additive. Excess stain is wiped off with a rag a few minutes after application, leaving only the stain that has penetrated the substrate. Usually, a surface stained with a transparent stain is subsequently coated with a clear finish such as varnish to bring out the color and figure of the wood and produce a durable, easily cleaned surface.

Semitransparent stains have more pigment and vehicle than transparent ones. They are not wiped after application. They are intended for exterior application in two coats and do not require a clear topcoat. Also self-sufficient are the solid stains, which are usually water based. These contain much more pigment and vehicle than the other two types of stains and resemble a dilute paint more than they do a true stain. They are intended for exterior use.

The *clear coatings* are high in vehicle and solvent content and contain little or no pigment. Their purpose is to protect the substrate, make it easier to keep clean, and bring out its inherent beauty, whether it be wood, metal, stone, or brick. *Lacquers* are clear coatings that dry extremely rapidly by solvent evaporation. They are based on nitrocellulose or acrylics and are employed chiefly in factories and shops for rapid finishing of cabinets and millwork. A slower-drying clear coating is known as a *varnish*. Varnishes may be either solvent- or water-based. Most harden either by oxidation of an oil vehicle or by moisture curing. Varnishes are useful for on-site finishing. Varnishes and lacquers are available in gloss, semigloss, and flat formulations.

Shellac is a clear coating for interior use that is made from secretions of a particular Asian insect that have been dissolved in alcohol. Shellac dries rapidly and gives a very fine finish, but it is highly susceptible to damage by water or alcohol.

There are many finishes, intended primarily for furniture and indoor woodwork, that are based on simple formulations of natural oils and waxes. A mixture of boiled linseed oil and turpentine, rubbed into wood in many successive coats, gives a soft, water-resistant finish that is attractive to sight, smell, and touch. Waxes such as beeswax and carnauba can be rubbed over sealed (and sometimes unsealed) surfaces of wood and masonry to give a pleasingly lustrous finish. Finishes based on waxes and oils usually require periodic reapplication in service to maintain the character of the surface.

There are countless specialized coatings formulated for particular purposes, such as the intumescent coatings that can add fire resistance to steel (see Chapter 11), or *high-performance coatings*, such as epoxies and urethanes, designed to provide greater degrees of resistance to physical wear, chemical attack, and corrosion than are possible with ordinary paints. There are also specialized impermeable coatings based on various polymer vehicles and on materials such as asphalt (for painting roofs) and portland cement (for painting masonry and concrete).

Many of the newest and most durable architectural coatings are designed to be applied in the factory, where controlled environmental conditions and customized machines permit the use of many types of materials and techniques that would be difficult or impossible to use in the field. These include powder coatings, which are sprayed on dry and fused into continuous films by the application of heat, two-component coating formulations such as fluoropolymer finishes (see Chapter 21), and many others.

Field Application of Architectural Coatings

No aspect of painting and finishing work is more important than surface preparation. Unless the substrate is clean, dry, smooth, and sound, no paint or clear coating will perform satisfactorily. Normal preparation of wood surfaces involves scraping and sanding to remove any previous coatings, patching and filling holes and cracks, and sanding the surface to make it smooth. Preparation of metals may involve solvent cleaning to remove oil and grease, scraping and wire brushing to remove corrosion and mill scale, sandblasting if the corrosion and scale are tenacious, and, on some metals, chemical etching of the bare metal to improve the paint bond. New masonry, concrete, and plaster surfaces generally require some aging before coating to ensure that the chemical curing reactions within the materials themselves are complete and that excess water has evaporated from the material.

A number of materials are designed specifically to prepare a surface to receive paints or clear coatings. Paste fillers are used to fill the small pores in open-grained woods such as oak, walnut, and mahogany prior to finishing. Various patching and caulking compounds serve to fill larger holes in substrates. A *primer* is a pigmented coating especially formulated to make a surface more paintable. A wood primer, for example, improves the adhesion of paint to wood. It also hardens the surface fibers of the wood so that it can be sanded smooth after priming. Other primers are designed as first coats for various metals, masonry materials, plaster, and gypsum board. Wood trim and casings in high-quality work are back primed by applying primer to their back surfaces before they are installed; this helps equalize the rate of moisture change on both sides of the wood during periods of changing humidity, which reduces cupping and other distortions. A *sealer* is a thin,

Paints and Coatings (continued)

unpigmented liquid that can be thought of as a primer for a clear coating. It seals the pores in the substrate so that the clear coating will not be absorbed.

To prevent premature drying, freshly painted surfaces should not be exposed to direct sunlight, and air temperatures during application should be between 50 and 90 degrees Fahrenheit (10–32°C). For exterior work, wind speeds should not exceed 15 miles per hour (7 m/s). The paint materials themselves should be at normal room temperature.

Paint and other coatings may be applied by brush, roller, pad, or spray. Brushing is the slowest and most expensive method; it is best for detailed work and for applying many types of stains and varnishes. Spraying is the fastest and least expensive, but also the most difficult to control. Roller application is economical and effective for large expanses of flat surface. Many painters prefer to apply transparent stains to smooth surfaces by rubbing with a rag that has been saturated with stain.

A single coat of paint or varnish is usually insufficient to cover the substrate and build the required thickness of film over it. A typical requirement for a satisfactory paint coating is one coat of primer plus two coats of finish material. Two coats of varnish are generally required over raw wood. The surface is lightly sanded after the first coat has dried to produce a smooth surface to which the final coat is applied.

Different coatings need curing periods that range from minutes for lacquers to days for most paints. During this time, environmental conditions similar to those required for the application of the coating need to be maintained to ensure that the coating cures properly.

Deterioration of Paints and Finishes

Coatings are the parts of a building that are exposed to the most wear and weathering, and they deteriorate with time, requiring recoating. The ultraviolet (UV) component of sunlight is particularly damaging, causing fading of paint colors and chemical decomposition of paint films. Clear coatings are especially susceptible to UV damage, often lasting no more than a year before discoloring and peeling, which is why they are generally avoided in exterior locations. Some clear coatings are manufactured with special UV-blocking ingredients so that they may be used on exterior surfaces. The other major force of destruction for paints and other coatings is water. Most peeling of paint is caused by water getting behind the paint film and lifting it off. The most common sources of this water in wood sidings are lumber that is damp at the time it is coated, rainwater leakage at joints, and water vapor migrating from damp interior spaces to the outdoors during the winter. Good construction practices and proper design of air barriers and vapor retarders can minimize these problems.

Other major forces that cause deterioration of architectural coatings are oxygen, air pollutants, fungi, dirt, degradation of the substrate through rust or decay, and mechanical wear. Most exterior paints are designed to "chalk" slowly in response to these forces, allowing the rain to wash the surface clean at frequent intervals.

Typical Architectural Coating Systems

The accompanying table summarizes some typical specifications of coating systems for new surfaces of buildings. Where alkyd coatings are indicated, conformance of product VOC emissions to applicable limits should be verified.

Standards for Architectural Coating Systems

The Master Painters Institute (MPI) sets standards for paints and painting methods. Its manual of coating systems standardizes and simplifies the specification of complete coating systems, that is, combinations of surface preparation, primer, and topcoats appropriate for any given substrate and meeting a variety of performance levels.

Green standards for paints and other architectural coatings continue to evolve. Green Seal's *GS-11 Environmental Standard for Paints* sets basic performance standards for hiding power (opacity) and cleanability, limits VOC emissions, and restricts the inclusion of certain toxic or harmful ingredients in paints. Compliance with GS-11 is currently recognized by LEED for credit for low-emitting architectural paints and coatings. MPI's *Green Performance Standard for Paints and Coatings* is an alternative standard that establishes criteria in the same categories of performance, emissions, and restricted ingredients, but in a manner that more closely coordinates with that organization's other product standards.

For more information on considerations of sustainability in architectural coatings, see page 250.

Some Typical Coating Specifications for New Surfaces

Exterior

Substrate	*Surface Preparation*	*Primer and Topcoats*
Wood siding, trim, window, doors	Sand smooth, spot-prime knots and pitch streaks. Fill and sand surface blemishes after applying the prime coat	Exterior latex (or alkyd primer where required by difficult substrate conditions) and two coats of latex paint; or two coats of semitransparent or solid stain
Concrete masonry	Must be clean and dry, at least 30 days old	Block filler primer and two coats of latex paint
Concrete walls	Must be clean and dry, at least 30 days old	Masonry primer where required and two coats of latex paint
Stucco	Must be clean and at least 7 days old	Masonry primer where required and two coats of latex paint
Iron and steel	Remove rust, mill scale, oil, and grease	General-purpose latex or corrosion-resistant metal primer and two coats of latex or alkyd enamel; or two coats of direct-to-metal latex enamel
Aluminum	Clean with a solvent to remove oil, grease, and oxide	Latex metal primer and two coats of latex or alkyd enamel; or two coats of direct-to-metal latex enamel

Interior

Substrate	*Surface Preparation*	*Primer and Topcoats*
Plaster	Conventional plaster must be 30 days old (veneer plaster may be coated with latex paint immediately after hardening)	Interior latex primer and two coats of latex paint
Gypsum board	Must be clean, dry, and free from dust	Interior latex primer and one or two coats of latex paint
Wood doors, windows, and trim	Sand smooth. Fill and sand small surface blemishes after applying the prime coat. Sand lightly between coats. Remove sanding dust before coating	Interior latex or alkyd enamel primer and one or two coats of latex or alkyd enamel
Concrete masonry	Must be clean and dry, at least 30 days old	Masonry primer where required and two coats of latex paint
Hardwood floors	Fill surface blemishes and sand before coating. Sand lightly between coats	Oil stain if color change is desired, and two coats of oil or polyurethane varnish

Siding

The exterior cladding material applied to the walls of a wood light frame building is referred to as *siding*. Many different types of materials are used as siding: wood boards with various profiles, applied either horizontally or vertically; plywood; wood shingles; metal or plastic sidings; fiber-cement; brick or stone; and stucco (Figures 6.16–6.34).

Board Sidings

Horizontally applied *board sidings*, made of solid wood, wood composition board, or fiber-cement, are usually nailed in such a way that the nails pass through the wall moisture barrier and sheathing and into the studs, giving a very secure attachment. This procedure also allows horizontal sidings to be applied directly over insulating sheathing materials without requiring a *nail-base sheathing*, a material such as plywood or OSB that is dense enough to hold nails itself. *Siding nails*, whose heads are intermediate in size between those of common nails and those of finish nails, are used to give the best compromise between holding power and appearance when attaching horizontal siding. Siding nails should be hot-dip galvanized, or made of aluminum or stainless steel, to prevent corrosion and staining. Ring-shank nails are preferred because of their high resistance to pulling out as the siding boards shrink and swell with changes in moisture content. Nailing is done in such a way that the individual pieces of siding may expand and contract freely without damage (Figure 6.16). Horizontal sidings are butted tightly to corner boards and window and door trim, usually with a sealant material applied to the joint during assembly.

More economical horizontal siding boards made of various types of wood composition materials have become readily available. Some of

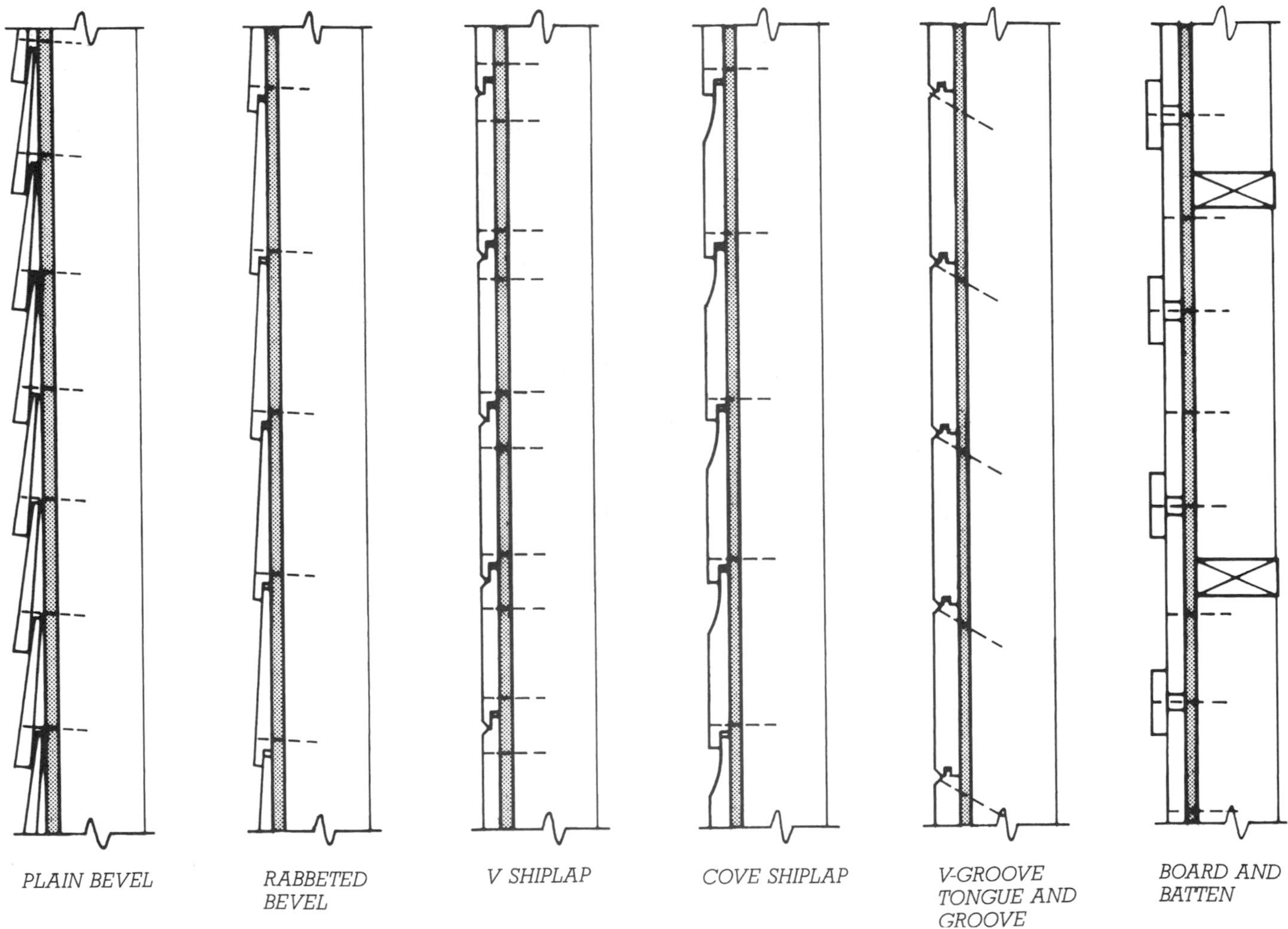

Figure 6.16

Six types of wood siding from among many. The four *bevel* and *shiplap sidings* are designed to be applied in a horizontal orientation. *Tongue-and-groove siding* may be used either vertically or horizontally. *Board-and-batten siding* may only be applied vertically. The nailing pattern (shown with broken lines) for each type of siding is designed to allow for expansion and contraction of the boards. Nail penetration into the sheathing and framing should be a minimum of 1½ inches (38 mm) for a satisfactory attachment.

these have experienced problems in service with excessive absorption of water, decomposition, or fungus growth, resulting in large class-action lawsuits and withdrawal of most such products from the market. Board sidings made from wood fiber in a portland cement binder have proven to perform well in service. They accept and hold paint well, do not decay or support fungus growth, and are dimensionally stable, highly resistant to fire, and very durable.

In North America, horizontal siding boards are customarily nailed tightly over the wall sheathing and housewrap. Siding may also be nailed to vertical wood spacers called *furring strips*, usually made from preservative-treated 1 × 3s (19 × 63 mm) or similar lengths of treated plywood or plastic, that are aligned over the studs (Figure 6.17). The space created behind the siding provides a free drainage path for leakage through the siding, permits rapid drying of the siding should it become soaked with water, and enhances the wall assembly's capability to expel water vapor that may accumulate within the insulated portions of the wall. Such construction is often referred to as *rainscreen siding*, even though it does not necessarily meet the requirements

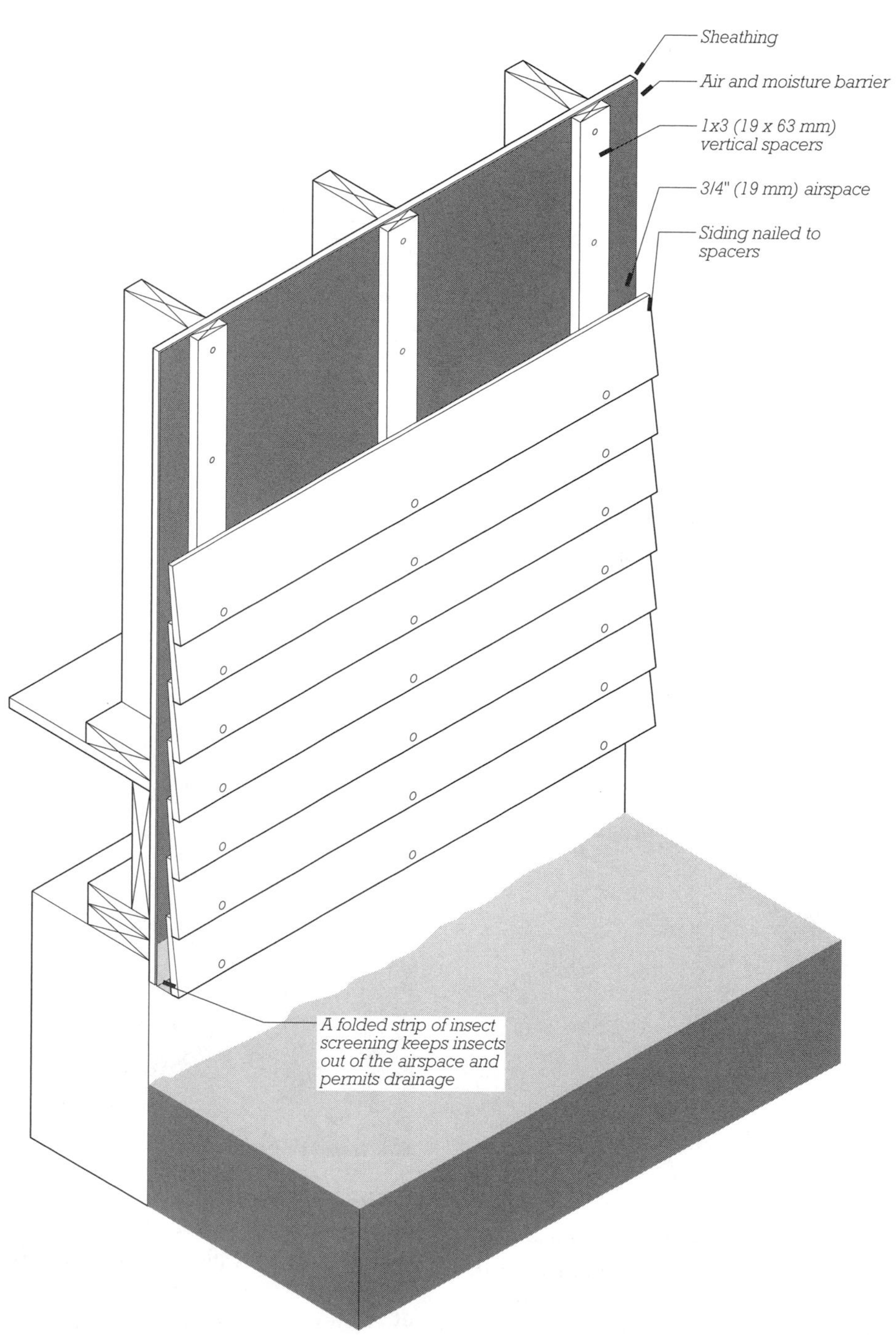

FIGURE 6.17
Rainscreen siding application. Any water that penetrates through joints or holes in the siding drains away before it reaches the sheathing.

of a fully pressure-equalized system (see Chapter 19). In such construction, special attention must be given to cornerboards and window and door casings to account for the additional thickness of the furred wall cladding. Alternatively, several manufacturers sell thin drainage mat materials, which can be sandwiched between the siding and the housewrap. These products provide some improvement in drainage and ventilation in comparison to conventional siding installations, but at lower cost and with less impact on the detailing of exterior trim in comparison to siding over vertical furring.

Vertically applied sidings (Figures 6.18 and 6.19) are nailed at the top

FIGURE 6.18
A carpenter applies V-groove tongue-and-groove redwood siding to an eave soffit, using a pneumatic nail gun. *(Courtesy of Senco Products, Inc.)*

FIGURE 6.19
A completed installation of tongue-and-groove siding, vertically applied in the foreground of the picture and diagonally in the area seen behind the chairs. The lighter-colored streaks are sapwood in this mixed heartwood/sapwood grade of redwood. The windows are framed in dark-colored aluminum. *(Architect: Zinkhan/Tobey. Photo by Barbeau Engh. Courtesy of California Redwood Association)*

and bottom plates of the wall framing and at one or more intermediate horizontal lines of wood blocking installed between the studs.

Heartwood redwood, cypress, and cedar sidings may be left unfinished if desired to weather to various shades of gray. The bare wood will erode gradually over a period of decades and will eventually have to be replaced. Other woods must be either stained or painted to prevent weathering and decay. If these coatings are renewed faithfully at frequent intervals, the siding beneath will last indefinitely.

Plywood Sidings

Plywood panel siding (Figures 6.20 and 6.21) is often chosen for its economy. The cost of the material per unit area of wall is usually somewhat less than for other siding materials, and labor costs tend to be relatively low because the large sheets of plywood are more quickly installed than equivalent areas of boards. In many cases, the sheathing can be eliminated from the building (with the wall moisture barrier applied directly to the studs) if plywood is used for siding, leading to further cost savings. All plywood

FIGURE 6.20
Grooved plywood siding is used vertically on this commercial building by Roger Scott Group, Architects. The horizontal metal flashings between sheets of plywood are purposely emphasized here with a special projecting flashing detail that casts a dark shadow line. *(Courtesy of APA–The Engineered Wood Association)*

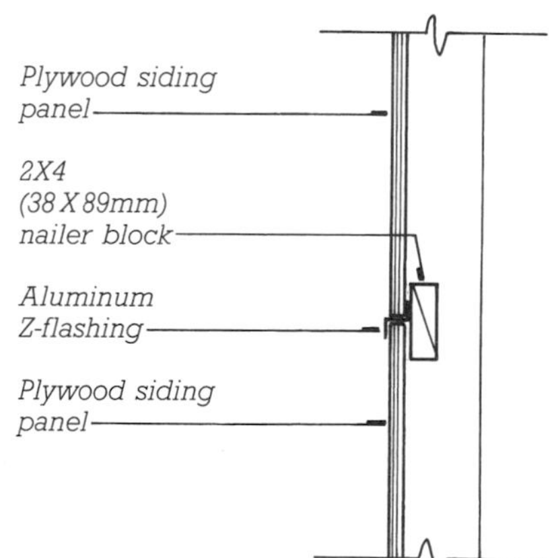

FIGURE 6.21
A detail of a simple Z-flashing, the device most commonly used to prevent water penetration at horizontal joints in plywood siding.

sidings must be painted or stained, even those made of decay-resistant heartwoods, because their veneers are too thin to withstand weather erosion for more than a few years. The most popular plywood sidings are those that are grooved to imitate board sidings and conceal the vertical joints between sheets.

The largest problem in using plywood sidings for multistory buildings is how to detail the horizontal end joints between sheets. A *Z-flashing* of aluminum (Figure 6.21) is the usual solution and is clearly visible. The designer should include in the construction drawings a sheet layout for the plywood siding that organizes the horizontal joints in an acceptable manner. This will help avoid a random, unattractive pattern of joints on the face of the building.

Shingle Sidings

Wood shingles and shakes (Figures 6.22–6.28) require a nail-base sheathing material such as OSB or plywood. Either corrosion-resistant box nails or gun-driven staples may be used for attachment. Most shingles are of cedar or redwood heartwood and do not need to be coated with paint or stain unless such a coating is desired for cosmetic reasons.

The application of wood shingle siding is labor intensive, especially around corners and openings, where many shingles must be cut and fitted. Figures 6.24 and 6.25 show two different ways of turning corners with wood shingles. Several manufacturers

FIGURE 6.22
Applying wood shingle siding over asphalt-saturated felt building paper. The corners are woven as illustrated in Figure 6.24. *(Photo by Edward Allen)*

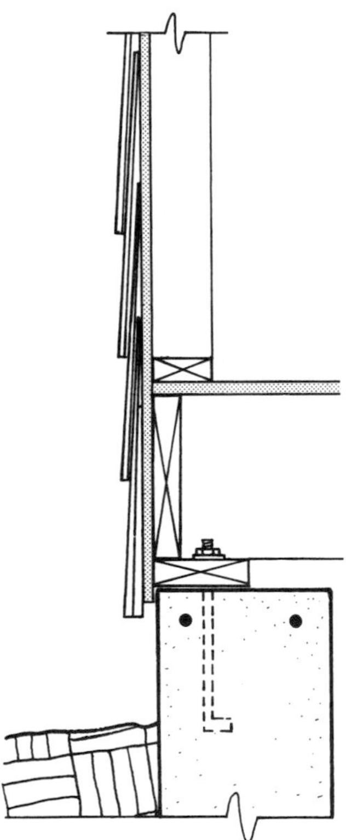

FIGURE 6.23
A detail of wood shingle siding at the sill of a wood platform frame building. The first course of shingles projects below the sheathing to form a drip, and is doubled so that all the open vertical joints between the shingles of the outside layer are backed up by the undercourse of shingles. Succeeding courses are single, but are laid so that each course covers the open joints in the course below.

produce shingle siding in panel form by stapling and/or gluing shingles to wood backing panels in a factory. A typical panel size is approximately 2 feet high and 8 feet long (600 × 2450 mm). Several different shingle application patterns are available in this form, along with prefabricated woven corner panels. The application of panelized shingles is much more rapid than that of individual shingles, which results in sharply lower on-site labor costs and, in most cases, a lower overall cost.

Metal and Plastic Sidings

Painted wood sidings are apt to deteriorate unless they are carefully scraped and repainted every 3 to 6 years. *Aluminum or vinyl siding,* formed of prefinished sheets of aluminum or molded of vinyl plastic, are usually designed to imitate wood sidings and are generally guaranteed against needing repainting for long periods, typically 20 years (Figures 6.29 and 6.30). Such sidings do have their own problems, however, including the poor resistance of aluminum sidings to denting and the tendency of plastic sidings to crack and occasionally shatter on impact, especially

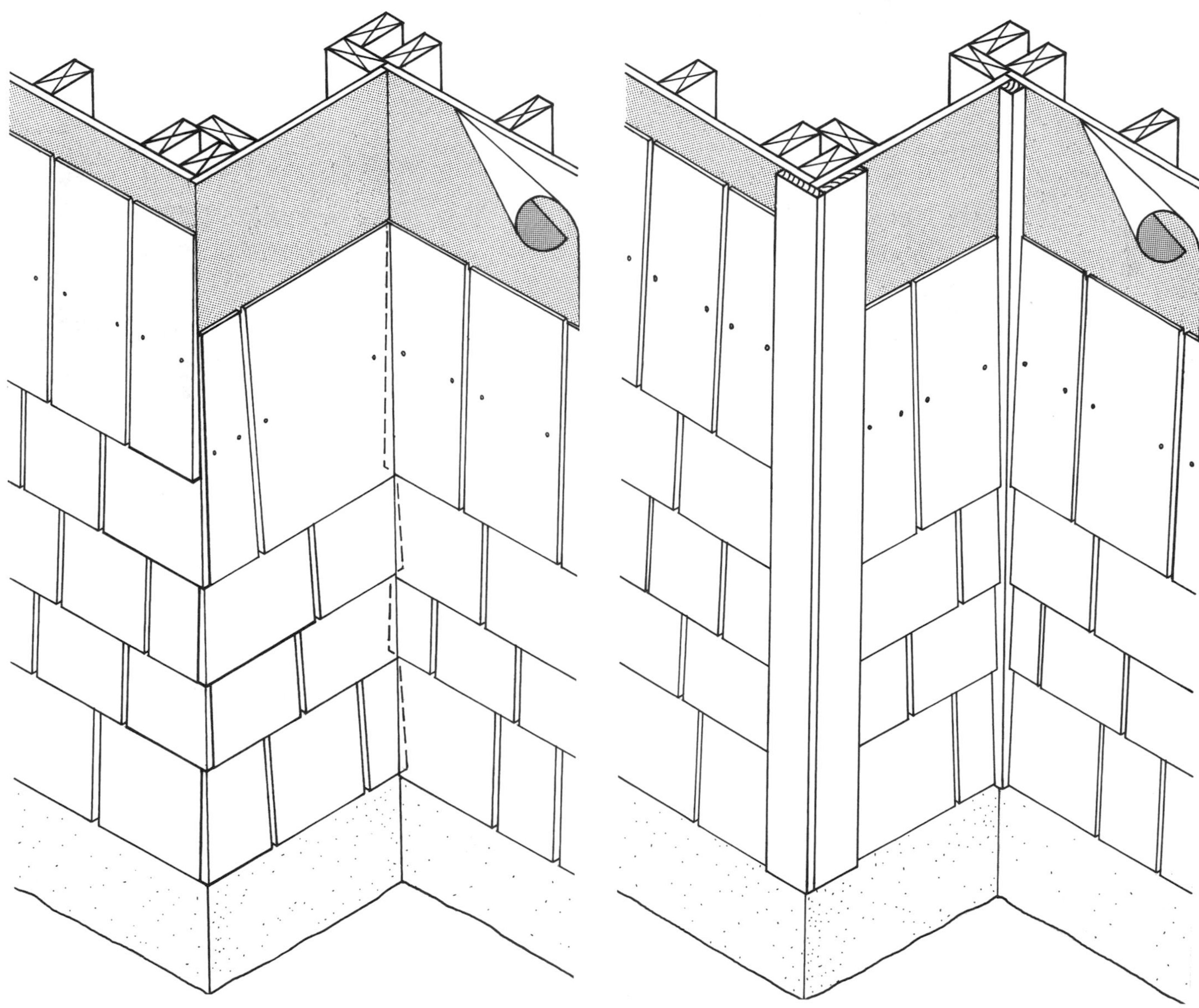

Figure 6.24
Wood shingles can be woven at the corners to avoid corner boards. Each corner shingle must be carefully trimmed to the proper line with a block plane, which is time-consuming and relatively expensive, but the result is a more continuous, sculptural quality in the siding.

Figure 6.25
Corner boards save time when shingling walls and become a strong visual feature of the building. Notice in this and the preceding diagram how the joints and nail heads in each course of shingles are covered by the course above.

FIGURE 6.26
Fancy cut wood shingles were often a featured aspect of shingle siding in the late 19th century. Notice the fish-scale shingles in the gable end, the serrated shingles at the lower edges of walls, and the sloping double-shingle course along the rakes. Corners are woven. *(Photo by Edward Allen)*

FIGURE 6.27
Fancy cut wood shingles stained in contrasting colors are used here on a contemporary restaurant. *(Courtesy of Shakertown Corporation)*

FIGURE 6.28
Both the roof and walls of this New England house by architect James Volney Righter are covered with wood shingles. Wall corners are woven. *(© Nick Wheeler/Wheeler Photographics)*

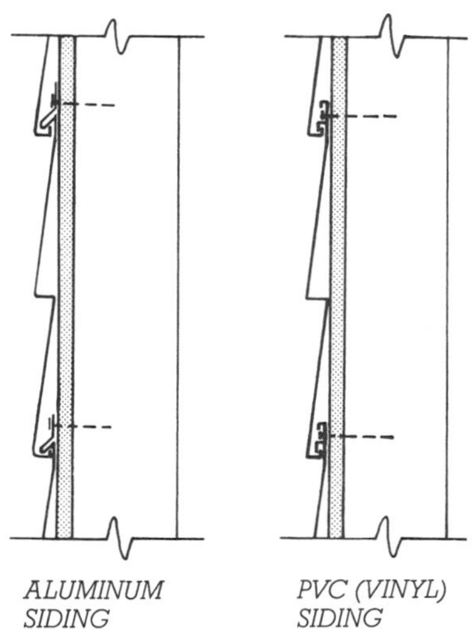

FIGURE 6.29
Aluminum and vinyl sidings are both intended to imitate wood horizontal bevel siding. Their chief advantage in either case is low maintenance. Nails are completely concealed in both systems.

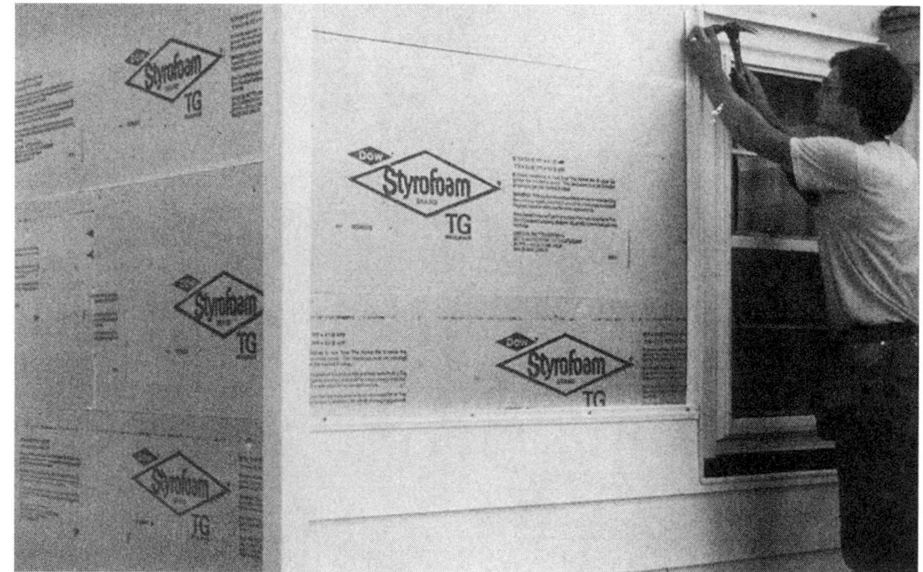

FIGURE 6.30
Retrofitting aluminum siding over insulating foam sheathing on an existing residence. Special aluminum pieces are provided for corner boards and window casings; each piece has a shallow edge channel to accept the cut ends of the siding. *(Photograph provided by The Dow Chemical Company)*

in cold weather. Although plastic and aluminum sidings bear a superficial similarity to the wood sidings that they mimic, their details around openings and corners in the wall are sufficiently different that they are usually inappropriate for use in historic restoration projects.

Stucco

Stucco, a *portland cement plaster*, is a strong, durable, economical, fire-resistant material for siding light frame buildings. It is normally applied in three coats over wire lath, either by hand or by spray apparatus (Figure 6.31). Despite its monolithic appearance, stucco is a porous material and prone to the development of hairline shrinkage cracks. When it is used in locations exposed to heavy wind and rain, significant quantities of water can pass through the material. In such circumstances, stucco

Figure 6.31
Applying exterior stucco over woven wire lath, often referred to as "chicken wire." The workers to the right hold a hose that sprays the stucco mixture onto the wall, while the man at the left levels the surface of the stucco with a straightedge. The small rectangular opening at the base of the wall is a crawlspace vent. *(Photo compliments Keystone Steel & Wire Co.)*

Figure 6.32
This small office building in Los Angeles demonstrates the plasticity of form that is possible with stucco siding. Vertical and horizontal joints in the stucco serve to minimize its tendency to shrink and crack as it dries out after curing. Although the building is supported from below on concrete columns and steel girders, it is actually a wood light frame building. *(Eric Owen Moss Architects. Photo © Tom Bonner)*

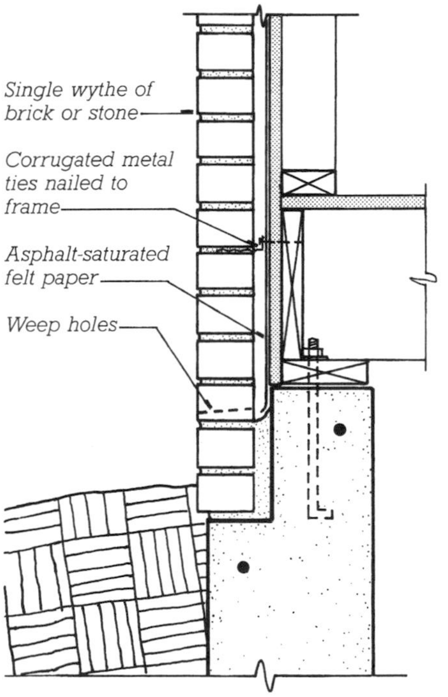

Single wythe of brick or stone
Corrugated metal ties nailed to frame
Asphalt-saturated felt paper
Weep holes

Figure 6.33
A detail of masonry veneer facing for a platform frame building. The weep holes drain any moisture that might collect in the cavity between the masonry and the sheathing. The cavity should be at least 1 inch (25 mm) wide. A 2-inch (50-mm) cavity is better because it is easier to keep free of mortar droppings that can clog the weep holes.

is best installed as a rainscreen cladding (Figure 6.17) or with a double-layer wall moisture barrier to provide good drainage and protect the wall construction from water.

Masonry Veneer

Light frame buildings can be faced with *masonry veneer,* a single wythe of brick or stone in the manner shown in Figure 6.33. The corrugated metal ties prevent the masonry from falling away from the building while allowing for differential vertical movement between the masonry and the frame. Like stucco, brick veneer is a porous material. The cavity behind the brick creates a rainscreen-like system that provides a path of free drainage to protect the wood-framed wall from water penetration. Masonry materials and detailing are covered in Chapters 8 and 9.

Artificial Stone

Artificial stone or *manufactured masonry* is made from mixtures of cement, sand, other natural aggregates, and mineral pigments such as iron oxide. It is cast into any of a great variety of shapes, textures, and colors simulating the appearance of traditional brick and natural stone products but applied as a thin facing. Artificial stone units range in thickness from approximately 1 inch to $2\frac{5}{8}$ inches (25–67 mm) and are typically applied over a base of metal lath and portland cement plaster.

Fiber-Cement Panel Siding

Fiber-cement panel siding is made of cement, sand, organic or inorganic fibers, and fillers manufactured into panels 4 feet × 8 feet (1219 mm × 2438 mm) in size and $\frac{1}{4}$ or $\frac{5}{16}$ inch (6 or 8 mm) thick (Figure 6.34). They

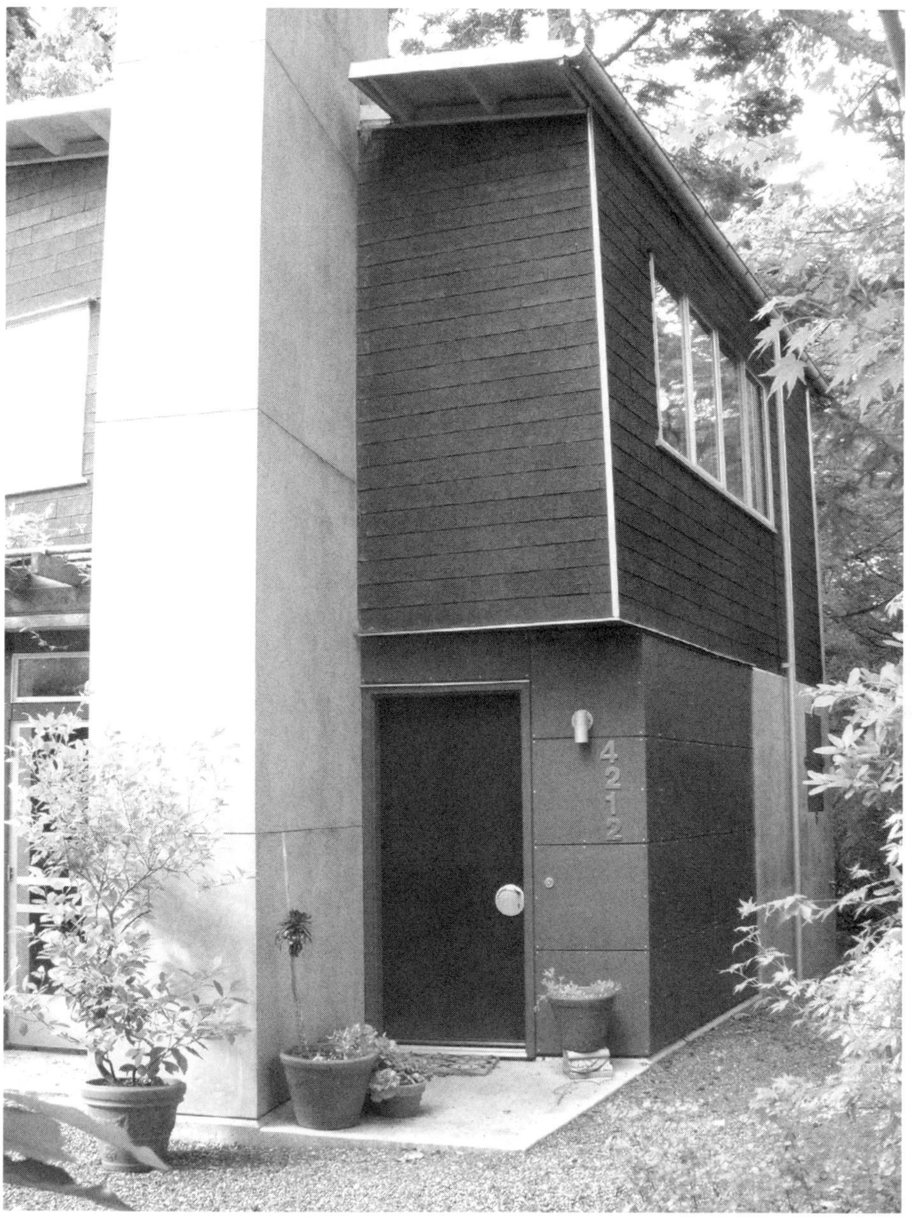

Figure 6.34
This wood-framed house illustrates the use of four siding materials. To the left, the chimney is clad in stucco. Note the horizontal expansion joints spaced at roughly 8 feet (2400 mm). The second story is clad in asphalt shingles. The area around the entry door is covered with fiber-cement panels. The panels were field-painted and fastened with stainless steel washer head screws that remain visible and make a distinctive pattern. To the right of the fiber-cement panels is a concrete wall with no applied cladding. *(Architect: Studio Ectypos, www.studioectypos.com. Photo by Joseph Iano)*

are manufactured with a smooth face or with various textures, and can be painted in the field or provided prefinished. They are attached with nails or screws driven into the wall studs or other solid framing behind the moisture barrier and wall sheathing. Gaps between panels can be sealed with joint sealant or filled with preformed plastic or aluminum trim.

Corner Boards and Exterior Trim

Most siding materials require vertical corner boards as well as trim boards around windows and doors. Roof edges require trim boards of various types, including fascias, soffits, frieze boards, and moldings, the exact patterns depending on the style and detailing of the building (Figure 6.35). For most sidings, the traditional choice of material for these exterior trim components is pine. For wood shingles, unpainted cedar or redwood may be preferred. Plastic and metal sidings are trimmed with special accessory strips of the same material. Fiber-cement siding may be trimmed either with wood or with fiber-cement boards made for the purpose.

A high-appearance grade of pine is usually chosen for wood trim boards. Traditionally, the boards are back primed before installation with the application of a prime coat of paint on the back side. This helps to reduce cupping and other distortions of the trim boards with changes of humidity. Fully preprimed trim boards are also available at most lumberyards.

Many historical styles require trim that is ornamentally shaped. The shaping generally takes place at a mill, so that the trim pieces come to the site ready to install.

As appearance grades of lumber have become more costly, substitutes have been introduced to the market, including boards made of high-density plastic foams, wood composites of various kinds, and lumber that has been cut into short pieces to

Figure 6.35
Painters finish the exterior of a house. The painter at the far left is standing on the top of a stepladder, which is a very unsafe practice. *(Courtesy of Energy Studies in Buildings Laboratory, Center for Housing Innovation, Department of Architecture, University of Oregon)*

FIGURE 6.36
Careful detailing is evident in every aspect of the exterior finishes of this commercial building. Notice especially the cleanly detailed window casings, the purposeful use of both vertical boards and wood shingles for siding, and the neat junction between the sidewall shingles and the rake boards. *(Woo & Williams, Architects. Photographer: Richard Bonarrigo)*

FIGURE 6.37
This house in a rainy climate is designed to shelter every window and door with a roof overhang. The base of the wall is sided with water-resistant cement board. Special attention to roof overhang proportions, exposed framing details, window muntin patterns, and railing details gives this house a uniquely appealing character. *(Photo by Rob Thallon)*

CONSIDERATIONS OF SUSTAINABILITY IN PAINTS AND OTHER ARCHITECTURAL COATINGS

- Paints and other coatings can be significant emitters of VOCs, and thus require care in selection and specification. Water-based acrylic latex paints generally have lower VOC emissions than solvent-based paints.
- The quality and durability of low- and zero-VOC paints continue to improve.
- Architectural paints and coatings achieving the Green Seal Organization's Green Seal Certification satisfy current LEED requirements for low-emitting materials.
- Unused paint can be reprocessed to make *recycled paint,* eliminating paint waste from the waste disposal stream.
- In addition to volatile organic compounds, paints and coatings can emit other toxic or unpleasant-smelling chemicals. Some manufacturers also report their products' emissions of such chemicals and offer products with low emissions in these categories.
- Paints that wear quickly and require frequent recoating may increase VOC emissions over the full life of a facility in comparison to others with higher emissions that are more durable and require less frequent recoating.

FIGURE 6.38
A carefully wrought exterior deck of redwood. The decking is spaced to allow water to drain through. *(Designer: John Matthias. Photo by Ernest Braun. Courtesy of California Redwood Association)*

eliminate knots, then reassembled end to end with finger joints and glue. These products tend to be more consistently straight and free of defects than natural, solid lumber. Some of them are not only produced as flat boards, but also molded to imitate traditional trim. They must be painted and are not suitable for transparent finishes. Many are also available pre-primed or prefinished.

Exterior Construction

Wood is widely used outdoors for porches, decks, stairs, stoops, and retaining walls (Figure 6.38). Decay-resistant heartwoods, wood that is pressure treated with preservatives, and moisture-resistant wood/plastic composite planks are suitable for these exposed uses. If nondurable woods are used, they will soon decay at the joints, where water is trapped and held by capillary action. Fasteners must be hot-dip galvanized or stainless steel to avoid corrosion. Wood decking that is exposed to the weather should always have open, spaced joints to allow for drainage of water through the deck and for expansion and contraction of the decking. Plastic composite planks are durable and attractive, although some are not as strong or stiff as wood, and may require more closely spaced joists for support.

Sealing Exterior Joints

After the completion of exterior carpentry work, exposed gaps between siding, trim, frames of doors and windows, and other exterior materials are sealed with *joint sealant* to protect against the entry of water. Exterior sealants must have good adhesion to the materials being sealed, remain permanently flexible, and be unaffected by exposure to sunlight. *Paintable sealants* can be installed before

CSI/CSC

MasterFormat Sections for Exterior Finishes for Wood Light Frame Construction

04 20 00	UNIT MASONRY
04 21 00	Clay Unit Masonry Brick Veneer Masonry
04 70 00	MANUFACTURED MASONRY
06 20 00	FINISH CARPENTRY Exterior Finish Carpentry
07 30 00	STEEP SLOPE ROOFING
07 31 00	Shingles and Shakes
07 32 00	Roof Tiles
07 40 00	ROOFING AND SIDING PANELS
07 46 00	Siding
07 60 00	FLASHING AND SHEET METAL
07 63 00	Sheet Metal Flashing and Trim
07 65 00	Flexible Flashing
08 10 00	DOORS AND FRAMES
08 14 00	Wood Doors Clad Wood Doors
08 15 00	Plastic Doors
08 16 00	Composite Doors
08 50 00	WINDOWS
08 51 00	Metal Windows
08 52 00	Wood Windows
08 53 00	Plastic Windows
08 54 00	Composite Windows
09 20 00	PLASTER AND GYPSUM BOARD
09 24 00	Portland Cement Plastering Portland Cement Stucco
09 90 00	PAINTING AND COATING
09 91 00	Painting Exterior Painting
09 93 00	Staining and Transparent Finishing Exterior Staining and Finishing
09 96 00	High-Performance Coatings

finish painting and then painted over to match adjacent surfaces. Other sealants, such as most silicones, cannot hold paint. These are selected from a range of available premixed colors and then applied after finish painting. Sealant materials and sealant joint design are discussed in detail in Chapter 19.

Exterior Painting, Finish Grading, and Landscaping

The final steps in finishing the exterior of a light frame building are painting or staining of exposed wood surfaces; finish grading of the ground around the building; installation of paving for drives, walkways, and terraces; and seeding and planting of landscape materials. By the time these operations take place, interior finishing operations are also usually well underway, having begun as soon as the roofing, sheathing, windows, and doors were in place.

Selected References

In addition to consulting the references listed at the end of Chapter 5, the reader should acquire current product information from manufacturers of residential windows, doors, roofing, and siding materials. Many building supply retailers also distribute unified catalogs of windows, doors, and millwork that can be an invaluable part of the designer's reference shelf.

1. National Roofing Contractors Association. *NRCA Steep-slope Roofing Manual.* Rosemont, IL, updated regularly.

This industry standard reference provides technical guidelines and recommended details for asphalt shingle roofing and other types of steep-slope roofing.

2. The Tauton Press, Inc. *Fine Homebuilding.* Newtown, CT, published monthly; and *Journal of Light Construction*, Williston, VT, published monthly.

These two magazines are excellent references on all aspects of wood light frame construction, including topics related to exterior carpentry and finishing.

3. R. Sam Williams. "Finishing of Wood," in Forest Products Laboratory. *Wood Handbook: Wood as an Engineering Material.* Madison, Wisconson, 1999.

This article provides a comprehensive introduction to the topic of wood finishing. It can be viewed online for free on the Forest Product Laboratory's web site, www.fpl.fs.fed.us. The complete handbook is also available in printed form from various publishers.

Web Sites

Exterior Finishes for Wood Light Frame Construction

Author's supplementary web site: **www.ianosbackfill.com/06_exterior_finishes_for_wood_light_frame_construction**

Roofing

National Roofing Contractors Association: **www.nrca.net**

Windows and Doors

Anderson Windows & Doors: **www.andersonwindows.com**
Fleetwood Windows & Doors: **www.fleetwoodusa.com**
Pella Windows and Doors: **www.pella.com**
Quantum Windows & Doors: **www.quantumwindows.com**

Siding

Cedar Shake & Shingle Bureau: **www.cedarbureau.org**
James Hardie Fiber Cement Siding: **www.jameshardie.com**
Portland Cement Association—Stucco (Portland Cement Plaster): **www.cement.org/stucco**
Vinyl Siding Institute: **www.vinylsiding.org**
Western Red Cedar Lumber Association: **www.wrcla.org**

Exterior Painting

Benjamin Moore Paints: **www.benjaminmoore.com**
Cabot Stains: **www.cabotstain.com**
Green Seal: **www.greenseal.org**

Master Painters Institute (MPI): **www.paintinfo.com**
MPI Specify Green: **www.specifygreen.com**
Olympic Paints and Stains: **www.olympic.com**
Sherwin-Williams Coatings: **www.sherwin-williams.com**

Key Terms and Concepts

roofing underlayment
building felt
synthetic roofing underlayment
moisture barrier, water-resistive barrier, weather-resistive barrier, air barrier
Grade D sheathing paper
housewrap
eave
rake
gutter
downspout, leader
external gutter
internal gutter, concealed gutter
splash block
dry well
ice dam
ventilated roof
soffit vent
gable vent
ridge vent
ice barrier
rubberized underlayment, ice and water shield
unventilated roof
asphalt shingle
flashing
sill pan
nail-fin window, flanged window
finless window, flangeless window
substrate (for painting)
vehicle, binder, film-former
solvent
pigment
solvent-based coating
water-based coating
volatile organic compound (VOC)
paint
enamel
stain
clear coating
lacquer
varnish
shellac
high-performance coating
primer
sealer
siding
board siding
nail-base sheathing
siding nail
bevel siding
shiplap siding
tongue-and-groove siding
board-and-batten siding
furring strip
rainscreen siding
plywood panel siding
Z-flashing
wood shingle
wood shake
aluminum siding
vinyl siding
stucco, portland cement plaster
masonry veneer
artificial stone, manufactured masonry
fiber-cement panel siding
joint sealant
paintable sealant
recycled paint

Review Questions

1. In what order are exterior finishing operations carried out on a platform frame building and why?

2. At what point in exterior finishing operations can interior finishing operations begin?

3. Which types of siding require a nail-base sheathing? What are some nail-base sheathing materials? Name some sheathing materials that cannot function as nail bases.

4. What are the reasons for the relative economy of plywood sidings? What special precautions are advisable when designing a building with plywood siding?

5. How does one make corners when siding a building with wood shingles?

6. Specify two alternative exterior coating systems for a building clad in wood bevel siding.

7. What are the usual reasons for premature paint failure on a wood-sided house?

Exercises

1. For a completed wood frame building, make a complete list of the materials used for exterior finishes and sketch a set of details of the eaves, rakes, corners, and windows. Are there ways in which each could be improved?

2. Visit a building materials supply store and look at all the alternative choices of sidings, windows, doors, trim lumber, and roofing. Study one or more systems of gutters and downspouts. Look at eave vents, gable vents, and ridge vents.

3. For a wood light frame building of your design, list precisely and completely the materials you would like to use for the exterior finishes. Sketch a set of typical details to show how these finishes should be applied to achieve the appearance you desire, with special attention to the roof edge details.

7

Interior Finishes for Wood Light Frame Construction

Architect Michael Craig Moore uses a rich palette of stone and wood for floors and millwork contrasting elegantly with light-colored walls and ceiling surfaces. In the foreground is slate flooring. Stairs and wood floors are heartwood only pine ("heart pine"). Wood cabinets and trim are vertical grain Douglas Fir. Cabinet counters and the window seat top (middleground of the image) are a composite sheet material made from compressed paper and phenolic resin. Cabinet pulls are bronze. The exposed structural wood column, to the right in the image, is also Douglas Fir. All wood surfaces are finished with clear polyurethane. Walls and ceilings are covered with standard gypsum wallboard finished with latex primer and top coats. *(Photograph by Michael Craig Moore, AIA)*

As the exterior roofing and siding of a platform frame building approach completion, the framing carpenters and roofers are joined by workers from a number of other building trades. Masons commence work on fireplaces and chimneys (Figures 7.1 and 7.2). Plumbers begin *roughing in* their piping ("roughing in" refers to the process of installing the components of a system that will not be visible in the finished building). First to be installed are the large *DWV (drain–waste–vent) pipes*, which drain by gravity and must therefore have first choice of space in the building; then the small *supply pipes*, which bring hot and cold water to the fixtures; and the gas piping (Figures 7.3–7.6). If the building will have central warm air heating and/or air conditioning, sheet metal workers install the furnace and *ductwork* (Figures 7.7–7.9). If the building is to have a *hydronic (forced hot water) heating system*, the plumbers put in the boiler and rough in the heating pipes and *convectors* at this time (Figure 7.10). A special variety of hydronic heating is the *radiant heating system* that warms the floors of the building by means of plastic piping, built into the floors, through which hot water is circulated (Figure 7.11). The electricians are usually the last of the mechanical and electrical trades to complete their roughing in because their wires are flexible and can generally be routed around the pipes and ducts without difficulty (Figure 7.12). When the plumbers, sheet metal workers, and electricians have completed their rough work, which consists of everything except installing the plumbing fixtures, electrical outlets, and air registers and grills, inspectors from the local building department check each of the systems for compliance with the plumbing, electrical, and mechanical codes, as well as to ensure that framing has not been damaged during the installation of these other components.

Once these inspections have been passed, connections are made to external sources of water, gas, electricity, and communications services, and to a means of sewage disposal, either a sewer main or a septic tank and leaching field. Thermal insulation and a vapor retarder are added to the exterior ceilings and walls.

Now a new phase of construction begins, the interior finishing operations, during which the inside of the building undergoes a succession of radical transformations. The elaborate tangle of framing members, ducts, pipes, wires, and insulation rapidly disappears behind the finish wall and ceiling materials. The interior millwork—doors, finish stairs, railings, cabinets, shelves, closet interiors, and door and window casings—is installed. The finish flooring materials are installed as late in the process as possible to save them from damage by the passing armies of workers, and carpenters follow behind the flooring installers to add the baseboards that cover the last of the rough edges in the construction.

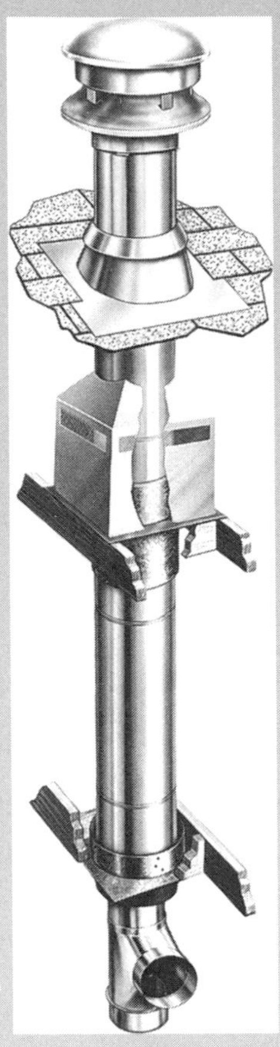

Finally, the building hosts the painters who prime, paint, stain, varnish, and paper its interior surfaces. The plumbers, electricians, and sheet metal workers make brief return appearances on the heels of the painters to install the plumbing fixtures; the electrical receptacles, switches, and lighting fixtures; and the air grills and registers. At last, following a final round of inspections and a last-minute round of repairs and corrections to remedy lingering defects, the building is ready for occupancy.

FIGURE 7.1
Insulated metal flue systems are often more economical than masonry chimneys (shown here) for furnaces, boilers, water heaters, package fireplaces, and solid-fuel stoves. *(Photo courtesy of Selkirk Metalbestos)*

FIGURE 7.2
A mason adds a section of clay flue liner to a chimney. The large flue is for a fireplace, and the three smaller flues are for a furnace and two wood-burning stoves. *(Photo by Edward Allen)*

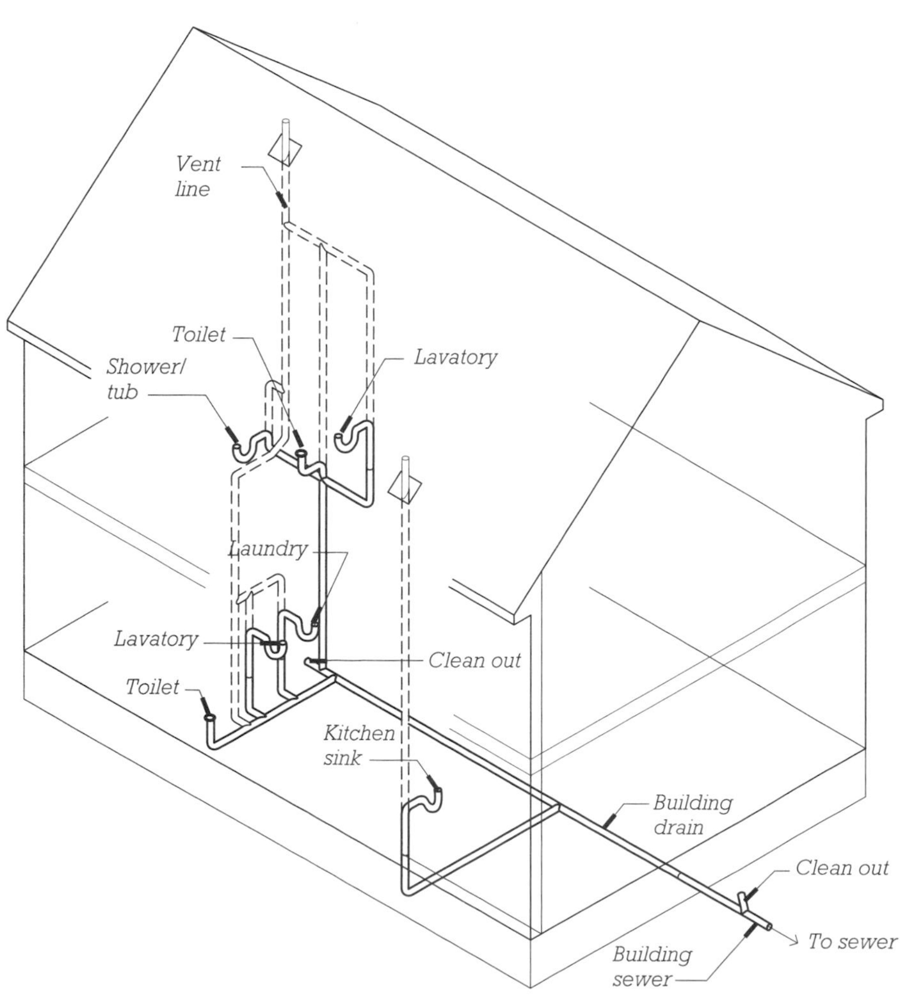

Figure 7.3
A typical residential wastewater system. All fixtures drain to the building drain through sloping or vertical branch lines. The waste pipe is vented to the exterior at each fixture through a network of vent pipes, shown with broken lines.

FIGURE 7.4
A typical residential water supply system. Water enters the house through a buried line and branches into two parallel sets of distribution lines. One is for cold water. The other passes through the water heater and supplies hot water.

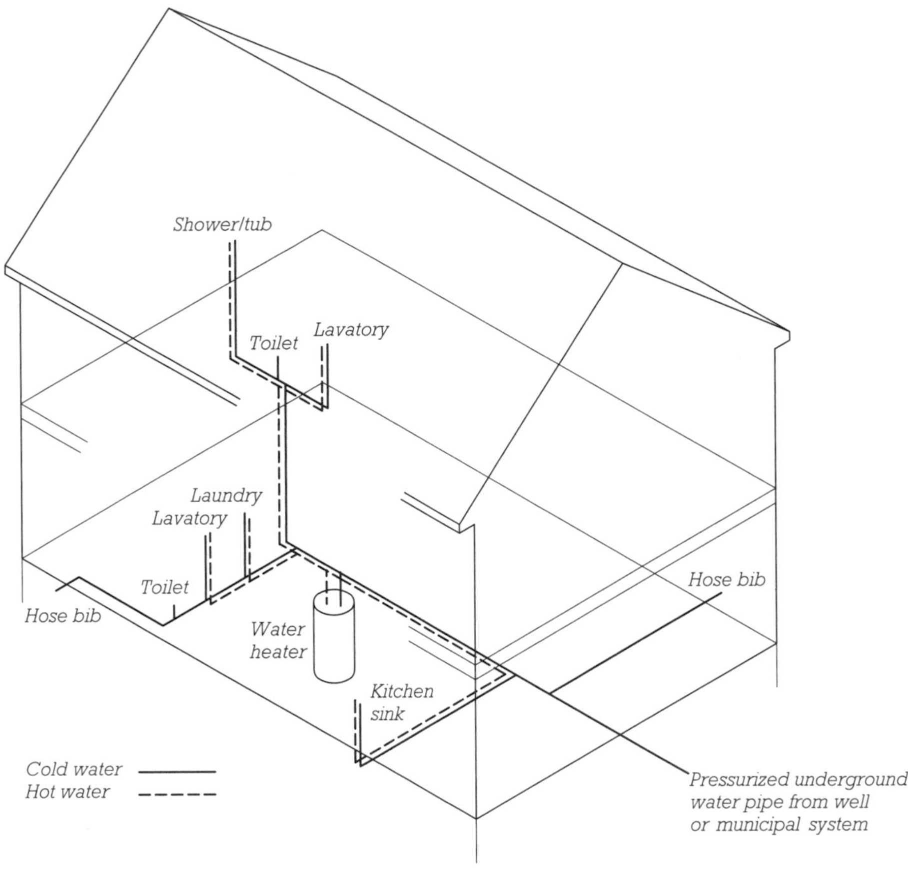

FIGURE 7.5
At the basement ceiling, the plumber installs the plastic waste pipes first to be sure that they are properly sloped to drain. The copper supply pipes for hot and cold water are installed next. *(Photo by Edward Allen)*

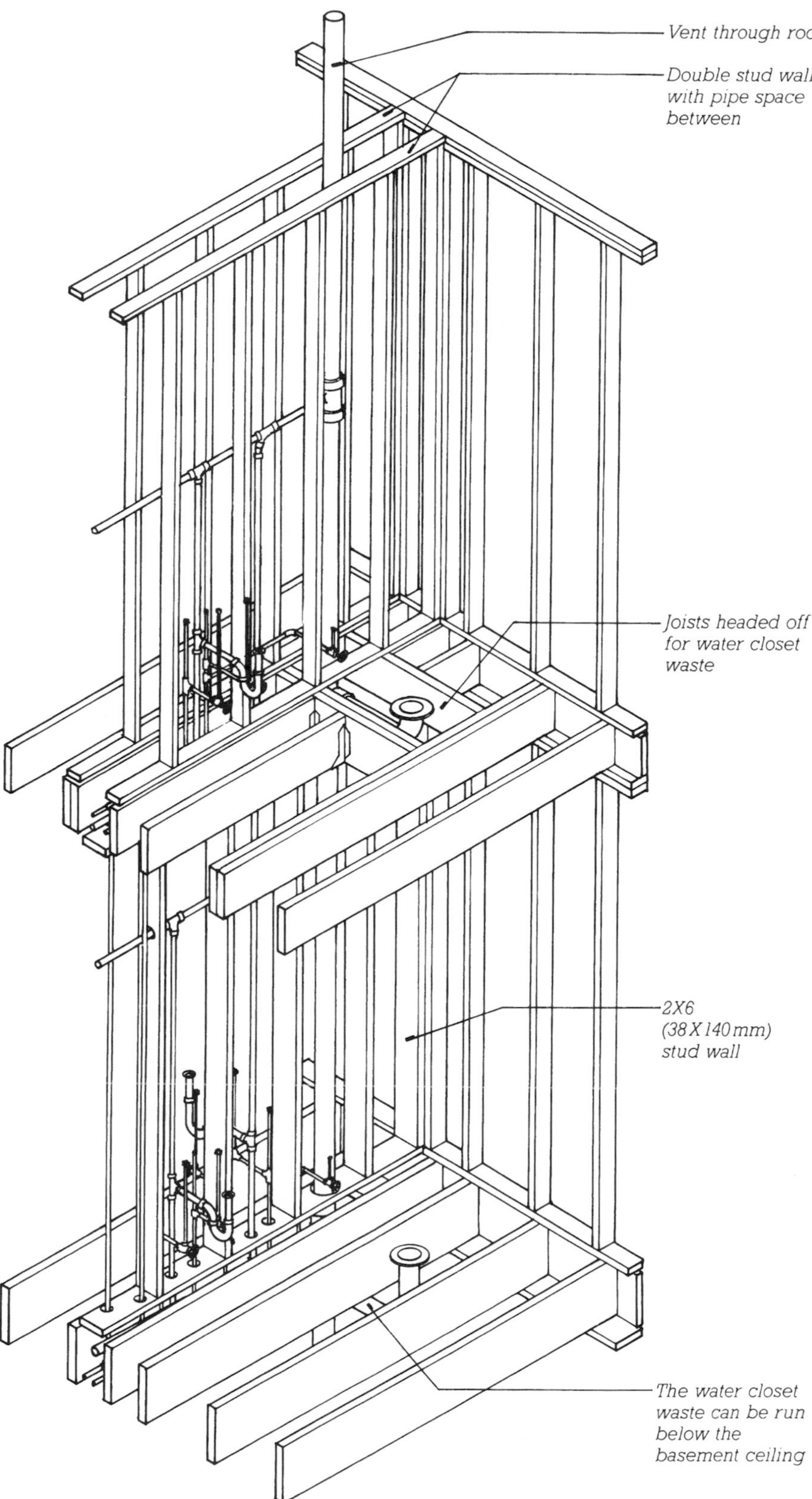

FIGURE 7.6
The plumber's work is easier and less expensive if the building is designed to easily accommodate piping. The "stacked" arrangement shown here, in which a second-floor bathroom and a back-to-back kitchen and bath on the first floor share the same vertical runs of pipe, is economical and easy to rough in, compared with plumbing that does not align vertically from one floor to the next. The double wall framing on the second floor allows plenty of space for the waste, vent, and supply pipes. The second-floor joists are located to provide a slot through which the pipes can pass at the base of the double wall, and the joists beneath the water closet (toilet) are headed off to house its waste pipe. The first floor shows an alternative type of wall framing using a single layer of deeper studs, which must be drilled to permit horizontal runs of pipe to pass through.

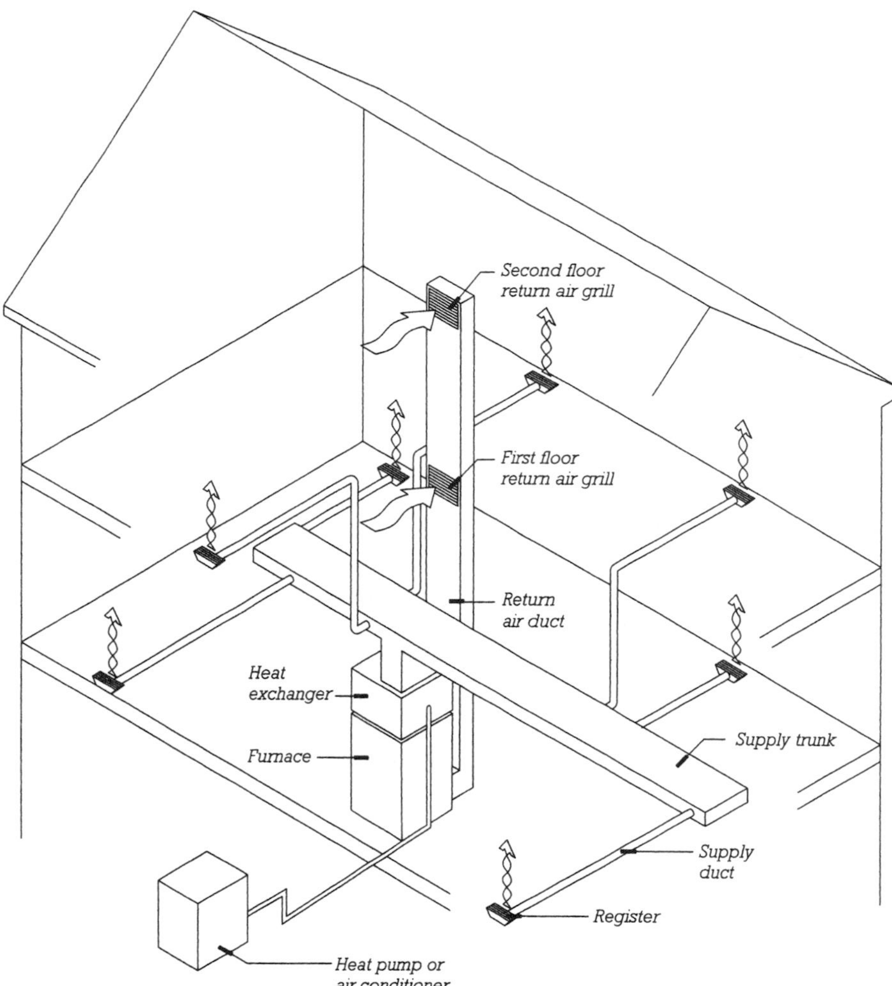

FIGURE 7.7
A forced-air system in a two-story building with a basement. The furnace is in the basement. It burns gas or oil, or uses electric resistance heating to warm a heat exchanger that creates warm air. It blows the warm air through sheet metal supply ducts to registers in the floor near the exterior walls. The air returns to the furnace through a centrally located return air duct that has a return air grill near the ceiling of each floor. With the addition of a heat pump or air conditioner, this system can deliver cool, dehumidified air during the warm months.

FIGURE 7.8
The installation of this hot air furnace and air conditioning unit is almost complete, needing only electrical connections. The metal pipe running diagonally to the left carries the exhaust gases from the oil burner to the masonry chimney. The ductwork is insulated to prevent moisture from condensing on it during the cooling season and to prevent excessive losses of energy from the ducts. *(Photo by Edward Allen)*

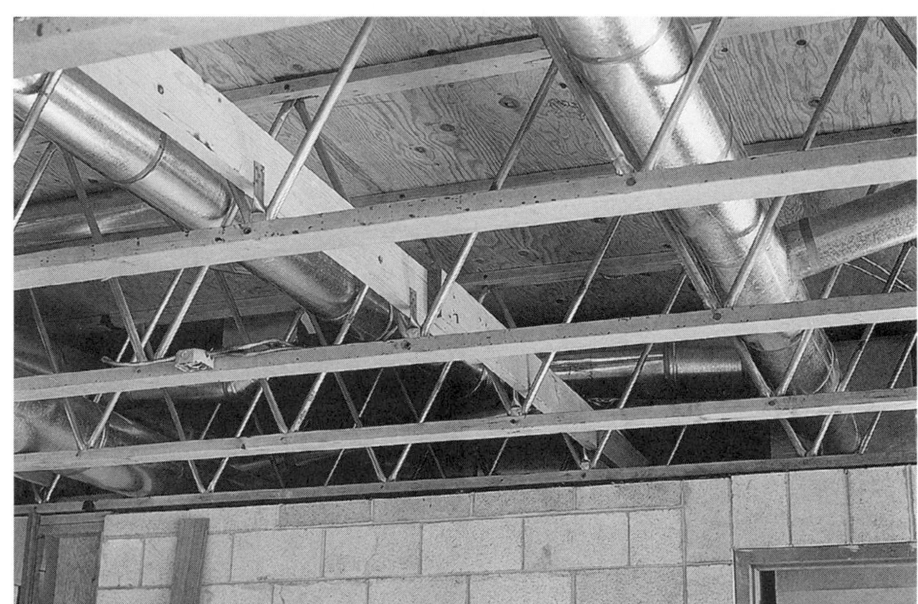

FIGURE 7.9
Ductwork and electrical wiring are installed conveniently through the openings in these floor trusses, making it easy to apply a finish ceiling if desired. The 2 × 6 that runs through the trusses in the center of the photograph is a bridging member that restrains the trusses from buckling. *(Courtesy of Trus Joist Corporation)*

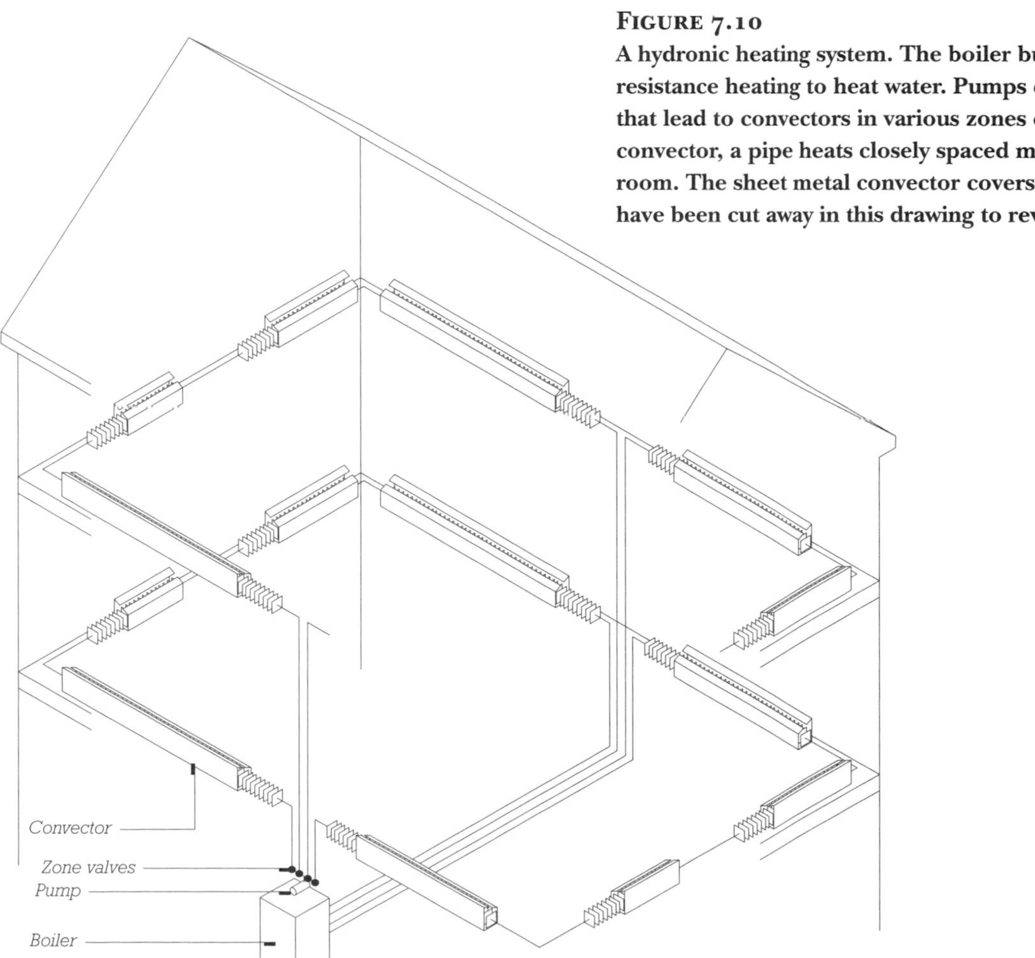

FIGURE 7.10
A hydronic heating system. The boiler burns gas or oil, or uses electric resistance heating to heat water. Pumps circulate the water through pipes that lead to convectors in various zones of the building. Inside each convector, a pipe heats closely spaced metal fins that warm the air in the room. The sheet metal convector covers are shown schematically and have been cut away in this drawing to reveal the metal fins.

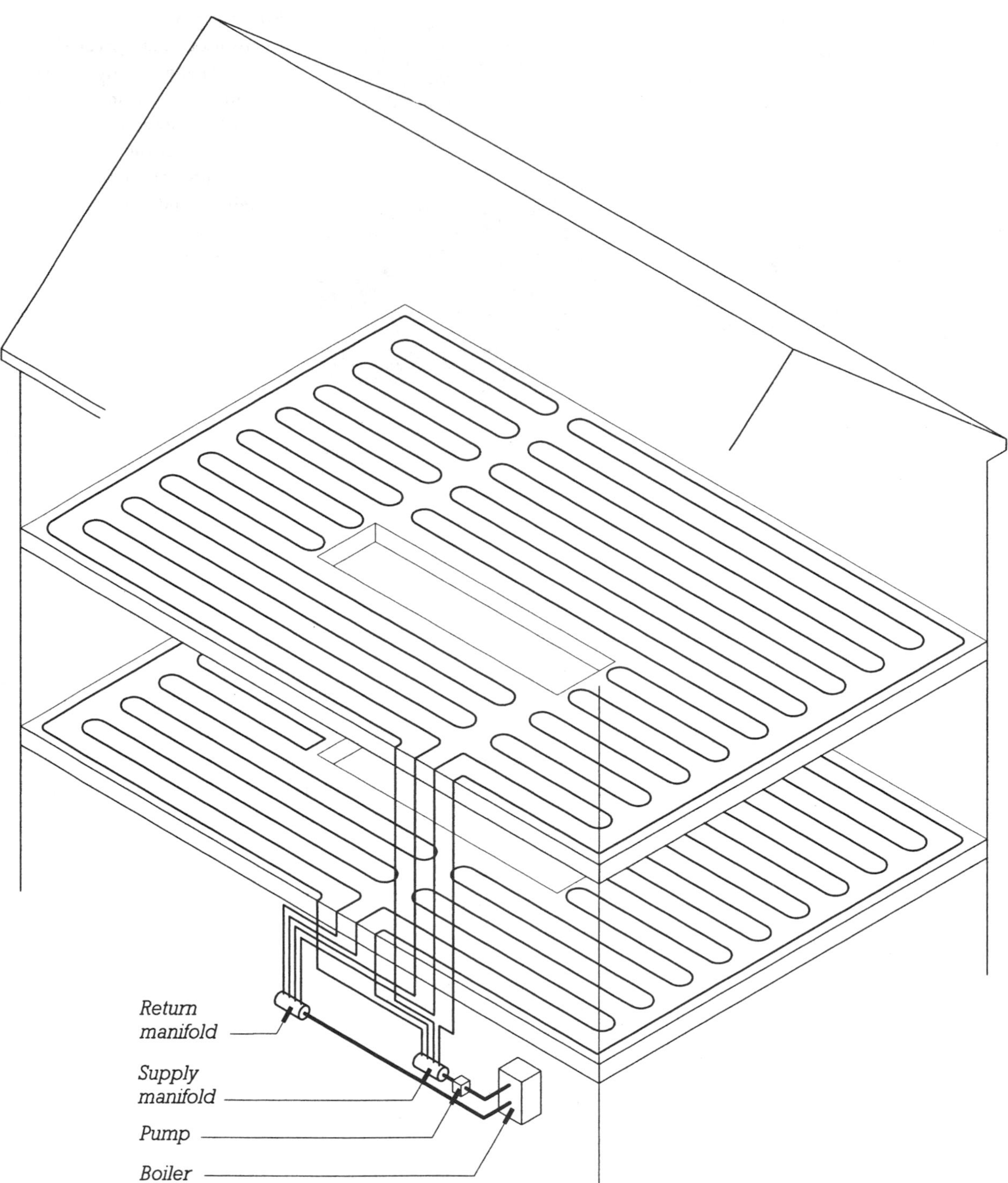

FIGURE 7.11
The radiant heating system in this two-story building delivers heat through hydronic tubes that warm the floors. There are four zones, two upstairs and two downstairs. Water is heated in a boiler and pumped through a supply manifold, where thermostatically controlled valves control the water flow to each of the zones.

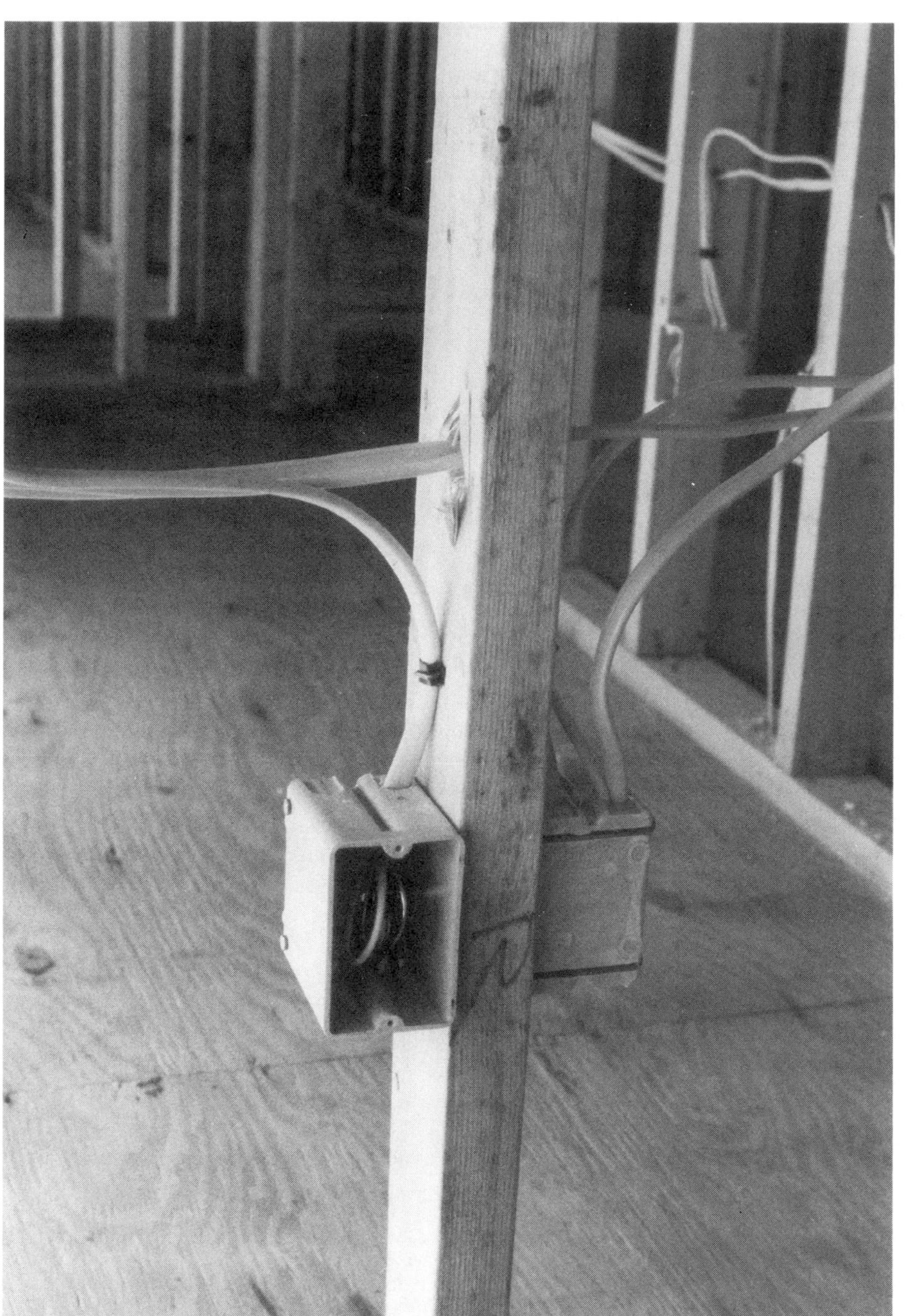

FIGURE 7.12
The electrician begins work by nailing plastic fixture boxes to the framing of the building in the locations shown on the electrical plan. Then holes are drilled through the framing, and the plastic-sheathed cable, which houses two insulated copper conductors and an uninsulated ground wire, is pulled through the holes and into the boxes, where it is held by insulated staples driven into the wood. After the interior wall materials are in place, the electrician returns to connect outlets, lighting fixtures, and switches to the wires and the boxes and to affix cover plates to finish the installation. *(Photo by Edward Allen)*

COMPLETING THE BUILDING ENCLOSURE

The walls, roofs, and other surfaces of a building that separate the indoors from the outdoors may be referred to as the *thermal envelope, building envelope,* or *building enclosure.* The building enclosure controls the flow of heat, air, and moisture between the interior and exterior of the building. Well-designed and carefully constructed enclosure assemblies help keep a building cooler in summer and warmer in winter by retarding the passage of heat through the exterior surfaces of the building. They help keep the occupants of the building more comfortable by moderating the temperatures of the interior surfaces of the building and reducing convective drafts. They reduce the energy consumption of the building for heating and cooling to a fraction of what it otherwise would be. And they prevent the harmful accumulation of condensation within the exterior walls, roof, and other enclosure assemblies of the building.

Insulating the Building Frame

Thermal insulating materials resist the conduction of heat. *Thermal insulation* is added to virtually all buildings to limit winter heat loss and reduce summer cooling loads. A material's resistance to the conduction of heat is measured as *thermal resistance,* abbreviated as *R-value,* or, in metric units, as *RSI-value* (or in many cases, also simply as R-value). A material with a higher R-value is a better insulator than one with a lower R-value. Figure 7.13 lists the most important types of thermal insulating materials used in wood light frame buildings and gives some of their characteristics. *Glass fiber batts* are the most popular type of insulation for use in wood light frame construction, but all of the materials listed find use. Examples of the installation of some of these materials are shown in Figures 6.30, and 7.14–7.16. For a more in-depth discussion of the role of thermal insulation in buildings, see pages 658–662.

Type	Materials	Method of Installation	R-Value[a]	Combustibility	Advantages and Disadvantages
Batt or blanket	Glass wool; rock wool	The batt or blanket is installed between framing members and is held in place either by friction or by a facing stapled to the framing	3.2–3.7 *22–26*	The glass wool or rock is noncombustible, but paper facings are combustible.	Low in cost, fairly high R-value, easy to install
High-density batt	Glass wool	Same as above	4.3 *30*	Same as above	Same as above
Loose fill	Glass wool; rock wool	The fill is blown onto attic floors, or into wall cavities through holes drilled in the siding	2.5–3.5 *17–24*	Noncombustible	Good for retrofit insulation in older buildings. May settle somewhat in walls.
Loose fibers with binder	Treated cellulose; glass wool	As the fill material is blown from a nozzle, a light spray of water activates a binder that adheres the insulation in place and prevents settlement	3.1–4.0 *22–28*	Noncombustible	Low in cost, fairly high R-value
Foamed in place	Polyurethane	The foam is mixed from two components and sprayed or injected into place, where it adheres to the surrounding surfaces	5–7 *35–49*	Combustible, gives off toxic gases when burned	High R-value, high cost, seals against air leakage, vapor impermeable
Foamed in place	Polyicynene	Two components are sprayed or injected into place, where they react chemically and adhere to the surrounding surfaces	3.6–4.0 *25–28*	Resistant to ignition, combustible, self-extinguishing	High R-value, seals against air leakage, vapor permeable
Rigid board	Expanded polystyrene foam (EPS)	The boards are applied over wall framing, on roof decks, around foundations, and under concrete slabs on grade.	3.6–4.2 *25–29*	Combustible, but self-extinguishing in most formulations	High R-value, low vapor permeance; can be used in contact with earth, moderate cost
Rigid board	Extruded polystyrene foam (XPS)	Same	5 *35*	Same as above	Same as EPS, but with higher compressive strength and lower moisture absorption
Rigid board	Polyisocyanurate foam	Primarily used as roof insulation applied over roof deck. Also used as wall insulation.	5.6–6 *39–42*	Combustible; gives off toxic gases when burned	High R-value, high cost
Rigid board	Mineral fiber	Same	4 *28*	Noncombustible	Moderate cost, vapor permeable
Rigid board	Cellulose (wood or vegetable) fiber	same	2.8 *19*	Combustible	Moderate cost, vapor permeable

[a]R-values are per unit thickness, expressed first in English units of ft²-hr-°F/Btu-in. and second in metric units (italicized) of m-°K/W.

FIGURE 7.13
Thermal insulation materials commonly used in light frame wood buildings. The R-values offer a direct means of comparing the relative effectiveness of the different types. For more information on rigid insulation boards commonly used with low-slope, see Figure 16.10.

(a)

(b)

(c)

(d)

(e)

FIGURE 7.14
(*a*) Installing a polyethylene vapor retarder over glass fiber batt insulation using a staple hammer, which drives a staple each time it strikes a solid surface. The batts are unfaced and stay in place between the studs by friction. (*b*) Stapling faced batts between roof trusses. The R-value of the insulation is printed on the facing. (*c*) Placing unfaced batts between ceiling joists in an existing attic. Vent spacers should be used at the eaves (see Figure 6.6). (*d*) Working from below to insulate a floor over a crawlspace. Batts in this type of installation are usually retained in place by pieces of stiff wire cut slightly longer than the distance between joists and sprung into place at frequent intervals below the insulation. (*e*) Insulating crawlspace walls with batts of insulation suspended from the sill. The header space between the joist ends has already been insulated. (See also Figure 5.8.) *(Courtesy of Owens-Corning Fiberglas Corporation)*

(a)

(b)

FIGURE 7.15
(*a*) Spraying a low-density polyicynene foam insulation between studs of a wood light frame building. At the time of spraying, the components are dense liquids, but they react immediately with one another to produce a low-density foam, as has already occurred at the bottom of the cavity. The foam in the cavity to the right has already been trimmed off flush with the studs. (*b*) Trimming off excess foam with a special power saw. *(Courtesy of Icynene, Inc., Toronto, Canada)*

Increasing Levels of Thermal Insulation

A wall framed with 2 × 4 (38 × 89 mm) studs and filled with glass fiber batt insulation can achieve a thermal resistance of approximately R-13 to R-15 (RSI-90 to 104). According to current North American energy code standards, this is adequate for a residential building in the southern portions of the United States and Hawaii but is insufficient for colder regions. To achieve a higher insulation value, the designer may either increase the thickness of the insulation or use an insulation material that has a higher R-value for an equal thickness. Figure 7.17 shows two commonly employed solutions, either of which achieves an insulation value of R-19 (RSI-132) or greater. Figure 7.20 illustrates two possible approaches to achieving even higher insulation levels in walls. Spray foam and rigid foam insulation materials can also be used, taking advantage

FIGURE 7.16
Blowing loose-fill glass fiber insulation into a ceiling below an attic. A vapor retarder was installed on the bottom side of the joists and then a gypsum board ceiling, which supports the insulation. Vent spacers were installed at the eaves to prevent the insulation from blocking eave vents. *(Courtesy of Owens-Corning Fiberglas Corporation)*

of the relatively high R-value per thickness of these materials, and in the case of spray foam, its ability to also reduce air leakage.

Figures 7.18 and 7.19 illustrate methods for reducing *thermal bridging* at corners and at headers over doors and windows. Thermal bridging occurs where solid framing members interrupt the thermal insulation layer, creating wall areas with a lower thermal resistance than surrounding areas and reducing the insulation efficiency of the wall as a whole. Use of advanced framing techniques (Figure 5.64) or insulating sheathing (Figure 7.17, right) are examples of other approaches to reducing thermal bridging in wall framing.

The insulation level can also be reduced where ceiling insulation under a sloped roof must be compressed somewhat in the diminishing space between the roof sheathing and the top of the exterior wall. This can be observed in Figures 7.17 and 7.20. A raised-heel roof truss (Figure 7.21) is one way to overcome this problem.

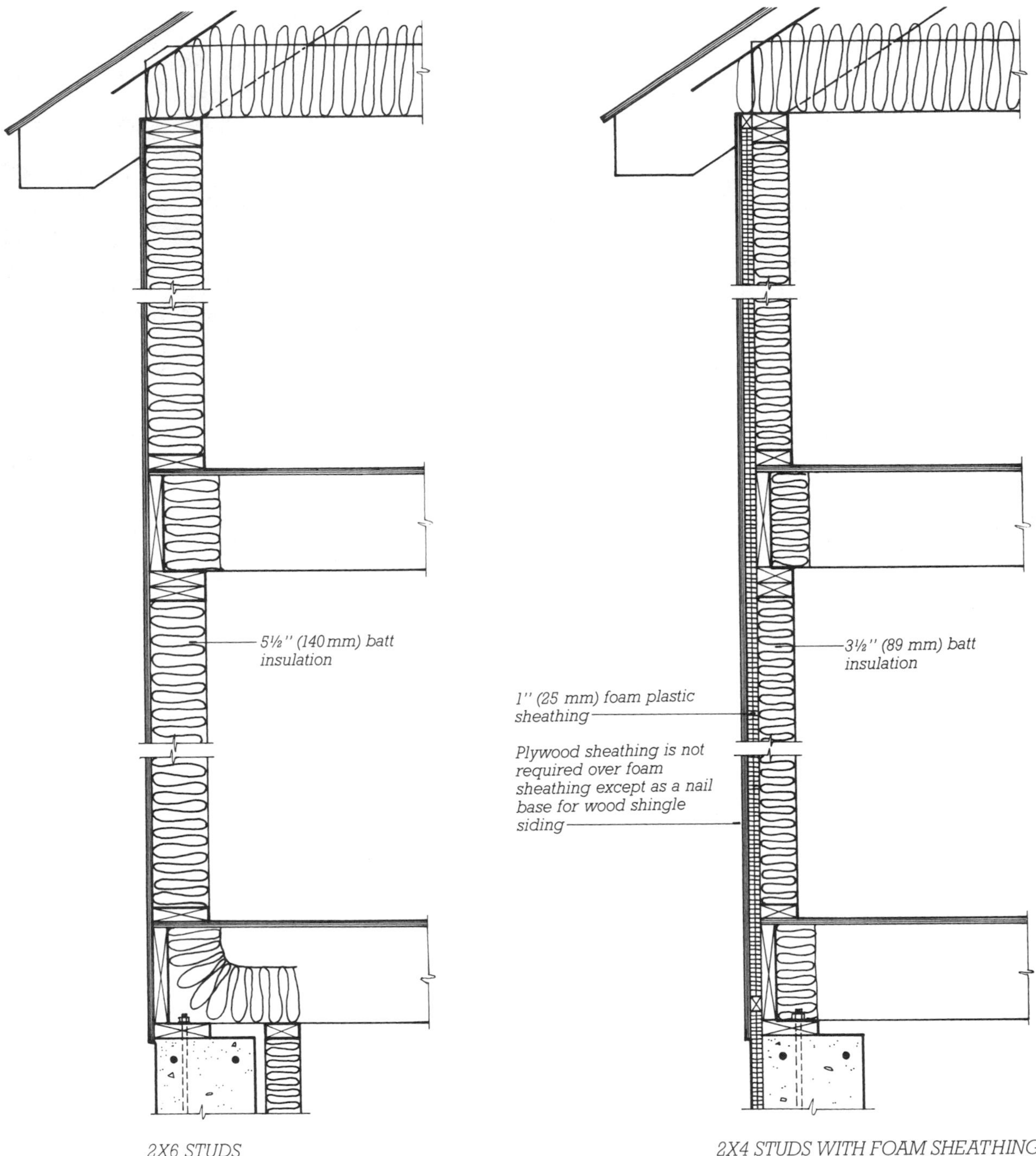

FIGURE 7.17
Insulation levels in walls of light frame buildings can be increased from the R13 to R-15 (RSI-90 to RSI-104) of a 2 × 4 stud wall to R19 (RSI-132) or more by using either 2 × 6 framing and thicker batt insulation (*left*) or 2 × 4 framing with plastic foam sheathing in combination with batt insulation (*right*). The foam sheathing insulates the wood framing members as well as the cavities between them but can complicate the process of installing some types of siding.

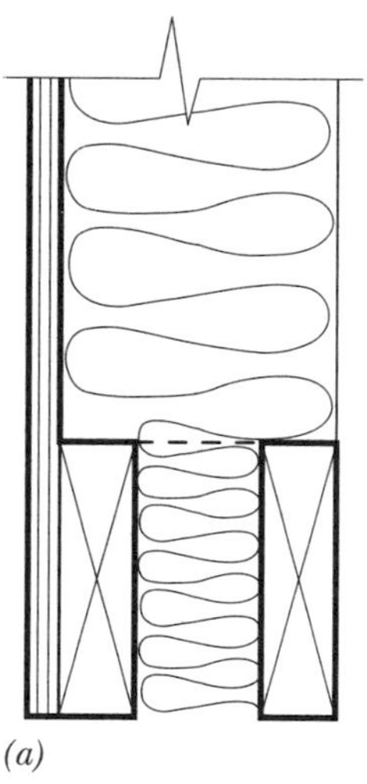

(a)

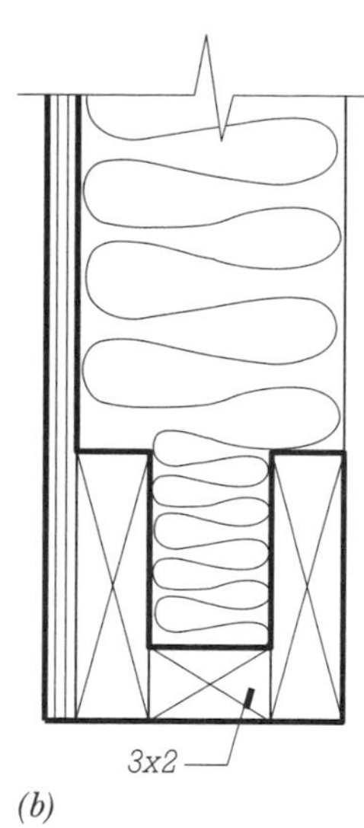

(b)

FIGURE 7.18
Window and door headers in 2 × 6 framing require special detailing. Two alternative header details are shown here in section view: (*a*) The header members are installed flush with the interior and exterior surfaces of the studs, with an insulated space between. This detail is thermally efficient but may not provide sufficient nailing for interior finish materials around the window. (*b*) A 3 × 2 spacer provides full nailing around the opening.

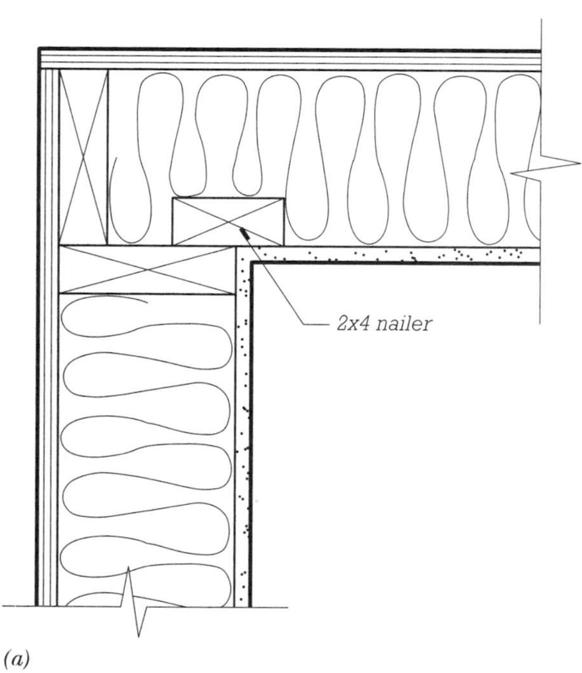

(a)

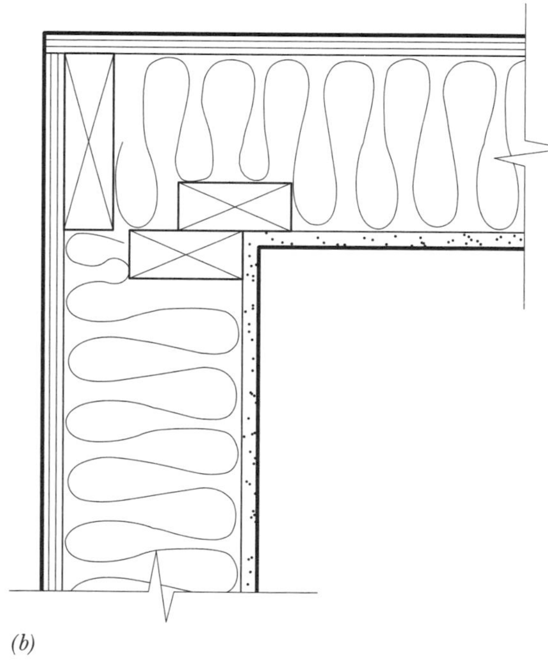

(b)

Figure 7.19
Two alternative corner post details for 2 × 6 stud walls, shown in plan view: (*a*) Each wall frame ends with a full 2 × 6 stud, and a 2 × 4 nailer is added to accept fasteners from the interior wall finish. (*b*) For maximum thermal efficiency, one wall frame ends with a 2 × 4 stud flush with the interior surface, which eliminates any thermal bridging through studs.

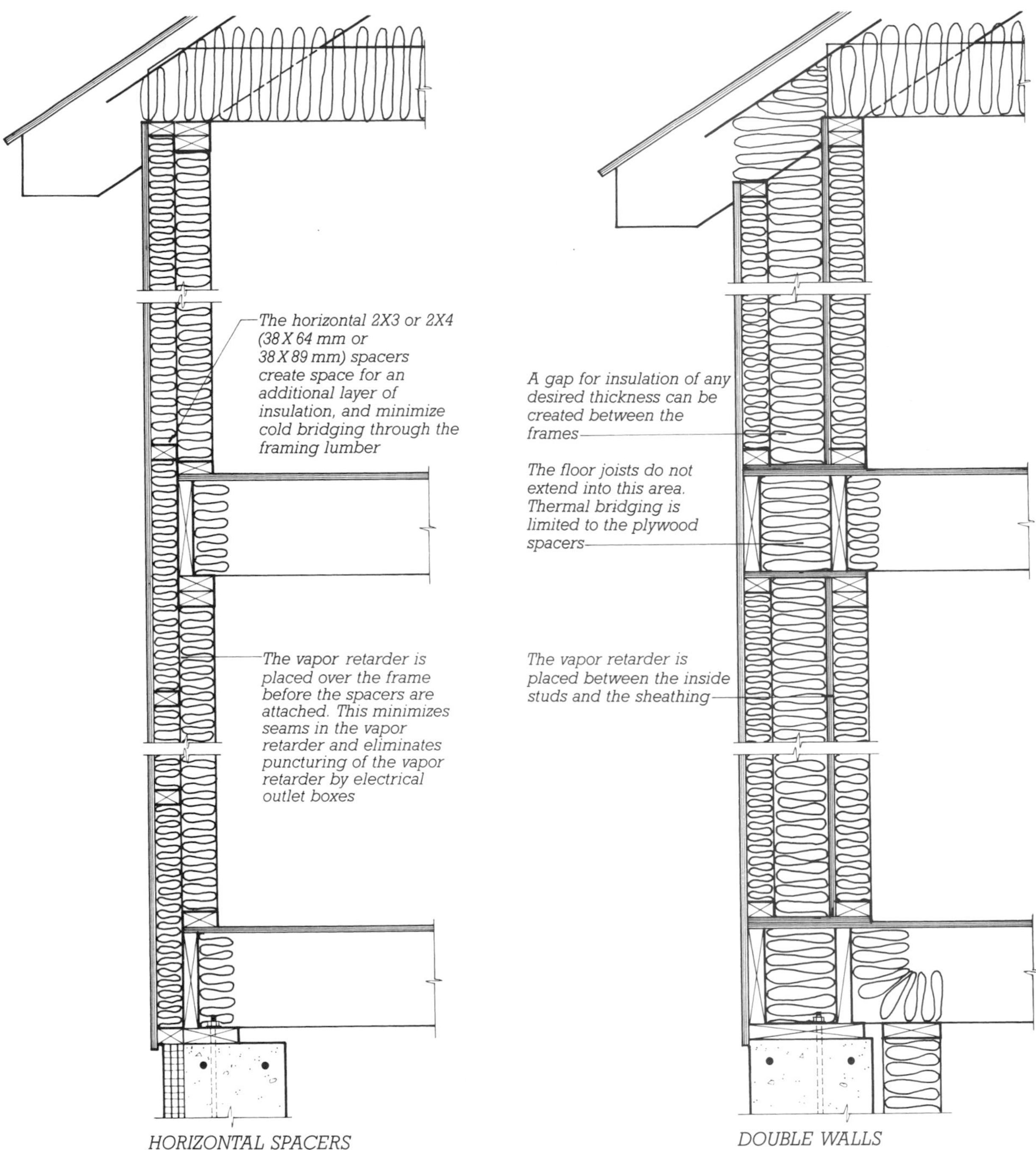
The horizontal 2X3 or 2X4 (38 X 64 mm or 38 X 89 mm) spacers create space for an additional layer of insulation, and minimize cold bridging through the framing lumber
The vapor retarder is placed over the frame before the spacers are attached. This minimizes seams in the vapor retarder and eliminates puncturing of the vapor retarder by electrical outlet boxes
HORIZONTAL SPACERS
A gap for insulation of any desired thickness can be created between the frames
The floor joists do not extend into this area. Thermal bridging is limited to the plywood spacers
The vapor retarder is placed between the inside studs and the sheathing
DOUBLE WALLS

FIGURE 7.20
With the two framing methods shown here, walls can be insulated to any desired level of thermal resistance.

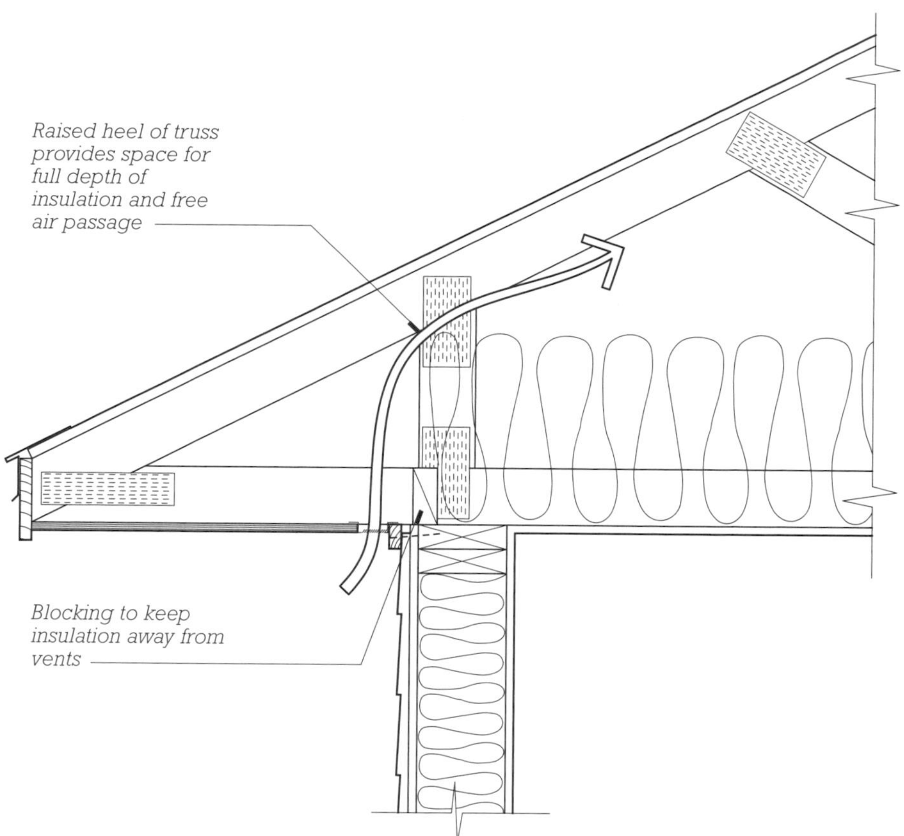
Raised heel of truss provides space for full depth of insulation and free air passage
Blocking to keep insulation away from vents

FIGURE 7.21
A raised-heel roof truss provides plenty of space for attic insulation at the eave.

Radiant Barriers

In warmer regions, *radiant barriers* may be used in roofs and walls to reduce the flow of solar heat into the building. These are thin sheets or panels faced with a bright metal foil that blocks the transmission of infrared (heat) radiation. Most types of radiant barriers are made to be installed over the rafters or studs and beneath the sheathing, so they must be put in place during the framing of the building. They are effective only if the bright surface of the barrier faces a ventilated airspace; this allows the reflective surface to work properly and provides for the convective removal of solar heat that has passed through the outer skin of the building. Some radiant barrier panels are configured with folds that provide this airspace automatically. Radiant barriers are used in combination with conventional insulating materials to achieve the desired overall thermal performance.

Vapor Retarders

A *vapor retarder* (often called, less accurately, a *vapor barrier*) is a membrane of metal foil, plastic, treated paper, or primer paint placed on the warm side of thermal insulation to prevent water vapor from entering the insulation and condensing into liquid. The function of a vapor retarder is explained in detail on pages 658–661. Its role increases in importance in colder climates, as thermal insulation levels increase, and for interior spaces such as pools or spas with high humidity levels.

Many batt insulation materials are furnished with a vapor retarder layer of treated paper or aluminum foil already attached. Most designers and builders in cold climates, however, prefer to use unfaced insulating batts and to apply a separate vapor retarder of polyethylene sheet, because a vapor retarder attached to batts has a seam at each stud that can leak significant quantities of air and vapor, while the separately applied sheet has fewer seams.

Air Barriers

Air barriers control the leakage of air through the building enclosure. They significantly reduce building energy consumption and help protect enclosure assemblies from moisture condensation by restricting the infiltration of humid air. The role of the air barrier in the building enclosure is discussed in more detail on pages 800–803.

Housewraps applied over exterior sheathing, discussed on page 222, are a common way to incorporate an air barrier into wood light frame buildings. Or, when plastic sheeting used as a vapor retarder is carefully sealed against air leakage, it too can function as an air barrier. Another method, called the *airtight*

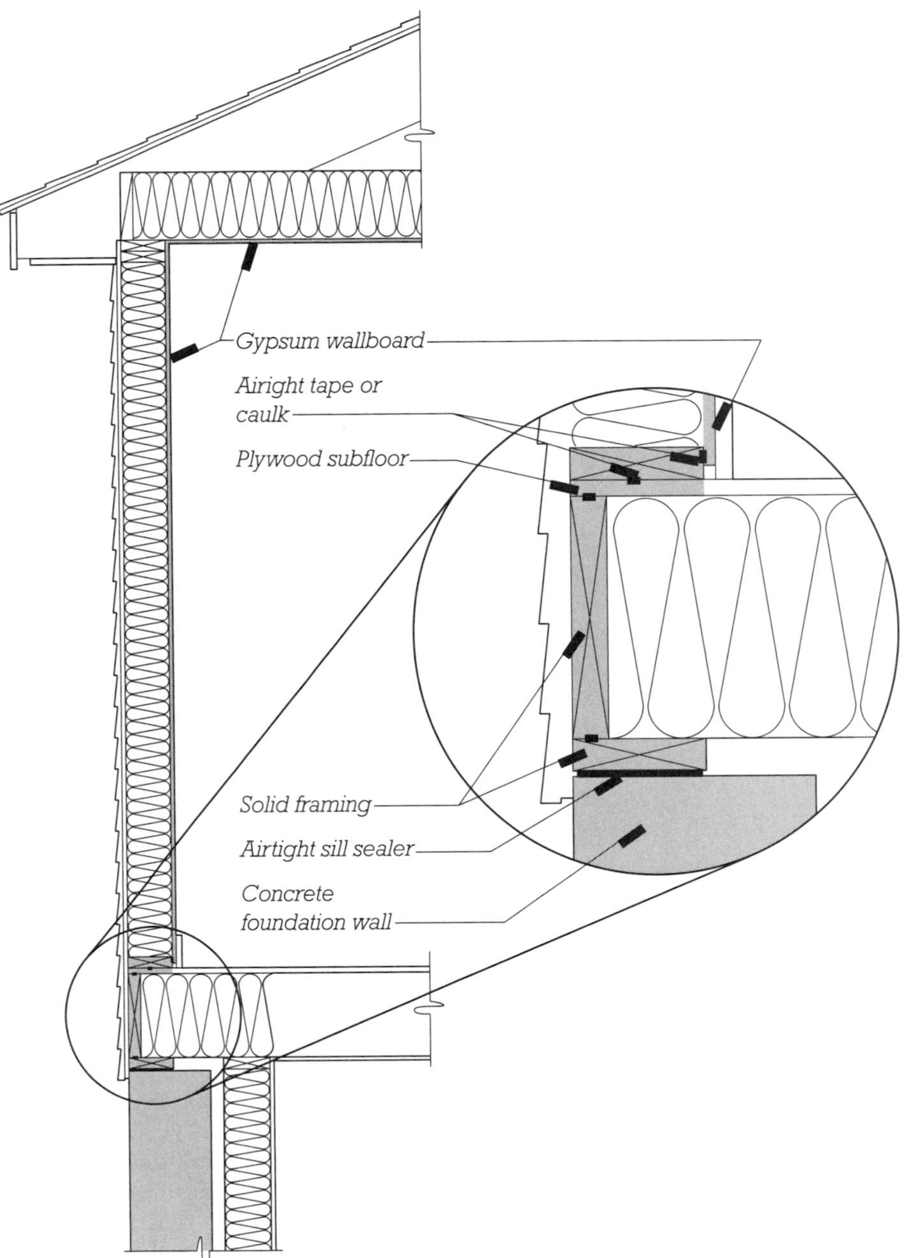

FIGURE 7.22
An air barrier system must form a continuous airtight boundary between the inside and outside of the building. In the ADA, the air barrier system is constructed of gypsum wallboard, wood framing members, and the concrete foundation wall (the shaded elements in the figure). Tapes and caulks (see the enlarged inset) are used to seal leakage paths between these components.

drywall approach (ADA), relies on the gypsum board (drywall) panels used to finish interior walls and ceilings to create the air barrier. In this method, meticulous attention is given to sealing joints between these panels and framing members so as to create a continuous air-impermeable boundary around interior conditioned spaces (Figure 7.22). The ADA requires careful attention to details of construction: Potential air leakage paths around the edges of gypsum panels and between abutting framing members must be sealed with compressible foam tape gaskets or joint sealants during the installation of these members; gypsum board is applied to all the interior surfaces of the outside walls before the interior partitions are framed, eliminating potential air leaks where these interior partitions join the outside walls; and gaskets, sealants, or special airtight boxes are used to seal air leakage paths around electrical fixtures and other penetrations.

The ADA requires careful attention to joint sealing during the construction of the building frame and the installation of drywall. A related system, called *simple caulk and seal,* relies on the strategic application of joint sealants after framing and drywalling are complete to achieve much the same result as the ADA. Simple caulk and seal is easier to coordinate, because the joint sealing occurs separately from the work of other trades. However, it may be less successful at eliminating air leakage paths that become concealed and inaccessible before the sealing work takes place.

The increased concern for controlling air leakage through the building enclosure has also led to higher standards of window and door construction and greater care in sealing around exterior electrical outlets and other penetrations of the exterior wall.

Air Infiltration and Ventilation

Reducing the flow of heat and air through the building enclosure reduces energy consumption. However, tightly constructed homes or apartment buildings may exchange so little air with the outdoors that if they are not adequately ventilated, indoor moisture, odors, and chemical pollutants can build up to intolerable or unhealthful levels. Opening a window to ventilate the dwelling is one way to introduce fresh air, but during heating or cooling seasons this wastes fuel. A better solution is a mechanical ventilation system designed to supply fresh air to interior spaces at a controlled and consistent rate. Such a ventilation system may be integrated with a forced-air heating or cooling system or it may be dedicated solely to satisfying ventilation needs. For greater energy efficiency, an *air-to-air heat exchanger*—a device that recovers most of the heat from the air exhausted from the building and adds it to the outside air that is drawn in—may be part of such a system. Nonresidential buildings of wood light frame construction generally require higher rates of ventilation than residential structures, and here, too, low-infiltration construction and appropriately designed ventilation systems are important to maintaining a healthful and comfortable interior environment while also reducing heating and cooling costs.

Wall and Ceiling Finish

Gypsum-based plaster and drywall finishes have always been the most popular for walls and ceilings in wood frame buildings. Their advantages include substantially lower installed costs than any other types of finishes, adaptability to either painting or wallpapering, and, importantly, a degree of fire resistance that offers considerable protection to a combustible frame. *Three-coat gypsum plaster* applied to *wood strip lath* was the prevalent wall and ceiling finish system until the Second World War, when *gypsum board* (also called *drywall*) came into increasing use because of its lower material cost, more rapid installation, and utilization of less skilled labor. More recently, *veneer plaster* systems have been developed that offer surfaces of a quality and durability superior to those of gypsum board, often at comparable prices.

Plaster, veneer plaster, and gypsum board finishes are presented in detail in Chapter 23. Gypsum board remains the favored material for small builders who do all the interior finish work in a building themselves because the skills and tools it requires are largely those of carpenters rather than plasterers. In geographic areas where there are plenty of skilled plasterers, veneer plaster captures a substantial share of the market. Almost everywhere in North America, there are subcontractors who specialize in gypsum board installation and finishing and who are able to finish the interior surfaces of larger projects, such as apartment buildings, retail stores, and rental office buildings, as well as individual houses, at highly competitive prices.

In most small buildings, all wall and ceiling surfaces are covered with plaster or gypsum board. Even wood paneling should be applied over a gypsum board backup layer for increased fire resistance. In buildings that require fire walls between dwelling units or fire separation walls between areas with different uses, a gypsum board wall of the required degree of fire resistance can be installed, eliminating the need to employ masons to put up a wall of brick or concrete masonry (Figures 5.67 and 5.68).

Millwork and Finish Carpentry

Millwork (so named because it is manufactured in a planing and molding mill) includes all the wood interior finish components of a building. Millwork is generally produced from much higher-quality wood than that used for framing: The softwoods used are those with fine, uniform grain structure and few defects, such as Sugar pine and Ponderosa pine. Flooring, stair treads, and millwork intended for transparent finish coatings such as varnish or shellac are customarily made of hardwoods such as Red and White oak, Cherry, Mahogany, or Walnut, or of similar quality hardwood plywoods.

Moldings are also produced from high-density plastic foams, parallel strand lumber, and medium-density fiberboard as a way of reducing costs and minimizing moisture expansion and contraction. All these materials must be painted.

The quality of millwork is regulated by the Architectural Woodwork Institute (AWI), which defines three grades: Economy, Custom, and Premium. *Economy Grade* cabinetry and millwork represents the minimum expectation of quality. *Custom Grade* provides a well-defined degree of control over the quality of materials, workmanship, and installation and is the grade to which most cabinets are built. *Premium Grade* is the most expensive, and is reserved for the very finest cabinets and millwork.

Proportioning Fireplaces

Ever since fireplaces were first developed in the Middle Ages, people have sought formulas for their construction to ensure that the smoke from a fire would go up the chimney and the heat into the room, rather than the reverse, which is too often the case. To this day, there is little scientific information on how fireplaces work and how to design them. What we do have are some measurements taken from fireplaces that seem to work reasonably well. These have been correlated and arranged into a table of dimensions (Figures B and C) that enables designers to reproduce the critical features of these fireplaces as closely as possible.

Several general principles are clear: The chimney should be as tall as possible. The cross-sectional area of the *flue* should be about one-tenth the area of the front opening of the fireplace. A *damper* should be installed to close off the chimney when no fire is burning and to regulate the passage of air through the *firebox* when the fire is burning (Figure A). A *smoke shelf* above the damper reduces fireplace malfunctions caused by cold downdrafts in the chimney. Splayed sides and a sloping back in the firebox reduce smoking and throw more heat into the room.

Starting from these general principles, two schools of thought have developed concerning how a fireplace should be shaped and proportioned. Most fireplaces in North America are built to the conventional standards that are tabulated here. Many designers, however, favor the rules for fireplace construction that were formulated by Count Rumford in the 1790s. These produce a fireplace with a taller opening and a shallower firebox than the conventional fireplace. The intention of the Rumford design is to attain a higher efficiency by throwing more radiant heat from the fire-warmed bricks at the back of the firebox to the occupants of the building.

Building codes place a number of restrictions on fireplace and chimney construction. Typically, these call for a 2-inch (51-mm) clearance between the wood framing and the masonry of a chimney or fireplace, and clearances to combustible finish materials around the opening of the fireplace as shown in Figure B. Also specified by code are the minimum thicknesses of masonry around the firebox and flue, the minimum size of the flue, the minimum extension of the chimney above the roof, and steel reinforcing for the chimney. A combustion air inlet must be provided to bring air from the outdoors to the base of the fire. For ready reference in proportioning fireplaces, use the values in the accompanying Figures B and C. In most cases, the designer need not detail the internal construction of the fireplace beyond the information given in these dimensions, because masons are well versed in the intricacies of assembling a fireplace.

There are several alternatives to the conventional masonry fireplace. One is a steel or ceramic fireplace liner that takes the place of the firebrick lining, damper, and smoke chamber. Many of these products have internal passages that draw air from the room, warm it with the heat of the fire, and return it to the room. The liner is set onto the underhearth and built into a masonry facing and chimney by the mason.

Another alternative is the "package" fireplace, a self-contained, fully insulated unit that needs no masonry whatsoever. It is usually set directly on the subfloor and fitted with an insulated chimney of prefabricated metal pipe. It may be faced with any desired ceramic or masonry materials. Many package fireplaces are made to burn gas rather than wood.

A third alternative is a freestanding metal stove that burns wood, coal, or other solid fuel. Stoves are available in hundreds of styles and sizes. Their principal advantage is that they provide more heat to the interior of the building per unit of fuel burned than a fireplace. A stove requires a noncombustible hearth and a fire-protected wall that are rather large in extent. The designer should consult the local building code at an early stage of design to be sure that the room is big enough to hold a stove of the desired dimensions.

Figure A
A cutaway view of a conventional masonry fireplace. Concrete masonry is used wherever it will not show, to reduce labor and material costs, but if hollow concrete blocks are used, they must be filled solid with grout. The damper and ash doors are prefabricated units of cast iron, as is a combustion air inlet in the hearth (not shown here) that connects with a sheet metal duct to an outdoor air intake. The flue liners are made of fired clay and are highly resistant to heat. The flue from the furnace or boiler in the basement slopes as it passes the firebox so as to adjoin the fireplace flue and keep the chimney as small as possible.

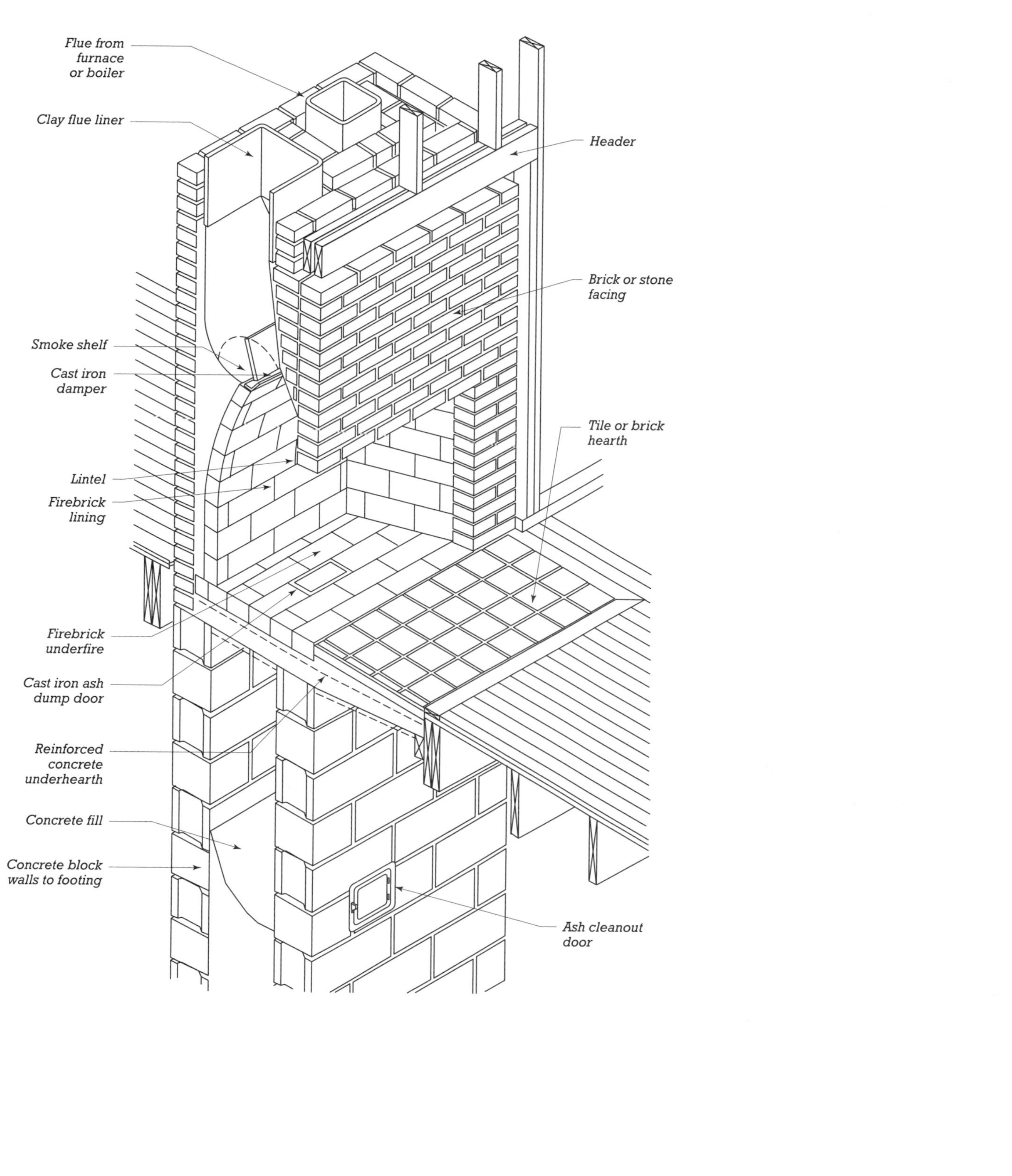
Flue from furnace or boiler
Clay flue liner
Header
Brick or stone facing
Smoke shelf
Cast iron damper
Tile or brick hearth
Lintel
Firebrick lining
Firebrick underfire
Cast iron ash dump door
Reinforced concrete underhearth
Concrete fill
Concrete block walls to footing
Ash cleanout door

PROPORTIONING FIREPLACES *(CONTINUED)*

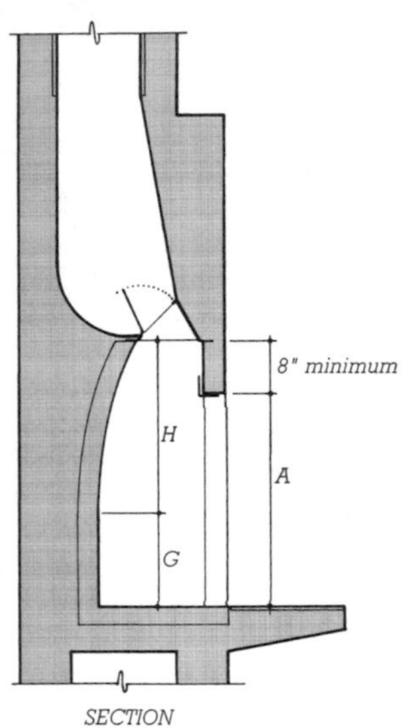

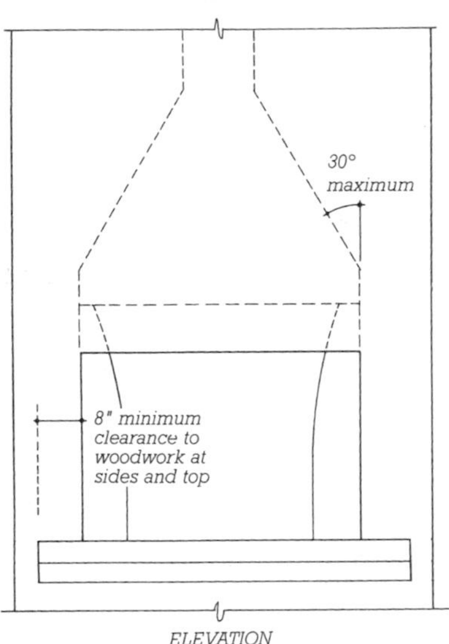

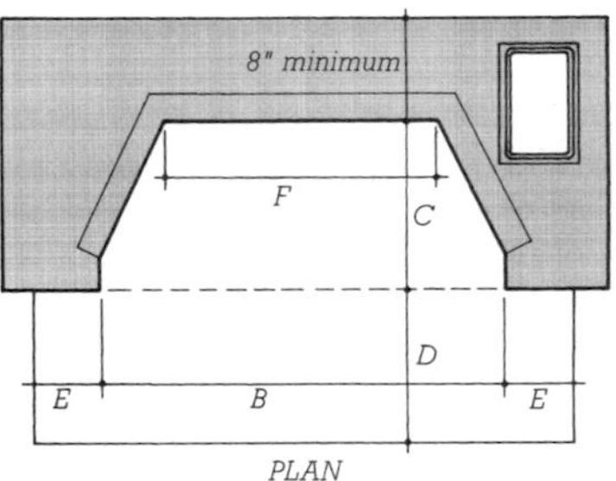

FIGURE B
The critical dimensions of a conventional masonry fireplace, keyed to the table in Figure C. Dimension D, the depth of the hearth, is commonly required to be 16 inches (405 mm) for fireplaces with openings up to 6 square feet (0.56 m^2) and 20 inches (510 mm) if the opening is larger. The side extension of the hearth, E, is usually fixed at 8 inches (200 mm) for smaller fireplace openings and 12 inches (305 mm) for larger ones.

FIGURE C
Recommended proportions for conventional masonry fireplaces, based largely on figures given in Ramsey/Sleeper, *Architectural Graphic Standards* (9th ed.), New York, John Wiley & Sons, Inc., 1994.

Fireplace Opening						Flue Lining	
Height (A)	Width (B)	Depth (C)	Minimum Backwall Width (F)	Vertical Backwall Height (G)	Inclined Backwall Height (H)	Rectangular (outside dimensions)	Round (inside diameter)
24″	28″	16 to 18″	14″	14″	16″	8½ × 13″	10″
(610 mm)	*(710 mm)*	*(405 to 455 mm)*	*(355 mm)*	*(355 mm)*	*(405 mm)*	*(216 × 330 mm)*	*(254 mm)*
28 to 30″	30	16 to 18″	16″	14″	18″	8½ × 13″	10″
(710 to 760 mm)	*(760 mm)*	*(405 to 455 mm)*	*(405 mm)*	*(355 mm)*	*(455 mm)*	*(216 × 330 mm)*	*(254 mm)*
28 to 30″	36″	16 to 18″	22″	14″	18″	8½ × 13″	12″
(710 to 760 mm)	*(915 mm)*	*(405 to 455 mm)*	*(560 mm)*	*(355 mm)*	*(455 mm)*	*(216 × 330 mm)*	*(305 mm)*
28 to 30″	42″	16 to 18″	28″	14″	18″	13 × 13″	12″
(710 to 760 mm)	*(1065 mm)*	*(405 to 455 mm)*	*(710 mm)*	*(355 mm)*	*(455 mm)*	*(330 × 330 mm)*	*(305 mm)*
32″	48″	18 to 20″	32″	14″	24″	13 × 13″	15″
(815 mm)	*(1220 mm)*	*(455 to 510 mm)*	*(815 mm)*	*(355 mm)*	*(610 mm)*	*(330 × 330 mm)*	*(381 mm)*

Millwork is manufactured and delivered to the building site at a very low moisture content, typically about 10 percent, so it is important to protect it from moisture and high humidity before and during installation to avoid swelling and distortions. The humidity within the building is frequently high at the conclusion of plastering or gypsum board work. The framing lumber, concrete work, masonry mortar, plaster, drywall joint compound, and paint are still diffusing large amounts of excess moisture into the interior air. As much of this moisture as possible should be ventilated to the outdoors before *finish carpentry* (the installation of millwork) commences. Windows should be left open for a few days, and in cool or damp weather the building's heating system should be turned on to raise the interior air temperature and help drive off excess water. In hot, humid weather, the air conditioning system should be activated to dry the air.

Interior Doors

Figure 7.23 illustrates five doors that fall into three general categories: Z-brace, panel, and flush. *Z-brace doors*, mostly built on site, are used infrequently because they are subject to distortions and large amounts of moisture expansion and contraction in the broad surface of boards whose grain runs perpendicular to the width of the door. *Panel doors* were developed centuries ago to minimize dimensional changes and distortions caused by the seasonal changes in the moisture content of the wood. They are widely available in ready-made form from millwork dealers. *Flush doors* are smooth slabs with no surface features except the grain of the wood. They may be either solid core or hollow core. *Solid-core* doors consist of two veneered faces glued to a solid core of wood blocks or bonded wood chips (Figure 7.24). They are much heavier, stronger, and more resistant to the passage of sound than hollow-core doors and are also more expensive. In residential buildings, their use is usually confined to entrance doors, but they are frequently installed throughout commercial and institutional buildings, where doors are subject to greater abuse. *Hollow-core doors* have two thin plywood faces separated by an airspace. The airspace is maintained by an interior grid of wood or paperboard spacers to which the veneers are bonded. Flush doors of either type are available in a variety of veneer species, the least expensive of which are intended to be painted.

For speed and economy of installation, most interior doors are furnished *prehung*, meaning that they have been hinged and fitted to frames at the mill. The carpenter on the site merely tilts the prehung door and frame unit up into the rough opening, plumbs it carefully with a spirit level, shims it with pairs of wood shingle wedges between the finish and rough jambs, and nails it to the studs with finish nails through the jambs (Figure 7.25). *Casings* are then nailed around the frame on both sides of the partition to close the ragged gap between the door frame and the wall finish (Figure 7.26). To save the labor of applying casings, door units can also be purchased with *split jambs* that enable the door to be cased at the mill. At the time of installation, each door unit is separated into halves, and the halves are installed from opposite sides of the partition to telescope snugly together before being nailed in place (Figure 7.27).

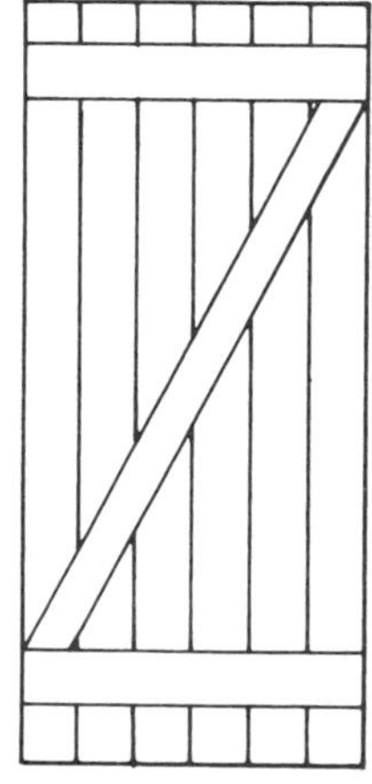

Z-BRACE

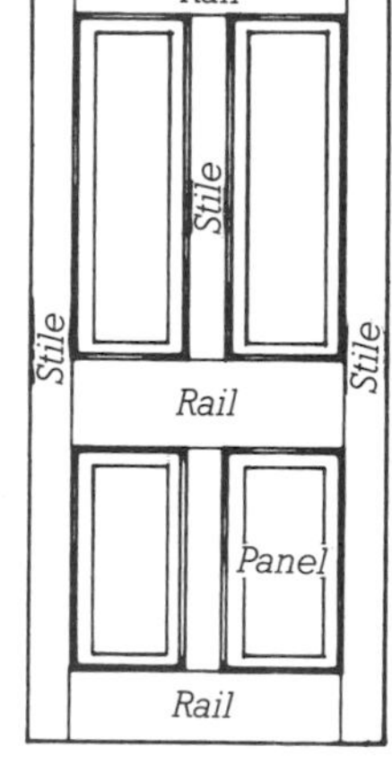

FOUR PANEL

SIX PANEL

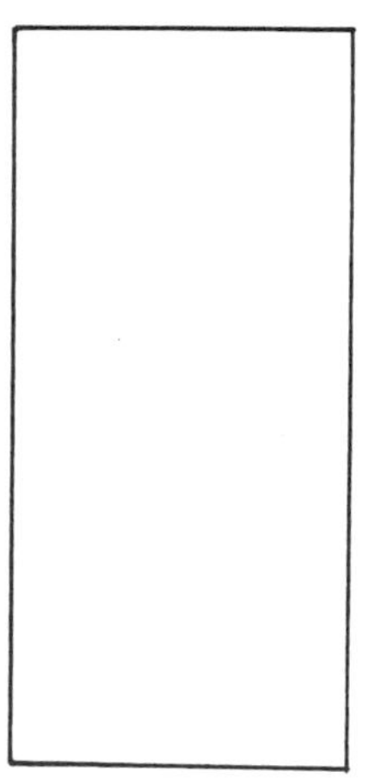

FLUSH SOLID CORE

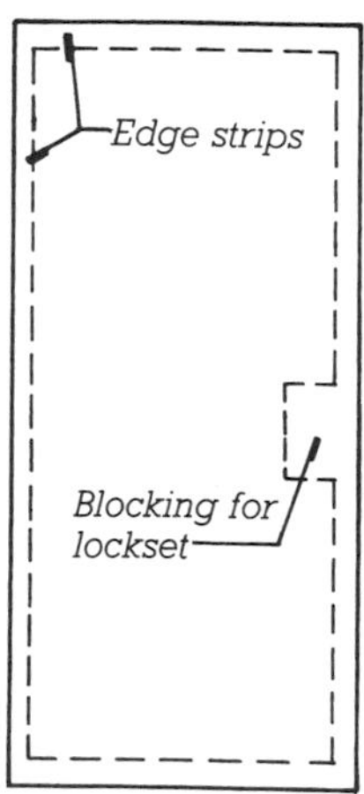

FLUSH HOLLOW CORE

FIGURE 7.23
Types of wood doors.

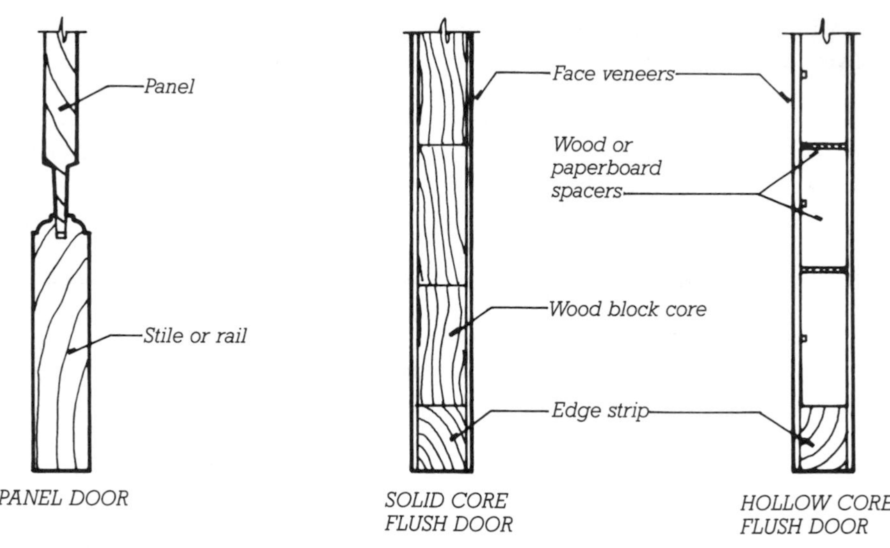

FIGURE 7.24
Edge details of three types of wood doors. The *panel* is loosely fitted to the *stiles* and *rails* in a panel door to allow for moisture expansion of the wood. The spacers and edge strips in hollow-core doors have ventilation holes to equalize air pressures inside and outside the door.

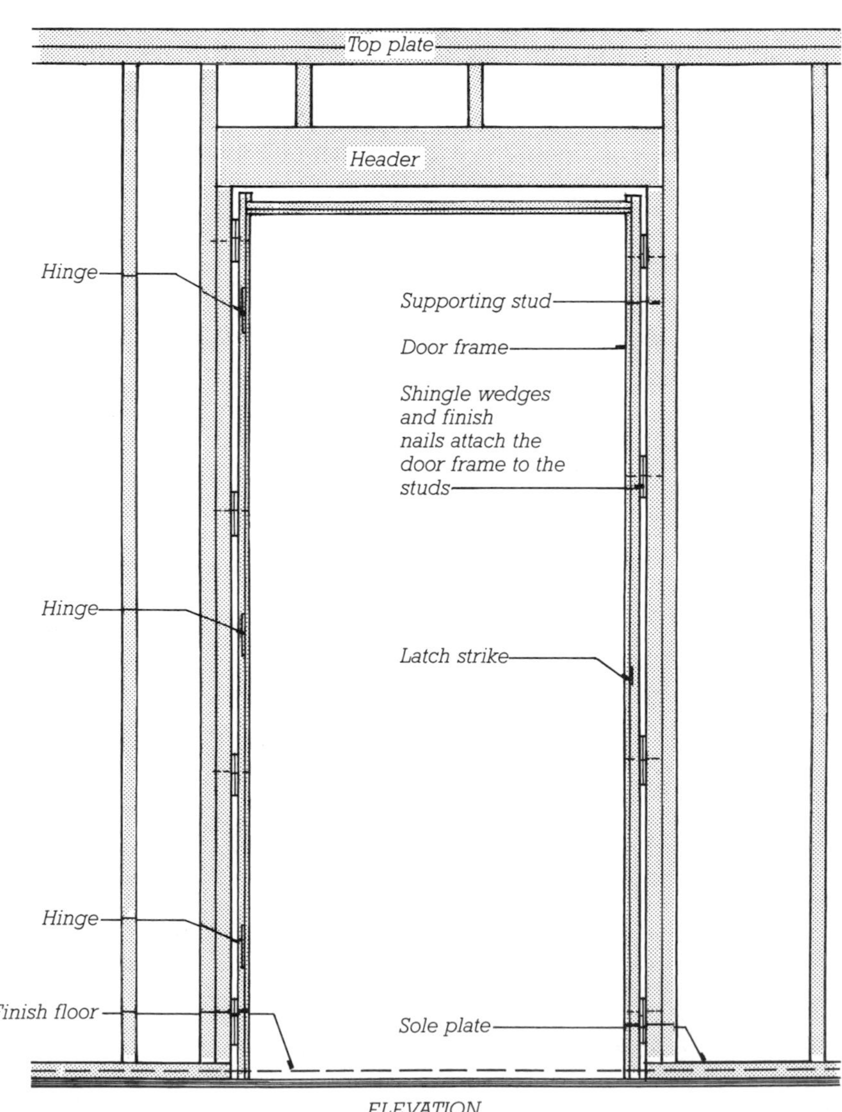

FIGURE 7.25
Installing a door frame in a rough opening. The shingle wedges at each nailing point are paired in opposing directions to create a flat, precisely adjustable shim to support the frame.

Window Casings and Baseboards

Windows are cased in much the same manner as doors (Figures 7.28 and 7.29). After the finish flooring is in place, *baseboards* are installed to cover the gap between the flooring and the wall finish and to protect the wall finish against damage by feet, furniture legs, and cleaning equipment (Figures 7.36K–N and 24.32).

When installing casings or baseboards, the carpenter recesses the heads of finish nails below the surface of the wood, traditionally using a hammer and a *nail set* (a hardened steel punch), or, as is more likely today, a powered finish nail gun. Later, the painters will fill these nail holes with a paste filler and sand the surface smooth after the filler has dried so that the holes will be invisible in the painted woodwork. For transparent wood finishes, nail holes are usually filled after the finishes have been applied, using wax-based fillers that are supplied in a range of colors to match the full range of wood species.

The nails in casings and baseboards must reach through the plaster or gypsum board to penetrate the framing members beneath in order to make a secure attachment. Eight- or ten-penny finish or casing nails are customarily used.

Cabinets

Cabinets for kitchens, bathrooms, bedrooms, workrooms, and other spaces may be either custom- or factory-fabricated. *Custom cabinets* are fabricated in specialty woodworking

(a)

(b)

(c)

(d)

FIGURE 7.26
Casing a door frame. (*a*) The heads of the finish nails in the frame are recessed below the surface of the wood with a steel nail set. (*b*) The top piece of casing, mitered to join the vertical casings, is ready to install, and glue is spread on the edge of the frame. (*c*) The top casing is nailed into place. (d) The nails are set below the surface of the wood, ready for filling. *(Photos by Joseph Iano)*

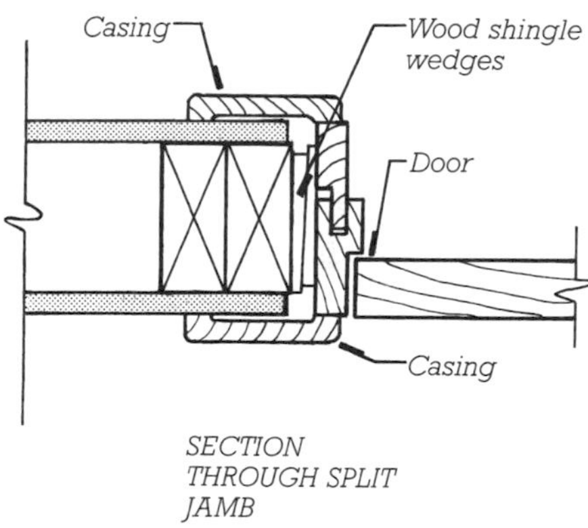

FIGURE 7.27
A split-jamb interior door arrives on the construction site prehung and precased. The halves of the frame are separated and installed from opposite sides of the partition.

(a) (b) (c) (d) (e) (f) (g) (h) (i)

Figure 7.28
Casing a window. (*a*) Marking the length of a casing. (*b*) Cutting the casing to length with a power miter saw. (*c*) Nailing the casing. (*d*) Coping the end of a molded edge of an apron with a coping saw so that the molding profile will terminate neatly at the end of the apron. (*e*) Planing the edge of the apron, which has been ripped (sawn lengthwise, parallel to the grain of the wood) from wider casing stock. (*f*) Applying glue to the apron. (*g*) Nailing the apron, which has been wedged temporarily in position with a stick. (*h*) The coped end of the apron, in place. (*i*) The cased window, ready for filling, sanding, and painting. *(Photos by Joseph Iano)*

shops according to drawings and specifications prepared individually for each project. Like quality millwork, custom cabinets are constructed to AWI specifications, usually Premium or Custom Grade. Less expensive, *prefabricated cabinets* are factory-manufactured to standard sizes and configurations. Both types of cabinets are usually delivered to the construction site fully finished. In project specifications, custom cabinets are specified as *architectural wood casework*, and factory-made cabinets are specified as *manufactured wood casework.*

On the construction site, cabinets are installed by shimming against wall and floor surfaces as necessary to make them level and screwing through the backs of the cabinet units into the wall studs (Figures 7.30 and 7.31). The tops are then attached with screws driven up from the cabinets beneath. Kitchen and bath *countertops* are cut out for built-in sinks and lavatories, which are subsequently installed by the plumber.

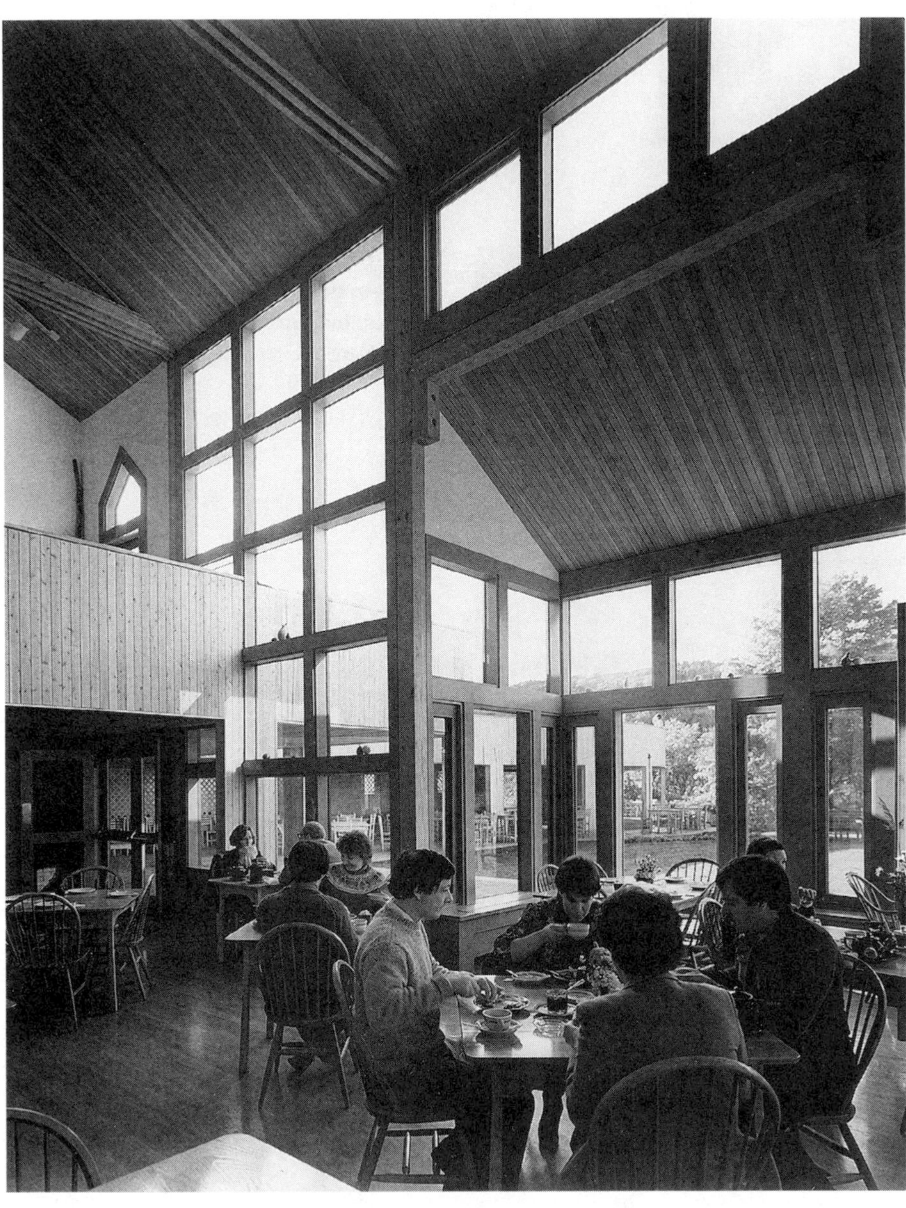

FIGURE 7.29
Simple but carefully detailed and skillfully crafted window casings in a restaurant.
(Woo & Williams, Architects. Photographer: Richard Bonarrigo)

FIGURE 7.30
Prepainted wood kitchen cabinets installed before the installation of shelves, drawers, doors, and countertops. *(Photo by Edward Allen)*

FIGURE 7.31
Custom-designed cabinets brighten a remodeled kitchen in an older house. *(Architects: Dirigo Design. Photo © Lucy Chen)*

Finish Stairs

Finish stairs are either constructed in place (Figures 7.32 and 7.33) or shop built (Figures 7.34 and 7.35). Shop-built stairs tend to be more tightly constructed and to squeak less in use, but site-built stairs can be fitted more closely to the walls and are more adaptable to special situations and framing irregularities. Stair treads are usually made of wear-resistant hardwoods such as Oak or Maple. Risers and stringers may be made of any reasonably hard wood, such as Oak, Maple, or Douglas fir.

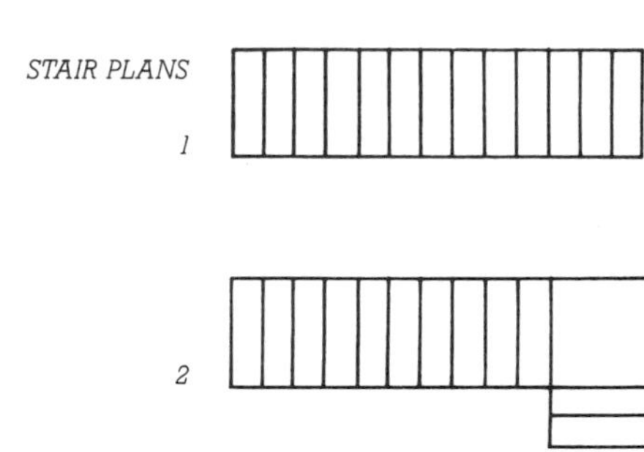

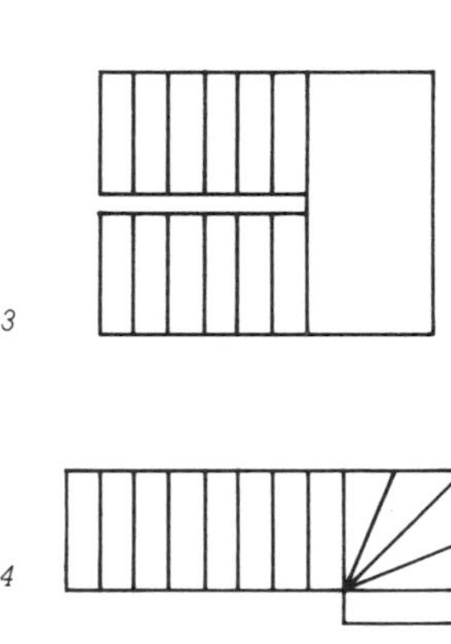

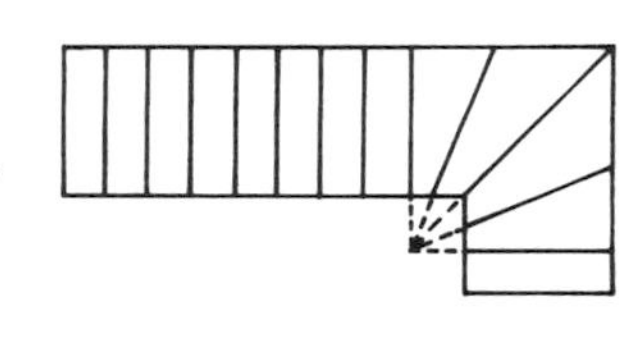

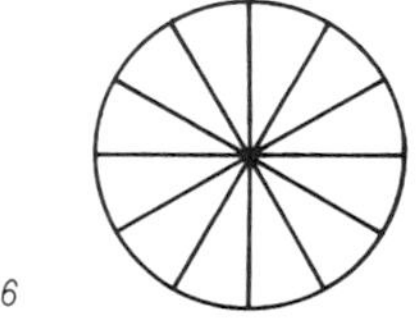

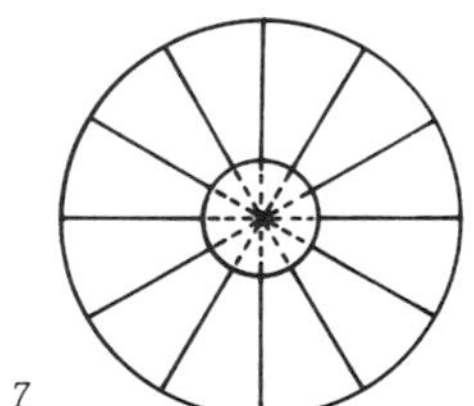

Figure 7.32
Left: stair terminology and clearances for wood frame residential construction. Right: types of stairs. *1.* Straight run. *2.* L-shaped stair with landing. *3.* 180-degree turn with landing. *4.* L-shaped stair with winders (triangular treads). Winders are helpful in compressing a stair into a much smaller space but are perilously steep where they converge, and their treads become much too shallow for comfort and safety. Building codes limit their minimum dimensions and restrict their use to within dwelling units. *5.* L-shaped stair whose winders have an offset center. The offset center can increase the minimum tread dimension to within legal limits. *6.* A spiral stair (in reality a helix, not a spiral) consists entirely of winders and is generally illegal for any use but a secondary stair in a single-family residence. *7.* A spiral or circular stair with an open center of sufficient diameter can have its treads dimensioned to legal standards.

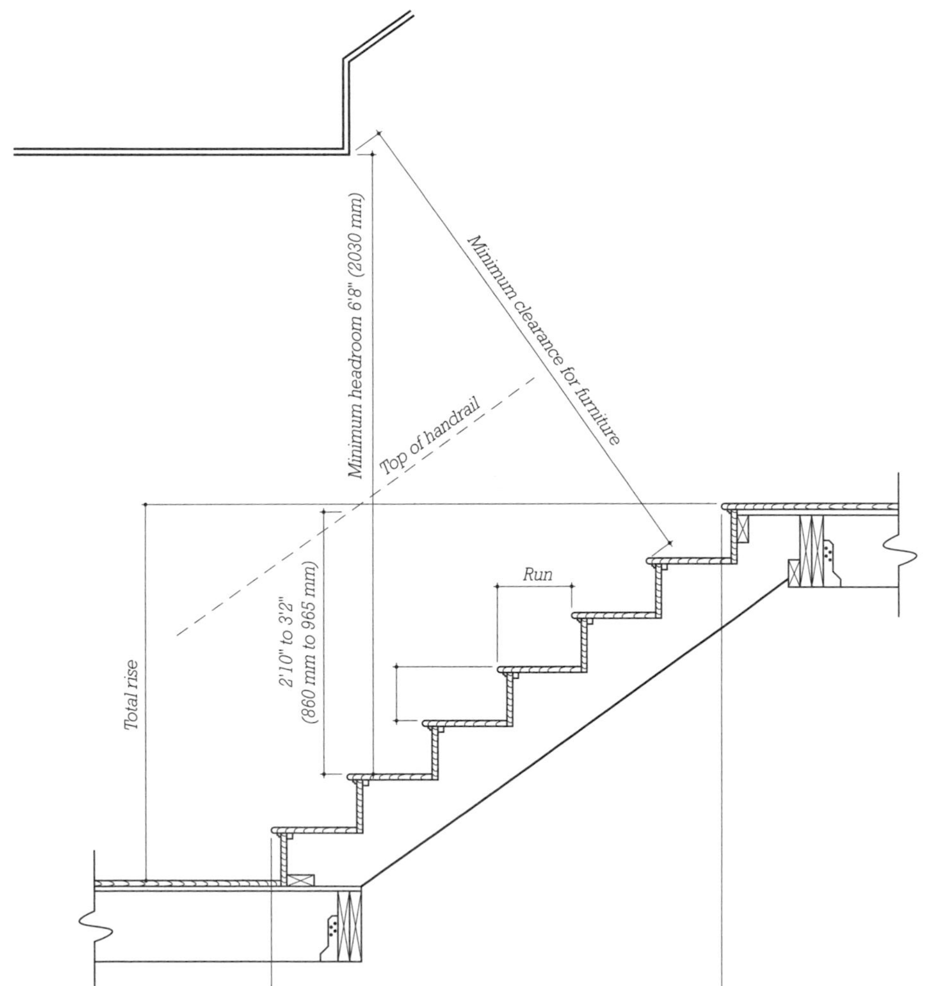

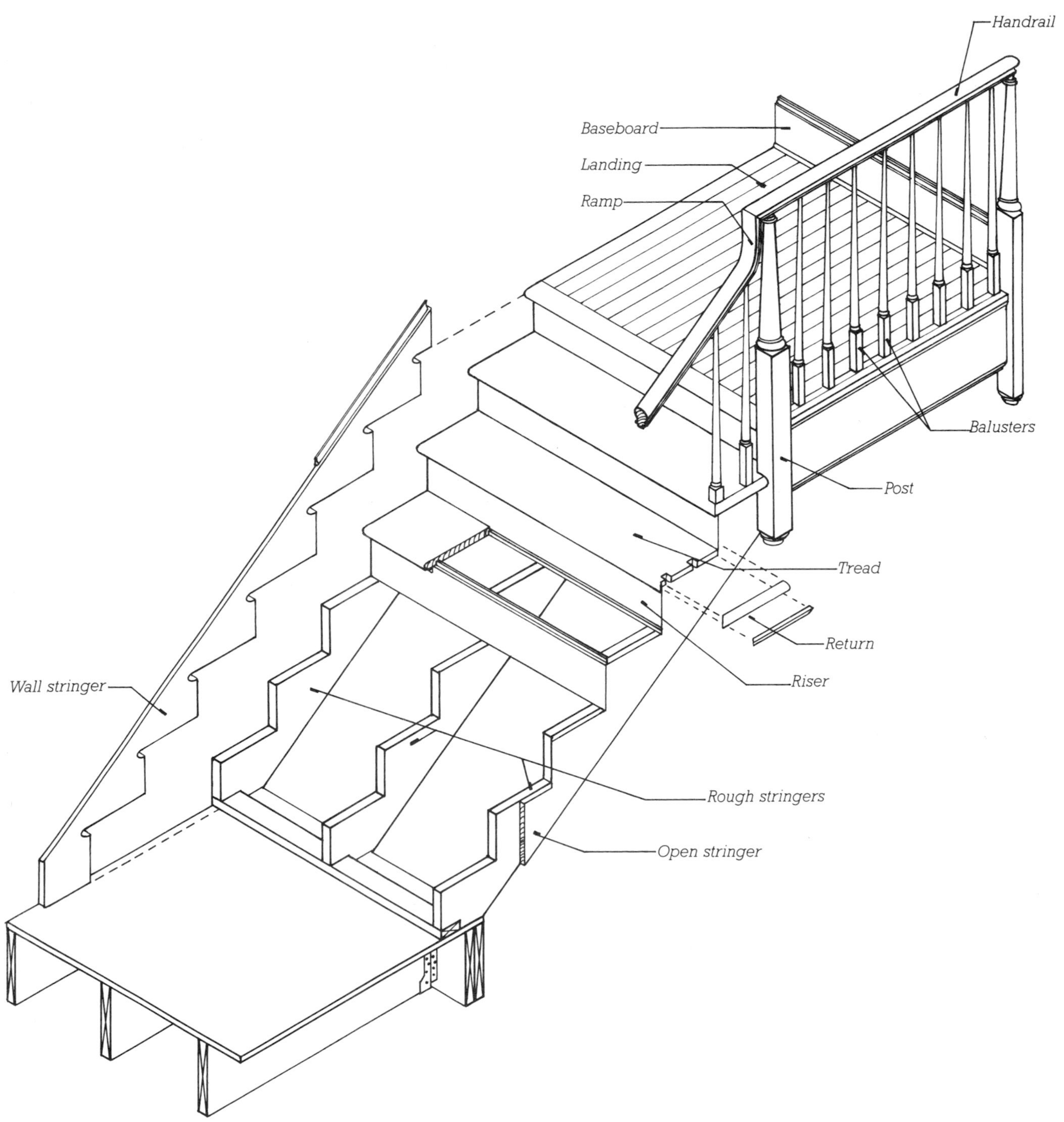

FIGURE 7.33
Constructing a finished stair in place. The joint between the riser and the open stringer is a miter. The balusters, posts, and handrail are purchased ready made from millwork suppliers and cut to fit.

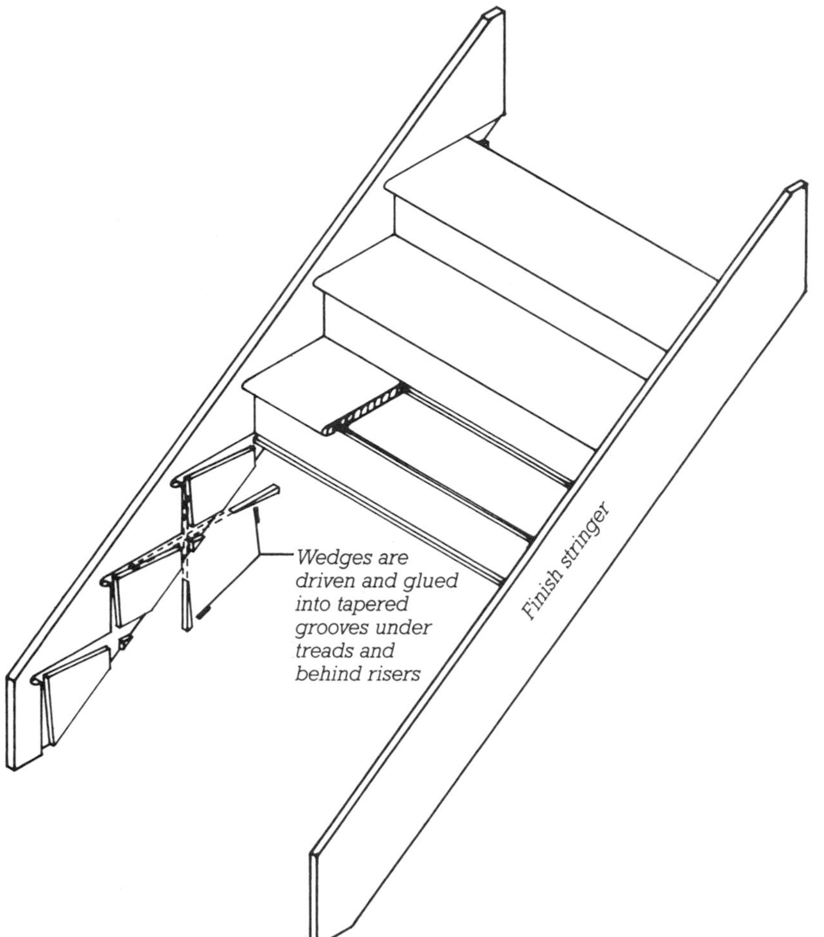
Wedges are driven and glued into tapered grooves under treads and behind risers
Finish stringer

FIGURE 7.34
A shop-built stair. All the components are glued firmly together in the shop, and the stair is installed as a single piece.

Three openings are required in stair-cases; the first is the door thro' which one goes up to the stair-case, which the less it is hid to them that enter into the house, so much the more it is to be commended. And it would please me much, if it was in a place, where before that one comes to it, the most beautiful part of the house was seen; because it makes the house (even tho' small) seem very large; but however, let it be manifest, and easily found. The second opening is the windows that are necessary to give light to the steps; they ought to be in the middle, and high, that the light may be spread equally every where alike. The third is the opening thro' which one enters into the floor above; this ought to lead us into ample, beautiful, and adorned places.

Andrea Palladio, *The Four Books of Architecture*, 1570.

Miscellaneous Finish Carpentry

Finish carpenters install dozens of miscellaneous items in the average building—closet shelves and poles, pantry shelving, bookshelves, wood paneling, chair rails, picture rails, ceiling moldings, mantelpieces, laundry chutes, folding attic stairs, access hatches, door hardware, weatherstripping, doorstops, and bath accessories (towel bars, paper holders, and so on). Many of these items are available ready-made from millwork and hardware suppliers (Figures 7.36 and 7.37), but others have to be crafted by the carpenter.

(a)

(b)

FIGURE 7.35
(*a*) A worker completes a highly customized curving stair in a shop. (*b*) Balusters, newel post, and handrail are finely finished in a historical style. *(Courtesy of Staircase & Millwork Co., Alpharetta, GA)*

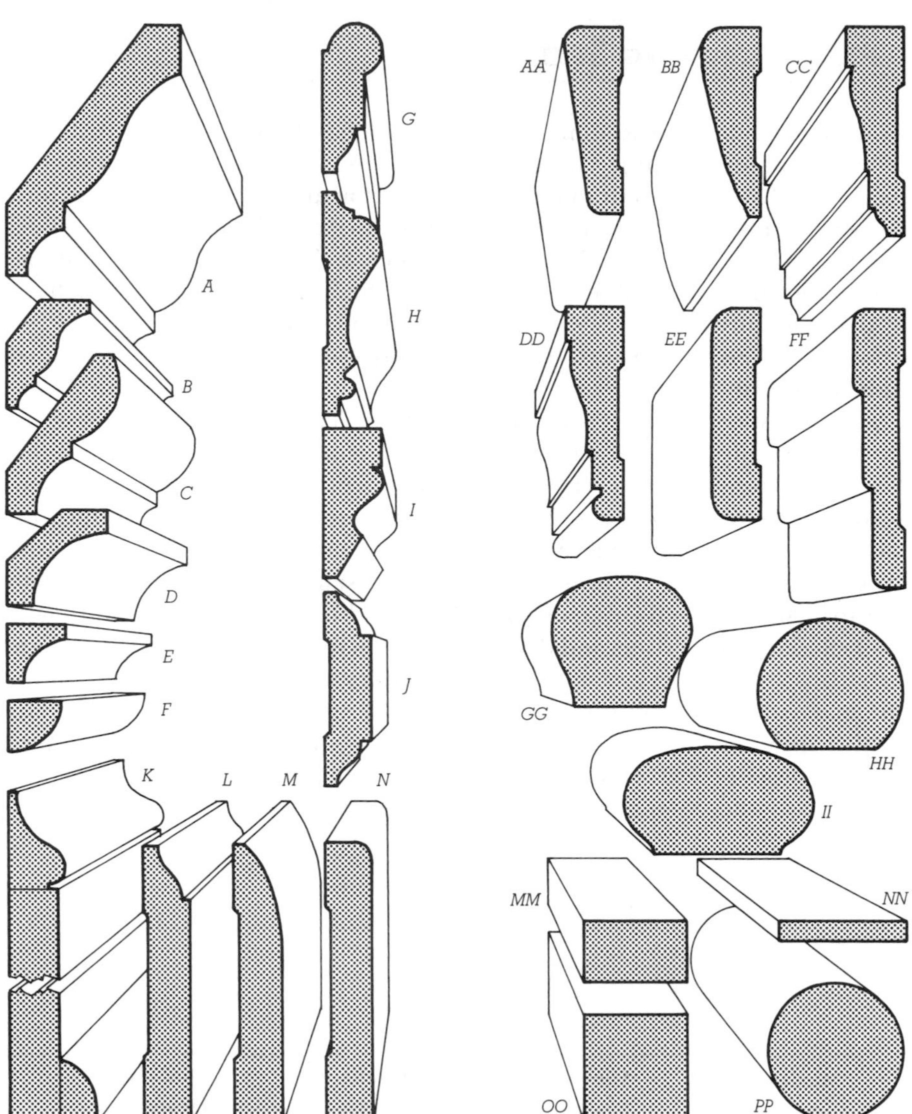

FIGURE 7.36
Some common molding patterns for wood interior trim. *A* and *I* are crowns, *C* is a bed, and *D* and *E* are coves. All are used to trim the junction of a ceiling and a wall. *F* is a quarter-round for general-purpose trimming of inside corners. Moldings *G–J* are used on walls. *G* is a picture molding, applied near the top of a wall so that framed pictures can be hung from it at any point on special metal hooks that fit over the rounded portion of the molding. *H* is a chair rail, installed around dining rooms to protect the walls from damage by the backs of chairs. *I* is a panel molding and *J* is a batten, both used in traditional paneled wainscoting. Baseboards (*K–N*) include three single-piece designs and one traditional design (*K*) using a separate cap molding and shoe in addition to a piece of S4S stock that is the baseboard itself (see also Figure 24.32). Notice the shallow groove, also called a "relieved back," in the single-piece baseboards and many other flat moldings on this page. This serves to reduce cupping forces on the piece and makes it easier to install even if it is slightly cupped.

Designs *AA–FF* are standard casings for doors and windows. *GG*, *HH*, and *II* are handrail stock. *MM* is representative of a number of sizes of S4S material available to the finish carpenter for miscellaneous uses. *NN* is lattice stock, also used occasionally for flat trim. *OO* is square stock, used primarily for balusters. *PP* represents several available sizes of round stock for balusters, handrails, and closet poles. Wood moldings are furnished in either of two grades: N Grade, for transparent finishes, must be of a single piece. P Grade, for painting, may be finger jointed or edge glued from smaller pieces of wood. P Grade is less expensive because it can be made up of short sections of lower-grade lumber with the defects cut out. Once painted, it is indistinguishable from N Grade. The shapes shown here represent a fraction of the moldings that are generally available from stock. Custom molding patterns can be easily produced because the molding cutters used to produce them can be ground quickly to the desired profiles, working from the architect's drawings.

Proportioning Stairs

Building codes ensure the design of safe stairs through a number of dimensional requirements. The limitations for stairs as given in the International Building Code (IBC) are summarized in Figure A. The required width of an exit stairway is also calculated according to the number of occupants served by the stair in accordance with formulas given in the code and may be wider than the minimums indicated in this figure. A stair may also not rise more than 12 feet (3660 mm) between landings. Landings contribute to the safety of a stair by providing a moment's rest to the legs between flights of steps. (Architects also generally avoid designing flights of less than three risers because short flights, especially in public buildings, sometimes go unnoticed, leading to dangerous falls.) The width of the landing must equal the width of the stair. The length of a landing must also equal the width of the stair for stairs up to 48 inches (1219 mm) in width but need not exceed this dimension for wider stairs.

Working within the dimensional limits of the IBC, combinations of tread and riser dimensions that are most comfortable underfoot can be found by using the proportional rule that twice the riser dimension added to the tread dimension should equal 24 to 25 inches (610–635 mm). This formula was derived in France two centuries ago from measurements of actual dimensions of comfortable stairs. Figure B gives an example of how this formula is used in designing a new stair, in this case for a single-family dwelling. Because the IBC does not allow variations greater than ⅜ inch (9.5 mm) between successive treads or risers, the floor-to-floor dimension should be divided equally into risers to an accuracy of 0.01 inch or 1 mm to avoid cumulative errors. The framing square used by carpenters in the United States to lay out stair stringers has a scale of hundredths of an inch, and riser dimensions should be given in these units rather than fractions to achieve the necessary accuracy.

Monumental outdoor stairs, such as those that lead to entrances of public buildings, are designed with lower risers and deeper treads than indoor stairs. Many designers relax the proportions of the $2R + T$ formula a bit for outdoor stairs, raising the sum to 26 or 27 inches (660 or 685 mm), but it is best to make a full-scale mockup of a section of such a stair to be sure that it is comfortable underfoot.

	Minimum Width	**Maximum Riser Height**	**Minimum Riser Height**	**Minimum Tread Depth**	**Minimum Headroom**
Stair within a residence	36″ (915 mm)	7¾″ (197 mm)	(none)	10″ (254 mm)	6′8″ (2032 mm)
Nonresidential stair, serving an occupant load of less than 50	36″ (915 mm)	7″ (178 mm)	4″ (102 mm)	11″ (279 mm)	6′8″ (2032 mm)
Nonresidential stair, serving an occupant load of 50 or more	44″ (1118 mm)	7″ (178 mm)	4″ (102 mm)	11″ (279 mm)	6′8″ (2032 mm)

Figure A
Dimensional limitations for stairs as established by the IBC.

Procedure	English	Metric
1. Determine the height (H) from finish floor to finish floor	$H = 9'4\frac{3}{8}''$, or 112.375″	H = 2854 mm
2. Divide H by the approximate riser height desired—in this example, 7″ or 180 mm—and round off to obtain a trial number of risers for the stair.	Approximate riser height = 7″ $\frac{H}{7''} = \frac{112.375''}{7''} = 16.05$ Try 16 risers	Approximate riser height = 180 mm $\frac{H}{7''} = \frac{2854\text{ mm}}{180\text{ mm}} = 15.9$ Try 16 risers
3. Divide H by the trial number of risers to obtain an exact riser height (R). Work to the nearest hundredth of an inch, or the nearest millimeter, to avoid any cumulative error that would result in one riser being substantially lower or higher than the rest. Check to make sure that this riser height falls within the limits set by the building code.	$R = \frac{H}{16} = \frac{112.975''}{16} = 7.02''$ R = 7.02″ 7.02″ = 7.75″ maximum; riser OK	$R = \frac{H}{16} = \frac{2854\ mm}{16} = 178\ mm$ R = 178 mm 178 mm = 179 mm maximum; riser OK
4. Substitute this riser height into the formula given for proportioning treads and risers and solve for the tread depth. The depth can be rounded down somewhat if desired, as long as $2R\ T = 24$. Check the tread depth against the code minimum.	$2R + T$ = 25″ 2(7.02″) + T = 25″ T = 25″ − 14.04″ T = 10.96″, say 10.9″ 10.9″ = 10″ code minimum; tread OK	$2R + T$ = 635 mm 2(178 mm) + T = 635 mm T = 635 mm − 356 mm T = 279 mm, say 275 mm 275 mm = 254 mm code minimum; tread OK
5. Summarize the results of these calculations. There is always one fewer tread than risers in a flight of stairs.	16 risers @ 7.02″ 15 treads @ 10.9″ Total run = (15) (10.9″) = 164.5″ = 13′7½″	16 risers @ 178 mm 15 treads @ 275 mm Total run = (15) (275 mm) = 4125 mm
6. If desired, a steeper or shallower stair can be tried as an alternative by subtracting or adding one tread and riser and recalculating dimensions. Reducing the number of treads and risers results in a steeper stair but also shortens the total run of the stair significantly, which is helpful when designing a stair for a limited amount of space. Adding risers and treads results in a stair that is less steep and occupies more space in the plan.	Try 15 risers: $R = \frac{H}{15} = \frac{112.375''}{15} = 7.49''$ 7.49″ = 7.75″ code maximum; riser OK 2(7.49″) + T = 25″ T = 10.02″, say 10.0″ 10″ = 10″ minimum; tread OK Summary: 15 risers @ 7.49″ 14 treads @ 10″ Total run = (14) (10″) = 11′8″ Subtracting one tread and riser shortens the stair run by almost 2 feet.	Try 15 risers: $R = \frac{H}{15} = \frac{2854\text{ mm}}{15} = 190\text{ mm}$ 190 mm = 197 code maximum; riser OK 2(190 mm) + T = 635 T = 255 mm 255 mm = 254 mm code minimum; tread OK Summary: 15 risers @ 190 mm 14 treads @ 255 mm Total run = (14) (255 mm) = 3570 mm Subtracting one tread and riser shortens the stair run by 555 mm.

FIGURE B
A sample calculation for proportioning a residential stair.

FIGURE 7.37
Fireplace mantels are available from specialized mills in a number of traditional and contemporary designs. Each mantel is furnished largely assembled but is detailed in such a way that it can easily be adjusted to fit any fireplace within a wide range of sizes. *(Courtesy of Mantels of Yesteryear, McCaysville, GA, www.mantelsofyesteryear.com)*

FLOORING AND CERAMIC TILE WORK

Before finish flooring can be installed, the subfloor is scraped free of plaster droppings and swept thoroughly. Underlayment panels of C–C Plugged plywood or particleboard (in areas destined for resilient flooring materials and carpeting) are glued and nailed over the subfloor, their joints staggered with those in the subfloor to eliminate weak spots. The thicknesses of the underlayment panels are chosen to make the finished floor surfaces as nearly equal in level as possible at junctions between different flooring materials.

In multistory wood light frame commercial and apartment buildings, a floor underlayment specially formulated of poured gypsum or lightweight concrete is often poured over the subfloor. This has a threefold function: It provides a smooth, level surface for finish floor materials; it furnishes additional fire resistance to the floor construction; and it reduces the transmission of sound through the floor to the apartment or office below. The gypsum or concrete is formulated with superplasticizer additives that make it virtually self-leveling as it is applied (Figure 7.38). A minimum thickness is $^{3}/_{4}$ inch (19 mm). Poured underlayments are also used to level floors in older buildings, to add fire and sound resistance to precast concrete floors, and to embed plastic tubing or electric resistance wires for in-floor radiant heat.

Floor finishing operations require cleanliness and freedom from traffic, so members of other trades are banished from the area as the flooring materials are applied.

Hardwood flooring is sanded level and smooth after installation, then vacuumed to remove the sanding dust. The finish coatings are applied in as dust-free an atmosphere as possible to avoid embedded specks. Resilient-flooring installers vacuum the underlayment meticulously so that particles of dirt will not become trapped beneath the thin flooring and cause bumps in the surface. The finished floors are often covered with sheets of heavy paper or plastic to protect them during the final few days of construction activity. Carpet installation is less sensitive to dust, and the installed carpets are less prone to damage than hardwood and resilient floorings, but temporary coverings are applied as necessary to protect the carpet from paint spills and water stains.

The application of ceramic tile to a portland cement plaster base coat over metal lath for a shower stall is illustrated in Figure 7.39, and finished ceramic tile work is shown in Figure 7.40. Cementitious backer board may be used as a less costly substitute for a cement plaster base

FIGURE 7.38
Workers apply a gypsum underlayment to an office floor. The gypsum is pumped through a hose and distributed with a straightedge tool. Because the gypsum seals against the bottoms of the interior partitions, it can also help reduce sound transmission from one room to another. *(© Gyp-Crete Corporation, Hamel, Minnesota)*

FIGURE 7.39
Installing sheets of ceramic tile in a shower stall. The base coat of portland cement plaster over metal lath has already been installed. Now the tilesetter applies a thin coat of tile adhesive to the base coat with a trowel and presses a sheet of tiles into it, taking care to align the tiles individually around the edges. A day or two later, after the adhesive has hardened sufficiently, the joints will be grouted to complete the installation. *(Photos by Joseph Ianor)*

FIGURE 7.40
Ceramic tile is used for the floor, countertops, and backsplash in this kitchen. The border was made by selectively substituting tiles of four different colors for the white tiles used for the field of the floor. *(Designer: Kevin Cordes. Courtesy of American Olean Tile)*

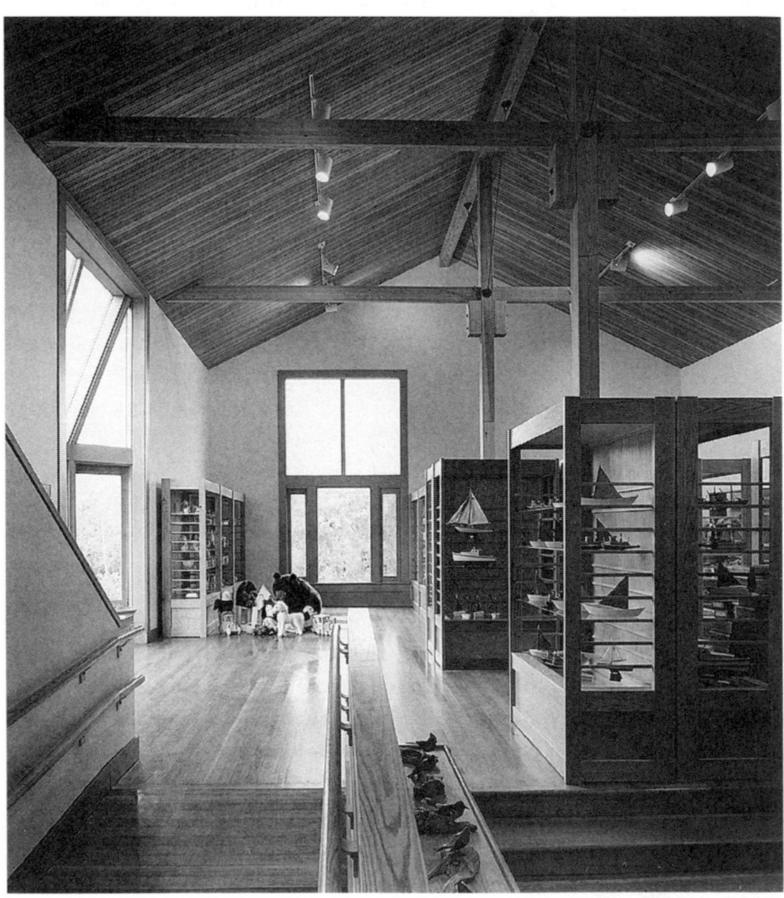

FIGURE 7.41
Varnished oak flooring, millwork, and casework. *(Woo & Williams, Architects. Photographer: Richard Bonarrigo)*

coat (Figure 24.29). Wood flooring installation is depicted in Figures 24.32 and 24.33. A finished hardwood floor is shown in Figures 7.41 and 24.34. The installation of ceramic tile and finish flooring materials is covered in more detail in Chapters 23 and 24.

Finishing Touches

When flooring and painting are finished, the plumbers install and activate the lavatories, water closets, tubs, sinks, and shower fixtures. Gas lines are connected to appliances and the main gas valve is opened. The electricians connect the wiring for the heating and air conditioning equipment and (if electric) water heater; mount the receptacles, switches, and lighting fixtures; and put metal or plastic cover plates on the switches and receptacles. The electrical circuits are energized and checked to be sure that they work. The smoke alarms and heat alarms, required by most codes in residential structures, are also connected and tested by the electricians, along with any communications, entertainment, and security system wiring. The heating and air conditioning system is completed with the installation of air grills and registers, or with the mounting of metal convector covers, then turned on and tested. Painted surfaces that have been scuffed or marred are touched up, and last-minute problems are identified and corrected through cooperative effort by the contractors, the owner of the building, and the architect. The building inspector is called in for a final inspection and issuance of an occupancy permit. After a thorough cleaning, the building is ready for use.

CSI/CSC

MasterFormat Sections for Interior Finishes for Wood Light Frame Construction

06 20 00	**FINISH CARPENTRY**
	Interior Finish Carpentry
06 22 00	**Millwork**
06 26 00	**Board Paneling**
06 40 00	**ARCHITECTURAL WOODWORK**
	Interior Architectural Woodwork
06 41 00	**Architectural Wood Casework**
06 42 00	**Wood Paneling**
06 43 00	**Wood Stairs Railings**
06 44 00	**Ornamental Woodwork**
06 46 00	**Wood Trim**
06 48 00	**Wood Frames**
07 20 00	**THERMAL PROTECTION**
07 21 00	**Thermal Insulation**
	Board Insulation
	Blanket Insulation
	Foamed-In-Place insulation
	Loose-Fill Insulation
	Blown Insulation
	Sprayed Insulation
07 25 00	**Weather Barriers**
07 26 00	**Vapor Retarders**
07 27 00	**Air Barriers**
08 10 00	**DOORS AND FRAMES**
08 14 00	**Wood Doors**
	Flush Wood Doors
	Stile and Rail Wood Doors
08 70 00	**HARDWARE**
08 71 00	**Door Hardware**
12 30 00	**CASEWORK**
12 32 00	**Manufactured Wood Casework**
12 36 00	**Countertops**

Selected References

1. Lstiburek, Joseph. *Builder's Guide to Cold Climates.* Westford, MA, Building Science Corporation, 2006.

This guide, along with companion guides for other major climate zones, explains the roles of insulation, vapor retarders, and air barriers in the performance of the building enclosure, and provides guidelines for the construction of energy-efficient and weather-resistant home.

2. Thallon, Rob. *Graphic Guide to Interior Details: For Builders and Designers.* Newtown, CT, Taunton Press, 2004.

Profusely illustrated, clearly written, and encyclopedic in scope, this book offers complete guidance on interior finishing of wood light frame buildings.

3. Dietz, Albert G. H. *Dwelling House Construction* (5th ed.). Cambridge, MA, MIT Press, 1990.

This classic text has extensive chapters with clear illustrations concerning chimneys and fireplaces, insulation, wallboard, lath and plaster, and interior finish carpentry.

4. Architectural Woodwork Institute. *AWI Quality Standards Illustrated.* Reston, VA, updated regularly.

Every detail of every grade of interior woodwork and cabinetry is illustrated and described in this thick volume.

Web Sites

Interior Finishes for Wood Light Frame Construction

Author's supplementary web site: **www.ianosbackfill.com/07_interior_finishes_for_wood_light_frame_construction**

Thermal Insulation and Vapor Retarder

Building Science Corporation: **www.buildingscience.com**
Dow Chemical rigid foam insulation: **www.dow.com/styrofoam**
Gypsum Association: **www.gypsum.org**
Icynene Corporation spray-foam insulation: **www.icynene.com**
Owens Corning insulation products: **www.owenscorning.com**

Wall and Ceiling Finish

USG gypsum products: **www.usg.com**

Millwork and Finish

Architectural Woodwork Institute: **www.awinet.org**
Hardwood Plywood & Veneer Association: **www.hpva.org**
Jeld-Wen Windows & Doors: **www.jeld-wen.com**
Window and Door Manufacturers Association: **www.wdma.org**

Proportioning Fireplaces

Buckley Rumford Company: **www.rumford.com**

Flooring and Ceramic Tile Work

American Olean Tile: **www.americanolean.com**
Tile Council of America: **www.tileusa.com**

Key Terms and Concepts

roughing in
DWV (drain–waste–vent) pipe
supply pipe
ductwork
hydronic heating system, forced hot water heating system
convector
radiant heating system
thermal envelope, building envelope, building enclosure
thermal insulation
thermal resistance
R-value, RSI-value
glass fiber batt
glass wool insulation
rock wool insulation
treated cellulose insulation
foamed in place polyurethane insulation
foamed in place polyicynene insulation
rigid board expanded polystyrene (EPS) foam insulation
rigid board extruded polystyrene (EXS) foam insulation
rigid board polyisocyanuarate foam insulation
rigid board mineral fiber insulation
rigid board cellulose (wood or vegetable) fiber insulation
thermal resistance
thermal bridging
radiant barrier
vapor retarder, vapor barrier
air barrier
airtight drywall approach (ADA)
simple caulk and seal
air-to-air heat exchanger
three-coat gypsum plaster
wood strip lath
gypsum board, drywall
veneer plaster
flue
damper
firebox
smoke shelf
millwork
molding
AWI Economy Grade, Custom Grade, Premium Grade
finish carpentry
Z-brace door
panel door
flush door
solid-core door
hollow-core door
panel
stile
rail
prehung door
casing
split jamb
baseboard
nail set
custom cabinet, architectural wood cabinet
prefabricated cabinet, manufactured wood cabinet
countertop

Review Questions

1. List the sequence of operations required to complete the interior of a wood light frame building and explain the logic of the order in which these operations occur.

2. What are some alternative ways of insulating the walls of a wood light frame building to R-values beyond the range normally possible with ordinary 2 × 4 (38 × 89 mm) studs?

3. Why are plaster and gypsum board so popular as interior wall finishes in wood frame buildings? List as many reasons as possible.

4. What is the level of humidity in a building at the time installation of the interior wall finishes is completed? Why? What should be done about this and why?

5. Summarize the most important things to keep in mind when designing a stair.

Exercises

1. Design and detail a fireplace for a building that you are designing, using the information provided on page 276 to work out the exact dimensions and the information in Chapter 8 to help in detailing the masonry.

2. Design and detail a stairway for a building that you are designing, using the information provided on pages 288 and 289 to calculate the dimensions.

3. Visit a wood frame building that you admire. Make a list of the interior finish materials and components, including finishes and species of wood where possible. How does each material and component contribute to the overall feeling of the building? How do they relate to one another?

4. Make measured drawings of millwork details in an older building that you admire. Analyze each detail to discover its logic. What woods were used and how were they sawn? How were they finished?

8

BRICK MASONRY

Flemish bond brickwork combines simply and directly with cut limestone lintels and sills in this townhouse on Boston's Beacon Hill. *(Photo by the Edward Allen)*

Masonry is the simplest of building techniques: The *mason* stacks pieces of material (bricks, stones, or concrete blocks, collectively called *masonry units*) atop one another to make walls. But masonry is also the richest and most varied of techniques, with its endless selection of colors, textures, and patterns. And because the pieces of which it is made are small, masonry can take any shape, from a planar wall to a sinuous surface that defies the distinction of roof from wall.

Masonry is the material of earth, taken from the earth and comfortably at home in foundations, pavings, and walls that grow directly from the earth. With modern techniques of reinforcing, however, masonry can rise many stories from the earth, and in the form of arches and vaults, masonry can take wing and fly across space.

The most ancient of our building techniques, masonry remains labor intensive, requiring the patient skills of experienced and meticulous artisans to achieve a satisfactory result. It has kept pace with the times and remains highly competitive technically and economically with other systems of structure and enclosure, the more so because one mason can produce in one operation a completely finished, insulated, loadbearing wall ready for use.

Masonry is durable. The designer can select masonry materials that are scarcely affected by water, air, or fire, ones with brilliant colors that will not fade, ones that will stand up to heavy wear and abuse, and make from them a building that will last for generations.

Masonry is a material of the small entrepreneur. One can set out to build a building of bricks with no more tools than a *trowel*, a shovel, a hammer, a measuring rule, a level, a square of scrap plywood, and a piece of string. Yet many masons can work together, aided by mechanized handling of materials, to put up projects as large as the human mind can conceive.

HISTORY

Masonry began spontaneously in the creation of low walls by stacking stones or pieces of caked mud taken from dried puddles. Mortar was originally the mud smeared into the joints of the rising wall to lend stability and weathertightness. Where stone lay readily at hand, it was preferred to bricks; where stone was unavailable, bricks were made from local clays and silts. Changes came with the passing millennia: People learned to quarry, cut, and dress stone with increasing precision. Fires built against mud brick walls brought a knowledge of the advantages of burned brick, leading to the invention of the kiln. Masons learned the simple art of turning limestone into lime, and lime mortar gradually replaced mud.

By the fourth millennium B.C., the peoples of Mesopotamia were building palaces and temples of stone and sun-dried brick. In the third millennium, the Egyptians erected

FIGURE 8.1
The Parthenon, constructed of marble, has stood on the Acropolis in Athens for more than 24 centuries. *(Photo by James Austin, Cambridge, England)*

the first of their stone temples and pyramids. In the last centuries prior to the birth of Christ, the Greeks perfected their temples of limestone and marble (Figure 8.1). Control of the Western world then passed to the Romans, who made the first large-scale use of masonry arches and roof vaults in their temples, basilicas, baths, palaces, and aqueducts.

Medieval civilizations in both Europe and the Islamic world brought masonry vaulting to a very high plane of development. The Islamic craftsmen built magnificent palaces, markets, and mosques of brick and often faced them with brightly glazed clay tiles. The Europeans directed their efforts toward fortresses and cathedrals of stone, culminating in the pointed vaults and flying buttresses of the great Gothic churches (Figures 8.2 and 8.3). In Central America, South America, and Asia, other civilizations were carrying on a simultaneous evolution of masonry techniques in cut stone.

During the Industrial Revolution in Europe and North America, machines were developed that quarried and worked stone, molded bricks, and sped the transportation of these heavy materials to the building site. Sophisticated mathematics were applied for the first time to the analysis of the structure of masonry arches and to the art of stonecutting. Portland cement mortar came into widespread use, enabling the construction of masonry buildings of greater strength and durability.

In the late 19th century, masonry began to lose its primacy among the materials of construction. The very tall buildings of the central cities required frames of iron or steel to replace the thick masonry bearing walls that had limited the heights to which one could build. Reinforced concrete, poured rapidly and economically into simple forms made of wood, began to replace brick and stone masonry in foundations and walls. The heavy masonry vault was supplanted by lighter floor and roof structures of steel and concrete that were faster to erect.

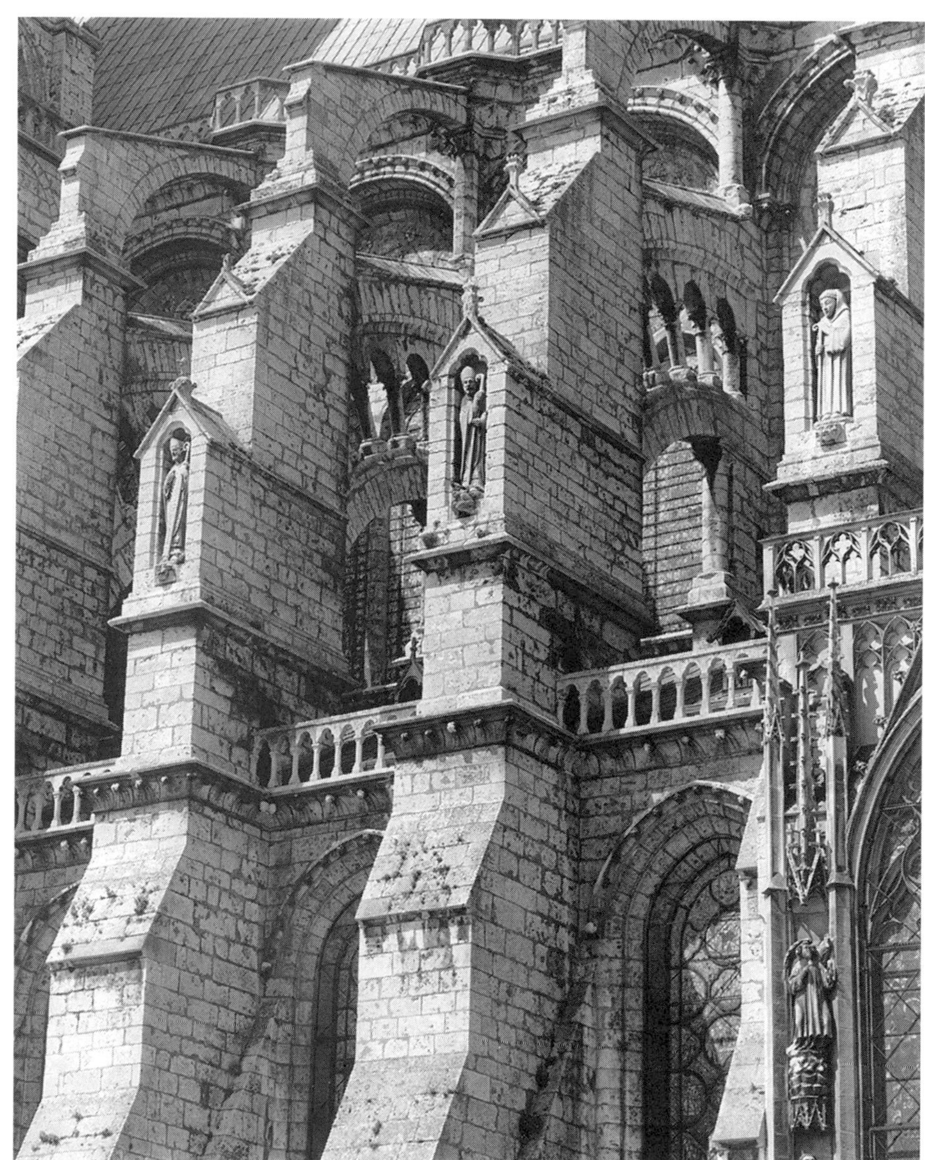

FIGURE 8.2
Construction in ashlar limestone of the magnificent Gothic cathedral at Chartres, France, was begun in A.D. 1194 and was not finished until several centuries later. Seen here are the flying buttresses that resist the lateral thrusts of the stone roof vaulting.
(Photo by James Austin, Cambridge, England)

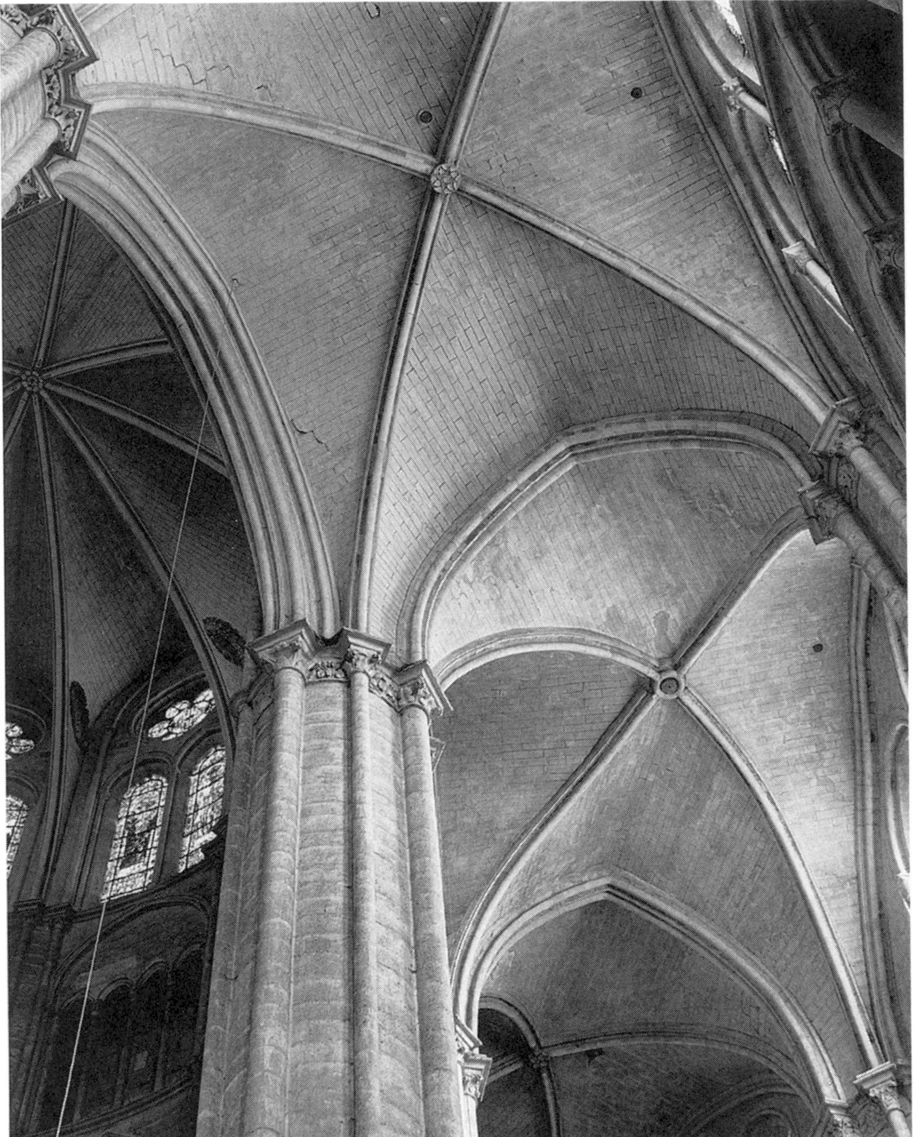

FIGURE 8.3
The Gothic cathedrals were roofed with lofty vaults of stone blocks. The ambulatory roof at Bourges (built 1195–1275) evidences the skill of the medieval French masons in constructing vaulting to cover even a curving floor plan. *(Photo by James Austin, Cambridge, England)*

FIGURE 8.4
Despite the steady mechanization of construction operations in general, masonry construction in brick, concrete block, and stone is still based on simple tools and the highly skilled hands that use them. *(Courtesy of International Masonry Institute, Washington, DC)*

The 19th-century invention of the hollow concrete block helped to avert the extinction of masonry as a craft. The concrete block was much cheaper than cut stone and required much less labor to lay than brick. It could be combined with brick or stone facings to make lower-cost walls that were still satisfactory in appearance. The brick cavity wall, an early-19th-century British invention, also contributed to the survival of masonry, for it produced a warmer, more watertight wall that was later to adapt easily to the introduction of thermal insulation when appropriate insulating materials became available in the mid-20th century.

Other 20th-century contributions to masonry construction include the development of techniques for steel-reinforced masonry, high-strength mortars, masonry units (both bricks and concrete blocks) that are higher in structural strength, and masonry units of many types that reduce the amount of labor required for masonry construction.

. . . and the smothered incandescence of the kiln: in the fabulous heat, mineral and chemical treasure baking on mere clay, to issue in all the hues of the rainbow, all the shapes of imagination that never yield to time . . . these great ovens would cast a spell upon me as I listened to the subdued roar deep within.

Frank Lloyd Wright, *In the Nature of Materials,* 1942.

If this book had been written as recently as 125 years ago, it would have had to devote little space to construction materials other than masonry and wood. Because other construction materials were so late in developing, most of the great works of architecture in the world, and many of the best-developed vernacular architectures, are built of masonry. We live amid a rich heritage of masonry buildings. There is scarcely a town in the world that is without a number of beautiful examples from which the serious student of masonry architecture can learn.

Mortar

Mortar is as vital a part of masonry as the masonry units themselves. Mortar serves to cushion the masonry units, giving them full bearing against one another despite their surface irregularities. Mortar seals between the units to keep water and wind from penetrating; it adheres the units to one another to bond them into a monolithic structural unit; and, inevitably, it is important to the appearance of the finished masonry wall.

Mortar Ingredients

The most characteristic type of mortar is *cement–lime mortar,* made of portland cement, hydrated lime, an inert aggregate, and water. The *aggregate,* sand, must be clean and must be screened to eliminate particles that are too coarse or too fine; ASTM specification C144 establishes standards for mortar sand. The *portland cement* is the bonding agent in the mortar. Its composition and manufacture are described in more detail beginning on page 518 of this book. (Only portland cement Types I, II, and III are recommended for use in masonry mortars.) Mortar made only with portland cement, however, is "harsh," meaning that it does not flow well on the trowel or under the brick, so lime is added to impart smoothness and workability. *Lime* is produced by burning limestone or seashells (calcium carbonate) in a kiln to drive off carbon dioxide and leave *quicklime* (calcium oxide). The quicklime is then slaked by allowing it to absorb as much water as it will hold, resulting in the formation of calcium hydroxide, called *slaked lime* or *hydrated lime.* The slaking process, which releases large quantities of heat, is usually carried out in the factory. The hydrated lime is subsequently dried, ground, and bagged for shipment. ASTM specification C207 governs the production of lime. Water is also an important ingredient in mortar because it is chemically involved in the curing of the cement and lime. Water used in mortar should be clean and free of acids, alkalis, and organic matter. Water that is potable is generally considered suitable for use in mortar.

Blended hydraulic cements, ASTM C595, are blends of portland cement with other cementitious materials, such as blast furnace slag, that may be used in place of ordinary portland cement alone in the cement–lime mortar mix.

Masonry cements and mortar cements are prepackaged cements that do not require the addition of lime by the mason on the job site. Their main advantages are convenience, consistency (since they are premixed), and good workability. *Masonry cements* are proprietary formulations that may contain portland cement or blended hydraulic cements, lime or other plasticizing ingredients, and other additives. Formulations vary from one manufacturer to another, but all must comply with ASTM C91. In order to achieve a workability equivalent to that of conventional cement–lime mortars, masonry cement mortars are formulated with *air-entraining admixtures* that result in a higher air content in the cured mortar than cement–lime mortar. This reduces the bond strength between the mortar and the masonry unit to about half that of conventional mortar, which means that the flexural and shear strength of the wall is reduced and the wall is more permeable to water. For these reasons, masonry cements should not be specified for masonry work that requires high strength and low permeability.

Mortar cements are also blends of portland cement, lime, and other additives. However, they are formulated according to ASTM C1329, with limits on air entrainment that allow them to meet bond strength requirements comparable to those of cement–lime mortars. Structural codes treat mortars made with mortar cement as equivalent to traditional cement–lime mortars.

Cements are available in a range of colors. The most common color is light gray, about the same color as ordinary concrete block. Cements are also available in white, as well as in a range of darker grays, all achieved by controlling the ingredients used in the manufacture of the cement itself. In the final mortar mix, a much broader range of colors can be produced either by adding pigments to the mortar at the time of mixing or by purchasing dry mortar mix that has been custom colored at the factory. Packaged mortar mixes can be obtained in shades ranging from pure white to pure black, including all the colors of the spectrum.

Because mortar makes up a considerable fraction of the exposed surface area of a brick wall, typically about 20 percent, the color of the mortar is extremely important in the appearance of a brick wall and is almost as important in the appearance of stone or concrete masonry walls. Small sample walls are often constructed before a major building goes under construction to view and compare different color combinations of brick and mortar and make a final selection.

Mortar Mixes

Mortars mixed with portland cement, blended hydraulic cement, masonry cement, or mortar cement are all specified according ASTM C270. Four basic *mortar types*, distinguished primarily by differences in strength, are defined:

- *Type N mortar* is a general-purpose mortar with a balance of good bonding capabilities and good workability. It is recommended for exterior veneers, nonloadbearing exterior walls, parapets, chimneys, and interior loadbearing walls.
- *Type S mortar* has a higher flexural bond strength than Type N mortar. It is recommended for exterior reinforced masonry, exterior loadbearing masonry walls, and veneers and walls subject to high wind forces or high seismic loads.
- *Type O mortar* is a low-strength mortar recommended mainly for interior nonloadbearing masonry and historic restoration work.
- *Type M mortar* is a high-strength mortar with less workability than Type S or N mortars. It is recommended for masonry construction below grade, masonry subject to high lateral or compressive loads, or masonry exposed to severe frost action.

Because lower-strength mortars are more workable than higher-strength mortars, as a general rule, the lowest-strength mortar that meets project requirements should be chosen. The majority of mortar for masonry work used in new construction is either Type N or Type S. As a memory aid, the letters used to designate mortar types, in order of decreasing strength, come from taking every other letter in the phrase MaSoN wOrK. (Type K mortar is a very-low-strength mortar used in historic preservation work that is no longer part of the ASTM C270 specification.)

According to ASTM C270, mortar mixes may be specified in one of two ways: either by *proportion specification*, in which the quantities of ingredients used to prepare the mix are specified, or by *property specification*, in which the compressive strength and other properties of the hardened mortar as determined by laboratory testing are defined. Proportion specification is the simpler (no laboratory testing is required) and more common method. On large jobs, however, property specification gives the mason more flexibility in the choice of mortar ingredients and can result in an overall cost savings, even after the costs of laboratory testing are considered. These two methods are summarized in Figures 8.5 and 8.6, respectively.

Lime Mortar

Modern masonry mortars are made with *hydraulic cements*, that is, cements that cure by chemical reaction with water, a process called "hydration" discussed in more detail below. Until the late 19th and early 20th centuries, however, mortar was made without portland cement, and the lime itself was the bonding agent. Such traditional *lime mortars*, made from a mix of lime, sand, and water, continue to find use principally in the restoration of historic structures. Unlike modern hydraulic cements, lime is a *nonhydraulic cement*, and mortars made with lime as the sole cementing ingredient cure through a reaction with carbon dioxide in the atmosphere. This process, called *carbonation*, occurs gradually and may continue for many years. Such mortars remain at least partially water soluble and retain some ability to self-heal in the event of hairline cracking caused by movement within the wall. By adding other cementitious materials, lime mortars with greater degrees of hydraulic properties can also be formulated.

Mortar Hydration

Hydraulic cement mortars cure by *hydration*, not by drying: A complex set of chemical reactions take up water and combine it with the constituents of the cement and lime to create a dense, strong crystalline structure that binds the sand particles and masonry units together. Once hydraulic cements harden, they become water insoluble.

Mortar Type	Parts by Volume of Portland Cement or Blended Hydraulic Cement	Parts by Volume of Mortar Cement or Masonry Cement	Parts by Volume Hydrated Lime	Aggregate, Measured in a Damp, Loose Condition
M	1		¼	Not less than 2¼ and not more than 3 times the sum of the volumes of cement and lime materials used
M	1	1 (Type N)		
M		1 (Type M)		
S	1		over ¼ to ½	
S	½	1 (Type N)		
S		1 (Type S)		
N	1		over ½ to 1¼	
N		1 (Type N)		
O	1		over 1¼ to 2½	
O		1 (Type N)		

FIGURE 8.5
Mortar types by proportion specification. The general principle is that the greater the proportion of cement to lime, the greater the compressive strength of the mortar. For each of the four mortar types, the first example is a cement–lime mortar mix consisting of portland cement or blended hydraulic cement, lime, and aggregate. The second and sometimes third examples for each type are mortars made with mortar cement or masonry cement. Note that no added lime is required for these mixes; the necessary lime and other plasticizers are included in the prepackaged ingredients. Note also that a Type M mortar can be made with a Type N cement by increasing the total cement content of the mix with the addition of portland cement or blended hydraulic cement on the job site. In the same way, Type S mortar can be made with Type N prepackaged mortar and added cement.

Mortar Type	Minimum Average Compressive Strength at 28 Days
M	2500 psi (17.2 MPa)
S	1800 psi (12.4 MPa)
N	750 psi (5.2 MPa)
O	350 psi (2.4 MPa)

FIGURE 8.6
Minimum compressive strength for mortar types by property specification. Not shown here, but also included in the ASTM C270 specification, are requirements for maximum air content, minimum water retention (a factor affecting the bond strength of the mortar), and volume ratio of aggregate to cement and lime. When mortar is specified by property, the mason is free to use any mix that meets the specified strength and other requirements as demonstrated through laboratory testing.

Mortar that has been mixed but not yet used can become too stiff for use, either by drying out or by commencing its hydration. If the mortar was mixed less than 90 minutes prior to its stiffening, it has merely dried and the mason can safely *retemper* it with water to make it workable again. If the unused mortar is more than 2½ hours old, it must be discarded because it has already begun to hydrate and cannot be retempered without reducing its final strength. On large masonry projects, an *extended-life admixture* is sometimes included in the mortar. This allows the mortar to be mixed in large batches and kept for as long as 72 hours before it must be discarded. Most masonry units should not be wetted before laying, but to prevent premature drying of mortar, which would weaken it, masonry units that are highly absorptive of water should be dampened slightly before laying.

Considerations of Sustainability in Brick Masonry

Brick Masonry Materials

• Mortar is made of minerals that are generally abundant in the earth. Portland cement and lime are energy-intensive products. (For more information about the sustainability of cement production, see Chapter 13.)

• Clay and shale, the raw materials for bricks, are plentiful. They are usually obtained from open pits, with the attendant disruption of drainage, vegetation, and wildlife habitat.

• Clay brick can include recycled brick dust, postindustrial wastes such as fly ash, and a variety of other waste products in their manufacture.

Brick Manufacturing

• Brick manufacturing plants are usually located close to the sources of their raw materials.

• Brick manufacturing produces few waste materials. Unfired clay is easily recycled into the production process. Fired bricks that are unusable are ground up and recycled into the production process or used as landscaping material.

• Brick manufacturing requires relatively large amounts of water. Water that doesn't evaporate can be reused many times. Little if any water need be discharged as waste.

• Because of the energy used in its firing, brick is a relatively energy-intensive product. Its embodied energy may range from about 1000 to 4000 BTU per pound (2.3-9.3 MJ/kg).

• The most common energy source for brick kilns is natural gas, although oil and coal are also used. Firing of clay masonry produces fluorine and chlorine emissions. Other types of air pollution can result from improperly regulated kilns.

• Most bricks are sold for use in regional markets close to their point of manufacture. This reduces the energy required for shipping and makes much brick eligible for credit as a regional material.

Brick Masonry Construction

• Relatively small amounts of waste are generated on a construction site during brick masonry work, including partial bricks, unsatisfactory bricks, and unused mortar. These wastes generally go into landfills or are buried on the site.

• Sealers applied to brick masonry to provide water repellency and protection from staining are potential sources of emissions. Solvent-based sealers generally have higher emissions than water-based products.

Brick Masonry Buildings

• Brick masonry is not normally associated with any indoor air quality problems, although in rare circumstances it can be a source of radon gas.

• The thermal mass effect of brick masonry can be a useful component of fuel-saving heating and cooling strategies such as solar heating and nighttime cooling.

• Brick masonry is a durable form of construction that requires relatively little maintenance and can last a very long time.

• Construction with brick masonry can reduce reliance on paint finishes, a source of volatile organic compounds.

• Brick masonry is resistant to moisture damage and mold growth.

• When a brick building is demolished, sound bricks may be cleaned of mortar and reused (once their physical properties have been verified as adequate for the new use). Brick waste can be crushed and used for landscaping. Brick and mortar waste can also be used as on-site fill. Much such waste, however, is disposed of off-site in landfills.

Brick Masonry

Among the masonry materials, brick is special in two respects: fire resistance and size. As a product of fire, it is the most resistant to building fires of any masonry unit type. Its size may account for much of the love that many people instinctively feel for brick: A traditional brick is shaped and dimensioned to fit the human hand. Hand-sized bricks are less likely to crack during drying or firing than larger bricks, and they are easy for the mason to manipulate. This small unit size makes brickwork very flexible in adapting to small-scale geometries and patterns and gives a pleasing scale and texture to a brick wall or floor.

Molding of Bricks

Because of their weight and bulk, which make them expensive to ship over long distances, *bricks* are produced by a large number of relatively small, widely dispersed factories from a variety of local clays and shales. The raw material is dug from pits, crushed, ground, and screened to reduce it to a fine consistency. Then it is tempered with water to produce a plastic clay ready for forming into bricks.

Three major methods are used today for forming bricks: the soft mud process, the dry-press process, and the stiff mud process. The oldest is the *soft mud process,* in which a relatively moist clay (20–30 percent water) is pressed into simple rectangular molds, either by hand or with the aid of molding machines (Figure 8.7). To keep the sticky clay from adhering to the molds, the molds may be dipped in water immediately before being filled, producing bricks with a relatively smooth, dense surface that are known as *water-struck bricks.* If the wet mold is dusted with sand just before forming the brick, *sand-struck* or *sand-mold bricks* are produced, with a matte-textured surface.

The *dry-press process* is used for clays that shrink excessively during drying. Clay mixed with a minimum of water (up to 10 percent) is pressed into steel molds by a machine working at a very high pressure.

The high-production *stiff mud process* is the one most widely used today. Clay containing 12 to 15 percent water is passed through a vacuum to remove any pockets of air, then extruded through a rectangular die (Figures 8.8 and 8.9). As the clay leaves the die, textures or thin mixtures of colored clays may be applied to its surface as desired. The rectangular column of moist clay is pushed by the pressure of extrusion across a cutting table, where automatic cutter wires slice it into bricks.

After molding by any of these three processes, the bricks are dried for 1 or 2 days in a low-temperature dryer kiln. They are then ready for transformation into their final form by a process known as *firing* or *burning.*

Firing of Bricks

Before the advent of modern kilns, bricks were most often fired by stacking them in a loose array called a *clamp,* covering the clamp with earth

Figure 8.7
A simple wooden mold produces seven water-struck bricks at a time. *(Photo by Edward Allen)*

Figure 8.8
A column of clay emerges from the die in the stiff mud process of molding bricks. *(Courtesy of Brick Industry Association)*

FIGURE 8.9
Rotating groups of parallel wires cut the column of clay into individual bricks, ready for drying and firing. *(Courtesy of Brick Industry Association)*

(*a*)

(b)

(c)

FIGURE 8.10
Three stages in the firing of water-struck bricks in a small factory. ***(a)*** **Bricks stacked on a kiln car ready for firing. The open passages between the bricks allow the hot kiln gases to penetrate to the interior of the stack. The bed of the kiln car is made of a refractory material that is unaffected by the heat of the kiln. The rails on which the car runs are recessed in the floor.** ***(b)*** **The cars of bricks are rolled into the far end of this gas-fired periodic tunnel kiln. When firing has been completed, the large door in the near end of the kiln is opened and the cars of bricks are rolled out on the rails that can be seen at the lower right of the picture.** ***(c)*** **After the fired bricks have been sorted, they are strapped into these "cubes" for shipping.** *(Photos by Edward Allen)*

or clay, building a wood fire under the clamp, and maintaining the fire for a period of several days. After cooling, the clamp was disassembled and the bricks were sorted according to the degree of burning each had experienced. Bricks adjacent to the fire (*clinker bricks*) were often overburned and distorted, making them unattractive and therefore unsuitable for use in exposed brickwork. Bricks in a zone of the clamp near the fire were fully burned but undistorted, suitable for exterior facing bricks with a high degree of resistance to weather. Bricks farther from the fire were softer and were set aside for use as backup bricks, while some bricks from around the perimeter of the clamp were not burned sufficiently for any purpose and were discarded. In the days before mechanized transportation, bricks for a building were often produced from clay obtained from the building site and were burned in clamps adjacent to the work.

Today, bricks are usually burned in either a periodic kiln or a continuous tunnel kiln. The *periodic kiln* is a fixed structure that is loaded with bricks, fired, cooled, and unloaded (Figure 8.10). For higher productivity, bricks are passed continuously through a long *tunnel kiln* on special railcars to emerge at the far end fully burned. In either type of kiln, the first stages of burning are *water smoking* and *dehydration*, which drive off the remaining water from the clay. The next stages are *oxidation* and *vitrification*, during which the temperature rises to 1800 to 2400 degrees Fahrenheit (1000–1300°C) and the clay is transformed into a ceramic material. This may be followed by a stage called *flashing*, in which the fire is regulated to create a reducing atmosphere in the kiln that develops color variations in the bricks. Finally, the bricks are cooled under controlled conditions to achieve the desired color and avoid thermal cracking. The cooled bricks are inspected, sorted, and packaged for shipment. The entire process of firing takes 40 to 150 hours and is monitored continuously to maintain product quality. Considerable shrinkage takes place in the bricks during drying and firing; this must be taken into account when designing the molds for the bricks. The higher the temperature is, the greater the shrinkage and the darker the color of the brick. Bricks are often used in a mixed range of colors, with the darker bricks inevitably being smaller than the lighter bricks. Even in bricks of uniform color, some size variation is to be expected, and bricks, in general, are subject to a certain amount of distortion from the firing process.

The color of a brick depends on the chemical composition of the clay or shale and the temperature and chemistry of the fire in the kiln. Higher temperatures, as noted in the previous paragraph, produce darker bricks. The iron that is prevalent in most clays turns red in an oxidizing fire and purple in a reducing fire. Other chemical elements interact in a similar way to the kiln atmosphere to make still other colors. For bright colors, the faces of the bricks can be glazed like pottery, either during the normal firing or in an additional firing.

Brick Sizes

There is no truly standard brick. The nearest thing in the United States is the *modular brick*, dimensioned to construct walls in modules of 4 inches (101 mm) horizontally and 8 inches (203 mm) vertically, but the modular brick has not found ready acceptance in some parts of the country and traditional sizes persist on a regional basis. Figure 8.11 shows the brick sizes that represent about 90 percent of all the bricks used in the United States. In practice, the designer, when selecting bricks for a building, usually views actual samples before completing the drawings for the building and dimensions the drawings in accordance with the size of the particular bricks selected (Figure 8.22). For most bricks in the traditional range of sizes, three courses of bricks plus the accompanying three mortar joints add up to a height of 8 inches (203 mm). Length dimensions must be calculated specifically for the brick selected and must include the thicknesses of the mortar joints.

The use of larger bricks can lead to substantial economies in construction. A *Utility brick*, for example, has the same face height as a standard modular brick, but because it is longer, its in-the-wall cost per square foot is about 25 percent lower and the compressive strength of the wall is about 25 percent higher because of the smaller proportion of mortar. The designer should also consider, however, that a wall built with larger bricks can deceive the viewer regarding the scale of the building.

Custom shapes and sizes of brick are often required for buildings with special details, ornamentation, or unusual geometries (Figures 8.12 and 8.13). These are readily produced by most brick manufacturers if sufficient lead time is given.

Name	Width	Length	Height
Modular	3½″ or 3⅝″ (90 mm)	7½″ or 7⅝″ (190 mm)	2¼″ (57 mm)
Standard	3½″ or 3⅝″ (90 mm)	8″ (200 mm)	2¼″ (57 mm)
Engineer Modular	3½″ or 3⅝″ (90 mm)	7½″ or 7⅝″ (190 mm)	2¾″ to 2 13/16″ (70 mm)
Engineer Standard	3½″ or 3⅝″ (90 mm)	8″ (200 mm)	2¾″ (70 mm)
Closure Modular	3½″ or 3⅝″ (90 mm)	7½″ or 7⅝″ (190 mm)	3½″ or 3⅝″ (90 mm)
Closure Standard	3½″ or 3⅝″ (90 mm)	8″ (200 mm)	3⅝″ (90 mm)
Roman	3½″ or 3⅝″ (90 mm)	11½″ or 11⅝″ (290 mm)	1⅝″ (40 mm)
Norman	3½″ or 3⅝″ (90 mm)	11½″ or 11⅝″ (290 mm)	2¼″ (57 mm)
Engineer Norman	3½″ or 3⅝″ (90 mm)	11½″ or 11⅝″ (290 mm)	2¾″ to 2 13/16″ (70 mm)
Utility	3½″ or 3⅝″ (90 mm)	11½″ or 11⅝″ (290 mm)	3½″ or 3⅝″ (90 mm)
King Size	3″ (75 mm)	9⅝″ (240 mm)	2⅝ or 2¾″ (70 mm)
Queen Size	3″ (75 mm)	7⅝″ or 8″ (190 mm)	2¾″ (70 mm)

FIGURE 8.11

Dimensions of bricks commonly used in North America, as established by the Brick Industry Association. This list gives an idea of the diversity of sizes and shapes available, and of the difficulty of generalizing about brick dimensions. Modular bricks are dimensioned so that three courses plus mortar joints add up to a vertical dimension of 8 inches (203 mm), and one brick length plus mortar joint has a horizontal dimension of 8 inches (203 mm). The alternative dimensions of each brick are calculated for ⅜″ (9.5-mm) and ½″ (12.7-mm) mortar joint thicknesses.

FIGURE 8.12
Bricks may be custom molded to perform particular functions. This rowlock water table course in an English bond wall was molded to an ogee curve. *(Photo by Edward Allen)*

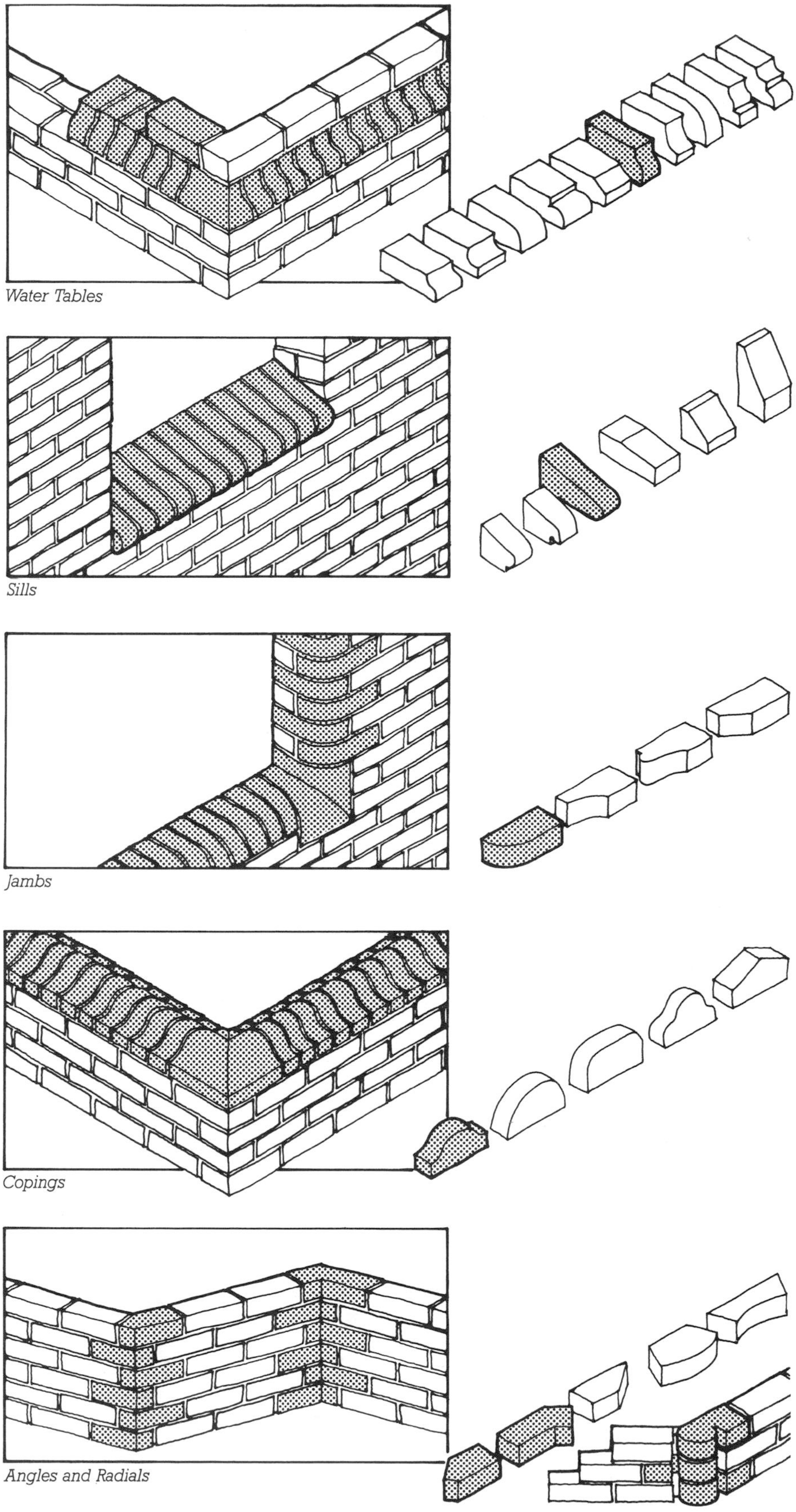

Water Tables

Sills

Jambs

Copings

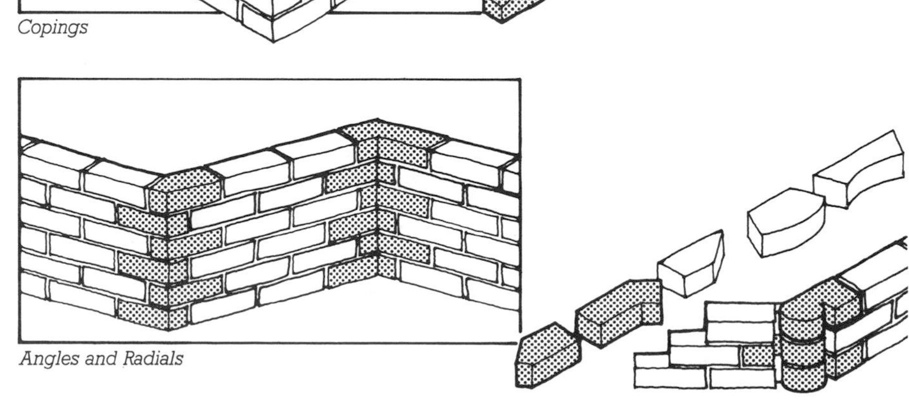

FIGURE 8.13
Some commonly used custom brick shapes. Notice that each water table, jamb, and coping brick shape requires special inside and outside corner bricks as well as the basic rowlock or header brick. The angle bricks are needed to make neat corners in walls that meet at other than right angles. The hinge brick at the extreme lower right of the drawing is a traditional shape that can be used to make a corner at any desired angle. Radial bricks produce a smoothly curved wall surface of any specified radius. Common shapes not pictured here include voussoirs for any desired shape and size of arch and rounded-edge bricks for stair treads.

Brick Classifications

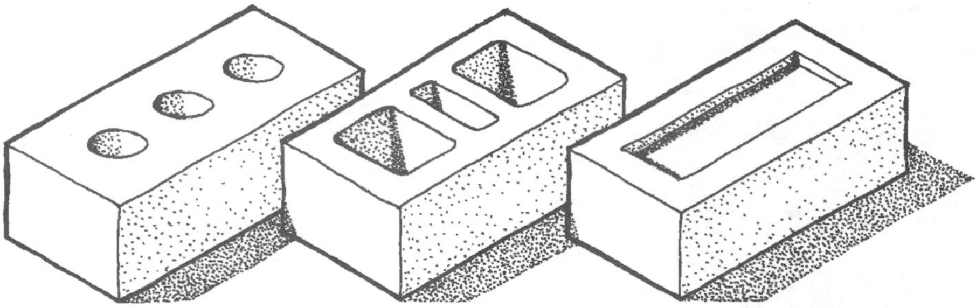

FIGURE 8.14
From left to right: cored, hollow, and frogged bricks. Cored and frogged bricks are considered solid, as long as they remain at least 75 percent solid. A hollow brick may be up to 60 percent void.

The most common bricks used in building construction are classified as facing brick (ASTM C216), building brick (ASTM C62), or hollow brick (ASTM C652). *Facing bricks* (also called *face brick*) are intended for both structural and nonstructural uses where appearance is important. *Building bricks* are used where appearance does not matter, such as in backup wythes of masonry that will be concealed in the finished work. Both facing bricks and building bricks are specified as *solid units.* Solid units may, in fact, be genuinely solid, or despite their name, they may be *cored* or *frogged* (Figure 8.14), as long as any plane measured parallel to the bearing surface of the brick is at least 75 percent solid. By reducing the volume and thickness of the clay, cores and frogs permit more even drying and firing of bricks, reduce fuel costs for firing, reduce shipping costs, and create bricks that are lighter and easier to handle. (Where genuinely solid bricks are required, specifications should call for 100 percent solid, uncored, unfrogged brick.) *Hollow bricks,* defined according to ASTM C652, may be up to 60 percent void and are used primarily to enable the insertion and grouting of steel reinforcing bars in single wythes of brickwork (Figure 8.14).

Paving bricks (ASTM C902) are used for the paving of walks, drives, and patios and must conform to special requirements not only for freeze-thaw resistance, but water absorption and abrasion resistance as well. *Firebricks* (ASTM C64) are used for the lining of fireplaces or furnaces. These are made from special clays, called *fireclays,* which produce bricks with refractory qualities (resistance to very high temperatures). Firebricks are laid in very thin joints of *fireclay mortar.*

Choosing Bricks

We have already considered three important qualities that the designer must consider in choosing the bricks for a particular building: molding process, color, and size. Several other qualities are also important. The ASTM standards for brick establish brick grades based on durability, and for brick that will be exposed to view, types, based on uniformity of shape and size (Figure 8.15). *Brick Grade* establishes minimum requirements for compressive strength and water absorption. The overall durability of the brick and its resistance to weathering can then be related to a map of weathering indices derived from data on winter rainfall and freeze-thaw cycles (Figure 8.16). Grade SW brick is recommended for exterior use in all regions on the map labeled as either severe or moderate weathering, as well as for all brick in contact with the earth. Grade MW brick is recommended for use only above grade, in areas labeled on the map as negligible weathering. Grade NW brick should be used only in exterior locations sheltered from the weather or indoors.

In addition to influencing durability, a brick's compressive strength is also of obvious importance when used in the construction of structural walls and piers. According to ASTM standards, minimum compressive strengths for building bricks and face bricks range, depending on the grade, from 1500 to 3000 pounds per square inch (psi) (10–21 MPa). However, higher-strength bricks are readily available. A compressive strength of 10,000 psi (69 MPa) is not uncommon for bricks used in structural masonry. In high-strength applications, brick strength may exceed 20,000 psi (138 MPa).

The strength of constructed brickwork depends not only on the strength of the bricks but also on the strength of the mortar and, if steel reinforced, on the strength and quantity of reinforcing. Typical working stresses in compression for low-strength, unreinforced brick walls range from 75 to 400 psi (0.52–2.8 MPa). With higher- strength masonry materials and the addition of steel reinforcing, significantly higher working stress values can be achieved.

Brick Type defines limits on the variation in size, distortion in shape, and *chippage* (extent of physical damage to face or visible corners) among brick units (Figure 8.15). Type FBS is considered a general-purpose face brick and is the most common type sold. Type FBX bricks have more stringent limits on appearance characteristics and are intended for masonry work with very thin joints or for bond patterns demanding very close dimensional tolerances. Type FBA bricks are characterized by significant variations in size and shape, as is typical of handmade brick or brick intentionally manufactured for such effect.

Durability of Facing Brick, Building Brick, and Hollow Brick	
Grade SW	Any weathering region, all brick in contact with the earth
Grade MW	Aboveground brick in regions indicated in Figure 8.16 as *Negligible weathering* only
Grade NW	Sheltered or indoor locations only
Appearance Uniformity of Facing Brick	
Type FBX	Least variation in size per unit, least distortion in shape, minimum chippage
Type FBS	General-purpose face brick, with wider variation in shape and size, greater chippage
Type FBA	Great nonuniformity in size and shape permitted, as defined by the manufacturer

Figure 8.15
Brick grade and type. Brick grades classify brick according to their durability and resistance to freeze-thaw action. The grades listed here apply to facing brick, building brick, and hollow brick. Paving brick are graded similarly, but using the designations, in decreasing order of weathering resistance, SX, MX, and NX. Brick types classify brick according to their uniformity of size and shape. The types listed here apply only to facing brick. Hollow brick are manufactured in types designated HBX, HBS, HBB, and HBA, in decreasing order of uniformity, and paving brick are manufactured in types PX, PS, and PA. Building brick, which are not visible in the finished construction, are not classified for appearance.

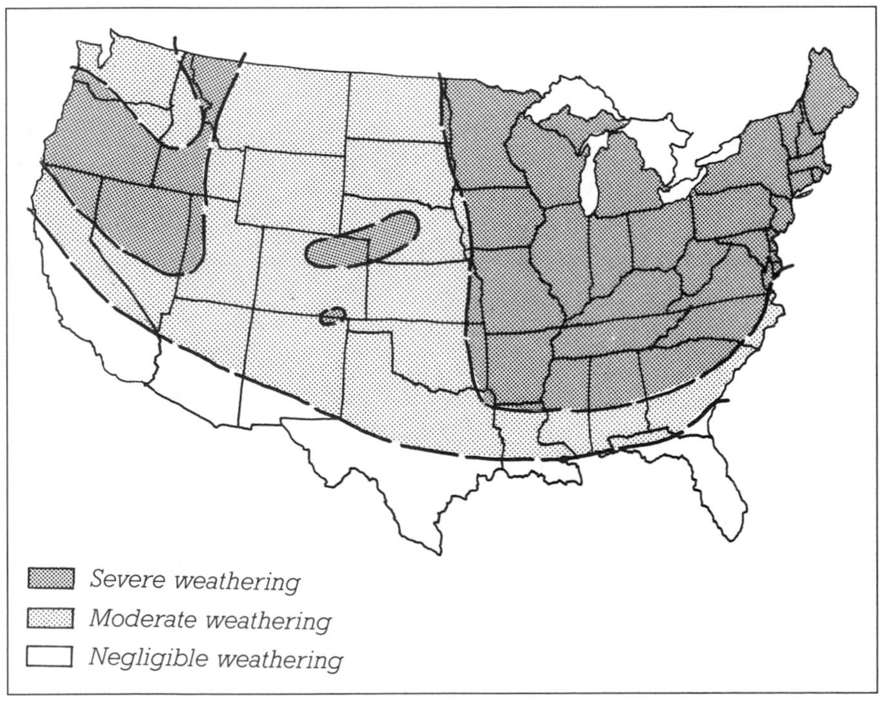

Figure 8.16
Weathering regions of the United States, as determined by winter rainfall and freezing cycles. Grade SW brick is recommended for all brick in contact with the earth and for exterior masonry in all but negligible weathering regions. *(Courtesy of Brick Industry Association)*

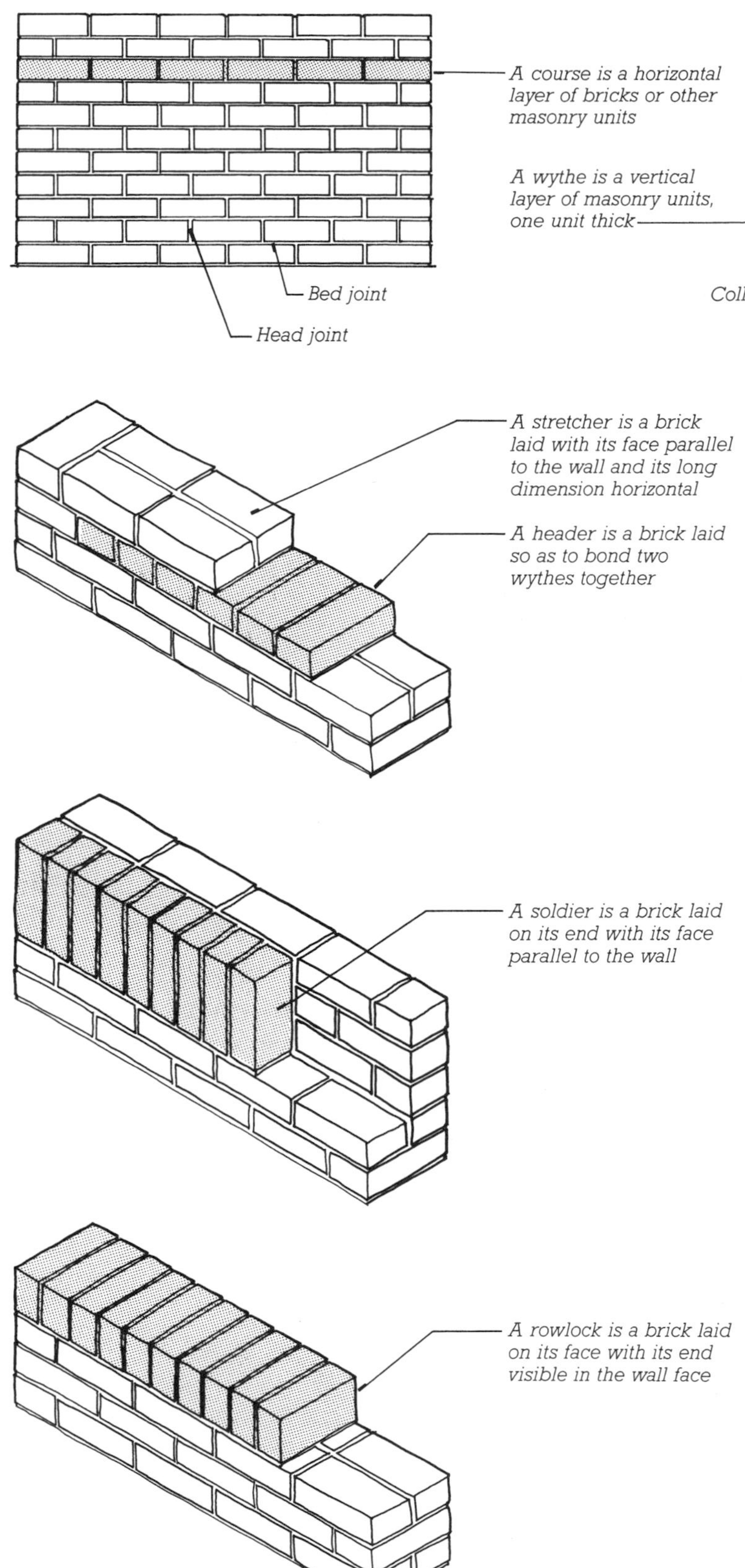
A course is a horizontal layer of bricks or other masonry units
A wythe is a vertical layer of masonry units, one unit thick
Bed joint
Head joint
Collar joint
A stretcher is a brick laid with its face parallel to the wall and its long dimension horizontal
A header is a brick laid so as to bond two wythes together
A soldier is a brick laid on its end with its face parallel to the wall
A rowlock is a brick laid on its face with its end visible in the wall face

Laying Bricks

Figure 8.17 shows the basic vocabulary of bricklaying. Bricks are laid in various positions for visual reasons, structural reasons, or both. The simplest brick wall is a single *wythe* of *stretcher courses.* For walls two or more wythes thick, *headers* are used to bond the wythes together into a structural unit. *Rowlock courses* are often used for caps on garden walls and for sloping sills under windows, although such

FIGURE 8.17
Basic brickwork terminology.

caps and sills are not durable in severe climates. Architects frequently employ *soldier courses* for visual emphasis in such locations as window lintels or tops of walls.

The problem of bonding multiple wythes of brick has been solved in many ways in different regions of the world, often resulting in surface patterns that are particularly pleasing to the eye. Figures 8.18 and 8.19 show some *structural bonds* for brickwork, among which *common bond, Flemish bond,* and *English bond* are the most popular. On the exterior of buildings the *cavity wall,* with its single outside wythe, offers the designer little excuse to use anything but *running bond.* Inside a building, safely out of the weather, one may use solid brick walls in any desired bond. For fireplaces and other very small brick constructions, however, it is often difficult to create a long enough stretch of unbroken wall to justify the use of bonded brickwork.

Figure 8.18
Frequently used structural bonds for brick walls. Partial closer bricks are necessary at the corners to make the header courses come out even while avoiding alignments of head joints in successive courses. The mason cuts the closers to length with a mason's hammer or diamond saw.

Running Bond consists entirely of stretchers

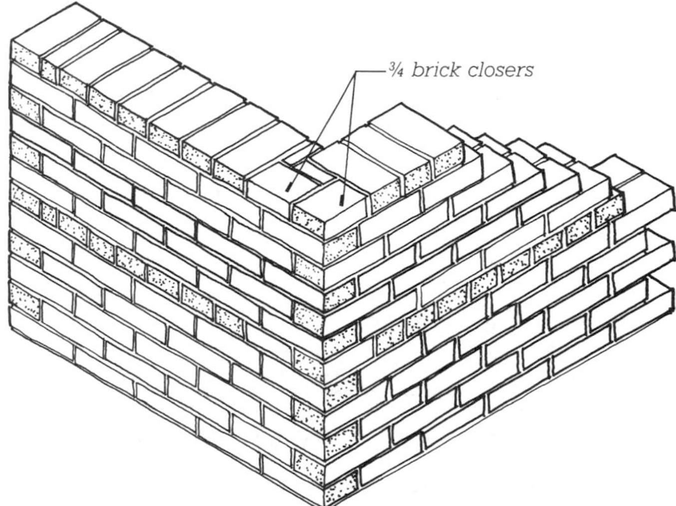

Common Bond (also known as American Bond) has a header course every sixth course. Notice how the head joints are aligned between the header and stretcher courses

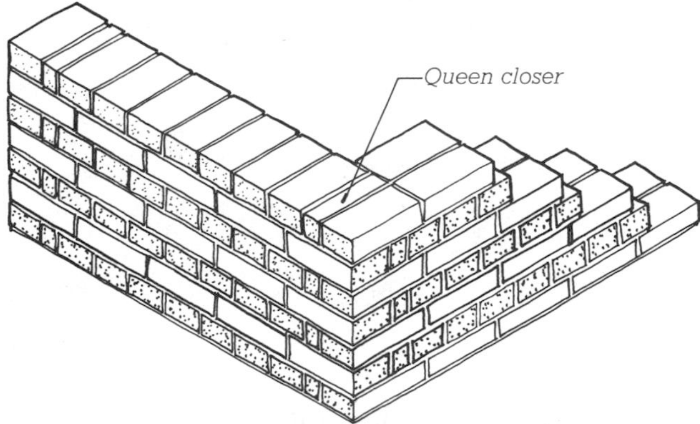

English Bond alternates courses of headers and stretchers

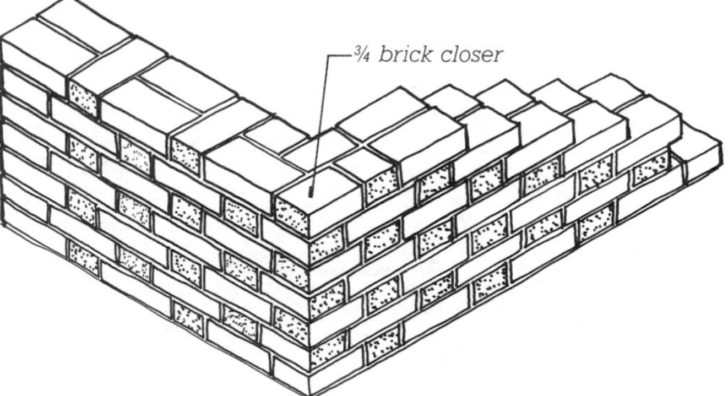

Flemish Bond alternates headers and stretchers in each course

(a) Running bond
(b) common bond
(c) english garden wall bond w/ flemish header courses
(d) English bond
(e) flemish bond
(f) monk bond

FIGURE 8.19
Photographs of some brick bonds. *(a)* Running bond, *(b)* common bond, and *(c)* English garden wall bond with Flemish header courses. In the right column, *(d)* English bond, *(e)* Flemish bond, and *(f)* monk bond, which is a Flemish bond with two stretchers instead of one between headers. The running bond example shown here is from the late 18th century. Its extremely thin joints require mortar made with very fine sand. Notice in the common bond wall (dating from the 1920s in this case) that the header course began to fall out of alignment with the stretcher courses, so the mason inserted a partial stretcher to make up the difference; such small variations in workmanship account for some of the visual appeal of brick walls. Flemish header courses, such as those used in the English garden wall bond, are often used with bricks whose length, including mortar joint, is substantially more than twice their width; the Flemish header course avoids the thick joints between headers that would otherwise result. The Flemish bond example is modern and is composed of modular sand-mold bricks. The monk bond shown here has unusually thick bed joints, approximately 3/4″ (19 mm) high; these joints are difficult for the mason to lay unless the consistency of the mortar is very closely controlled. *(Photos by Edward Allen)*

The process of bricklaying is summarized in Figures 8.20 and 8.21. While conceptually simple, bricklaying requires both extreme care and considerable experience to produce a satisfactory result, especially where a number of bricklayers working side by side must produce identical work on a major structure. Yet speed is essential to the economy of masonry construction. The work of a skilled mason is impressive both for its speed and for its quality. This level of expertise takes time and hard work to acquire, which is why the apprenticeship period for brickmasons is both long and demanding.

The laying of *leads* (pronounced "leeds") is relatively labor intensive. A mason's rule or a *story pole* that is marked with the course heights is used to establish accurate course heights in the leads. The work is checked frequently with a spirit level to ensure that surfaces are flat and plumb and courses are level. When the leads have been completed, a mason's line (a heavy string) is stretched between the leads, using L-shaped *line blocks* at each end to locate the end of the line precisely at the top of each course of bricks.

The laying of the infill bricks between the leads is much faster and easier because the mason needs only a trowel in one hand and a brick in the other to *lay to the line* and create a perfect wall. It follows that leads are expensive compared to the wall surfaces between them, so where economy is important, the designer should seek to minimize the number of corners in a brick structure.

Bricks may be cut as needed, either with sharp, well-directed blows of the chisel-pointed end of a mason's hammer or, for greater accuracy and more intricate shapes, with a power saw that utilizes a water-cooled diamond blade (Figure 9.25). Cutting of bricks slows the process of bricklaying considerably, however, and ordinary brick walls should be dimensioned to minimize cutting (Figure 8.22).

I remember the masons on my first house. I was not much older than the apprentice whom I found choking back tears of frustration with the clumsiness of his work and the rebukes of his boss.

Nearly thirty years later, we still collaborate on sometimes difficult masonry walls, fireplaces and paving patterns. I work with bricklayers...whose years of learning their craft paralleled my years of trying to understand my profession. We are friends and we talk about our work like pilgrims on a journey to the same destination.

Henry Klein, architect, at his acceptance of the Louis Sullivant Award for Architecture, 1981, quoted in *Blueprints Magazine,* 1985.

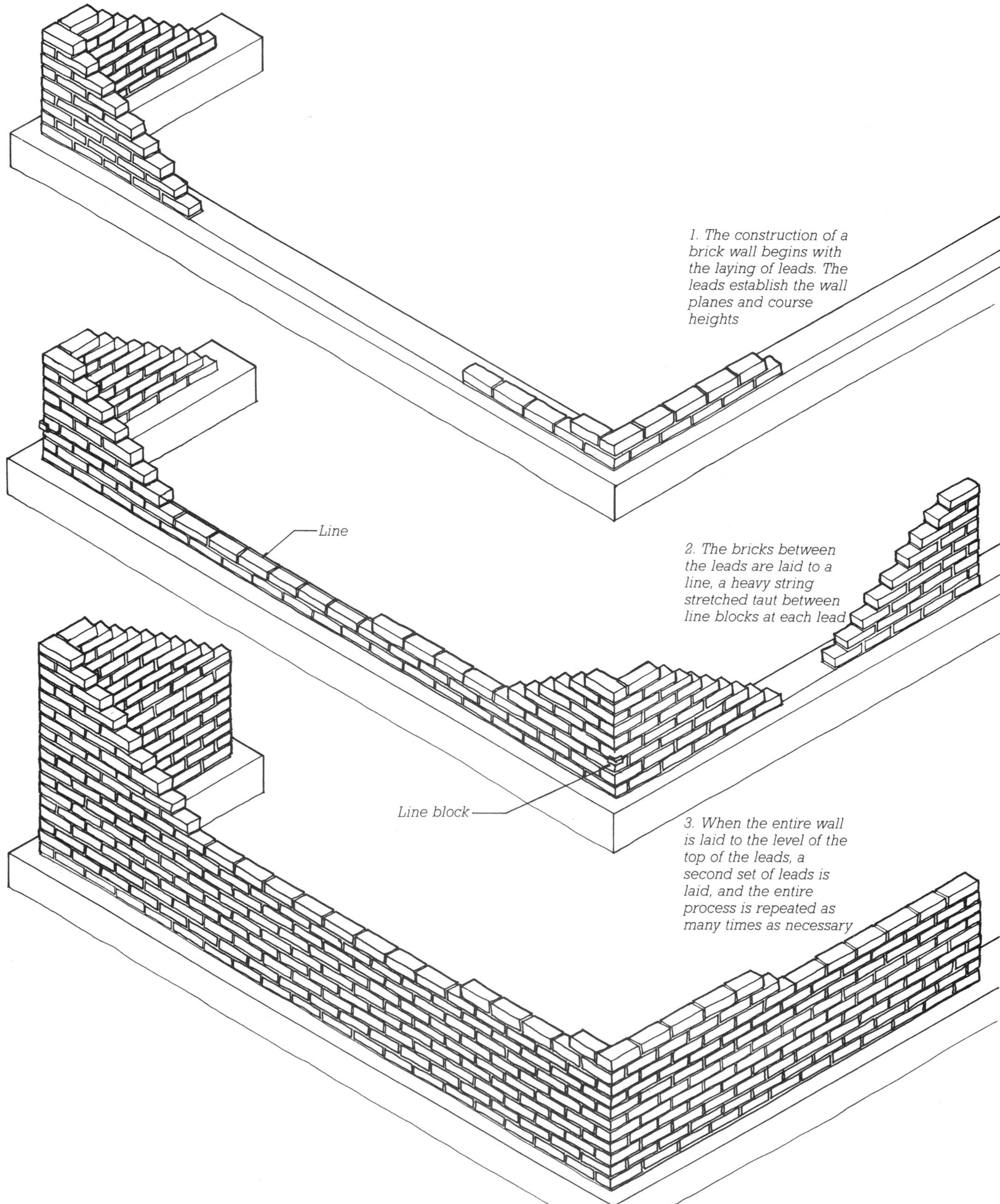

1. The construction of a brick wall begins with the laying of leads. The leads establish the wall planes and course heights
Line
2. The bricks between the leads are laid to a line, a heavy string stretched taut between line blocks at each lead
Line block
3. When the entire wall is laid to the level of the top of the leads, a second set of leads is laid, and the entire process is repeated as many times as necessary

FIGURE 8.20
The procedure for building brick walls. This example is a single wythe of running bond.

(a) (b) (c) (d) (e) (f)

FIGURE 8.21
Laying a brick wall. ***(a)*** **The first course of bricks for a lead is bedded in mortar, following a line marked on the foundation.** ***(b, c, d)*** **As each lead is built higher, the mason uses a spirit level to make sure that each course is level, straight, plumb, and in the same plane as the rest of the lead. A mason's rule or a story pole is also used to check the heights of the courses.** ***(e)*** **A finished lead.** ***(f)*** **A mason lays brick to a line stretched between two leads. When laying to a line, there is no need to use a level or rule.** *(Courtesy of International Masonry Institute, Washington, DC)*

Mortar joints can vary in thickness from about 1/4 inch (6 mm) to more than 1/2 inch (13 mm). Thin joints work only when the bricks are identical to one another within very small tolerances and the mortar is made with a fine sand. Very thick joints require a stiff mortar that is difficult to work with. Mortar joints are usually standardized at 3/8 inch (10 mm), which is easy for the mason and allows for considerable distortion and unevenness in the bricks. One-half inch (13 mm) joints are also common.

The joints in brickwork are *tooled* an hour or two after laying as the mortar begins to harden, to give a neat appearance and to compact the mortar into a profile that meets the visual and weather-resistive requirements of the wall (Figures 8.23 and 8.24). Outdoors, the *vee joint* and *concave joint* shed water and resist freeze-thaw damage better than the others. Indoors, a *raked* or *stripped joint* can be used if desired to accentuate the pattern of bricks in the wall and deemphasize the mortar.

After joint tooling, the face of the brick wall is swept with a soft brush to remove the dry crumbs of mortar left by the tooling process. If the mason has worked cleanly, the wall is now finished, but most brick walls are later given a final cleaning by scrubbing with *muriatic acid* (hydrochloric acid) and rinsing with water to remove mortar stains from the faces of the bricks. Light-colored bricks can be stained by acids and should be cleaned by other means.

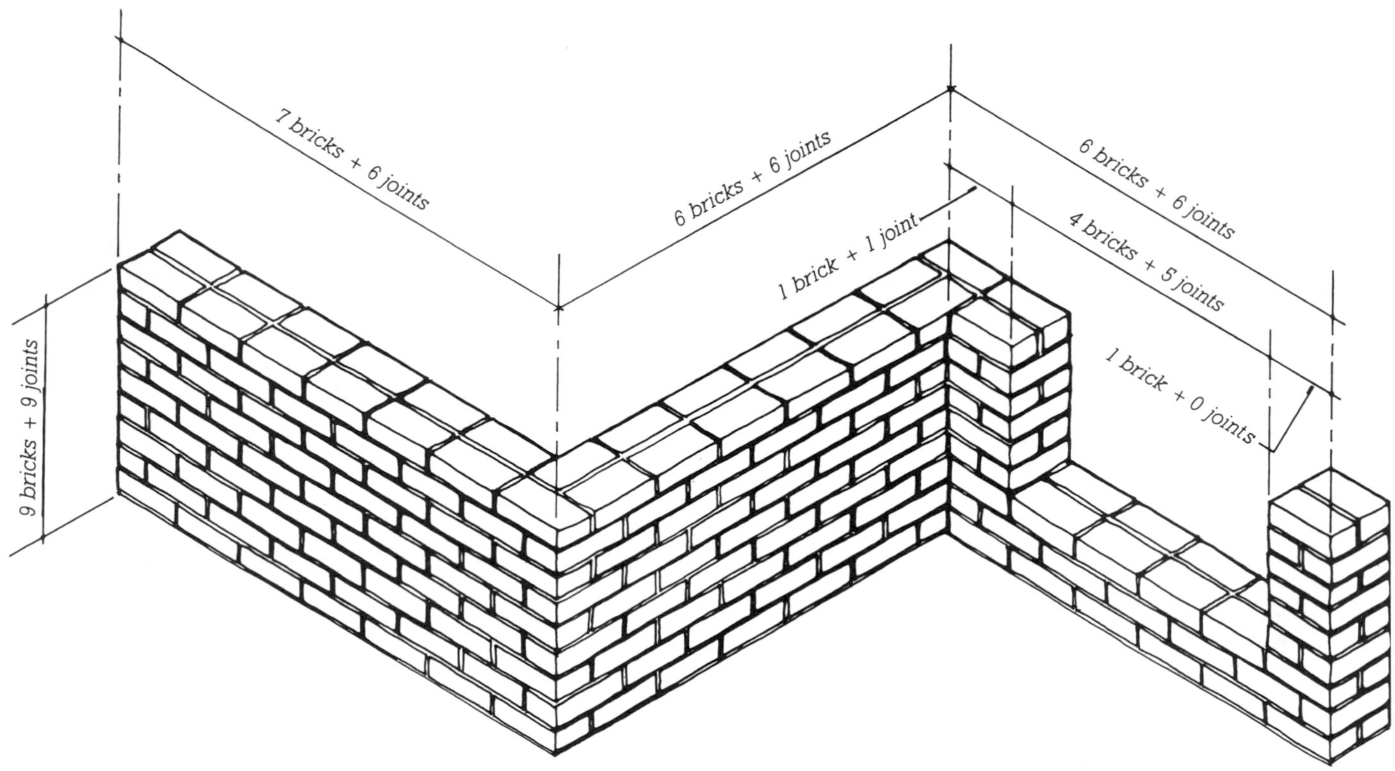

FIGURE 8.22
Dimensions for brick buildings are worked out in advance by the architect based on the actual dimensions of the bricks and mortar joints to be used in the building. Bricks and mortar joints are carefully counted and converted to numerical dimensions for each portion of the wall.

FIGURE 8.23
***(a)* Tooling horizontal joints to a concave profile. *(b)* Tooling vertical joints to a concave profile. The excess mortar squeezed out of the joints by the tooling process will be swept off with a brush, leaving a finished wall. *(c)* Raking joints with a common nail held in a skate-wheel joint raker. The head of the nail digs out the mortar to a preset depth.** *(Courtesy of Brick Industry Association)*

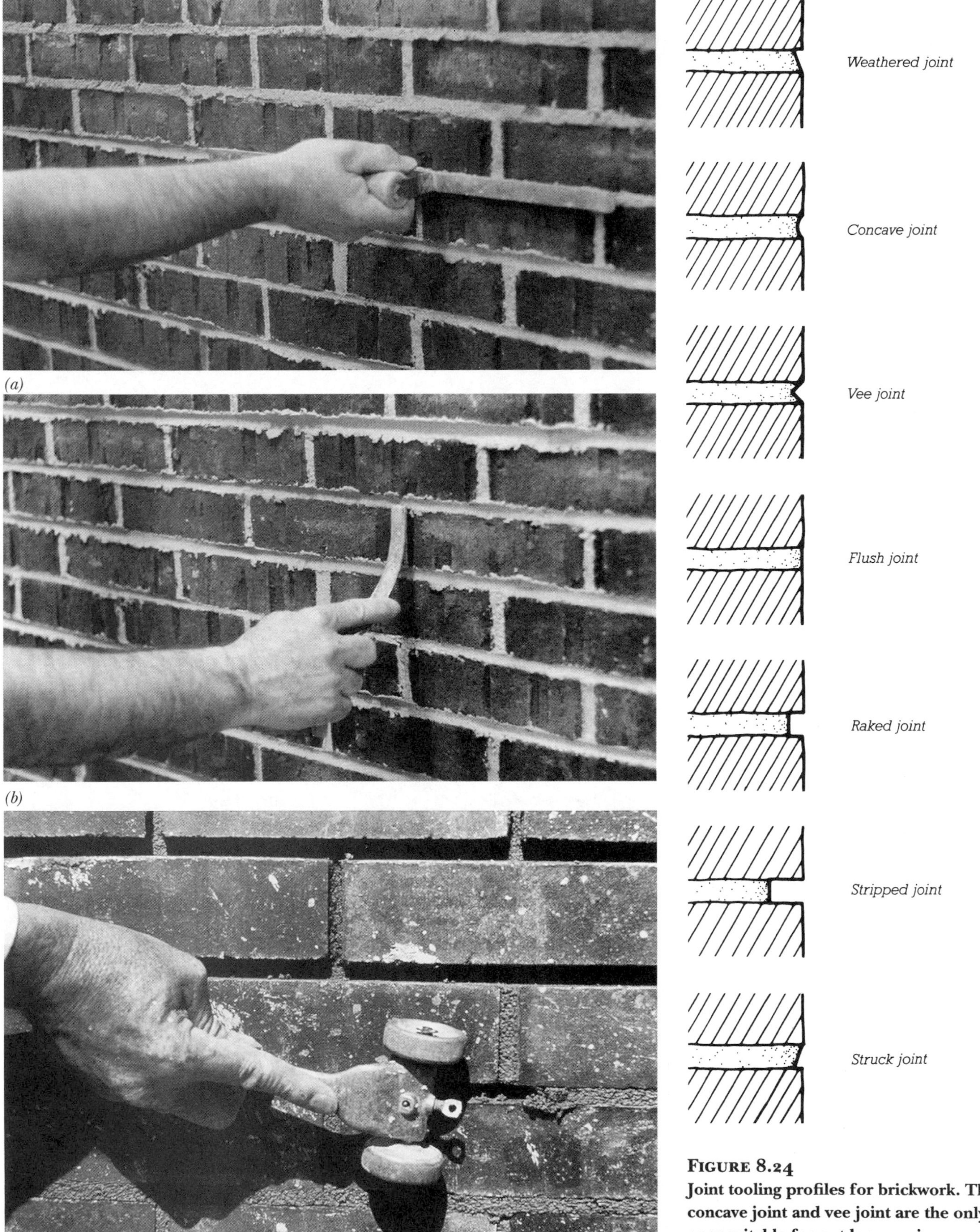

FIGURE 8.24
Joint tooling profiles for brickwork. The concave joint and vee joint are the only ones suitable for outdoor use in severe climates.

Spanning Openings in Brick Walls

Brick walls must be supported above openings for windows and doors. *Lintels* of reinforced concrete, reinforced brick, or steel angles (Figures 8.25 and 8.26) are all equally satisfactory from a technical standpoint. The near invisibility of the steel lintel is a source of delight to some designers but dissatisfies those who prefer that a building express visually its means of support. Wood is no longer used for lintels because of its tendency to burn, to decay, and to shrink and allow the masonry above to settle and crack.

The *corbel* is an ancient structural device of limited spanning capability, one that may be used for small openings in brick walls, for beam brackets, and for ornament (Figures 8.27–8.29). A good rule of thumb for designing corbels is that the projection of each course should not exceed half the course height; this results in a corbel angle of about 60 degrees to the horizontal and minimizes flexural stress in the bricks.

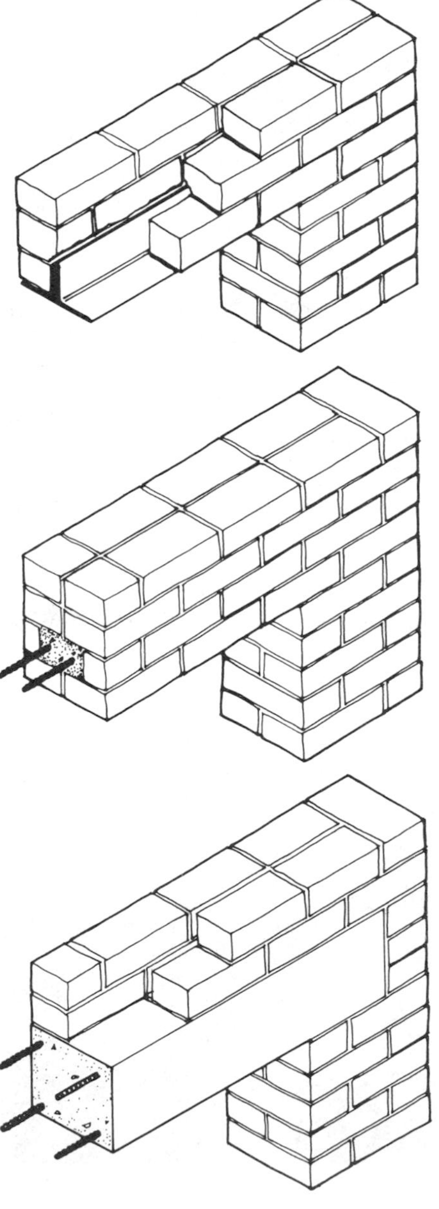

Figure 8.25
Three types of lintels for spanning openings in brick walls. The double-angle steel lintel (top) is scarcely visible in the finished wall. The reinforced brick lintel (center) works in the same manner as a reinforced concrete beam and gives no visible clues as to what supports the bricks over the opening. The precast reinforced concrete lintel (bottom) is clearly visible. For short spans, cut stone lintels without reinforcing can be used in the same manner as concrete lintels.

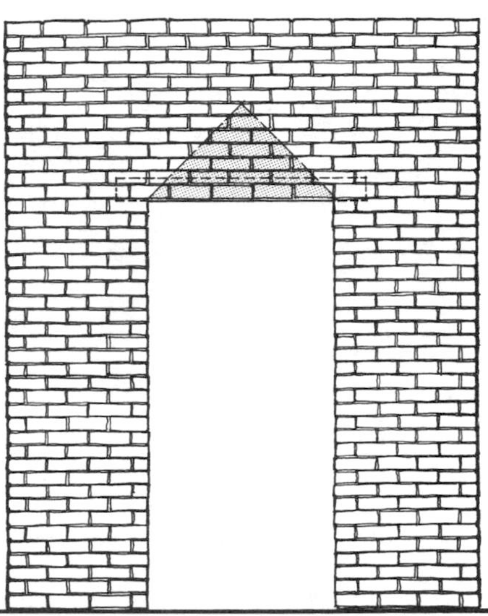

Figure 8.26
Because of corbelling and arching action in the bricks, a lintel is considered to carry only the triangular area of brickwork indicated by the shaded portion of this drawing. The broken line indicates a concealed steel angle lintel.

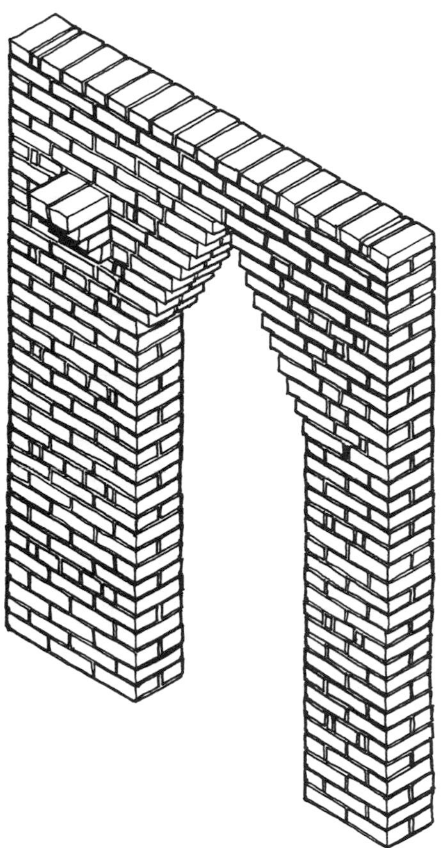

FIGURE 8.27
Corbelling has many uses in masonry construction. It is used in this example both to span a door opening and to create a bracket to support a beam.

FIGURE 8.28
Corbelling creates a transition from the cylindrical tower to a hexagonal roof. Cut limestone is used for window sills, lintels, arch intersections, and grotesquely carved rainwater spouts. The building is the Gothic cathedral in Albi, France. *(Photo by Edward Allen)*

FIGURE 8.29
All the skills of the 19th-century mason were called into play to create the corbels and arches of this brick cornice in Boston's Back Bay. *(Photo by Edward Allen)*

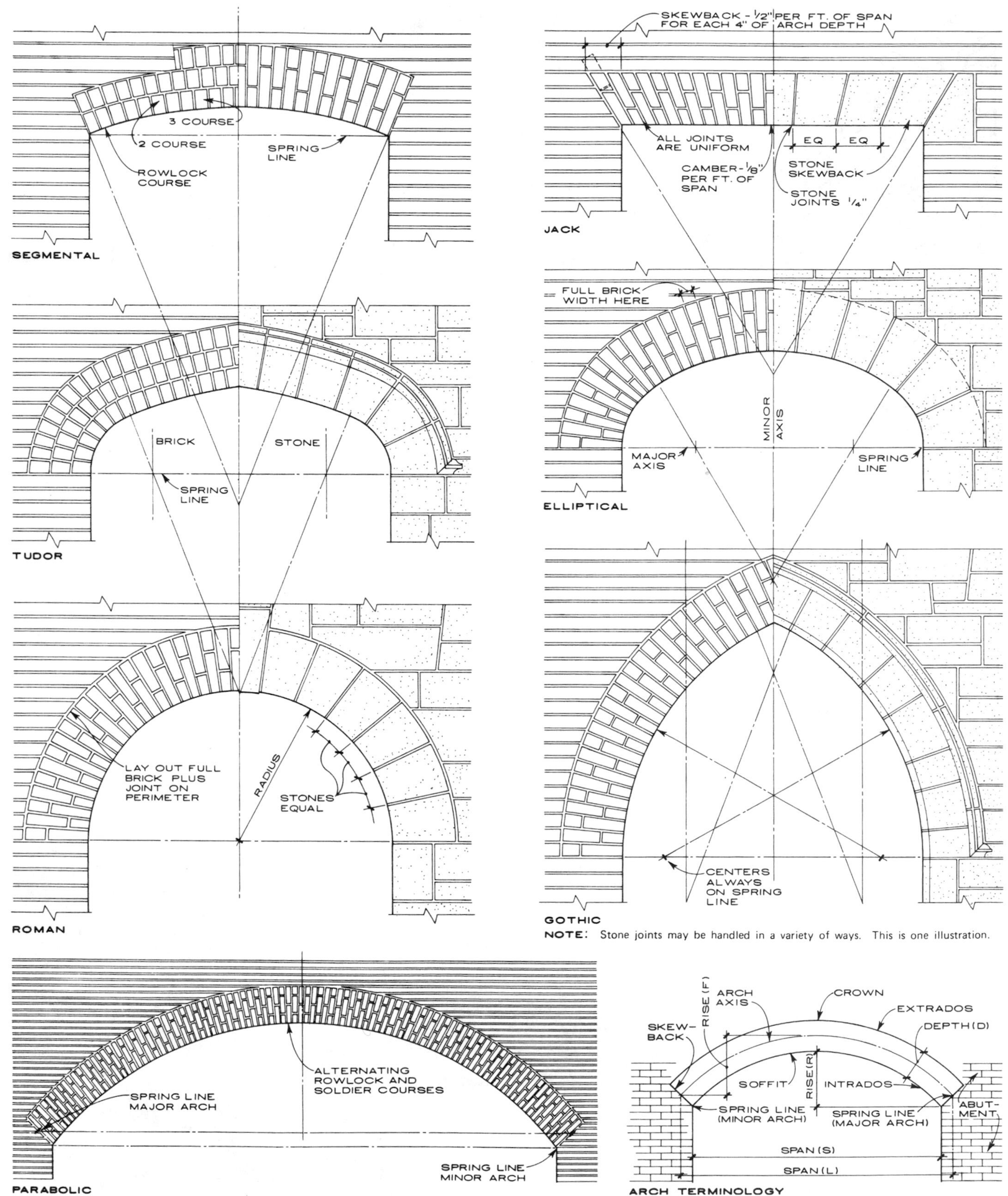

NOTE: Stone joints may be handled in a variety of ways. This is one illustration.

FIGURE 8.30
Arch forms and arch terminology in brick and cut stone. The spandrel is the area of wall that is bounded by the extrados of the arch. *(Reprinted by permission of John Wiley & Sons, Inc., from Ramsey/Sleeper, Architectural Graphic Standards (7th ed.), Robert T. Packard, A.I.A., Editor, copyright 1981 by John Wiley & Sons, Inc.)*

The brick *arch* is a structural form so widely used and so powerful, both structurally and symbolically, that entire books have been devoted to it (Figure 8.30). Given a *centering* of wood or steel (Figure 8.31), a mason can lay a brick arch very rapidly, although the *spandrel*, the area of flat wall that adjoins the arch, is slow to construct because many of its bricks must be cut to fit. In an arch of *gauged brick*, each brick is rubbed

(a)

(b)

(c)

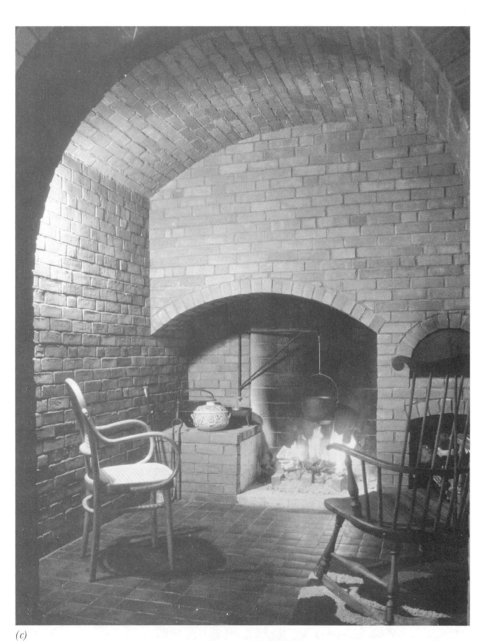

FIGURE 8.31
***(a)* Two rough brick arches under construction, each on its wooden centering. *(b)* The brick locations were marked on the centering in advance to be sure that no partial bricks or unusual mortar joint thicknesses will be required to close the arch. This was done by laying the centering on its side on the floor and placing bricks around it, adjusting their positions by trial and error to achieve a uniform spacing. Then the location of each brick was marked with pencil on the curved surface of the centering. *(c)* The brick arches whose construction is illustrated in the previous two photographs span a fireplace room that is roofed with a brick barrel vault. The firebox is lined with firebrick and the floor is finished with quarry tiles.** *(Photos by Edward Allen)*

to the required wedge shape on an abrasive stone, which is laborious and expensive. The *rough arch*, which depends on wedge-shaped mortar joints for its curvature, is therefore much more usual in today's buildings (Figures 8.32 and 8.33). A number of brick manufacturers will mold to order sets of tapered bricks for arches of any shape and span.

An arch translated along a line perpendicular to its plane produces a *barrel vault*. An arch rotated about its vertical centerline becomes a *dome*. From various intersections of these two basic roof shapes comes the infinite vocabulary of vaulted masonry construction. Brick vaults and domes, if their lateral thrusts are sufficiently tied or *buttressed*, are strong, stable forms. In parts of the world where labor is inexpensive, they continue to be built on an everyday basis (Figures 8.34 and 8.46). In North America and most of Europe, where labor is more costly, they have been replaced almost entirely by less expensive, more compact spanning elements such as beams and slabs of wood, steel, or concrete.

FIGURE 8.32
A deep semicircular brick arch frames a view of a brick arcade and the harbor beyond at the Federal Courthouse in Boston, Massachusetts. *(Architects: Pei, Cobb, Freed and Partners. Photo © 1998 Steve Rosenthal)*

FIGURE 8.33
A rough *jack arch* (also called a *flat arch*) in a wall of Flemish bond brickwork.
(Photo by Edward Allen)

(a)

(b)

FIGURE 8.34
(a) **Masons in Mauritania, drawing on thousands of years of experience in masonry vaulting, build a dome for a patient room of a new hospital. The masonry is self-supporting throughout the process of construction; only a simple radius guide is used to maintain a constant diameter. The dome is double with an airspace between to insulate the room from the sun's heat.** ***(b)*** **The walls are buttressed with stack bond brick headers to resist the outward thrust of the domes.** *(Courtesy of ADAUA, Geneva, Switzerland)*

Reinforced Brick Masonry

Reinforced brick masonry (RBM) is analogous to reinforced concrete construction. The same deformed steel reinforcing bars used in concrete are placed in thickened collar joints to strengthen a brick wall or lintel. A reinforced brick wall (Figure 8.35) is created by constructing two wythes of brick 2 to 4 inches (50–100 mm) apart, placing the reinforcing steel in the cavity, and filling the cavity with *grout*. Grout is a mixture of portland cement, aggregate, and water. ASTM C476 specifies the proportions and qualities of grout for use in filling masonry loadbearing walls. It is important that grout be fluid enough to flow readily into the narrow cavity and fill it completely. The excess water in the grout that is required to achieve this fluidity is quickly absorbed by the bricks and does not detract from the eventual strength of the grout, as it would from concrete poured into formwork. Highly flowable *self-consolidating grout*, analogous to self-consolidating concrete as described in Chapter 13, can also be used.

There are two methods for grouting reinforced brick walls: low lift and high lift. In *low-lift grouting*, the masonry is constructed to a height not greater than 4 feet (1200 mm) before grouting, taking care to keep the cavity free of mortar squeezeout and droppings, which might interfere with the placement of the reinforcing and grout. The vertical reinforcing bars are inserted into the cavity and are left projecting at least 30 bar diameters above the top of the brickwork to transfer their loads to the steel in the next lift. The cavity is then filled with grout to within 1½ inches (38 mm) of the top, and the process is repeated for the next lift.

In the *high-lift grouting* method, the wall is grouted a story at a time. The cleanliness of the cavity is ensured by temporarily omitting some of the bricks in the lowest course of masonry to create *cleanout holes*. As the bricklaying progresses, the cavity is flushed periodically from above with water to drive debris down and out through the cleanouts. To resist the hydrostatic pressure of the wet grout, the wythes are held together by galvanized steel wire *ties* laid into the bed joints and across the cavity, usually at intervals of 24 inches (600 mm) horizontally and 16 inches (400 mm) vertically. After the cleanouts have been filled with bricks and the mortar has cured for at least 3 days, the reinforcing bars are placed and grout is pumped into the cavity from above in increments not more than 4 feet (1200 mm) high. To minimize pressure on the brickwork, each increment is allowed to harden for an hour or so before the next increment is poured above it.

The low-lift method is generally easier for small work where the grout is poured by hand. Where grout pumping equipment must be rented, the high-lift method is preferred because it minimizes rental costs.

Although unreinforced brick walls are adequate for many structural purposes, RBM walls are much stronger against vertical loads, flexural loads from wind or earth pressure, and shear loads. With RBM, it is possible to build bearing wall buildings to heights formerly possible only with steel and concrete frames, and to do so with surprisingly thin walls (Figures 8.36 and 8.37). RBM is also used for brick piers, which are analogous to concrete columns, and, less commonly, for structural lintels (Figure 8.25), beams, slabs, and retaining walls.

Reinforced brickwork may also be created at a smaller scale by inserting reinforcing bars and grout into the cores of hollow bricks. This technique is especially useful for single-family residential construction and for single-wythe prefabricated curtain wall panels (Chapter 20).

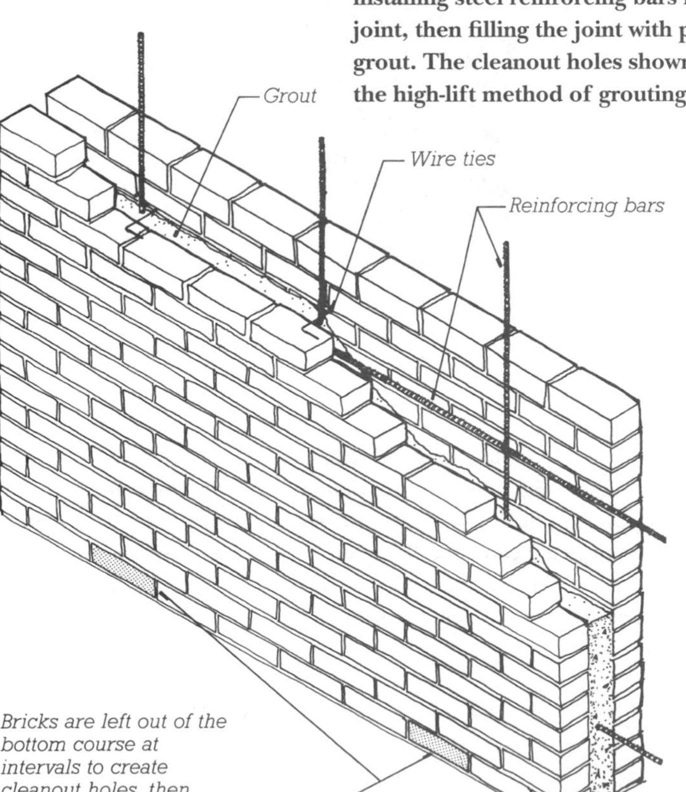
Grout
Wire ties
Reinforcing bars
Bricks are left out of the bottom course at intervals to create cleanout holes, then inserted before grouting

FIGURE 8.35
A reinforced brick loadbearing wall is built by installing steel reinforcing bars in a thickened collar joint, then filling the joint with portland cement grout. The cleanout holes shown here are used in the high-lift method of grouting.

FIGURE 8.36
Twelve-inch (300-mm) reinforced brick walls of constant thickness bear the concrete floor and roof structures of a hotel.
(Photo by Edward Allen)

MASONRY WALL CONSTRUCTION

Bricks, stones, and concrete masonry units are mixed and matched in both loadbearing and nonloadbearing walls of many types. Stone and concrete masonry are discussed in the next chapter, and the more important types of masonry wall constructions are presented in Chapter 10.

FIGURE 8.37
Because loads in a building accumulate from top to bottom, the unreinforced brick walls of the 16-story Monadnock Building, built in Chicago in 1891, are 18 inches (460 mm) thick at the top and 6 feet (1830 mm) thick at the base of the building.
(Architects Burnham and Root. Photo by William T. Barnum. Courtesy of Chicago Historical Society, ICHi-18292)

FIGURE 8.38
Ornamental corbelled brickwork in an 18th-century New England chimney. Step flashings of lead sheet waterproof the junction between the chimney and the wood shingles of the roof. *(Photo by Edward Allen)*

Figure 8.39
In the gardens he designed at the University of Virginia, Thomas Jefferson used unreinforced serpentine walls of brick that are only a single wythe thick. The shape of the wall makes it extremely resistant to overturning despite its thinness. *(Photo by Wayne Andrews)*

FIGURE 8.40
Cylindrical bays of brick with stone lintels front these Boston rowhouses.
(Photo by the Edward Allen)

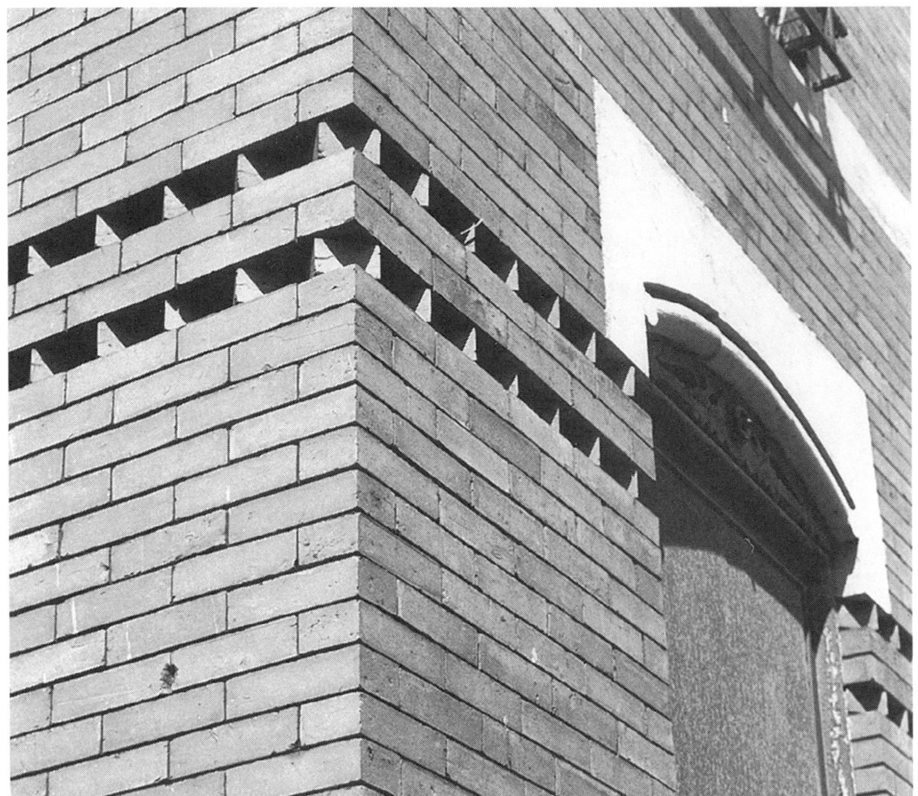

FIGURE 8.41
Bricks were laid diagonally in two of the courses to create this mouse-tooth pattern. The window is spanned with a segmental arch of cut limestone. *(Photo by Edward Allen)*

FIGURE 8.42
***Quoins* originated long ago as cut stone blocks used to form strong corners on walls of weak masonry materials such as mud bricks or round fieldstones. In more recent times, quoins (pronounced "coins") have been used largely for decorative purposes. At the left, cut limestone quoins and a limestone *water table* dress up a common bond brick wall. The mortar joints between quoins are finished in a protruding beaded profile to emphasize the pattern of the stones. At the right, brick quoins are used to make a graceful termination of a concrete masonry wall at a garage door opening. Notice that three brick courses match perfectly one block course.** *(Photos by Edward Allen)*

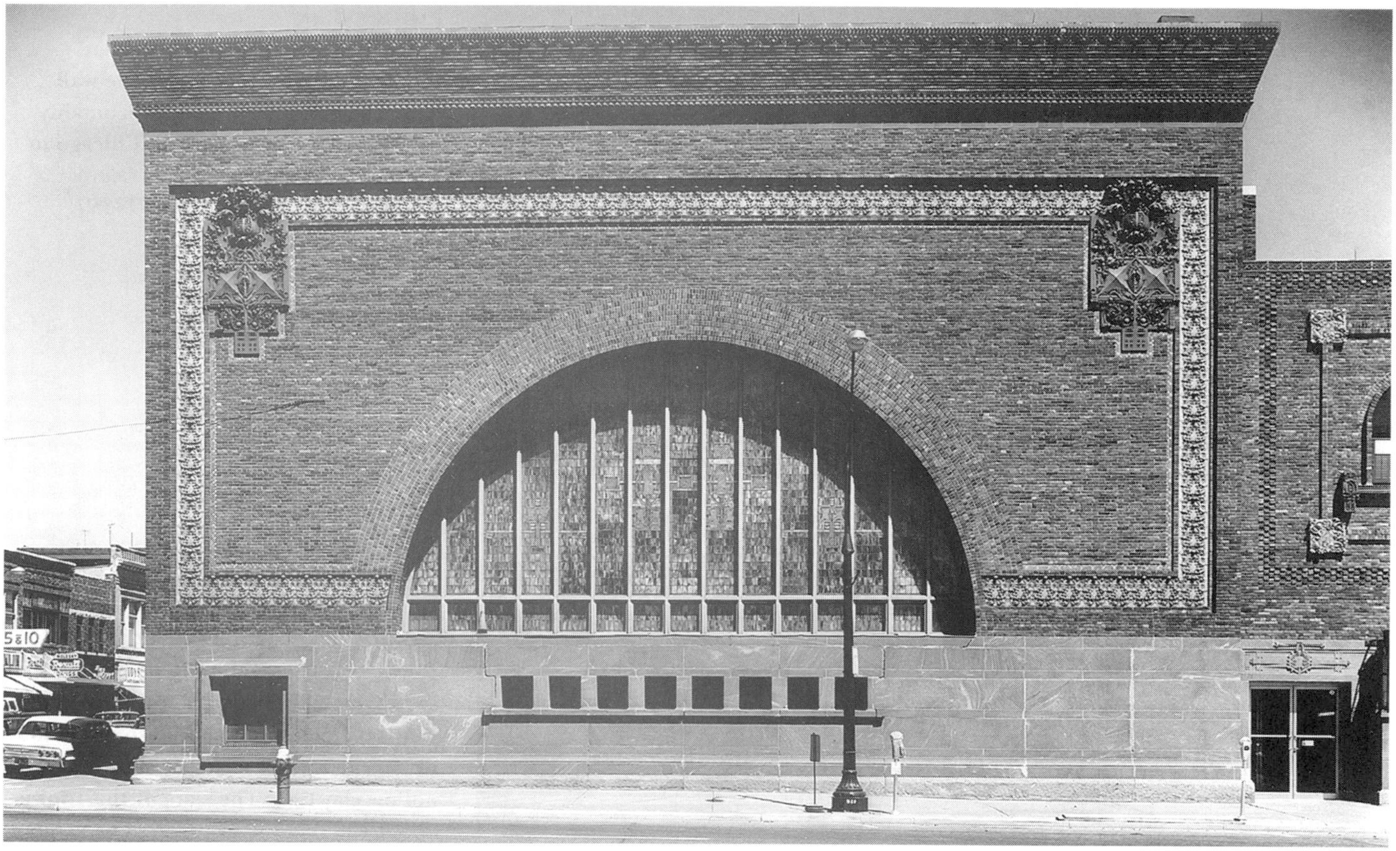

FIGURE 8.43
Louis Sullivan's National Farmers' Bank in Owatonna, Minnesota, completed in 1908, rises from a red sandstone base. Enormous rowlock brick arches span the windows in the two street facades. Bands of glazed terra-cotta ornament in rich blues, greens, and browns outline the walls, and a flaring cornice of corbelled brick and terra cotta caps the building. *(Photo by Wayne Andrews)*

FIGURE 8.44
Architects Pei, Cobb, Freed employed brick half-domes to mark the entrances to courtrooms in the Boston Federal Courthouse. *(Photo © 1998 Steve Rosenthal)*

FIGURE 8.45
Frank Lloyd Wright used long, flat Roman bricks and cut limestone wall copings to emphasize the horizontality of the Robie House, built in Chicago in 1906. *(Photo by Mildred Mead. Courtesy of Chicago Historical Society, IChi-14191)*

FIGURE 8.46
During the second half of the 20th century, Uruguayan engineer Eladio Dieste constructed hundreds of industrial buildings with long-span roof vaults of reinforced clay masonry, either bricks or, like the example shown here, hollow clay tiles. The span of this roof is about 100 feet (30 m). ***(Left)*** **The vaults are long strips with clerestory windows between for daylighting.** ***(Right)*** **Each strip is S-shaped in cross section. This stiffens the vault to prevent buckling and also provides the openings for the windows.** *(Photos by Edward Allen)*

CSI/CSC	
MasterFormat Sections for Brick Masonry	
04 00 00	MASONRY
04 05 00	Common Work Results for Masonry Masonry Grouting Masonry Reinforcing Bars
04 20 00	UNIT MASONRY
04 21 00	Clay Unit Masonry Brick Masonry
04 50 00	REFACTORY MASONRY
05 52 00 04 57 00	Combustion Chamber Masonry Masonry Fireplaces

Selected References

1. Beall, Christine. *Masonry Design and Detailing for Architects, Engineers, and Builders* (5th ed.). New York, McGraw-Hill, 2003.

This 500-page book is an excellent general design reference on brick, stone, and concrete masonry.

2. Brick Industry Association. *BIA Technical Notes of Brick Construction.* McLean, VA, various dates.

This collection of more than 50 bulletins is available in ring binder format or as free downloads from the Brick Industry Association's web site. Included is up-to-date information on every aspect of bricks and brick masonry.

3. Brick Industry Association. *Principles of Brick Masonry.* Reston, VA, 1989.

The 70 pages of this booklet present a complete curriculum in clay masonry construction for the student of building construction.

Web Sites

Brick Masonry

Author's supplementary web site: **www.ianosbackfill.com/08_brick_masonry**
Brick Industry Association: **www.bia.org**
General Shale Brick: **www.generalshale.com**
Glen-Gery Brick: **www.glengerybrick.com**
International Masonry Institute: **www.imiweb.org**

Key Terms and Concepts

mason
masonry unit
trowel
mortar
cement–lime mortar
aggregate
portland cement
lime
quicklime
slaked lime, hydrated lime
blended hydraulic cement
masonry cement
air-entraining admixture
mortar cement
mortar type
Type N mortar
Type S mortar
Type O mortar
Type M mortar
proportion specification
property specification
hydraulic cement
lime mortar
nonhydraulic cement
carbonation
hydration
retemper
extended-life admixture
brick
soft mud process
water-struck brick
sand-struck brick, sand-mold brick
dry-press process
stiff mud process
firing, burning
clamp
clinker brick
periodic kiln
tunnel kiln
water smoking
dehydration
oxidation
vitrification
flashing
modular brick
Utility brick
facing brick, face brick
building brick
solid unit
cored unit
frogged unit
hollow brick
paving brick
firebrick
fireclay
fireclay mortar
brick Grade
brick Type
chippage
wythe
stretcher course
header
rowlock course
soldier course
coursebed joint
head joint
collar joint
stretcher
header
soldier
rowlock
structural bond
common bond
Flemish bond
English bond
cavity wall
running bond
lead
story pole
line block
lay to the line
tooled joint
weathered joint
concave joint
vee joint
flush joint
raked joint
stripped joint
struck joint
muriatic acid
lintel
corbel
arch
centering
spandrel
gauged brick
rough arch
jack arch, flat arch
barrel vault
dome
buttress
reinforced brick masonry (RBM)
grout
self-consolidating grout
low-lift grouting
high-lift grouting
cleanout hole
tie
quoin
water table

Review Questions

1. How many syllables are in the word *masonry*? (Hint: There cannot be more syllables in a word than there are vowels. Many people, even masons and building professionals, mispronounce this word.)

2. What are the most common types of masonry units?

3. What are the molding processes used in manufacturing bricks? How do they differ from one another?

4. List the functions of mortar.

5. What are the ingredients of mortar? What is the function of each ingredient?

6. Why are mortar joints tooled? Which tooling profiles are suitable for a brick wall in a severe climate?

7. What is the function of a structural brick bond such as common or Flemish bond? Draw the three most popular brick bonds from memory.

Exercises

1. What is the exact height of a brick wall that is 44 courses high when three courses of brick plus their three mortar joints are 8 inches (203.2 mm) high?

2. What are the inside dimensions of a window opening in a wall of modular bricks with 3/8 inch (9.5-mm) mortar joints if the opening is 6 1/2 bricks wide and 29 courses high?

3. Obtain sand, hydrated lime, several hundred bricks, and basic bricklaying tools from a masonry supply house. Arrange for a mason to help everyone in your class learn a bit of bricklaying technique. Use lime mortar (hydrated lime, sand, and water), which hardens so slowly that it can be retempered with water and used again and again for many weeks. Lay small walls in several different structural bonds. Make simple wooden centering and construct an arch. Construct a dome about 4 feet (1.2 m) in diameter without using centering, as it is done in Figure 8.34. Dismantle what you build at the end of each day, scrape the bricks clean, stack them neatly for reuse, and retemper the mortar with water, covering it with a sheet of plastic to keep it from drying out before it is used again.

4. Design a brick fireplace for a house that you are designing. Select the size and color of brick and the color of mortar. Proportion the fireplace according to the guidelines in Chapter 7. Dimension the fireplace so that it uses only full and half-bricks. Draw every brick and every mortar joint in each view of the fireplace. Use rowlocks, soldiers, corbels, and arches as desired for visual effect. How will you span the fireplace opening?

9

Stone and Concrete Masonry

Stone offers a wide range of expressive possibilities to the architect. In this detail of a 19th-century church, the columns to the left are made of polished granite and rest on bases of carved limestone. The limestone blocks to the right, squared and dressed by hand, have rough-pointed faces and tooth-axed edges. (*Architects: Cummings and Sears. Photo by Edward Allen*)

Stone masonry and concrete masonry are similar in concept to brick masonry. Both involve the stacking of masonry units in the same mortar that is used for brick masonry. However, there are important differences: Whereas bricks are molded to shape, building stone must be wrested from quarries in rough blocks, then cut and carved to the shapes that we want. We can control the physical and visual properties of bricks to some extent, but we cannot control the properties of stone, so we must learn to select from the bountiful assortment provided by the earth the type and color that we want and to work with it as nature provides it to us. Concrete masonry units, like bricks, are molded to shape and size, and their properties can be closely controlled. Most concrete masonry units, however, are much larger than bricks, and, like stone, they require slightly different techniques for laying.

STONE MASONRY

Types of Building Stone

Building stone is obtained by taking rock from the earth and reducing it to the required shapes and sizes for construction. It is a natural, richly diverse material that can vary greatly in its chemistry, structure, physical properties, and appearance. Geologically, stone can be classified into three types according to how it was formed:

- *Igneous rock* is rock that was deposited in a molten state.
- *Sedimentary rock* is rock that was deposited by the action of water and wind.
- *Metamorphic rock* was formerly either igneous or sedimentary rock. Subsequently, its properties were transformed by heat and pressure.

For commercial purposes, ASTM C119 classifies stone used in building construction into six groups: Granite, Limestone, Quartz-Based Stone, Slate, Marble, and Other.

Granite Group

Granite is the igneous rock most commonly quarried for construction in North America. It is a mosaic of mineral crystals, principally feldspar and quartz (silica), and can be obtained in a range of colors that includes gray, black, pink, red, brown, buff, and green. Granite is nonporous, hard, strong, and durable, the most nearly permanent of building stones, suitable for use in contact with the ground or in locations where it is exposed to severe weathering. Its surface can be finished in any of a number of textures, including a mirrorlike polish. In North America, it is quarried chiefly in the East and the upper Midwest. Various granites are also imported from abroad, chiefly from Brazil, China, India, and Italy. Domestic granites are classified according to whether they are fine-grained, medium-grained, or coarse-grained. Requirements for granite dimension stone are defined in ASTM C615.

Basalt, like granite, is a very dense and durable igneous rock. It is usually found only in a dark gray color, and is one of a group of stones that may be collectively referred to as "black granites." It is generally used in the form of rubble and is seldom machined.

FIGURE 9.1
Austin Hall at Harvard University (1881–1884), designed by Henry Hobson Richardson, is a virtuoso performance in stone masonry. Notice the intricate carving of the yellow Ohio sandstone capitals and arch components. The spandrels above the arches are a mosaic of two colors of Longmeadow sandstone blocks. The depth of the arches is intentionally exaggerated to impart a feeling of massiveness to the wall at the entrance to the building. (*Photo by Steve Rosenthal*)

FIGURE 9.2
The loadbearing walls of the Cistercian Abbey Church in Irving, Texas, are made of 427 rough blocks of Big Spring, Texas, limestone, each 2 by 3 by 6 feet (0.6 × 0.9 × 1.8 m) and weighing about 5000 pounds (2300 kg). The stones were brought directly from the quarry to the site without milling; drill holes are visible in many of the stones. Each stone was bedded in Type S mortar 1 inch (25 mm) thick to allow for irregularities in its horizontal faces. Darker stones are grouped in bands for visual effect. The columns were turned from limestone. Cunningham Architects, designing for a Catholic order that originated 900 years ago in Europe, wanted to build a church that would last for another 900 years. (*Photo by James F. Wilson*)

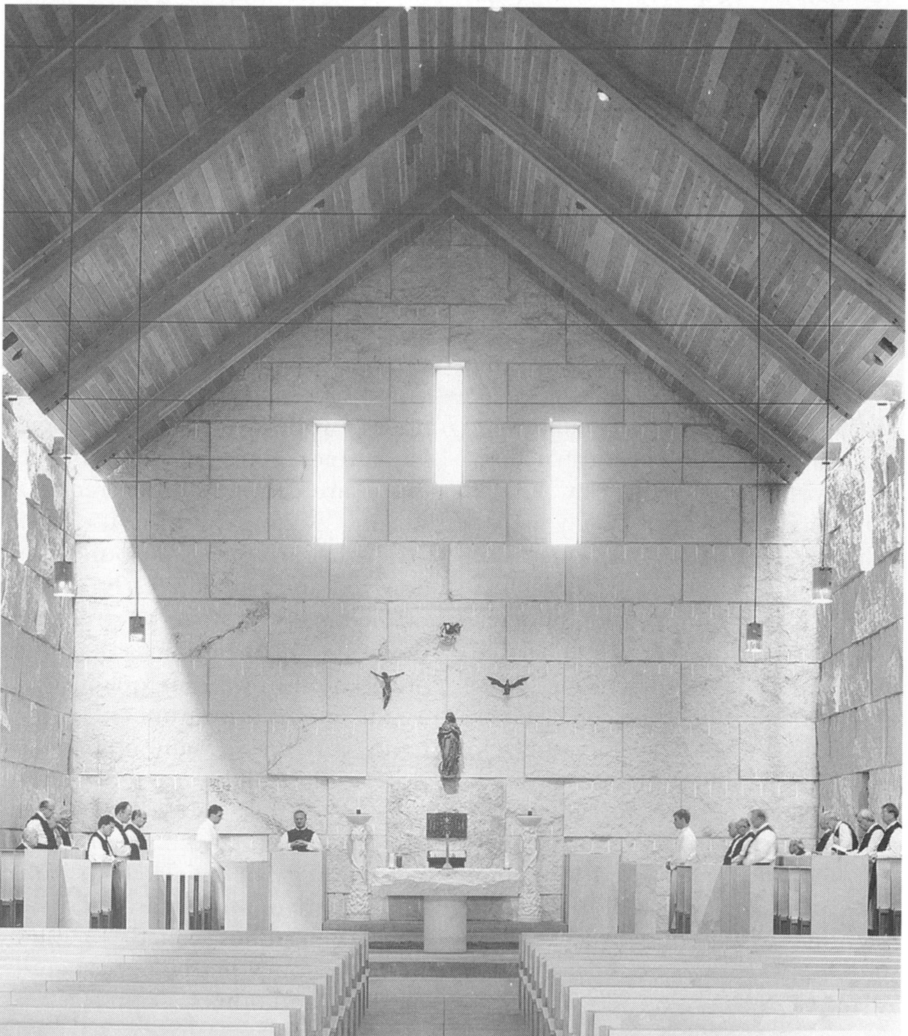

FIGURE 9.3
The heavy timber roof of the Cistercian Abbey Church, flanked by continuous skylights, appears to float above the simple stone volume of the nave. (*Photo by James F. Wilson*)

Limestone Group

Limestone is one of the two principal sedimentary rock types used in construction. It may be found in a strongly stratified form or in deposits that show little stratification (*freestone*). It is quarried throughout North America, with major quarries for large-dimension stone found in Missouri and Indiana. France, Germany, Italy, Spain, Portugal, and Croatia are major suppliers of imported limestone. Requirements for limestone dimension stone are specified in ASTM C568.

Limestone may be composed either of calcium carbonate (*oolitic limestone*) or of a mixture of calcium and magnesium carbonates (*dolomitic limestone*). Both types were formed long ago from the skeletons or shells of marine organisms. Colors range from almost white through gray and buff to iron oxide red. Limestone is porous and contains considerable groundwater (*quarry sap*) when quarried. While still saturated with quarry sap, most limestones are easy to work but are susceptible to frost damage. After seasoning in the air to evaporate the quarry sap, the stone becomes harder and is resistant to frost damage. Some dense limestones can be polished (and may be classified as marbles), but most are produced with varying degrees of surface texture.

According to ASTM C568, limestones are classified as I Low-Density, II Medium-Density, or III High-Density. Higher density classifications generally are stronger and less porous than those with lower densities. All three classifications are suitable for building applications.

The Indiana Limestone Institute classifies limestone into two colors, buff and gray, and four grades. Select grade consists of fine- to average-grained stone with a minimum of natural flaws. Standard-grade stone may include fine- to moderately large-grained stone with an average number of natural flaws. Rustic grade may have fine- to very-coarse-grained stone with an above-average number of natural flaws. The fourth grade, variegated, consists of an unselected mixture of the first three grades and permits both gray- and buff-colored stone.

Quartz-Based Dimension Stone Group

Sandstone is the second major sedimentary rock type used in building construction. Like limestone, it may be found in either a strongly stratified form or as more homogeneous freestone. Sandstone was formed in ancient times from deposits of quartz sand (silicon dioxide). Its color and physical properties vary significantly with the material that cements the sand particles, which may consist of silica, carbonates of lime, or iron oxide. Sandstone is quarried principally in New York, Ohio, and Pennsylvania. Two of its more familiar forms are *brownstone*, widely used in wall construction, and *bluestone*, a highly stratified, durable stone especially suitable for paving and wall copings. It is quarried in the northeastern United States. Sandstone will not accept a high polish. Requirements for quartz-based dimension stone are specified in ASTM C616.

Slate Group

Slate is one of the two metamorphic stone groups utilized in construction. Slate was formed from clay. It is a dense, hard stone with closely spaced planes of cleavage, along which it is easily split into sheets, making it useful for paving stones, roof shingles, and thin wall facings. It is quarried in Vermont, Virginia, New York, Pennsylvania, and parts of Canada, in a variety of colors including black, gray, purple, blue, green, and red. China and India are the largest foreign suppliers of slate to North America, with Italy, the United Kingdom, and Spain also providing significant quantities. Requirements for slate dimension stone are specified in ASTM C629.

Marble Group

Marble is the second of the major metamorphic rock groups. In its true geologic form it is a recrystallized form of limestone. It is easily carved and polished and occurs in white, black, and nearly every color, often with beautiful patterns of veining. The marbles used in North America come chiefly from Alabama, Tennessee, Vermont, Georgia, Missouri, and Canada. Marbles are also imported to this continent from all over the world, primarily from Turkey, Lebanon, Mexico, Italy, China, and France. The physical properties and appearance of marble vary greatly, depending on the chemistry of the original limestone from which it was formed and, even more so, on the processes by which it was metamorphosed. Requirements for marble dimension stone are specified in ASTM C503.

The Marble Institute of America has established a four-step grading system for marbles, in which Group A includes sound marbles and stone with uniform and favorable working qualities. Group B marbles have working qualities somewhat less favorable than those of Group A, and may have some natural faults that require sticking (cementing together) and waxing (filling of voids with cements, shellac, or other materials). Group C marbles may have further variations in working qualities. Geological flaws, voids, veins, and lines of separation are common, often requiring sticking, waxing, and the use of liners, which are pieces of sound stone doweled and cemented to the backs of the sheets of stone to strengthen them. Group D permits a maximum of variation in working qualities and a still larger proportion of natural faults and flaws. Many of the most prized, most highly colored marbles belong to this lowest group. The Marble Group also includes other stones that can take a high polish but that are not true marbles, such as dense limestones, called "limestone marble," "onyx marble," "serpentine marble," and others.

Other Group

The ASTM C119 Other Group includes a variety of less frequently used building stones. *Travertine* is a relatively rare, partially crystallized, and richly patterned calcite (having a chemistry similar to that of limestone) rock deposited by ancient springs. It is marblelike in its physical qualities. Requirements for travertine dimension stone are specified in ASTM C1527. Also included in the ASTM C119 Other Group are alabaster, greenstone, schist, serpentine, and soapstone.

Quarrying and Milling of Stone

The construction industry uses stone in many different forms. *Fieldstone* is rough building stone obtained from riverbeds and rock-strewn fields. *Rubble* consists of irregular quarried fragments that have at least one good face to expose in a wall. *Dimension stone* is stone that has been quarried and cut into rectangular form; large slabs are often referred to as *cut stone*, and small rectangular blocks are called *ashlar*. *Flagstone* consists of thin slabs of stone, either rectangular or irregular in outline, that are used for flooring and paving. Crushed and broken stone are useful in site work as freely draining fill material, as base layers under concrete slabs and pavings, as surfacing materials, and as aggregates in concrete and asphalt. Stone dust and powder are used by landscapers for walks, drives, and mulch.

The maximum sizes and minimum thicknesses of sheets of cut stone vary from one type of stone to another. Granite, the strongest stone, may be utilized in sheets as thin as ⅜ inch (9.5 mm) in some applications. Marble is not generally cut thinner than ¾ inch (19 mm). In both granite and marble, somewhat greater thicknesses than these are advisable in most applications. Limestone, the weakest building stone, is never cut thinner than 2 inches (51 mm), and 3 inches (76 mm) is the preferred thickness for conventionally set stone. In a 6-inch (152-mm) thickness, limestone may be handled in sheets as large as 5 × 18 feet (1.5 × 5.5 m). A Group A marble has a maximum recommended sheet size of 5 × 7 feet (1.5 × 2.1 m).

Dimension stone, whether granite, limestone, sandstone, or marble, is cut from the quarry in blocks. The methods for doing this vary somewhat with the type of stone, and quarrying technology is changing rapidly. The most advanced machines for cutting marble and limestone in the quarry are chain saws and belt saws that utilize diamond blades (Figures 9.4 and 9.5). Granite is much harder than other stones, so it is quarried either by drilling and blasting or by the use of a *jet burner* that combusts fuel oil with compressed air. The hot flame at the end of the jet burner lance induces

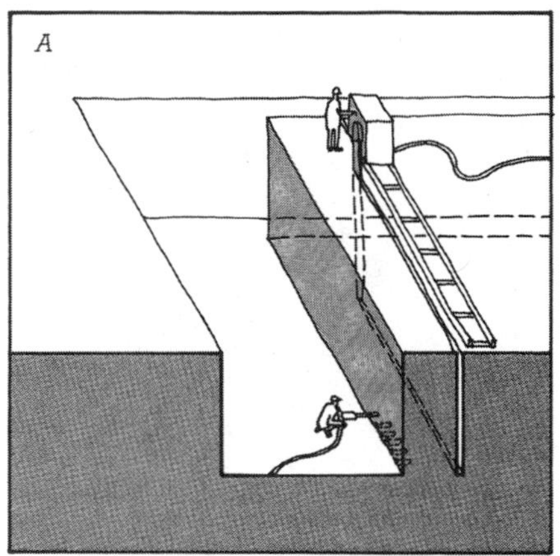

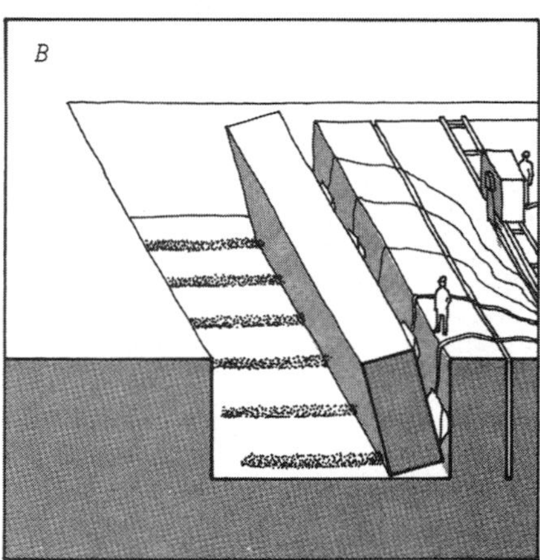

Figure 9.4
A typical procedure for quarrying limestone. (*a*) A diamond belt saw divides the limestone bedrock into long cuts, each about 50 feet (15 m) long and 12 feet (3.6 m) high. Multiple horizontal drill holes create a plane of weakness under the outermost cut. (*b*) Rubber air bags are inflated in the saw kerf to "turn the cut," breaking it free of the bedrock and tipping it over onto a prepared bed of stone chips that cushions its fall. (*c*) Quarry workers use sledgehammers to drive steel wedges into shallow drilled holes to split the cut into blocks. A front end loader removes the blocks and stockpiles them ready for trucking to the mill.

(*a*)

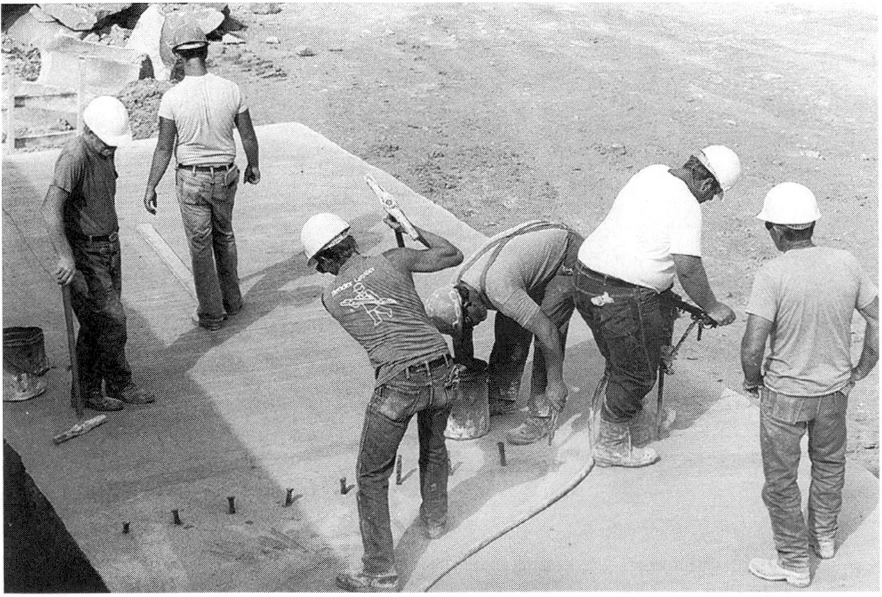

(*b*)

Figure 9.5
(*a*) The long blade of a diamond belt saw, only the shank of which is visible here, cuts limestone full depth in one pass, using water to lubricate the saw and flush away the stone dust. The saw advances automatically on its portable rails at a maximum rate of about 2½ inches (65 mm) per minute. A chain saw is similar to the belt saw shown here, but it uses a chain of linked, rigid teeth, whereas the belt saw uses a narrow, flexible belt of steel-reinforced polyethylene with diamond cutting segments. (*b*) After the cut has been turned, quarry workers split it into transportable blocks. The man at the top of the photograph is laying out the splitting pattern with a measuring tape and straightedge, working from a list of the sizes of blocks that the mill requires for the specific jobs it is working on. At the right, a worker drills shallow holes in the stone. He is followed by a second worker who places steel wedges in the holes and a third who drives the wedges until the cut splits. Each split takes just several minutes to accomplish. (*Photos by the Edward Allen*)

local thermal stress and *spalling* (cracking or flaking of the surface) in the granite. With repeated passes of the lance, the operator gradually excavates a deep, narrow trench to isolate each block of granite from the bedrock. Certain types of granite are quarried successfully with a diamond wire technique, in which a long, taut loop of diamond-studded wire is run at high speed against the stone to make the cut (pages 593–597).

When preparing cut stone for a building, the stone producer works from the architect's drawings to make a set of shop drawings that show the shape and dimension of each individual stone in the building. After these drawings have been checked by the architect, they are used to guide the work of the mill in producing the stones. Rough blocks of stone are selected in the yard, brought into the mill, and sawed into slabs. The slabs may be sawed into smaller pieces, edged, planed flat or to a molding profile, turned on a lathe, or carved, as required. Automated equipment is often used to cut and carve repetitive pieces. After the desired surface finish has been created, holes for *lewises* and anchors are drilled as needed (Figures 9.6, 9.7, and 9.11). Each finished piece of stone is marked to correspond to its position in the building as indicated on the shop drawings before being shipped to the construction site.

His powers continually increased, and he invented ways of hauling the stones up to the very top, where the workmen were obliged to stay all day, once they were up there. Filippo had wineshops and eating places arranged in the cupola to save the long trip down at noon. . . . He supervised the making of the bricks, lifting them out of the ovens with his own hands. He examined the stones for flaws and hastily cut model shapes with his pocket knife in a turnip or in wood to direct the men. . . .

Giorgio Vasari, writing of Filippo Brunelleschi (1377–1446), the architect of the great masonry dome of the Cathedral of Santa Maria del Fiore in Florence, in *The Lives of the Artists*, 1550.

Selecting Stone for Buildings

The selection of building stone is complicated by several factors. Common names for stone do not necessarily correspond to their geologic origins, mineral composition, or physical properties. For example, some stones labeled commercially as marble are devoid of carbonates, the essential mineral group in true marbles, and they differ markedly in their dimensional stability. Commercially applied names for stone often differ regionally as well. In North America, for example, the name "basalt" is applied to a granitelike igneous rock, whereas elsewhere around the globe this term may be used to describe several varieties of sedimentary sandstones or siltstones. Even within one region, the physical characteristics of a particular stone type can vary, sometimes significantly, from one quarry to another, or even within single batches of stone extracted from the same locale.

When stone is obtained from an established source, its properties and fitness for use can usually be extrapolated from past experience. Where a particular stone has a history of successful performance as exterior cladding, for example, it is generally safe to assume that new supplies of this stone from the same source will continue to be suitable for this type of application. However, where stone comes from an unproven source or will be used in a manner for which it does not have a history of past performance, it should be tested in the laboratory to determine its physical properties and verify its suitability for the proposed use. Stone intended for masonry may undergo *petrographic analysis* (microscopic examination of the stone's mineral content and structure), as well as testing for water absorption, density, compressive strength, dimensional stability, frost resistance, and resistance to attack from salt or other chemicals. Stone panels used as external cladding, as discussed in Chapter 20, may also be tested for bending strength, modulus of rupture, thermal expansion and contraction, and anchor capacity (the load capacity of the proposed metal anchor system). Examples of physical property requirements for some common building stone types are listed in Figure 9.8.

The stone industry is international in scope and operation. Architects and building owners have tended to select stone primarily on the basis of appearance, durability, and cost, often with little regard to national origin. The most technically advanced machinery for stoneworking, used all over the world, is designed and manufactured in Italy and Germany. A number of Italian companies have earned reputations for cutting and finishing stone to a very high standard at a reasonable cost. As a result of these factors and the low cost of ocean freight relative to the value of stone, it has been common to choose, for example, a granite that is quarried in Finland, to have the quarry blocks shipped to Italy for cutting and finishing, and to have the finished stone shipped to a North American port, where it is transferred to railcars or trucks for delivery to the building site. Furthermore, because of the unique character of many domestic stones, both U.S. and Canadian quarries and mills ship millions of tons of stone to foreign countries each year, in addition to the even larger amount that is quarried, milled, and erected at home. With a heightened awareness of sustainability factors, however, there is an increasing emphasis on the use of local stone and local stone mills.

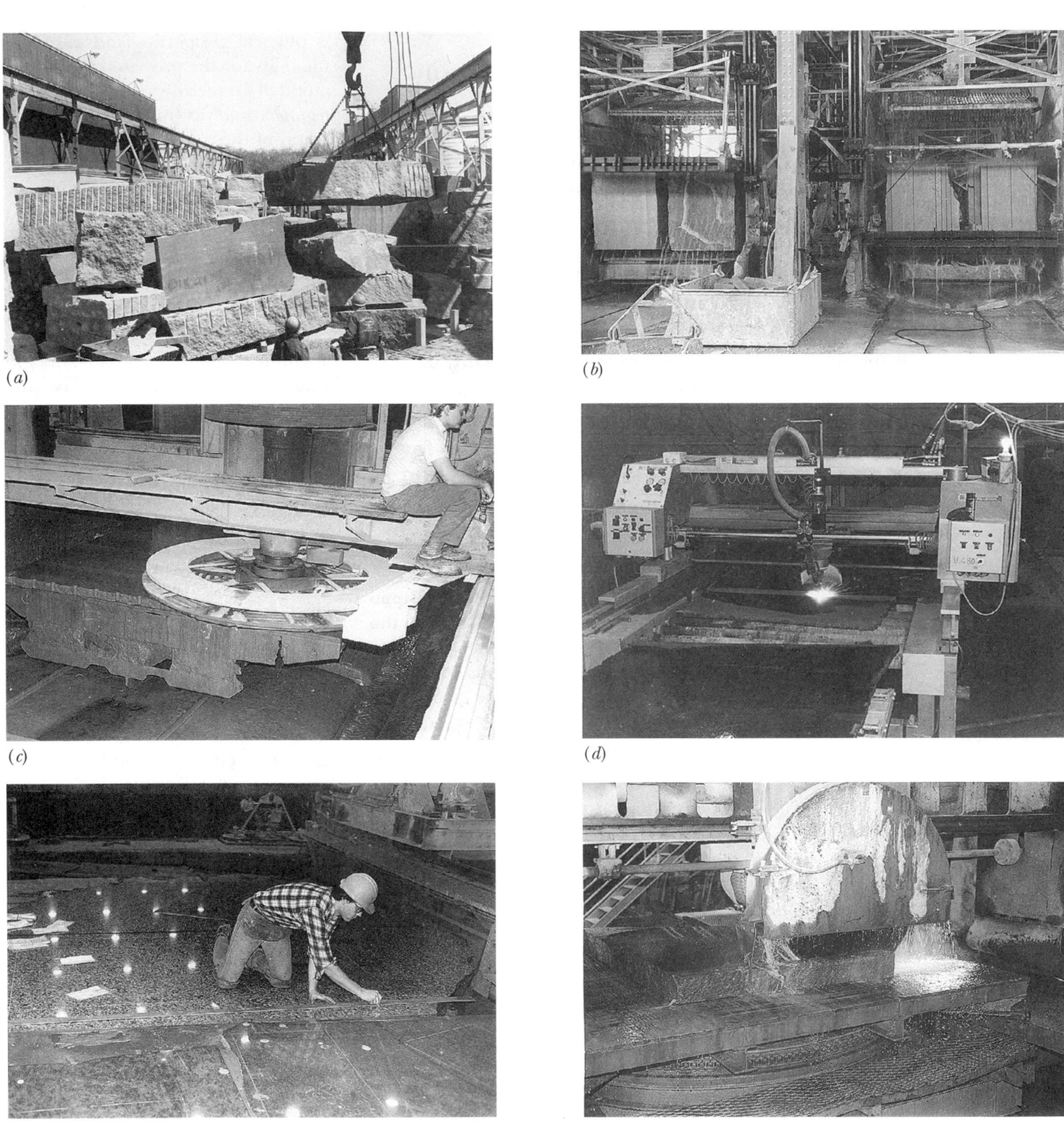

FIGURE 9.6
Stone milling operations, showing a composite of techniques for working granite and limestone. *(a)* An overhead crane lifts a rough block of granite from the storage yard to transport it into the plant. The average block in this yard weighs 60 to 80 tons (55 to 75 metric tons). *(b)* Two reciprocating gang saws slice blocks of limestone into slabs. The saw at the right has just completed its cuts, while the one at the left is just beginning. The intervals between the parallel blades are set by the saw operator to produce the desired thicknesses of slabs. Water cools the diamond blades of the saws and flushes away the stone dust. *(c)* A slab of granite is ground to produce a flat surface prior to polishing operations. *(d)* If a textured thermal finish is desired on a granite slab, a propane–oxygen torch is passed across the slab under controlled conditions to cause small chips to explode off the surface. *(e)* A layout specialist, working from shop drawings for a specific building, marks a polished slab of granite for cutting. *(f)* The granite slab is cut into finished pieces with a large diamond circular saw that is capable of cutting 7 feet (2.1 m) per minute at a depth of 3 inches (76 mm).

(*g*) Small pneumatic chisels with carbide-tipped bits are used for special details in granite. (*h*) Hand polishers are used to finish edges of granite that cannot be finished by automatic machinery. (*i*) Cylindrical components of limestone are turned on a lathe. (*j*) A cylindrical column veneer is ground to its true radius. (*k*) Linear shapes of Indiana limestone, a relatively soft stone, can be formed by a profiled silicon carbide blade in a planer. The piece of stone is clamped to the reciprocating bed, which passes it back and forth beneath the blade. The blade's pressure and depth are controlled by the operator's left hand pressing on the wooden lever. Here the planer is making a stepped profile for a cornice. (*l*) Working freehand with a vibrating pneumatic chisel, a carver finishes a column capital of Indiana limestone. Different shapes and sizes of interchangeable chisel bits rest in the curl of the capital. *(Photos a, c–h, and j courtesy of Cold Spring Granite Company; photo i courtesy of Indiana Limestone Institute; photos b, k, and l by Edward Allen)*

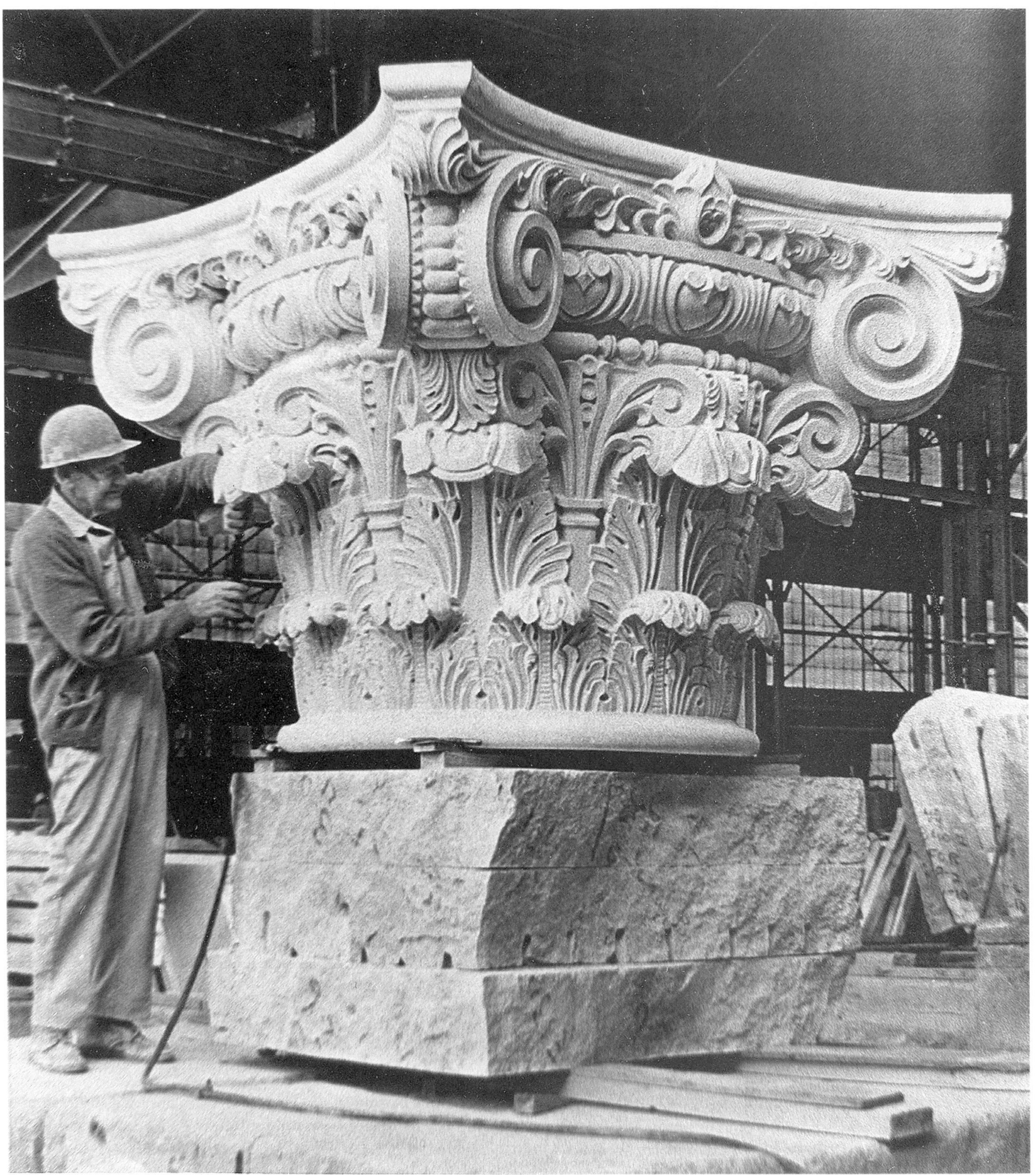

FIGURE 9.7
This 9-ton (8-metric-ton) Corinthian column capital was carved from a single 30-ton (27-metric-ton) block of Indiana limestone. Rough cutting took 400 hours and carving another 500. Eight of these capitals were manufactured for a new portico on an existing church. *(Architect: I. M. Pei & Partners. Courtesy Indiana Limestone Institute)*

	Water Absorption by Weight, Maximum	Density, Minimum	Compressive Strength, Minimum	Modulus of Rupture, Minimum	Flexural Strength, Minimum
Granite ASTM C615	0.40%	160 lb/ft^3 *2560 kg/m^3*	19,000 psi *131 MPa*	1500 psi *10.34 MPa*	1200 psi *8.27 MPa*
Limestone ASTM C568	3–12%	110–160 lb/ft^3 *1760–2560 kg/m^3*	1800–8000 psi *12–55 MPa*	400–1000 psi *2.9–6.9 MPa*	
Sandstone ASTM C616	1–8%	125–160 lb/ft^3 *2003–2560 kg/m^3*	4000–20,000 psi *27.6–137.9 MPA*	350–2000 psi *2.4–13.9 MPa*	
Marble ASTM C503	0.20%	144–162 lb/ft^3 *2305–2595 kg/m^3*	7500 psi *52 MPa*	1000 psi *7 MPa*	1000 psi *7 MPa*

Figure 9.8
Minimum property requirements for some common stone types, according to ASTM standards. Moisture absorption is a good indicator of stone durability. The less absorptive the stone, generally the less susceptible it is to freeze-thaw damage or chemical deterioration. Higher moisture levels within stone can also cause accelerated corrosion of metal anchors or reinforcing within the constructed masonry assembly. Stone density is also a good general measure of durability, correlating with higher strength and lower absorption. Compressive strength is especially important for stone used in loadbearing walls, where the stone must withstand the gravity loads imposed upon it. Modulus of rupture is a measure of a stone's resistance to shear and tension forces, a property particularly relevant to the performance of metal anchors used to attach stone to the building. Flexural strength is important in determining the wind load resistance of relatively thin stone panels.

Stone Masonry

Stone is used in two fundamentally different ways in buildings: It may be laid in mortar, much like bricks or concrete blocks, to make walls, arches, and vaults, a method of construction referred to as *stone masonry*. Or it may be mechanically attached to the structural frame or walls of a building as a facing, called *stone cladding*. This chapter deals only with stone masonry laid in mortar; the detailing and installation of stone cladding are covered in Chapters 20 and 23.

There are two simple distinctions that are useful in classifying patterns of stone masonry (Figures 9.9 and 9.10):

- Rubble masonry is composed of unsquared pieces of stone, whereas ashlar is made up of squared pieces.
- *Coursed stone masonry* has continuous horizontal joint lines, whereas *uncoursed* or *random stone masonry* does not.

Rubble can take many forms, from rounded, river-washed stones to broken pieces from a quarry. It may be either coursed or uncoursed. Ashlar masonry may be coursed or uncoursed and may be made up of blocks that are all the same size or several different sizes. The terms are obviously very general in their meaning, and the reader will find some variation in their use even among people experienced in the field of stone masonry.

Rubble stonework is laid very much like brickwork, except that the irregular shapes and sizes of the stones require the mason to select each stone carefully to fit the available space and, occasionally, to trim a stone with a mason's hammer or chisel. Ashlar stonework, though also similar in many ways to brickwork, frequently uses stone pieces that are too heavy to lift manually, and therefore relies on hoisting equipment to assist the mason in positioning the stones. This, in turn, requires a means of attaching the hoisting ropes to the sides or top of the stone block so as not to interfere with the mortar joint, and several types of devices are commonly used for this purpose (Figure 9.11). Both ashlar and rubble are usually laid with the *quarry bed* or *grain* of the stone running in the horizontal direction because stone is both stronger and more weather resistant in this orientation.

Stone masonry is frequently combined with concrete masonry to reduce costs. The stone is used where it can be seen, but concrete masonry, which is less expensive, quicker to erect, and more easily reinforced, is used for concealed parts of the wall. One example of such techniques, an exterior stone veneer with a concrete masonry backup wall, is illustrated in Figure 9.12.

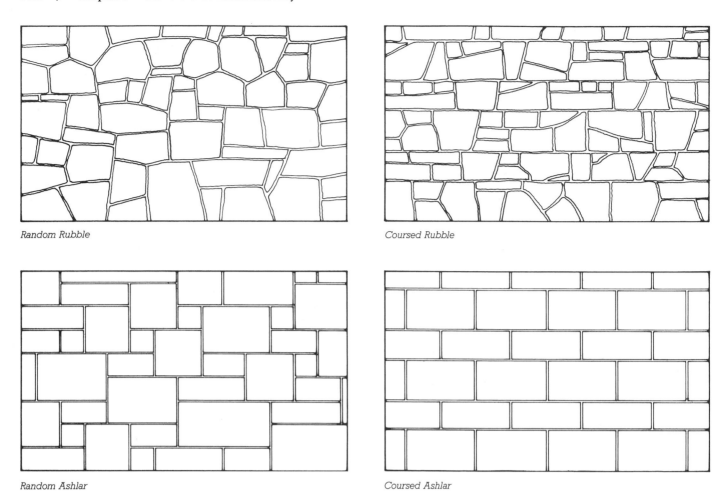

Random Rubble

Coursed Rubble

Random Ashlar

Coursed Ashlar

FIGURE 9.9
Rubble and ashlar stone masonry, coursed and random.

(*a*)

(*b*)

FIGURE 9.10
Random granite rubble masonry (*left*) and random ashlar limestone (*right*). (*Photos by Edward Allen*)

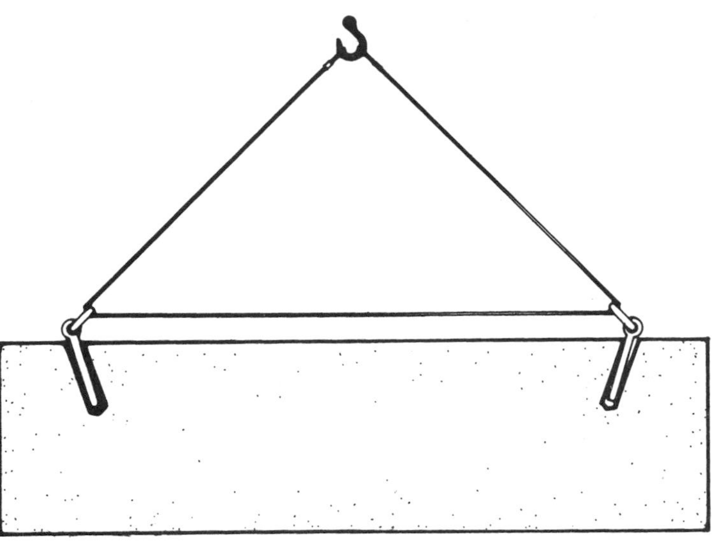
Lewis Pins

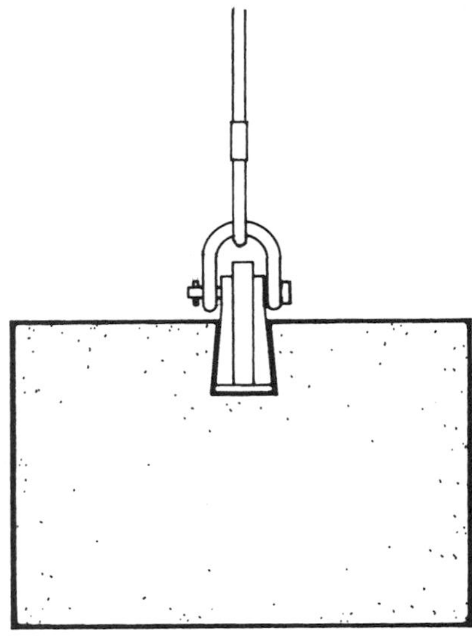
Box Lewis

FIGURE 9.11
Lewises permit the lifting and placing of blocks of building stone without interfering with the bed joints of mortar.

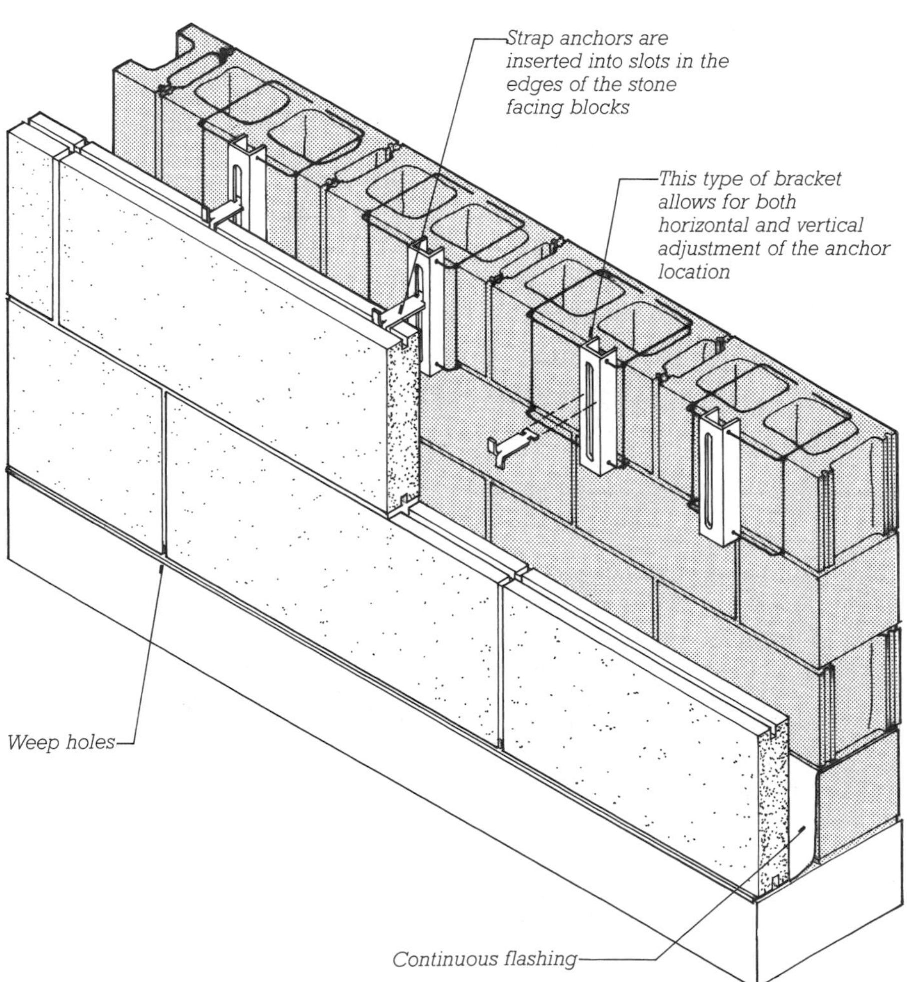
Strap anchors are inserted into slots in the edges of the stone facing blocks
This type of bracket allows for both horizontal and vertical adjustment of the anchor location
Weep holes
Continuous flashing

FIGURE 9.12
A conventional method of anchoring blocks of cut stone facing to a concrete masonry backup wall. The strap anchors, which are in direct contact with the stone, should be made only of highly corrosion-resistant stainless steel. (When steel corrodes, it expands. Corrosion of the anchors where they engage with the stone could quickly lead to damage to the stone itself.) The brackets attached to the concrete masonry backup wall may be stainless steel or less expensive galvanized steel. The brackets are built into the concrete masonry backup wall as it rises. The airspace, or cavity, between the stone and concrete masonry creates a drainage plane that conducts water penetrating the stone facing to the bottom of the cavity. The waterproof flashing at the base of the cavity forces the water to exit the wall back to the exterior through small drains in the facing called "weep holes."For more information about cavity wall construction, see the next chapter. For methods of attaching larger panels of stone to a building, see Chapter 20.

Considerations of Sustainability in Stone and Concrete Masonry

Stone and Concrete Masonry Materials

- Stone is a plentiful but finite resource. It is usually obtained from open pits, with the attendant disruption of drainage, vegetation, and wildlife habitat.
- The detrimental impacts of stone quarrying can long outlive the buildings for which the stone was extracted.
- Quarry reclamation practices, such as revegetation, land reshaping, and habitat restoration, can mitigate some of the adverse environmental impacts of stone quarrying and convert exhausted quarry sites to other beneficial uses.
- Concrete used in the manufacture of masonry units may include recycled materials such as fly ash, crushed glass, slag, and other postindustrial wastes. For more information regarding the sustainability of concrete, see Chapter 13.
- Mortar used for stone and concrete masonry is made from minerals that are generally abundant in the earth. However, portland cement and lime are energy-intensive products to manufacture. For more information about the sustainability of cement production, see Chapter 13.

Stone and Concrete Masonry Processing and Manufacturing

- Stone is heavy. It is expensive and energy intensive to transport. Stone may originate from local quarries or from sources in many places around the world. Fabrication may take place close to the source of the stone, close to the building site, or in some other location remote from both the stone's source and destination. Where uniquely sourced stones are desired or where specialized fabrication processes or skills are required, shipping over long distances may be required.
- The cutting, shaping, and polishing operations that take place during stone fabrication use large quantities of water that becomes contaminated with stone residue, lubricants, and abrasives. Water filtration and recycling systems can prevent contaminants from entering the wastewater stream and minimize water consumption.
- As much as one-half of quarried stone may become waste during fabrication. Depending on the type of stone, waste may be crushed and used as fill material on construction sites or as aggregate in concrete or asphalt. Stone with a strong color or other unique appearance qualities may be processed into aggregate for use in the manufacture of terrazzo, architectural concrete masonry units, or synthetic stone products. Much stone waste, however, is disposed of as landfill.
- The embodied energy of building stone can vary significantly with the source of the stone, fabrication processes, and distances and methods of shipping. Stone that is easily quarried and fabricated, and that is used locally, may have an embodied energy of as little as 300 to 400 BTU per pound (0.7–0.9 MJ/kg). On the other hand, stone that requires more effort and energy to extract and fabricate, and that is transported over long distances before arriving at the building site, may have an embodied energy 10 or even 20 times greater.
- Most concrete masonry units are manufactured in regional plants relatively close to their final end-use destinations.
- The use of lightweight concrete masonry reduces transportation-related costs and energy consumption.

Mortar joints in stone masonry are frequently *raked* (excavated) to a depth ranging from ½ inch to 1 inch or more (13-25 mm or more) after setting of the stones. This prevents uneven settling of stones or spalling of the stone edges that can occur when mortar at the face of the wall dries out and hardens faster than mortar deeper in the joint. After the remaining mortar has fully cured, the masons return to *point* the wall by filling the joints out to the face with *pointing mortar* and tooling them to the desired profile. The primary function of the pointing mortar is to form a good weather seal at the face of the stone. For this reason, low-strength mortar with good workability and adhesion characteristics, such as Types O or N mortar, is used. To ensure that the pointing mortar does not itself lead to concentrated stresses and possible spalling at the face of the masonry, it should never be of higher strength than the mortar deeper in the joint. (See the previous chapter for a discussion of mortar types.) Using pointing mortar also affords the opportunity to use a mortar of different color from the setting mortar. Alternatively, raked joints can be pointed with elastomeric joint sealant. High- quality sealant that does not stain the stone and has the appropriate elastic capabilities should be used. Joint sealant material types and their application are described more fully in Chapter 19.

Some building stones, especially limestone and marble, deteriorate rapidly in the presence of acids. This

• The embodied energy of concrete masonry units is slightly higher than that of the concrete from which they are made, due to the additional energy consumed in the curing of the units. Ordinary concrete masonry units have an embodied energy of approximately 250 BTU per pound (0.6 MJ/kg).

Stone and Concrete Masonry Construction

• Relatively small amounts of waste are generated on a construction site during stone and concrete masonry construction, including, for example, stone cutoffs, partial blocks, and unused mortar. These wastes generally go into landfills or are buried on the site.

• Sealers applied to stone and concrete masonry to provide water repellency and protection from staining are potential sources of emissions. Solvent-based sealers generally emit more air pollutants than water-based products.

Stone and Concrete Masonry Buildings

• Stone and concrete masonry are not normally associated with indoor air quality problems. In rare instances, stone aggregate in concrete or stone used in stone masonry has been found to be a source of radon gas emissions.

• The thermal mass effect of stone and concrete masonry can be a useful component of fuel-saving heating and cooling strategies such as solar heating and nighttime cooling.

• Stone and concrete masonry are dense materials that can effectively reduce sound transmission between adjacent spaces.

• Stone and concrete masonry construction are non-combustible. Lightweight concrete masonry units are especially effective for construction of fire resistance rated assemblies.

• Lightweight concrete masonry units have greater thermal resistance than more dense concrete units, stone, or brick.

• Construction with stone or concrete masonry can reduce reliance on paint finishes, a source of volatile organic compounds.

• Stone and concrete masonry are durable forms of construction that require relatively little maintenance and can last a very long time.

• Stone and concrete masonry are resistant to moisture damage and mold growth.

• When a building with stone or concrete masonry is demolished, the stone or masonry units can be crushed and recycled for use as on-site fill or as aggregate for paving. Some building stone can be salvaged for new construction.

Concrete Masonry Sitework

• Concrete masonry permeable pavers can facilitate on-site capture of storm water.

• Light-colored concrete pavers can lessen urban heat island effects.

• Interlocking concrete masonry units used in earth retaining walls are easily disassembled and reused

restricts their outdoor use in regions whose air is heavily polluted, and it also prevents their being cleaned with acid, as is often done with bricks. Exceptional care is taken during construction to keep stonework clean: Nonstaining mortars are used, high standards of workmanship are enforced, and the work is kept covered as much as possible. Flashings in stone masonry must be of plastic or nonstaining metal. Stonework normally may be cleaned only with mild soap, water, and a soft brush. Once the stonework is clean and dry, it may be treated with a *clear sealer*, also called a *water repellent*, a breathable and virtually invisible coating that reduces water absorption by the stonework, helping to protect it from staining and weathering.

From thousands of years of experience in building with stone, we have inherited a rich tradition of styles and techniques (Figures 9.13–9.20). This tradition is all the richer for its regional variations that are nurtured by the locally abundant stones: warm, creamy Jerusalem limestone in Israel; pristine, white Pentelic marble in Greece; stern gray granite and gray-streaked white Vermont marble in the northeastern United States; joyously colorful marbles and wormy travertine in Italy; golden limestone and gray Aswan granite in Egypt; red and brown sandstones in New York; cool, carveable limestone in France; gray-black basalt and granite in Japan. Though the stone industry is now global in character, its greatest glories are often regional and local.

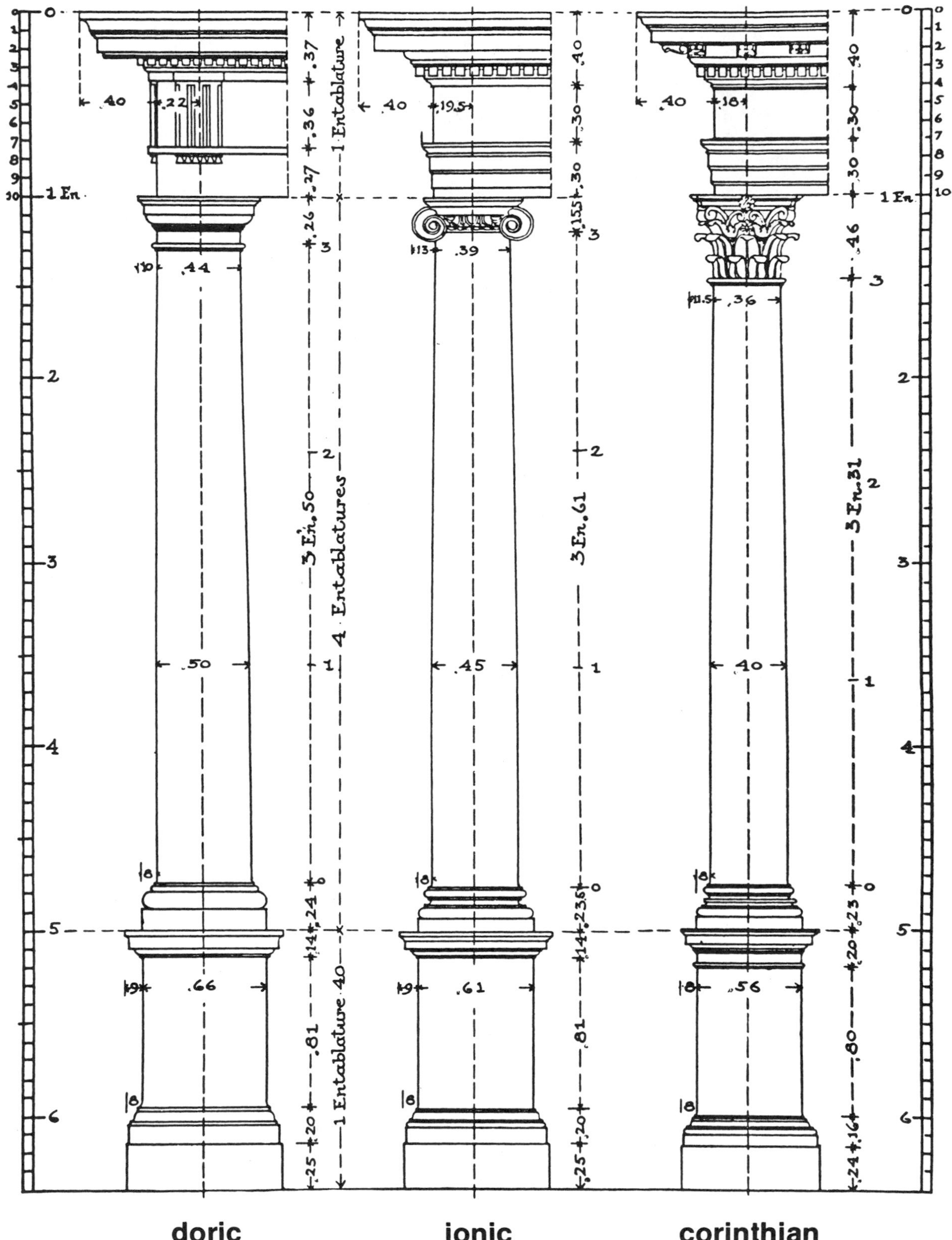

FIGURE 9.13
Proportioning rules for the classical orders of architecture. (*Courtesy Indiana Limestone Institute*)

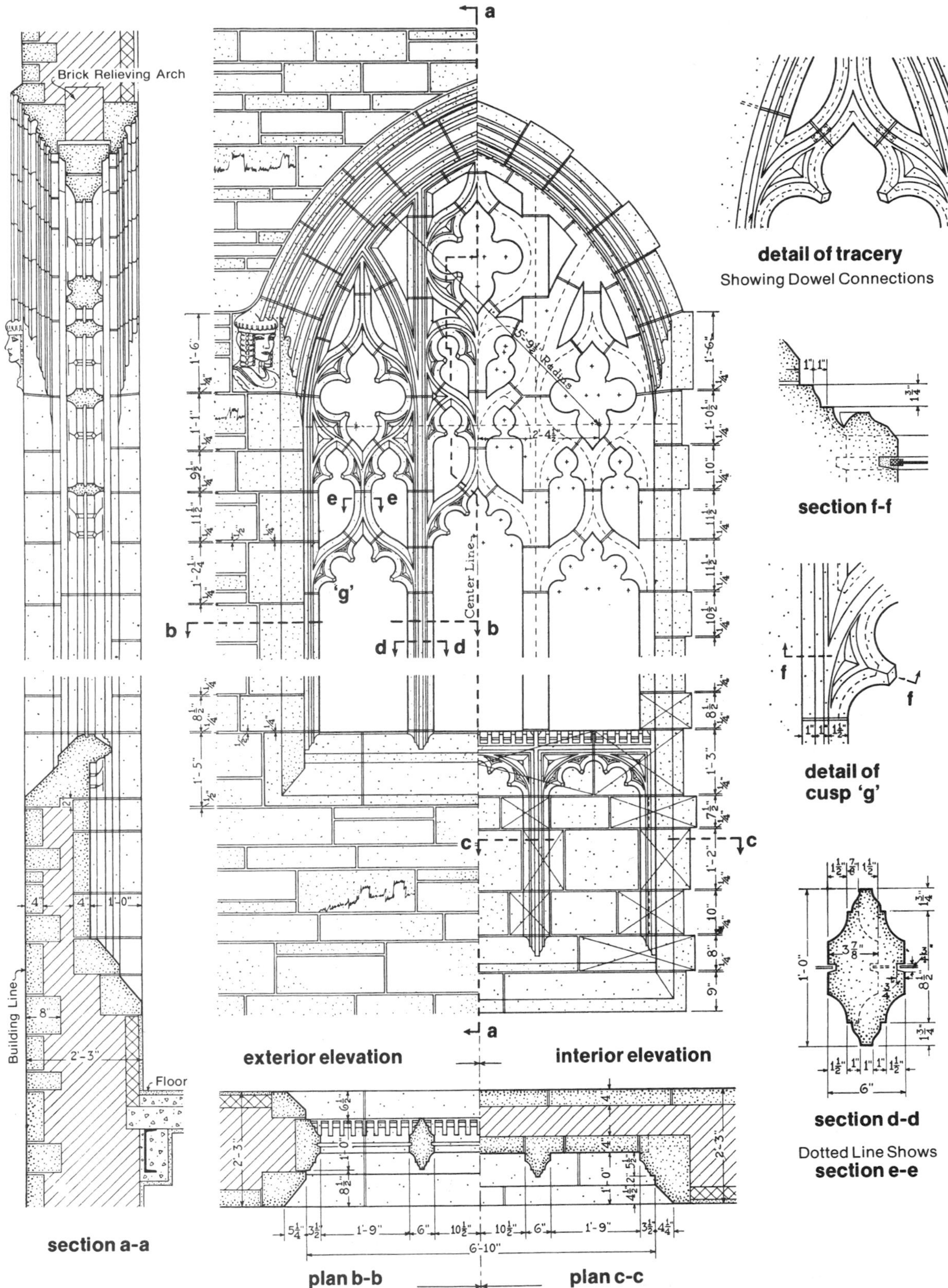

FIGURE 9.14
Details of Gothic window framing and tracery in limestone. (*Courtesy Indiana Limestone Institute*)

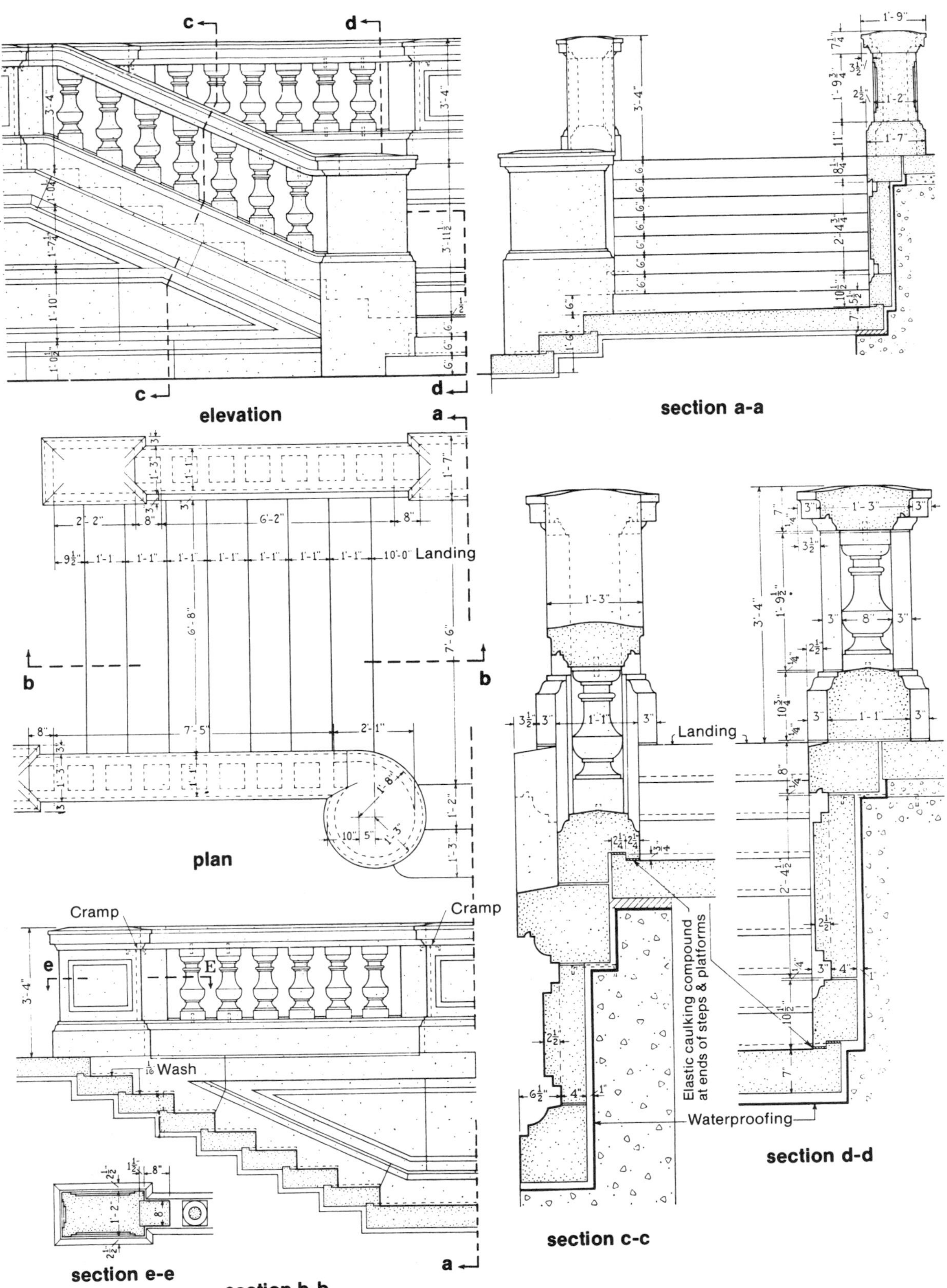

FIGURE 9.15
Details for a limestone stair and balustrade in the classical manner. The balusters are turned in a lathe, as shown in Figure 9.6*i*. (*Courtesy Indiana Limestone Institute*)

FIGURE 9.16
The Marshall Field Wholesale Store, built in 1885 in Chicago, rested on a two-story base of red granite. Its upper walls were built of red sandstone, and its interior was framed with heavy timber. *(Architect: H. H. Richardson. Courtesy of Chicago Historical Society, IChi-01688)*

FIGURE 9.17
Granite and sandstone details from the rectory of H. H. Richardson's Trinity Church in Boston, 1872–1877. *(Photo by Edward Allen)*

(a) (b) (c) (d) (e) (f)

FIGURE 9.18

Cut stone detailing. ***(a)*** **Random ashlar of broken face granite on H. H. Richardson's Trinity Church in Boston (1872–1877) still shows the drill holes of its quarrying.** ***(b)*** **Dressed stonework in and around a window of the same church contrasts with the rough ashlar of the wall.** ***(c)*** **Stonework on a 19th-century apartment building grows more refined as it distances itself from the ground and meets the brick walls above.** ***(d)*** **The base of the Boston Public Library (1888–1895, McKim, Mead, and White, Architects) is constructed of pink granite, with strongly rusticated blocks between the windows.** ***(e)*** **A college chapel is simply detailed in limestone.** ***(f)*** **A contemporary college library is clad in bands of limestone.** *(Architects: Shepley, Bulfinch, Richardson and Abbott. Photos by Edward Allen)*

FIGURE 9.19
Ashlar limestone loadbearing walls at the Wesleyan University Center for the Arts, designed by architects Kevin Roche, John Dinkeloo, and Associates. (*Courtesy Indiana Limestone Institute*)

FIGURE 9.20
The East Building of the National Gallery of Art in Washington, D.C., is clad in pink marble. (*Architects: I. M. Pei & Partners. Photo by Ezra Stoller, © ESTO*)

Concrete Masonry

Concrete masonry units (CMUs) are manufactured in three basic forms: solid bricks, larger hollow units that are commonly referred to as *concrete blocks*, and, less commonly, larger solid units.

Manufacture of Concrete Masonry Units

Concrete masonry units are manufactured by vibrating a stiff concrete mixture into metal molds, then immediately turning out the wet blocks or bricks onto a rack so that the mold can be reused at the rate of 1000 or more units per hour. The racks of concrete masonry units are cured at an accelerated rate by subjecting them to steam, either at atmospheric pressure or, for faster curing, at higher pressure (Figure 9.21). After steam curing, the units are bundled on wooden pallets for shipping to the construction site.

Concrete masonry units are made in a variety of sizes and shapes (Figures 9.22 and 9.23). They are also made with different densities of concrete, some of which use cinders, pumice, blast furnace slag, or expanded lightweight aggregates rather than crushed stone or gravel. Many colors and surface textures are available. Special shapes are relatively easy to produce if a sufficient number of units will be produced to amortize the expense of the mold. The major ASTM standards under which concrete masonry units are manufactured are C55 for concrete bricks, C90 for loadbearing units, and C129 for nonloadbearing units.

ASTM C90 establishes three weights of loadbearing concrete masonry units, as shown in Figure 9.24. Although all three weight classifications require the same minimum compressive strength, heavier blocks are denser and typically yield higher compressive strength than lighter blocks. Heavier blocks are also less expensive to manufacture, absorb moisture less readily, have better resistance to sound transmission, and are more resistant to abuse. But their

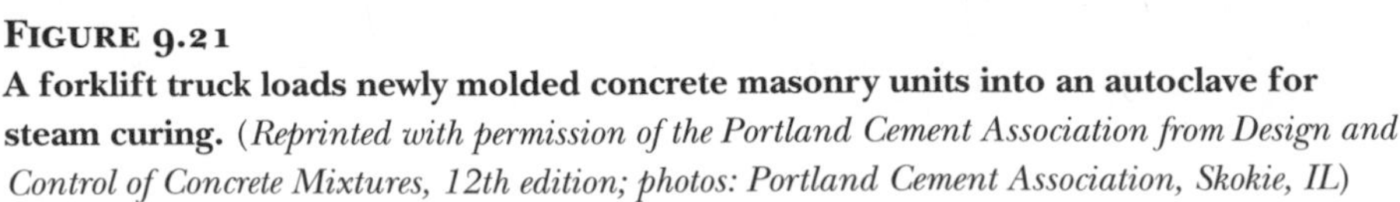

FIGURE 9.21
A forklift truck loads newly molded concrete masonry units into an autoclave for steam curing. (*Reprinted with permission of the Portland Cement Association from Design and Control of Concrete Mixtures, 12th edition; photos: Portland Cement Association, Skokie, IL*)

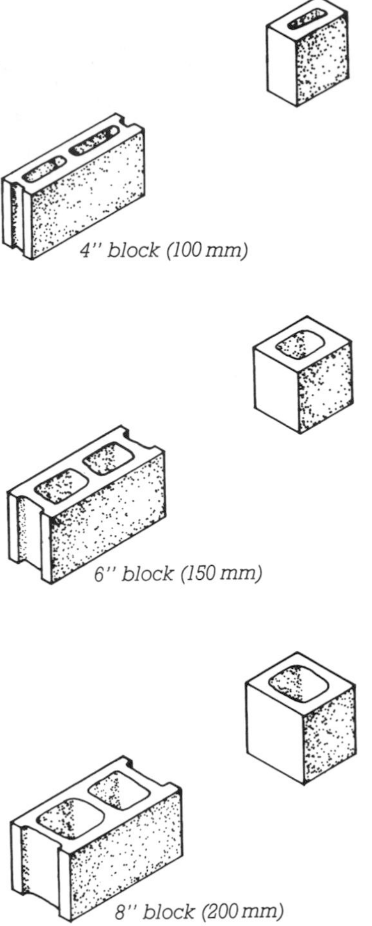

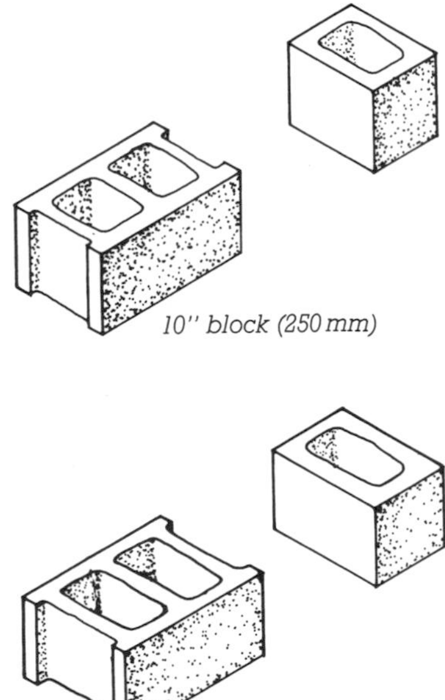

FIGURE 9.22
American standard concrete blocks and half-blocks. Each full block is nominally 8 inches (200 mm) high and 16 inches (400 mm) long.

FIGURE 9.23
Other concrete masonry shapes. Concrete bricks are interchangeable with modular clay bricks. Header units accept the tails of a course of headers from a brick facing. The use of control joint units is illustrated in Figure 10.21. Bond beam units have space for horizontal reinforcing bars and grout and are used to tie a wall together horizontally. They are also used for reinforced block lintels. A-blocks are used to build walls with vertical reinforcing bars grouted into the cores in situations where there is insufficient space to lift the blocks over the tops of the projecting bars; one such situation is a concrete masonry backup wall that is built within the frame of a building, as seen in Figure 20.1.

ASTM C90 Weight Classification	Density of Concrete (dry)	Typical Weights of Individual Units
Normal weight (sometimes also referred to as "Heavyweight")	At least 125 pcf (2000 kg/m^3)	33–39 lb (15–18 kg)
Medium weight	From 105 pcf to less than 125 pcf (1680–2000 kg/m^3)	28–32 lb (13–15 kg)
Lightweight	Less than 105 pcf (1680 kg/m^3)	20–27 lb (9–12 kg)

FIGURE 9.24
Specified densities and typical weights of hollow concrete masonry units.

FIGURE 9.25
Concrete blocks and bricks can be cut very accurately with a water-cooled, diamond-bladed saw. For rougher sorts of cuts, a few skillful blows from the mason's hammer will suffice. (*Reprinted with permission of the Portland Cement Association from Design and Control of Concrete Mixtures, 12th edition; photos: Portland Cement Association, Skokie, IL*)

greater weight also makes heavier blocks more expensive to ship and more labor intensive and expensive for masons to lay up in comparison to lighter-weight blocks. The greater density of heavier blocks also gives them lower thermal resistance and lower fire resistance. Medium-weight blocks are probably the most common weight classification specified, although the availability of blocks of different weights varies with regional differences in building practices.

Single-wythe exterior concrete masonry walls tend to leak in wind-driven rains. Except in dry climates, they should be painted on the outside with masonry paint or a breathable elastomeric coating, manufactured with integral water repellent additives, or cavity wall construction should be used instead, as described in Chapter 10.

Although innumerable special sizes, shapes, and patterns are available, the dimensions of American concrete masonry units are based on an 8-inch (203-mm) cubic module. The most common block is nominally 8 × 8 × 16 inches (203 × 203 × 406 mm). The actual size of the block is 7⅝ × 7⅝ × 15⅝ inches (194 × 194 × 397 mm), which allows for a mortar joint ⅜ inch (9 mm) thick. This standard block is designed to be lifted and laid conveniently with two hands (compared to a brick, which is designed to be laid with one). Its double-cube proportions work well for running bond stretchers, for headers, and for corners.

Although concrete masonry units can be cut with a diamond-bladed power saw (Figure 9.25), it is more economical and produces better results if the designer lays out buildings of concrete masonry in dimensional units that correspond to the module of the block (Figure 9.26). Nominal 4-inch, 6-inch, and 12-inch (102-mm, 152-mm, and 305-mm) block thicknesses are also common, as is a solid concrete brick that is identical in size and proportion to a modular clay brick. A handy feature of the standard 8-inch (203-mm) block height is that it corresponds exactly to three courses of ordinary clay or concrete brickwork, or two courses of oversize bricks, making it easy to interweave blockwork and brickwork in composite walls.

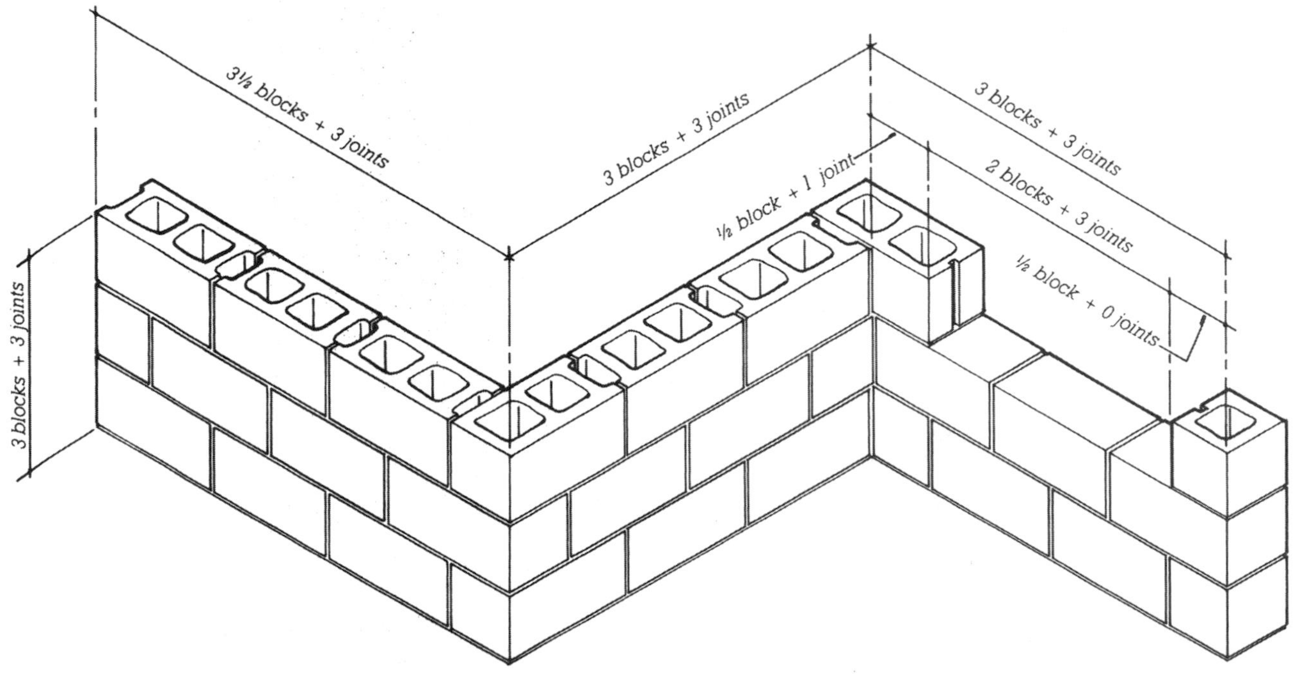

FIGURE 9.26
Modular dimensioning of concrete masonry construction. Concrete masonry buildings should be dimensioned to use uncut blocks except under special circumstances.

Laying Concrete Blocks

The accompanying photographic sequence (Figure 9.27) illustrates the technique for laying concrete block walls. The mortar is identical to that used in brick walls, but in most walls only the face shells of the block are mortared, with the webs left unsupported (Figure 9.27*c*, *j*, and *k*).

Concrete masonry is often reinforced with steel to increase its loadbearing capacity, resistance to cracking, or resistance to seismic forcers. *Horizontal reinforcing* is usually

(*a*) (*b*) (*c*) (*d*)

FIGURE 9.27
Laying a concrete masonry wall. *(a)* A bed of mortar is spread on the footing. *(b)* The first course of blocks for a lead is laid in the mortar. Mortar for the head joint is applied to the end of each block with the trowel before the block is laid. *(c)* The lead is built higher. Mortar is normally applied only to the face shells of the block and not to the webs. *(d)* As each new course is started on the lead, its height is meticulously checked with either a folding rule or, as shown here, a story pole marked with the height of each course.

inserted in the form of *joint reinforcing*, welded grids of small-diameter steel rods that are laid into the mortar bed joints at the desired vertical intervals (Figure 9.28). If stronger horizontal reinforcing is required, bond beam blocks (Figure 9.23) or special blocks with channeled webs (Figure 9.29) allow heavier reinforcing bars to be placed in the horizontal direction. The horizontal bars may be embedded in grout before the next course is laid, with the grout contained in the cores of the reinforced course by a strip of metal mesh that

(e) (f) (g) (h)

(e, f) **Each new course is also checked with a spirit level to be sure that it is level and plumb. Time expended in making sure that the leads are accurate is amply repaid in the accuracy of the wall and the speed with which blocks can be laid between the leads.** ***(g)*** **The joints of the lead are tooled to a concave profile.** ***(h)*** **A soft brush removes mortar crumbs after tooling.**

continued

(*i*)

(*j*)

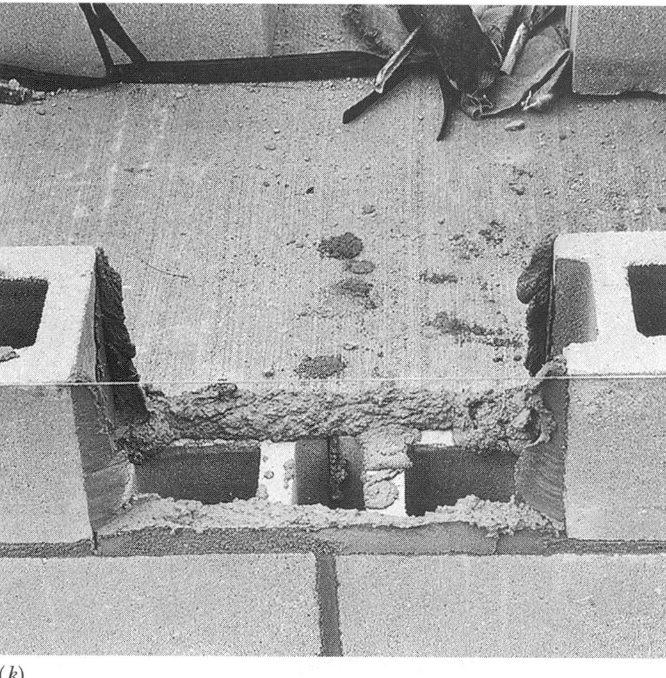

(*k*)

(*l*)

FIGURE 9.27 *continued*
(*i*) A mason's line is stretched taut between the leads on line blocks. (*j*) The courses of blocks between the leads are laid rapidly by aligning each block with the stretched line; no story pole or spirit level is necessary. The mason has laid bed joint mortar and "buttered" the head joints for a number of blocks. (*k*) The last block to be installed in each course of infill blocks, the closer, must be inserted between blocks that have already been laid. The bed and head joints of the already laid blocks are buttered. (*l*) Both ends of the closer blocks are also buttered with mortar, and the block is lowered carefully into position. Some touching up of the head joint mortar is often necessary. (*Reprinted with permission of the Portland Cement Association from* Design and Control of Concrete Mixtures, *12th edition; photos: Portland Cement Association, Skokie, IL*)

was previously laid into the bed joint beneath the course to bridge across the core openings. Alternatively, the horizontal bars may be grouted simultaneously with the vertical bars.

Vertical block cores are easily reinforced by inserting bars and grouting, using either the low-lift or high-lift technique, as described in Chapter 8. In most cases only those cores that contain reinforcing bars are grouted, but sometimes all the vertical cores are filled, whether or not they contain bars, for added strength (Figure 9.30).

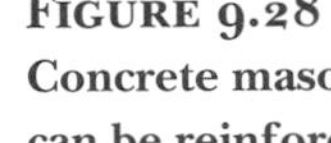

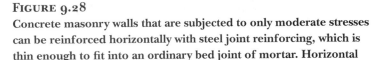

FIGURE 9.28
Concrete masonry walls that are subjected to only moderate stresses can be reinforced horizontally with steel joint reinforcing, which is thin enough to fit into an ordinary bed joint of mortar. Horizontal joint reinforcing is available in both a "truss" pattern, as illustrated, and a "ladder" pattern. Both are equally satisfactory. Vertical reinforcing is done with ordinary reinforcing bars grouted into the cores of the blocks.

FIGURE 9.29
In this proprietary system for building more heavily reinforced concrete masonry walls, the webs are grooved to allow the insertion of horizontal reinforcing bars into the wall. The cores of the blocks are then grouted to embed the bars. (*Courtesy of G. R. Ivany and Associates, Inc.*)

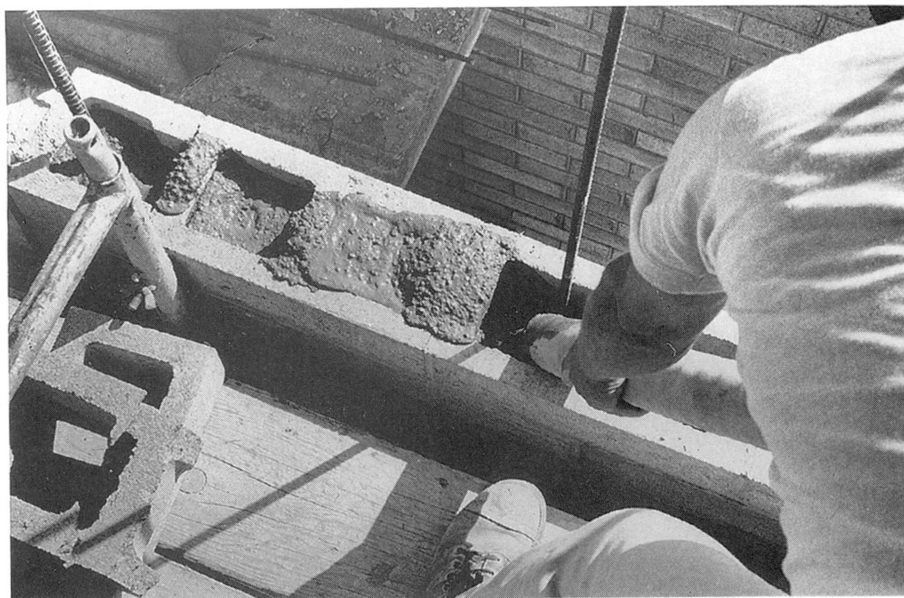

FIGURE 9.30
Grout is deposited in the cores of a reinforced concrete masonry wall using a grout pump and hose. (*Reprinted with permission of the Portland Cement Association from* Design and Control of Concrete Mixtures, *12th edition; photos: Portland Cement Association, Skokie, IL*)

Lintels for concrete block walls may be made of steel angles, combinations of rolled steel shapes, reinforced concrete, or bond beam blocks with grouted horizontal reinforcing (Figure 9.31).

Surface bonding of concrete masonry walls has found application in certain low-rise buildings where the cost or availability of skilled labor is a problem. The blocks are laid without mortar, course upon course, to make a wall. Then a thin layer of a special cementitious compound containing short fibers of alkali-resistant glass is applied to each side of the wall with plastering tools. After it has cured, this surface bonding compound joins the blocks securely to one another both in tension and in compression. It also serves as a surface finish whose appearance resembles stucco.

Decorative Concrete Masonry Units

Decorative concrete masonry units, also called architectural concrete masonry units, are easily and economically manufactured in an unending variety of surface patterns, textures, and colors intended for exposed use in exterior and interior walls. A few such units are diagrammed in Figure 9.33, and some of the resulting surface textures are depicted in Figures 9.34–9.37. Mold costs for producing special units are low when spread across the number of units required for a medium-sized to large building. Many of the textured concrete masonry units that are now considered standard originated as special designs created by architects for particular buildings.

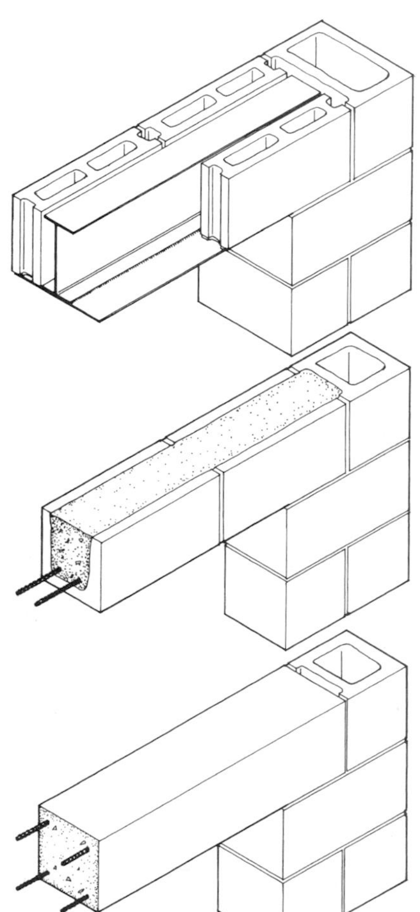

FIGURE 9.31
Lintels for openings in concrete masonry walls. At the top, a steel lintel for a broad opening is made up of a wide-flange section welded to a plate. Steel angle lintels are used for narrower openings. In the middle, a reinforced block lintel is composed of bond beam units. At the bottom, a precast reinforced concrete lintel is seen.

FIGURE 9.32
A small house by architects Clark and Menefee is cleanly detailed in concrete masonry. A reinforced masonry retaining wall helps to link the building visually to the site. (*Photo by Jim Rounsevell*)

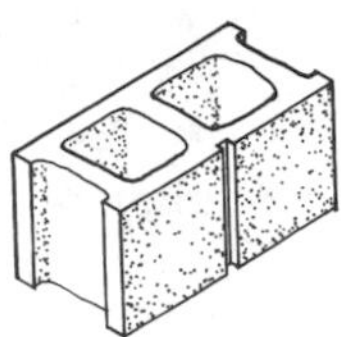

Scored-face unit

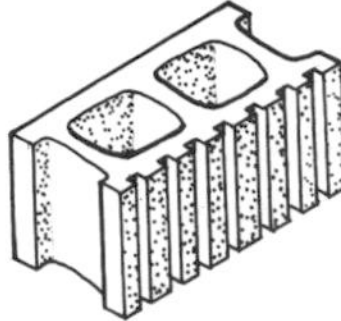

Ribbed-face unit

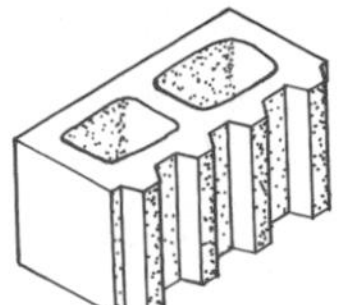

Ribbed-face unit

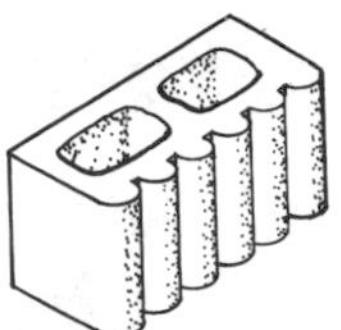

Fluted-face unit

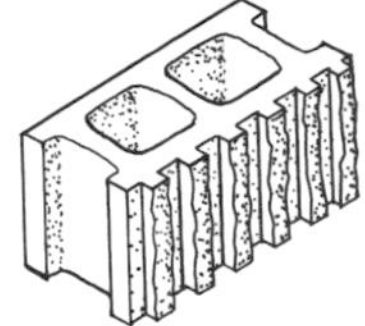

Ribbed split-face unit

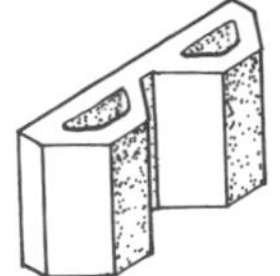

Angular-face unit

FIGURE 9.33
Some decorative concrete masonry units, representative of literally hundreds of designs currently in production. The scored-face unit, if the slot in the face is filled with mortar and tooled, produces a wall that looks as if it were made entirely of half-blocks. The ribbed split-face unit is produced by casting "Siamese twin" blocks that are joined at the ribs, then shearing them apart.

FIGURE 9.34
A facade of split-face concrete masonry. (*Architects: Paderewski, Dean, Albrecht & Stevenson. Courtesy of National Concrete Masonry Association*)

(a)

(b)

(c)

(d)

(e)

FIGURE 9.35
Some walls of decorative concrete masonry. ***(a)*** **Split block.** ***(b)*** **Slump block, which is molded from relatively wet concrete and allowed to sag slightly after molding and before curing.** ***(c)*** **Split blocks of varying sizes laid in a random ashlar pattern.** ***(d)*** **Ribbed split-face blocks.** ***(e)*** **Striated blocks.** (*Courtesy of National Concrete Masonry Association*)

FIGURE 9.36
Split-face blocks are used indoors in this high-school auditorium. (*Architects: Marcel Breuer and Associates. Courtesy of National Concrete Masonry Association*)

The Economy of Concrete Masonry Construction

Concrete masonry is a versatile building material, and walls built from it are usually more economical than comparable ones made of brick or stone masonry: The concrete blocks themselves are cheaper on a volumetric basis and are made into a wall much more quickly because of their larger size (a single standard concrete block occupies the same volume as 12 modular bricks). Concrete blocks can be produced to required degrees of strength, and because their hollow cores allow for the easy insertion of reinforcing steel and grout, they are widely used in bearing wall construction. Concrete blocks are often used for the backup wythe behind a brick or stone facing. Block walls also accept plaster, stucco, or tile work directly, without the need for metal lath. In one recent year, nearly 8 billion concrete blocks were manufactured in North America alone.

Other Types of Masonry Units

Bricks, stones, and concrete blocks are the most commonly used types of masonry units. In the past, hollow tiles of cast gypsum or fired clay were often used for partition construction (Figure 23.39). Both have been supplanted in the United States by concrete blocks, though hollow clay tiles are still widely used in other parts of the world. *Structural glazed facing tiles* of clay remain in use, especially for partitions, where their durable, easily cleaned surfaces are advantageous, as in public corridors, toilet rooms, institutional kitchens, locker and shower rooms, and industrial plants (Figure 23.40).

Structural terra cotta, glazed or unglazed molded decorative units of fired clay, was widely used until the mid-20th century and is often seen on the facades of late-19th-century masonry buildings in the United States (Figure 8.43). Terra cotta had almost disappeared from the U.S. materials market a few decades ago, but restoration work on older buildings created a demand for it and generated a renewed desire to use it in new buildings, rescuing the last manufacturers from the brink of obsolescence.

Glass blocks are available in many textures and in clear, heat-absorbing, and reflective glass (Figures 9.38 and 9.39). Glass blocks are nonabsorbent. When glass masonry walls are constructed, the mortar stiffens more slowly than it does with more absorbent units of clay or concrete, so temporary spacers are inserted between units to maintain proper spacing until the mortar sets up.

Autoclaved aerated concrete (AAC), though it has been manufactured and used in Europe for many years, has found only limited application in North America. Its ingredients are sand, lime, water, and a small amount of aluminum powder. These materials are reacted with steam to produce an aerated concrete that consists primarily of calcium silicate hydrates. AAC is available in solid blocks that are laid in mortar like other concrete masonry units. It is also made in such forms as unreinforced wall panels and steel-reinforced lintels and floor/ceiling panels. Because of its trapped gas bubbles, which are created by the reaction of the aluminum powder with the lime, the density of AAC is similar to that of wood, and it is easily sawed, drilled, and shaped. It has moderately good thermal insulating properties. It is not nearly as strong as normal-density concrete, but it is sufficiently strong to serve as loadbearing walls, floors, and roofs in low-rise construction. Walls of AAC are too porous to be left exposed; they are usually stuccoed on the exterior and plastered on the interior.

Masonry Wall Construction

Bricks, stones, and concrete masonry units are mixed and matched in both loadbearing and nonloadbearing walls of many types. The more important of these wall constructions are presented in the next chapter.

FIGURE 9.37
High-rise apartments constructed with reinforced bearing walls of fluted-face units. Specially cast blocks were used to produce the curved balcony fronts.
(Architect: Paul Rudolph. Courtesy of National Concrete Masonry Association)

FIGURE 9.38
Glass blocks used to enclose a conference room in a corporate office. (*Courtesy of Pittsburgh Corning Corporation*)

FIGURE 9.39
Exterior use of glass blocks in the wall of a stairwell. (*Architect: Gwathmey/Siegel. Courtesy of Pittsburgh Corning Corporation*)

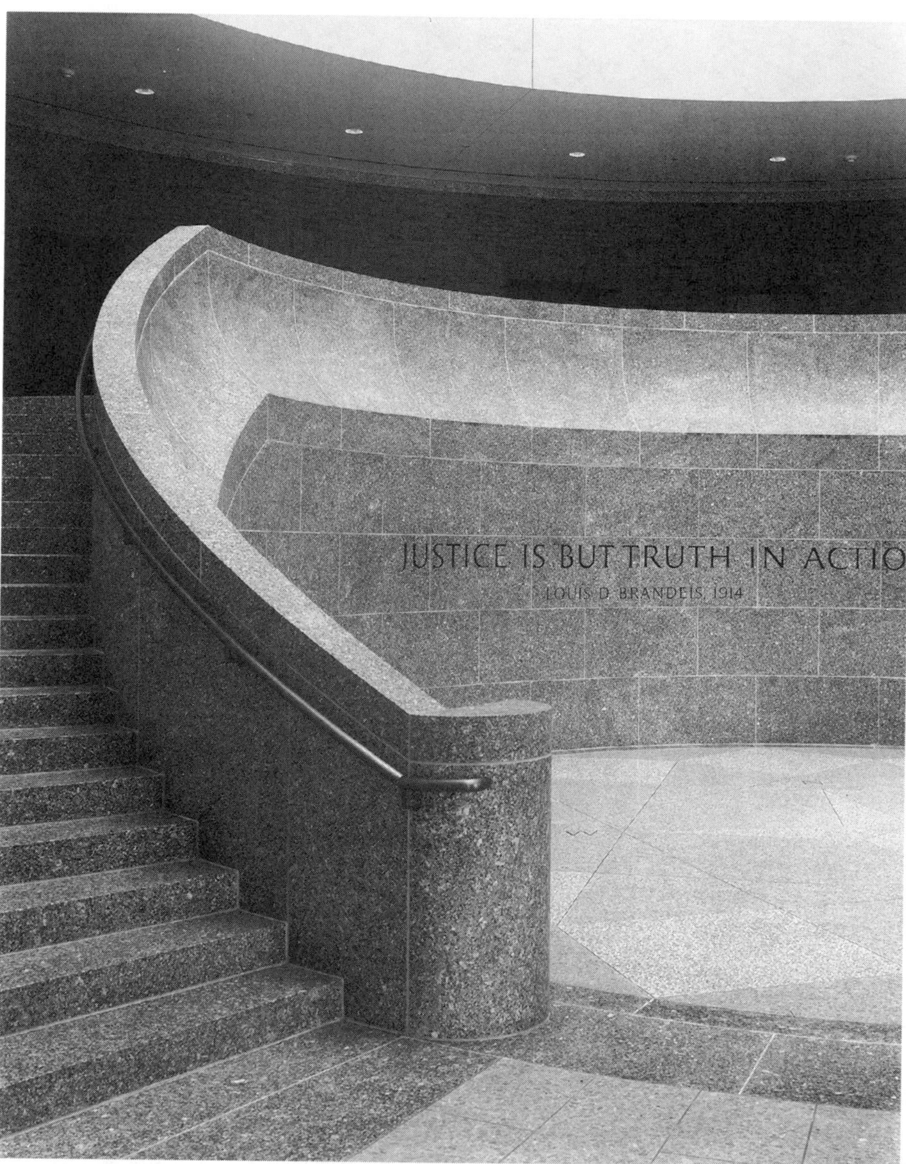

FIGURE 9.40
Dark granite, simply and beautifully detailed, creates a feeling of quiet dignity for the main stair of the Federal Courthouse in Boston. The floor is made of terrazzo with matching and contrasting colors of stone chips. (*Photo © Steve Rosenthal*)

FIGURE 9.41
Some masonry paving patterns in brick and granite. All six of these examples are laid without mortar in a bed of sand. Outdoor pavings of masonry may also be laid in mortar over reinforced concrete slabs. (*Photos by Edward Allen*)

CSI/CSC

MasterFormat Sections for Stone and Concrete Masonry

04 00 00	MASONRY
04 05 00	Common Work Results for Masonry
	Masonry Grouting
	Masonry Anchorage and Reinforcing
	Continuous Joint Reinforcement
	Masonry Reinforcing Bars
	Stone Anchors
04 20 00	UNIT MASONRY
04 21 00	Clay Unit Masonry
	Glazed Structural Clay Tile Masonry
	Terra Cotta Masonry
04 22 00	Concrete Unit Masonry
	Surface-Bonded Concrete Unit Masonry
	Architectural Concrete Unit Masonry
04 23 00	Glass Unit Masonry
04 40 00	STONE ASSEMBLIES
04 43 00	Stone Masonry

Selected References

1. Indiana Limestone Institute of America, Inc. *ILI Handbook.* Bedford, IN, updated regularly.

This is the definitive reference on the use of limestone in building. It includes the history and provenance of Indiana limestone, recommended standards and details for its use, and architectural case histories.

2. Marble Institute of America. *Dimensional Stone Design Manual.* Cleveland, OH, updated regularly.

Standards, specifications, and details for dimensional stonework are included in this large looseleaf volume. Though all stone types are included, the emphasis is on marble.

3. Panarese, W., S. Kosmatka, and F. Randall. *Concrete Masonry Handbook for Architects, Engineers, and Builders* (6th ed.). Skokie, IL, Portland Cement Association, 2008.

This is a clearly written, beautifully illustrated guide to every aspect of concrete masonry.

4. National Concrete Masonry Association. *Annotated Design and Construction Details for Concrete Masonry.* Herndon, VA, 2003.

A treasury of typical concrete masonry details.

5. National Concrete Masonry Association. *TEK Manual for Concrete Masonry Design and Construction.* Herndon, VA, update regularly.

This comprehensive collection of technical bulletins covers every major aspect of concrete masonry products and design. The complete TEK bulletin series is also available free online, through links from the NCMA web site.

Web Sites

Stone and Concrete Masonry

Author's supplementary web site: **www.ianosbackfill.com/09_stone_and_concrete_masonry**

Autoclaved Aerated Concrete Products Association: **www.aacpa.org**

Building Stone Institute: **www.buildingstoneinstitute.org**

Bybee Stone Company: **www.bybeestone.com**

Genuine Stone: **www.genuinestone.com**

Indiana Limestone Institute: **www.iliai.com**

Marble Institute of America: **www.marble-institute.com**

National Concrete Masonry Association: **www.ncma.org**

Portland Cement Association: **www.cement.org**

Trenwyth Architectural Masonry Units: **www.trenwyth.com**

Key Terms and Concepts

building stone
igneous rock
sedimentary rock
metamorphic rock
granite
basalt
limestone
freestone
oolitic limestone
dolomitic limestone
quarry sap
sandstone
brownstone
bluestone
slate
marble
travertine
fieldstone
rubble
dimension stone
cut stone
ashlar
flagstone
jet burner
spalling
lewis
petrographic analysis
stone masonry
stone cladding
coursed stone masonry
uncoursed stone masonry, random stone masonry
quarry bed, grain
raked mortar joint
pointed mortar joint
pointing mortar
clear sealer, water repellent
concrete masonry unit (CMU)
concrete block
horizontal reinforcing
joint reinforcing
surface bonding
decorative concrete masonry unit, architectural concrete masonry unit
structural glazed facing tile
structural terra cotta
glass block
autoclaved aerated concrete (AAC)

Review Questions

1. What are the major types of stone used in construction? How do their properties differ?

2. What sequence of operations would be used to produce rectangular slabs of polished marble from a large quarry block?

3. In what ways is the laying of stone masonry different from the laying of bricks?

4. What are the advantages of concrete masonry units over other types of masonry units?

5. How long is a wall that is made up of 22 concrete masonry units, each nominally 8 × 8 × 16 inches (203 × 203 × 406 mm), joined with ⅜-inch (10 mm) mortar joints?

6. How may horizontal and vertical steel reinforcement be introduced into a CMU wall?

Exercises

1. Design a stone masonry facade for a downtown bank building that is 32 feet (9.8 m) wide and two stories high. Draw all the joints between stones on the elevation. Draw a detail section to show how the stones are attached to a concrete masonry backup wall.

2. Design a masonry gateway for one of the entrances to a college campus with which you are familiar. Choose whatever type of masonry you feel is appropriate, and make the fullest possible use of the decorative and structural potentials of the material. Show as much detail of the masonry on your drawings as you can.

3. Visit a masonry supply company and view all the types of concrete masonry units that available. What is the function of each?

4. Design a simple concrete masonry house that a student could build for his or her own use, keeping the floor area to 400 square feet (37 m^2).

5. Design a decorative CMU that can be used to build a richly textured wall.

BICYCLES

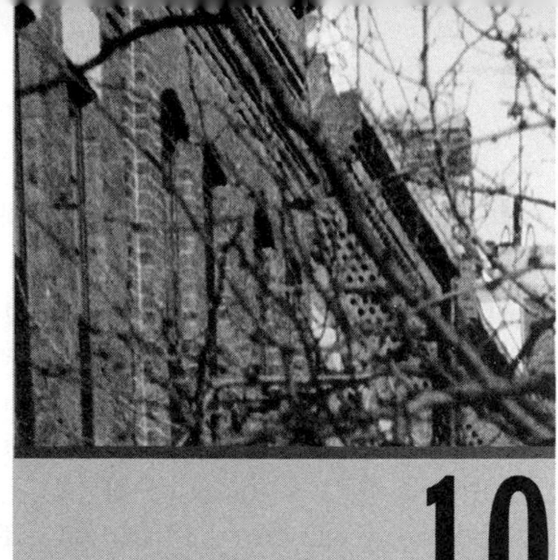

10

Masonry Wall Construction

America's rich fabric of 19th-century downtown buildings is made up largely of buildings whose construction is defined by building codes as ordinary construction: exterior masonry loadbearing walls with floors and roof spanned by wood joists. (*Photo by Edward Allen*)

The world's most enduring buildings are constructed of *solid masonry*. Such buildings' thick monolithic walls can stand for centuries, their great mass providing strength, durability, and moderation of the flow of heat and moisture between indoors and outdoors. Traditional masonry construction methods survive to the present day in such buildings as the Washington National Cathedral (Washington, D.C.), constructed primarily of solid limestone masonry. However, most contemporary construction combines the essential materials of masonry—brick, stone, concrete block, and mortar—with more modern materials and technological innovations to create walls that are thinner, lighter in weight, faster to erect, and better able to control the flow of heat and moisture through the wall.

Figure 10.2
Masonry ties and joint reinforcing. These are only a few examples from among dozens of types available. Corrugated ties *(a)* have little strength or stiffness and are recommended only for anchoring low-rise veneer to wood framed backup walls. Z-ties *(b)* are stiffer than corrugated ties. Adjustable ties *(c)* allow for irregularities in course height between wythes. Adjustable stone ties *(d)* have excessive play, unless the collar joint is filled with mortar. The ladder and truss ties *(e–g)*, combine horizontal joint reinforcement with the masonry tie. Truss-type horizontal reinforcing *(g)* can develop composite action between wythes, resulting in a stiffer wall. It is suitable for solid walls but should not be used with cavity walls, where differential rates of expansion and contraction between the separate wythes can lead to bowing of the wall. (In cases where these illustrations apply to composite masonry walls, the mortar fill between the wythes has been omitted for clarity.)

Types of Masonry Walls

Composite Masonry Walls

For an optimum balance between appearance and economy, solid masonry walls of more than one wythe may be constructed as *composite masonry walls,* with an outer wythe of stone or face brick and a backup wythe of hollow concrete masonry. The two wythes are bonded together either by steel horizontal joint reinforcing (Figure 10.2*g*) or steel ties (Figure 10.2*d*), and the space between the two wythes is filled with mortar. In traditional composite masonry construction, wythes may be bonded with header bricks from the outer wythe that engage with the backup wythe. The headers may penetrate completely through the backup wythe or they may interlock with courses of header blocks (Figure 9.23). Composite masonry walls may be either loadbearing or nonloadbearing.

Figure 10.1
Traditional, unreinforced solid masonry walls rely on their mass for their strength and durability. However, such construction lacks tensile strength and, as a consequence, is vulnerable to seismic forces. This building was severely damaged in Seattle's 2001 Nisqually earthquake and subsequently underwent extensive reconstruction. Steel crossbracing, part of an internal structural framework inserted for added resistance to lateral forces, can be seen through the center street-level window. Efflorescence (white powderlike deposite) just below the parapet corbelling is evidence of newly constructed masonry. This phenomenon is explained later in this chapter. *(Photo by Joseph Iano)*

A. Corrugated Tie

B. Z-Tie

C. Adjustable Tie

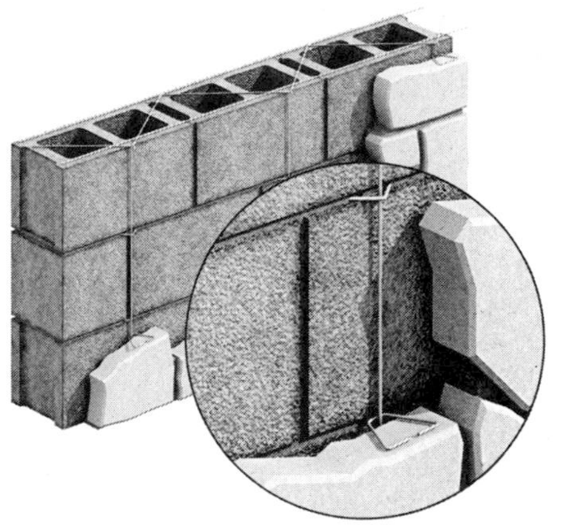

D. Adjustable Stone Tie

E. Two-Wire Ladder Tie

F. Ladder Loop Tie

G. Three-Wire Truss Tie

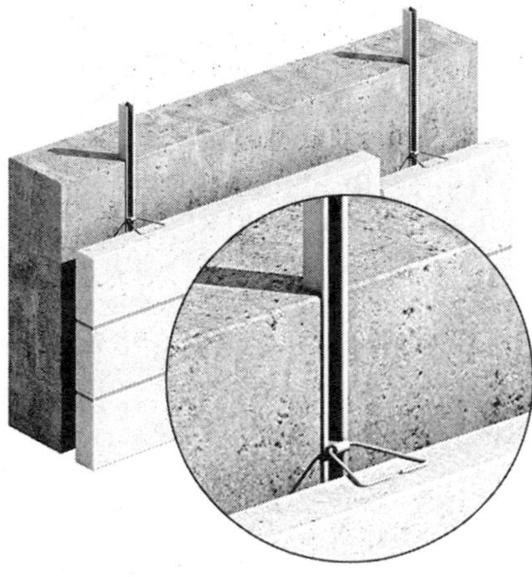

H. Dovetail Anchors for Concrete Backup

I. Steel Column Anchor

In a composite masonry wall, the separate wythes are intimately bonded and behave as a single mass. When designing such walls, the designer must be sure that the differences in the thermal and moisture expansion characteristics of the two masonry materials are not so great as to cause bowing or cracking of the wall as they expand or contract at different rates. In loadbearing composite walls, the different strengths and elasticities of the two materials must also be taken into account to ensure the proper placement of reinforcing and for calculating the deformations the wall will experience under load.

Masonry Cavity Walls

Every masonry wall is porous to some degree. Some water will find its way through even new masonry if the wall is wetted for a sustained period. Older masonry walls and walls that are imperfectly constructed will allow even more water to pass. A *cavity wall* prevents water from reaching the interior of the building by interposing a hollow space between the outside and inside wythes of the wall. The two wythes are separated by a continuous airspace that is spanned only by ties made of corrosion-resistant galvanized steel or stainless steel that hold the wythes together (Figure 10.2*b*, *c*, *e*, and *f*; Figures 10.3, 10.8 and 10.9). When water penetrates the outer wythe and reaches the cavity, it has no place to go but down. When it reaches the bottom of the cavity, it is caught by a thin, impervious membrane called a *flashing* and drained through *weep holes* to the exterior of the building.

To further protect against water penetration, a water-repellent coating, or *dampproofing*, may be applied to the cavity side of the interior wythe of the wall. In the event that water does bridge the cavity, this coating discourages seepage into the interior wythe. Membranes or coatings that can also function as air barriers may be applied to this surface, as illustrated in Figures 20.1 and 20.3. (See Chapter 19 for an explanation of air barriers.) To control the flow of heat though a cavity wall, rigid foam plastic insulation boards may be inserted into the cavity (Figure 10.3*d*-*f*).

The minimum recommended separation between wythes of a cavity wall is 2 inches (50 mm). This provides sufficient space for the masons to keep the cavity free of mortar obstructions while the wall is being built. Where insulation boards are inserted into the cavity, the remaining clear space between the face of the insulation and the back of the outer wythe should not be reduced to less than 1 inch (25 mm). Weep holes should be installed not more than 24 inches (600 mm) apart horizontally in brick and 32 inches (800 mm) apart in concrete masonry, and they should lie immediately on top of the flashing in the wall to keep the bottom of the cavity as dry as possible. A minimum diameter for a weep hole is ¼ inch (6 mm). A weep hole may be created in any of several ways: with a short piece of rope laid in the mortar joint and later pulled out (Figures 10.12*b* and 10.13*b*), with a plastic or metal tube laid in the mortar, or by simply leaving a head joint unmortared. Plastic and metal weep hole accessories with insect screens are available for installation in unmortared head joints to prevent bees or other insects from taking up residence within the cavity.

Figure 10.3
Photographs of joint reinforcing and ties. *(a)* Ladder reinforcing for a single wythe of concrete masonry. *(b)* Truss reinforcing. *(c)* The double rods of seismic ladder reinforcing offer greater strength than the single rods of normal ladder reinforcing. *(d)* Ladder-type side rod reinforcing reaches across an insulated cavity to reinforce the face wythe of brick masonry and tie it to the backup wythe of concrete masonry. *(e)* Truss-type eye-and-pintle tie reinforcing allows for some adjustment in course heights. *(f)* Ladder-type seismic tab tie reinforcing allows the construction of the face wythe to lag behind that of the concrete masonry backup wythe, and still provides reinforcement and positive tying of the face wythe to the backup.
(Courtesy of Dur-O-Wal Corporation)

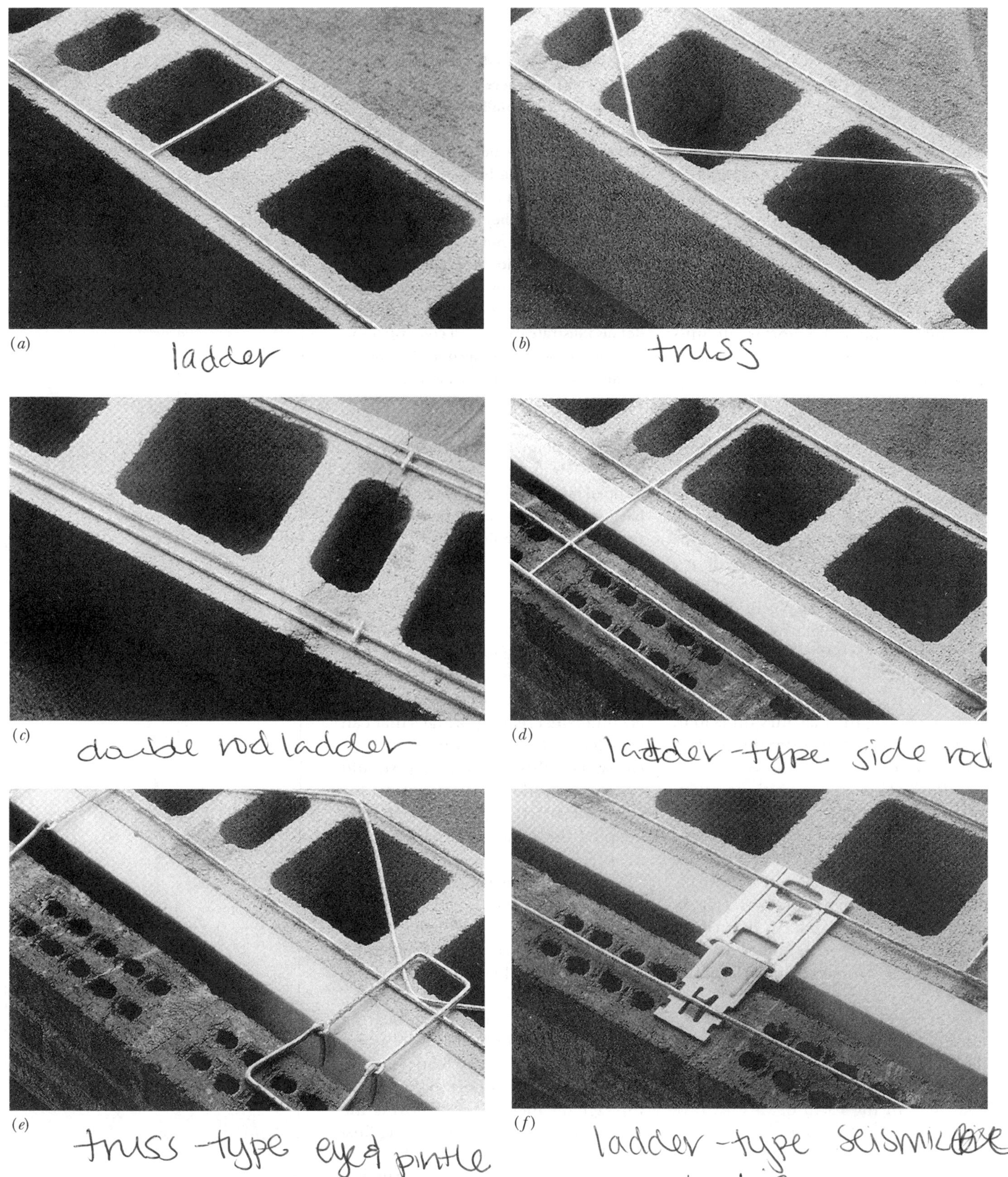
(a)
ladder
(b)
truss
(c)
double rod ladder
(d)
ladder-type side rod
(e)
truss-type eye & pintle
(f)
ladder-type seismic
tab tie

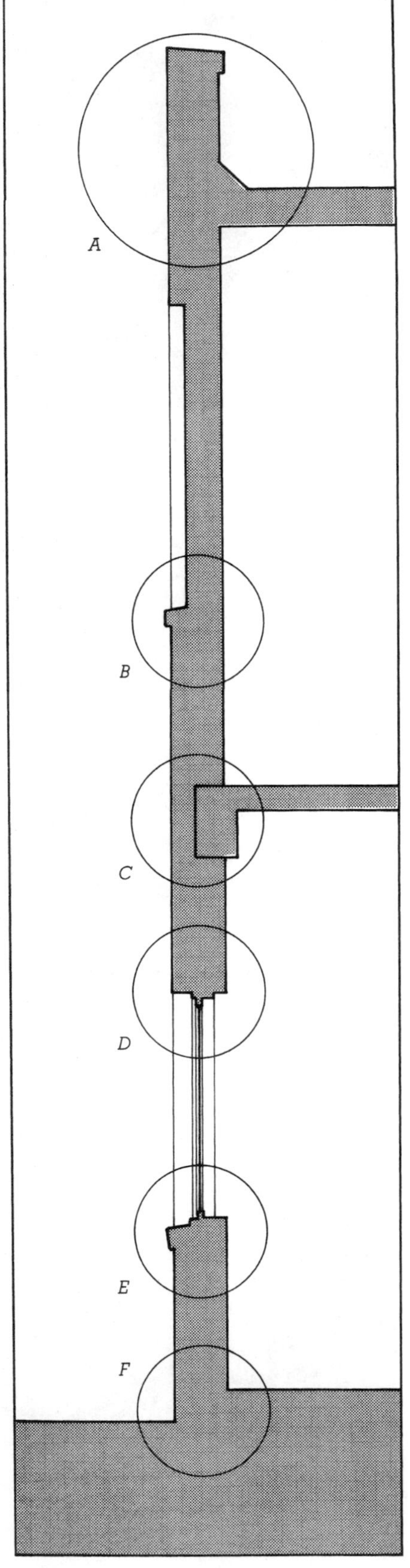

FIGURE 10.4
Typical flashings in masonry walls. Flashings are required at the top of the wall and anywhere the cavity is obstructed horizontally. The letters on the full-height wall section on this page refer to the large-scale details on the facing page. The flashings in a parapet (*a*) may be thought of as external flashings because they do not drain the cavity, and they do not usually have weep holes associated with them. The flashing shown for a recess or projection in (*b*) is for a composite masonry wall. All other details illustrate cavity wall construction. The reglet used in the second shelf angle detail (*c*) is a length of formed metal or molded plastic that is attached to the inside of the formwork before the concrete is poured. The end dam shown in the center drawing prevents water trapped by the flashing from draining back into the cavity rather than out through the weep holes.

During construction, the cavity must be kept free of mortar droppings, brick chips, and other debris, which can clog the weep holes or form bridges that can transport water across the cavity to the inner wythe. This requires good construction practices and careful inspection. When a masonry unit is pushed into position in a bed joint of mortar, some mortar is squeezed out of the joint on each side. Normally, the mason cuts off this extruded mortar with a trowel and removes it from the cavity, but inevitably some of the mortar falls into the cavity. Eventually, enough mortar may accumulate at the bottom of the cavity to block drainage holes. One solution to this problem is to place a strip of wood on the steel ties in the cavity to catch such mortar droppings, then to pull out this strip and scrape the mortar off it before placing the next row of ties. However, this method is an added inconvenience for the mason and is prone to spilling mortar as the wood strip is lifted from the cavity. An alternative is for the mason to bevel the bed joint with the trowel, making it thinner at the cavity face and thicker at the outside face, before placing the masonry units in it. This minimizes the squeezing out of mortar into the cavity.

Regardless of precautions taken during construction of the wall, the cavity should be inspected before it is closed to be sure that the weep holes have not been obstructed by mortar droppings or other debris. It was once common practice to specify that a layer of ⅜-inch-diameter (9-mm-diameter) gravel be deposited in the base of the cavity to prevent such obstruction. However, research has shown that this is not only ineffectual, but actually tends to trap water even if the cavity is clean. More recently, *cavity drainage materials*, in the form of various proprietary free-draining, woven, or matted products have come into use. These materials are inserted within the cavity, either at the bottom of the cavity or continuously throughout its full height, to catch mortar droppings so that they do not obstruct drainage (Figure 10.5). Such products, though helpful in keeping weep holes open, may still permit mortar droppings to accumulate, creating bridges that can transfer heat and moisture across the cavity. For this reason, they should not be considered a substitute for good construction practices aimed at keeping the cavity as free of mortar droppings as possible.

In a loadbearing cavity wall, the backup wythe normally carries the structural loads while the outer wythe serves as a nonstructural facing or *veneer.* (Less commonly, both wythes can participate in carrying structural loads.) In nonloadbearing walls, the backup wythe does not carry structural gravity loads but still provides critical support to the veneer through the metal ties, anchoring the veneer in place and bracing it against wind and seismic forces. Additional examples of cavity wall construction can be found elsewhere in this text: Brick veneer curtain wall with concrete masonry backup is illustrated in

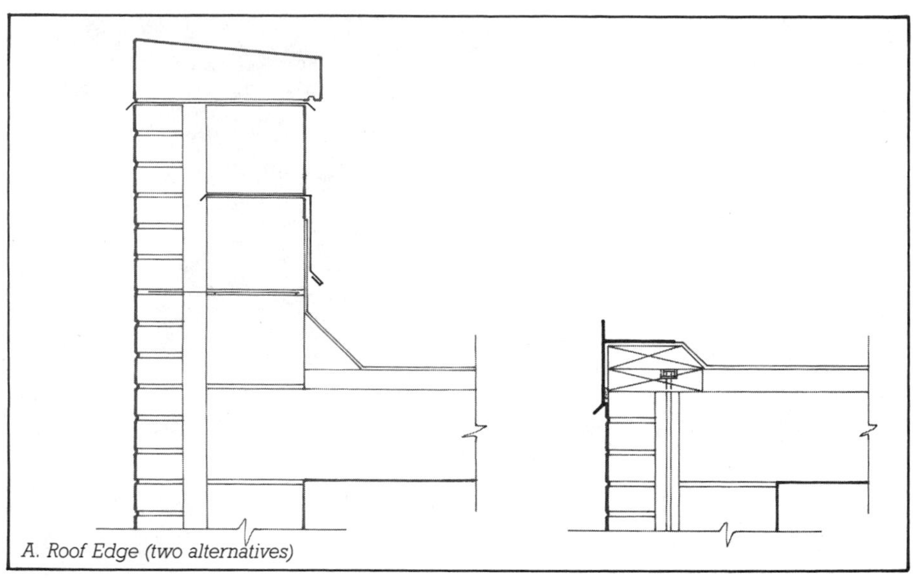
A. Roof Edge (two alternatives)

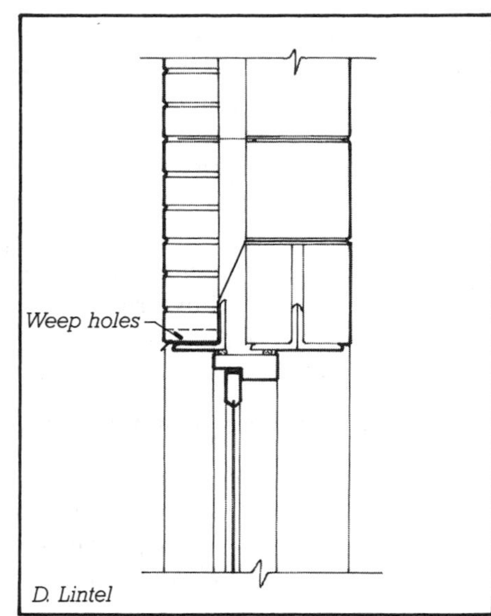

D. Lintel

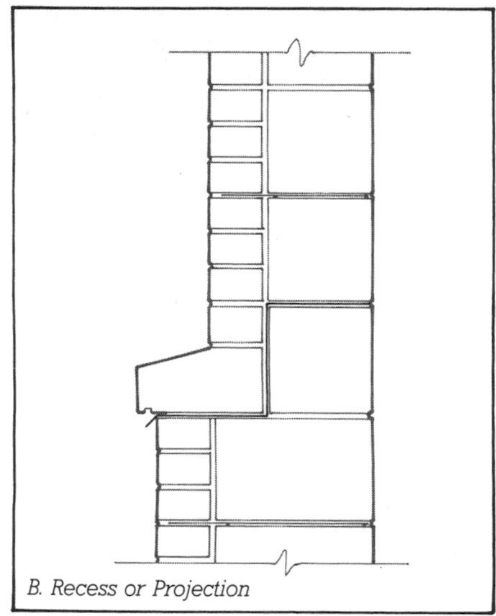
B. Recess or Projection

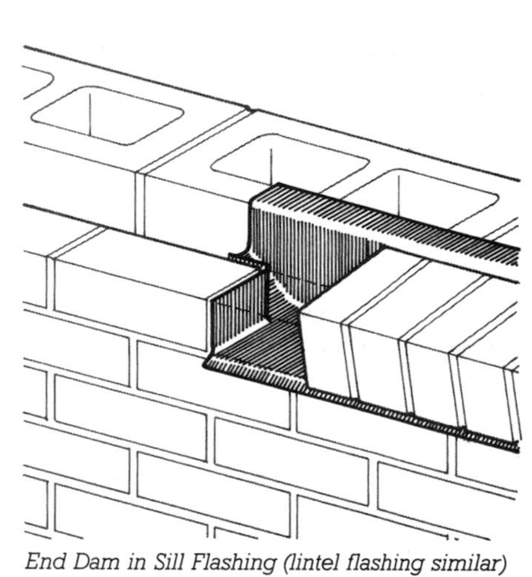
End Dam in Sill Flashing (lintel flashing similar)

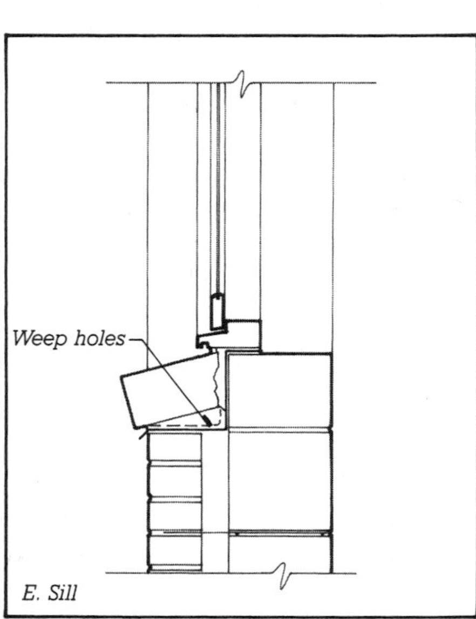

E. Sill

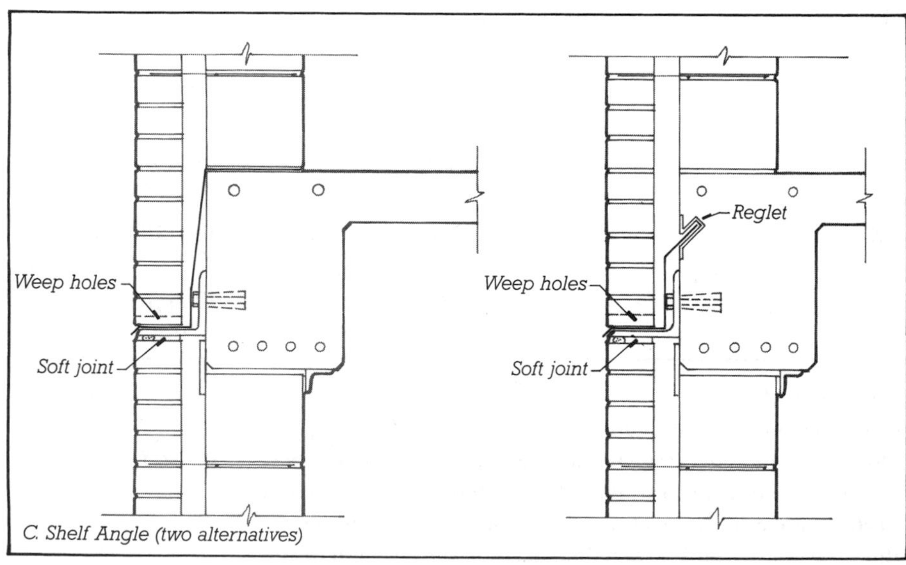

C. Shelf Angle (two alternatives)

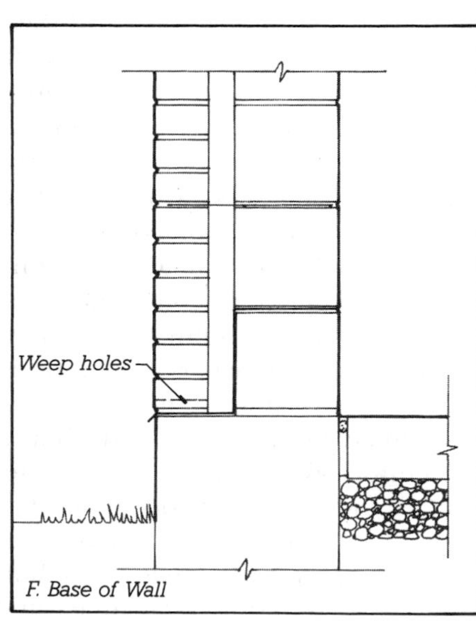

F. Base of Wall

Figures 20.1 through 20.3 and is discussed in detail in the accompanying captions and text. Brick veneer with a steel stud backup wall is shown in Figure 20.5, a case study of brick veneer cladding is presented on pages 810–817, and brick veneer with wood light framing is illustrated in Figure 6.33. An example of exterior stone facing is shown in Figure 9.12.

Masonry Loadbearing Walls

Walls constructed of brick, stone, or concrete masonry can be used to support roof and floor structures of wood light framing, heavy timber framing, steel, sitecast concrete, precast concrete, or masonry vaulting. Because these *masonry loadbearing walls* (usually called simply *bearing walls*) do double duty by serving also as exterior walls or interior partitions, they are often a very economical system of construction compared to systems that carry their structural loads on columns of wood, steel, or concrete but require separate systems of construction to fill in between these members.

Reinforced Masonry Walls

Loadbearing masonry walls may be built with or without reinforcing. Unreinforced masonry walls, however, cannot carry such high compressive stresses as reinforced walls, and they have little ability to resist tension forces. This makes them unsuitable for use in regions with seismic risk, or for walls subject to strong lateral forces such as from heavy winds or earth pressures. Steel-reinforced masonry loadbearing walls can also be constructed thinner than comparable unreinforced walls, resulting, especially in taller walls, in savings in materials, construction labor, and floor area consumed by the wall. In the past, unreinforced walls were used in the United States to support buildings as tall as 16 stories (Figure 8.37). In contemporary construction, all but the smallest and simplest loadbearing masonry walls are reinforced.

FIGURE 10.5
One manufacturer's proprietary cavity drainage material. Strips of tangled matting cut with a keystone pattern are inserted into the bottom of the wall cavity, where they prevent accumulations of mortar from obstructing weep holes. As noted in the text, mortar droppings in the cavity have other detrimental effects and should be avoided, regardless of the use of such materials. (*©Copyright 2007, MotarNet® USA, Ltd. All rights reserved.*)

	Ultimate Compressive Strength	Density
Concrete masonry units	1700–6000 psi *12–41 MPa*	75–145 lb/ft^3 *1200–2320 kg/m^3*
Bricks	2000–20,000 psi *14–140 MPa*	100–140 lb/ft^3 *1600–2240 kg/m^3*
Limestone	2600–33,000 psi *18–230 MPa*	130–170 lb/ft^3 *2080–2720 kg/m^3*
Sandstone	4000–35,000 psi *28–240 MPa*	130–165 lb/ft^3 *2080–2640 kg/m^3*
Marble	7500–27,000 psi *502–190 MPa*	165–170 lb/ft^3 *2640–2720 kg/m^3*
Granite	19,000–45,000 psi *130–310 MPa*	165–170 lb/ft^3 *2640–2720 kg/m^3*

FIGURE 10.6
Common ranges of strength and density for bricks, concrete blocks, and building stone to allow comparison between these materials. In practice, the allowable compressive stresses used in structural calculations for masonry are much lower than the values given, both to take into account the strength of the mortar and to provide a substantial safety factor.

The engineering design of masonry loadbearing walls, both reinforced and unreinforced, is governed by *Building Code Requirements for Masonry Structures, ACI 530/ASCE 5/TMS 402,* a standard jointly prepared by the American Concrete Institute, the American Society of Civil Engineers, and The Masonry Society. This document establishes requirements for both masonry materials (masonry units, reinforcing materials, mortar, metal ties and accessories, and grout) and methods of construction. It also sets forth the engineering procedures by which the strength and stiffness of masonry structural elements are calculated. Methods of reinforcing brick and concrete masonry walls have been described in Chapters 8 and 9, and are further illustrated in this chapter in Figures 10.2, 10.3, 10.7–10.10, and 10.15. Reinforcing requirements for relatively short, lightly loaded bearing walls often can be read from standard design tables. However, for larger or more heavily loaded walls, the number, locations, and sizes of the steel reinforcing bars must be determined individually for each structural condition by a structural engineer.

Posttensioned Masonry Walls

Masonry walls may be posttensioned utilizing high-strength steel threaded rods or flexible tendons rather than conventionally reinforced with ordinary vertical reinforcing bars (Figure 10.7). These posttensioning elements are anchored into the foundation and run vertically through the masonry wall, either in a cavity between wythes or in the cores of concrete masonry units. After the wall has been completed and the mortar has cured, each rod or tendon is tensioned (stretched very tightly) and anchored in its tensioned condition to a horizontal steel plate at the top of the wall. Posttensioning rods are threaded so that each may be tensioned by tightening a nut against a steel plate. Tendons are stretched

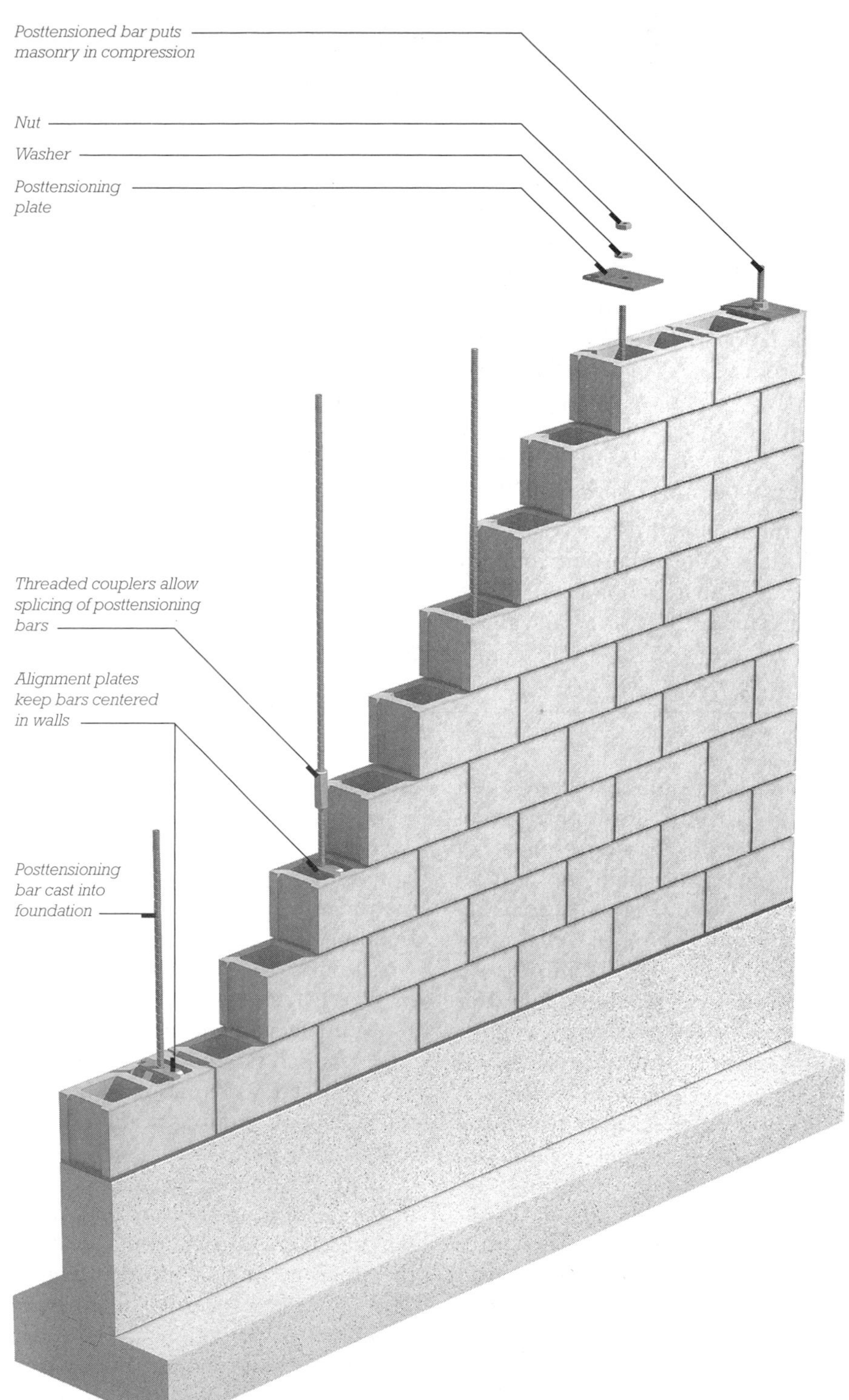

Posttensioned bar puts masonry in compression
Nut
Washer
Posttensioning plate
Threaded couplers allow splicing of posttensioning bars
Alignment plates keep bars centered in walls
Posttensioning bar cast into foundation

FIGURE 10.7
One system for posttensioning a concrete masonry wall. Short sections of round, threaded high-strength steel bar are joined with threaded couplers as the wall rises. At the base of the wall, the bar is anchored to a threaded insert that has been epoxied into a hole drilled in the concrete foundation. At the top, the bar passes through a steel plate. When the nut at the top end of the bar is tightened, the masonry wall is placed in compression. A load indicator washer tells when the bar is applying sufficient compression to the wall. Load indicator washers are shown in Figure 11.17.

For Preliminary Design of a Loadbearing Masonry Structure

- To estimate the size of a reinforced brick masonry column, add up the total roof and floor area supported by the column. A 12-inch (300-mm) square column can support up to about 2000 square feet (185 m^2) of area, a 16-inch (400-mm) column 3000 square feet (280 m^2), a 20-inch (500-mm) column 5000 square feet (465 m^2), and a 24-inch (600-mm) column 7000 square feet (650 m^2).
- To estimate the size of a reinforced concrete masonry column, add up the total roof and floor area supported by the column. A 12-inch (300-mm) square column can support up to about 1000 square feet (95 m^2) of area, a 16-inch (400-mm) column 2000 square feet (185 m^2), a 20-inch (500-mm) column 3000 square feet (280 m^2), and a 24-inch (600-mm) column 4000 square feet (370 m^2).
- To estimate the thickness of a reinforced brick masonry loadbearing wall, add up the total width of floor and roof decks that contribute load to a 1-foot (305-mm) length of wall. An 8-inch (200-mm) wall can support up to approximately 650 feet (200 m) of deck and a 12-inch (300-mm) wall about 800 feet (245 m).
- To estimate the thickness of a reinforced concrete masonry loadbearing wall, add up the total width of floor and roof decks that contribute load to a 1-foot (305-mm) length of wall. An 8-inch (200-mm) wall can support up to approximately 400 feet (120 m) of deck, a 12-inch (300-mm) wall 700 feet (215 m).

These approximations are valid only for purposes of preliminary building layout and must not be used to select final member sizes. They apply to the normal range of building occupancies, such as residential, office, commercial, and institutional buildings, and parking garages. For manufacturing and storage buildings, use somewhat larger members.

For more comprehensive information on preliminary selection and layout of a structural system and sizing of structural members, see Edward Allen and Joseph Iano, *The Architect's Studio Companion* (4th ed.), New York, John Wiley & Sons, Inc., 2007.

by a special hydraulic jack and then anchored in their tensioned condition with the aid of a steel chuck that grips the wires of which the tendon is made. In either case, this tensioning of the reinforcing places the entire wall under a vertical compressive prestress that is considerably higher than would be created by the weights of the masonry and the floors and roofs that it supports. The effect of posttensioning is to strengthen the wall against forces that would normally induce tension in the wall, such as from wind or seismic loads. This allows the use of thinner walls with fewer grouted cores, which saves material and labor. The concept of posttensioning is discussed and illustrated in more detail in Chapter 13.

Spanning Systems for Masonry Bearing Wall Construction

Ordinary Joisted Construction

So-called *Ordinary construction*, in which the floors and roof are framed with wood joists and rafters and supported at the perimeter on masonry walls, is the fabric of which American center cities largely were built in the 19th century. It still finds use today in a small percentage of new buildings and is listed as Type III construction in the building code table in Figure 1.2. Ordinary construction is essentially balloon framing (Figure 5.2) in which the outer walls of wood are replaced with masonry bearing walls. Balloon framing of the interior loadbearing partitions is used instead of platform framing because it minimizes the sloping of floors that might be caused by wood shrinkage along the interior lines of support. Figure 10.8 shows the essential features of Ordinary construction. Notice two very important details: the firecut ends of the joists and the metal anchors used to tie the wood framing and the masonry wall together.

Heavy Timber or Mill Construction

Heavy Timber or *Mill construction*, listed as Type IV-HT in the code table (Figure 1.2), like Ordinary construction, combines masonry exterior walls with a wood frame interior. However, it uses heavy timbers rather than light joists, rafters, and studs, and thick timber decking rather than thin wood panel sheathing and subflooring. Because heavy timbers are slower to catch fire and burn than nominal 2-inch (38-mm) framing members, somewhat larger floor areas and greater building heights are permitted with Mill construction than with Ordinary construction. Mill construction is discussed in detail in Chapter 4 and a contemporary example is illustrated in Figure 4.13.

Steel and Concrete Decks with Masonry Bearing Walls

Spanning systems of structural steel, sitecast concrete, and precast concrete are frequently used in combination with masonry bearing walls. Figures 10.9 and 10.10 show representative details of two of these combinations. Depending on the degree of fire resistance of the spanning elements, these constructions may be classified under Type I or Type II in the table in Figure 1.2.

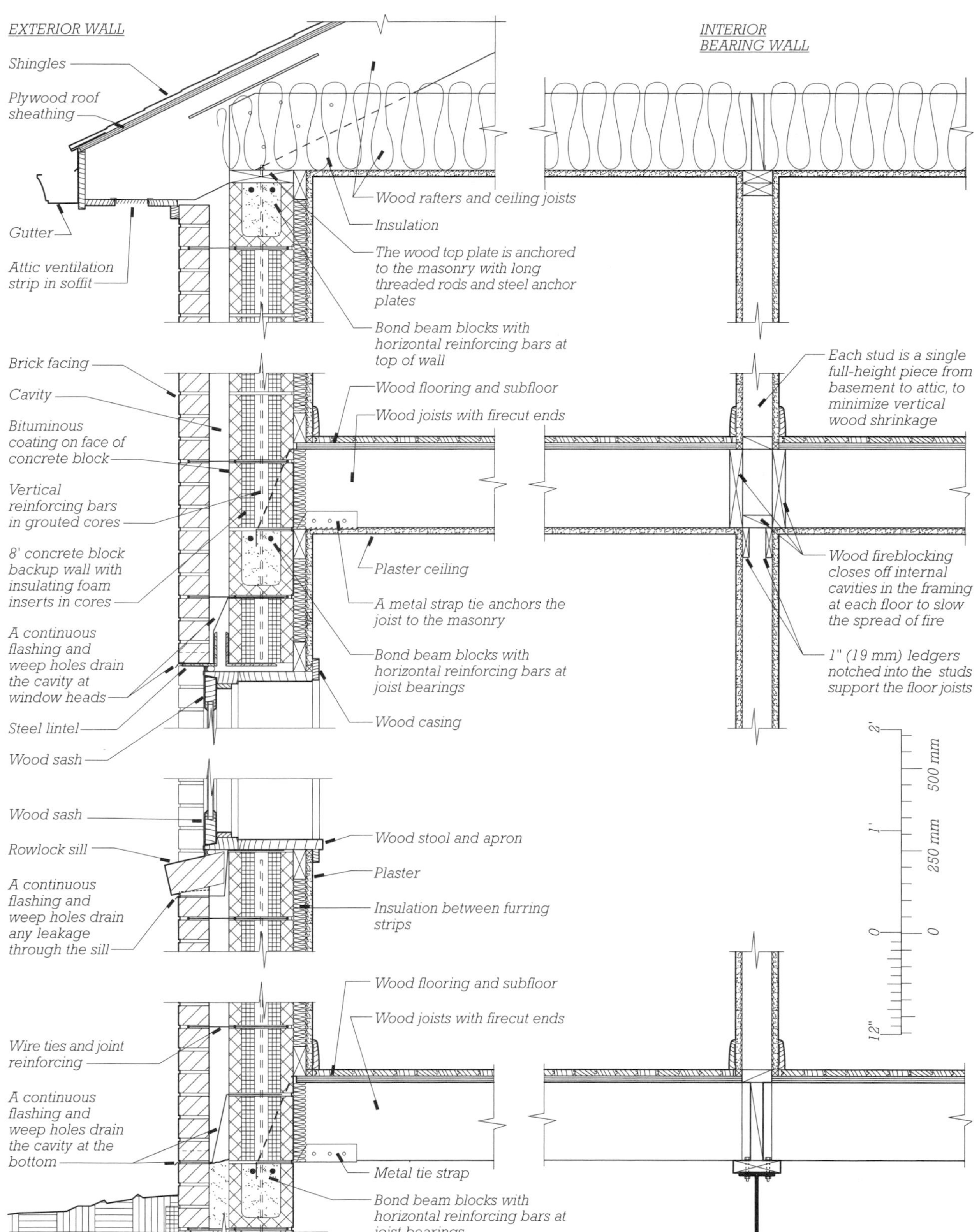

FIGURE 10.8
Ordinary construction, shown here with a cavity wall of brick with a concrete masonry loadbearing wythe. Thermal insulation is installed between wood furring strips on the interior side. The interior of a building of Ordinary construction is framed with balloon framing of wood studs and joists. At the perimeter of the building, the joists are supported by the masonry wall. The rowlock window sill detailed here is similar to the one whose construction is shown in Figure 10.13.

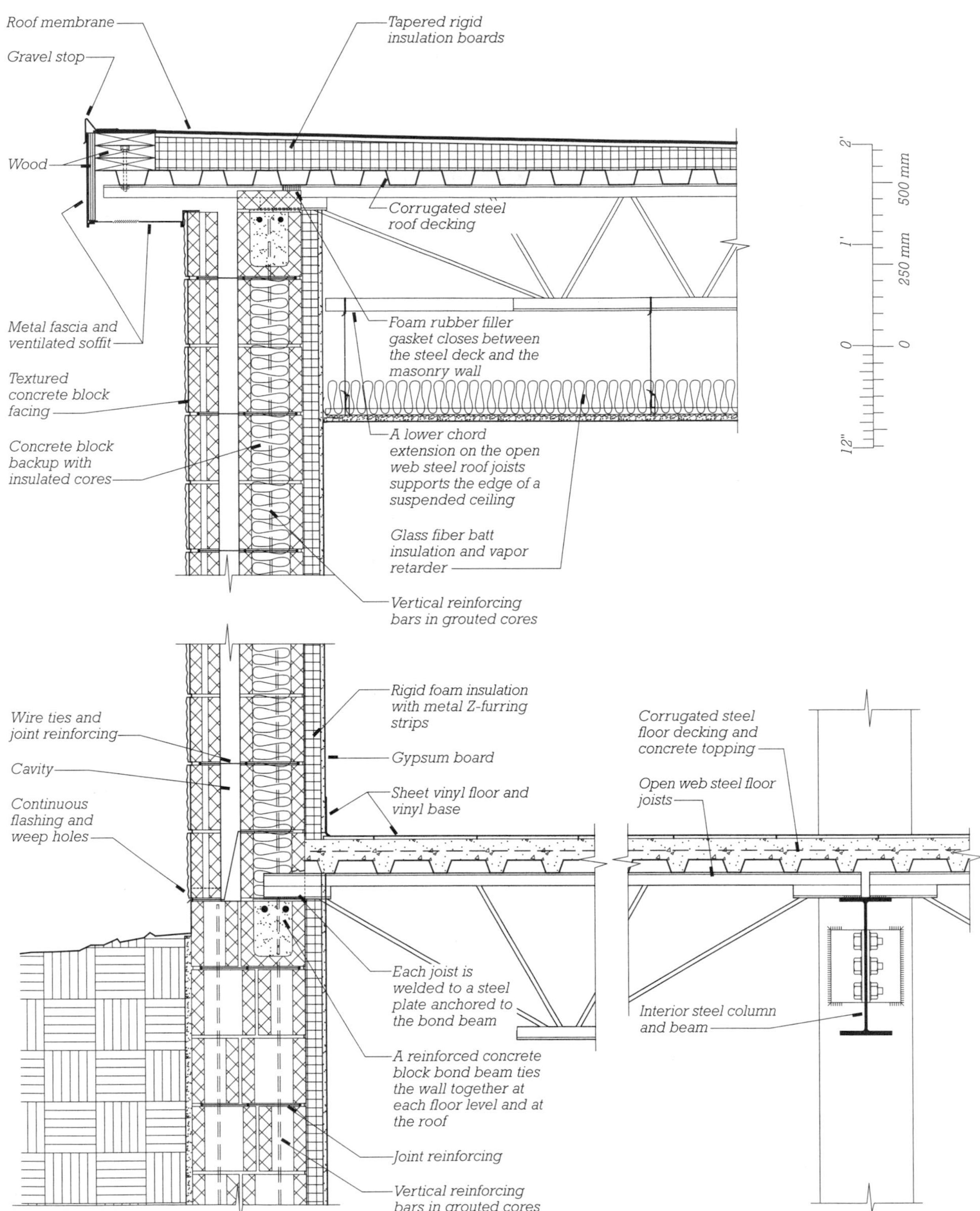

FIGURE 10.9
An example of a concrete masonry exterior bearing wall with a roof and floor of steel open-web joists and corrugated steel decking. In regions with very severe winters, insulation in the wall cavity or on the inside surface of the wall would offer better cold-weather performance than insulation within the cores, as shown here. With suitable fire protection for the steel structure, this type of construction could be built many stories high.

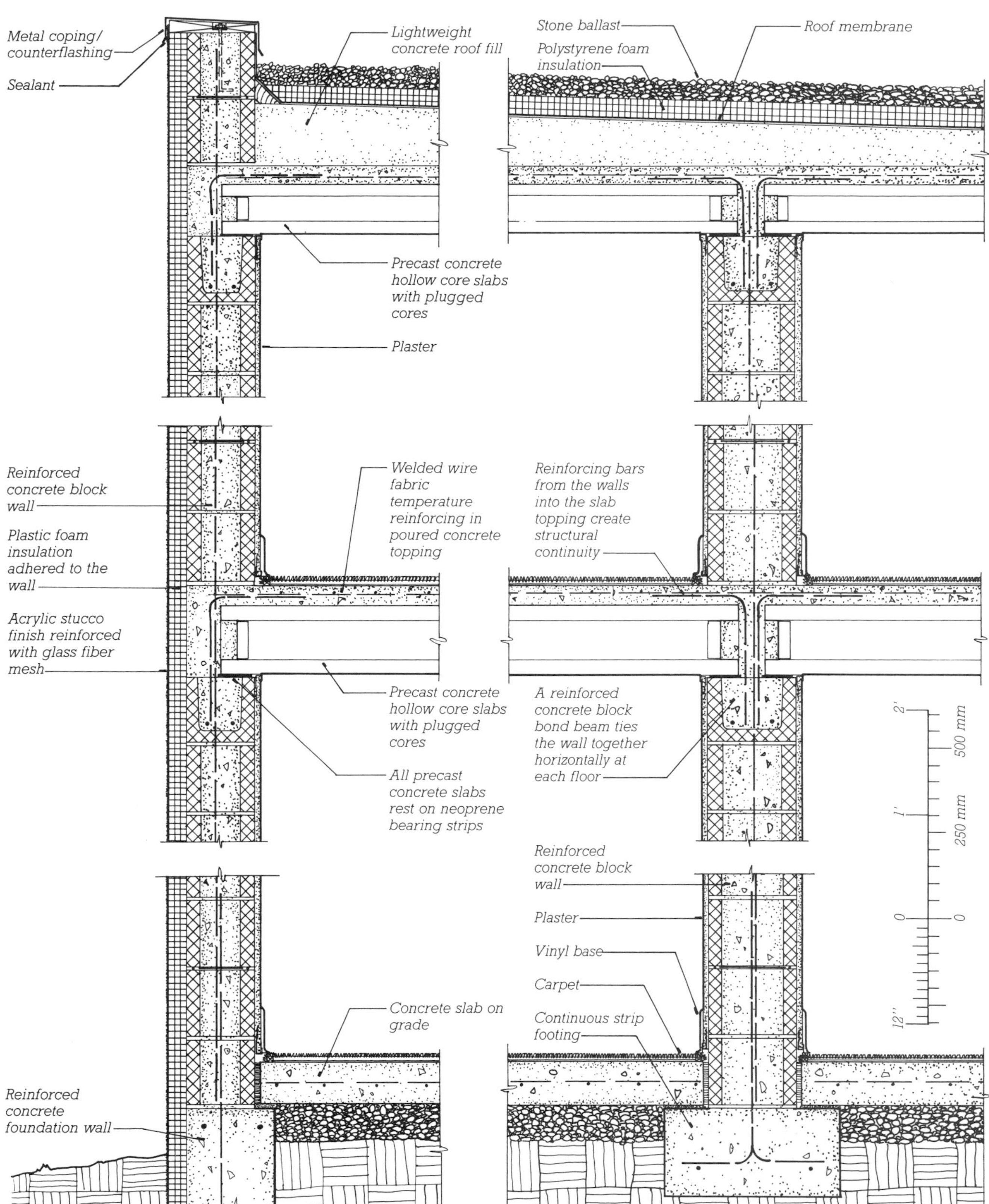

FIGURE 10.10
An example of concrete block exterior and interior bearing walls with precast concrete hollow-core slabs spanning the roof and floors. This system can be used for multistory buildings. Full wall reinforcing is shown, with EIFS applied to the outside of the building. Details of this exterior finishing system are further discussed later in this chapter.

Detailing Masonry Walls

Flashings and Drainage

Two general types of flashing are used in masonry construction: *External flashings* prevent moisture from penetrating into the masonry wall at its exposed top or where it intersects the roof. *Internal flashings* (also known as *concealed* or *through-wall flashings*) catch water that has penetrated a masonry wall and drain it through weep holes back to the exterior.

The external flashing at the intersection of a flat roof and a wall parapet is usually constructed in two overlapping parts, a *base flashing* and a *counterflashing* or *cap flashing*. This makes installation easier and allows for some movement between the wall and roof components. Often the base flashing is formed by the roof membrane itself. The base flashing is normally turned up for a height of at least 8 inches (200 mm). The counterflashing is embedded in the masonry wall above the base flashing and extends downward, lapping over the base flashing. Counterflashings are often made in two interlocking pieces, making them easier to install and easier to remove when the roof membrane must be replaced (Figures 10.4*a*, 10.11, 16.29).

Internal flashings are installed by masons as they construct the wall. We have already discussed the internal flashing that is placed at the bottom of the wall cavity. Additional flashings are required at every location where the cavity is interrupted: at heads of windows and doors, at window sills, at shelf angles, and at spandrel beams. Where an internal flashing crosses the wall cavity, it should be turned up 6 to 9 inches (150–225 mm) at the back face of the cavity and penetrate the inner wythe by at least 2 inches (50 mm). In this manner, water draining down the cavity is intercepted by the flashing and directed toward the exterior of the wall. Examples of this internal flashings can be seen in Figure 10.4–*b-f*. If the cavity is backed up by a concrete beam or wall, it may terminate in a *reglet*, a horizontal slot formed in the face of the concrete, as shown on the right-hand side of Figure 10.4*c*. At the outside face of the wall, the flashing should be carried at least ¾ inch (19 mm) beyond the face of the wall and turned down at a 45-degree angle so that water draining from the flashing drips free of the

Figure 10.11
Copper flashings in a parapet. The single-ply roof membrane is turned up to self-flash under the copper counterflashing. (*Courtesy of Copper Development Association Inc.*)

wall, rather than being drawn under the flashing by capillary action and returning into the wall. It is a common but dangerous practice to terminate the flashing just inside the face of the wall in order to hide the flashing; this can cause significant amounts of water to be reabsorbed by the wall underneath the flashing.

Flashings may be made of sheet metal, modified asphalt membranes, plastics, rubbers, or composite sheets. Sheet metal flashings are the most durable and the most expensive. Copper and stainless steel are best; galvanized steel eventually rusts and disintegrates. Aluminum and lead are unsuitable for flashings in masonry walls because they react chemically with mortar. Asphaltic flashings are made of polymer-modified asphalts laminated to plastic backings. Most are manufactured with preapplied adhesive on one side, for which reason they are frequently referred to as *self-adhered flashings*. In comparison to metal flashings, they can be more easily shaped and sealed at corners and laps. They are often used in combination with sheet metal. The sheet metal can support the flexible asphaltic membrane where it spans the wall cavity and can extend beyond the outer face of the wall to form the recommended drip. (Asphaltic flashings are too flexible to form a projecting drip, and this material cannot be permanently exposed to sunlight.) Composite (laminated) flashings combine layers of various materials and are intermediate in price. Most consist of a heavy foil of copper or lead laminated with polyester film, glass fiber mesh, bitumen-coated fabric, or waterproofed kraft paper. Many composite flashings are very durable. Flashings of plastic and rubber are the least expensive. Some are durable, but others deteriorate too rapidly for use in masonry walls. The designer should specify only those types that have been proven suitable for the purpose.

Flashings within a wall are almost impossible to replace if they fail in service. Even the most expensive flashing materials cost only a very small fraction of the total price of a masonry wall, so there is little reason to use cheap ones in a misguided effort to save money.

Careful supervision is necessary to ensure that flashings are properly installed. At corners and other junctions where separate pieces of flashing material meet, the pieces should be lapped at least 6 inches (150 mm) and either soldered together or sealed with a suitable mastic (Figure 10.12). Head and sill flashings should extend beyond the jambs of the opening and terminate in folded *end dams* (Figures 10.4 and 10.13*a*). The end dams ensure that water trapped by the flashing drains back to the exterior of the wall rather than spilling off the end of the flashing within the cavity. End dams should also be formed wherever else a flashing is interrupted horizontally, such as at exposed columns or expansion and control joints.

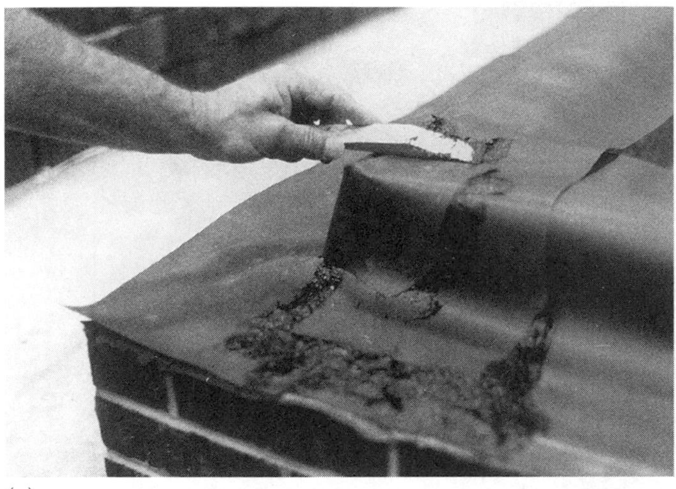

(*a*)

(*b*)

FIGURE 10.12

Flashing the base of a cavity wall with composite flashing sheet. *(a)* The mason uses a mastic adhesive to seal the lap joints in the flashing at the corners. *(b)* A piece of rope forms a weep hole. Because this flashing material is sensitive to sunlight and is too flexible to project beyond the face of the wall, it should be trimmed back slightly inside the face and cemented to the top of a strip of sheet copper or stainless steel that projects beyond the face and turns down to form a drip. (*Courtesy of Brick Institute of America*)

(*a*)

(*b*)

(*c*)

FIGURE 10.13
Flashing a rowlock brick window sill with composite flashing sheet. ***(a)*** **The flashing sheet is turned up to form a dam at the end. As in Figure 10.12, the projecting portion should be made of sheet metal.** ***(b)*** **Pieces of rope form weep holes in the bed joint beneath the sloping rowlock sill bricks. These will be pulled out after the mortar has stiffened.** ***(c)*** **The finished sill.** (*Courtesy of Brick Institute of America*)

Thermal Insulation of Masonry Walls

A solid masonry wall is a good conductor of heat, which is another way to say that it is a poor insulator. In many hot, dry climates, the capacity of an uninsulated masonry wall to store heat and retard its passage keeps the inside of the building cool during the hot day and warm during the cold night. But in climates with sustained cold or hot seasons, measures must be taken to improve the thermal resistance of masonry walls. The introduction of an empty cavity into a wall improves its thermal insulating properties considerably, but not to a level fully sufficient for cold climates.

There are three general ways of insulating masonry walls: on the outside face, within the wall, and on the inside face. Insulation on the outside face is a relatively recent development. It is usually accomplished by means of an *exterior insulation and finish system (EIFS)*, which consists of panels of plastic foam that are attached to the masonry and covered with a thin, continuous layer of polymeric stucco reinforced with glass fiber mesh. The appearance is that of a stucco building. The masonry is completely concealed, and can be of inexpensive materials and unrefined workmanship. EIFS is frequently used for insulating existing masonry buildings in cases where the exterior appearance of the masonry does not need to be retained. An advantage of EIFS is that the masonry is protected from temperature extremes and can function effectively to stabilize the interior temperature of the building. Disadvantages are that the thin stucco coatings are usually not very resistant to denting or penetration damage and that the EIFS is combustible. Furthermore, most EIFSs have no internal flashings or drainage, so puncture damage and lapses in workmanship, especially around edges and openings, can lead to substantial moisture leakage into the walls and the building. Figure 10.10 details a building insulated in this manner, and photographs of the installation of exterior insulation and finish systems are shown in Figure 20.23.

Insulation within the wall can take several forms. If the cavity in a wall is made sufficiently wide, the masons can insert slabs of plastic foam insulation against the inside wythe of masonry as the wall is built (Figure 10.3*d–f*). The hollow cores of a concrete block wall can be filled with loose granular insulation (Figure 10.14) or with special molded-to-fit liners of foam plastic (Figure 10.15). Insulating the cores of concrete blocks does not retard the passage of heat through the webs of the blocks, however, and is most effective when it is coupled with an unbroken layer of insulation in the cavity or on one face of the wall.

The inside surface of a masonry wall can be insulated by attaching

FIGURE 10.14
Insulating the cores in a concrete block wall with a dry fill insulation. The insulation is noncombustible, inorganic, nonsettling, and treated to repel water that might be present in the block cores from condensation or leakage. (*Courtesy of W. R. Grace & Co.*)

FIGURE 10.15
This proprietary concrete masonry system is designed with polystyrene foam inserts that provide a high degree of thermal insulation. The webs of the blocks are minimal in area to reduce thermal bridging where they pass through the foam. Vertical reinforcing bars can be placed as shown, and horizontal bars can be laid in the grooved webs. The cores must be grouted where reinforcing is used. *(Courtesy of Korfil, Inc., P.O. Box 1000, West Brookfield, MA 01585)*

FIGURE 10.16
A worker installs foam plastic insulation on the interior of a concrete masonry wall, using a patented system of steel furring strips in which only isolated clips, visible in this photograph, contact the masonry wall to minimize thermal conduction through the metal. The furring strips serve as a base to which interior finish panels such as gypsum board can be attached. Other furring systems for masonry walls are shown in Figures 23.4–23.6. *(Courtesy of W. R. Grace & Co.)*

wood or metal *furring strips* to the inside of the wall with masonry nails or powder-driven metal fasteners (Figures 10.8, 10.9, 10.16, and 23.4–23.6). Furring strips may be of any desired depth to house the necessary thickness of fibrous or foam insulation. The gypsum wallboard or other interior finish material is then fastened to the furring strips. Furring strips can also solve another chronic problem of masonry construction by creating a space in which electrical wiring and plumbing can easily be concealed.

Some Special Problems of Masonry Construction

Expansion and Contraction

Masonry walls expand and contract slightly in response to changes in both temperature and moisture content. Thermal movement is relatively easy to quantify (Figure 10.17). Moisture movement is more difficult: New clay masonry units tend to absorb water and expand under normal atmospheric conditions. New concrete masonry units usually shrink somewhat as they give off excess water following manufacture. Expansion and shrinkage in masonry materials are small compared to the moisture movement in wood or the thermal movement in plastics or aluminum, but they must be taken into account in the design of the building by providing surface divider joints to avoid an excessive buildup of forces that could crack or spall the masonry (see the accompanying sidebar, "Movement Joints in Buildings").

Three different kinds of surface divider joints are used in masonry. Expansion joints are intentionally created slots that can close slightly to accommodate expansion of surfaces made of brick or stone masonry. Control joints are intentionally created cracks that can open to accommodate shrinkage in surfaces made of concrete masonry. Abutment joints, sometimes called "construction joints" or "isolation joints," are placed at junctions between masonry and other materials, or between new masonry and old masonry, to accommodate differences in movement. Figures 10.18–10.21 illustrate the use of movement joints in masonry walls. Movement joints are also critical in masonry facings applied over multistory structural frames of steel or concrete to prevent fracture of the masonry when the frame deflects under load, as discussed in Chapter 20.

Joint reinforcing must be interrupted at movement joints so that it does not restrain the opening or closing of the joint. To prevent out-of-plane displacements of the wall, various kinds of vertically interlocking details are often used, as seen in Figure 10.21. Most movement joints in masonry walls are closed with flexible sealants to prevent air and water from passing through.

Masonry is a massive material, taking forms permitted by the law of gravity. Our vocabulary of masonry forms was developed in buildings which became essays about gravity—great weights piled high, buttresses braced against the thrust of arched vaulting. Long after the internal steel frame relieved the need for such forms, they still retain meaning for us through their historical references and in their familiarity. The basic masonry forms have become symbols.

Michael Shellenbarger, *Landmarks: A Tradition of Portland Masonry Architecture*, 1984.

Material	Coefficient of Linear Thermal Expansion	
	in./in.-°F	mm/mm-°C
Clay or shale brick masonry	3.6×10^{-6}	6.5×10^{-6}
Lightweight concrete masonry	4.3×10^{-6}	7.7×10^{-6}
Limestone	4.4×10^{-6}	7.9×10^{-6}
Granite	4.7×10^{-6}	8.5×10^{-6}
Normal-weight concrete masonry	5.2×10^{-6}	9.4×10^{-6}
Marble	7.3×10^{-6}	13.1×10^{-6}
Normal-weight concrete	5.5×10^{-6}	9.9×10^{-6}
Structural steel	6.5×10^{-6}	11.7×10^{-6}

Figure 10.17
Average linear coefficients of thermal expansion for some masonry materials.

Movement Joints in Buildings

Building materials and buildings are constantly in motion. Many of these motions are cyclical and never-ending. Some are caused by temperature changes: All materials shrink as they grow cooler and expand as they become warmer, each material doing so at its own characteristic rate (see Figure 10.17 and the table of coefficients of thermal expansion in the Appendix). Some are caused by changes in moisture content: Most porous materials grow larger when wetted by water or humid air and smaller when they dry out, also at rates that vary from one material to another. These cyclical motions caused by temperature and moisture can occur on a seasonal basis (warm and moist in summer, cold and dry in winter), and they can also occur in much shorter cycles (warm days, cool nights; warm when the sun is shining on a surface, cool when a cloud covers the sun). Some motions are caused by structural deflections under load, such as the slight sagging of beams, girders, joists, and slabs. These motions can be very long term for dead loads and for floors supporting stored materials in a warehouse. Deflections can be medium term for snow on a roof and very short term for walls resisting gusting winds. Some motions are one-time phenomena: Concrete and stucco shrink as they cure and dry out, whereas gypsum plaster expands upon curing. Clay bricks expand slightly over time as they absorb atmospheric moisture. Concrete columns shorten slightly, and concrete beams and slabs sag a bit due to plastic creep of the material during the first several years of a building's life; then they stabilize. Posttensioned slabs and beams grow measurably shorter as they take up the compression induced by the stretched steel tendons. Soil compresses under the pressure of the foundations of a new building and then, in most cases, stops moving.

Chemical processes can cause movement in building components: If a steel reinforcing bar rusts, it expands, cracking the concrete around it. Solvent release sealants shrink as they cure. Some plastics shrink and crack upon prolonged exposure to sunlight. Motion can also be caused by the freezing expansion of water, as happens in the upward heaving of insufficiently deep footings during a cold winter, or in the spalling of concrete and masonry surfaces exposed to wetting and freezing.

Most of these motions are small in magnitude. They are largely unpreventable. But they occur in buildings of every size and every material, and if they are ignored in design and construction, they can tear the building apart, cracking brittle materials and applying forces in unanticipated ways to both structural and nonstructural building components that can cause these elements to fail.

We accommodate these inevitable movements in buildings in two different ways: In some cases, we strengthen a material to enable it to resist the stress that will be caused by an anticipated movement, as we do in adding shrinkage–temperature steel to a concrete slab. In most cases, however, we install movement joints in the fabric of the building that are designed to allow the movements to occur without damaging the building. We locate these joints in places where we anticipate maximum potential distress from expected movements. We also place them at regular intervals in large surfaces and assemblies to relieve movement-caused stresses before they can cause damage. A building that is not provided suitably with movement joints will make its own joints by cracking and spalling at points of maximum stress, creating a situation that is unsightly at best and sometimes dangerous or catastrophic.

Types of Building Joints

The usual terminology applied to joints in buildings is confusing and often contradictory. The term

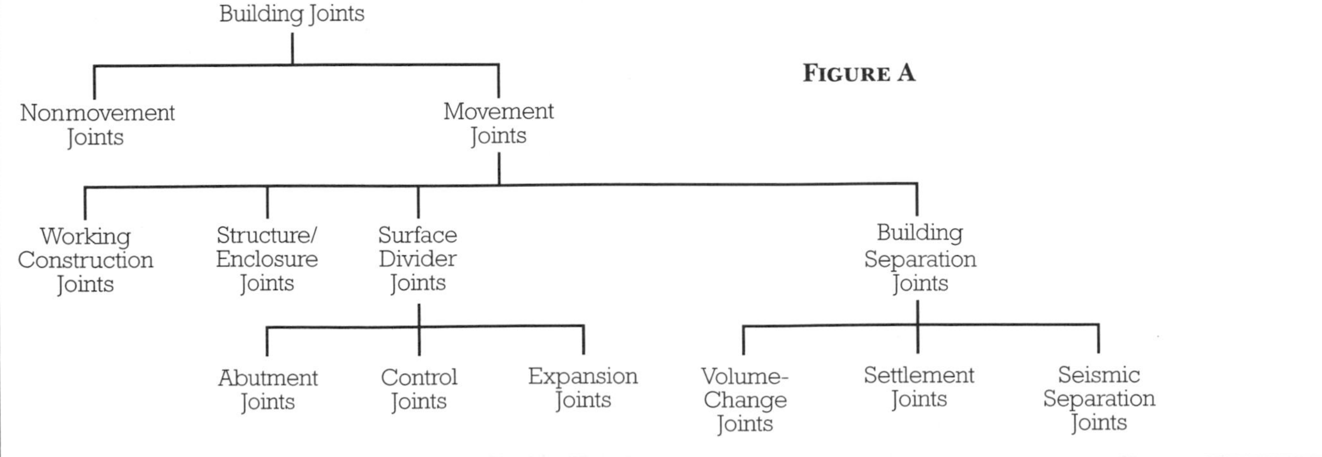

Figure A

"expansion joint," for example, is commonly and erroneously applied to almost any type of movement joint.

A more logical system of terminology is proposed in Figure A. This establishes two broad classifications, movement joints and nonmovement joints. *Nonmovement joints* include most types of joints that are used to connect pieces of material in a building, joints such as the nailed connections in the wooden frame of a house, mortar joints between masonry units, welded and bolted connections in a steel frame, and joints between pours of concrete. A nonmovement joint can be made to move only by overloading the joint, as in the pulling apart of a nailed connection, the slipping of steel members in a bolted connection, or the cracking of a weld, a mortar joint, or a concrete slab.

Movement joints are of many different kinds. What they have in common is a designed-in ability to adjust to expected amounts of motion without distress.

- The simplest movement joints are *working construction joints,* which are designed into various building materials and created in the normal process of assembling a building. An excellent example is the ordinary shingled roof, which is made up of small units of material that are applied in an overlapping pattern so that small amounts of thermal or moisture movement in the underlying roof structure or in the shingles themselves can be tolerated without distress (Figure 16.40). Other examples are wood bevel siding (Figure 6.16), which is nailed in such a way that moisture expansion and contraction are provided for; the metal clips and pans from which a sheet metal roof is assembled (Figure 16.51), which slip as necessary to allow for thermal movement; and most types of sealant joints and glazing joints (Chapters 17 and 19).
- *Structure/enclosure joints* separate structural from nonstructural elements so that they will act independently. A simple example is the sealant joint used at the top of an interior partition (Figure 23.22*a*), which ensures that the partition will bear no structural load even if the floor above sags. An important example of a structure/enclosure joint is the "soft joint" that is placed just beneath a shelf angle that supports a masonry veneer (Figures 10.4*c* and 20.1); like the joint at the top of a partition, it prevents a nonstructural element (in this case, a brick facing) from being subjected to a structural load for which it is not designed. Many other cladding attachment details shown in Chapters 19–21 are designed to allow the structural frame of the building and the exterior skin to move independently of one another, and these attachments are always associated with soft sealant joints in the skin panels. Yet another example of the structure/enclosure joint is the joint that is provided around the edge of a basement floor slab to allow for separate movement of the loadbearing wall and the slab (Figure 5.6). This type of joint is often called an *isolation joint* because it isolates the adjacent components from each other so that they can move independently.
- *Surface divider joints,* as their name implies, are used to accommodate movement in the plane of a floor, wall, ceiling, or roof. Surface divider joints can be further classified as abutment joints, control joints, and expansion joints.
- *Abutment joints* separate new construction from old construction. They are used when an existing building is altered or enlarged, to allow normal amounts of one-time movement to take place in the new materials without disturbing the original construction. If an existing brick wall is extended horizontally with new masonry, for example, a vertical abutment joint should be provided between the old work and the new rather than trying to interlock the new courses of brick with the old. The last drawing in Figure 10.21 is an abutment joint. Abutment joints are sometimes called "construction joints" or "isolation joints."
- *Control joints* are deliberately created lines of weakness along which cracking will occur as a surface of brittle material shrinks, relieving the stresses that would otherwise cause random cracking. The regularly spaced grooves across concrete sidewalks are control joints; they serve to channel the cracking tendency of the sidewalk into an orderly pattern of straight lines rather than random jagged cracks. Elsewhere in this book, control joint designs are shown for concrete floor slabs (Figure 14.3*c*), concrete masonry walls (Figure 10.21), and plaster (Figure 23.12).
- *Expansion joints* are open seams that can close slightly to allow expansion to occur in adjacent areas of material. Expansion joints in brick walls permit the bricks to expand slightly under moist conditions (Figure 10.21). Expansion joints in aluminum curtain wall mullions (Figure 21.14) allow the elements of the wall to increase in size when warmed by sunlight.

Control joints and expansion joints should be located at geometric discontinuities such as corners, changes in the height or width of a surface, or openings. In long or large surfaces, they should also be spaced at intervals that will relieve the expected stresses in the material before those stresses rise to levels that can cause damage.

- *Building separation joints* divide a large or geometrically complex

Movement Joints in Buildings (continued)

building mass into smaller, discrete structures that can move independently of one another. Building separation joints can be classified into three types: volume-change joints, settlement joints, and seismic separation joints.

- Large-scale effects of expansion and contraction caused by temperature and moisture are relieved by *volume-change joints.* These are generally placed at horizontal or vertical discontinuities in the massing of the building, where cracking would be most likely to occur (Figure B). They are also located at intervals of 150 to 200 feet (40–60 m) in very long buildings, the exact dimension depending on the nature of the materials and the rate at which dimensional changes occur.
- *Settlement joints* are designed to avoid distress caused by different rates of anticipated foundation settlement between different portions of a building, as between a high-rise tower and a connected low-rise wing, or between portions of a building that bear on different soils or have different types of foundations.
- *Seismic separation joints* are used to divide a geometrically complex building into smaller units that can move independently of one another during an earthquake. (Buildings in seismic zones should also be detailed with structure/enclosure joints that permit the frame of the building to deform during an earthquake without damaging brittle cladding or partition elements.)

Building separation joints are created by constructing independent structures on either side of the plane of the joint, usually with entirely separate foundations, columns, and slabs (Figure B). Each of these independent structures is small enough and compact enough in its geometry that it is reasonable to believe that it will move as a unit in response to the forces that are expected to act upon it.

Detailing Movement Joints

The first imperative in detailing a movement joint is to determine what type or types of movement the joint must be designed to accommodate. This is not always simple. Often the same joint is called upon to perform simultaneously in several of the ways that are outlined here—as a volume-change joint, a settlement joint, and a seismic joint, for example. A joint in a composite masonry wall may serve as both an expansion joint for the brick facing and a control joint for the concrete masonry backup. Once the function or functions of a joint have been determined, the expected character

Figure B

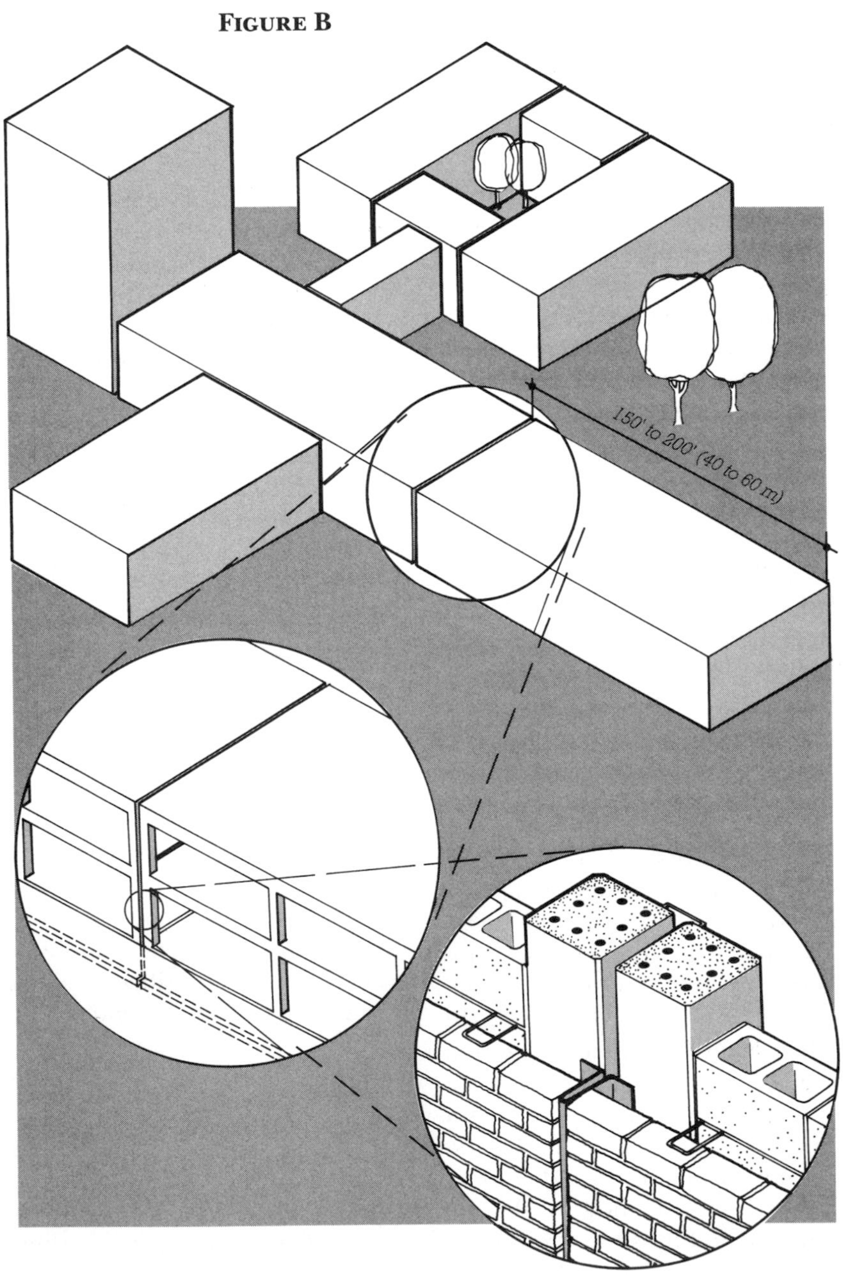

and magnitude of motion can be estimated with the aid of standard technical reference works, and the joint can then be designed accordingly.

It is important that any structural materials that would restrict movement be discontinued at a movement joint. Reinforcing bars or welded wire fabric should not extend through a control joint. Expanded metal lath is interrupted at control joints in plaster or stucco. The primary loadbearing frame of a building is interrupted at building separation joints. At the same time, it is often important to detail a movement joint so that it will maintain a critical alignment of one sort or another. Figure 10.21 shows several expansion and control joints that use interlocking masonry units or hard rubber gaskets to avoid out-of-plane movement across the joint. In concrete slabs, smooth, greased, closely spaced steel dowels are often inserted across a control joint at the midheight of the slab; these permit the joint to open up while ensuring that the slab will remain at the same level on both sides of the joint. The curtain wall mullion in Figure 21.14 allows for movement along one axis while maintaining alignment along the two other axes.

Joints must be designed to stop the passage of heat, air, water, light, sound, and fire. Some must carry traffic, as in the case of joints in floors or bridge pavements. All must be durable and maintainable, while simultaneously adjusting to movement and maintaining an acceptable appearance. Each joint must be detailed to allow for the expected direction and extent of movement: Some joints will have to operate only in a push–pull manner, while others are expected to accommodate a shearing motion and even a twisting motion as well. The exterior joint closure is usually obtained by means of a bellows of metal or synthetic rubber (Figure B). Some typical interior joint closures are shown in Figure 22.4. A perusal of manufacturers' catalogs classified under Section 07 95 00, Expansion Control, of the CSI/CSC MasterFormat system will reveal hundreds of different joint closure devices for every purpose.

Every designer of buildings must develop a sure sense of where movement joints are needed in buildings and a feel for how to design them. This is neither quick nor easy to do, for the topic is large and complex, and authoritative reference material is widely scattered. Numerous buildings are built each year by designers who have not acquired this intuition. Many of these buildings are filled with cracks even before they have been completed. This brief essay and the related illustrations throughout the book are intended to create an awareness of the problem of movement in buildings and to establish a logical framework that the reader can fill in with more detailed information over time.

References

In addition to the references listed at the ends of the chapters of this book, many of which treat movement problems, a good general reference is *Cracks, Movements and Joints in Buildings*, Ottawa, National Research Council of Canada, 1976 (NRCC 15477). For a practical summary, see pages 91–102 of Edward Allen and Patrick Rand, *Architectural Detailing: Function, Constructibility, Aesthetics*, Hoboken, NJ, John Wiley & Sons, Inc., 2007.

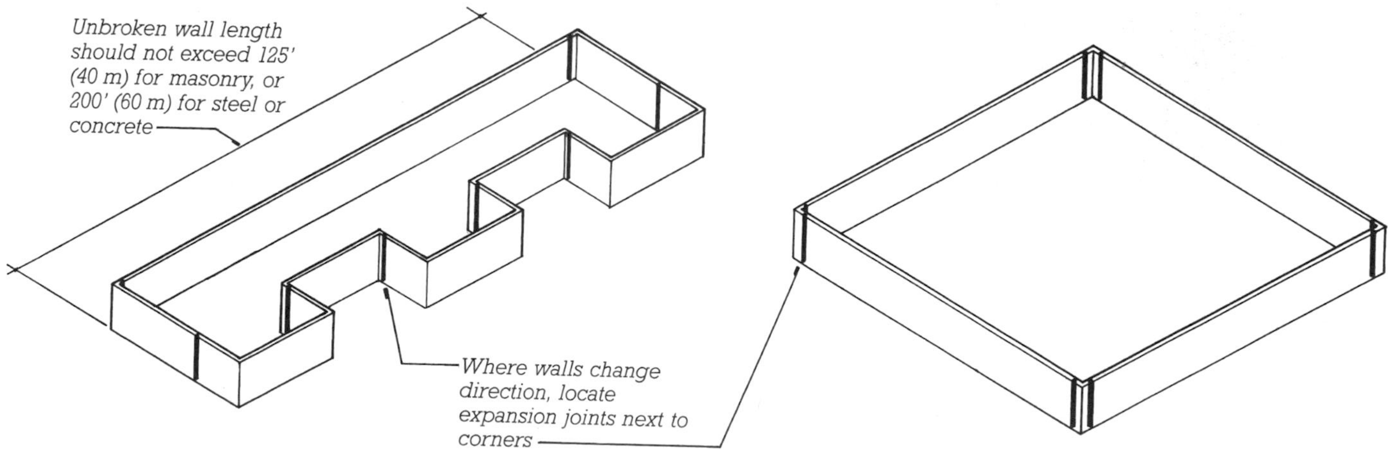
Unbroken wall length should not exceed 125' (40 m) for masonry, or 200' (60 m) for steel or concrete
Where walls change direction, locate expansion joints next to corners

FIGURE 10.18
Movement joints in masonry walls should be located near changes of direction in the walls.

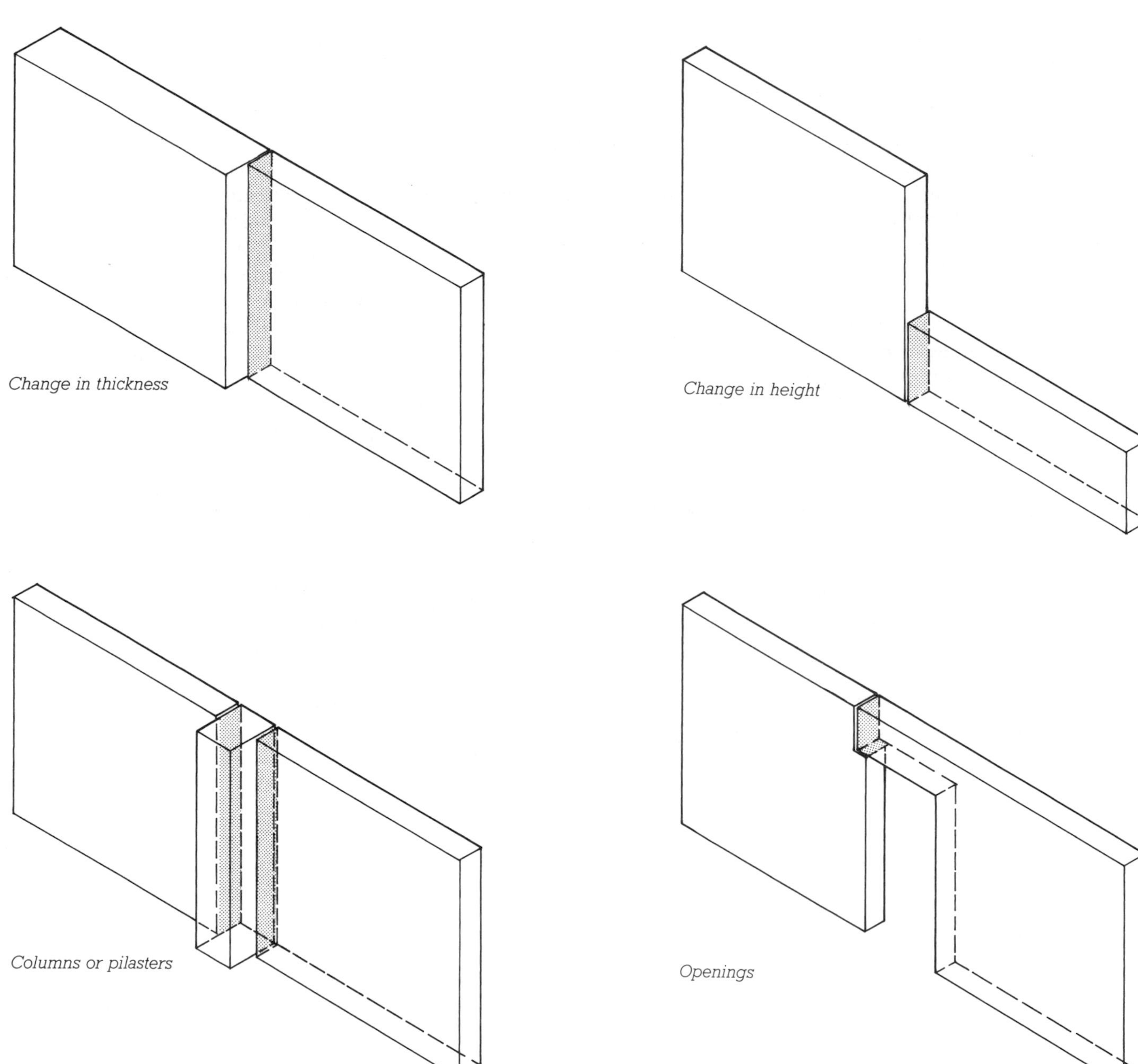

FIGURE 10.19
Movement joints in masonry walls should also be located at discontinuities in the wall, where cracks tend to form.

FIGURE 10.20
A window head and expansion joint in a brick veneer wall. The soldier course is supported on a steel angle lintel. (*Photo by Edward Allen*)

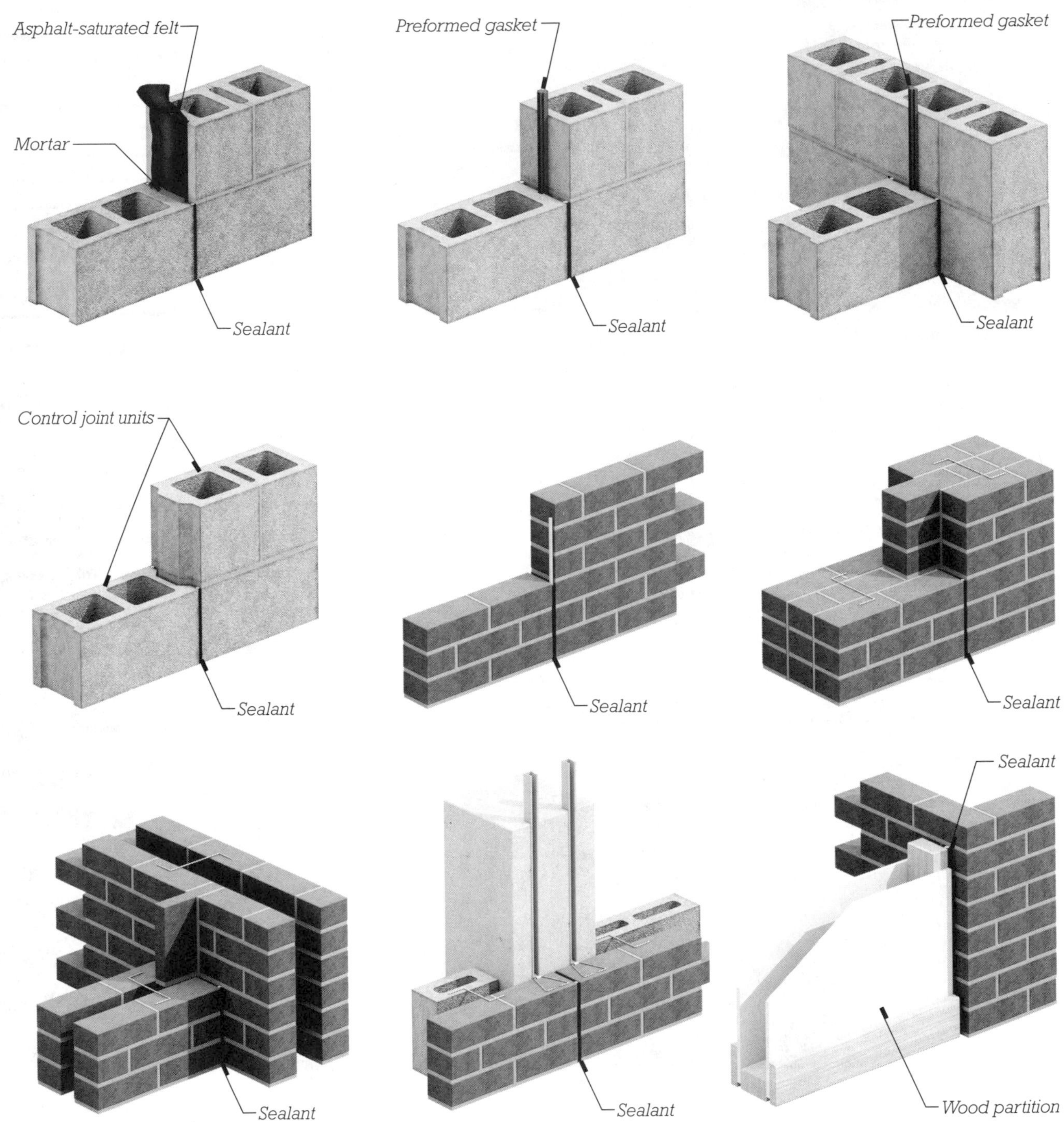

FIGURE 10.21
Some ways of making movement joints in masonry. The joints in concrete masonry walls are control joints to control shrinkage cracking. Those in brick walls are expansion joints to allow for moisture expansion of the bricks. The joint at the lower right is an abutment joint. Notice that in many of these details the masonry units interlock to prevent out-of-plane movement of the walls. Detailing of sealant joints is covered in Chapter 19.

FIGURE 10.22
Efflorescence on a wall of Flemish bond brickwork. (*Photo by Edward Allen*)

Efflorescence

Efflorescence is a fluffy crystalline powder, usually white, that sometimes appears on the surface of a wall of brick, stone, or concrete masonry (Figures 10.1, 10.22). It consists of one or more water-soluble salts that were originally present either in the masonry units or in the mortar. These were brought to the surface and deposited there by water that seeped into the masonry, dissolved the salts, and then migrated to the surface and evaporated. Efflorescence can usually be avoided by choosing masonry units that have been shown by laboratory testing not to contain water-soluble salts, by using clean ingredients in the mortar, and by minimizing water intrusion into the masonry construction. Most types of efflorescence form soon after the completion of construction and are easily removed with water and a brush. Although efflorescence is likely to reappear after such a washing, it will normally diminish and finally disappear with time as the salt is gradually leached out of the wall. Efflorescence that forms for the first time after a period of years is an indication that water has only recently begun to enter the wall, and is best controlled by investigating and correcting the source of leakage.

Mortar Joint Deterioration

Mortar joints are the weakest link in most masonry walls. Water running down a wall tends to accumulate in the joints, where cycles of freezing and thawing weather can gradually *spall* (split off flakes of) the mortar in an accelerating process of destruction that eventually creates water leaks and loosens the masonry units. To forestall this process as long as possible, a suitably weather-resistant mortar formulation must be used, and joints must be well filled and tightly compacted with a concave or vee tooling at the time the masonry is laid. Even with all these precautions, a masonry wall in a severe climate will show substantial joint deterioration after many years of weathering and may require *repointing*, sometimes called *tuckpointing*, a process of raking and cutting out the defective mortar and replacing it with fresh mortar.

Moisture Resistance of Masonry

Most masonry materials, including mortar, are porous and can transmit water from the outside of the wall to the inside. Water can also enter through cracks between the masonry units and the mortar and through defects in the mortar joints. To prevent water from entering a building through a masonry wall, the designer should begin by specifying appropriate types of masonry units, mortar, and joint tooling. Cavity wall construction should be utilized rather than solid or composite wall construction. The construction process should be supervised closely to be sure that all mortar joints are free of voids, that flashings and weep holes are properly installed, and that cavities are kept clean. Masonry walls should be protected against excessive wetting of the exterior surface of the wall to the extent practical through proper roof drainage and roof overhangs. Beyond these measures, consideration may also be given to coating the wall with stucco, paint, or *clear water repellent.* It is important that any exterior coating be highly permeable to water vapor to avoid blistering and rupture of the coating from outward vapor migration (see pages 658 and 659). Masonry primer/sealers and paints based on portland cement fill the pores of the wall without obstructing the outward passage of water vapor. Most other masonry paints are based on latex formulations and are also permeable to water vapor. When exterior walls are constructed of solid masonry, *integral water repellents* should be added to the mortar, and if concrete masonry units are used, to the concrete from which these are made as well.

Below grade, masonry should first be parged (plastered on the outside) with two coats of Type M mortar to a total thickness of ½ inch (13 mm) to seal cracks and pores. After the parging has cured and dried, it can be coated with a bituminous dampproofing compound or, if a truly watertight wall is required below grade, it can be covered with a waterproofing layer, as discussed in Chapter 2.

Cold- and Hot-Weather Construction

Mortar cannot be allowed to freeze before it has cured; otherwise, its strength and watertightness may be seriously damaged. In cold climates, special precautions are necessary if masonry work is carried on during the winter months. These include such measures as keeping masonry units and sand dry, protecting them from freezing temperatures prior to use, warming the mixing water (and sometimes the sand as well) to produce mortar at an optimum temperature for workability and curing, using a Type III (high early strength) cement to accelerate the curing of the mortar, and mixing the mortar in smaller quantities so that it does not cool excessively before it is used. The masons' workstations should be protected from wind with temporary enclosures. They should also be heated if temperatures inside the enclosures do not remain above freezing. The finished masonry must be protected against freezing for at least 2 to 3 days after it is laid, and the tops of walls should be protected from rain and snow. Chemical accelerators and so-called "antifreeze" admixtures are, in general, harmful to mortar and reinforcing steel and should not be used.

In hot weather, mortar may dry excessively before it cures. Some types of masonry units may have to be dampened before laying so that they do not absorb too much water from the mortar. It is also helpful to keep the masonry units and mortar ingredients, as well as the masons' workstations, in the shade.

Masonry and the Building Codes

Because masonry construction is noncombustible, its use is permitted in any code-defined construction type, examples of which have been discussed in earlier sections of this chapter. Walls made of masonry are also effective barriers to the passage

Wall Type	Fire Resistance (hours)	STC
4″ (100-mm) brick	1	45
6″ (150-mm) brick	2	51
8″ (200-mm) brick	2–4	52
10″, 12″ (250-mm, 300-mm) brick	4	59
4″ concrete block	½–1	43–47[a]
6″ concrete block	1	44–51[a]
8″ concrete block	1–2	45–55[a]
10″ concrete block	3–4	46–59[a]

[a]Coarse-textured block must be painted or plastered on both sides.

Figure 10.23
Rule-of-thumb fire resistance ratings and Sound Transmission Class (STC) ratings for some masonry partitions. Fire resistance of masonry construction varies with the density of the masonry units (less dense units conduct heat more slowly and can achieve higher ratings), the total mass of the wall (greater mass absorbs more heat with less rise in temperature and can achieve a higher rating), and other factors, such as whether the wall is solid or has a cavity, whether combustible members are framed into the wall, and the presence of applied finishes, such as plaster or gypsum board, that can contribute to fire resistance. STC ratings also vary with wall density, mass, and the presence of coatings.

of fire, making them suitable for use as fire walls and other types of fire-resistance-rated separations between spaces within a building or between adjacent buildings. Figure 10.23 gives rule-of-thumb values for fire resistance ratings for brick and concrete masonry walls of various thicknesses.

Due to their mass, masonry walls are also effective in limiting the transmission of sound from one space to another. Sound Transmission Class (STC) ratings for brick and concrete masonry walls, also listed in Figure 10.23, allow comparison of the acoustical properties of these walls with those of other partition systems discussed in Chapter 23. (See Chapter 22 for more information about acoustic criteria for walls.)

Uniqueness of Masonry

Masonry is often chosen as a construction material for its association in people's minds with beautiful buildings and architectural styles of the past, and with the qualities of permanence and solidity. It is often chosen for its unique colors, textures, and patterns; for its fire resistance; and for its easy compliance with building code requirements for noncombustibility and fire resistance. Masonry is often chosen, too, because it is economical. Although it is labor intensive, it can create a high-performance, long-lasting structure and enclosure in a single operation by a single trade, bypassing the difficulties that are frequently encountered in managing the numerous trades and subcontractors needed to erect a comparable building of other materials.

Masonry construction, like wood light frame construction, is carried out with small, relatively inexpensive tools and machines on the construction site. Unlike steel and concrete construction, it does not require (except in the case of ashlar stonework) a large and expensively equipped shop to fabricate the major materials prior to erection. It shares with site-cast concrete construction the long construction schedule that requires special precautions and can encounter delays during periods of very hot, very cold, or very wet weather. In general, however, it does not require an extensive period of preparation and fabrication in advance of the beginning of construction because it uses standardized units and materials that are put into final form as they are placed in the building.

From the beginning of human civilization, masonry has been the medium from which we have created our most nearly permanent, most carefully crafted, most highly prized buildings. It has given us the massiveness of the Egyptian pyramids, the inspirational elegance of the Parthenon, and the light-filled loftiness of the great European cathedrals, as well as the reassuring coziness of the fireplace, the brick cottage, and the walled garden. Masonry can express our highest aspirations and our deepest yearnings for rootedness in the earth. It reflects both the tiny scale of the human hand and the boundless power of that hand to create.

Material	Working Strength in Tension[a]	Working Strength in Compression[a]	Density	Modulus of Elasticity
Wood (framing lumber)	300–1000 psi 2.1–6.9 MPa	600–1700 psi 4.1–12 MPa	30 pcf 480 kg/m^3	1,000,000–1,900,000 psi 6900–13,000 MPa
Brick masonry (including mortar, unreinforced)	0	250–1300 psi 1.7–9.0 MPa	120 pcf 1900 kg/m^3	700,000–3,700,000 psi 4800–25,000 MPa
Structural steel	24,000–43,000 psi 170–300 MPa	24,000–43,000 psi 170–300 MPa	490 pcf 7800 kg/m^3	29,000,000 psi 200,000 MPa
Concrete (unreinforced)	0	1000–4000 psi 6.9–28 MPa	145 pcf 2300 kg/m^3	3,000,000–4,500,000 psi 21,000–31,000 MPa

[a]Allowable stress or approximate maximum stress under normal loading conditions.

Figure 10.24
Comparative physical properties for four common structural materials: wood, brick masonry (shaded row), steel, and concrete. Brick masonry itself has no useful tensile strength, but its compressive strength is considerable, and when combined with steel reinforcing, it can be used for a wide range of structural types. The ranges of values of strength and stiffness reflect variations among available bricks and mortar types.

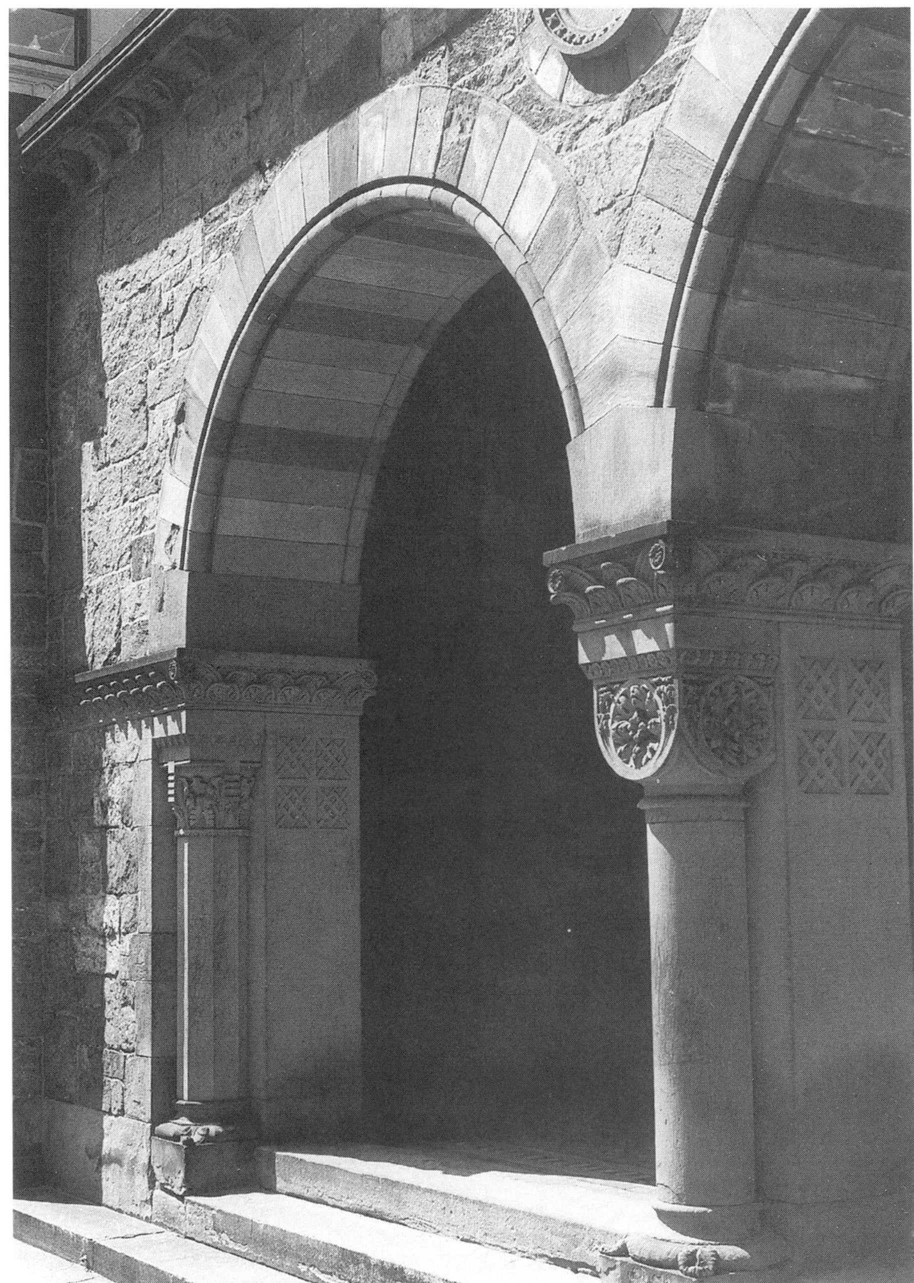

FIGURE 10.25
A detail of the porch of H. H. Richardson's First Baptist Church, Boston, built in 1871. *(Photo by Edward Allen)*

CSI/CSC

MasterFormat Sections for Masonry Wall Construction

04 00 00	MASONRY
04 05 00	Common Work Results for Masonry Masonry Grouting Masonry Anchorage and Reinforcing Masonry Accessories Masonry Control and Expansion Joints Masonry Embedded Flashing Masonry Cavity Drainage, Weepholes, and Vents
04 20 00	UNIT MASONRY
04 21 00	Clay Unit Masonry
04 22 00	Concrete Unit Masonry
04 27 00	Multiple-Wythe Unit Masonry Composite Unit Masonry Cavity Wall Unit Masonry
04 40 00	STONE ASSEMBLIES
04 43 00	Stone Masonry
07 10 00	DAMPPROOFING AND WATERPROOFING
07 11 00	Dampproofing
07 19 00	Water Repellents
07 60 00	FLASHING AND SHEET METAL
07 62 00	Sheet Metal Flashing and Trim
07 65 00	Flexible Flashing Laminated Sheet Flashing Modified Bituminous Sheet Flashing Plastic Sheet Flashing Rubber Flashing Self-Adhering Sheet Flashing

Selected References

In addition to the references listed in Chapters 8 and 9, the reader is referred to:

1. The Masonry Society. *Masonry Designers' Guide*. Boulder, CO, updated regularly.

This comprehensive design guide is based on *Building Code Requirements for Masonry Structures, ACI 530/ASCE 5/TMS 402* (discussed in this chapter) and the related *Specifications for Masonry Structures*. It also includes extensive commentary, illustrations, and design examples relating to the design and construction of masonry structures.

Web Sites

Author's supplementary web site:
www.ianosbackfill.com/10_masonry_loadbearing_wall_construction
Dur-O-Wall: **www.dur-o-wal.com**
Grace Construction Products: **www.na.graceconstruction.com** (select "Masonry Products")
Hyload Flashings: **www.hyloadflashing.com**
The Masonry Society: **www.masonrysociety.org**
Mortar Net: **www.mortarnet.com**
Thermadrain: **www.thermadrain.com**

Key Terms and Concepts

solid masonry, monolithic masonry
composite masonry wall
cavity wall
flashing
weep hole
dampproofing
cavity drainage material
veneer
masonry loadbearing wall, bearing wall
ACI 530/ASCE 5/TMS 402
Ordinary construction
Heavy Timber construction, Mill construction
external flashing
internal flashing, concealed flashing, through-wall flashing
base flashing
counterflashing, cap flashing
reglet
self-adhered flashing
end dam
exterior insulation and finish system (EIFS)
furring strip
nonmovement joint
movement joint
working construction joint
structure/enclosure joint
isolation joint
surface divider joint
abutment joint
control joint
expansion joint
building separation joint
volume-change joint
settlement joint
seismic separation joint
efflorescence
spall
repointing
tuckpointing
clear water repellent
integral water repellent

Review Questions

1. Describe how a cavity wall works and sketch its major construction features. What aspects of cavity wall construction are most critical to its success in preventing water leakage?

2. Where should flashings be installed in a masonry wall? What is the function of the flashing in each of these locations?

3. Where should weep holes be provided? Describe the function of a weep hole and indicate several ways in which it may be constructed.

4. What are the differences between Ordinary construction and Mill construction? What features of each are related to fire resistance?

5. What types of movement joints are required in a concrete masonry wall? In a brick masonry wall? Where should these joints be located?

6. What are some ways of insulating masonry walls?

7. Why is balloon framing used rather than platform framing in Ordinary construction?

8. What precautions should be taken when constructing masonry walls in Minneapolis in the winter?

Exercises

1. What are the allowable height and floor area for a restaurant (Occupancy Group A-2) in a building of Mill construction (Type IV)? How do these figures change if unprotected Ordinary construction (Type IIIB) is used instead? (Assume that sprinklers are not required.) What if unprotected steel joists are substituted for the wood joists? Or precast concrete plank floors with a 2-hour fire rating?

2. Examine the masonry walls of some new buildings in your area. Where have movement joints been placed in these walls? What type of joint is each? Why is it placed where it is? Do you agree with this placement?

3. Look at weep holes and flashings in these same buildings. How are they detailed? Can you improve on these details?

4. Design a masonry gateway for one of the entrances to a college campus with which you are familiar. Choose whatever type of masonry you feel is appropriate, and make the fullest possible use of the decorative and structural potentials of the material.

11

Steel Frame Construction

Ironworkers place open-web steel joists on a frame of steel wide-flange beams as a crane lowers bundles of joists from above. *(Photo by Balthazar Korab. Courtesy Vulcraft Division of Nucor)*

Steel, strong and stiff, is a material of slender towers and soaring spans. Precise and predictable, light in proportion to its strength, it is also well suited to rapid construction, highly repetitive building frames, and architectural details that satisfy the eye with a clean, precise elegance. Among the metals, it is uniquely plentiful and inexpensive. If its weaknesses—a tendency to corrode in certain environments and a loss of strength during severe building fires—are held in check by intelligent construction measures, it offers the designer possibilities that exist with no other material.

History

Prior to the beginning of the 19th century, metals had little structural role in buildings except in connecting devices. The Greeks and Romans used hidden cramps of bronze to join blocks of stone, and architects of the Renaissance countered the thrust of masonry vaults with wrought iron chains and rods. The first all-metal structure, a cast iron bridge, was built in the late 18th century in England and still carries traffic across the Severn River more than two centuries after its construction. Cast iron, produced from iron ore in a blast furnace, and wrought iron, iron that has been purified by beating it repeatedly with a hammer, were used increasingly for framing industrial buildings in Europe and North America in the first half of the 19th century, but their usefulness was limited by the unpredictable brittleness of cast iron and the relatively high cost of wrought iron.

Until that time, steel had been a rare and expensive material,

Figure 11.1
Landscape architect Joseph Paxton designed the Crystal Palace, an exposition hall of cast iron and glass, which was built in London in 1851. *(Bettmann Archive)*

Figure 11.2
Allied Bank Plaza, designed by Architects Skidmore, Owings, and Merrill. *(Permission of American Institute of Steel Construction)*

C O

produced only in small batches for such applications as weapons and cutlery. Plentiful, inexpensive steel first became available in the 1850s with the introduction of the *Bessemer process,* in which air was blown into a vessel of molten iron to burn out the impurities. By this means, a large batch of iron could be made into steel in about 20 minutes, and the structural properties of the resulting metal were vastly superior to those of cast iron. Another economical steelmaking process, the *open-hearth method,* was developed in Europe in 1868 and was soon adopted in America. By 1889, when the Eiffel Tower was built of wrought iron in Paris (Figure 11.3), several steel frame skyscrapers had already been erected in the United States (Figure 11.4). A new material of construction had been born.

The Material Steel

Steel

Steel is any of a range of alloys of iron that contain less than 2 percent carbon. Ordinary structural steel, called *mild steel,* contains less than three-tenths of 1 percent carbon, plus traces of beneficial elements such as manganese and silicon, and of detrimental impurities such as phosphorus, sulfur, oxygen, and nitrogen. In contrast, ordinary *cast iron* contains 3 to 4 percent carbon and greater quantities of impurities than steel, while *wrought iron* contains even less carbon than most steel alloys. Carbon content is a crucial determinant of the properties of any *ferrous* (iron-based) *metal:* Too much carbon makes a hard but brittle metal (like cast iron), while too little produces a malleable, relatively weak material (like wrought iron). Thus, mild steel is iron whose properties

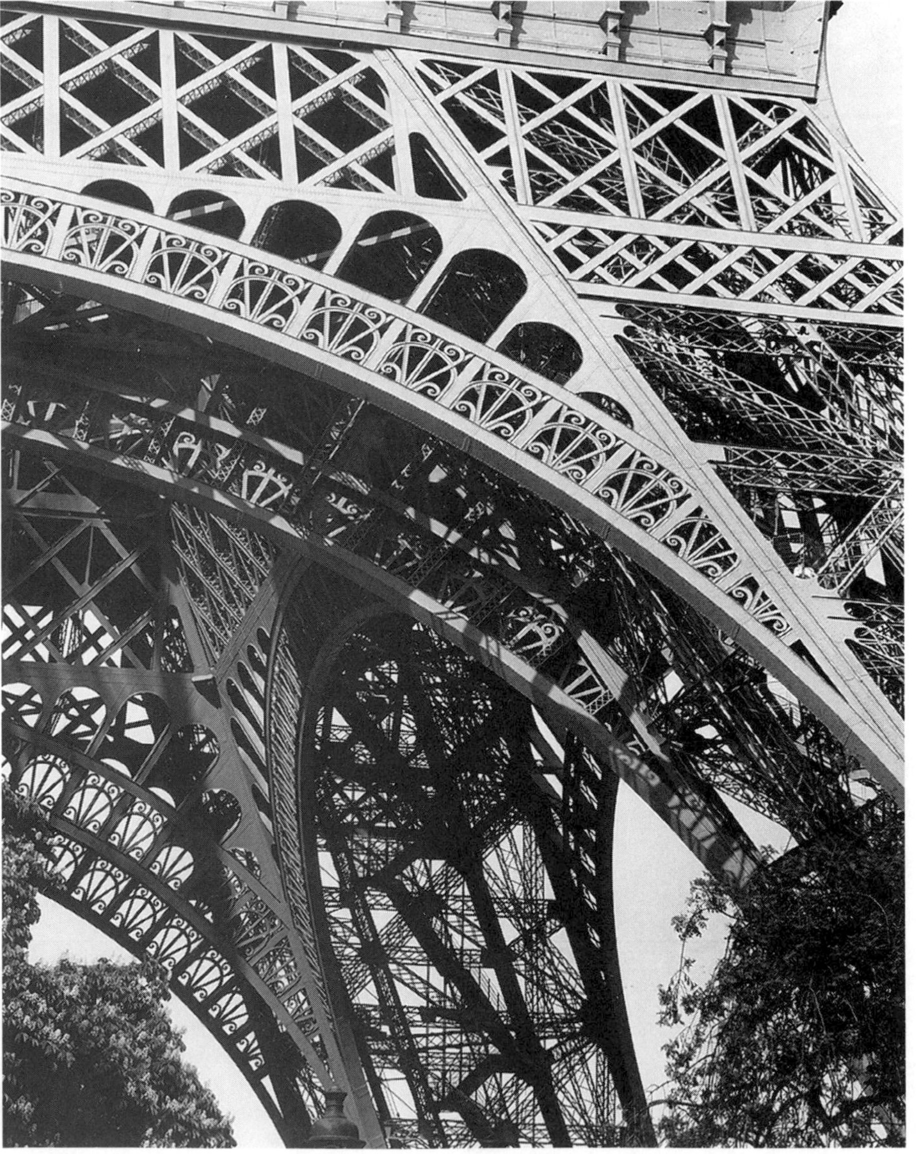

Figure 11.3
Engineer Gustave Eiffel's magnificent tower of wrought iron was constructed in Paris from 1887 to 1889. *(Photo by James Austin, Cambridge, England)*

The gap between stone and steel-and-glass was as great as that in the evolutionary order between the crustaceans and the vertebrates.

Lewis Mumford, *The Brown Decades,* New York, Dover Publications, Inc., 1955, pp. 130–131.

have been optimized for structural purposes by controlling the amounts of carbon and other elements in the metal.

The process of converting iron ore to steel begins with the smelting of ore into cast iron. Cast iron is produced in a blast furnace charged with alternating layers of *iron ore* (oxides of iron), *coke* (coal whose volatile constituents have been distilled out, leaving only carbon), and crushed limestone (Figure 11.6). The coke is burned by large quantities of air forced into the bottom of the furnace to produce carbon monoxide, which reacts with the ore to reduce it to elemental iron. The limestone forms a slag with various impurities, but large amounts of carbon and other elements are inevitably incorporated into the iron. The molten iron is drawn off at the bottom of the furnace and held in a liquid state for processing into steel.

Most steel that is converted from iron is manufactured by the *basic oxygen process* (Figure 11.5), in which a hollow, water-cooled lance is lowered into a container of molten iron and recycled steel scrap. A stream of pure oxygen at very high pressure is blown from the lance into the metal to burn off the excess carbon and impurities. A flux of lime and fluorspar is added to the metal to react with other impurities, particularly phosphorus, and forms a slag that is discarded. New metallic elements may be added to the container at the end of the process to adjust the composition of the steel as desired: Manganese gives resistance to abrasion and impact, molybdenum gives strength, vanadium imparts strength and toughness, and nickel and chromium give corrosion resistance, toughness, and stiffness. The entire process takes place with the aid of careful sampling and laboratory analysis techniques to ensure the finished quality of the steel and takes less than an hour from start to finish.

Today, most structural steel for frames of buildings is produced from scrap steel in so-called "mini-mills," utilizing *electric arc furnaces.* These mills are miniature only in comparison to the conventional mills that they have replaced; they are housed in enormous buildings and roll structural shapes up to 40 inches (1 m) deep. The scrap from which structural steel is made comes mostly from defunct automobiles, one mini-mill alone consuming 300,000 junk cars in an average year. Through careful metallurgical testing and control, these are recycled into top-quality steel.

FIGURE 11.4
The Home Insurance Company Building, designed by William LeBaron Jenney and built in Chicago in 1893, was among the earliest true skyscrapers. The steel framing was fireproofed with masonry, and the exterior masonry facings were supported on the steel frame. *(Photo by Wm. T. Barnum. Courtesy of Chicago Historical Society ICHi-18293)*

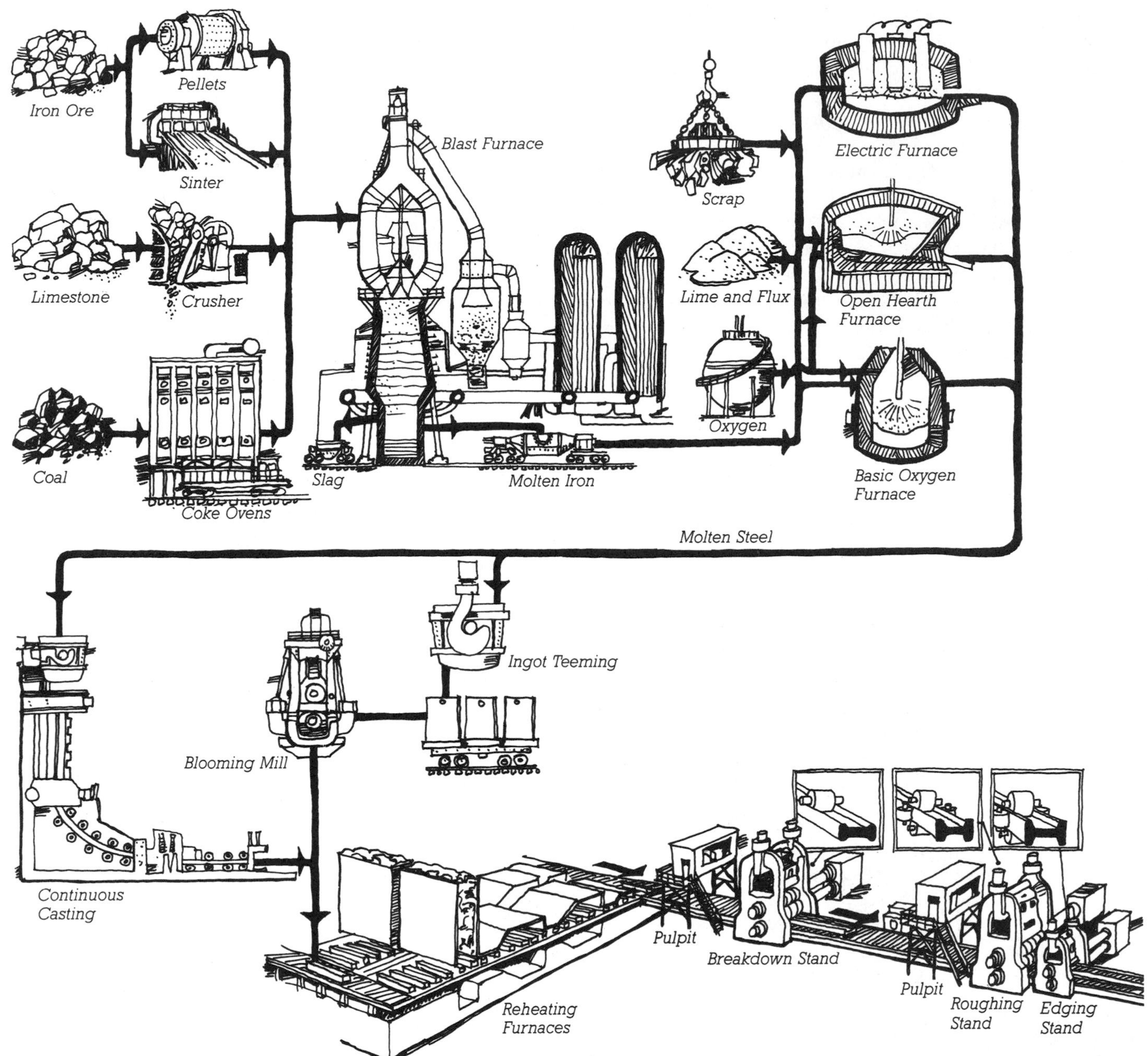

FIGURE 11.5
The steelmaking process, from iron ore to structural shapes. Notice particularly the steps in the evolution of a wide-flange shape as it progresses through the various stands in the rolling mill. Today, most structural steel in the United States is made from steel scrap in electric furnaces. *(Adapted from "Steelmaking Flowlines," by permission of the American Iron and Steel Institute)*

For Preliminary Design of a Steel Structure

- Estimate the depth of **corrugated steel roof decking** at 1⁄40 of its span. Standard depths are 1, 1½, 2, and 4 inches (25, 38, 50, and 100 mm).
- Estimate the overall depth of **corrugated steel floor decking plus concrete topping** at 1⁄24 of its span. Typical overall depths range from 2½ to 7 inches (65–180 mm).
- Estimate the depth of **open-web steel joists** at 1⁄20 of their span for heavily loaded floors or widely spaced joists and at 1⁄24 of their span for roofs, lightly loaded floors, or closely spaced joists. The spacing of joists depends on the spanning capability of the decking material. Typical joist spacings range from 2 to 10 feet (0.6–3.0 m). Standard joist depths are given elsewhere in this chapter.
- Estimate the depth of **steel beams** at 1⁄20 of their span and the depth of **steel girders** at 1⁄15 of their span. The width of a beam or girder is usually 1⁄3 to 1⁄2 of its depth. For composite beams and girders, use the same ratios but apply them to the overall depth of the beam or girder, including the floor deck and concrete topping. Standard depths of steel wide-flange shapes are given elsewhere in this chapter.
- Estimate the depth of **triangular steel roof trusses** at 1⁄4 to 1⁄5 of their span. For rectangular trusses, the depth is typically 1⁄8 to 1⁄12 of their span.
- To estimate the size of a **steel column**, add up the total roof and floor area supported by the column. A W8 column can support up to about 4000 square feet (370 m^2) and a W14 column 30,000 square feet (2800 m^2). Very heavy W14 shapes, which are substantially larger than 14 inches (355 mm) in dimension, can support up to 50,000–100,000 square feet (4600– 9300 m^2). Steel column shapes are usually square or nearly square in proportion.

These approximations are valid only for purposes of preliminary building layout and must not be used to select final member sizes. They apply to the normal range of building occupancies such as residential, office, commercial, and institutional buildings and parking garages. For manufacturing and storage buildings, use somewhat larger members.

For more comprehensive information on preliminary selection and layout of a structural system and sizing of structural members, see Edward Allen and Joseph Iano, *The Architect's Studio Companion* (4th ed.), New York, John Wiley & Sons, Inc., 2007.

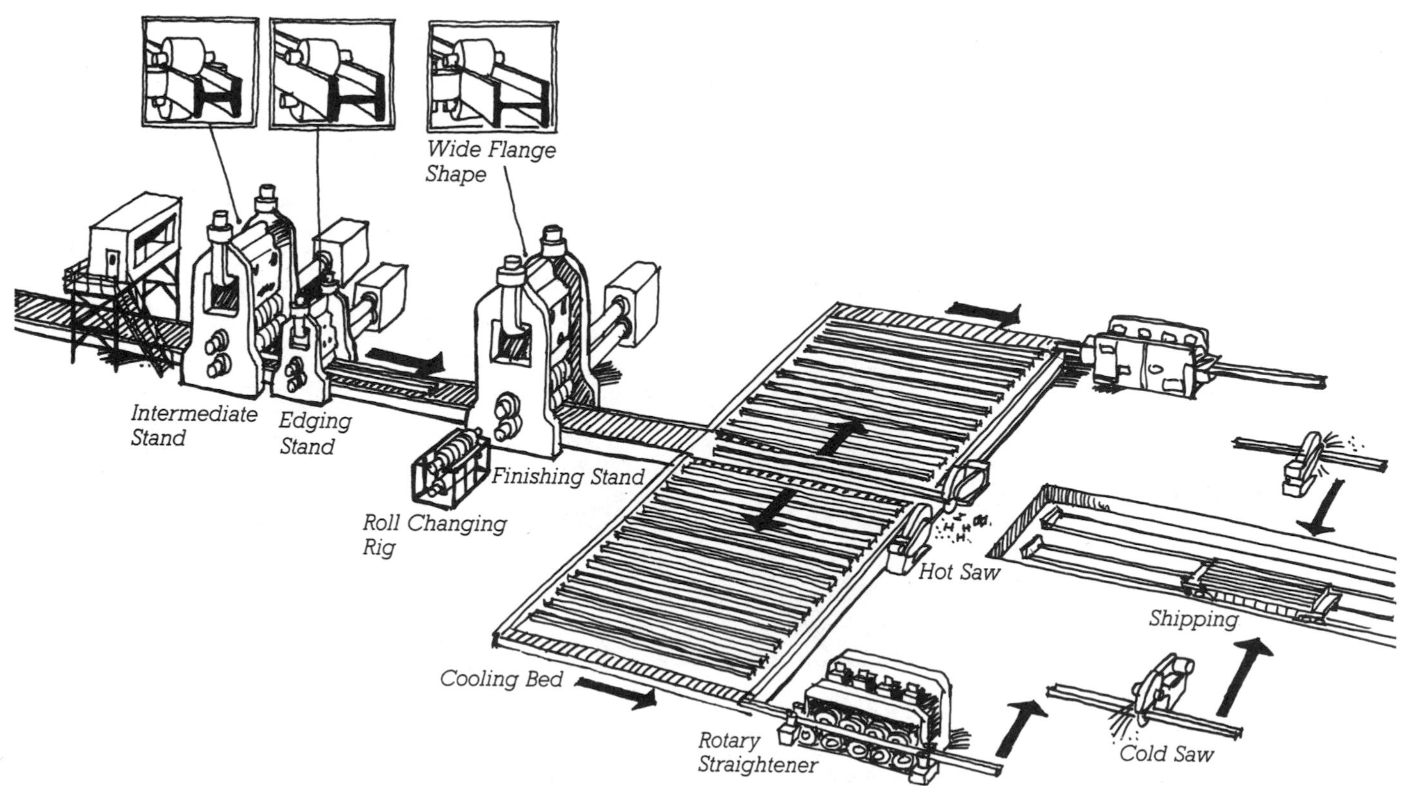

FIGURE 11.6
Molten iron is poured into a crucible to begin its conversion to steel in the basic oxygen process. *(Courtesy of U.S. Steel Corp.)*

Regardless of the particular steelmaking process, finished steel is cast continuously into *beam blanks* or *blooms*, very thick approximations of the desired final shape, which are then rolled into final form, as described later in this chapter.

Steel Alloys

By adjusting the mix of metallic elements used in the production of steel, its strength and other properties can be manipulated. Mild structural steel, known by its ASTM designation of A36, was for decades the predominant steel type used in building frames. But today's mini-mills, using scrap as their primary raw material, routinely produce stronger, less expensive *high-strength, low-alloy steels*, such as those designated ASTM A992 or ASTM A572. ASTM A992 steel is the preferred steel type for standard wide-flange structural shapes, while ASTM A36 steel, or, where higher strength is needed, ASTM A572 steel, are specified for angles, channels, plates, and bars. (For an explanation of standard steel shapes, see the discussion below.)

Where steel without any protective finish will remain exposed to exterior conditions in the completed construction, *weathering steel* (ASTM A588) may be specified. This steel alloy develops a tenacious oxide coating when exposed to the atmosphere that, once formed, protects against further corrosion and eliminates the need for paint or another protective coating. While mostly used for highway and bridge construction where it reduces maintenance costs, weathering steel also finds occasional use in buildings, where the deep, warm hue of the oxide coating can be exploited as an aesthetic feature. With the addition of nickel and chromium to steel, various grades of *stainless steel* (ASTM A240 and A276) with even greater corrosion resistance and costing significantly more than conventional structural steel, can be produced. Steel can also be protected from corrosion by galvanizing, the application of a zinc coating, which is discussed further on pages 507–508.

Production of Structural Shapes

In the *structural mill* or *breakdown mill*, the beam blank is reheated as necessary and then passed through a succession of rollers that squeeze the metal into progressively more refined approximations of the desired shape and size (Figure 11.7). The finished shape exits from the last set of rollers as a continuous length that is cut into shorter segments by a *hot saw* (Figure 11.8). These segments are cooled on a *cooling bed* (Figure 11.9). Then a *roller straightener* corrects any residual crookedness. Finally, each piece is cut to length and labeled with its shape designation and the number of the batch of steel from which it was rolled. Later, when the

FIGURE 11.7
A glowing steel wide-flange shape emerges from the rolls of the finishing stand of the rolling mill. *(Photo by Mike Engestrom. Courtesy of Nucor-Yamato Steel Company)*

Figure 11.8
A hot saw cuts pieces of wide-flange stock from a continuous length that has just emerged from the finishing stand in the background. Workers in the booth control the process. *(Courtesy of U.S. Steel Corp.)*

Figure 11.9
Wide-flange shapes are inspected for quality on the cooling bed. *(Photo by Mike Engestrom. Courtesy of Nucor-Yamato Steel Company)*

piece is shipped to a fabricator, it will be accompanied by a certificate that gives the chemical analysis of that particular batch, as evidence that the steel meets standard structural specifications.

The roller spacings in the structural mill are adjustable; by varying the spacings between rollers, a number of different shapes with the same nominal dimensions can be produced (Figure 11.10). This provides the architect and the structural engineer with a finely graduated selection of shapes from which to select each structural member in a building, thereby avoiding the waste of steel through the specification of shapes that are larger than required.

Wide-flange shapes are used for most beams and columns, superseding the older *American Standard (I-beam)* shapes (Figure 11.11). American Standard shapes are less efficient structurally than wide flanges because the roller arrangement that produces them is incapable of increasing the amount of steel

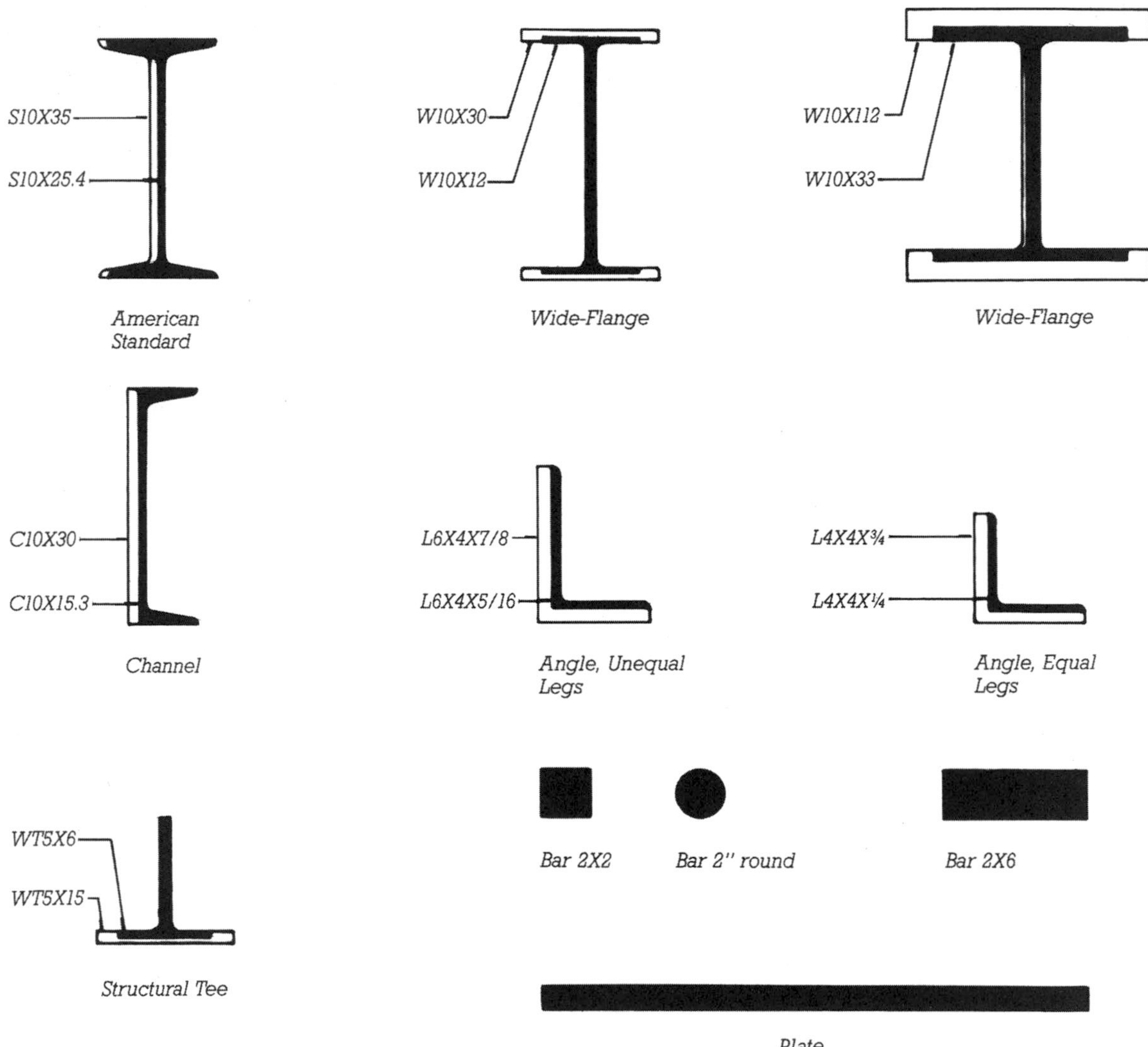

FIGURE 11.10
Examples of the standard shapes of structural steel. Where two shapes are superimposed, they illustrate different weights of the same section, produced by varying the spacing of the rollers in the structural mill. Structural steel shapes and their general requirements are defined in ASTM A6. *Bars* are round, rectangular, and hexagonal solid shapes generally not greater than 8 inches (203 mm) in any cross-sectional dimension. Wider solid shapes are called *plate* or *sheet*, depending on their thickness in relation to their width. Plate is thicker than sheet.

in the flanges without also adding steel to the web, where it does little to increase the load-carrying capacity of the member. Wide flanges are available in a vast range of sizes and weights. The smallest available depth in the United States is nominally 4 inches (100 mm), and the largest is 44 inches (1117 mm). Weights per linear foot of member range from 9 to 730 pounds (13–1080 kg/m), the latter for a nominal 14-inch (360-mm) shape with flanges nearly 5 inches (130 mm) thick. Some producers construct heavier wide-flange sections by welding together flange and web plates rather than rolling, a procedure that is also used for producing very deep, long-span plate girders (Figure 11.79).

Wide flanges are manufactured in two basic proportions: tall and narrow for beams and squarish for columns and foundation piles. The accepted nomenclature for wide-flange shapes begins with the letter W, followed by the nominal depth of

Shape	Sample Designation	Explanation	Range of Available Sizes
Wide-flange	W21 × 83	W denotes a wide-flange shape. The first number is the nominal depth in inches and the second number is the weight in pounds per foot of length.	Nominal depths from 4 to 18″ in 2″ increments, from 18 to 36″ in 3″ increments, and from 36 to 44″ in 4″ increments.
American Standard beam	S18 × 70	S denotes an American Standard beam. The first number is the nominal depth in inches, and the second number is the weight in pounds per foot of length.	Nominal depths of 3″, 4″, 5″, 6″, 8″, 10″, 12″, 15″, 18″, 20″, and 24″
Channel	MC10 × 33.6	MC denotes a channel. The first number is the nominal depth in inches, and the second number is the weight in pounds per foot of length.	Nominal depths of 6″, 7″, 8″, 9″, 10″, 12″, 13″, and 18″.
American Standard channel	C6 × 13	C denotes an American Standard channel. The first number is the nominal depth in inches, and the second number is the weight in pounds per foot of length.	Nominal depths of 3″, 4″, 5″, 6″, 7″, 8″, 9″, 10″, 12″, and 15″.
Structural tee	WT13.5 × 47	WT denotes a tee made by splitting a W shape. The first number is the nominal depth in inches, and the second number is the weight in pounds per foot. (The example tee listed here was made from a W27 × 94.) Tees split from American Standard beams are designated ST rather than WT.	See the available sizes of wide-flange and American Standard beams listed above, and divide by 2 to arrive at available depths for structural tees made from these shapes.
Angle	L4 × 3 × 3⁄8	L denotes an angle. The first two numbers are the nominal depths in inches of the two legs, and the last number is the thickness in inches of the legs.	Leg depths of 2″, 2½″, 3″, 3½″, 4″, 5″, 6″, 7″, and 8″. Leg thicknesses from 1⁄8″ to 11⁄8″.
HSS Square, Rectangular, Round, or Elliptical	HSS10 × 8 × ½	HSS denotes a hollow structural section. The first two numbers are the nominal size in inches of the two sides of a square, rectangular, or elliptical shape. For round tubes, a single number indicates nominal diameter. The last number is the thickness in inches of the wall of the tube.	For square or rectangular shapes, nominal depths from 1″ to 48″ and wall thickness from 1⁄8″ to 5⁄8″. For round shapes, nominal diameters from 1.66 to 20″ and wall thicknesses from 0.109 to 0.625″.

FIGURE 11.11
Commonly used steel shapes.

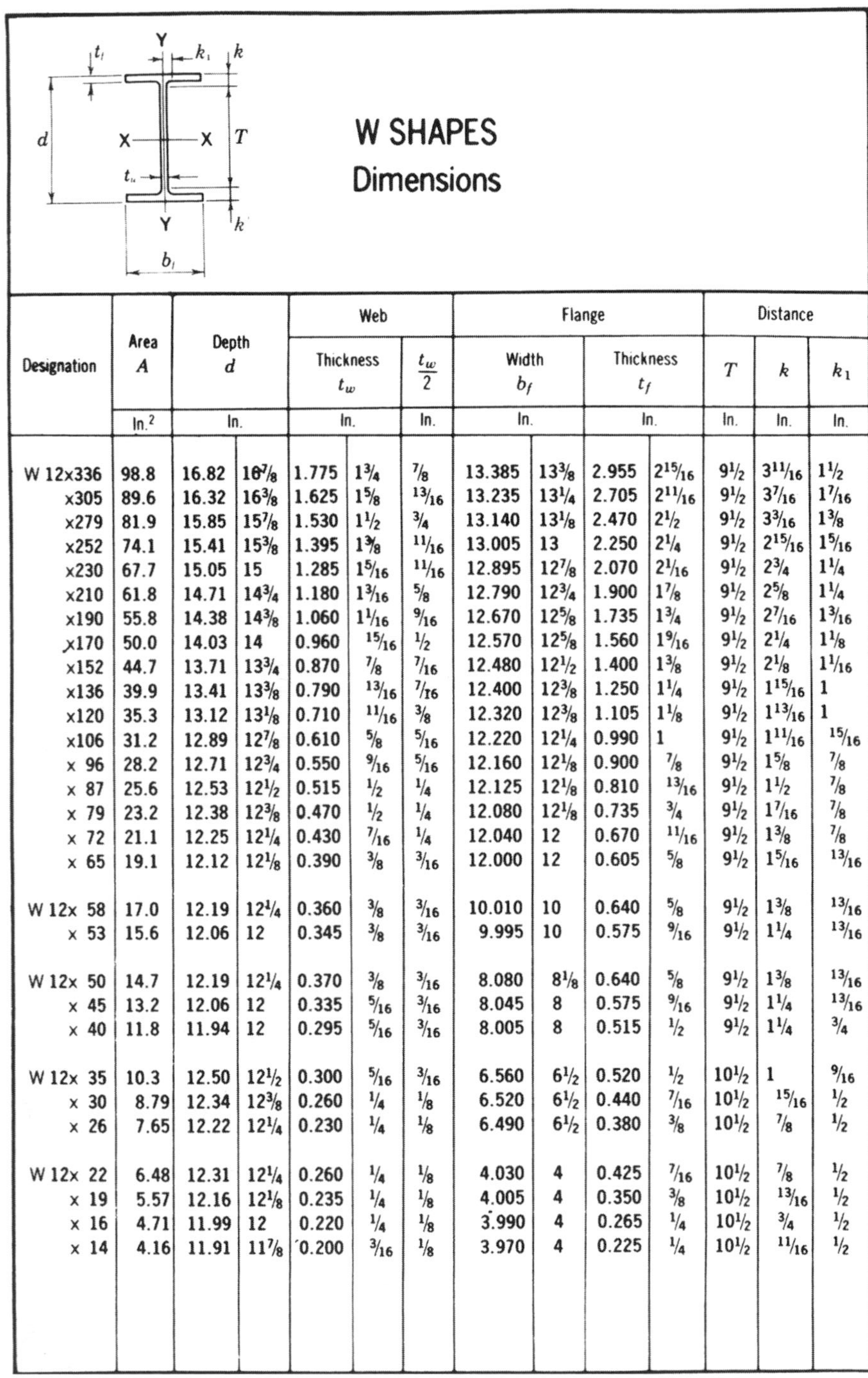

W SHAPES Dimensions

Designation	Area A	Depth d		Web Thickness t_w		$\frac{t_w}{2}$	Flange Width b_f		Flange Thickness t_f		Distance T	k	k_1
	In.^{2}	In.		In.		In.	In.		In.		In.	In.	In.
W 12x336	98.8	16.82	16 7/8	1.775	1 3/4	7/8	13.385	13 3/8	2.955	2 15/16	9 1/2	3 11/16	1 1/2
x305	89.6	16.32	16 3/8	1.625	1 5/8	13/16	13.235	13 1/4	2.705	2 11/16	9 1/2	3 7/16	1 7/16
x279	81.9	15.85	15 7/8	1.530	1 1/2	3/4	13.140	13 1/8	2.470	2 1/2	9 1/2	3 3/16	1 3/8
x252	74.1	15.41	15 3/8	1.395	1 3/8	11/16	13.005	13	2.250	2 1/4	9 1/2	2 15/16	1 5/16
x230	67.7	15.05	15	1.285	1 5/16	11/16	12.895	12 7/8	2.070	2 1/16	9 1/2	2 3/4	1 1/4
x210	61.8	14.71	14 3/4	1.180	1 3/16	5/8	12.790	12 3/4	1.900	1 7/8	9 1/2	2 5/8	1 1/4
x190	55.8	14.38	14 3/8	1.060	1 1/16	9/16	12.670	12 5/8	1.735	1 3/4	9 1/2	2 7/16	1 3/16
x170	50.0	14.03	14	0.960	15/16	1/2	12.570	12 5/8	1.560	1 9/16	9 1/2	2 1/4	1 1/8
x152	44.7	13.71	13 3/4	0.870	7/8	7/16	12.480	12 1/2	1.400	1 3/8	9 1/2	2 1/8	1 1/16
x136	39.9	13.41	13 3/8	0.790	13/16	7/16	12.400	12 3/8	1.250	1 1/4	9 1/2	1 15/16	1
x120	35.3	13.12	13 1/8	0.710	11/16	3/8	12.320	12 3/8	1.105	1 1/8	9 1/2	1 13/16	1
x106	31.2	12.89	12 7/8	0.610	5/8	5/16	12.220	12 1/4	0.990	1	9 1/2	1 11/16	15/16
x 96	28.2	12.71	12 3/4	0.550	9/16	5/16	12.160	12 1/8	0.900	7/8	9 1/2	1 5/8	7/8
x 87	25.6	12.53	12 1/2	0.515	1/2	1/4	12.125	12 1/8	0.810	13/16	9 1/2	1 1/2	7/8
x 79	23.2	12.38	12 3/8	0.470	1/2	1/4	12.080	12 1/8	0.735	3/4	9 1/2	1 7/16	7/8
x 72	21.1	12.25	12 1/4	0.430	7/16	1/4	12.040	12	0.670	11/16	9 1/2	1 3/8	7/8
x 65	19.1	12.12	12 1/8	0.390	3/8	3/16	12.000	12	0.605	5/8	9 1/2	1 5/16	13/16
W 12x 58	17.0	12.19	12 1/4	0.360	3/8	3/16	10.010	10	0.640	5/8	9 1/2	1 3/8	13/16
x 53	15.6	12.06	12	0.345	3/8	3/16	9.995	10	0.575	9/16	9 1/2	1 1/4	13/16
W 12x 50	14.7	12.19	12 1/4	0.370	3/8	3/16	8.080	8 1/8	0.640	5/8	9 1/2	1 3/8	13/16
x 45	13.2	12.06	12	0.335	5/16	3/16	8.045	8	0.575	9/16	9 1/2	1 1/4	13/16
x 40	11.8	11.94	12	0.295	5/16	3/16	8.005	8	0.515	1/2	9 1/2	1 1/4	3/4
W 12x 35	10.3	12.50	12 1/2	0.300	5/16	3/16	6.560	6 1/2	0.520	1/2	10 1/2	1	9/16
x 30	8.79	12.34	12 3/8	0.260	1/4	1/8	6.520	6 1/2	0.440	7/16	10 1/2	15/16	1/2
x 26	7.65	12.22	12 1/4	0.230	1/4	1/8	6.490	6 1/2	0.380	3/8	10 1/2	7/8	1/2
W 12x 22	6.48	12.31	12 1/4	0.260	1/4	1/8	4.030	4	0.425	7/16	10 1/2	7/8	1/2
x 19	5.57	12.16	12 1/8	0.235	1/4	1/8	4.005	4	0.350	3/8	10 1/2	13/16	1/2
x 16	4.71	11.99	12	0.220	1/4	1/8	3.990	4	0.265	1/4	10 1/2	3/4	1/2
x 14	4.16	11.91	11 7/8	0.200	3/16	1/8	3.970	4	0.225	1/4	10 1/2	11/16	1/2

W SHAPES Properties

Nominal Wt. per Ft.	Compact Section Criteria $\frac{b_f}{2t_f}$	F_y'	$\frac{d}{t_w}$	F_y'''	r_T	$\frac{d}{A_f}$	Axis X-X I	Axis X-X S	Axis X-X r	Axis Y-Y I	Axis Y-Y S	Axis Y-Y r	Torsional constant J	Plastic Modulus Z_x	Z_y
Lb.		Ksi		Ksi	In.		In.^{4}	In.^{3}	In.	In.^{4}	In.^{3}	In.	In.^{4}	In.^{3}	In.^{3}
336	2.3	—	9.5	—	3.71	0.43	4060	483	6.41	1190	177	3.47	243	603	274
305	2.4	—	10.0	—	3.67	0.46	3550	435	6.29	1050	159	3.42	185	537	244
279	2.7	—	10.4	—	3.64	0.49	3110	393	6.16	937	143	3.38	143	481	220
252	2.9	—	11.0	—	3.59	0.53	2720	353	6.06	828	127	3.34	108	428	196
230	3.1	—	11.7	—	3.56	0.56	2420	321	5.97	742	115	3.31	83.8	386	177
210	3.4	—	12.5	—	3.53	0.61	2140	292	5.89	664	104	3.28	64.7	348	159
190	3.7	—	13.6	—	3.50	0.65	1890	263	5.82	589	93.0	3.25	48.8	311	143
170	4.0	—	14.6	—	3.47	0.72	1650	235	5.74	517	82.3	3.22	35.6	275	126
152	4.5	—	15.8	—	3.44	0.79	1430	209	5.66	454	72.8	3.19	25.8	243	111
136	5.0	—	17.0	—	3.41	0.87	1240	186	5.58	398	64.2	3.16	18.5	214	98.0
120	5.6	—	18.5	—	3.38	0.96	1070	163	5.51	345	56.0	3.13	12.9	186	85.4
106	6.2	—	21.1	—	3.36	1.07	933	145	5.47	301	49.3	3.11	9.13	164	75.1
96	6.8	—	23.1	—	3.34	1.16	833	131	5.44	270	44.4	3.09	6.86	147	67.5
87	7.5	—	24.3	—	3.32	1.28	740	118	5.38	241	39.7	3.07	5.10	132	60.4
79	8.2	62.6	26.3	—	3.31	1.39	662	107	5.34	216	35.8	3.05	3.84	119	54.3
72	9.0	52.3	28.5	—	3.29	1.52	597	97.4	5.31	195	32.4	3.04	2.93	108	49.2
65	9.9	43.0	31.1	—	3.28	1.67	533	87.9	5.28	174	29.1	3.02	2.18	96.8	44.1
58	7.8	—	33.9	57.6	2.72	1.90	475	78.0	5.28	107	21.4	2.51	2.10	86.4	32.5
53	8.7	55.9	35.0	54.1	2.71	2.10	425	70.6	5.23	95.8	19.2	2.48	1.58	77.9	29.1
50	6.3	—	32.9	60.9	2.17	2.36	394	64.7	5.18	56.3	13.9	1.96	1.78	72.4	21.4
45	7.0	—	36.0	51.0	2.15	2.61	350	58.1	5.15	50.0	12.4	1.94	1.31	64.7	19.0
40	7.8	—	40.5	40.3	2.14	2.90	310	51.9	5.13	44.1	11.0	1.93	0.95	57.5	16.8
35	6.3	—	41.7	38.0	1.74	3.66	285	45.6	5.25	24.5	7.47	1.54	0.74	51.2	11.5
30	7.4	—	47.5	29.3	1.73	4.30	238	38.6	5.21	20.3	6.24	1.52	0.46	43.1	9.56
26	8.5	57.9	53.1	23.4	1.72	4.95	204	33.4	5.17	17.3	5.34	1.51	0.30	37.2	8.17
22	4.7	—	47.3	29.5	1.02	7.19	156	25.4	4.91	4.66	2.31	0.847	0.29	29.3	3.66
19	5.7	—	51.7	24.7	1.00	8.67	130	21.3	4.82	3.76	1.88	0.822	0.18	24.7	2.98
16	7.5	—	54.5	22.2	0.96	11.3	103	17.1	4.67	2.82	1.41	0.773	0.10	20.1	2.26
14	8.8	54.3	59.6	18.6	0.95	13.3	88.6	14.9	4.62	2.36	1.19	0.753	0.07	17.4	1.90

FIGURE 11.12
A portion of the table of dimensions and properties of wide-flange shapes from the *Manual of Steel Construction* of the American Institute of Steel Construction. One inch equals 25.4 mm. *(By permission of American Institute of Steel Construction)*

the shape in inches, a multiplication sign, and the weight of the shape in pounds per foot. Thus, a W12 × 26 is a wide-flange shape nominally 12 inches (305 mm) deep that weighs 26 pounds per foot of length (38.5 kg/m). More information about this shape is contained in a table of dimensions and properties in the *Manual of Steel Construction* published by the American Institute of Steel Construction (Figure 11.12): Its actual depth is 12.22 inches (310.4 mm), and its flanges are 6.49 inches (164.9 mm) wide. These proportions indicate that the shape is intended mainly for use as a beam or girder and not as a column or foundation pile. Reading across the table column by column, the designer can learn everything there is to know about this section, from its thicknesses and the radii of its fillets to various quantities that are useful in computing its structural behavior under load. At the upper end of the portion of the table dealing with 12-inch (305-mm)-wide-flanges, we find shapes weighing up to 336 pounds per foot (501 kg/m), with actual depths of almost 17 inches (432 mm). These heavier shapes have flanges nearly as wide as the shapes are deep, suggesting that they are intended for use as columns. U.S. producers manufacture steel shapes only in conventional units of measurement, inches and pounds. In other parts of the world, a standard range of metric sizes is used. The United States has adopted a "soft" conversion to metric sizes, merely tabulating metric dimensions for shapes that are produced in conventional units.

Steel *angles* (Figure 11.13) are extremely versatile. They can be used for very short beams supporting small

ANGLES
Equal legs and unequal legs
Properties for designing

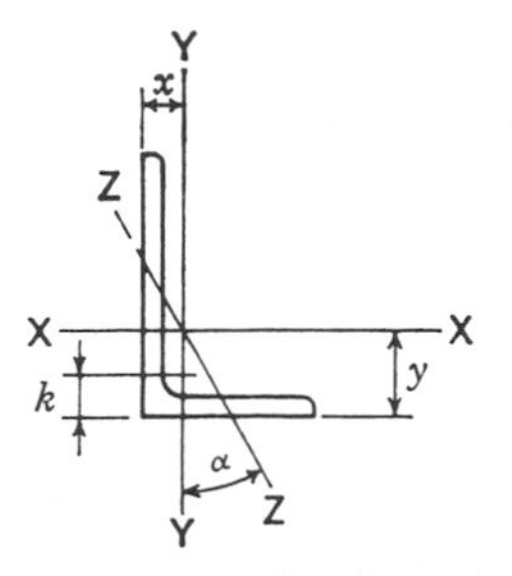

Size and Thickness	k	Weight per Foot	Area	AXIS X-X				AXIS Y-Y				AXIS Z-Z	
				I	*S*	*r*	*y*	*I*	*S*	*r*	*x*	*r*	Tan
In.	In.	Lb.	In.^{2}	In.^{4}	In.^{3}	In.	In.	In.^{4}	In.^{3}	In.	In.	In.	*α*
L 4 x3 x 5/8	1 1/16	13.6	3.98	6.03	2.30	1.23	1.37	2.87	1.35	0.849	0.871	0.637	0.534
1/2	15/16	11.1	3.25	5.05	1.89	1.25	1.33	2.42	1.12	0.864	0.827	0.639	0.543
7/16	7/8	9.8	2.87	4.52	1.68	1.25	1.30	2.18	0.992	0.871	0.804	0.641	0.547
3/8	13/16	8.5	2.48	3.96	1.46	1.26	1.28	1.92	0.866	0.879	0.782	0.644	0.551
5/16	3/4	7.2	2.09	3.38	1.23	1.27	1.26	1.65	0.734	0.887	0.759	0.647	0.554
1/4	11/16	5.8	1.69	2.77	1.00	1.28	1.24	1.36	0.599	0.896	0.736	0.651	0.558
L 3 1/2x3 1/2x 1/2	7/8	11.1	3.25	3.64	1.49	1.06	1.06	3.64	1.49	1.06	1.06	0.683	1.000
7/16	13/16	9.8	2.87	3.26	1.32	1.07	1.04	3.26	1.32	1.07	1.04	0.684	1.000
3/8	3/4	8.5	2.48	2.87	1.15	1.07	1.01	2.87	1.15	1.07	1.01	0.687	1.000
5/16	11/16	7.2	2.09	2.45	0.976	1.08	0.990	2.45	0.976	1.08	0.990	0.690	1.000
1/4	5/8	5.8	1.69	2.01	0.794	1.09	0.968	2.01	0.794	1.09	0.968	0.694	1.000
L 3 1/2x3 x 1/2	15/16	10.2	3.00	3.45	1.45	1.07	1.13	2.33	1.10	0.881	0.875	0.621	0.714
7/16	7/8	9.1	2.65	3.10	1.29	1.08	1.10	2.09	0.975	0.889	0.853	0.622	0.718
3/8	13/16	7.9	2.30	2.72	1.13	1.09	1.08	1.85	0.851	0.897	0.830	0.625	0.721
5/16	3/4	6.6	1.93	2.33	0.954	1.10	1.06	1.58	0.722	0.905	0.808	0.627	0.724
1/4	11/16	5.4	1.56	1.91	0.776	1.11	1.04	1.30	0.589	0.914	0.785	0.631	0.727
L 3 1/2x2 1/2x 1/2	15/16	9.4	2.75	3.24	1.41	1.09	1.20	1.36	0.760	0.704	0.705	0.534	0.486
7/16	7/8	8.3	2.43	2.91	1.26	1.09	1.18	1.23	0.677	0.711	0.682	0.535	0.491
3/8	13/16	7.2	2.11	2.56	1.09	1.10	1.16	1.09	0.592	0.719	0.660	0.537	0.496
5/16	3/4	6.1	1.78	2.19	0.927	1.11	1.14	0.939	0.504	0.727	0.637	0.540	0.501
1/4	11/16	4.9	1.44	1.80	0.755	1.12	1.11	0.777	0.412	0.735	0.614	0.544	0.506
L 3 x3 x 1/2	13/16	9.4	2.75	2.22	1.07	0.898	0.932	2.22	1.07	0.898	0.932	0.584	1.000
7/16	3/4	8.3	2.43	1.99	0.954	0.905	0.910	1.99	0.954	0.905	0.910	0.585	1.000
3/8	11/16	7.2	2.11	1.76	0.833	0.913	0.888	1.76	0.833	0.913	0.888	0.587	1.000
5/16	5/8	6.1	1.78	1.51	0.707	0.922	0.865	1.51	0.707	0.922	0.865	0.589	1.000
1/4	9/16	4.9	1.44	1.24	0.577	0.930	0.842	1.24	0.577	0.930	0.842	0.592	1.000
3/16	1/2	3.71	1.09	0.962	0.441	0.939	0.820	0.962	0.441	0.939	0.820	0.596	1.000

Angles in shaded rows may not be readily available. Availability is subject to rolling accumulation and geographical location, and should be checked with material suppliers.

FIGURE 11.13
A portion of the table of dimensions and properties of angle shapes from the *Manual of Steel Construction* of the American Institute of Steel Construction. One inch equals 24.5 mm. *(By permission of American Institute of Steel Construction)*

loads and are frequently found playing this role as lintels spanning door and window openings in masonry construction. In steel frame buildings, their primary role is in connecting wide-flange beams, girders, and columns, as we will see shortly. They also find use as diagonal braces in steel frames and as members of steel trusses, where they are paired back to back to connect conveniently to flat *gusset plates* at the joints of the truss (Figure 11.82). *Channel* sections are also used as truss members and bracing, and for short beams, lintels, and stringers in steel stairs. *Tees, plates, bars,* and *sheets* all have their various roles in a steel frame building, as shown in the diagrams that accompany this text.

The structural properties of steel can also be adjusted after rolling, using various so-called "thermo-mechanical processes." For example, immediately after rolling, ASTM A913 steel is subjected to a process of *quenching* (rapid cooling) and then *tempering* (partial reheating) to give the steel an optimized balance of strength, toughness, and weldability characteristics.

Cast Steel

While the vast majority of structural steel is produced as rolled shapes, structural shapes can also be produced as *cast steel,* that is, by pouring molten steel directly into molds and allowing the steel to cool. Although steel castings are, pound for pound, more expensive than rolled steel shapes, they have other advantages: Because cast parts are produced in small quantities, they can economically utilize specialized steel alloys selected on the basis of a part's unique requirements. Because they are cast in discrete molds rather than formed through a continuous rolling process, cast steel parts can be nonuniform in section, they can readily incorporate curves or complex geometries, and their shapes can be carefully tailored to the particular requirements of the part. Cast steel is especially well suited for the production of custom-shaped connections for steel structures that are stronger, lighter, and more attractive than possible with those fabricated from conventional rolled steel.

Cold-Worked Steel

Steel can be *cold-worked* or *cold-formed* (rolled or bent) in a so-called "cold" state (at room temperature). Cold working causes the steel to gain considerable strength through a realignment of its crystalline structure. Light-gauge (thin) steel sheet is formed into C-shaped sections to make short-span framing members that are frequently used to frame partitions and exterior walls of larger buildings and floor structures of smaller buildings (see Chapter 12). Steel sheet stock is also rolled into corrugated configurations utilized as floor and roof decking in steel-framed structures (Figures 11.59–11.64).

Heavier sheet or plate stock may be cold-formed into square, rectangular, round, and elliptical hollow shapes that are then welded along the longitudinal seam to form *hollow structural sections (HSS)* (Figures 11.11, 11.14, and 11.84). Also called "structural tubing," they are often used for columns and for members of welded steel trusses and space trusses. Their hollow shape makes them especially suitable for members that are subjected to torsional (twisting) stresses or to buckling associated with compressive loads.

The normal range of wide-flange shapes is too large to be cold-rolled, but cold rolling is used to produce small-section steel rods and steel components for open-web joists, where the higher strength can be utilized to good advantage. Steel is also cold-drawn through dies to produce the very high strength wires used in wire ropes, bridge cables, and concrete prestressing strands.

Open-Web Steel Joists

Among the many structural steel products fabricated from hot- and cold-rolled shapes, the most common is the *open-web steel joist (OWSJ)*, a mass-produced truss used in closely spaced arrays to support floor and roof decks (Figure 11.14). According to Steel Joist Institute (SJI) specifications, open-web joists are produced in three series: K series joists are for spans up to 60 feet (18 m) and range in depth from 8 to 30 inches (200–760 mm). LH series joists are designated as "Longspan" and can span as far as 96 feet (29 m). Their depths range from 18 to 48 inches (460–1220 mm). The DLH "Deep Longspan" series of open-web joists are 52 to 72 inches deep (1320–1830 mm) and can span up to 144 feet (44 m). Most buildings that use open-web joists utilize K series joists that are less than 2 feet (600 mm) deep to achieve spans of up to 40 feet (13 m). The spacings between joists commonly range from 2 to 10 feet (0.6–3 m), depending on the magnitude of the applied loads and the spanning capability of the decking. Some joist manufacturers also produce proprietary open-web steel joists types capable of longer spans than trusses designed to SJI specifications.

Joist girders are prefabricated steel trusses designed to carry heavy loads, particularly bays of steel joists (Figure 11.14). They range in depth from 20 to 72 inches (500–1800 mm). They can be used instead of wide-flange beams and girders in roof and floor structures where their greater depth is not objectionable. Open-web joists and joist girders are invariably made of high-strength steel.

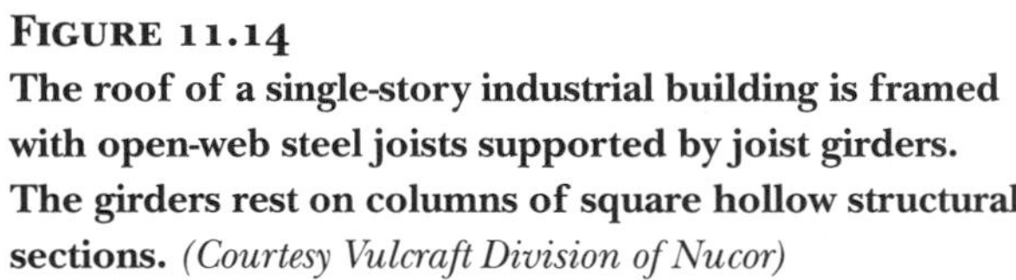

FIGURE 11.14
The roof of a single-story industrial building is framed with open-web steel joists supported by joist girders. The girders rest on columns of square hollow structural sections. *(Courtesy Vulcraft Division of Nucor)*

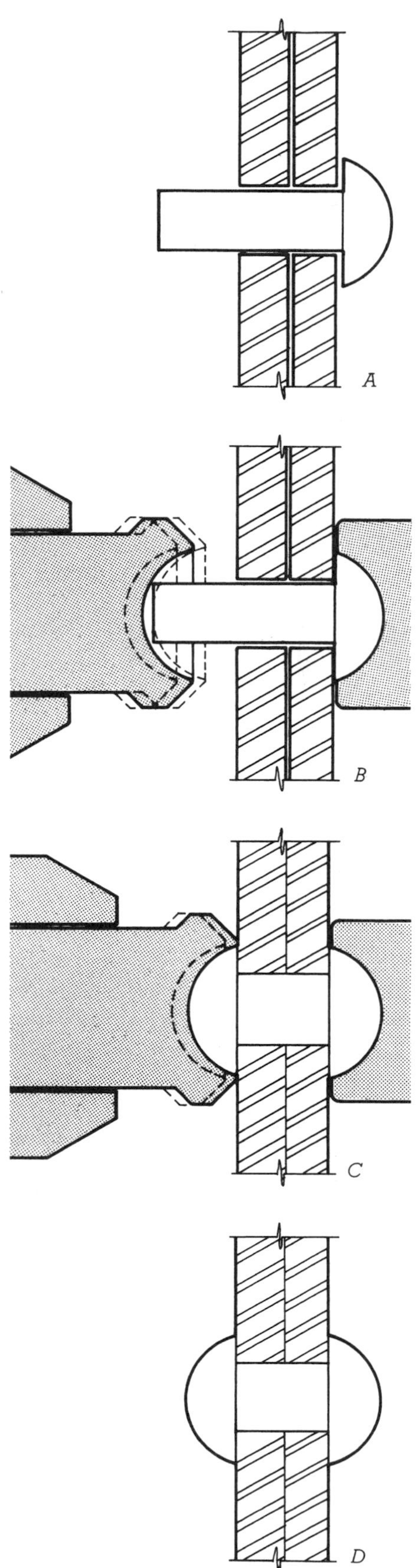

FIGURE 11.15
How riveted connections are made. (*a*) A hot steel rivet is inserted through holes in the two members to be joined. (*b, c*) Its head is placed in the cup-shaped depression of a heavy, hand-held hammer. A pneumatic hammer drives a rivet set repeatedly against the body of the rivet to form the second head. (*d*) The rivet shrinks as it cools, drawing the members tightly together.

Joining Steel Members

Rivets

Steel shapes can be joined into a building frame with any of three fastening techniques—rivets, bolts, or welds—and by combinations of these. A *rivet* is a steel fastener consisting of a cylindrical body and a formed head. It is brought to a white heat in a forge, inserted through holes in the members to be joined, and hot-worked with a pneumatic hammer to produce a second head opposite the first (Figure 11.15). As the rivet cools, it shrinks, clamping the joined pieces together and forming a tight joint. Riveting was for many decades the predominant fastening technique used in steel frame buildings, but it has been completely replaced in contemporary construction by the less labor-intensive techniques of bolting and welding.

Bolts

Bolts used in steel frame construction may be either *high-strength bolts* (ASTM A325 and A490), which are heat treated during manufacturing to develop their greater strength, or lower-strength *carbon steel bolts* (ASTM A307). In contemporary steel frame construction, bolted structural connections rely almost exclusively on high-strength bolts. Carbon steel bolts (also called "unfinished" or "common" bolts) find only limited use, such as in the fastening of minor framing elements or temporary connections.

The manner in which a bolted structural steel connection derives its strength depends on how the bolts are installed. In a *bearing-type connection*, bolts need only be installed to a *snug-tight* condition. In this case, movement between the joined members is resisted by the bolts themselves as the sides of the bolt holes in the connected members bear against the bodies of the bolts. In a *slip-critical* (or *friction type*) *connection*, bolts are *preloaded* (tightened during installation) to such an extent that friction between the adjoining faces of the steel members (the *faying surfaces*) resists movement between the members. Under normal load conditions, bolts in bearing-type connections are stressed primarily in shear, while those in slip-critical connections are stressed in tension.

When a bearing-type connection is first loaded, a slight slippage of the joint occurs as the sides of the bolt holes in the joined members achieve full bearing against the bodies of the bolts. In contrast, a slip-critical connection will reach its full design capacity with virtually no initial slippage. For this reason, only slip-critical connections are used where the small changes in alignment that can occur with bearing-type connections could be detrimental to the performance of a structure. For example, column splices and beam-to-column connections in tall buildings must be designed as slip-critical, as must connections that experience load reversals.

In a typical connection, bolts are inserted into holes $\frac{1}{16}$ inch (2 mm) larger than the diameter of the bolt. Depending on a variety of factors, hardened steel washers may be inserted under one or both ends of the fastener. Washers are required with slotted or oversized holes to ensure that the bolt head and nut have adequate contact with the surfaces of the joined members. When installing preloaded bolts, washers may be required to prevent damage, such as *galling* (tearing), to the surfaces of the joined members. Many bolt tension verification methods, discussed below, also require washers under at least one end of the bolt to ensure consistent tensioning results.

FIGURE 11.16
An ironworker tightens high-strength bolts with a pneumatic impact wrench. *(Courtesy of Bethlehem Steel Corporation)*

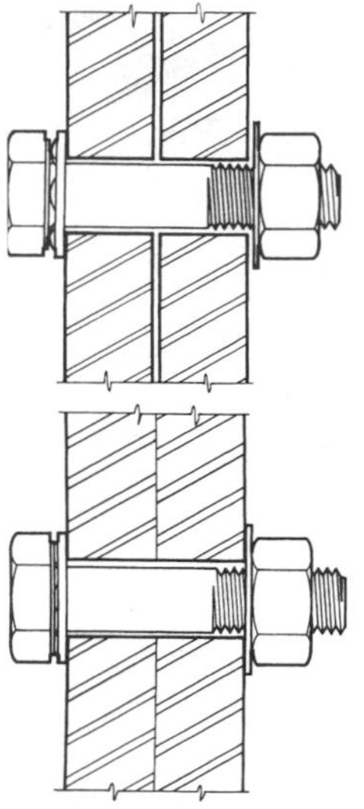

FIGURE 11.17
***Top*: An untightened high-strength bolt with a load indicator washer under the head. *Bottom*: The bolt and washer after tightening; notice that the protrusions on the load indicator washer have flattened.**

FIGURE 11.18
A splined tension control bolt. *(Courtesy of LeJeune Bolt Company)*

FIGURE 11.19
The compact design of the electric shear wrench used for tightening tension control bolts makes it easy to reach bolts in tight situations. *(Courtesy of LeJeune Bolt Company)*

Bolts are usually tightened using a pneumatic or electric *impact wrench* (Figure 11.16). In a bearing-type connection, the amount of tension in the bolts is not critical. In a slip-critical connection, bolts must be tightened reliably to at least 70 percent of their ultimate tensile strength.

A major problem in the assembly of slip-critical connections is how to verify that the necessary tension has been achieved in each bolt. This can be accomplished in any of several ways. In the *turn-of-nut method*, each bolt is tightened snug, then turned a specified additional fraction of a turn. Depending on bolt length, bolt alloy, and other factors, the additional tightening required will range from one-third of a turn to a full turn.

In another method, a *load indicator washer*, also called a *direct tension indicator (DTI)*, is placed under the head or nut of the bolt. As the bolt is tightened, protrusions on the washer are progressively flattened in proportion to the tension in the bolt (Figure 11.17). Inspection for proper bolt tension then becomes a relatively simple matter of inserting a feeler gauge to determine whether the washer has flattened sufficiently to indicate the required tension. One washer manufacturer has made inspection even easier by attaching tiny dye capsules to the washer; when the protrusions on the washer flatten sufficiently, the capsules squirt a highly visible dye onto the surface of the washer.

Less frequently used to verify bolt tension is the *calibrated wrench method*, in which a special torque control wrench is used to tighten the bolts. The torque setting of the wrench is carefully calibrated for the particular size and type of fasteners being installed so as to achieve the required bolt tension. A washer under the turned end of the

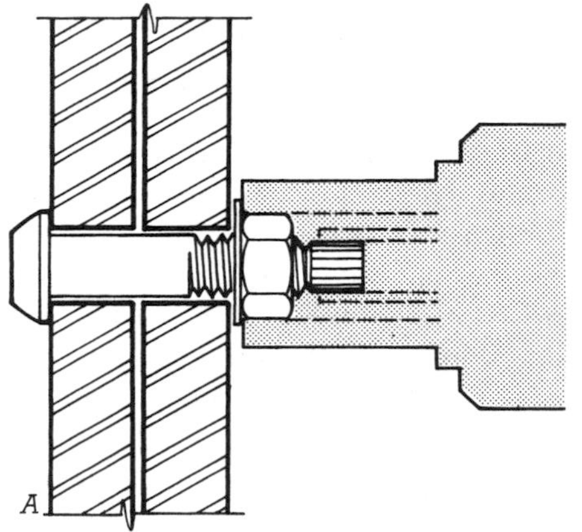

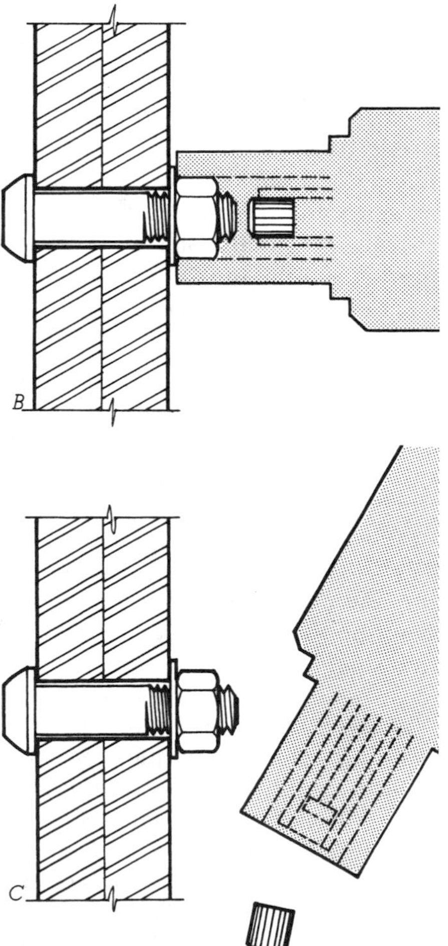

FIGURE 11.20
Tightening a tension control bolt. (*a*) The wrench holds both the nut and the splined body of the bolt and turns them against one another to tighten the bolt. (*b*) When the required torque has been achieved, the splined end twists off in the wrench. (*c*) A plunger inside the wrench discharges the splined end into a container. *(Courtesy of LeJeune Bolt Company)*

bolt minimizes friction and ensures a consistent relationship between the tightening force applied and the tension achieved in the bolt.

Yet another method of bolt tension verification employs *tension control bolts.* These have protruding splined ends that extend beyond the threaded portion of the body of the bolt (Figure 11.18). The nut is tightened by a special power-driven *shear wrench* (Figure 11.19) that grips both the nut and the splined end simultaneously, turning the one against the other. The splined end is formed in such a way that when the required torque has been reached, the end twists off (Figure 11.20). Verification of adequate bolt tensioning in installed bolts then becomes a simple matter of visually checking for the absence of splines. Another advantage of this fastener type is that it is installed by a single worker, unlike conventional bolts, which require a second worker with a wrench to prevent the other end of the bolt assembly from turning during tightening.

An alternative to the high-strength bolt is the *lockpin and collar fastener* or *swedge bolt,* a boltlike steel pin with annular rings that relies on a steel collar in lieu of a conventional nut to hold the pin. The swedge bolt is installed using a special power tool to hold the pin under high tension while cold forming (swaging, a crimpinglike action) the collar around its end to complete the connection. As the installation process is completed, the tail of the lockpin breaks off, furnishing visual evidence that the necessary tension has been achieved in the fastener. Like the tension control bolt, the swedge bolt can be installed by a single worker.

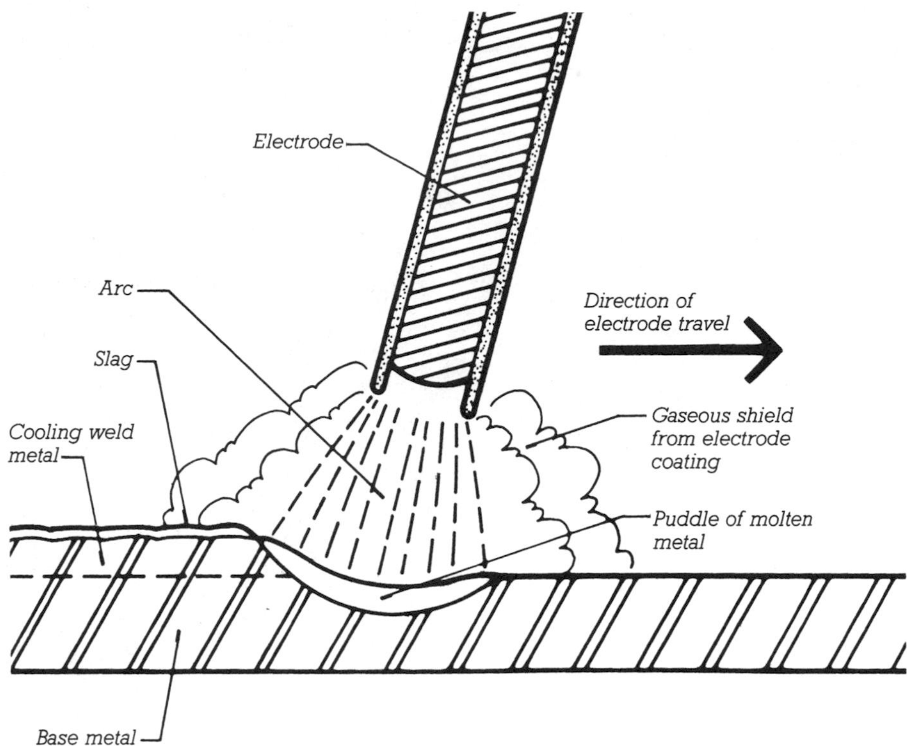

FIGURE 11.21
Close-up diagram of the electric arc welding process.

Welding

Welding offers a unique and valuable capability to the structural designer: It can join the members of a steel frame as if they were a monolithic whole. Welded connections, properly designed and executed, are stronger than the members they join in resisting both shear and moment forces. Although it is possible to achieve this same performance with high-strength bolted connections, such connections are often cumbersome compared to equivalent welded joints. Bolting, on the other hand, has its own advantages: It is quick and easy for field connections that need only to resist shearing forces, and it can be accomplished under conditions of adverse weather or difficult physical access that would make welding impossible. Often welding and bolting are combined in the same connections to take advantage of the unique qualities of each: Welding may be used in the fabricator's shop for its inherent economies and in the field for its structural continuity, whereas bolting is often employed in the simpler field connections and to hold connections in alignment prior to welding. The choice of bolting, welding, or combinations of the two is often dictated by the designer, but it may also be influenced by considerations such as the fabricator's and erector's equipment and expertise, availability of electric power, climate, and location.

Electric arc welding is conceptually simple. An electrical potential is established between the steel pieces to be joined and a metal *electrode* held either by a machine or by a person. When the electrode is held close to the seam between the steel members, a continuous electric arc is established that generates sufficient heat to melt both a localized area of the steel members and the tip of the electrode (Figure 11.21). The molten steel from the electrode merges with that of the members to form a single puddle. The electrode is drawn slowly along the seam, leaving behind a continuous bead of metal that cools and solidifies to form a continuous connection between the members. For small members, a single pass of the electrode may suffice to make the connection. For larger members, a number of passes are made in order to build up a weld of the required depth.

In practice, welding is a complex science. The metallurgy of the structural steel and the welding electrodes must be carefully coordinated. Voltage, amperage, and polarity of the electric current are selected to achieve the right heat and penetration for the weld. Air must be kept away from the electric arc to prevent rapid oxidation of the liquid steel; this is accomplished in simple welding processes either by a thick coating on the electrode that melts to create a liquid and gaseous shield around the arc or by a core of vaporizing flux in a tubular steel electrode. It may also be done by means of a continuous flow of inert gas around the arc, or with a dry flux that is heaped over the end of the electrode as it moves across the work.

FIGURE 11.22
Standard weld symbols, as used on steel connection detail drawings.

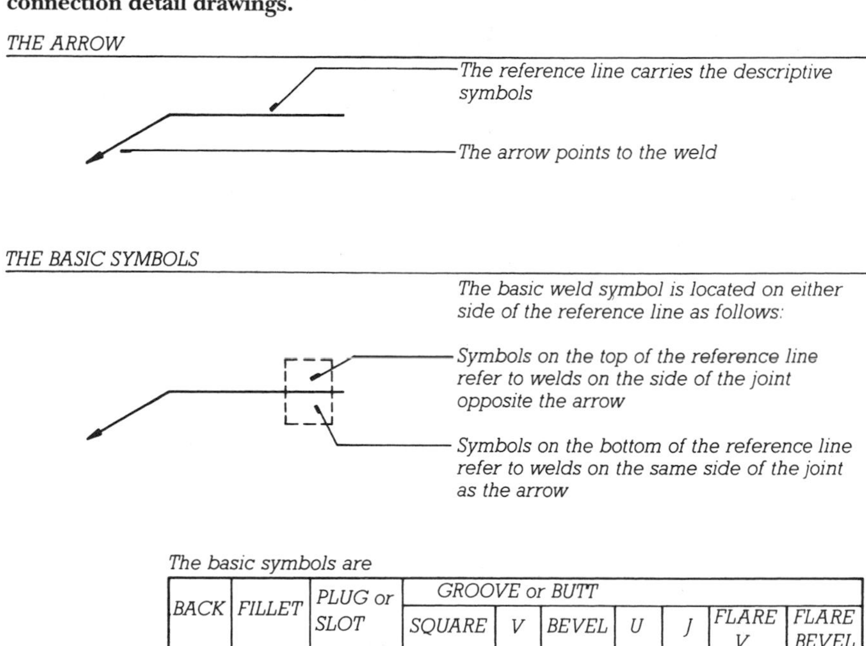

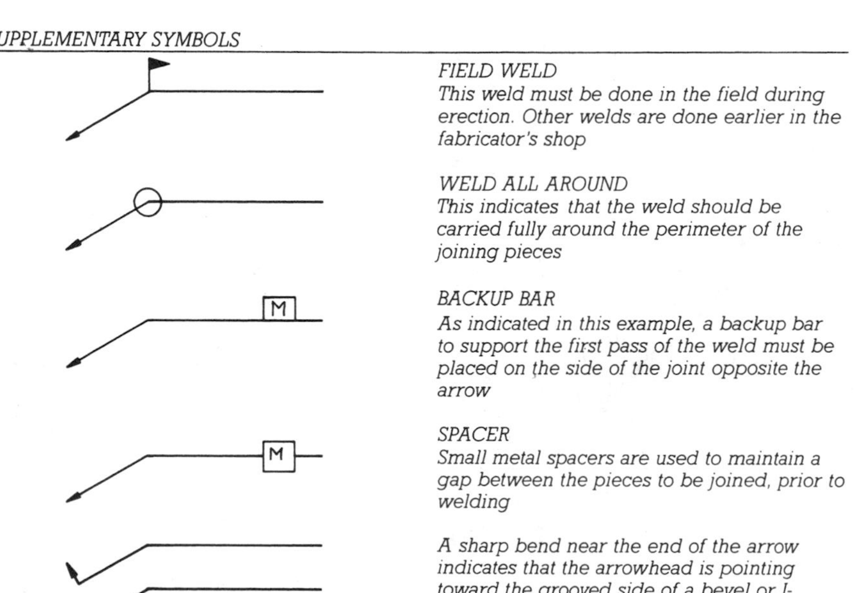

The required thickness and length of each weld are calculated by the designer to match them to the forces to be transmitted between the members, and are indicated on fabrication drawings using standardized *weld symbols* (Figure 11.22). For deep welds, such as the full-penetration welds shown in Figure 11.23, the edges of the members are beveled to create a groove that permits access of the electrode to the full thickness of the piece. Small strips of steel called *backup bars* or *backing bars* are welded beneath the groove prior to beginning the actual weld to prevent the molten metal from flowing out the bottom of the groove. The weld then is deposited in a number of passes of the electrode until the groove is fully filled. In some cases, *runoff bars* are required at the ends of a groove weld to facilitate the formation of a full thickness of weld metal at the edges of the member (Figure 11.46).

Workers who do structural welding are methodically trained and periodically tested to ensure that they have the required level of skill and knowledge. When an important weld is completed, it is inspected to make sure that it is of the required size and quality; this often involves sophisticated magnetic particle, dye penetrant, ultrasonic, or radiographic testing procedures that search for hidden voids and flaws within each weld.

Welds in structural connections that may be subjected to very high stresses during a seismic event and that are critical to maintaining the stability of the building structure are termed *demand-critical welds.* They must meet special requirements related to their design, materials, installation, and inspection to ensure their reliability under these conditions.

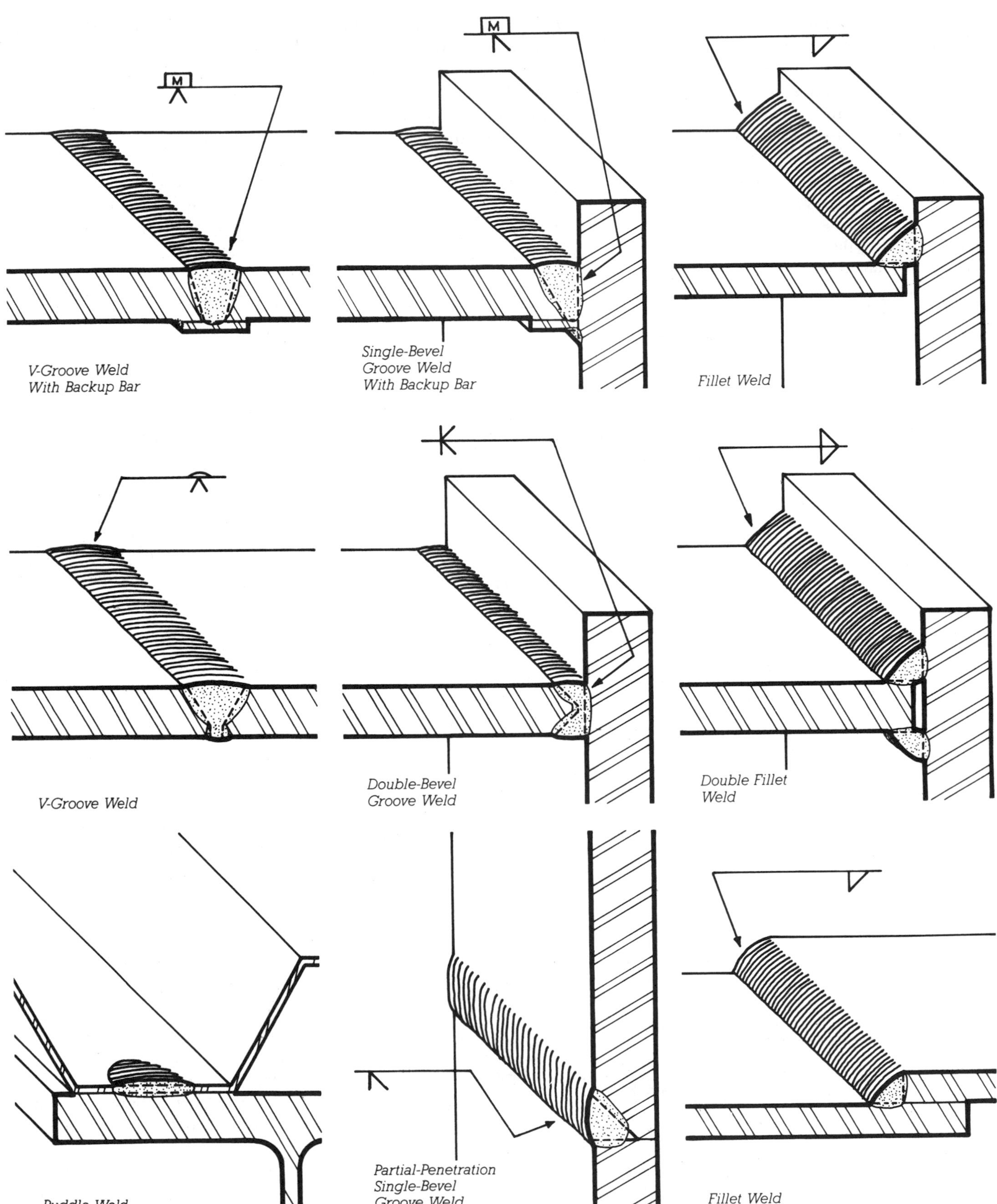

FIGURE 11.23
Typical welds used in steel frame construction. Fillet welds are the most economical because they require no advance preparation of the joint, but full-penetration groove welds are stronger. The standard symbols used here are explained in Figure 11.22.

Details of Steel Framing

Typical Connections

Most steel frame connections use angles, plates, or tees as transitional elements between the members being connected. A simple bolted beam-to-column-flange connection requires two angles and a number of bolts (Figures 11.24–11.27). The angles are cut to length, and the holes are made in all the components prior to assembly. The angles are usually bolted to the web of the beam in the fabricator's shop. The bolts through the flange of the column are added as the beam is erected on the construction site. This type of connection, which joins only the web of the beam, but not the flanges, to the column is known as a *shear connection*. It is capable of transmitting vertical forces (*shear*) from a beam to a column. However, because it does not connect the beam flanges to the column, it is of no value in transmitting bending forces (*bending moment*) from one to the other.

To produce a *moment connection*, one capable of transmitting bending forces between a beam and column, it is necessary to connect the beam flanges strongly across the joint, most commonly by means of *full-penetration groove welds* (Figures 11.28 and 11.29). If the column flanges are insufficiently strong to accept the forces transmitted from the beam flanges, *stiffener plates* must be installed inside the flanges of the column to better distribute these forces into the body of the column. (Though much less common, it is also possible to design moment-transmitting connections that rely solely on bolting.)

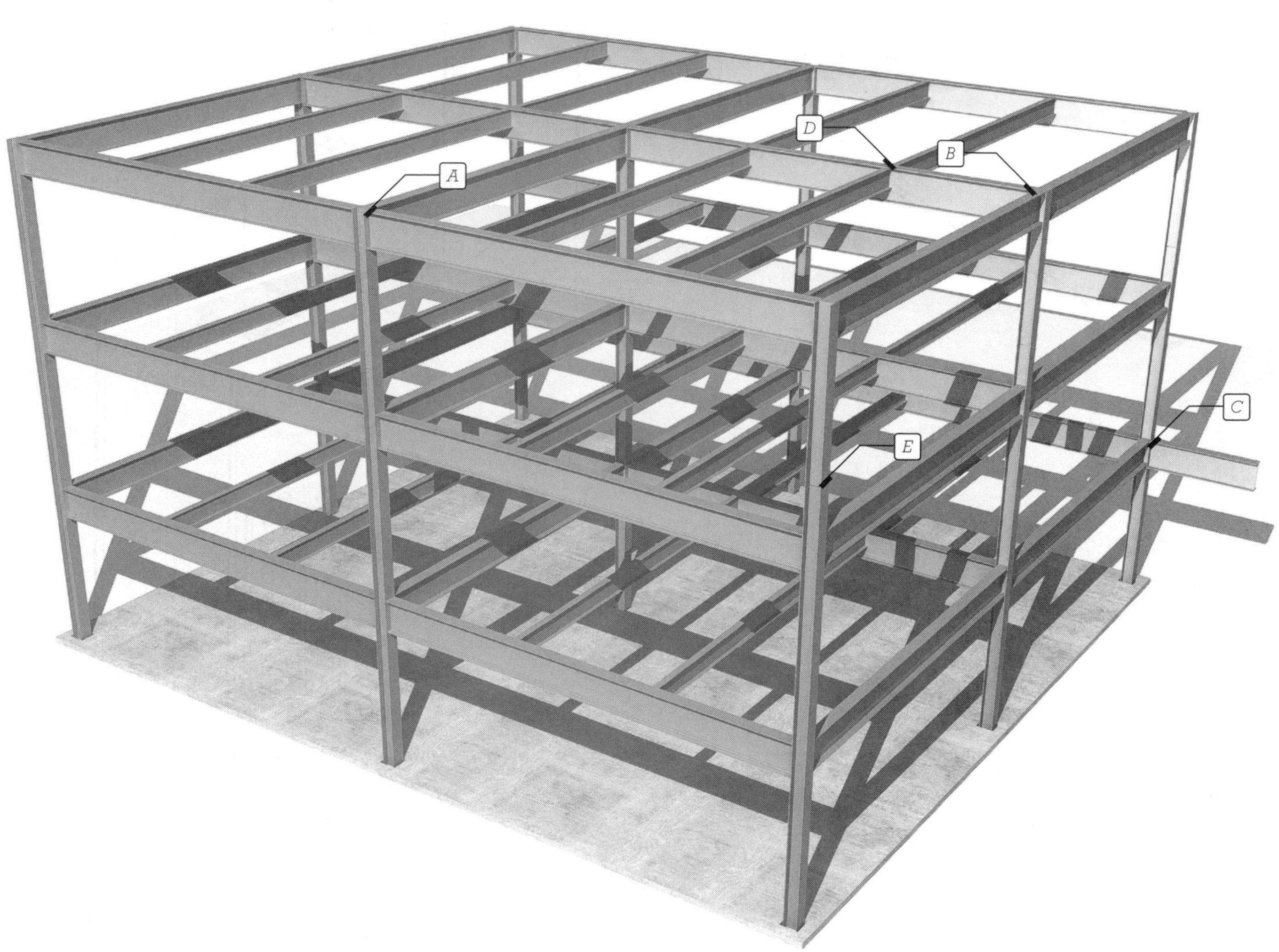

FIGURE 11.24
A generic steel building frame. The letters are keyed to the connection details in the figures that follow.

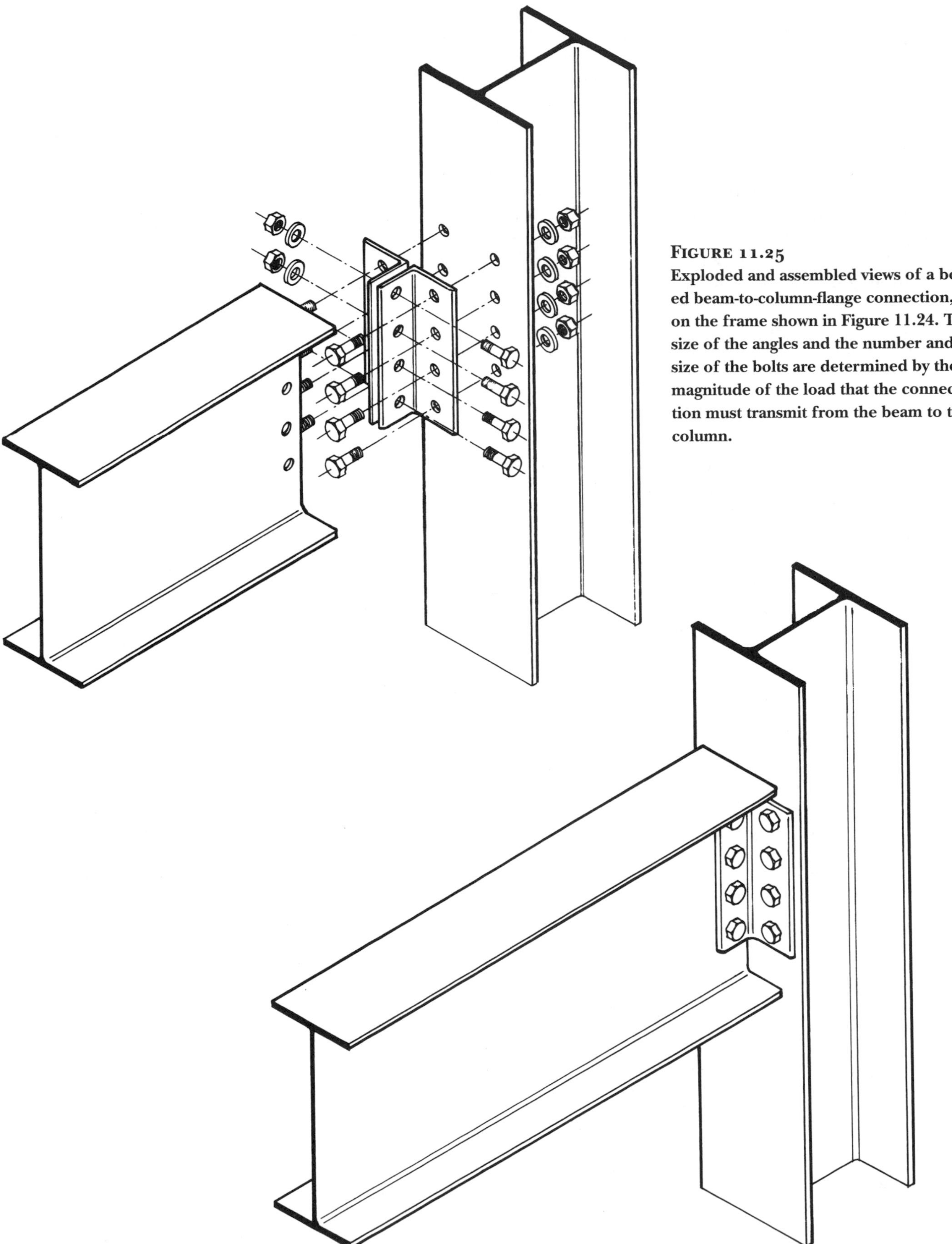

FIGURE 11.25
Exploded and assembled views of a bolted beam-to-column-flange connection, *A* on the frame shown in Figure 11.24. The size of the angles and the number and size of the bolts are determined by the magnitude of the load that the connection must transmit from the beam to the column.

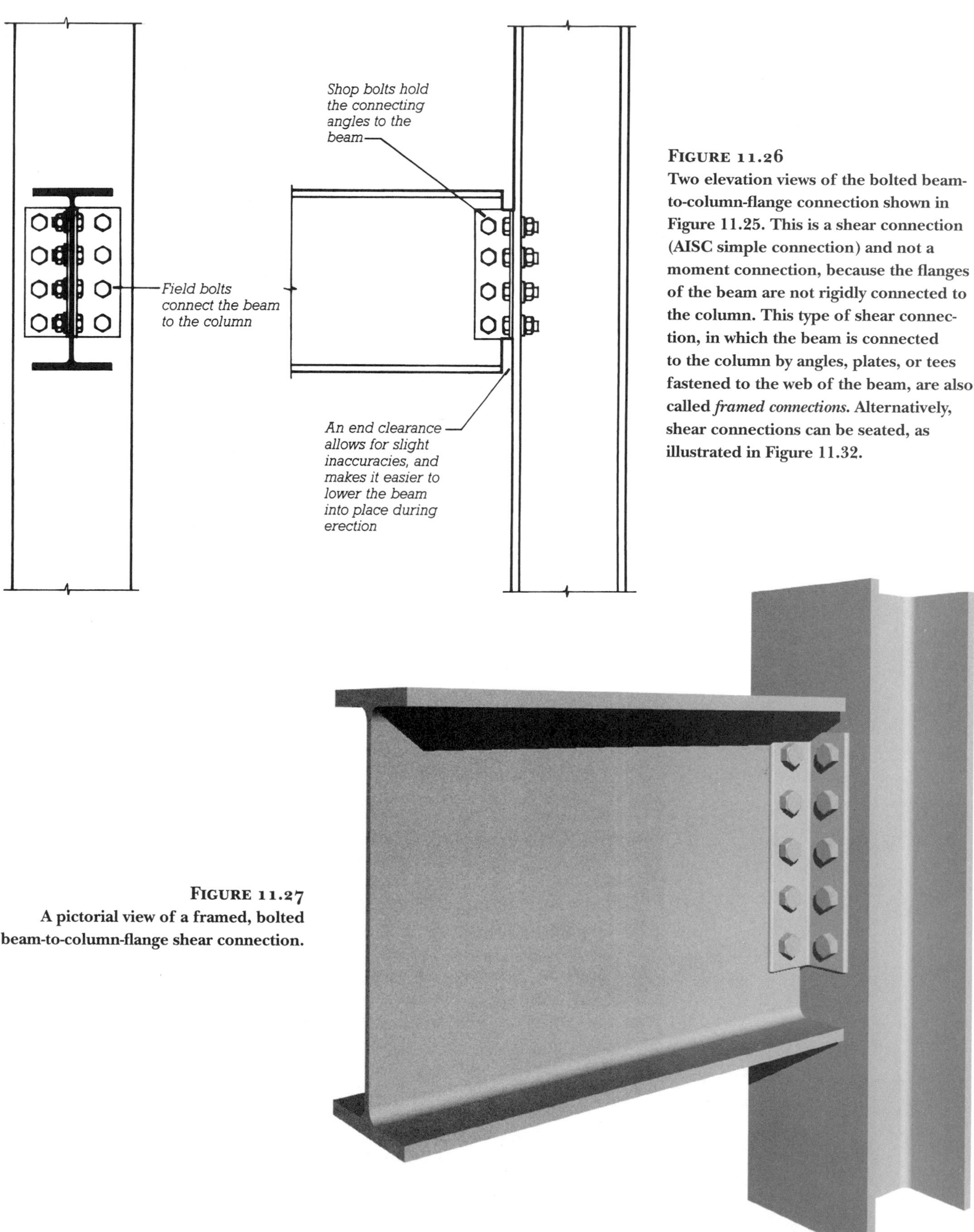
Shop bolts hold the connecting angles to the beam
Field bolts connect the beam to the column
An end clearance allows for slight inaccuracies, and makes it easier to lower the beam into place during erection

FIGURE 11.26
Two elevation views of the bolted beam-to-column-flange connection shown in Figure 11.25. This is a shear connection (AISC simple connection) and not a moment connection, because the flanges of the beam are not rigidly connected to the column. This type of shear connection, in which the beam is connected to the column by angles, plates, or tees fastened to the web of the beam, are also called *framed connections*. Alternatively, shear connections can be seated, as illustrated in Figure 11.32.

FIGURE 11.27
A pictorial view of a framed, bolted beam-to-column-flange shear connection.

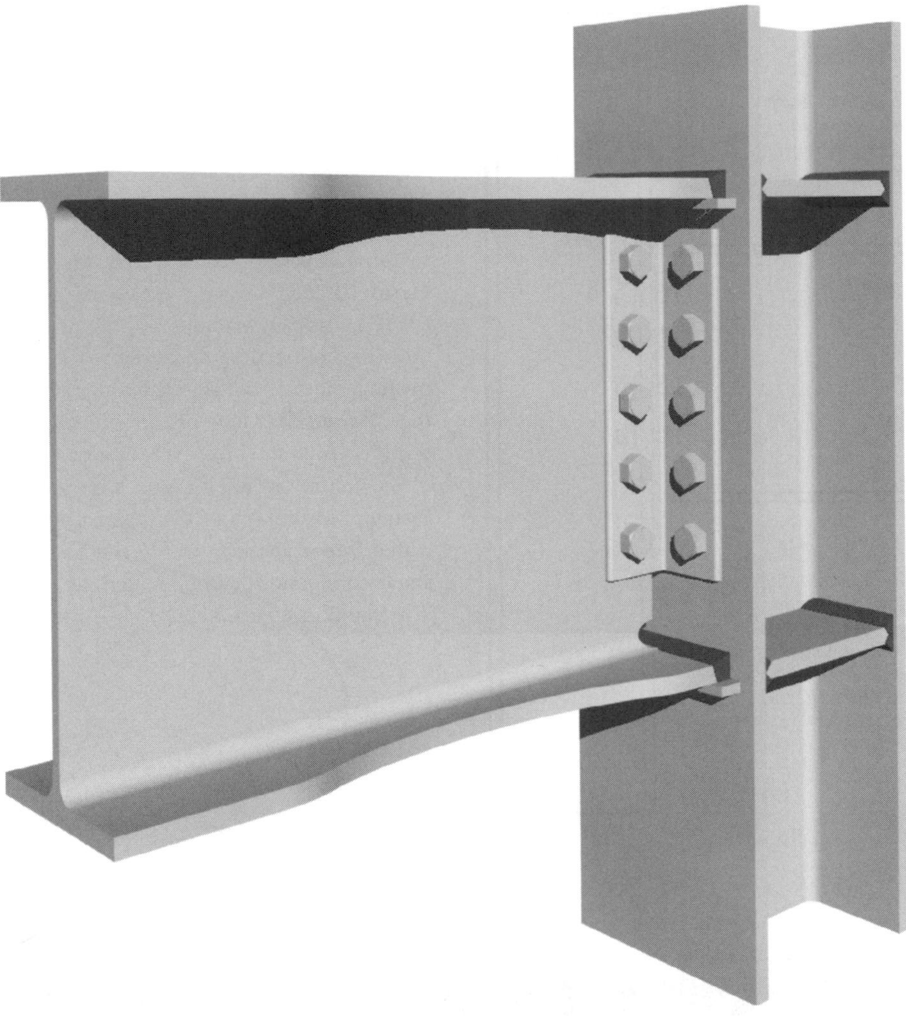

FIGURE 11.28
A welded moment connection (AISC Fully Restrained) for joining a beam to a column flange. This is the type of connection that would be used instead of the shear connection at location *A* in Figure 11.24 if a moment connection were required. The bolts hold the beam in place for welding and also provide shear resistance. Small rectangular backup bars are welded beneath the end of each beam flange to prevent the welding arc from burning through. A clearance hole is cut from the top of the beam web to permit the backup bar to pass through. A similar clearance hole at the bottom of the beam web allows the bottom flange to be welded entirely from above for greater convenience. The groove welds develop the full strength of the flanges of the beam, allowing the connection to transmit moments between the beam and the column. If the column flanges are not stiff enough to accept the moments from the beam, stiffener plates are welded between the column flanges as shown here. The flanges of the beam are cut to a "dog bone" configuration to create a zone of the beam that is slightly weaker in bending than the welded connection itself. During a violent earthquake, the beam will deform permanently in this zone while protecting the welded connection against failure.

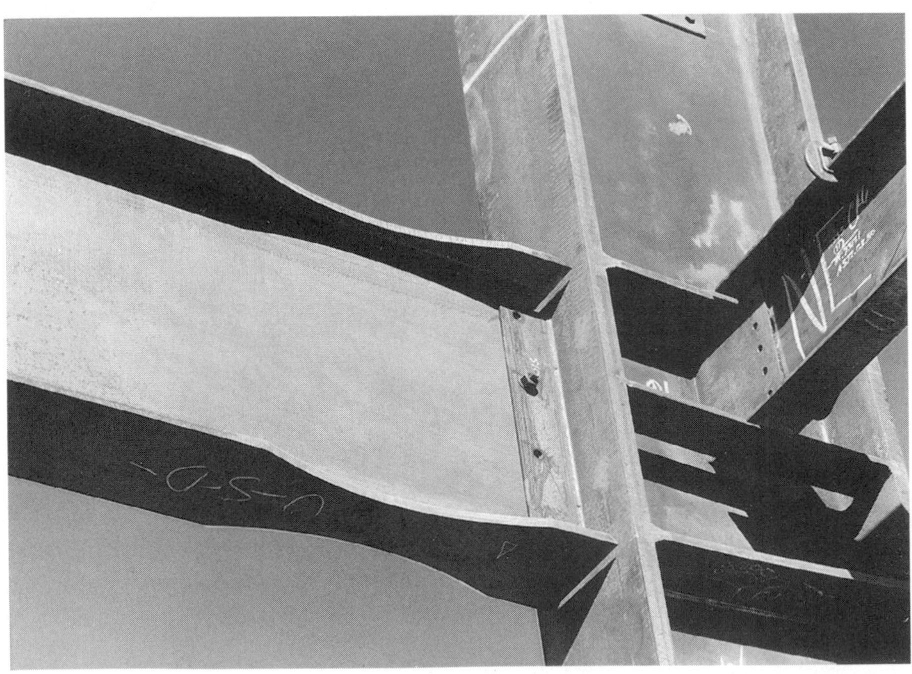

FIGURE 11.29
Photograph of a moment connection similar to the one shown in Figure 11.28. The beam has just been bolted to a shear tab that is welded to the column. Next, backup bars will be welded to the column just under the beam flanges, after which the flanges will be welded to the column. *(Permission of American Institute of Steel Construction)*

Stabilizing the Building Frame

In order to understand the respective roles of shear connections and moment connections in a building frame, it is necessary to understand the means by which buildings may be made stable against the lateral forces of wind and earthquake. Three basic stabilizing mechanisms are commonly used: braced frames, shear walls, and moment-resisting frames (Figure 11.30). A *braced frame* works by creating stable triangular configurations, or *diagonal bracing*, within the otherwise unstable rectilinear geometry of a steel building frame. The connections between beams and columns within a braced frame need not transmit moments (bending forces); they can behave like pins or hinges, which is another way of saying that they can be shear connections such as the one in Figure 11.27. (Though it may not be readily apparent, this type of connection is capable of the small rotations necessary for it to behave essentially as if it is hinged.)

A special case of the braced frame is the *eccentrically braced frame*. Because the ends of the diagonal braces are offset some distance from each other, where they connect to the beams, the structure as a whole is more resilient than a frame with conventional bracing. Eccentric bracing is used primarily as a way of causing a building frame to absorb energy during an earthquake and thus to protect against collapse. Like conventional braced frames, eccentrically braced frames can rely exclusively on shear connections between beams and columns.

Shear walls are stiff walls made of steel, concrete, or reinforced concrete masonry. They serve the same purpose as the diagonal bracing within a braced frame structure and, like the braced frame, moment connections between beams and columns are not required.

Moment-resisting frames have neither diagonal bracing nor shear walls to provide lateral stability. Rather, they rely on moment connections between beams and columns that are resistant to rotation and thereby capable of stabilizing the frame against lateral forces. Depending on the configuration of the structure and the magnitude of the forces involved, not all of the connections in a moment-resisting frame necessarily need be moment connections. Since moment connections are more costly to make than shear connections, they are used only to the extent required, with the remainder of the frame relying on simpler, less costly shear connections.

There are two common methods of arranging stabilizing elements within the frame of a tall building (Figure 11.31). One is to provide a *rigid core* in the center of the building. The core, which is the area that contains the elevators, stairs, mechanical chases, and washrooms, is structured as a stiff tower, using diagonal bracing, shear walls, or moment connections. The remainder of the building frame may then be constructed with shear connections

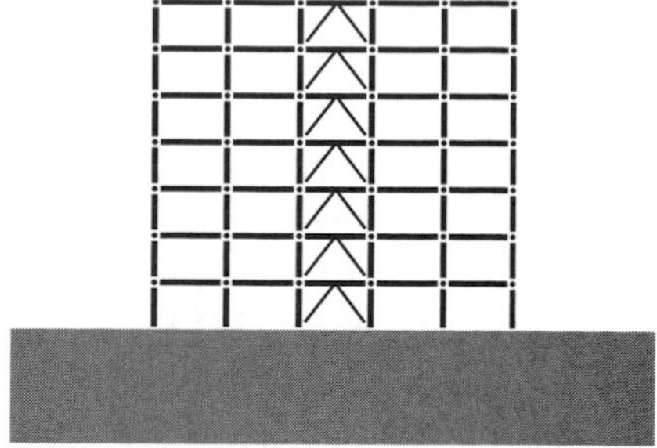

Braced Frame

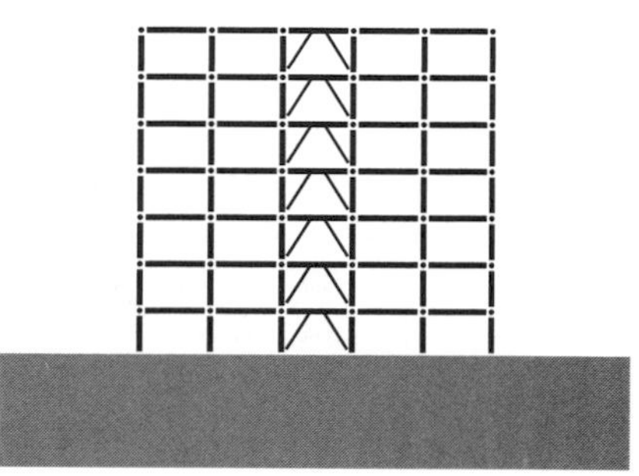

Eccentrically Braced Frame

Moment-Resisting Frame

Shear walls

FIGURE 11.30
Elevation views of the basic means for imparting lateral stability to a structural frame. Connections made with dots are shear connections, and solid intersections indicate moment connections. The braced frame (*top*) is illustrated here with *Chevron* (or *inverted V*) *bracing*. *Crossbracing*, in which paired diagonals run from opposite corners of the braced bay, are also common. Eccentric bracing (*second from top*) is used in situations where it is advantageous for the frame to absorb seismic energy during an earthquake. The connections in the moment-resisting frame (*third from top*), called "moment connections," are sufficiently resistant to rotation to stabilize the structure against lateral forces without diagonal bracing or shear walls. Moment-resisting frames are also sometimes called "rigid frames," although this name can be misleading. While the connections in such frames are relatively rigid, whole frames that rely only on such connections for lateral stability are typically somewhat less stiff than those stabilized with diagonal bracing or shear walls.

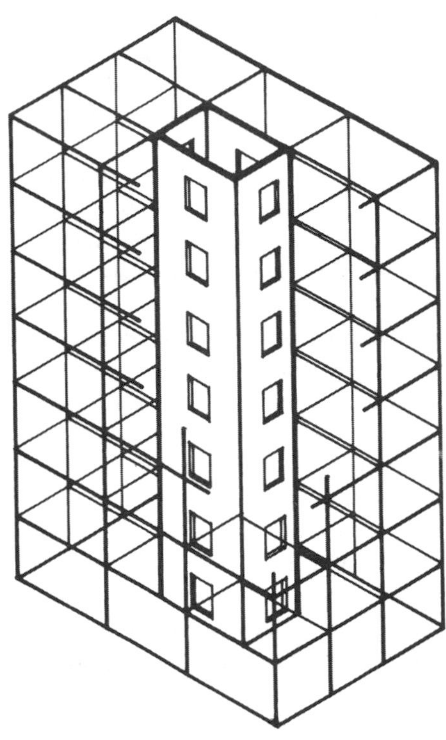

Rigid Core

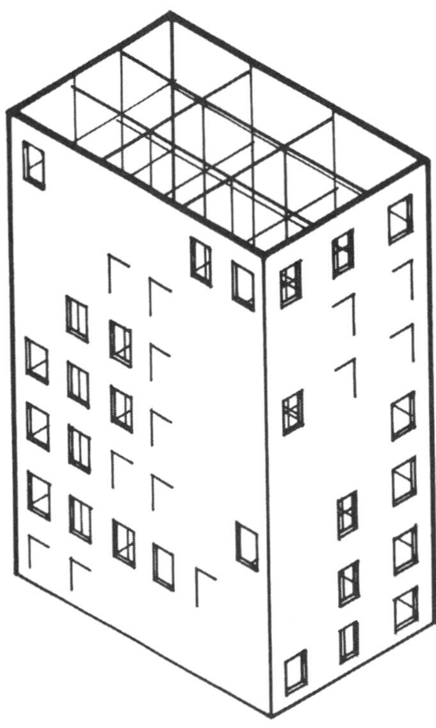

Rigid Perimeter

FIGURE 11.31
Rigid core versus rigid perimeter.

and stabilized by the *diaphragm action* (the rigidity possessed by a thin plate of material such as a welded steel deck with a concrete topping) of the floors and roof that connect these outer bays to the rigid core. Or, where additional resistance to lateral forces or greater stiffness is required, beam-to-column moment connections may be introduced into some portions of the building frame as well.

A second arrangement for achieving stability is to provide a *rigid perimeter*, again by using diagonal bracing, shear walls, or moment connections. When this is done, the entire interior of the structure can be assembled with shear connections, relying on diaphragm action in the floor and roof plates to impart stability to these portions of the structure.

In summary, shear connections between beams and columns are sufficient to transmit vertical loads through the building frame, but they are not, on their own, capable of providing resistance to lateral forces. Lateral force resistance may be provided by the introduction of diagonal bracing (braced frame), shear walls, beam-to-column moment connections (moment-resisting frame), or some combination of these elements into portions of the frame. Because shear connections are easier and less expensive to construct than moment connections, moment connections are used only to the extent necessary and often in combination with other stabilizing mechanisms.

Shear Connections and Moment Connections

The American Institute of Steel Construction (AISC) defines three types of beam-to-column connections, classified according to their moment-resisting capability. *Fully-Restrained* (*FR*) *moment connections* (formerly *AISC Type 1*) are sufficiently rigid that the geometric angles between members will remain virtually unchanged under normal loading. *Partially-Restrained* (*PR*) *moment connections* (formerly *AISC Type 3*) are not as rigid as FR connections, but nonetheless possess a dependable and predictable moment-resisting capacity that can be used to stabilize a building frame. FR and PR moment connections are also sometimes referred to as "rigid" and "semirigid" connections, respectively. Both connection types can be used to construct moment-resisting building frames. *Simple connections* (formerly *AISC Type 2*), otherwise known as shear connections, are considered to be capable of unrestrained rotation under normal loading conditions and to have negligible moment-resisting capacity. Buildings framed solely with simple connections must depend on diagonal bracing or shear walls for lateral stability.

A series of simple, fully bolted shear (AISC simple) connections are shown as a beginning basis for understanding steel connection details (Figures 11.25–11.27, 11.32, and 11.37). These are interspersed with a corresponding series of welded moment (AISC Fully- and Partially-Restrained) connections (Figures 11.28, 11.29, 11.33, and 11.36). Welding is also widely used for making shear connections, two examples of which are shown in Figures 11.34 and 11.35. A series of column connections is illustrated in Figures 11.38–11.41.

In practice, there are a number of different ways of making any of these connections, using various kinds of connecting elements and different combinations of bolting and welding. The object is to choose a method of stabilization and designs for individual connections that will result in the greatest possible economy of construction for the building as a whole. For standard joint conditions in simple structures, the choice of which connection to use may be left to the fabricator, who has firsthand knowledge of the safest, most erectable methods that will utilize the company's labor and equipment most efficiently. For more complex structures or for unique joining conditions, the structural engineer or architect may dictate a specific connection detail.

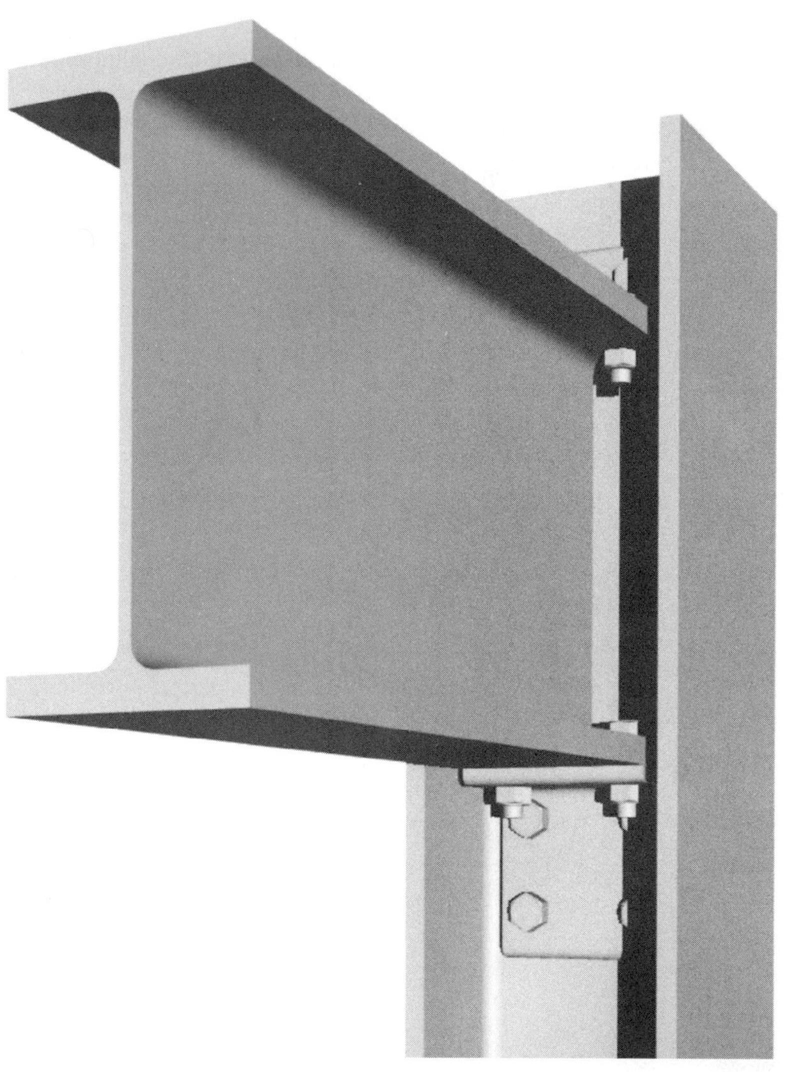

FIGURE 11.32
A *seated* beam-to-column-web *connection*, location *B* on the frame in Figure 11.24. Although the beam flanges are connected to the column by a seat angle below and a stabilizing angle above, this is a shear (AISC simple) connection, not a moment connection, because the two bolts are incapable of developing the full strength of the beam flange. This seated connection is used rather than a framed connection, as illustrated in Figure 11.27, to connect to a column web because there is usually insufficient space between the column flanges to insert a power wrench to tighten all the bolts in a framed connection.

FIGURE 11.33
A welded beam-to-column-web moment(AISC Fully Restrained) connection, used at location *B* on Figure 11.24 when a rigid connection is required. A vertical *shear tab*, welded to the web of the column at its centerline, serves to receive bolts that join the column to the beam web and hold the beam in place during welding. The horizontal stiffener plates that are welded inside the column flanges are thicker than the beam flanges and extend out beyond the column flanges to reduce concentrations of stress at the welds.

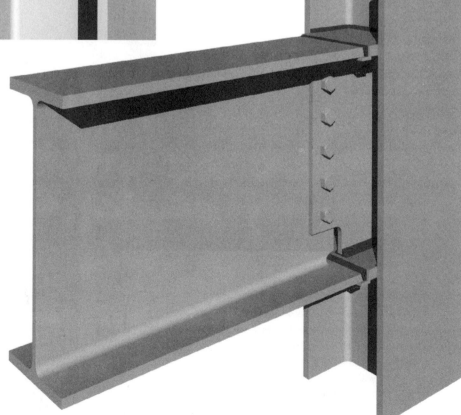

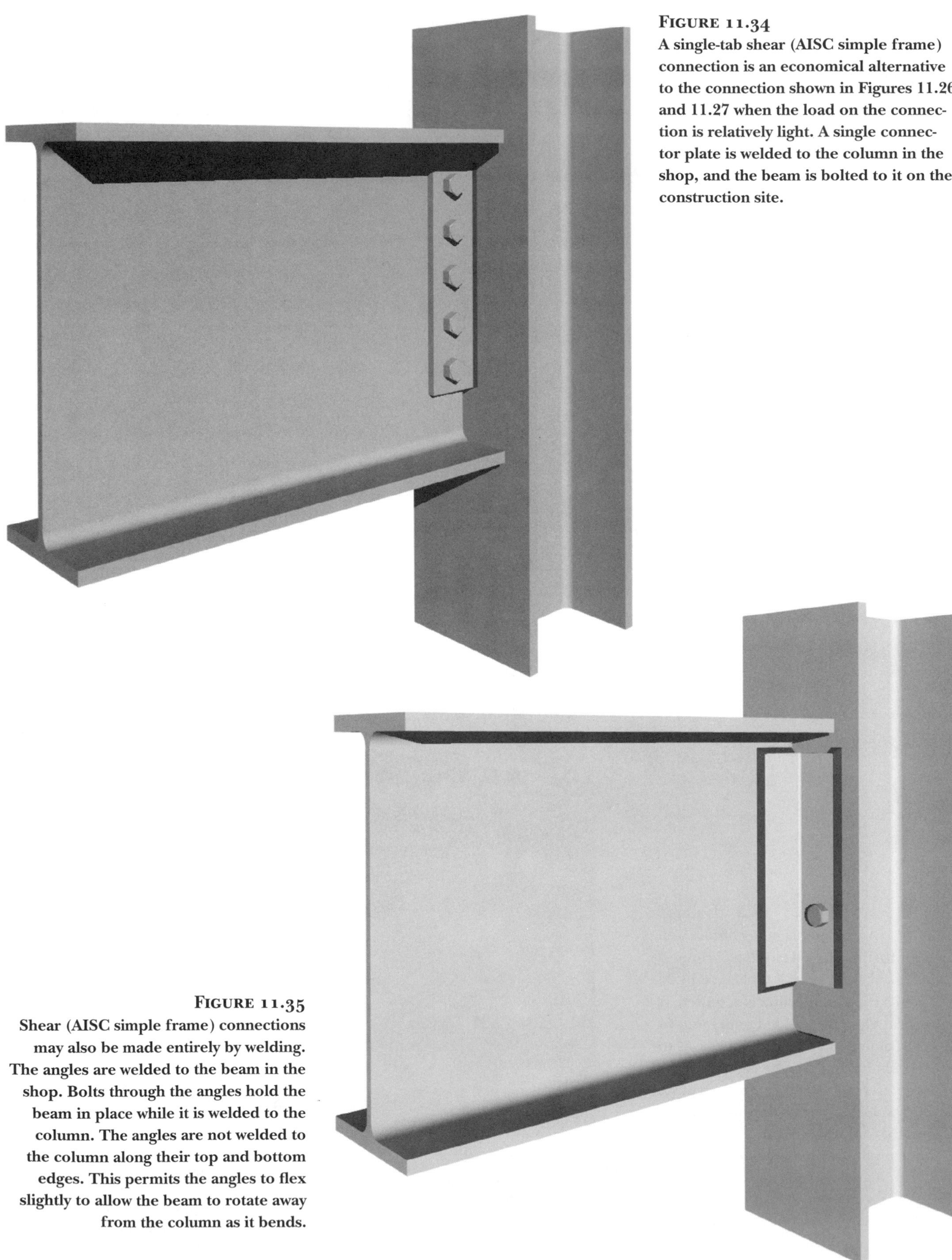

FIGURE 11.34
A single-tab shear (AISC simple frame) connection is an economical alternative to the connection shown in Figures 11.26 and 11.27 when the load on the connection is relatively light. A single connector plate is welded to the column in the shop, and the beam is bolted to it on the construction site.

FIGURE 11.35
Shear (AISC simple frame) connections may also be made entirely by welding. The angles are welded to the beam in the shop. Bolts through the angles hold the beam in place while it is welded to the column. The angles are not welded to the column along their top and bottom edges. This permits the angles to flex slightly to allow the beam to rotate away from the column as it bends.

FIGURE 11.36
A welded/bolted *end plate* beam–column *connection.* As shown, this is a semi-rigid (AISC Partially-Restrained) connection. With more bolts, this can become a rigid, AISC Fully-Restrained connection, and could be used to support a short cantilever beam such as the one at location *C* in Figure 11.24. The plate is welded to the end of the beam in the shop and bolted to the column on the building site.

FIGURE 11.37
A coped beam–girder shear (AISC simple) connection, used at location *D* in Figure 11.24. A girder is a beam that supports other beams. This connection may also be made with single tabs rather than angles if the load is not too great. The top flanges of the beams are cut away (coped) so that the tops of the beams and the girder are all level with one another, ready to receive the floor or roof decking. Bending moments at the ends of a beam are normally so small that the flanges may be coped without compromising the strength of the beam.

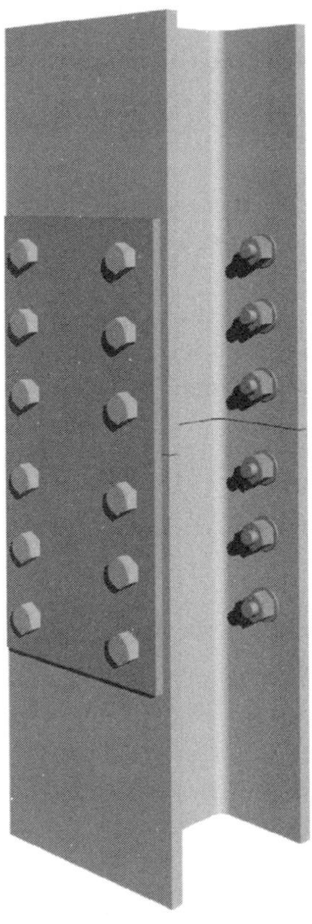

FIGURE 11.38
A bolted column–column connection for columns that are the same size. The plates are bolted to the lower section of the column in the shop and to the upper section on the site. All column connections are made at waist height above the floor, location *E* in Figure 11.24.

FIGURE 11.39
Column sizes diminish as the building rises, requiring frequent use of shim plates at connections to make up for differences in flange thicknesses.

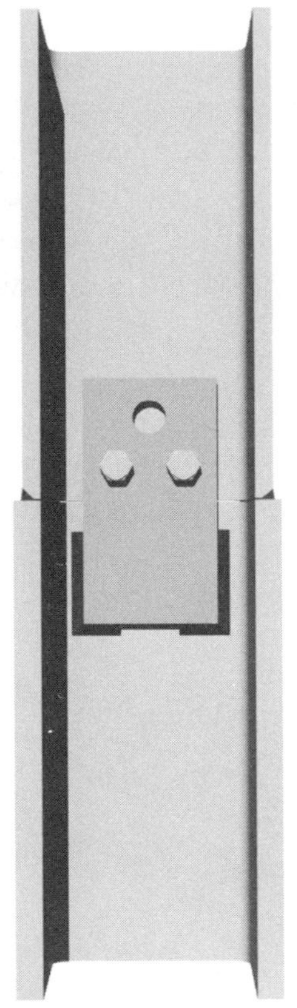

FIGURE 11.40
Column connections may be welded rather than bolted. The connector plate is welded to the lower column section in the fabricator's shop. The hole in the connector plate is used to attach a lifting line during erection. The bolts hold the column sections in alignment, while the flanges are connected in the field with partial-penetration welds in bevel grooves. Partial-penetration welding allows one column to rest on the other prior to welding.

FIGURE 11.41
A welded butt plate connection is used where a column changes from one nominal size of wide flange to another. The thick butt plate, which is welded to the lower column section in the shop, transfers the load from one section of column to the other. The partial-penetration weld at the base of the upper column is made on the site.

The Construction Process

A steel building frame begins as a rough sketch on the drafting board of an architect or engineer. As the building design process progresses, the sketch evolves through many stages of drawings and calculations to become a finished set of structural drawings. These show accurate column locations, the shapes and sizes of all the members of the frame, and all the loads of the members, but they do not give the exact length to which each member must be cut to mate with the members it joins, and they do not give details of the more routine connections of the frame. These are left to be worked out by a subsequent recipient of the drawings, the *fabricator.*

The Fabricator

The fabricator's job is to deliver to the construction site steel components that are ready to be assembled without further processing. This work begins with the preparation in the fabricator's shop of detailed drawings that show exactly how each piece will be made and what its precise dimensions will be. The fabricator designs connections to transmit the loads indicated by the engineer's drawings. Within the limits of accepted engineering practice, the fabricator is free to design the connections to be made as economically as possible, using various combinations of welding and bolting that best suit available equipment and expertise. Drawings are also prepared by the fabricator to show the general contractor exactly where and how to install foundation anchor bolts to connect to the columns of the building and to guide the erector in assembling the steel frame on the building site. When completed, the fabricator's *shop drawings* are submitted to the engineer and the architect for review and approval to be sure that they conform exactly to the intentions of the design team. Meanwhile, the fabricator places an order with a producer of steel for the stock from which the structural steel members will be fabricated. (The major beams, girders, and columns are usually ordered cut to exact length by the mill.) When the approved shop drawings, with corrections and comments, are returned to the fabricator by the design team, revisions are made as necessary, and full-size templates of cardboard or wood are prepared as required to assist the shop workers in laying out the various connections on the actual pieces of steel.

Plates, angles, and tees for connections are brought into the shop and cut to size and shape with gas-fueled cutting torches, power shears, and saws. With the aid of the templates, bolt hole locations are marked. If the plates and angles are not unusually thick, the holes may be made rapidly and economically with a punching machine. In very thick stock, or in pieces that will not fit conveniently into the punching machines, holes are drilled rather than punched.

Pieces of steel stock for the beams, girders, and columns are brought into the fabricator's shop with an overhead traveling crane or conveyor system. Each piece is stenciled or painted with a code that tells which building it is intended for and exactly where it will go in the building. With the aid of the shop drawings, each piece is measured and marked for its exact length and for the locations of all holes, stiffeners, connectors, and other details. Cutting to length, for those members not already cut to length at the mill, is done with a power saw or a flame cutting torch. The ends of column sections that must bear fully on base plates or on one another are squared and made perfectly flat by sawing, milling, or facing. In cases where the columns will be welded to one another, and for beams and girders that are to be welded, the ends of the flanges are beveled as necessary. Beam flanges are *coped* as required. Bolt holes are punched or drilled (Figure 11.43). *Plasma* (high-temperature ionized gas) *cutting* and *laser*

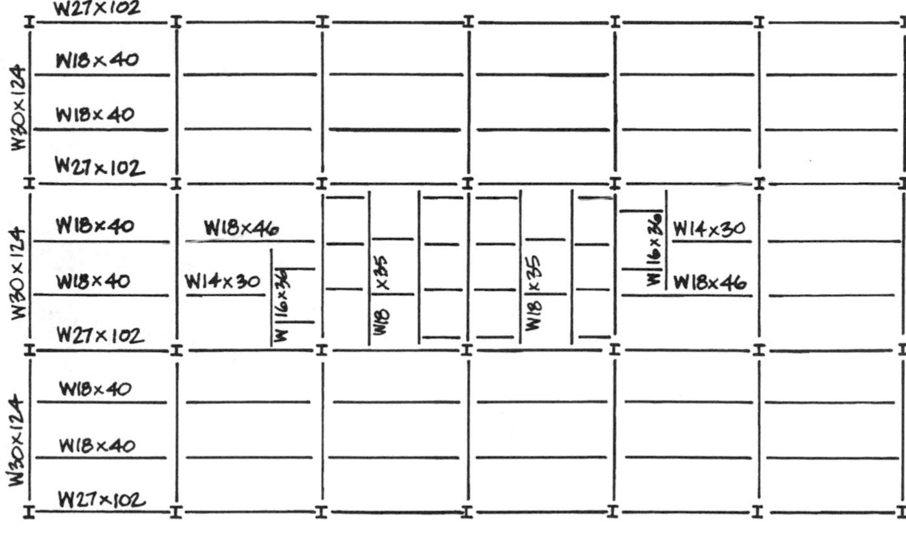

FIGURE 11.42
A typical framing plan for a multistory steel-framed building, showing size designations for beams and girders. Notice how this frame requires beam-to-column-flange connections where the W30 girders meet the columns, beam-to-column-web connections where the W27 beams meet the columns, and coped beam-girder connections where the W18 beams meet the W30 girders. The small squares in the middle of the building are openings for elevators, stairways, and mechanical shafts. An architect's or engineer's framing plan would also give dimensions between centers of columns, and would indicate the magnitudes of the loads that each joint must transfer, to enable the fabricator to design each connection.

cutting are also finding increased use in steel fabrication. Both of these types of cutting devices can be driven by machines that allow the fully automated cutting and shaping of parts from digitally prepared models.

Where called for, beams and girders are *cambered* (curved slightly in an upward direction) so that they will deflect into a straight line under load. Cambering may be accomplished by a hydraulic ram that bends the beam enough to force a permanent deformation. Steel shapes can also be bent to a smooth radius with a large machine that passes the shape through three rollers that flex it sufficiently to impart a permanent curvature (Figure 11.44). An older, much more costly means of cambering involves heating local areas of one flange of the member with a large oxyacetylene torch. As each area is heated to a cherry-red color, the metal softens, expands, and deforms to make a slight bulge in the width and thickness of the flange because the surrounding steel, which is cool, prevents the heated flange from lengthening. As the heated flange cools, the metal contracts, pulling the member into a slight bend at that point. By repeating this process at several points along the beam, a camber of the desired shape and magnitude is produced.

As a last step in fabricating beams, girders, and columns, stiffener plates are arc welded to each piece as required, and connecting plates, angles, and tees are welded or bolted at the appropriate locations (Figure 11.45). As much connecting as possible is done in the shop, where tools are handy and access is easy. This saves time and money during erection, when tools and working conditions are less optimal and total costs per worker-hour are higher.

Plate girders, built-up columns, trusses, and other large components are assembled in the shop in units as large as can practically be transported to the construction site, whether by truck, railway, or barge (Figure 11.46). Intricate assemblies such as large

FIGURE 11.43
Punching bolt holes in a wide-flange beam. *(Courtesy of W. A. Whitney Corp.)*

FIGURE 11.44
Rectangular hollow structural sections for this frame were bent into curves by the fabricator. Wide-flange shapes can also be bent. *(Photo by Eliot Goldstein, The Goldstein Partnership, Architects)*

FIGURE 11.45
Welders attach connector plates to an exceptionally heavy column section in a fabricator's shop. The twin channels bolted to the end of the column will be used to attach a lifting line for erection, after which they will be removed and reused. *(Courtesy of U.S. Steel Corp.)*

FIGURE 11.46
Machine welding plates together to form a box column. The torches to the left preheat the metal to help avoid thermal distortions in the column. Mounds of powdered flux around the electrodes in the center indicate that this is the submerged arc process of welding. The small steel plates tacked onto the corners of the column at the extreme left are runoff bars, which are used to allow the welding machine to go past the end of the column to make a complete weld. These will be cut off as soon as welding is complete. *(Courtesy of U.S. Steel Corp.)*

trusses are usually preassembled in their entirety in the shop, to be sure that they will go together smoothly in the field, then broken down again into transportable components.

As the members are completed, each is straightened, cleaned and prime painted as necessary, and inspected for quality and for conformance to the job specifications and shop drawings. The members are then taken from the shop to the fabricator's yard by crane, conveyor, trolley, or forklift, where they are organized in stacks according to the order in which they will be needed on the building site.

As an alternative to the traditional process in which the fabricator produces the final design for the steel connections and submits these designs for review by the structural engineer, a structural engineer may use three-dimensional modeling software to design the steel connections and supply digital data to the fabricator to drive the fabricator's automated equipment. While this method requires the engineer to assume more responsibility for the final design of the steel connection details, it also can shorten the time required for steel to arrive on site and can improve the coordination between the structural steel and other building systems.

There are 175,000 ironworkers in this country . . . and apart from our silhouettes ant-size atop a new bridge or skyscraper, we are pretty much invisible.

Mike Cherry, *On High Steel: The Education of an Ironworker,* New York, Quadrangle/The New York Times Book Co., 1974, p. xiii

FIGURE 11.47
Three typical column base details. *Upper left*: A small column with a welded baseplate set on a steel leveling plate. *Upper right*: A larger column with a welded baseplate set on leveling nuts. *Below*: A heavy column field welded to a loose baseplate that has been previously leveled and grouted.

The Erector

Where the fabricator's job ends, the *erector's* job begins. Some companies both fabricate and erect, but more often the two operations are done by separate companies. The erector is responsible for assembling into a frame on the building site the steel components furnished by the fabricator. The erector's workers, by tradition, are called *ironworkers.*

Erecting the First Tier

Erection of a multistory steel building frame starts with assembly of the first two-story *tier* of framing. Lifting of the steel components is begun with truck-mounted or crawler-mounted mobile crane. In accordance with the erection drawings prepared by the fabricator, the columns for the first tier, usually furnished in sections two stories high, are picked up from organized piles on the site and lowered carefully over the anchor bolts and onto the foundation, where the ironworkers bolt them down.

Foundation details for steel columns vary (Figure 11.47). Steel *baseplates,* which distribute the concentrated loads of the steel columns across a larger area of the concrete foundation, are shop welded to all but the largest columns. The foundations and anchor bolts were put in place previously by the general contractor, following the plan prepared by the fabricator. The contractor may, if requested, provide thin steel *leveling plates* that are set perfectly level at the proper height on a bed of *grout* atop each concrete foundation. The baseplate of the column rests upon the leveling plate and is held down with the protruding anchor bolts. Alternatively, especially for larger baseplates with four anchor bolts, the leveling plate is omitted. The column is supported at the proper elevation on stacks of steel shims inserted between the baseplate and the foundation, or on leveling nuts placed beneath the baseplate on the anchor bolts. After the first tier of framing is plumbed up as described below, the baseplates are grouted and the anchor bolts tightened. For very large, heavy columns, baseplates are shipped independently of the columns (Figure 11.48). Each is leveled in place with shims, wedges, or shop-attached leveling screws, then grouted prior to column placement.

After the first tier of columns has been erected, the beams and girders for the first two stories are bolted in place (Figures 11.50–11.55). First, a *raising gang,* working with a crane, positions the components and inserts enough bolts to hold them together temporarily. A gang of bolters follows behind, inserting bolts in all the holes and partially tightening them. The two-story tier of framing is then *plumbed up* (straightened and squared) with diagonal cables and turnbuckles while checking the alignment with plumb bobs, transits, or laser levels. When the tier is plumb, connections are tightened, baseplates are grouted if necessary, welds are made, and permanent diagonal braces, if called for, are rigidly attached. Ironworkers scramble back and forth, up and down on the columns and beams, protected from falling by safety harnesses that are connected to steel cable lifelines (Figure 11.50).

At the top of each tier, a temporary working surface of 2- or 3-inch (50- or 75-mm) wood planks or corrugated steel decking may be laid over the steel framing. Similar platforms will be placed every second story as the frame rises, unless safety nets are used instead or the permanent floor decking is installed as erection

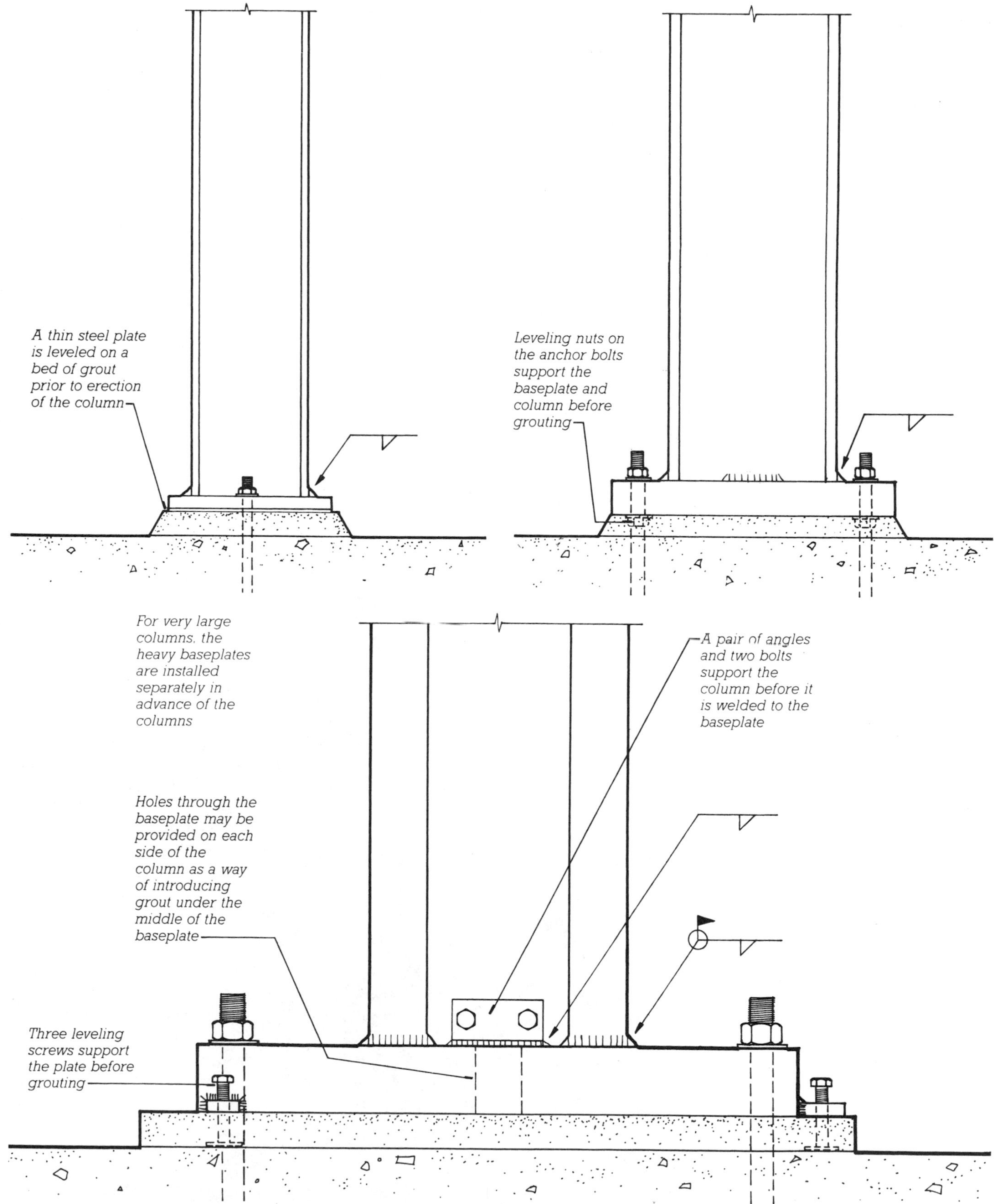
A thin steel plate is leveled on a bed of grout prior to erection of the column
Leveling nuts on the anchor bolts support the baseplate and column before grouting
For very large columns, the heavy baseplates are installed separately in advance of the columns
A pair of angles and two bolts support the column before it is welded to the baseplate
Holes through the baseplate may be provided on each side of the column as a way of introducing grout under the middle of the baseplate
Three leveling screws support the plate before grouting

progresses. The platforms protect workers on lower levels of the building from falling objects. They also furnish a convenient working surface for tools, materials, and derricks. Column splices are made at waist level above this platform, both as a matter of convenience and as a way of avoiding conflict between the column splices and the beam-to-column connections. Columns are generally fabricated in two-story lengths, a transportable size that also corresponds to the two-story spacing of the plank surfaces.

Erecting the Upper Tiers

Erection of the second tier proceeds much like that of the first. Two-story column sections are hoisted into position and connected by splice plates to the first tier of columns. The beams and columns for the two floors are set, the tier is plumbed and tightened up, and another layer of planks, decking, or safety netting is installed.

If the building is not too tall, the mobile crane will do the lifting for the entire building. For a taller building, the mobile crane does the work until it gets to the maximum height to which it can lift a *tower crane* (Figure 11.49). The tower crane builds itself an independent tower as the building rises, either alongside the building or within an elevator shaft or a vertical space temporarily left open in the frame.

FIGURE 11.48
Ironworkers guide the placement of a very heavy column fabricated by welding together two rolled wide-flange sections and two thick steel plates. It will be bolted to its baseplate through the holes in the small plate welded between the flanges on either side. *(Courtesy of Bethlehem Steel Corporation)*

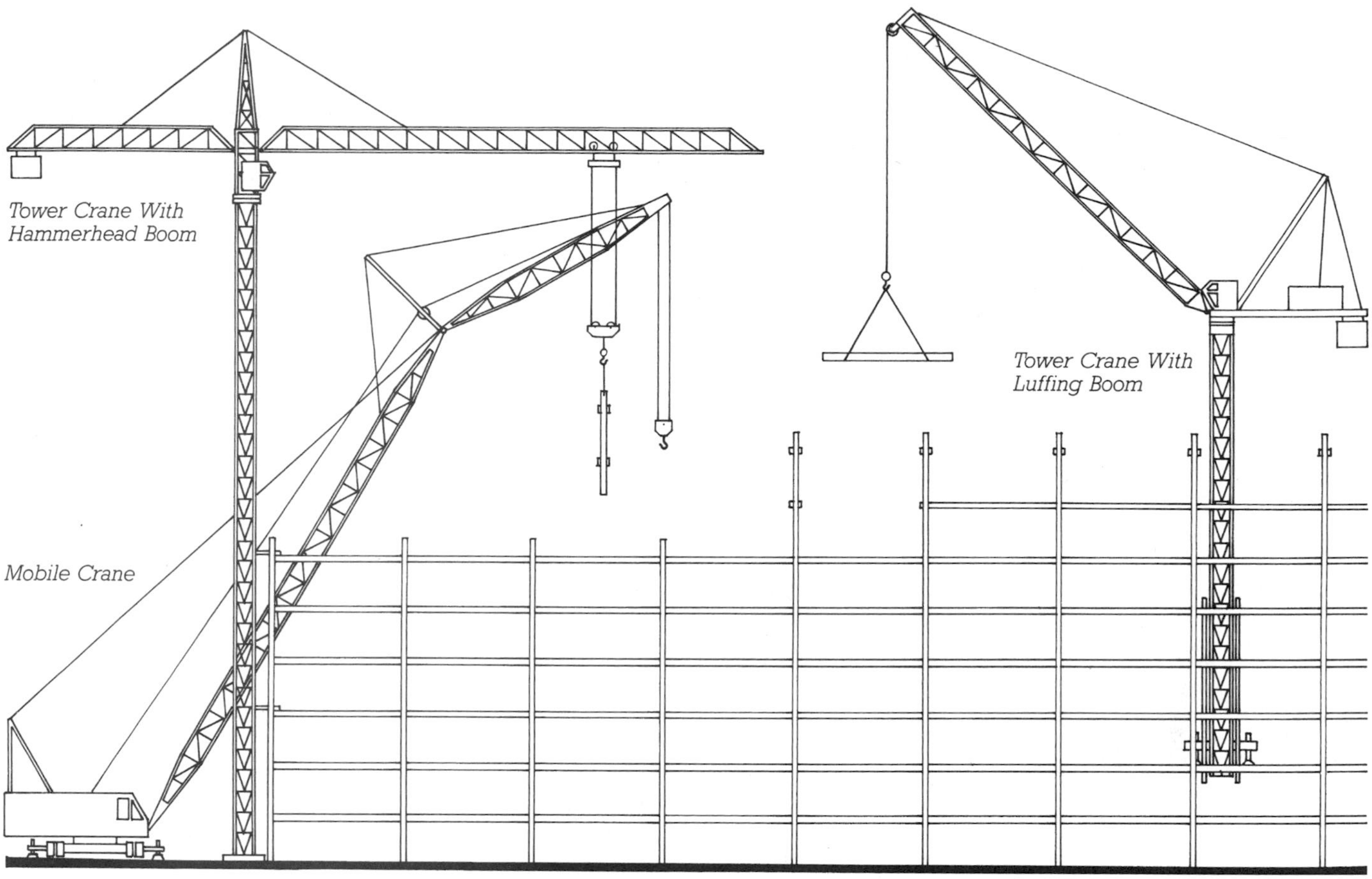

Tower Crane With Hammerhead Boom

Tower Crane With Luffing Boom

Mobile Crane

FIGURE 11.49
Two common types of tower cranes and a mobile crane. The *luffing-boom crane* can be used in congested situations where the movement of the *hammerhead boom crane* would be limited by obstructions. Both can be mounted on either external or internal towers. The internal tower is supported in the frame of the building, whereas the external tower is supported by its own foundation and braced by the building. Tower cranes climb as the building rises by means of self-contained hydraulic jacks.

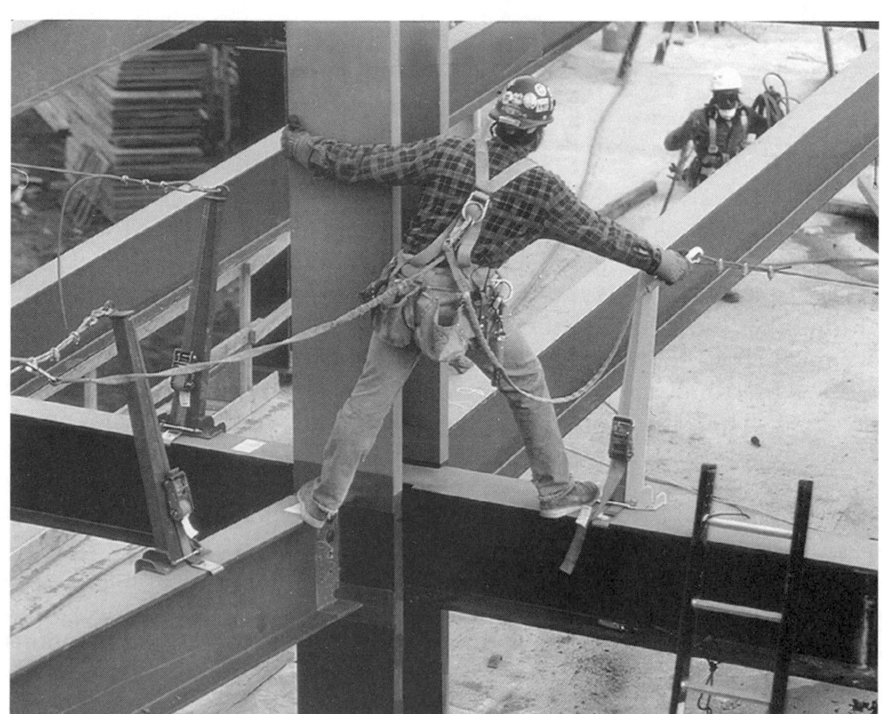

FIGURE 11.50
An ironworker clips his body harness to a safety line as he moves around a column.
(Photo by James Digby. Courtesy of LPR Construction Company)

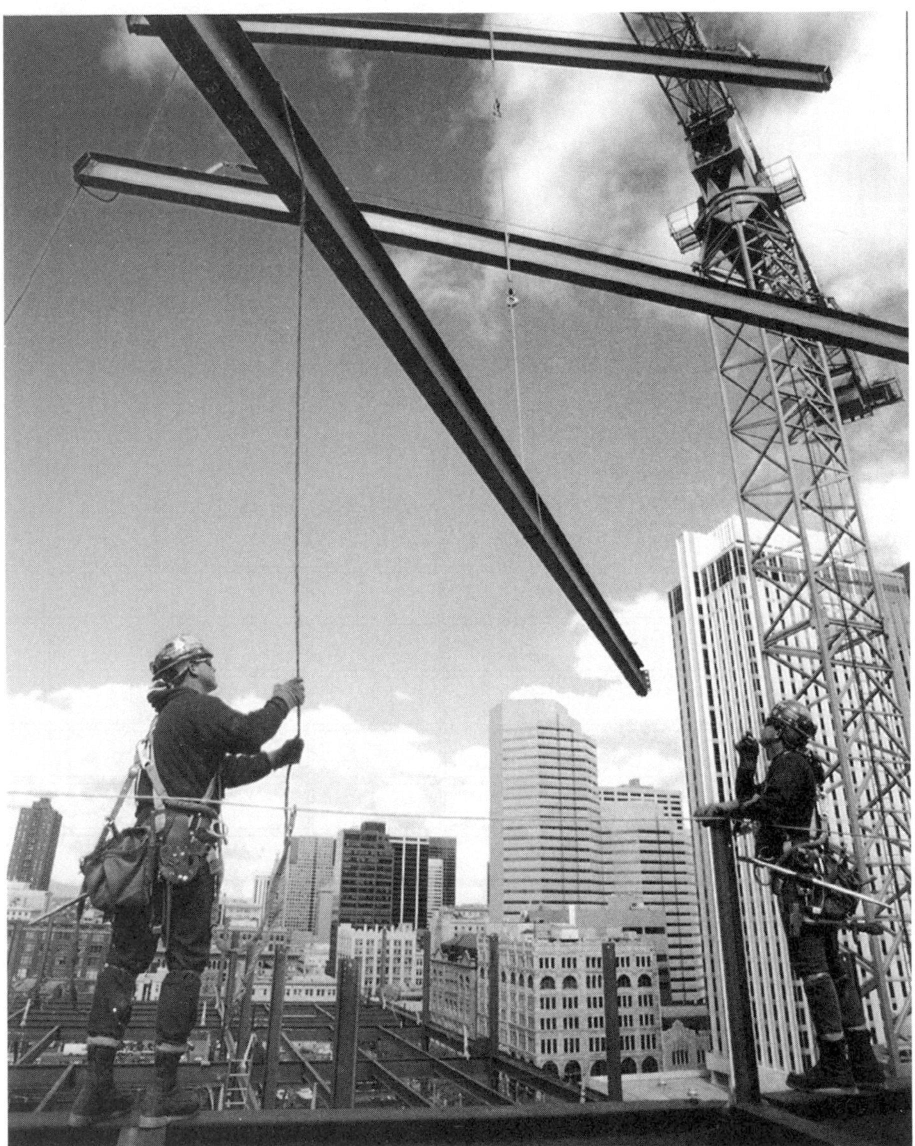

FIGURE 11.51
A tower crane lowers a series of beams to ironworkers. The worker to the left uses a tagline to maneuver the beam into the proper orientation. *(Photo by James Digby. Courtesy of LPR Construction Company)*

As each piece of steel is lowered toward its final position in the frame, it is guided by an ironworker who holds a rope called a *tagline*, the other end of which is attached to the piece. Other ironworkers in the raising gang guide the piece by hand as soon as they can reach it, until its bolt holes align with those in the mating pieces (Figure 11.53). Sometimes crowbars or hammers must be used to pry, wedge, or drive components until they fit properly, and bolt holes may, on occasion, have to be reamed larger to admit bolts through slightly misaligned pieces. When an approximate alignment has been achieved, tapered steel *drift pins* from the ironworker's tool belt are shoved into enough bolt holes to hold the pieces together until a few bolts can be inserted. The bolters follow behind the raising gang, filling the remaining holes with bolts from leather carrying baskets and tightening them first with hand wrenches and then with impact wrenches. Field-welded connections are initially held in alignment with bolts, then welded when the frame is plumb.

The last beam is placed at the top of the building with a degree of ceremony appropriate to the magnitude of the building. At the very least, a small evergreen tree, a national flag, or both are attached to the beam before it is lifted (Figure 11.56). For major buildings, assorted dignitaries are likely to be invited to a building-site *topping-out* party that includes music and refreshments. After the party, work goes on as usual, for although the frame is complete, the building is not. Roofing, cladding, and finishing operations will continue for many months.

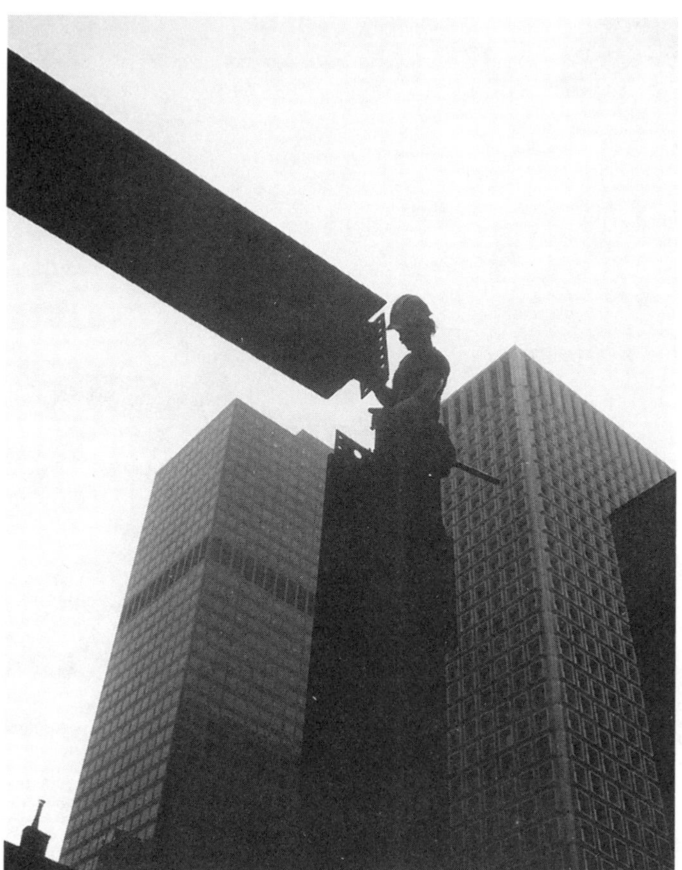

FIGURE 11.52
Connecting a beam to a column. *(Courtesy of Bethlehem Steel Corporation)*

FIGURE 11.53
Ironworkers attach a girder to a box column. Each worker carries two combination wrench–drift pin tools in a holster on his belt and inserts the tapered drift pins into each connection to hold it until a few bolts can be added. Bundles of corrugated steel decking are ready to be opened and distributed over the beams to make a floor deck. *(Courtesy of Bethlehem Steel Corporation)*

FIGURE 11.54
Bolting joist girders to a column.
(Courtesy Vulcraft Division of Nucor)

FIGURE 11.55
Welding open-web steel joists to a wide-flange beam. *(Courtesy Vulcraft Division of Nucor)*

FIGURE 11.56
Topping out: The last beam in a steel frame is special. *(Courtesy of U.S. Steel Corp.)*

FIGURE 11.57
A 10-story steel frame nears completion. The lower floors have already been decked with corrugated steel decking. *(Courtesy Vulcraft Division of Nucor)*

If nobody plumbed-up, all the tall buildings in our cities would lean crazily into each other, their elevators would scrape and bang against the shaft walls, and the glaziers would have to redress all the windowglass into parallelograms. . . . What leaned an inch west on the thirty-second floor is sucked back east on the thirty-fourth, and a column that refused to quit leaning south on forty-six can generally be brought over on forty-eight, and by the time the whole job is up the top is directly over the bottom.

Mike Cherry, *On High Steel: The Education of an Ironworker,* New York, Quadrangle/The New York Times Book Co., 1974, pp. 110–111.

Floor and Roof Decking

If plank decks are used during erection of the frame, they must be replaced with permanent floor and roof decks of incombustible materials. In early steel frame buildings, shallow arches of brick or tile were often built between the beams, tied with steel tension rods, and filled over with concrete to produce level surfaces (Figure 11.58). These were heavier than the metal deck systems commonly used today and required larger framing members to carry their weight. They were also much more labor intensive.

Metal Decking

Metal decking at its simplest is a thin sheet of steel that has been corrugated to increase its stiffness. The spanning capability of the deck is determined mainly by the thickness of the sheet from which it is made and the depth and spacing of the corrugations. It also depends on whether the decking sheets are single or cellular. Single corrugated sheets are commonly used for *roof decking,* where concentrated loads are not expected to be great and deflection criteria are not as stringent as in floors. They are also used as permanent formwork for concrete floor decks, with a reinforced concrete slab supported by the steel decking until the slab can support itself and its live loads (Figures 11.59–11.61). *Cellular decking* is manufactured by welding together two sheets, one corrugated and one flat. It can be made sufficiently stiff to support normal floor loads without structural assistance from the concrete topping that is poured over it. Cellular decking can offer the important side benefit of providing spaces for electrical and communications wiring, as illustrated in Chapter 24.

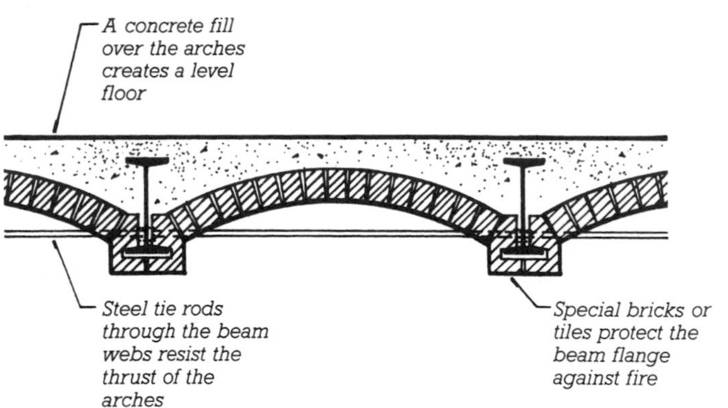
A concrete fill over the arches creates a level floor
Steel tie rods through the beam webs resist the thrust of the arches
Special bricks or tiles protect the beam flange against fire

FIGURE 11.58
Tile or brick arch flooring is found in many older steel frame buildings.

FIGURE 11.59
Workers install corrugated steel form decking over a floor structure of open-web steel joists supported by joist girders. *(Photo by Balthazar Korab. Courtesy Vulcraft Division of Nucor)*

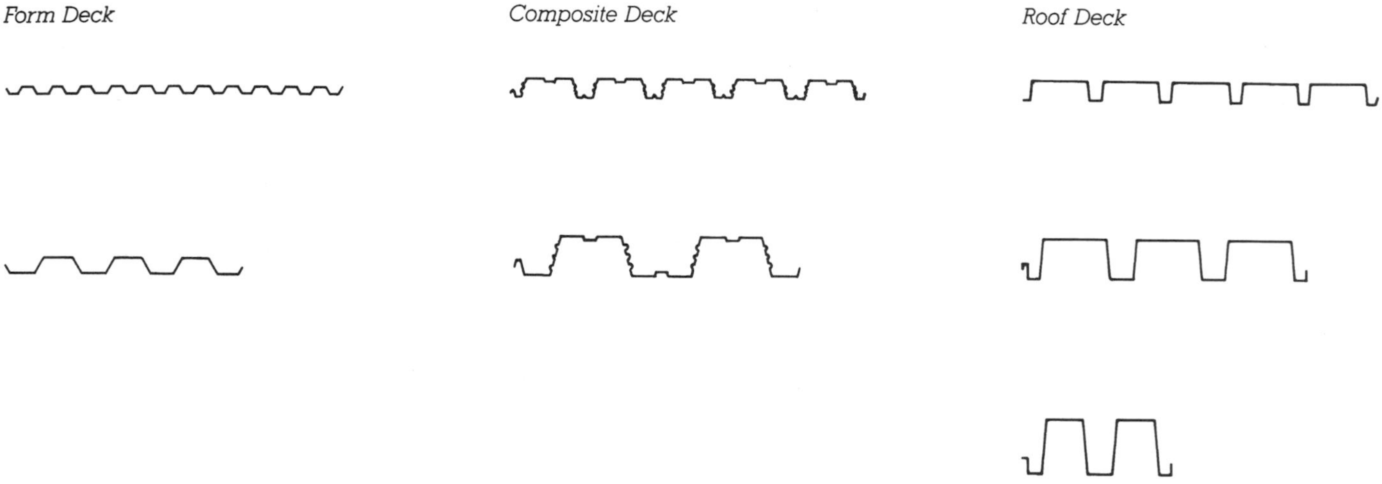

Form Deck

Composite Deck

Roof Deck

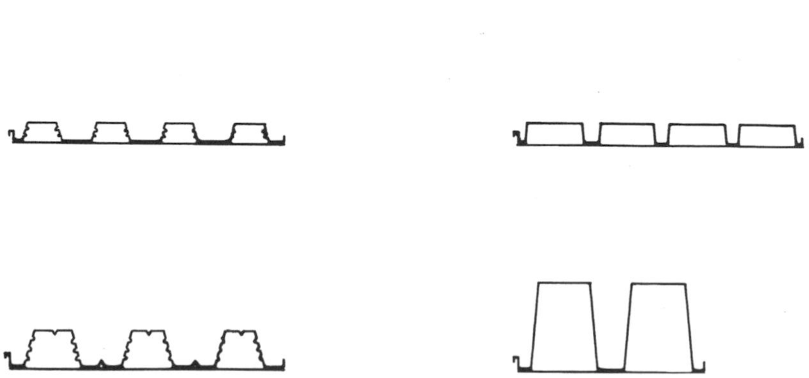

FIGURE 11.60
Typical profiles of corrugated steel decking. Standard depths of form decking range from ½ to 2½ inches (13–64 mm). Composite decking depths lie between 1½ and 3 inches (38 and 76 mm). Roof decking is available in depths of 1½ to 7 inches (38–178 mm). The bottom two examples of composite and roof decking are cellular.

Metal decking is usually *puddle welded* to the joists, beams, and girders by melting through the decking to the supporting members below with a welding electrode. Self-drilling, self-tapping screws or powder-driven pins may also be used for decking attachment. If the deck is required to act as a diaphragm, the longitudinal edges of the decking panels must be connected to one another at frequent intervals with screws or welds.

Composite Construction

Composite metal decking (Figures 11.60–11.62) is designed to work together with the concrete floor topping to make a stiff, lightweight, economical deck. The metal decking serves as

FIGURE 11.61
Samples of corrugated steel decking. The second sample from the bottom achieves composite action by keying of the concrete topping to the deformations in the decking. The bottom sample has a closed end, which is used at the perimeter of the building to prevent the concrete topping from escaping during pouring. *(Courtesy of Wheeling Corrugating Company Division, Wheeling-Pittsburgh Steel Corp.)*

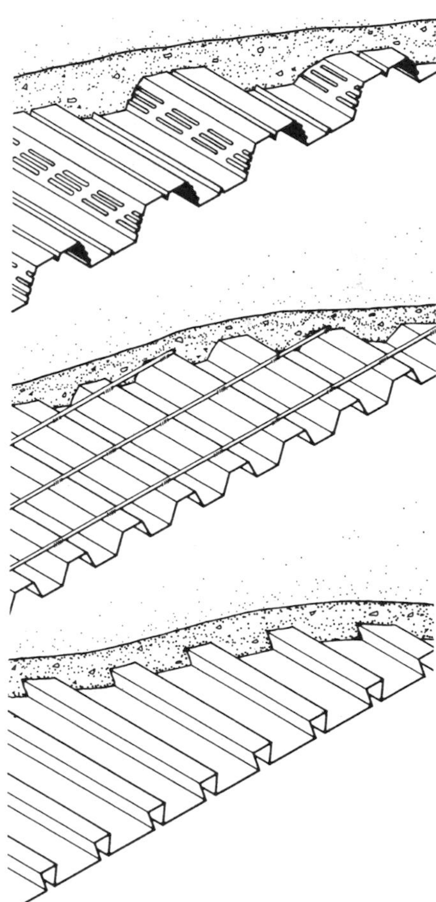

FIGURE 11.62
Composite decking acts as steel reinforcing for the concrete topping installed over it. The top example bonds to the concrete with deformed ribs and the middle example with welded steel rods. The bottom type makes an attractive ceiling texture if left exposed and furnishes dovetail channels for the insertion of special fastening devices to hang ductwork, piping, conduits, and machinery from the ceiling.

tensile reinforcing for the concrete, to which it bonds by means of special rib patterns in the sheet metal or by small steel rods or wire fabric welded to the tops of the corrugations.

Composite construction is often carried a step beyond the decking to include the beams of the floor. Before the concrete is poured over the metal deck, *shear studs* are welded every few inches to the top of each beam, using a special electric welding gun (Figures 11.63 and 11.64). It would be more economical to attach the shear studs in the shop rather than in the field, but the danger that ironworkers would trip on the studs during the erection process delays their installation until the steel decking is in place. The purpose of the studs is to create a strong shear connection between the concrete slab and the steel beam. A strip of the slab can then be assumed to act together with the top flange of the steel shape to resist compressive forces. The result of the composite action of the two materials is a steel member whose loadbearing capacity has been greatly enhanced at relatively low cost by taking advantage of the unused strength of the concrete topping that must be present in the construction anyway. The payoff is a stiffer, lighter, less expensive frame.

Concrete Decks

Concrete floor and roof slabs are often used in steel building frames instead of metal decking and concrete fill. Concrete may be poured in place over removable plywood forms, or it may be erected in the form of precast concrete planks lifted into place much like the metallic elements of the building (Figure 11.65). Precast concrete decks are relatively light in weight and are quick to erect, even under weather conditions that would preclude the pouring of concrete, but they usually require the addition of a thin, poured-in-place concrete topping to produce a smooth floor.

Roof Decking

For roofs of low steel-framed buildings, many different types of decks are available. Corrugated metal may be used, with or without a concrete fill; many types of rigid insulation

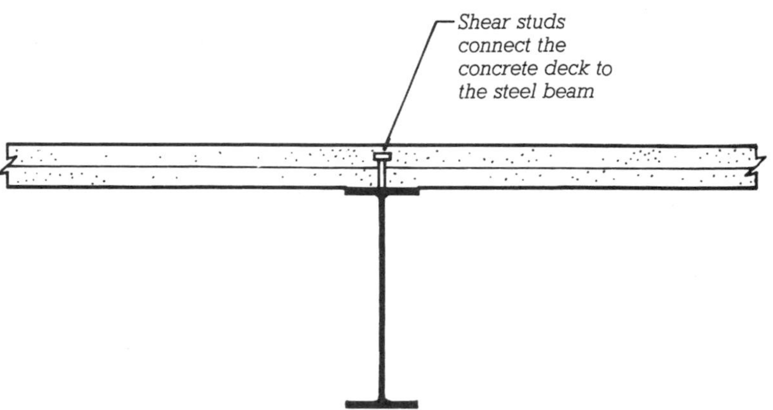
Shear studs connect the concrete deck to the steel beam

Figure 11.63
Composite beam construction.

Figure 11.64
Pouring a concrete fill on a steel roof deck, using a concrete pump to deliver the concrete from the street below to the point of the pour. Shear studs are plainly visible over the lines of the beams below. The welded wire reinforcing strengthens the concrete against cracking. *(Courtesy of Schwing America, Inc.)*

FIGURE 11.65
A tower crane installs precast concrete hollow-core planks for floor decks in an apartment building. Precast concrete is also used for exterior cladding of the building. The steel framing is a design known as the "staggered truss" system, in which story-height steel trusses at alternate levels of the building support the floors. The trusses are later enclosed with interior partitions. An advantage of the staggered truss system is that the floor structure is very thin, allowing overall floor-to-floor heights as small as $8^{2}/_{3}$ feet (2.6 m). *(Courtesy of Blakeslee Prestress, Inc.)*

boards are capable of spanning the corrugations to make a flat surface for the roof membrane. Some corrugated decks are finished with a weather-resistant coating that allows them to serve as the water-resistive surface of the roof. Many different kinds of insulating deck boards are produced of wood fibers or glass fibers bonded with portland cement, gypsum, or organic binders. Many of these insulating boards are designed as permanent formwork for reinforced, poured slabs of gypsum or lightweight concrete. In this application, they are usually supported on steel *subpurlins* (Figures 11.66 and 11.67). Heavy timber decking, or even wood joists and plywood sheathing, are also used over steel framing in situations where building codes permit combustible materials.

Corrugated steel sheets are also often used for siding of industrial buildings, where they are supported on *girts*, which are horizontal Z-shapes or channels that span between the outside columns of the building (Figure 11.80).

Architectural Structural Steel

Where structural steel members will remain exposed in the finished building and a high standard of appearance quality is desired, steel may be specified as *architecturally exposed structural steel (AESS)*. AESS specifications may include special requirements for dressing and finishing of welds, closer tolerances in connections between members, removal of marks made on the steel during fabrication, application of high-quality finishes, and other considerations.

FIGURE 11.66
Welding truss-tee subpurlins to open-web steel joists for a roof deck. *(Courtesy of Keystone Steel & Wire Co.)*

FIGURE 11.67
Installing insulating formboard over truss-tee subpurlins. A light wire reinforcing mesh and poured gypsum fill will be installed over the formboards. The trussed top edges of the subpurlins become embedded in the gypsum slab to form a composite deck. *(Courtesy of Keystone Steel & Wire Co.)*

Fireproofing of Steel Framing

Building fires are not hot enough to melt steel, but are often able to weaken it sufficiently to cause structural failure (Figures 11.68 and 11.69). For this reason, building codes generally limit the use of exposed steel framing to buildings of one to five stories, where escape in case of fire is rapid. For taller buildings, it is necessary to protect the steel frame from heat long enough for the building to be fully evacuated and the fire extinguished or allowed to burn out on its own.

Fireproofing ("fire protection" might be a more accurate term) of steel framing was originally done by encasing steel beams and columns in brick masonry or poured concrete (Figures 11.70 and 11.71). These heavy encasements were effective, absorbing heat into their great mass and dissipating some of it through dehydration of the mortar and concrete, but their weight added considerably to the load that the steel frame had to bear. This added, in turn, to the weight and cost of the frame. The search for lighter-weight fireproofing led first to thin enclosures of metal lath and plaster around the steel members (Figures 11.70–11.72). These derive their effectiveness from the large amounts of heat needed to dehydrate the water of crystallization from the gypsum plaster. Plasters based on lightweight aggregates such as vermiculite instead of sand have come into use to further reduce the weight and increase the thermal insulating properties of the plaster.

FIGURE 11.68
An exposed steel structure following a prolonged fire in the highly combustible contents of a warehouse. *(Courtesy of National Fire Protection Association, Quincy, Massachusetts)*

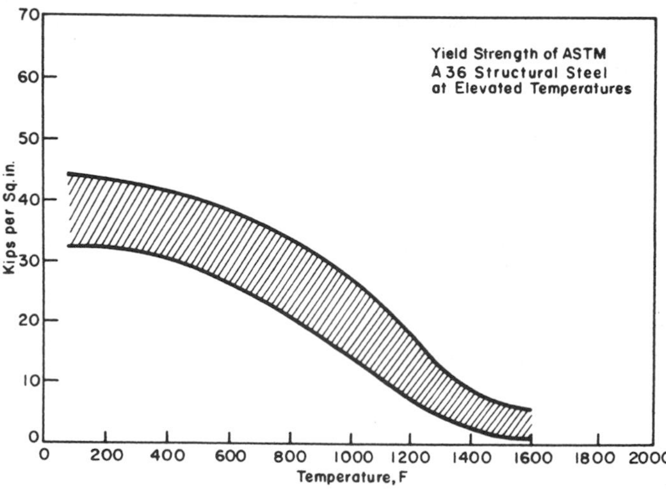

FIGURE 11.69
The relationship between temperature and strength in structural steel. *(Courtesy of American Iron and Steel Institute)*

Today's designers can also choose from a group of fireproofing techniques that are lighter still. Plaster fireproofing has largely been replaced by beam and column enclosures made of boards or slabs of gypsum or other fire-resistive materials (Figures 11.70–11.75). These are fastened mechanically around the steel shapes, and in the case of gypsum board fireproofing, they can also serve as the finished surface on the interior of the building.

Where the fireproofing material need not serve as a finished

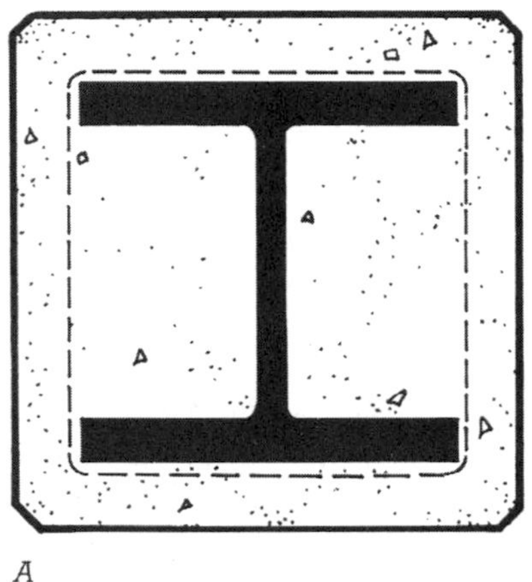

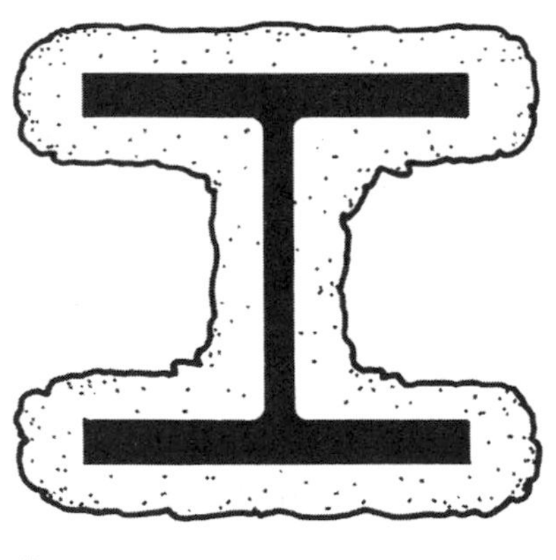

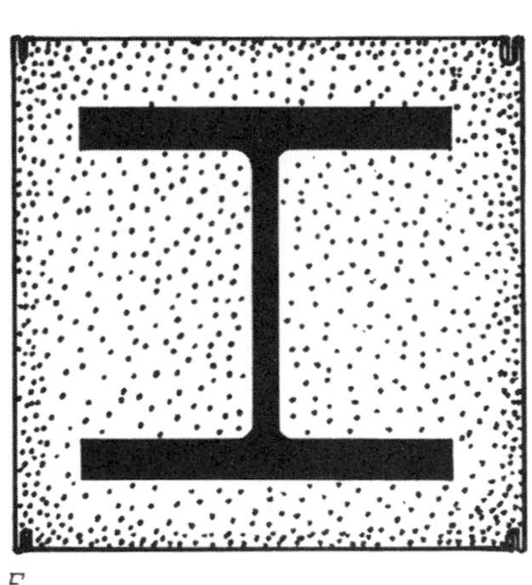

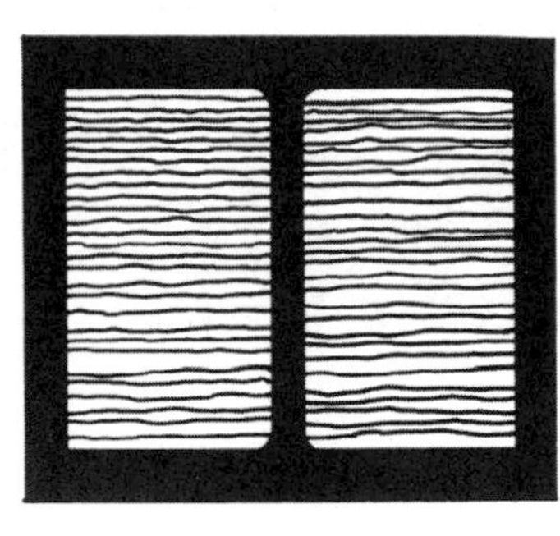

FIGURE 11.70
Some methods for fireproofing steel columns. (*a*) Encasement in reinforced concrete. (*b*) Enclosure in metal lath and plaster. (*c*) Enclosure in multiple layers of gypsum board. (*d*) Spray-on fireproofing. (*e*) Loose insulating fill inside a sheet metal enclosure. (*f*) Water-filled box column made of a wide-flange shape with added steel plates.

surface, *spray-applied fire-resistive materials (SFRM)*, commonly referred to as "spray-applied fireproofing," have become the most prevalent type. These generally consist of either a fiber and a binder or a cementitious mixture, and are sprayed over the steel to the required thickness (Figure 11.76). These products are available in weights of about 12 to 40 pounds per cubic foot (190–640 kg/m^3). The lighter materials are fragile and must be covered with finish materials. The denser materials are generally more durable. All spray-applied materials act primarily by insulating the steel from high temperatures for long periods of time. They are usually the least expensive form of steel fireproofing. Spray-applied fireproofing is most commonly applied in the

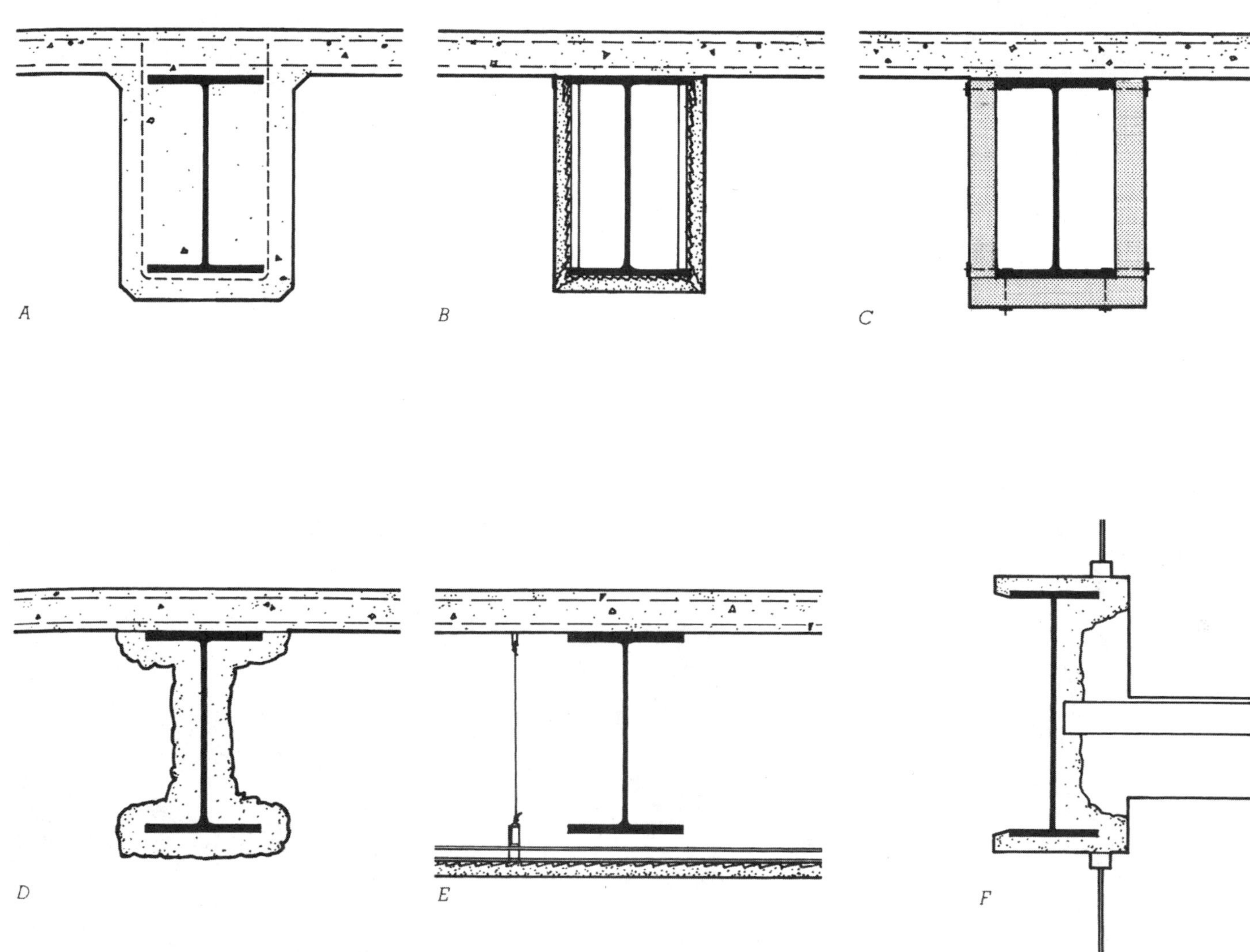

FIGURE 11.71
Some methods for fireproofing steel beams and girders. (*a*) Encasement in reinforced concrete. (*b*) Enclosure in metal lath and plaster. (*c*) Rigid slab fireproofing. (*d*) Spray-on fireproofing. (*e*) Suspended plaster ceiling. (*f*) Flame-shielded exterior spandrel girder with spray-on fireproofing inside.

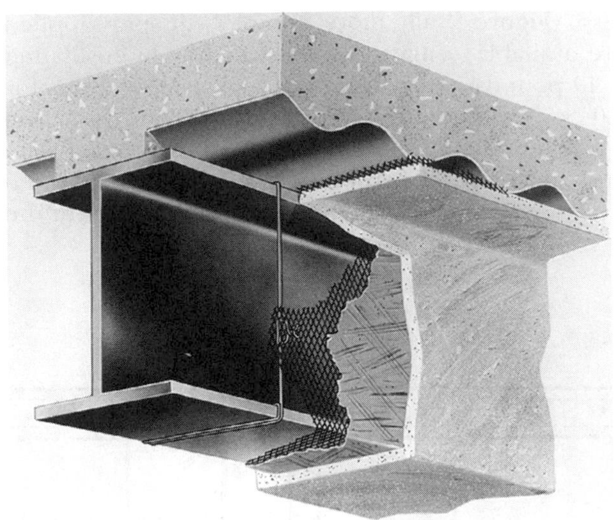

FIGURE 11.72
Lath-and-plaster fireproofing around a steel beam.
(Courtesy of United States Gypsum Company)

FIGURE 11.73
Gypsum board fireproofing around a steel column. The gypsum board layers are screwed to the four cold-formed steel C-channels at the corners of the column, and finished with steel corner bead and drywall compound on the corners. *(Courtesy of United States Gypsum Company)*

FIGURE 11.74
Attaching slab fireproofing made of mineral fiber to a steel column, using welded attachments. *(Courtesy of United States Gypsum Company)*

FIGURE 11.75
Slab fireproofing on a steel beam.
(Courtesy of United States Gypsum Company)

field after the steel has been erected and the connections between members are completed. It can also be applied in the fabrication shop, where controlled environmental conditions and easier access to the steel members can result in faster and more consistent-quality application.

When terrorists flew airliners into Manhattan's World Trade Center towers in 2001, the fireproofing on steel structural components of the buildings was dislodged by the impact of the airplanes. This left these structures vulnerable to the heat of the fires that followed the crashes and is believed to be a primary cause of the eventual collapse of both buildings. In response to this failure, the International Building Code now requires higher bond strengths for spray-applied fireproofing used in buildings with occupied floors greater than 75 feet (23 m) above the level of firefighter access.

The newest generation of fireproofing techniques for steel offers new possibilities to the designer. *Intumescent mastics* and *intumescent paints* are thin coatings that allow steel structural elements to remain exposed to view in situations of low to moderate fire risk. They expand when exposed to fire to form a thick, stable char that insulates the steel from the heat of the fire for varying lengths of time, depending on the thickness of the coating. Most intumescent coatings are available in an assortment of colors. They can also serve as a base coat under ordinary paints if another color is desired.

A rather specialized technique, applicable only to steel box or tube columns exposed on the exterior of buildings, is to fill the columns with water and antifreeze (Figure 11.70*f*). Heat applied to a region of a column by a fire is dissipated throughout the column by convection in the liquid filling.

Mathematical and computer-based techniques have been developed for calculating temperatures that will be reached by steel members in various situations during a fire. These allow the designer to experiment with a variety of ways of protecting the members, including metal flame shields that allow the component to be left exposed on the exterior of a building (Figure 11.71*f*).

FIGURE 11.76
Applying spray-on fireproofing to a steel beam, using a gauge to measure the depth. *(Courtesy of W. R. Grace & Co.)*

Longer Spans in Steel

Standard wide-flange beams are suitable for the range of structural spans normally encountered in offices, schools, hospitals, apartment buildings hotels, retail stores, warehouses, and other buildings in which columns may be brought to earth at intervals without obstructing the activities that take place within. For many other types of buildings—athletic buildings, certain types of industrial buildings, aircraft hangars, auditoriums, theaters, religious buildings, transportation terminals—longer spans are required than can be accomplished with wide-flange beams. A rich assortment of longer-span structural devices is available in steel for these uses.

Improved Beams

One general class of longer-span devices might be called "improved beams." The *castellated beam* (Figures 11.77 and 11.78) is produced by flame cutting the web of a wide-flange section along a zigzag path, then reassembling the beam by welding its two halves point to point, thus increasing its depth without increasing its weight. This greatly augments the spanning potential of the beam, provided that the superimposed loads are not exceptionally heavy.

For long-span beams tailored to any loading condition, *plate girders* are custom designed and fabricated. Steel plates and angles are assembled by bolting or welding in such a way as to put the steel exactly where it is needed: The flanges are often made thicker in the middle of the span where bending forces are higher, more web stiffeners are provided near the ends where web stresses are high, and areas around the supports are specially reinforced. Almost any depth can be manufactured as needed, and very long spans are possible, even under heavy loads (Figure 11.79). These members are often tapered, having greater depth where the bending moment is largest.

Rigid steel frames are efficiently produced by welding together steel wide-flange sections or plate girders. They may be set up in a row to roof a rectangular space (Figure 11.80) or arrayed around a vertical axis to cover a circular area. Their structural action lies midway between that of a rectilinear frame and that of an arch. Like an arch, they may sometimes require steel tie rods at the base to resist lateral thrust, in which case these rods are usually concealed within the floor slab.

Castellated beams, plate girders, and rigid steel frames share the characteristic that because they are long, slender elements, they frequently must be braced laterally by purlins, girts, decking, or diagonal bracing to prevent them from buckling.

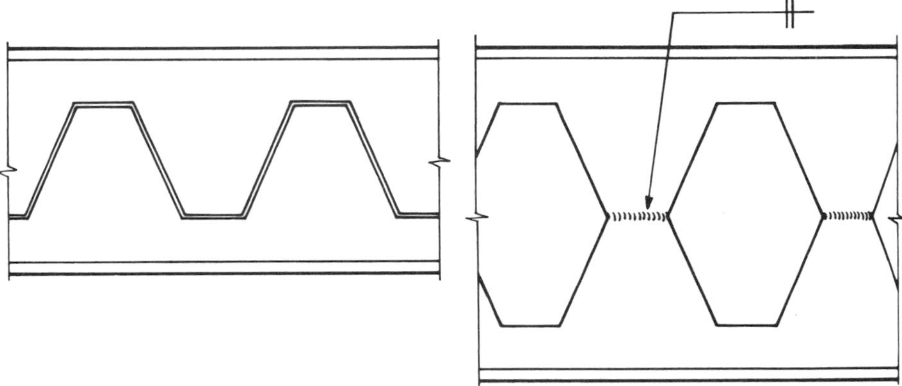

Figure 11.77
Manufacture of a castellated beam.

Figure 11.78
Castellated beams and girders frame into a wide-flange column.
(Courtesy of Castelite Steel Products, Midlothian, Texas)

FIGURE 11.79
Erecting a welded steel plate girder. Notice how the girder is custom made with cutouts for the passage of pipes and ductwork. The section being erected is 115 feet (35 m) long, 13 feet (4 m) deep, and weighs 192,000 pounds (87,000 kg). *(Courtesy of Bethlehem Steel Corporation)*

FIGURE 11.80
The steel rigid frames of this industrial building carry steel purlins that will support the roof deck and girts to support the wall cladding. The depth of each frame varies with the magnitude of the bending forces and is greatest at the eave connections, where these forces are at a maximum. *(Courtesy of Metal Building Manufacturers Association)*

Trusses

Steel trusses (Figures 11.81–11.84) are triangulated arrangements of steel members that are generally deeper and lighter than improved beams and can span correspondingly longer distances. They can be designed to carry light or heavy loads. Earlier in this chapter, one class of steel trusses, open-web joists and joist girders, was presented. These are standardized light members for light loadings that are capable of fairly long spans, and they are usually less expensive than custom-made trusses. Custom-made roof trusses for light loadings are most often made up of steel tee or paired-angle top and bottom *chords* with paired-angle internal members. The angles within each pair are spaced just far enough

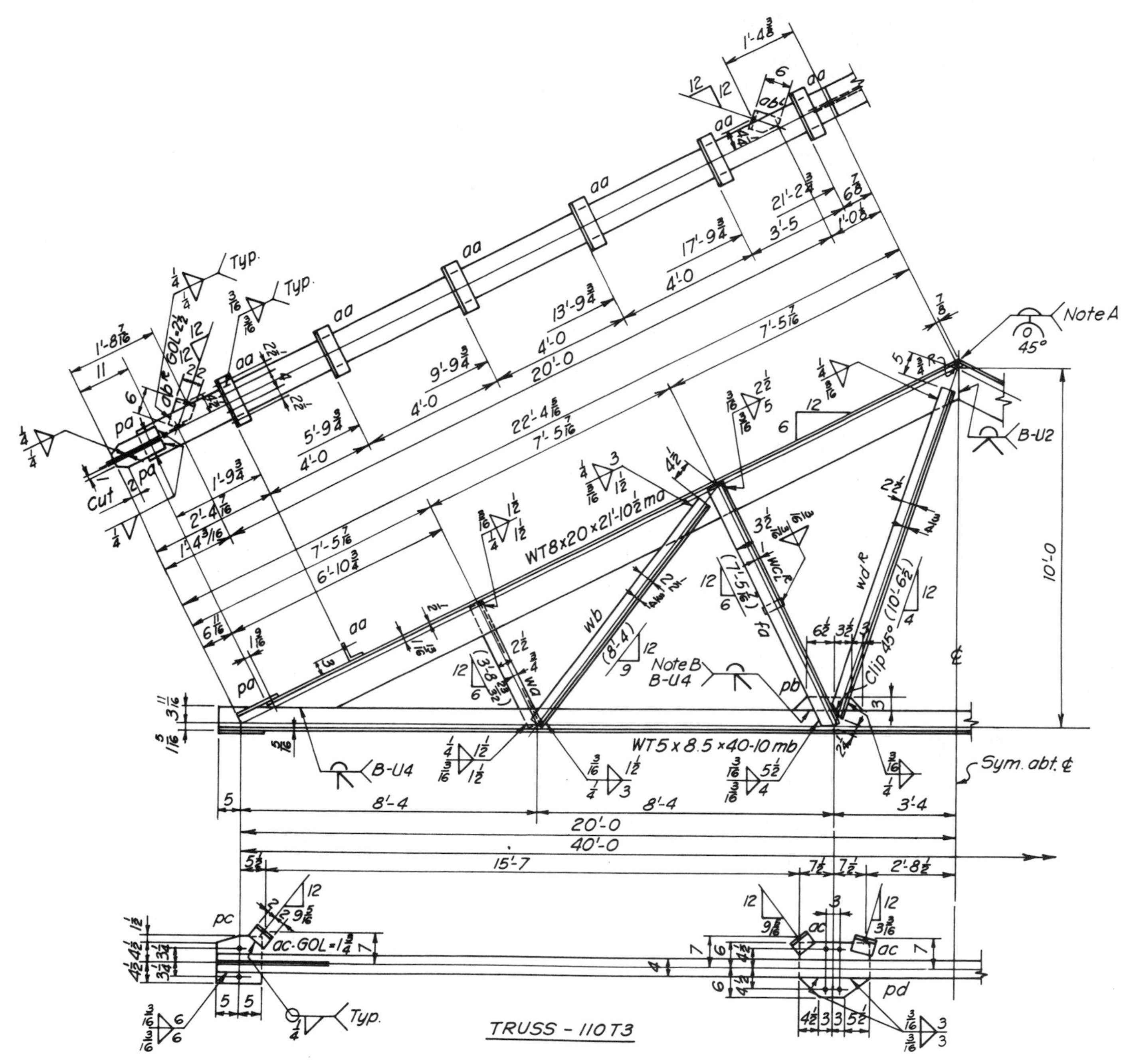

FIGURE 11.81
A fabricator's shop drawing of a welded steel roof truss made of tees and paired-angle diagonals. *(From* Detailing for Steel Construction, *Chicago, AISC, 1983. By permission of American Institute of Steel Construction)*

apart to leave space between for the steel gusset plate connectors that join them to the other members of the truss. They may be either welded or bolted to the gusset plates. Trusses for heavier loadings, such as the transfer trusses that are used in some building frames to transmit column loads from floors above across a wide meeting room or lobby in a building, can be made of wide-flange or tubular shapes.

A steel *space truss* (more popularly called a *space frame*) is a truss made three dimensional (Figures 11.85 and 11.86). It carries its load by bending along both of its axes, much like a two-way concrete slab (Chapter 13). It must be supported by columns that are spaced more or less equally in both directions.

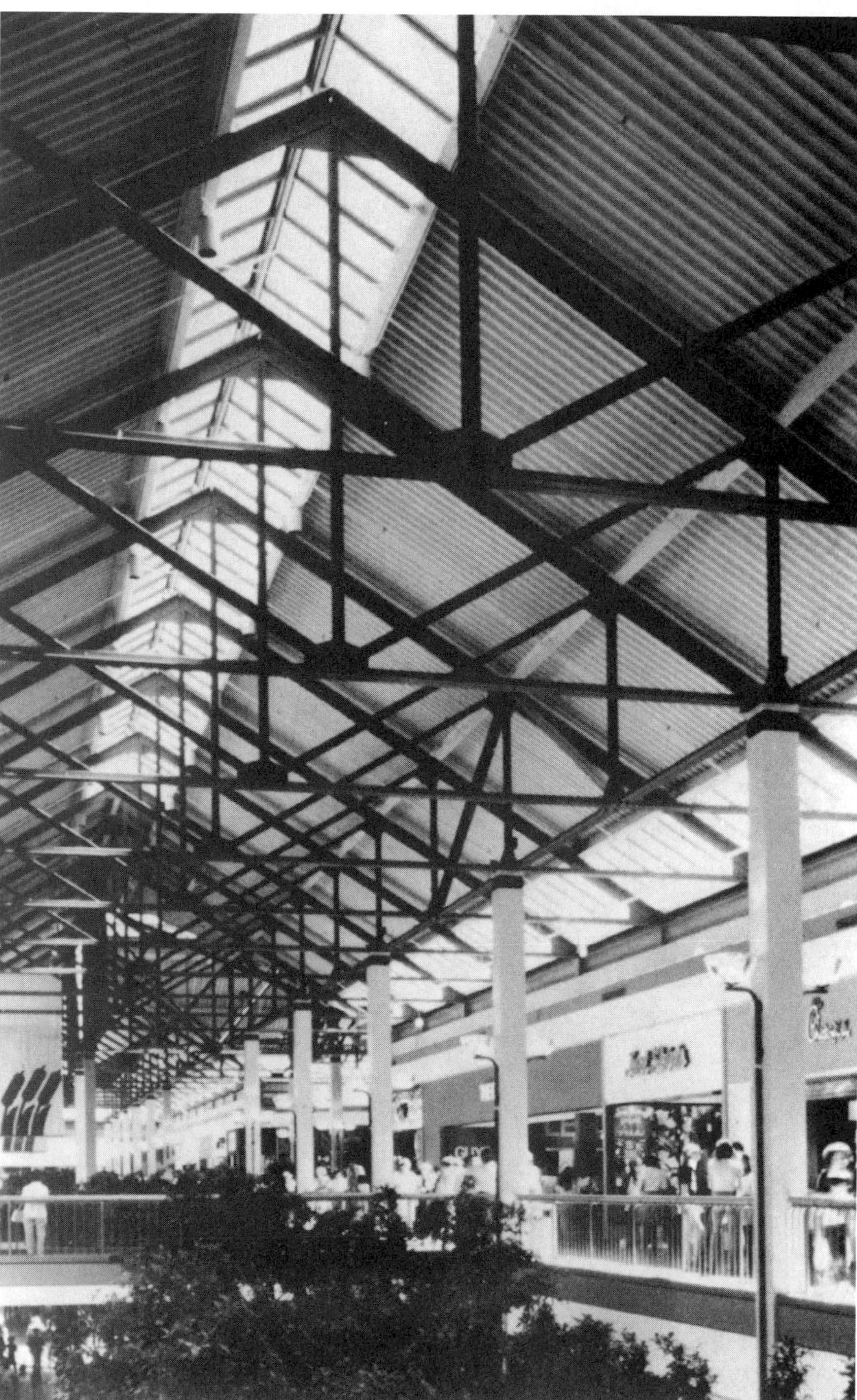

FIGURE 11.82
Bolted steel roof trusses over a shopping mall support steel purlins that carry the corrugated steel roof deck. *(Permission of American Institute of Steel Construction)*

FIGURE 11.83
Ironworkers seat the end of a heavy roof truss made of wide-flange sections. *(Permission of American Institute of Steel Construction)*

FIGURE 11.84
Tubular steel trusses support the roof of a convention center. *(Permission of American Institute of Steel Construction)*

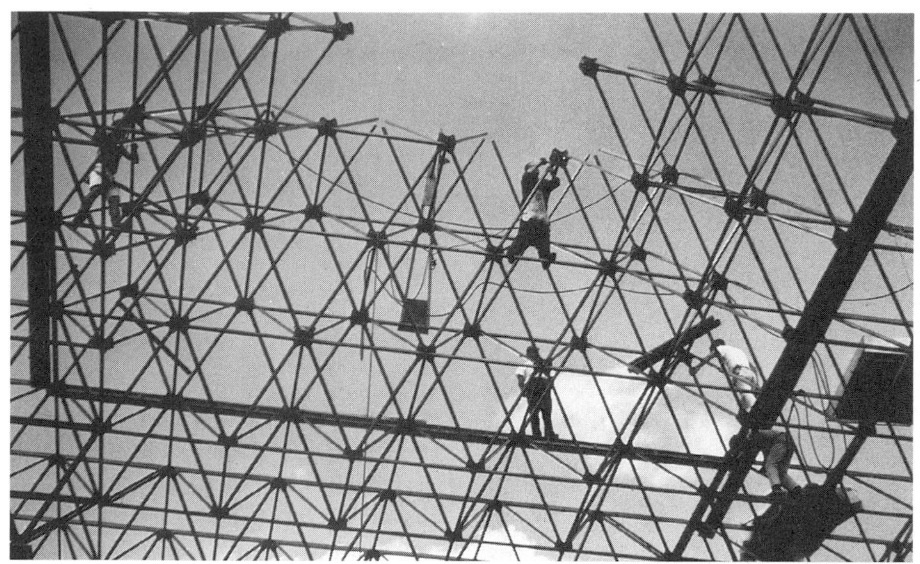

FIGURE 11.85
Assembling a space truss. *(Courtesy of Unistrut Space-Frame Systems, GTE Products Corp.)*

Arches

Steel *arches*, produced by bending standard wide-flange shapes or by joining plates and angles, can be made into cylindrical roof vaults or circular domes of considerable span (Figure 11.87). For greater spans still, the arches may be built of steel trusswork. Lateral thrusts are produced at the base of an arch and must be resisted by the foundations or by a tie rod. At the time of this writing, the longest single-span roof structure in the world is reported to be a retractable football stadium roof, currently under construction in Arlington, Texas, supported on a pair of steel box trusses, spanning 1225 ft (373 m). The trusses are each 35 feet (11 m) deep and weigh 3255 tons (2955 metric tons).

FIGURE 11.86
A space truss carries the roof of a ferry terminal. *(Architects: Braccia/DeBrer/Heglund. Structural engineer: Kaiser Engineers. Photo by Barbeau Engh. Permission of American Institute of Steel Construction)*

FIGURE 11.87
Erecting the steel dome at Disney World.
(© Walt Disney Productions. Permission of American Institute of Steel Construction)

FIGURES 11.88 AND 11.89
The Olympic Stadium roof in Munich, Germany, is made of steel cables and transparent acrylic plastic panels. For scale, notice the worker seen through the roof at the upper left of Figure 11.89.
(Architects: Frei Otto, Ewald Bubner, and Benisch and Partner. Courtesy of Institute for Lightweight Structures, Stuttgart)

(continued)

Tensile Structures

High-tensile-strength wires of cold-drawn steel, made into cables, are the material for a fascinating variety of tentlike roofs that can span very long distances (Figures 11.88 and 11.89). With *anticlastic* (saddle-shaped) *curvature, cable stays,* or other means of restraining the cable net, hanging roofs are fully rigid against wind uplift and flutter. For smaller spans, fabrics can do most of the work, supported by steel cables along the edges and at points of maximum stress, as presented in the accompanying sidebar.

FIGURES 11.88 AND 11.89
(continued)

The spider web is a good inspiration for steel construction.

Frank Lloyd Wright, "In the Cause of Architecture: The Logic of the Plan," ***Architectural Record*** **(January 1928).**

Fabric Structures

Fabric structures are not new: People have constructed tents since the earliest days of human civilization. During the last several decades, however, new, durable fabrics and computerized methods for finding form and forces have helped to create a new construction type: a permanent, rigid, stable fabric structure that will last for 20 years or more.

Types of Fabric Structures

Fabric structures are either tensile or pneumatic (Figure A). A *tensile fabric structure* is a membrane supported by masts or other rigid structural elements such as frames or arches. The membrane usually consists of a woven textile fabric and is generally reinforced with steel cables along the main lines of stress. The fabric and cables transmit external loads to the rigid supports and ground anchors by means of tensile forces.

Pneumatic structures depend on air pressure for their stability and their capacity to carry snow and wind loads. The most common type of pneumatic structure is the *air-supported structure*, in which an airtight fabric, usually reinforced with steel cables, is held up by pressurizing the air in the inhabited space below it. The fabric and cables in an air-supported structure are stressed in tension.

Fabrics for Permanent Structures

Nearly all fabric structures are made of woven cloth that has been coated with a synthetic material. The cloth provides structural strength to resist the tensile forces in the structure, and the coating makes the fabric airtight and water resistant. The most widely used fabric is polyester cloth that has been laminated or coated with polyvinyl chloride (vinyl, PVC). Two other frequently used, longer-lasting fabrics are based on glass fiber cloth: One is coated with polytetrafluorethylene (PTFE, the most common brand of which is known as Teflon); the other is coated with silicone. The polyester/PVC fabric is the most economical of the three, but it does not meet U.S. building code requirements for a noncombustible material and is used predominantly for smaller structures. The glass/PTFE and glass/silicone fabrics are classified as noncombustible. Though it is more expensive, the glass/PTFE fabric remains clean longer than the other two. All three fabrics are highly resistant to such forces of deterioration as ultraviolet light, oxidation, and fungi.

Fabrics may be white, colored, or imprinted with patterns or graphics. A fabric may be totally opaque, obstructing all passage of light, or it may be translucent, allowing a controlled percentage of light to pass through.

Though a single layer of fabric has little resistance to the flow of heat, a properly designed fabric structure can achieve substantial energy savings over conventional enclosures through selective use of translucency and reflectivity. Translucency can be used to provide natural illumination, gather solar heat in the winter, and cool the space at night in the summer. A highly reflective fabric can reduce solar heat gain and conserve artificial illumination. With the addition of a second layer, a fabric liner that is suspended about a foot (300 mm) below the structural fabric, the thermal resistance of the structure can be improved. An acoustic inner liner can help to control internal sound reflection, which is especially important in air-supported structures, which tend to focus sound. Tensile structures, because of their anticlastic curvature, tend to disperse sound rather than focus it.

Figure A
Simple tensile and pneumatic (air-supported) fabric structures. *(Sketch by Edward Allen)*

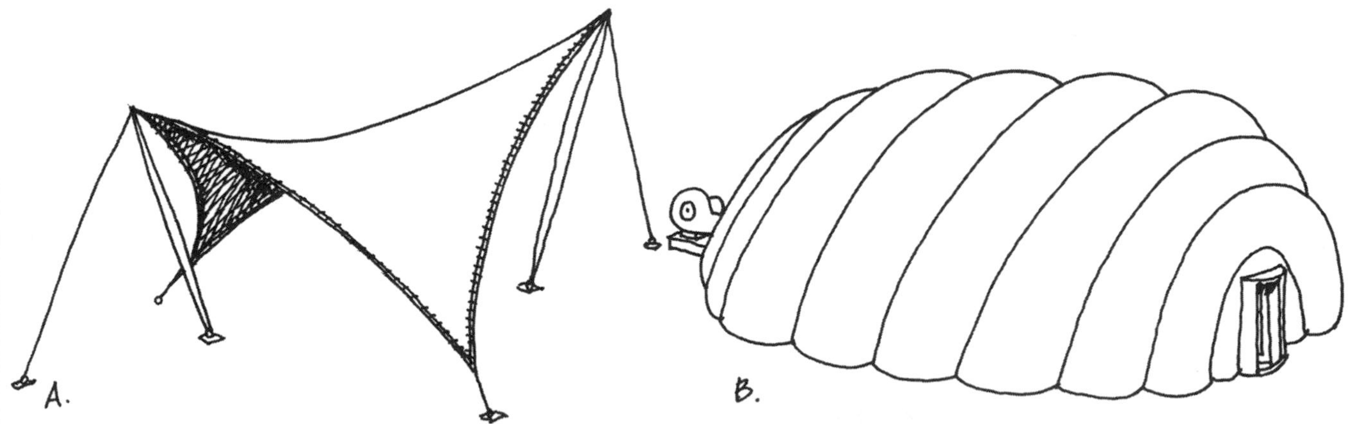

FIGURE B
The world's largest roof structure covers the Haj Terminal in Jeddah, Saudi Arabia, a colossal airport facility that is used to facilitate the travel of vast numbers of Muslim faithful during a short period of annual pilgrimage. The roof is made up of radial shapes. *(Architects: Skidmore, Owings & Merrill. Roof designer and structural engineer: Geiger Berger Associates. Photographs for Figures B–G are furnished by the courtesy of the photographer, Horst Berger.)*

FIGURE C
The fabric of the Haj Terminal is PTFE-coated glass fiber cloth.

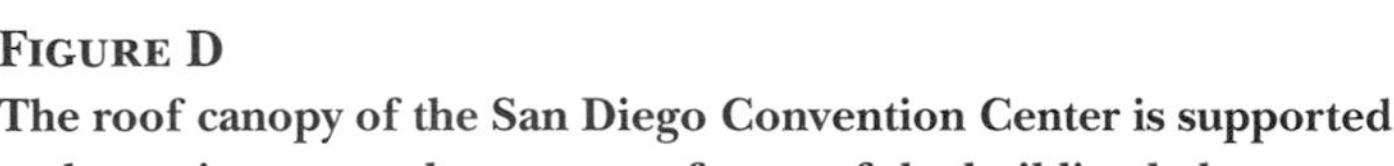

FIGURE D
The roof canopy of the San Diego Convention Center is supported at the perimeter on the concrete frame of the building below.

FIGURE E
The San Diego Convention Center's roof is raised at the middle on rigid steel struts that rest on cables supported by the concrete frame. *(Architects: Arthur Erickson Associates. Roof designer and structural engineer: Horst Berger Partners)*

Fabric Structures *(continued)*

Figure F
The form of the Denver International Airport roof combines saddle and radial shapes to echo the forms of the surrounding mountains. *(Architects: C. W. Fentriss, J. H. Bradburn & Associates. Roof designer and structural engineer: Severud Associates, Horst Berger, Principal Consultant)*

Figure G
Like the petals of a giant flower, tensile structures arranged in a huge circle shade the grandstand of King Fahd Stadium in Riyadh, Saudi Arabia. The masts are 190 feet (58 m) high. *(Architects: Ian Fraser, John Roberts & Partners. Roof designer and structural engineer: Geiger Berger Associates)*

Tensile Structures

Tensile structures are stabilized by anticlastic curvature and *prestress.* Anticlastic curvature means that the fabric is curved simultaneously in two opposite directions. Two basic geometries may be used to create anticlastic curvature: One is the saddle shape (Figures D, E, and G), the other the radial tent (Figures B and C). It is from combinations and variations of these geometries, as in Figure F, that all tensile structures are shaped.

Prestress is the introduction of permanent tension into the fabric in two opposing directions. Without anticlastic curvature and prestress, the fabric would flutter in the wind and destroy itself within a short time. The amount of curvature and the amount of prestressing force must both be sufficient to maintain stability under expected wind and snow conditions. If the curvature is too flat or if the prestressing tension is too low, excessive deflection or flutter will occur.

The design of a tensile structure usually begins by experimenting with simple physical models. These often are made of pantyhose material or stretch fabric, either of which is easily stretched and manipulated. After a general shape has been established with the model, a computer is used to find the exact equilibrium shape, determine the stresses in the fabric and supporting members under wind and snow loadings, and generate cutting patterns for the fabric. The design process is referred to as *form finding,* because a tensile structure cannot be made to take any arbitrary shape. Just as a hanging chain will always take a form that places its links in equilibrium with one another, a tensile structure must take a form that maintains proportionate amounts of tension in all parts of the fabric under all expected loading conditions. The designer's task is to find such a form.

A good design for a tensile structure employs short *masts* to minimize buckling problems. The fabric generally cannot come to a peak at the mast, but must terminate in a cable ring that is attached to the mast in order to avoid high tensile stresses in the fabric. The perimeter edges of the fabric usually terminate along curving steel cables. To make a stable structure, these cables must have adequate curvature and must be anchored to foundations that offer firm resistance to uplift forces. The fabric may be attached to the cables by sleeves sewn into the edges of the fabric or clamps that grip the fabric and pull it toward the cable.

Air-Supported Structures

Air-supported structures are pressurized by the fans that are used to heat, cool, and ventilate the building. The required air pressures are so low that they are scarcely discernible by people entering or leaving the building, but they are high enough (5–10 pounds per square foot, or 0.25–0.50 kPa) to prevent ordinary swinging doors from opening. For this reason, revolving doors, whose operation is unaffected by internal pressure and that maintain a continual seal against loss of air, are usually used for access.

The fabric of an air-supported structure is prestressed by its internal air pressure to prevent flutter. For low-profile roof shapes, a cable net is employed to resist the high forces that result from the flat curvature. The fabric spans between the cables. The fabric and cables pull up on the foundations with a total force that is equal to the internal air pressure multiplied by the area of ground covered by the roof. The supporting elements and foundations must be designed to resist this force.

Wind causes suction forces to occur on many areas of an air-supported structure, which results in additional tension in the fabric and cables. The downward forces from wind or snow load on an air-supported structure must be resisted directly by the internal air pressure pushing outward against them. In geographic areas where snow loads are larger than acceptable internal pressures, snow must be removed from the roof. Failure to do so has led to unplanned deflations of several air-supported roofs.

In theory, air-supported structures are not limited in span. In practice, flutter and perimeter uplift forces restrict their span to a few hundred meters, but this is sufficient to house entire football stadiums. For safety, the outer edges of most air-supported roofs terminate at a level that is well above the floor level within. Thus, if the roof deflates because of fan failure, inadequate snow removal, or air leakage, the roof fabric will hang in suspension at a height well above the floor of the building (Figure H).

For further information, see Horst Berger, *Light Structures; Structures of Light,* Basel, Birkhäuser Verlag, 1996.

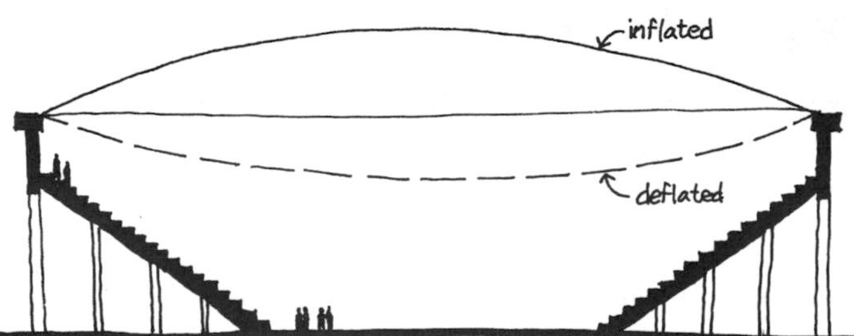

FIGURE H
Most air-supported structures are designed so that if air pressure fails, the membrane will hang at a safe level above the heads of the occupants. *(Sketch by Edward Allen)*

Composite Columns

Columns that combine the strength of structural steel shapes and sitecast concrete have been used in buildings for many years. One type of composite column surrounds a steel wide-flange column with sitecast reinforced concrete. Another type consists of a steel pipe that is filled with concrete. In a third type, a wide-flange column is inserted within the pipe before the concrete is added to create a higher loadbearing capacity.

Several recent high-rise buildings use very large steel pipe columns filled with very-high-strength concrete to carry a major portion of both vertical and lateral loads. These columns enable reductions of as much as 50 percent in the overall quantity of steel required for the building (Figure 11.90). In one such building, a 720-foot (200-m) office tower, four 10-foot-diameter (3-m) pipe columns filled with 19,000-psi (131-MPa) concrete carry 40 percent of the gravity loads and a large proportion of the wind loads. There is no reinforcing or other steel inside the pipes except at certain connections that carry very heavy loads. The potential advantages of composite columns in tall buildings include reduced steel usage, greater rigidity of the building against wind forces, and simplified beam-column connections.

Industrialized Systems in Steel

Steel adapts well to industrialized systems of construction. The two most successful and most economical prefabrication systems in the United States are probably the manufactured home (often referred to as a "mobile home") and the package industrial building. The manufactured home, built largely of wood, is made possible by a rigid undercarriage (chassis) welded together from rolled steel shapes. The package building is most commonly based on a structure of welded steel rigid frames supporting an enclosure of corrugated metal sheets. The manufactured home is

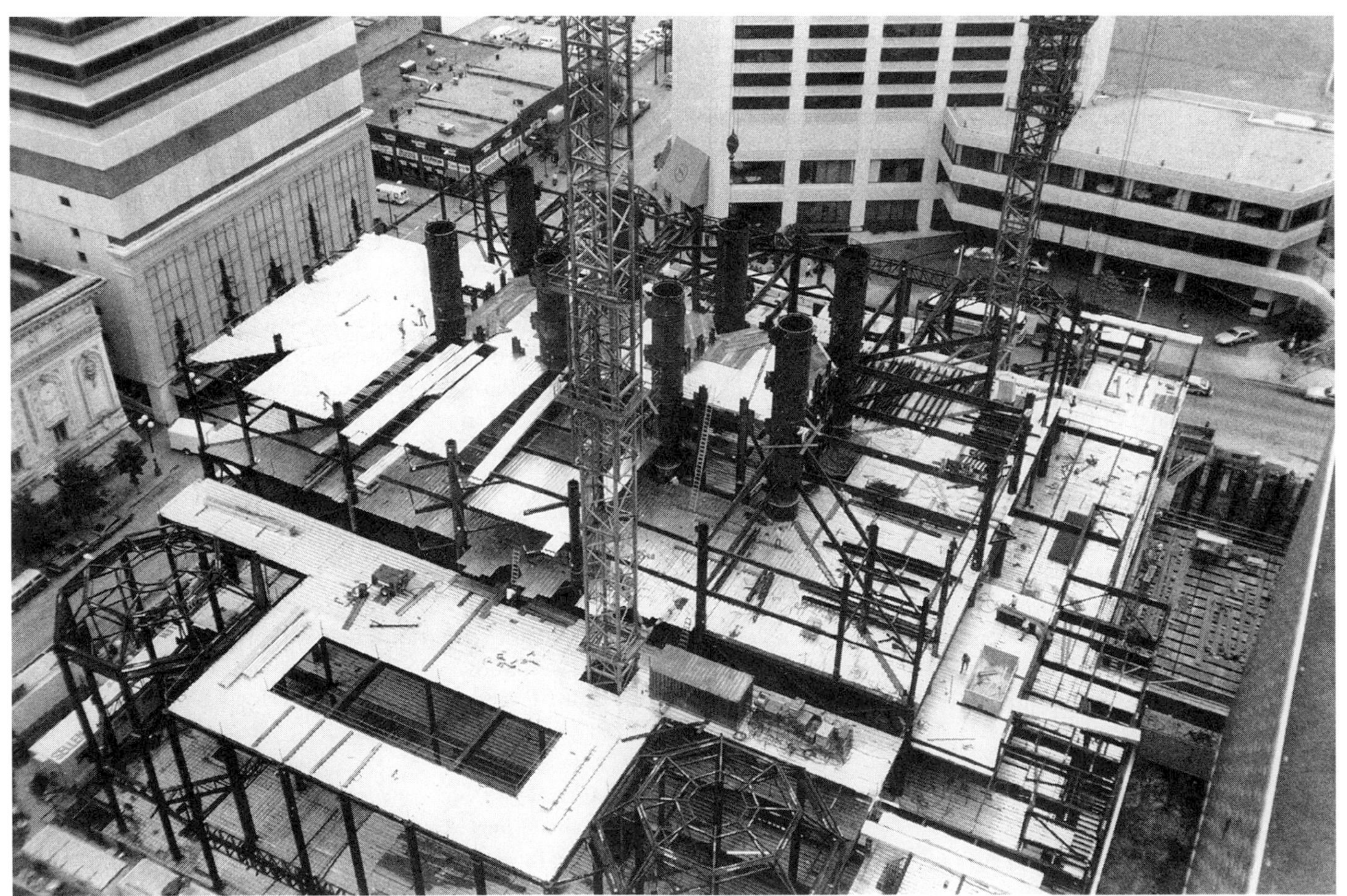

Figure 11.90
A core structure of eight large composite columns, each a concrete-filled pipe 7½ feet (2.3 m) in diameter, carries the majority of gravity and wind loads in this 44-story Seattle office building. The perimeter of the building is supported by smaller-diameter composite pipe columns. *(Courtesy of Skilling Ward Magnusson Barkshire, Inc., Seattle, Washington)*

Considerations of Sustainability in Steel Frame Construction

Manufacture

- The raw materials for steel are iron ore, coal, limestone, air, and water. The ore, coal, and limestone are minerals whose mining and quarrying cause disruption of land and loss of wildlife habitat, often coupled with pollution of streams and rivers. Coal, limestone, and low-grade iron ore are plentiful, but high-grade iron ore has been depleted in many areas of the earth.
- The steel industry has worked hard to reduce pollution of air, water, and soil, but much work remains to be done.
- Supplies of some alloying metals, such as manganese, chromium, and nickel, are becoming depleted.
- The manufacture of a ton of steel from iron ore by the basic oxygen process consumes 3170 pounds (1440 kg) of ore, 300 pounds (140 kg) of limestone, 900 pounds (410 kg) of coke (made from coal), 80 pounds (36 kg) of oxygen, and 2575 pounds (1170 kg) of air. In the process, 4550 pounds (2070 kg) of gaseous emissions are given off, and 600 pounds (270 kg) of slag and 50 pounds (23) of dust are generated. Further emissions emanate from the process of converting coal to coke.
- The embodied energy of steel produced from ore by the basic oxygen process is about 14,000 BTU per pound (33 MJ/kg). In modern facilities, scrap steel is typically added as an ingredient during this process, resulting in recycled materials content of 25 to 35 percent.
- Today, most structural steel in North America is made from recycled scrap by the electric arc furnace process; its embodied energy is approximately 4000 BTU per pound (9.3 MJ/kg), less than one-third that of steel made from ore. The recycled materials content of steel made by this process is 90 percent or higher.
- In North America, virtually all hot-rolled structural steel shapes are manufactured by the electric arc furnace process. Steel plate and sheet, used in the manufacture, for example, of light gauge steel members, decking, and hollow structural sections, may be produced by either the electric arc furnace or basic oxygen processes.
- Ninety-five percent or more of all structural steel used in North American building construction is eventually recycled or reused, which is a very high rate. In a recent one-year period, 480 million tons (430 million metric tons) of scrap steel were consumed worldwide.
- Scrap used in the production of structural steel in mini-mills usually comes from sources within approximately 300 miles (500 km) of the mill. When the steel produced in such mills is then used for the construction of buildings not too far from the mill, the steel is potentially eligible for credit as a regionally extracted, processed, and produced material. This is most likely for the most commonly used steel alloys that are produced in the greatest number of mills. However, some less commonly produced steel alloys are only available from a limited number of mills or, in some cases, are produced solely overseas, and are not eligible for such a credit except for projects located fortuitously close to the mills where these particular types of steel are produced.

Construction

- Steel fabrication and erection are relatively clean, efficient processes, although the paints and oils used on steel members can cause air pollution.
- Steel frames are lighter in weight than concrete frames that would do the same job. This means that a steel building generally has smaller foundations and requires less excavation work.
- Some spray-on fireproofing materials can pollute the air with stray fibers.

In Service

- Steel framing, if protected from water and fire, will last for many generations with little or no maintenance.
- Steel exposed to weather needs to be repainted periodically unless it is galvanized, given a long-lasting polymer coating, or made of more expensive stainless steel.
- Steel framing members in building walls and roofs should be thermally broken or insulated in such a way that they do not conduct heat between indoors and outdoors.
- When a steel building frame is demolished, its material is almost always recycled.
- Steel seldom causes indoor air quality problems, although surface oils and protective coatings sometimes outgas and cause occupant discomfort.

founded on steel because of steel's matchless stiffness and strength. The package building depends on steel for these qualities, for the repeatable precision with which components can be produced, and for the ease with which the relatively light steel components can be transported and assembled. It is but a short step from the usual process of steel fabrication and erection to the serial production of repetitive building components.

Steel and the Building Codes

Steel frame construction appears in the typical building code tables in Figures 1.2 and 1.3 as six different construction types—I, II, and III—the exact classification depending on the degree of fireproofing treatment applied to the various members of the frame. With a high degree of fireproofing, especially on members supporting more than one floor, unlimited building heights and areas are permitted for most occupancy groups. With no fireproofing whatsoever of steel members, building heights and areas are severely limited, but many occupancy groups can easily meet these restrictions.

Uniqueness of Steel

Among the common structural materials for fire-resistant construction—masonry, concrete, and steel—steel alone has useful tensile strength, which, along with compressive strength, it possesses in great abundance (Figure 11.91). A relatively small amount of steel can do a structural job that would take a much greater amount of another material. Thus, steel, the most dense structural material, is also the one that produces the lightest structures and those that span the greatest distances.

The infrastructure needed to bring steel shapes to a building site—the mines, the mills, the fabricators, and the scrap metal industry—is vast and complex. An elaborate sequence of advance planning and preparatory activities is required for making a steel building frame. Once on the site, however, a steel

Material	Working Strength in Tension[a]	Working Strength in Compression[a]	Density	Modulus of Elasticity
Wood (framing lumber)	300–1000 psi 2.1–6.9 MPa	600–1700 psi 4.1–12 MPa	30 pcf 480 kg/m^3	1,000,000–1,900,000 psi 6900–13,000 MPa
Brick masonry (including mortar, unreinforced)	0	250–1300 psi 1.7–9.0 MPa	120 pcf 1900 kg/m^3	700,000– 3,700,000 psi 4800–25,000 MPa
Structural steel	24,000–43,000 psi 170–300 MPa	24,000–43,000 psi 170–300 MPa	490 pcf 7800 kg/m^3	29,000,000 psi 200,000 MPa
Concrete (unreinforced)	0	1000–4000 psi 6.9–28 MPa	145 pcf 2300 kg/m^3	3,000,000–4,500,000 psi 21,000–31,000 MPa

[a]Allowable stress or approximate maximum stress under normal loading conditions.

Figure 11.91
Comparative physical properties for four common structural materials: wood, brick masonry, steel (shaded row), and concrete. Steel is many times stronger and stiffer than these other structural materials. The ranges of values of strength and stiffness reflect differences among structural steel alloys.

frame goes together quickly, and with relatively few tools, in an erection process that is rivaled for speed and all-weather reliability only by certain precast concrete systems. With proper design and planning, steel can frame almost any shape of building, including irregular angles and curves. Ultimately, of course, structural steel produces only a frame. Unlike masonry or concrete, it does not lend itself easily to forming a total building enclosure except in certain industrial applications. This is of little consequence, however, because steel mates easily with glass, masonry, and panel systems of enclosure and because steel does its own job, that of carrying loads high and wide with apparent ease, so very well.

FIGURE 11.92
This elegantly detailed house in southern California was an early example of the use of structural steel at the residential scale. *(Architect: Pierre Koenig, FAIA. Photo: Julius Shulman, Hon. AIA)*

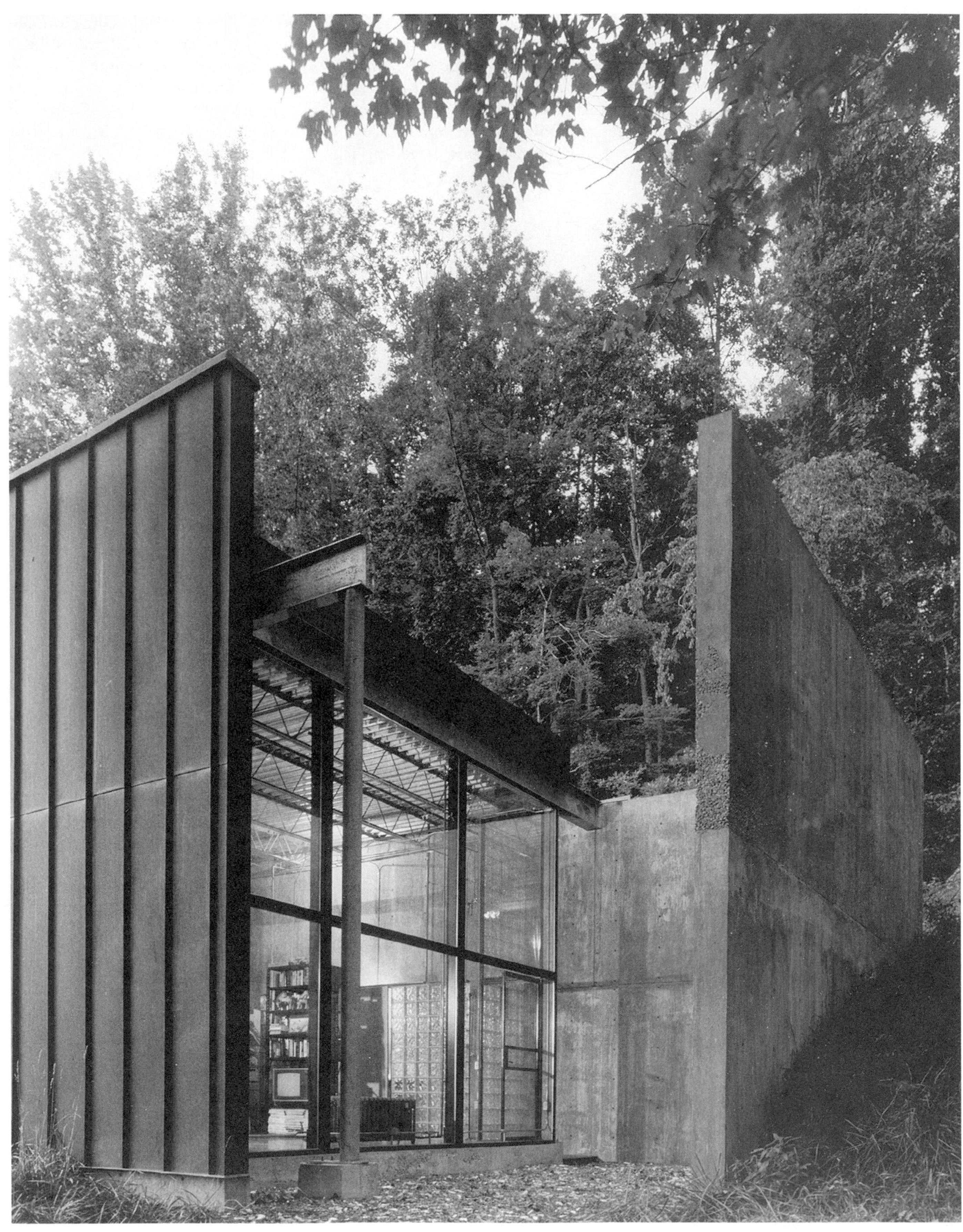

Figure 11.93
Architect Peter Waldman utilized steel pipe columns, wide-flange beams, open-web steel joists, and corrugated steel roof decking for his own house in Charlottesville, Virginia. *(Photo by Maxwell McKenzie)*

Figure 11.94
The Chicago Police Training Center expresses elegantly the logic and simplicity of a straightforward steel frame. *(Architect: Jerome R. Butler, Jr. Engineer: Louis Koncza. Permission of American Institute of Steel Construction)*

FIGURE 11.95
Architect Suzane Reatig structured the roof of a Washington, D.C., church with trusses made of steel angles. The ribs of the roof decking add a strong texture to the ceiling. *(Photo by Robert Lautman)*

FIGURE 11.96
The United Airlines Terminal at Chicago's O'Hare Airport is a high-tech wonderland of steel framing and fritted glass. *(Architect: Murphy-Jahn. Photo by Edward Allen)*

FIGURE 11.97
Chicago is famous for its role in the development of the steel frame skyscraper (see Figure 11.4). One of the tallest in the United States is the Sears Tower, seen in the foreground of this photograph. *(Architect and engineer: Skidmore, Owings and Merrill. Photo by Chicago Convention and Tourism Bureau, Inc. Permission of the American Institute of Steel Construction.)*

CSI/CSC MasterFormat Sections for Steel Frame Construction	
05 10 00	STRUCTURAL METAL FRAMING
05 12 00	Structural Steel Framing
05 16 00	Structural Cabling
05 20 00	METAL JOISTS
05 21 00	Steel Joist Framing
05 30 00	METAL DECKING
05 31 00	Steel Decking
05 35 00	Raceway Decking Assemblies
05 36 00	Composite Metal Decking
05 50 00	METAL FABRICATIONS
05 56 00	Metal Castings

SELECTED REFERENCES

1. American Institute of Steel Construction, Inc. *Steel Construction Manual.* Chicago, updated regularly.

This is the bible of the steel construction industry in the United States. It contains detailed tables of the dimensions and properties of all standard rolled steel sections, data on standard connections, and specifications and code information.

2. American Iron and Steel Institute. *Specification for Structural Steel Buildings.* Washington, DC, 2005.

This specification, included in the Steel Construction Manual, can also be viewed for free on the American Institute of Steel Construction's web site, www.aisc.org.

3. American Iron and Steel Institute. *Designing Fire Protection for Steel Beams* and *Designing Fire Protection for Steel Trusses.* Washington, DC, 1984 and 1991, respectively.

The problem of fireproofing steel building elements is discussed, and a range of fireproofing details is illustrated in these concise booklets.

4. Ambrose, James, and Patrick Tripeny. *Simplified Design of Steel Structures* (8th ed.). Hoboken, NJ, John Wiley & Sons, Inc., 2007.

This is an excellent introduction to the calculation of steel beams, columns, and connections.

5. Geoffrey L. Kulak, John W. Fisher, and John H. A. Struik. *Guide to Design Criteria for Bolted and Riveted Joints* (2nd ed.). Chicago, AISC, 2001.

This 300-plus page guide provides detailed engineering guidelines for the design of bolted and riveted steel connections. This guide can be viewed for free on the Research Council on Structural Connections web site, www.boltcouncil.org.

6. Steel Joist Institute. *Catalog of Standard Specifications and Load Tables for Steel Joists and Joist Girders.* Myrtle Beach, SC, updated regularly.

Load tables, sizes, and specifications for open-web joists are given in this booklet.

Web Sites

Steel Frame Construction

Author's supplementary web site: **www.ianosbackfill.com/11_steel_frame_construction**

The Material Steel

American Institute of Steel Construction (AISC): **www.aisc.org**
American Iron and Steel Institute: **www.steel.org**
Chaparral Steel: **www.chaparralsteel.com**
Jacob Stainless Steel Fittings and Wire: **www.jakobstainlesssteel.com**
Lincoln Electric Welding: **www.lincolnelectric.com**
Nucor Steel: **www.nucor.com**
Nucor–Vulcraft Group: **www.vulcraft.com**
Research Council on Structural Connections web site: **www.boltcouncil.org**
Steel Joist Institute (SJI): **www.steeljoist.org**
Steel Recycling Institute: **www.recycle-steel.org**

Longer Spans in Steel

Birdair Tensioned Membrane and Lightweight Structures: **www.birdair.com**

Key Terms and Concepts

Bessemer process
open-hearth method
steel
mild steel
cast iron wrought iron
ferrous metal
iron ore
coke
basic oxygen process
electric arc furnace
beam blank, bloom
high-strength, low-alloy steel
weathering steel
stainless steel
structural mill, breakdown mill
hot saw
cooling bed
roller straightener
bar
plate
sheet
wide-flange shape
American Standard shape, I-beam
angle
gusset plate
channel
tee
plate
bar
sheet
quenching
tempering
cast steel
cold-worked steel, cold-formed steel
hollow structural section (HSS)
open-web steel joist (OWSJ)
joist girder
rivet
high-strength bolt
carbon steel bolt
bearing-type connection
sung-tight
slip-critical connection, friction connection
preloaded
faying surface
galling
impact wrench
turn-of-nut method
load indicator washer, direct tension indicator (DTI)
calibrated wrench method
tension-control bolt
shear wrench
lockpin and collar fastener, swedge bolt
electric arc welding
electrode
weld symbols
backup bar, backing bar
runoff bar
demand-critical weld
shear connection
shear
bending moment
framed connection
moment connection
full-penetration groove weld
stiffener plate
braced frame
diagonal bracing
eccentrically braced frame
Chevron bracing, inverted V bracing
cross bracing
shear wall
moment-resisting frame
rigid core
diaphragm action
rigid perimeter
Fully-Restrained moment connection, AISC Type 1 connection
Partially-Restrained moment connection, AISC Type 3), connection
Simple connection. AISC Type 2 connection
seated connection
shear tab
end plate connection

fabricator
shop drawing
coped flange
plasma cutting
laser cutting
camber
erector
ironworker
tier
baseplate
leveling plate
grout
luffing-boom crane
hammerhead boom crane
raising gang
plumbing up
tower crane
tagline
drift pin
topping out
metal decking
roof decking
cellular decking
puddle weld
composite metal decking
shear stud
subpurlin
girt
architecturally exposed structural steel (AESS)
fireproofing
spray-applied fire-resistive materials (SFRM)
intumescent mastic
intumescent paint
castellated beam
plate girder
rigid steel frame
steel truss
chord
space truss, space frame
arch
anticlastic curvature
cable stay
tensile fabric structure
pneumatic structure
air-supported structure
prestress
form finding
mast

Review Questions

1. What is the difference between iron and steel? What is the difference between wrought iron and cast iron?

2. By weight, what is the major raw material used in the making of cast iron?

3. How are steel structural shapes produced? How are the weights and thicknesses of a shape changed?

4. How does the work of the fabricator differ from that of the erector?

5. Explain the designation W21 × 68.

6. How can you tell a shear connection from a moment connection? What is the role of each?

7. Why might a beam be coped?

8. What is the advantage of composite construction?

9. Explain the advantages and disadvantages of a steel building structure with respect to fire. How can the disadvantages be overcome?

10. List three different structural systems in steel that might be suitable for the roof of an athletic fieldhouse.

Exercises

1. For a simple multistory office building of your design:

    a. Draw a steel framing plan for a typical floor.

    b. Draw an elevation or section showing a suitable method of making the building stable against lateral forces—wind and earthquake.

    c. Make a preliminary determination of the approximate sizes of the decking, beams, and girders, using the information in the box on page 417.

    d. Sketch details of the typical connections in the frame, using actual dimensions from the *Manual of Steel Construction* (reference 1) for the member size you have determined and work to scale.

2. Select a method of fireproofing, and sketch typical column and beam fireproofing details for the building in Exercise 1.

3. What fire-resistance ratings in hours are required for the following elements of a steel framed department store, three stories in height, unsprinklered, with 21,000 square feet of area per floor? (The necessary information is found in Figures 1.2 and 1.3.)

    a. Lower-floor columns

    b. Floor beams

    c. Roof beams

    d. Interior nonbearing walls and partitions

4. Find a steel building frame under construction. Observe the connections carefully and figure out why each is detailed as it is. If possible, arrange to talk with the structural engineer of the building to discuss the design of the frame.

12

Light Gauge Steel Frame Construction

Driving self-drilling, self-tapping screws with electric screw guns, framers add diagonal bracing straps to a wall frame made from light gauge steel studs and runner channels. (*Courtesy of United States Gypsum Company*)

To manufacture the members used in *light gauge steel* frame construction, sheet steel is fed from continuous coils through machines at room temperature that cold-work the metal (see Chapter 11) and fold it into efficient structural shapes, producing linear members that are stiff and strong. Thus, these members are referred to as *cold-formed metal framing* to differentiate them from the much heavier hot-rolled shapes that are used in structural steel framing. The term "light gauge" refers to the relative thinness (gauge) of the steel sheet from which the members are made.

The Concept of Light Gauge Steel Construction

Light gauge steel construction is the noncombustible equivalent of wood light frame construction. The external dimensions of the standard sizes of light gauge members correspond closely to the dimensions of the standard sizes of nominal 2-inch (38-mm) framing lumber. These steel members are used in framing as closely spaced studs, joists, and rafters in much the same way as wood light frame members are used, and a light gauge steel frame building may be sheathed, insulated, wired, and finished inside and out in the same manner as a wood light frame building.

The steel used in light gauge members is manufactured to ASTM standard A1003 and is metallic-coated with zinc or aluminum-zinc alloy to provide long-term protection against corrosion. The thickness of the metallic coating can be varied, depending on the severity of the environment in which the members will be placed. For studs, joists, and rafters, the steel is formed into C-shaped *cee sections* (Figure 12.1). The webs of cee members are punched at the factory to provide holes at 2-foot (600-mm) intervals; these are designed to allow wiring, piping, and bracing to pass through studs and joists without the necessity of drilling holes on the construction site. For top and bottom wall plates and for joist headers, *channel sections* are used. The strength and stiffness of a member depend on the shape and depth of the section and the *gauge* (thickness) of the steel sheet from which it is made. A standard range of depths and gauges is available from each manufacturer. Commonly used metal thicknesses for loadbearing members range from 0.097 to 0.033 inch (2.46–0.84 mm) and are as thin as 0.018 inch (0.45 mm) for nonloadbearing members (Figure 12.2).

At least one manufacturer produces nonloadbearing light gauge steel members by passing steel sheet through rollers with mated patterned surfaces, producing a dense array of dimples in the metal of the formed members. The additional cold working of the metal that occurs during the forming process and the finished

Figure 12.1
Typical light gauge steel framing members. To the left are the common sizes of cee studs and joists. In the center are channel studs. To the right are runner channels.

Considerations of Sustainability in Light Gauge Steel Framing

In addition to the sustainability issues raised in the previous chapter, which also apply here, the largest issue concerning the sustainability of light gauge steel construction is the high thermal conductivity of the framing members. If a dwelling framed with light gauge steel members is framed, insulated, and finished as if it were framed with wood, it will lose heat in winter at about double the rate of the equivalent wood structure. To overcome this limitation, energy codes now require light gauge steel framed buildings constructed in cold regions, including most of the continental United States, to be sheathed with plastic foam insulation panels in order to eliminate the extensive thermal bridging that can otherwise occur through the steel framing members.

Even with insulating sheathing, careful attention must be given to avoid undesired thermal bridges. For example, on a building with a sloped roof, a significant thermal bridge may remain through the ceiling joist-rafter connections, as seen in Figure 12.4*b*. Foam sheathing on the inside wall and ceiling surfaces is one possible way to avoid this condition, but adding insulation to the inside of the metal framing exposes the studs and stud cavities to greater temperature extremes and increases the risk of condensation. It also still allows thermal bridging through the screws used to fasten interior gypsum wallboard to the framing. Though small in area, these thermal bridges can readily conduct heat and result in spots of condensation on interior finish surfaces in very cold weather.

patterned surface result in members made from thinner sheet stock that are equal in and strength and stiffness to conventionally formed members produced from heavier gauge material.

For large projects, members may be manufactured precisely to the required lengths. Otherwise, they are furnished in standard lengths. Members may be cut to length on the construction job site with power saws or special shears. A variety of sheet metal angles, straps, plates, channels, and miscellaneous shapes are manufactured as accessories for light gauge steel construction (Figure 12.3).

Light gauge steel members are usually joined with *self-drilling, self-tapping screws*, which drill their own holes and form helical threads in the holes as they are driven. Driven rapidly by hand-held electric or pneumatic tools, these screws are plated with cadmium or zinc to resist corrosion, and they are available in an assortment of diameters and lengths to suit a full range of connection situations. Welding is often employed to assemble panels of light gauge steel framing that are prefabricated in a factory, and it is sometimes used on the building site where particularly strong connections are needed. Other fastening techniques that are widely used include hand-held clinching devices that join members without screws or welds and pneumatically driven pins that penetrate the members and hold by friction.

FIGURE 12.2
Minimum thicknesses of base sheet metal (not including the metallic coating) for light gauge steel framing members. Traditional gauge designations are also included (note how lower gauge numbers correspond to greater metal thickness). Gauge numbers are no longer recommended for specification of sheet metal thickness due to lack of a uniform standard for the translation between these numbers and actual metal thickness. Sheet metal thickness may also be specified in mils, or thousandths of an inch. For example, a thickness of 0.033 inch can be expressed as 33 mils.

	Minimum Thickness of Steel Sheet	
Gauge	Loadbearing Light Gauge Steel Framing	Nonloadbearing Light Gauge Steel Framing
12	0.097″ (2.46 mm)	
14	0.068″ (1.73 mm)	
16	0.054″ (1.37 mm)	0.054″ (1.37 mm)
18	0.043″ (1.09 mm)	0.043″ (1.09 mm)
20	0.033″ (0.84 mm)	0.030″ (0.75 mm)
22		0.027″ (0.69 mm)
25		0.018″ (0.45 mm)

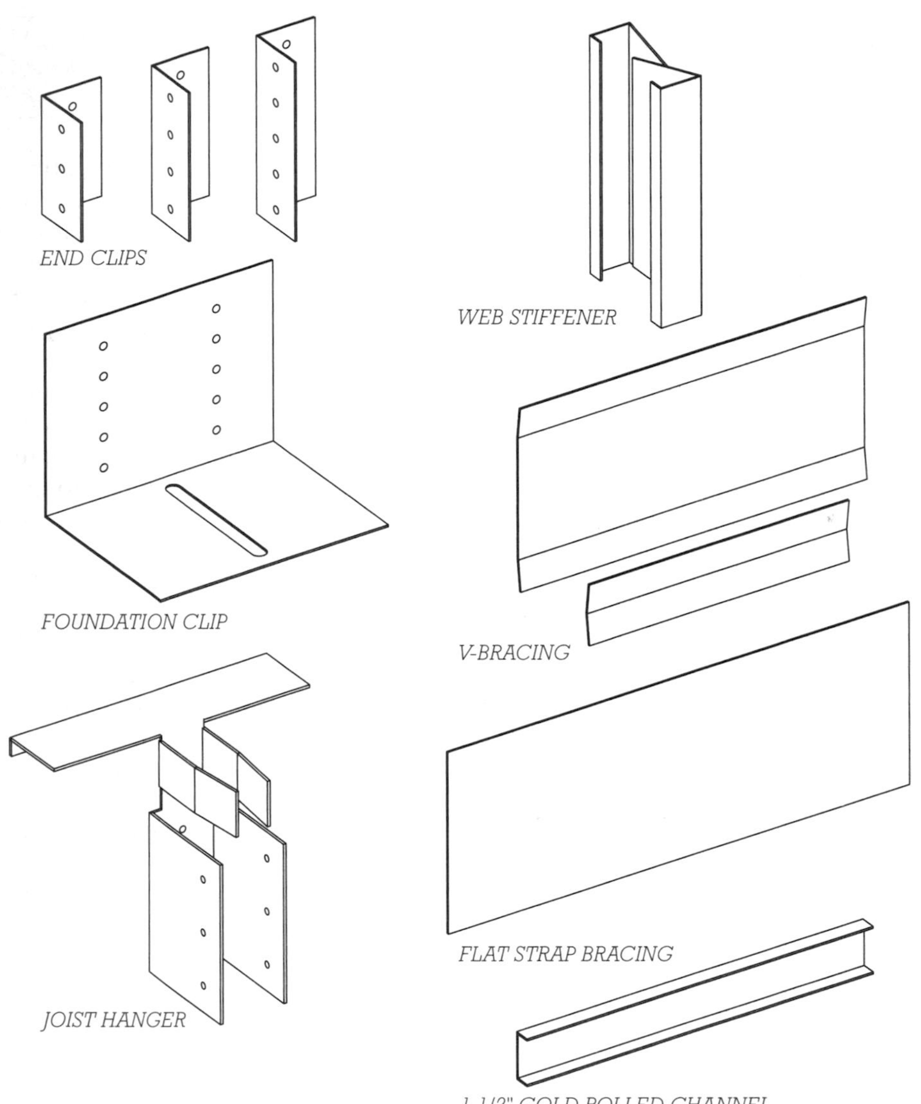

FIGURE 12.3
Standard accessories for light gauge steel framing. End clips are used to join members that meet at right angles. Foundation clips attach the ground-floor platform to anchor bolts embedded in the foundation. Joist hangers connect joists to headers and trimmers around openings. The web stiffener is a two-piece assembly that is inserted inside a joist and screwed to its vertical web to help transmit wall loads vertically through the joist. The remaining accessories are used for bracing.

FRAMING PROCEDURES

The sequence of construction for a building that is framed entirely with light gauge steel members is essentially the same as that described in Chapter 5 for a building framed with nominal 2-inch (38-mm) wood members (Figure 12.4). Framing is usually constructed platform fashion: The ground floor is framed with steel joists. Mastic adhesive is applied to the upper edges of the joists, and wood panel subflooring is laid down and fastened to the upper flanges of the joists with screws. Steel studs are laid flat on the subfloor and joined to make wall frames. The wall frames are sheathed either with wood panels or, for noncombustible construction, with *gypsum sheathing panels*, which are similar to gypsum wallboard but with glass mat faces and a water-resistant core formulation. The wall frames are tilted up, screwed down to the floor frame, and braced. The upper-floor platform is framed, then the upper-floor walls. Finally, the ceiling and roof are framed in much the same way as in a wood-framed house. Prefabricated trusses of light gauge steel members that are screwed or welded together are often used to frame ceilings and roofs (Figures 12.15 and 12.16). It is possible, in fact, to frame any building with light gauge steel members that can be framed with nominal 2-inch (38-mm) wood members. To achieve a more fire-resistive construction type under the building code, floors of corrugated steel decking with a concrete topping are sometimes substituted for wood panel subflooring.

Openings in floors and walls are framed analogously to openings in wood light frame construction, with doubled members around each opening and strong headers over doors and windows (Figures 12.5–12.9). Joist hangers and right-angle clips of

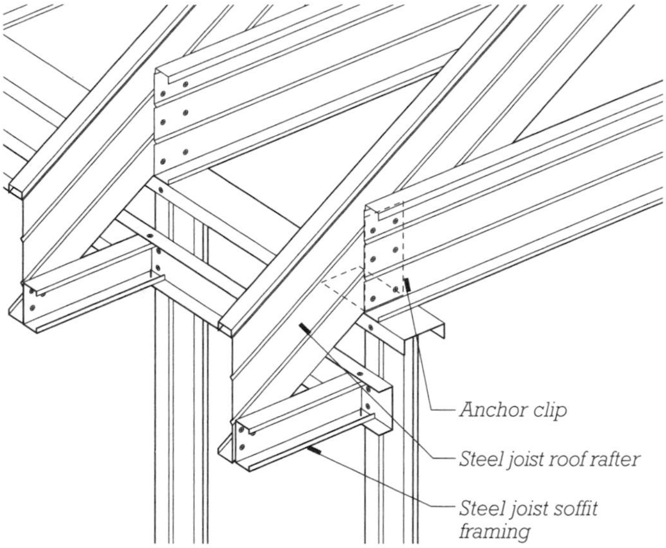

Anchor clip

Steel joist roof rafter

Steel joist soffit framing

B EAVE

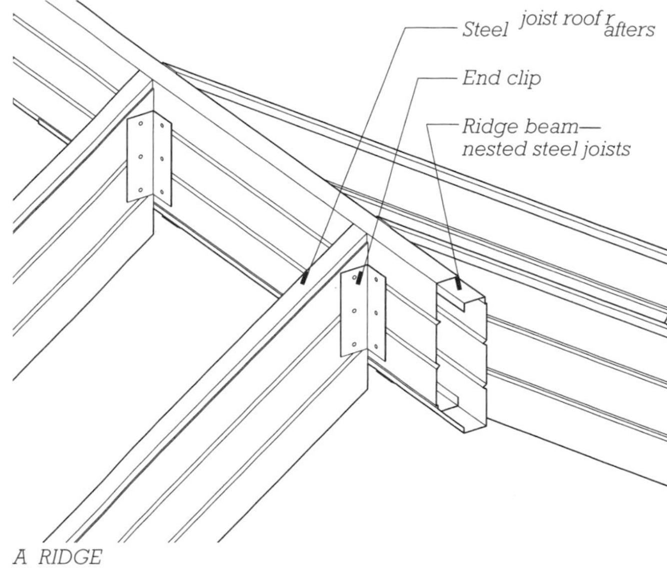

A RIDGE

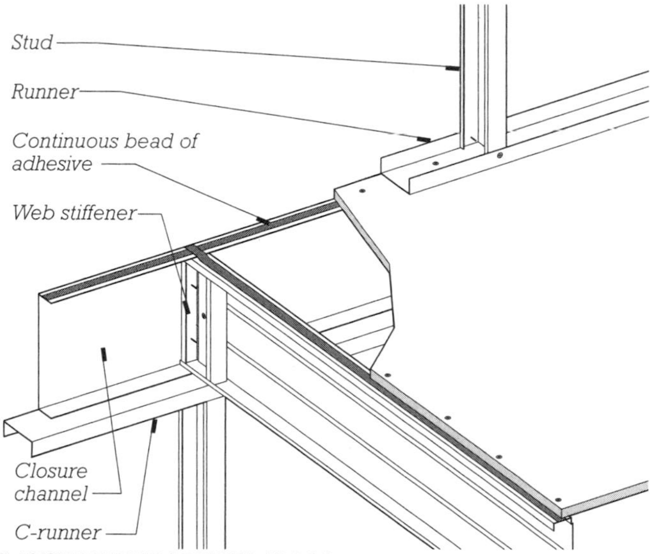

Stud

Runner

Continuous bead of adhesive

Web stiffener

Closure channel

C-runner

C JOIST BEARING AT UPPER FLOOR

Figure 12.4
Typical light gauge framing details. Each detail is keyed by letter to a circle on the whole-building diagram in the center of the next page to show its location in the frame. (*a*) A pair of nested joists makes a boxlike ridge board or ridge beam. (*b*) Anchor clips are sandwiched between the ceiling joists and rafters to hold the roof framing down to the wall. (*c*) A web stiffener helps transmit vertical forces from each stud through the end of the joist to the stud in the floor below. Mastic adhesive cushions the joint between the subfloor and the steel framing. (*d*) Foundation clips anchor the entire frame to the foundation. (*e*) At interior joist bearings, joists are overlapped back to back and a web stiffener is inserted.

(continued)

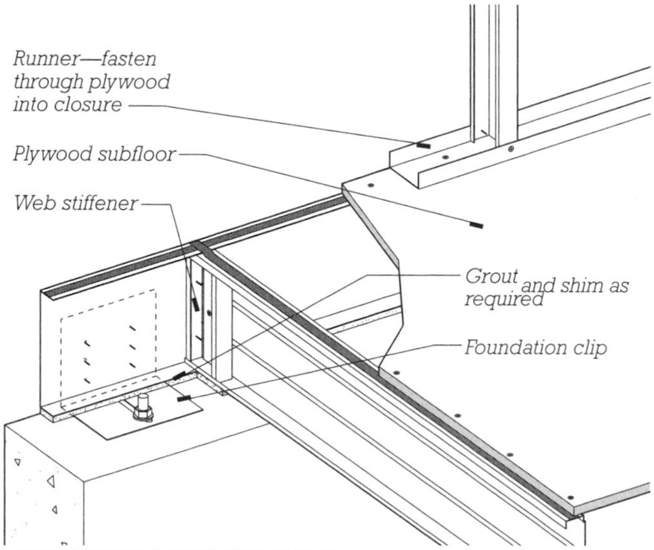

Runner—fasten through plywood into closure

Plywood subfloor

Web stiffener

Grout and shim as required

Foundation clip

D JOIST BEARING AT FOUNDATION

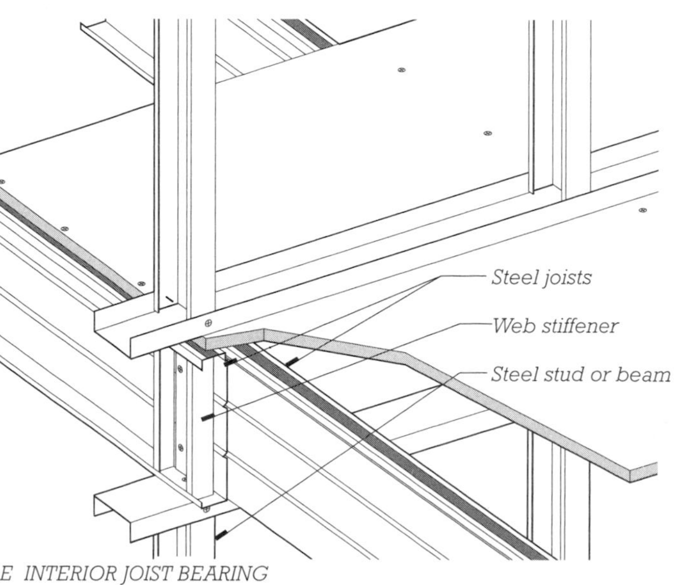

E INTERIOR JOIST BEARING

H GABLE END FRAMING

G JOIST PARALLEL TO END WALL

F JOIST PARALLEL TO FOUNDATION

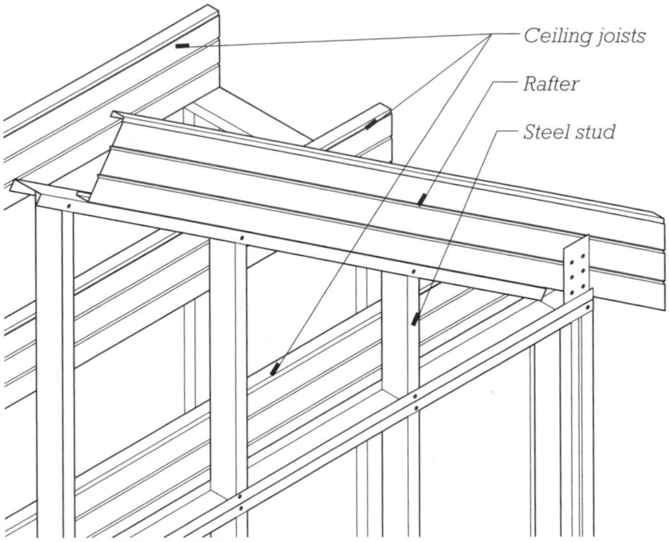

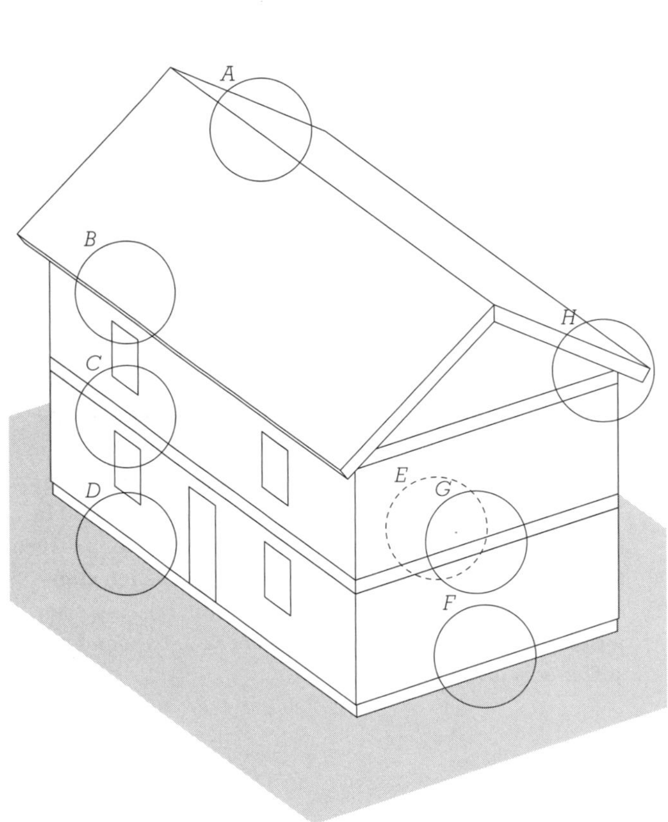

FIGURE 12.4 *(continued)*
(*f, g*) Short crosspieces brace the last joist at the end of the building and help transmit stud forces through to the wall below. (*h*) Like all these details, the gable end framing is directly analogous to the corresponding detail for a wood light frame building as shown in Chapter 5.

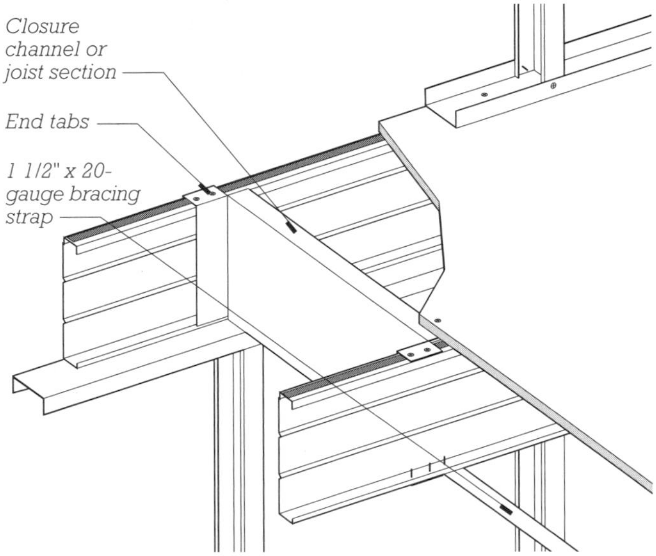

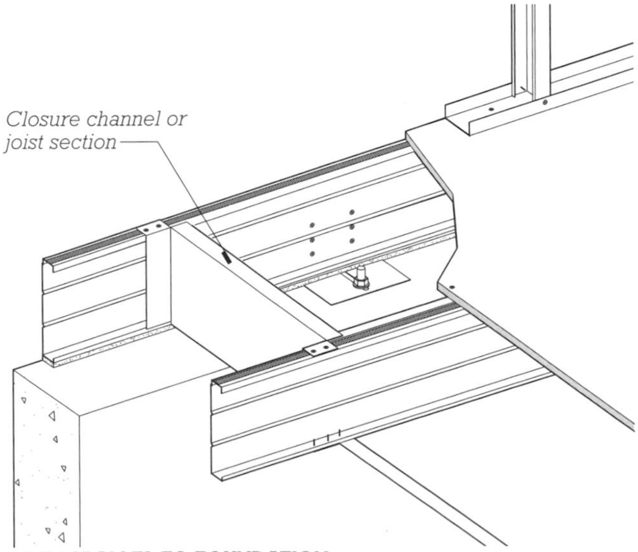

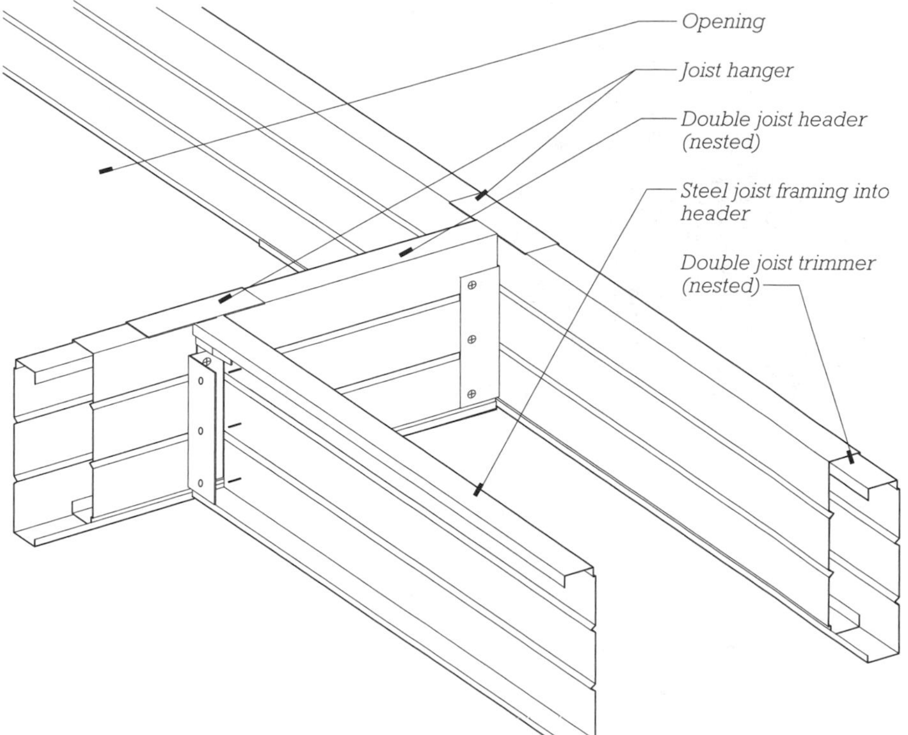

FIGURE 12.5
Headers and trimmers for floor openings are doubled and nested to create a strong, stable box member. Only one vertical flange of the joist hanger is attached to the joist; the other flange would be used instead if the web of the joist were oriented to the left rather than the right.

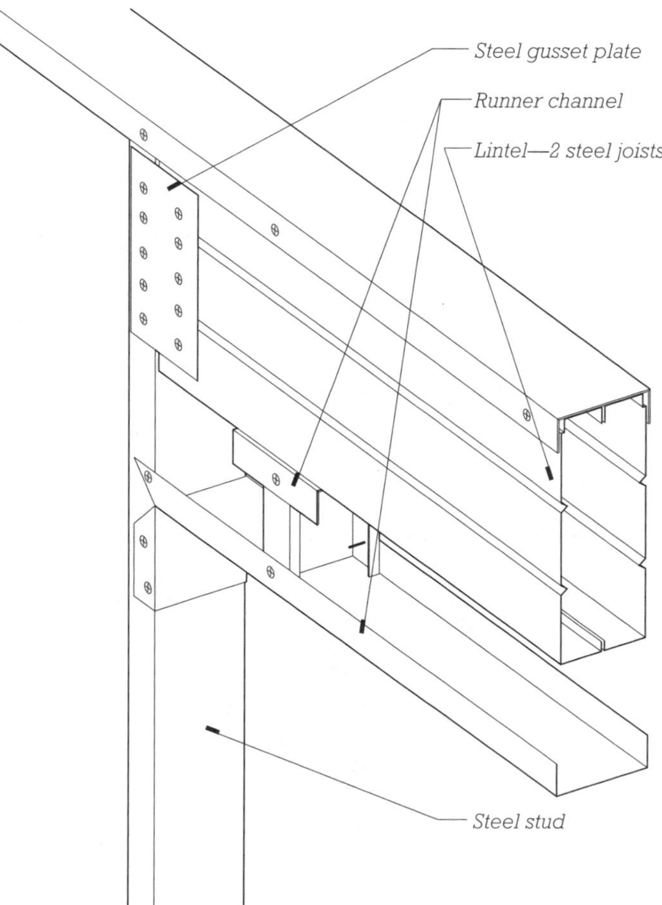

FIGURE 12.6
A typical window or door head detail. The header is made of two joists placed with their open sides together. The top plate of the wall, which is a runner channel, continues over the top of the header. Another runner channel is cut and folded at each end to frame the top of the opening. Short studs are inserted between this channel and the header to maintain the rhythm of the studs in the wall.

FIGURE 12.7
Diagonal strap braces stabilize upper-floor wall framing for an apartment building. *(Courtesy of United States Gypsum Company)*

FIGURE 12.8
Temporary braces support the walls at each level until the next floor platform has been completed. Cold-rolled channels pass through the web openings of the studs; they are welded to each stud to help stabilize them against buckling. *(Courtesy of Unimast Incorporated—www.unimast.com)*

FIGURE 12.9
A detail of a window header. Because a supporting stud has been inserted under the end of the header, a large gusset plate such as the one shown in Figure 12.6 is not required. *(Courtesy of Unimast Incorporated—www.unimast.com)*

FIGURE 12.10
Ceiling joists in place for an apartment building. A brick veneer cladding has already been added to the ground floor. *(Courtesy of United States Gypsum Company)*

sheet steel are used to join members around openings. Light gauge members are designed so that they can be *nested* to form a tubular configuration that is especially strong and stiff when used for a ridge board or header (Figures 12.4*a* and 12.5).

Because light gauge steel members are much more prone than their wood counterparts to twisting or buckling under load, somewhat more attention must be paid to their bracing and bridging. The studs in tall walls are generally braced at 4-foot (1200-mm) intervals, either with steel straps screwed to the edges of the studs or with 1½-inch (38-mm) cold-formed steel channels passed through the punched openings in the studs and welded or screwed to an angle clip at each stud (Figure 12.8). Floor joists are bridged with cee-joist blocking between and steel straps screwed to their top and bottom edges. In locations where large vertical forces must pass through floor joists (as occurs where loadbearing studs sit on the edge of a floor platform), steel *web stiffeners* are screwed to the thin webs of the joists to prevent them from buckling (Figure 12.4*c,e*). Wall bracing consists of diagonal steel straps screwed to the studs (chapter-opening photo, Figure 12.7). Permanent resistance to buckling, twisting, and lateral loads such as wind and earthquake is imparted largely and very effectively by subflooring, wall sheathing, and interior finish materials.

FIGURE 12.11
A detail of eave framing. *(Courtesy of Unimast Incorporated—www.unimast.com)*

FIGURE 12.12
A power saw with an abrasive blade cuts quickly and precisely through steel framing members. *(Courtesy of Unimast Incorporated—www.unimast.com)*

Other Common Uses of Light Gauge Steel Framing

Light gauge steel members are used to construct many components of fire-resistant buildings whose structures are made of structural steel, concrete, or masonry. These components include interior walls and partitions (Chapter 23), suspended ceilings (Chapter 24), and fascias, parapets, and backup walls for such exterior claddings as masonry veneer, exterior insulation and finish system (EIFS), glass-fiber-reinforced concrete (GFRC), metal panels, and various thin stone cladding systems (Chapters 19 and 20; see also Figures 12.13 and 12.14). Light gauge steel members used for framing interior partitions and other nonloadbearing applications are properly referred to and specified as *nonstructural metal framing*, as distinct from cold-formed metal framing, the latter term reserved for light gauge steel members used in structural applications and exterior wall cladding systems (even though both types of members are, in fact, cold-formed).

Light gauge steel studs can be combined with concrete to produce thin, but relatively stiff, wall panel systems. Both loadbearing and nonloadbearing panels can be made that are suitable for use in residential and light commercial buildings. A variety of production methods are possible that generally involve casting an approximately 2-inch (50-mm)-thick concrete facing onto a framework of steel studs. The concrete may be sitecast (on the building site) or precast (in a factory). The concrete-to-steel bond may be created by a variety of devices welded or screwed to the studs that then become embedded in the concrete, such as stud anchors, sheet metal shear strips, welded wire reinforcing, or expanded metal. In loadbearing applications, the concrete panels provide shear resistance while the steel studs provide most of the resistance to gravity loads and to wind loads acting perpendicular to the face of the panel.

Figure 12.13
Light gauge steel stud walls frame the exterior walls of a building whose floors and roof are framed with structural steel. *(Courtesy of Unimast Incorporated—www.unimast.com)*

FIGURE 12.14
The straightness of steel studs is apparent in these tall walls that enclose a building framed with structural steel.
(Courtesy of Unimast Incorporated—www.unimast.com)

FIGURE 12.15
A worker tightens the last screws to complete a connection in a light gauge steel roof truss. The truss members are held in alignment during assembly by a simple jig made of plywood and blocks of framing lumber. *(Courtesy of Unimast Incorporated—www.unimast.com)*

FIGURE 12.16
Installing steel roof trusses. *(Courtesy of Unimast Incorporated—www.unimast.com)*

For Preliminary Design of a Light Gauge Steel Frame Structure

- Estimate the depth of rafters on the basis of the horizontal (not slope) distance from the outside wall of the building to the ridge board in a gable or hip roof and the horizontal distance between supports in a shed roof. Estimate the depth of a rafter at 1⁄24 of this span, rounded up to the nearest 2-inch (50-mm) dimension.
- The depth of light gauge steel **roof trusses** is usually based on the desired roof pitch. A typical depth is one-quarter of the width of the building, which corresponds to a 6⁄12 pitch.
- Estimate the depth of light gauge steel **floor joists** as 1⁄20 of the span, rounded up to the nearest 2-inch (50-mm) dimension.
- For **loadbearing studs**, add up the total width of floor and roof slabs that contribute load to the stud wall. A 3⅝-inch (92-mm) or 4-inch (102-mm) stud wall can support a combined width of approximately 60 feet (18 m), and a 6-inch (152-mm) or 8-inch (203-mm) stud wall can support a combined width of approximately 150 feet (45 m).
- For **exterior cladding backup walls**, estimate that a 3⅝-inch (92-mm) stud may be used to a maximum height of 12 feet (3.7 m), a 6-inch (150-mm) stud to 19 feet (5.8 m), and an 8-inch (100-mm) stud to 30 feet (9.1 m). For brittle cladding materials such as brick masonry, select a stud that is 2 inches (50 mm) deeper than these numbers would indicate.

All framing members are usually spaced at 24 inches (600 mm) o.c.

These approximations are valid only for purposes of preliminary building layout and must not be used to select final member sizes. They apply to the normal range of building occupancies such as residential, office, commercial, and institutional buildings. For manufacturing and storage buildings, use somewhat larger members.

For more comprehensive information on preliminary selection and layout of structural members, see Edward Edward and Joseph Iano, *The Architect's Studio Companion* (4th ed.), New York, John Wiley & Sons, Inc., 2007.

In situations where noncombustibility is not a requirement, metal and wood light framing are sometimes mixed in the same building. Some builders find it economical to use wood to frame exterior walls, floors, and roof, with steel framing for interior partitions. Sometimes all walls, interior and exterior, are framed with steel, and floors are framed with wood. Steel trusses made of light gauge members may be applied over wood frame walls. In such mixed uses, special care must be taken in the details to ensure that wood shrinkage will not create unforeseen stresses or damage to finish materials. Steel framing also may be used in lieu of wood where the risk of damage from termites is very high.

Advantages and Disadvantages of Light Gauge Steel Framing

Light gauge steel framing shares most of the advantages of wood light framing: It is versatile and flexible; requires only simple, inexpensive tools; furnishes internal cavities for utilities and thermal insulation; and accepts an extremely wide range of exterior and interior finish materials. Additionally, steel framing may be used in buildings for which noncombustible construction is required by the building code, thus extending its use to larger buildings and those whose uses require a higher degree of resistance to fire.

Steel framing members are significantly lighter in weight than the wood members to which they are structurally equivalent, an advantage that is often enhanced by spacing steel studs, joists, and rafters at 24 inches (600 mm) o.c. rather than 16 inches (400 mm) o.c. Light gauge steel joists and rafters can span slightly longer distances than nominal 2-inch (50-mm) wood members of the same depth. Steel members tend to be straighter and more uniform than wood members, and they are much more stable dimensionally because they are unaffected by changing humidity. Although they may corrode if exposed to moisture over an extended period of time, particularly in oceanfront locations, steel framing members cannot fall victim to termites or decay.

Compared to walls and partitions of masonry construction, equivalent walls and partitions framed with steel studs are much lighter in weight, easier to insulate, and accept electrical wiring and pipes for plumbing and heating much more readily. Steel framing, because it is a dry process, may be carried out under wet or cold weather conditions that would make masonry construction difficult. Masonry walls tend to be much stiffer and more resistant to the passage of sound than steel-framed walls, however.

The thermal conductivity of light gauge steel framing members is much higher than that of wood. In cold regions, light gauge steel framing should be detailed with *thermal breaks*, that is, materials with high resistance to the flow of heat, such as foam plastic sheathing or insulating edge spacers between studs and sheathing, to prevent the rapid loss of heat through the steel members. Without such measures, the thermal performance of the wall or roof is greatly reduced, energy losses increase substantially, and moisture condensation within the framing cavity or on interior building surfaces may occur, with attendant damage to materials, growth of mold and mildew, and discoloration of surface finishes. Special attention must be given to designing details to block excessive heat flow in every area of the frame. At the eave of a steel-framed house, for instance, the ceiling joists readily conduct heat from the warm interior ceiling along their length to the cold eave unless insulating edge spacers or foam insulation boards are used between the ceiling finish material and the joists.

Light Gauge Steel Framing and the Building Codes

Although light gauge steel framing members will not burn, they will lose their structural strength and stiffness rapidly if exposed to the heat of fire. They must therefore be protected from fire in accordance with building code requirements. With suitable protection provided by gypsum sheathing and gypsum wallboard or plaster, light gauge steel construction may be classified as either Type I or Type II Construction in the building code table shown in Figure 1.2, enabling its use for a wide range of building types and sizes.

In its International Residential Code for One- and Two-Family Dwellings, the International Code Council has incorporated *prescriptive requirements* for steel-framed residential construction. In many cases, these requirements, with their structural tables and standard details, allow builders to design and construct light gauge steel-framed houses without having to employ an engineer or architect, just as they are able to do with wood light frame construction.

Finishes for Light Gauge Steel Framing

Any exterior or interior finish material that is used in wood light frame construction may be applied to light gauge steel frame construction. Whereas finish materials are often fastened to a wood frame with nails, only screws may be used with a steel frame. Wood trim components are applied with special finish screws, analogous to finish nails, which have very small heads.

Figure 12.17
Gypsum sheathing panels have been screwed onto most of the ground-floor walls of this large commercial building. *(Courtesy of Unimast Incorporated—www.unimast.com)*

FIGURE 12.18
Waferboard (a wood panel product similar to OSB) sheaths the walls of a house framed with light gauge steel studs, joists, and rafters. *(Courtesy of Unimast Incorporated—www.unimast.com)*

Metals in Architecture

Metals are dense, lustrous materials that are highly conductive of heat and electricity. They are generally *ductile*, meaning that they can be hammered thin or drawn into wires. They can be liquefied by heating and will resolidify as they cool. Most metals corrode by oxidation. Metals include the strongest building materials presently in common use, although stronger materials based on carbon or aramid fibers are beginning to appear more frequently in building construction applications.

Most metals are found in nature in the form of oxide ores. These ores are refined by processes that involve heat and reactant materials or, in the case of aluminum, electrolysis.

Metals may be classified broadly as either ferrous, meaning that they consist primarily of iron, or nonferrous (all other metals). Because iron ore is an abundant mineral and is relatively easy to refine, ferrous metals tend to be much less expensive than nonferrous ones. The ferrous metals are also the strongest, but most have a tendency to rust. Nonferrous metals in general are considerably more expensive on a volumetric basis than ferrous metals, but unlike ferrous metals, most of them form thin, tenacious oxide layers that protect them from further corrosion under normal atmospheric conditions. This makes many of the nonferrous metals valuable for finish components of buildings. Many of the nonferrous metals are also easy to work and attractive to the eye.

Modifying the Properties of Metals

A metal is seldom used in its chemically pure state. Instead, it is mixed with other elements, primarily other metals, to modify its properties for a particular purpose. Such mixtures are called *alloys*. An alloy that combines copper with a small amount of tin is known as "bronze." A very small, closely controlled amount of carbon mixed with iron makes steel. In both of these example, the alloy is stronger and harder than the metal that is its primary ingredient. Several alloys of iron (several different steels, to be more specific) are mentioned in Chapter 11. Some of these steel alloys have higher strengths and some form self-protecting oxide layers because of the influence of the alloying elements they contain. Similarly, there are many alloys that consist primarily of aluminum; some are soft and easy to form, others are very hard and springy, still others are very strong, and so on.

The properties of many metals can also be changed by *heat treatment*. Steel that is *quenched*, that is, heated red-hot and then plunged in cold water, becomes much harder but very brittle. Steel can be *tempered* by heating it to a moderate degree and cooling it more slowly, making it both hard and strong. Steel that is brought to a very high temperature and then cooled very slowly, a process called *annealing*, will become softer, easier to work, and less brittle. Many aluminum alloys can also be heat treated to modify their characteristics.

Cold working is another way of changing the properties of a metal. When steel is beaten or rolled thinner at room temperature, its crystalline structure is changed in a way that makes it much stronger and somewhat more brittle. The highest-strength metals used in construction are steel wires and cables used to prestress concrete. Their high strength (about four times that of normal structural steel) is the result of drawing the metal through smaller and smaller orifices to produce the wire, a process that subjects the metal to a high degree of cold working. Cold-rolled steel shapes with substantially higher strengths than hot-rolled structural steel are used as reinforcing and as components of open-web joists. The effects of cold working are easily reversed by annealing. Hot rolling, which is, in effect, a self-annealing process, does not increase the strength of metal.

To change the appearance of metal or to protect it from oxidation, it can be coated with a thin layer of another metal. Steel is often *galvanized* by coating it with zinc to protect against corrosion, as described below. *Electroplating* is widely used to coat metals such as chromium and cadmium onto steel to improve its appearance and protect it from oxidation. An electrolytic process is used to *anodize* aluminum, adding a thin oxide layer of controlled color and consistency to the surface of the metal. To protect them and enhance their appearance, metals are frequently finished with nonmetallic coatings such as paints, lacquers, high-performance organic coatings, porcelain enamel, and thermosetting powders.

Fabricating Metals

Metals can be shaped in many different ways. *Casting* is the process of pouring molten metal into a shaped mold; the metal retains the shape of the mold as it cools. *Rolling*, which may be done either hot or cold, forms the metal by squeezing it between a series of shaped rollers. *Extrusion* is the process of squeezing heated but not molten metal through a shaped die to produce a long metal piece with a shaped profile matching the cutout in the die. *Forging* involves heating a piece of metal until it becomes soft, then beating it into shape. Forging was originally done by hand with a blacksmith's forge, hammer, and anvil, but most forging is now done with powerful hydraulic machinery that forces the metal into shaped dies. *Stamping* is the process of squeezing sheet metal between two matching dies to give it a desired shape or texture. *Drawing* produces wires by pulling a metal rod through

METALS IN ARCHITECTURE *(CONTINUED)*

a series of progressively smaller orifices in hardened steel plates until the desired diameter is reached. These forming processes have varying effects on the strength of the resulting material: Cold drawing and cold rolling will harden and strengthen many metals. Forging imparts a grain orientation to the metal that closely follows the shape of the piece for improved structural performance. Casting tends to produce somewhat weaker metal than most other forming processes, but it is useful for making elaborate shapes (like lavatory faucets) that could not be manufactured economically in any other way. Recent developments in steel casting enable the production of castings that are as strong as rolled steel shapes.

Metals can also be shaped by *machining*, which is a process of cutting unwanted material from a piece of metal to produce the desired shape. Among the most common machining operations is *milling*, in which a rotating cutting wheel is used to cut metal from a workpiece. To produce cylindrical shapes, a piece of metal is rotated against a stationary cutting tool in a *lathe*. Holes are produced by *drilling*, which is usually carried out either in a *drill press* or a lathe. Screw threads may be produced in a hole by the use of a helical cutting tool called a *tap*, and the external threads on a steel rod are cut with a *die*. (The threads on mass-produced screws and bolts are formed at high speed by special rolling machines.) Grinding and polishing machines are used to create and finish flat surfaces. Sawing, shearing, and punching operations, described in Chapter 11, are also common methods of shaping metal components.

An economical method of cutting steel of almost any thickness is with a *flame cutting torch* that combines a slender, high-temperature gas flame with a jet of pure oxygen to burn away the metal. *Plasma cutting* with a tiny supersonic jet of superheated gas that blows away the metal can give more precise cuts at thicknesses of up to 2 inches (50 mm), and *laser cutting* gives high-quality results in thin metal plates.

Sheet metal is fabricated with its own particular set of tools. Shears are used to cut metal sheets, and folds are made on large machines called *brakes*.

Joining Metal Components

Metal components may be joined either mechanically or by fusion. Most mechanical fastenings require drilled or punched holes for the insertion of screws, bolts, or rivets. Some small-diameter screws that are used with thin metal components are shaped and hardened so that they are capable of drilling and tapping as they are driven. Many sheet metal components, especially roofing sheet and ductwork, are joined primarily with interlocking, folded connections.

High-temperature fusion connections are made by *welding*, in which a gas flame or electric arc melts the metal on both sides of the joint and allows it to flow together with additional molten metal from a welding rod or consumable electrode. *Brazing* and *soldering* are lower-temperature processes in which the parent metal is not melted. Instead, a different metal with a lower melting point (bronze or brass in the case of brazing and a lead-tin alloy in the most common type of solder) is melted into the joint and bonds to the pieces that it joins. A fully welded connection is generally as strong as the pieces it connects. A soldered connection is not as strong, but it is easy to make and works well for connecting copper plumbing pipes and sheet metal roofing. As an alternative to welding or soldering, adhesives are occasionally used to join metals in certain nonstructural applications.

Common Metals Used in Building Construction

The ferrous metals include cast iron, wrought iron, steel, and stainless steel. **Cast iron** contains relatively large amounts of carbon and impurities. It is the most *brittle* (subject to sudden failure) ferrous metal. **Wrought iron** is produced by hammering semimolten iron to produce a metal with long fibers of iron interleaved with long fibers of slag. It has very low iron content, making it stronger in tension and much less brittle than cast iron. Both cast iron and wrought iron found significant use in early metal structures. But with the introduction of economical steel-making processes, the roles of both of these earlier metals were largely taken over by steel. Even the ornamental metalwork that we refer to today as "wrought iron" is frequently made of mild steel. **Steel** is discussed in some detail in Chapter 11, and its many uses are noted throughout this book. In general, all these ferrous metals are very strong, relatively inexpensive, easy to form and machine, and must be protected from corrosion.

Stainless steel, made by alloying steel with other metals, primarily chromium and nickel, forms a self-protecting oxide coating that makes it highly resistant to corrosion. It is harder to form and machine than mild steel and is more costly. It is available in attractive finishes that range from matte textures to a mirror polish. Stainless steel is frequently used in the manufacture of fasteners, roofing and flashing sheet, hardware, railings, and other ornamental metal items.

Stainless steel is available in different alloys distinguished, most importantly, by their level of corrosion resistance. *Type 304 stainless steel* is the type most commonly specified and provides adequate corrosion resistance for most applications. Type 304 stainless steel may also be referred to as Type 18-8, the two numbers referring to the percentages of chromium and nickel, respectively, in this alloy. *Type 316 stainless steel*, with higher nickel content and the addition of small amounts of molybdenum, is more corrosion resistant than Type 304. It is frequently specified for use in marine environments where salt-laden air can lead to the accelerated corrosion of less resistant stainless steel alloys. *Type 410 stainless steel* has a lower chromium content and is less corrosion resistant than the 300 series alloys. However, this alloy also has a different metallic crystal structure that, unlike the 300 series alloys, allows it to be hardened through heat treatment. Self-drilling, self-tapping stainless steel fasteners, whose threads must be tough enough to cut through structural steel or concrete, are frequently made of hardened Type 410 stainless steel.

Aluminum (spelled and pronounced *aluminium* in the British Commonwealth) is the nonferrous metal most often used in construction. Its density is about one-third that of steel and it has moderate to high strength and stiffness, depending on which of a multitude of alloys is selected. It can be hardened by cold working, and some alloys can be heat treated for increased strength. It can be hot- or cold-rolled, cast, forged, drawn, and stamped, and is particularly well adapted to extrusion (see Chapter 21). Aluminum is self-protecting from corrosion, easy to machine, and has thermal and electrical conductivities that are almost as high as those of copper. It is easily made into thin foils that find wide use in thermal insulating and vapor-retarding materials. With a mirror finish, aluminum in foil or sheet form reflects more heat and light than any other architectural material. Typical uses of aluminum in buildings include roofing and flashing sheet, ductwork, curtain wall components, window and door frames, grills, ornamental railings, siding, hardware, electrical wiring, and protective coatings for other metals, chiefly steel. Aluminum powder is used in metallic paints, and aluminum oxide is used as an abrasive in sandpaper and grinding wheels.

Copper and copper alloys are widely used in construction. Copper is slightly more dense than steel and is bright orange-red in color. When it oxidizes, it forms a self-protecting coating that ranges in color from blue-green to black, depending on the contaminants in the local atmosphere. Copper is moderately strong and can be made stronger by alloying or cold working, but it is not amenable to heat treatment. It is ductile and easy to fabricate. It has the highest thermal and electrical conductivity of any metal used in construction. It may be formed by casting, drawing, extrusion, and hot or cold rolling. The primary uses of copper in buildings are roofing and flashing sheet, piping and tubing, and wiring for electricity and communications. Copper is an alloying element in certain corrosion-resistant steels, and copper salts are used as wood preservatives.

Copper is the primary constituent of two versatile alloys, bronze and brass. **Bronze** is a reddish-gold metal that traditionally consists of 90 percent copper and 10 percent tin. Today, however, the term "bronze" is applied to a wide range of alloys that may also incorporate such metals as aluminum, silicon, manganese, nickel, and zinc. These various bronzes are found in buildings in the form of statuary, bells, ornamental metalwork, door and cabinet hardware, and weatherstripping. **Brass** is formulated of copper and zinc plus small amounts of other metals. It is usually lighter in color than bronze, more of a straw yellow, but in contemporary usage the line between brasses and bronzes has become rather indistinct, and the various brasses occur in a wide range of colors, depending on the formulation. Brass, like bronze, is resistant to corrosion. It can be polished to a high luster. It is widely used in hinges and doorknobs, weatherstripping, ornamental metalwork, screws, bolts, nuts, and plumbing faucets (where it is usually plated with chromium). On a volumetric basis, brass, bronze, and copper are expensive metals, but they are often the most economical materials for applications that require their unique combination of functional and visual properties. For greater economy, they are frequently plated electrolytically onto steel for such uses as door hinges and locksets.

Zinc is a blue-white metal that is low in strength, relatively brittle, and moderately hard. Zinc alloy sheet is used for roofing and flashing. Alloys of zinc are also used for casting small hardware parts such as doorknobs, cabinet pulls and hinges, bathroom accessories, and components of electrical fixtures. These *die castings*, which are usually electroplated with another metal such as chromium for appearance, are not especially strong, but they are economical and they can be very finely detailed.

The most important use of zinc in construction is for galvanizing, the application of a zinc coating to prevent steel from rusting. The zinc itself forms a self-protecting gray oxide coating, and even if the zinc is accidentally scratched through to the steel beneath, the zinc interacts electrochemically with the exposed steel to continue to protect the steel from corrosion—a phenomenon called *galvanic protection*. *Hot-dip galvanizing*, in which the steel parts are submerged in a molten zinc bath to produce a thick coating, is the most durable form of galvanizing. Much less durable is the thin coating produced by *electrogalvanizing*. Threaded steel fasteners and other small parts may be *mechanically galvanized*, in which zinc is fused to the steel at room temperature in a tumbler that contains zinc dust, impact media (such as ball bearings, for example), and other materials. Mechanical galvanizing produces a coating that

Metals in Architecture (continued)

is especially uniform and consistent in thickness. Steel sheet for architectural roofing is also frequently coated with an aluminum-zinc alloy coating. The aluminum provides a superior protective oxide coating, and the zinc provides galvanic protection if the coating becomes damaged and the base steel exposed. (For a more detailed discussion of galvanic action, see pages 698–700.)

Tin is a soft, ductile silvery metal that forms a self-protecting oxide layer. The ubiquitous "tin can" is actually made of sheet steel with an internal corrosion-resistant coating of tin. Tin is found in buildings primarily as a constituent of terne metal, an alloy of 80 percent lead and 20 percent tin that was used in the past as a corrosion-resistant coating for steel or stainless steel roofing sheet. Today, zinc-tin alloy coated steel and stainless steel sheets are available for use as roofing metals that are close in appearance and durability to traditional terne metal.

Chromium is a very hard metal that can be polished to a brilliant mirror finish. It does not corrode in air. It is often electroplated onto other metals for use in ornamental metalwork, bathroom and kitchen accessories, door hardware, and plumbing and lighting fixtures. It is also a major alloying ingredient in stainless steel and many other metals, to which it imparts hardness, strength, and corrosion resistance. Chromium compounds are used as colored pigments in paints and ceramic glazes.

Magnesium is a strong, remarkably lightweight metal (less than one-quarter the density of steel) that is much used in aircraft but remains too costly for general use in buildings. It is found on the construction site as a material for various lightweight tools and as an alloying element that increases the strength and corrosion resistance of aluminum.

Titanium is also low in density, about half the weight of steel, very strong, and one of the most corrosion-resistant of all metals. It is a constituent in many alloys, and its oxide has replaced lead oxide in paint pigments. Titanium is also a relatively expensive metal and has only recently begun to appear on the construction in the form of roofing sheet metal.

CSI/CSC

MasterFormat Sections for Light Gauge Steel Frame Construction

05 40 00	COLD-FORMED METAL FRAMING
05 41 00	Structural Metal Stud Framing
05 42 00	Cold-Formed Metal Joist Framing
04 44 00	Cold-Formed Metal Trusses
06 10 00	ROUGH CARPENTRY
06 16 00	Sheathing
	Gypsum Sheathing
09 20 00	PLASTERING AND GYPSUM BOARD
09 22 16	Non-Structural Metal Framing

Selected References

1. American Iron and Steel Institute. AISI Cold-Formed Steel Design Manual. 1996, Chicago.

This is an engineering reference work that contains structural design tables and procedures for light gauge steel framing.

2. International Code Council. International Residential Code for One- and Two-Family Dwellings. Falls Church, VA, 2002.

This code incorporates full design information and other code provisions, applicable throughout most of the United States, for light gauge steel frame residential construction.

Web Sites

Light Gauge Steel Frame Construction

Author's supplementary web site: **www.ianosbackfill.com/12_light_gauge_steel_frame_construction**

Center for Cold-Formed Steel Structures: **web.umr.edu/~ccfss/research&abstracts.html**

Dietrich Metal Framing: **www.dietrichindustries.com**

Steel Framing Alliance: **www.steelframingalliance.com**

United States Gypsum: **www.usg.com**

Key Terms and Concepts

light gauge steel
cold-formed metal framing
cee section
channel section
gauge
self-drilling, self-tapping screw
gypsum sheathing panel
nested member
web stiffener
nonstructural metal framing
thermal break
prescriptive requirement
ductile
alloy
heat treatment
quench
temper
anneal
cold working
galvanize
electroplating
anodize
casting
rolling
extrusion
forging
stamping
drawing
machining
milling
lathe
drilling
drilling
drill press
tap
die
flame cutting torch
plasma cutting
laser cutting
brake
welding
brazing
soldering
brittle
Types 304, 316, 410 stainless steel
die casting
galvanic protection
hot-dip galvanizing
electrogalvanizing
mechanical galvanizing

Review Questions

1. How are light gauge steel framing members manufactured?

2. How do the details for a house framed with light gauge steel members differ from those for a similar house with wood platform framing?

3. What special precautions should you take when detailing a steel-framed building to avoid excessive conduction of heat through the framing members?

4. If a building framed with light gauge steel members must be totally noncombustible, what materials would you use for subflooring and wall sheathing?

5. What is the advantage of a prescriptive building code for light gauge steel framing?

6. Compare the advantages and disadvantages of wood light frame construction and light gauge steel frame construction.

Exercises

1. Convert a set of details for a wood light frame house to light gauge steel framing.

2. Visit a construction site where light gauge steel studs are being installed. Grasp an installed stud that has not yet been sheathed at chest height and twist it clockwise and counterclockwise. How resistant is the stud to twisting? How is this resistance increased as the building is completed?

3. On this same construction site, make sketches of how electrical wiring, electric fixture boxes, and pipes are installed in metal framing.

PROJECT: Camera Obscura at Mitchell Park, Greenport, New York

ARCHITECT: SHoP/Sharples Holden Pasquarelli

The *camera obscura* is an ancient device—a room-sized projector used to display views of the room's surroundings within the camera, where these images may be viewed by the camera's occupants. In undertaking the Camera Obscura at Mitchell park, SHoP Studio accepted the nostalgic theme of the client's program, and added to it its own interests in developing cutting-edge design and construction methods.

SHoP designed and documented the Camera Obscura entirely in the form of a three-dimensional digital model. Beyond facilitating the project's unconventional geometry, the use of

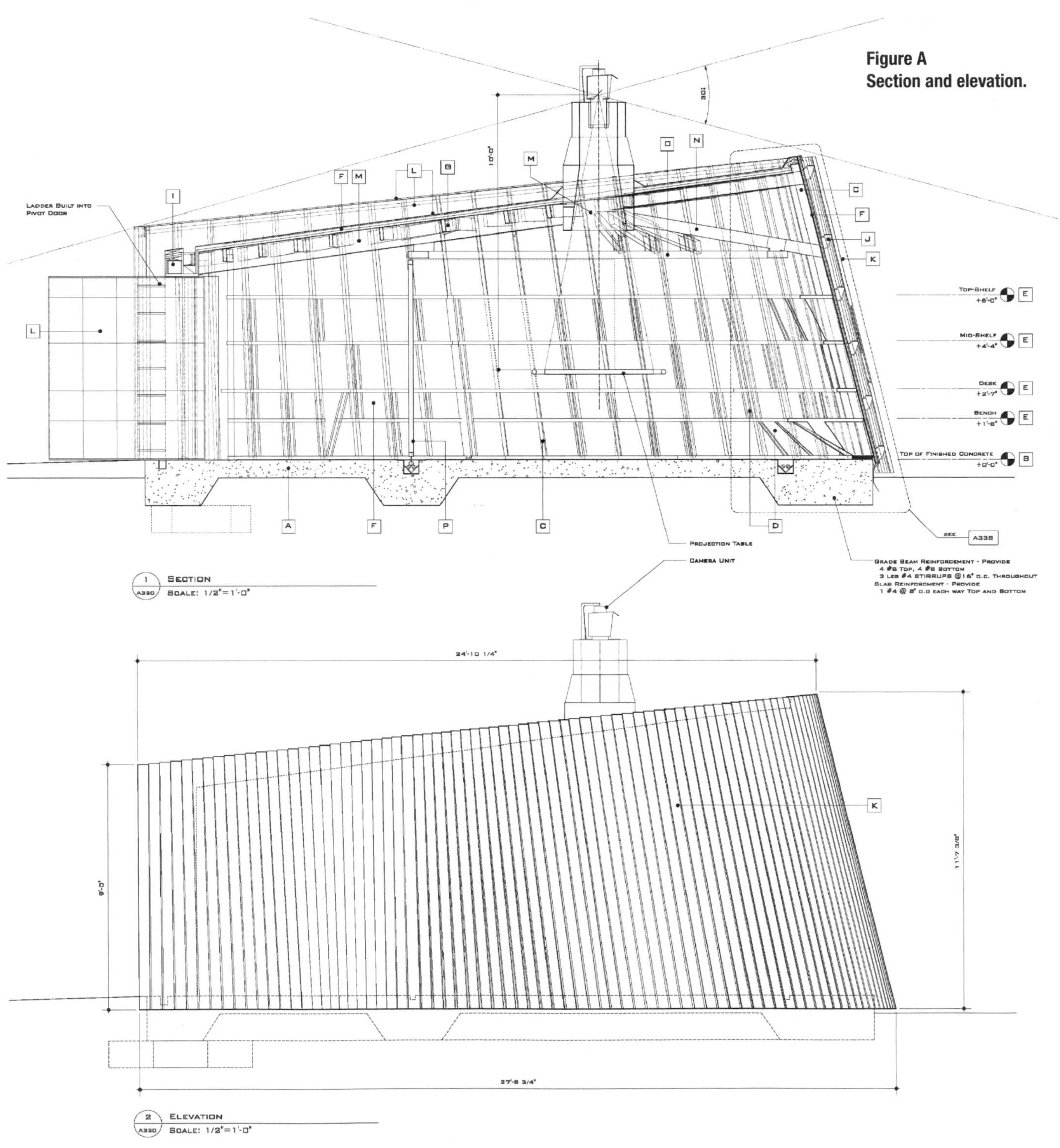

Figure A
Section and elevation.

digital modeling created significant opportunities for changing the way in which this project would be built and altering the architect's contribution to that process.

For example, as a consequence of the digital model, much of the traditional construction-phase shop drawing preparation process was stood on its head in this project. Instead of the fabricator preparing drawings for the review of the architect/engineer team, the model created by SHoP for the project design was used to generate templates that are supplied by the architect in digital form to the fabricator. The fabricator used these templates to drive automated machinery that transformed raw materials stock to cut, formed, and drilled components. Pieces were delivered to the construction site individually prelabeled, ready for assembly in the final structure.

Figure B
Cutting templates derived from the digital model.

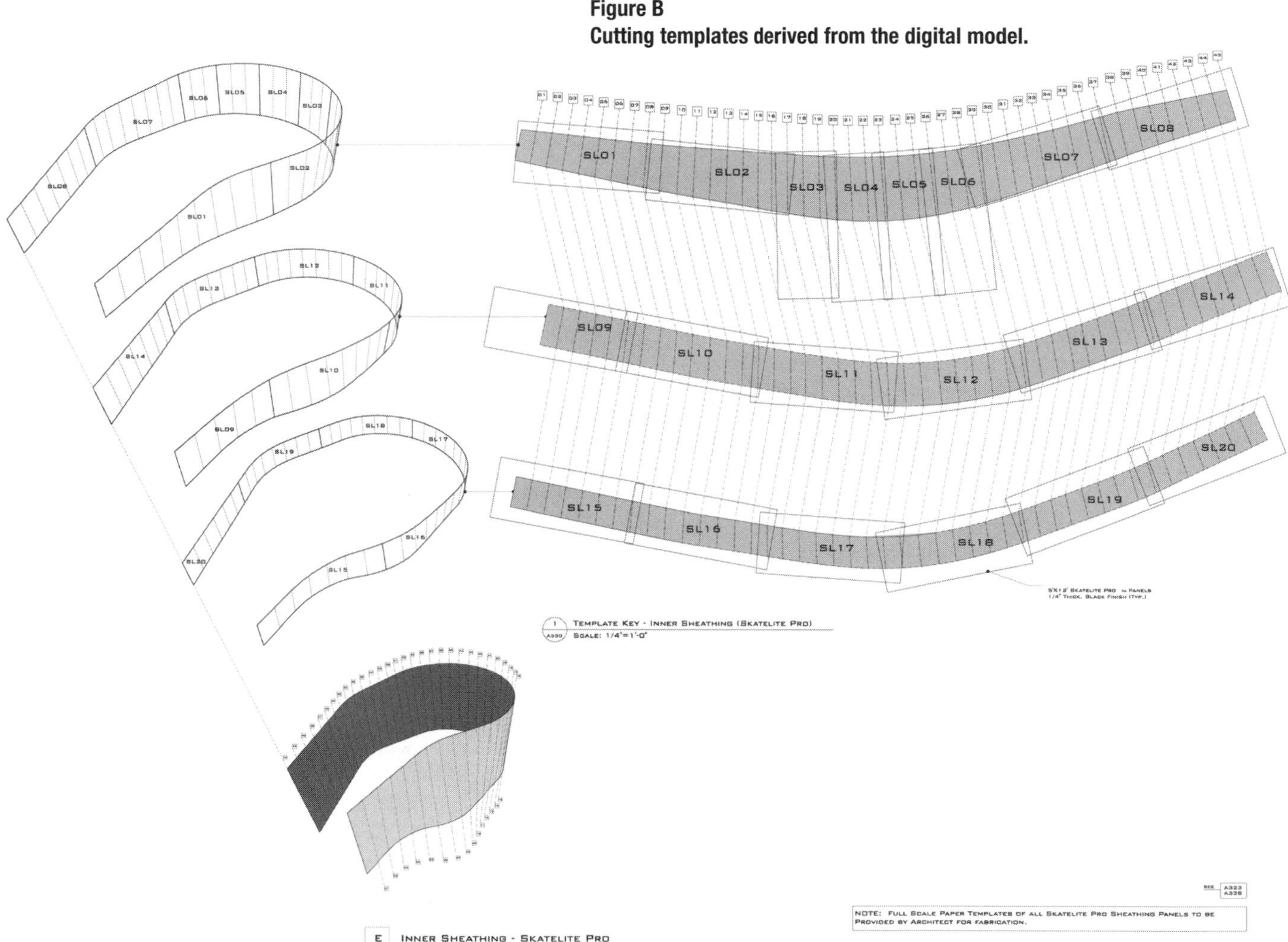

Figure C
Individually sized and shaped aluminum fins.

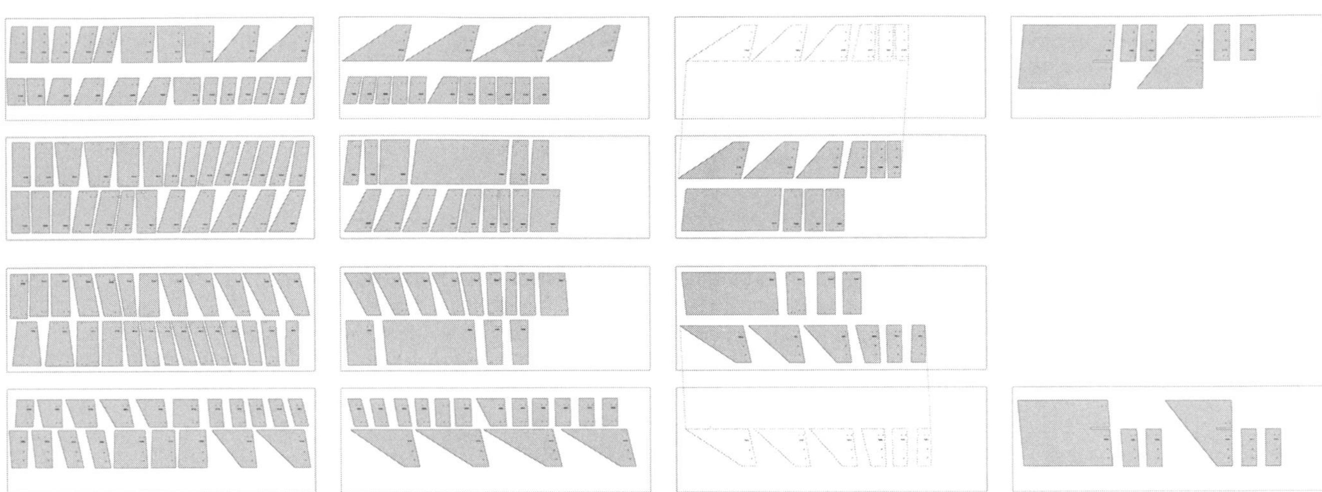

The digital building model also allowed SHoP to explore the possibilities of customization beyond what is practical with more conventional design methods. In the Camera Obscura, many of the building pieces were unique in shape and were intended for use in a single predetermined location within the building. If this proposition were undertaken using conventional construction methods, it would imply significant cost premiums. By capitalizing on the descriptive capabilities of the model coupled with automated fabrication, the costs to produce these items and to organize their assembly can be made competitive with traditional construction.

SHoP also used the digital building model to generate construction drawings that communicate how the building would be assembled in the field. For example, exploded assembly diagrams were used to study and illustrate the sequences in which systems were constructed. Cutting patterns were organized to minimize cutting time and material waste. Templates were plotted full size on paper and delivered to the building site to assist with construction layout.

SHoP's interest in creatively exploring the means of construction carried with it additional responsibilities. Because SHoP provided templates for forming various components, it assumed greater responsibility for ensuring that these components would fit properly when assembled in the field. As a consequence, SHoP worked closely with fabricators and suppliers to educate themselves regarding both the potential capabilities and the limitations of the materials with which they designed. In some instances, material properties, such as the practical bend radii of metals of various gauges, were built into the parameters of the digital model itself. Full-size mockups could be constructed on-site to verify assembly concepts and tolerances prior to fabricating the bulk of the project's components. And as with any design firm committed to improving its professional capabilities, the lessons SHoP learned from completed work are conscientiously applied to new projects.

SHoP's goal was to connect the tools of design with the techniques of construction. Note that all images shown here are taken from actual construction drawings, for a project awarded through a competitive, public bid process. With the innovative application of new design tools and a willingness to challenge the conventional professional boundaries, SHoP aims to open up new architectural possibilities. These efforts are still new, and their full potential is perhaps is not yet realized. Yet they already demonstrate how the exploration of materials and techniques of construction can be an integral part of a creative design practice.

Special thanks to SHoP/Sharples Holden Pasquarelli, and William Sharples, Principal, for assistance with the preparation of this case study.

Figure D
Assembly diagram.

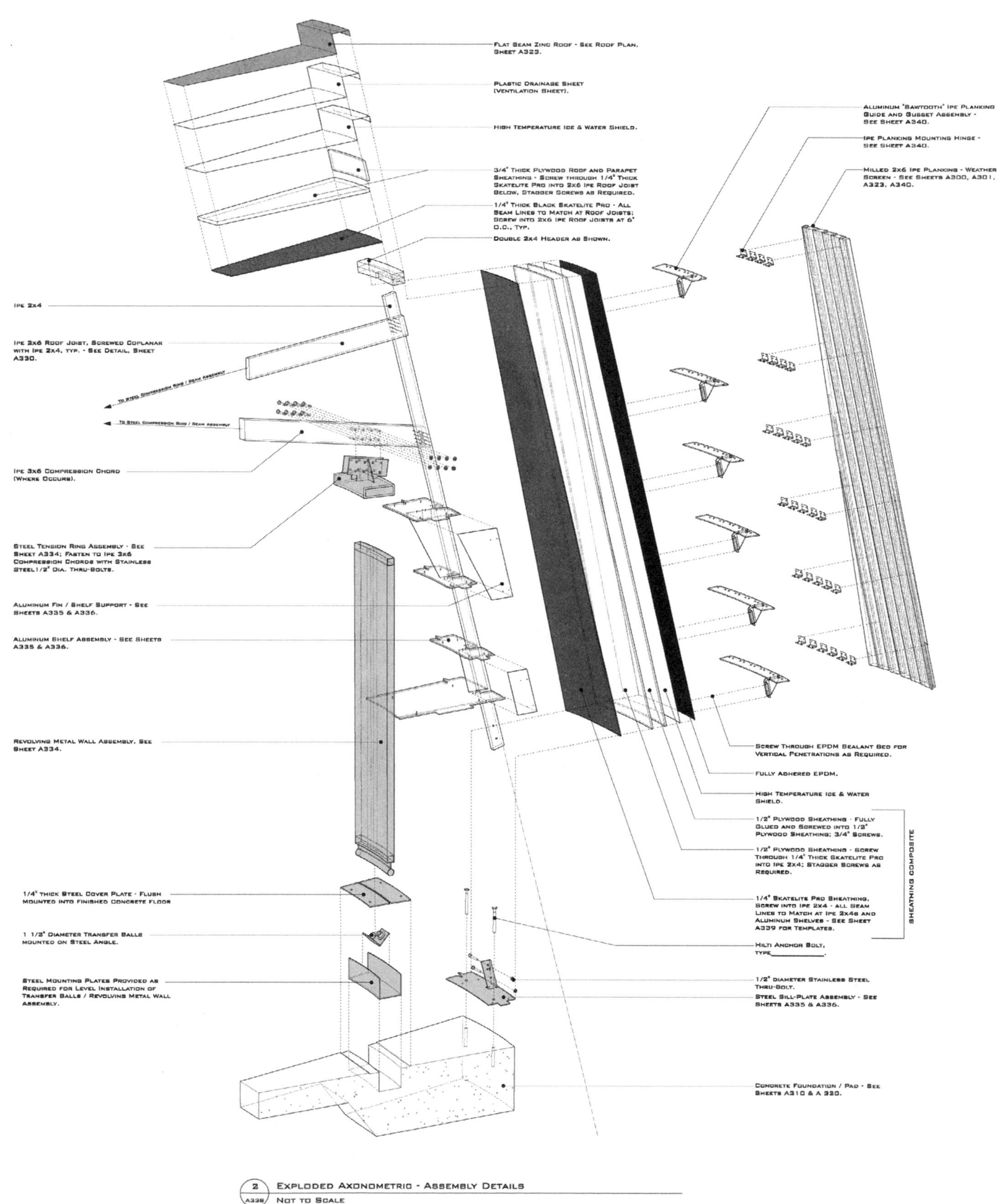

Flat Seam Zinc Roof - See Roof Plan, Sheet A323.
Plastic Drainage Sheet (Ventilation Sheet).
High Temperature Ice & Water Shield.
3/4" Thick Plywood Roof and Parapet Sheathing - Screw through 1/4" Thick Skatelite Pro into 2x6 Ipe Roof Joist Below, Stagger Screws as Required.
1/4" Thick Black Skatelite Pro - All Seam Lines to Match at Roof Joists; Screw into 2x6 Ipe Roof Joists at 6" O.C., Typ.
Double 2x4 Header as Shown.
Aluminum "Sawtooth" Ipe Planking Guide and Gusset Assembly - See Sheet A340.
Ipe Planking Mounting Hinge - See Sheet A340.
Milled 2x6 Ipe Planking - Weather Screen - See Sheets A300, A301, A323, A340.
Ipe 2x4
Ipe 2x6 Roof Joist, Screwed Coplanar with Ipe 2x4, typ. - See Detail, Sheet A330.
To Steel Compression Ring / Beam Assembly
To Steel Compression Ring / Beam Assembly
Ipe 3x6 Compression Chord (Where Occurs).
Steel Tension Ring Assembly - See Sheet A334; Fasten to Ipe 3x6 Compression Chords with Stainless Steel1/2" Dia. Thru-Bolts.
Aluminum Fin / Shelf Support - See Sheets A335 & A336.
Aluminum Shelf Assembly - See Sheets A335 & A336.
Revolving Metal Wall Assembly, See Sheet A334.
Screw Through EPDM Sealant Bed for Vertical Penetrations as Required.
Fully Adhered EPDM.
High Temperature Ice & Water Shield.
1/2" Plywood Sheathing - Fully Glued and Screwed into 1/2" Plywood Sheathing; 3/4" Screws.
1/2" Plywood Sheathing - Screw Through 1/4" Thick Skatelite Pro into Ipe 2x4; Stagger Screws as Required.
1/4" Skatelite Pro Sheathing, Screw into Ipe 2x4 - all Seam Lines to Match at Ipe 2x4s and Aluminum Shelves - See Sheet A339 for Templates.
Sheathing Composite
1/4" thick Steel Cover Plate - Flush mounted into Finished Concrete Floor
1 1/2" Diameter Transfer Balls mounted on Steel Angle.
Hilti Anchor Bolt, type________.
Steel Mounting Plates Provided as Required for Level Installation of Transfer Balls / Revolving Metal Wall Assembly.
1/2" diameter Stainless Steel Thru-Bolt.
Steel Sill-Plate Assembly - See Sheets A335 & A336.
Concrete Foundation / Pad - See Sheets A310 & A 320.
2
A336
Exploded Axonometric - Assembly Details
Not to Scale

13

CONCRETE CONSTRUCTION

A physical sciences center at Dartmouth College, built in a highly irregular space bounded by three existing buildings, typifies the potential of reinforced concrete to make expressive, highly individual buildings. *(Architects: Shepley Bulfinch Richardson and Abbott. Photograph: Ezra Stoller © ESTO)*

Concrete is the universal material of construction. According to the World Business Council for Sustainable Development, concrete is, after water, the most widely used material on earth. The raw ingredients for its manufacture are readily available in almost every part of the globe, and concrete can be made into buildings with tools ranging from a primitive shovel to a computerized precasting plant. Concrete does not rot or burn; it is relatively low in cost; and it can be used for every building purpose, from lowly pavings to sturdy structural frames to handsome exterior claddings and interior finishes.

But concrete is the only major structural material commonly manufactured on site, it has no form of its own, and it has no useful tensile strength. Before its limitless architectural potential can be realized, the designer and builder must learn to produce concrete of consistent and satisfactory quality, to combine concrete skillfully with steel reinforcing to bring out the best structural characteristics of each material, and to mold and shape it to forms appropriate to its qualities and to our building needs.

History

The ancient Romans, while quarrying limestone for mortar, accidentally discovered a silica- and alumina-bearing mineral on the slopes of Mount Vesuvius that, when mixed with limestone and burned, produced a cement that exhibited a unique property: When mixed with water and sand, it produced a mortar that could harden underwater as well as in the air. In fact, it was stronger when it hardened underwater. This mortar was also harder, stronger, much more adhesive, and cured much more quickly than the ordinary lime mortar to which they were accustomed. In time, it not only became the preferred mortar for use in all their building projects, but it also began to alter the character of Roman construction. Masonry of stone or brick came to be used to build only the surface layers of piers, walls, and vaults, and the hollow interiors were filled entirely with large volumes of the new type of mortar (Figure 13.2). We now know that this mortar contained all the essential ingredients of modern portland cement and that the Romans were the inventors of concrete construction.

Knowledge of concrete construction was lost with the fall of the Roman Empire, not to be regained until the latter part of the 18th century, when a number of English inventors began experimenting with both natural and artificially produced cements. Joseph Aspdin, in 1824, patented an artificial cement that he named *portland cement,* after English Portland limestone, whose durability as a building stone was legendary. His cement was soon in great demand, and the name "Portland" remains in use today.

Reinforced concrete, in which steel bars are embedded to resist tensile forces, was developed in the 1850s by several people simultaneously. Among them were the Frenchman J. L. Lambot, who built several reinforced concrete boats in Paris in 1854, and an American, Thaddeus Hyatt, who made and tested a number of reinforced concrete beams. But the combination of steel and concrete did not come into widespread use until a

Figure 13.1
At the time concrete is placed, it has no form of its own. This bucket of fresh concrete was filled on the ground by a transit-mix truck and hoisted to the top of the building by a crane. The worker at the right has opened the valve in the bottom of the bucket to discharge the concrete into the formwork. *(Reprinted with permission of the Portland Cement Association from* Design and Control of Concrete Mixtures, *12th edition; Photos: Portland Cement Association, Skokie, IL)*

French gardener, Joseph Monier, obtained a patent for reinforced concrete flower pots in 1867 and went on to build concrete water tanks and bridges of the new material. By the end of the 19th century, engineering design methods had been developed for structures of reinforced concrete and a number of major structures had been built. By this time, the earliest experiments in prestressing (placing the reinforcing steel under tension before the structure supports a load) had also been carried out, although it remained for Eugene Freyssinet in the 1920s to establish a scientific basis for the design of prestressed concrete structures.

Cement and Concrete

Concrete is a rocklike material produced by mixing coarse and fine *aggregates*, portland cement, and water and allowing the mixture to harden. *Coarse aggregate* is normally gravel or crushed stone, and *fine aggregate* is sand. Portland cement, hereafter referred to simply as "cement," is a fine gray powder. During the hardening, or *curing*, of concrete, the cement combines chemically with water to form strong crystals that bind the aggregates together, a process called *hydration*. During this process, considerable heat, called *heat of hydration*, is given off, and, especially as excess water evaporates from the concrete, the concrete shrinks slightly, a phenomenon referred to as *drying shrinkage*. The curing process does not end abruptly unless it is artificially interrupted. Rather, it tapers off gradually over long periods of time, though, for practical purposes, concrete is normally considered fully cured after 28 days.

In properly formulated concrete, the majority of the volume consists of coarse and fine aggregate, proportioned and graded so that the fine particles completely fill the spaces between the coarse ones (Figure 13.3). Each particle is completely coated with a paste of cement and water that bonds it fully to the surrounding particles.

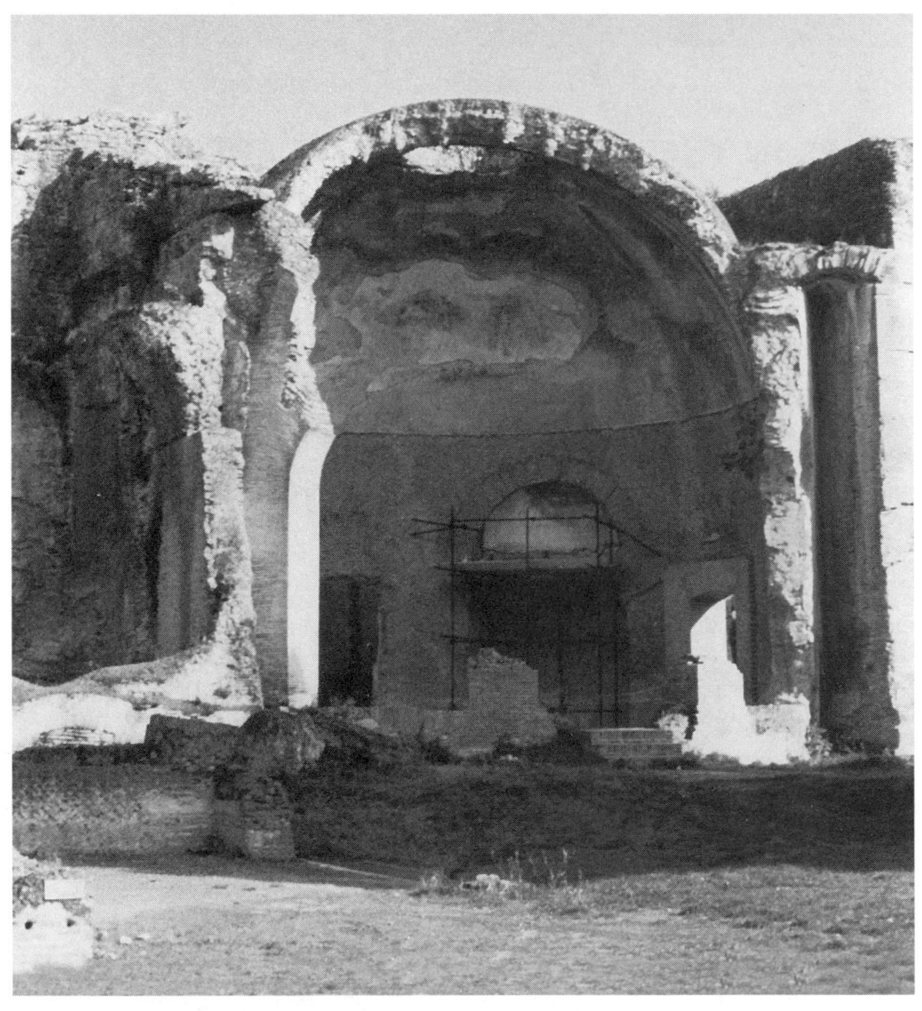

Figure 13.2
Hadrian's Villa, a large palace built near Rome between A.D. 125 and 135, used unreinforced concrete extensively for structures such as this dome.
(Photo by Edward Allen)

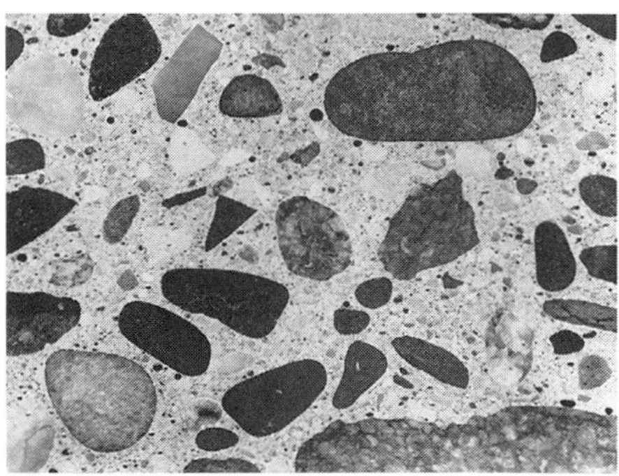

Figure 13.3
Photograph of a polished cross section of hardened concrete, showing the close packing of coarse and fine aggregates and the complete coating of every particle with cement paste. *(Reprinted with permission of the Portland Cement Association from* Design and Control of Concrete Mixtures, *12th edition; Photos: Portland Cement Association, Skokie, IL)*

Cement

Portland cement may be manufactured from any of a number of raw materials, provided that they are combined to yield the necessary amounts of lime, iron, silica, and alumina. Lime is commonly furnished by limestone, marble, marl, or seashells. Iron, silica, and alumina may be provided by clay or shale. The exact ingredients depend on what is readily available, and the recipe varies widely from one geographic region to another, often including slag or flue dust from iron furnaces, chalk, sand, ore washings, bauxite, and other minerals. To make portland cement, the selected constituents are crushed, ground, proportioned, and blended. Then they are conducted through a long, rotating kiln at temperatures of 2600 to 3000 degrees Fahrenheit (1400–1650°C) to produce *clinker* (Figures 13.4 and 13.5). After cooling, the clinker is pulverized to a powder finer than flour. Usually at this stage a small amount of gypsum is added to act as a retardant during the eventual concrete curing process. This finished powder, portland cement, is either packaged in bags or shipped in bulk. In the United States, a standard bag of cement contains 1 cubic foot (0.09 m^2) of volume and weighs 94 pounds (43 kg).

The quality of portland cement is established by ASTM C150, which identifies eight different types:

Type I	Normal
Type IA	Normal, air entraining
Type II	Moderate resistance to sulfate attack
Type IIA	Moderate sulfate resistance, air entraining
Type III	High early strength
Type IIIA	High early strength, air entraining
Type IV	Low heat of hydration
Type V	High resistance to sulfate attack

Type I cement is used for most purposes in construction. Types II

STEPS IN THE MANUFACTURE OF PORTLAND CEMENT

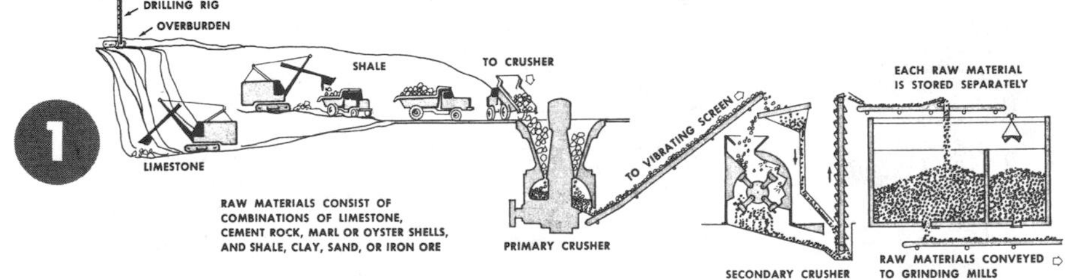

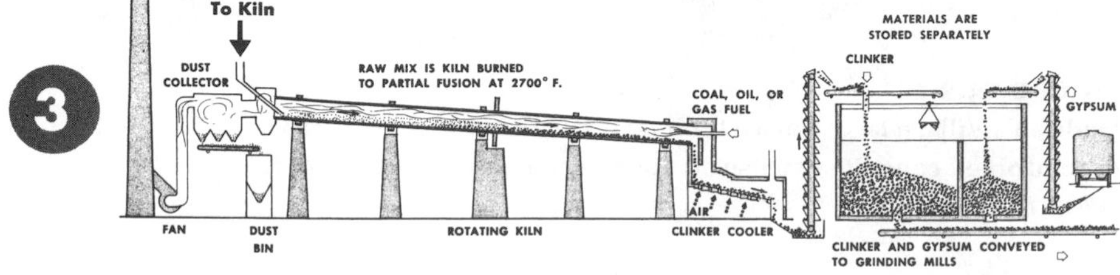

FIGURE 13.4
A rotary kiln manufacturing cement clinker. *(Reprinted with permission of the Portland Cement Association from* Design and Control of Concrete Mixtures, *12th edition; Photos: Portland Cement Association, Skokie, IL)*

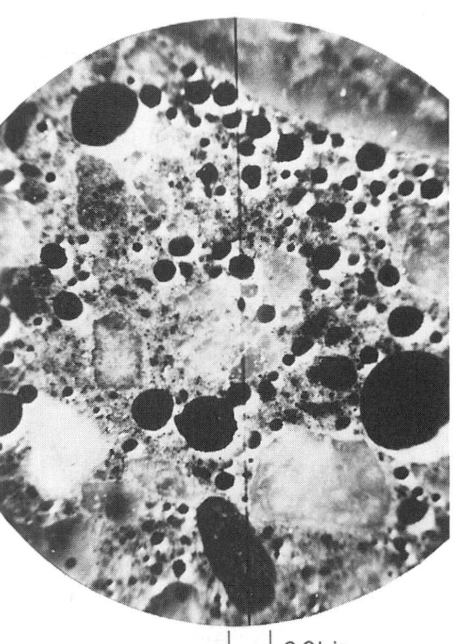

FIGURE 13.6
A photomicrograph of a small section of air-entrained concrete shows the bubbles of entrained air (0.01 inch equals 0.25 mm). *(Reprinted with permission of the Portland Cement Association from* Design and Control of Concrete Mixtures, *12th edition; Photos: Portland Cement Association, Skokie, IL)*

FIGURE 13.5
Steps in the manufacture of portland cement. *(Reprinted with permission of the Portland Cement Association from* Design and Control of Concrete Mixtures, *12th edition; Photos: Portland Cement Association, Skokie, IL)*

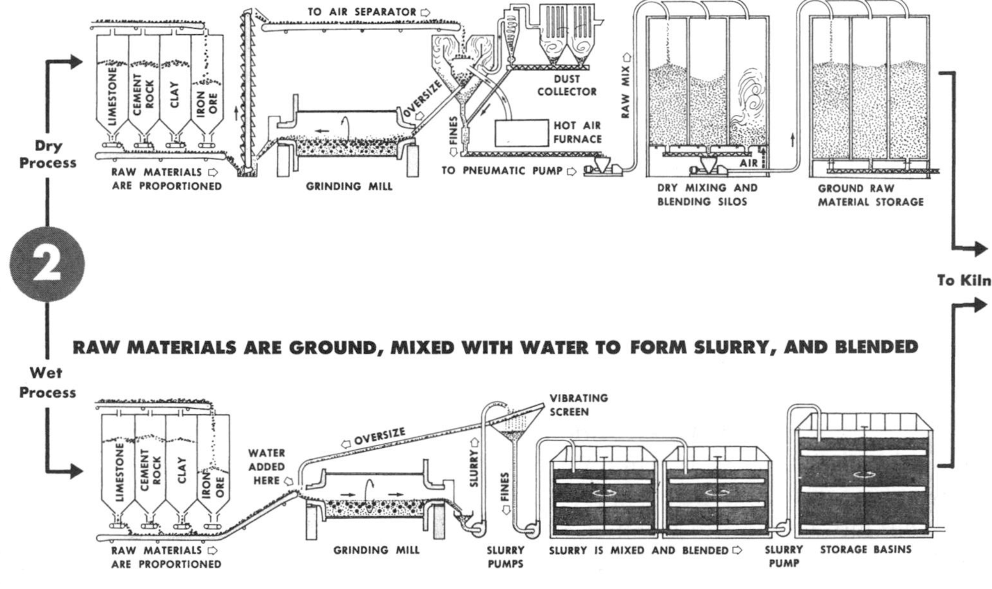

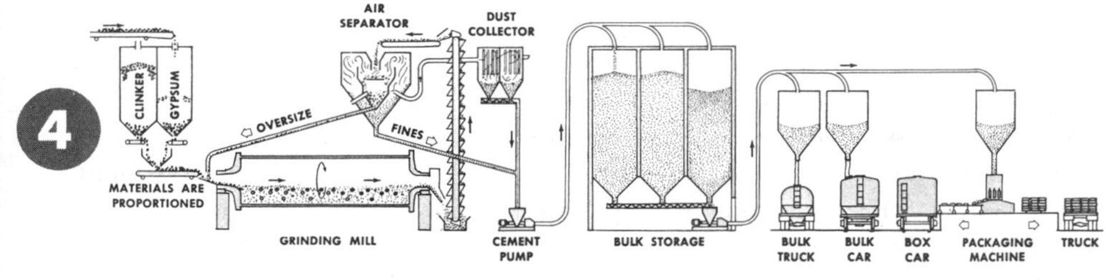

and V are used where the concrete will be in contact with water that has a high concentration of sulfates. Type III hardens more quickly than the other types and is employed in situations where a reduced curing period is desired (as may be the case in cold weather), in the precasting of concrete structural elements, or when the construction schedule must be accelerated. Type IV is used in massive structures such as dams, where the heat emitted by curing concrete may raise the temperature of the concrete to damaging levels.

Recent changes to the ASTM C150 standard allow the inclusion of ground limestone in portland cement (as an additive in the finished cement, distinct from its use as a raw ingredient in the manufacture of clinker). This will provide economic and environmental benefits, reducing consumption of raw materials and energy as well as lessening emissions of carbon dioxide and cement kiln dust.

Air-entraining cements contain ingredients that cause microscopic air bubbles to form in the concrete during mixing (Figure 13.6). These bubbles, which usually comprise 2 to 8 percent

Considerations of Sustainability in Concrete Construction

- Worldwide each year, the making of concrete consumes 1.6 billion tons (1.5 billion metric tons) of portland cement, 10 billion tons (9 billion metric tons) of sand and rock, and 1 billion tons (0.9 billion metric tons) of water, making the concrete industry the largest user of natural resources in the world.
- The quarrying of the raw materials for concrete in open pits can result in soil erosion, pollutant runoff, habitat loss, and ugly scars on the landscape.
- Concrete construction also uses large quantities of other materials—wood, wood panel products, steel, aluminum, plastics—for formwork and reinforcing.
- The total energy embodied in a pound of concrete varies, especially with the design strength. This is because higher-strength concrete relies on a greater proportion of portland cement in its mix, and the energy required to produce portland cement is very high in comparison to concrete's other ingredients. For average-strength concrete, the embodied energy ranges from about 200 to 300 BTU per pound (0.5-0.7 MJ/kg).
- There are various useful approaches to increasing the sustainability of concrete construction:
    - Use waste materials from other industries, such as fly ash from power plants, slag from iron furnaces, copper slag, foundry sand, mill scale, sandblasting grit, and others, as components of cement and concrete.
    - Use concrete made from locally extracted materials and local processing plants to reduce the transportation of construction materials over long distances.
    - Minimize the use of materials for formwork and reinforcing.
    - Reduce energy consumption, waste, and pollutant emissions from every step of the process of concrete construction, from quarrying of raw materials through the eventual demolition of a concrete building.
    - In regions where the quality of the construction materials is low, improve the quality of concrete so that concrete buildings will last longer, thus reducing the demand for concrete and the need to dispose of demolition waste.

Portland Cement

- The production of portland cement is by far the largest user of energy in the concrete construction process, accounting for about 85 percent of the total energy required. Portland cement production also accounts for roughly 5 percent of all carbon dioxide gas generated by human activities worldwide and about 1.5 percent of such emissions in North America.
- Since 1970, the North American cement industry has reduced the amount of energy expended in cement production by one-third, and the industry continues to work toward further reductions.
- The manufacture of cement produces large amounts of air pollutants and dust. For every ton of cement clinker produced, almost a ton of carbon dioxide, a greenhouse gas, is released into the atmosphere. Cement production accounts for approximately 1.5 percent of carbon dioxide emissions in the United States and 5 percent of carbon dioxide emissions worldwide.
- In the past 35 years, the emission of particulates from cement production has been reduced by more than 90 percent.
- The cement industry is committed to reducing greenhouse gas emissions per ton of product by 10 percent from 1990 levels by the year 2020. According to the Portland Cement Association, over concrete's lifetime, it reabsorbs roughly half of the carbon dioxide released during the original cement manufacturing process.
- The amount of portland cement used as an ingredient in concrete, and as a consequence, the energy required to produce the concrete, can be substantially reduced by the addition of certain industrial waste materials with cementing properties to the concrete mix. Substituting such supplementary cementitious materials, including fly ash, silica fume, and blast furnace slag, for up to half the portland cement in the concrete, can result in reductions in embodied energy of as great as one-third.
- When added to concrete, fly ash is most commonly substituted for portland cement at rates of between 15 and 25 percent. Mixes with even higher replacement rates, called *high-volume-fly-ash (HVFA) concrete,* are also finding increased acceptance. Concrete mixed with fly ash as an ingredient gains other benefits as well: It needs less water than normal concrete, its heat of hydration is lower, and it shrinks less, all characteristics that lead to a denser, more durable product. Research is underway to develop concrete mixes in which fly ash completely replaces all portland cement.
- Waste materials from other industries can also be used as cementing agents—wood ash and rice-husk ash are two examples. Used motor oil and used rubber vehicle tires can be employed as fuel in cement kilns. And while consuming waste products from other industries, a cement manufacturing plant can, if efficiently operated, generate virtually no solid waste itself.

Aggregates and Water

- Sand and crushed stone come from abundant sources in many parts of the world, but high-quality aggregates are becoming scarce in some countries.
- In rare instances, aggregate in concrete has been found to be a source of radon gas. Concrete itself is not associated with indoor air quality problems.
- Waste materials such as crushed, recycled glass, used foundry sand, and crushed, recycled concrete can substitute for a portion of the conventional aggregates in concrete.
- Water of a quality suitable for concrete is scarce in many developing countries. Concretes that use less water by using superplasticizers, air entrainment, and fly ash could be helpful.

Wastes

- A significant percentage of fresh concrete is not used because the truck that delivers it to the building site contains more than is needed for the job. This concrete is often dumped on the site, where it hardens and is later removed and taken to a landfill for disposal. An empty transit-mix truck must be washed out after transporting each batch, which produces a substantial volume of water that contains portland cement particles, admixtures, and aggregates. These wastes can be recovered and recycled as aggregates and mixing water, but more concrete suppliers need to implement schemes for doing this.

Formwork

- Formwork components that can be reused many times have a clear advantage over single-use forms, which represent a large waste of construction material.
- Form release compounds and curing compounds should be chosen for low volatile organic compound content and biodegradability.
- Insulating concrete forms eliminate most temporary formwork and produce concrete walls with high thermal insulating values.

Reinforcing

- In North America, reinforcing bars are made almost entirely from recycled steel scrap, primarily junked automobiles. This reduces resource depletion and energy consumption significantly.

Demolition and Recycling

- When a concrete building is demolished, its reinforcing steel can be recycled.
- In many if not most cases, fragments of demolished concrete can be crushed, sorted, and used as aggregates for new concrete. At present, however, most demolished concrete is buried on the site, used to fill other sites, or dumped in a landfill.

Green Uses of Concrete

- Pervious concrete, made with coarse aggregate only, can be used to make porous pavings that allow stormwater to filter into the ground, helping to recharge aquifers and reduce stormwater runoff.
- Concrete is a durable material that can be used to construct buildings that are long-lasting and suitable for adaptation and reuse, thereby reducing the environmental impacts of building demolition and new construction.
- In brownfield development, concrete fill materials can be used to stabilize soils and reduce leachate concentrations.
- Where structured parking garages (often constructed of concrete) replace surface parking, open space is preserved.
- Concrete's thermal mass can be exploited to reduce building heating and cooling costs by storing excess heat during overheated periods of the day or week and releasing it back to the interior of the building during underheated periods.
- Lighter-colored concrete paving reflects more solar radiation than darker asphalt paving, leading to lower paving surface temperatures and reduced urban heat island effects.
- Interior concrete slabs made with white concrete can improve illumination, visibility, and worker safety within interior spaces without the expense or added energy consumption of extra light fixtures or increasing the light output from existing fixtures. White concrete is made with white cement and white aggregates.
- Photocatalytic agents can be added to concrete used in the construction of roads and buildings. In the presence of sunlight, the concrete chemically breaks down carbon monoxide, nitrogen oxide, benzene, and other air pollutants.

of the volume of the finished concrete, improve workability during placement of the concrete and, more importantly, greatly increase the resistance of the cured concrete to damage caused by repeated cycles of freezing and thawing. Air-entrained concrete is commonly used for pavings and exposed architectural concrete in cold climates. With appropriate adjustments in the formulation of the mix, air-entrained concrete can achieve the same structural strength as normal concrete.

White portland cement is produced by controlling the quantities of certain minerals, such as oxides of iron and manganese, found in the ingredients of cement, that contribute to cement's usual gray color. White portland cement is used for architectural applications to produce concrete that is lighter and more uniform in color or, when combined with other coloring agents, to enhance the appearance of integrally colored concrete.

Aggregates and Water

Because aggregates make up roughly three-quarters of the volume of concrete, the structural strength of a concrete is heavily dependent on the quality of its aggregates. Aggregates for concrete must be strong, clean, resistant to freeze-thaw deterioration, chemically stable, and properly graded for size distribution. An aggregate that is dusty or muddy will contaminate the cement paste with inert particles that weaken it, and an aggregate that contains any of a number of chemicals from sea salt to organic compounds can cause problems ranging from corrosion of reinforcing steel to retardation of the curing process and ultimate weakening of the concrete. A number of standard ASTM laboratory tests are used to assess the various qualities of aggregates.

Size distribution of aggregate particles is important because a range of sizes must be included and properly proportioned in each concrete mix to achieve close packing of the particles. A concrete aggregate is graded for size by passing a sample of it through a standard assortment of sieves with diminishing mesh spacings, then weighing the percentage of material that passes through each sieve. This test makes it possible to compare the particle size distribution of an actual aggregate with that of an ideal aggregate. Size of aggregate is also significant because the largest particle in a concrete mix must be small enough to pass easily between the most closely spaced reinforcing bars and to fit easily into the formwork. In general, the maximum aggregate size should not be greater than three-fourths of the clear spacing between bars or one-third the depth of a slab. For very thin slabs and toppings, a ⅜-inch (9-mm) maximum aggregate diameter is often specified. A ¾-inch or 1½-inch (19-mm or 38-mm) maximum size is common for much slab and structural work, but aggregate diameters up to 6 inches (150 mm) are used in dams and other massive structures. Producers of concrete aggregates sort their product for size using a graduated set of screens and can furnish aggregates graded to order.

Lightweight aggregates are used instead of sand and crushed stone for various special types of concrete. *Structural lightweight aggregates* are made from minerals such as shale. The shale is crushed to the desired particle sizes, then heated in an oven to a temperature at which the shale becomes plastic in consistency. The small amount of water that occurs naturally in the shale turns to steam and "pops" the softened particles like popcorn. Concrete made from this *expanded shale aggregate* has a density about 20 percent less than that of normal concrete, yet it is nearly as strong. Nonstructural lightweight concretes are made for use in insulating roof toppings that have densities only one-fourth to one-sixth that of normal concrete. The aggregates in these concretes are usually expanded mica (*vermiculite*) or expanded volcanic glass (*perlite*), both produced by processes much like that used to make expanded shale. However, both of these aggregates are much less dense than expanded shale, and the density of the concretes in which they are used is further reduced by admixtures that entrain large amounts of air during mixing.

FIGURE 13.7
Taking a sample of coarse aggregate from a crusher yard for testing.
(Reprinted with permission of the Portland Cement Association from Design and Control of Concrete Mixtures, *12th edition; Photos: Portland Cement Association, Skokie, IL)*

ASTM standard C1602 defines the requirements for mixing water for concrete. Generally, water must be free of harmful substances, especially organic material, clay, and salts such as chlorides and sulfates. Water that is suitable for drinking has traditionally been considered suitable for making concrete.

Supplementary Cementitious Materials

Various mineral products, called *supplementary cementitious materials (SCMs),* may be added to concrete mixtures as a substitute for some portion of the portland cement to achieve a range of benefits. Supplementary cementitious materials are classified as either pozzolans or hydraulic cements.

Pozzolans are materials that react with the calcium hydroxide in wet concrete to form cementing compounds. They include:

- *Fly ash,* a fine powder that is a waste product from coal-fired power plants, increases concrete strength, decreases permeability, increases sulfate resistance, reduces temperature rise during curing, reduces mixing water, and improves pumpability and workability of concrete. Fly ash also reduces concrete drying shrinkage.
- *Silica fume,* also known as *microsilica,* is a powder that is approximately 100 times finer than portland cement, consisting mostly of silicon dioxide. It is a byproduct of electronic semiconductor chip manufacturing. When added to a concrete mix, it produces extremely high-strength concrete that also has very low permeability.
- *Natural pozzolans,* mostly derived from shales or clays, are used for purposes such as reducing the internal temperature of curing concrete, reducing the reactivity of concrete with aggregates containing sulfates, or improving the workability of concrete. *High reactivity metakaolin* is a unique white-colored natural pozzolan that enhances the brilliance of white or colored concrete while also improving the material's workability, strength, and density. These characteristics make it especially well suited as an ingredient in exposed architectural concrete applications where appearance and finish quality are critical.

Blast furnace slag (also called *slag cement*), a byproduct of iron manufacture, is a *hydraulic cement,* meaning that, like portland cement, it reacts directly with water to form a cementitious compound. It may be added to concrete mixes to improve workability, increase strength, reduce permeability, reduce temperature rise during curing, and improve sulfate resistance.

Supplementary cementitious materials may be added to portland cement during the cement manufacturing process, in which case the resulting product is called a *blended cement,* or they may be added to the concrete mix at the batch plant. The use of supplementary cementitious materials also enhances the sustainability of concrete by reducing reliance on more energy-intensive portland cement and, in many cases, by making productive use of waste products from other industrial manufacturing processes. Half or more of the concrete produced in North America includes some supplementary cementitious materials in its mix.

Admixtures

Ingredients other than cement and other cementitious materials, aggregates, and water, broadly referred to as *admixtures,* are often added to concrete to alter its properties in various ways:

- *Air-entraining admixtures* increase the workability of the wet concrete, reduce freeze-thaw damage, and, when used in larger amounts, create very lightweight nonstructural concretes with thermal insulating properties.
- *Water-reducing admixtures* allow a reduction in the amount of mixing water while retaining the same workability, which results in a higher-strength concrete.
- *High-range water-reducing admixtures,* also known as *superplasticizers,* are organic compounds that transform a stiff concrete mix into one that flows freely into the forms. They are used either to facilitate placement of concrete under difficult circumstances or to reduce the water content of a concrete mix in order to increase its strength.
- *Accelerating admixtures* cause concrete to cure more rapidly, and *retarding admixtures* slow its curing to allow more time for working with the wet concrete.
- *Workability agents* improve the plasticity of wet concrete to make it easier to place in forms and finish. They include pozzolans and air-entraining admixtures, along with certain fly ashes and organic compounds.
- *Shrinkage-reducing admixtures* reduce drying shrinkage and the cracking that results.
- *Corrosion inhibitors* are used to reduce rusting of reinforcing steel in structures that are exposed to road deicing salts or other corrosion-causing chemicals.
- *Freeze protection admixtures* allow concrete to cure satisfactorily at temperatures as low as 20 degrees Fahrenheit (7°C).
- *Extended set-control admixtures* may be used to delay the curing reaction in concrete for any period up to several days. They include two components: The stabilizer component, added at the time of initial mixing, defers the onset of curing indefinitely; the activator component, added when desired, reinitiates the curing process.
- *Coloring agents* are dyes and pigments used to alter and control the color of concrete for building components whose appearance is important.

Making and Placing Concrete

The quality of cured concrete is measured by any of several criteria, depending on its end use. For structural columns, beams, and slabs, compressive strength and stiffness are important. For pavings and floor slabs, flatness, surface smoothness, and abrasion resistance are also important. For pavings and exterior concrete walls, a high degree of weather resistance is required. Watertightness is important in concrete tanks, dams, and walls. Regardless of the criterion to which one is working, however, the rules for making high-quality concrete are much the same: Use clean, sound ingredients; mix them in the correct proportions; handle the wet concrete properly to avoid segregating its ingredients; and cure the concrete carefully under controlled conditions.

Proportioning Concrete Mixes

The design of concrete mixtures is a science that can be described here only in its broad outlines. The starting point of any mix design is to establish the desired workability characteristics of the wet concrete, the desired physical properties of the cured concrete, and the acceptable cost of the concrete, keeping in mind that there is no need to spend money to make concrete better than it needs to be for a given application. Concretes with ultimate compressive strengths as low as 2000 psi (13.8 MPa) are satisfactory for some foundation elements. Concretes with ultimate compressive strengths of 20,000 psi (140 MPa) and more, produced with the aid of silica fume, fly ash, and superplasticizer admixtures, are currently being employed in the columns of some high-rise buildings. Acceptable workability is achievable at any of these strength levels.

Given a proper gradation of satisfactory aggregates, the strength of cured concrete is primarily dependent on the amount of cement in the mix and on the *water–cement (w-c) ratio.* Although water is required as a reactant in the curing of concrete, much more water must be added to a concrete mix than is needed for the hydration of the cement to give the wet concrete the necessary fluidity and plasticity for placing and finishing. The extra water eventually evaporates from the concrete, leaving microscopic voids that reduce the strength and surface qualities of the concrete (Figure 13.8). For common concrete applications, absolute water–cement ratios range from about 0.45 to 0.60 by weight, meaning that the weight of the water in the mix does not exceed 45 to 60 percent of the weight of the portland cement. Relatively high water–cement ratios are often favored by concrete workers because they produce a fluid mixture that is easy to place in the forms, but the resulting concrete is likely to be deficient in strength and surface qualities. Lower water–cement ratios make concrete that is denser and strongr and that shrinks less during curing. But unless air-entraining or water-reducing admixtures are included in a low water–cement ratio mix to improve its workability, the concrete will not flow easily into the forms, it will have large voids, and it will finish poorly. It is important that concrete be formulated with the right quantity of water for each situation, enough to ensure workability but not enough to adversely affect the properties of the cured material.

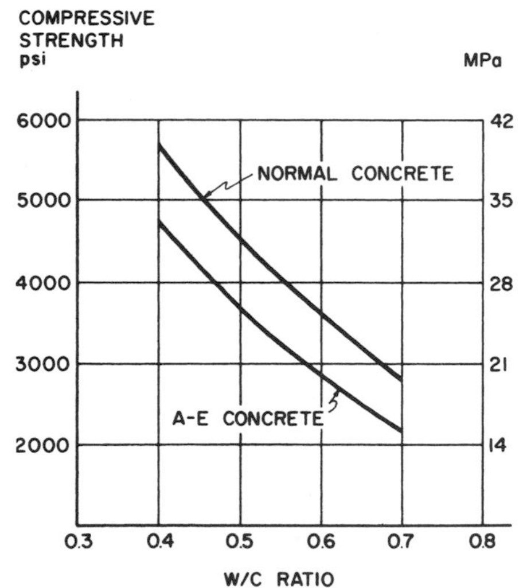

Figure 13.8
The effect of the water–cement ratio on the strength of concrete. "A-E concrete" on the graph refers to air-entrained concrete. *(Reprinted with permission of the Portland Cement Association from* Design and Control of Concrete Mixtures, *12th edition; Photos: Portland Cement Association, Skokie, IL)*

Most concrete in North America is proportioned at central batch plants, using laboratory equipment and engineering knowledge to produce concrete of the proper quality for each project. The concrete is *transit mixed* en route in a rotating drum on the back of a truck so that it is ready to pour by the time it reaches the job site (Figures 13.9 and 13.10). For very small jobs, concrete may be mixed at the job site, either in a small power-driven mixing drum or on a flat surface with shovels. For these small jobs, where the quality of the finished concrete generally does not need to be precisely controlled, proportioning is usually done by rule of thumb. Typically, the dry ingredients are measured volumetrically, using a shovel as a measuring device, in proportions such as one shovel of cement to two of sand to three of gravel, with enough water to make a wet concrete that is neither soupy nor stiff.

Each load of transit-mixed concrete is delivered with a certificate from the batch plant that lists its ingredients and their proportions. As a further check on quality, a *slump test* may be performed at the time of pouring to determine if the desired degree of workability has been achieved without making the concrete too wet (Figures 13.11 and 13.12). For structural concrete, standard test cylinders are also poured from each truckload. Within 48 hours of pouring, the cylinders are taken to a testing laboratory, cured for a specified period under

FIGURE 13.9
Charging a transit-mix truck with measured quantities of cement, aggregates, admixtures, and water at a central batch plant. *(Reprinted with permission of the Portland Cement Association from* Design and Control of Concrete Mixtures, *12th edition; Photos: Portland Cement Association, Skokie, IL)*

FIGURE 13.10
A transit-mix truck discharges its concrete, which was mixed en route in the rotating drum, into a truck-mounted concrete pump, which forces it through a hose to the point in the building at which it is being poured. *(Reprinted with permission of the Portland Cement Association from* Design and Control of Concrete Mixtures, *12th edition; Photos: Portland Cement Association, Skokie, IL)*

standard conditions, and tested for compressive strength (Figure 13.13). If the laboratory results are not up to the required standard, test cores are drilled from the actual members made from the questionable batch of concrete. If the strength of these core samples is also deficient, the contractor may be required to cut out the defective concrete and replace it. Frequently, test cylinders are also cast and cured on the construction site under the same conditions as the concrete in the forms; these may then be tested to determine when the concrete is strong enough to allow removal of forms and temporary supports.

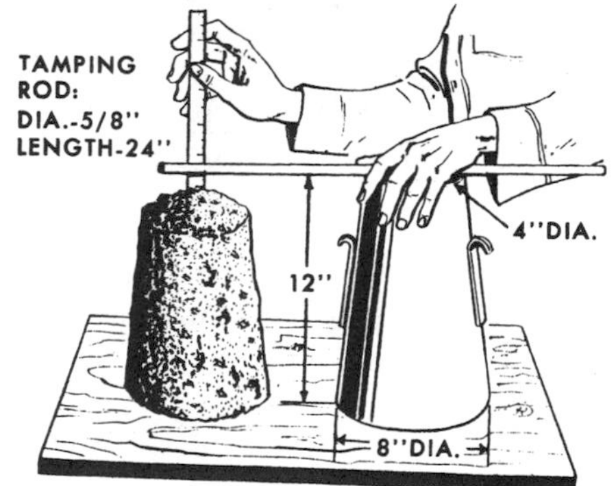

FIGURE 13.11
Illustration of the concrete slump test. The hollow metal cone is filled with concrete and tamped with the rod according to a standard procedure. The cone is carefully lifted off, allowing the wet concrete to sag, or slump, under its own weight. The slump in inches is measured in the manner shown. *(From the U.S. Department of Army, Concrete, Masonry, and Brickwork)*

FIGURE 13.12
Photograph of slump being measured. *(Reprinted with permission of the Portland Cement Association from* Design and Control of Concrete Mixtures, *12th edition; Photos: Portland Cement Association, Skokie, IL)*

FIGURE 13.13
Inserting a standard concrete test cylinder into a structural testing machine, where it will be crushed to determine its strength. *(Reprinted with permission of the Portland Cement Association from* Design and Control of Concrete Mixtures, *12th edition; Photos: Portland Cement Association, Skokie, IL)*

FIGURE 13.14
Concrete being placed in a basement floor slab with the aid of a concrete pump. Concrete can be pumped for long horizontal distances and many stories into the air. Note also the rather substantial rakers that brace the wall of the excavation. *(Reprinted with permission of the Portland Cement Association from* Design and Control of Concrete Mixtures, *12th edition; Photos: Portland Cement Association, Skokie, IL)*

Handling and Placing Concrete

Freshly mixed concrete is not a liquid but a *slurry*, a semistable mixture of solids suspended in liquid. If it is vibrated excessively, moved horizontally for long distances in the forms, or dropped through constrained spaces, it is likely to *segregate*, which means that the coarse aggregate works its way to the bottom of the form and the water and cement paste rise to the top. The result is concrete of nonuniform and generally unsatisfactory properties. Segregation is prevented by depositing the concrete, fresh from the mixer, as close to its final position as possible. If concrete must be moved a large horizontal distance to reach inaccessible areas of the formwork, it should be pumped through hoses (Figure 13.14) or conveyed in buckets or buggies, rather than pushed across or through the formwork. If concrete must be dropped a distance of more than 3 to 5 feet (1 m or so), care must be taken to ensure that the concrete can fall freely, without obstruction, so that segregation will not occur, or it should be deposited through *dropchutes* that break the fall of the concrete.

Concrete must be *consolidated* in the forms to eliminate trapped air and to fill completely the space around the reinforcing bars and in all corners of the formwork. This may be done by repeatedly thrusting a rod, spade, or immersion-type vibrator into the concrete at closely spaced intervals throughout the formwork. Excessive agitation of the concrete must be avoided, however, or segregation will occur.

Self-consolidating concrete (SCC), a concrete that fills forms completely without requiring vibration or any other method of consolidation, is a more recent development. It is formulated with more fine aggregates than coarse ones, a reversal of the usual proportions, and it includes special superplasticizing admixtures based on polycarboxylate ethers and, in some cases, other viscosity-modifying agents. The result is a concrete that flows freely, yet does not allow its coarse aggregate to sink to the bottom of the mix. Self-consolidating concrete may be used where forms are crowded with steel reinforcing, making consolidation of conventional concrete difficult. The consistent surface characteristics and crisp edges produced by self-consolidating concrete make it well suited to the production of high-finish-quality architectural concrete. By eliminating the separate consolidation step and allowing more rapid placement, self-consolidating concrete can improve productivity in precast concrete and large-volume sitecast concrete operations. However, formwork costs for self-consolidating concrete may be higher than those for conventional concrete, as the greater fluid pressures exerted by the freely flowing material require forms that are especially stiff and strong.

Curing Concrete

Because concrete cures by hydration, the chemical bonding of the water and cement, and not by simple drying, it is essential that it be kept moist until its required strength is achieved. The curing reaction takes place over a very long period of time, but concrete is commonly designed on the basis of the strength that it reaches after 28 days. If it is allowed to dry out at any point during this time period, the strength of the resulting concrete will be reduced, and its surface hardness and durability are likely to be adversely affected (Figure 13.15). Concrete cast in formwork is protected from dehydration on most surfaces by the formwork, but the top surfaces must be kept moist by repeatedly spraying or flooding with water, by covering with moisture-resistant sheets of paper or film, or by spraying on a curing compound that seals the surface of the concrete against loss of moisture. These measures are even more important for concrete slabs, whose large surface areas make them especially susceptible to premature drying. Such drying is a particular danger when slabs are poured in hot or windy weather, which can cause the surface of the pour to dry out and crack even before the concrete begins to cure. Temporary windbreaks may have to be erected, shade may have to be provided, evaporation retarders may be added to the concrete, and frequent fogging of the air directly over the surface of the slab with a fine spray of water may be required until the slab is hard enough to be finished and covered or sprayed with curing compound.

At low temperatures, the curing reaction in concrete proceeds much more slowly. If concrete reaches subfreezing temperatures while curing, the curing reaction stops

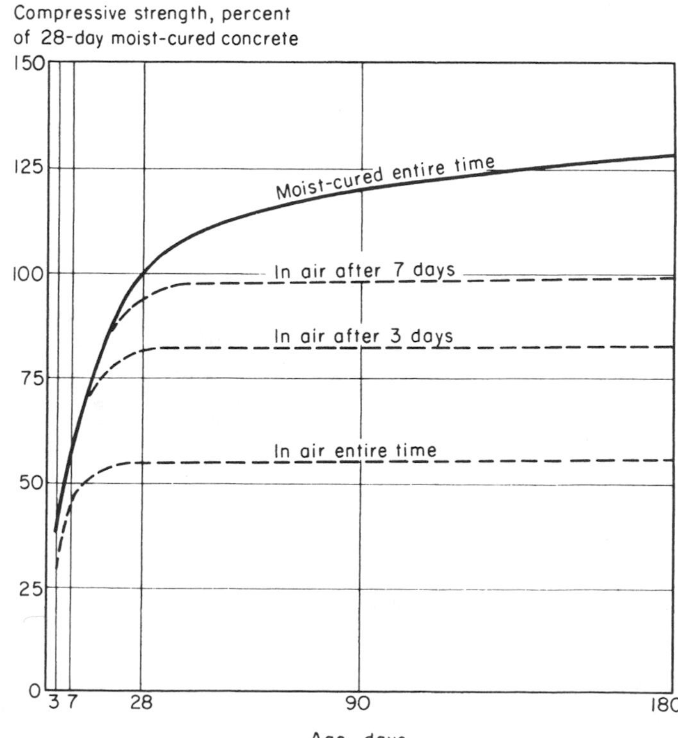

FIGURE 13.15
The growth of compressive strength in concrete over time. Moist-cured concrete is still gaining strength after 6 months, whereas air-dried concrete virtually stops gaining strength altogether. *(Reprinted with permission of the Portland Cement Association from* Design and Control of Concrete Mixtures, *12th edition; Photos: Portland Cement Association, Skokie, IL)*

completely until the temperature of the concrete rises above the freezing mark. It is important that the concrete be protected from low temperatures and especially from freezing until it is fully cured. If freshly poured concrete is covered and insulated, its heat of hydration is often sufficient to maintain an adequate temperature in the concrete even at fairly low air temperatures. Under more severe winter conditions, the ingredients of the concrete may have to be heated before mixing, and both a temporary enclosure and a temporary source of heat may have to be provided during placing and curing.

In very hot weather, the hydration reaction is greatly accelerated, and concrete may begin curing before there is time to place and finish it. This tendency can be controlled by using cool ingredients and, under extreme conditions, by replacing some of the mixing water with an equal quantity of crushed ice, making sure that the ice has melted fully and the concrete has been thoroughly mixed before placing. Another method of cooling concrete is to bubble liquid nitrogen through the mixture at the batch plant.

FORMWORK

Because concrete is put in place as a shapeless slurry with no physical strength, it must be shaped and supported by *formwork* until it has cured sufficiently to support itself. Formwork is usually made of braced panels of wood, metal, or plastic. It is constructed as a negative of the shape intended for the concrete. Formwork for a beam or slab serves as a temporary working surface during the construction process and as the temporary means of support for reinforcing bars. Formwork must be strong enough to support the considerable weight and fluid pressure of wet concrete without excessive deflection, which often requires temporary supports that are major structures in themselves. During curing, the formwork helps to retain the necessary water of hydration in the concrete. When curing is complete, the formwork must pull away cleanly from the concrete surfaces without damage either to the concrete or to the formwork, which is usually used repeatedly as a construction project progresses. This means that the formwork should have no reentrant corners that will trap or be trapped by the concrete. Any element of formwork that must be withdrawn directly from a location in which it is surrounded on four or more surfaces by concrete, such as a joist pan (Figures 14.23 and 14.24), must be tapered. Formwork surfaces that are in contact with concrete are also usually coated with a *form release compound*, which is an oil, wax, or plastic that prevents adhesion of the concrete to the form.

FIGURE 13.16
Casting concrete on the building site requires the construction of a complete temporary structure that will be removed once the concrete has been placed and cured.

The quality of the concrete surfaces can be no better than the quality of the forms in which they are cast, and the requirements for surface quality and structural strength of formwork are rigorous. Top-grade wooden boards and plastic-overlaid plywoods are frequently used to achieve high-quality surfaces. The ties and temporary framing members that support the boards or plywood are spaced closely to avoid bulging of the forms under the high pressure of the wet concrete.

In a sense, formwork constitutes an entire temporary building that must be erected and then demolished in order to produce a second, permanent building of concrete (Figure 13.16). The cost of this formwork accounts for a major portion—often one-half or more—of the overall cost of a concrete building frame. This cost is one of the factors that has led to the development of *precasting*, a process in which concrete is cast into reusable forms at an industrial plant. Rigid, fully cured structural units from the plant are then transported to the job site, where they are hoisted into place and connected much as if they were structural steel shapes. The alternative to precasting, and the more usual way of building with concrete, is *sitecasting*, also called *cast-in-place construction*, in which concrete is poured into forms that are erected on the job site. In the two chapters that follow, formwork is shown for both sitecast and precast concrete.

Reinforcing

The Concept of Reinforcing

Concrete has no useful tensile strength (Figure 13.17). Historically, its structural uses were limited until the concept of steel reinforcing was developed. The compatibility of steel and concrete is a fortuitous accident. If the two materials had grossly different coefficients of thermal expansion, a reinforced concrete structure would tear itself apart during seasonal cycles of temperature variation. If the two materials were chemically incompatible, the steel would corrode or the concrete would be degraded. If concrete did not adhere to steel, a very different and more expensive configuration of reinforcing would be necessary. Concrete and steel, however, change dimension at nearly the same rate in response to temperature changes; steel is protected from corrosion by the alkaline chemistry of concrete; and concrete bonds strongly to steel, providing a convenient means of adapting brittle concrete to structural elements that must resist not only compression, but tension, shear, and bending as well.

The basic theory of *reinforced concrete* is extremely simple: Put the reinforcing steel where there are tensile forces in a structural member, and let the concrete resist the compression. This accounts fairly precisely for the location of most of the reinforcing steel that is used in a concrete structure. However, there are some important exceptions: Steel is used to resist a share of the compression in concrete columns and in beams whose height must be reduced for architectural reasons. It is used in the form of column ties, discussed below, to prevent buckling of vertical reinforcing in columns. It is used to resist cracking that might otherwise be caused by curing shrinkage, and by thermal expansion and contraction in slabs and walls.

Material	Working Strength in Tension[a]	Working Strength in Compression[a]	Density	Modulus of Elasticity
Wood (framing lumber)	300–1000 psi 2.1–6.9 MPa	600–1700 psi 4.1–12 MPa	30 pcf 480 kg/m^3	1,000,000– 1,900,000 psi 6900–13,000 MPa
Brick masonry (including mortar, unreinforced)	0	250–1300 psi 1.7–9.0 MPa	120 pcf 1900 kg/m^3	700,000– 3,700,000 psi 4800–25,000 MPa
Structural steel	24,000–43,000 psi 170–300 MPa	24,000–43,000 psi 170–300 MPa	490 pcf 7800 kg/m^3	29,000,000 psi 200,000 MPa
Concrete (unreinforced)	0	1000–4000 psi 6.9–28 MPa	145 pcf 2300 kg/m^3	3,000,000– 4,500,000 psi 21,000–31,000 MPa

[a]Allowable stress or approximate maximum stress under normal loading conditions.

Figure 13.17
Comparative physical properties of four common structural materials: wood, brick masonry, steel, and concrete (shaded row). Concrete, like masonry, has no useful tensile strength, but its compressive strength is considerable, and when combined with steel reinforcing, it can be used for every type of structure. The ranges of values in strength and stiffness reflect variations in concrete mix properties. (Specially formulated concretes are capable of substantially higher strengths than those listed in this table.)

Steel Bars for Concrete Reinforcement

Steel reinforcing bars ("rebar") for concrete construction are hot-rolled in much the same way as structural shapes. They are round in cross section and *deformed* with surface ribs for better bonding to concrete (Figures 13.18 and 13.19). At the end of the rolling line in the mill, the bars are cut to a standard length (commonly 60 feet, or 18.3 m, in the United States), bundled, and shipped to local fabricating shops.

Reinforcing bars are rolled in a limited number of standard diameters. In the United States, bars are specified by a simple numbering system in which the number corresponds to the number of eighths of an inch (3.2 mm) of bar diameter (Figure 13.20). For example, a number 6 reinforcing bar is $\frac{6}{8}$ or $\frac{3}{4}$ inch (19 mm) in diameter, and a number 8 is $\frac{8}{8}$ or 1 inch (25.4 mm) in diameter. Bars larger than number 8 vary slightly from these nominal diameters in order to correspond to convenient cross-sectional areas of steel. For the increasing volume of work in the United States that is carried out in SI units, a "soft" conversion of units is used: The bars are exactly the same, but a different numbering system is used, corresponding roughly to the diameter of each bar in millimeters. This avoids the expensive process of converting rolling mills to produce a slightly different set of bar sizes. In most countries other than the United States, a "hard metric" range of reinforcing bar sizes is standard (Figure 13.21).

In selecting reinforcing bars for a given beam or column, the structural engineer knows from calculations the required cross-sectional area of steel that is needed in a given location. This area may be achieved with a larger number of smaller bars, or a smaller number of larger bars, in any of several combinations. The final bar arrangement is based on the physical space available in the concrete

FIGURE 13.18
Glowing strands of steel are reduced to reinforcing bars as they snake their way through a rolling mill. *(Courtesy of Bethlehem Steel Company)*

Figure 13.19
The deformations rolled onto the surface of a reinforcing bar help it to bond tightly to concrete. *(Photo by Edward Allen)*

ASTM Standard Reinforcing Bars

Bar Size		Nominal Dimensions					
		Diameter		Cross-Sectional Area		Weight	
American	Metric	in.	mm	in.^{2}	mm^2	lb/ft	kg/m
#3	#10	0.375	9.5	0.11	71	0.376	0.560
#4	#13	0.500	12.7	0.20	129	0.668	0.944
#5	#16	0.625	15.9	0.31	199	1.043	1.552
#6	#19	0.750	19.1	0.44	284	1.502	2.235
#7	#22	0.875	22.2	0.60	387	2.044	3.042
#8	#25	1.000	25.4	0.79	510	2.670	3.973
#9	#29	1.128	28.7	1.00	645	3.400	5.060
#10	#32	1.270	32.3	1.27	819	4.303	6.404
#11	#36	1.410	35.8	1.56	1006	5.313	7.907
#14	#43	1.693	43.0	2.25	1452	7.65	11.38
#18	#57	2.257	57.3	4.00	2581	13.6	20.24

FIGURE 13.20
American standard sizes of reinforcing bars. These sizes were originally established in conventional units of inches and square inches. More recently, "soft metric" designations have also been given to the bars without changing their sizes. Notice that the size designations of the bars in both systems of measurement correspond very closely to the rule-of-thumb values of 1/8 inch or 1 mm per bar size number.

Metric Reinforcing Bars

Size Designation	Nominal Mass, kg/m	Nominal Dimensions	
		Diameter, mm	Cross Sectional Area, mm^2
10M	0.785	11.3	100
15M	1.570	16.0	200
20M	2.355	19.5	300
25M	3.925	25.2	500
30M	5.495	29.9	700
35M	7.850	35.7	1000
45M	11.775	43.7	1500
55M	19.625	56.4	2500

FIGURE 13.21
These "hard metric" reinforcing bar sizes are used in most countries of the world.

member, the required depth of concrete that must cover the reinforcing, the clear spacing required between bars to allow passage of the concrete aggregate, and the sizes and number of bars that will be most convenient to fabricate and install.

Most reinforcing bars are manufactured according to ASTM standard A615 and are available in grades 40, 60, and 75, corresponding to steel with yield strengths of 40,000, 60,000, and 75,000 psi (280, 420, and 520 MPa), respectively (Figure 13.22). Grade 60 is generally the most economical and readily available of the three, although grade 75 is finding increasing use in column reinforcing. Reinforcing bars conforming to ASTM A706, made with low-alloy steel meeting special ductility requirements, are used where concrete structures must meet special seismic design criteria or where extensive welding of reinforcing is required. In structures with especially heavy reinforcing requirements, reinforcing bars conforming to ASTM A1035 with strengths as high as 120,000 psi (830 MPa) may be used. With such high-strength bars, bar size may be reduced and the spacing between the bars may be increased in comparison to designs with conventional-strength reinforcing. This reduces *rebar congestion*, making it easier to place and consolidate the concrete around the reinforcing.

Reinforcing bars in concrete structures that are exposed to salts such as deicing salts or those in seawater are prone to rust. Galvanized reinforcing bars and epoxy-coated reinforcing bars are often used in marine structures, highway structures, and parking garages to resist this corrosion. Stainless steel bars, stainless-steel-clad bars, zinc-and-polymer-coated bars, and proprietary corrosion-resistant alloy bars are newer types of corrosion-resistant reinforcing. Still in the experimental stage or newest to market are nonmetallic reinforcing bars made from high-strength fibers of carbon, aramid, or glass embedded in a polymeric matrix.

As an alternative to conventional reinforcing bars, reinforcing steel for slabs is produced in sheets or rolls of *welded wire reinforcing (WWR)*, also called *welded wire fabric (WWF)*, a grid of wires or round bars spaced 2 to 12 inches (50–300 mm) apart (Figure 13.23). The lighter styles of welded wire fabric resemble cattle fencing and are used to reinforce concrete slabs on grade and certain precast concrete elements. The heavier styles find use in concrete walls and structural slabs. The principal advantage of welded wire fabric over individual bars is economy of labor in placing the reinforcing, especially where a large number of small bars can be replaced by a single sheet of material. The size and spacing of the wires or bars for a particular application are specified by the structural engineer or architect of the building.

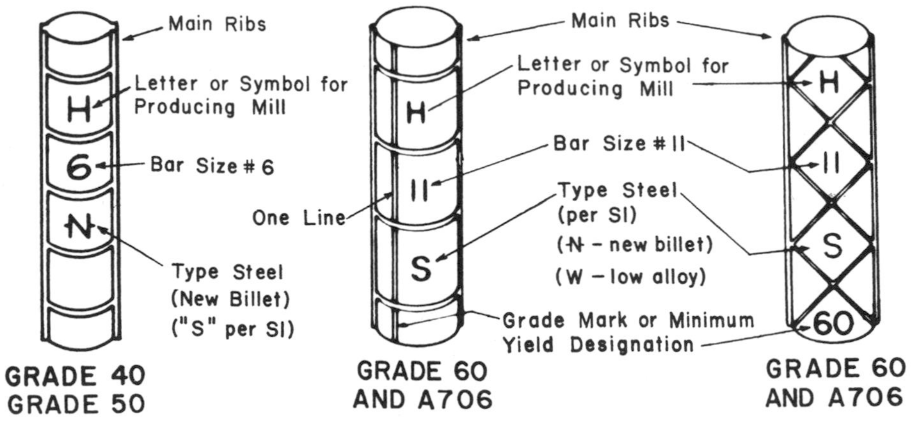

FIGURE 13.22
Reinforcing bars are manufactured with identification marks, denoting the mill that produced the bars, bar size, type of steel, and steel grade. Steel grade is indicated either with a number, such as "60" for Grade 60 steel, or with short bars (grade lines), in which no bars indicates Grade 40 or 50 steel, one bar indicates Grade 60 steel, and two bars indicates grade 75 steel. *(Courtesy of Concrete Reinforcing Steel Institute)*

Sectional Area and Weight of Welded Wire Reinforcing

Wire Size Number		Nominal Diameter, in.	Nominal Weight, lb/lin ft	Area in Square Inches per Foot of Width for Various Spacings						
				Center-Center Spacing						
Smooth	Deformed			2″	3″	4″	6″	8″	10″	12″
W20	D20	0.505	.680	1.20	.80	.60	.40	.30	.24	.20
W18	D18	0.479	.612	1.08	.72	.54	.36	.27	.216	.18
W16	D16	0.451	.544	.96	.64	.48	.32	.24	.192	.16
W14	D14	0.422	.476	.84	.56	.42	.28	.21	.168	.14
W12	D12	0.391	.408	.72	.48	.36	.24	.18	.144	.12
W11	D11	0.374	.374	.66	.44	.33	.22	.165	.132	.11
W10.5		0.366	.357	.63	.42	.315	.21	.157	.126	.105
W10	D10	0.357	.340	.60	.40	.30	.20	.15	.12	.10
W9.5		0.348	.323	.57	.38	.285	.19	.142	.114	.095
W9	D9	0.338	.306	.54	.36	.27	.18	.135	.108	.09
W8.5		0.329	.289	.51	.34	.255	.17	.127	.102	.085
W8	D8	0.319	.272	.48	.32	.24	.16	.12	.096	.08
W7.5		0.309	.255	.45	.30	.225	.15	.112	.09	.075
W7	D7	0.299	.238	.42	.28	.21	.14	.105	.084	.07
W6.5		0.288	.221	.39	.26	.195	.13	.097	.078	.065
W6	D6	0.276	.204	.36	.24	.18	.12	.09	.072	.06
W5.5		0.265	.187	.33	.22	.165	.11	.082	.066	.055
W5	D5	0.252	.170	.30	.20	.15	.10	.075	.06	.05
W4.5		0.239	.153	.27	.18	.135	.09	.067	.054	.045
W4	D4	0.226	.136	.24	.16	.12	.08	.06	.048	.04
W3.5		0.211	.119	.21	.14	.105	.07	.052	.042	.035
W3		0.195	.102	.18	.12	.09	.06	.045	.036	.03
W2.9		0.192	.099	.174	.116	.087	.058	.043	.035	.029
W2.5		0.178	.085	.15	.10	.075	.05	.037	.03	.025
W2.1		0.162	.070	.126	.084	.063	.042	.031	.025	.021
W2		0.160	.068	.12	.08	.06	.04	.03	.024	.02
W1.5		0.138	.051	.09	.06	.045	.03	.022	.018	.015
W1.4		0.134	.048	.084	.056	.042	.028	.021	.017	.014

FIGURE 13.23
Standard configurations of welded wire reinforcing. The heaviest "wires" are more than ½ inch (13 mm) in diameter, making them suitable for structural slab reinforcing. Welded wire reinforcing is specified by first indicating the spacing of the wires and then the wire types. For example, the designation *6 × 12-W12 × W5* indicates welded wire reinforcing with W12 longitudinal wires spaced at 6 inches (150 mm) and W5 transverse wires spaced at 12 inches (300 mm). *(Courtesy of Concrete Reinforcing Steel Institute)*

STANDARD HOOKS

All specific sizes recommended by CRSI below meet minimum requirements of ACI 318

RECOMMENDED END HOOKS
All Grades

D=Finished bend diameter

Bar Size	180° HOOKS			90° HOOKS
	D	A or G	J	A or G
# 3	2¼	5	3	6
# 4	3	6	4	8
# 5	3¾	7	5	10
# 6	4½	8	6	1-0
# 7	5¼	10	7	1-2
# 8	6	11	8	1-4
# 9	9½	1-3	11¾	1-7
#10	10¾	1-5	1-1¼	1-10
#11	12	1-7	1-2¾	2-0
#14	18¼	2-3	1-9¾	2-7
#18	24	3-0	2-4½	3-5

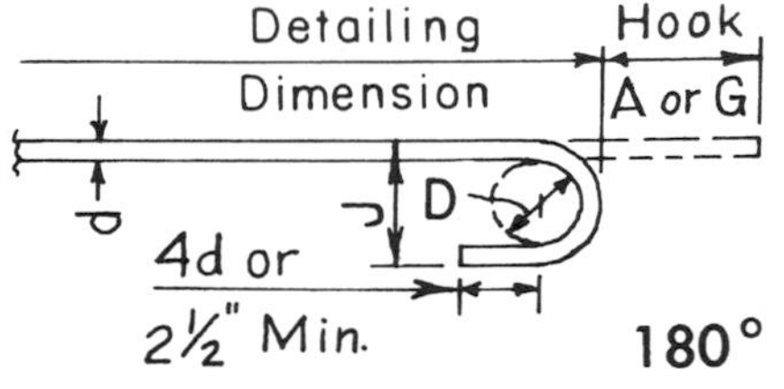

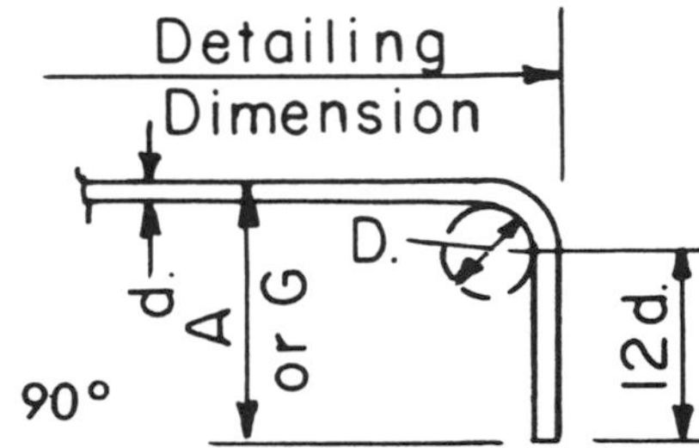

STIRRUP AND TIE HOOKS

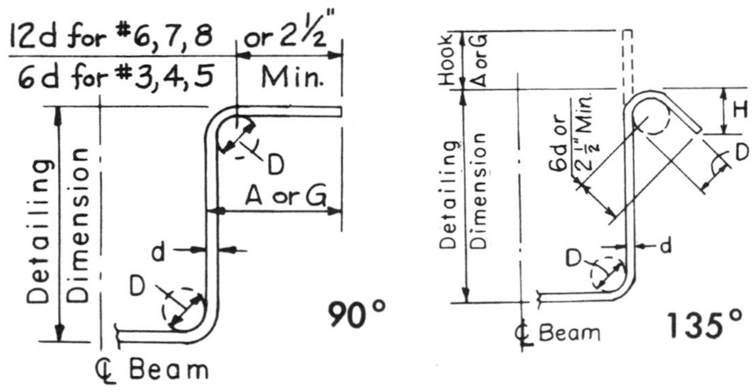

STIRRUPS
(TIES SIMILAR)

135° SEISMIC STIRRUP/TIE HOOKS

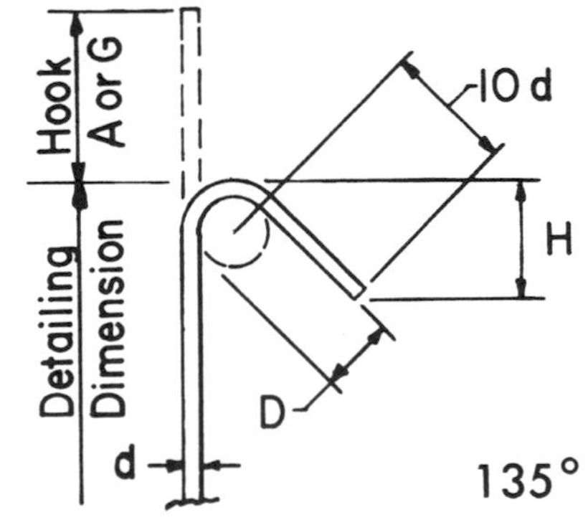

STIRRUP AND TIE HOOK DIMENSIONS
Grades 40-50-60 ksi

Bar Size	D (in.)	90° Hook	135° Hook	
		Hook A or G	Hook A or G	H Approx.
#3	1½	4	4	2½
#4	2	4½	4½	3
#5	2½	6	5½	3¾
#6	4½	1-0	7¾	4½
#7	5¼	1-2	9	5¼
#8	6	1-4	10¼	6

135° SEISMIC STIRRUP/TIE HOOK DIMENSIONS
Grades 40-50-60 ksi

Bar Size	D (in.)	135° Hook	
		Hook A or G	H Approx.
#3	1½	5	3½
#4	2	6½	4½
#5	2½	8	5½
#6	4½	10¾	6½
#7	5¼	1-0½	7¾
#8	6	1-2¼	9

FIGURE 13.24
The bending of reinforcing bars is done according to precise standards in a fabricator's shop. *(Courtesy of Concrete Reinforcing Steel Institute)*

Fabrication and Erection of Reinforcing Bars

The fabrication of reinforcing steel for a concrete construction project is analogous to the fabrication of steel shapes for a steel frame building (Chapter 11). The fabricator receives the engineering drawings for the building from the contractor and prepares shop drawings for the reinforcing bars. After the shop drawings have been checked by the engineer or architect, the fabricator sets to work cutting the reinforcing bar stock to length, making the necessary bends (Figure 13.24) and tying the fabricated bars into bundles that are tagged to indicate their destination in the building. The bundles are shipped to the building site as needed. There they are broken down, lifted by hand or hoisted by crane, and wired (or occasionally welded) together in the forms to await pouring of the concrete. The wire has a temporary function only, which is to hold the reinforcement in position until the concrete has cured. Any transfer of load from one reinforcing bar to another in the completed building is done by the concrete. Where two bars must be spliced, they are overlapped a specified number of bar diameters (typically 30), and the loads are transferred from one to the other by the surrounding concrete. The one common exception occurs in heavily reinforced columns where there is insufficient space to overlap the bars; there they are often spliced end to end rather than overlapped, and loads are transferred through welds or sleevelike mechanical splicing devices (Figure 13.25).

The composite action of concrete and steel in reinforced concrete structural elements is such that the reinforcing steel is usually loaded axially in tension or compression, and occasionally in shear, but never in bending. The bending stiffness of the reinforcing bars themselves is of no consequence in imparting strength to the concrete.

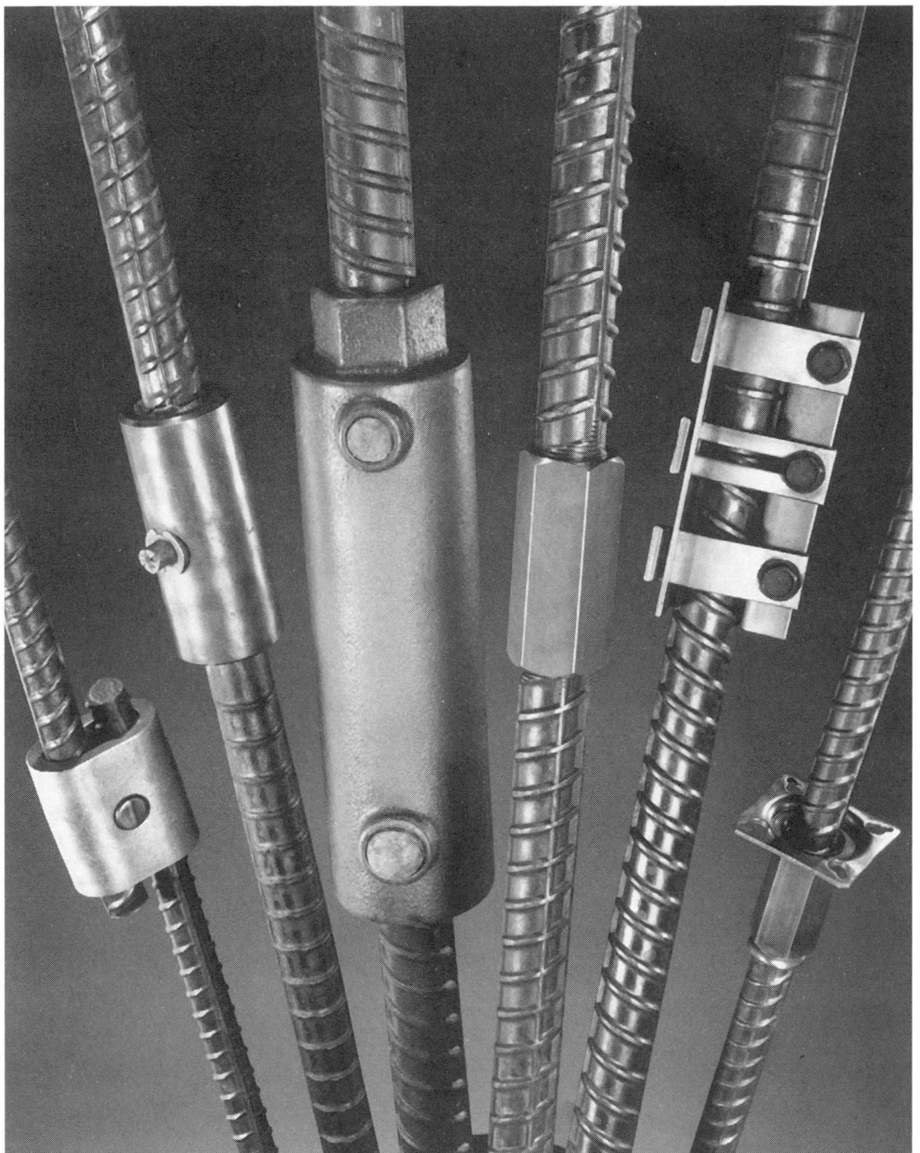

Figure 13.25
Some mechanical devices for splicing reinforcing bars. From left to right: A lapped, wedged connection, used primarily to connect new bars to old ones when adding to an existing structure. A welded connector, very strong and tough. A grouted sleeve connector for joining precast concrete components: One bar is threaded and screwed into a collar at one end of the sleeve, and the other bar is inserted into the remainder of the sleeve and held there with injected grout. A threaded sleeve, with both bars threaded and screwed into the ends of the sleeve. A simple clamping sleeve that serves to align compression bars in a column. A flanged coupler for splicing bars at the face of a concrete wall or beam: The coupler is screwed onto the threaded end of one bar, and its flange is nailed to the inside face of the formwork. After the formwork has been stripped, the other bar is threaded and screwed through a hole in the flange and into the coupler. *(Photo courtesy of Erico, Inc.)*

Reinforcing a Simple Concrete Beam

In an ideal, simply supported beam under uniform loading, compressive (squeezing) forces follow a set of archlike curves that create a maximum compressive stress in the top of the beam at midspan, with progressively lower compressive stresses toward either end. A mirrored set of curves correspond to paths of tensile (stretching) force, with stresses again reaching a maximum at the middle of the span (Figure 13.26). In an ideally reinforced concrete beam, steel reinforcing bars would be bent to follow these lines of tension, and the bunching of the bars at midspan would serve to resist the higher stresses at that point. It is difficult, however, to bend bars into these curves and to support the curved bars adequately in the formwork, so a simpler rectilinear arrangement of reinforcing steel is substituted.

This arrangement consists of a set of bottom bars and stirrups. The *bottom bars* are placed horizontally near the bottom of the beam, leaving a specified amount of concrete below and to the sides of the rods as *cover* (Figure 13.27). The concrete cover provides a full embedment for the reinforcing bars and protects them against fire and corrosion. The bars are most heavily stressed at the midpoint of the beam span, with progressively smaller amounts of stress

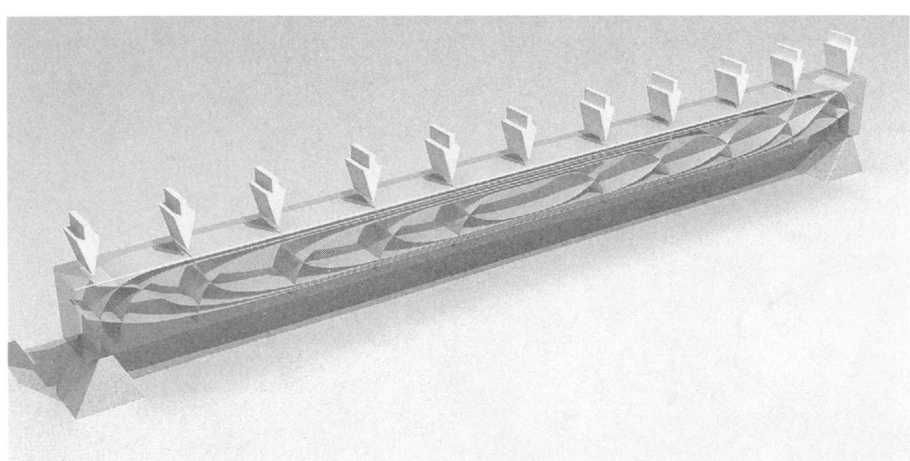

(*a*)

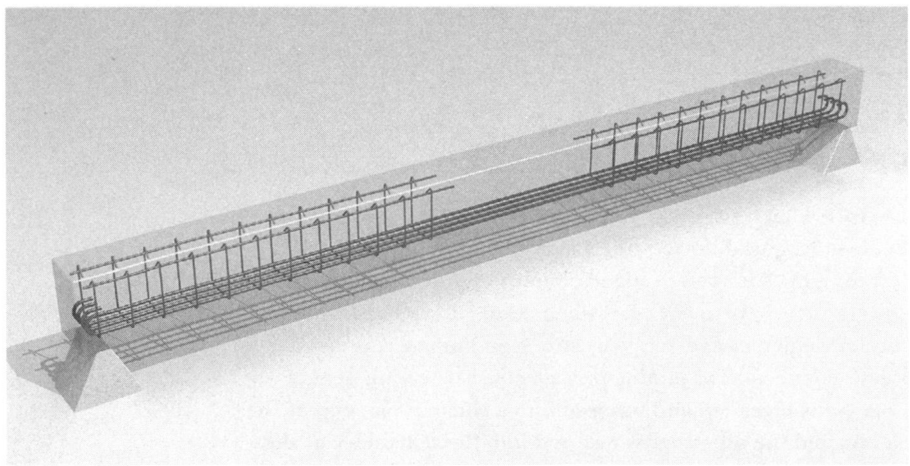

(*b*)

FIGURE 13.26
(*a*) The directions of force in a simply supported (supported at the ends only) beam under uniform loading. The archlike lines represent compression, and the cablelike lines represent tension. Near the ends of the beam, the lines of strongest tensile force move upward diagonally through the beam. (*b*) Steel reinforcing for a simply supported beam under uniform loading. The concrete resists compressive forces. The horizontal bars near the bottom of the beam resist the major tensile forces. The vertical stirrups resist the lesser diagonal tensile forces near the ends of the beam.

Concrete cover protects the steel against fire and corrosion
Bar spacings must be large enough that coarse aggregate can pass through easily

FIGURE 13.27
A cross section of a rectangular concrete beam showing cover and bar spacing.

toward each of the supports. The differences in stress are dissipated from the bars into the concrete by means of *bond* forces, the adhesive forces between the concrete and the steel, aided by the ribs on the surface of the bars. At the ends of the beam, some stress remains in the steel, but there is no further length of concrete into which the stress can be dissipated. This problem is solved by bending the ends of the bars into *hooks*, which are semicircular bends of standard dimensions.

The bottom bars do the heavy tensile work in the beam, but some lesser tensile forces occur in a diagonal orientation near the ends of the beam. These are resisted by a series of *stirrups* (Figure 13.26). The stirrups may be either open *U-stirrups*, as shown, or *closed stirrup-ties*, which are full rectangular loops of steel that wrap all the way around the longitudinal bars. U-stirrups are less expensive to make and install and are sufficient for many situations, but stirrup-ties are required in beams that will be subjected to torsional (twisting) forces or to high compressive forces in the top or bottom bars. In either case, the stirrups furnish vertical tensile reinforcing to resist the cracking forces that run diagonally across them. A more efficient use of steel would be to use diagonal stirrups oriented in the same direction as the diagonal tensile forces, but they would be difficult to install.

When the simple beam of our example is formed, the bottom steel is supported at the correct cover height by *chairs* made of heavy steel wire or plastic (Figures 13.28 and 13.29). In a broad beam or slab, bars are supported by long chairs called *bolsters*. Chairs and bolsters remain in the concrete after pouring, even though their work is finished, because there is no way to get them out. In outdoor concrete work, the feet of the chairs and bolsters sometimes rust where they come in contact with the face of the beam or slab unless plastic or plastic-capped steel chairs are used. Where reinforced concrete is poured in direct contact with the soil, concrete bricks or small pieces of concrete are used to support the bars instead of chairs to prevent rust from forming under the feet of the chairs and spreading up into the reinforcing bars.

The stirrups in the simple beam that we have been examining are supported by wiring them to the bottom bars and by tying their tops to horizontal #3 top bars (the smallest standard size) that have no function in the beam other than to keep the stirrups upright and properly spaced until the concrete has been poured and cured.

FIGURE 13.28
A two-piece plastic bar support, called a "tower chair," supports a steel reinforcing bar for a structural concrete slab. To the left of the chair, a small concrete brick supporting a second bar in a position closer to the bottom of the slab is also partially visible. *(Photo by Joseph Iano)*

SYMBOL	BAR SUPPORT ILLUSTRATION	BAR SUPPORT ILLUSTRATION PLASTIC CAPPED OR DIPPED	TYPE OF SUPPORT	SIZES
SB	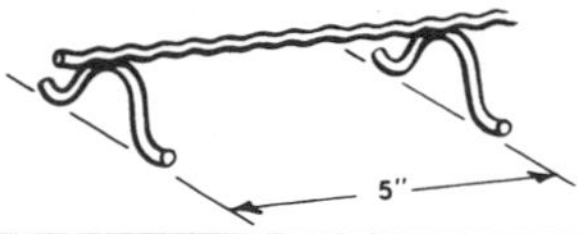 5"	CAPPED 5" 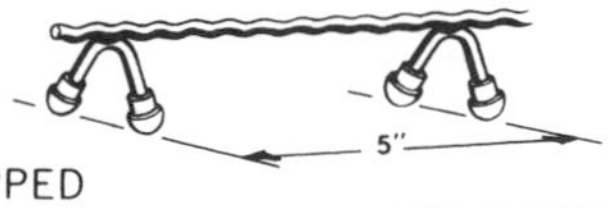	Slab Bolster	¾, 1, 1½, and 2 inch heights in 5 ft. and 10 ft. lengths
SBU	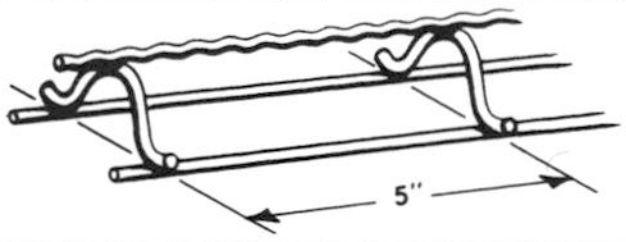 5"		Slab Bolster Upper	Same as SB
BB	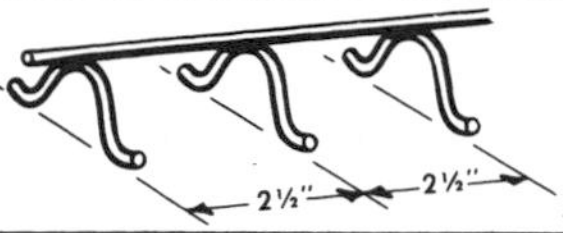 2½" 2½"	CAPPED 2½" 2½" 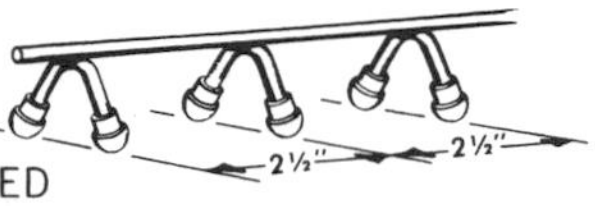	Beam Bolster	1, 1½, 2, over 2" to 5" heights in increments of ¼" in lengths of 5 ft.
BBU	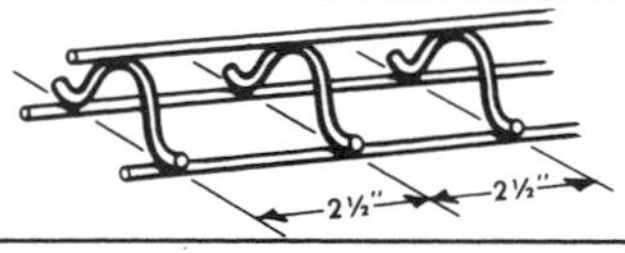 2½" 2½"		Beam Bolster Upper	Same as BB
BC	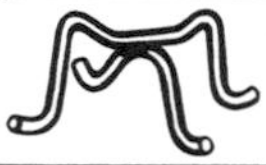	DIPPED 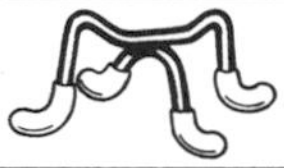	Individual Bar Chair	¾, 1, 1½, and 1¾" heights
JC	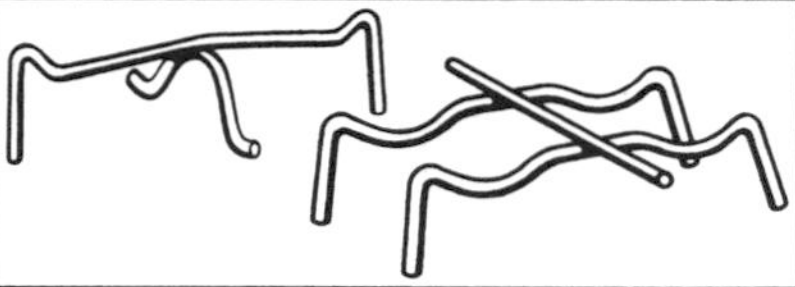	DIPPED 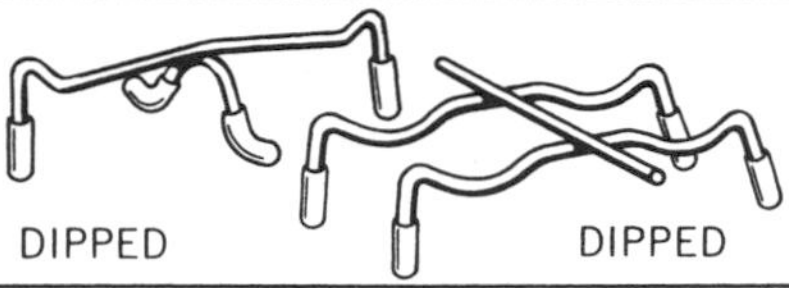 DIPPED	Joist Chair	4, 5, and 6 inch widths and ¾, 1 and 1½ inch heights
HC		CAPPED 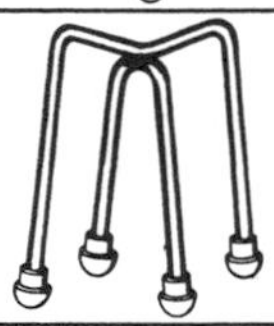	Individual High Chair	2 to 15 inch heights in increments of ¼ inch
HCM	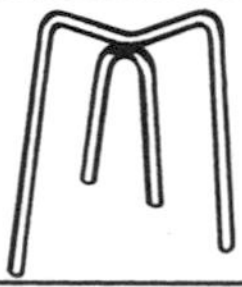		High Chair for Metal Deck	2 to 15 inch heights in increments of ¼ in.
CHC	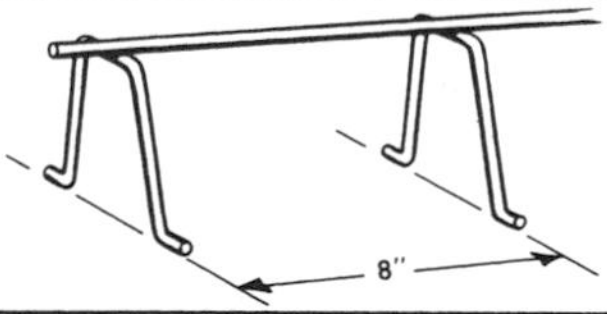 8"	CAPPED 8" 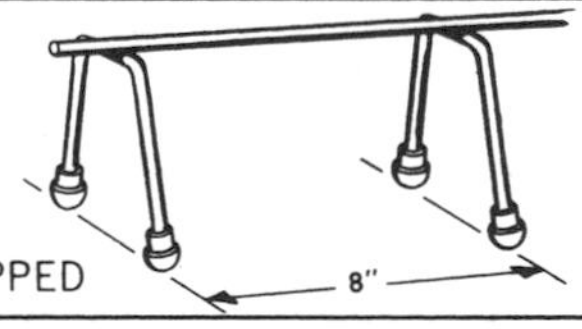	Continuous High Chair	Same as HC in 5 foot and 10 foot lengths
CHCU	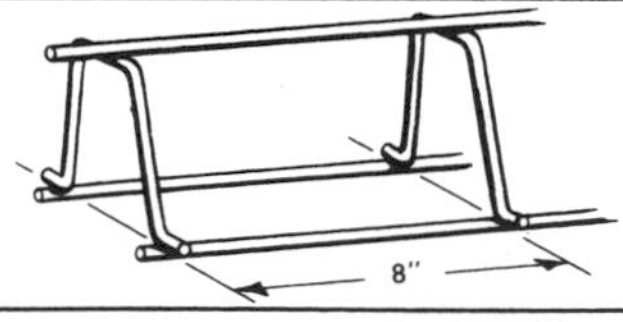 8"		Continuous High Chair Upper	Same as CHC
CHCM	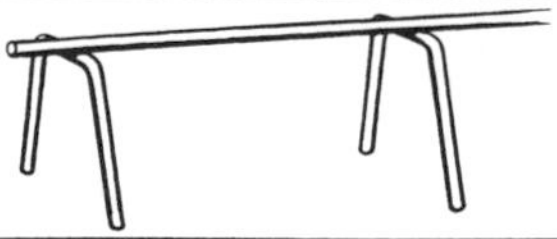		Continuous High Chair for Metal Deck	Up to 5 inch heights in increments of ¼ in.
JCU	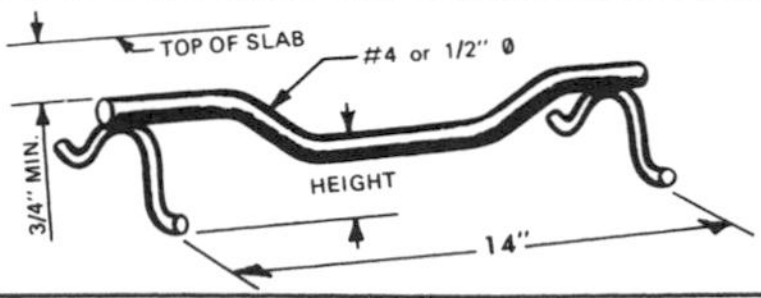 TOP OF SLAB #4 or 1/2" ø 3/4" MIN. HEIGHT 14"	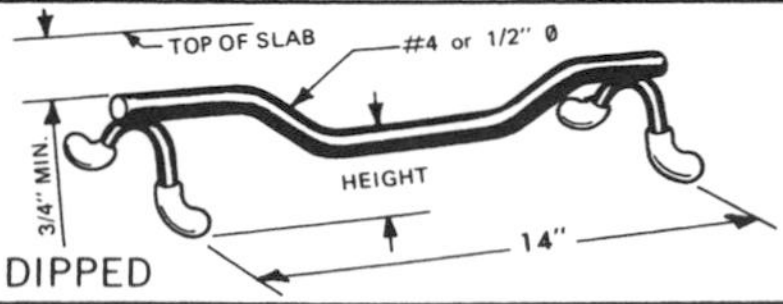 TOP OF SLAB #4 or 1/2" ø 3/4" MIN. HEIGHT 14" DIPPED	Joist Chair Upper	14" span. Heights −1" thru +3½" vary in ¼" increments

FIGURE 13.29
Chairs and bolsters for supporting reinforcing bars in beams and slabs. Bolsters and continuous chairs are made in long lengths for use in slabs. Chairs support only one or two bars each. *(Courtesy of Concrete Reinforcing Steel Institute)*

Reinforcing a Continuous Concrete Beam

Most sitecast concrete beams are not of this simple type because concrete lends itself most easily to one-piece structural frames with a high degree of structural continuity from one beam span to the next. In a continuous structure, the bottom of the beam is in tension at midspan, and the top of the beam is in tension at supporting girders, columns, or walls. This means that top bars must be provided over the supports, and bottom bars in midspan, along with the usual stirrups, as illustrated in Figure 13.30. Until a few decades ago, it was common practice to bend some of the horizontal bars up or down at the points of bending reversal in continuous concrete beams so that the same bars could serve both as bottom steel at midspan and as top steel over the columns, but this has largely been abandoned in favor of the simpler practice of using only straight bars for horizontal reinforcing.

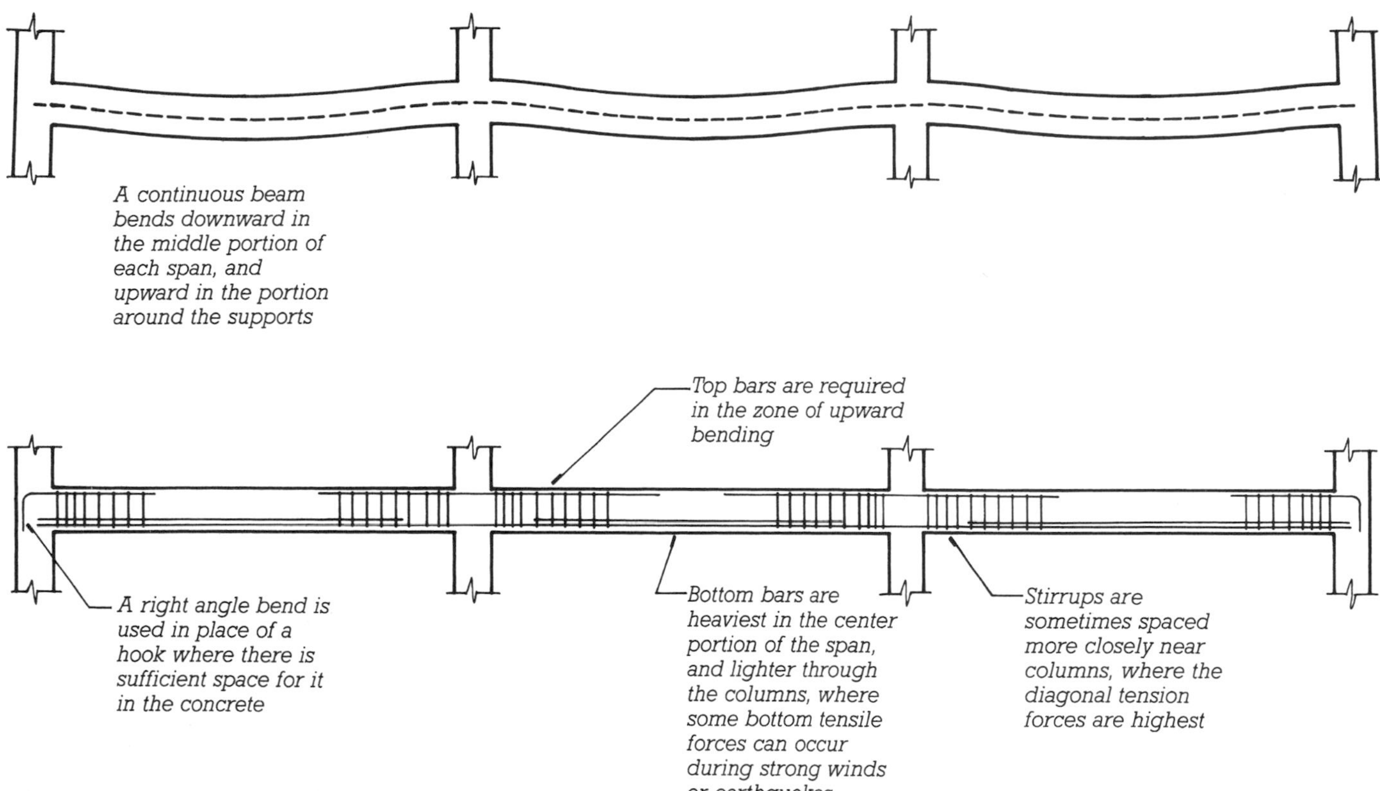

A continuous beam bends downward in the middle portion of each span, and upward in the portion around the supports
Top bars are required in the zone of upward bending
A right angle bend is used in place of a hook where there is sufficient space for it in the concrete
Bottom bars are heaviest in the center portion of the span, and lighter through the columns, where some bottom tensile forces can occur during strong winds or earthquakes
Stirrups are sometimes spaced more closely near columns, where the diagonal tension forces are highest

FIGURE 13.30
Reinforcing for a concrete beam that is continuous across several spans. The upper diagram shows in exaggerated form the shape taken by a continuous beam under uniform loading; the broken line is the centerline of the beam. The lower diagram shows the arrangement of bottom steel, top steel, and stirrups conventionally used in this beam. The bottom bars are usually placed on the same level, but they are shown on two levels in this diagram to demonstrate the way in which some of the bottom steel is discontinued in the zones near the columns. There is a simple rule of thumb for determining where the bending steel must be placed in a beam: Draw an exaggerated diagram of the shape the beam will take under load, as in the top drawing of this illustration, and put the bars as close as possible to the convex edges.

Reinforcing Structural Concrete Slabs

A concrete slab that spans across parallel beams (*one-way action*) is, in effect, a very wide beam. The reinforcing pattern for such a slab is similar to the reinforcing pattern in a beam, but with a larger number of smaller top and bottom bars distributed evenly across the width of the slab. Because the slab is wide, it has a large cross-sectional area of concrete that can usually resist the relatively weak diagonal tension forces near its supports without the aid of stirrups.

One-way slabs must be provided with *shrinkage–temperature steel*, a set of small-diameter reinforcing bars set at right angles to, and on top of, the primary reinforcing in the slab. Their function is to prevent cracks from forming parallel to the primary reinforcing because of concrete shrinkage, temperature-induced stresses, or miscellaneous forces that may occur in the building (Figure 13.31).

Two-Way Slab Action

A structural economy unique to concrete frames is realized through the use of *two-way action* in floor and roof slabs. Two-way slabs, which work best for bays that are square or nearly square, are reinforced equally in both directions and share the bending forces equally between the two directions. This allows two-way slabs to be somewhat shallower than one-way slabs, to use less reinforcing steel, and thus to cost less. Figure 13.32 illustrates the concept of two-way action. Several different two-way concrete framing systems will be shown in detail in the next chapter.

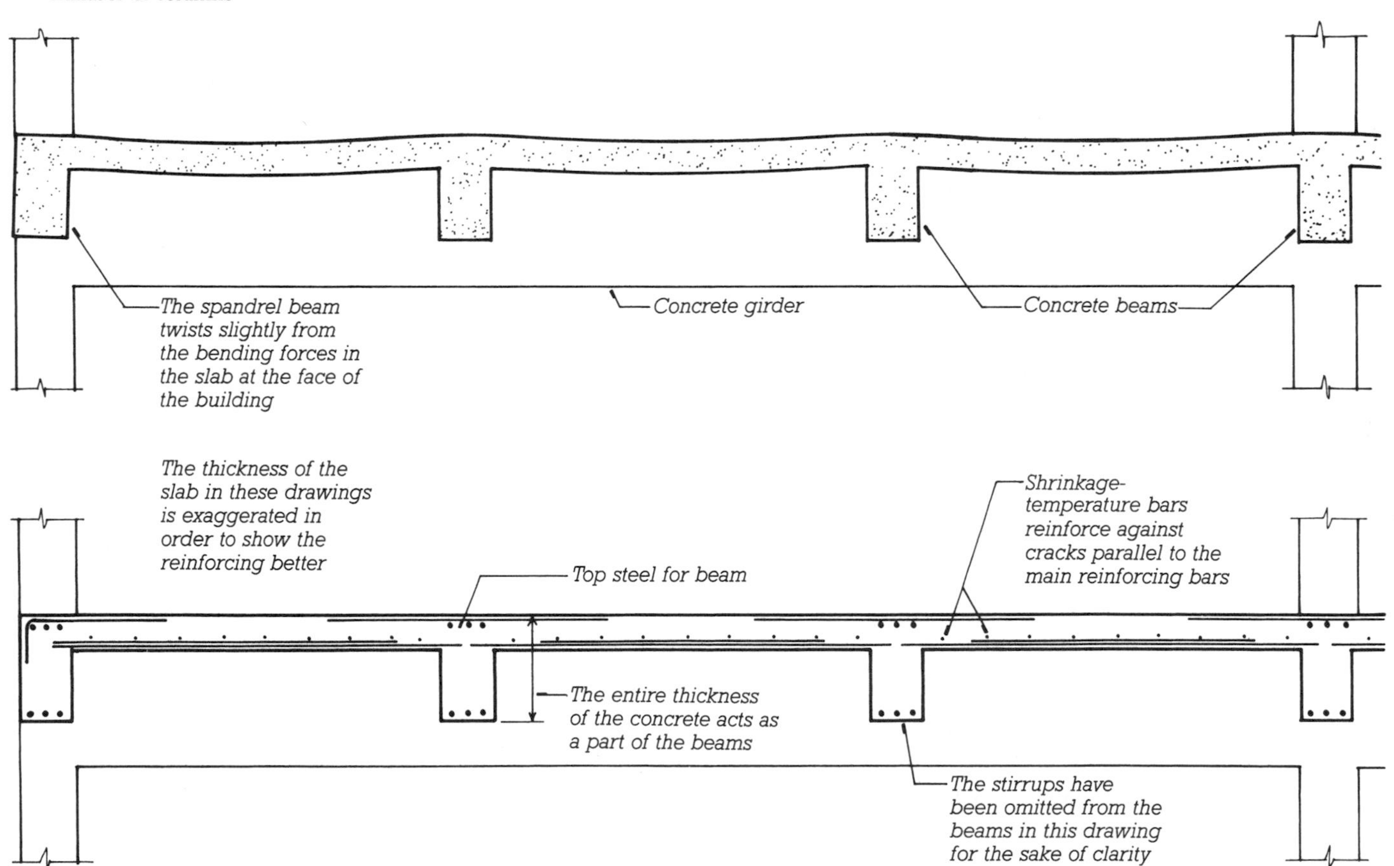
A concrete slab supported by a number of beams bends in the same pattern as a concrete beam supported by a number of columns
The spandrel beam twists slightly from the bending forces in the slab at the face of the building
Concrete girder
Concrete beams
The thickness of the slab in these drawings is exaggerated in order to show the reinforcing better
Top steel for beam
Shrinkage-temperature bars reinforce against cracks parallel to the main reinforcing bars
The entire thickness of the concrete acts as a part of the beams
The stirrups have been omitted from the beams in this drawing for the sake of clarity

FIGURE 13.31
Reinforcing for a one-way concrete slab. The reinforcing is similar to that for a continuous beam, except that stirrups are not usually required in the slab, and shrinkage–temperature bars must be added in the perpendicular direction. The slab does not sit on the beams; rather, the concrete around the top of a beam is part of both the beam and the slab. A concrete beam in this situation is considered to be a T-shaped member, with a portion of the slab acting together with the stem of the beam, resulting in greater structural efficiency and reduced beam depth.

The country . . . near Taliesin my home and workshop, is the bed of an ancient glacier drift. Vast busy gravel pits abound there, exposing heaps of yellow aggregate once and still everywhere near, sleeping beneath the green fields. Great heaps, clean and golden, are always waiting there in the sun. And I never pass . . . without an emotion, a vision of the long dust-whitened stretches of the cement mills grinding to impalpable fineness the magic powder that would "set" my vision all to shape; I wish both mill and gravel endlessly subject to my will. . . . Materials! What a resource.

Frank Lloyd Wright *Architectural Record, October 1928.*

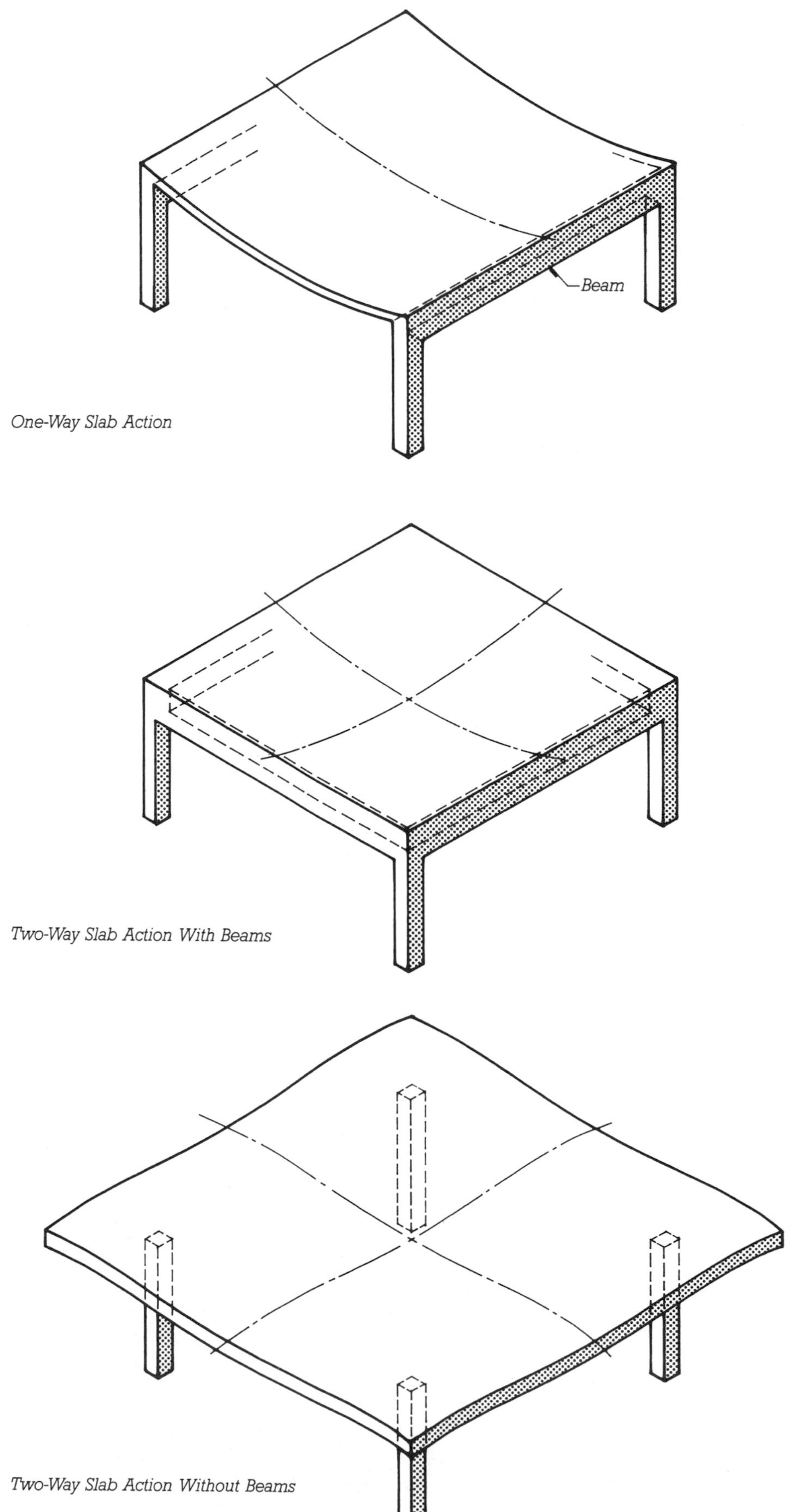

FIGURE 13.32
One-way and two-way slab action, with deflections greatly exaggerated.

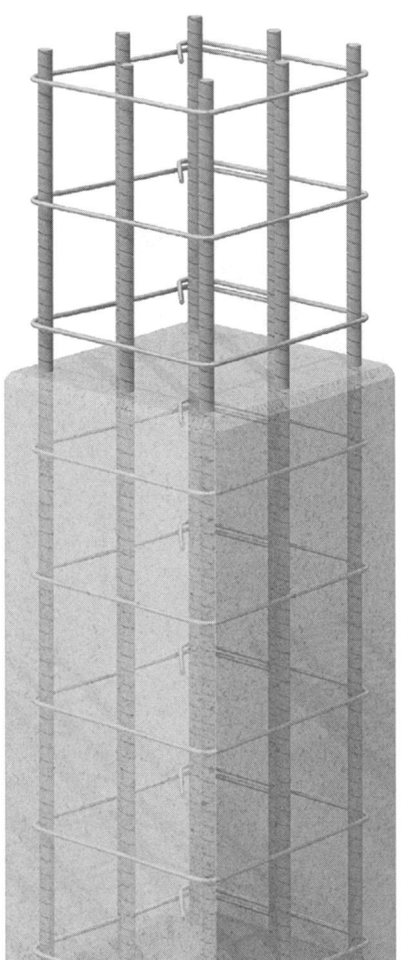

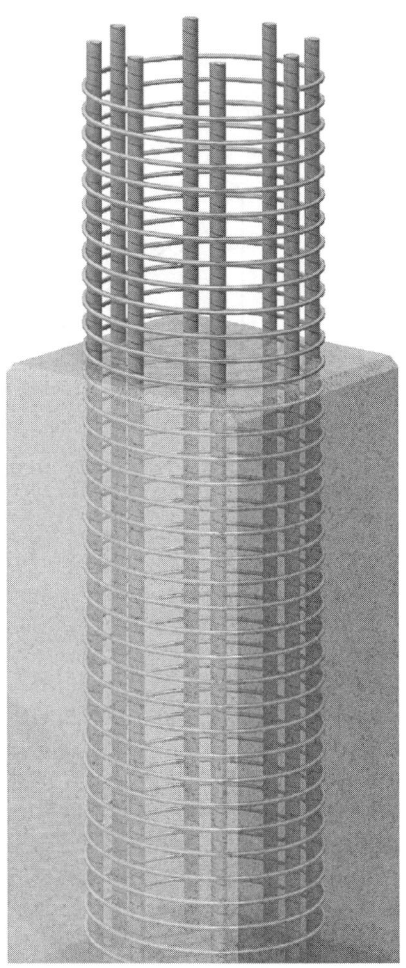

FIGURE 13.33
Reinforcing for concrete columns. To the left is a column with a rectangular arrangement of vertical bars and column ties. To the right is a circular arrangement of vertical bars with a column spiral. Either arrangement may be used in either a round or a square column.

Figure 13.34
Column spirals. Each double circle of vertical bars will be embedded in a single rectangular column. *(Courtesy of Concrete Reinforcing Steel Institute)*

Reinforcing Concrete Columns

Columns contain two types of reinforcing: *Vertical bars* (also called *column bars*) are large-diameter reinforcing bars that share the compressive loads with the concrete and resist the tensile stresses that occur in columns when a building frame is subjected to wind or earthquake forces. *Ties* of small-diameter steel bars wrapped around the vertical bars help to prevent them from buckling under load: Inward buckling is prevented by the concrete core of the column and outward buckling by the ties (Figures 13.33 and 13.34). The vertical bars may be arranged either in a circle or in rectangular patterns. The ties may be either of two types: *column ties* or *column spirals*. Spirals are shipped to the construction site as tight coils of rod that are expanded accordion fashion to the required spacing and wired to the vertical bars. They are generally used only for square or circular arrangements of vertical bars. For rectangular arrangements of vertical bars, discrete column ties must be wired to the vertical bars one by one. Each corner bar and alternate interior bars must be contained inside a bend of each tie, so two or more column ties are often attached at each level (Figure 13.35). A circular arrangement of vertical bars is often more economical than a rectangular one because it avoids the need to enclose bars in the corners of ties. Column ties are generally more economical than spirals, so even columns with circular bar arrangements are often tied with discrete circular ties. The sizes and spacings of column ties and spirals are determined by the structural engineer. To minimize labor costs on the job site, the ties and vertical bars for each column are usually wired together in a horizontal position at ground level, and the finished column cage is lifted into its final position with a crane.

Fibrous Reinforcing

Fibrous reinforcing is composed of short fibers of glass, steel, or polypropylene that are added to a concrete mix. *Microfiber reinforcing* is added in relatively low dosages and is intended to reduce *plastic shrinkage cracking*, which frequently occurs while the concrete is still in a plastic state, during the earliest stages of curing. Microfiber reinforcing makes little if any contribution to the properties of fully cured concrete. *Macrofiber reinforcing*, usually of polypropylene or a steel-polypropylene blend, not only protects against plastic shrinkage but also resists longer-term cracking due to drying shrinkage and thermal stresses. Macrofiber reinforcing is added to concrete at dosages 10 or 20 times greater than those of microfiber reinforcing and, in some cases, fully replaces the usual shrinkage–temperature steel in concrete slabs. Macrofiber reinforcing can also improve concrete's resistance to impact, abrasion, and shock. Glass fibers are also added to concrete to produce glass-fiber-reinforced concrete (GFRC), which is used for cladding panels (see Chapter 20).

FIGURE 13.35
Multiple column ties at each level are arranged so that the four corner bars and alternate interior bars are contained in the corners of the ties. *(Courtesy of Concrete Reinforcing Steel Institute)*

Concrete Creep

In addition to plastic and drying shrinkage, concrete is subject to long-term *creep*. When placed under sustained compressive stress from its own weight, the weight of other permanent building components, or the force of prestressing (as described later in this chapter), concrete will gradually and permanently shorten over a period of months or years. In some circumstances, this dimensional change is of sufficient magnitude that it must be accounted for in the design and detailing of building systems. For example, when a brick veneer cladding system (see Chapter 20) is supported on a concrete building frame, the shrinkage of the concrete combined with other factors affecting movement of the masonry require that horizontal movement joints be designed into the cladding system to accommodate the differential movement between the cladding and the building structure. If these joints are not provided, or if they are too narrow and have insufficient capacity to absorb movement, the cladding system can fail as it becomes compressed, in part, by the shortening of the concrete structure. As a rule of thumb, sitecast concrete building frames can be expected to shorten in height under the influence of their own weight and other dead loads at the rate of 1⁄16 inch for every 10 feet (½ mm per meter) of building height.

Prestressing

When a beam supports a load, the compression side of the beam is squeezed slightly and the tension side is stretched. In a reinforced concrete beam, the stretching tendency is resisted by the reinforcing steel but not by the brittle concrete. When the steel elongates (stretches) under tension, the concrete around it forms cracks that run from the edge of the beam to the horizontal plane in the beam, above which compressive forces occur. This cracking is visible to the unaided eye in reinforced concrete beams that are loaded to (or beyond) their full load-carrying capacity. In effect, over half of the concrete in the beam is doing no useful work except to hold the steel in position and protect it from fire and corrosion (Figure 13.36).

If the reinforcing bars could be stretched to a high tension before the beam is loaded and then released against the concrete that surrounds them, they would place the concrete in compression. If a load were subsequently put on the beam, the tension in the stretched steel would increase further, and the compression in the concrete surrounding the steel would diminish. If the initial tension or *prestress* in the steel bars were of sufficient magnitude, however, the surrounding concrete would never be subjected to tension, and no cracking would occur. Furthermore, the

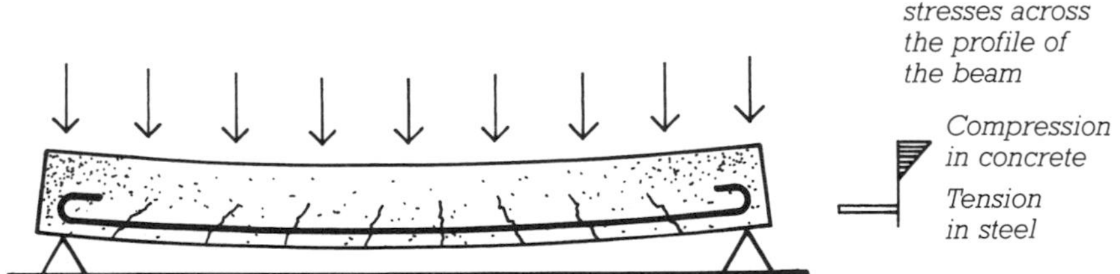

In a reinforced concrete beam, less than half the concrete is in compression, and cracks will appear in the bottom of the beam under full load

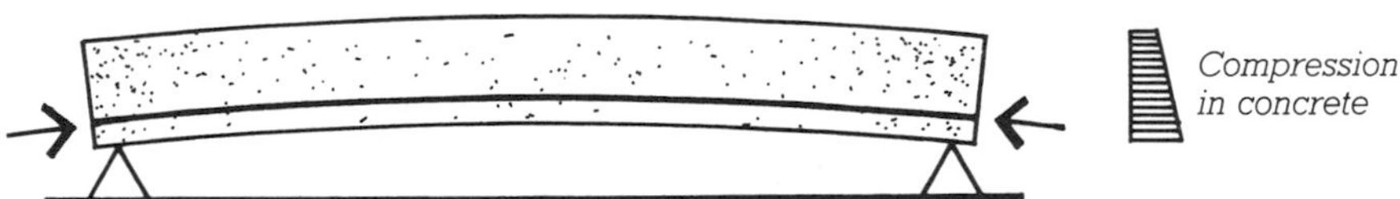

When a concrete beam is prestressed, all the concrete acts in compression. The off-center location of the prestressing steel causes a camber in the beam

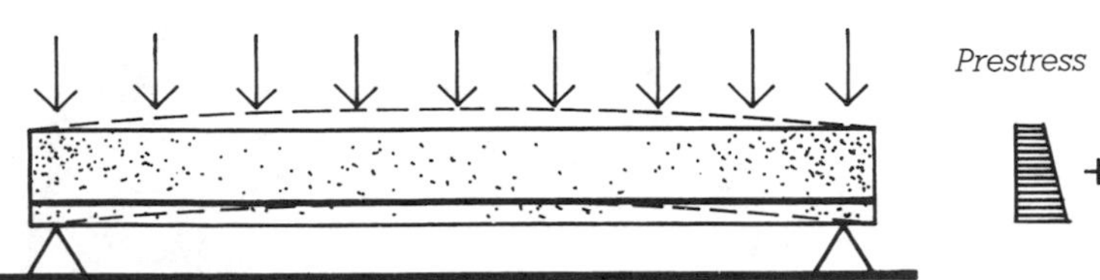

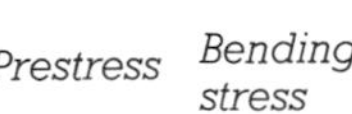

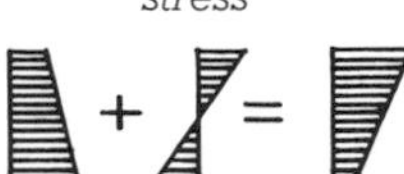

Under loading, the prestressed beam becomes flatter, but all the concrete still acts in compression, and no cracks appear

Figure 13.36
The rationale for prestressing concrete. In addition to the absence of cracks in the prestressed beam, the structural action is more efficient than that of a reinforced beam. Therefore, the prestressed beam uses less material. The small diagrams to the right indicate the distribution of stresses across the vertical cross section of each of the beams at midspan.

beam would be capable of carrying a much greater load with the same amounts of concrete and steel than if it were merely reinforced in the conventional manner. This is the rationale for *prestressed concrete.* Prestressed members, particularly those designed to work in bending, contain less concrete than reinforced members of equivalent strength and thus are, in general, less expensive. Their lighter weight also pays off by making precast, prestressed concrete members easier and cheaper to transport. For this reason, structural precast concrete used for slabs, beams, and girders (and in some cases columns as well) is typically prestressed.

In practice, ordinary reinforcing bars are not sufficiently strong to serve as prestressing steel. Prestressing is practical only with extremely high-strength steel strands that are manufactured for the purpose. These are made of cold-drawn steel wires that are formed into small-diameter cables.

Pretensioning

Prestressing is accomplished in two different ways. *Pretensioning* is used with precast concrete members: High-strength steel strands are stretched tightly between abutments in a precasting plant, and the concrete member (or, more commonly, a series of concrete members arranged end to end) is cast around the stretched steel (Figure 15.8). The curing concrete adheres to the strands along their entire length. After the concrete has cured to a specified minimum compressive strength, the strands are cut off at either end of each member. This releases the external tension on the steel, allowing it to recoil slightly, which squeezes all of the concrete of the member into compression. If, as is usually the case, the steel is placed as close as possible to the tension side of the member, the member takes on a decided *camber* (lengthwise arching) at the time the steel strands are cut (Figure 13.37). Much or all of this camber disappears later when the member is subjected to loads in a building.

The strong abutments needed to hold the tensioned strands prior to the pouring of concrete are very expensive to construct except in a single fixed location where many concrete members can be created within the same set of abutments. For this reason, pretensioning is useful only for concrete members cast in precasting plants.

Posttensioning

Unlike pretensioning, which is always done in a factory, *posttensioning* is done almost exclusively in place on the building site. The high-strength steel strands (called *tendons*) are covered with a steel or plastic tube to prevent them from bonding to the concrete and are not tensioned until the concrete has cured. Each tendon is anchored to a steel plate embedded in one end of the beam or slab.

1. The first step in pretensioning is to stretch the steel prestressing strands tightly across the casting bed

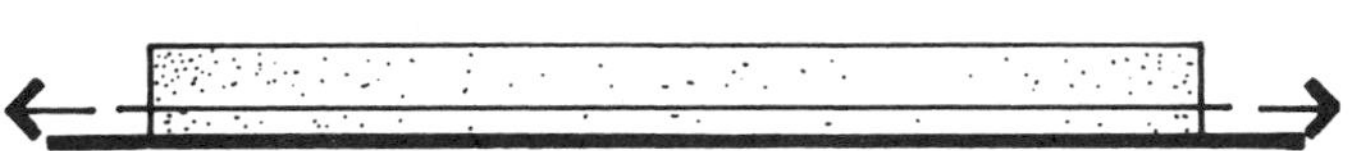

2. Concrete is cast around the stretched strands and cured. The concrete bonds to the strands

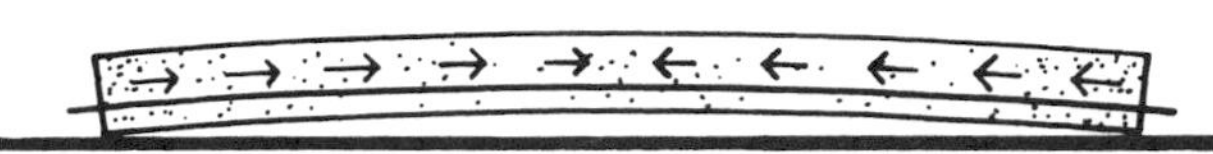

3. When the strands are cut the concrete goes into compression and the beam takes on a camber

FIGURE 13.37
Pretensioning. Photographs of pretensioned steel strands for a beam are shown in Chapter 15.

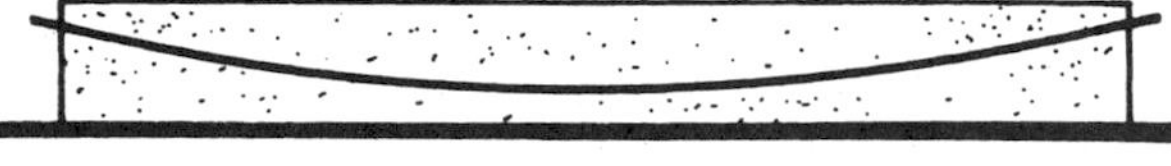

1. In posttensioning, the concrete is not allowed to bond to the steel strands during curing

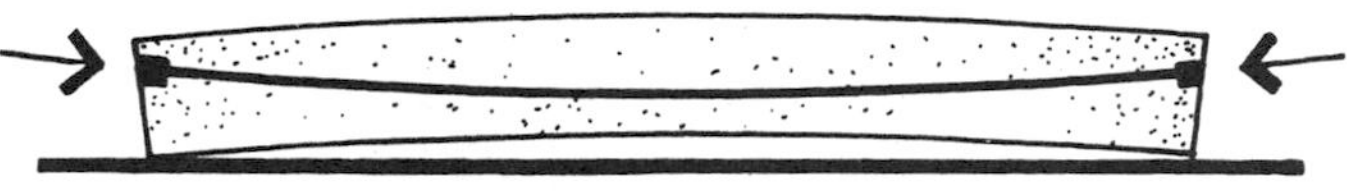

2. After the concrete has cured, the strands are tensioned with a hydraulic jack and anchored to the ends of the beam. If the strands are draped, as shown here, higher structural efficiency is possible than with straight strands

FIGURE 13.38
Posttensioning, using draped strands to more nearly approximate the flow path of tensile forces in the beam.

A hydraulic jack is inserted between the other end of the tendon and a similar steel plate in the other end of the member. The jack applies a large tensile force to the tendon while compressing the concrete with an equal but opposite force that is applied through the plate. The stretched tendon is anchored to the plate at the second end of the member before the jack is removed (Figures 13.38–13.40). (For very long members, the tendons are jacked from both ends to be sure that frictional losses in the tubes do not prevent uniform tensioning.)

The net effect of posttensioning is essentially identical to that of pretensioning. The difference is that in posttensioning, abutments are not needed because the member itself provides the abutting force needed to tension the steel. When the posttensioning process is complete, the tendons may be left unbonded, or, if they are in a steel tube, they may be bonded by injecting grout to fill the space between the tendons and the tube. Bonded construction is common in bridges and other heavy structures, but most posttensioning in buildings is done with unbonded tendons. These are made up of seven cold-drawn steel wires and are either 0.5 or 0.6 inch (12.7 or 15.2 mm) in diameter (Figures 13.40 and 13.41). The tendon is coated with a lubricant and covered with a plastic sheath at the factory.

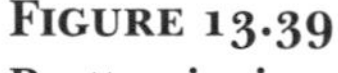

FIGURE 13.39
Posttensioning draped tendons in a large concrete beam with a hydraulic jack. Each tendon consists of a number of individual high-strength steel strands. The bent bars projecting from the top of the beam will be embedded in the concrete slab that the beam will support to allow them to act together as a composite structure. *(Reprinted with permission of the Portland Cement Association from* Design and Control of Concrete Mixtures, *12th edition; Photos: Portland Cement Association, Skokie, IL)*

FIGURE 13.40
Most beams and slabs in buildings are posttensioned with plastic-sheathed, unbonded tendons. The pump and hydraulic jack (also called a "ram") are small and portable. *(Courtesy of Constructive Services, Inc., Dedham, Massachusetts)*

(*a*)

(*b*)

(*c*)

FIGURE 13.41
An end anchorage for small posttensioning tendons is ingeniously simple. (*a*) A steel anchor plate and a plastic pocket former are nailed to the inside of the concrete formwork, with the larger circular face of the pocket former placed against the vertical face of the formwork. The end of the tendon extends through a hole drilled in the formwork. (*b*) After the concrete has cured, the formwork is stripped and the pocket former is retracted, leaving a neat pocket in the edge of the slab for access to the anchor plate, which lies recessed below the surface of the concrete. Two conical wedges with sharp ridges inside are inserted around the tendon and into the conical hole in the anchor plate. (*c*) The ram presses against the wedges and draws the tendon through them until the gauge on the pump indicates that the required tension has been achieved. When the ram is withdrawn, the wedges are drawn into the conical hole, grip the tendon, and maintain the tension. After all the tendons have been tensioned, the excess length of tendon is cut off and the pocket is grouted flush with the edge of the slab. *(Photos by Edward Allen)*

Even higher structural efficiencies are possible in a prestressed beam or slab if the steel strands are placed to follow as closely as possible the lines of tensile force that are diagrammed in Figure 13.26. In a posttensioned beam or slab, this is done by using chairs of varying heights to support the tendons along a curving line that traces the center of the tensile forces in the member. Such a tendon is referred to as being *draped* (Figures 13.38 and 13.39). Draping is impractical in pretensioned members because the tendons would have to be pulled down at many points along their length. But pretensioned strands can be *harped*, that is, pulled up and down in the formwork to make a downward-pointing or flattened V shape in each member that approximates very roughly the shape of a draped tendon (Figure 13.42).

Because it is always highly compressed by prestressing force, the concrete in a prestressed member is subject to long-term progressive shortening (creep). The steel strands also stretch slightly over time and lose some of their prestressing force. Initial prestressing forces must be increased slightly above their theoretically correct values to make up for these long-term movements. Further increases in initial tension are needed to accommodate the slight curing shrinkage that takes place in concrete, small, short-term movements caused by elastic shortening of the concrete during structural loading, and frictional losses and initial slippage or "set" of the strand anchors in posttensioned members.

Succeeding chapters will discuss prestressed concrete, both pretensioned and posttensioned, in greater detail, showing its application to various standard precast and cast-in-place systems of construction.

Innovations in Concrete Construction

The basic materials, concrete and steel, are constantly undergoing research and development, leading to higher allowable strengths and decreased weight of structures. Structural lightweight concretes are being used more widely to reduce loads still further. Shrinkage-compensating cements have been developed for use in concrete structures that cannot be allowed to shrink during curing. Self-consolidating concrete is being used to improve productivity and reduce labor costs. Improved coatings for reinforcing steel, steel alloys with greater corrosion resistance, and nonmetallic reinforcing are being used to extend the life of structures exposed to the weather, marine environments, and road salts.

Concretes with compressive strengths as high as 30,000 psi (200 MPa) and exhibiting even usable tensile strength are called *ultra-high performance concretes (UHPCs)*. They are formulated from portland cement, silica fume, silica or quartz flour (extremely finely ground silica or quartz), fine sand, high-range water reducer, water, and steel fibrous reinforcing. There is no large aggregate. The resulting concrete is characterized by a dense packing of the fine aggregate and finely ground mineral powders within the cementitious matrix, resulting in concrete that is stronger, less permeable, and more durable than conventional high-strength mixes. The addition of steel macrofiber reinforcing imparts tensile strength and ductility (toughness). Ultra-high performance concrete can be used in the casting of remarkably thin architectural elements. Precast exterior cladding panels and arched canopies, for example, both no thicker than ¾ inch (20 mm) and with no conventional reinforcing, have been produced from this material (Figures 13.43 and 13.44). Ultra-high performance concrete is suitable for structural applications where very high load capacity, low weight, and high durability are required. By reducing the volume of concrete required in such applications, its use also reduces green house gas emissions resulting from the manufacture of concrete in comparison to the use of conventional, lower-strength concretes.

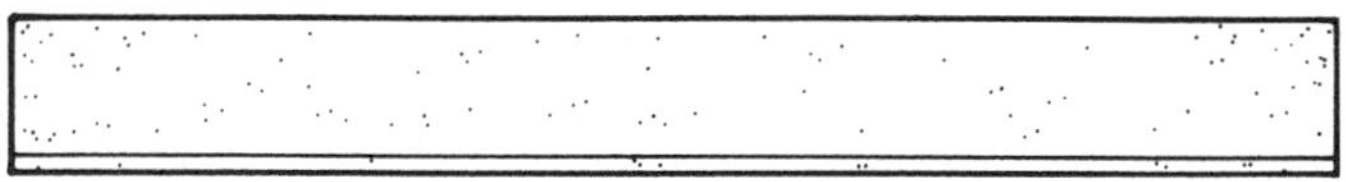

Straight pretensioning strands

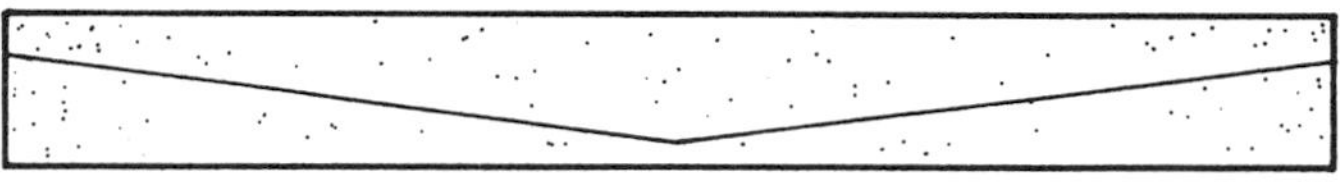

Depressed pretensioning strands

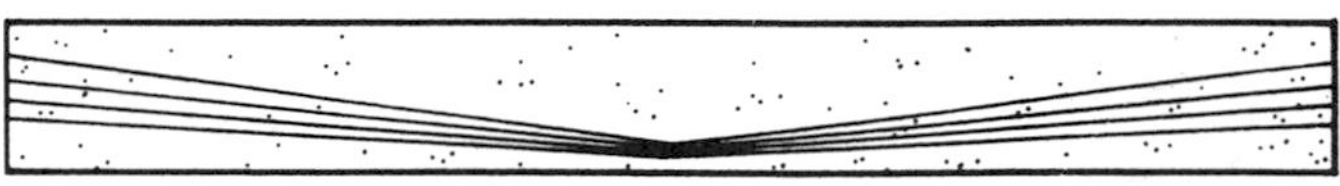

Harped pretensioning strands

Figure 13.42
Shaping pretensioning strands to improve structural efficiency. Examples of depressed and harped strands are shown in Chapter 15.

FIGURE 13.43
The Shawnessy LRT Station in Calgary, Alberta, designed by Stantec Architecture, Ltd. The canopy shells were cast from ultra-high-performance concrete and are only ¾ inch (19 mm) thick. *(Courtesy of Lafarge North America, Inc. Photography by Tucker Photography)*

Light transmitting concrete is made from precast concrete blocks or panels with embedded optic fibers or fabrics that allow light to pass through the material while retaining the strength and durability of the concrete. This recently formulated material is finding application in nonstructural partitions, countertops, and other architectural elements.

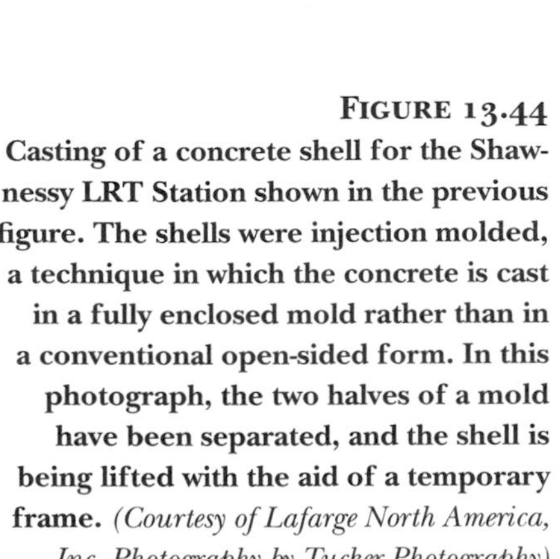

FIGURE 13.44
Casting of a concrete shell for the Shawnessy LRT Station shown in the previous figure. The shells were injection molded, a technique in which the concrete is cast in a fully enclosed mold rather than in a conventional open-sided form. In this photograph, the two halves of a mold have been separated, and the shell is being lifted with the aid of a temporary frame. *(Courtesy of Lafarge North America, Inc. Photography by Tucker Photography)*

ACI 301

In North America, concrete structures are built according to the requirements of *ACI 301: Specifications for Structural Concrete for Buildings*, a publication of the American Concrete Institute. This is a comprehensive document that covers every aspect of concrete work: formwork, reinforcement, chairs and bolsters, concrete mixtures, handling and placing of concrete, lightweight concrete, prestressing, and the use of concrete in exposed architectural surfaces. It is a standard that is familiar to architects, engineers, contractors, and building inspectors, and it furnishes the common basis for everyone who works in designing and constructing a concrete building.

CSI/CSC

MasterFormat Sections for Concrete Construction

03 10 00	CONCRETE FORMING AND ACCESSORIES
03 20 00	CONCRETE REINFORCING
03 21 00	Reinforcing Steel
03 22 00	Welded Wire Reinforcing
03 23 00	Stressing Tendons
03 24 00	Fibrous Reinforcing
03 30 00	CAST-IN-PLACE CONCRETE
03 40 00	PRECAST CONCRETE

Selected References

1. Portland Cement Association. *Design and Control of Concrete Mixtures* (14th ed). Skokie, IL, 2002.

The 372 pages of this book summarize clearly and succinctly, with many explanatory photographs and tables, the state of current practice in making, placing, finishing, and curing concrete.

2. Concrete Reinforcing Steel Institute. *Manual of Standard Practice.* Schaumburg, IL, updated regularly.

Specifications for reinforcing, steel, welded wire fabric, bar supports, detailing, fabrication, and installation are standardized in this booklet.

3. American Concrete Institute. *Field Reference Manual: Standard Specifications for Structural Concrete ACI 301 with Selected ACI References.* Farmington Hills, MI, ACI International, updated regularly.

This manual includes ACI 301, the detailed standard specification for every aspect of structural concrete, as well as numerous other ACI references relevant to cast-in-place concrete construction.

4. Mehta, P. Kumar, and Paulo J. M. Monteiro. *Concrete: Microstructure, Properties, and Materials* (3rd ed.). New York, McGraw-Hill, 2006.

For the reader who wishes to explore further the science and mechanics of concrete, this text provides an in-depth treatment of concrete materials, formulation, and behavior.

Web Sites

Concrete Construction

Author's supplementary web site: **www.ianosbackfill.com/13_concrete_construction**
American Concrete Institute International: **www.concrete.org**
Portland Cement Associaton: **www.cement.org**

Reinforcing

Concrete Reinforcing Steel Institute **www.crsi.org**

Key Terms and Concepts

portland cement
concrete
aggregate
coarse aggregate
fine aggregate
curing
hydration
heat of hydration
drying shrinkage
clinker
air-entraining cement
white portland cement
high-volume-fly-ash concrete (HVFA concrete)
lightweight aggregate
structural lightweight aggregate
expanded shale aggregate
vermiculite
perlite
supplementary cementitious material (SCM)
pozzolan
fly ash
silica fume, microsilica
natural pozzolan
high reactivity metakaolin
blast furnace slag, slag cement
hydraulic cement
blended cement
concrete admixture
air-entraining admixture
water-reducing admixture
high-range water-reducing admixture, superplasticizer
accelerating admixture
retarding admixture
workability agent
shrinkage-reducing admixture
corrosion inhibitor
freeze protection admixture
extended set-control admixture
coloring agent
water–cement ratio, w-c ratio
transit-mixed concrete
slump test
slurry
segregation
dropchute
consolidation
self-consolidating concrete (SCC)
formwork
form release compound
precasting
sitecasting, cast-in-place construction
reinforced concrete
steel reinforcing bars
deformed reinforcing bars
rebar congestion
welded wire reinforcing (WWR), welded wire fabric (WWF)
bottom bar
cover
bond
hook
stirrup
U-stirrup
closed stirrup-tie
chair
bolster
one-way action
shrinkage–temperature steel
two-way action
vertical bar, column bar
tie
column tie
column spiral
fibrous reinforcing
microfiber reinforcing
plastic shrinkage cracking
macrofiber reinforcing
creep
prestress
prestressed concrete
pretensioning
camber
posttensioning
tendon
draped tendon
harped tendon
ultra-high performance concrete (UHPC)
light transmitting concrete
ACI 301

Review Questions

1. What is the difference between cement and concrete?
2. List the conditions that must be met to make a satisfactory concrete mix.
3. List the precautions that should be taken to cure concrete properly. How do these change in very hot, very windy, and very cold weather?
4. What problems are likely to occur if concrete has too low a slump? Too high a slump? How can the slump be increased without increasing the water content of the concrete mixture?
5. Explain how steel reinforcing bars work in concrete.
6. Explain the role of stirrups in beams.
7. Explain the role of ties in columns.
8. What does shrinkage-temperature steel do? Where is it used?
9. Explain the differences between reinforcing and prestressing and the relative advantages and disadvantages of each.
10. Under what circumstances would you use pretensioning, and under what circumstances would you use posttensioning?
11. Explain the advantages of using higher strength reinforcing bars in concrete that requires very heavy reinforcing.

Exercises

1. Design a simple concrete mixture. Mix it and pour some test cylinders for several water–cement ratios. Cure and test the cylinders. Plot a graph of concrete strength versus water–cement ratio.
2. Sketch from memory the pattern of reinforcing for a continuous concrete beam. Add notes to explain the function of each feature of the reinforcing.
3. Design, form, reinforce, and cast a small concrete beam, perhaps 6 to 12 feet (2–4 m) long. Get help from a teacher or professional, if necessary, in designing the beam.
4. Visit a construction site where concrete work is going on. Examine the forms, reinforcing, and concrete work. Observe how concrete is brought to the site, transported, placed, compacted, and finished. How is the concrete supported after it has been poured? For how long?

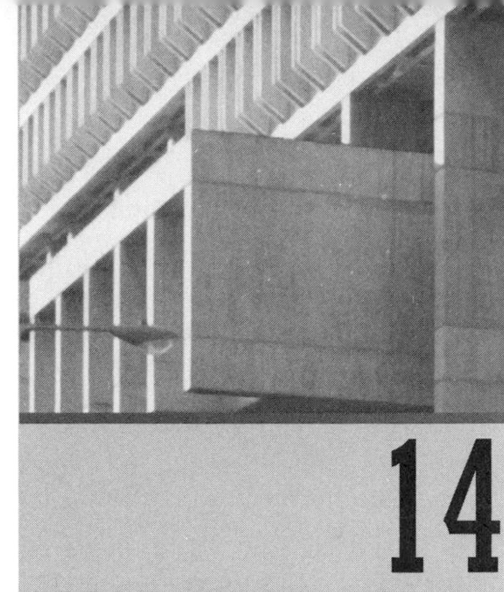

14

SITECAST CONCRETE FRAMING SYSTEMS

Boston City Hall makes bold use of sitecast concrete in its structure, facades, and interiors. Its base is faced with brick masonry. *(Architects: Kallmann, McKinnell, and Wood. Photograph: Ezra Stoller © ESTO)*

Concrete that is cast into forms on the building site offers almost unlimited possibilities to the designer. Any shape that can be formed can be cast, with any of a limitless selection of surface textures, and books on modern architecture are filled with graphic examples of the realization of this extravagant promise. Certain types of concrete elements cannot be precast, but can only be cast on the site—foundation caissons and spread footings, slabs on grade, structural elements too large or too heavy to transport from a precasting plant, elements so irregular or special in form as to rule out precasting, slab toppings over precast floor and roof elements, and many types of structures with two-way slab action or full structural continuity from one member to another. In many cases where sitecast concrete could be replaced with precast, sitecast remains the method of choice simply because of its more massive, monolithic architectural character.

Sitecast concrete structures tend to be heavier than most other types of structures, which can lead to the selection of a precast concrete or structural steel frame instead if foundation loadings are critical. Sitecast buildings are also relatively slow to construct because each level of the building must be formed, reinforced, poured, cured, and stripped of formwork before construction of another level can begin. In effect, each element of a sitecast concrete building is manufactured in place, often under variable weather conditions, whereas the majority of the work on steel or precast concrete buildings is done in factories and shops where worker access, tooling, materials-handling equipment, and environmental conditions are generally superior to those on the job site. The technology of sitecasting has evolved rapidly, however, in response to its own inherent limitations, with streamlined methods of materials handling, systems of reusable formwork that can be erected or taken down almost instantaneously, extensive prefabrication of reinforcing elements, and mechanization of finishing operations. The rapid pace of this evolution has kept sitecast concrete among the construction techniques most favored by building owners, architects, and engineers.

FIGURE 14.1
Unity Temple in Oak Park, Illinois, was constructed by architect Frank Lloyd Wright in 1906. Its structure and exterior surfaces were cast in concrete, making it one of the earliest buildings in the United States to be built primarily of this material. *(Photo by John McCarthy. Courtesy of Chicago Historical Society ICHi-18291)*

Casting a Concrete Slab on Grade

A concrete *slab on grade* is a level surface of concrete that lies directly on the ground. Slabs on grade are used for roads, sidewalks, patios, airport runways, and basements or ground floors of buildings. A slab-on-grade floor usually experiences little structural stress except a direct transmission of compression between its superimposed loads and the ground beneath, so it furnishes a simple example of the operations involved in the sitecasting of concrete (Figure 14.2).

To prepare for the placement of a slab on grade, the unstable topsoil is scraped away to expose the subsoil beneath. If the exposed subsoil is too soft, it is compacted or replaced with more stable material. Next, a layer of approximately ¾-inch-diameter (19-mm) crushed stone at least 4 inches (100 mm) deep, also referred to as a *capillary break*, is compacted over the subsoil as a drainage layer to keep water away from the underside of the slab. Where the slab is not being cast within surrounding walls, a simple edge form—a strip of wood or metal fastened to stakes driven into the ground—is constructed around the perimeter of the area to be poured and is coated with a form release compound to prevent the concrete from sticking (Figure 14.3). Where walls surround the slab to be poured, a compressible joint filler material is placed to create an isolation joint between the slab and the surrounding walls, as explained later in this chapter. The top edge of the form is carefully leveled. The thickness of the slab may range from 3 inches (100 mm) for a residential floor to 6 or 8 inches (150 or 200 mm) for an industrial floor. It may reach a foot or more (300 mm) for an airport runway, which must carry very large, concentrated loads from airplane wheels and diffuse them into the ground. For interior floor slabs on grade but not exterior slabs, a *moisture barrier* (also called a *vapor retarder*), usually consisting of a heavy plastic sheet, is laid over the crushed stone to prevent water vapor from rising through the slab from the ground beneath.

For some interior floor slab construction, a layer of fine crushed stone or sand, 2 to 4 inches (50 to 100 mm) thick, may be placed over the vapor retarder. Prior to placing concrete, this layer of fine aggregate protects the vapor retarder sheet from damage. Once concrete is placed, this layer absorbs excess water from the

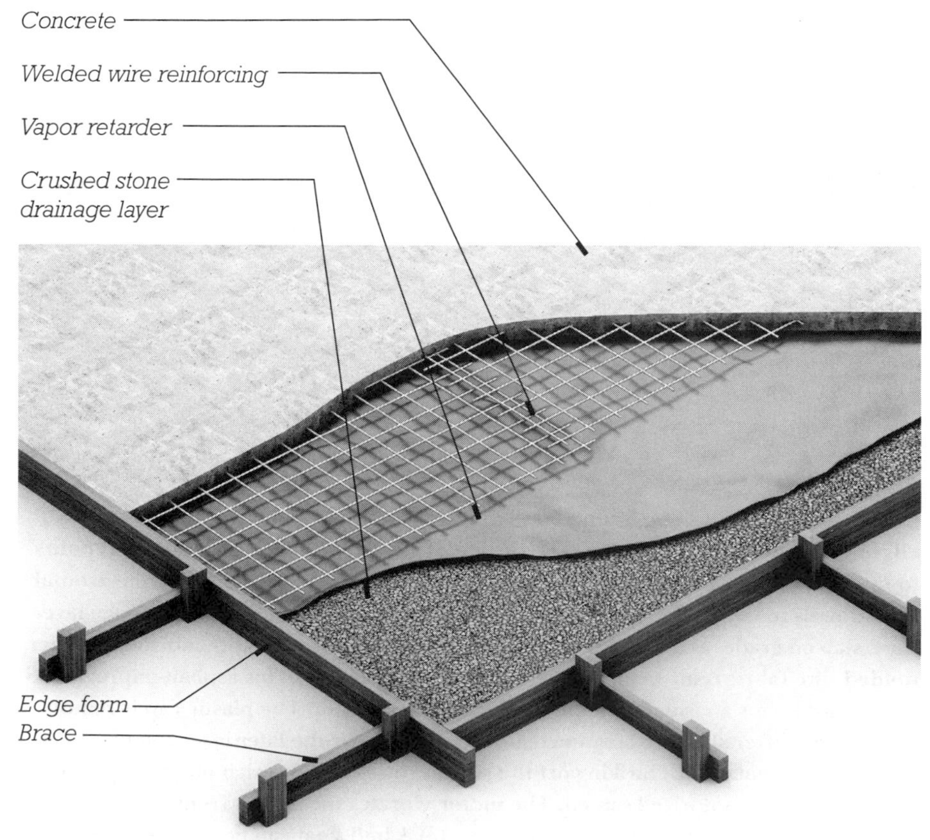

Figure 14.2
The construction of a concrete slab on grade. Notice how the wire reinforcing fabric is overlapped where two sheets of fabric join. As explained in the accompanying text, a layer of fine crushed stone or sand is sometimes added over the vapor retarder to protect the vapor retarder from damage before the concrete is poured and to promote uniform curing once the concrete is placed.

(*a*) (*b*) (*c*) (*d*) (*e*) (*f*) (*g*) (*h*) (*i*) (*j*)

FIGURE 14.3
Constructing and finishing a concrete slab on grade. (*a*) Attaching a proprietary slab edge form to a supporting stake. The profile of the edge form causes adjacent pours to interlock. Hole knockouts in the metal form allow for the placement of horizontal steel dowels to tie the pours together. (*b*) To the right, a crushed-stone drainage layer for a slab on grade, and to the left, a slab section ready to pour, with moisture barrier, welded wire fabric reinforcing, and edge forms in place. (*c*) This asphalt-impregnated fiberboard forms an isolation joint when cast into the slab. The plastic cap is removed immediately after slab finishing to create a clean slot for the later insertion of an elastomeric joint sealant. (*d*) Striking off the surface of a concrete slab on grade just after pouring, using a motorized screed. The motor vibrates the screed from end to end to work the wet concrete into a level surface. (*e*) A bull float can be used for preliminary smoothing of the surface immediately after screeding. (*f*) Hand floating brings cement paste to the surface and produces a plane surface.

concrete, helping to prevent curling (warping) of the slab that can occur during curing when the top of the slab loses moisture more rapidly than the bottom. However, this practice also runs the risk of creating a reservoir for the storage of water under the slab that can lead to moisture-related problems after the slab is in service, especially when the slab is covered with floor coverings that are sensitive to moisture, such as resilient sheet flooring or vinyl composition tiles. Particularly with the advent of tougher vapor retarder sheets that are less vulnerable to damage during construction operations, the practice of placing an aggregate layer over the vapor retarder has become less prevalent.

A reinforcing mesh of welded wire reinforcing (also called "welded wire fabric"), cut to a size just a bit smaller than the dimensions of the slab, is laid over the moisture barrier or crushed stone. The reinforcing most commonly used for lightly loaded slabs, such as those in houses, is 6 × 6-W1.4 × W1.4, which has a wire spacing of 6 inches (150 mm) in each direction and a wire diameter of 0.135 inch (3.43 mm—see Figure 13.23 for more information on welded wire fabric). For slabs in factories, warehouses, and airports, a fabric made of heavier wires or a grid of reinforcing bars may be used instead. The grid of wires or bars helps protect the slab against cracking that might be caused by concrete shrinkage, temperature stresses, concentrated loads, frost heaving, or settlement of the ground beneath. Fibrous (macrofiber) reinforcing, discussed in the previous chapter, can also be used in place of wire reinforcing for general crack control in slabs.

(*g*) Floating can be done by machine instead of by hand. (*h*) Steel troweling after floating produces a dense, hard, smooth surface. (*i*) A section of concrete slab on grade finished and ready for curing. The dowels inserted through the edge form will connect to the sections of slab that will be poured next. (*j*) One method of damp curing a slab is to cover it with polyethylene sheet to retain moisture in the concrete. *(Photos a, b, c, and i courtesy of Vulcan Metal Products, Inc., Birmingham, Alabama; photos d, e, f, g, h, and j reprinted with permission of the Portland Cement Association from Design and Control of Concrete Mixtures, 12th edition, Portland Cement Association, Skokie, IL)*

Pouring and Finishing the Slab on Grade

Pouring (casting) of the slab commences with placing concrete in the formwork. This may be done directly from the chute of a transit-mix truck, or with wheelbarrows, concrete buggies, a large crane-mounted concrete bucket, a conveyor belt, or a concrete pump and hoses. The method selected will depend on the scale of the job and the accessibility of the slab to the truck delivering the concrete. The concrete is spread by workers using shovels or rakes until the form is full. Then the same tools are used to agitate the concrete slightly, especially around the edges, to eliminate air pockets. As the concrete is placed, the concrete masons reach into the wet concrete with metal hooks and raise the welded wire fabric to approximately the midheight of the slab, the position in which it is best able to resist tensile forces caused by forces acting either upward or downward on the slab.

The first operation in finishing the slab is to *strike off* or *screed* the concrete by drawing a stiff plank of wood or metal across the top edges of the formwork to achieve a level concrete surface (Figure 14.3*d*). This is done with an end-to-end sawing motion of the plank that avoids tearing the projecting pieces of coarse aggregate from the surface of the wet concrete. A bulge of concrete is maintained in front of the screed as it progresses across the slab, so that when a low point is encountered, concrete from the bulge will flow in to fill it.

Immediately after striking off the concrete, the slab receives its initial *floating*. This step is usually performed by hand, using flat-surfaced tools, typically 4 to 10 feet (1.2 to 3 m) in length, called *bull floats* or *darbies* (Figure 14.3*e*). These are drawn across the concrete to flatten and consolidate its surface. After this initial floating, the top of the slab is level but still rather rough. If a concrete topping will later be poured over the slab, or if a floor finish of terrazzo, stone, brick, or quarry tile will be applied, the slab may be left to cure without further finishing.

If a smoother surface is desired, additional finishing operations proceed after a period of time during which the concrete begins to stiffen and the watery sheen, called *bleed water*, evaporates from the surface of the slab. First, specially shaped hand tools may be used to form neatly rounded edges around the perimeter of the slab and control joints in the interior. Next, the slab is floated a second time to further consolidate its surface. At this stage, small slabs may be floated by hand (Figure 14.3*f*), but for larger slabs, rotary power floats are used (Figure 14.3*g*). The working surfaces of floats are made of wood or of metal with a slightly rough surface. As the float is drawn across the surface, the friction generated by this roughness vibrates the concrete gently and brings cement paste to the surface, where it is smoothed over the coarse aggregate and into low spots by the float. If too much floating is done, however, an excess of paste and free water rises to the surface and forms puddles, making it almost impossible to get a good finish. Experience on the part of the mason is essential to floating, as it is to all slab finishing operations, to know just when to begin each operation and just when to stop. The floated slab has a lightly textured surface that is appropriate for outdoor walks and pavings without further finishing.

For a completely smooth, dense surface, the slab must also be *troweled.*

This is done immediately after the second floating, either by hand, using a smooth, rectangular *steel trowel* (Figure 14.3*h*), or with a *rotary power trowel.* If the concrete mason cannot reach all areas of the slab from around the edges, squares of plywood called *knee boards,* two per mason, are placed on the surface of the concrete. These distribute the mason's weight sufficiently that he or she can kneel on the surface without making indentations. Any marks left by the knee boards are removed by the trowel as the mason works backward across the surface from one edge to the other. If a nonslip surface is required, a stiff-bristled janitor's pushbroom is drawn across the surface of the slab after troweling to produce a striated texture called a *broom finish.*

Where a concrete slab must meet narrow floor flatness limits, it may be restraightened after each floating or troweling operation. *Restraightening* is commonly performed with a rectangular flat-bottomed *straightedge,* roughly 10 feet (3 m) in length, which is drawn across the concrete slab surface to remove minor undulations produced during floating or troweling.

Shake-on hardeners are sometimes sprinkled over the surface of a slab between the screeding and floating operations. These dry powders react with the concrete to form a very hard, durable surface for such heavy-wear applications as warehouses and factories.

When the finishing operations have been completed, the slab should be cured under damp conditions for at least a week; otherwise, its surface may crack or become dusty from premature drying. Damp curing may be accomplished by covering the slab with an absorbent material such as sawdust, earth, sand, straw, or burlap and maintaining the material in a damp condition for the required length of time. Or an impervious sheet of plastic or waterproof paper may be placed over the slab soon after troweling to prevent the escape of moisture from the concrete (Figure 14.3*j*). The same effect may be obtained by spraying the concrete surface with one or more applications of a liquid *curing compound,* which forms an invisible moisture barrier membrane over the slab.

No concrete floor is perfectly flat. The normal finishing process produces a surface that undulates, usually imperceptibly, between low and high areas that go unnoticed in everyday use. Traditionally, the flatness of concrete slabs is specified as the maximum gap size, typically in the range of ⅛ to ⅜ inch (3 to 10 mm), permitted under a 10-foot (3-m) straightedge placed anywhere on the floor. Industrial warehouses that use high-rise forklift trucks, however, require floors whose flatness is controlled to within more narrow tolerances. These *superflat floors,* as well as other floors where close control over flatness is desired, are specified according to a more complicated system of index numbers, called *F-numbers,* that correspond to the degrees of flatness (waviness) and levelness (conformity to a horizontal plane) that are required, and are produced using special finishing equipment and techniques. Because of its extreme accuracy, a laser-guided automatic straightedging machine (Figure 14.4) is often used in the creation of superflat floors. This device produces a slab surface that is flat and level to within consistently small tolerances, and does so at a very rapid rate.

FIGURE 14.4
Guided by a laser beam, the motorized straightedging device on this machine can strike off 240 square feet (22 m²) of slab surface per minute to an extremely exacting standard of flatness. The worker to the right smooths the surface with a bull float. *(Photo by Wironen, Inc. Courtesy of the Laser Screed Company, Inc., New Ipswich, New Hampshire)*

Controlling Cracking in Concrete Slabs on Grade

Because concrete slabs on grade are relatively thin in relation to their horizontal dimensions and usually are relatively lightly reinforced, they are particularly prone to cracking. The stresses that cause cracking may originate from the shrinkage that is a normal part of the concrete curing process, from thermal expansion and contraction of the slab, or from differential movement between the slab and abutting building elements. If such cracks are allowed to occur randomly, they can be unsightly and can compromise the functionality of the slab.

Most commonly, cracking in concrete slabs on grade is managed by introducing an organized system of joints into the slab that allow stresses to be relieved without compromising the appearance or performance of the slab. *Control joints,* also called *contraction joints,* are intentionally weakened sections through a concrete slab where the tensile forces caused by concrete drying shrinkage can be relieved without disfiguring the slab. They are usually formed as grooves that extend at least one-quarter of the depth of the slab, created either by running a special trowel along a straightedge while the concrete is still plastic or by sawing partially through the concrete shortly after it begins to harden using a diamond or abrasive saw blade in a power circular saw. To provide a further inducement for cracks to occur at control joint locations rather than elsewhere in the slab, reinforcing in the slab may be partially discontinued where it crosses these joints. Control joint spacing recommendations vary with the thickness of the slab and the shrinkage rate of the concrete. For example, for slabs 4 to 8 inches (100–200 mm) thick made with ordinary concrete, joints spacings from 11 feet 6 inches to 17 feet 6 inches (3.6–5.3 m) are recommended (the thinner the slab, the closer the joint spacing) Where control joints run in perpendicular directions, they should be spaced more or less equally in each direction to create panels that are square or close to square in proportion.

Reinforced concrete made "pilotis" possible. The house is in the air, away from the ground; the garden runs under the house, and it is also above the house, on the roof. . . . Reinforced concrete is the means which makes it possible to build all of one material. . . . Reinforced concrete brings the free plan into the house! Floors no longer have to stand simply one on top of the other. They are free. . . . Reinforced concrete revolutionizes the history of the window. Windows can run from one end of the facade to the other . . .

Le Corbusier and P. Jeanneret, *Oeuvre Complète 1910–1929,* 1956.

Isolation joints, sometimes also called *expansion joints,* are formed by casting full-depth preformed joint materials, typically ⅜ to ¾ inch (10–20 mm) in width, into the slab (Figure 14.3*c*), completely separating the slab from adjacent elements. Isolation joints relieve potential stresses by allowing freedom of movement of the slab with respect to other building parts or other portions of the slab—movements that may occur due to thermal expansion and contraction, structural loading, or differential settlement. Isolation joints are commonly provided where the edge of a concrete slab abuts adjacent walls or curbs, as well as around elements, such as columns or loadbearing walls, which pass through the slab within its perimeter. Isolation joints are also used to divide large or irregularly shaped slabs into smaller, more simply shaped rectangular areas that are less prone to stress accumulation.

Concrete itself can be manipulated to reduce cracking: Shrinkage-reducing chemical admixtures and some supplementary cementitious materials, such as fly ash, reduce drying shrinkage. Lowering the water–cement ratio of the concrete mix both reduces drying shrinkage and results in finished concrete that is stronger and more crack resistant. Specially formulated *shrinkage-compensating cements* can completely nullify drying shrinkage, allowing the casting of large slabs on grade entirely free of contraction joints. The amount of steel reinforcing in the slab can be increased or fibrous reinforcing can be added to the concrete mix to improve a concrete slab's resistance to tensile forces. Protecting a freshly poured concrete slab from premature drying during the damp curing process reduces cracking while the concrete hardens and ensures that the concrete attains its full design strength.

Under certain circumstances, it is advantageous to posttension a slab on grade, using level tendons in both directions at the midheight of the slab rather than welded wire fabric. Posttensioning places the entire slab under sufficient compressive stress so that it will never experience tensile stress under any anticipated conditions. Posttensioning makes floors resistant to cracking under concentrated loads, eliminates the need to make control joints, and often permits the use of a thinner slab. It is especially effective for slabs over unstable or inconsistent soils and for superflat floors.

Casting a Concrete Wall

A reinforced concrete wall at ground level usually rests on a poured concrete strip footing (Figures 14.5–14.7). The footing is formed and poured much like a concrete slab on grade. Its cross-sectional dimensions and its reinforcing, if any, are determined by the structural engineer. A *key*, a groove that will form a mechanical connection to the wall, is sometimes formed in the top of the footing with strips of wood that are temporarily embedded in the wet concrete. Vertically projecting *dowels* of steel reinforcing bars are usually installed in the footing before pouring; these will later be overlapped with the vertical bars in the walls to form a strong structural connection. After pouring, the top of the footing is screeded; no further finishing operations are required. The footing is left to cure for at least a day before the wall forms are erected.

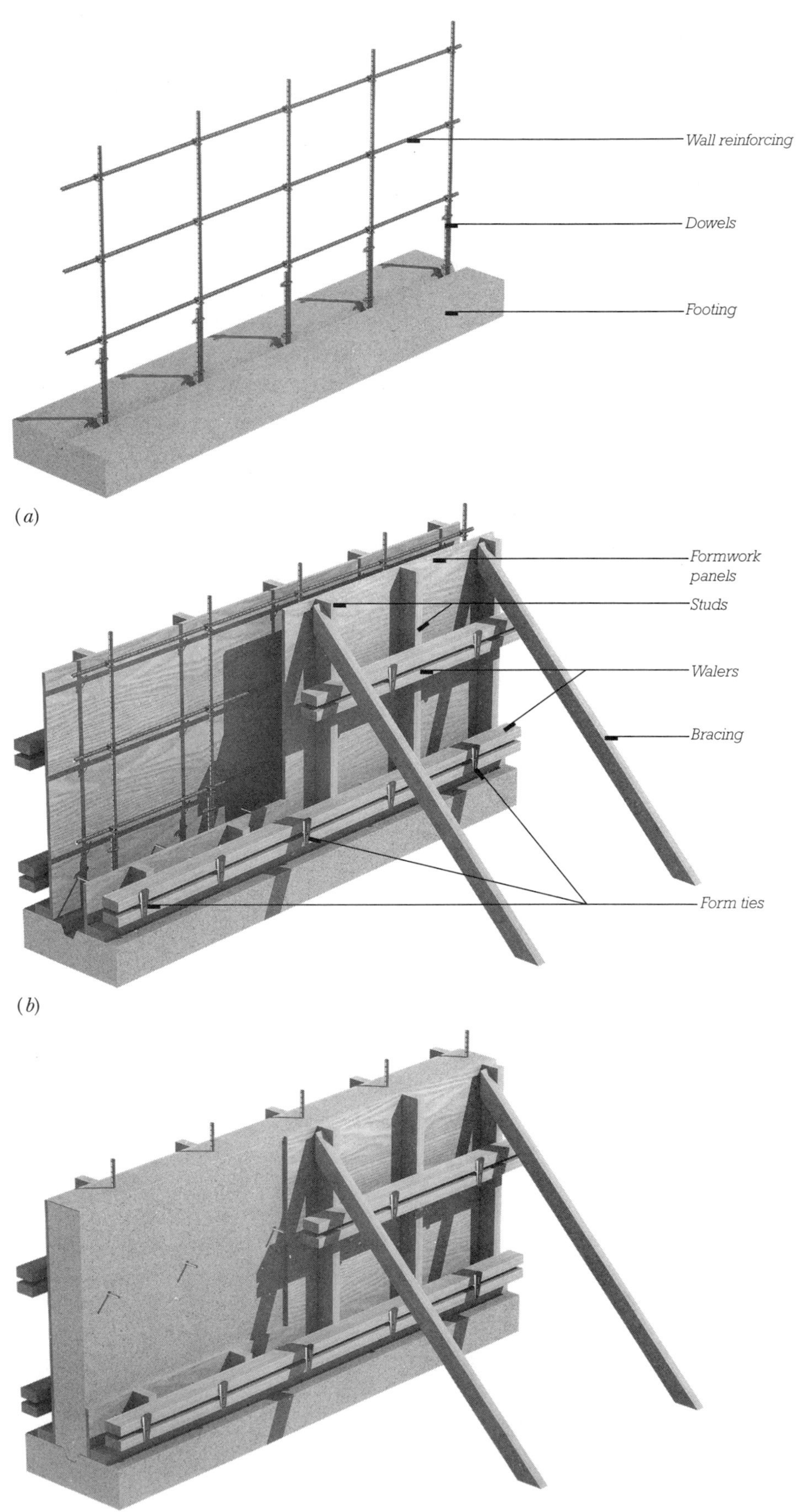

Figure 14.5
The process of casting a concrete wall. (*a*) Vertical reinforcing bars are wired to the dowels that project from the footing, and horizontal bars are wired to the vertical bars. (*b*) The formwork is erected. Sheets of plywood form the faces of the concrete. They are supported by vertical wood studs. The studs are supported against the pressure of the wet concrete by horizontal walers. The walers are supported by steel rod ties that pass through holes in the plywood to the walers on the other side. The ties also act as spreaders to maintain a spacing between the plywood walls that is equal to the desired thickness of the wall. Diagonal braces keep the whole assembly plumb and straight. (*c*) After the concrete has been poured, consolidated, and cured, the wedges that secure the walers to the form ties are driven out and the formwork is pulled off the concrete. The projecting ends of the form ties are broken off.

FIGURE 14.6
Protected against falling by a safety harness, a worker stands on the reinforcing bars for a concrete wall to wire another horizontal bar in position. *(Photo courtesy of DBI/SALA, Red Wing, Minnesota)*

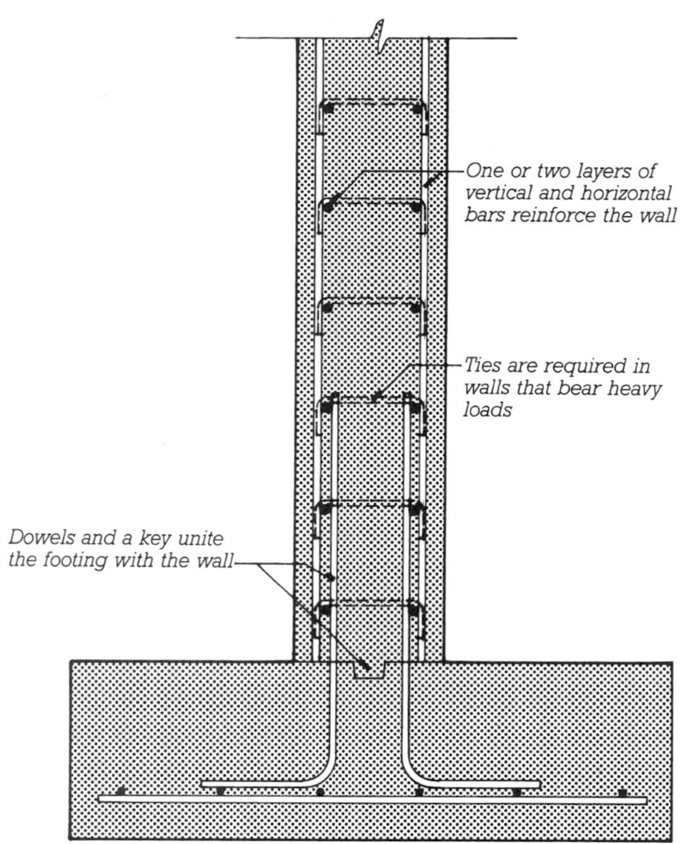
One or two layers of vertical and horizontal bars reinforce the wall
Ties are required in walls that bear heavy loads
Dowels and a key unite the footing with the wall

FIGURE 14.7
Section through a reinforced concrete wall, with two layers of horizontal and vertical reinforcing bars for greater strength.

The wall reinforcing, either in one vertical layer of horizontal and vertical bars at the center of the wall or two vertical layers near the faces of the wall, as specified by the structural engineer, is installed next, with the bars wired to one another at the intersections. The vertical bars are overlapped with the corresponding dowels projecting from the footing. L-shaped horizontal bars are installed at wall corners to maintain full structural continuity between the two portions of the wall. If the top of the wall will connect to a concrete floor or another wall, rods are left projecting from the formwork. These will be embedded in the later pour of concrete to form a continuous connection.

Wall forms may be custom built of lumber and plywood for each job, but it is more usual to use standard reusable prefabricated formwork panels. The panels for one side of the form are coated with a form release compound, set on the footing, aligned carefully, and braced. The *form ties*, which are small-diameter steel rods specially shaped to hold the formwork together under the pressure of the wet concrete, are inserted through holes provided in the formwork panels and secured to the back of the form by devices supplied with the form ties. Both ties and fasteners vary in detail from one manufacturer to another (Figures 14.8 and 14.9). The ties will pass straight through the concrete wall from one side to the other and remain embedded permanently in the wall after it is poured. This may seem like an odd way to hold wall forms together, but the pressures of the wet, heavy concrete on the forms are so large that there is no other economical way of dealing with them.

When the ties are in place and the reinforcing has been inspected, the formwork for the second side of the wall is erected, the *walers* and *braces* are added (Figure 14.5), and the forms are inspected to be sure that they are straight, plumb, correctly aligned, and adequately tied and braced. A surveyor's transit or laser leveling device is used to establish the exact height to which the concrete will be poured, and this height is marked all around the inside of the forms. Pouring may then proceed.

Concrete is brought to the site, test cylinders are made, and a slump test is performed to check for the proper pouring consistency. Concrete is then transported to the top of the wall by a large crane-mounted bucket or by a concrete pump and hose. Workers standing on planks at the tops of the forms deposit the concrete in the forms, consolidating it with a vibrator to eliminate air pockets (Figure 14.10). When the form has been filled and consolidated up

FIGURE 14.8
Detail of a form tie assembly. Two plastic cones just inside the faces of the form maintain the correct wall thickness. Tapered, slotted wedges at the ends transmit force from the tie to the walers. After the forms have been stripped, the cones will be removed from the concrete and the ties snapped off inside the voids left by the cones. The conical holes may be left open, filled with mortar, or plugged with conical plastic plugs.

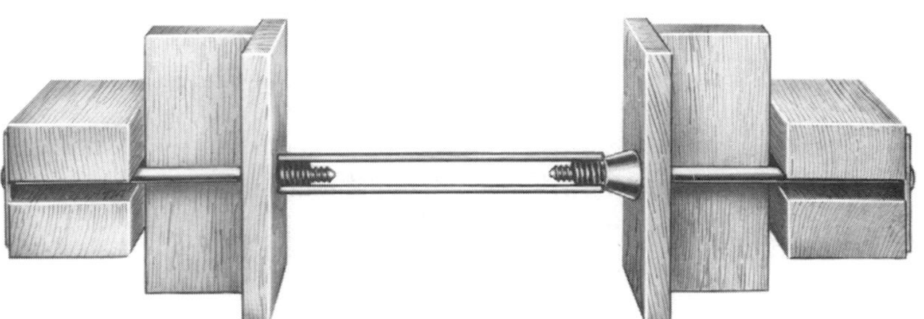

FIGURE 14.9
Detail of a heavy-duty form tie. This assembly is tightened with special screws that engage a helix of heavy wire welded into the tie. The wire components remain in the concrete, but the screws and the plastic cone on the right are removed and reused after stripping. The purpose of the cone is to give a neatly finished hole in the exposed surface of the concrete. *(Courtesy of Richmond Screw Anchor Co., Inc., 7214 Burns St., Fort Worth, TX 76118)*

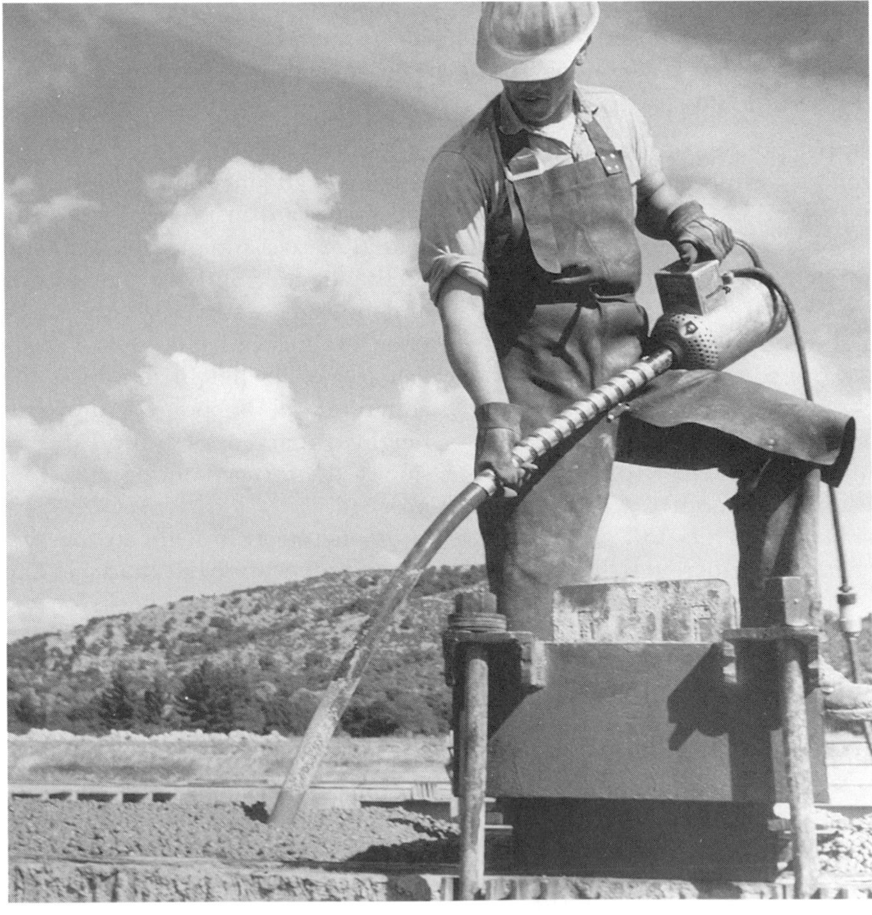

FIGURE 14.10
Consolidating wet concrete after pouring, using a mechanical vibrator immersed in the concrete. *(Reprinted with permission of the Portland Cement Association from* Design and Control of Concrete Mixtures, *12th edition; Photos: Portland Cement Association, Skokie, IL)*

FIGURE 14.11
Three stages in the construction of a reinforced concrete wall on a strip footing. In the foreground, the reinforcing bars have been wired to the dowels that project from the footing, ready for erection of the formwork. In the center, a section of wall has been poured in proprietary steel formwork that is tied through the wall with small steel straps secured by wedges. In the background, the forms have been stripped, and some of the ties have been snapped off. *(Reprinted with permission of the Portland Cement Association from* Design and Control of Concrete Mixtures, *12th edition; Photos: Portland Cement Association, Skokie, IL)*

to the level that was marked inside the formwork, hand floats are used to smooth and level the top of the wall. The top of the form is then covered with a plastic sheet or canvas, and the wall is left to cure.

After a few days of curing, the bracing and walers are taken down, the connectors are removed from the ends of the form ties, and the formwork is *stripped* from the wall (Figure 14.11). This leaves the wall bristling with projecting ends of form ties. These are twisted off with heavy pliers, and the *form tie holes* that they leave in the surfaces of the wall are carefully filled with grout. If required, major defects in the wall surface caused by defects in the formwork, inadequate filling of the forms with concrete, or poor consolidation of the concrete can be repaired at this time. The wall is now complete.

(*a*)

(*c*)

FIGURE 14.12
Forming a concrete wall with insulating concrete forms. (*a*) The forms are manufactured as interlocking blocks. The inner and outer halves of the blocks are tied together by steel mesh webs that connect to sheet metal strips on the inner and outer surfaces. These strips later serve to receive screws that fasten interior and exterior finish materials to the wall. (*b*) Workers stack the blocks to form all the exterior walls of a house. Openings for doors and windows are formed with dimension lumber. The worker to the right is cutting a formwork block to length with a hand saw. (*c*) This sample wall, from which some of the blocks have been removed, shows that the completed wall contains a continuous core of reinforced concrete with thermal insulation inside and out. *(Courtesy of American Polysteel Forms)*

(*b*)

Insulating Concrete Forms

An alternative way of casting a concrete wall, particularly one that will be an exterior wall of a building, is to use *insulating concrete forms (ICFs)* that serve both to form the concrete and to remain in place permanently as thermal insulation (Figure 14.12). The forms are manufactured in slightly differing configurations by dozens of companies. The most common form is interlocking hollow blocks of polystyrene foam, but some manufacturers produce planks or panels of foam with plastic ties that join them to make wall forms. Whatever the exact configuration, these systems weigh so little and are so accurately made that they go together almost as easily and quickly as a child's plastic building blocks. The tops of the blocks are channeled in such a way that horizontal reinforcing bars may be laid in the top of each course and vertical bars may be inserted into the vertical cores. The wall forms must be braced strongly to prevent them from moving during pouring. Concrete is usually deposited in the cores of the foam blocks from the hose of a concrete pump. The full height of a wall cannot be cast in one operation because the pressure of so great a depth of wet concrete would blow out the sides of the fragile blocks. The normal procedure is to deposit the concrete in several "lifts" of limited height, working all the way around the structure with each lift so that by the time the second lift is begun, the first lift has had an hour or two to harden somewhat, relieving the pressure at the bottom of the forms. Interior and exterior finish materials must be applied to the foam plastic faces to protect them from sunlight, mechanical damage, and fire. The thermal insulating value of the finished wall is usually R17 to R22, which is generally sufficient to meet current code requirements.

Casting a Concrete Column

A column is formed and cast much like a wall, with a few important differences. The footing is usually an isolated column footing, a pile cap, or a caisson rather than a strip footing (Figure 14.13). The dowels are sized and spaced in the footing to match the vertical bars in the column. The cage of column reinforcing is assembled with wire ties and hoisted into place over the dowels. If space is too tight in the region where the vertical

(*a*)

(*b*)

Figure 14.13
(*a*) A column footing almost ready for pouring but lacking dowels. The reinforcing bars are supported on pieces of concrete brick. (*b*) Column footings poured with projecting dowels to connect to both round and rectangular columns. *(Photos by Edward Allen)*

bars and dowels overlap, the bars may be spliced end to end with welds or mechanical connectors instead (Figure 13.25). The column form may be a rectangular box of plywood or composite panels, a cylindrical steel or plastic tube bolted together in halves so that it can later be removed, or a waxed cardboard tube that is stripped after curing by unwinding the layers of paper that make up the tube (Figures 14.14 and 14.15). Unless a rectangular column is very broad and wall-like, form ties through the concrete are not required. The vertical bars project from the top of the column to overlap or splice to the bars in the column for the story above, or they are bent over at right angles to splice into the roof structure. Where vertical bars overlap, the tops of the bars from the column below are offset (bent inward) by one bar diameter to avoid interference.

FIGURE 14.14
In the foreground, a square column form is tied with pairs of L-shaped steel brackets. In the background, a worker braces a round column form made of sheet steel. *(Courtesy of the Ceco Corporation, Oakbrook Terrace, Illinois)*

FIGURE 14.15
Round columns may also be formed with single-use cardboard tubes. Notice the density of the steel shoring structure that is being erected to support the slab form, which will carry a very heavy load of wet concrete. *(Courtesy of Sonoco Products Company)*

ONE-WAY FLOOR AND ROOF FRAMING SYSTEMS

One-Way Solid Slab System

A *one-way solid slab* (Figures 14.16–14.19) spans across parallel lines of support furnished by walls and/or beams. The walls and columns are poured prior to erecting the formwork for a one-way slab, but the forms for the girders and beams are nearly always built continuously with those for the slab, and girders, beams, and slab are poured simultaneously as a single piece.

The girder and beam forms are erected first, then the slab forms. The forms are supported on temporary joists and beams of metal or wood, and the temporary beams are supported on temporary *shores* (adjustable-length columns). The weight of uncured concrete that must be supported is enormous, and the temporary beams and shoring must be both strong and closely spaced. Formwork is, in fact, designed by a contractor's structural engineers just as carefully as it would be if it were a permanent building, because a structural failure in formwork is an intolerable risk to workers and property.

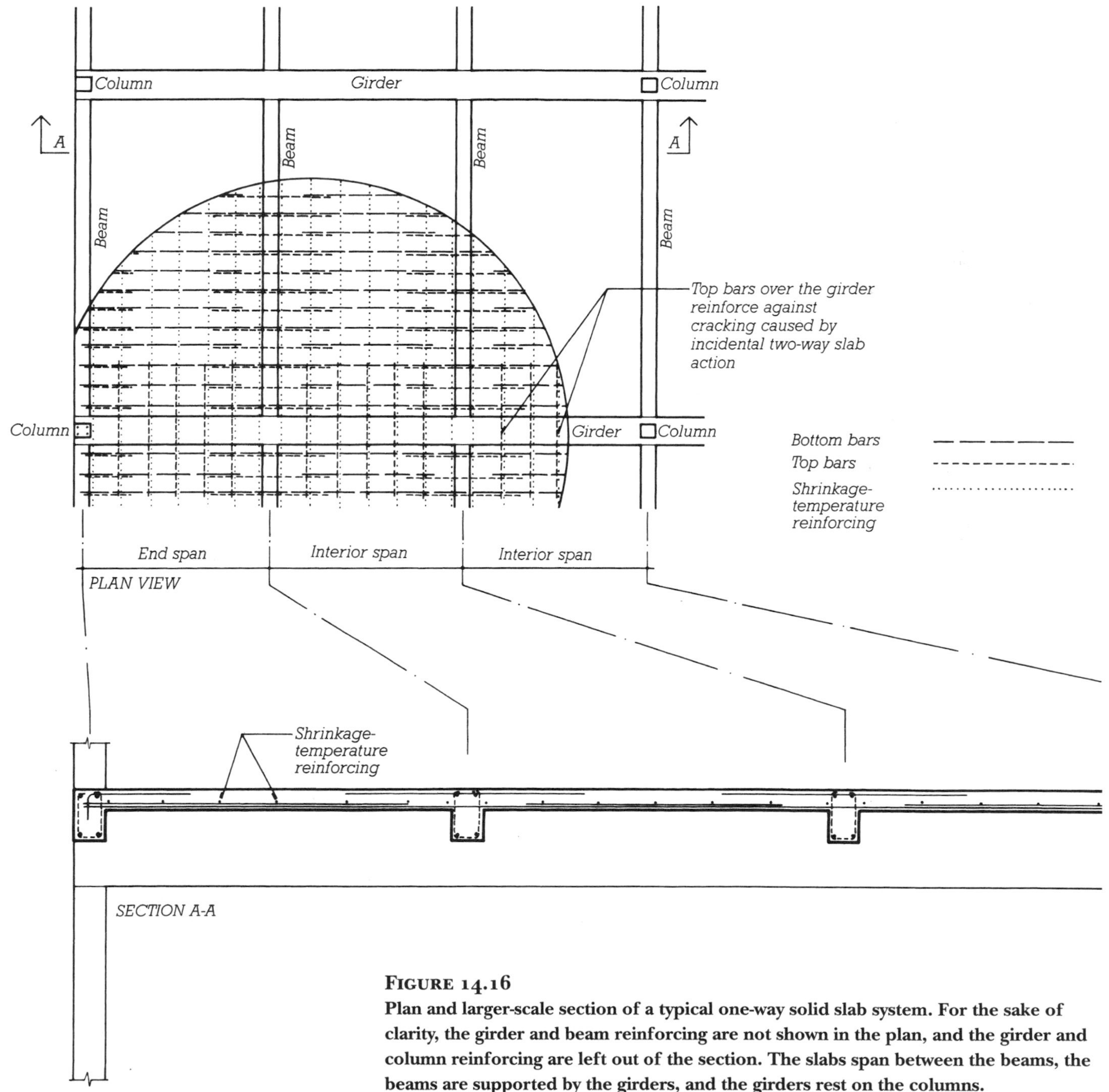

FIGURE 14.16
Plan and larger-scale section of a typical one-way solid slab system. For the sake of clarity, the girder and beam reinforcing are not shown in the plan, and the girder and column reinforcing are left out of the section. The slabs span between the beams, the beams are supported by the girders, and the girders rest on the columns.

Edges of concrete structural elements are beveled or rounded by inserting shaped strips of wood or plastic into the corners of the formwork to produce the desired profile. This is done because sharp edges of concrete often break off during form stripping to leave a ragged edge that is almost impossible to patch. In service, sharp edges are easily damaged by, and are potentially damaging to, people, furniture, and vehicles.

A form release compound is applied to all formwork surfaces that will be in contact with concrete. Then, in accordance with reinforcing diagrams and schedules prepared by the structural engineer, the girder and beam reinforcing—bottom bars, top bars, and stirrups—is installed in the forms, supported on chairs and bolsters to maintain the required cover of concrete. Next, the slab reinforcing—bottom bars, top bars, and shrinkage–temperature bars—is placed on bolsters. After the reinforcing and formwork have been inspected, the girders, beams, and slab are poured in a single operation, with sample cylinders being made for later testing to be sure that the concrete is strong enough. One-way slab depths are typically 4 to 10 inches (100–250 mm), depending on the span and loading intensity. The top of the slab is finished in the same manner as a slab on grade, usually to a steel trowel finish, and the slab is

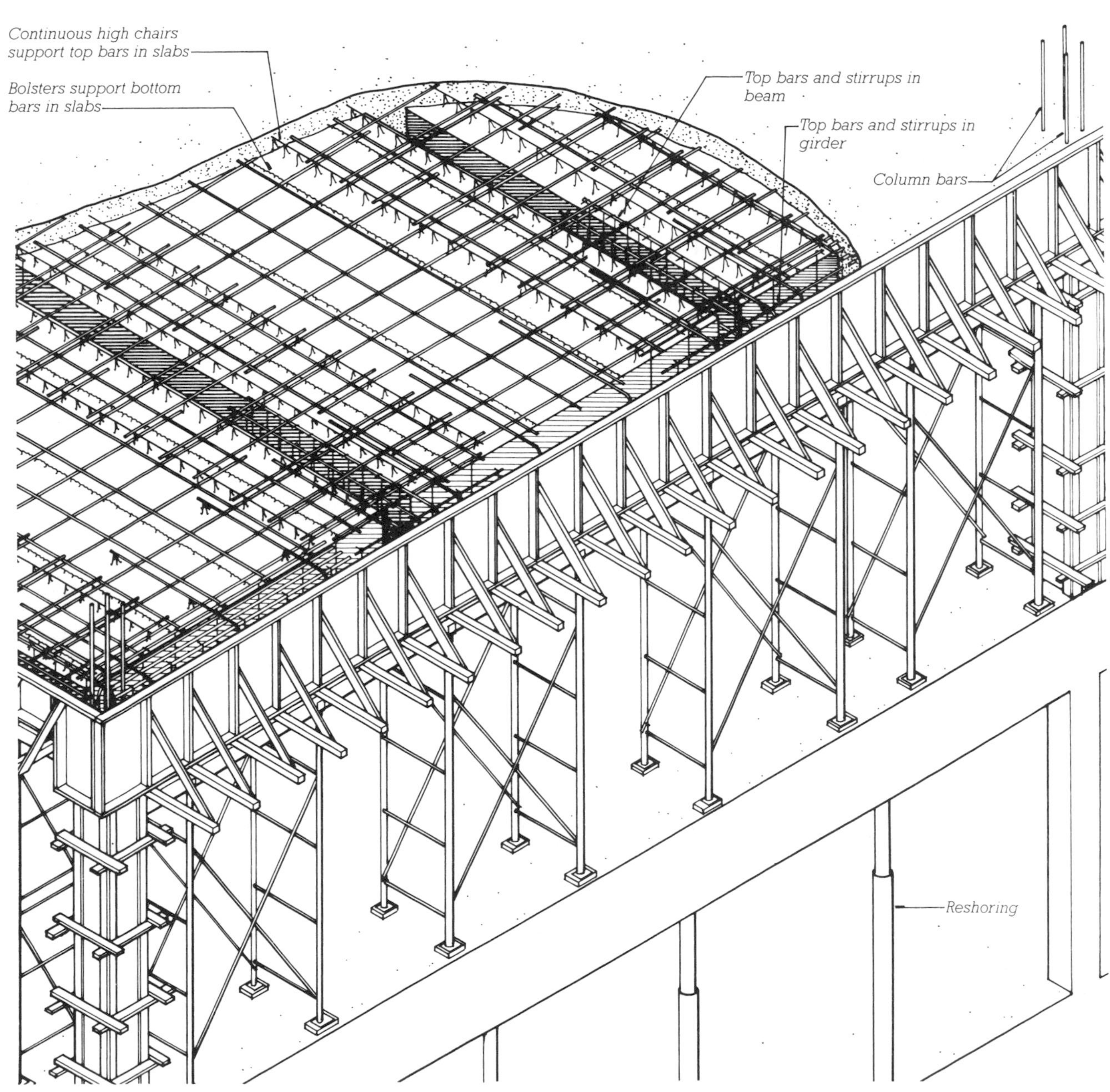

FIGURE 14.17
Isometric view of a one-way solid slab system under construction. The slab, beams, and girders are created in a single pour.

sealed or covered for damp curing. The only components left projecting above the slab surface at this stage are the offset column bars, which are now ready to overlap with, or splice to, the column bars for the floor above.

When the slab and beams have attained enough strength to support themselves safely, the formwork is stripped and the slabs and beams are *reshored* with vertical props to relieve them of loads until they have reached full strength, which will take several more weeks. Meanwhile, the formwork and the remainder of the shoring are cleaned and moved up a level above the slab and beams just poured, where the cycle of forming, reinforcing, pouring, and stripping is repeated (Figure 14.19).

Ordinarily, the most efficient and economical concrete beam is one whose depth is twice or three times its breadth. One-way solid slabs are often supported, however, by beams that are several times as broad as they are deep. These are called *slab bands* (Figure 14.20). Banded slab construction offers two kinds of economy: The width of the slab band reduces the span of the slab, which can result in a reduced thickness for the slab and consequent savings of concrete and reinforcing steel. Also, the reduced depth of the slab band compared to a more conventionally proportioned concrete beam allows for reduction of the story height of the building, with attendant economies in columns, cladding, partitions, and vertical runs of piping and ductwork.

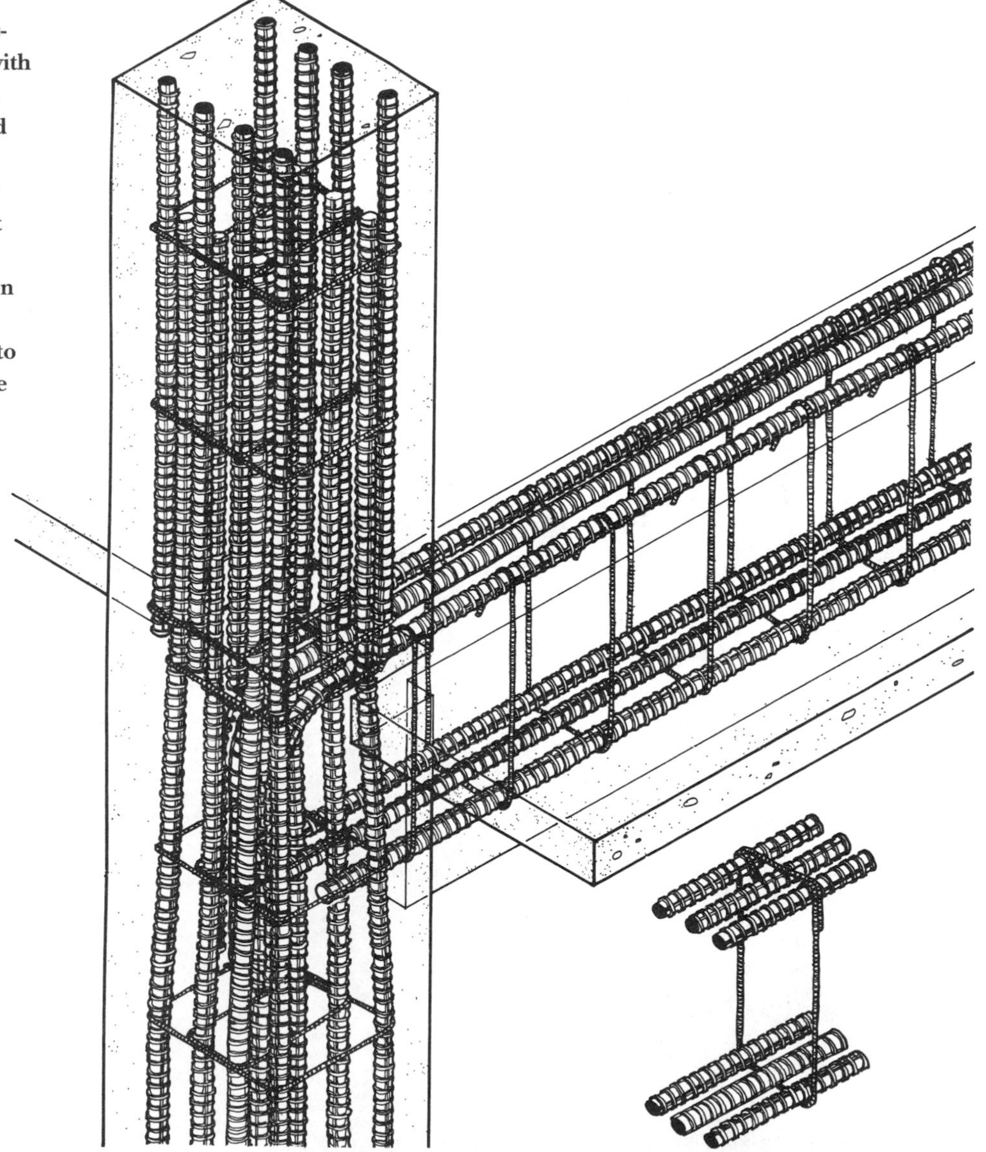

FIGURE 14.18
An example of a beam–column connection in a one-way solid slab structure, with the slab reinforcing omitted for clarity. Notice how the column bars are spliced by overlapping them just above floor level. The bars from the column below are offset at the top so that they lie just inside the bars of the column above at the splice. Structural continuity between the beam and column is established by running the top bars from the beam into the column. U-stirrups are shown in the beam; closed stirrup ties, shown in the inset detail, are often used instead.

FIGURE 14.19
Reshoring in concrete slab construction. The three uppermost floors visible below the formwork system in this concrete tower have been reshored. The formwork is a climbing system that lifts itself up the structure as each floor is completed. The solid panels surrounding the outside of the formwork protect the work and the workers from the wind and rain. The concrete system is a two-way flat plate, a floor framing system described later in this chapter. *(Photo by Joseph Iano)*

FIGURE 14.20
Banded slab construction. Compare the depth and breadth of the slab band in the center of the building to those of the conventional concrete beams around the perimeter. *(Reprinted with permission of the Portland Cement Association from Design and Control of Concrete Mixtures, 12th edition; Photos: Portland Cement Association, Skokie, IL)*

FIGURE 14.21
This helical ramp is a special application of one-way solid slab construction. The formwork is made of overlaid plywood for a smooth surface finish. *(Courtesy of APA–The Engineered Wood Association)*

One-Way Concrete Joist System (Ribbed Slab)

As one-way solid slab spans increase, a progressively thicker slab is required. Beyond a certain span, the slab becomes so thick that the weight of the slab itself is an excessive burden, unless a substantial portion of the nonworking concrete in the lower part of the slab can be eliminated to lighten the load. This is the rationale for the *one-way concrete joist system* (Figures 14.22–14.25), also called a *ribbed slab*. The bottom steel is concentrated in spaced ribs or joists. The thin slab that spans across the top of the joists is reinforced only by shrinkage–temperature bars. There is little concrete in this system that is not working, with the result that a one-way concrete joist system can efficiently and economically span considerably longer distances than a one-way solid slab. Each joist is reinforced as a small beam, except that stirrups are not usually used in concrete joists because of the restricted space in the narrow joist. Instead, the ends of the joists are broadened sufficiently that the concrete itself can resist the diagonal tension forces.

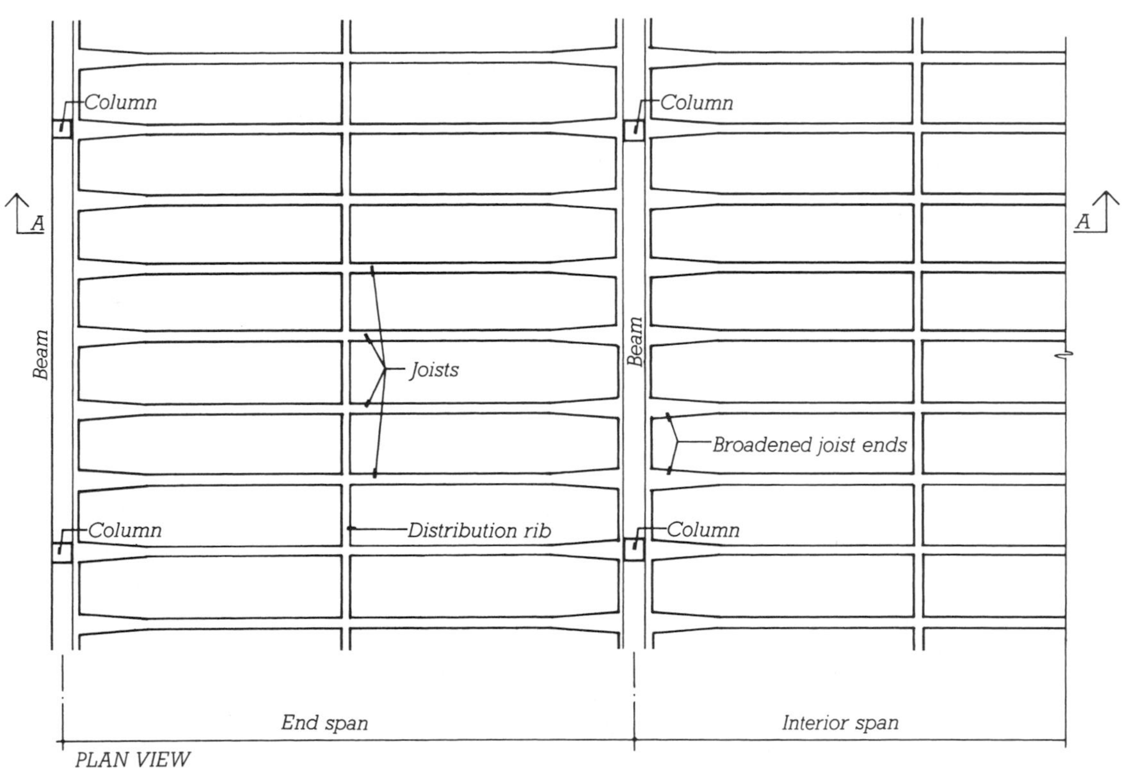

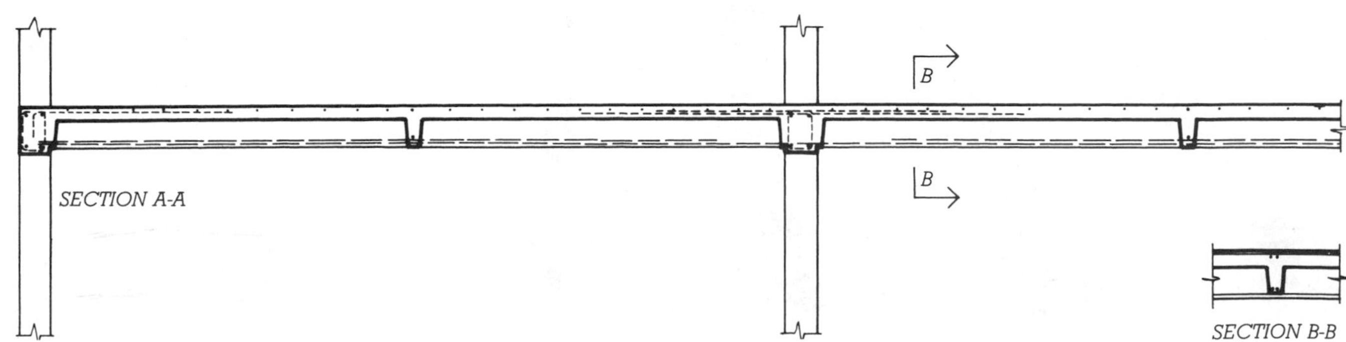

FIGURE 14.22
Plan and larger-scale section of a typical one-way concrete joist system. For the sake of clarity, no reinforcing is shown in the plan, and the column reinforcing is not shown on the section. All bottom and top reinforcing occurs in the ribs, and all shrinkage–temperature bars are placed in the slab.

FIGURE 14.23
Standard steel form dimensions for one-way concrete joist construction. (One inch equals 25.4 mm.) *(Courtesy of the Ceco Corporation, Oakbrook Terrace, Illinois)*

Filler widths (10" and 15") are available for filling nonstandard spaces only.

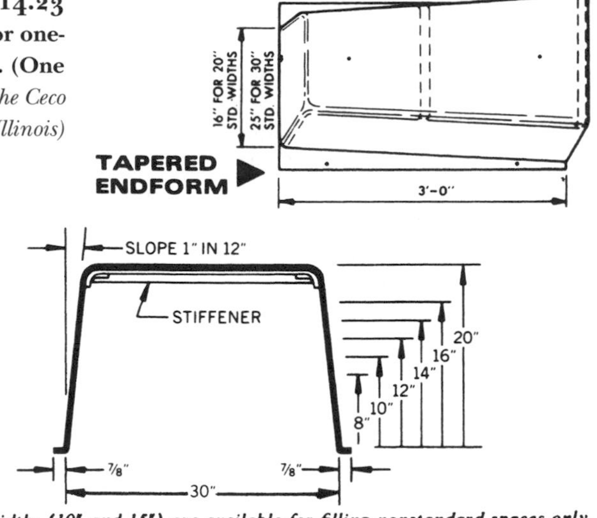

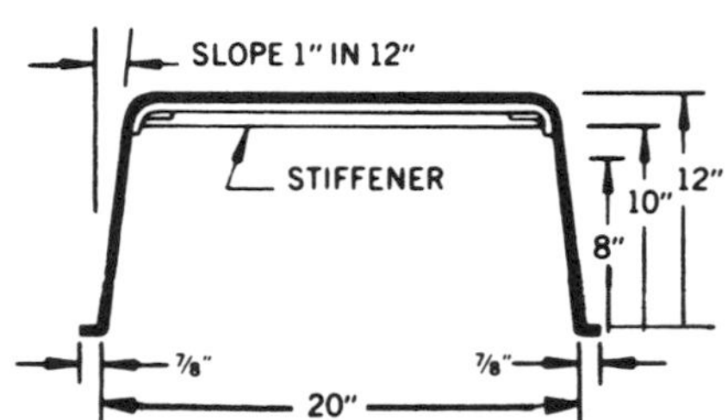

FIGURE 14.24
Reinforcing being placed for a one-way concrete joist floor. Electrical conduits and boxes have been put in place, and welded wire fabric is being installed as shrinkage–temperature reinforcing. Both the tapered end pans and the square endcaps for the midspan distribution rib are clearly visible. *(Courtesy of the Ceco Corporation, Oakbrook Terrace, Illinois)*

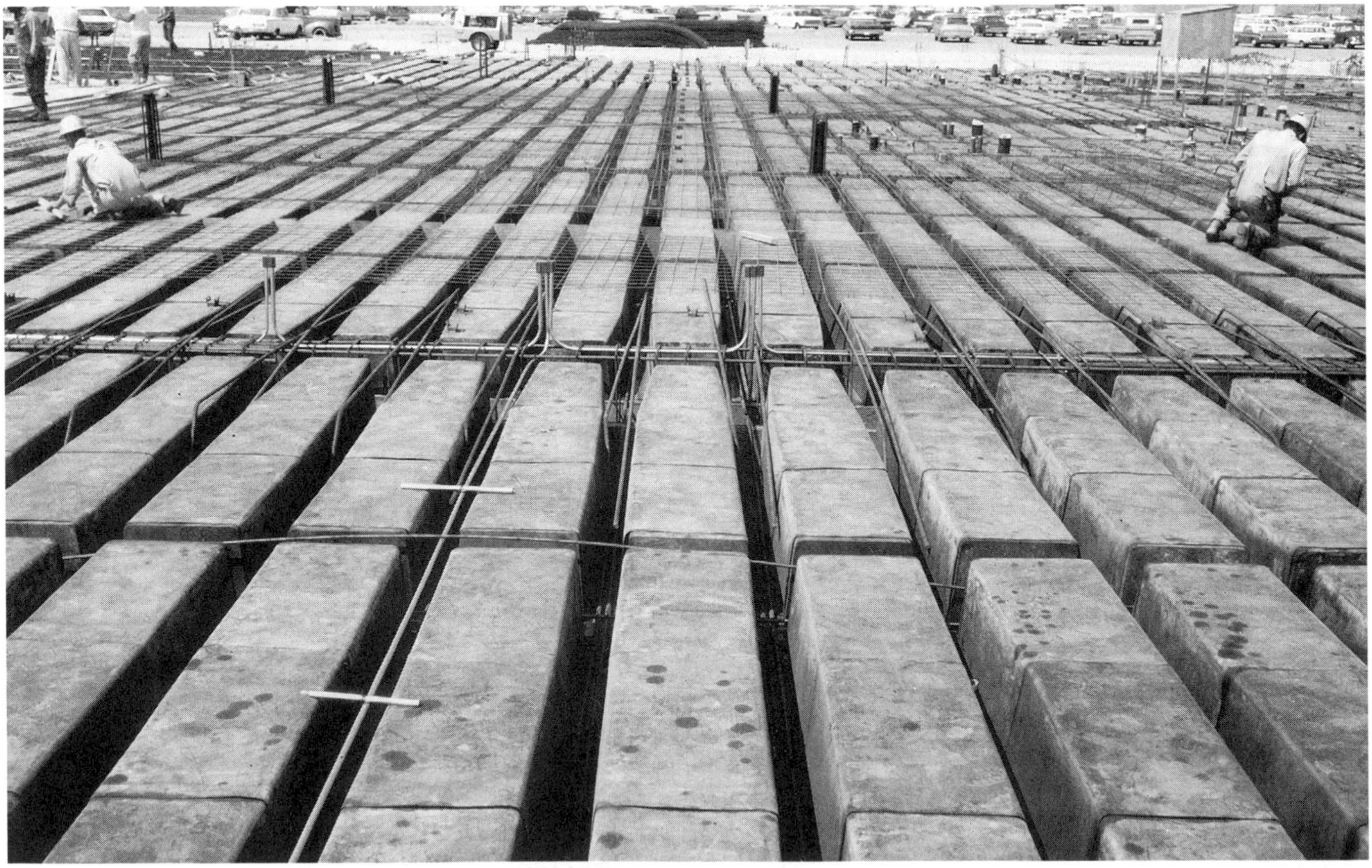

The joists are formed with metal or plastic *pans* supported on a temporary plywood deck. Pans are available in two standard widths, 20 inches (508 mm) and 30 inches (762 mm), and in depths ranging up to 20 inches (508 mm), as shown in Figure 14.23. (Larger pans are also available for forming floors designed as one-way solid slab systems.) The sides of the pans taper from bottom to top to allow them to drop easily from the hardened concrete during stripping. The joist width can be varied by placing the rows of pans closer together or farther apart. The bottom of each joist is formed by the wood deck. The joist ends are broadened with standard end pans whose width tapers. A *distribution rib* is sometimes formed across the joists at midspan to distribute concentrated loads to more than one joist. After application of a form release compound, the beam and joist reinforcing are placed, the shrinkage–temperature bars are laid crosswise on bolsters over the pans, and the entire system is poured and finished (Figures 14.24 and 14.25).

One-way concrete joists are usually supported on *joist bands*, which are broad beams that are only as deep as the joists. Although a deeper beam would be more efficient structurally, a joist band is formed by the same plywood deck that supports the pans, which eliminates expensive beam formwork entirely and produces a simpler underside of slab profile with a uniform floor-to-ceiling height throughout.

Wide-Module Concrete Joist System

When fire-resistance requirements of the building code dictate a slab thickness of 4½ inches (115 mm) or more, the slab is capable of spanning a much greater distance than the normal space between joists in a one-way concrete joist system. This has led to the development of the *wide-module concrete joist system*, also called the *skip-joist system*, in which the joists are placed 4 to 6 feet (1220–1830 mm) apart. The name "skip-joist" arose from the original practice of achieving this wider spacing by laying strips of wood over alternate joist cavities in conventional joist pan formwork to block out the concrete. Pans are now specially produced for

FIGURE 14.25
A one-way concrete joist system after stripping of the formwork, showing broadened joist ends at the lower edge of the photograph and a distribution rib in the foreground. The dangling wires are hangers for a suspended finish ceiling (see Chapter 24). *(Reprinted with permission of the Portland Cement Association from* Design and Control of Concrete Mixtures, *12th edition; Photos: Portland Cement Association, Skokie, IL)*

wide-module construction (Figures 14.26 and 14.27).

Because wide-module joists must each carry about double the weight carried by conventionally spaced joists, stirrups are generally required near the ends of each joist. If conventional U-stirrups are used, they must be installed on a diagonal as seen from above, in order to fit into the narrow joist, or single-leg stirrups may be used instead.

FIGURE 14.26
Formwork for a wide-module concrete joist system. These pans have been placed over a flat plywood deck, which will also serve to form the bottoms of the joist band beams. *(Courtesy of the Ceco Corporation, Oakbrook Terrace, Illinois)*

FIGURE 14.27
The underside of the finished wide-module joists, joist bands, and slab. *(Courtesy of the Ceco Corporation, Oakbrook Terrace, Illinois)*

Two-Way Floor and Roof Framing Systems

Two-Way Flat Slab and Two-Way Flat Plate Systems

Two-way concrete framing systems are generally more economical than one-way systems in buildings where the columns can be spaced in bays that are square or nearly square in proportion. A *two-way solid slab* is a rarely seen system, occasionally used for very heavily loaded industrial floors, in which the slab is supported by a grid of beams running in both directions over the columns. Most two-way floor and roof framing systems, however, even for heavy loadings, are made without beams. The slab is reinforced in such a way that the varying stresses in the different zones of the slab are accommodated within a uniform thickness of concrete. This simplifies formwork construction and reinforcing bar patterns considerably.

The *two-way flat slab* (Figure 14.28), a system suited to heavily loaded buildings such as storage and industrial buildings, illustrates this concept.

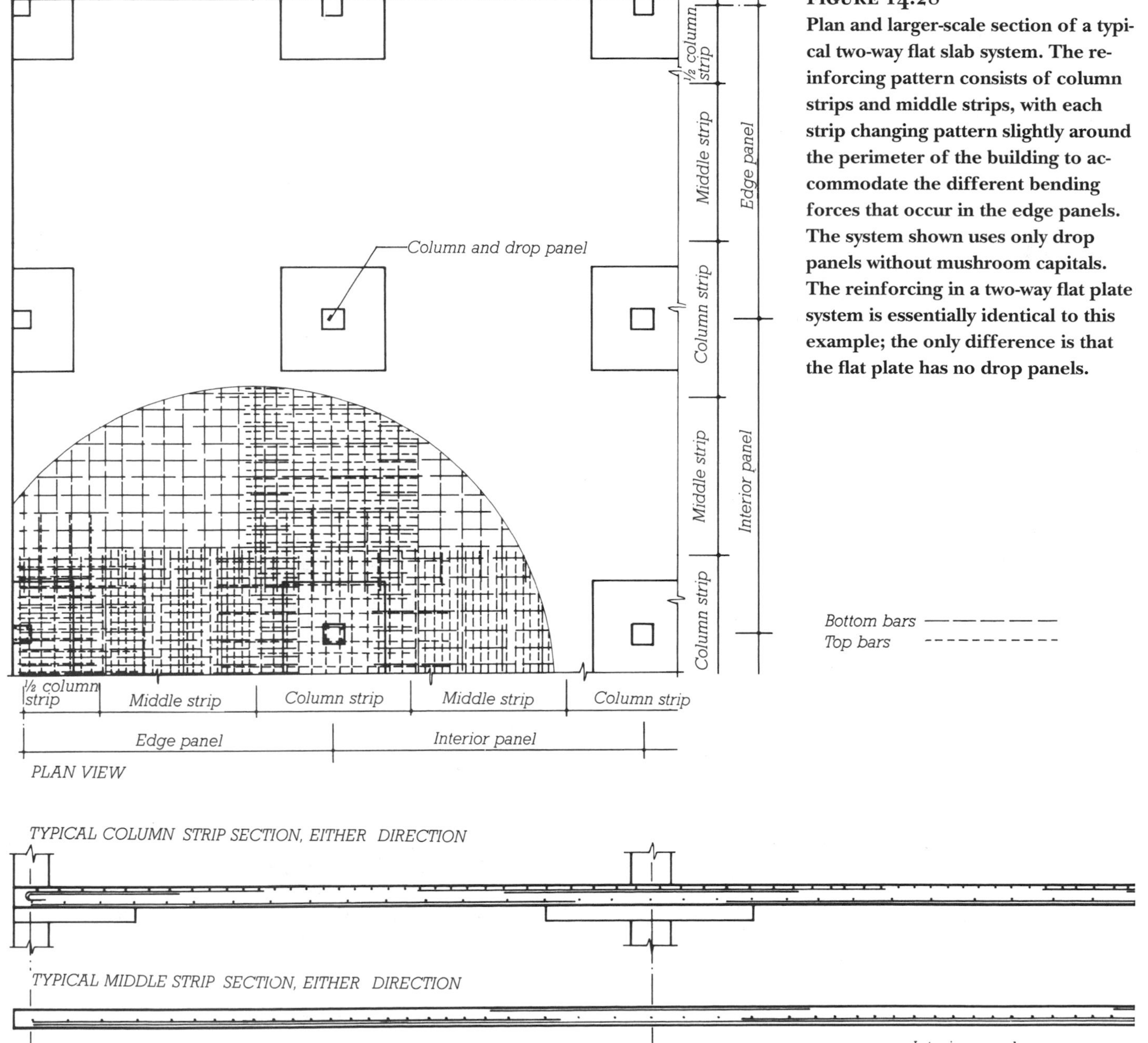

FIGURE 14.28
Plan and larger-scale section of a typical two-way flat slab system. The reinforcing pattern consists of column strips and middle strips, with each strip changing pattern slightly around the perimeter of the building to accommodate the different bending forces that occur in the edge panels. The system shown uses only drop panels without mushroom capitals. The reinforcing in a two-way flat plate system is essentially identical to this example; the only difference is that the flat plate has no drop panels.

The formwork is completely flat except for a thickening of the concrete to resist the high shear forces around the top of each column. Traditionally, this thickening was accomplished with both a funnel-shaped *mushroom capital* and a square *drop panel*, but today the capital is usually eliminated to reduce the formwork cost, leaving a drop panel to do the work alone

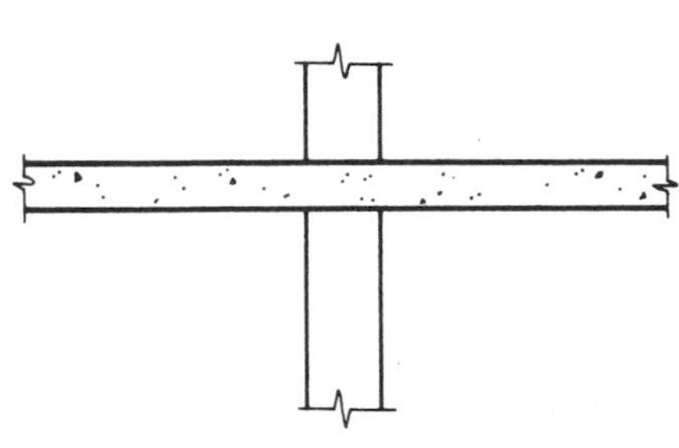

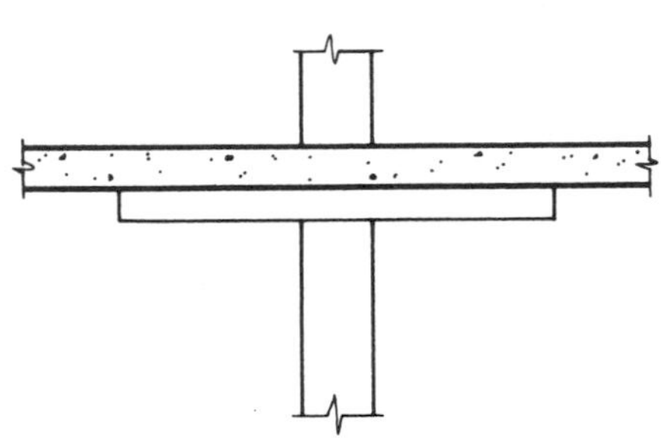

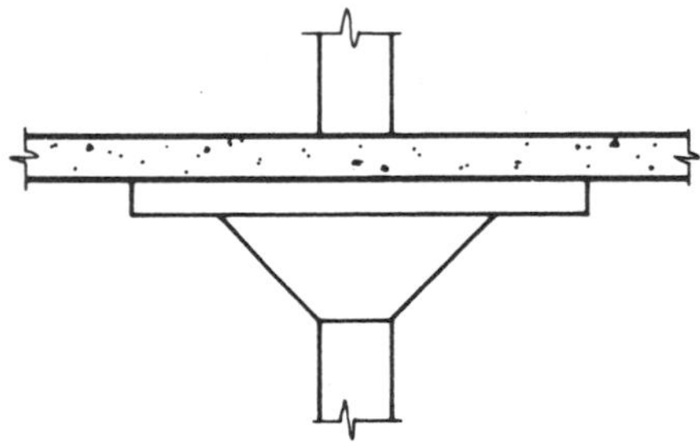

FIGURE 14.29
Column capitals for two-way concrete framing systems. For slabs bearing heavy loads, shear stresses around the column are reduced by means of a mushroom capital and drop panel or a drop panel alone. For lighter loads, no thickening of the slab is required.

FIGURE 14.30
A two-way flat plate floor of a high-rise apartment building is poured with the aid of a concrete pump at the base of the building. The concrete is delivered up a telescoping tower to a nozzle at the end of an articulated boom. The column dowels in the center of the newly poured area demonstrate the ability of flat plate construction to adapt to irregular patterns and spacings of columns. *(Courtesy of Schwing America, Inc.)*

FIGURE 14.31
The underside of a two-way flat plate floor. Pipes have been installed for an automatic sprinkler fire suppression system. *(Reprinted with permission of the Portland Cement Association from* Design and Control of Concrete Mixtures, *12th edition; Photos: Portland Cement Association, Skokie, IL)*

(Figure 14.29). Typical depths for the slab itself range from 6 to 12 inches (150 to 300 mm).

The reinforcing is laid in both directions in half-bay-wide strips of two fundamental types: *Column strips* are designed to carry the higher bending forces encountered in the zones of the slab that cross the columns. *Middle strips* have a lighter reinforcing pattern. Shrinkage–temperature steel is not needed in two-way systems because the concrete must be reinforced in both directions to resist bending. The drop panel and capital (if any) have no additional reinforcing beyond that provided by the column strip; the greater thickness of concrete furnishes the required shear resistance.

In more lightly loaded buildings, such as hotels, hospitals, dormitories, and apartment buildings, the slab need not be thickened at all over the columns. This makes the formwork extremely simple and even allows some columns to be moved off the grid a bit if it will facilitate a more efficient floor plan arrangement (Figure 14.30). The completely flat ceilings of this system allow room partitions to be placed anywhere with equal ease. Because there are no beams and girders, only a thin slab, the story heights of the building may be kept to an absolute minimum, which reduces the cost of exterior cladding (Figure 14.31). Typical slab depths for this *two-way flat plate* system range from 5 to 12 inches (125–305 mm).

The zones along the exterior edges of both the two-way flat slab system and the two-way flat plate system require special attention. To take full advantage of structural continuity, the slabs should be cantilevered beyond the last row of columns a distance equal to about 30 percent of the interior span. If such a cantilever is impossible, additional reinforcing must be added to the slab edges to carry the higher stresses that will result.

Because a two-way flat plate has no drop panel, it requires additional reinforcing bars in the slab at the top of each column to resist the high shear stresses that occur in this region. Alternatively, a proprietary system of vertical steel studs may be installed in the formwork at each column head to act as stirrups and replace a much larger volume of horizontal bars (Figures 14.32 and 14.33).

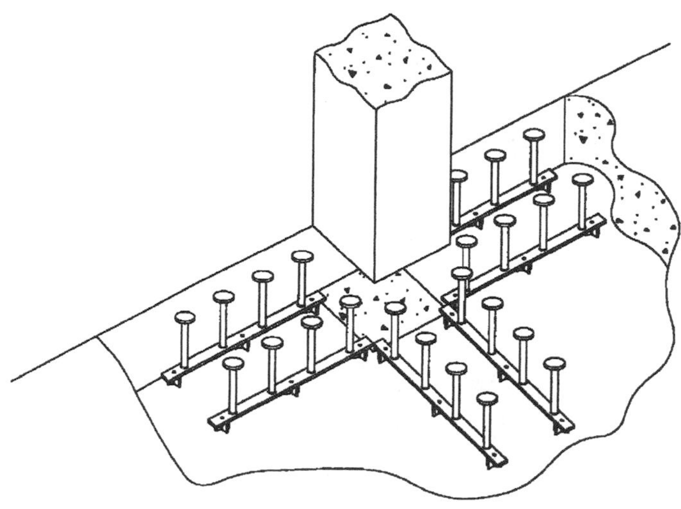

FIGURE 14.32
Shear reinforcement around columns in a two-way flat plate can be simplified by using Studrails®, a proprietary system of steel studs factory welded to horizontal rails. *(U.S. and Canada patents Nos. 4406103 and 1085642, respectively. Licensee: Deha, represented by Decon, 105C Atsion Rd., P.O. Box 1575, Medford, NJ 08055-6675 and 35 Devon Road, Bramton, Ontario L6T 5B6)*

FIGURE 14.33
Studrails nailed to the formwork around column bars, ready for installation of the top and bottom bars for the two-way flat plate. *(U.S. and Canada patents Nos. 4406103 and 1085642, respectively. Licensee: Deha, represented by Decon, 105C Atsion Rd., P.O. Box 1575, Medford, NJ 08055-6675 and 35 Devon Road, Bramton, Ontario L6T 5B6)*

Two-Way Waffle Slab System

The *waffle slab*, or *two-way concrete joist system* (Figure 14.34), is the two-way equivalent of the one-way concrete joist system. Metal or plastic pans called *domes* are used as formwork to eliminate nonworking concrete from the slab, allowing considerably longer spans than are feasible in the two-way flat plate system. The standard domes form joists 6 inches (152 mm) wide on 36-inch (914-mm) centers, or 5 inches (127 mm) wide on 24-inch (610-mm) centers, in a variety of depths up to 20 inches (500 mm), as shown in Figures 14.35–14.38. Special domes are also available in larger sizes. Solid concrete *heads* are created around the tops of the columns by leaving the domes out of the formwork in these areas. A head serves the same function as a drop panel in the two-way flat slab system. If a waffle slab cannot be cantilevered at the perimeter of the building, a perimeter beam must be provided. Stripping of the domes is facilitated in many cases by a compressed air fitting at the inside top of each dome. This allows the domes to be popped out of the concrete with the application of a puff of compressed air. The waffle slab system is suited to longer-span, heavily loaded applications, and its coffered underside presents

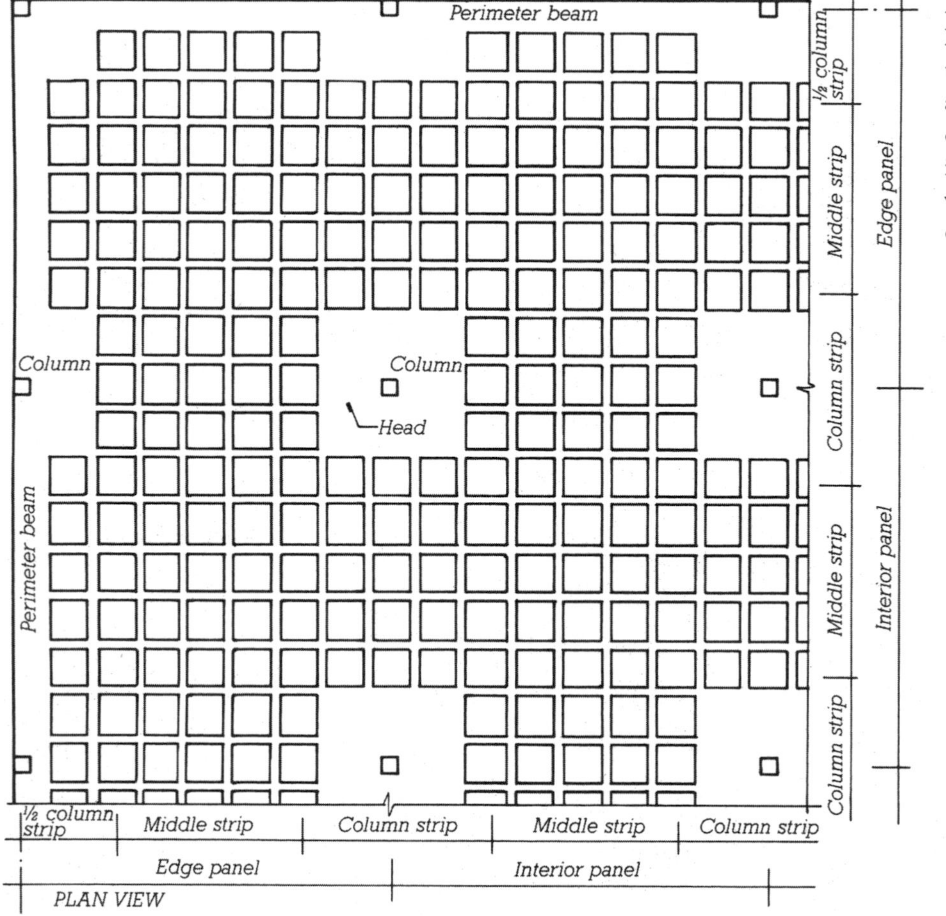

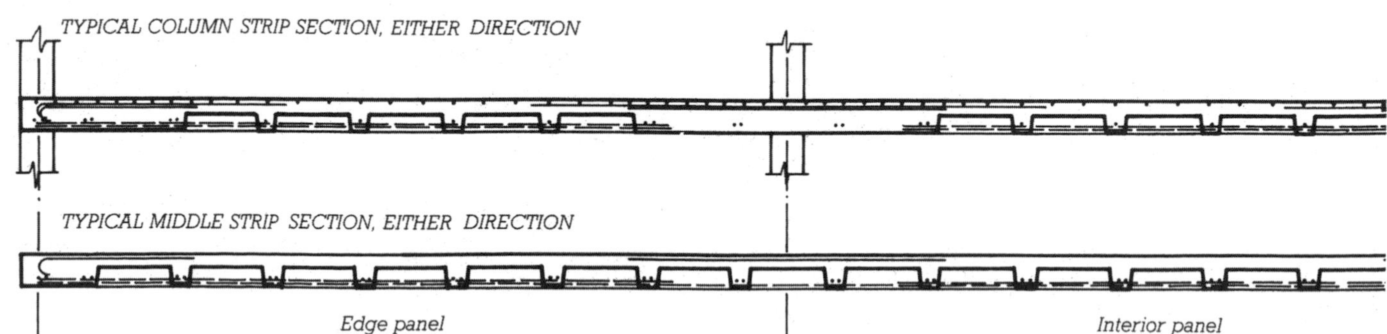

FIGURE 14.34
Plan and larger-scale section of a typical two-way concrete joist system, also known as a waffle slab. No reinforcing is shown on the plan drawing for the sake of clarity, and the section does not show the welded wire fabric that is spread over the entire form before pouring.

2′-0″ MODULE
(19″ x 19″ Dome System)

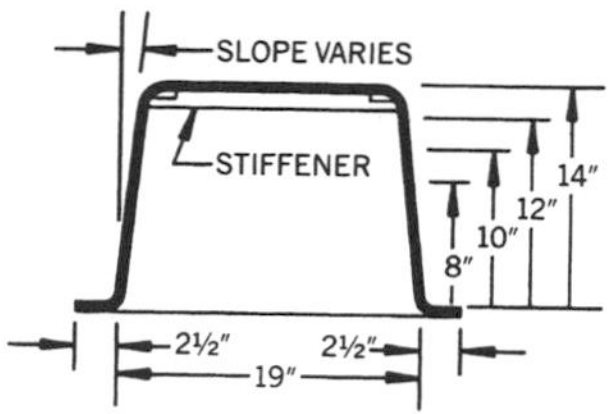

3′-0″ MODULE
(30″ x 30″ Dome System)

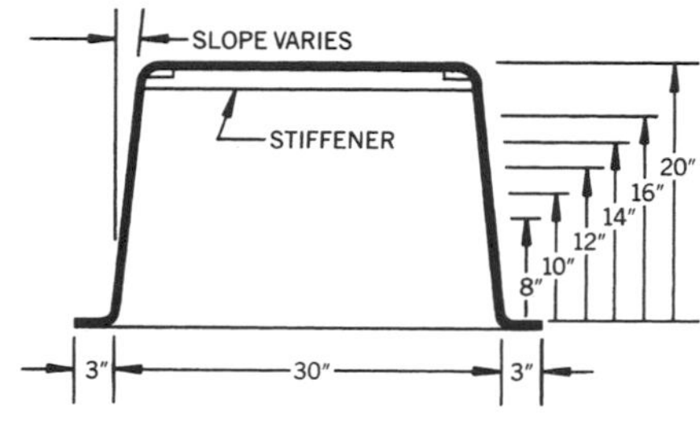

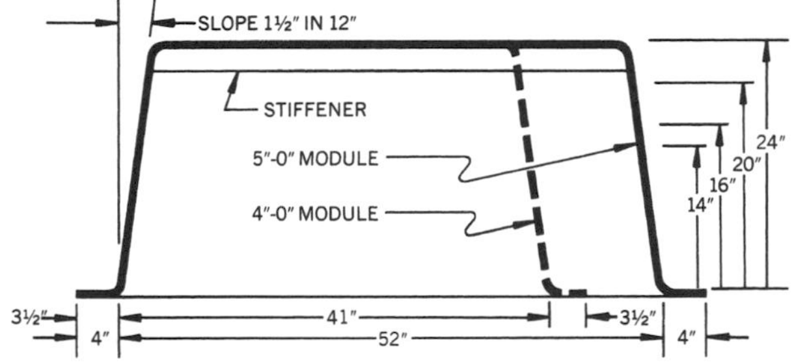

FIGURE 14.35

Standard steel dome forms for two-way concrete joist construction. A 2½-foot (760-mm) module, utilizing forms 24 inches (610 mm) square, is also available from some manufacturers. (One inch equals 25.4 mm.) *(Courtesy of the Ceco Corporation, Oakbrook Terrace, Illinois)*

FIGURE 14.36

Steel domes being placed on a temporary plywood deck to form a two-way concrete joist floor. Pans are omitted around columns to form solid concrete heads. *(Courtesy of the Ceco Corporation, Oakbrook Terrace, Illinois)*

FIGURE 14.37
Plastic dome formwork being assembled for a two-way concrete joist floor. Notice the electrical conduit and junction box in the foreground, ready to be embedded in the slab. *(Courtesy of Molded Fiber Glass Concrete Forms Company)*

FIGURE 14.38
Stripping plastic domes after removal of the temporary plywood deck. *(Courtesy of Molded Fiber Glass Concrete Forms Company)*

Figure 14.39
The underside of a two-way concrete joist floor. Notice how the joists are cantilevered beyond the column line for maximum structural efficiency. *(Courtesy of the Ceco Corporation, Oakbrook Terrace, Illinois)*

rich architectural opportunities as an exposed ceiling (Figure 14.39). However, the complexity of waffle slab formwork typically renders this system less economical than other systems with comparable span and load-carrying capability, such as one-way joist.

Concrete Stairs

A concrete stair (Figure 14.40) may be thought of as an inclined one-way solid slab with additional concrete added to make risers and treads. The underside of the formwork is planar. The top is built with riser forms, which are usually inclined to provide greater toe space and make the stair more comfortable to users. The concrete is poured in one operation, and the treads are tooled to a steel trowel finish.

Sitecast Posttensioned Framing Systems

Posttensioning can be applied to any of the sitecast concrete framing systems. It is used in beams, girders, and slabs, both one-way and two-way, to reduce member sizes, reduce deflection, and extend spanning capability.

Two-way flat plate structures are very commonly posttensioned, especially when spans are long or zoning restrictions on the height of the building require minimal slab depths. The tendon layout, however, is quite different from the conventional reinforcing layout shown in Figure 14.28. Instead of being placed identically in both directions, the *draped tendons* are evenly distributed in one direction and banded closely together over the line of columns in the other direction (Figures 14.41 and 14.42). This arrangement functions better structurally in posttensioned slabs because it balances the maximum upward force from the banded tendons against the maximum downward force from the distributed tendons. It is also much easier to install than distributed, draped tendons running in both directions. If the structural bay is square, the same number of tendons is used in each direction. The prestressing force from the banded tendons becomes evenly distributed throughout the width of the slab within a short distance of the end anchorages because of corbelling action in the concrete.

As with any prestressed concrete framing system, both short-term and long-term losses of prestressing force must be anticipated. The short-term losses in posttensioning are caused by elastic shortening of the concrete, friction between the tendons and the concrete, and initial movements (*set*) in the anchorages. The long-term losses are caused by concrete shrinkage, concrete creep, and steel relaxation. The structural engineer calculates the total of these expected losses and specifies an additional amount of initial posttensioning force to compensate for them.

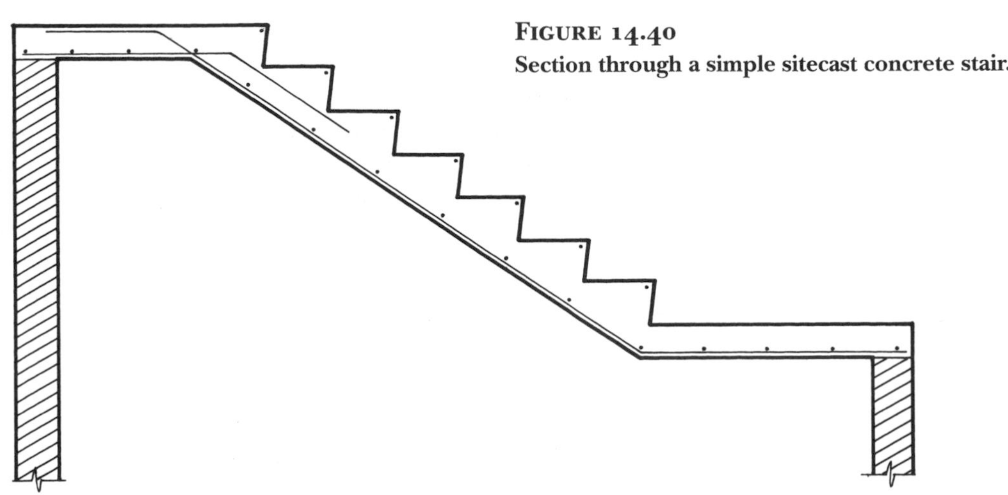

Figure 14.40
Section through a simple sitecast concrete stair.

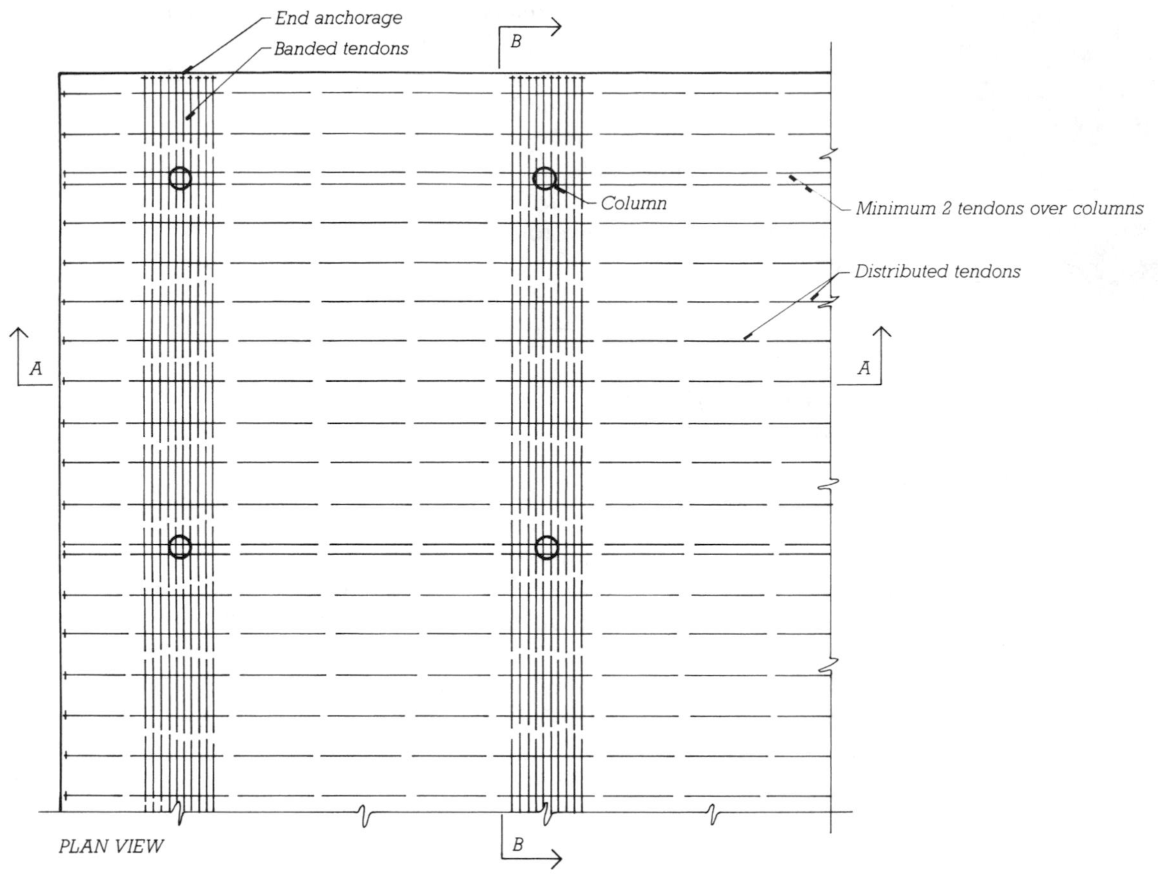

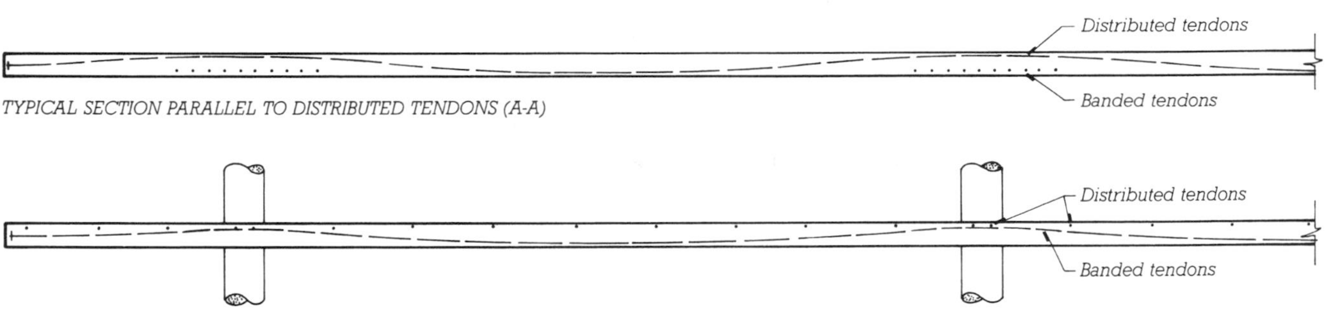

FIGURE 14.41
A plan and two larger-scale sections of the tendon layout in a two-way flat plate floor with banded posttensioning. The number of tendons running in each of the two directions is identical, but those in one direction are concentrated into bands that run over the tops of the columns. The draping of the tendons is evident in the two section drawings. Building codes require that at least two distributed tendons run directly over each column to help reinforce against shear failure of the slab in this region. In addition to the tendons, conventional steel reinforcing is used around the columns and in midspan, but this has been omitted from these drawings for the sake of clarity.

FIGURE 14.42
Banded tendons run directly through the concrete column of this flat plate floor. A substantial amount of conventional reinforcing is used here for shear reinforcing. Notice the end anchorage plates nailed to the vertical surface of the formwork at the upper right; see also Figure 13.40. (*Courtesy of Post-Tensioning Institute*)

SELECTING A SITECAST CONCRETE FRAMING SYSTEM

Preliminary factors to be considered in the selection of a sitecast concrete framing system for a building include the following (Figures 14.43 and 14.44):

1. Are the bays of the building square or nearly square? If so, a two-way system will probably be more economical than a one-way system.

2. How long are the spans? Spans up to 25 or 30 feet (7.6 or 9.1 m) are usually accomplished most economically with a two-way flat plate or flat slab system because of the relative simplicity of the formwork. For longer spans, a one-way joist system may be a good choice. Posttensioning extends significantly the economical span range of any of these systems.

3. How heavy are the loads? Heavy industrial loadings are borne better by thicker slabs and larger beams than they are by light joist construction. Ordinary commercial, institutional, and residential loadings are carried easily by flat plate or joist systems.

4. Will there be a finish ceiling beneath the slab? If not, flat plate and one-way slab construction have smooth, paintable undersides that can serve as ceilings.

5. Does the lateral stability of the building against wind and seismic loads have to be provided by the rigidity of the concrete frame? Flat plate floors may not be sufficiently rigid for this purpose, which would favor a one-way system with its deeper beam-to-column connections.

INNOVATIONS IN SITECAST CONCRETE CONSTRUCTION

The development of sitecast concrete construction continues along several lines. The basic materials, concrete and steel, continue to undergo innovations, as described in Chapter 13. The continuing evolution of high-strength, high-stiffness concrete, along with improvements in concrete forming systems and concrete pumping technology, have enabled sitecast concrete construction to remain economically competitive with structural steel for buildings of virtually any type or size. The world's tallest building at the time of this writing, the Burj Dubai, in Dubai, United Arab Emirates, is being constructed for most of its height as a steel-reinforced sitecast concrete structure.

Formwork generally accounts for more than half the cost of sitecast concrete construction. Efforts to reduce this cost have led to many innovations, including new types of formwork panels that are especially smooth, durable, and easy to clean after they have been stripped. These can be reused dozens of times before they wear out.

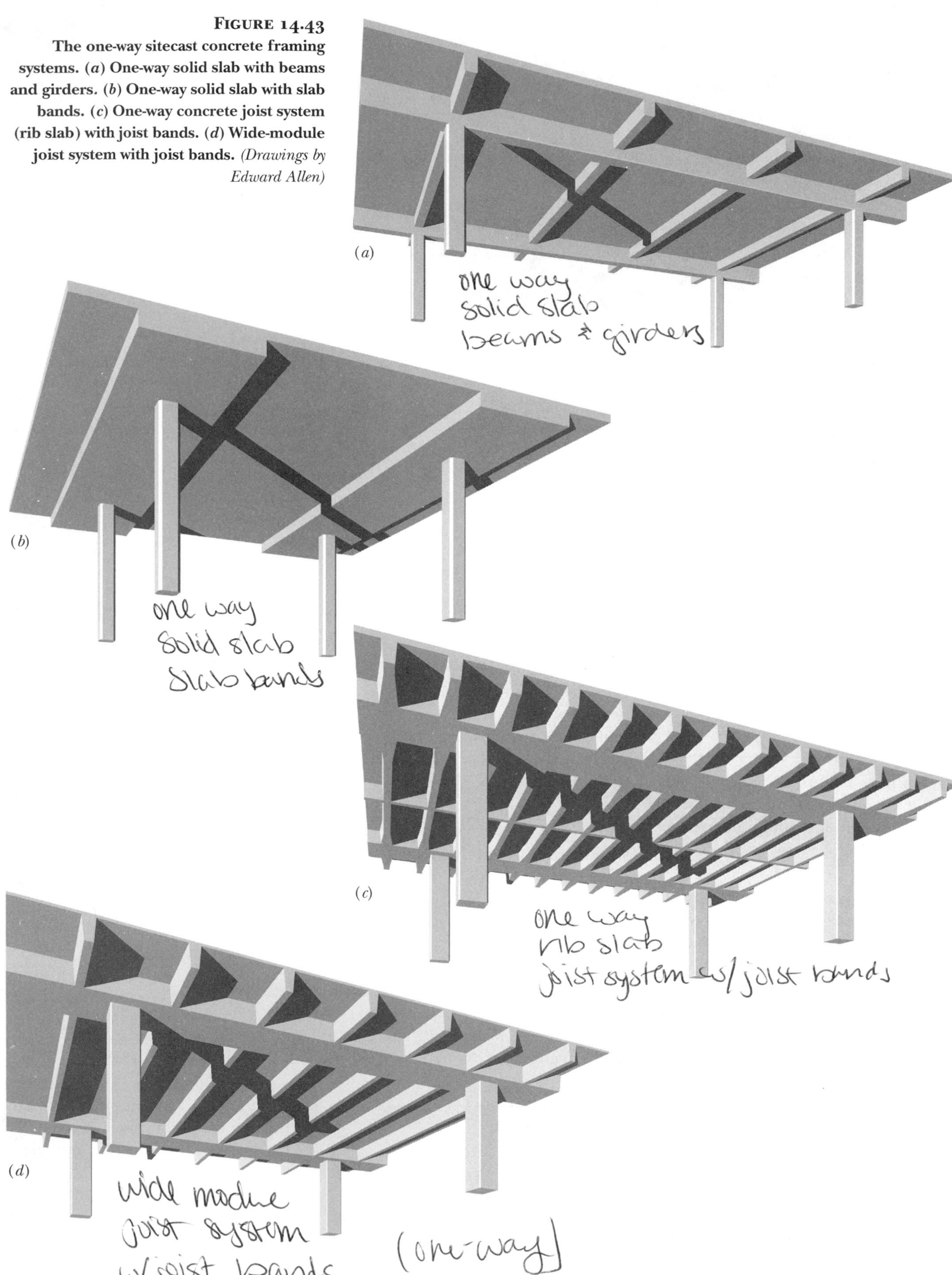

FIGURE 14.43
The one-way sitecast concrete framing systems. (*a*) One-way solid slab with beams and girders. (*b*) One-way solid slab with slab bands. (*c*) One-way concrete joist system (rib slab) with joist bands. (*d*) Wide-module joist system with joist bands. *(Drawings by Edward Allen)*

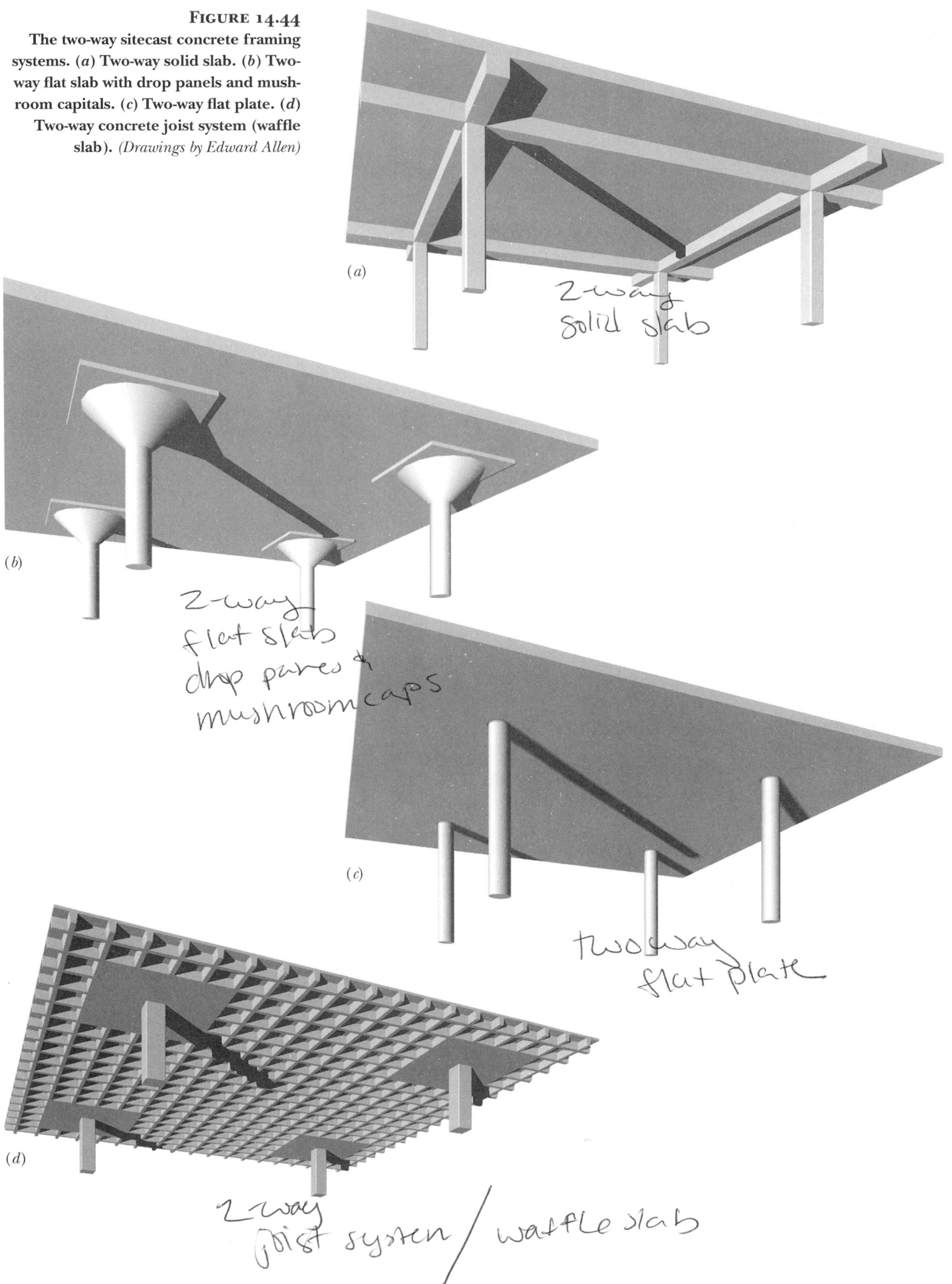

FIGURE 14.44
The two-way sitecast concrete framing systems. (*a*) Two-way solid slab. (*b*) Two-way flat slab with drop panels and mushroom capitals. (*c*) Two-way flat plate. (*d*) Two-way concrete joist system (waffle slab). *(Drawings by Edward Allen)*

For Preliminary Design of a Sitecast Concrete Structure

- Estimate the depth of a **one-way solid slab** at 1/22 of its span if it is conventionally reinforced or 1/40 of its span if it is posttensioned. Depths range typically from 4 to 10 inches (100–250 mm).
- Estimate the total depth of a **one-way concrete joist system** or **wide-module system** at 1/18 of its span if it is conventionally reinforced or 1/36 of its span if it is posttensioned. For standard sizes of the pans used to form these systems, see Figure 14.23. To arrive at the total depth, a slab thickness of 3 to 4½ inches (75–115 mm) must be added to the depth of the pan that is selected.
- Estimate the depth of **concrete beams** at 1/16 of their span if they are conventionally reinforced or 1/24 of their span if they are posttensioned. For concrete girders, use ratios of 1/12 and 1/20, respectively.
- Estimate the depth of **two-way flat plates** and **flat slabs** at 1/30 of their span if they are conventionally reinforced or 1/45 of their span if they are posttensioned. Typical depths are 5 to 12 inches (125–305 mm). The minimum column size for a flat plate is approximately twice the depth of the slab. The width of a drop panel for a flat slab is usually one-third of the span, and the projection of the drop panel below the slab is about one-half the thickness of the slab.
- Estimate the depth of a **waffle slab** at 1/24 of its span if it is conventionally reinforced or 1/35 of its span if it is posttensioned. For standard sizes of the domes used to form waffle slabs, see Figure 14.35. To arrive at the total depth, a slab thickness of 3 to 4½ inches (75–115 mm) must be added to the depth of the dome that is selected.
- To estimate the size of a **concrete column** of normal height, add up the total roof and floor area supported by the column. A 12-inch (300-mm) column can support up to about 2000 square feet (190 m^2) of area, a 16-inch (400-mm) column 4000 square feet (370 m^2), a 20-inch (500-mm) column 6000 square feet (560 m^2), a 24-inch (600-mm) column 9000 square feet (840 m^2), and a 28-inch (700-mm) column 10,500 square feet (980 m^2). These sizes are greatly influenced by the strength of the concrete used and the ratio of reinforcing steel to concrete. Columns are usually round or square.
- To estimate the thickness of a **concrete loadbearing wall**, add up the total width of floor and roof slabs that contribute load to the wall. An 8-inch (200-mm) wall can support approximately 1200 feet (370 m) of slab, a 10-inch (250-mm) wall 1500 feet (460 m), a 12-inch (300-mm) wall 1700 feet (520 m), and a 16-inch (400-mm) wall 2200 feet (670 m). These thicknesses are greatly influenced by the strength of the concrete used and the ratio of reinforcing steel to concrete.

These approximations are valid only for purposes of preliminary building layout and must not be used to select final member sizes. They apply to the normal range of building occupancies, such as residential, office, commercial, and institutional buildings, and parking garages. For manufacturing and storage buildings, use somewhat larger members.

For more comprehensive information on preliminary selection and layout of a structural system and sizing of structural members, see Edward Allen and Joseph Iano, *The Architect's Studio Companion* (4th ed.), Hoboken, John Wiley & Sons, Inc., 2007.

Lift-slab construction, used chiefly with two-way flat plate structures, virtually eliminates formwork. The floor and roof slabs of a building are cast in a stack on the ground. Then hydraulic jacks are used to lift the slabs up the columns to their final elevations, where they are welded in place using special cast-in-place steel slab collars (Figure 14.45).

Ganged forms for wall construction are large units made up of a number of panels that are supported by the same set of walers. These are handled by cranes and are often more economical than conventional small panels that are maneuvered by hand. For floor slabs that are cast in place, *flying formwork* is fabricated in large sections that are supported on deep metal trusses. The sections are moved from one floor to the next by crane, eliminating much of the labor usually expended on stripping and reerecting formwork (Figure 14.46).

Slip forming is useful for tall-walled structures such as elevator shafts, stairwells, and storage silos. A ring of formwork is pulled steadily upward by jacks supported on the vertical reinforcing bars, while workers add concrete and horizontal reinforcing in a continuous process. Manufacturers of concrete formwork have developed more sophisticated systems of self-climbing formwork that offer many advantages over conventional slip forming (Figures 14.19 and 14.47).

In *tilt-up construction* (Figure 14.48), a floor slab is cast on the ground and reinforced concrete wall panels are poured over it in a horizontal position. When curing is complete, the panels are tilted up into a vertical orientation and hoisted into position by a crane, then grouted together. The elimination of most of the usual wall formwork results in formwork costs that are typically less than 5 percent of the cost of the overall concrete system, making tilt-up construction often economical for single-story buildings. Although most tilt-up

Figure 14.45
Lift-slab construction in progress. The paired steel rods silhouetted against the sky to the right are part of the lifting jacks seen at the tops of the columns. In North America, this form of construction is used infrequently, in part due to a history of past construction accidents. *(Reprinted with permission of the Portland Cement Association from* Design and Control of Concrete Mixtures, *12th edition; Photos: Portland Cement Association, Skokie, IL)*

Figure 14.46
Flying formwork for a one-way concrete joist system being moved from one floor to the next in preparation for pouring. Stiff metal trusses allow a large area of formwork to be handled by a crane as a single piece. *(Courtesy of Molded Fiber Glass Concrete Forms Company)*

panel construction is for walls no taller than 45 feet (13.7 m), panels for walls approaching heights as great as 100 feet (30 m) are feasible. Tilt-up is probably the most widely used of the innovative concrete construction methods described here.

Shotcrete (*pneumatically placed concrete*) is sprayed into place from the nozzle of a hose by a stream of compressed air. Because of its very low slump, even walls with vertical sides can be placed with little in the way of conventional formwork, though some kind of solid surface to spray against is required. Shotcrete is used for foundation walls, stabilization of steep slopes, repairing damaged concrete on the faces of beams and columns, seismic retrofits, and the production of free-form structures such as swimming pools and playground structures.

Even greater savings in formwork costs can be realized by casting concrete in reusable molds in a precasting plant, or by combining

FIGURE 14.47
A proprietary system of self-climbing formwork is being used to form these sitecast concrete elevator shafts for a tall building. The top level is a working surface from which reinforcing bars are handled and the concrete is poured. The outer panels of the formwork are mounted on overhead tracks just beneath the top level. The panels can be rolled back to the outside of the perimeter walkway after each pour, allowing workers to clean the formwork and install the reinforcing for the next pour. The entire two-story apparatus raises itself a story at a time with built-in hydraulic jacks. *(Courtesy of Patent Scaffolding Company, Fort Lee, New Jersey)*

FIGURE 14.48
Tilt-up construction. The exterior wall panels were reinforced and cast flat on the floor slab. Using special lifting rings that were cast into the panels and a lifting harness that exerts equal force on each of the lifting rings, a crane tilts up each panel and places it upright on a strip foundation at the perimeter of the building. Each erected panel is braced temporarily with diagonal steel struts until the roof structure has been completed. *(Reprinted with permission of the Portland Cement Association from* Design and Control of Concrete Mixtures, *12th edition; Photos: Portland Cement Association, Skokie, IL)*

sitecast concrete with precast forming elements, both of which are the subject of the next chapter.

Advances in reinforcing for sitecast concrete, other than the adoption of posttensioning, include a move to higher-strength steels and a trend toward increased prefabrication of reinforcing bars prior to installation in the forms. With developments in welding and fabricating machinery, the concept of welded wire reinforcing is expanding beyond the familiar grid of heavy wire to include complete cages of column reinforcing and entire bays of slab reinforcing. As lighter-weight reinforcing bars based on carbon fiber and aramid fiber come into more common use, other new forms of reinforcing are sure to evolve.

Concrete pumps are becoming the universal means of moving wet concrete from delivery trucks to the point where it is poured. Pumping has many advantages, among them that it can deliver concrete to places that could not possibly be reached by a crane-mounted bucket or even a wheelbarrow. For most projects, concrete is pumped through a flexible hose, but in some very large projects, fixed, rigid pipes are installed for the duration of the construction process to carry the concrete over long distances. The concrete mixture should be designed with the participation of the pumping subcontractor to be sure that it will not clog the line when it is put under pressure by the pump. Concrete can be pumped to astonishing heights and horizontal distances: For the Burj Dubai tower, concrete was pumped more than 1970 vertical feet (600 m).

Architectural Concrete

Concrete that is intended as finished interior or exterior surfaces, and is specified with highly prescribed finish characteristics, is known as *architectural concrete*. Most formed concrete surfaces, although structurally sound, have too many blemishes and irregularities to be visually attractive. A vast amount of thought and effort has been expended to develop handsome surface finishes for architectural concrete (Figures 14.49–14.52). *Exposed aggregate finishes* involve the scrubbing and hosing of concrete surfaces shortly after the initial set of the concrete to remove the cement paste from the surface and reveal the aggregate. This process is often aided by chemicals that retard the set of the cement paste; these are either sprayed on the surface of a slab or used as a coating inside formwork. Because

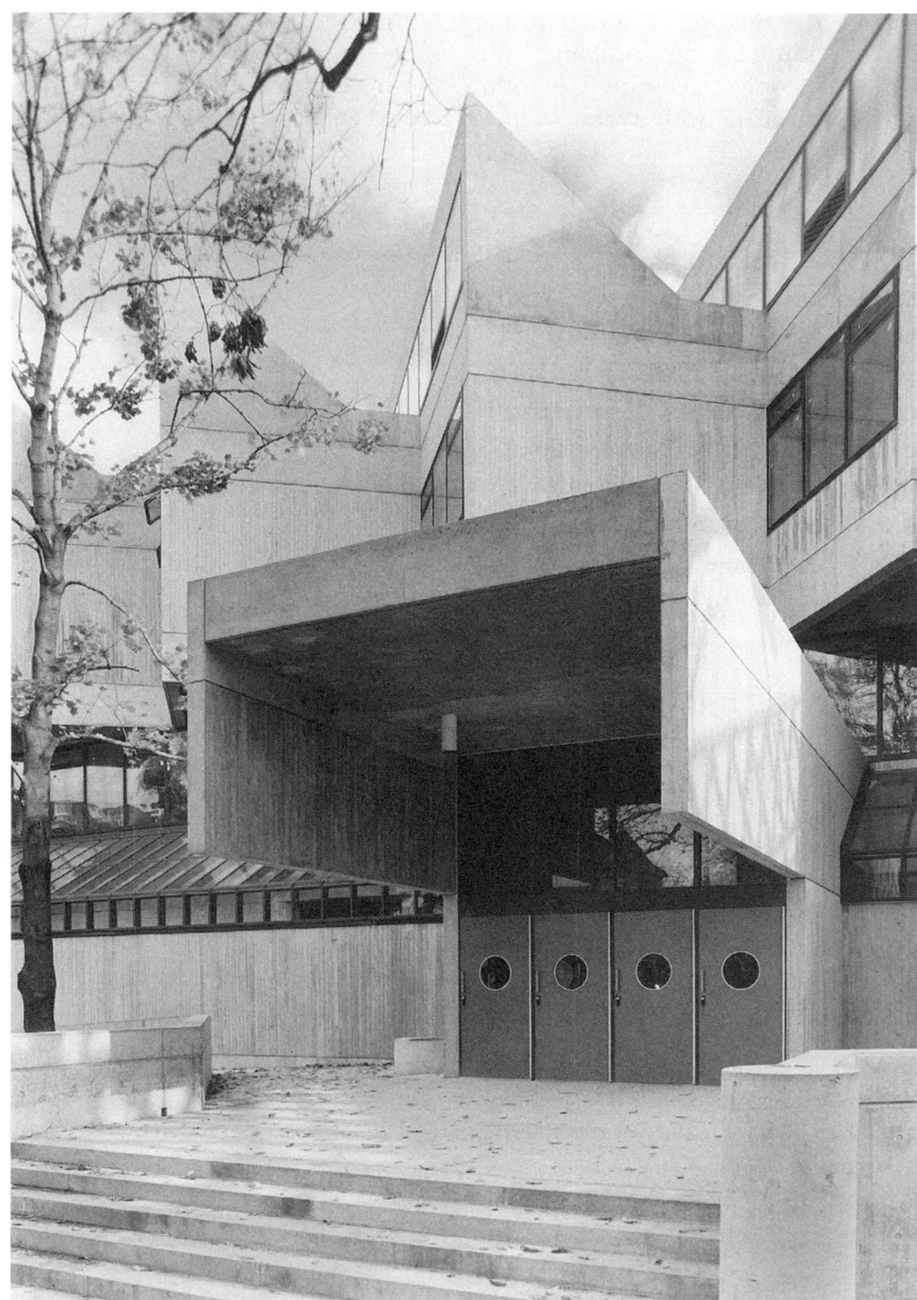

Figure 14.49
Exposed wall surfaces of sitecast concrete. Narrow boards were used to form the walls, and form tie locations were carefully worked out in advance. *(Architect: Eduardo Catalano. Photo by Erik Leigh Simmons. Courtesy of the architect)*

concrete can take on almost any texture that can be imparted to the surface of formwork, much work has gone into developing formwork surfaces of wood, wood panel products, metal, plastic, and rubber to produce textures that range from almost glassy smooth to ribbed, veined, board textured, and corrugated. After partial curing, other steps can be taken to change the texture of concrete, including sandblasting, rubbing with abrasive stones, grinding smooth, and hammering with various types of flat, pointed, or toothed masonry hammers. Many types of pigments, dyes, paints, and sealers can be used to add color or gloss to concrete surfaces and to give protection against weather, dirt, and wear.

Exposed wall surfaces of concrete need special attention from the designer and contractor (Figure 14.51). Chairs and bolsters need to be selected that will not create rust spots on exterior concrete surfaces. Form tie locations in exposed concrete walls should be patterned to harmonize with the layout of the walls themselves, and the holes left in the concrete surfaces by snapped-off ties must be patched or plugged securely to prevent rusting through. Joints between pours can be concealed gracefully with recesses called *rustication strips* in the face of the concrete. The formulation of the concrete needs to be closely controlled for color consistency from one batch to the next. In cold climates, air entrainment is advisable to prevent freeze-thaw damage of exterior wall surfaces.

If we were to train ourselves to draw as we build, from the bottom up . . . stopping our pencil to make a mark at the joints of pouring or erecting, ornament would grow out of our love for the expression of method.

Louis I. Kahn, quoted in Vincent Scully, Jr., *Louis I. Kahn,* 1962

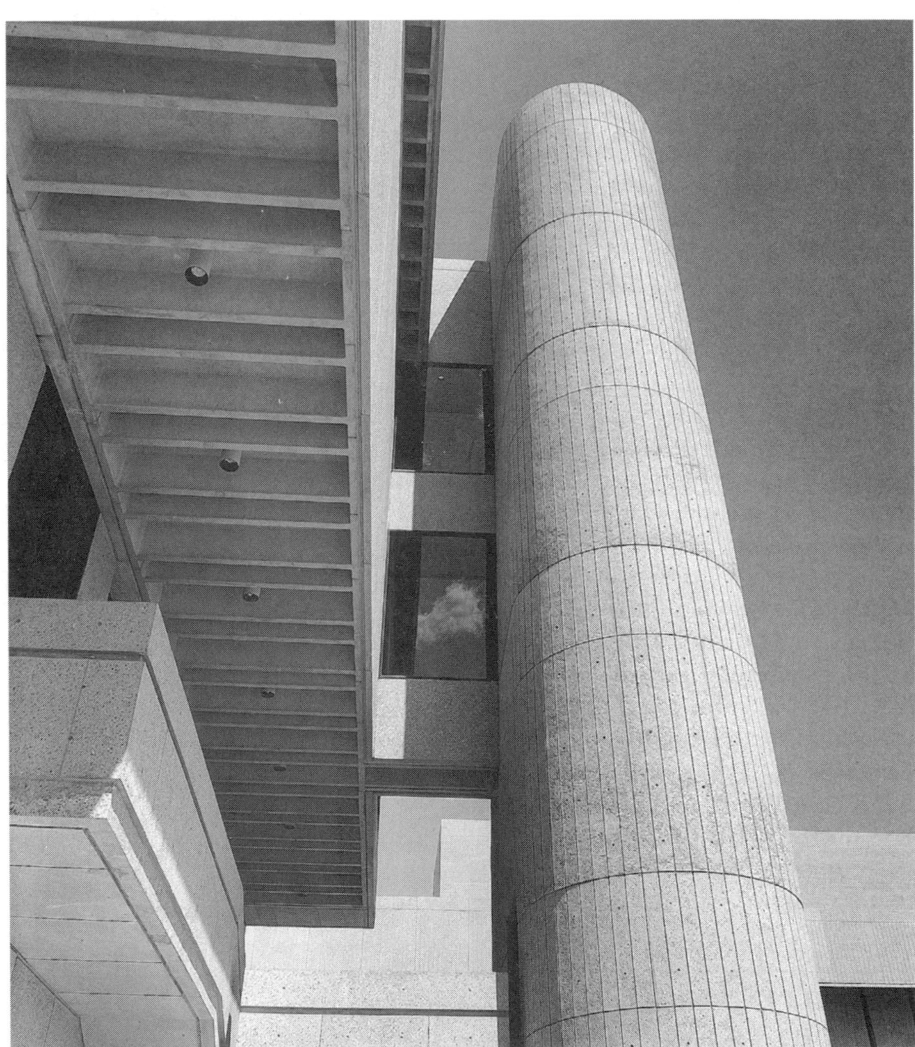

FIGURE 14.50
Exposed wall surfaces of concrete, sandblasted to expose the aggregate. Note the regular spacing of the form tie holes, which was worked out by the architect as an integral feature of the building design. *(Architect: Eduardo Catalano. Photo by Gordon H. Schenck, Jr. Courtesy of the architect)*

FIGURE 14.51
Standards specified by the architect to ensure satisfactory visual quality in the exposed concrete walls of the buildings illustrated in Figures 14.49 and 14.50. *(Courtesy of Eduardo Catalano, Architect)*

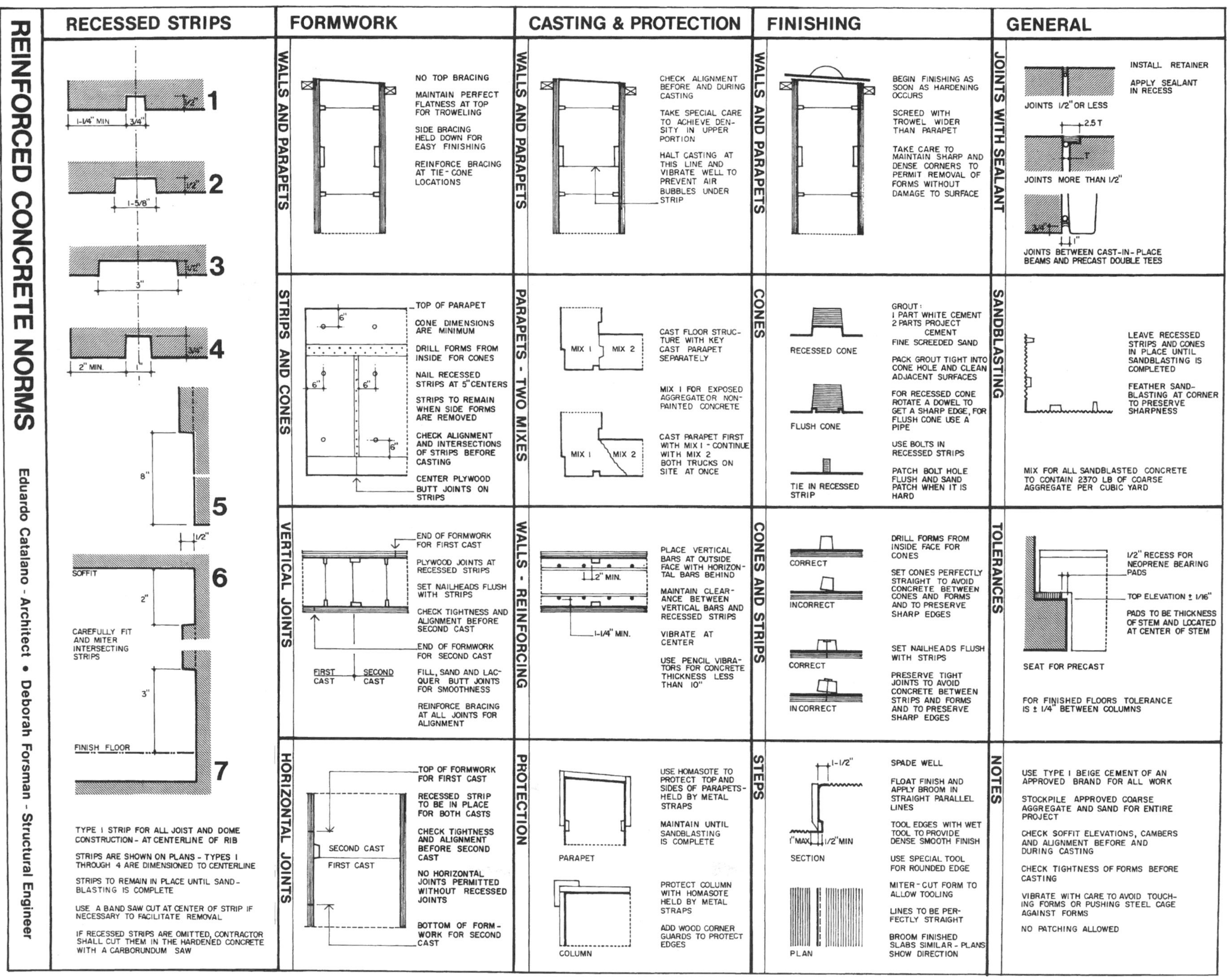

REINFORCED CONCRETE NORMS
Eduardo Catalano - Architect • Deborah Forsman - Structural Engineer
RECESSED STRIPS
1
2
3
4
5
6
7
1-1/4" MIN
3/4"
1/2"
1-5/8"
3"
2" MIN.
1"
8"
2"
SOFFIT
CAREFULLY FIT AND MITER INTERSECTING STRIPS
FINISH FLOOR
TYPE I STRIP FOR ALL JOIST AND DOME CONSTRUCTION - AT CENTERLINE OF RIB
STRIPS ARE SHOWN ON PLANS - TYPES I THROUGH 4 ARE DIMENSIONED TO CENTERLINE
STRIPS TO REMAIN IN PLACE UNTIL SANDBLASTING IS COMPLETE
USE A BAND SAW CUT AT CENTER OF STRIP IF NECESSARY TO FACILITATE REMOVAL
IF RECESSED STRIPS ARE OMITTED, CONTRACTOR SHALL CUT THEM IN THE HARDENED CONCRETE WITH A CARBORUNDUM SAW
FORMWORK
WALLS AND PARAPETS
NO TOP BRACING
MAINTAIN PERFECT FLATNESS AT TOP FOR TROWELING
SIDE BRACING HELD DOWN FOR EASY FINISHING
REINFORCE BRACING AT TIE-CONE LOCATIONS
STRIPS AND CONES
TOP OF PARAPET
CONE DIMENSIONS ARE MINIMUM
DRILL FORMS FROM INSIDE FOR CONES
NAIL RECESSED STRIPS AT 5" CENTERS
STRIPS TO REMAIN WHEN SIDE FORMS ARE REMOVED
CHECK ALIGNMENT AND INTERSECTIONS OF STRIPS BEFORE CASTING
CENTER PLYWOOD BUTT JOINTS ON STRIPS
6"
VERTICAL JOINTS
END OF FORMWORK FOR FIRST CAST
PLYWOOD JOINTS AT RECESSED STRIPS
SET NAILHEADS FLUSH WITH STRIPS
CHECK TIGHTNESS AND ALIGNMENT BEFORE SECOND CAST
END OF FORMWORK FOR SECOND CAST
FILL, SAND AND LACQUER BUTT JOINTS FOR SMOOTHNESS
REINFORCE BRACING AT ALL JOINTS FOR ALIGNMENT
FIRST CAST
SECOND CAST
HORIZONTAL JOINTS
TOP OF FORMWORK FOR FIRST CAST
RECESSED STRIP TO BE IN PLACE FOR BOTH CASTS
CHECK TIGHTNESS AND ALIGNMENT BEFORE SECOND CAST
NO HORIZONTAL JOINTS PERMITTED WITHOUT RECESSED JOINTS
BOTTOM OF FORMWORK FOR SECOND CAST
SECOND CAST
FIRST CAST
CASTING & PROTECTION
WALLS AND PARAPETS
CHECK ALIGNMENT BEFORE AND DURING CASTING
TAKE SPECIAL CARE TO ACHIEVE DENSITY IN UPPER PORTION
HALT CASTING AT THIS LINE AND VIBRATE WELL TO PREVENT AIR BUBBLES UNDER STRIP
PARAPETS - TWO MIXES
MIX I
MIX 2
CAST FLOOR STRUCTURE WITH KEY CAST PARAPET SEPARATELY
MIX I FOR EXPOSED AGGREGATE OR NON-PAINTED CONCRETE
CAST PARAPET FIRST WITH MIX I - CONTINUE WITH MIX 2 BOTH TRUCKS ON SITE AT ONCE
WALLS - REINFORCING
2" MIN.
1-1/4" MIN.
PLACE VERTICAL BARS AT OUTSIDE FACE WITH HORIZONTAL BARS BEHIND
MAINTAIN CLEARANCE BETWEEN VERTICAL BARS AND RECESSED STRIPS
VIBRATE AT CENTER
USE PENCIL VIBRATORS FOR CONCRETE THICKNESS LESS THAN 10"
PROTECTION
PARAPET
COLUMN
USE HOMASOTE TO PROTECT TOP AND SIDES OF PARAPETS - HELD BY METAL STRAPS
MAINTAIN UNTIL SANDBLASTING IS COMPLETE
PROTECT COLUMN WITH HOMASOTE HELD BY METAL STRAPS
ADD WOOD CORNER GUARDS TO PROTECT EDGES
FINISHING
WALLS AND PARAPETS
BEGIN FINISHING AS SOON AS HARDENING OCCURS
SCREED WITH TROWEL WIDER THAN PARAPET
TAKE CARE TO MAINTAIN SHARP AND DENSE CORNERS TO PERMIT REMOVAL OF FORMS WITHOUT DAMAGE TO SURFACE
CONES
RECESSED CONE
FLUSH CONE
TIE IN RECESSED STRIP
GROUT: I PART WHITE CEMENT 2 PARTS PROJECT CEMENT
FINE SCREEDED SAND
PACK GROUT TIGHT INTO CONE HOLE AND CLEAN ADJACENT SURFACES
FOR RECESSED CONE ROTATE A DOWEL TO GET A SHARP EDGE, FOR FLUSH CONE USE A PIPE
USE BOLTS IN RECESSED STRIPS
PATCH BOLT HOLE FLUSH AND SAND PATCH WHEN IT IS HARD
CONES AND STRIPS
CORRECT
INCORRECT
CORRECT
INCORRECT
DRILL FORMS FROM INSIDE FACE FOR CONES
SET CONES PERFECTLY STRAIGHT TO AVOID CONCRETE BETWEEN CONES AND FORMS AND TO PRESERVE SHARP EDGES
SET NAILHEADS FLUSH WITH STRIPS
PRESERVE TIGHT JOINTS TO AVOID CONCRETE BETWEEN STRIPS AND FORMS AND TO PRESERVE SHARP EDGES
STEPS
1-1/2"
1" MAX
1/2" MIN
SECTION
PLAN
SPADE WELL
FLOAT FINISH AND APPLY BROOM IN STRAIGHT PARALLEL LINES
TOOL EDGES WITH WET TOOL TO PROVIDE DENSE SMOOTH FINISH
USE SPECIAL TOOL FOR ROUNDED EDGE
MITER-CUT FORM TO ALLOW TOOLING
LINES TO BE PERFECTLY STRAIGHT
BROOM FINISHED SLABS SIMILAR - PLANS SHOW DIRECTION
GENERAL
JOINTS WITH SEALANT
INSTALL RETAINER
APPLY SEALANT IN RECESS
JOINTS 1/2" OR LESS
2.5 T
T
JOINTS MORE THAN 1/2"
3/4"
1"
JOINTS BETWEEN CAST-IN-PLACE BEAMS AND PRECAST DOUBLE TEES
SANDBLASTING
LEAVE RECESSED STRIPS AND CONES IN PLACE UNTIL SANDBLASTING IS COMPLETED
FEATHER SANDBLASTING AT CORNER TO PRESERVE SHARPNESS
MIX FOR ALL SANDBLASTED CONCRETE TO CONTAIN 2370 LB OF COARSE AGGREGATE PER CUBIC YARD
TOLERANCES
1/2" RECESS FOR NEOPRENE BEARING PADS
TOP ELEVATION ± 1/16"
PADS TO BE THICKNESS OF STEM AND LOCATED AT CENTER OF STEM
SEAT FOR PRECAST
FOR FINISHED FLOORS TOLERANCE IS ± 1/4" BETWEEN COLUMNS
NOTES
USE TYPE I BEIGE CEMENT OF AN APPROVED BRAND FOR ALL WORK
STOCKPILE APPROVED COARSE AGGREGATE AND SAND FOR ENTIRE PROJECT
CHECK SOFFIT ELEVATIONS, CAMBERS AND ALIGNMENT BEFORE AND DURING CASTING
CHECK TIGHTNESS OF FORMS BEFORE CASTING
VIBRATE WITH CARE TO AVOID TOUCHING FORMS OR PUSHING STEEL CAGE AGAINST FORMS
NO PATCHING ALLOWED

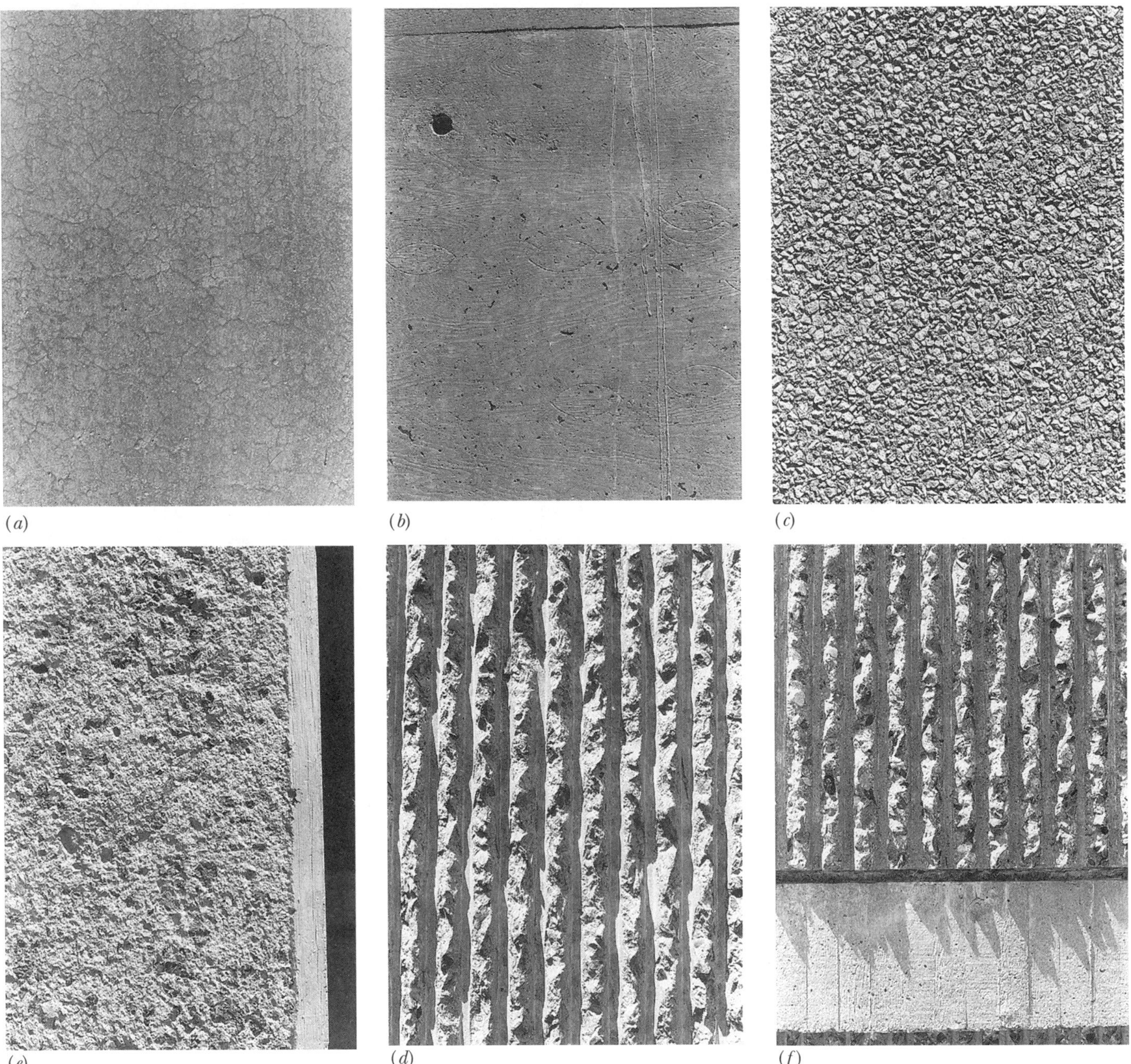

(*a*) (*b*) (*c*) (*e*) (*d*) (*f*)

FIGURE 14.52
Close-up photographs of some surface textures for exposed concrete walls. (*a*) Concrete cast against overlaid plywood to obtain a very smooth surface shows a crazing pattern of hairline cracks after 10 years of service. (*b*) The boat-shaped patches and rotary-sliced grain figure of A-veneered plywood formwork are mirrored faithfully in this surface. A neatly plugged form tie hole is seen at the upper left, and several lines of overspill from a higher pour have dribbled over the surface. (*c*) This exposed aggregate surface was obtained by coating the formwork with a curing retarder and scrubbing the surface of the concrete with water and a stiff brush after stripping the formwork. (*d*) The bush-hammered surface of this concrete column is framed by a smoothly formed edge. (*e, f*) Architect Paul Rudolph developed the techniques of casting concrete walls against ribbed formwork, then bush hammering the ribs to produce a very heavily textured, deeply shadowed surface. In the example to the right, the ribbed wall surface is contrasted to a board-formed slab edge, with a recessed rustication strip between. *(Photos by Edward Allen)*

Cutting Concrete, Stone, and Masonry

It is often necessary to cut hard materials, both in the course of obtaining and processing construction materials and during the construction process itself. The quarrying and milling of stone require many cutting operations. Precast concrete elements are frequently cut to length in the factory. Masonry units often need to be cut on the construction site, and masonry walls sometimes require the cutting of fastener holes and utility openings. Concrete cutting has become an industry in itself because of the need for utility openings, fastener holes, control joints, and surface grinding and texturing. Cutting and drilling are required to create new openings and remove unwanted construction during the renovation of masonry and concrete buildings. Core drilling is used to obtain laboratory test specimens of concrete, masonry, and stone. Finally, cutting is sometimes necessary to remove incorrect work and to perform building demolition operations.

In preindustrial times, hard materials were cut with hand tools, such as steel saws that employed an abrasive slurry of sand and water beneath the blade, and hardened steel drills and chisels that were driven with a heavy hammer. Wedges and explosives in drilled holes were used to split off large blocks of material. These techniques and mechanized variations of them are still used to some extent, but diamond cutting tools are rapidly taking over the bulk of the tough cutting chores in the construction industry. Diamond tools are expensive in first cost, but they cut much more rapidly than other types of tools, they cut more cleanly, and they last much longer, so they are usually more economical overall. Furthermore, diamond tools can sometimes do things that conventional tools cannot, such as precision sawing marble and granite into very thin sheets for floor and wall facings.

Diamonds cut hard materials efficiently because they are the hardest known material. Most of the industrial diamonds that go into cutting tools are synthetic. They are produced by subjecting graphite and a catalyst to extreme heat and pressure, then sorting and grading the small diamonds that result. Although some natural diamonds are still used in industry, synthetic diamonds are preferred for most tasks because of their more consistent behavior in use.

To manufacture a cutting tool, the diamonds are first embedded in a metallic bonding matrix and the mixture is formed into small cutting segments. The choice of diamonds and the exact composition of the bonding matrix are governed by the type of material that is to be cut. The cutting segments are brazed to steel cutting tools—circular saw blades, gangsaw blades, core drill cylinders—and the cutting tools are mounted in the machines that drive them. Some tools are designed to cut dry, but most are used with a spray of water that cools the blade and washes away the cut material. The accompanying illustrations (Figures A–J) show a variety of machines that use diamond cutting tools. Another example is shown in Figure 9.25. Diamonds are also made into grinding wheels, which are used for everything from sharpening tungsten carbide tools to flattening out-of-level concrete floors and polishing granite.

Cutting tools based on materials other than diamonds are still common on the construction job site. Tungsten carbide is used for the tips of small-diameter masonry and concrete drills and for the teeth of woodworking saws.

(Continues)

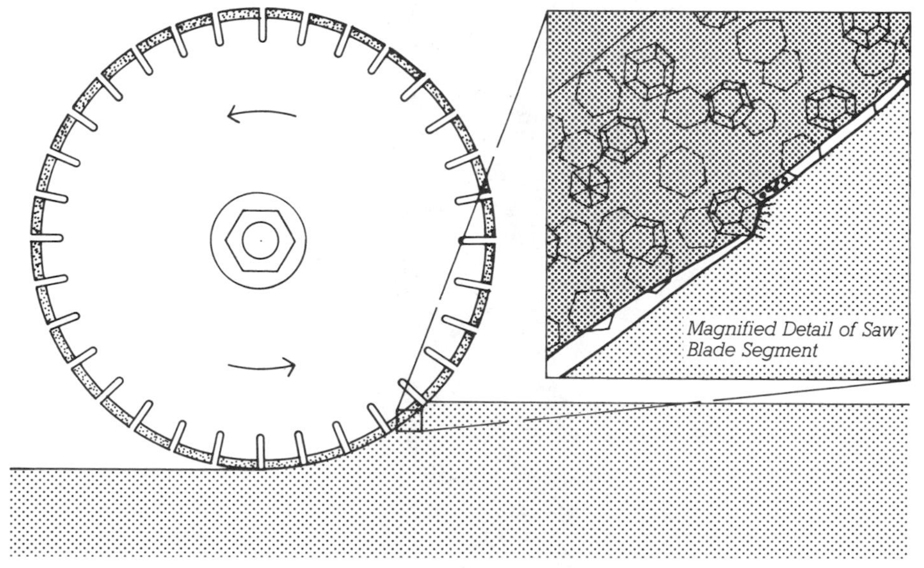

Figure A
A *diamond saw* blade is made up of cutting segments brazed to a steel blade core. Each cutting segment consists of diamond crystals embedded in a metallic bonding matrix. The diamonds in the cutting segment fracture chips from the material being cut. In doing so, each diamond gradually becomes chipped and worn and finally falls out of the bonding matrix altogether. The bonding matrix wears at a corresponding rate, exposing new diamonds to take over for those that have fallen out.

Cutting Concrete, Stone, and Masonry *(continued)*

Low-cost circular saw blades composed of abrasives such as silicon carbide are useful for occasional cutting of metals, concrete, and masonry, but the cutting action is slow and the blades wear very quickly. Less precise tools such as the pneumatic jackhammer, hydraulic splitters, and the traditional sledgehammer also have their uses. For large-volume precision cutting and drilling, however, there is no substitute for the industrial diamond.

Figure B
A hand-held pneumatically powered diamond circular saw cuts excess length from a concrete pile. *(Courtesy of Sinco Products, Inc.)*

Figure C
A pneumatically powered diamond circular saw slices a control joint in a concrete pavement. *(Courtesy of Partner Industrial Products, Itasca, Illinois)*

Figure D
Cutting a new opening in a masonry wall with a diamond circular saw. *(Courtesy of Partner Industrial Products, Itasca, Illinois)*

FIGURE E
Although a conventional circular saw can cut to a depth of only about a third of its blade diameter, this edge-driven ring saw can cut hard materials to a depth of almost three-quarters of its diameter. *(Courtesy of Partner Industrial Products, Itasca, Illinois)*

FIGURE F
A hydraulically driven chain saw with diamond teeth cuts a concrete masonry wall. *(C-150 Hydracutter, photo courtesy of Reimann & George Construction, Buffalo New York)*

Cutting Concrete, Stone, and Masonry (*continued*)

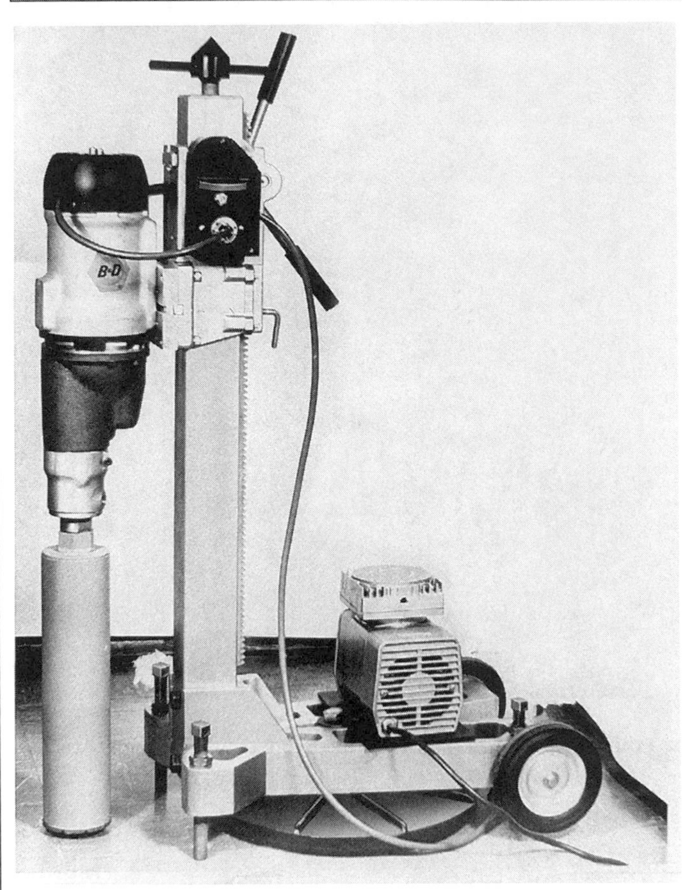

Figure G
A diamond core drill is used for cutting round holes. *(Photo courtesy of Sprague & Henwood, Inc., Scranton, Pennsylvania)*

Figure H
Using a technique called "stitch drilling," a core drill cuts an opening in a very thick wall. *(Courtesy of GE Superabrasives)*

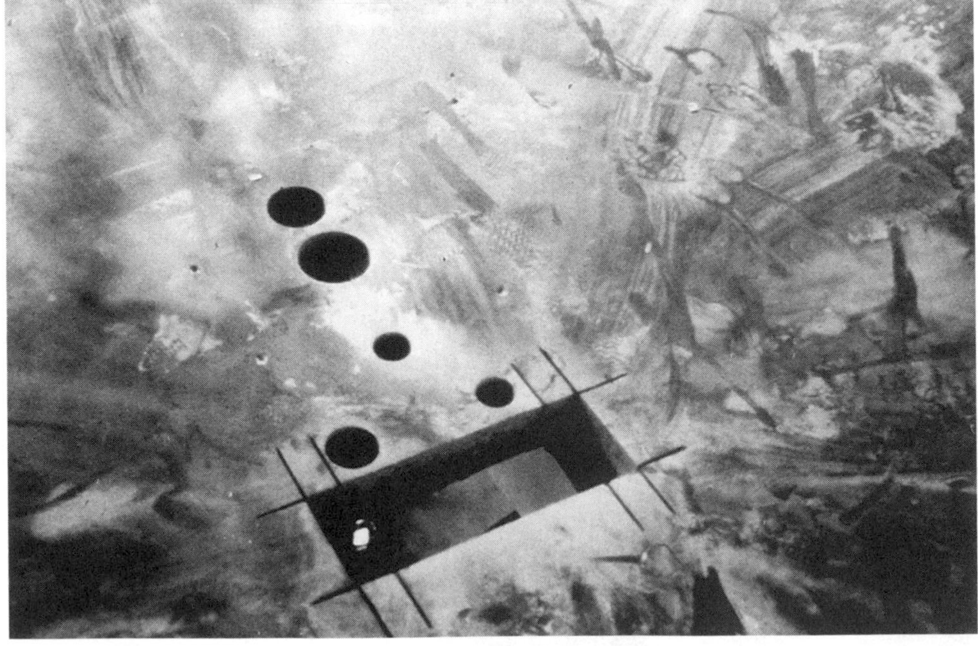

Figure I
Sawed and drilled openings for utility lines in a concrete floor slab. *(Courtesy of GE Superabrasives)*

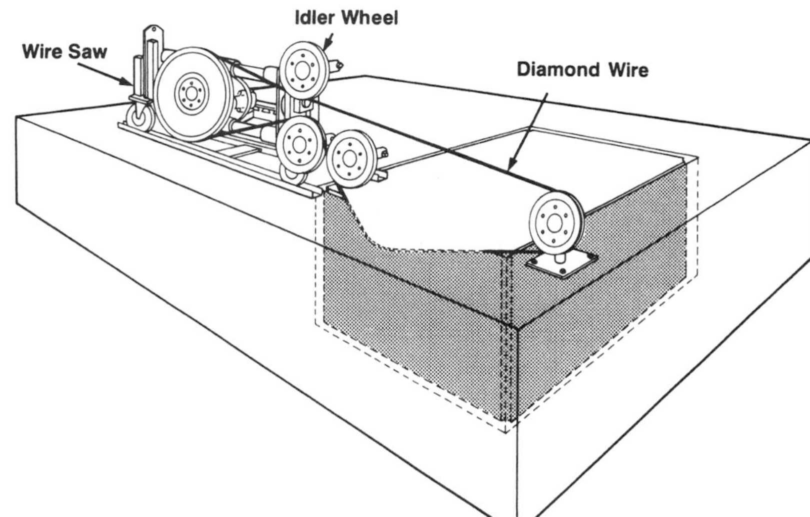

FIGURE J
Wire saws are capable of cutting depths and thicknesses of material that cannot be cut with any other kind of tool. Holes are first drilled at the corners of the cut, and the diamond cutting wire is strung through the holes and around the wheels of the sawing equipment. The equipment maintains a constant tension on the wire as it is pulled at high speed through the material. The wire gradually cuts its way out, leaving smooth planar surfaces. The wire is actually a steel cable on which steel beads with diamonds embedded in them are strung. *(© 1988 Cutting Technologies, Inc., Cincinnati, Ohio. All rights reserved)*

FIGURE K
This hotel was created from 36 concrete grain silos originally built in 1932. The 7-inch (180-mm) walls of the silos were cut with diamond circular saw blades ranging in diameter from 24 to 42 inches (610–1070 mm) to create the window and door openings. *(Courtesy of GE Superabrasives)*

Basically there are two approaches to the problem of producing a good surface finish on concrete. One is to remove the cement that is the cause of the blemishes and expose the aggregate. The other is to superimpose a pattern or profile that draws attention from the blemishes.

Henry Cowan, *Science and Building: Structural and Environmental Design in the Nineteenth and Twentieth Centuries,* New York, John Wiley & Sons, Inc., 1978, p. 283

Longer Spans in Sitecast Concrete

The ancient Romans built unreinforced concrete vaults and domes as roofs for temples, baths, palaces, and basilicas (Figure 13.2). Impressive spans were constructed, including a dome over the Pantheon in Rome, still standing, that approaches 150 feet (45 m) in diameter. Today, the arch, dome, and vault remain favorite devices for spanning long distances in concrete because of concrete's suitability to structural forms that work entirely in compression (Figures 14.53–14.55). Through folding or scalloping of vaulted forms, or through the use of warped geometries such as the hyperbolic paraboloid, the required resistance to buckling can be achieved with a surprisingly thin layer of concrete, often proportionally thinner than the shell of an egg (Figure 14.55*b*).

Long-span beams and trusses are possible in concrete, including post-tensioned beams and girders and reinforced deep girders analogous to steel plate girders and rigid frames. Concrete trusses and space frames are not common but are built from time to time. By definition, a truss includes strong tensile forces as well as compressive forces and is heavily dependent on steel reinforcing or prestressing.

Barrel shells and *folded plates* (Figures 14.54, 14.55*c*, and 14.56) derive their stiffness and strength from the folding or scalloping of a thin concrete plate to increase its rigidity and structural depth without adding material. Each of these forms depends

Figure 14.53
The same wooden centering was used four times to form this concrete arch bridge. *(Courtesy of Gang-Nail Systems, Inc.)*

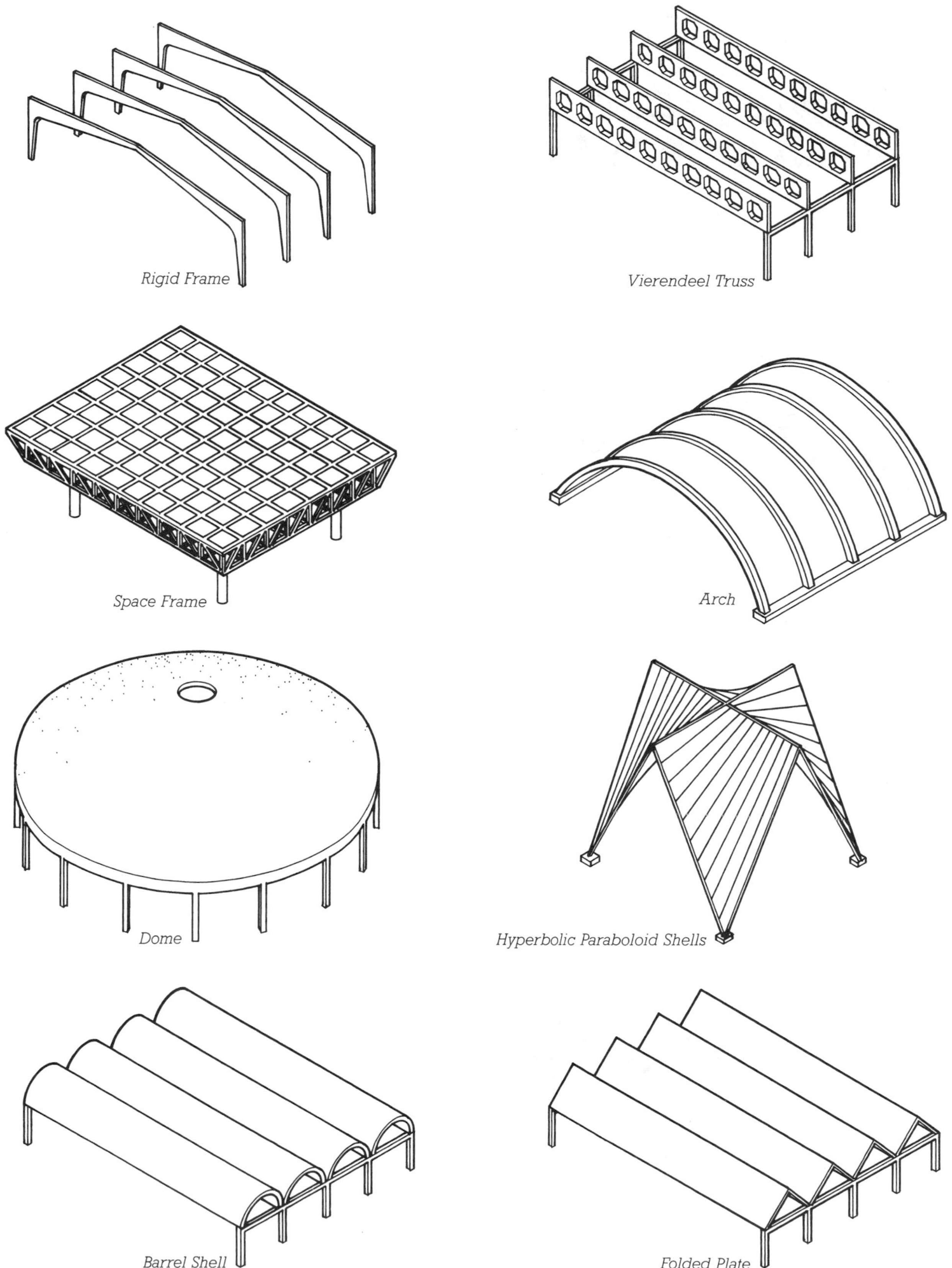

FIGURE 14.54

Examples of eight types of longer-span structures in concrete. Each is a special case of an infinite variety of forms. All can be sitecast, but the rigid frame, space frame, and Vierendeel truss are more likely to be precast for most applications.

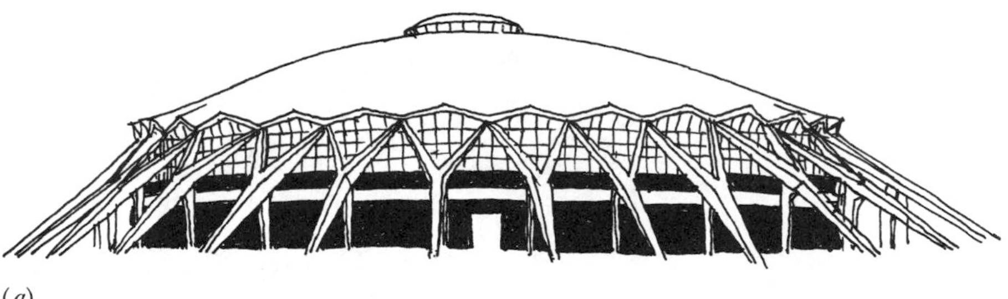

(*a*)

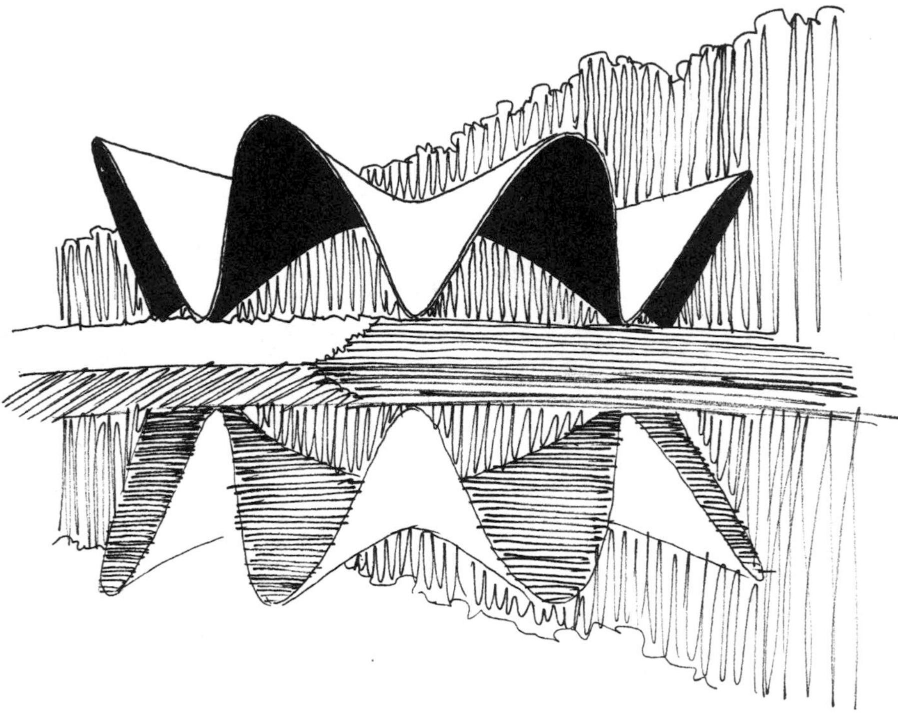

(*b*)

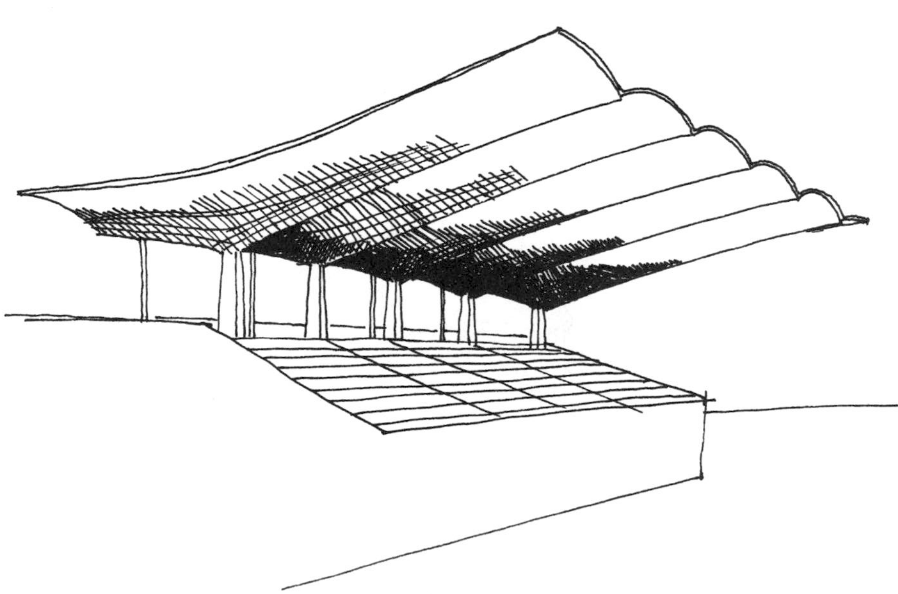

(*a*)

FIGURE 14.55
Three concrete shell structures by 20th-century masters of concrete engineering. (*a*) A domed sports arena by Pier Luigi Nervi. (*b*) A lakeside restaurant of hyperbolic paraboloid shells by Felix Candela. (*c*) A racetrack grandstand roofed with cantilevered concrete barrel shells by Eduardo Torroja. *(Drawings by Edward Allen)*

FIGURE 14.56
Flying formwork is removed from a bay of a folded plate concrete roof for an air terminal. *(Architects: Thorshov and Cerny. Photo courtesy of APA–The Engineered Wood Association)*

on reinforcing or posttensioning to resist the tensile forces that it may experience.

Designing Economical Sitecast Concrete Buildings

The cost of a concrete building frame can be broken down into the costs of the concrete, the reinforcing steel, and the formwork. Of the three, the cost of concrete is usually the least significant in North America and the cost of formwork is the most significant. Accordingly, simplification and standardization of formwork are the first requirements for an economical concrete frame. Repetitive, identical column spacings and bay sizes allow the same formwork to be used again and again without alterations. Flat plate construction is often the most economical, simply because its formwork is so straightforward. Joist band construction is usually more economical than joist construction that uses beams proportioned more efficiently for their structural requirements, because enough is saved on formwork costs to more than compensate for the added concrete and reinforcing steel in the beams. This same reasoning applies if column and beam dimensions are standardized throughout the building, even though loads may vary; the amount of reinforcing and the strengths of the concrete and reinforcing steel can be changed to meet the varying structural requirements (Figure 14.57).

Sitecast Concrete and the Building Codes

Concrete structures are inherently fire resistant. When fire attacks concrete, the water of hydration is gradually driven out and the concrete loses strength, but this deterioration is slow because considerable heat is needed to raise the temperature of the mass of concrete to the point where dehydration begins, and a large additional quantity of heat is required to vaporize the water. The steel reinforcing bars or prestressing strands are buried beneath a concrete cover that protects them for an extended period of time. Except under unusual circumstances, such as a prolonged fire fueled by stored

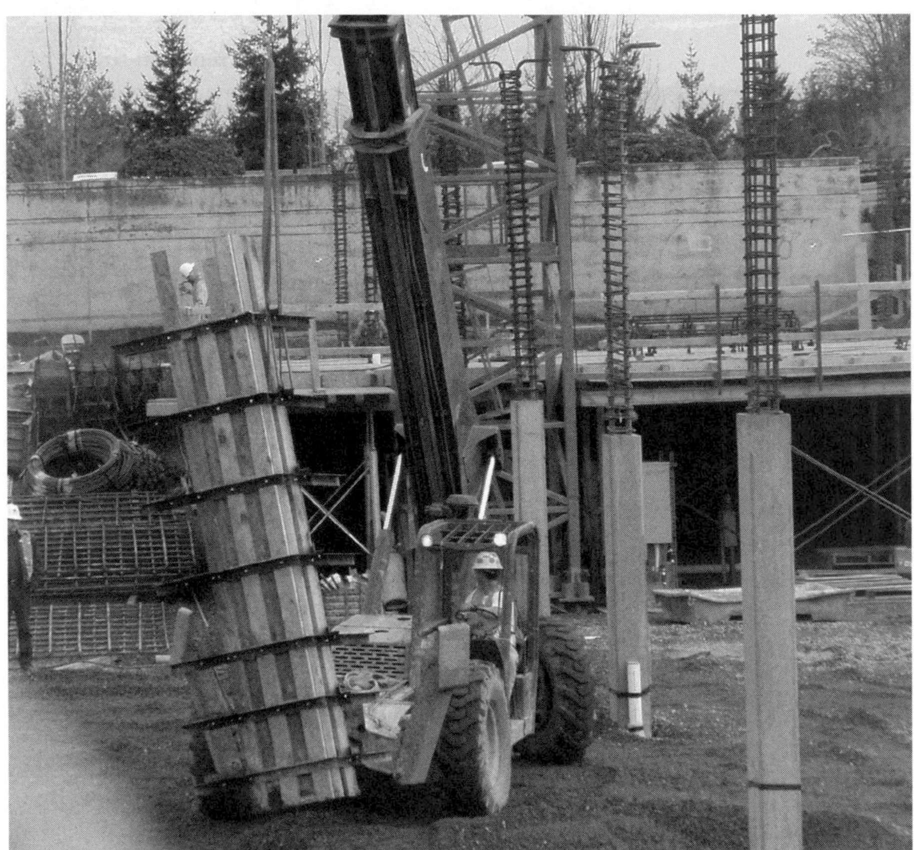

FIGURE 14.57
By keeping sizes of columns, beams, and other formed elements as consistent as possible, significant savings can be achieved in the cost of formwork for sitecast concrete construction. Here, a column form, having been stripped from a recently cast column, is being carried to a new location for reuse. *(Photo by Joseph Iano)*

petroleum products, concrete structures usually survive fires with only cosmetic damage and are repaired with relative ease.

Concrete structures with adequate cover over reinforcing and adequate slab thicknesses are classified as Type I buildings under the International Building Code. Slab thickness requirements for the several construction types are complex, depending on the type of aggregate used in the concrete and whether or not a given structural member is restrained from movement by surrounding construction. Fire resistance requirements for the highest construction types can be met in joist and waffle systems either by increasing their slab thickness beyond what is structurally necessary or by applying fireproofing materials to the lower surfaces of the floor structures.

Sitecast concrete buildings have rigid joints and in many cases need no additional structural elements to achieve the necessary resistance to wind and seismic forces. More restrictive seismic design provisions in the building codes have, however, increased the attention paid by structural engineers to column ties and beam stirrups, particularly in the zones where beams and columns meet, to be sure that vertical bars in columns and horizontal bars in beams are adequately restrained against the unusually strong forces that can occur in these zones under seismic loadings. The joints between flat plate floors and columns may not be sufficiently rigid to brace a building of more than modest height, unless drop panels or beams are added to stiffen the slab-to-column junction.

Uniqueness of Sitecast Concrete

Concrete is a shapeless material that must be given form by the designer. For economy, the designer can adopt a standard system of concrete framing. For excitement, one can invent new shapes and textures, a route taken by many of the leading architects. Some have pursued its sculptural possibilities, others its surface patterns and textures, still others its structural logic. From each of these routes have come masterpieces—Le Corbusier's chapel at Ronchamp (Figure 14.62),

FIGURE 14.58
Concrete work nears the 1475-foot (450-m) summit of, at the time of their construction, the world's tallest buildings, the twin Petronas Towers in Kuala Lumpur, Malaysia. Each tower is supported by a perimeter ring of 16 cylindrical concrete columns and a central core structure, also made of concrete. The columns vary in diameter from 8 feet (2400 mm) at the base of the building to 4 feet (1200 mm) at the top. For speed of construction, the floors are framed with steel and composite metal decking. Concrete with strengths as high as 11,600 psi (80 MPa) was used in the columns. The architect was Cesar Pelli & Associates, Inc. The structural engineers were Thornton-Tomasetti and Rahnill Bersekutu Sdn Bhd. The U.S. partner in the joint venture team that constructed the towers was J. A. Jones Construction Co., Charlotte, North Carolina. *(Photograph by Uwe Hausen, J. A. Jones, Inc.)*

Wright's Unity Temple (Figure 14.1), and the elegant structures of Torroja, Candela, and Nervi, examples of which are sketched in Figure 14.55. Many of these masterpieces, especially from the latter three designers, were also constructed with impressive economy. Sitecast concrete can do almost anything, be almost anything, at almost any scale, and in any type of building. It is a potent architectural material, and therefore a material both of spectacular architectural achievements and dismal architectural failures. A material so malleable demands skill and restraint from those who would build with it, and a material so commonplace requires imagination if it is to rise above the mundane.

As designers, we have learned to express the internal composition of concrete by exposing its aggregates at the surface, or to show the beauties of the formwork in which it was cast by leaving the marks of the ties and the textures of the formboards. But we have yet to discover how to reveal in the finished structure the lovely and complex geometries of the steel bars that constitute half of the structural partnership that makes concrete buildings stand up.

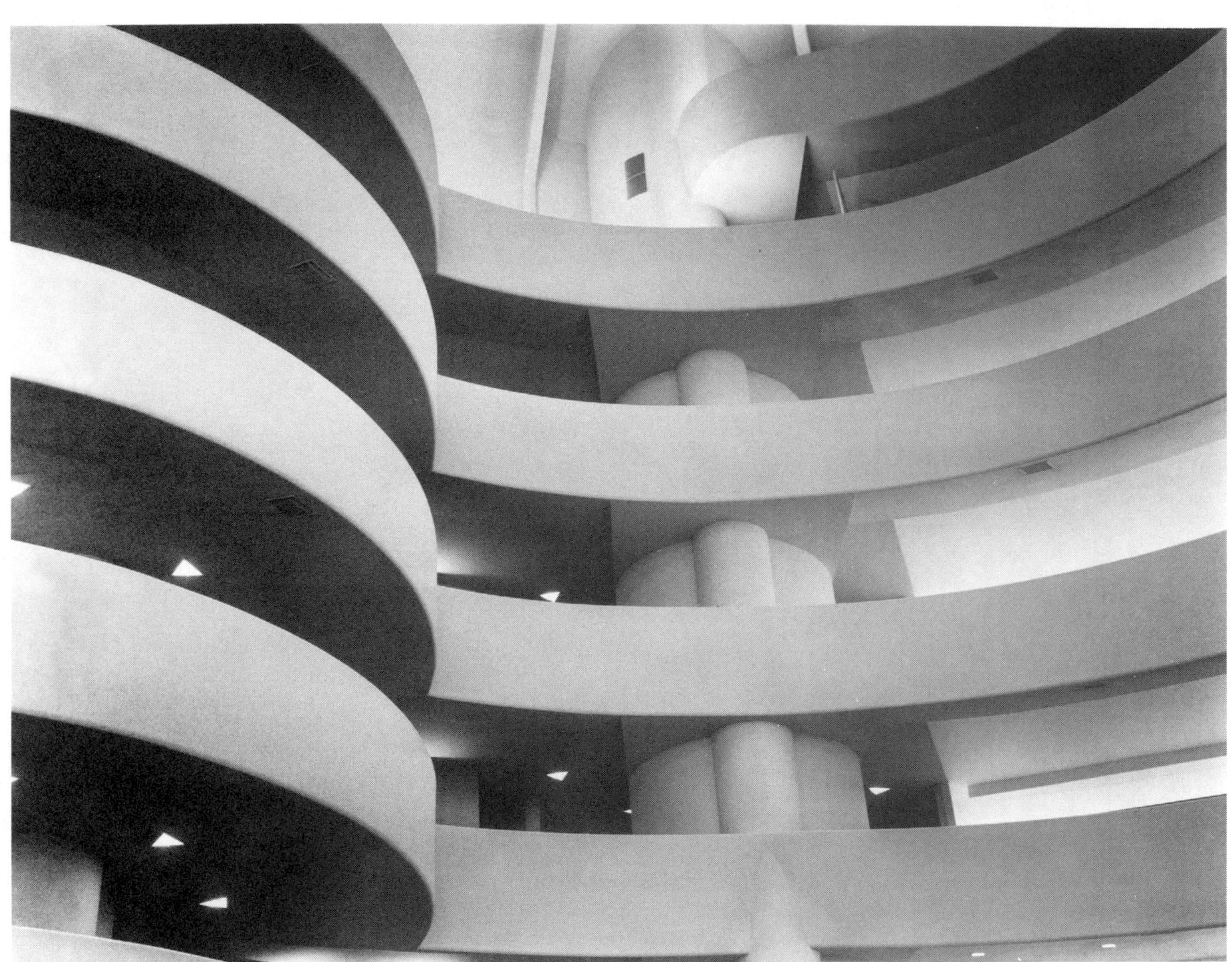

FIGURE 14.59
The plastered surfaces of Frank Lloyd Wright's Guggenheim Museum (1943–1956) cover a helical ramp of cast-in-place concrete. *(Photo by Wayne Andrews)*

Figure 14.60
The Chapel of St. Ignatius, Seattle University, designed by architect Steven Holl, is a tilt-up concrete structure. *(Photo by Joseph Iano)*

Figure 14.61
A sitecast concrete house in Lincoln, Massachusetts. *(Architects: Mary Otis Stevens and Thomas F. McNulty)*

FIGURE 14.62
Le Corbusier's most sculptural building in his favorite material, concrete: the chapel of Notre Dame de Haut at Ronchamp, France (1950–1955). *(Drawing by Edward Allen)*

FIGURE 14.63
The TWA Terminal at John F. Kennedy Airport, New York, 1956–1962. *(Architect: Eero Saarinen. Photo by Wayne Andrews)*

CSI/CSC	
MasterFormat Sections for Concrete Construction	
03 30 00	**CAST-IN-PLACE CONCRETE**
03 31 00	**Structural Concrete**
03 33 00	**Architectural Concrete**
03 34 00	**Low Density Concrete**
03 35 00	**Concrete Finishing**
03 37 00	**Specialty Placed Concrete**
	Shotcrete
	Pumped Concrete
03 39 00	**Concrete Curing**

Selected References

1. American Concrete Institute. *ACI 318: Building Code Requirements for Structural Concrete and Commentary*. Farmington Hills, MI, updated regularly.

This booklet establishes the basis for the engineering design and construction of reinforced concrete structures in the United States.

2. Concrete Reinforcing Steel Institute. *CRSI Design Handbook*. Schaumburg, IL, Concrete Reinforcing Steel Institute, updated regularly.

Structural engineers working in concrete use this handbook, which is based on the ACI Code (reference 1) as their major reference. It contains examples of engineering calculation methods and hundreds of pages of tables of standard designs for reinforced concrete structural elements.

3. Concrete Reinforcing Steel Institute. *Placing Reinforcing Bars*. Schaumburg, IL, updated regularly.

Written as a handbook for those engaged in the business of fabricating and placing reinforcing steel, this small volume is clearly written and beautifully illustrated with diagrams and photographs of reinforcing for all the common concrete framing systems.

4. Hurd, M. K. *Formwork for Concrete* (7th ed.). Farmington Hills, MI, American Concrete Institute, 2005.

Profusely illustrated, this book is the bible on formwork design and construction for sitecast concrete.

5. American Concrete Institute. *ACI 303R: Guide to Cast-in-Place Architectural Concrete Practice*. Farmington Hills, MI, updated regularly.

This is a comprehensive handbook on how to produce attractive surfaces in concrete.

6. Post-Tensioning Institute. *Post-Tensioning Manual*. Phoenix, AZ, updated regularly.

This heavily illustrated volume is both an excellent introduction to posttensioning for the beginner and a basic engineering manual for the expert.

8. Concrete Reinforcing Steel Institute. *Structural System Selection: Guide to Structural System Selection* and *Workbook for Evaluating Concrete Structures*. Schaumburg, IL, 1997.

These companion volumes make it easy to select an appropriate concrete framing system for a building and assign approximate sizes to its members.

9. See also the Selected References listed in Chapter 13.

Web Sites

Sitecast Concrete Framing Systems

Author's supplementary web site: **www.ianosbackfill.com/14_sitecast_concrete_framing_systems**

Dayton Superior formwork accessories: **www.daytonconcreteacc.com**

Dywidag-Systems International posttensioning systems: **www.dsiamerica.com**

Molded Fiber Glass Construction Products Co. formwork: **www.mfgcp.com**

Symons Corp. formwork: **www.symons.com**

Key Terms and Concepts

slab on grade
capillary break
moisture barrier, vapor retarder
strike off, screed
floating
bull float
darby
bleed water
troweled slab
steel trowel
rotary power trowel
knee board
broom finish
restraightening
straightedge
shake-on hardener
curing compound
superflat concrete floor
F-number
control joint, contraction joint
isolation joint, expansion joint
shrinkage-compensating cement
key
dowel
form tie
waler
brace
formwork stripping
form tie hole
insulating concrete form (ICF)
one-way solid slab
shore
reshoring
slab band
one-way concrete joist system, ribbed slab
pan
distribution rib
joist band
wide-module concrete joist system, skip-joist system
two-way solid slab
two-way flat slab
mushroom capital
drop panel
column strip
middle strip
two-way flat plate
waffle slab, two-way concrete joist system
dome
head
draped tendon
set
lift-slab construction
ganged form
flying formwork
slip forming
tilt-up construction
shotcrete, pneumatically placed concrete
architectural concrete
exposed aggregate finish
rustication strip
diamond saw
barrel shell
folded plate

Review Questions

1. Draw from memory a detail of a typical slab on grade and list the steps in its production. Why can't the surface be finished in one operation, instead of waiting for hours before final finishing?

2. What are control joints and isolation joints? Explain the purpose and typical locations for each in a concrete slab.

3. List the steps that are followed in forming and pouring a concrete wall.

4. Distinguish one-way concrete framing systems from two-way systems. Are steel and wood framing systems one-way or two-way? Is one-way construction more efficient structurally than two-way construction?

5. List the common one-way and two-way concrete framing systems and indicate the possibilities and limitations of each.

6. Why posttension a concrete structure rather than merely reinforce it?

Exercises

1. Propose a suitable reinforced concrete framing system for each of the following buildings and determine an approximate thickness for each:

   a. An apartment building with a column spacing of about 16 feet (5 m) in each direction

   b. A newsprint warehouse, column spacing 20 feet by 22 feet (6 m x 6.6 m)

   c. An elementary school, column spacing 24 feet by 32 feet (7.3 m x 9.75 m)

   d. A museum, column spacing 36 feet by 36 feet (11 m x 11 m)

   e. A hotel where overall building height must be minimized in order to build as many stories as possible within a municipal height limit

2. Look at several sitecast concrete buildings. Determine the type of framing system used in each and explain why you think it was selected. If possible, talk to the designers of the building and find out if you were right.

3. Observe a concrete building under construction. What is its framing system? Why? What types of forms are used for its columns, beams, and slabs? In what form are the reinforcing bars delivered to the site? How is the concrete mixed? How is it raised and deposited into the forms? How is it consolidated in the forms? How is it cured? Are samples taken for testing? How soon after pouring are the forms stripped? Are the forms reused? How long are the shores kept in place? Keep a diary of your observations over a period of a month or more.

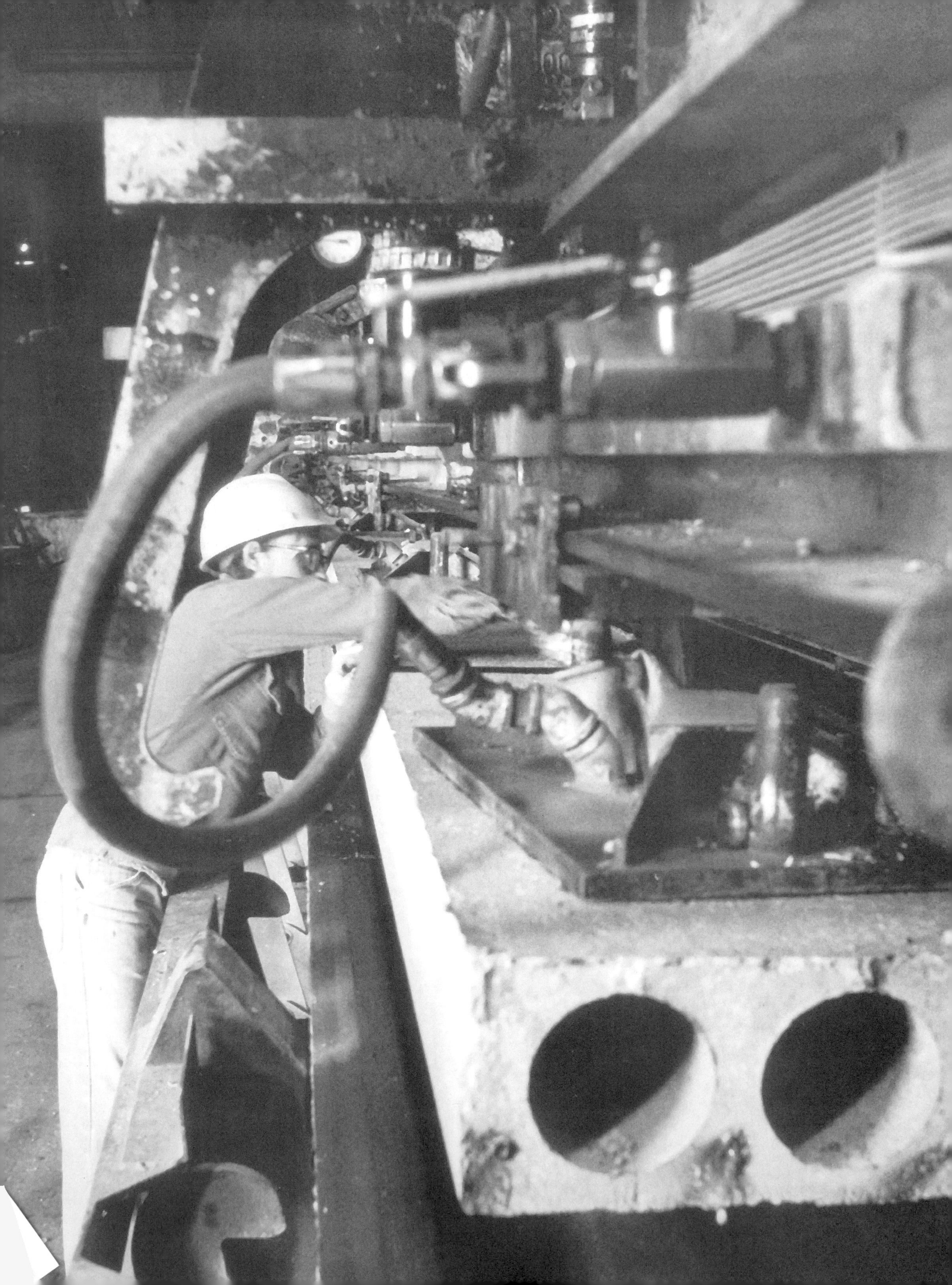

15

Precast Concrete Framing Systems

Suction devices lift a precast concrete hollow-core slab from the casting bed where it was manufactured. *(Courtesy of The Flexicore Co., Inc.)*

Structural *precast concrete* elements—slabs, beams, girders, columns, and wall panels—are cast and cured in factories, transported to the construction job site, and erected as rigid components. Precasting offers many potential advantages over sitecasting of concrete: The production of precast elements is carried out conveniently at ground level. The mixing and pouring operations are often highly mechanized, and frequently, especially in difficult climates, they are carried out under shelter. Control of the quality of materials and workmanship is generally better than on the construction job site. The concrete is cast in permanent forms made of steel, concrete, glass-fiber-reinforced plastic, or wood panels with smooth overlays, whose excellent surface properties are mirrored in the high-quality surfaces of the finished precast elements that they produce. The forms may be reused hundreds or thousands of times before they have to be renewed, so formwork costs per unit of finished concrete are low. The forms are equipped to pretension the steel in the precast elements for greater structural efficiency, which translates into longer spans, lesser depths, and lower weights than for comparable reinforced concrete elements. Concrete and steel of superior strength are used in precast elements, typically 5000-psi (35-MPa) concrete and 270,000-psi (1860-MPa) prestressing steel.

For the fastest possible curing, the concrete for precast concrete elements is made with Type III portland cement, high early strength. And the elements are usually *steam cured*. Steam furnishes heat to accelerate the hardening of the concrete and moisture for full hydration. Thus, a precasting plant is able to produce fully cured structural elements, from laying of the prestressing strands to removal of the finished elements from the casting bed, on a 24-hour cycle. (Self-consolidating concrete, described in Chapter 13, is also finding increased use in precast concrete production, where it has the potential to increase productivity and produce architectural concrete elements with better finish quality.)

When the elements produced by this expeditious technique are delivered to the construction job site, further advantages are realized: The erection process is similar to that of structural steel, but it is often faster because most precast concrete systems include a deck as an integral part of the major spanning elements, without the need for placing additional joist or decking components (Figures 15.1 and 15.2). Erection is much faster than that of sitecast concrete because there is no formwork to be erected and stripped, and little or no waiting for concrete to cure. And erection of precast structures can take place under adverse weather conditions, such as extremely high or low temperatures, which would not permit the sitecasting of concrete.

When choosing between precast and sitecast concrete, the designer must weigh the potential advantages of precasting against some potential disadvantages. Precast structural elements, although lighter in weight than similar elements of sitecast concrete, are nevertheless heavy and bulky to transport over the roads and hoist into place. This restricts somewhat the size and proportions of most precast elements: They can be rather long, but only as wide as the maximum legal vehicle width of 12 to 14 feet (3.66–4.27 m). This restricted width usually precludes utilization of the efficiencies of two-way structural action in precast slabs. And the three-dimensional sculptural possibilities of sitecast concrete are largely absent in precast concrete.

Although whole walls or rooms are sometimes precast as single units in concrete, this chapter will focus on the standard precast elements that are commonly mass-produced as structural components.

FIGURE 15.1
A fully precast building frame under construction. A poured concrete topping will cover the hollow-core slabs and the beams to create a smooth floor and tie the precast elements together. *(Courtesy of The Flexicore Co., Inc.)*

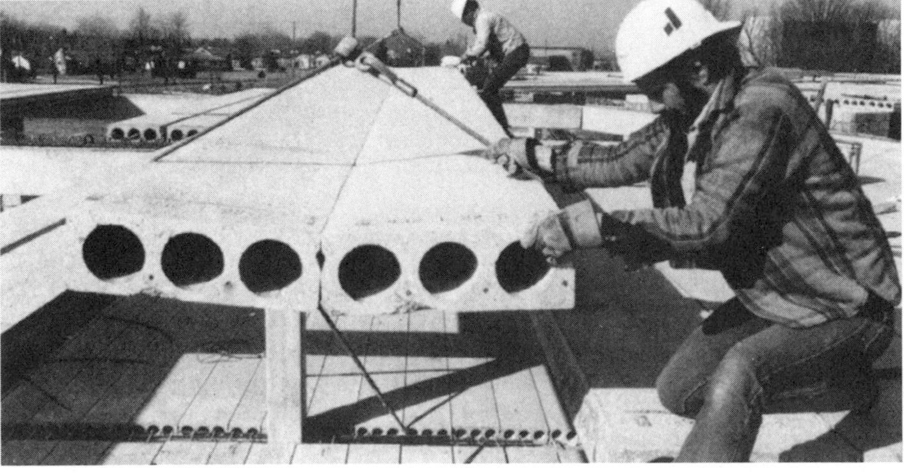

FIGURE 15.2
Workers guide two hollow-core slabs, lowered by a crane in wire rope slings, onto the precast concrete beams that will support them. *(Courtesy of The Flexicore Co., Inc.)*

Precast, Prestressed Concrete Structural Elements

Precast Concrete Slabs

The most fully standardized precast concrete elements are those used for making floor and roof slabs (Figure 15.3). These may be supported by bearing walls of precast concrete or masonry or by frames of steel, sitecast concrete, or precast concrete. Four kinds of precast slab elements are commonly produced: For short spans and minimum slab depths, *solid slabs* are appropriate. For longer spans, deeper elements must be used, and precast solid slabs, like their sitecast counterparts, become inefficient because they contain too much deadweight of nonworking concrete. In *hollow-core slabs*, precast elements suitable for intermediate spans, internal longitudinal voids replace much of the nonworking concrete. For the longest spans, still deeper elements are required, and *double tees* and *single tees* eliminate still more nonworking concrete.

For most applications, precast slab elements of any of the four types are manufactured with a rough top surface. After the elements have been erected, a concrete *topping* is poured over them and finished to a smooth surface. The topping, usually 2 inches (50 mm) thick, bonds during curing to the rough top of the precast elements and becomes a working part of their structural action. The topping also helps the precast elements to act together as a structural unit rather than as individual planks in resisting concentrated loads and diaphragm loads, and it conceals the slight differences in camber that often occur in prestressed components. Structural continuity across a number of spans can be achieved by casting reinforcing bars into the topping over the supporting beams or walls. Underfloor electrical conduits may also be embedded in the topping. Smooth-top precast slabs are sometimes used without topping, as is discussed later in this chapter.

Either normal-density concrete or structural lightweight concrete may be selected for use in any of the precast slab elements. Lightweight concrete, approximately 20 percent less dense than normal concrete, reduces the load on the frame and foundations of a building but is more expensive than normal-density concrete.

There is considerable overlapping of the economical span ranges of the different kinds of precast slab elements, allowing the designer some latitude in choosing which to use in a particular situation. Solid slabs and hollow-core slabs save on overall building height in multistory structures, and their smooth undersides can be painted and used as finish ceilings in many applications. For longer spans, double tees are generally preferred to the older single-tee design because they do not need to be supported against tipping during erection.

Precast Concrete Beams, Girders, and Columns

Precast concrete beams and girders are made in several standard shapes (Figure 15.4). The projecting *ledgers* on *L-shaped beams* and *inverted tees* provide direct support for precast slab elements. They conserve headroom in a building by supporting slabs near the bottoms of the beams, compared to rectangular beams without ledgers, where slab elements must rest on top. *AASHTO* (American Association of State Highway and Transportation Officials) *girders* were designed originally as efficient shapes for bridge structures, but they are sometimes used in buildings as well. Precast concrete columns are usually square or rectangular in section and may be prestressed or simply reinforced.

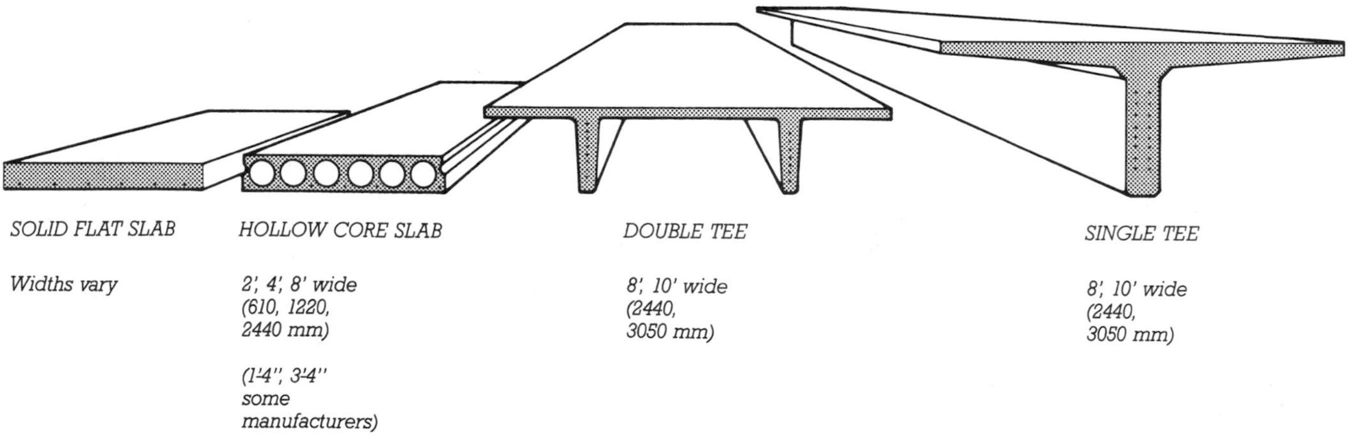

Figure 15.3
The four major types of precast concrete slab elements. Hollow-core slabs are produced by different companies in a variety of cross-sectional patterns, using several different processes. Single tees are much less commonly used than double tees because they need temporary support against tipping until they are permanently fastened in place.

For Preliminary Design of a Precast Concrete Structure

- Estimate the depth of a **precast solid slab** at 1⁄40 of its span. Depths typically range from 3½ to 8 inches (90–200 mm).
- An 8-inch (200-mm) **precast hollow-core slab** can span approximately 25 feet (7.6 m), a 10-inch (250-mm) slab 32 feet (9.8 m), and a 12-inch (300-mm) slab 40 feet (12 m).
- Estimate the depth of **precast concrete double tees** at 1⁄28 of their span. The most common depths of double tees are 12, 14, 16, 18, 20, 24, and 32 inches (300, 350, 400, 460, 510, 610, and 815 mm). Some manufacturers can provide double tees that are 48 inches (1220 mm) deep.
- A **precast concrete single tee** 36 inches (915 mm) deep spans approximately 85 feet (926 m) and a 48-inch (1220-mm) tee 105 feet (32 m).
- Estimate the depth of **precast concrete beams and girders** at 1⁄15 of their span for light loadings and 1⁄12 of their span for heavy loadings. These ratios apply to rectangular, inverted-tee, and L-shaped beams. The width of a beam or girder is usually about one-half of its depth. The projecting ledgers on inverted-tee and L-shaped beams are usually 6 inches (150 mm) wide and 12 inches (300 mm) deep.
- To estimate the size of a **precast concrete column**, add up the total roof and floor area supported by the column. A 10-inch (250-mm) column can support up to about 2300 square feet (215 m^2) of area, a 12-inch (300-mm) column 3000 square feet (280 m^2), a 16-inch (400-mm) column 5000 square feet (465 m^2), and a 24-inch (600-mm) column 9000 square feet (835 m^2). These values may be interpolated to columns in 2-inch (50-mm) increments. Columns are usually square.

These approximations are valid only for purposes of preliminary building layout and must not be used to select final member sizes. They apply to the normal range of building occupancies, such as residential, office, commercial, and institutional buildings, and parking garages. For manufacturing and storage buildings, use somewhat larger members.

For more comprehensive information on preliminary selection and layout of structural system and sizing of structural members, see Edward Allen and Joseph Iano, *The Architect's Studio Companion* (4th ed.), New York, John Wiley & Sons, Inc., 2007.

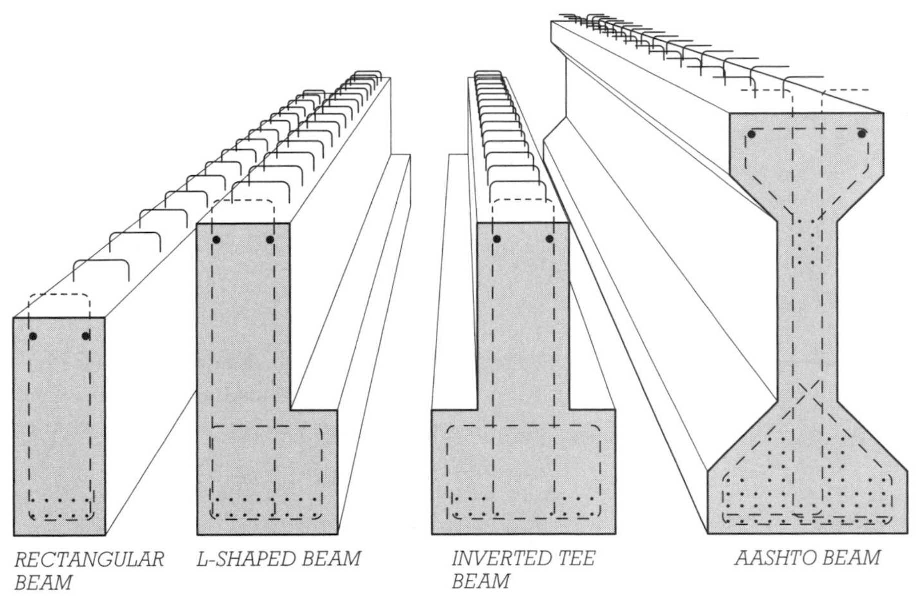

Figure 15.4
Standard precast concrete beam and girder shapes. The larger dots represent mild steel reinforcing bars, and the smaller dots represent high-strength prestressing strands. The broken lines show mild steel stirrups. Stirrups usually project above the top of the beam, as shown, to bond to a sitecast concrete topping for composite structural action.

Precast Concrete Wall Panels

Precast concrete panels, either prestressed or conventionally reinforced, are commonly used as loadbearing wall panels in many types of low-rise and high-rise buildings. Solid panels typically range from 3½ to 10 inches (90 to 250 mm) in thickness and can span one or two stories in height. When prestressed, strands are located in the vertical midplane of the wall panels to strengthen the panels against buckling and to eliminate camber. Ribbed or hollow-core panels, or sandwich panels with integral rigid foam insulation, may be as deep as 12 to 24 inches (305 to 610 mm) and can span up to four stories (Figure 15.12).

A number of manufacturers produce proprietary precast concrete wall panels for residential foundation construction. Vertical ribs, or so-called "concrete studs," spaced at 24 inches (610 mm), provide surfaces with sheet metal or wood strips for the attachment of conventional interior finishes and create cavity space that can accommodate insulation or the routing of building services. Some manufacturers integrate rigid foam insulation or reinforced footings into the design of the panels.

Nonloadbearing precast concrete wall panels are discussed in Chapter 20.

Assembly Concepts for Precast Concrete Buildings

Figure 15.5 shows a building whose precast slab elements (double tees in this example) are supported on a skeleton frame of L-shaped precast girders and precast columns. The slab elements in Figure 15.6 are supported on precast loadbearing wall panels. Figure 15.7 illustrates a building whose slabs are supported on a combination of wall panels and girders. These three fundamental ways of supporting precast slabs—on a precast concrete skeleton, on precast loadbearing wall panels, and on a combination of the two—occur in endless variations in buildings. The skeleton may be one bay or many bays deep; the loadbearing walls are often constructed of reinforced masonry or of various configurations of precast concrete; the slab elements

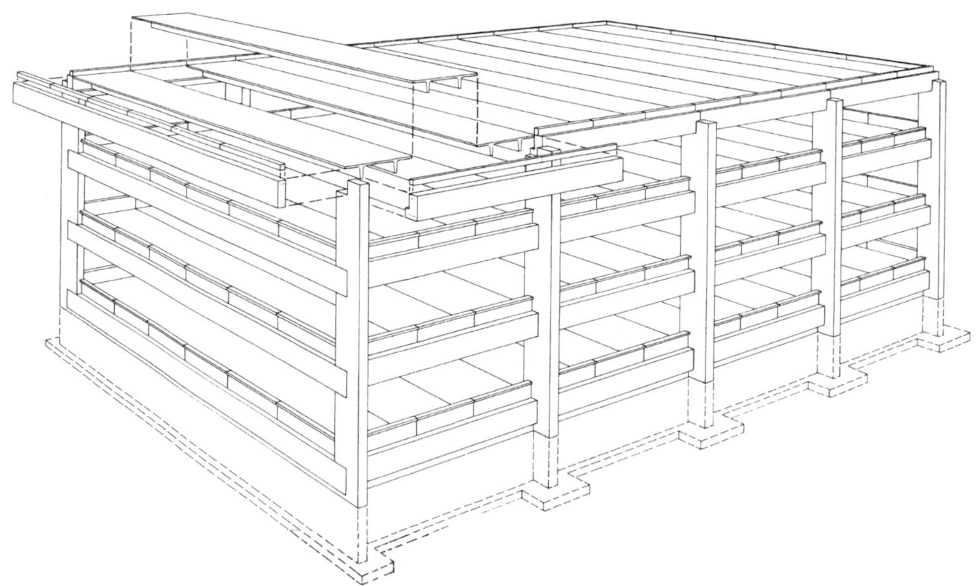

Figure 15.5
Double-tee slab elements supported on a frame of precast columns and L-shaped girders.
(Courtesy of Precast/Prestressed Concrete Institute)

Figure 15.6
Hollow-core slab elements supported on precast concrete loadbearing wall panels.
(Courtesy of Precast/Prestressed Concrete Institute)

may be solid, hollow-core, or double tee, topped or untopped. One of the principal virtues of precast concrete as a structural material is that it is locally manufactured to order and is easily customized to an individual building design, usually at minimal additional cost.

Manufacture of Precast Concrete Structural Elements

Casting Beds

Most precast concrete elements are produced in permanent forms called *casting beds.* Casting beds average 400 feet (125 m) in length but extend 800 feet (250 m) or more in some plants (Figure 15.8). A cycle of precasting usually begins in the morning, as soon as the elements that were cast the previous day have been lifted from the beds. High-strength steel reinforcing strands are strung between the abutments at the extreme ends of the bed. The strands are pretensioned with hydraulic jacks, during which process they stretch considerably. Once the strands are fully tensioned, transverse bulkhead separators may be placed along the bed at the required intervals to divide the individual elements from one another. (For solid slabs, cored slabs, and wall panels, the bulkheads are often omitted; the cured slab or panel is simply sawed into the required lengths before it is removed from the bed.)

Figure 15.7
Double-tee slab elements supported on a perimeter of precast concrete loadbearing wall panels and an interior structure of precast columns and inverted tee beams. *(Courtesy of Precast/Prestressed Concrete Institute)*

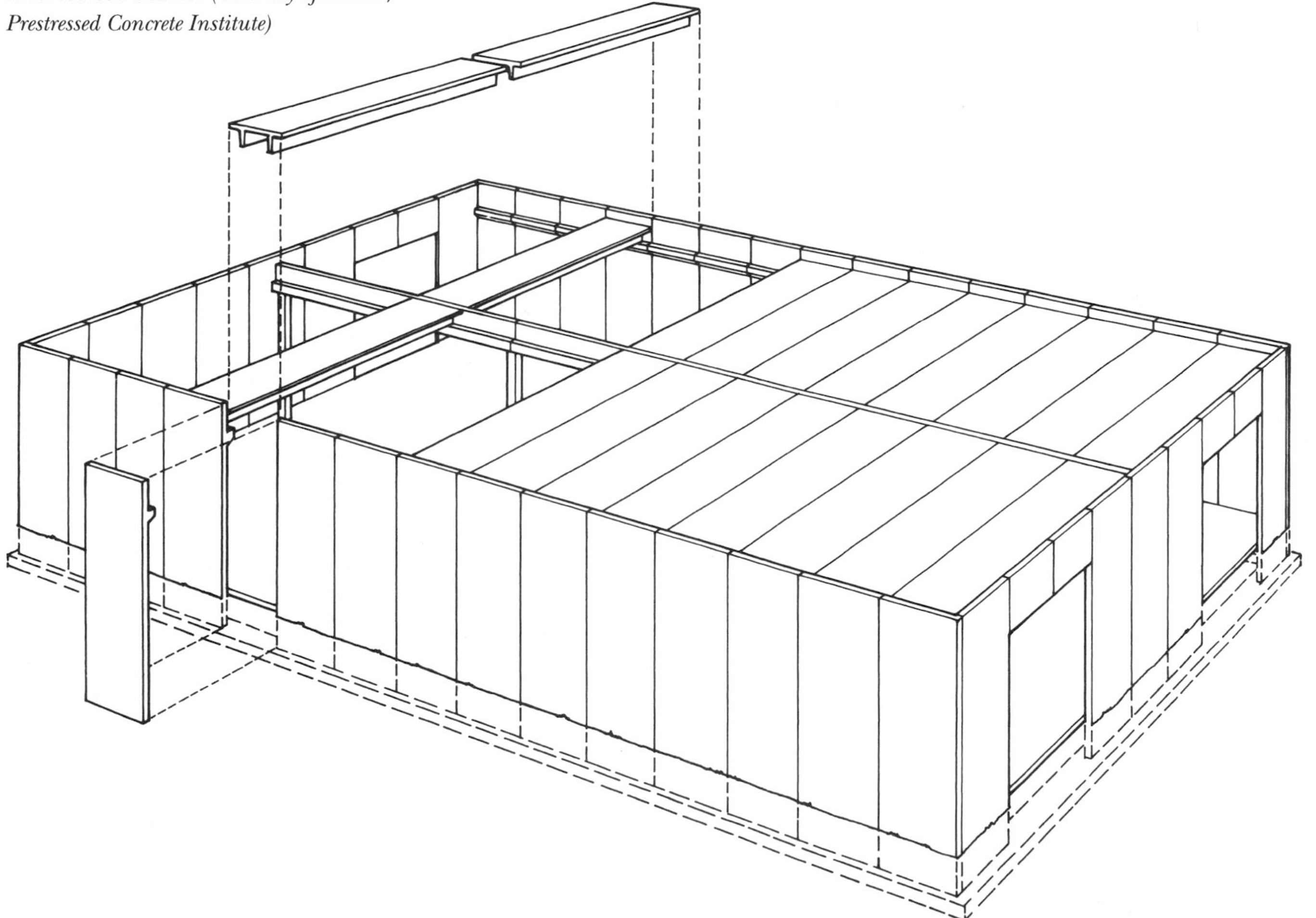

(*a*)

FIGURE 15.8
Manufacturing double-tee slabs. *(a)* A worker inserts weld plates with their V-shaped anchors of reinforcing bar in the casting bed prior to pouring the concrete. The prestressing strands and wire fabric shear reinforcement have been installed in the stems of the double tees, and a portion of the welded wire fabric for the top slab can be seen in the foreground. Notice the great length of the casting bed; many elements can be cast end to end at the same time. *(b)* The top surface of the concrete is straightedged by machine. *(c)* The next morning, following overnight steam curing, a worker cuts the prestressing strands between bulkhead separators with an oxyacetylene cutting torch. The welded wire fabric is exposed at the ends of the elements because they will be used as untopped slabs (see Figure 15.20). *(d)* Using lifting loops cast into the ends, the slabs are lifted from the bed and will be stockpiled outside. The dapped stems will rest on inverted-tee and L-shaped girders; the notched corner will fit around a column. *(e)* Loading the double tees for trucking. *(Photos by Alvin Ericson)*

(c)

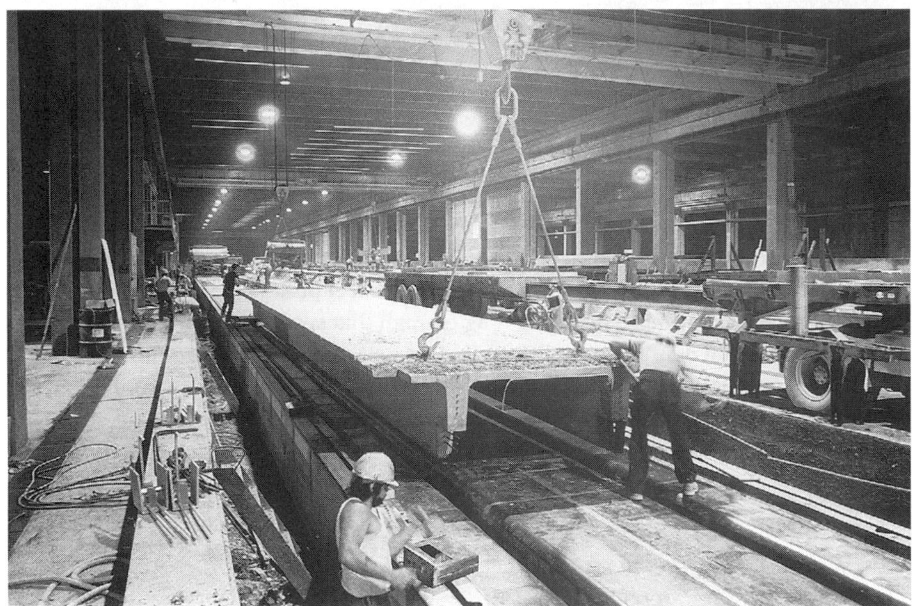

(d)

(e)

When pretensioning and separator placement have been completed, mild steel reinforcing bars and welded wire fabric are placed as required. *Weld plates* and other embedments are installed. Then the concrete is placed in the bed, vibrated to eliminate voids, and struck off level. If the slabs are to be used without topping, the top surface is finished further with floats and trowels. Live steam or radiant heat is then applied to the concrete to accelerate its curing. Ten to twelve hours after pouring, the concrete has reached a compressive strength of 2500 to 4000 psi (24–28 MPa) and has bonded to the steel strands. The next morning, after the strength of the concrete has been verified by testing cylinders in the laboratory, the exposed strands are cut between the bulkhead separators, releasing the external force in the strands, which spring back to prestress the concrete. Asymmetrically prestressed slab and beam elements immediately arch up from the casting bed to take on a pronounced camber as the prestressing force is released into them. When the elements have been separated from one another, they are hoisted off the bed and stockpiled ready for shipment. Then a new cycle of casting begins.

Prestressing and Reinforcing Steel

Solid slabs, hollow-core slabs, and wall panels are cast around horizontal strands. Tees, double tees, beams, and girders are often cast around depressed or harped strands for greater efficiency of structural action (Figures 13.42 and 15.9).

Ordinary mild steel reinforcing is also cast into prestressed concrete elements for various purposes: Beams or slabs that will cantilever beyond their supports are given top reinforcing bars over the cantilever points. Welded wire reinforcing is used to reinforce the flanges of tees and double tees and for general reinforcing of wall panels. Where stirrups are required in the stems of beams and single or double tees, they are made of either mild steel reinforcing bars or welded wire reinforcing (Figure 15.10). Additional reinforcing may be installed to add strength around *dapped* (notched) *ends* and openings in panels or slabs for pipes, ducts, columns, and hatchways. Weld plates and other metal connecting devices are cast into the elements as required. Projecting steel loops are cast into many types of elements as crane attachments for lifting.

FIGURE 15.9
A precasting bed being readied for the pouring of a very long AASHTO girder. Side forms for the mold can be seen in the background to the right. The depressed strands are held down in the center of the beam by steel pulleys that will be left in the concrete after pouring. The bed is long enough that several girders are being cast end to end, with the depressed strands pulled up and down as required. Mild steel reinforcing bars are used for stirrups. The projecting tops of the stirrups will bond to the sitecast topping. Vertical twists of prestressing strand near the end of the girder will serve as lifting loops. *(Courtesy of Blakeslee Prestress, Inc.)*

Carbon Fiber Reinforcing

Carbon fiber reinforcing is gaining increasing use as a substitute for mild steel reinforcing (such as shear stirrups and temperature steel) in precast concrete products including wall panels, double tees, and deck panels for floors and roofs. Because carbon fiber does not require protection from corrosion (unlike steel), less concrete cover is required than for steel reinforcing, significantly reducing the overall thickness and weight of carbon fiber–reinforced components. The low thermal con-[illegible]uctivity of carbon fiber, combined [illegible] the thinner concrete sections [illegible]e with this type of reinforcing, permit the casting of insulated panels with superior thermal performance. The much higher tensile strength and stiffness of carbon fiber in comparison to mild steel, and the innovative ways in which grids of carbon fiber reinforcing can be integrated into precast concrete components, yield improvements in structural efficiency as well.

Carbon fiber composite cable substitutes for high-strength steel posttensioning strands are finding use in experimental precast concrete highway structures. It is likely that these materials will eventually enter the building construction market as well.

Hollow-Core Slab Production

The longitudinal voids in hollow-core slabs can be formed by a number of processes. In the *extruded process*, extrusion devices squeeze an extremely dry, stiff concrete mix through a moving extrusion die to produce the voided shape directly. This method has the disadvantage that vertical openings and weld plates cannot easily be cast in; where openings are required in extruded slabs, they must be cut out of the stiff but still wet concrete just after extrusion or sawed after curing. Weld plates are added to the slabs by hand before the concrete has completely cured. In the *wet-cast*

FIGURE 15.10
Workers install side forms for inverted-tee beams in an outdoor casting bed. Mild steel reinforcing bars are used extensively for stirrups. *(Courtesy of Blakeslee Prestress, Inc.)*

process, a bottom layer of wet concrete is deposited in the casting bed; then a second layer of concrete, with collapsible tubes, dry crushed stone, or lightweight aggregate carefully positioned to form the voids, is placed. Special forms may be easily placed in the bed to make openings as required, and weld plates may be cast in by this process. The tubes or aggregate are removed after the concrete has cured. In the *slip-form process,* a moving hopper deposits a zero-slump concrete mix in the casting bed. Tubes that form the slab cores move with the hopper and are pulled out of the slab as the casting process proceeds from one end of the slab to the other. This process falls somewhere between the wet-cast and extrusion processes in terms of the relative ease or difficulty of placing embeds and forming special openings.

After curing, the slabs are conveyed to an aggregate recovery area, where the dry stone is poured out of the voids and saved for reuse (Figure 15.11). In a third type of process, air-inflated tubes are used to form the voids. In some plants, hollow-core slabs are cast atop one another in stacks rather than in a single layer and are wet cured for 7 days rather than steam cured overnight.

Column Production

Precast concrete columns may be reinforced with ordinary mild steel or with pretensioned strands. Pretensioned columns are often made and shipped in multistory sections with *corbels* (Figures 15.16 and 15.18) to support beams or slabs. Columns with corbels on one side or on two opposing sides are easily cast in flat beds. If corbels are required on three sides or on two adjacent sides, box forms are set atop the upper side of the column form as it lies in the casting bed. For corbels on the fourth side, steel plate inserts are cast into the bottom face of the column in the bed, to which reinforcing bars are welded after the column is removed from the bed. The corbels on the fourth side are then cast around the reinforcing bars in a separate operation.

A

B

C

D

F

FIGURE 15.11
Steps in the manufacture of Span-Deck, a proprietary hollow-core slab. ***(a)*** **A thin bottom slab of low-slump concrete is deposited into the casting bed with a traveling hopper.** ***(b)*** **A ridged roller compacts the bottom slab and makes indentations to key to the concrete webs.** ***(c)*** **Four prestressing strands are pulled onto the bed; these will be pretensioned with jacks.** ***(d)*** **After pretensioning, an extrusion device travels the length of the bed, forming the webs and the top of the slab and filling the cores with dry, lightweight aggregate that serves as temporary support for the fresh concrete.** ***(e)*** **The next day, after curing of the slab has been completed, a saw cuts it into the required lengths. Each length is then lifted from the bed and transported by an overhead crane to the aggregate recovery area, where the dry aggregate is poured out and saved for reuse.** ***(f)*** **Finished Span-Deck slabs stockpiled, ready for transportation to the construction site. The wads of paper in the cores prevent concrete from filling the cores during pouring of the topping.** *(Courtesy of Blakeslee Prestress, Inc.)*

FIGURE 15.12
A tilting table being used for casting a foam-insulated concrete wall panel. To the left, a concrete panel face with welded wire fabric reinforcing is being cast. Sheets of rigid foam insulation with wire ties are seen in the center, and the welded wire fabric for the second panel face is seen at the right, with a pair of vertical bars and a pocket at the bottom being cast in as part of a system of connections. Notice the pipes at the left edge of the table for heating the mold to accelerate curing. *(Courtesy of Blakeslee Prestress, Inc.)*

Joining Precast Concrete Elements

Figures 15.13–15.22 and 15.27 show some frequently used connection details for precast concrete construction. Bolting, welding, and grouting are all commonly employed in these connections. Connections can be posttensioned to produce continuous beam action at points of support (Figure 15.16). Exposed metal connectors that are not covered by topping are usually *dry packed* with portland cement grout (the grout is stiff but not actually dry at the time that it is installed) after being joined to protect them from fire and corrosion.

The simplest joints in precast concrete construction are those that rely on gravity by placing one element atop another, as is done where slab elements rest on a bearing wall or beam or where a beam rests on the corbel of a column. Column-to-footing or column-to-column joints are usually grouted, to allow full bearing between sections (Figures 15.13–15.15). Bearing walls are joined with slabs in a similar manner (Figure 15.27). With spanning members, *bearing pads* are usually inserted at points of contact. These serve to avoid the grinding, concrete-to-concrete contact that might create points of high stress (Figures 15.16–15.20) while also allowing for movement caused by expansion and contraction or structural deflection of the members. For solid and hollow-core slabs, these pads are strips of high-density plastic. Under elements with higher point loadings such as tees and beams, pads of synthetic rubber are used. For resistance to seismic and wind forces, the members in these simple joints must be tied together laterally. Slab elements are often joined over supports by reinforcing bars that are cast into the topping or, where no topping is used, into the grout keys between the slabs (Figures 15.18–15.22). Beams, single tees, and double tees may additionally be connected by welding together steel plate inserts that have been embedded in the elements in the plant (Figures 15.18–15.20). The stems (lower portions) of single and double tees and the bottoms of beams are never welded to the supports but are left free to move on their bearing pads as the slab elements deflect under load.

Fastening to Concrete

As a concrete frame is finished, many things need to be fastened to it, including exterior wall panels and facings; interior partitions; hangers for pipes, ducts, and conduits; suspended ceilings; stair railings; cabinets; and machinery. Drawings *A–H* are examples of fastening systems that are cast into the concrete. *A* is a familiar *anchor bolt. B* is a steel plate welded to a bent rod or strap anchor; this weld plate or *embed plate* furnishes a surface to which steel components can be welded. The steel angle in *C* has a threaded stud welded to it so that another component can be attached by bolting. *D* is an adjustable insert of malleable iron that is nailed to the formwork through the slots in the ears on either side. A special nut twists and locks into the slot to accept a bolt or threaded rod from below. A slightly different form of this type of insert is shown in the detail of a masonry shelf angle attachment in Chapter 20. *E* and *F* are two different designs of threaded inserts that are cast into the concrete. The sheet steel *dovetail slot* in *G* is used with special anchor straps as shown to tie masonry facings to a sitecast concrete frame or wall. *H* is simply a dovetailed wood nailer strip cast into the concrete, a detail that is risky because the wood may absorb moisture, swell, and crack the concrete, or dry out, shrink, and become loose. *A–F* are heavy-duty devices that are available in capacities sufficient to anchor heavy building components and machinery.

Details *I–P* depict fastening devices that are inserted in holes drilled into the cured concrete. *I* shows the steel post of a railing anchored into an oversized hole in a concrete slab using grout, poured lead, or epoxy; this detail is also used to fasten bolts to concrete and can carry heavy loads if properly designed and installed. *J* is a plastic sleeve, *K* a wood or fiber plug, and *L* a lead sleeve; all three are inserted into drilled holes and expand to grip the sides of the hole when a screw is driven into them. *M* is a similar type of metal sleeve but has a special nail that

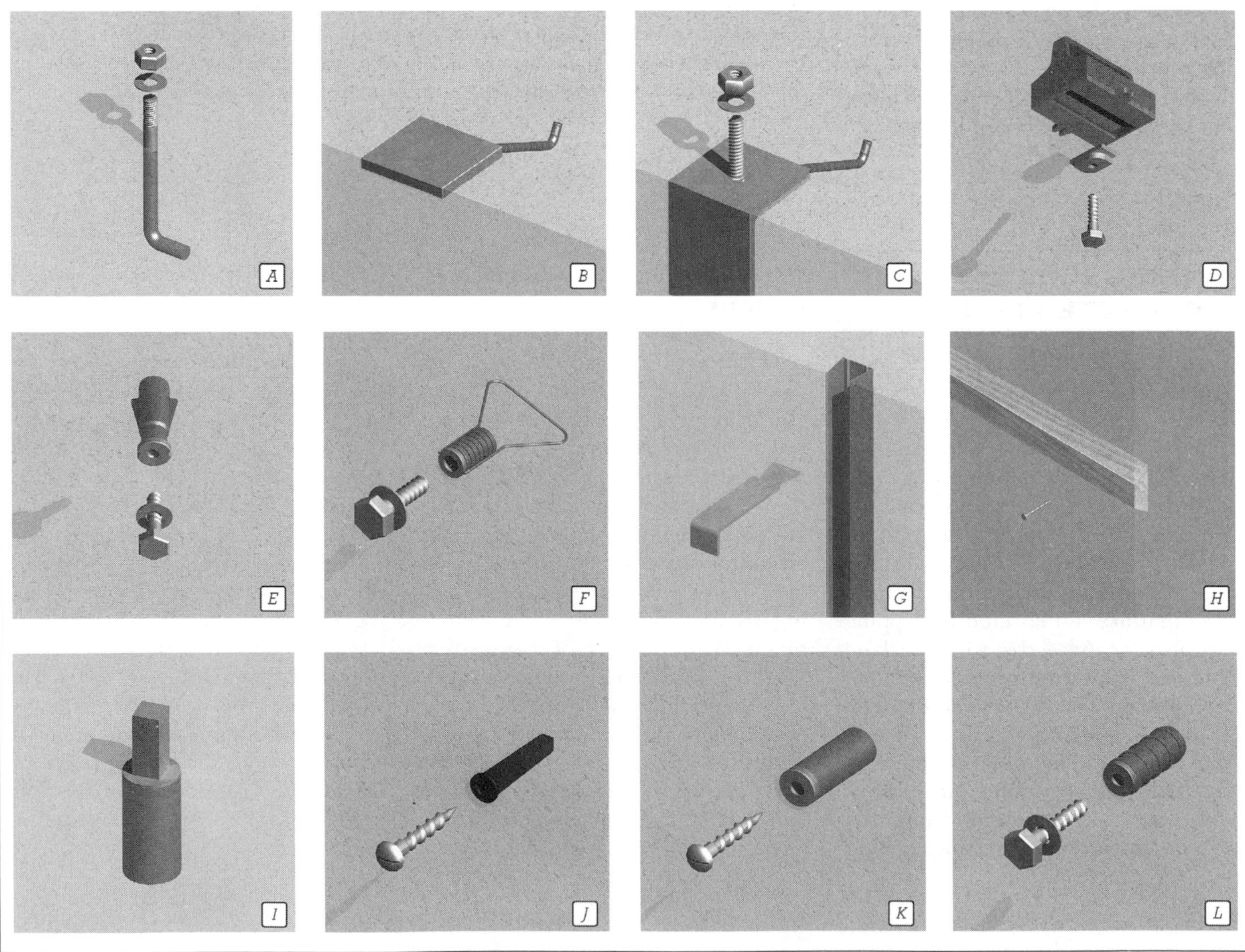

expands the sleeve as it is driven. *N* is a special bolt with a steel sleeve over a tapered shank at the inner end. The sleeve catches against the concrete as the bolt is driven into the hole and is expanded by the taper as the bolt is tightened. *O* is a special screw and *P* is a special nail, both designed to grip tightly when inserted into drilled holes of the correct diameter. *J–P* are light- to-medium-duty fasteners, with the exception of *L* and *N*, which can carry rather heavy loads.

Q and *R* are driven anchors. *Q* is the familiar concrete nail or masonry nail, made of hardened steel. If driven through a strip of wood with a few blows of a very heavy hammer or inserted with a nail gun, it will penetrate concrete just enough to provide some shear resistance for furring strips and sleepers, but it has a tendency to loosen, particularly if driven with too many blows. Shown in *R* are three examples of *powder-driven* (often called *powder-actuated*) *fasteners*, which are driven into steel or concrete by an exploding cartridge of gunpowder. The first fastener is a simple pin used for attaching wood or sheet metal components to a wall or slab. The middle one is threaded to accept a nut. The eye on the fastener to the right allows a wire, such as a hanger wire for a suspended ceiling, to be attached. Powder-driven fasteners are rapidly installed, economical, and have a moderately high load-carrying capacity.

S typifies devices whose perforated metal plates adhere securely with a mastic-type adhesive to surfaces of concrete or masonry. The fastener shown here has a thin sheet metal spike, over which a panel of foam plastic insulation can be impaled. The tip of the spike is then bent across the face of the insulation panel to hold it in place. Roof edge systems employing perforated plates and adhesives to fasten them to the building are illustrated in Chapter 16.

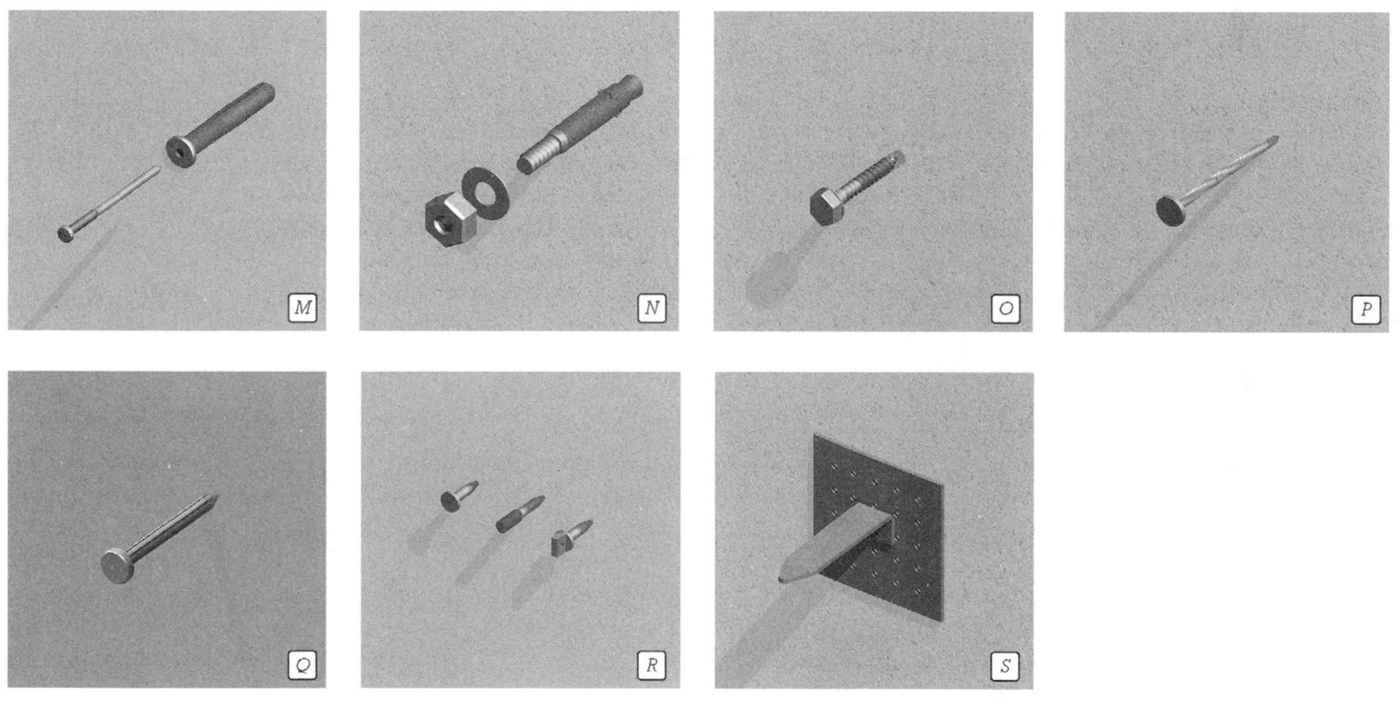

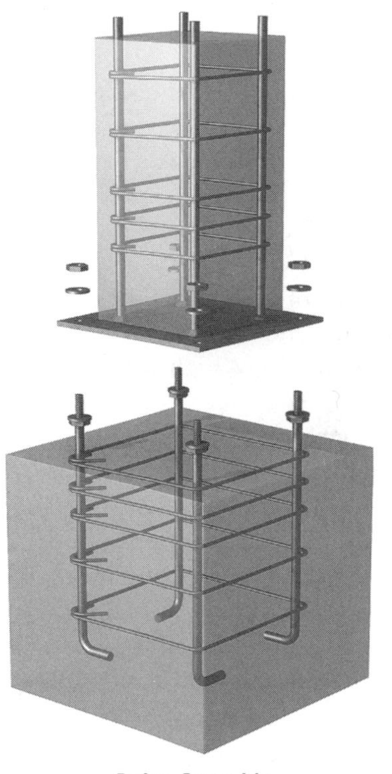
Before Assembly

Assembled

Grouted

FIGURE 15.13
A simple base detail for precast concrete columns. Four anchor bolts project from the top of the sitecast foundation. Nuts and washers are placed on these bolts to support the column temporarily. The column, which was cast around steel dowels welded to a baseplate, is lowered by crane. Workers guide the column so that the anchor bolts come through the holes in the baseplate. Washers and nuts are added to the anchor bolts on top of the baseplate. The eight nuts are used to adjust the height of the column and to make it plumb. When this has been accomplished, the nuts are tightened and stiff grout is dry packed under the baseplate. *(All photorealistic art in this chapter is by Lon Grohs)*

In some cases, precast slab elements, especially solid slabs, require temporary shoring at midspan to help support the weight of the topping until it has cured. After curing, the topping becomes part of the slabs and increases their strength and stiffness. For construction economy, smooth-topped precast slab elements are sometimes used without topping, especially for roofs, where any unevenness between elements is bridged by rigid thermal insulation. Untopped slabs may also be used for floors that will be finished with a pad and carpet and for parking garages. Untopped slabs require special connection details that do not rely on reinforcing bars placed in the topping (Figures 15.18 and 15.20).

Posttensioning can be used to combine large precast elements into even larger ones on the site. This is done to assemble precast concrete box segments into very long, deep girders for bridges (Figures 15.23 and 15.24) and to create tall shear walls from story-high precast panels in multistory buildings. In either case, ducts for the posttensioning tendons are placed accurately in the sections before casting so that they will mate perfectly end to end when the sections are assembled on the site. After assembly, tendons are inserted into the ducts, horizontally in the case of girders or vertically in the case of shear walls, tensioned with portable hydraulic jacks, then grouted if required.

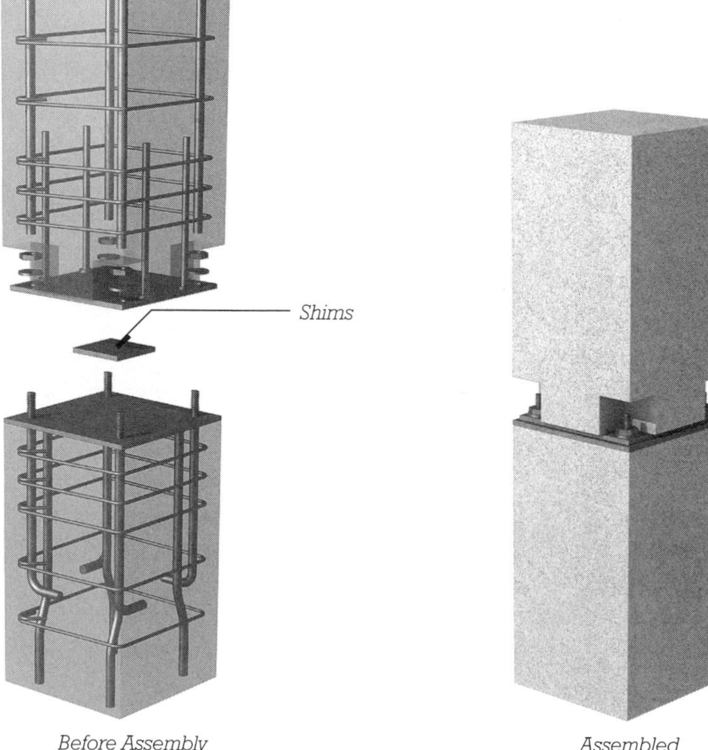

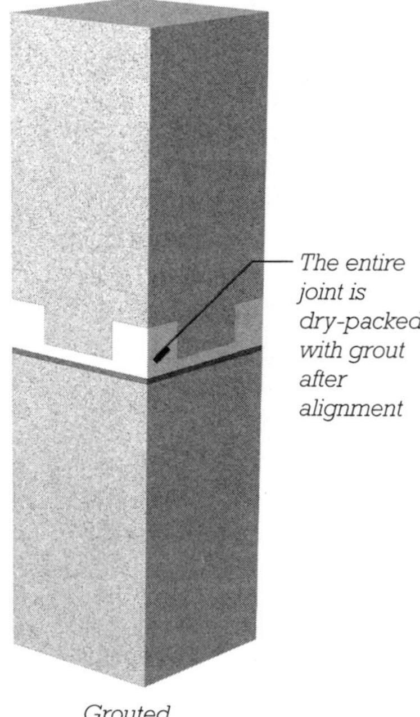

FIGURE 15.14
Similar details serve both as an alternative way of placing a column on its foundation and for column-to-column connections. Metal shims support the upper column sections at the proper height until the grout cures. The open corners are dry packed with stiff grout after the column has been aligned and bolted; this protects the metal parts of the connection from fire and corrosion.

Joint design is the area in which precast concrete technology is developing most rapidly. Historically, precast concrete building construction has been limited in regions of high seismic risk due to uncertainty regarding this structure type's performance under seismic loadings. More recently, innovations in the joining of precast concrete members have led to connection systems that can more reliably absorb the energy imparted into the structure during a seismic event. Unbonded posttensioning strands that allow the structure to respond elastically to seismic displacements, hybrid joint connections that combine the ductility of mild steel reinforcing with the high strength of prestressed strands, ungrouted or "open" joint systems that permit controlled movements between members, joint friction dampers that limit frame movements while absorbing seismic energy, and other techniques have come into use or are under development.

Other development efforts focus on improving the ease and economy of joining precast members. New joining systems are patented each year, and as reinforcing techniques, grouts, and adhesives develop further, there will be additional simplifications and improvements of many kinds in precast concrete framing details.

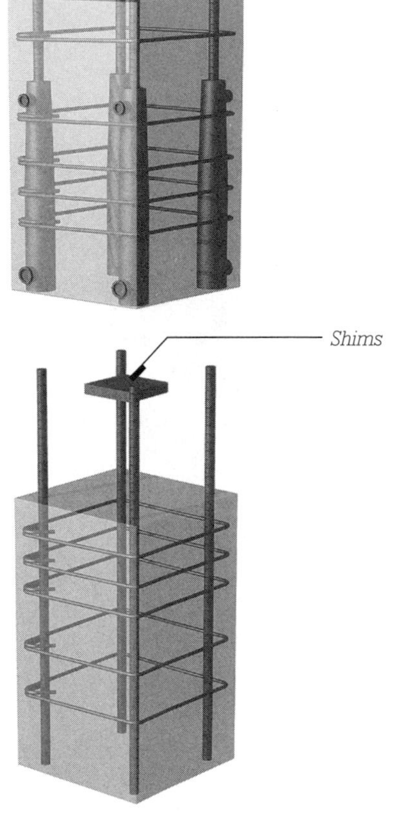

Before Assembly

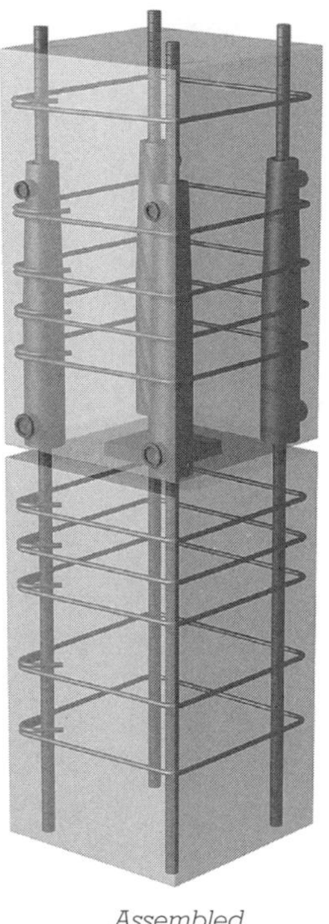

Assembled

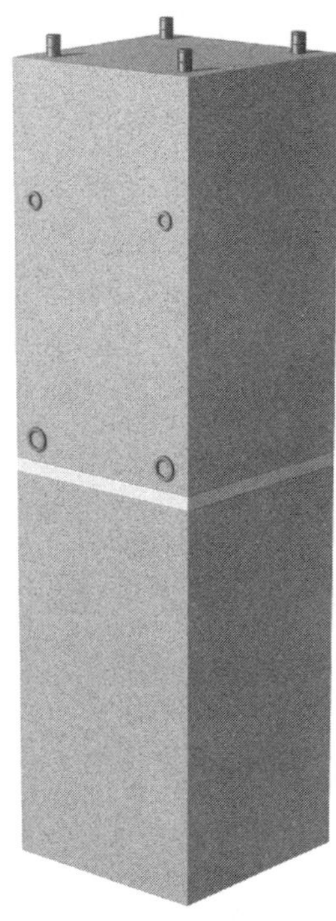

Grouted

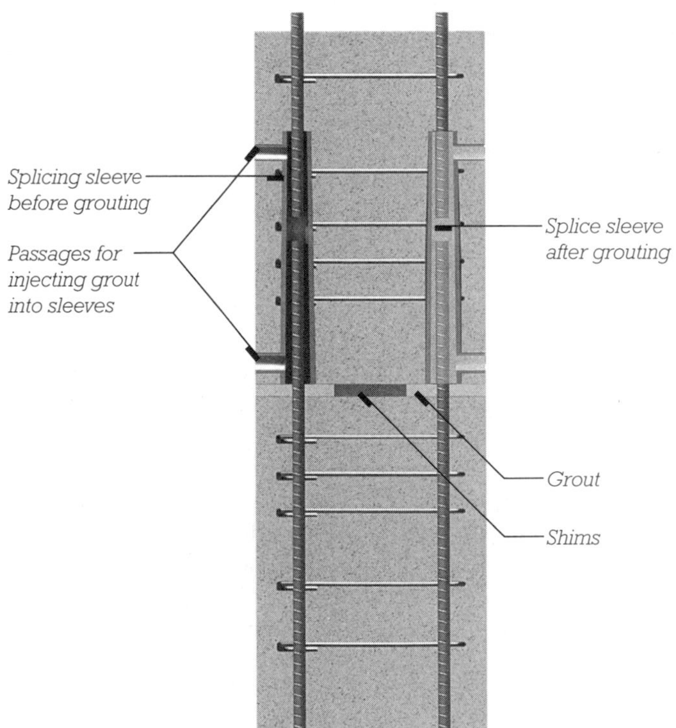

Section

FIGURE 15.15

This column-to-column connection uses proprietary sleeves that are cast into the lower end of the upper column section. Before the sections are assembled (upper left), the lower ends of the vertical bars from the upper column section, which reach down to the midheight of each sleeve, are the only contents of the sleeves. Assembly of the column sections starts with the placement of a stack of steel shims in the center of the top of the lower section. These shims serve to adjust the height of the column and to maintain a space for grouting between the two sections.

In the next two drawings, the sections have been assembled by lowering the upper section onto the lower one. The sleeves mate with projecting reinforcing bars from the lower column section. After the upper column section has been shimmed to exactly the right height and plumbed up, a fluid grout is injected into each sleeve, where it cures and serves to connect the reinforcing bars. A stiff grout is dry packed between the ends of the columns. The grouted sleeves develop the full strength of the reinforcing bars that they connect.

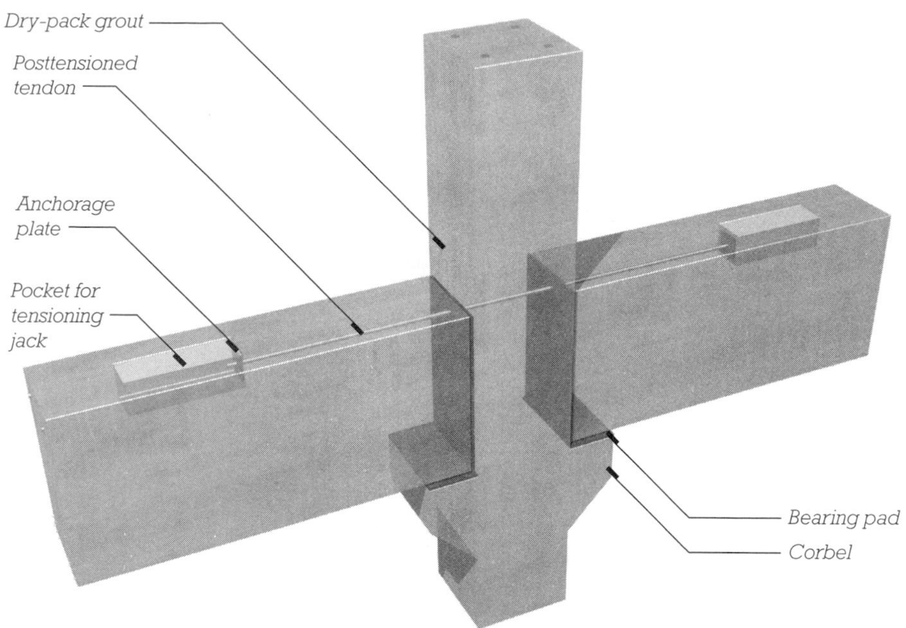

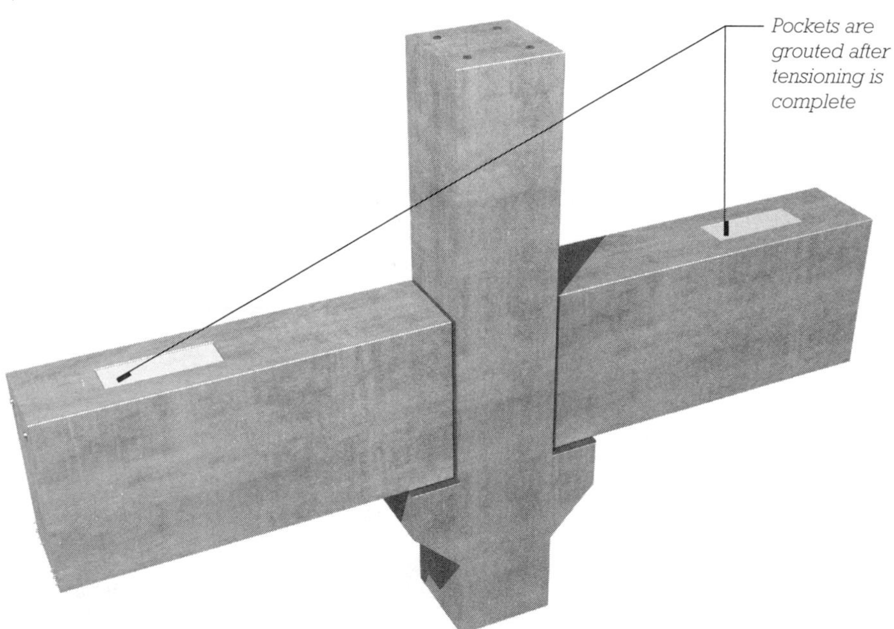

FIGURE 15.16
A posttensioned, structurally continuous beam–column connection may be created by passing a tendon from a pocket in the top of one beam, through the column, to a pocket in the top of the other beam. The tendon is anchored to a plate in one pocket as it is tensioned by a jack in the other pocket.

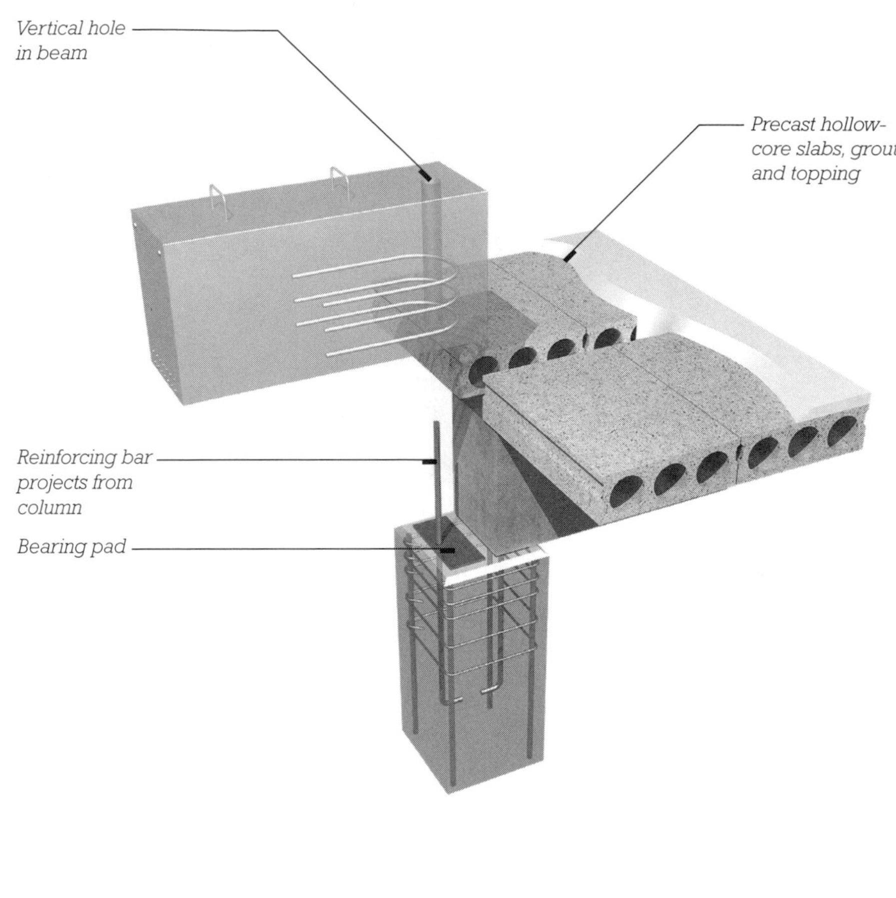
Vertical hole
in beam
Precast hollow-
core slabs, grout,
and topping
Reinforcing bar
projects from
column
Bearing pad

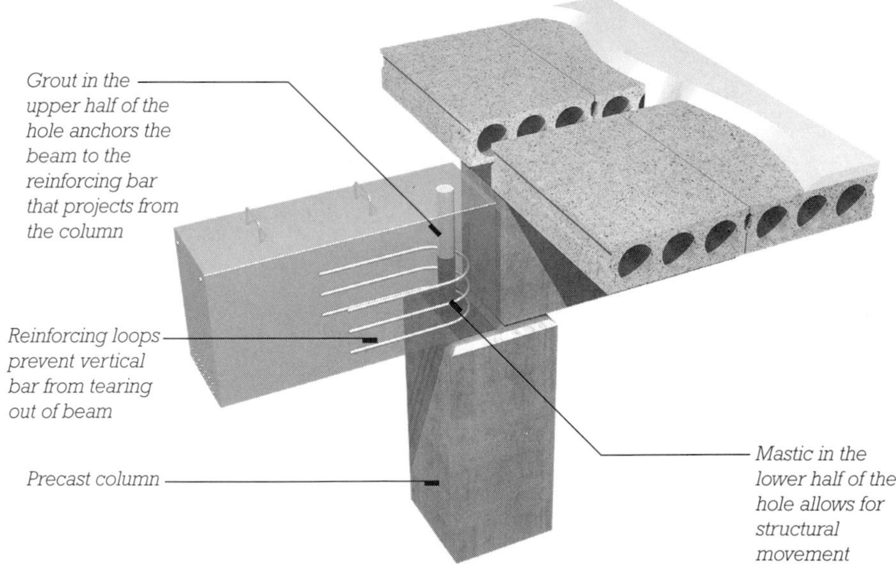
Grout in the
upper half of the
hole anchors the
beam to the
reinforcing bar
that projects from
the column
Reinforcing loops
prevent vertical
bar from tearing
out of beam
Precast column
Mastic in the
lower half of the
hole allows for
structural
movement

FIGURE 15.17
Topped hollow-core roof slabs supported on beams are joined to a column with vertical rods. A similar connection can be used for floor beams resting on corbels.

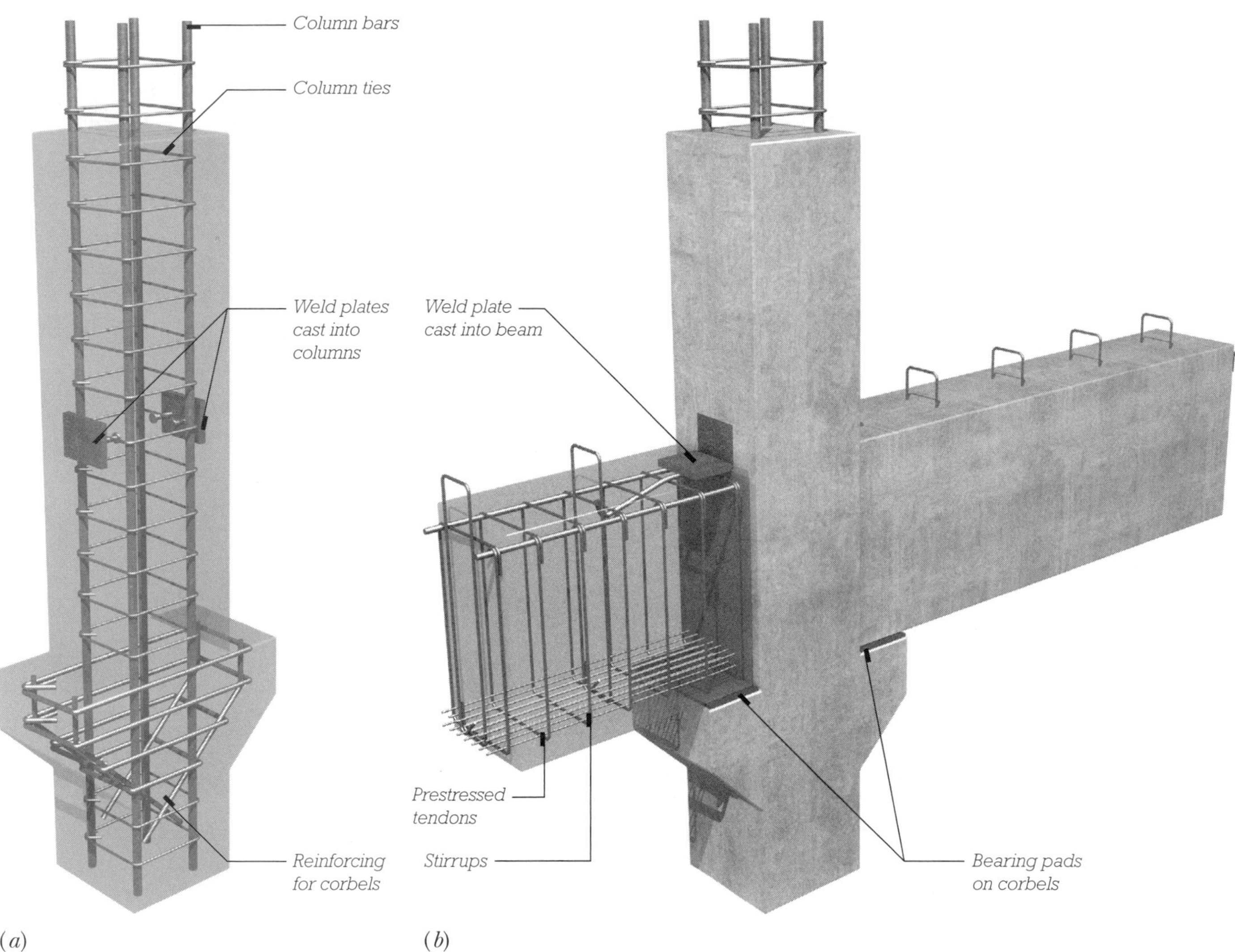

FIGURE 15.18
The beams in this system of framing rest on concrete corbels that are integrally cast with the column. The smooth-topped hollow-core slabs are detailed for use without topping. *(a)* Weld plates are cast into the column. *(b)* Beams are placed on bearing pads on the corbels. There is a weld plate cast into the top of each beam at the end.

(*continued*)

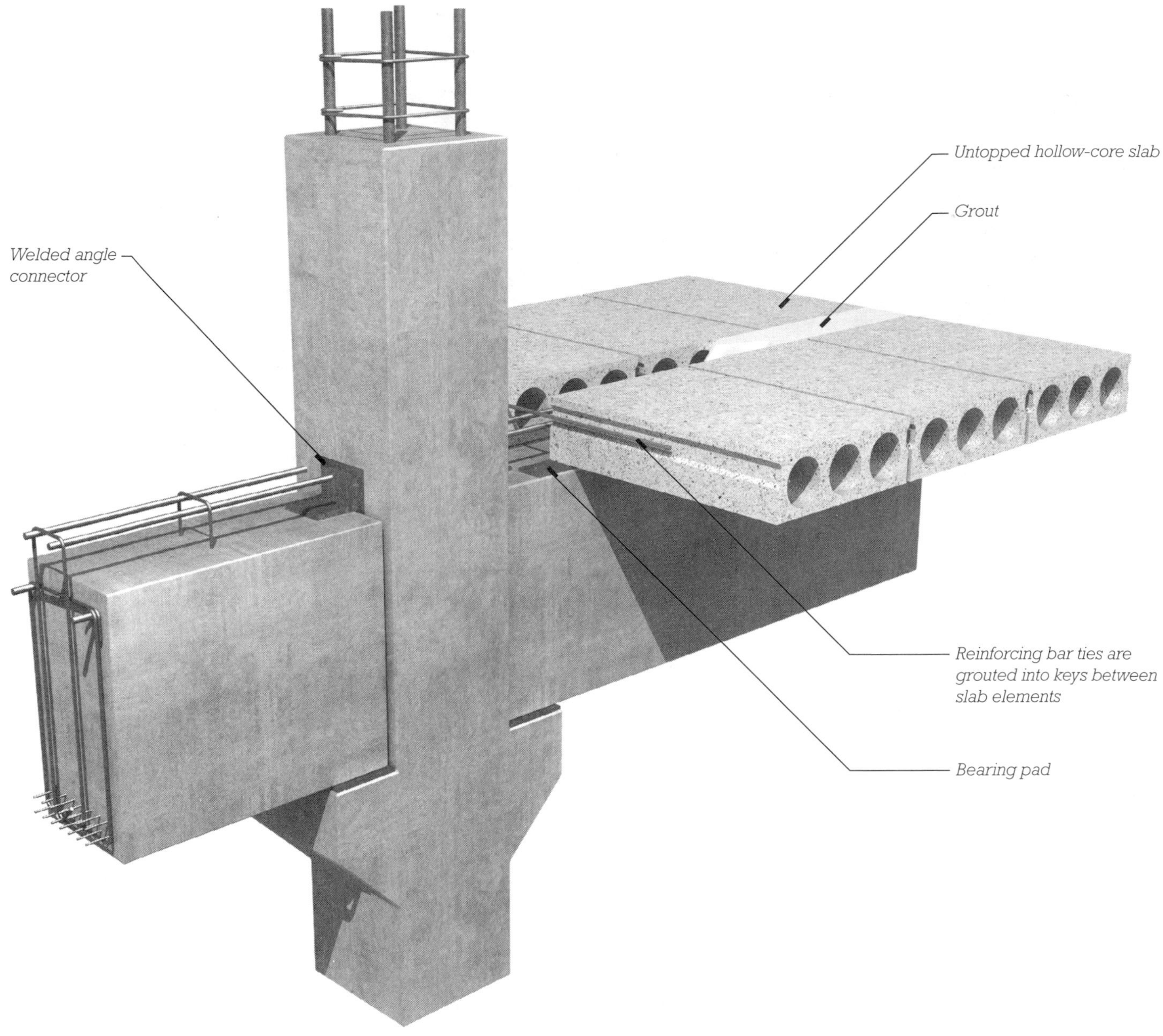

(c)

FIGURE 15.18 *(continued)*
***(c)* Short pieces of steel angle are welded to the plates to join the beams to the columns. Smooth-top hollow-core precast concrete planks are placed on bearing pads on top of each beam. Grout is poured into the gap between the ends of the planks to unite loops of reinforcing that project from the tops of the beams, reinforcing bars that are inserted through the loops, and lateral pieces of reinforcing bar that are grouted into the keys between planks. The end result is a tightly connected assembly that supports an untopped precast concrete floor or roof.**

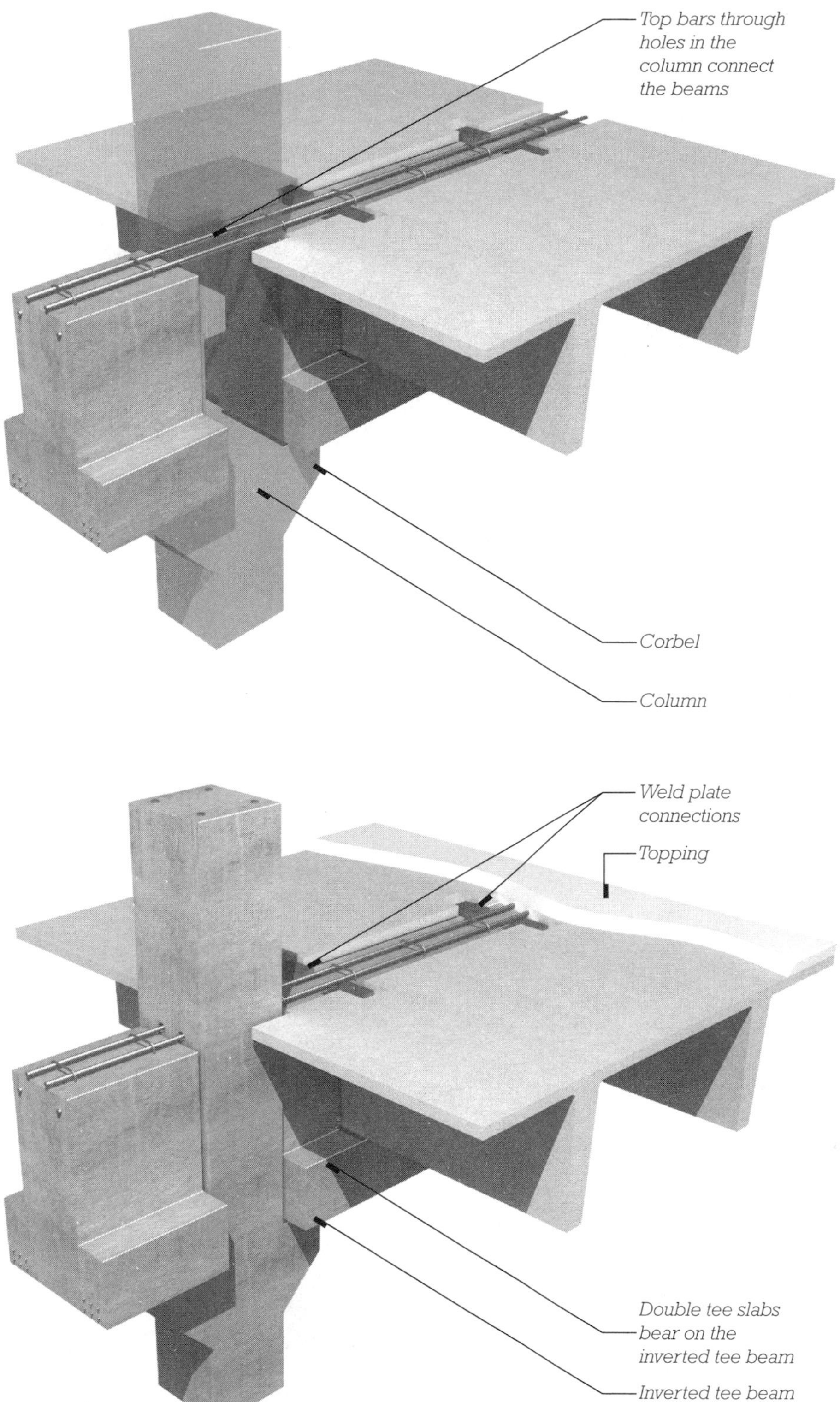
Top bars through holes in the column connect the beams
Corbel
Column
Weld plate connections
Topping
Double tee slabs bear on the inverted tee beam
Inverted tee beam

FIGURE 15.19
Topped double-tee floor slabs are supported by inverted-tee beams in this detail. Reinforcing bars that pass through hollow tubes cast into the column connect the beams and column. The sitecast topping ties all the components together and gives a smooth, level surface.

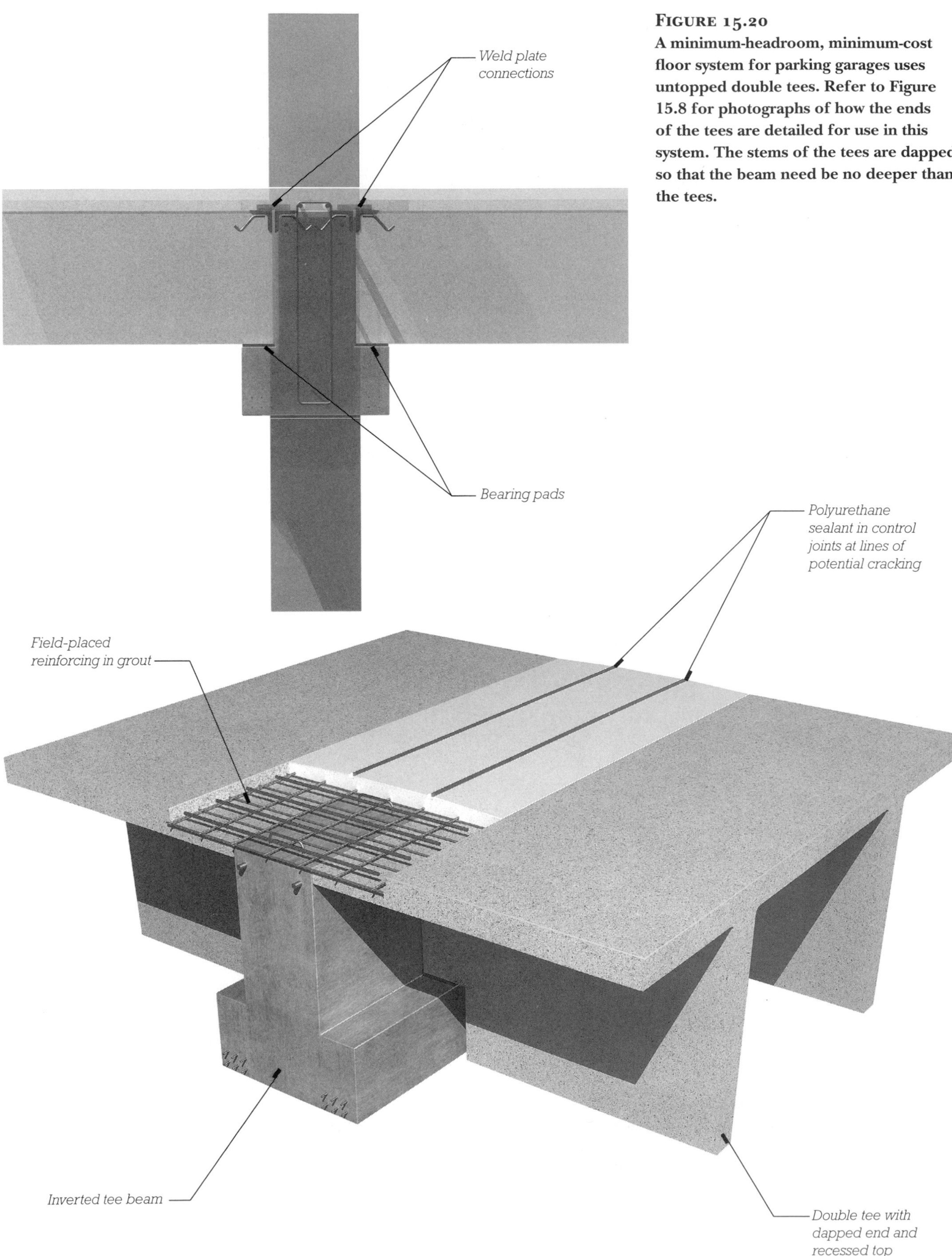

FIGURE 15.20
A minimum-headroom, minimum-cost floor system for parking garages uses untopped double tees. Refer to Figure 15.8 for photographs of how the ends of the tees are detailed for use in this system. The stems of the tees are dapped so that the beam need be no deeper than the tees.

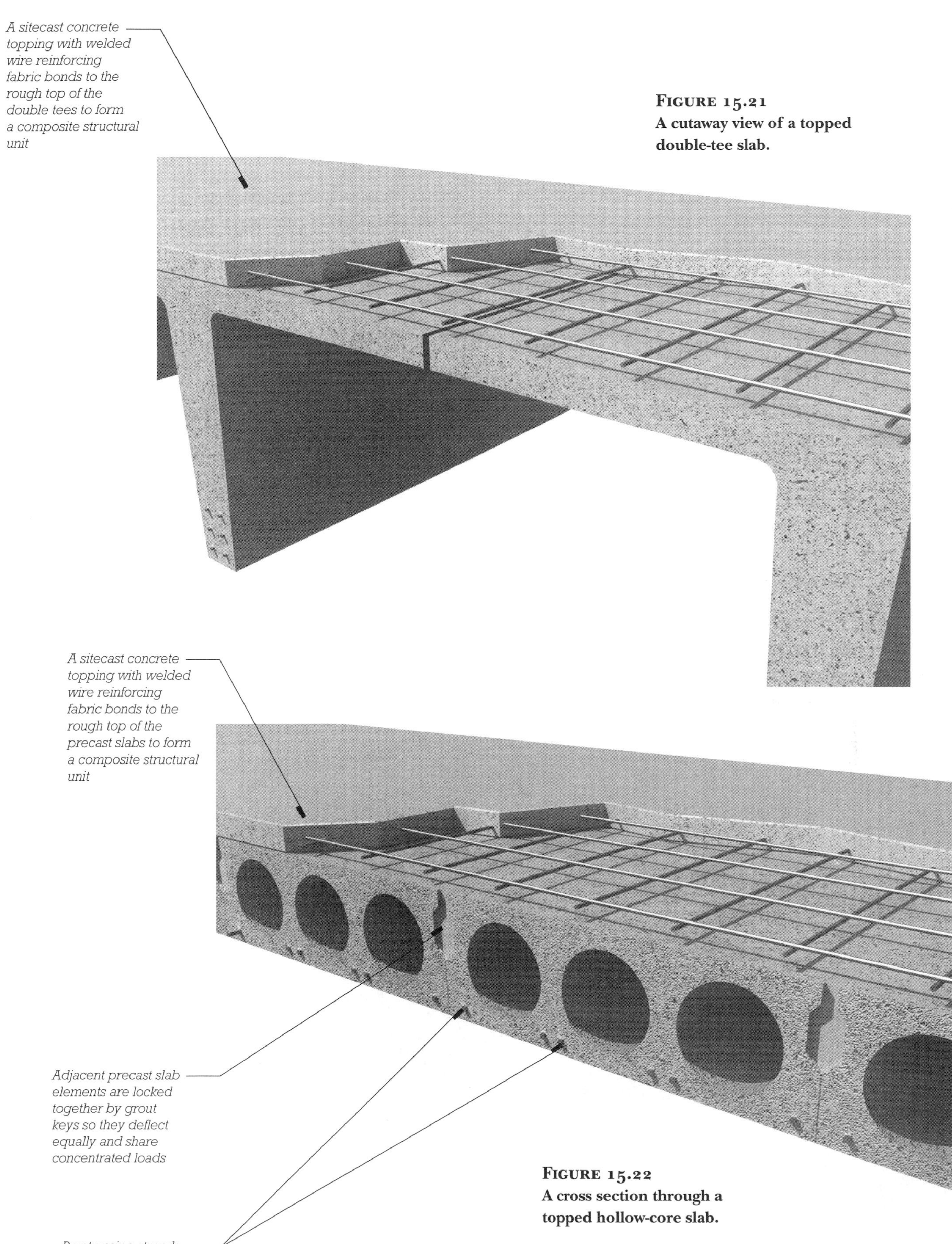

A sitecast concrete topping with welded wire reinforcing fabric bonds to the rough top of the double tees to form a composite structural unit
A sitecast concrete topping with welded wire reinforcing fabric bonds to the rough top of the precast slabs to form a composite structural unit
Adjacent precast slab elements are locked together by grout keys so they deflect equally and share concentrated loads
Prestressing strands

FIGURE 15.21
A cutaway view of a topped double-tee slab.

FIGURE 15.22
A cross section through a topped hollow-core slab.

FIGURE 15.23
The Linn Cove Viaduct in Linnville, North Carolina, was built of short precast segments that were posttensioned together as they were placed to form a continuous box girder deck. A section is being installed by the derrick at the extreme right. The maximum clear span is 180 feet (55 m). *(Engineer: Figg and Muller Engineers, Inc. Courtesy of Precast/Prestressed Concrete Institute)*

FIGURE 15.24
The Linn Cove Viaduct was constructed with very little temporary shoring to disrupt the natural landscape below as little as possible. The precast hollow box sections formed a girder that could cantilever for long distances during construction. The box profile is highly resistant to the torsional forces that occur in a curving beam. *(Courtesy of Precast/Prestressed Concrete Institute)*

FIGURE 15.25
In the filigree precast concrete system, precast concrete elements serve as formwork for sitecast concrete. In this photograph, precast units for a filigree slab and beam system have been erected and the placement of steel reinforcing is in progress. Later, concrete will be sitecast on top of this formwork.

Once the construction of this system is complete, the precast units and sitecast concrete work together as a unified structural system: Prestressed strands in the precast units provide bottom reinforcing. Top bars are added on site as conventional reinforcing. Lightweight bar joist reinforcing, cast into the precast units so as to remain partially exposed, will form an intimate structural bond between the precast and sitecast concrete portions of the system, creating composite structural action in a manner very similar to that achieved with shear studs in structural steel composite construction (Figures 11.63 and 11.64). Note also the shallow slab band form running from lower left to upper right in this photograph. *(Courtesy of Midstate Filigree Systems, www.filigreeinc.com)*

Composite Precast/Sitecast Concrete Construction

In the *filigree precast concrete system,* relatively thin precast elements that are either conventionally reinforced or prestressed are used as the formwork for site casting of beams and slabs. Once the process is complete, composite structural action between the sitecast concrete and precast units results in a unified, structurally efficient system. Since the precast units remain in place as part of the finished system, formwork costs are much less in comparison to conventional sitecast concrete construction methods (Figure 15.25).

The Construction Process

The construction process for precast concrete framing is directly parallel to that for steel framing. The structural drawings for the building are sent to the precasting plant, where engineers and drafters prepare shop drawings that show all the dimensions and details of the individual elements and how they are to be connected. These drawings are reviewed by the engineer and architect for conformance with their design intentions and corrected as necessary. Then the production of the precast components proceeds, beginning with construction of any special molds that are required and fabrication of reinforcing cages, then continuing through cycles of casting, curing, and stockpiling as previously described. The finished elements, marked to designate their positions in the building, are transported to the construction site as needed and placed by crane in accordance with erection drawings prepared by the precasting plant.

Precast Concrete and the Building Codes

The fire resistance of precast concrete building frames and bearing wall panels is dependent on whether they are made of structural lightweight concrete or normal concrete and on the amount of concrete cover that protects the prestressing strands and reinforcing bars. The building code table shown in Figure 1.3 indicates the fire resistance ratings in hours required for components of Type I and II construction. When the construction type has been determined by the architect or engineer, the precaster can assist in determining how the necessary degrees of fire resistance can be achieved in each component of the building. Slab elements are readily available in 1- and 2-hour fire resistance ratings, and beams and columns in ratings ranging from 1 to 4 hours. The fire resistance ratings of precast concrete slab elements may be increased by adding a topping or by increasing the thickness of a required topping. Solid and hollow-core slabs may achieve ratings as high as 3 hours by this means. Single and double tees require the addition of applied fireproofing material or a membrane ceiling fire protection beneath in order to achieve a fire resistance rating higher than 2 hours.

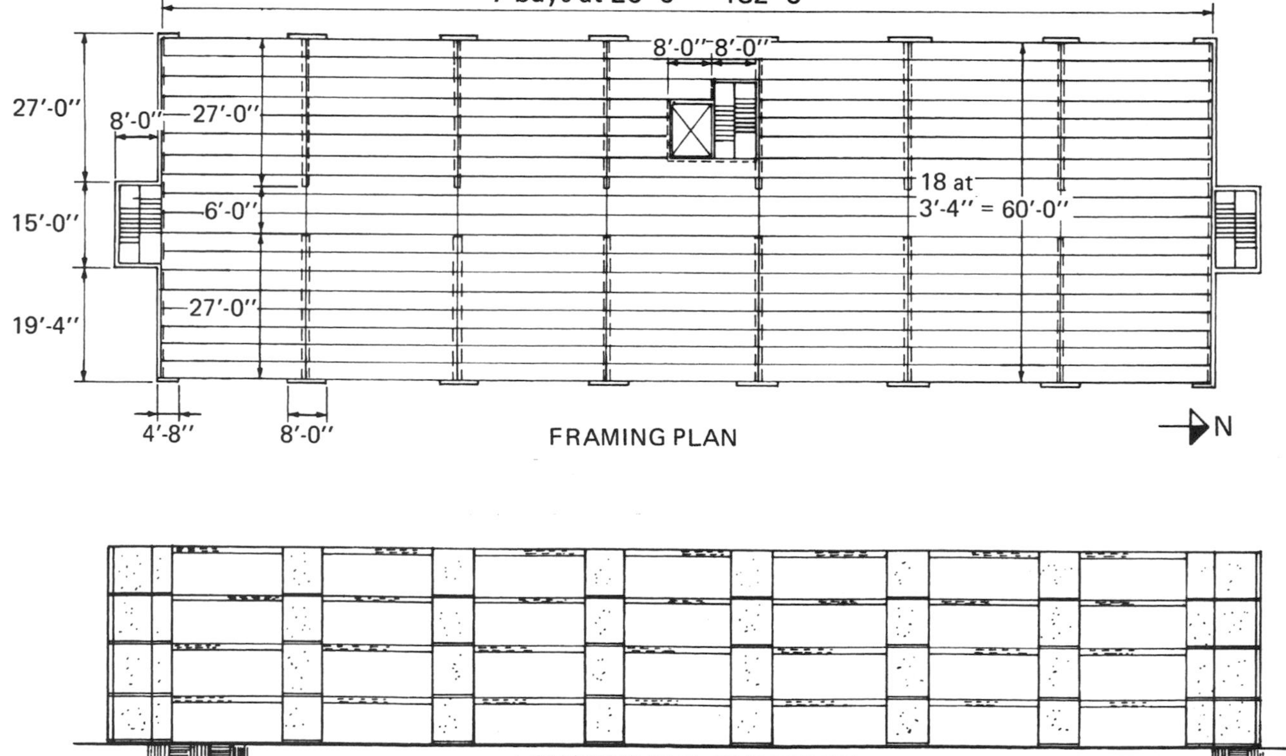

Figure 15.26
A framing plan and elevation of a simple four-story building made of loadbearing precast concrete wall panels and hollow-core slab elements. *(Courtesy of Precast/Prestressed Concrete Institute)*

Considerations of Sustainability in Precast Concrete Construction

In addition to the issues of sustainability of concrete construction that were raised in Chapter 13, there are issues that pertain especially to precast concrete construction:

- Because of the higher-strength concrete mixes typically used in the production of precast concrete, its embodied energy is higher on a pound-for-pound basis than that of conventional concrete, generally falling in the range of 500 to 600 BTU per pound (1.1-1.4 MJ/kg).
- Precast concrete production encourages the reuse of formwork, reducing waste. Wood and fiberglass forms can be used up to 50 times without major maintenance. Concrete and steel forms can be reused hundreds or thousands of times.
- Because precast concrete is manufactured in a controlled, factory-like setting, raw materials are used more efficiently and less waste is produced. Gray water used in various production processes, sand used in finishing, and large aggregate used to create voids in hollow planks can all be readily reused.
- In many cases, the optimized design of precast concrete results in elements that use less material than comparable sitecast concrete systems.
- Precast concrete elements with high-quality architectural finishes reduce the need for volatile organic compound–emitting paints or other finish coatings. Concrete is not easily damaged by moisture and does not support the growth of mold.
- Precast concrete wall panels with properly sealed joints have low permeability to air leakage, reducing building heating and cooling costs and contributing to good indoor air quality.
- Precast concrete wall panels can be reused when buildings are altered.

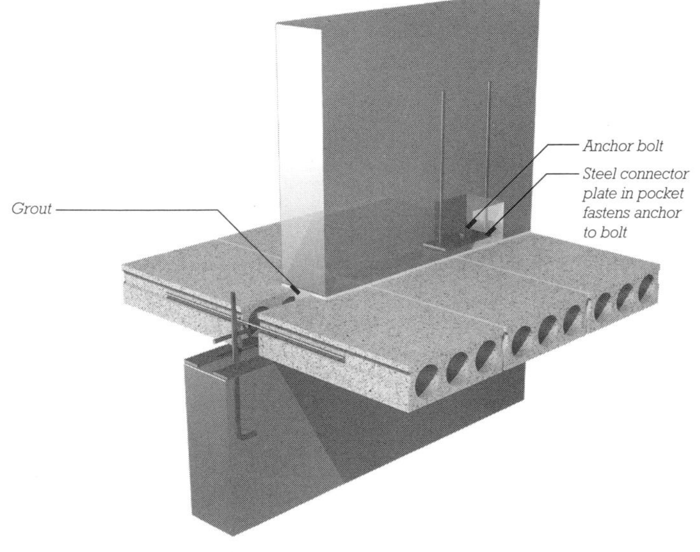

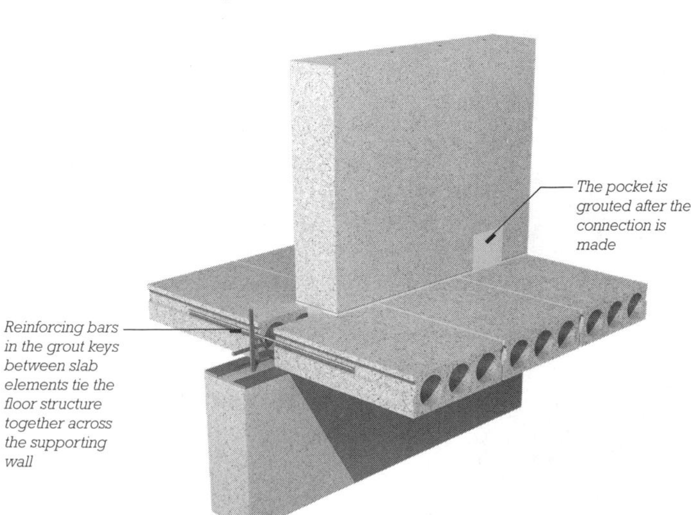

Figure 15.27
A typical detail for the slab–wall junctions in the structure shown in Figure 15.26. The reinforcing in the wall panels and the prestressing steel in the slabs have been omitted from these drawings for the sake of clarity.

FIGURE 15.28
A view of the construction of a building that uses precast concrete loadbearing wall panels. The rectangular pockets at the lower edges of the wall panels are for bolted connections of the type shown in Figure 15.27. Steel pipe braces support the panels until all the connections have been made and the structure becomes self-stable. *(Courtesy of Blakeslee Prestress, Inc.)*

FIGURE 15.29
Exterior loadbearing wall panels are often made of specially colored concrete, with surface textures cast in. *(Courtesy of Blakeslee Prestress, Inc.)*

FIGURE 15.30
A crane hoists a single-piece precast stair for a high-rise building of loadbearing precast concrete wall panels. *(Courtesy of Blakeslee Prestress, Inc.)*

FIGURE 15.31
A precast column section is lifted from the ground by a crane. *(Courtesy of Blakeslee Prestress, Inc.)*

FIGURE 15.32
Erecting an upper-story column, using a connection of the type illustrated in Figure 15.14. *(Courtesy of Blakeslee Prestress, Inc.)*

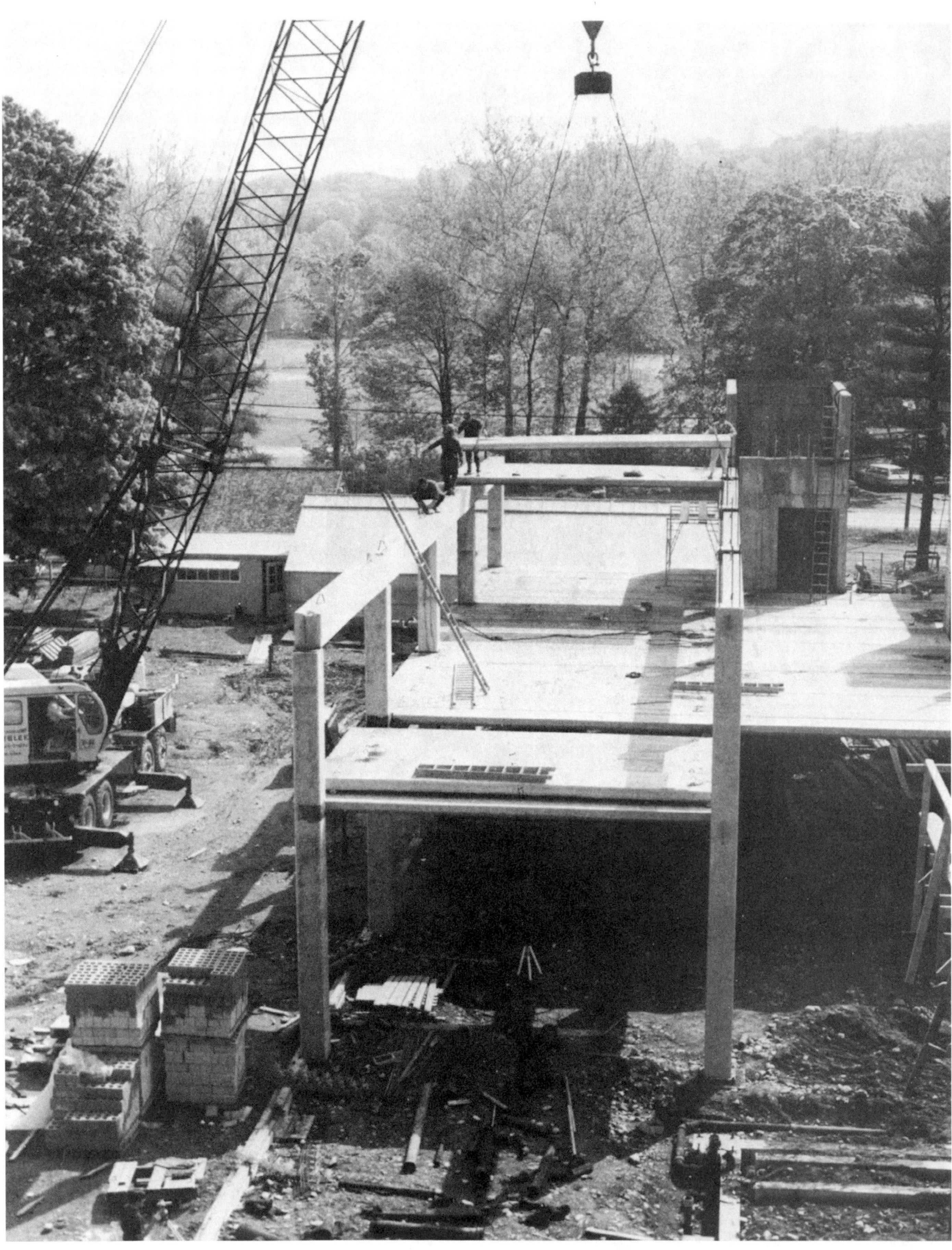

FIGURE 15.33
Placing a hollow-core slab deck on a precast frame. *(Courtesy of Blakeslee Prestress, Inc.)*

UNIQUENESS OF PRECAST CONCRETE

Precast, prestressed concrete structural elements are crisp, slender in relation to span, precise, repetitive, and highly finished. They combine the rapid all-weather erection of structural steel framing with the self-fireproofing of sitecast concrete framing to offer economical framing for many kinds of buildings. Because precast concrete is the newest and least developed of the major framing materials for buildings, its architectural aesthetic is still maturing. Solid and hollow-core slabs have become an accepted part of our structural vocabulary in schools, hotels, apartment buildings, and hospitals, where they are ideal both functionally and economically. Engineers and architects have long been comfortable with precast concrete in longer-span building types, especially parking structures, warehouses, and industrial plants, where its unique structural potential and efficient serial production of identical elements can be fully utilized and openly expressed. Now we are becoming increasingly successful in creating public buildings of the highest architectural quality that consist of precast concrete both inside and out (Figures 15.34–15.39). It is reasonable to expect that many innovative buildings in the coming years will be built of this sleek, sinewy, rapidly developing new material of construction.

(*a*)

FIGURE 15.34
***(a)* A crane lifts a column section from a flatbed truck to begin erection of a tennis stadium in New Haven, Connecticut.**

(continued)

(*b*)

(*d*)

(*c*)

(*e*)

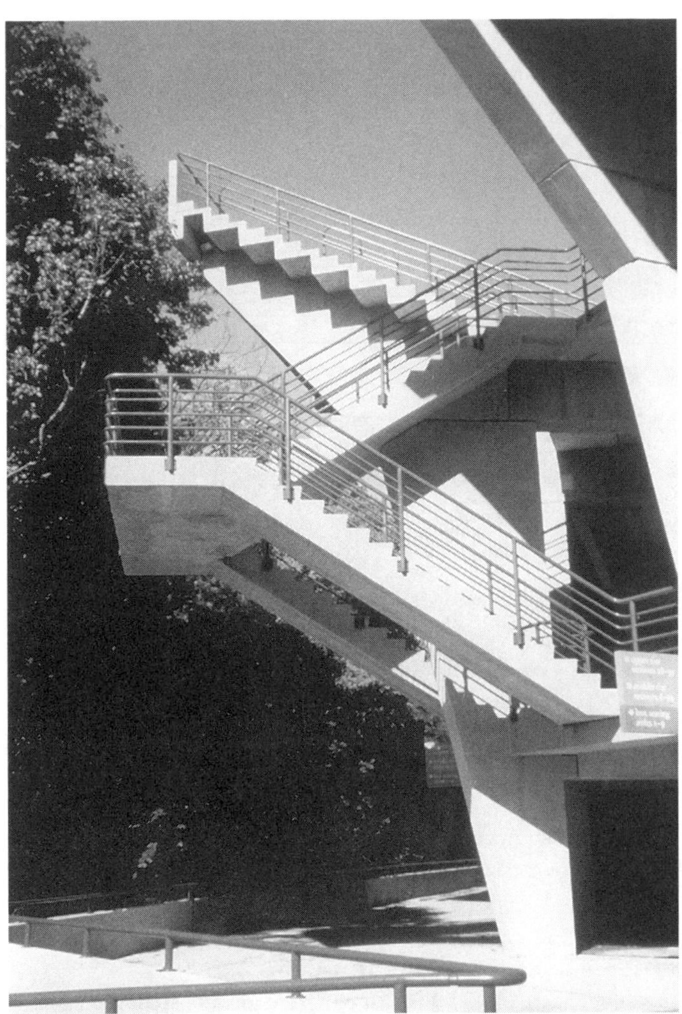

(f)

(g)

FIGURE 15.34 ***(continued)***
(b) **Grouted sleeve connectors of the type shown in Figure 15.15 were used to join the precast sections of the bents that support the grandstand seating.** ***(c)*** **Stepped sections of grandstand floor await lifting.** ***(d)*** **A precast stair section is placed.** ***(e)*** **Precast hollow-core planks will be used for miscellaneous areas of floor.** ***(f)*** **The relationship of the stairs, bents, and stepped floor sections is evident in this photograph of the finished stadium.** ***(g)*** **The stadium in use, after a construction process that took only 11 months. Seating capacity is 15,000.** *(Structural engineer: Spiegel Zamecnik & Shah, Inc. Photos by Clark Broadbent. Compliments of Blakeslee Prestress, Inc., P.O. Box 510, Branford, CT 06405)*

Figure 15.35
A low-rise office building in Texas shows a carefully detailed frame of precast concrete. *(Architects: Omniplan Architects. Courtesy of Precast/Prestressed Concrete Institute)*

Figure 15.36
Precast concrete girders span the roof of a paper mill in British Columbia. *(Engineer: Swan Wooster Engineering Co., Ltd. Courtesy of Precast/Prestressed Concrete Institute)*

FIGURE 15.37
The precast walls and slabs of these condominium apartments were erected during the winter in the New Mexico mountains. *(Architect: Antoine Predock. Courtesy of Precast/Prestressed Concrete Institute)*

FIGURE 15.38
Highly customized precast concrete framing is used in this courthouse by architect Eduardo Catalano. *(Photo by Gordon H. Schenk, Jr. Courtesy of the architect)*

FIGURE 15.39
The 39 stories of the Paramount Building rise 420 feet (128 m) on a precast concrete frame. The building is located in San Francisco, California, in the zone of highest seismic risk in the United States. Beam–column connections are posttensioned to resist seismic forces. The precast concrete cladding panels act to resist lateral forces on the building. *(Architects: Elkus Manfredi and Kwan Henmi. Structural engineers: Robert Engelkirk Consulting Engineers, Inc. Photograph © Bernard André)*

CSI/CSC
MasterFormat Sections for Concrete Construction

03 40 00	PRECAST CONCRETE
03 41 00	Precast Structural Concrete Precast Concrete Hollow Core Planks Precast Concrete Slabs Precast Structural Pretensioned Concrete
03 45 00	Precast Architectural Concrete

Selected References

1. Precast/Prestressed Concrete Institute. *PCI Design Handbook.* Chicago, updated regularly.

This is the major reference handbook for those engaged in designing precast, prestressed concrete buildings. It includes basic building assembly concepts, load tables for standard precast elements, engineering design methods, and a few suggested connection details.

2. Precast/Prestressed Concrete Institute. *Design and Typical Details of Connections for Precast and Prestressed Concrete* (2nd ed.). Chicago, 1988.

This engineering design manual includes a large collection of drawings of standard connection details.

3. Precast/Prestressed Concrete Institute. *Erector's Manual: Standards and Guidelines for the Erection of Precast Concrete Products.* Chicago, 1999.

The 96 pages of this manual describe and illustrate the best ways of hoisting and assembling the elements of a precast concrete building.

4. Precast/Prestressed Concrete Institute. *Architectural Precast Concrete.* Chicago, 1989.

This design manual provides technical and design guidelines for achieving high-quality architectural finishes in precast concrete, and includes hundreds of images and line drawings.

Web Sites

Precast Concrete Framing Systems

Author's supplementary web site: **www.ianosbackfill.com/15_precast_concrete_framing_systems**

Altus Group carbon reinforced precast concrete: **www.altusprecast.com**

Dayton Superior: **www.daytonconcreteacc.com**

Precast/Prestressed Concrete Institute (PCI): **www.pci.org**

Spancrete: **www.spancrete.com**

Key Terms and Concepts

precast concrete
steam curing
solid slab
hollow-core slab
double tee
single tee
topping
ledger
L-shaped beam
inverted tee
AASHTO girder
casting bed
weld plate
dapped end
carbon fiber reinforcing
hollow core slab extruded process
hollow core slab wet-cast process
hollow core slab slip-form process
corbel
anchor bolt
embed plate
dovetail slot
powder-driven fastener, powder-actuated fastener
dry pack
bearing pad
filigree precast concrete system

Review Questions

1. Under what circumstances might a designer choose a precast concrete framing system over a sitecast system? Under what circumstances might a sitecast system be favored?

2. Why are precast concrete structural elements usually cured with steam?

3. Explain several methods of producing hollow-core slabs.

4. Diagram from memory several different ways of connecting precast concrete beams to columns.

5. Diagram from memory a method of connecting a pair of untopped double-tee slabs to an inverted-tee beam. Then work out a similar way of connecting a double tee to an L-shaped beam, as would occur around the perimeter of the same building.

6. Explain the construction process for a filigree precast concrete system. What are this system's unique advantages over other concrete systems?

Exercises

1. Design a simple two-story rectangular warehouse, 90 by 180 feet (27 by 54 m), using precast concrete for the floor and roof structure and for the walls as well. Use the preliminary structural design information given in this chapter to help determine the column spacing, the types of elements to use, and the depths of the elements. Draw a framing plan and typical connections for the building.

2. Locate a concrete precasting plant in your area and arrange a visit to view the production process. If possible, arrive at the plant early in the morning, when the strands are being cut and the elements are being lifted from the molds.

3. Learn from the management of the precasting plant where a precast concrete building is being erected; then visit the building site. Trace a typical precast concrete structural element from raw materials through precasting, transporting, and erecting. Are there ways in which this process could be made more efficient? Sketch a few of the typical connections being used in the project.

16

ROOFING

- **Low-Slope Roofs**
  - Roof Decks
  - Thermal Insulation and Vapor Retarder
  - BUILDING ENCLOSURE ESSENTIALS: THERMAL INSULATION AND VAPOR RETARDER
  - Rigid Insulating Materials for Low-Slope Roofs
  - Vapor Retarders for Low-Slope Roofs
  - Membranes for Low-Slope Roofs
  - Ballasting and Traffic Decks
  - Edge and Drainage Details for Low-Slope Roofs
  - Structural Panel Metal Roofing for Low-Slope Roofs
- **Steep Roofs**
  - Shingles
  - Architectural Sheet Metal Roofing
  - CONSIDERATIONS OF SUSTAINABILITY IN ROOFING
- **Sustainable Roofing**
  - Cool Roofs
  - Green Roofs
  - Photovoltaic Roofing
- **Roofing and the Building Codes**
  - BUILDING ENCLOSURE ESSENTIALS: DISSIMILAR METALS AND THE GALVANIC SERIES

Architect Tom Kundig, of Olson Sundberg Kundig Allen Architects, uses noncombustible, corrugated steel for the roof of a hilltop residence in a fire-prone area of Southern California. The simple roof form is lifted above the main living space of the house, shading it from the sun and allowing naturally cool offshore breezes to circulate freely around and through it. *(Photograph by Tim Bies, courtesy of Olson Sundberg Kundig Allen Architects)*

A building's roof is its first line of defense against the weather, protecting it from rain, snow, and sun. The roof helps to insulate the building from extremes of heat and cold and to control the accompanying problems of air leakage and water vapor condensation. And like any frontline defender, it must itself take the brunt of the attack: A roof is subject to the most intense solar radiation of any part of a building. At midday, the sun broils a roof with radiated heat and ultraviolet light. On clear nights, a roof radiates heat to the blackness of space and becomes colder than the surrounding air. From noon to midnight of the same day, it is possible for the surface temperature of a roof to vary from near boiling to below freezing. In cold climates, snow and ice cover a roof after winter storms, and cycles of freezing and thawing gnaw at the materials of the roof. A roof is vital to the sheltering function of a building, yet it is singularly vulnerable to the destructive forces of nature.

Roofs can be covered with many different materials. These can be organized conveniently into two groups: those that work on *steep roofs* and those that work on *low-slope roofs*, those that are nearly flat. The distinction is important: A steep roof drains itself quickly of water, giving wind and gravity little opportunity to push or pull water through the roofing material. Therefore, steep roofs can be covered with roofing materials that are fabricated and applied in small, overlapping units—shingles of wood, slate, or artificial composition; tiles of fired clay or concrete; or even tightly wrapped bundles of reeds, leaves, or grasses (Figures 16.1 and 16.2). There are several advantages to these materials: Many of them are inexpensive. The small, individual units are easy to handle and install. Repair of localized damage to the roof is easy. The effects of thermal expansion and contraction, and of movements in the structure that supports the roof, are minimized by the ability of the small roofing units to move with respect to one another. Water vapor vents itself easily from the interior of the building through the loose joints in the roofing material. And a steep roof of well-chosen materials skillfully installed can be a delight to the eye.

Low-slope roofs have none of these advantages. Water drains relatively slowly from the surfaces, and small errors in design or construction can cause them to trap puddles of standing water. The membranes that cover low-slope roofs must be absolutely watertight. Even small punctures, tears, or gaps in seams, caused by defects in construction, physical wear and tear, or movements within the building structure, can allow large quantities of water to enter into the building structure and its interior, with potentially disastrous results. Or, water vapor pressure from within the building can blister and rupture the membrane. But low-slope roofs also have overriding advantages: A low-slope roof can cover a building of any horizontal dimension, whereas a steep roof becomes uneconomically tall when used on a very broad building. A building with a low-slope roof has a much simpler geometry that is often less expensive to construct. And low-slope roofs, when appropriately detailed, can serve as balconies, decks, patios, and even landscaped gardens or parks.

FIGURE 16.1
A steep roof can be made waterproof with any of a variety of materials. This thatched roof is being constructed by fastening bundles of reeds to the roof structure in overlapping layers in such a way that only the butts of the reeds are left exposed to the weather.

FIGURE 16.2
The finished thatched roof has gently rounded contours and a pleasing surface texture. The decorative pattern of the ridge cap is the unique signature of the thatcher who made the roof. *(Photos of thatched roofs courtesy of Warwick Cottage Enterprises, Anaheim, California)*

Low-Slope Roofs

A low-slope roof (often referred to, inaccurately, as a "flat" roof) is usually defined as one whose slope is less than 2:12, or 17 percent. A low-slope roof is a highly interactive assembly made up of multiple components. The *roof deck* is the structural surface that supports the roof. *Thermal insulation* is installed to slow the passage of heat into and out of the building. An *air barrier* restricts the leakage of air through the roof assembly, and a *vapor retarder* is essential in colder climates or when enclosing humid spaces to prevent moisture vapor from condensing within it. The *roof membrane* is the impervious sheet of material that keeps water out of the building. *Drainage* components, such as roof drains, gutters, and downspouts, remove the water that runs off the membrane. Around the membrane's edges and wherever it is penetrated by pipes, vents, expansion joints, electrical conduits, or roof hatches, special *flashings* and details must be designed and installed to prevent water penetration.

Roof Decks

Previous chapters of this book have presented the types of structural decks ordinarily used under low-slope roofs: wood panels over wood joists, solid wood decking over heavy timber framing, corrugated steel decking, panels of wood fiber bonded together with portland cement, sitecast concrete slab, and precast concrete slab. For a durable low-slope roof installation, it is important that the deck be adequately stiff under expected roof loadings and fully resistant to wind uplift forces. The deck must slope toward drainage points at an inclination sufficient to drain reliably despite the effects of structural deflections. A slope of at least ¼ inch per foot of run (1 in 50, or 2 percent) is normally required by the building code and by most manufacturers of low-slope roof membranes. To create the slopes in a low-slope roof, the beams that support the deck may be sloped by varying the height of the supporting columns. Or the deck may be constructed level and the slopes created by a layer of thermal insulation of varying thickness installed on the top of the deck. This layer may consist of lightweight insulating concrete or a system of tapered rigid insulation boards.

If the slope of a low-slope roof is too shallow, *ponding* occurs: the formation of puddles of water that stand for extended periods of time, leading to premature deterioration of the roofing materials in those areas.

If water accumulates in low spots caused by structural deflections, progressive structural collapse becomes a possibility, with deepening puddles attracting more and more water during rainstorms and becoming heavier and heavier, until the beams or joists become loaded to the point of failure (Figure 16.3).

The roof membrane must be laid over a smooth surface. A wood deck that is to receive a roof membrane should have no large gaps, knotholes, or protruding fasteners. A sitecast concrete deck should be troweled smooth, and a precast concrete plank deck, if not topped with a concrete fill, must be grouted at junctions between planks to fill the cracks and form a smooth surface. A corrugated steel deck must be covered with *substrate board,* thin panels of rigid insulation, wood, or gypsum board that can bridge the flutes in the deck and create a continuous, smooth surface.

It is extremely important that the deck be dry at the time roofing operations commence to avoid later problems with water vapor trapped under the roof membrane. A deck should not be roofed when rain, snow, or frost is present in or on the deck material. Concrete decks and insulating fills must be fully cured and thoroughly air dried.

If a deck is large in extent, the roofing system should be provided with enough movement joints so that expansion and contraction or other movement within the deck does not excessively stress the overlying membrane. Where building separation joints occur within a building structure, these joints must carry through the roof membrane system as well (Figure 16.27). Where such joints do not occur or are too far apart to satisfy the separation requirements of the membrane, *area dividers,* which are much like building separation joints but do not extend below the surface of the roof deck, may be installed (Figure 16.28).

Thermal Insulation and Vapor Retarder

Thermal insulation for a low-slope roof may be installed in any of three positions: below the structural deck, between the deck and the membrane, or above the membrane (Figure 16.4).

Insulation Below the Deck

Below the deck, batt insulation of mineral fiber or glass fiber is installed either between framing members or

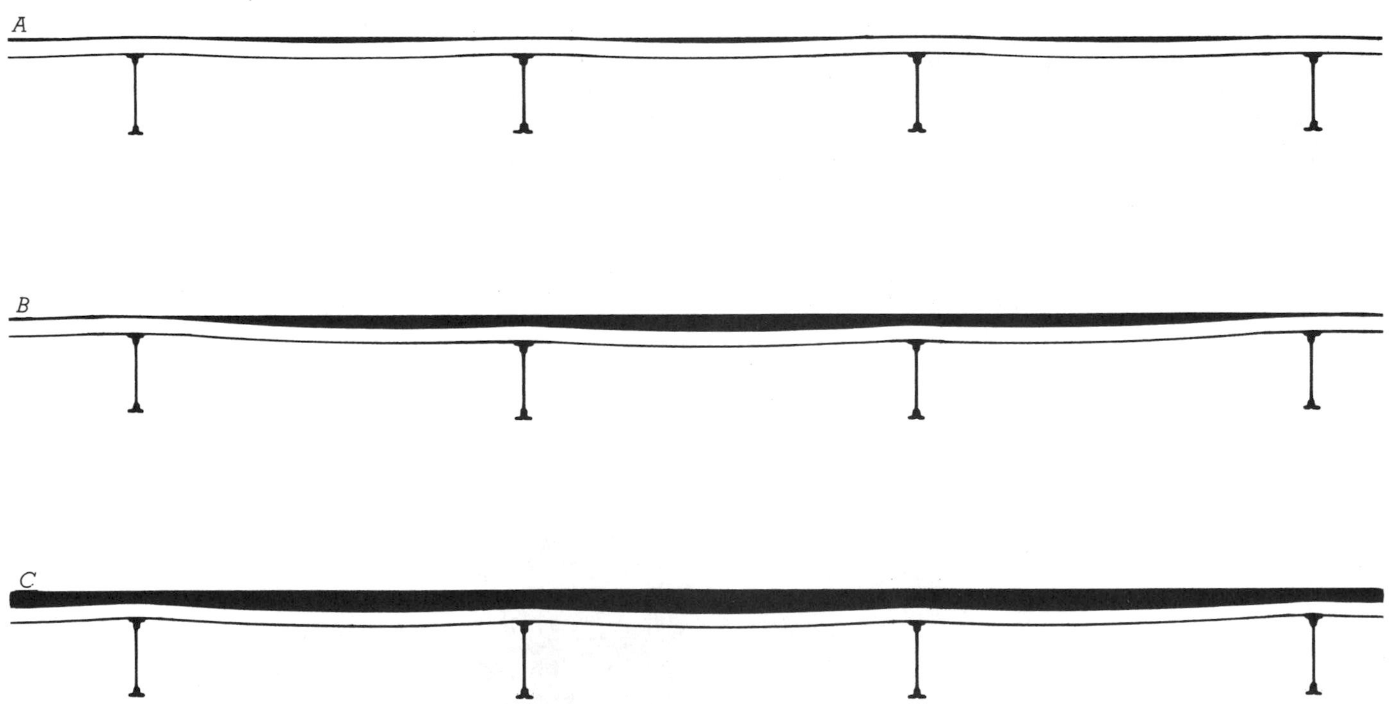

FIGURE 16.3
A low-slope roof with insufficient slope toward a drain is subject to ponding, and possible structural failure through progressive collapse, as demonstrated in this sequence of cross sections. *(a)* Water stands on the roof in puddles, its weight causing slight deflections of the roof deck between supporting beams or joists. *(b)* If heavy rainfall continues, the puddles grow and join, and the accumulating weight of the water begins to cause substantial deflections in the supporting structural elements. The deflections encourage water from a broader area of the roof to run into the puddle. *(c)* As structural deflections increase, the depth of the puddle increases more and more rapidly until the overloaded structure collapses.

on top of a suspended ceiling assembly. The building code normally requires a ventilated airspace between the insulation and the underside of the deck to dissipate excess water vapor. Below-deck insulation is relatively economical and trouble-free, but it leaves the deck and the membrane exposed to the full range of outdoor temperature fluctuations. In cold climates or when enclosing humid spaces, a vapor retarder is recommended on the warm, conditioned side of the roof insulation to control the diffusion of water vapor into the insulated portions of the roof where condensation could occur.

Spray foam insulation can also be installed below a roof deck. When air-impermeable foam insulation is used, the risk of moisture accumulation within the assembly is reduced, and ventilation between the deck and the insulation may not be required.

Insulation Between the Deck and the Membrane

The traditional position for low-slope roof insulation is between the deck and the roof membrane. Insulation in this position must be in the form of low-density rigid boards or lightweight concrete in order to support the membrane. The insulation protects the deck from temperature extremes and is itself protected from the weather by the membrane.

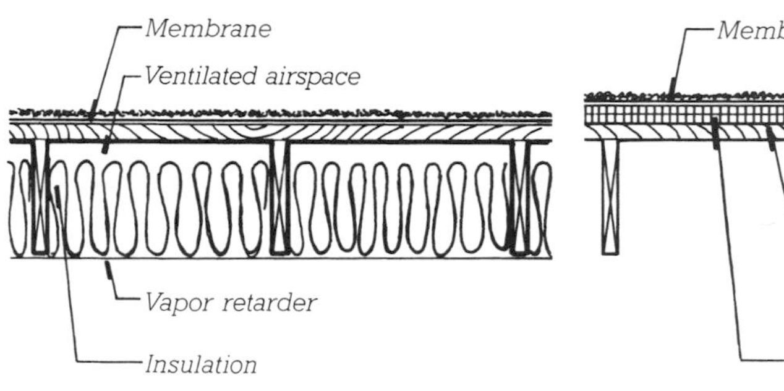

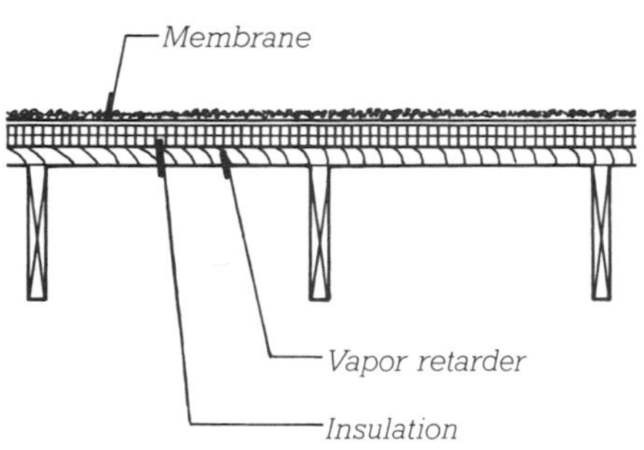

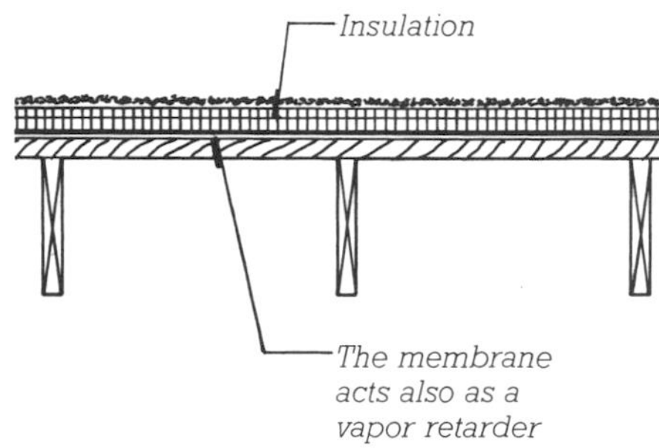

FIGURE 16.4
Low-slope roofs with thermal insulation in three different positions, shown here with a wood joist roof deck. At the left, insulation is located below the deck, with a vapor retarder on the warm side of the insulation. In the center, insulation is placed between the deck and the membrane, with a vapor retarder on the warm side of the insulation. At the right, a protected membrane roof, insulation is placed above the membrane and no separate vapor retarder is required. When insulation is installed below the roof deck, a space that is continuously ventilated with outside air is usually required between the insulation and the deck to prevent water vapor accumulation.

FIGURE 16.5
Topside roof vents are being installed in this built-up roof membrane to release vapor pressure that may accumulate beneath it. *(Courtesy of Manville Corporation)*

But the roof membrane in this type of installation remains exposed to extreme temperature variations. Additionally, any moisture that may accumulate in the insulation is trapped beneath the membrane, which can lead to decay of the insulation and roof deck, as well as blistering and eventual rupture of the membrane from vapor pressure. (See the discussion on pages 658–661 for a more in-depth explanation of insulation and vapor retarders.)

When installing insulation between the deck and the membrane in cold climates, two precautions may be taken: a vapor retarder installed below the insulation and ventilation within the insulation to allow the escape of any moisture that reaches there. Ventilation is accomplished through the installation of *topside vents*, one per 1000 square feet (100 m^2), that allow water vapor to escape upward through the membrane (Figures 16.5 and 16.6). Topside vents are most effective with a loose-laid membrane (a membrane that is not adhered to the underlying surface), which allows trapped moisture to easily work its way toward the vents from any part of the insulating layer.

Insulation Above the Membrane: The Protected Membrane Roof

In a *protected membrane roof (PMR)* system, insulation is installed above the roof membrane. This offers two advantages: The membrane is protected from extremes of heat and cold, and the membrane is on the warm side of the insulation, where it is immune to vapor blistering problems. Because the insulation itself is exposed to water when placed above the membrane, it must be made of a material that retains its insulating value when wet and does not decay or disintegrate. Extruded polystyrene foam is the one material that has these qualities (Figure 16.7). The panels of polystyrene are either embedded in a coat of hot asphalt to adhere them to the membrane below or are laid loose. They are held down and protected from sunlight (which disintegrates polystyrene) by a layer of *ballast*, which may consist of crushed stone or gravel, a thin concrete layer laminated at the factory to the upper surface of the insulating board, or interlocking concrete blocks (Figures 16.8, 16.9,

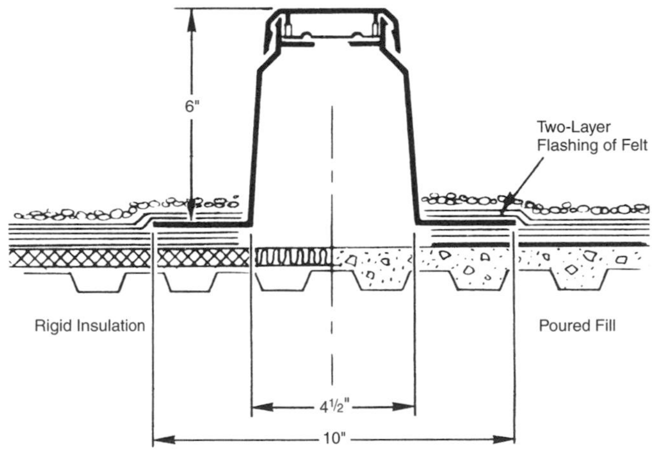

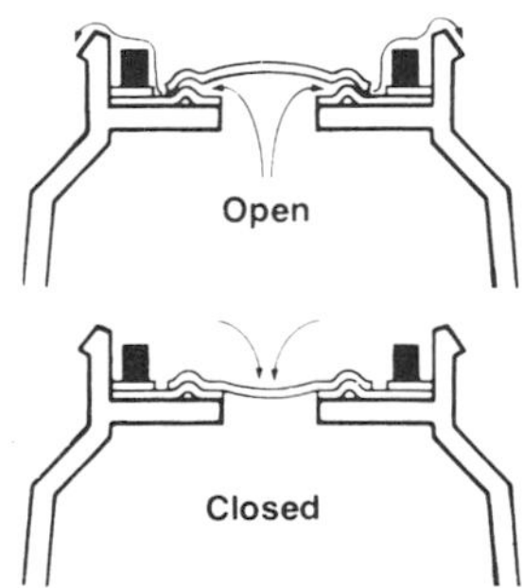

FIGURE 16.6
This proprietary topside roof vent, made of molded plastic with a synthetic rubber valve, allows moisture vapor to escape from beneath the membrane but closes to prevent water or air from entering. *(Courtesy of Manville Corporation)*

FIGURE 16.7
Installing extruded polystyrene foam insulation over a roof membrane to create a protected membrane roof. *(Photograph provided by the Dow Chemical Company)*

and 16.23). Because the membrane in a PMR system is shielded from sunlight and temperature extremes by the insulation and ballast above, it can be expected to last roughly twice as long as in an assembly where it remains directly exposed to these elements. However PMR systems do have potential disadvantages. When the roof insulation is exposed to precipitation, it may absorb moisture and lose some of its resistance to the flow of heat. When performing energy calculation for PMR systems, R-values for roof insulation may be slightly reduced from their standard values to account for such losses. PMR systems may not be appropriate for climates with extended periods of rainy, cold weather, as cool water continuously flowing around and under the insulation boards may negate much of their insulating value. And repairs to protected membrane roofs, if required, are more costly and time-consuming because accessing the roof membrane requires removal of the layers of material above it.

FIGURE 16.8
A cutaway detail of a proprietary type of protected membrane roof shows, from bottom to top, the roof deck, the membrane, polystyrene foam insulation, a polymeric fabric that separates the ballast from the insulation, and the ballast. *(Photograph provided by The Dow Chemical Company)*

FIGURE 16.9
This proprietary system of 2-inch (50-mm)-thick polystyrene foam insulation for a protected membrane roof is topped with a $^{3}/_{8}$-inch (9-mm) layer of latex-modified concrete. The concrete protects the foam from sunlight and wear and also ballasts it to prevent it from lifting off the roof in high winds. *(Photo courtesy of T. Clear Protected Membrane Roof System)*

Building Enclosure Essentials: Thermal Insulation and Vapor Retarder

Thermal Insulation

Thermal insulation is material added to a building assembly to slow the conduction of heat through the assembly. In North America, insulation is almost always installed in new roof and exterior wall assemblies, in floors over unheated spaces, around foundations and concrete slabs on grade, and other areas where heated or cooled interior space comes in contact with unconditioned space, the earth, or the outdoors. A well-insulated building enclosure increases occupant comfort and reduces the energy required to heat or cool the building.

A material's effectiveness in resisting the conduction of heat is called its *thermal resistance*, abbreviated as *R* and expressed as square foot-hour-degree Fahrenheit per BTU (ft^2-hr-°F/BTU). In the metric system, thermal resistance is abbreviated as *RSI* (or sometimes also simply as R), expressed as square meter-degree Kelvin per watt (m^2-°K/W). An insulating value of R-1 is equivalent to RSI 0.176. For a list of common insulating materials and their properties, see Figures 7.13 and 16.10. The higher a material's R-value, the higher its resistance to heat flow and the better its performance as a thermal insulator.

The thermal performance of a complete building assembly depends on the sum of the thermal resistances of the materials from which it is made. Every component of an assembly contributes in some measure to the assembly's overall thermal resistance, the amount of the contribution depending on the type of material and its thickness. Metals are poor insulators, and concrete and masonry are only slightly better. Wood has a substantially higher thermal resistance, but not nearly as high as that of commonly used insulating materials. In conventional wall and roof assemblies, most of the thermal resistance comes from the insulation materials themselves.

In wintertime, it is warm inside a heated building and cold outside and the inside surface of a wall or roof assembly is warm, and the outside surface is cold. Between the two surfaces, the temperature varies according to the thermal resistances of the various layers of the assembly. Because most of the insulating value of the assembly is in the insulation itself, most of the temperature change within the assembly occurs across the thickness of the insulating material (Figure A).

Water Vapor and Condensation

Water exists in three physical states, depending on its temperature and pressure: solid (ice), liquid, and vapor. Air always contains some water in the form of *water vapor*, an invisible gas. The higher the temperature of the air, the more water vapor it can contain. At a given temperature, the amount of water vapor the air actually contains, in proportion to the maximum amount of water vapor it can contain, is the *relative humidity* of the air. For example, air at 50 percent relative humidity contains half as much water vapor as it is capable of holding at that particular temperature.

If a mass of air is cooled, its relative humidity rises. The amount of water vapor in the air mass has not changed, but the ability of the air mass to hold water vapor has diminished because the air has become cooler. If the cooling of the air mass continues, a temperature will be reached at which the humidity is 100 percent. This is the temperature at which the air is fully saturated with water vapor, also known as the *dew point*. The dew point

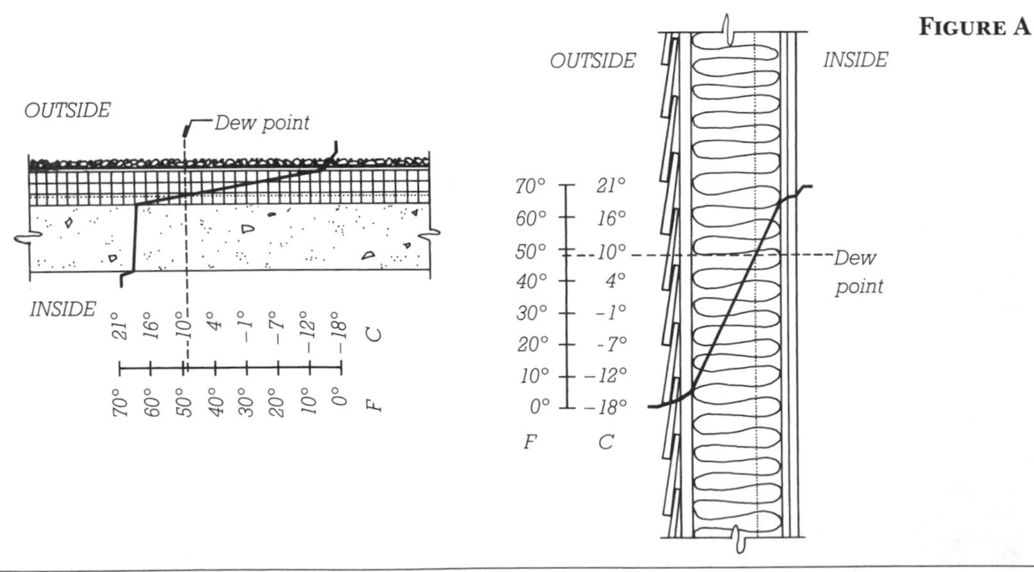

Figure A

is different for every air mass. A roomful of very humid air has a high dew point, which is another way of saying that the air in the room would not have to be cooled very much before it reached 100 percent humidity. In comparison, a roomful of dry air has a lower dew point and can be cooled to a lower temperature before it reaches saturation.

When a mass of air is cooled below its dew point, it can no longer contain all its water vapor and some of the vapor turns to liquid. The further the air mass is cooled below this point, the less water vapor it can hold and the more vapor turns to liquid. This process of converting water vapor to liquid by cooling is called *condensation.*

Condensation takes place in buildings in many different ways. In winter, room air circulating against a cold pane of glass may be cooled to below its dew point, and water droplets will form on the glass. If the air is very humid, the droplets will grow in size, then run down the glass to accumulate in puddles on the window sill. If the glass is very cold, the condensate may freeze into patterns of ice crystals on the glass. In a similar fashion, on a hot, humid summer day, water vapor in the air may condense on the surface of a cold water pipe, a cool basement wall, or an ice-cold glass of lemonade. Though hidden from view, condensation can also occur within the walls, roofs, and other assemblies that enclose a building.

Water Vapor in Building Assemblies

Water vapor is a gas and exerts pressure, called *vapor pressure,* on the surfaces that contain it. The more water vapor an air mass contains, the greater the vapor pressure. Under wintertime heating conditions, the air inside a building is at a higher temperature and contains more water vapor than the air outside. This is especially true in areas of the building with many occupants or where cooking, wet industrial processes, bathing, or washing take place. The result is a net difference in vapor pressure acting from inside to outside, causing water vapor to diffuse outward directly through the various material layers of the enclosing assembly. If the rate of diffusion is high enough and the drop in temperature within the assembly great enough, water vapor will reach its dew point and condense within the assembly. If moisture accumulates, insulation may loose its effectiveness and materials may be damaged by rusting, mildew, decay, freeze-thaw, or other harmful processes. If water runs out of the assembly, finishes or building contents may be harmed. (Water vapor can also be carried through an assembly by air leaking through gaps in materials. This is a different mechanism than the diffusion of water vapor through materials and is discussed under the topic of air barriers on pages 800–803.)

Under summertime cooling conditions in hot, humid weather, the diffusion of water vapor through building assemblies is reversed. Water vapor is driven from the warm, damp outside air toward the cooler, drier air within. In most of North America, this summer condition is not as severe as that in winter: Differences in summer temperature and humidity between indoors and outdoors are not as great as those in winter, and the cooling season is short compared to the heating season. Where heating conditions predominate, control of water vapor in building assemblies focuses primarily on the flow of vapor from interior to exterior, though conditions of reverse, inward flow may also warrant some consideration. In areas of the American South and in the Hawaiian Islands, however, the summer condition is the more severe, and the flow of water vapor from exterior to interior is the predominant problem to be solved.

Vapor Retarder

A *vapor retarder* (often called, inexactly, a *vapor barrier*) is a material used to slow the diffusion of water vapor through a building assembly. Vapor retarders are continuous sheets or coatings made of plastic, metal foil, coated paper, or any other material resistant to the passage of water vapor.

Vapor retarders are located toward the warmer side of the insulation in a building assembly. In this position, they can restrict the diffusion of water vapor into the assembly from the side of higher vapor pressure, limiting chances for dew point conditions and condensation to occur within the assembly's cooler portions. For buildings in most parts of North America, where winter heating conditions predominate, vapor retarders are placed toward the interior, heated side of insulation in the assembly. In humid regions where warm-weather cooling predominates within buildings, the vapor retarder should be located toward the exterior side of insulation. In relatively mild or balanced climates, or where assemblies are designed to minimize condensation conditions, a vapor retarder may not be necessary at all.

The better a material's resistance to water vapor diffusion, that is, the lower its *vapor permeance,* the more effective it is as a vapor retarder. Permeance is measured in *perms,* defined by ASTM E96 as the passage of one grain of water vapor per hour through 1 square foot of material at a pressure differential of 1 inch of mercury between the two sides of the material (grains/hr-ft^2-in. Hg). In metric units, a perm is measured in grams per second per square meter per Pascal of pressure difference (g/s-m^2-Pa). One

Building Enclosure Essentials: Thermal Insulation and Vapor Retarder (continued)

U.S. perm equals 5.72 × 10^{-8} metric perms. To arrive at more workable units, metric perms may also be calculated with nanograms (billionths of a gram) in place of grams. In this case, one U.S. perm equals 57.2 ng/s-m^2-Pa. Model building codes generally define materials used as vapor retarders as having a U.S. perm rating of 1 (metric perm rating of 60 ng/s-m^2-Pa) or less.

When evaluating vapor retarders, vapor permeance should not be confused with *vapor permeability*. Vapor permeability is defined as a material's vapor permeance for a unit of thickness. For example, the vapor permeability of a particular rigid insulation may be given as 0.75 perm-inch. The vapor permeance of an insulation board made of this material is then found by dividing the material's permeability by the actual thickness of the board used. For example, if the insulation board is ½ inch thick, its permeance is 1.5 perms (0.75 perm-in./ 0.5 inch), or if 2 inches thick, 0.375 perm (0.75 perm-in./2 inches). In metric units, vapor permeability is calculated as ng-m/ Pa-s-m^2, which reduces to ng/Pa-s-m.

The vapor permeance of some materials varies. Materials that absorb liquid moisture, such as plywood or the kraft paper facing found on much glass fiber batt insulation, tend to increase in permeance when damp. At least one manufacturer's proprietary vapor retarder membrane is engineered purposely to increase in permeance under conditions of high humidity as well. Such materials can provide benefits, inhibiting vapor diffusion into the assembly under normal conditions but becoming more permeable and enhancing the ability of the assembly to dry outward if moisture accumulates within it.

Commonly used vapor retarder materials are polyethylene plastic sheet, kraft paper facing on glass fiber batt insulation, aluminum foil facing on various types of insulation, and special paint primers with low water vapor permeability. Some foam insulation materials, depending on their formulation and thickness, can also act as vapor retarders. In low-slope roof construction, where vapor retarders are frequently installed as part of the roof membrane system, vapor retarders are often made of roofing felts layered in hot asphalt or of adhered rubberized asphalt sheets (Figure B).

Vapor Retarder Usage

Vapor retarders are used in insulated building assemblies to prevent water vapor condensation within the assembly. This is most likely where an assembly is exposed to large temperature differences from one side to the other and the relative humidity is high on the warmer side. In the International Energy Conservation Code (2006), requirements for vapor retarders differ between residential and commercial building types and with building location. Generally speaking, vapor retarders are required in exterior wall and ceiling assemblies for buildings in regions of the United States where winter heating conditions predominate, that is, in roughly the upper two-thirds of the continental United States as well as Alaska. The Building Code of Canada (2005) requires vapor retarders in most insulated building assemblies.

Where vapor retarders are used, the cooler side of the assembly, opposite the vapor retarder, must be "breathable," that is, designed so that any moisture that does find its way into the assembly can be dispersed by means of ventilation or diffusion through vapor-permeable materials. Examples of this strategy include ventilated roof systems, ventilated rainscreen cladding, and the use of vapor- permeable house wraps. Assemblies with multiple vapor-impermeable layers should be approached with caution: They can trap moisture and provide no means for the moisture to escape. A rule of thumb sometimes used for the selection of materials on the opposite side of an assembly from a vapor retarder is that they have a vapor permeance at least 10 times that of the vapor retarder itself.

Additional factors to consider in the use of vapor retarders include the following:

- *The temperature and humidity conditions to which a building assembly is exposed over time are not static.* A wall system designed exclusively for winter heating conditions may perform poorly when vapor migrates from the exterior inward during summer months with air conditioning in operation. Solar heating of a rain-soaked exterior cladding system can cause strong inward water vapor movement at any time of the year.
- *Building assemblies must be able to dry out when wet.* Water introduced into a building assembly, such as from penetrating rain or from construction with wet building materials, must have a means to escape so that the assembly can dry.
- *Air barriers are also important to the control of water vapor.* Under many circumstances, air leakage can transport many times greater amounts of water vapor into an assembly than can vapor diffusion. Air barriers are discussed on pages 800–803.
- *Design for water vapor control should consider all components of an assembly.* The permeability of exterior sheathing, cladding or roofing, provisions for ventilation, and the

location and types of insulating materials within an assembly can all influence the choice and placement of vapor retarders.

For more information on the design of vapor retarders in insulated building assemblies, see the references listed at the end of this chapter.

	Vapor Permeance	
	U.S. Perms, grains/hr-ft^2-in. Hg	Metric Perms, ng/s-m^2-Pa
Aluminum foil, 1 mil (0.025 mm)	~0	~0
Double-ply roofing felt with hot asphalt	0.005	0.29
Polyethylene sheet, 4 mil (0.08 mm)	0.08	4.6
Paint, exterior alkyd (oil), three coats over wood	0.3–1.0	17–57
Kraft paper facing for glass fiber batt insulation		
Low humidity	0.4–1.0	23–57
High humidity	7.5	430
Paint, latex vapor barrier primer	0.43	25
EPDM single-ply roofing, 45 mil	0.43	25
High-density spray polyurethane foam, various thicknesses	0.43–2	25–110
Plywood, exterior glue, ½ in. (12 mm) thick		
Low humidity	0.5	29
High humidity	3–10	170–570
Brick masonry, 4 in. (100 mm) thick	0.8–1.1	46–63
Proprietary variable-permeability vapor retarder membrane		
Low humidity	<1	<57
High humidity	10–35	570–2000
Paint, interior alkyd (oil), interior primer plus one flat finish coat over plaster	1.6–3.0	92–172
Gypsum wallboard, ½ in. (12 mm)		
Painted, interior latex	2–3	110–172
Unpainted	25	1400
Building paper, asphalt-impregnated felt	3.3	190
Housewraps	5–50	290–2900
Low-density spray polyurethane foam insulation, 5 in. (125 mm)	10	570
Plaster on metal lath, ¾ in. (19 mm) thick	15	860

FIGURE B
Vapor permeance of some common building materials. Shaded rows indicate materials exceeding building code defined limits for vapor retarders.

	R-Value[a]	Composition	Advantages	Disadvantages
Cellulose fiber board	2.8 *19*	A rigid, low-density board of wood or vegetable fibers and a binder, which may be coated or impregnated with asphalt	Economical; compatible with hot bitumens	Lower insulating efficiency than plastic foams; susceptible to absorption of moisture and decay
Perlite board	2.8 *19*	Granules of expanded volcanic glass and a binder pressed into a rigid board	Inert; fire resistant; compatible with hot bitumens; dimensionally stable	Much lower insulating efficiency than plastic foams
Cellular glass board	2.9 *20*	A rigid, low-density board of heat-fused, closed-glass cells with kraft paper facers	Inert; fire resistant; compatible with hot bitumens; dimensionally stable	Lower insulating efficiency than plastic foams
Mineral fiber board	4 *28*	A rigid, low-density board of rock, slag, or glass fibers and binder, usually with glass fiber mat facers	Inert; fire resistant; dimensionally stable; compatible with hot bitumens; ventilates moisture freely	
Expanded polystyrene foam board (EPS)	3.6–4.2 *25–29*	A closed-cell rigid foam of polystyrene plastic manufactured through a molding process	Less expensive than extruded polystyrene foam; compatible with asphalt at low temperatures	Combustible; high coefficient of thermal expansion; greater moisture absorption than extruded polystyrene foam; not compatible with hot bitumens
Extruded polystyrene foam board (XPS)	5 *35*	A closed-cell rigid foam of polystyrene plastic manufactured through an extrusion process	Resistant to moisture; suitable for protected membrane roof systems; compatible with asphalt at low temperatures; available in high densities suitable for roofs subject traffic loads or heavy overburdens	Combustible; high coefficient of thermal expansion; not compatible with hot bitumens
Polyisocyanurate foam board	5.6 *39*	A closed-cell rigid foam of polyisocyanurate, with facings of aluminum foil, glass fiber mats, or cellulosic mats	High insulating efficiency; compatible with hot bitumens; moderate fire resistance	Expensive; rated R-value may decrease over time
Composite insulating boards	Varies	Sandwich layers of foam plastic and other materials such as perlite board, mineral fiber board, and saturated felt	Combine the high insulating efficiency and moisture resistance of foam plastics with the fire resistance, structural rigidity, and/or bitumen compatibility of other materials	
Lightweight insulating concrete	1–1.5 *6.9–10*	Concretes made from lightweight mineral aggregates (perlite or vermiculite) or foaming, air-entraining agents	Inert; can easily produce a tapered insulation layer for positive roof drainage; fire resistant; dimensionally stable; often combined with embedded foam plastic insulation to achieve higher insulating efficiency	Residual moisture from mixing water can cause blistering of membrane

[a]R-values are per unit thickness, expressed first in English units of ft^2-hr-°F/Btu-in. and second in metric units (italicized) of m-°K/W.

FIGURE 16.10
A comparative summary of some rigid insulating materials for low-slope roofs.

Rigid Insulating Materials for Low-Slope Roofs

An insulating material for low-slope roofs should have high thermal resistance, adequate resistance to compression, denting, gouging, moisture decay, and fire, and, if part of a hot-applied system, high resistance to melting or dissolving when hot bitumens are mopped onto it. No single material has all these virtues. Some rigid insulating materials commonly used on low-slope roofs in North America are listed in Figure 16.10, along with a summary of their advantages and disadvantages. The best choice is often a combination of materials, or a composite board that combines two or more materials into one product, to exploit the best qualities of each. A composite insulating board for installation under a built-up bituminous roof membrane might include, for example, a bottom layer of polyisocyanurate foam with high insulating value and a top layer of perlitic board resistant to hot bitumens.

If rigid insulating boards are located below the roof membrane, they may be adhered to the deck with hot asphalt or adhesives, or fastened to the deck mechanically with screws or any of a variety of fasteners made especially for the purpose. Mechanical fasteners are favored by insurance companies because they are more secure against wind uplift (Figures 16.11 and 16.12).

Lightweight insulating concrete is an economical insulating material that also creates a nailable roof deck. Formulated with lightweight aggregates or foaming, air entraining agents, this material has densities ranging from 20 to 40 lb/ft^3 (320 to 640 kg/m^3) compared to 145 lb/ft^3 (2320 kg/m^3) for conventional concrete. Lightweight concrete may be applied directly to corrugated steel decking or over rough concrete decks and can easily be tapered during installation to slope toward points of roof drainage. Thermal resistance per inch is not as high for this material as for most other types of roof insulation. However, plastic foam boards may be embedded in the insulating concrete to achieve higher insulating values within reasonable thicknesses. Lightweight concrete fill insulation contains large amounts of free water at the time it is placed. It must be cured and dried as thoroughly as possible before application of the membrane, and some form of venting to allow the escape of moisture vapor from the insulation during the life of the roof, via either topside vents or bottomside slotted metal roof decking, is usually advisable. Poured-in-place gypsum, another decking material popular in the past for forming lightweight, nailable sloping roof decks, is no longer used in new construction.

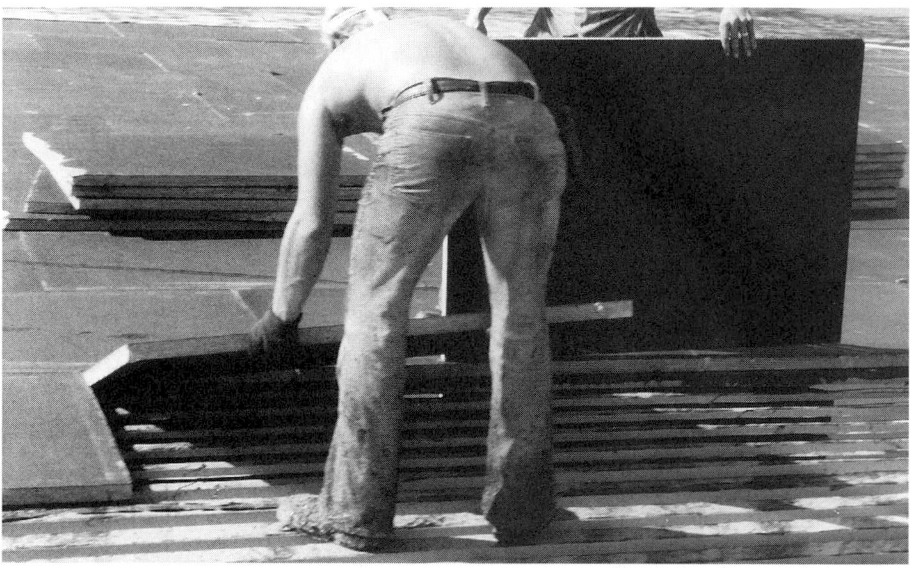

FIGURE 16.11
Workers bed rigid insulation boards in strips of hot asphalt over a corrugated metal roof deck. *(Courtesy of GAF Corporation)*

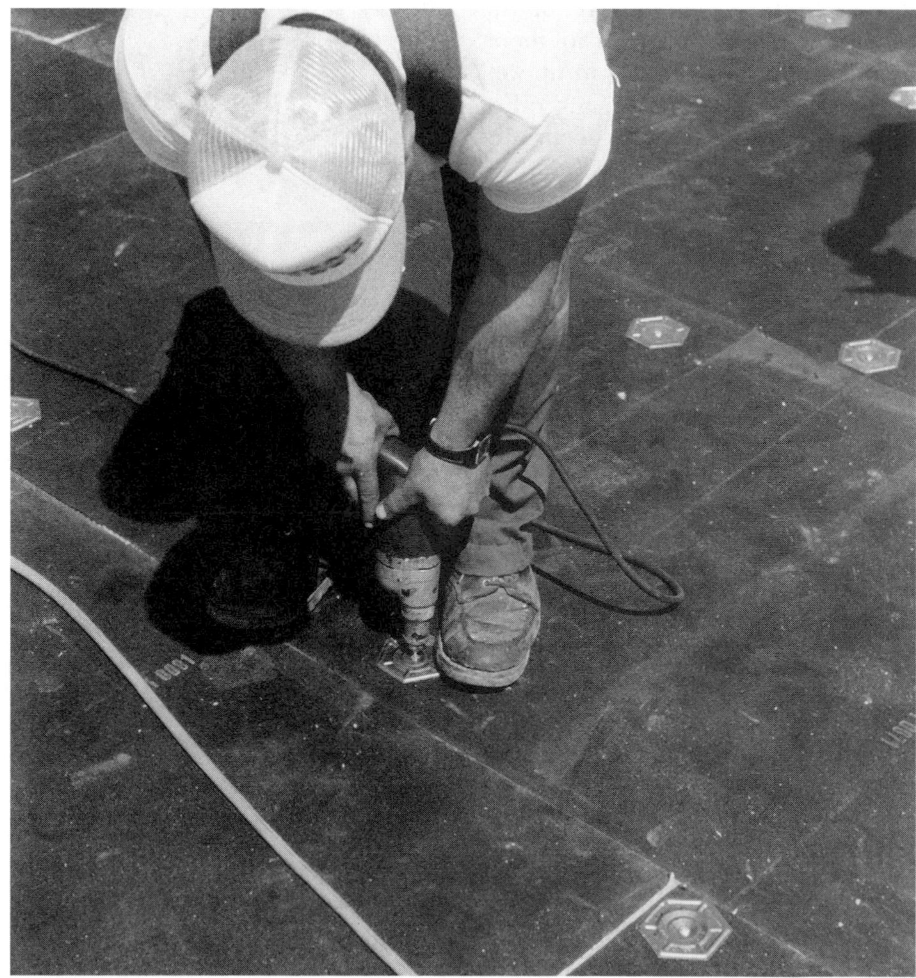

FIGURE 16.12
Screws and large sheet metal washers attach insulation more securely to a metal deck than can hot asphalt. *(Courtesy of GAF Corporation)*

Vapor Retarders for Low-Slope Roofs

The membrane in a protected membrane roof serves also as the vapor retarder. However, when insulation is located below the roof membrane, a separate vapor retarder is recommended in cold climates or when enclosing high-humidity interior spaces.

The most common type of vapor retarder for a low-slope roof consists of two layers of asphalt-saturated roofing felt bonded together and adhered to the roof deck with hot asphalt. Vapor retarder sheets made from factory-manufactured self-adhering bituminous membranes are also common. Polyethylene sheeting, used as a vapor retarder in many other types of construction, is seldom used in low-slope roofs because it melts at the application temperature of the hot bitumen used in many roof membranes, and it does not stand up well to the rigors of foot traffic and other construction activities that occur during roofing installation.

When a vapor retarder is included in a low-slope roof assembly, it must be located within the assembly such that it will always be warmer than the dew point of the interior air under common conditions of use. This usually means putting the vapor retarder below the insulation. However, a vapor retarder should not be installed directly over a corrugated steel deck, where it would have to bridge across the corrugations and would be vulnerable to damage until it was covered by insulation. In such cases, substrate board (thin panels of wood, gypsum board, or foam insulation) is first laid over the deck, followed by the vapor retarder and then the insulation boards. When some portion of the roof insulation is located below the vapor retarder, the designer must carefully calculate the dew point location in this assembly to be sure that the vapor retarder lies below it.

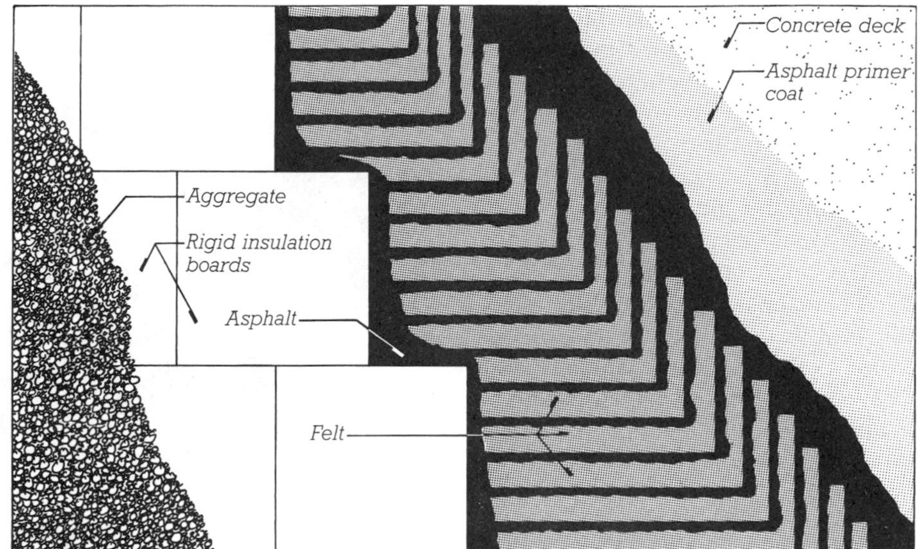

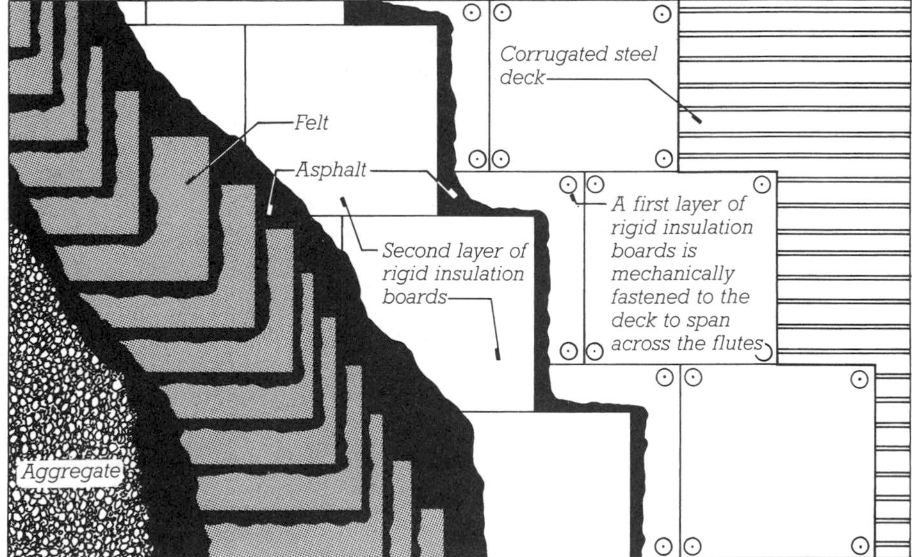

FIGURE 16.13
Two typical built-up roof constructions, as seen from above. The top diagram is a cutaway view of a protected membrane roof over a sitecast concrete roof deck. The membrane is made from plies of felt overlapped in such a way that at no location is it less than four plies thick. Rigid foam insulation boards are bedded in hot asphalt over the membrane and ballasted with stone aggregate to keep them in place and protect them from sunlight. The bottom diagram shows how rigid insulation boards are attached to a corrugated steel roof deck in two staggered layers to provide a firm, smooth base for application of the membrane. A three-ply membrane is shown. In cold climates, a vapor retarder should be installed between the layers of insulation or under the insulation over a substrate board fastened to the steel roof deck.

Membranes for Low-Slope Roofs

The membranes used for low-slope roofing fall into three categories: bituminous roof membranes, single-ply roof membranes, and fluid-applied roof membranes.

Bituminous Roof Membranes

Bituminous roof membranes are of two types, built-up or modified bitumen. A *built-up roof (BUR) membrane* is assembled in place from multiple layers of asphalt-impregnated roofing felt bedded in additional layers of bitumen (Figures 16.13–16.15). The felt, made from cellulose, glass, or synthetic fibers, is saturated with asphalt at the factory and delivered to the site in rolls. The bitumen is usually asphalt derived from the distillation of petroleum, but for dead-level or very low slope roofs, coal tar pitch is used instead because of its greater resistance to standing water. Polymer-modified asphalts, as described below for modified bitumen roofs, may also be used. Both asphalt and coal tar pitch are applied hot in order to merge with the saturant bitumens in the felt and form a unified, multi-ply membrane. The felt is laminated in overlapping layers (plies) to form a membrane that is two to four plies thick. The more plies used, the more durable the roof. To protect the membrane from sunlight and physical wear, a layer of crushed stone or other mineral granule aggregate is embedded in the top surface. Less commonly, a built-up roof may be made from felt plies bedded in *cold-applied mastics* (*solvent-based asphalts*), that is, compounds of asphalt and other substances applied by spray or brush at ambient temperatures and then cured through solvent evaporation.

A *modified bitumen roof membrane* is made from factory-manufactured sheets of polymer-modified bitumens. Modified bitumens are asphalt materials to which compounds such as

FIGURE 16.14
A base sheet of asphalt-saturated felt is installed over rigid insulation, using a machine that unrolls the felt and presses it into a layer of hot asphalt. *(Courtesy of the Celotex Corporation)*

FIGURE 16.15
Overlapping layers of roofing felt are hot-mopped with asphalt to create a four-ply membrane. *(Courtesy of Manville Corporation)*

atactic polypropylene (APP) or *styrene-butadiene-styrene (SBS)* have been added in order to increase the material's flexibility, cohesion, toughness, and resistance to flow. Modified bitumen roof membrane sheets are also reinforced with plastic or glass fibers or fibrous mats. Sheet thickness typically ranges from 0.040 to 0.160 inch (1.0–4.0 mm).

Like a built-up roof, modified bitumen sheets are assembled in place in overlapping layers to form a multi-ply system, usually two or three plies thick. The sheets are bonded to one another in a number of possible ways: With a *torch-applied* membrane, as a roofing sheet is unrolled, an open-flame apparatus is used to thermally fuse the underside of the sheet to the top surface of the substrate or underlying sheet. A *hot-mopped* membrane relies on the application of hot asphalt to bond the sheets, a *cold process* or *cold-applied adhesive* membrane uses liquid adhesives, and a *self-adhered* membrane relies on factory-applied adhesives (Figures 16.16).

The top or *cap sheet* in a modified bitumen roof is surfaced with mineral granules, thin metallic laminates, or asphaltic or elastomeric coatings for greater resistance to ultraviolet deterioration, wear, and fire (Figure 16.17). Cap sheets with reflective white coatings that comply with cool roof standards are also available. In comparison to built-up roofs, modified bitumen roofs combine the toughness and redundancy of multi-ply field application with the improved material qualities of factory-manufactured sheets. Built-up and modified bitumen systems may

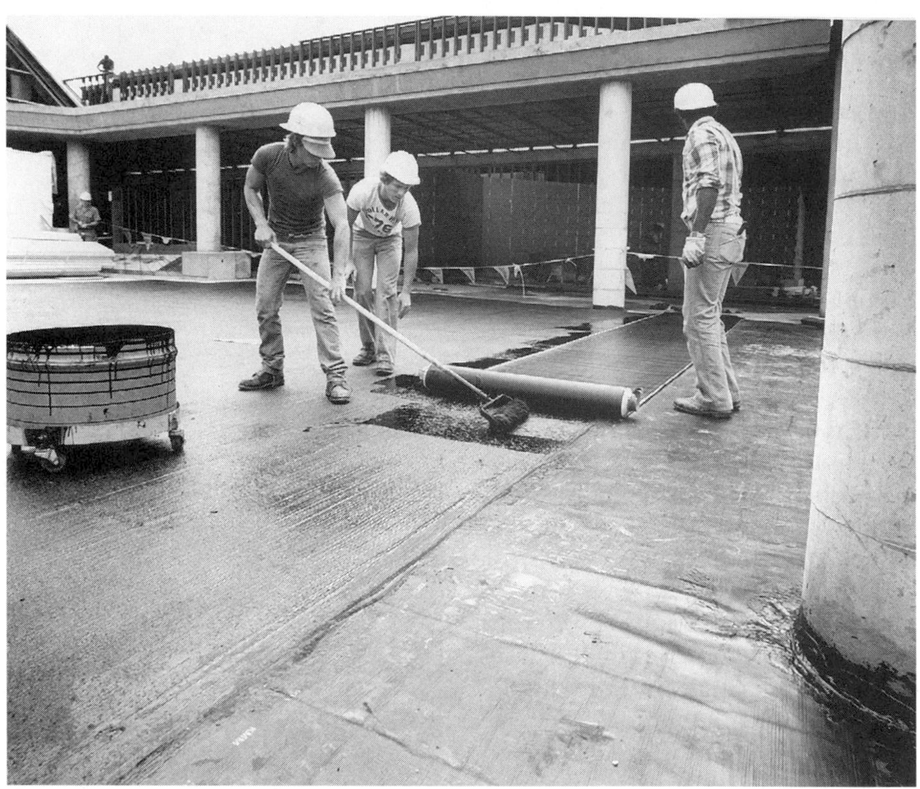

Figure 16.16
Roofers bond a polymer-modified bitumen membrane to a concrete deck with a cold-applied adhesive. The seams will be heat-fused together. *(Courtesy of Koppers Company, Inc.)*

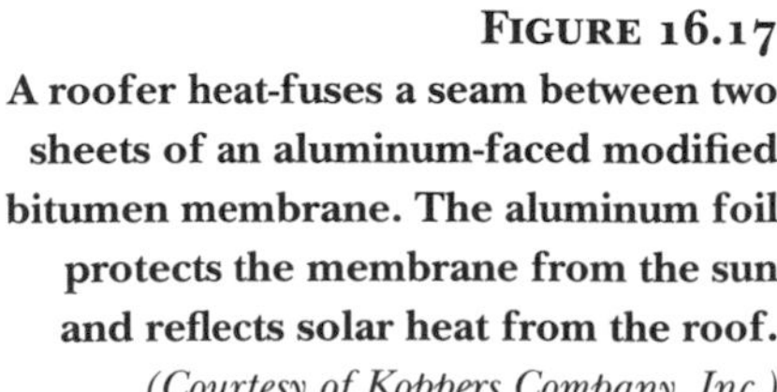

Figure 16.17
A roofer heat-fuses a seam between two sheets of an aluminum-faced modified bitumen membrane. The aluminum foil protects the membrane from the sun and reflects solar heat from the roof. *(Courtesy of Koppers Company, Inc.)*

also be combined with a modified bitumen cap sheet applied over several plies of built-up roofing to create a *hybrid membrane bituminous roof.* Bituminous roofing systems account for approximately 40 percent of the market for low-slope roofing membranes, with the larger portion of this share belonging to modified bitumen systems.

Single-Ply Roof Membranes

Single-ply roof membranes are a diverse and seemingly ever-evolving group of sheet materials that are applied to the roof in a single layer (Figure 16.18). Compared to bituminous roof membranes, they require less on-site labor, and especially in comparison to built-up roof membranes, they are more elastic and therefore less prone to cracking or tearing as they age. Common membrane thicknesses vary from 0.035 to 0.120 inch (0.9 to 3.0 mm), depending on the membrane material type and the requirements of the roofing application. Single-ply membranes may be affixed to the roof deck by adhesives, the weight of ballast, fasteners concealed in the seams between sheets, or, if the sheets are sufficiently flexible, with the use clever mechanical attachments that need not penetrate the membrane (Figures 16.19–16.21).

The materials used for single-ply membranes fall into two groups: thermoplastic and thermosetting. *Thermoplastic materials* can be softened by the application of heat and may be joined at the seams by heat (or solvent) welding. This welding process, which fully fuses one sheet to another, results in seams between sheets that are reliable, strong, and long-lasting. *Thermosetting materials,* also sometimes referred to as *elastomerics,* have a more tightly linked molecular structure and cannot be softened by heat. Thermosetting sheets must be joined at the seams by liquid adhesives or pressure-sensitive tapes, which have not always proved as reliable as the welded seams in thermoplastic membranes. Single-ply mem-

Figure 16.18
Workers unfold a large single-ply roof membrane. *(Courtesy of Carlisle SynTec Systems)*

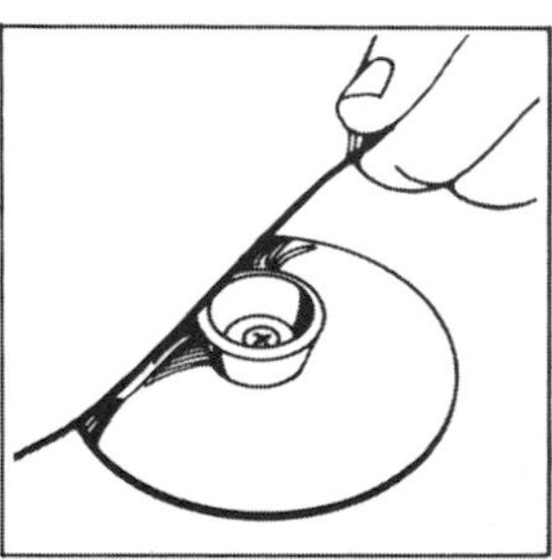

1. Roll membrane over knobbed base plate

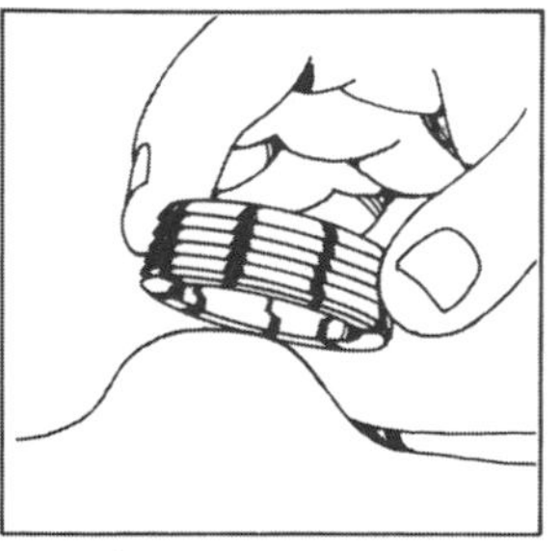

2. Roll and snap on white retainer clip

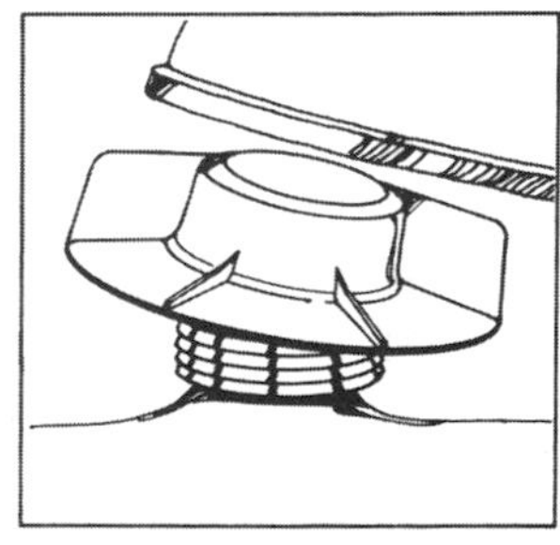

3. Snap and screw on threaded black cap

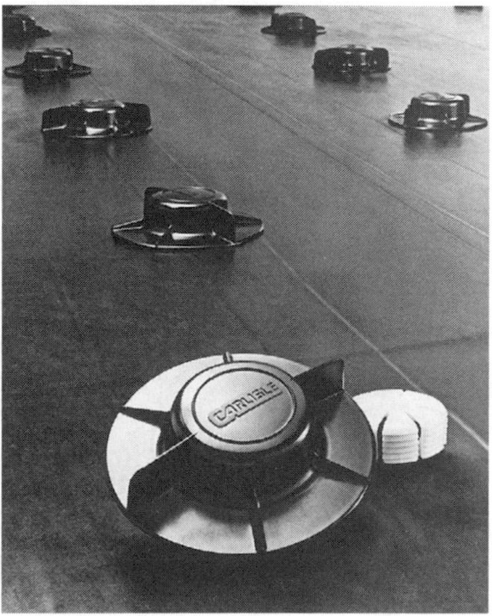

Figure 16.19 and 16.20
A proprietary nonpenetrating attachment system for a single-ply roof membrane. *(Courtesy of Carlisle SynTec Systems)*

branes, unlike multi-ply bituminous membranes, have no redundancy. Even a small defect in the joining of sheets can result in significant leakage through the membrane. (See pages 758–761 for more information about the difference between thermoplastic and thermosetting plastics.)

The most commonly used thermoplastic roof membrane materials are *polyvinyl chloride (PVC)* and *thermoplastic polyolefin (TPO)*. PVC roof membranes, made of PVC resins, plasticizers, stabilizers, and reinforcing fibers or fabrics, have a track record of successful performance established over many decades. They are available in various colors, including reflective white for cool roofs. However, concerns over the production of toxic chemicals (in particular, dioxin, a known carcinogen) during the manufacture and disposal of PVC have led to debate over the appropriateness of PVC for use as a construction material. No consensus has yet emerged, and as PVC manufacturers continue to improve their manufacturing processes and institute materials recycling programs, the pros and cons of this issue are likely to continue to evolve for the foreseeable future. TPO roof membranes are relatively new to the North American low-slope roofing market. They are made from blends of polyethylene, polypropylene, and ethylene-propylene rubber polymers reinforced with fibers or fabrics. TPO membranes exhibit good resistance to heat and ultraviolet (UV) radiation, characteristics more commonly associated with thermosetting membranes, but as thermoplastics, their seams can be heat welded. They are also available in a broad range of colors. PVC and TPO together account for approximately 20 percent of the market for low-slope roofing materials. Other, less widely used thermoplastic roof membrane materials include ketone ethylene ester (KEE) and a class of materials referred to as "PVC alloys" or "PVC compounded thermoplastics," made from various blends of PVC and other polymers.

The most commonly installed thermosetting roof membrane material is *ethylene propylene diene monomer (EPDM)*, a synthetic rubber that may or may not include fiber or fabric reinforcing. EPDM has a highly stable chemical structure with excellent resistance to ozone, heat, UV radiation, and weathering. It is most commonly black in color but is also available in cooler white from some manufacturers. Since EPDM cannot be heat welded, seaming is performed with tapes or adhesives. Like PVC, EPDM has a many-decades-long track record of successful performance. It is the most widely used material in North America for low-slope roofing of any type, accounting on its own for more than one-third of the market for such applications. Other thermosetting roof membrane materials include chorosulfonated polyethylene (CSPE) and polyisobutylene (PIB).

Fluid-Applied Roof Membranes

Fluid-applied roof membranes are used primarily for domes, shells, and other complex shapes that are difficult to roof by conventional means. Such shapes are often too flat on top for shingles but too steep on the sides for built-up roof membranes, and if doubly curved, they are difficult to cover with preformed sheets. Fluid-applied membranes are installed in liquid form with a roller or spray gun, usually in several coats, and cure to form a rubbery membrane. Materials applied by this method include neoprene (with a protective layer of CSPE), silicone, polyurethane, butyl rubber, and asphalt emulsion.

Fluid-applied membranes are also used as a waterproofing layer over sprayed-on polyurethane foam insulation in proprietary roofing systems designed for surfaces that are hard to fit with flat sheets of insulation and

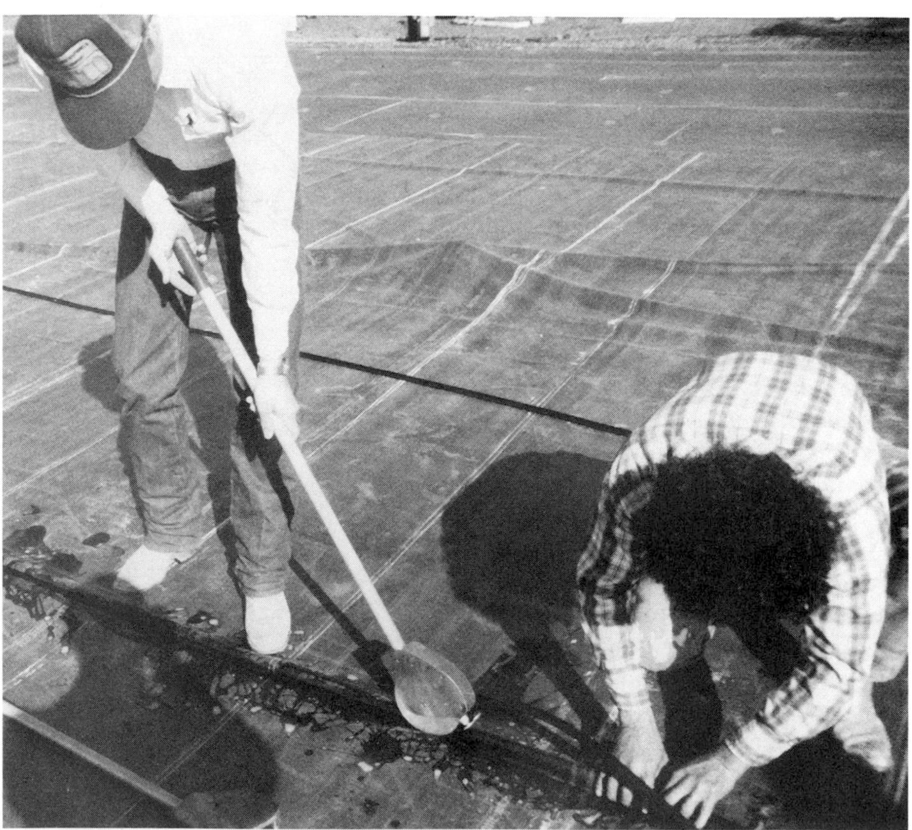

FIGURE 16.21
Another proprietary nonpenetrating attachment system folds the membrane into a continuous slot, where it is held by a synthetic rubber spline that is inserted with a wheeled tool. *(Courtesy of Firestone)*

membrane. These systems are also a convenient means for adding thermal insulation and a new roof membrane over existing deteriorated built-up roofs on any shape of building.

Membranes for low-slope roofing can have life expectancies ranging from 10 to 30 years, depending on the membrane material and thickness, the geographic location of the roof, its exposure to extremes of temperature and UV radiation, and the quality of the roof's installation and maintenance.

Ballasting and Traffic Decks

Roof membranes may be covered after installation with a *ballast* of loose, rounded stone aggregate ranging in size from 1½ to 2½ inches (38 to 64 mm) in diameter or with precast concrete paver blocks (Figures 16.8 and 16.22). The ballast serves to hold the membrane down against wind uplift, and it protects the membrane from ultraviolet light and physical wear. It may also contribute to the fire resistance of the roof covering.

Traffic decks are installed over flat roof membranes for walks, roof terraces, and sometimes driveways or parking surfaces. Two different details are used: In one, low blocks of plastic or concrete are set on top of the roof membrane to support the corners of heavy square paving stones or slabs with open joints (Figure 16.23). In the other, a drainage layer of gravel or no-fines concrete (a very porous concrete whose aggregate consists solely of coarse stone of a single size) is leveled over the membrane, and open-jointed paving blocks are installed on top. In either detail, water falls through the joints in the paving and is caught and drained away by the membrane below. Notice that the membrane is not pierced in either detail.

FIGURE 16.22
This ballasting system uses special concrete blocks that are joined by tubular plastic splines. The grooves in the bottoms of the blocks are designed to facilitate drainage of water from the membrane. The blocks are made with lightweight aggregate, so they contribute to the thermal insulation of the roof. *(Insulating roof ballast with variable interlock feature © National Concrete Masonry Association, 1988)*

FIGURE 16.23
A proprietary system for supporting stone or precast concrete paving blocks over a low-slope roof membrane permits the use of a low-slope roof as an outdoor terrace. Each high-density polyethylene pedestal supports the adjacent corners of four paving blocks. The vertical spacer fins on the pedestal provide a uniform drainage space between the blocks. Matching polyethylene leveling plates (not shown) can be placed over the pedestal to compensate for irregularities in the roof surface. *(Photo courtesy of Envirospec, Inc., Buffalo, New York)*

Edge and Drainage Details for Low-Slope Roofs

Some typical details of low-slope roofs are presented in Figures 16.24–16.33. All are shown with built-up roof membranes, but details for single-ply membranes are similar in principal.

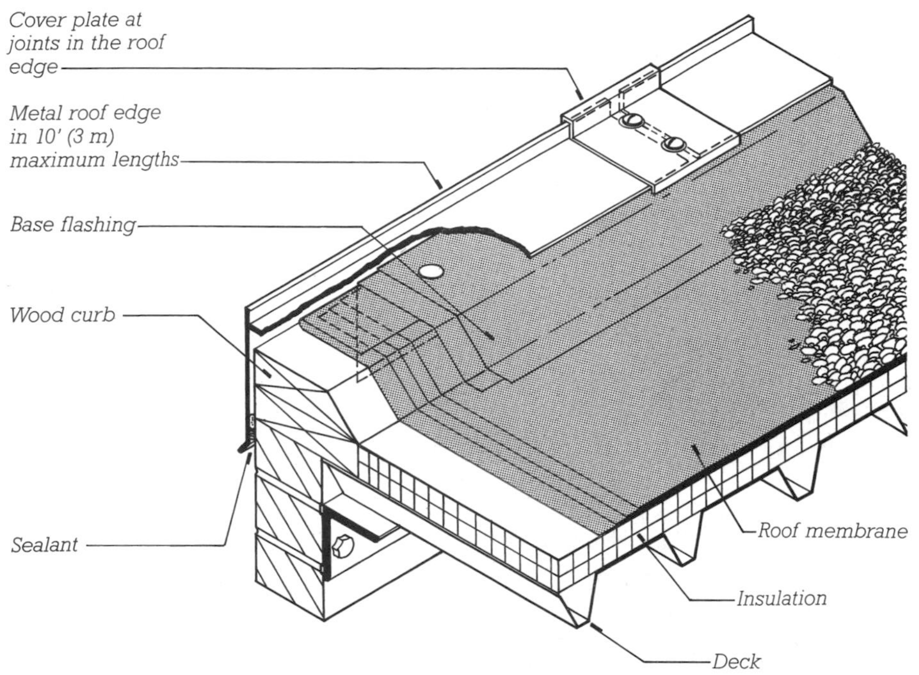

FIGURE 16.24
A roof edge for a conventional built-up roof. The membrane consists of four plies of felt bedded in asphalt with a gravel ballast. The *base flashing* is composed of two additional plies of felt that seal the edge of the membrane and reinforce it where it bends over the curb. The curb directs water toward interior drains or scuppers rather than allowing it to spill over the edge. The exposed vertical face of the metal roof edge is called a *fascia*.

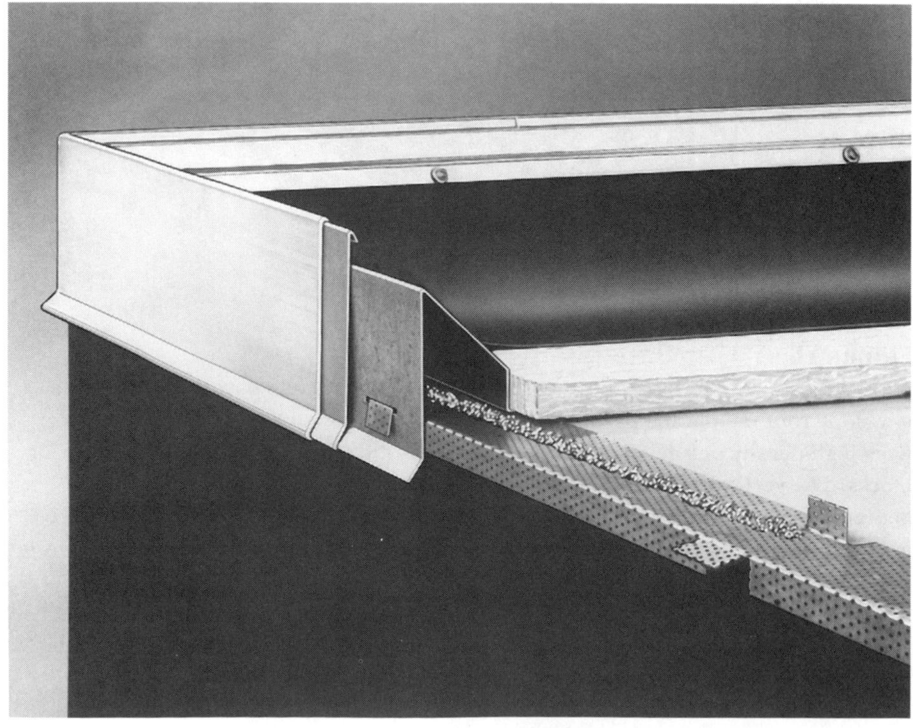

FIGURE 16.25
A proprietary roof edge system for low-slope roofs. The perforated metal strip is fastened to the roof with a mastic adhesive that oozes through the perforations to create a tighter bond. When the adhesive has hardened, a galvanized steel curb is fastened in place with the tabs of perforated metal and an aluminum roof edge is hooked on, with a backup piece at the end joints as shown to prevent leakage. Lastly, the roof edge and the membrane are locked in place simultaneously by installing a clamping strip that engages the hook on the top of the aluminum roof edge. The clamping strip is held in place by screws that pass through the edge of the membrane into the galvanized curb, as seen at the top of the photograph. *(Product of W. P. Hickman Company, Asheville, North Carolina)*

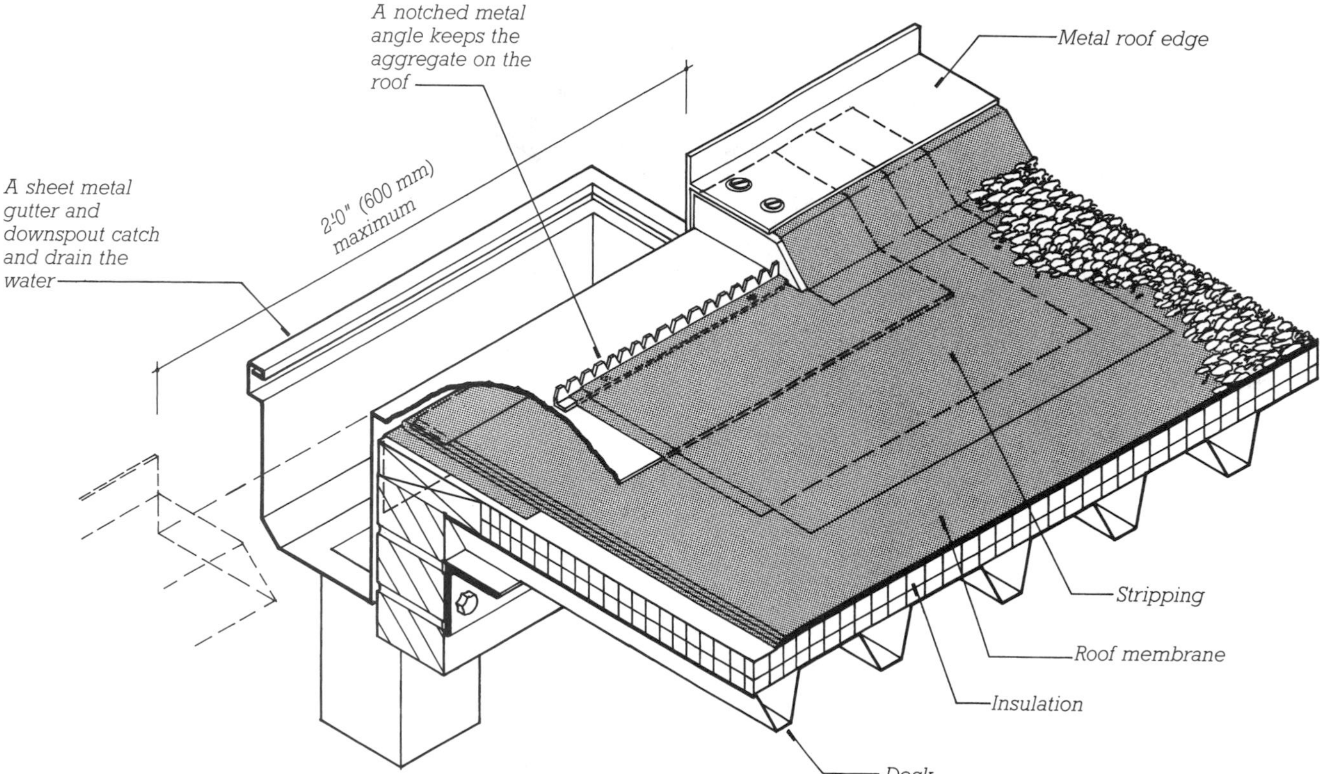
A notched metal angle keeps the aggregate on the roof
Metal roof edge
A sheet metal gutter and downspout catch and drain the water
2'-0" (600 mm) maximum
Stripping
Roof membrane
Insulation
Deck

FIGURE 16.26
Detail of a *scupper*. The curb is discontinued to allow water to spill off the roof into a gutter and downspout. Additional layers of felt, called *stripping*, seal around the sheet metal components. Most roofs use interior drains (Figure 16.31) as their primary means of drainage, with scuppers more frequently used as secondary drainage to limit ponding in the case of a clog in the primary drain.

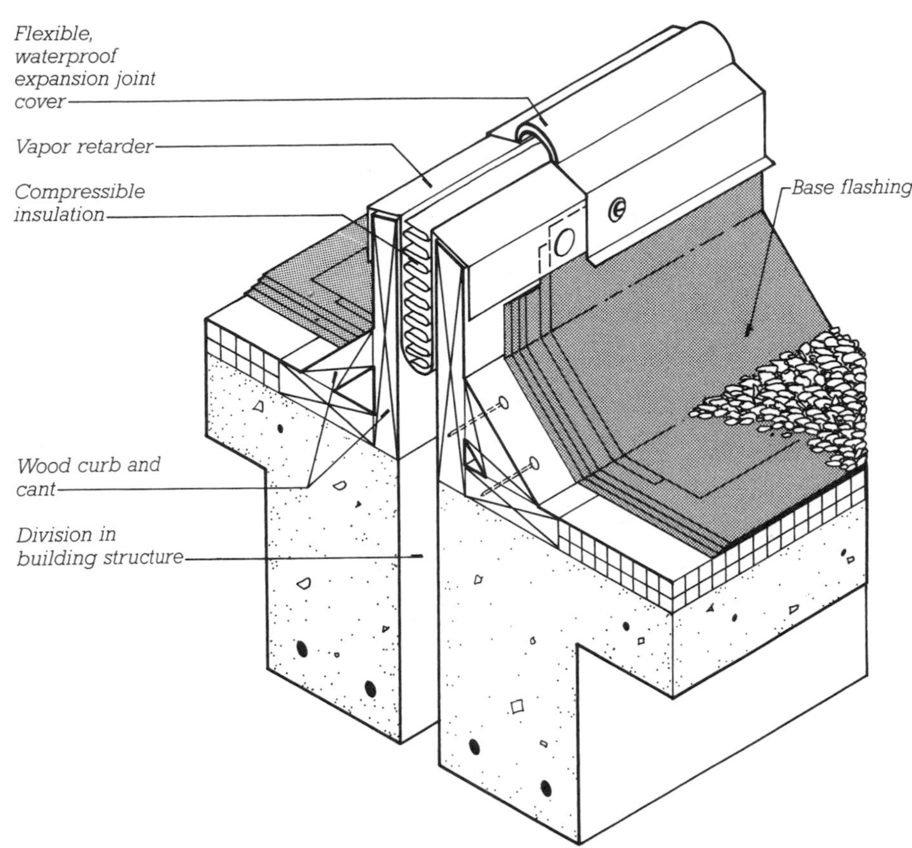

FIGURE 16.27
A building separation joint in a low-slope roof. Large differential movements between the adjoining parts of the structure can be tolerated with this type of joint because of the ability of the flexible joint cover to adjust to movement without tearing. High curbs keep standing water away from the edge of the membrane, which is sealed with a two-ply base flashing.

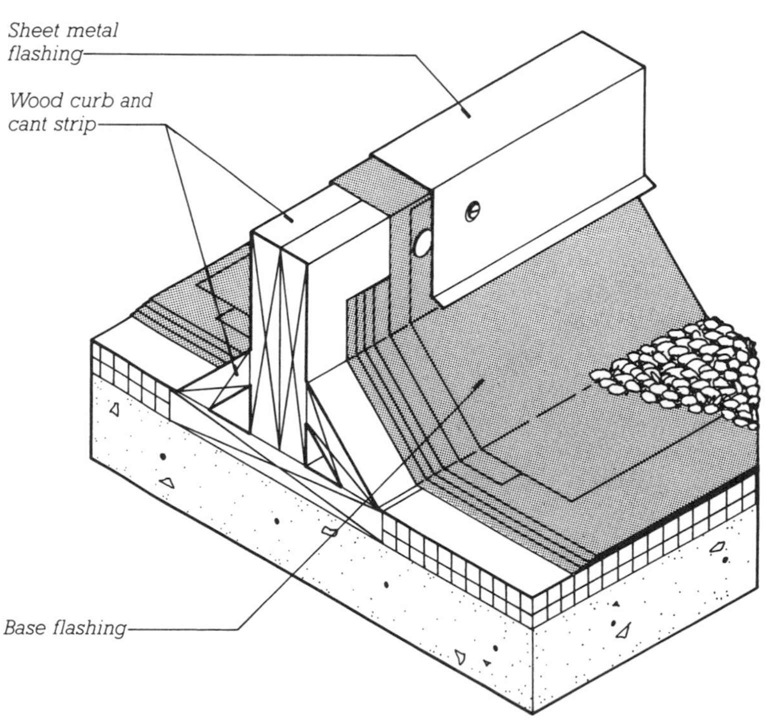

FIGURE 16.28
An area divider is designed to allow for some movement only in the membrane itself, not in the entire structure. It is used to subdivide a very large membrane to allow for thermal movement.

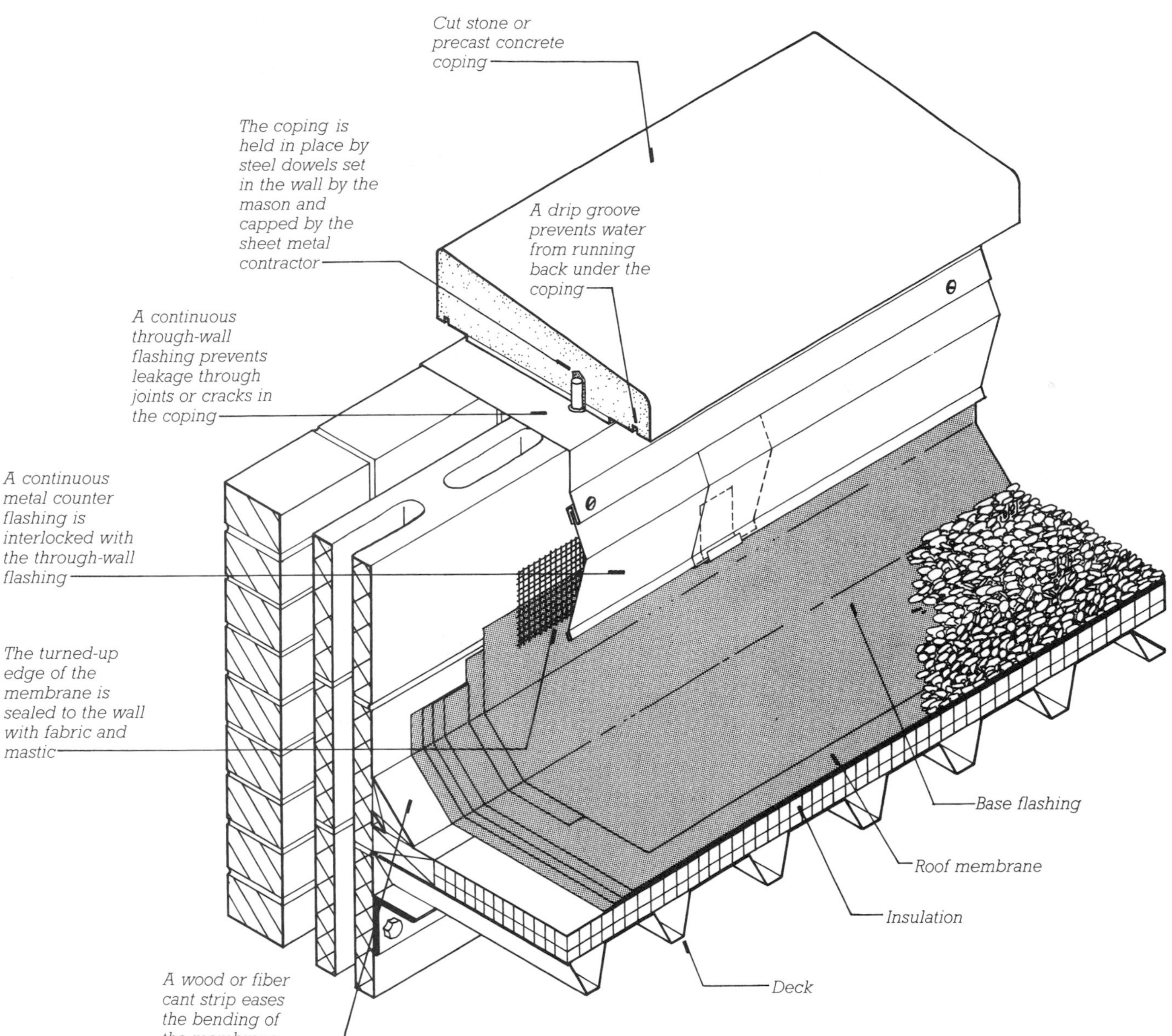
Cut stone or precast concrete coping
The coping is held in place by steel dowels set in the wall by the mason and capped by the sheet metal contractor
A drip groove prevents water from running back under the coping
A continuous through-wall flashing prevents leakage through joints or cracks in the coping
A continuous metal counter flashing is interlocked with the through-wall flashing
The turned-up edge of the membrane is sealed to the wall with fabric and mastic
Base flashing
Roof membrane
Insulation
Deck
A wood or fiber cant strip eases the bending of the membrane

FIGURE 16.29
A *parapet*—a low wall that projects above the roof edge—with conventional *counterflashing* and *coping*. *Cant* strips, as seen in this and many other illustrated details in this chapter where the roof membrane turns up onto a vertical surface, are commonly used with bituminous roofing, in which the less flexible membranes and multiple plies cannot easily make a sharp right-angle bend. Single-ply membranes typically do not require cant strips under such conditions.

FIGURE 16.30
A proprietary parapet coping system. The perforated metal channel is fastened to the masonry with a mastic adhesive. Sections of metal coping are snapped over the channel, with a special pan element beneath the joints to drain leakage. *(Product of W. P. Hickman Company, Asheville, North Carolina)*

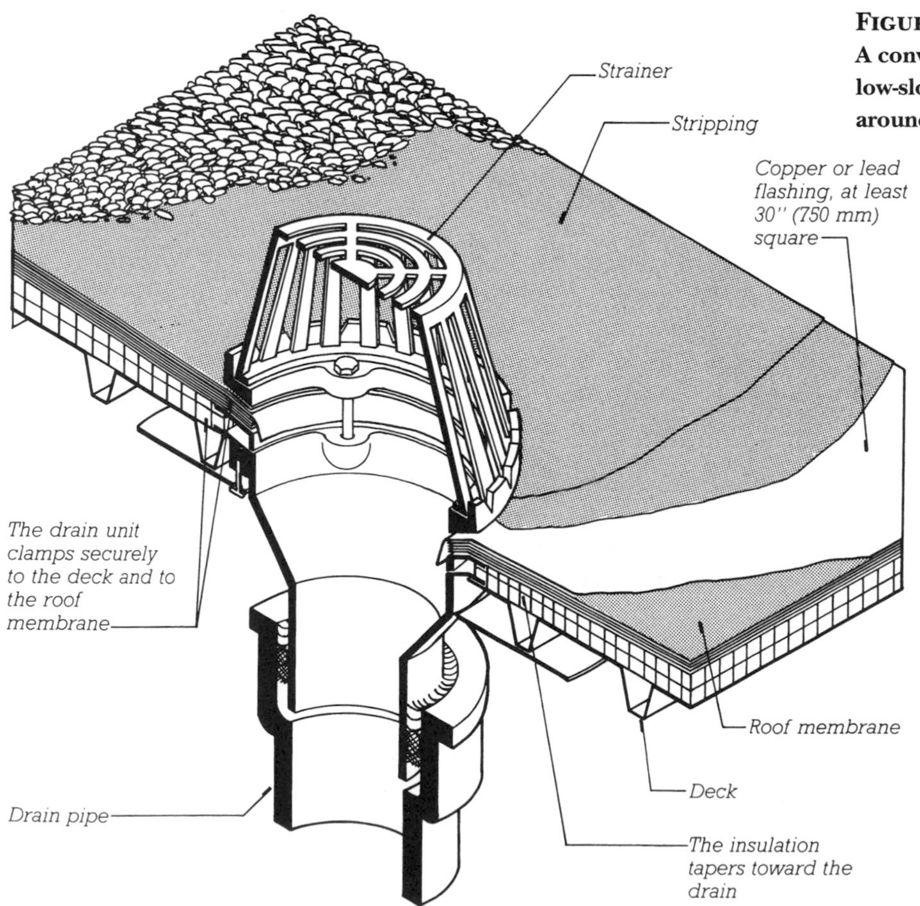

Strainer
Stripping
Copper or lead flashing, at least 30" (750 mm) square
The drain unit clamps securely to the deck and to the roof membrane
Roof membrane
Deck
Drain pipe
The insulation tapers toward the drain

FIGURE 16.31
A conventional cast iron interior *roof drain* for a low-slope roof. Two plies of felt stripping seal around the sheet metal flashing.

FIGURE 16.32
A proprietary single-piece roof drain made of molded plastic. *(Product of W. P. Hickman Company, Asheville, North Carolina)*

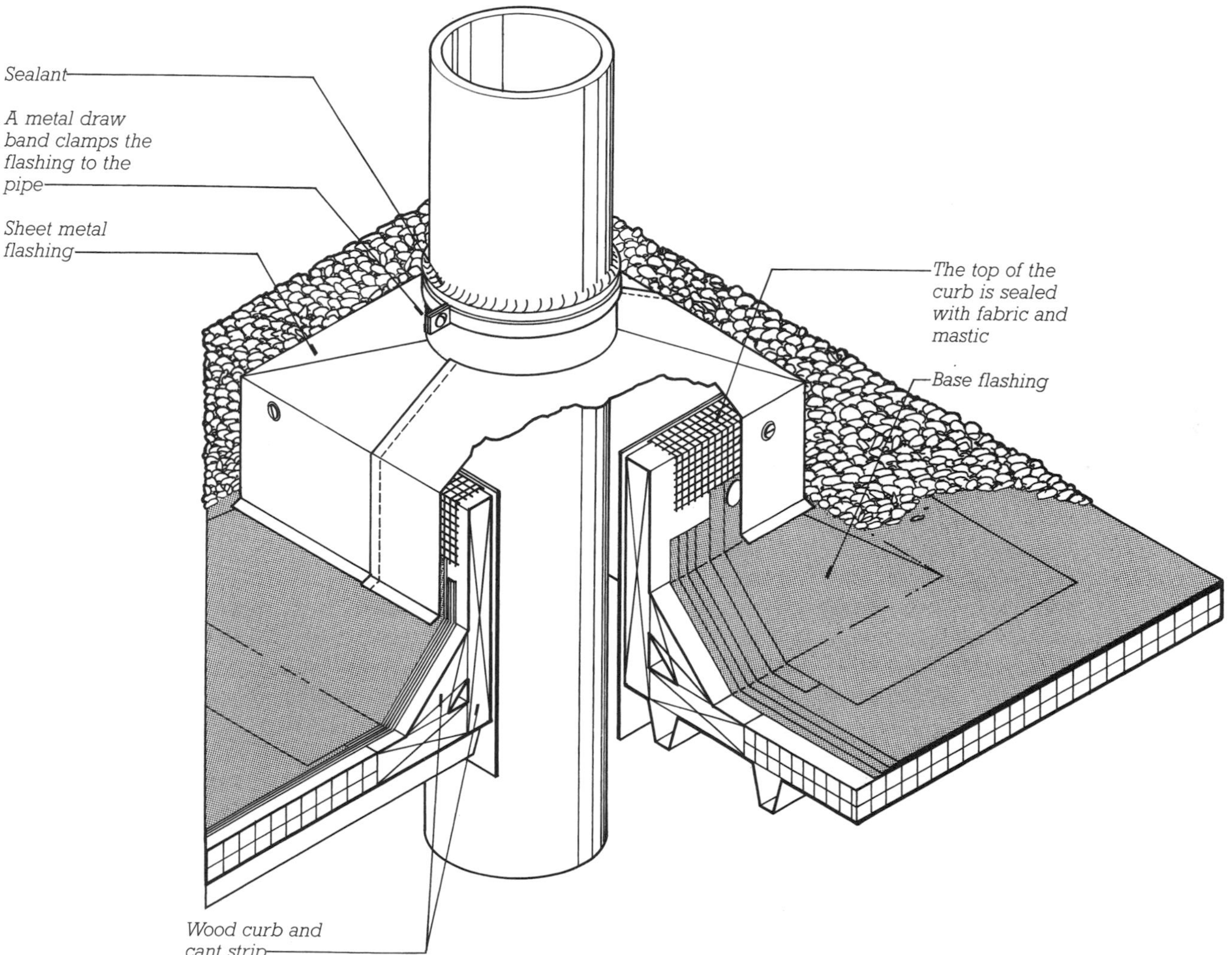
Sealant
A metal draw band clamps the flashing to the pipe
Sheet metal flashing
The top of the curb is sealed with fabric and mastic
Base flashing
Wood curb and cant strip

FIGURE 16.33
A *roof penetration* for a plumbing vent stack. Notice how this and all the previous edge and penetration details for a flat roof use the curb, cant strip, and stripping to keep standing water away from the edge of the membrane.

Structural Panel Metal Roofing for Low-Slope Roofs

Manufacturers of prefabricated metal building systems have developed proprietary systems of metal roofing panels that can be used as low-slope roofs at pitches as shallow as ¼:12 (1 in 48, or about 2 percent). These systems can be applied not only to prefabricated metal buildings but to buildings constructed of other materials as well (Figures 16.34–16.36). They are called *structural panel metal roofing* because the folded shape of the metal roofing gives it sufficient stiffness to support itself and normal snow loads between purlins without the need for a structural deck beneath. This name also distinguishes it from architectural sheet metal roofing, the traditional forms of metal roofing that are not self-supporting. *Architectural sheet metal roofing* is utilized largely on steep roofs and is described later in this chapter.

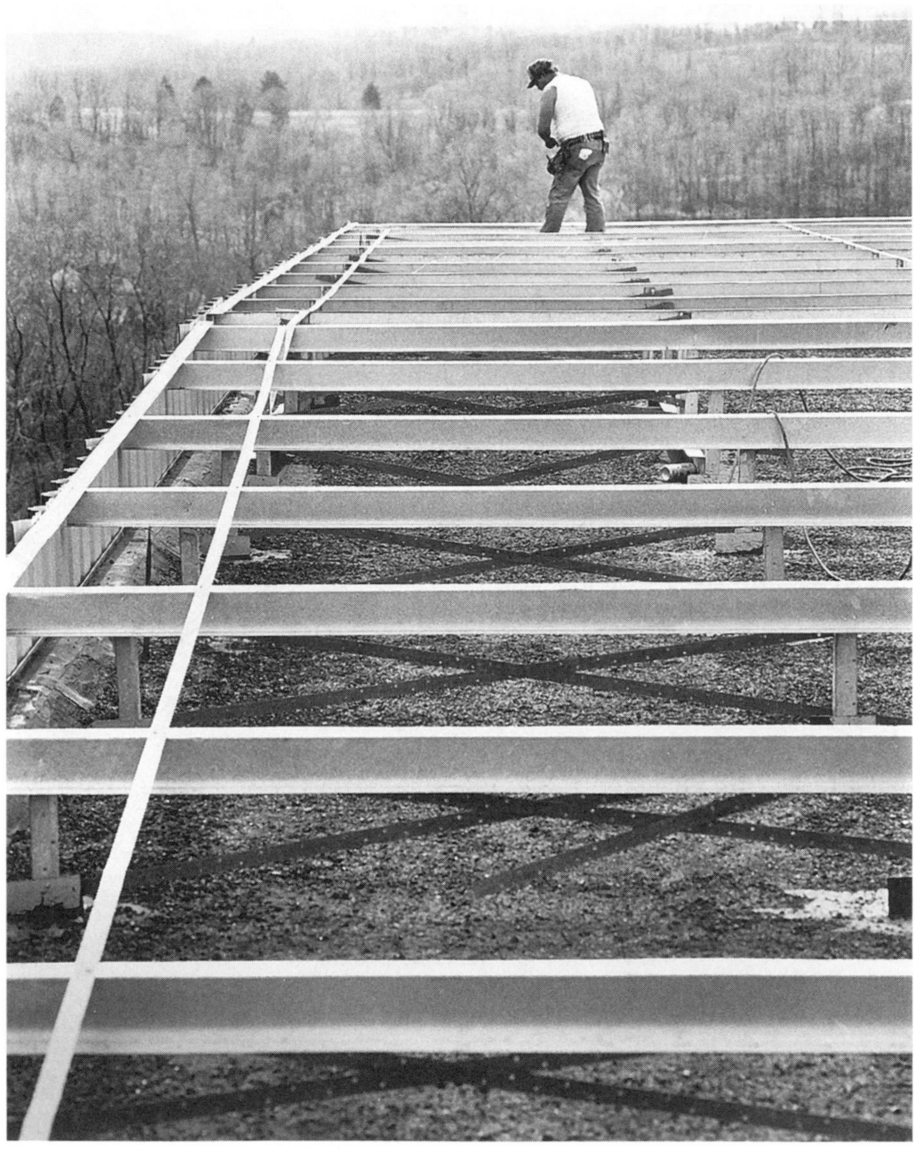

FIGURE 16.34
As a first step in reroofing a building with a structural metal roof, steel Z-purlins are erected over the old roof on tubular metal posts. *(Courtesy of Metal Building Manufacturers Association)*

Figure 16.35
A proprietary metal clip is used to fasten the metal roofing sheets to the Z-purlins while allowing for thermal movement in the sheets. The plastic foam collar avoids thermal bridging. *(Courtesy of Metal Building Manufacturers Association)*

Figure 16.36
The completed structural metal roof has a slope of only ¼:12 (1:48). The numerous penetrations for plumbing vents, air vents, and ductwork are typical of low-slope roofs. *(Courtesy of Metal Building Manufacturers Association)*

Steep Roofs

Roofs with a pitch of 2:12 (17 percent) or greater are referred to as "steep" roofs. Roof coverings for steep roofs fall into three general categories: thatch, shingles, and architectural sheet metal. *Thatch*, an attractive and effective roofing, consisting of bundles of reeds, grasses, or leaves (Figures 16.1 and 16.2), is highly labor-intensive and is rarely used today. Shingle and sheet metal roofs of many types are common to every type of building and range in price from the most economical roof coverings to the most expensive.

The insulation and vapor retarder in most steep roofs are installed below the roof sheathing or deck. Typical details of this practice are shown in Chapters 6 and 7. Where the underside of the deck is to be left exposed as a finish surface, a vapor retarder (if needed) and rigid insulation boards are applied above the deck, just below the roofing. A layer of plywood or OSB is then nailed over the insulation boards as a *nail base* for fastening the shingles or sheet metal, or special composite insulation boards with an integral nail base layer can be used.

In cold climates, steep roofs may have a tendency to form ice dams at the eaves under wintertime conditions. Where the risk of such ice damming is high, building codes require installation of a rubberized underlayment or other ice barrier material along the eaves to prevent trapped water from entering the building, as described on pages 227–228.

Shingles

The word *shingle* is used here in a generic sense to include wood shingles and shakes, asphalt shingles, slates, clay tiles, and concrete tiles. What these materials have in common is that they are applied to the roof in small units and in overlapping layers with staggered vertical joints.

Wood shingles are thin, tapered slabs of wood sawn from short pieces of tree trunk with the grain of the wood running approximately parallel to the face of the shingle (Figures 16.37 and 16.58). *Wood shakes* are split from the wood, rather than sawn, and exhibit a much rougher face texture than wood shingles (Figures 16.38 and 16.39). Most wood shingles and shakes in North America are made from Red

Figure 16.37
Applying Red cedar shingles, in this example as reroofing over asphalt shingles. Small corrosion-resistant nails are driven near each edge at the midheight of the shingle. Each succeeding course covers the joints and nails in the course below. *(Courtesy of Red Cedar Shingle and Handsplit Shake Bureau)*

FIGURE 16.38
Application of handsplit Red cedar shakes over an existing roof of asphalt shingles. Compare the shape, thickness, and surface texture of these shakes with those of the cedar shingles in the preceding illustration. Each course is interleaved with strips of heavy asphalt-saturated felt 18 inches (460 mm) wide as extra security against the passage of wind and water between the highly irregular and therefore loosely mated shakes. The nail stripper hung around the roofer's neck speeds his work by holding the nails and aligning them with points down, ready for driving. *(Courtesy of Red Cedar Shingle and Handsplit Shake Bureau)*

FIGURE 16.39
Shake application over a new roof deck using air-driven heavy-duty staplers for greater speed. The strips of asphalt-saturated felt have all been placed in advance with their lower edges unfastened. Each course of shakes is laid out, slipped up under its felt strip, then quickly fastened by roofers walking across the roof and inserting staples as fast as they can pull the trigger. *(Courtesy of Senco Products, Inc.)*

cedar, White cedar, or Redwood because of the natural decay resistance of these woods. Wood shingles and shakes are frequently recommended to be installed over spaced sheathing or a *breather mat* (a wiry plastic mat that creates a continuous airspace) to permit airflow under the shingles and prevent the accumulation of moisture and the accelerated deterioration of the shingles or shakes. Wood roof coverings are moderately expensive and are not highly resistant to fire unless the shakes or shingles have been pressure treated with fire-retardant chemicals. They eventually fail from erosion of the wood fibers and may be expected to last 15 to 25 years under average conditions.

Asphalt or *composition shingles* are die-cut from heavy sheets of asphalt-impregnated felt faced with mineral granules that act as a wearing layer and decorative finish. Most felts are based on glass fibers, but some still retain the older cellulose composition. The most common type of asphalt shingle, which covers probably 90 percent of the single-family houses in North America, is 12 inches by 36 inches (305 mm by 914 mm) in size. (A metric-size shingle 337 mm by 1000 mm is also widely marketed.) In the most popular pattern, each shingle is slotted twice to produce a roof that looks as though it were made of smaller shingles (Figures 16.40–16.43). Other sizes and many other styles are available, including thicker shingles that are laminated from several layers of material. Asphalt shingles are inexpensive to buy, quick to install, moderately fire resistant, and have a life expectancy of 15 to 25 years, depending on their composition.

The same sheet material from which asphalt shingles are cut is also manufactured in rolls 3 feet (914 mm) wide as asphalt *roll roofing*. Roll roofing is very inexpensive and is used primarily on storage and agricultural buildings. Its chief drawbacks are that thermal expansion of the roofing or shrinkage of the wood deck can cause unsightly ridges to

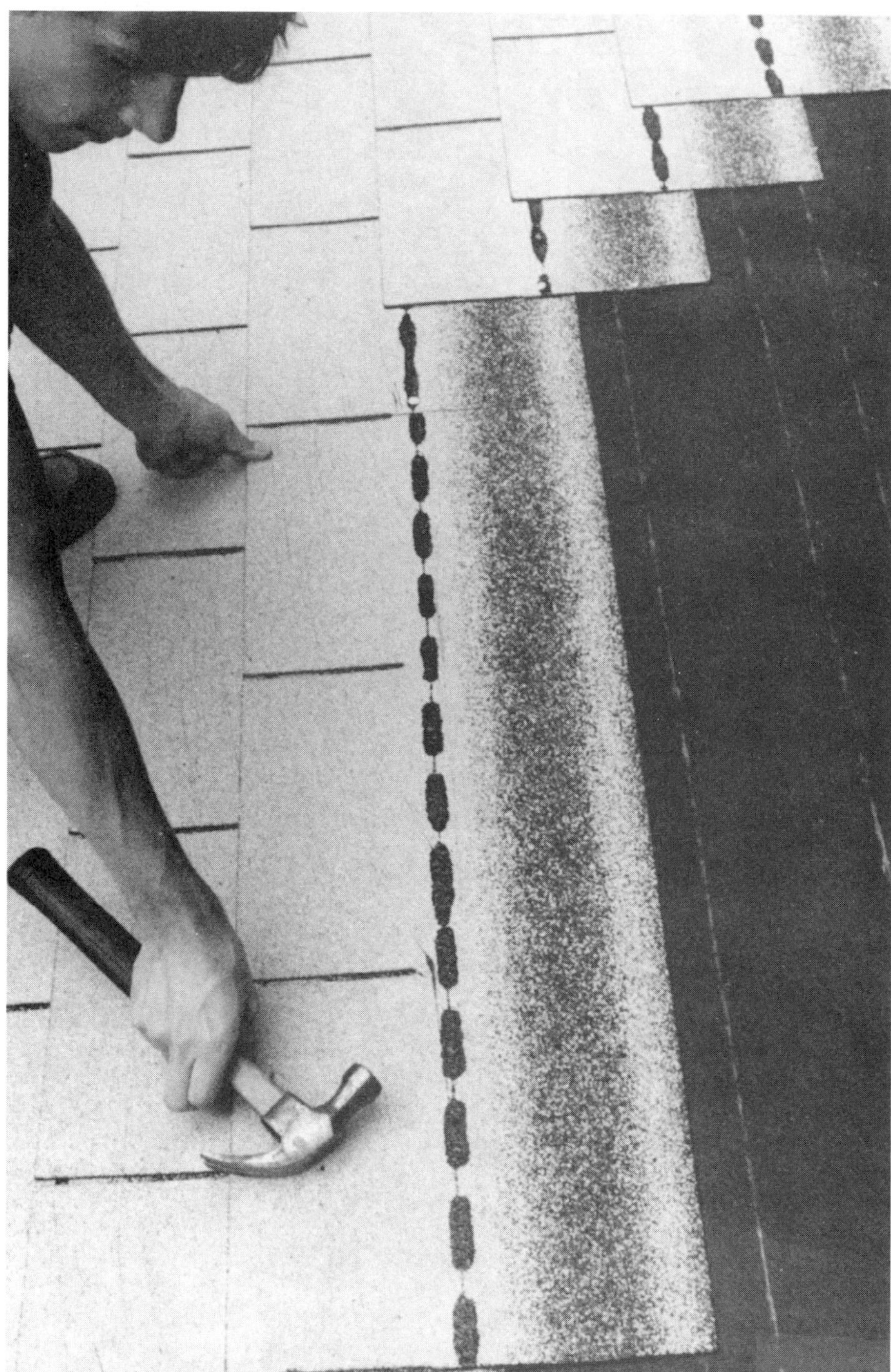

FIGURE 16.40
Installing asphalt shingles. To give a finer visual scale to the roof, the slots make each shingle appear to be three smaller shingles when the roof is finished. Many different patterns of asphalt shingles are available, including ones that do not have slots. *(Photo by Edward Allen)*

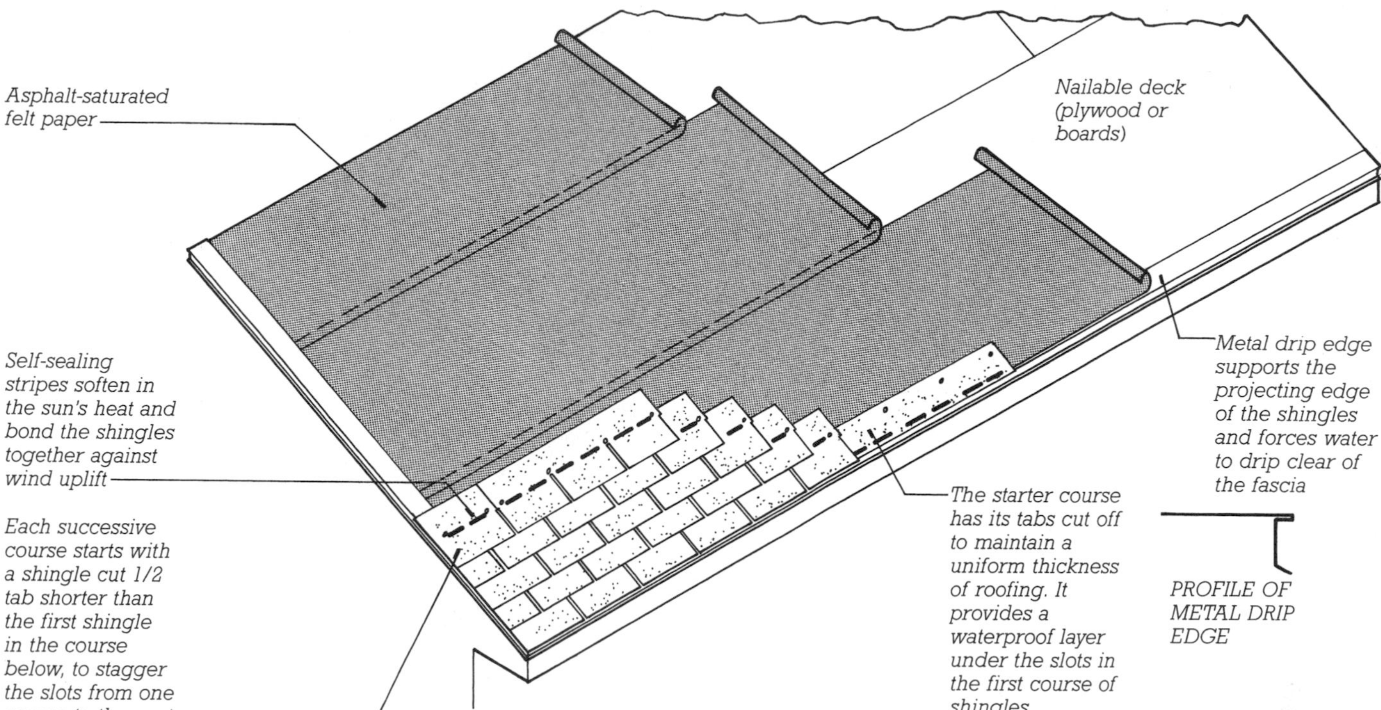

Asphalt-saturated felt paper
Nailable deck (plywood or boards)
Self-sealing stripes soften in the sun's heat and bond the shingles together against wind uplift
Metal drip edge supports the projecting edge of the shingles and forces water to drip clear of the fascia
Each successive course starts with a shingle cut 1/2 tab shorter than the first shingle in the course below, to stagger the slots from one course to the next
The starter course has its tabs cut off to maintain a uniform thickness of roofing. It provides a waterproof layer under the slots in the first course of shingles
PROFILE OF METAL DRIP EDGE

FIGURE 16.41
Starting an asphalt shingle roof. As explained in Chapter 6, building codes require the installation of an ice barrier beneath the shingles along the eave in regions with cold winters prone to ice damming. Where required, the ice barrier would replace the lowest course of asphalt-saturated felt paper shown in this illustration.

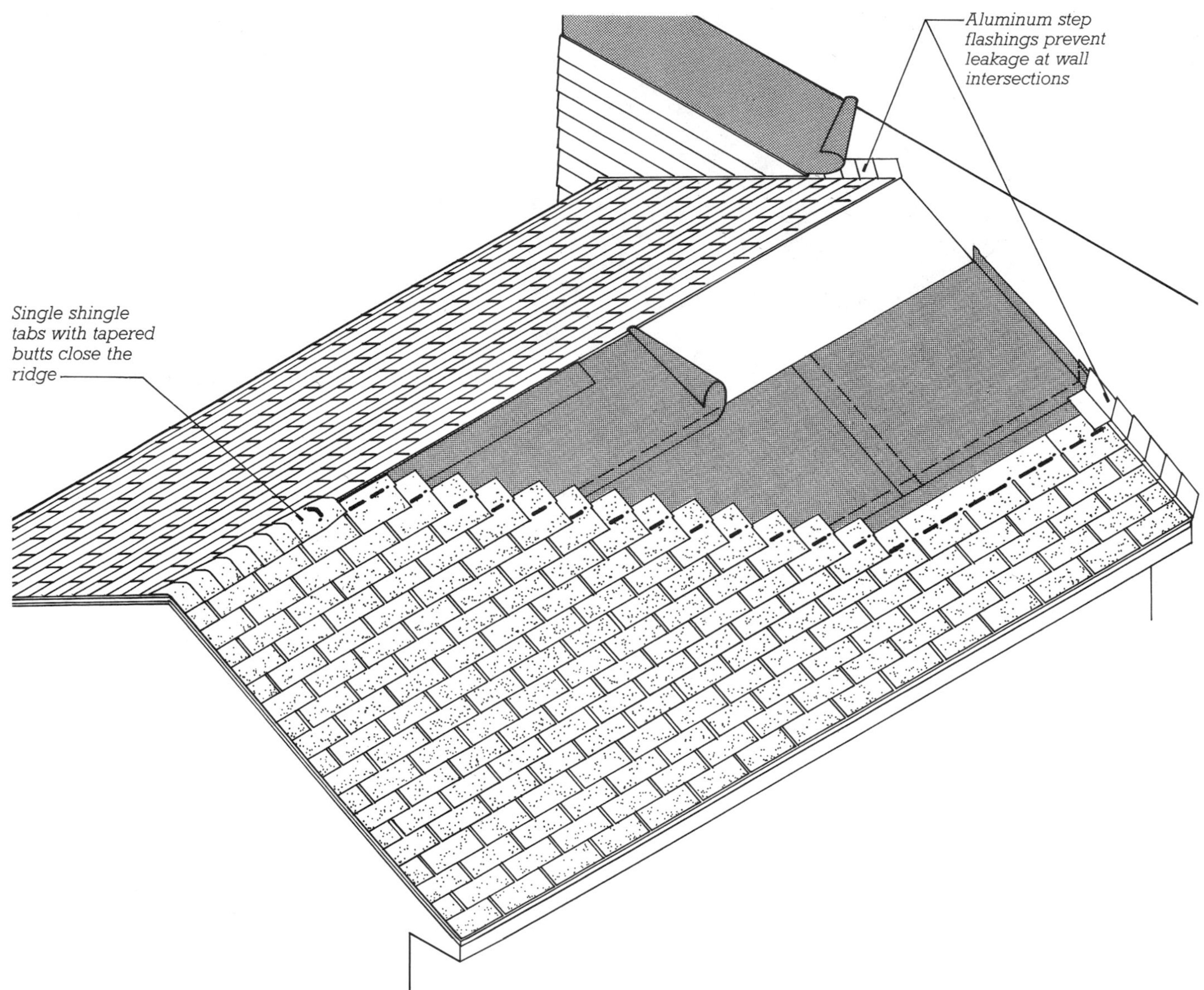
Aluminum step flashings prevent leakage at wall intersections
Single shingle tabs with tapered butts close the ridge

FIGURE 16.42
Completing an asphalt shingle roof. A metal or plastic ridge vent strip (see Chapter 6) is often substituted for the shingle tabs on the ridge to provide an outlet for ventilation under the roof sheathing.

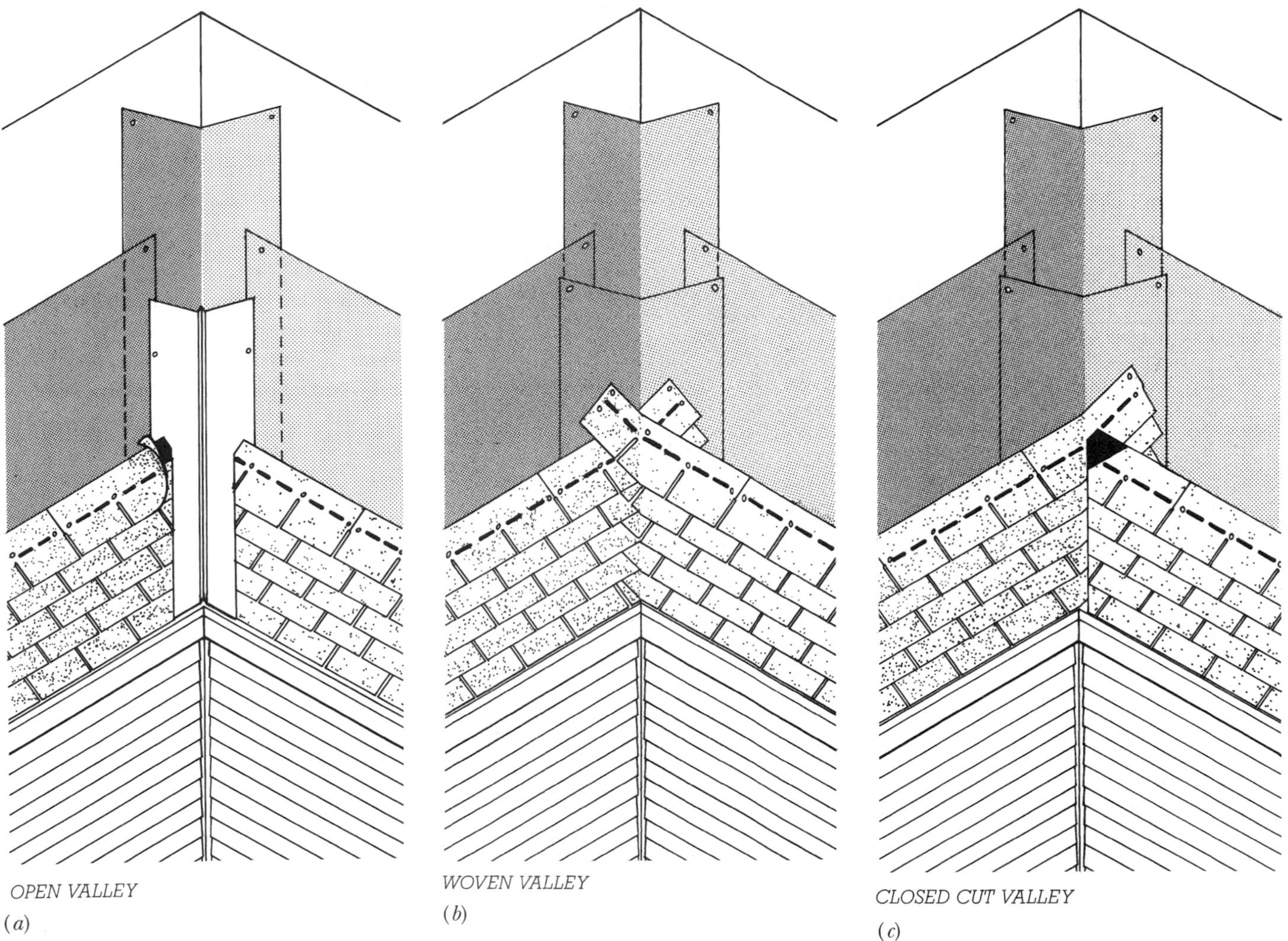

FIGURE 16.43
A *valley* is formed in a roof where two sloping roof planes meet above an inside corner of the building. Three alternative methods of making a valley in an asphalt shingle roof are shown here. *(a)* The *open valley* uses a sheet metal flashing; the ridge in the middle of the flashing helps prevent water that is coming off one slope from washing up under the shingles on the opposite slope. The *woven valley (b)* and *cut valley (c)* are favorites of roofing contractors because they require no sheet metal. The solid black areas on shingles in the open and cut valleys indicate areas to which asphaltic roofing cement is applied to adhere shingles to each other.

form in the roofing and that thermal contraction can tear it.

Slate shingles for roofing are delivered to the site split, trimmed to size, and punched or drilled for nailing (Figures 16.44 and 16.45). They form a fire-resistant, long-lasting roof that is suitable for buildings of the highest quality. Its first cost is high, but a slate roof typically lasts 60 to 80 years.

Clay tiles have been used on roofs for thousands of years. It is said that the tapered barrel tiles traditional to the Mediterranean region (similar to the mission tiles in Figure 16.46) were originally formed on the thighs of the tilemakers. Many other patterns of clay tiles are now available, both glazed and unglazed. *Concrete tiles* are generally less expensive than those of clay and are available in some of the same pat-

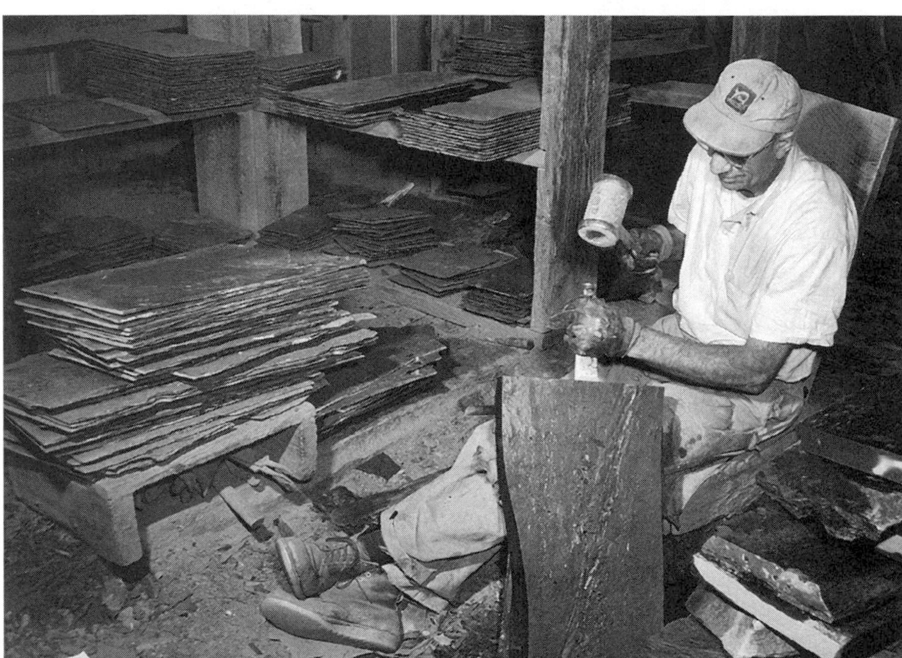

FIGURE 16.44
Splitting slate for roofing. The thin slates in the background will next be trimmed square and to dimension, after which nail holes will be punched in them. *(Photo by Flournoy. Courtesy of Buckingham-Virginia Slate Corporation)*

FIGURE 16.45
A slate roof during installation. *(Courtesy of Buckingham-Virginia Slate Corporation)*

terns. Tile roofs in general are heavy, durable, highly resistant to fire, and relatively expensive in first cost. Expected lifetimes range from 30 to 75 years, depending on climate and the resistance of the tiles to water absorption.

Other materials used for roof shingles include sheet metal, rubber, fiber-reinforced cement, and plastic. Each type of shingle must be laid on a roof deck that slopes sufficiently to ensure leakproof performance. Minimum slopes for each material are specified by the manufacturer and the building codes. For example, the minimum slope for a standard asphalt shingle roof is usually 4:12 (33 percent), although with specially protective underlayments, slopes as low as 2:12 (17 percent) may be acceptable in some circumstances. For any shingle type, slopes greater than the minimum should be used in locations where water is likely to be driven up the roof surface by heavy storm winds.

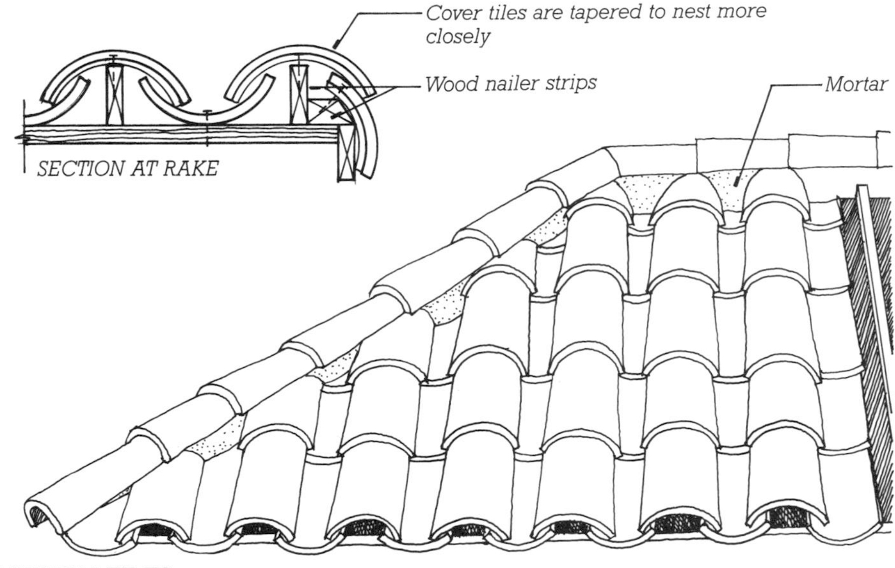

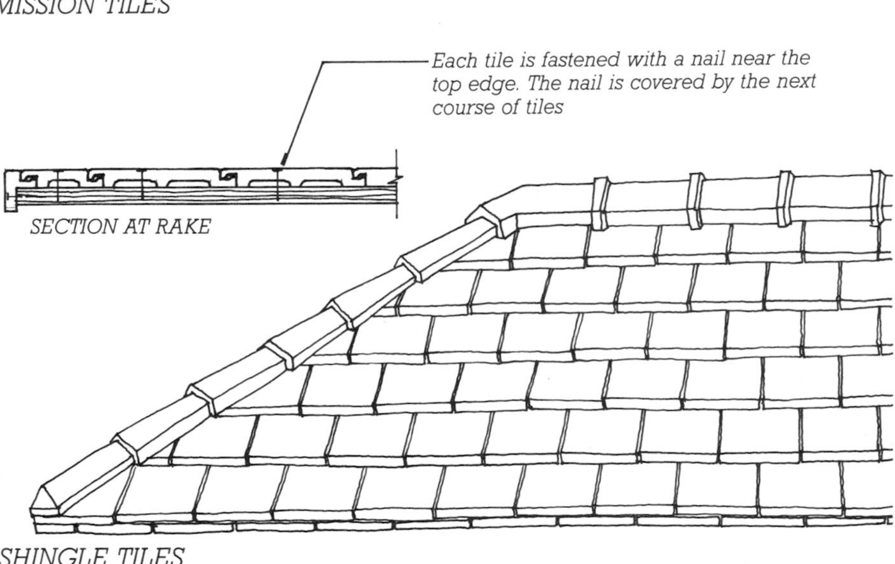

Figure 16.46
Two styles of clay tile roofs. The mission tile has ancient origins.

Figure 16.47
Protective devices keep workers from falling long distances off steep roofs.
(Courtesy of DBI/SALA, Red Wing, Minnesota)

Architectural Sheet Metal Roofing

Thin sheets of metal have been used for roofing since ancient times and remain a popular roofing material to this day. They are installed using ingenious systems of joining and fastening to maintain watertightness (Figures 16.48–16.52). Seams between sheets must be spaced close enough to secure the sheets adequately against wind uplift and to absorb expansion and contraction between sheets due to changes in temperature. These seams also create strong visual patterns that can be manipulated by the designer to emphasize the qualities of the roof shape. Architectural sheet metal roofing is relatively high in first cost but, when properly installed, can be expected to last for many decades.

Various types of metal may be used in the production of architectural sheet metal roofs:

- *Lead* is a soft, easily formed, very long-lasting metal that oxidizes over time to a dull white color. It is also a toxic material that requires special health precautions during its handling.
- *Copper* is a relatively soft metal that turns a beautiful blue-green in clean air and a dignified black in an industrial atmosphere; various chemical treatments can be used to obtain and preserve the desired color.
- *Lead-coated copper* sheet is sometimes used to combine the greater strength of copper with the gray-white color of lead and to avoid the staining of wall materials by oxides of copper that can occur with rainwater runoff from uncoated copper roofs.
- *Zinc* metal roofing is a long-lasting roofing metal made of zinc alloyed with small amounts of copper and titanium to improve its workability. It normally ages to a dark gray color and can also be treated in various ways to alter or preserve its appearance.
- *Stainless steel* and *titanium* are strong and long-lasting but less easily worked than other roofing metals. Both are silvery-white in color.

FIGURE 16.48
Standing-seam copper roofs. *(Designer: Emil Hanslin. Courtesy of Copper Development Association, Inc.)*

FIGURE 16.49
An automatic roll seamer, moving under its own power, locks standing seams in a copper roof. A cleat is just visible at the lower right. *(Courtesy of Copper Development Association, Inc.)*

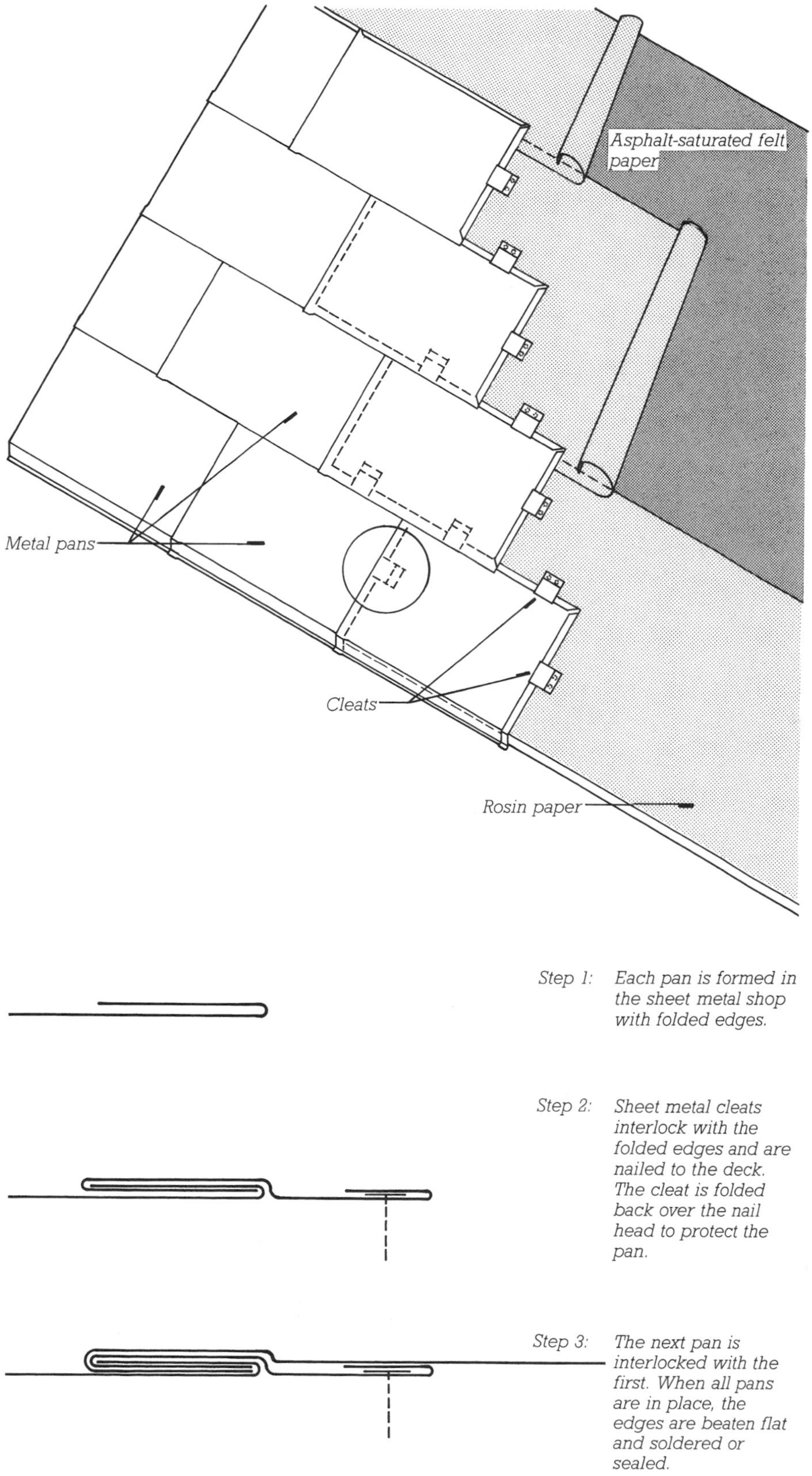

Asphalt-saturated felt paper

Metal pans

Cleats

Rosin paper

Step 1: *Each pan is formed in the sheet metal shop with folded edges.*

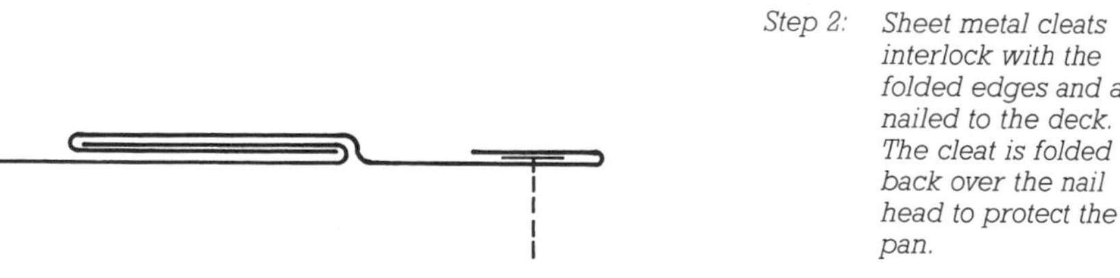

Step 2: *Sheet metal cleats interlock with the folded edges and are nailed to the deck. The cleat is folded back over the nail head to protect the pan.*

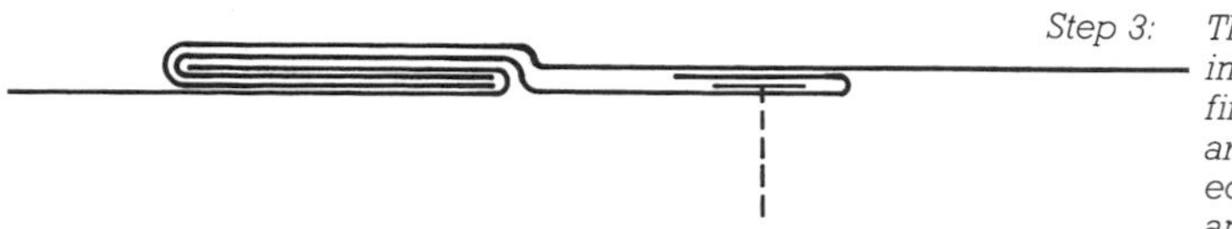

Step 3: *The next pan is interlocked with the first. When all pans are in place, the edges are beaten flat and soldered or sealed.*

FIGURE 16.50
Installing a *flat-seam* metal roof. The three diagrams at the bottom of the illustration show the three steps in creating the seam, viewed in cross section. The cleats, which fasten the roofing to the deck, are completely concealed when the roof is finished.

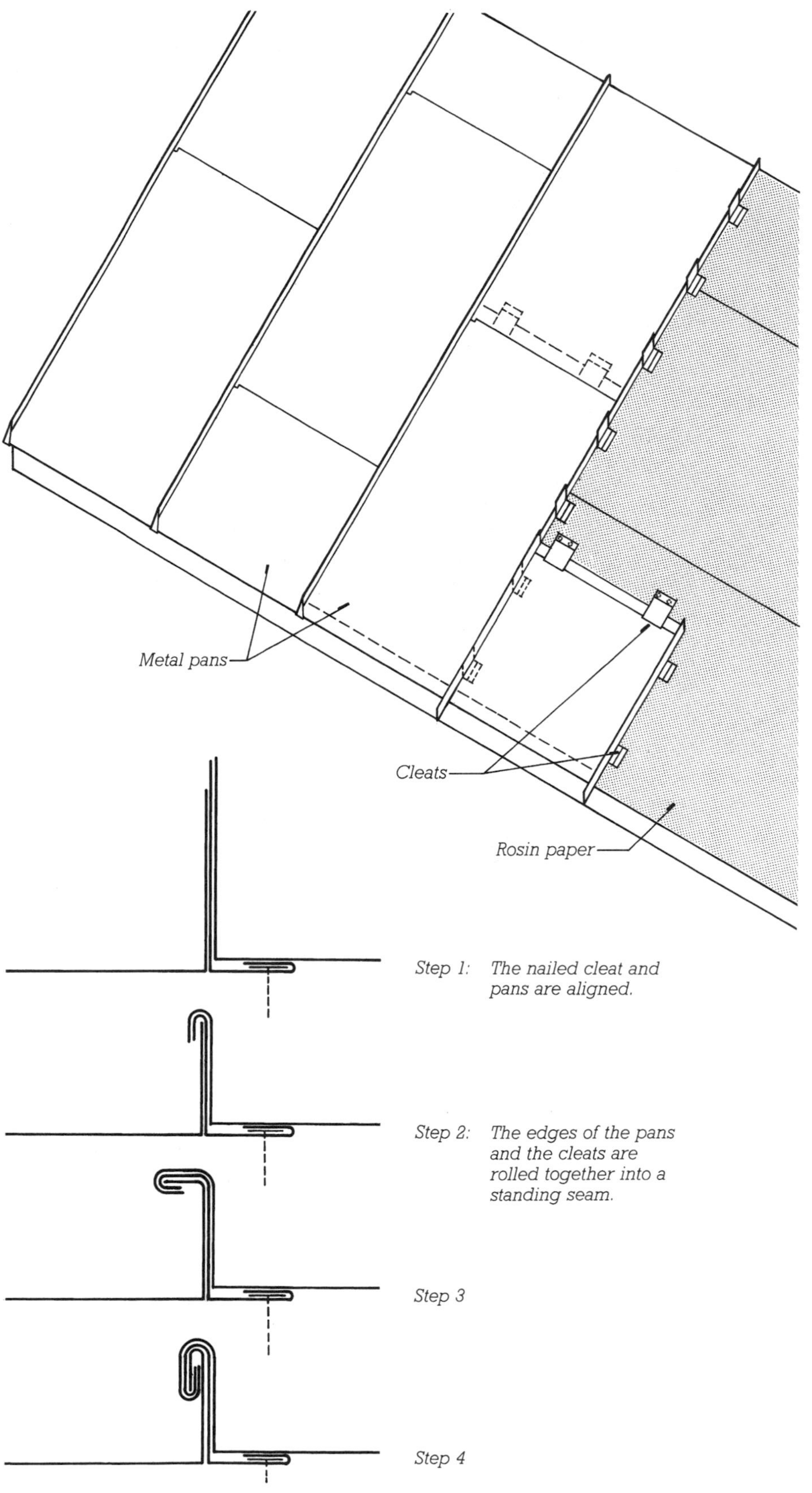
Metal pans
Cleats
Rosin paper
Step 1: The nailed cleat and pans are aligned.
Step 2: The edges of the pans and the cleats are rolled together into a standing seam.
Step 3
Step 4

FIGURE 16.51
Installing an architectural *standing-seam* metal roof.

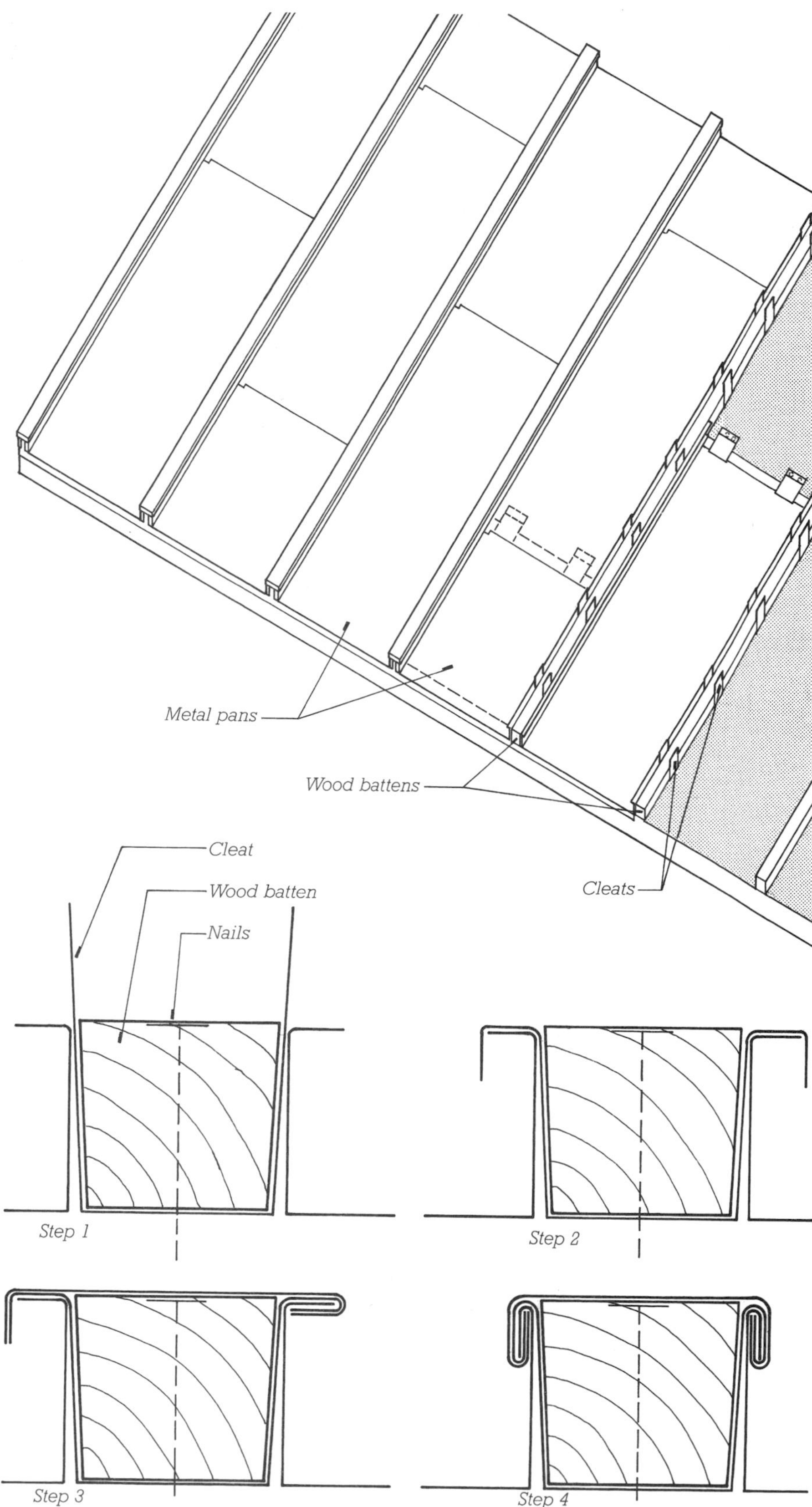

FIGURE 16.52
Installing a *batten-seam* metal roof. The battens are tapered in cross section to allow for expansion of the roofing metal.

• *Zinc-tin alloy coated stainless steel* has a darker, duller appearance than uncoated stainless steel. This material is very similar in appearance to—and is sometimes confused with—a lead-tin alloy coated metal called *terne-coated stainless steel* that is no longer manufactured but may still be found on older buildings.

The metals listed above all form self-protecting oxide coatings (patinas) that provide long-lasting resistance to corrosion. They are usually installed uncoated and allowed to patinate naturally. Other, less expensive metals, which are not as long-lasting in an uncoated condition, are commonly factory coated with high-performance organic (paint-like) coatings that extend their life expectancy and provide a wide range of color choice. These include *aluminum, metallic-coated steel* (steel coated with alloys of zinc or zinc-aluminum), or *zinc-tin alloy coated steel,* and even, occasionally, *ferrous steel* without any protective, metallic coating. (For more about architectural metals, see pages 505–508.)

The thickness of steel sheet traditionally has been specified by *gauge* (also spelled *gage*), a system of whole numbers in which lower numbers correspond to greater metal thickness. However, due to the absence of a uniform standard for the translation between gauge and actual metal thickness, ASTM standards discourage the use of gauge numbers to specify sheet metal thickness and instead recommend reliance on decimal or fractional inches to indicate thickness. The thickness of copper sheet is normally specified by weight, expressed in ounces per square foot (0.092 m^2). Aluminum is specified simply in decimal inches. In general, thicker metal sheets are longer-lasting, less prone to waviness or "oil canning," often more difficult to form into shapes, and more expensive than thinner

		Nominal Thickness	
Steel Sheet	**Gauge**	**Metallic-Coated Steel**	**Stainless Steel**
	18	0.052 in. *1.32 mm*	0.050 in. *1.27 mm*
	20	0.040 in. *1.02 mm*	0.038 in. *0.95 mm*
	22	0.034 in. *0.86 mm*	0.031 in. *0.79 mm*
	24	0.028 in. *0.71 mm*	0.025 in. *0.64 mm*
	25	0.025 in. *0.64 mm*	0.022 in. *0.56 mm*
	26	0.022 in. *0.56 mm*	0.019 in. *0.48 mm*
Copper Sheet	**Weight per square foot (0.092 m^2)**		
	24 oz.	0.032 in. *0.82 mm*	
	20 oz.	0.027 in. *0.68 mm*	
	16 oz.	0.022 in. *0.55 mm*	
Aluminum Sheet			
		0.060 in *1.52 mm*	
		0.050 in. *1.27 mm*	
		0.040 in. *1.02 mm*	
		0.032 in. *0.81 mm*	

Figure 16.53
Thicknesses of common sheet metals used in architectural sheet metal roofing.

sheets. Figure 16.53 lists typical thicknesses for some common sheet metal roofing materials.

Metal roofing may be fabricated and supplied in two distinct ways. When a roofing installer purchases unformed sheet metal stock and custom-forms the panels to the required shapes, the roofing is specified as *sheet metal roofing*. Alternatively, roofing panels may be factory-formed into a family of shapes that can be selected from a manufacturer's catalog. In this case, the roofing is specified as *metal roof panels* (Figure 16.54). Metal roof panels are most commonly made from aluminum or metallic-coated steel with factory-applied coatings. They may rely on interlocking seams with concealed fastener systems that imitate the appearance of traditional site-fabricated standing seam or batten seam sheet metal roofing, or they may consist of simpler corrugated or folded shapes fastened with exposed screws and rubber washers. Metal roof panels are generally less expensive than traditional, custom-formed sheet metal roofing. They may also be used in low-slope roofing applications as discussed earlier this chapter.

The roof plays a primal role in our lives. The most primitive buildings are nothing but a roof. If the roof is hidden, if its presence cannot be felt around the building, or if it cannot be used, then people will lack a fundamental sense of shelter.

Christopher Alexander et al., *A Pattern Language*, New York, Oxford University Press, 1977, p. 570.

Minimum recommended slopes for sheet metal roofing vary with the type of seam, the manner in which the seams are sealed, and the type of roofing underlayment. For premanufactured metal roof panels, consult the manufacturer's recommendations. For custom-fabricated sheet metal roofing, consult the appropriate references listed at the end of this chapter.

Protection from Corrosion Between Dissimilar Metals

The ideal way to avoid the corrosion that can occur between dissimilar metals is to use the same metal for every component of a sheet metal roof system, including its fasteners, anchor clips, roofing sheets, flashings, gutters, and downspouts. Alternatively, where this is not practical, metals must be mixed with an understanding of the reactions that can occur between them. For example, sheet metal roofs of copper, lead, or zinc may generally be safely anchored with fasteners and anchor clips made of stronger, harder stainless steel, because the stainless steel is electrochemically noble (passive) in relation to these other metals. For a more detailed discussion of galvanic action and corrosion between dissimilar metals, see pages 698–700.

FIGURE 16.54
The raised seams in the architectural panel metal roof accentuate the complex roof geometries of the International Center in Brattleboro, Vermont.
(Architect: William A. Hall Partnership. Photo by Stanley Jesudowich)

Considerations of Sustainability in Roofing

The roof can contribute in many ways to the sustainability of a building:

- A roof can capture rainwater and snowmelt and conduct them to a cistern, tank, or pond for use as domestic water, industrial water, or irrigation.
- A properly proportioned overhang can shade south-facing windows from the high summer sun but admit warming light from the low winter sun.
- A light-colored roof covering, if kept clean, can reflect half or more of the solar radiation striking its surface, improving occupant comfort and significantly reducing the heating load on the occupied space below. Even darker-hued roof materials, when coated with specially formulated cool color pigments, can reflect 25 percent or more of solar radiation. Such cool roofs can reduce cooling energy costs for buildings by 10 to 25 percent and extend the life of the roofing materials.
- Reflective roofs that reduce the absorption of solar heat can decrease the elevation of air temperatures in densely built areas, thereby reducing a building's indirect contributions to smog, degraded air quality, environmental discomfort, and other heat island effects.
- In a hot climate, a shading layer above a roof, with a freely ventilated space between, can eliminate most solar heat gain through the roof surfaces. The shading layer might consist of latticework, fabric, or corrugated metal; the exact material is less important than providing both shading and ventilation.
- A roof surface can support flat-plate solar heat collectors used to reduce building heating costs or arrays of photovoltaic cells to provide electrical power. Electrical power for building use can also be produced from thin-film photovoltaic materials laminated directly onto a variety of conventional roof coverings.

Roof Insulation

Thermal insulating materials in roofs and walls are probably the most cost-effective, planet-saving materials used in buildings. They increase occupant comfort by moderating the radiant temperatures of ceilings and walls. They reduce heating and cooling energy requirements to a fraction of what they would otherwise be. They pay for themselves through energy savings in a very short period of time.

Other environmental implications of thermal insulating materials are more complex:

- Cellulose insulation is the most environmentally friendly thermal insulating material. It can be specified to contain not less than 85 percent recycled material, primarily newspapers. Its embodied energy is only about 150 BTU per pound (0.35 MJ/kg). The borate compound used to make it fire resistant is generally not harmful to human beings. The binder used to hold the cellulose fibers in place is usually a simple glue that does not outgas. Installers should wear breathing protection, but the fibers are not known to be carcinogenic.
- Glass wool and mineral wool can be manufactured primarily from waste materials, recycled glass and blast furnace slag, respectively. The embodied energy in glass wool is about 13,000 BTU per pound (30 MJ/kg), but a pound of it goes a long way. The binders used in some glass fiber insulations give off small quantities of formaldehyde, but nonoutgassing products are available. In a large number of scientific studies, glass and mineral wools have not been shown to be carcinogenic. However, they do irritate and congest the lungs, and installers must wear breathing protection.
- Polystyrene foam is made from petroleum and has an embodied energy of about 50,000 BTU per pound (120 MJ/kg). It can be made from recycled styrene and polystyrene. Expanded polystyrene foam (bead foam) is made using pentane gas, which is not an ozone-destroying substance, as a blowing agent.
- Extruded polystyrene, polyisocyanurate, and polyurethane foams were formerly manufactured using ozone-depleting blowing agents. All are now manufactured using pentane or other chemicals less harmful to the atmosphere.

Low-Slope Membranes

Low-slope roof membranes have varying impacts on the environment:

- Bituminous roofing is largely based on asphaltic compounds derived from coal and petroleum.
- Roofing operations with hot asphalt and pitch give off plentiful quantities of fumes that are decidedly unpleasant and potentially unhealthful, but once the roof has cooled, these emissions subside.
- Most roofing felts today are made with cellulose or glass fibers, but very old felts in buildings being demolished or reroofed may contain asbestos, a proven carcinogen.
- The various rubber and plastic formulations used in single-ply membranes generally utilize petroleum as the primary raw material. Each has its own characteristics with regard to embodied energy and outgassing.
- Adhesive bonding, solvent welding, and heat welding of seams may give off volatile organic compounds (VOCs). Some single-ply membrane materials also emit VOCs.
- Materials from demolition of built-up roof membranes are generally incinerated or taken to landfills. Thermoplastic single-ply membranes can be recycled, although most are not at the present time. Thermosetting membranes cannot be recycled.

• Low-slope roofs may be developed as green roofs, in which the membrane is covered with soil and plants. These living constituents of the roof consume carbon dioxide, generate oxygen, shield the membrane and the inhabited space below from temperature extremes, cool the roof surface by evaporation and transpiration, delay the passage of stormwater into sewers and reduce its volume, and create a pleasant roofscape.

Steep Roof Materials

Steep roof materials—shingles, slates, tiles, sheet metal—come from a variety of sources with different environmental impacts:

• Asphalt shingles consist mostly of petroleum.

• The environmental impact of wood shingles is much the same as that of other wood products; see pages 90 and 91.

• Slates are quarried stone, and concrete tiles are, of course, concrete (pages 520 and 521).

• Copper, stainless steel, aluminum, terne, and lead roofing materials originate as ores that must be mined, refined, and manufactured into finished form, all at various environmental costs.

• Metal roofing materials can be recycled. There is little recycling of the other steep roof materials. There are available, however, proprietary shingles made almost entirely of recycled tires or other recycled materials.

Sustainable Roofing

Cool Roofs

Roofs are exposed to solar radiation daily, and as that radiation is absorbed and converted to heat, the temperature of the roof covering rises. Depending on the intensity of the radiation and the portion of it retained by the covering, roof surfaces may routinely reach temperatures of 150 degrees Fahrenheit (65°C) or higher. High roof temperatures can lead to overheating of interior spaces, reduced comfort for building occupants, increased building energy consumption, the need for larger, more expensive cooling equipment, shortened lifespan of roofing materials, and an increased contribution to urban heat island effects through elevation of the surrounding air temperature. Selecting a *cool roof* covering that minimizes such heating can significantly reduce these effects.

Solar heating of roofs is principally affected by two properties of the roofing material. A material's *solar reflectance*, or *albedo*, is a measure of its tendency to reflect solar radiation rather than absorb it. Solar reflectance is measured on a unitless scale from 0 to 1, where 1 represents a material that reflects all solar radiation and 0 represents one that absorbs all solar radiation. A higher solar reflectance corresponds to a cooler roof. *Thermal emittance* is a measure of a material's capacity to radiate infrared heat energy and cool itself as its temperature rises. Like solar reflectance, thermal emittance is measured on a scale of 0 to 1 and a higher thermal emittance implies a cooler roof.

Cool roof criteria differ among energy conservation standards and green building programs. Requirements for the U.S. Environmental Protection Agency's (EPA) Energy Star program are based solely on a roof covering's solar reflectance, measured both when the covering is new and after it has weathered. Requirements for the U.S. Green Building Council's LEED for New Buildings are based on a roof covering's *solar reflective index (SRI)*. SRI is a measure of solar heating potential, derived according to ASTM E1980, that accounts for a material's reflective and emittive properties, as well as for its ability to lose heat through thermal conductance to the surrounding air. Two roofing materials with the same SRI are expected to achieve the same surface temperature under comparable exposures. Higher SRI values correspond to cooler roof coverings, with an SRI value of 0 corresponding to a standard reference black surface and a value of 100 corresponding to a standard reference white surface (Figure 16.55).

Comparative solar heating properties for common roofing materials are listed in Figure 16.56. Product-specific data can be obtained from the manufacturer's product literature or from the Cool Roof Rating Council, an independent organization that maintains a roof material rating program and publishes the properties of tested products (see the list of web sites at the end of this chapter). In comparison to traditional dark-colored EPDM or bituminous membranes, highly reflective cool membranes on low-slope roofs can reduce roof surface temperatures by as much as 50 to 75 degrees Fahrenheit (25–40°C) and cut building cooling costs by an estimated 15 to 25 percent. Cool roofing materials on steep roofs have the potential to save an estimated 5 to 10 percent of building cooling costs.

	EPA Energy Star	USGBC LEED[a]
Low-Slope Roofs	Minimum solar reflectance: 0.65 unaged 0.50 aged 3 years	Minimum SRI: 78
Steep Roofs	Minimum solar reflectance: 0.25 unaged 0.15 aged 3 years	Minimum SRI: 29

[a]LEED NC 2.2, Sustainable Sites Credit 7.2 Heat Island Effect: Roof.

FIGURE 16.55
Cool roof requirements for low-slope and steep roofs. For EPA Energy Star programs, aged values are based on testing of material samples taken from installed, weathered roofs or from samples that have undergone accelerated aging to simulate natural weathering.

	Solar Reflectance	Thermal Emissivity	Solar Reflective Index (SRI)	Roof Surface Temperature Rise
Low-Slope Roofing				
White thermoplastic	0.85	0.90	105	10°F (5°C)
Bituminous membrane with reflective elastomeric coating or reflective film cap sheet	0.80	0.90	100	15°F (8°C)
White EPDM	0.75	0.90	95	20°F (11°C)
Bituminous membrane with white-coated granules	0.65	0.90	80	30°F (16°C)
Bituminous membrane with light granules	0.35	0.90	35	60°F (30°C)
Bituminous membrane with dark granules	0.25	0.90	25	70°F (39°C)
Black EPDM	0.10	0.85	5	87°F (48°C)
Steep Roofing				
Metal panel with cool white coating	0.70	0.85	85	25°F (14°C)
Aluminum, uncoated	0.60	0.05	40	56°F (31°C)
Red clay tile	0.35	0.90	35	60°F (33°C)
Metal panel with cool color coating	0.30	0.85	30	66°F (37°C)
Asphalt shingle with cool color granulated surface	0.25	0.90	25	70°F (39°C)
Unpainted concrete tile	0.25	0.90	25	70°F (39°C)
Gray asphalt shingle	0.22	0.90	20	73°F (41°C)
Black asphalt shingle	0.05	0.90	0	90°F (50°C)

FIGURE 16.56
Solar heating properties for some example roof materials. Materials in shaded rows do not meet the criteria for cool roofs listed in the previous figure. (Figures are approximate and should not be used to document compliance with cool roof rating programs.)

FIGURE 16.57
A green roof over a sloping roof deck. This extensive planted roof has a 6-inch (150-mm)-deep soil layer and supports native and ornamental of grasses. *(Courtesy of Steve Grim)*

Cool Color Roof Coverings

Cool color roofing materials are nonwhite in color but nevertheless reflect a significant portion of the sun's radiation. Cool colors are formulated with pigments that are selectively reflective to different portions of the solar spectrum. They are highly reflective of *near-infrared (NIR) radiation,* an invisible component of solar radiation that accounts for over half of the total heat energy radiated by the sun, while they remain absorptive in the visible light spectrum, which accounts for their apparent color. Cool color pigments can be applied to aggregate granules used to coat asphalt shingles, as well as to sheet metal, clay or concrete tile, fiber-cement shingles, and other roofing materials to produce products meeting cool roof standards for steep roofs. As the formulation of cool color pigments continues to evolve, smooth-surfaced roofing materials with reflectance values as high as 0.45 and granule-surfaced materials with values as high as 0.30 are anticipated.

Green Roofs

Green roofs, also called *eco roofs* or *vegetated roofs,* are roofing systems covered with vegetation and additional materials needed to support plant growth. Like protected membrane roofs, green roofs extend the life of the roof membrane by shielding it from UV radiation and extremes of temperature. Green roofs also reduce heating and cooling costs by moderating temperature swings in the roof assembly. They reduce the transmission of noise through the roof system and decrease the reflection of exterior noise. They reduce stormwater runoff and provide habitat for birds. By supporting plant growth and reducing heat island effects, green roofs improve air quality. They provide aesthetic value and, in some cases, create pleasant, usable space.

Extensive green roofs are relatively shallow, with soil depths of 2 to 6 inches (25–150 mm). They are planted with herbs, grasses, mosses, sedums, or other drought-tolerant plants that do not require irrigation or frequent maintenance. *Intensive green roofs* may have soils as deep as 30 inches (750 mm) and are designed to support a broader variety of plant types and shrubs. Intensive roofs typically require irrigation and regular maintenance such as weeding, trimming, pest management, and fertilization. Planning for the structural loads of soil and plant materials is an important part of green roof design. Because of their lesser depth, extensive roofs are relatively light in weight, imposing loads, when saturated with water, ranging from 12 to 35 psf (0.57 to 1.7 kPa) on the supporting roof structure. Intensive roofs impose loads of 50 psf (2.4 kPa) or more. While most green roofs are essentially flat, with special soil retention measures, extensive roofs with slopes as great as 12:12 (100 percent) are technically feasible (Figure 16.57).

From the top down, typical components of a green roof system include the following:

- **Plant materials** may be selected on the basis of hardiness, climate, depth of soil, maintenance expectations, and appearance.
- The **growth medium** (soil) must provide long-lasting, optimal growing conditions. Particulars of its formulation vary with the depth of soil and the types of vegetation supported. Soil stability, drainage properties, and, in extensive roofs, drought resistance are important considerations.
- A geotextile **filter fabric** prevents soil particles from being washed out of the growth medium and clogging the drainage layer below.
- **Soil restraints** made from perforated plastic or metal allow the free flow of water while confining the growth medium at the roof perimeter and around drains or scuppers.
- **Drainage layer** materials such as a molded plastic panel or an entangled plastic filament mat are used to provide efficient drainage and aeration beneath the soil. Some products also provide **water retention**, benefiting the plants' subsoil environment. Crushed rock or other aggregate material may also be used, although at the cost of added weight in comparison to the use of synthetic materials.
- Rigid foam **insulation** boards may be positioned above or below the roof membrane. Where insulation is placed above the membrane, extruded polystyrene is used because it retains its insulating value when wet.
- Depending on the membrane system, one or more **protection layers** to prevent root invasion and to relieve physical stress may be laid over the membrane.
- Since access to the **waterproofing membrane** becomes difficult once it is covered with the green roof components, it should be chosen with consideration of its long-term performance in a buried, continuously damp environment. Especially for intensive green roofs, robust waterproofing systems are recommended in place of lighter-duty, conventional roof membranes. Though many membrane materials may be used, hot rubberized asphalt, PVC, and multi-ply modified bitumen have the longest track records of successful installation. Where modular green roof systems are used (see below) and the membrane will remain more easily accessible, less expensive conventional roof membrane types may also be suitable. Prior to being covered, green roof membranes should be *flood tested*, that is, tested for watertightness by placing the membrane under continuously submerged conditions for a period of hours or days, to check for leaks prior to placing the overlying components.
- Requirements for the **vapor retarder** and **air barrier** are no different for green roofs than for conventional low-slope roof assemblies.
- The **roof deck** and **supporting structure** must be engineered to carry the additional loads imposed by green roof components.

With *modular green roof systems*, all the components of the green roof system above the membrane are preassembled in easily transported trays or modules. These trays, typically 2 to 4 feet (600–1200 mm) in plan dimension and 2 to 8 inches (50–200 mm) in depth, are preplanted and arrive on the construction site ready to be placed directly over the roof membrane. Modular green roof systems are lightweight, easy to specify, easy to assemble on site, and easy to remove or adjust at a later date.

Photovoltaic Roofing

Photovoltaic (PV) roofing materials are metal panels or shingles of asphalt, slate, fiber-cement, or metal laminated with thin-film semiconductor materials and capable of converting solar radiation into electrical energy. When they are installed as part of a *building-integrated photovoltaic (BIPV)* system, the electricity produced by these materials can be used to power a building's own equipment and systems or, where permitted by the public power utility, can be sold back to the utility to offset power consumption charges. Electrical power from BIPV systems is produced without pollution or the depletion of natural resources.

The essential components of a BIPV system are an array of PV modules (the PV roofing material), controller equipment to regulate the current generated by the modules, and an inverter to convert the direct current (DC) output of the modules to alternating current (AC) compatible with conventional building power. *Stand-alone* or *off-grid BIPV* systems—systems not connected to the public utility power grid—also require a power storage system, usually in the form of rechargeable batteries, and often also include a backup power system, such as diesel generators.

Estimating the potential power contribution of a BIPV system and evaluating its potential economic payback for a particular project require an analysis of the project's geographic location, solar exposure, projected power needs, utility costs, and additional factors. See the list of web sites at the end of this chapter for more information.

Roofing and the Building Codes

Manufacturing standards, minimum slopes, permitted underlayment materials, and installation requirements for roofing materials are specified by building codes. The codes also regulate a roof's required level of resistance to flame spread and fire penetration, tested according to standards ASTM E108 or UL 790 and rated as *Class A, B, or C roof coverings* (listed here in decreasing order of resistance). The International Building Code requires that roofs on buildings of Construction Types I, II-A, III-A, IV, or V-A meet at least Class B requirements and those on buildings of Construction Types II-B, III-B, and V-B meet at least Class C requirements (see Figure 1.3 and related text for an explanation of construction types). Roofs for single-family homes and other small residential or utility buildings generally may be nonclassified, except that where portions of such roofs are located close to property lines, a minimum Class C rating is required. Property insurance policies or local building regulations such as may apply in dense urban areas or in areas prone to wildfire may also impose roof class rating requirements.

Roof class rating tests apply to whole roof assemblies, including the membrane, shingles or other covering, underlayments, insulation, decking, and ballast, if any. Where a rated roof is required, the manufacturer of the roof covering should be consulted to determine the precise materials and construction requirements for meeting the classification. Broadly speaking, most low-slope roof membranes and noncombustible tiles (such as concrete or clay) can meet Class A requirements, as can some metal roof coverings and asphalt shingles made of glass felts; asphalt shingles made of organic felts and other metal roof coverings can meet Class B requirements; and fire-retardant wood shingles and shakes can meet Class C requirements.

Figure 16.58
A house is both roofed and sided with Red cedar shingles to feature its sculptural qualities. *(Architect: William Isley. Photo by Paul Harper. Courtesy of Red Cedar Shingle and Handsplit Shake Bureau)*

Building Enclosure Essentials: Dissimilar Metals and the Galvanic Series

When different metals come into contact in the presence of moisture or some other conducting medium, electric current flows between them. Such metal pairs, or *galvanic couples*, form the basis, for example, for the storage of electrical energy in batteries. In building construction, galvanic couples are important for the manner in which the exchange of electrons affects the corrosion rates of the metals involved—a phenomenon that, if applied properly, can be used to positive effect or, if ignored, can lead to the premature deterioration of critical building components.

The Galvanic Series

In any galvanic couple, the metal donating electrons, called the *anode*, experiences an accelerated rate of corrosion; conversely, the metal receiving electrons, called the *cathode*, experiences corrosion at a reduced rate. This effect is relatively great for pairs of metals with large differences between their electrochemical potentials and proportionally smaller for those with lesser potential difference.

To facilitate the prediction of the corrosion potential of metal pairs, metals may be listed in a *galvanic series*, that is, in order of their relative electrochemical potential, with the most anodic, or *active*, metals at one end of the list and the most cathodic, *noble* or *passive*, ones at the other. Metal pairs positioned far apart in the series have greater corrosion potential than pairs closer to each other. Any number of galvanic series can be compiled for a group of metals in different conducting mediums and under different environmental conditions. For building construction, a galvanic series based on metals immersed in flowing sea water is used (Figure A).

Applying the Galvanic Series to Problems in Building Construction

Corrosion in the Anode and the Cathode

In Figure A, any metal listed higher in the series behaves as the anode when paired with any other metal listed lower in the series, which behaves as the cathode. Furthermore, in any such galvanic couple, the anode experiences accelerated corrosion, while the cathode's rate of corrosion is reduced.

A common application of galvanic couples is in the use of anodic metals as protective coatings for other, more cathodic metals. For example, consider *galvanized* (zinc-coated) *steel* fasteners. Zinc appears higher in the galvanic series and is therefore the anode in this pair of metals. Even if the zinc coating is damaged and the steel itself becomes directly exposed to moisture and air, the steel can remain protected. As long as there is sufficient surrounding zinc to sustain galvanic action between the metals, only the zinc will corrode, in effect sacrificing itself to preserve the steel. For the same reason, zinc and zinc-aluminum alloys are frequently applied as protective metallic coatings to steel sheet used for roofing and flashing. These alloys form sacrificial, protective coatings for the underlying steel.

Environmental Conditions

Corrosion rates of metals are affected by the environment. Exterior metals exposed to marine, salt-laden air or to atmospheres laden with industrial pollutants will corrode at a faster rate than metals in less aggressive or more protected environments. The more severe the environment, the more important it is to avoid contact between dissimilar metals. Where dissimilar metals must be combined in such environments, metal pairs with the least electrochemical potential difference, that is, as close to each other as possible in the galvanic series, are preferred.

Relative Surface Area of Anode to Cathode

The corrosive effects within a galvanic couple are proportional to the relative surface areas of the metals involved. For example, consider copper roofing fastened with stainless steel fasteners. In this metal pair, copper is the anode and at risk of corrosion. However, the relative surface area of anode to cathode, that is, the surface area of the copper roofing, is very large in relation to the surface area of the stainless steel fasteners. This means that the electrochemical effects are diluted over a very large area of the copper and are therefore relatively weak. In fact, the use of stainless steel fasteners for the attachment of most types of metal roofing, including copper, is quite common. On the other hand, consider the use of galvanized steel fasteners for the attachment of the same copper roofing. In this case, the fasteners are the anode in the galvanic couple, and the ratio of surface areas of anode (fastener) to cathode (roof metal) is very low. This condition is likely to lead to greatly accelerated corrosion of the fasteners and is not recommended. As a general rule, when dissimilar metals are paired, the ratio of surface areas of anode to cathode should be kept as large as possible.

Cathodic Fasteners

As a corollary to the previous rule, where fasteners of one metal are used to attach metals of another type, the fasteners should be the cathodic metal in the pair. Since

Most Anodic or Active
Zinc
Aluminum
Cadmium
Steel, iron
Aluminum bronze
Naval brass, yellow brass, red brass
Tin
Copper
Lead-tin solder
Admiralty brass, aluminum brass
Manganese bronze, silicone bronze, tin bronze
Types 410 and 416 stainless steel (passive)
Nickel silver
High copper content nickel-copper alloys
Lead
Nickel-aluminum-bronze
Nickel
Silver
Monel, high nickel content nickel-copper alloys
Types 304 and 316 stainless steel (passive)
Titanium
Most Cathodic or Noble

FIGURE A

A galvanic series lists metals in order of electrochemical potential for a given conducting medium. The series shown here is for metals immersed in flowing sea water, adapted from ASTM G82. When paired as galvanic couples, metals positioned higher on the list will sacrificially corrode while protecting metals positioned lower on the list. The horizontal lines in the list separate metals into groups with relatively equal potential difference within each group. For example, although aluminum bronze and high copper content nickel-copper alloys are separated by eight other metals in this list, their electrochemical difference, and their relative corrosion potential as a galvanic couple, are roughly only as great as those between lead and titanium (separated by five other metals) or between cadmium and steel (which appear directly after each other).

Some metals, for example stainless steel, can appear in different locations within the series, depending on their electrochemical condition and the environment in which they are placed. In this series, stainless steels are listed in their so-called passive states, their expected state under conditions of normal architectural usage.

Building Enclosure Essentials: Dissimilar Metals and the Galvanic Series (continued)

fasteners are typically small in surface area in comparison to the materials they attach, this approach ensures a relatively large anode-to-cathode surface area ratio. In practice, stainless steel fasteners are normally considered suitable for fastening construction metals of virtually all other types. This is a safe practice since stainless steel is the most cathodic metal normally found on the building site.

Water Flowing from One Metal to Another

Where rainwater flows from one metal type to another, the first metal should be anodic in relation to the second. Consider again a copper roof, in this case draining into a stainless steel gutter. As rainwater washes off the roof metal, it will carry dissolved copper ions into the gutter. Since the stainless steel is cathodic in relation to the copper, the gutter metal itself will remain unharmed. Conversely, consider the same roof with a galvanized steel gutter. In this case, the galvanized steel gutter metal is anodic in relation to the dissolved copper washing through it, and it may experience accelerated corrosion.

Insulating Dissimilar Metals

For galvanic action to occur, electric current must flow between the dissimilar metals. Thus, by insulating dissimilar metals from each other, galvanic corrosion can be prevented. For example, a layer of felt paper or rubberized asphalt membrane may be used to separate dissimilar sheet metal materials used in roofing and flashing work.

Paints, aluminum anodizing, and other coatings can also frequently act as insulators between one metal and another. However, when using coatings to separate dissimilar metals, the metals to be coated should be chosen carefully. Where an insulating coating is applied only to the anodic metal in a metal pair, that metal is at risk of accelerated corrosion if the coating becomes damaged. This is because if the coating fails in only limited areas, the exposed area of anodic metal is relatively small, and the galvanic action becomes concentrated in these limited areas (in other words, the anode-to-cathode surface area ratio is very small). For this reason, where coatings are used to separate dissimilar metals, it is safer to coat only the cathodic metal in the pair or to coat both metals.

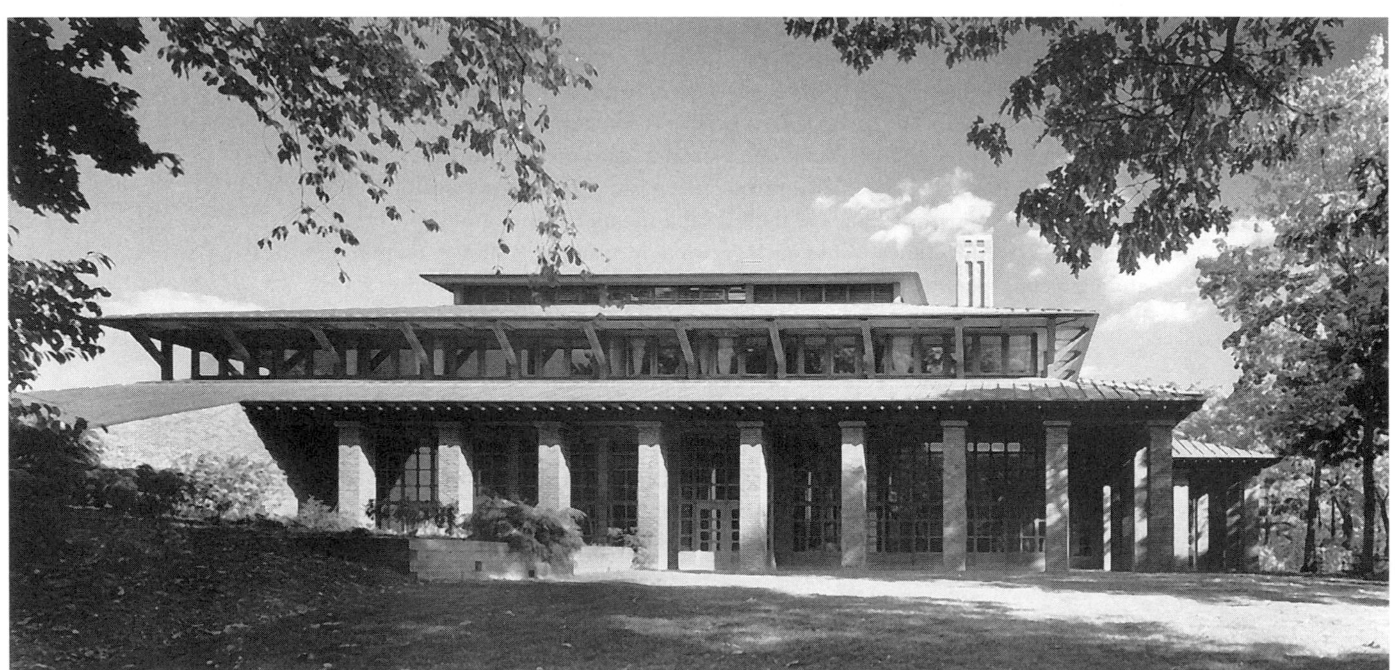

Figure 16.59
A standing-seam metal roof with beautifully detailed overhangs, designed by architects Kallmann and McKinnell.
(Photo by Steve Rosenthal)

CSI/CSC

MasterFormat Sections for Roofing

07 20 00	THERMAL PROTECTION
07 22 00	Roof and Deck Insulation
07 30 00	STEEP SLOPE ROOFING
07 31 00	Shingles and Shakes
	Asphalt Shingles
	Slate Shingles
	Wood Shingles and Shakes
07 32 00	Roof Tiles
	Clay Roof Tiles
	Concrete Roof Tiles
07 33 00	Natural Roof Coverings
	Sod Roofing
	Thatched Roofing
	Vegetated Roofing
07 40 00	ROOFING AND SIDING PANELS
07 41 00	Roof Panels
	Metal Roof Panels
07 50 00	MEMBRANE ROOFING
07 51 00	Built-Up Bituminous Roofing
07 52 00	Modified Bituminous Membrane Roofing
07 53 00	Elastomeric Membrane Roofing
	Ethylene-Propylene-Diene-Monomer Roofing
07 54 00	Thermoplastic Membrane Roofing
	Polyvinyl-Chloride Roofing
	Thermoplastic-Polyolefin Roofing
07 55 00	Protected Membrane Roofing
07 56 00	Fluid-Applied Roofing
07 57 00	Coated Foamed Roofing
	Sprayed Polyurethane Foam Roofing
07 58 00	Roll Roofing
07 60 00	FLASHING AND SHEET METAL
07 61 00	Sheet Metal Roofing
	Standing Seam Sheet Metal Roofing
	Batten Seam Sheet Metal Roofing
	Flat Seam Sheet Metal Roofing
	Tinplate and Ternplate Roofing
07 62 00	Sheet Metal Flashing and Trim
07 70 00	ROOF AND WALL SPECIALTIES AND ACCESSORIES
07 71 00	Roof Specialties
07 72 00	Roof Accessories
07 76 00	Roof Pavers

Selected References

1. National Roofing Contractors Association. *The NRCA Roofing Manual: Membrane Roof Systems, The NRCA Steep-Slope Roofing Manual, The NRCA Architectural Sheet Metal Manual, The NRCA Spray Polyurethane Foam-Based Roofing Manual,* and *The NRCA Green Roof Systems Manual.* Rosemont, IL.

These manuals, all updated regularly, are the most comprehensive guides to current U.S. practice for both low-slope and steep-slope roofing systems. The treatment is exhaustive, and both diagrams and text are excellent. These products are updated regularly and may also be available in combined editions.

2. Sheet Metal and Air Conditioning Contractors National Association. *Architectural Sheet Metal.* Chantilly, VA, updated regularly.

Architectural sheet metal roofs are copiously detailed in this excellent reference, along with every conceivable flashing, fascia, gravel stop, and gutter for flat and shingled roofs.

3. Zahner, L. William. *Architectural Metals: A Guide to Selection, Specification, and Performance* (3rd ed.). John Wiley & Sons, Hoboken, NJ, 1995.

This is one of the most comprehensive treatments of architectural metal types and their uses in construction available.

3. ASHRAE. *ASHRAE Handbook – Fundamentals.* American Society of Heating, Refrigerating, and Air-Conditioning Engineers, Atlanta, GA, updated regularly.

This handbook provides a definitive treatment of the physics of heat and moisture transfer through building assemblies and the application of these principals to building construction methods.

Web Sites

Roofing

Author's supplementary web site: **www.ianosbackfill.com/16_roofing**

Low-Slope Roofs

Asphalt Roofing Manufacturers Association: **www.asphaltroofing.org**
Carlisle SynTec: **www.carlisle-syntec.com**
Firestone Building Products: **www.firestonebpco.com**
National Roofing Contractors Association: **www.nrca.net**
Polyisocyanurate Insulation Manufacturers Association: **www.polyiso.org**
Whole Building Design Guide, Roofing Systems: **www.wbdg.org/design/env_roofing.php**

Steep Roofs

Asphalt Roofing Manufacturers Association: **www.asphaltroofing.org**
Cedar Shingle and Shake Bureau **www.cedarbureau.org**
Certainteed Roofing Products **www.certainteed.com**
Copper Development Association **www.copper.org**
GAF Roofing Products **www.gaf.com**
Metal Construction Association: **www.metalconstruction.org**
Sheet Metal and Air Conditioning Contractors National Association: **www.smacna.org**
Umicore Building Products (zinc roofing): **www.vmzinc-us.com**

Cool Roofs

Cool Metal Roofing Coalition: **www.coolmetalroofing.org**
Cool Roof Rating Council: **www.coolroofs.org**
Cool Roofing Materials Database: **eetd.lbl.gov/coolroofs**

Green Roofs

Green Roofs for Healthy Cities: **www.greenroofs.org**

Photovoltaic Roofs

Renewable Resource Data Center (RReDC): **rredc.nrel.gov**

Key Terms and Concepts

steep roof
low-slope roof
roof deck
thermal insulation
air barrier
vapor retarder
roof membrane
drainage
flashing
ponding
substrate board
area divider
topside vent
protected membrane roof (PMR)
ballast
thermal resistance (R, RSI)
water vapor
relative humidity
dew point
condensation
vapor pressure
vapor retarder, vapor barrier
vapor permeance
perm
vapor permeability
lightweight insulating concrete
bituminous roof membrane
built-up roof (BUR) membrane
cold-applied mastic, solvent-based asphalt
modified bitumen roof membrane
atactic polypropylene (APP)
styrene-butadiene-styrene (SBS)
torch-applied modified bitumen membrane
hot-mopped modified bitumen membrane
cold process applied modified bitumen membrane, cold-applied adhesive modified bitumen membrane
self-adhered modified bitumen membrane
cap sheet
hybrid membrane bituminous roof
single-ply roof membrane
thermoplastic material
thermosetting material, elastomeric
polyvinyl chloride (PVC)
thermoplastic polyolefin (TPO)
ethylene propylene diene monomer (EPDM)
fluid-applied roof membrane
ballast
base flashing
fascia
scupper
stripping
parapet
counterflashing
coping
cant
roof drain
roof penetration
traffic deck
structural panel metal roofing
architectural sheet metal roofing
thatch
nail base
shingle
wood shingle
wood shake
breather mat
valley
asphalt shingle, composition shingle
open valley
woven valley
cut valley
roll roofing
slate shingle
clay tile
concrete tile
standing seam
batten-seam
flat seam
lead
copper
lead-coated copper
zinc
stainless steel
titanium
zinc-tin alloy coated stainless steel
terne-coated stainless steel
aluminum
metallic-coated steel
zinc-tin alloy coated steel
ferrous steel
gauge, gage
sheet metal roofing
metal roofing panel
cool roof
solar reflectance, albedo
thermal emittance
solar reflective index (SRI)
cool color
near-infrared (NIR) radiation
green roof, eco roof, vegetated roof
extensive green roof
intensive green roof
flood test
modular green roof system
photovoltaic (PV) roofing
building-integrated photovoltaic (BIPV) system
stand-alone BIPV, off-grid BIPB
galvanic couple
anode
cathode
galvanic series
active metal
noble metal, passive metal
galvanized steel
Class A, B, C roof coverings

Review Questions

1. What are the major differences between a low-slope roof and a steep roof? What are the advantages and disadvantages of each type?

2. Discuss the three positions in which thermal insulation may be installed in a low-slope roof, and the advantages and disadvantages of each.

3. Explain in precise terms the function of a vapor retarder in an exterior wall or roof assembly.

4. Compare a bituminous roof membrane to a single-ply roof membrane.

5. What is the difference between cedar shingles and cedar shakes?

6. What metals are used for architectural sheet metal roofing? What are the strengths and drawbacks of each?

7. What are the benefits of a cool roof? What properties of a roofing material affect its solar heating and how?

8. List the major components of a green roof system and describe their functions.

9. Which of the following metal combinations are generally safe, and why?

a. Copper sheet metal roofing with stainless steel fasteners.

b. Copper sheet metal roofing with copper fasteners.

c. Copper sheet metal roofing with galvanized steel fasteners.

d. Galvanized steel sheet metal roofing with stainless steel fasteners.

e. Galvanized steel sheet metal roofing with galvanized steel fasteners.

Exercises

1. For a low-slope-roofed university classroom building with a masonry bearing wall, steel interior frame, corrugated steel roof deck, and parapet:

   a. Show two ways of achieving a 1:50 roof slope on structural bays 36 feet (11 m) square.

   b. Sketch a set of details of the parapet edge, building separation joint, area divider, and roof drain for a low-slope roof system of your choice. Show insulation, vapor retarder (if any), roof membrane, and flashings.

2. Sketch a fascia detail for a low-slope roof system of your choice, assuming that the wall below is made of precast concrete panels and the roof deck of precast concrete slab elements.

3. Find a low-slope roof system being installed and take notes on the process until the roof is completed. Ask questions of the roofers, the architect, or your instructor about anything you don't understand.

4. Examine a number of low-slope roofs around your campus or neighborhood, looking for problems such as cracking, blistering, tearing, and leaking. Attempt to explain the reasons for each problem that you discover.

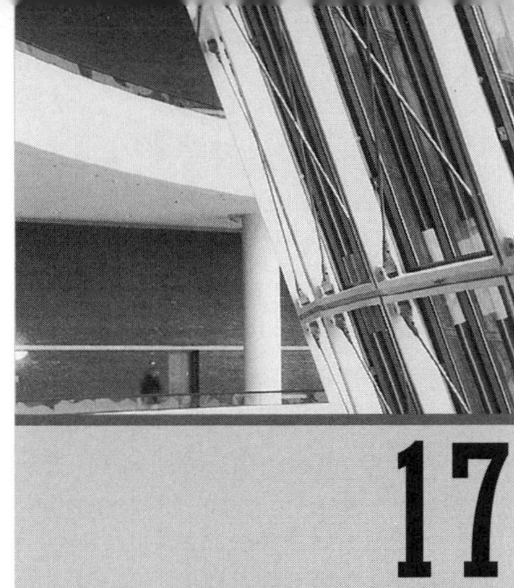

17

GLASS AND GLAZING

The lobby of the Federal Courthouse embraces a view of Boston Harbor with a dramatically curving wall of laminated glass. *(Architects: Pei, Cobb, Freed. © 1998 Steve Rosenthal)*

Glass plays many roles and takes many forms in buildings—Gothic church windows made of thousands of jewel-like pieces of colored glass; breathtaking expanses of smooth, uninterrupted glass that fill whole walls of today's buildings; Elizabethan casement windows with tiny diamond panes set in lead; skyscrapers that shimmer in facets of reflective glass mirroring the sky; cozy windows; comfortable windows; windows that bring soft, natural light; windows that frame spectacular views; windows that welcome winter sunlight to warm a room; windows that dissolve the boundaries between inside and outside. But glass can also form windows that make privacy impossible; windows that admit a harsh, glaring light; winter-cold surfaces that chill the body and tax the heating system; windows that broil a room in summer afternoon sunlight. Glass skillfully used in building contributes strongly to our enjoyment of architecture, but glass thoughtlessly used can make a building unattractive, uneconomical, and uncomfortable to inhabit.

History

The origins of glass are lost in prehistory. Initially a material for colored beads and small bottles, glass was first used in windows in Roman times. The largest known piece of Roman glass, a crudely cast sheet used for a window in a public bath at Pompeii, was nearly 3 by 4 feet (800 by 1100 mm) in size.

By the 10th century A.D., the Venetian island of Murano had become the major center of glassmaking, producing *crown glass* and *cylinder glass* for windows. Both the crown and cylinder processes began by blowing a large glass sphere. In the crown process, the heated glass sphere was adhered to an iron rod called a *punty* opposite the blowpipe. The blowpipe was then removed, leaving a hole opposite the punty. Next, the sphere was reheated, whereupon the glassworker spun the punty rapidly, causing centrifugal force to open the sphere into a large disk, or crown, 30 inches (750 mm) or more in diameter (Figure 17.2). When the crown was cut into panes, one pane always contained the "bullseye" where the punty was attached before being cracked off. In the cylinder process, the sphere, heated to a molten state, was swung back and forth, pendulum fashion, on the end of the blowpipe to elongate it into a cylinder. The hemispherical ends were cut off and the remaining cylinder was slit lengthwise, reheated, opened, and flattened into a rectangular sheet of glass that was later cut into panes of any desired size (Figure 17.3). Prior

Figure 17.1
An entire wall of the Baltimore Convention Center is made of low-emissivity double glazing supported by a tubular steel substructure. Adjustable stainless steel fittings with rubber washers connect the sheets of glass to the substructure. *(Architects: Cochran, Stevenson, & Donkervoet. Photo of Pilkington Planar System courtesy of W&W Glass Systems, Inc.)*

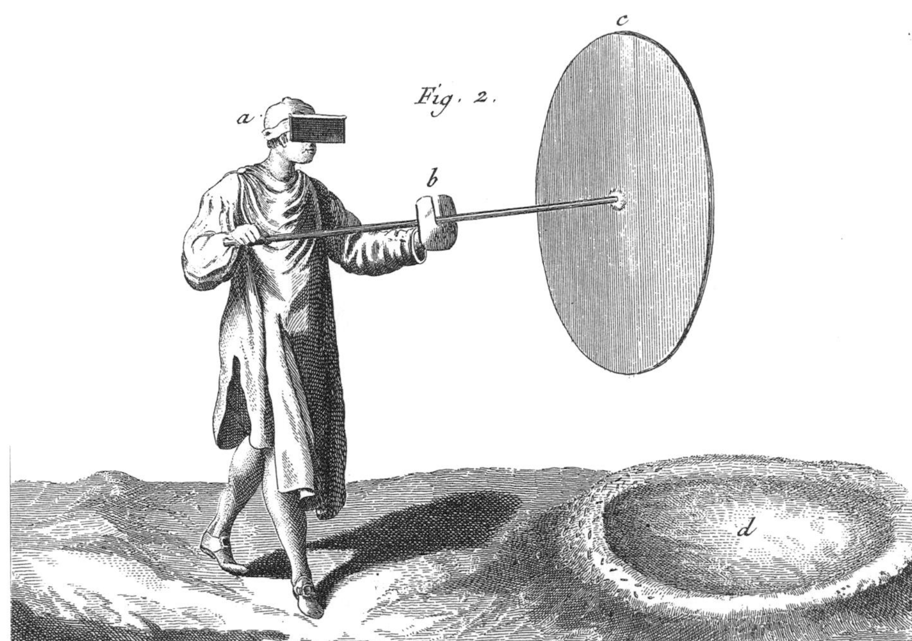

FIGURE 17.2
The glassworker in this old engraving wore a face shield *(a)* and hand shield *(b)* to protect against the heat of the large glass crown *(c)* that he had just spun on the end of a punty. After cooling, the crown was cut into small lights of window glass. *(Courtesy of the Corning Museum of Glass, Corning, New York)*

FIGURE 17.3
Making cylinder glass in the 19th century, Pittsburgh, Pennsylvania. Elongated glass bottles were blown by swinging the blowpipe back and forth in the pit in front of the furnace *(center)*. As each bottle solidified *(left)*, it was brought to another area where the ends were cut off to produce cylinders *(right)*. The cylinders were reheated and flattened into sheets from which window glass was cut. *(Courtesy of the Corning Museum of Glass, Corning, New York)*

to the introduction of modern glassmaking techniques, crown glass was favored over cylinder glass for its surface finish, which was smooth and brilliant because it was formed without contacting another material. Cylinder glass, though more economical to produce, was limited in surface quality by the texture and cleanliness of the surface on which it was flattened.

Neither crown glass nor cylinder glass was of sufficient optical quality for the fine mirrors desired by the 17th-century nobility. For this reason, *plate glass* was first produced, in France, in the late 17th century. Molten glass was cast into frames, spread into sheets by rollers, cooled, then ground flat and polished with abrasives, first on one side and then the other. The result was a costly glass of near-perfect optical quality in sheets of unprecedentedly large size. Mechanization of the grinding and polishing operations in the 19th century reduced the price of plate glass to a level that allowed it to be used for storefronts in both Europe and America.

In the 19th century, the cylinder process evolved into a method of drawing cylinders of molten glass vertically from a crucible. This made possible the routine, economical production of cylinders 40 to 50 feet (12–15 m) long. In 1851, the Crystal Palace in London (Figure 11.1) was glazed with 900,000 square feet (84,000 m^2) of cylinder glass supported on a cast iron frame.

In the early years of the 20th century, cylinder glass production was gradually replaced by processes that pulled flat sheets of *drawn glass* directly from a container of molten glass. Highly mechanized production lines for the grinding and polishing of plate glass were established, with rough glass sheets entering the line continuously at one end and finished sheets emerging at the other.

In 1959, the English firm of Pilkington Brothers Ltd. started production of *float glass*, which has since been licensed to other glassmakers and has become the worldwide standard, replacing both drawn glass and plate glass. In this process, a ribbon of molten glass is floated across a bath of molten tin, where it hardens before touching a solid surface (Figures 17.4–17.6). The resulting sheets of glass have parallel surfaces, high optical quality (virtually indistinguishable from that of plate glass), and a brilliant surface finish. Float glass has been produced in America since 1963 and now accounts for nearly all of domestic flat glass production.

The terminology associated with glass developed early in this long history. The term *glazing* as it applies to building refers to the installing of glass in an opening or to the transparent material (usually glass) in a glazed opening. The installer of glass is known as a *glazier*. Individual pieces of glass are known as *lights*, or often, to avoid confusion with visible light, *lites*.

The Material Glass

The major ingredient of glass is sand (silicon dioxide). Sand is mixed with soda ash (sodium hydroxide or sodium carbonate), lime, and small amounts of alumina, potassium oxide, and various elements to control color, then heated to form glass. The finished material, while seemingly crystalline and convincingly solid, is actually a supercooled liquid, for it has no fixed melting point and an open, noncrystalline microstructure.

When drawn into small fibers, glass is stronger than steel, though not nearly as stiff. In larger pieces, the microscopic imperfections that are an inherent characteristic of glass reduce its useful strength to significantly lower levels, particularly in tension. When a surface of a sheet of glass is placed in sufficient tension, as happens when an object strikes the glass, cracks propagate from an imperfection near the point of maximum tension and the glass shatters.

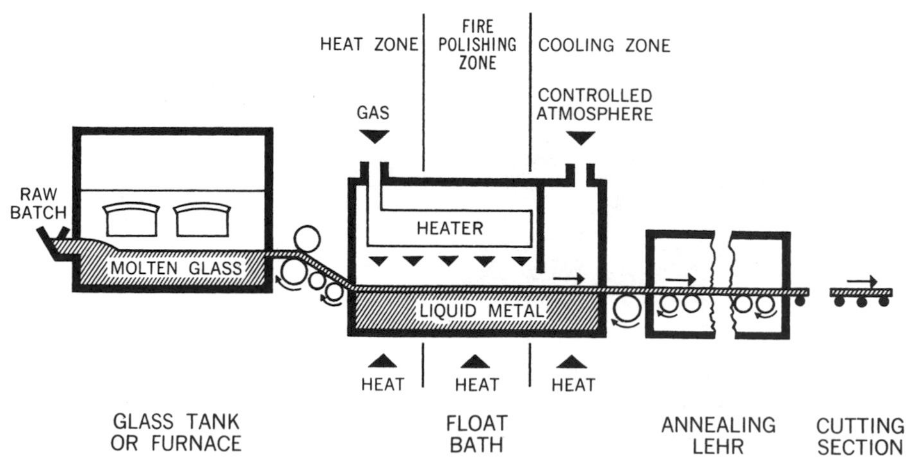

Figure 17.4
In the float glass process, molten glass from the furnace is floated on a bath of liquid tin to form a continuous sheet of glass. The annealing *lehr* cools the glass at a controlled rate to avoid internal stresses, after which it is cut into smaller sheets. *(Courtesy of PPG Industries)*

FIGURE 17.5
The superior flatness and bright surface finish of float glass are readily seen in the reflections on the glass ribbon emerging from the annealing lehr. *(Photo courtesy of LOF Glass, a Libby-Owens-Ford Company)*

FIGURE 17.6
Track-mounted cutting devices score the ribbon of cooled float glass as part of a computer-controlled cutting operation that automatically produces the glass sizes ordered by customers. *(Courtesy of PPG Industries)*

Considerations of Sustainability Relating to Glass

Glass Production

- The major raw materials for glass—sand, limestone, and sodium carbonate—are finite but abundant minerals.
- The high embodied energy of glass manufactured using traditional methods, roughly 7000 BTU per pound (16 MJ/kg), can be reduced by as much as 30 to 65 percent as new, more energy efficient manufacturing technologies are introduced.
- Some glass production involves the generation of potentially unhealthful or pollution-causing waste materials. Traditional mirror glass manufacturing, for example, generates an acidic waste effluent with high concentrations of copper or lead. However, recently, mirror glass manufactured with more environmentally friendly production techniques has become available.
- Although glass bottles and containers are recycled into new containers at a high rate, there is little recycling of flat glass at the present time. Most old glass goes to landfills.
- Efforts are underway to find new uses for waste glass. For example, vitrified glass aggregate (glass that has been melted and rapidly quenched to trap heavy metals and other contaminants) can be reused in asphalt, concrete, construction backfill, roofing shingles, and ceramic tiles.

Uses of Glass

- If it is not broken by accident or improper installation, glass lasts for a very long time with little degradation of quality, often much longer than most other building components.
- Glass is inert and does not affect indoor air quality. It is easily kept clean and free of molds and bacteria.
- The impact of glass on energy consumption can be very detrimental, very beneficial, or anything in between, depending on how intelligently it is used.
- If badly used, glass can contribute to summertime overheating from unwanted solar gain, excessive wintertime heat losses due to inherently low R-values, visual glare, wintertime discomfort caused by radiant heat loss from the body to cold glass surfaces, and condensation of moisture that can damage other building components.
- Well used, glass can bring solar heat into a building in winter and exclude it in summer, with attendant savings in heating and cooling energy. It can bring daylight into a building without glare, reducing both the use of electricity for lighting and the cooling load produced by that lighting.
- These benefits accrue over the entire life of the building, and the payoffs can be huge. Thus, glass is a key component of every energy-efficient building and a chief accomplice of the ill-informed designer in most energy-wasting buildings.

Thicknesses of Glass

Glass is manufactured in a series of thicknesses typically ranging from 3/32 inch (2.5 mm), also called *single-strength*, through 1/8 inch (3 mm), called *double-strength*, up to as much as 1 inch (25.4 mm), depending on the manufacturer. Glass thickness for a particular window is determined by the size of the light and the expected maximum wind loads on the glass. For low buildings with relatively small windows, glass 1/8 inch thick is usually sufficient. For larger windows and for windows in tall buildings, where wind velocities are high at higher altitudes, thicker glass is generally required, along with increased attention to how the glass is supported in its frame. (It has become standard practice for architects and structural engineers to order extensive wind tunnel testing of models of tall buildings during the design process to establish the expected maximum wind pressures and suctions on the windows.)

Because of unavoidable manufacturing defects in the glass, as well as the probability of damage to the glass during installation and while it is in service, a certain amount of breakage must always be anticipated in a large building. ASTM E1300 establishes standard procedures for evaluating the structural stability and probability of breakage in glass. These are used to determine the glass thickness that will result in an acceptably low probability of breakage for a window of given dimensions, support conditions, and wind pressure.

During its manufacture, ordinary window glass is *annealed*, meaning that it is cooled slowly under controlled conditions to avoid locked-in thermal stresses that might cause it to behave unpredictably in use. But other types of glass have come into use for particular purposes in buildings.

Heat-Treated Glass

Heat-treated glass is produced by reheating annealed glass in an oven to approximately 1150 degrees Fahrenheit (620°C) and then cooling (*quenching*) both of its surfaces rapidly with blasts of air while its core cools much more slowly. This process induces permanent compressive stresses in the edges and faces of the glass and tensile stresses in the core. The resulting glass is stronger in bending than annealed glass and more resistant to thermal stress and impact. These properties make heat-treated glass useful for windows exposed to heavy wind pressures, impact, or intense heat or cold. By adjusting the quenching process, greater or lesser degrees of residual stress may be introduced into the glass, producing products referred to as either "tempered" or "heat-strengthened glass."

Tempered Glass

Tempered glass has higher residual stresses than heat-strengthened glass and is about four times as strong in bending as annealed glass. If it does break, the sudden release of its internal stresses reduces tempered glass instantaneously to small, square-edged granules rather than long, sharp-edged shards. This characteristic, combined with its high strength, qualifies it for use as safety glazing (discussed below), that is, in situations of possible occupant impact. Tempered glass is also used for all-glass doors that have no frame at all (Figure 17.7), for whole walls of squash and handball courts, for hockey rink enclosures, and for basketball backboards.

Tempered glass is more costly than annealed glass. It often has noticeable optical distortions created by the tempering process. In addition, all cutting to size, drilling, and edging must be done before the heat treatment of the glass because any such operations after tempering will release the stresses in the glass and cause it to disintegrate. Tempered glass is also sometimes referred to as *fully tempered glass* to distinguish it more clearly from heat-strengthened glass.

Heat-Strengthened Glass

For many applications, lower-cost *heat-strengthened glass* may be used instead of tempered glass. The induced compressive stresses in the surface and edges of heat-strengthened glass are about one-third as high as those in fully tempered glass (typically 5000 psi compared to 15,000 psi for tempered glass, or 34 MPa versus 104 MPa). Heat-strengthened glass is about twice as strong in bending as annealed glass and is more resistant to thermal stress. It usually has fewer distortions than tempered glass. Its breakage behavior is more like that of annealed glass than tempered glass. For that reason, it cannot be used where safety glazing is required except in laminated form (laminated glass is discussed below).

Laminated Glass

Laminated glass is made by sandwiching a transparent *polyvinyl butyral (PVB) interlayer* between sheets of glass and bonding the three layers together

FIGURE 17.7
Tempered glass is used for strength and breakage safety in both the doors and windows of this store in a downtown shopping mall. *(Photo by Edward Allen)*

under heat and pressure. Laminated glass is not as strong as annealed glass of the same thickness, but when laminated glass breaks, the soft interlayer holds the shards of glass in place rather than allowing them to fall out of the frame of the window. This makes laminated glass useful for skylights and overhead glazing, because it reduces the risk of injury to people below in case of breakage (Figures 17.8 and 17.9). The PVB interlayer may be colored or patterned to produce a wide range of visual effects in laminated glass. Because laminated glass does not create dangerous, loose shards of glass when it breaks, it also qualifies as safety glazing.

Laminated glass is a better barrier to the transmission of sound than solid glass. It is used to glaze windows of residences, classrooms, hospital rooms, and other rooms that must be kept quiet in the midst of noisy environments. It is especially effective when installed in two or more layers with airspaces between. In comparison to solid glass, laminated glass also reduces the transmission of ultraviolet (UV) radiation, a component of sunlight that contributes significantly to fading and the degradation of interior finishes, furnishings, and fabrics.

Security glass, used for drive-in banking windows and other facilities that need to be resistant to burglary, is made of multiple layers of glass and PVB, and is available in a range of thicknesses to stop any desired caliber of bullet. Laminated glass is also used in blast-resistant and windborne debris-resistant glazing systems, which are described in more detail in the next chapter.

Chemically Strengthened Glass

Chemically strengthened glass is produced by an ion exchange process that takes place when annealed glass is immersed in a molten salt bath. As smaller sodium ions in the glass are replaced with larger potassium ions from the salt solution, the faces of

FIGURE 17.8
The entrance canopy of the Newport, Rhode Island Hospital is made of laminated glass supported by stainless steel spider fittings that transmit the weight of the roof to cantilevered steel beams. The heads of the bolts that fasten the glass to the fittings are recessed unobtrusively within the thickness of the glass. *(Taylor & Partners, Architects. Photo of Pilkington Planar System courtesy of W&W Glass Systems, Inc.)*

the glass are put into compression relative to the core, and the glass is prestressed in a manner similar to the one that occurs with heat treating. However, because the temperatures involved in chemical strengthening are lower, chemically strengthened glass does not experience the optical distortions or warping that are common with heat-treated glass. Depending on the particulars of the treatment process, the strength and toughness of chemically strengthened glass can exceed those of tempered glass.

Unlike tempered glass, chemically strengthened glass can be cut after strengthening, although its strength is diminished along the cut edges. When chemically strengthened glass breaks, it produces large, hazardous shards. So, like heat-strengthened glass, it cannot be used where safety glazing is required unless it is laminated. Chemical strengthening is used for pieces of glass that are not easily heat treated, such as those that are small, thin, or oddly shaped. It is also used in some fire-rated glass products (discussed below) and in laminated form for security glass, blast-resistant glass, and windborne debris-resistant glass.

Fire-Rated Glass

Fire-rated glass in fire doors, fire windows, and fire resistance rated walls must maintain its integrity as a barrier to the passage of smoke and flames even after it has been exposed to heat for a period of time. Some tempered or laminated glass products can achieve test ratings of up to

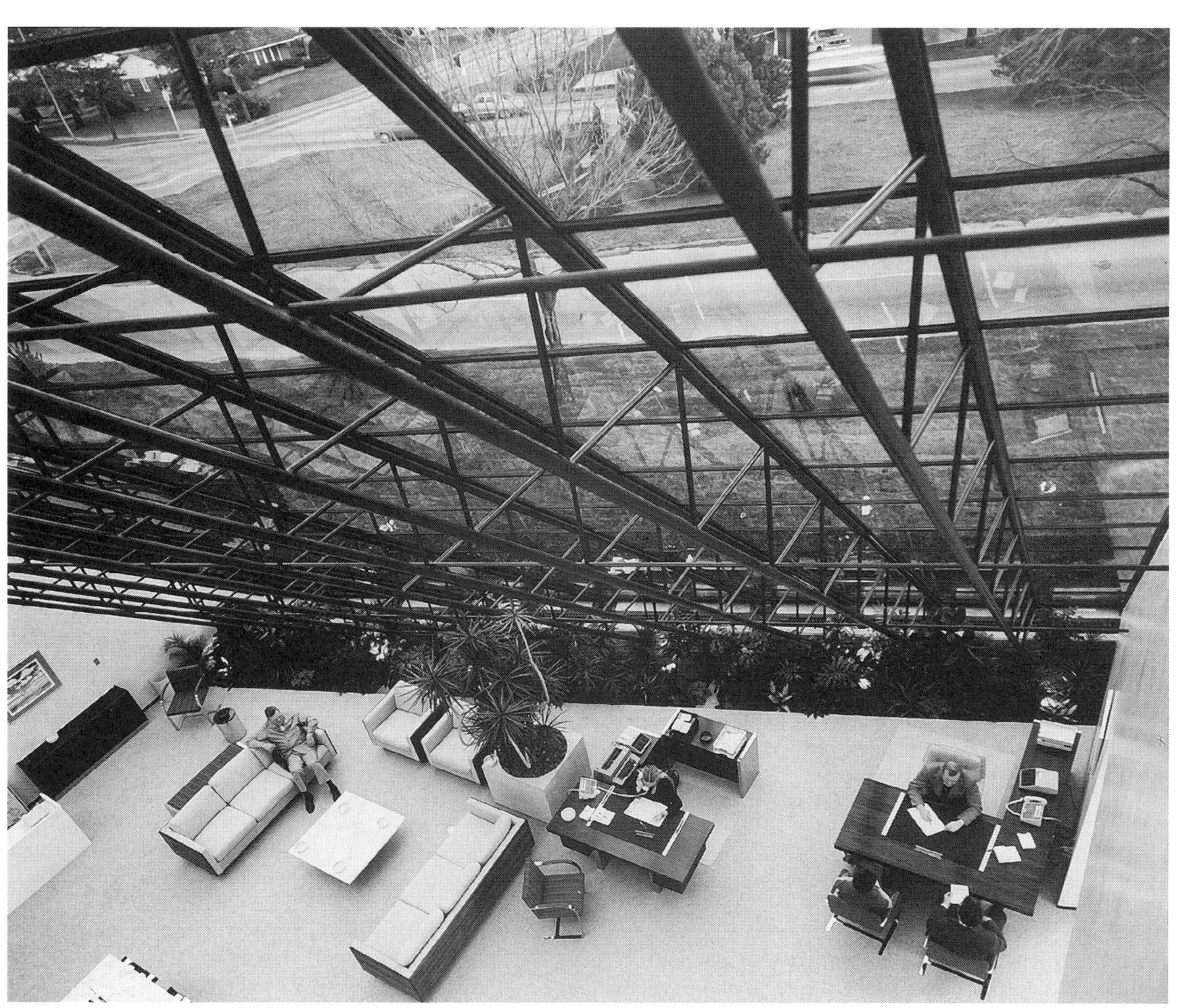

FIGURE 17.9
Laminated glass provides safety against falling shards in an overhead sloped glazing installation. *(Courtesy of PPG Industries)*

20 minutes of fire resistance. *Wired glass* is produced by rolling a mesh of small wires into a sheet of hot glass. When wired glass breaks from thermal stress, the wires hold the sheets of glass in place so that the glass continues to act as a fire barrier. It carries a fire resistance rating of 45 minutes. *Optical-quality ceramic* is more stable against thermal breakage than any type of glass. It looks and feels like glass and can achieve fire resistance ratings ranging from 20 minutes to 3 hours.

Two other fire-rated glass types are fire-retardant filled double glazing and intumescent interlayer laminated glazing. *Fire retardant filled double glazing* consists of a clear, heat-absorbing polymer gel contained between two sheets of tempered glass. *Intumescent interlayer laminated glazing* is made of thin layers of transparent intumescent material sandwiched between multiple layers of annealed glass. When either glass type is heated by fire, the gel or intumescent interlayers react to form opaque, insulating layers. As a result, these products not only resist the passage of flame and smoke, they also limit the rise in surface temperature of the glass on the side opposite the fire and prevent the transfer of radiant heat through the glass. These added protective properties make these glass types suitable for use in larger sizes and in a broader range of applications than other types of fire-rated glass. Fire resistance ratings of up to 2 hours can be achieved.

According to the International Building Code fire-rated glass must meet the fire endurance requirements of one of three tests—NFPA 252, NFPA 257, or ASTM E119—depending on whether the glass is part of a fire door, fire window, or fire-rated wall assembly, respectively. Fire-rated glass tested for use in doors and windows is limited in the maximum size of individual lights permitted. However, glass products that can pass the more stringent requirements of ASTM E119 for wall assemblies, including fire-retardant filled double glazing and intumescent interlayer laminated glazing, have no such size limits and can be used as full substitutes for fire resistance–rated wall construction. To distinguish between products that can meet all the requirements of fire-rated wall construction and those that are suitable only for use in fire doors and windows, the terms *glass fire walls* or *fire resistive glazing* may be applied to the former and *fire protective glazing* to the latter.

Due to its frequent use in doors and other hazardous locations, fire-rated glass must often also meet the impact resistance and breakage safety requirements of safety glazing. To meet these requirements, optical-quality ceramic is either laminated or film-faced to protect it from shattering dangerously when used in such locations. Ordinary annealed wired glass does not meet safety glazing requirements. While historically its use was permitted in fire doors and windows due to a lack of suitable alternatives, this is no longer the case. Where wired glass is now used in hazardous locations, it also is either film-faced or laminated. Fire-retardant filled glazing and intumescent laminated glazing are both capable of meeting safety glazing requirements.

Fritted Glass

A number of producers are equipped to imprint the surface of glass with silk-screened patterns of ceramic-based paints. The paints consist primarily of pigmented glass particles called *frit*. After the frit has been printed on the glass, the glass is dried and then fired in a tempering furnace, transforming the frit into a hard, permanent ceramic coating. Many colors are possible in both translucent and opaque finishes. Typical patterns for *fritted* or *silkscreened glass* are various dot and stripe motifs (Figure 17.10), but custom-designed patterns and even text are easily reproduced. Fritted glass is often used to control the penetration of solar light and heat into a space.

Spandrel Glass

Frits are used to create special opaque glasses for covering spandrel areas (the bands of wall around the edges of floors) in glass curtain wall construction (Figure 17.11). A uniform coating of frit is applied to what will be the interior surface of the glass. Some *spandrel glasses* are made as similar as possible in exterior appearance to the glass that will be used for the windows on a specific project. It is very difficult, however, even with reflective coated glass, to make the spandrels indistinguishable from the windows under all lighting conditions. Most spandrel glasses are made to contrast with the windows of the building. Many suppliers can apply thermal insulation on the interior of the glass, complete with vapor retarder. Spandrel glass is usually tempered or heat strengthened to resist the thermal stresses that can be caused by accumulation of solar heat behind the spandrel.

Tinted and Reflective Coated Glass

Solar heat buildup can be problematic in buildings with large areas of glass, especially during the warm part of the year. Fixed sun-shading devices outside the windows are the best ways of blocking unwanted sunlight, but glass manufacturers have also developed tinted and reflective glasses that reduce glare and cut down on solar heat gain.

Tinted Glass

The transparency of glass to visible light is called its *visible light transmittance (VT)*. It is measured as the ratio of visible light that passes through the glass relative to the amount of light striking the glass. Clear glasses have visible light transmittance in the range of 0.80 to 0.90, meaning that 80 to 90 percent of the visible light striking the glass passes through to the building interior. The remaining 10 to 20 percent is either reflected or absorbed by the glass and converted to heat.

By tinting glass, its visible light transmittance is reduced. *Tinted glass* is made by adding small amounts of selected chemical elements to the molten glass mixture to produce the desired hue and intensity of color in grays, bronzes, blues, greens, and golds. The visible light transmittance of commercially available tinted glasses ranges from about 0.75 in the lightest tints to 0.10 for dark gray. The overall reduction in solar heat gain is often significantly less, however, because the solar radiation absorbed by the glass and converted to heat must go somewhere, and a substantial portion of it is conducted or reradiated to the interior of the building (Figure 17.12).

To evaluate the effectiveness of glass in reducing heat gain from solar radiation, a measure called the *solar heat gain coefficient (SHGC)* is used; it is the ratio of solar heat admitted through a particular glass to the total heat energy striking the glass. SHGC accounts for the solar radiation that passes through glass, as well as for heat that is conducted or radiated into the space due to heating of the glass itself. Clear glasses have solar heat gain coefficients ranging from about 0.90 to 0.70, depending on the clarity and thickness of the glass. Solar heat gain coefficients for tinted glasses range from about 0.70 to 0.35, meaning that these glasses allow 70 to 35 percent of the solar heat energy striking the glass to pass through. Generally speaking, for buildings dominated by a heating load, glass with a high SHGC is desirable to take advantage of passive solar heat gains. In buildings dominated by cooling, glass with a low SHGC is preferable to minimize unwanted solar heating.

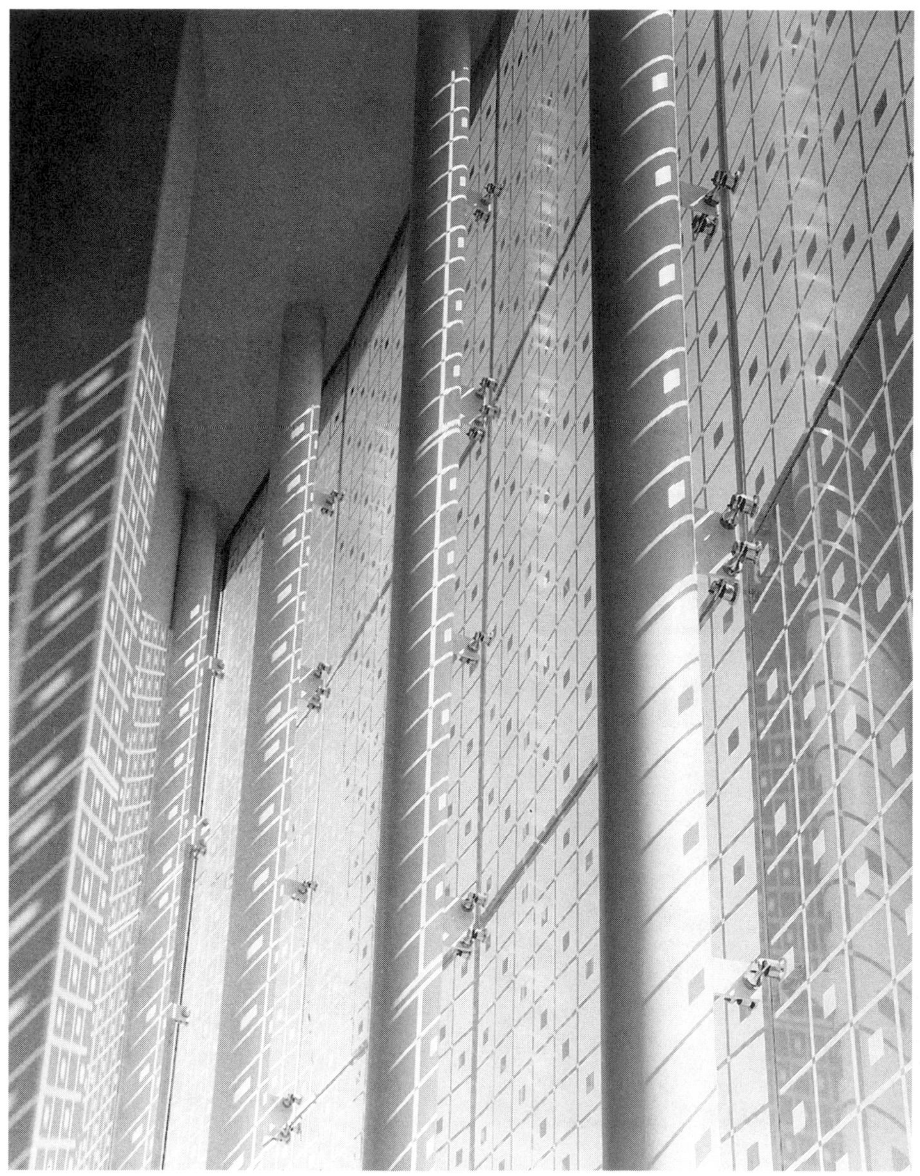

Figure 17.10
Fritted patterns modulate the sunlight that enters a theater lobby. *(Photo of Pilkington Planar System courtesy of W&W Glass Systems, Inc.)*

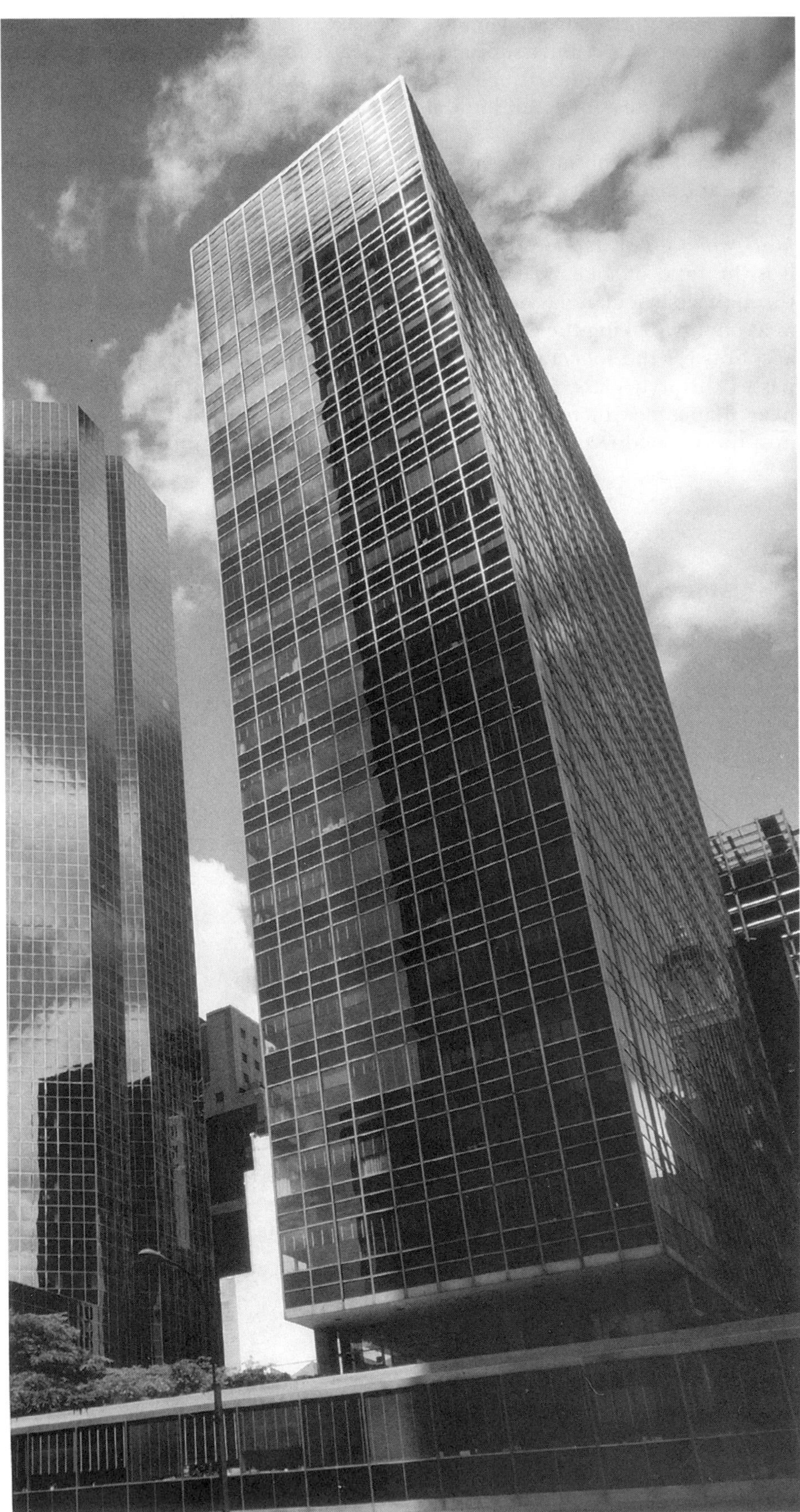

FIGURE 17.11
Lever House in New York, an early glass curtain wall building designed by architects Skidmore, Owings and Merrill, uses dark-green glass for the spandrels and lighter-green glass for the windows.
(Courtesy of PPG Industries)

(*Shading coefficient*, a measure similar to SHGC, is an older measure of reduction in solar heat gain that has been mostly replaced by SHGC.)

Visible transmittance and solar heat gain coefficient can be combined to determine the *light to solar gain (LSG) ratio*, a useful measure of the overall energy-conserving potential of glass. The LSG ratio is defined as the visible light transmittance divided by the solar heat gain coefficient. A glass with high LSG admits a relatively large portion of visible light in comparison to the amount of solar heat admitted, combining the greatest daylighting potential with the least solar heating potential. Green- and blue-tinted glasses tend to have high LSG ratios values, while those of bronze, gold, and gray tints tend to be lower.

Reflective Coated Glass

Thin, durable films of metal or metal oxide can be deposited on a surface of either clear or tinted glass sheets under closely controlled conditions to make *reflective coated glass*, also called *solar control glass*. Depending on its composition, the film may be applied to either the inside of the glass or the outside. In double glazing, it may also be applied to either of the surfaces that face the space between the layers of glass. While remaining thin enough to see through, the film reflects a substantial portion of the incident visible light. Visible

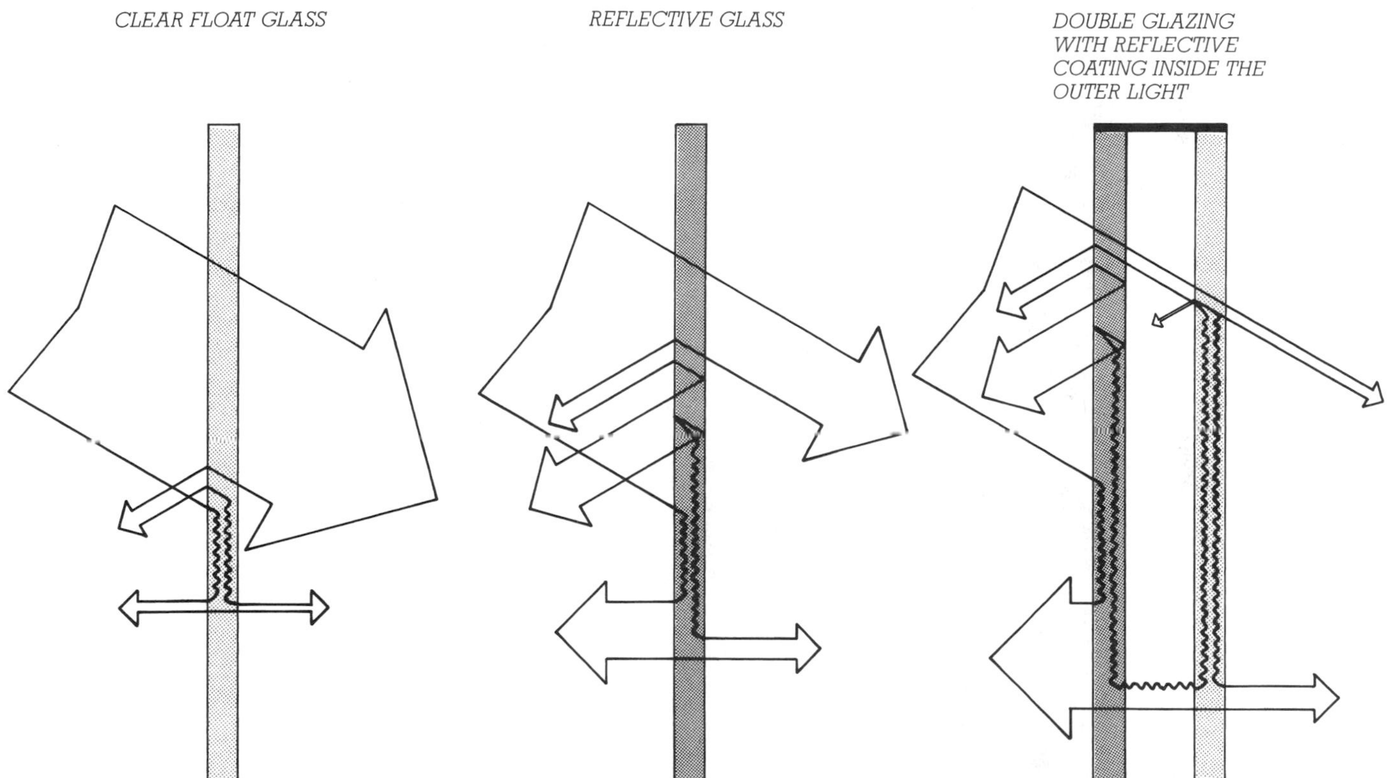

FIGURE 17.12
A schematic representation of the effect of three different glazing assemblies on incoming sunlight. Outdoors is to the left. The relative widths of the arrows indicate the relative percentages of the incoming light transmitted, reflected, and absorbed. In clear float glass, to the left, most of the light is transmitted, with small quantities reflected, absorbed, and reradiated as heat. Reflective coated glass, at the center, bounces a large proportion of the light back to the outdoors, and also absorbs and reradiates a significant amount. In double glazing, many different combinations of types of glass are possible; the one shown to the right of this diagram utilizes glass with a reflective coating on the inner side of the outer light.

light transmittance and solar heat gain coefficients (SHGC) for reflective coated glasses vary significantly, depending on the density of the metallic coating and the tinting of the glass to which it is applied. Reflective coated glasses appear as mirrors from the outside on a bright day and are often chosen by architects for this property alone (Figure 17.13). At night, with lights on inside the building, they appear as dark but transparent glass.

The sunlight reflected by a building glazed with reflective coated glass can be helpful in some circumstances by lighting an otherwise dark urban street space. It can also create problems in other situations by bouncing solar heat and glare into neighboring buildings and onto the street.

Who when he first saw the sand and ashes . . . would have imagined that in this shapeless lump lay concealed so many conveniences in life . . . by some such fortuitous liquefaction was mankind taught to procure a body at once in a high degree solid and transparent; which might admit the light of the sun, and exclude the violence of the wind; which might extend the sight of the philosopher to new ranges of existence. . . .

Dr. Samuel Johnson, writer and lexicographer, *The Rambler,* April 17, 1750.

FIGURE 17.13
Reflective coated windows with subtly different reflective coated glass spandrels. *(Architects: Paul Rudolph and 3D International. Photo courtesy of PPG Industries)*

Insulating Glass

Window glass is a poor thermal insulator. A single sheet of glass (*single glazing*) conducts heat about 5 times as fast as 1 inch (25 mm) of polystyrene foam insulation and 20 times as fast as a well-insulated wall. A second sheet of glass applied to a window with an airspace between the two sheets (*double glazing*) cuts this rate of heat loss in half, and a third sheet with its additional airspace (*triple glazing*) reduces the rate of heat loss to about a third of the rate through a single sheet. A triple-glazed window, however, still loses heat about six times as fast as the wall in which it is placed. Continuing to add additional sheets of glass and airspaces adds weight, bulk, and expense to the glazing unit and the frame that holds it, making double and triple glazing the practical maximums for normal building glazing applications.

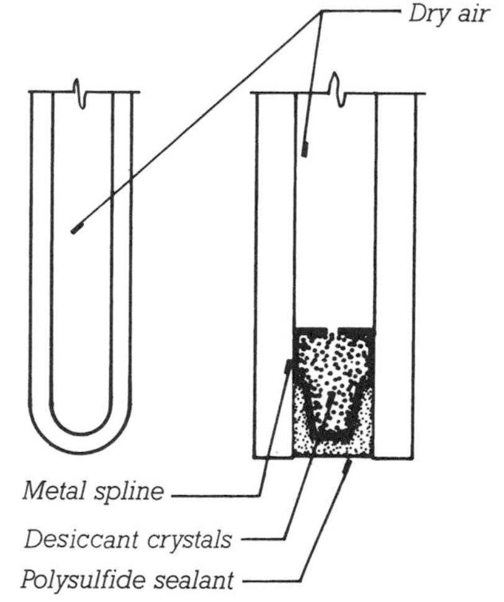

FIGURE 17.14
Two methods of sealing the edge of double glazing: fused glass edges, to the left, and a metal spline and organic sealant, to the right. Desiccant crystals in the spline (edge spacer) absorb any residual moisture in the airspace. Splines also improve the thermal performance of the insulated units in the areas close to the unit edges in comparison to fused glass edges.

To prevent moisture condensation within the airspace of double or triple glazing (also called *insulating glass units* or IGUs), the units are usually hermetically sealed at the time of manufacture with dry air inserted in the space between the glass lights. Originally, for small lights of double glass, the edges of the two sheets were simply fused together (Figure 17.14). However, this detail is seldom used now because the fused glass edge is highly conductive of heat. Instead, a hollow metal *edge spacer* (also called a *spline*) is inserted between the edges of the sheets of glass, and the edges are closed with an organic sealant compound. A small amount of a chemical drying agent, or *desiccant*, is left inside the spacer to remove any residual moisture from the trapped air. The air is always inserted at atmospheric pressure to avoid structural pressures on the glass. (When insulating glass units do exhibit internal condensation, it is a sign of failure of the edge seal, and the unit must be replaced.)

The thickness of the airspace in insulating glass units is less critical to the units' insulating value than the mere presence of the airspace: From ⅜ inch (9 mm) up to about 1 inch (25 mm) of thickness, the insulating value of the airspace increases somewhat, but above that thickness little additional benefit is gained. A standard overall thickness for large lights of double glazing is 1 inch (25.4 mm), which results in an airspace ½ inch (13 mm) thick if ¼-inch (6-mm) glass is used.

For slightly improved thermal performance, stainless steel, which is less conductive of heat, may be used instead of aluminum for the spacer, and a sealant material may be placed between the glass and the spacer as a thermal break. For even better thermal performance, so-called *warm edge spacers* made of thermally broken aluminum or extruded rubber may be used.

The thermal performance of insulated glazing units can also be improved by introducing gases with greater density and lower thermal conductivity than that of ordinary air between the sheets of glass. Depending on the gas used and the thickness of the space between the glass sheets, improvements in thermal performance of 12 to 18 percent are possible. Argon and krypton are the gases most commonly used.

The performance of glazing as a thermal insulator is quantified as its *U-Factor.* U-Factor is expressed as BTUs per square foot-hour-degree Fahrenheit (BTU/ft^2-hr-°F) or, in metric units, as watts per square meter-degree Kelvin (W/m^2-°K). U-Factor is the mathematical reciprocal of R (see page 658), and as such, lower values represent improved thermal performance. Some examples glazing configurations and their U-Factors are listed in Figure 17.15. For more detailed information, glass manufacturers' product literature should be consulted.

Insulated glazing products that rely on the evacuation of most of the air from the space between the glass sheets are also under development. When combined with low-emissivity coatings (see below), such *vacuum-insulated glazing units* are predicted to achieve U-Factors as low as 0.080 BTU/ft^2-hr-°F (0.45 W/m^2-°K) in units not more than ½-inch (12 mm) in total thickness.

Low-Emissivity Coated Glass

The thermal performance of glazing can be improved substantially by the use of glass with a *low-emissivity* (*low-e*) *coating.* Low-e coatings are ultrathin, virtually transparent, and almost colorless metallic coatings that selectively reflect solar radiation of different wavelengths. They have a high visible light transmittance and, depending on the particular coating, a low transmittance for some or all types of infrared radiation (heat).

Low-e coated glass is most commonly used as one of the two lights in double glazing, where it offers several benefits: By reducing the radiant

	VT	SHGC	LSG	U[a]	R[b]
Single glazing, clear	0.90	0.85	1.3	1.1 *6.3*	0.91 *0.16*
Double glazing, clear	0.79	0.70	1.1	0.47 *2.7*	2.1 *0.37*
Double glazing, medium gray tinted	0.40	0.45	0.9	0.47 *2.7*	2.1 *0.37*
Triple glazing, clear	0.53	0.52	1.0	0.34 *1.9*	2.9 *0.52*
Double glazing, clear, low-e (low SHGC)	0.64	0.27	2.4	0.28 *1.6*	3.6 *0.63*
Double glazing, clear, low-e (high SHGC), argon filled	0.78	0.63	1.2	0.27 *1.5*	3.7 *0.65*

[a]U-Factor: Btu/ft^2-hr-°F followed by *W/m^2-°K.*
[b]R-value: ft^2-hr-°F/Btu followed by *m^2-°KW.*

FIGURE 17.15
Comparative properties of some example glass types. Note the high light to solar gain (LSG) ratio possible with low-e double glazing. LSG is the quotient of visible transmittance (VT) over solar heat gain coefficient (SHGC). Higher LSG values indicate better overall energy efficiency of the glazing unit (in buildings where cooling loads dominate). U-Factors and R-values are center-of-glass values and, for the insulating glass units listed, do not account for reduced thermal performance around the edges of the units due to the greater conductivity of the spacers. The lower the U-Factor (higher the R-value), the better the thermal performance of the unit.

transfer of heat between individual lights, the overall thermal transmittance of the glazing unit is reduced to the extent that low-e double glazing can meet or exceed the thermal performance of ordinary triple glazing. By reflecting the majority of the infrared component of solar radiation, low-e double glazing can simultaneously provide high visible light transmittance with low solar heat gain, allowing such units to achieve the highest light to solar gain ratios of any insulated glass type.

By varying the properties of the low-e coating and by combining it with different types of tinted glass, the performance characteristics of the glazing unit can be tailored to meet different needs. For buildings dominated by wintertime heating, low-e units with high U-Factors (to minimize heat loss) and high solar heat gain coefficients (to promote wintertime solar heat gains) may be selected. For buildings dominated by cooling loads, units with low solar heat gain coefficients (to minimize solar heating) and lower visible light transmittance are used (Figure 17.15). Like laminated glass, low-e coated glass also has low UV radiation transmittance, a benefit to interior finishes and furnishings. Though less common, low-e coated glass may also be used in single or triple glazing to improve the thermal performance of these glass types.

When specifying glass with any type of coating (low-e glass or reflective coated glass), it is necessary to specify on which glass surface the coating is to be located. By convention, glass surfaces are numbered starting from the exterior side of a glazing unit and working inward. In single glazing, the outward face is *surface number* 1 and the inward face is surface number 2. In double glazing, the outward face of the outer glass light is surface number 1 and the inward face of this light is surface number 2; the outward face of the inner glass light is surface number 3, and its inward face is surface number 4. In low-e double glazing, the low-e coating is most commonly located on the number 2 surface, although, where a high solar heat gain coefficient is desired, it may be located on the number 3 surface instead.

Low-e coatings can also be applied to very thin membranes of transparent plastic. One or two of these plastic films can be installed within the center of the airspace or gas space of a double glazed unit, stretched tight, parallel to the sheets of glass, where they act as virtually weightless additional glazing elements. Combined with selective properties of the low-e coating, thermal performance values ranging from R-6 to R-20 are claimed by manufacturers of these films.

Self-Cleaning Glass

Glass tends to attract dirt, and must be washed periodically both inside and out to maintain its transparency. *Self-cleaning glass* is coated with titanium oxide on its exterior surface. This coating acts as a catalyst that enables sunlight to convert organic dirt to carbon dioxide and water. It also causes rainwater to run down the surface in sheets rather than to bead up. Nonorganic dirt, such as sand, is unaffected by the catalyst, but the sheets of water are more effective at removing such matter than beaded water. The coating is applied only to the outside of the glass; therefore, the interior surface of the glass must be washed manually.

Glass That Changes Its Properties

Glass that can change its optical properties is called *chromogenic glass. Thermochromic glass* becomes darker when it is warmed by the sun. *Photochromic glass* becomes darker when exposed to bright light. Both types are potentially valuable as passive devices to reduce cooling loads in buildings.

Electrochromic glass changes its transparency in response to the passage of electric current. Also called *switchable glass,* it can be actively controlled by building occupants or automated systems, allowing, in comparison to passive technologies, more precise response to requirements for control of solar heat gain, daylighting, or occupant privacy. Currently available electrochromic glass products mostly rely on solid state liquid crystal technology, similar to that used in electronic flat panel displays. These products are limited to interior applications where control over transparency and privacy is desired, but they are not suitable to the conditions to which exterior glazing is subjected. Other technologies currently under development are expected to result in products that are suitable for both exterior as well interior exposures, that can selectively control portions of the solar spectrum (such as infrared radiation), and that can switch between transparent and reflective states.

Gasochromic glass is another developing switchable technology for altering the light transmittance of glass, in which the transparency of a reactive coating on the number 2 surface of an insulated glass unit is altered by the pumping of gas into or out of the interstitial space of the unit.

Other Types of Glass

Glass can be produced with an amazing range of physical properties and variations in appearance, and new products with unique characteristics continue to be developed. Fins of *structural glass* function as beams to resist wind loads in very tall or wide curtain walls, and one glass manufacturer has produced hollow glass cylinders prestressed with steel wire running through its center axis that it claims can take the place of concrete or steel elements in resisting structural compressive loads.

Antireflective glass minimizes residual reflections that normally occur when light levels differ significantly on the opposite sides of glass. It is used for glazing in showrooms, display areas, sports stadiums, artwork framing, and other applications where the highest possible optical quality is desired. Mirrors are made of *mirror glass,* which has a thin silver-based coating on its back side. A thin layer of copper applied over the silver prevents corrosion, and a second layer of backing paint provides additional protection. *Patterned glass,* hot glass rolled into sheets with many different surface patterns and textures, is used where light transmission is desired but vision must be obscured for privacy. Glass manufactured with a high percentage of lead oxide can be used as *radiation-shielding glass. Photovoltaic glass* is coated with a thin film of amorphous silicon that generates electricity from sunlight. It allows a building with a large glazed area to create at least some of the electric energy that it uses for lights and machinery. Traditional *stained glass* and contemporary *colored glass,* formulated with ingredients that alter the color of the glass, can be used in a wide range of artistic and architectural applications. Glass may be blown, molded, fused, and colored to produce an infinite variety of types of *art glass* used for decorative and sculptural purposes.

Plastic Glazing Sheets

Transparent plastic sheet materials are often used instead of glass for specialized glazing applications. The two most common plastic glazing materials are *acrylic* and *polycarbonate.* Both are more expensive than ordinary float glass. Both have very high coefficients of thermal expansion, which cause them not merely to expand and contract with temperature changes, but also to bow visibly toward the warm side when subjected to high indoor–outdoor temperature differentials. This, in turn, requires that plastic sheet materials be installed in their frames with relatively expensive glazing details that allow for plenty of linear movement and rotation. Both polycarbonate and acrylic are soft and easily scratched, although more scratch-resistant formulations are available.

Plastic glazing is most commonly used where glass is inappropriate: Plastics can be cut to shapes with inside corners (L-shapes and T-shapes, for example) that are likely to crack if cut from glass. They can be bent easily to fit in curved frames. They can be heat-formed into domed glazing for skylights. And the plastics, especially polycarbonate, which is literally impossible to break under ordinary conditions, are widely used for windows in buildings where vandalism is a problem or high impact resistance is required. Polycarbonate plastics can be manufactured in a variety of colors and varying degrees of transparency. They can also be manufactured in a double-walled configuration, called *cellular polycarbonate glazing,* creating hollow panels, roughly ¼ inch to 1½ inches (6–40 mm) thick, with greater stiffness and better thermal performance than solid sheets. Plastic glazing sheets can also meet safety requirements for glazing used in areas subject to human impact.

Translucent but nontransparent plastic sheets reinforced with glass fibers (*fiberglass-reinforced polyester glazing*) are also used in buildings. Corrugated sheets are used for industrial skylights and residential patio roofs. Thin, flat sheets of a special formulation with a high translucency to solar energy are used for skylights and low-cost solar collector glazing.

Aerogel-Filled Glazing

Aerogel, a silicon-based foam that is 99.8 percent air, can be used to fill the cavity in double-glazed glass or plastic products. Though aerogel was invented many decades ago, its commercialization was delayed by its fragility and high cost of manufacture—problems that have only recently been solved.

Aerogel is milky in color, not fully transparent, and has a visible transmittance that varies with its thickness. Aerogel-filled glazing has a good light to solar gain ratio, making it an efficient source of diffuse, low-contrast, natural daylight. Currently available aerogel products can achieve insulating values of R-8 per inch (RSI-1.4 per 25 mm), more than twice that of glass fiber insulation. Products under development that rely on nanotechnology to improve thermal performance are claimed to have insulating values as high as R-40 per inch (RSI-7 per 25 mm).

Glazing

Glazing Small Lights

Small lights of glass are subjected neither to large wind force stresses nor to large amounts of thermal expansion and contraction. They may be glazed by very simple means (Figure 17.16). In traditional wood sash, the glass is first held in place by small metal *glazier's points* and then sealed on the outside with *glazing putty*, a simple compound of linseed oil and pigment that gradually hardens by oxidation of the oil. Putty must be protected from the weather by subsequent painting, and tends to harden and become brittle with age.

As an alternative to glazing putty, newer latex and silicone caulking compounds that can be applied more quickly and need not be painted may be used for field glazing. Improved, more adhesive, and more elastic putties or *glazing compounds* are employed for factory-glazed sash as well.

Glazing Large Lights

Large lights of glass (those over 6 square feet or 0.6 m^2 in area) require more care in glazing. Wind load stresses on each light of glass are higher, and the glass must span farther between its supporting edges. Any irregularities in the frame of the window may result in distortion of the glass, highly concentrated pressures on small areas of the glass, or glass-to-frame contact, any of which can lead to abrasion or fracture of the glass. In large lights, thermal expansion and contraction can also cause stresses to build up in the glass.

The design objectives for a large-light glazing system are:

1. To support the weight of the glass in such a way that the glass is not subjected to intense or abnormal stress patterns
2. To support the glass against wind pressure and suction
3. To isolate the glass from the effects of structural deflections in the frame of the building and in the smaller framework of *mullions* that supports the glass
4. To allow for expansion and contraction of both glass and frame without damage to either
5. To avoid contact of the glass with the frame of the window or with any other material that could abrade or stress the glass

The weight of the glass is supported in the frame by synthetic rubber *setting blocks*, normally two per light, located at the quarter points of the bottom edge of the light. For support against wind loads, a specified amount of *bite* (depth of grip on the edge of the glass) is provided by the supporting mullions. If the bite is too little, the glass may pop out under wind loading; if too much, the glass

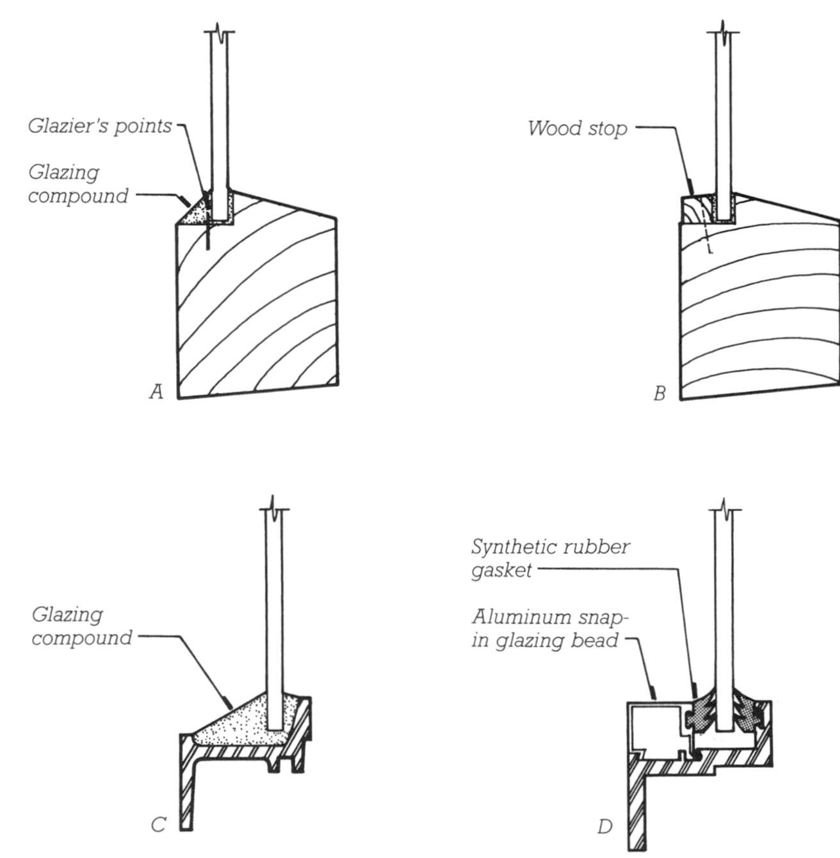

Figure 17.16
Alternative methods of single glazing lights of glass. Glass is traditionally mounted in wood sash using glazier's points and glazing compound *(a)* or a wood stop nailed to the sash *(b)*. Metal sash was once glazed in the manner shown in *(c)*, but most metal sash is now glazed with snap-in beads and synthetic rubber gaskets, as exemplified in *(d)*.

may not be able to deflect enough under heavy wind loads without being stressed at the edge. The mullions, of course, must be stiff enough to transmit the wind loads from the glass to the frame of the building without deflecting so far as to overstress the glass. The resilient glazing material used to seal the glass-to-mullion joint must be of sufficient dimension and elasticity to allow for any anticipated thermal movement and for possible irregularities in the mullions.

The glazing materials that are most commonly used between the mullions and the glass include *wet glazing components* and *dry glazing components.* The wet components are mastic sealants and glazing compounds. The dry components are rubber or elastomeric gaskets. Wet glazing, with good workmanship, is more effective in sealing against penetration of water and air. Dry glazing is faster, easier, and less dependent on workmanship than wet glazing. The two types are often used in combination to utilize the best properties of each.

In the large-light glazing systems shown in Figure 17.17, rubber setting

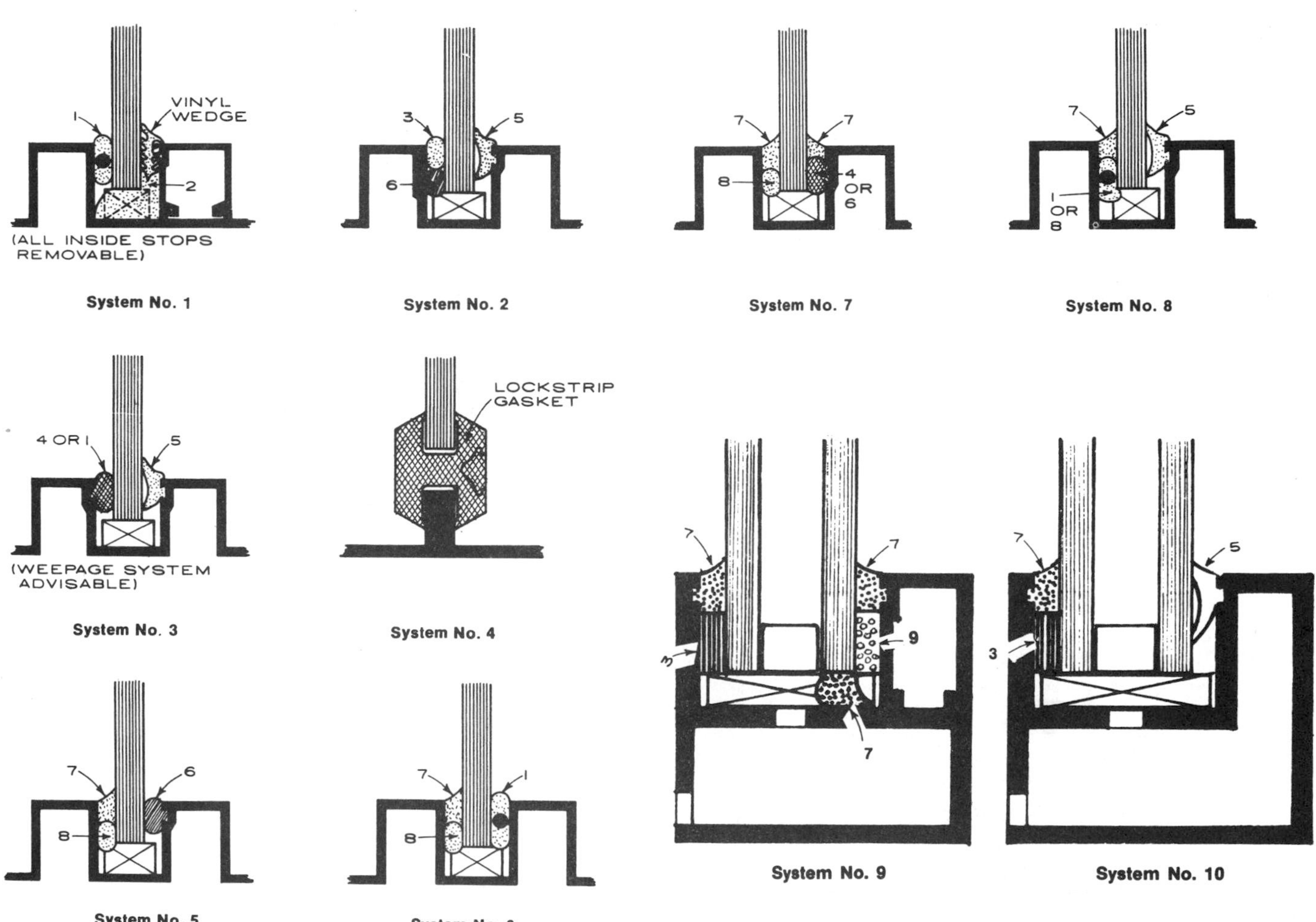

Outdoor side at left in all cases. System numbers have no significance as to order of preference.

KEY TO MATERIALS

1 Pre-shimmed butyl or polyisobutylene tape
2 Acrylic or bulyl sealant
3 Butyl or polyisobutylene tape
4 Cellular neoprene
5 Dense neoprene roll-in gasket
6 Dense neoprene gasket
7 High range sealants (polysulfide, silicone or acrylic sealant) (see Note)
8 Dense neoprene roll-in rod
9 Cellular Tape

Note: Choice of cap sealant depends on opening size and other conditions; consult sealant mfr.

FIGURE 17.17

An assortment of typical large-light glazing details, with the outdoor side to the left.

(American Architectural Manufacturers Association Window Selection Guide—1988)

Figure 17.18
Inserting the lockstrip into a lockstrip gasket to expand the gasket and seal it against the glass. *(Courtesy of Standard Products Company)*

Figure 17.19
A finished lockstrip gasket glazing installation. *(Courtesy of Standard Products Company)*

blocks are indicated by the rectangles with X's in them under the lower edge of the glass. Systems 1, 2, 3, 6, 9, and 10 use *preformed solid tape sealant*, a thick ribbon of very sticky synthetic that is adhered by pressure to the glass and the mullions. Several different formulations (butyl, polyisobutylene) are usually lumped together under the name *polybutene glazing tape*. The tape sealant material exerts an extremely strong hold on the glass and stays plastic indefinitely to allow for movement in the glazing system.

Systems 1, 2, 3, 8, and 10 utilize *wedge* or *roll-in gaskets*, strips of elastomeric material that are simply pushed into the gap between the glass and the mullion on the interior side to wedge the assembly tightly together and seal against air leakage. Systems 5–10 seal the outside gap with a wet glazing material. System 4 comprises a two-piece *lockstrip gasket* that is a completely self-contained glazing system. Figures 17.18 and 17.19 show applications of lockstrip gasket glazing.

The properties that solid tape sealants, mastic sealants, and compression gaskets have in common are that they possess the required degree of resiliency, they can be installed in the thickness required to cushion the glass against all expected movements, and they form a watertight seal against both glass and frame. To guard against possible leakage and moisture condensation, however, *weep holes* should be provided to drain water from the horizontal mullions to the exterior of the window frame, as seen directly below the bottom edge of the glass in Systems 9 and 10 in Figure 17.17.

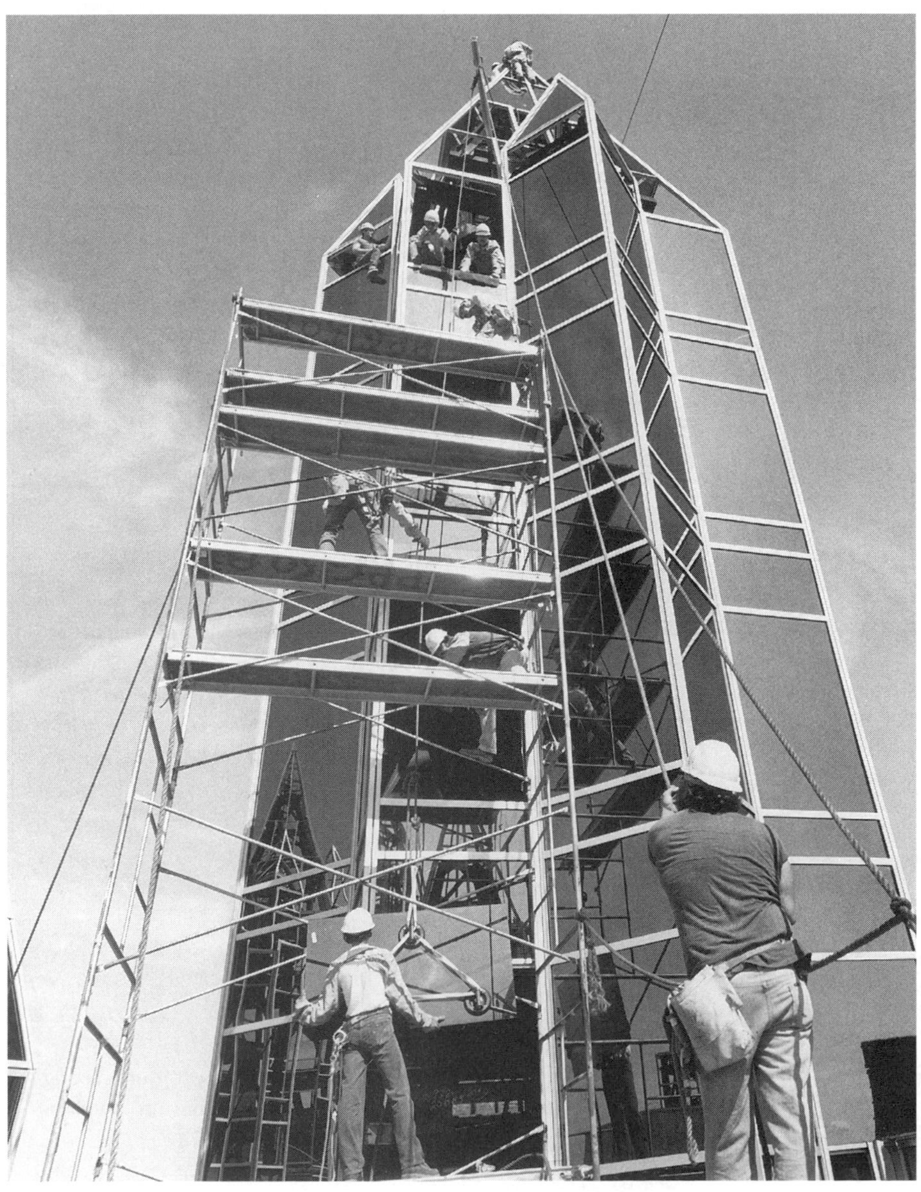

FIGURE 17.20
Glaziers install lights of reflective glass that weigh as much as 125 pounds (57 kg) each in a spire atop an office building. The glass is attached to the hoisting rope by the usual means, suction cups, as seen in front of the worker at the lower left.
(John Burgee Architects with Philip Johnson. Photo courtesy of PPG Industries)

Advanced Glazing Systems

In their quest to design ever more minimal details for buildings, architects have encouraged the development of systems of glazing that seem, in varying degrees, to defy gravity. In *butt-joint glazing*, the head and sill of the glass sheets are supported conventionally in metal frames, but vertical mullions are eliminated, the vertical joints between sheets of glass being made by the injection of a colorless silicone sealant. This gives a strong effect of unbroken horizontal bands of glass wrapping continuously around the building (Figures 17.21–17.23).

In *structural glazing*, the metal mullions lie entirely inside the glass, with the glass adhered to the mullions with *structural silicone sealant* or, more

FIGURE 17.21
Mullionless butt-joint glazing uses only a bead of colorless silicone sealant at the vertical joints in the glass. The glass in this example is single glazing ¾ inch (19 mm) thick. *(Photo courtesy of LOF Glass, a Libby-Owens-Ford Company)*

FIGURE 17.22
Another mullionless butt-joint glazing installation, as seen from the outside. *(Architects: Neuhaus & Taylor. Photo courtesy of PPG Industries)*

recently, *acrylic foam structural glazing tape*. Structural glazing allows the outside skin of the building to be completely flush, unbroken by protruding mullions (Figures 17.24–17.26). Notice in Figure 17.24 that the critical silicone sealant work is done in a factory, not on a construction site. The lights of glass are transported to the site already adhered to the small aluminum channels that will bind them to the mullions. In comparison to structural silicone glazing, acrylic foam structural glazing tape exhibits superior elastic properties, it can be applied more quickly and with less waste, it develops its adhesive bond more quickly, and it produces a cleaner visual appearance.

Most striking of all are the *suspended glazing systems*, used primarily for high walls of glass around building lobbies and enclosed grandstands. In the *glass mullion system*, the tempered glass sheets are suspended from above on special clamps and are stabilized against wind pressure by perpendicular stiffeners, also made of tempered glass, or by systems of tension cables. Where a single sheet of glass spans from the top of the window to the bottom, the glass and stiffeners are joined only by sealant. To create walls that are taller than a single sheet, metal fittings are used to join sheets at the corners and edges (Figures 17.1, 17.10, and 17.27). Stainless steel cables and fittings may be used to support large expanses of glass in roofs (Figures 17.28–17.30 and 17.35). The structure around the perimeter of the opening must be very stiff and strong to resist the pull of the cables that sustain the glass.

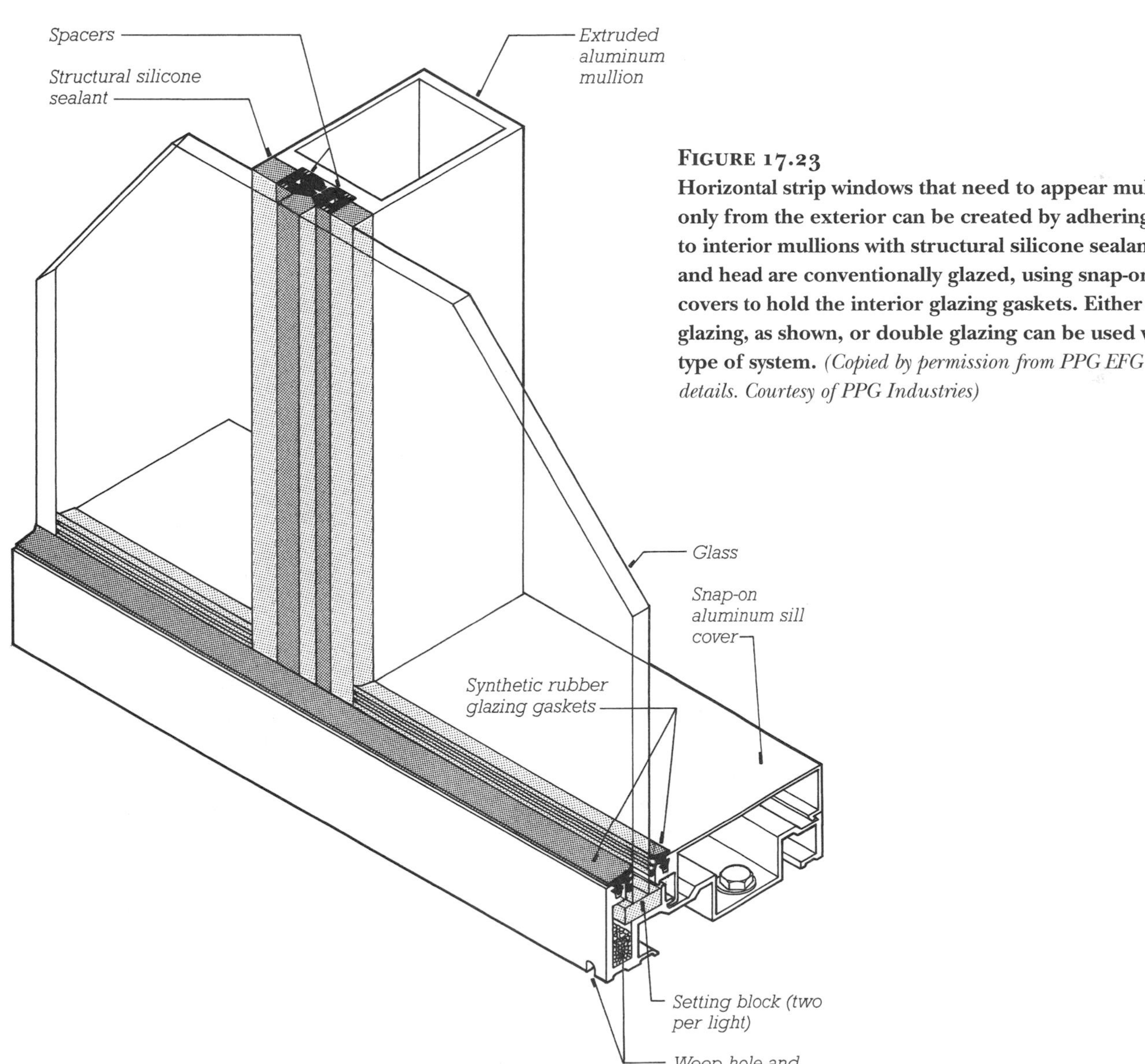

FIGURE 17.23
Horizontal strip windows that need to appear mullionless only from the exterior can be created by adhering the glass to interior mullions with structural silicone sealant. The sill and head are conventionally glazed, using snap-on aluminum covers to hold the interior glazing gaskets. Either single glazing, as shown, or double glazing can be used with this type of system. *(Copied by permission from PPG EFG System 401 details. Courtesy of PPG Industries)*

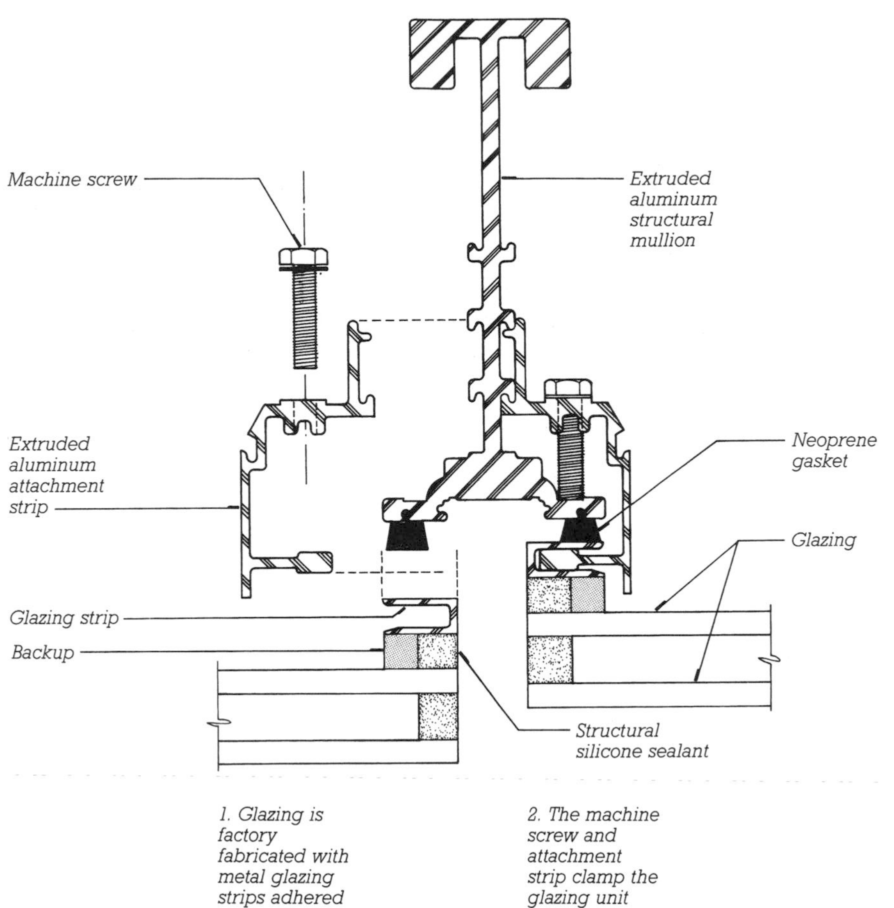
Machine screw
Extruded aluminum structural mullion
Extruded aluminum attachment strip
Neoprene gasket
Glazing
Glazing strip
Backup
Structural silicone sealant
1. Glazing is factory fabricated with metal glazing strips adhered to the glass using structural silicone sealant.
2. The machine screw and attachment strip clamp the glazing unit into place.

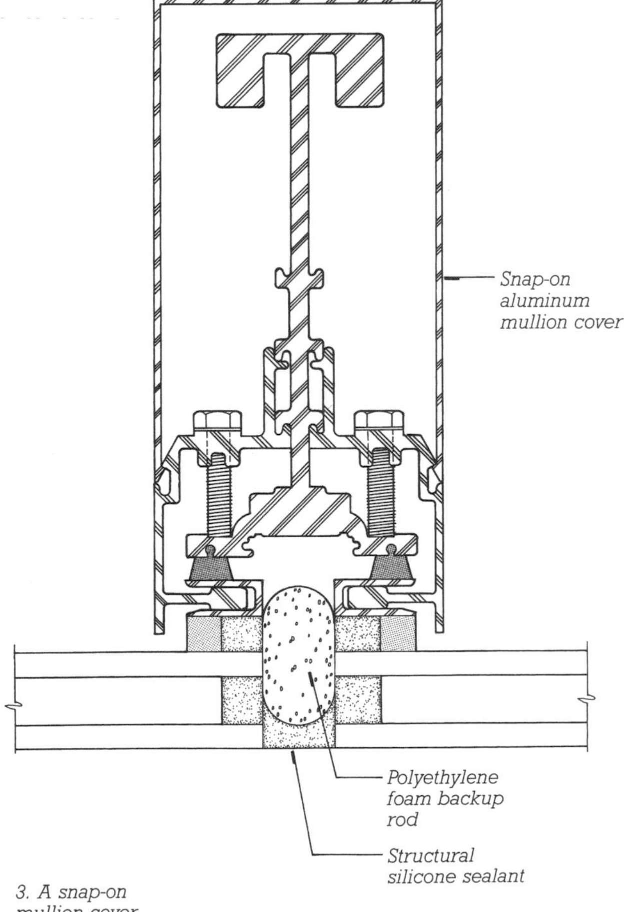
Snap-on aluminum mullion cover
Polyethylene foam backup rod
Structural silicone sealant
3. A snap-on mullion cover and an exterior sealant joint complete the assembly.

FIGURE 17.24

Steps in the assembly of a mullion for a four-side structural silicone exterior flush glazing system. This system is used to construct multistory glass walls with no metal exposed on the exterior of the building. The adhesive action of structural silicone sealant is the sole means by which the glass is held in place. This system is applicable either to double glazing, as shown, or to single glazing. Notice that the internal complexities of the aluminum components and sealant are completely concealed when the installation is finished. From the inside, one sees only a simple rectangular aluminum mullion, and from the outside, only glass and a thin bead of silicone sealant. *(Copied by permission from PPG EFG System 712 details. Courtesy of PPG Industries)*

FIGURE 17.25
Reflective glass mounted with a four-side structural silicone glazing system shows no metal on the exterior of this Texas office building, only thin lines of sealant. *(Architects: Haldeman, Miller, Bregman & Haman. Photo courtesy of PPG Industries)*

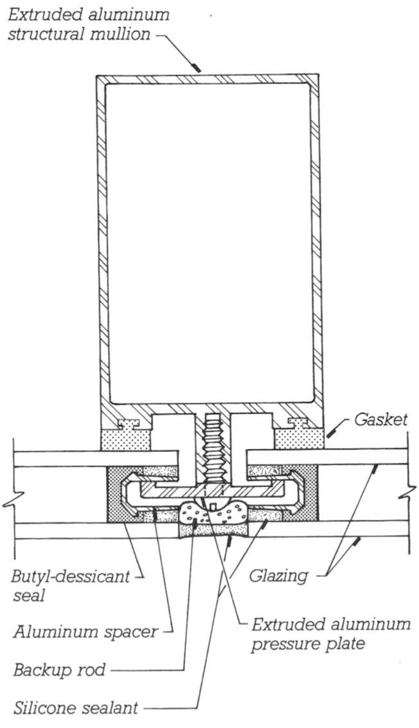

FIGURE 17.26
Structural spacer glazing is a patented system of flush glazing that provides a more positive attachment of the double-glazed units to the building. The glass is fastened to the mullion with an aluminum pressure plate that engages a slot in the spacer strip between the sheets of glass. The desiccant required to remove residual moisture from the airspace is mixed with the butyl sealant material in this system.

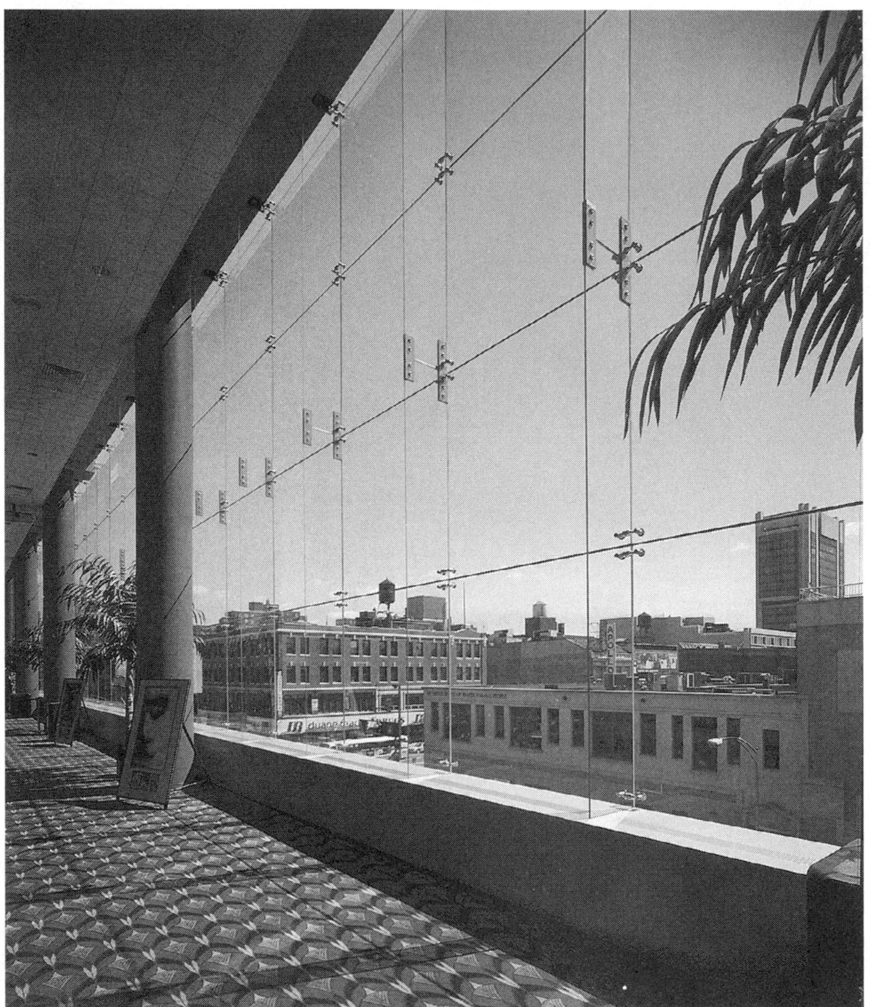

FIGURE 17.27
The *low-iron glass* in the Magic Johnson Theater in New York City is highly transparent, an effect that is accentuated by suspending the glass from above, using only vertical glass stiffeners to resist wind loads. Stainless steel fittings join the components of the wall. *(Photo of Pilkington Planar System courtesy of W&W Glass Systems, Inc.)*

FIGURE 17.28
Airside 2 Terminal at Orlando International Airport, designed by HOK Architects, features large areas of insulating laminated glass units. Laminated glass fins serve as beams to conduct the weight of the roof to vertical stainless steel rods that are supported by a stainless steel cable tensile structure. The downward-hanging cables transmit the weight to the stiff steel pipe trusses around the perimeter, which also resist the inward pull of the cables. The upward-arching cables hold the glass surface down against possible suction forces exerted by wind. Frit was used to diminish reflected glare on glass facing the control tower, and low-e coatings were applied to the insulating glass. *(Photo of Pilkington Planar System courtesy of W&W Glass Systems, Inc.)*

FIGURE 17.29
A detail of the glass roof supports. Notice the four-point and two-point stainless steel spider fittings that attach the glass components to each other and to the metal supporting structure. *(Photo of Pilkington Planar System courtesy of W&W Glass Systems, Inc.)*

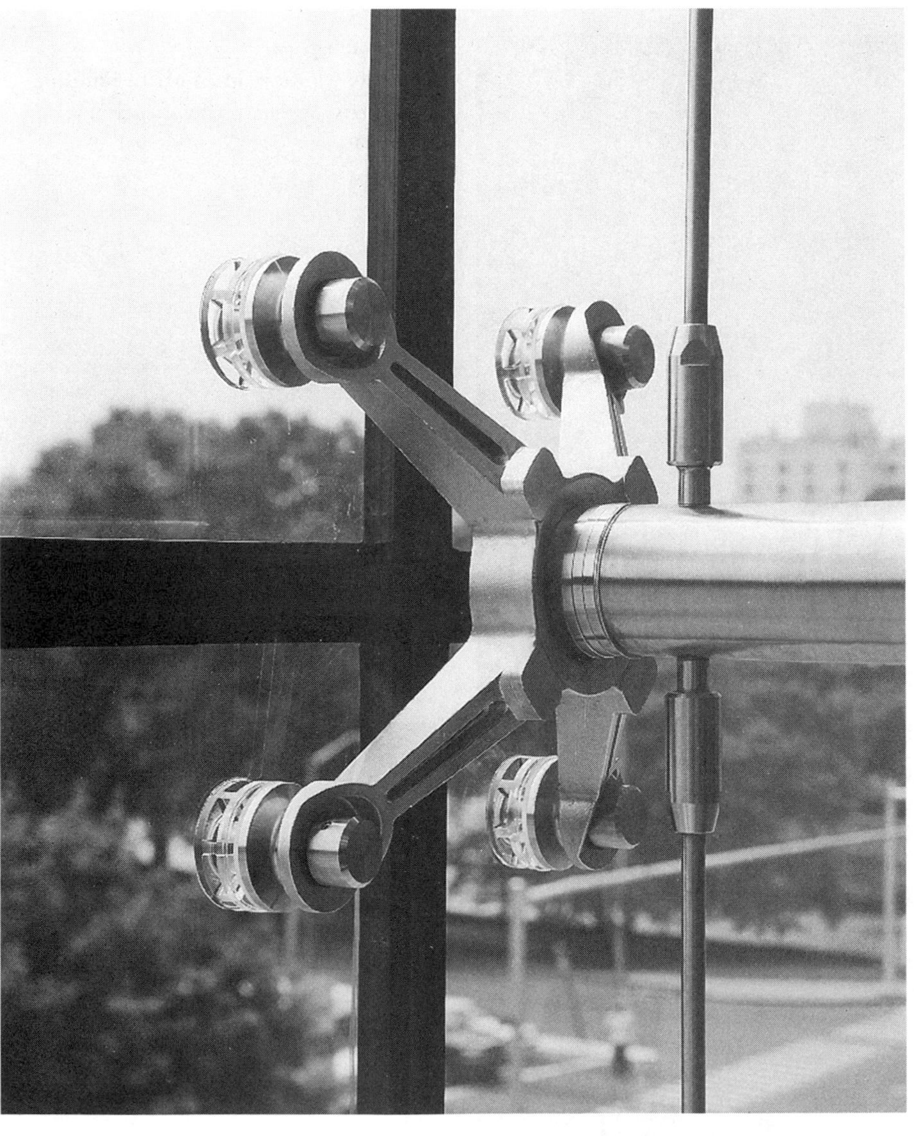

FIGURE 17.30
A close-up view of a four-point spider fitting that holds corners of four individual pieces of insulating glass. Vertical stainless steel rods in adjustable fittings carry the weight of the glass to the structure above. *(Photo of Pilkington Planar System courtesy of W&W Glass Systems, Inc.)*

Imagine a city iridescent by day, luminous by night, imperishable! Buildings, shimmering fabrics, woven of rich glass; glass all clear or part opaque and part clear, patterned in color or stamped to harmonize with the metal tracery that is to hold it all together, the metal tracery to be, in itself, a thing of delicate beauty consistent with slender steel construction. . . .

Frank Lloyd Wright, in *Architectural Record*, April 1928

FIGURE 17.31
Leaded diamond-pane windows of handmade rolled glass in an Elizabethan English bay window. *(Photo by Edward Allen)*

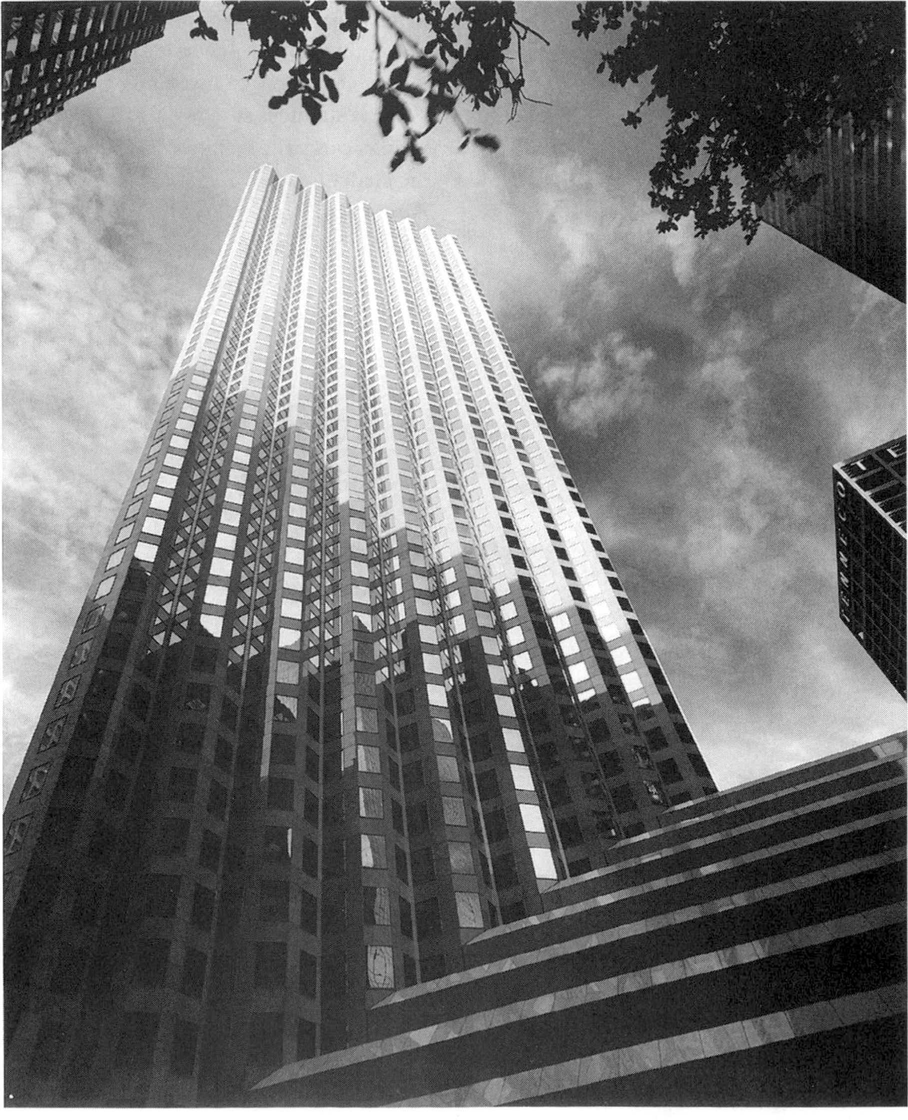

FIGURE 17.32
Discrete windows in an office building. *(Architects: Skidmore, Owings and Merrill. Photo courtesy of PPG Industries)*

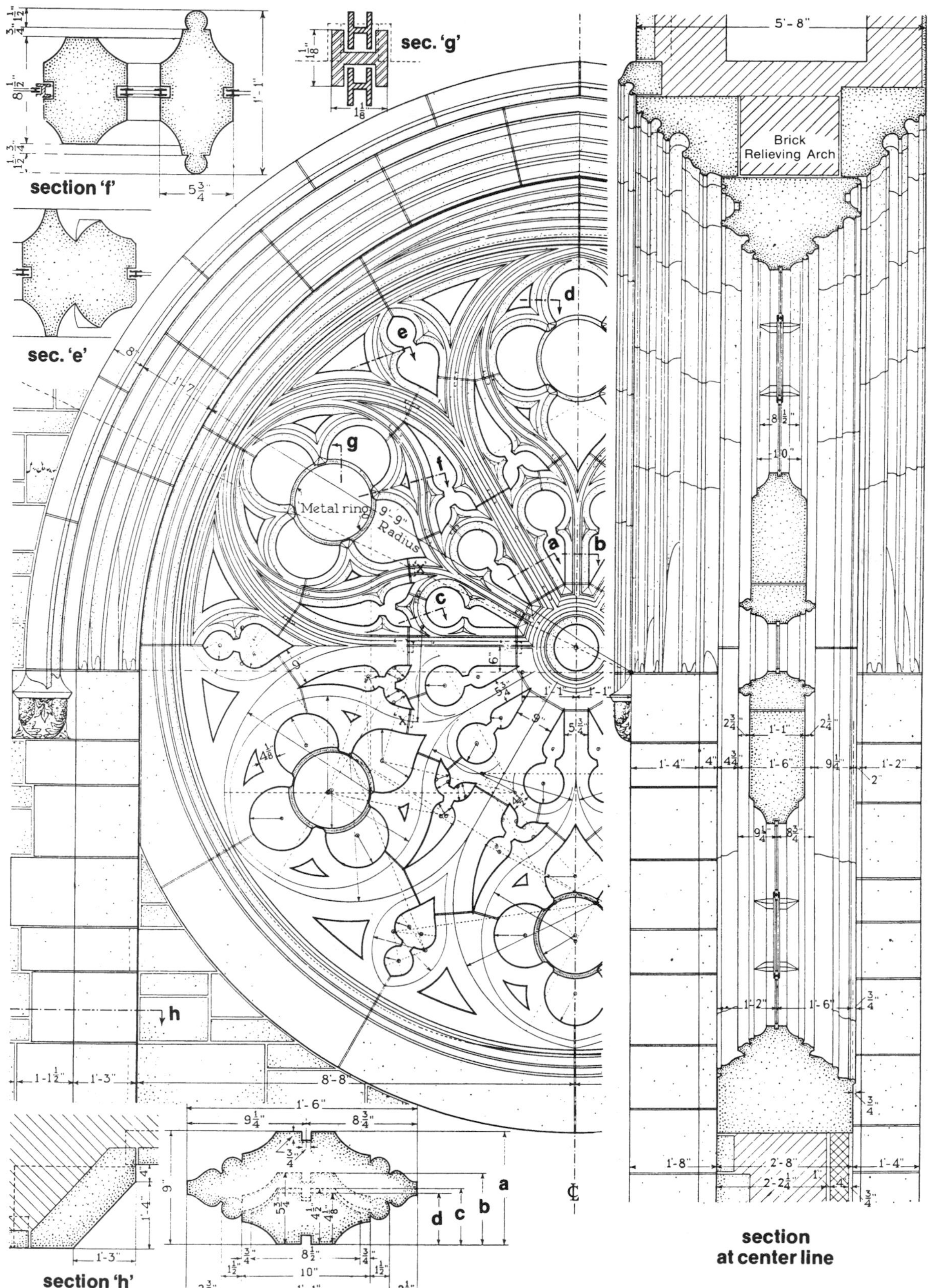

FIGURE 17.33
Limestone and metal mullions for a Gothic church window. *(Courtesy Indiana Limestone Institute)*

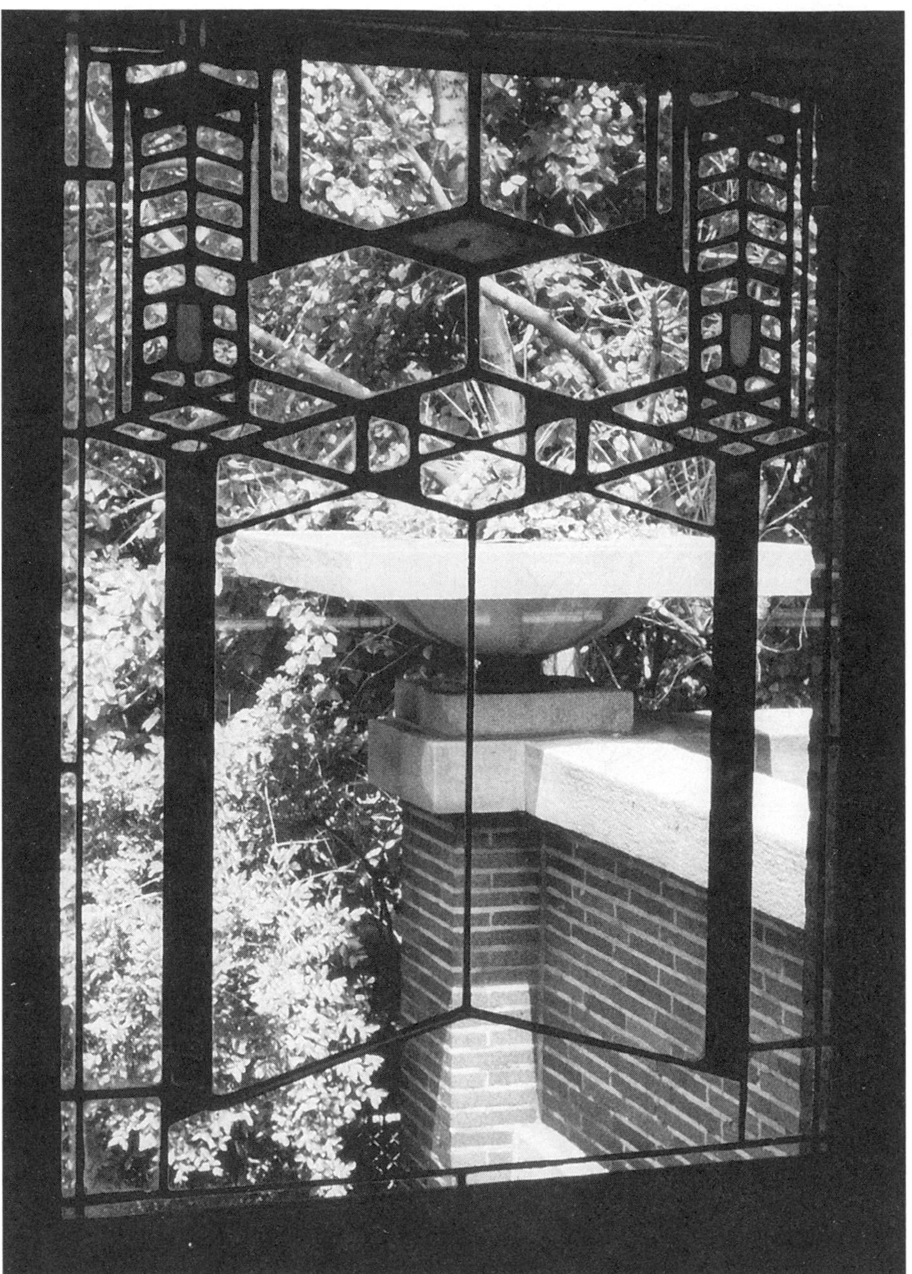

Winter dining rooms and bathrooms should have a southwestern exposure, for the reason that they need the evening light, and also because the setting sun, facing them in all its splendor but with abated heat, lends a gentler warmth to that quarter in the evening. Bedrooms and libraries ought to have an eastern exposure, because their purposes require the morning light . . . . Dining rooms for Spring and Autumn to the east; for when the windows face that quarter, the sun, as he goes on his career from over against them to the west, leaves such rooms at the proper temperature at the time when it is customary to use them.

Marcus Vitruvius Pollio, Roman architect, *The Ten Books of Architecture*, 1st century B.C.

FIGURE 17.34
Leaded stained glass in the Robie House, Chicago, 1906.
(Architect: Frank Lloyd Wright. Photo by Edward Allen)

GLASS AND ENERGY

Glass is a two-way pipeline for the flow of both conducted and radiated heat. As noted previously, glass, even when doubled or tripled, conducts heat rapidly into or out of a building. It can also collect and trap large amounts of solar heat inside a building.

In residential buildings, the conduction of heat through glass should be minimized in the extremely hot or cold seasons of the year. Double glazing, low-e coatings, low-conductivity gas fills, and snug curtains or shutters are desirable features for residential windows. Warming sunlight is welcome in winter but highly undesirable in summer, which leads the conscientious designer of residences to orient major windows toward the south, with overhangs or sunshades above to protect them against the high summer sun. Large east or west windows can cause severe overheating in summer and should be avoided unless they are shaded by nearby trees.

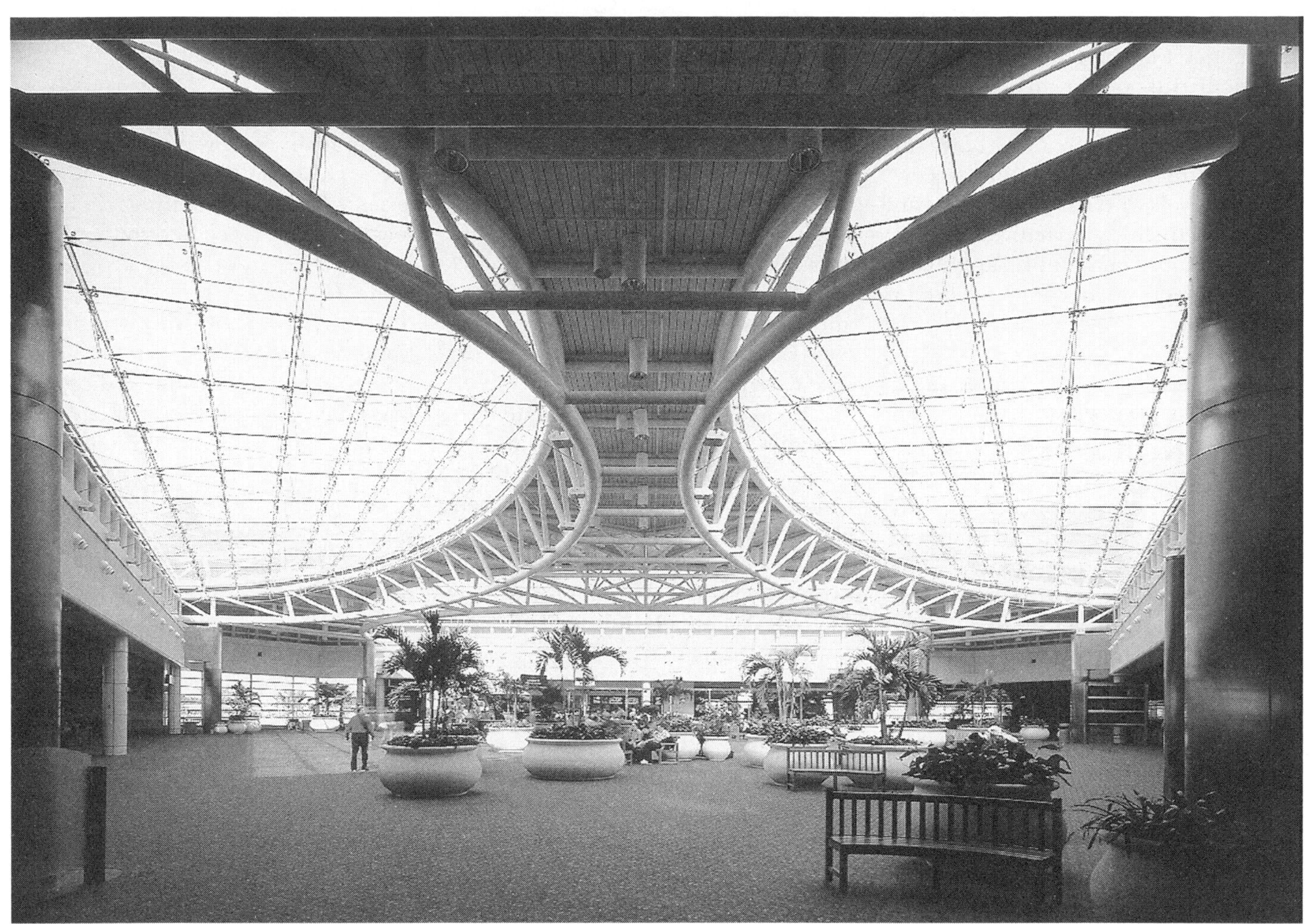

FIGURE 17.35
Airside 2 Terminal at Orlando International Airport. *(Photo of Pilkington Planar System courtesy of W&W Glass Systems, Inc.)*

In nonresidential buildings, heat generated within the building by lights, people, and machinery is often sufficient to maintain comfort throughout much of the winter. In warmer weather, this heat, along with any solar heat that has entered the windows, must be removed from the building by a cooling system. In this situation, north-facing windows contribute least to the problem of cooling the building, and south-facing windows with horizontal sunshades above allow the entry of solar heat only in the winter. East and west windows are problematic, as they contribute strongly to summertime overheating and are very difficult to shade. Shades or blinds inside the glass are helpful in eliminating the glare from such windows, but they do little to keep out the heat because once sunlight strikes them, its heat is already inside the building and little of it will escape.

Tinted and reflective glasses are of obvious value in controlling the entry of solar heat into buildings, to the point that they might be perceived as encouraging the designer to pay little attention to window size and orientation. But a growing number of larger-scale buildings are characterized by different glazing schemes for the different sides of the building, each designed to create an optimal flow of heat into and out of the building for that orientation, and each making creative use of the available types of glass for this purpose. The results, as measured in occupant comfort and energy savings, are generally impressive, and the aesthetic possibilities are intriguing.

This last statement could apply equally well to the role of glass in admitting light to a building. Electric lighting is often the major consumer of energy in a commercial building, especially when the heat generated by the lights must be removed from the building by a cooling system. Daylight shining through windows and skylights, distributed throughout a space by reflecting and diffusing surfaces, can reduce or eliminate the need for electric lights under many circumstances and is often more pleasant than artificial illumination. Low-cost computer models make it easy to predict the levels of daylighting that can be achieved with alternative designs, enabling more and more architects and engineers to become expert in this field.

Glass and the Building Codes

Building codes are concerned with several functional aspects of glass: its structural adequacy against wind and impact loads; its role in providing natural light in habitable rooms; its breakage safety; its safety in preventing the spread of fire through a building; and its role in determining the energy consumption of a building.

The International Building Code provides structural criteria for determining the necessary thicknesses of glass to resist wind and other structural loads. In coastal regions where hurricanes are common, the code also requires windows or window coverings to meet requirements for resistance to the impact of objects that may be blown against glazed areas by high winds.

The International Residential Code requires all habitable rooms to have a net exterior glazed area equal to at least 8 percent of their floor area. The International Building Code does not generally require exterior windows or glazing (with the exception that some bedrooms are required to have emergency escape windows or doors) and permits spaces to be lit by artificial illumination alone. The use of natural light to provide interior illumination and the provision of views to the exterior for building occupants are recognized components of healthy, energy-efficient buildings and are encouraged by LEED for New Construction and other sustainable design programs.

Breakage safety is regulated in skylights and overhead glazing to avert accidental injury that might be caused by falling shards of broken glass. Laminated glass and plastic glazing sheets, because they will not drop out of the skylight if broken, are the only skylight glazings that are permitted without restriction.

Breakage safety is also important in glazing that people might run into and break with their bodies. To avoid severe injury in such an occurrence, building codes mandate that sheets of glass in hazardous locations must be some type of *safety glazing*, that is, glass or plastic that does not create large, sharp, potentially lethal spears when it breaks. Examples of such locations include areas in and around exterior doors where people may accidentally bump against the glass, floor-to-ceiling sheets of glass that are often walked into by people who mistake them for openings in the wall, shower enclosures, and windows having a glazed area in excess of 9 square feet (0.9 m^2) whose lowest edge is within 18 inches (450 mm) of the floor. Glazing materials that meet safety glazing requirements include tempered glass, laminated glass, and plastic sheet.

Fire-rated glass must be used for openings in required fire doors and fire separation walls. The maximum areas of glazed openings in these locations are specified by building codes. The International Building Code (IBC) also requires that windows aligned above one another in buildings over three stories in height be separated vertically by fire-resistive spandrels of a specified minimum height, usually 36 inches (914 mm). The intent of this provision is to restrict the spread of fire from one floor of a building to the floors above. If a glass spandrel is used, it must be backed up inside with a material that offers the necessary fire resistance.

Code provisions relating to glass and energy consumption generally offer several approaches to the designer of a building: Prescriptive requirements may be followed that spell out clearly the maximum amount of glass that may be used, expressed as a percentage of the overall wall or floor area, and the minimum thermal resistance of the glass. Alternative approaches allow the designer to trade off thermal performance between different parts of the building or to perform a detailed energy analysis of the entire building using approved methods, demonstrating in either case that the overall energy performance of the building is equal or superior to that of the same building designed in conformance to prescriptive requirements.

CSI/CSC	
MasterFormat Sections for Roofing	
08 80 00	GLAZING
08 81 00	Glass Glazing
08 83 00	Mirrors
08 84 00	Plastic Glazing
08 85 00	Glazing Accessories
08 88 00	Special Function Glazing
	Hurricane-Resistant Glazing
	Cable-Suspended Glazing
	Pressure-Resistant Glazing
	Radiation-Resistant Glazing
	Security Glazing
	Ballistics-Resistant Glazing

Selected References

1. The most current information on glass will be found in manufacturers' product literature, available either in hard copy or on the Web.

2. Glass Association of North America. *Glazing Manual.* Topeka, KS, updated regularly.

This handbook summarizes current practice in the production and use of glass in buildings.

3. Schittich, Christian, et. al. *Glass Construction Manual.* Munich, Birkhauser Verlag, 2007.

This is a beautifully produced and comprehensive treatment of modern uses of glass in architecture, including the material properties of glass, glazing and building energy consumption, and the details of glazing systems.

Web Sites

Glass and Glazing

Author's supplementary web site: **www.ianosbackfill.com/17_glass_and_glazing**

Cardinal Glass: **www.cardinalcorp.com**

Corning Museum of Glass: **www.cmog.org**

Glass Association of North America: **www.glasswebsite.com**

Nathan Allan Glass Studios: **www.nathanallan.com**

National Glass Association: **www.glass.org**

Pilkington: **www.pilkington.com**

Pilkington Planar: **www.pilkington.com/planar**

PPG Industries: **corporateportal.ppg.com/NA/IdeaScapes/GlassOverview.htm**

TGP Fire Rated Glass and Framing: **www.fireglass.com**

Whole Building Design Guide, Glazing: **www.wbdg.org/design/env_fenestration_glz.php**

Key Terms and Concepts

crown glass
cylinder glass
punty
plate glass
drawn glass
float glass
lehr
glazing
glazier
light, lite
single-strength glass
double-strength glass
annealed glass
heat-treated glass
quenching
tempered glass, fully tempered glass
heat-strengthened glass
laminated glass
polyvinyl butyral (PVB) interlayer
security glass
chemically strengthened glass
fire-rated glass
wired glass
optical-quality ceramic
fire-retardant filled double glazing
intumescent interlayer laminated glazing
glass fire wall, fire resistive glazing
fire protective glazing
frit
fritted glass, silkscreened glass
spandrel glass
visible light transmittance (VT)
tinted glass
solar heat gain coefficient (SHGC)
shading coefficient
light to solar gain (LSG) ratio
reflective coated glass, solar control glass
single glazing
double glazing
triple glazing
insulating glass unit (IGU)
edge spacer, spline
desiccant
warm edge spacer
U-Factor
vacuum-insulated glazing units
low-emissivity (low-e) coating
surface number
self-cleaning glass
chromogenic glass
thermochromic glass
photochromic glass
electrochromic glass, switchable glass
gasochromic glass
structural glass
antireflective glass
mirror glass
patterned glass
radiation shielding glass
photovoltaic glass
stained glass, colored glass
art glass
acrylic glazing
polycarbonate glazing
cellular polycarbonate glazing
fiberglass-reinforced polyester glazing
aerogel
glazier's points
glazing putty
glazing compound
mullion
setting block
bite
wet glazing component
dry glazing component
preformed solid tape sealant
polybutene glazing tape
wedge gasket, roll-in gasket
lockstrip gasket
weep hole
butt-joint glazing
structural glazing
structural silicone sealant
acrylic foam structural glazing tape
suspended glazing system
glass mullion system
low-iron glass
safety glazing

Review Questions

1. What are the advantages of float glass over drawn glass? Over plate glass?

2. Name two situations in which you might use each of the following types of glass: (a) tempered glass, (b) laminated glass, (c) wired glass, (d) patterned glass, (e) reflective glass, (f) polycarbonate plastic glazing sheet.

3. What are the design objectives for a large-light glazing system?

4. Discuss the role of glass that faces each of the principal directions of the compass in adding solar heat to an air-conditioned office building in summer. How should windows be treated on each of these facades to minimize summertime solar heat gain?

5. In what ways does a typical building code regulate the use of glass and why?

6. What is a low-e coating, and what are its benefits?

Exercises

1. Examine the ways in which glass is mounted in several actual buildings and sketch a detail of each. Explain why each detail was used in its situation, and why you agree or disagree with the detail used.

2. Find a book on passive solar heating of houses, and in it, a table of solar heat gains for your area. For windows facing each of the four directions of the compass, plot a curve that represents the solar heat coming through a square foot of clear glass on an average day in each month of the year. Which window orientation maximizes wintertime heat and minimizes summertime heat? Is there an undesirable orientation that maximizes summertime heat and minimizes wintertime heat?

PROJECT: Skating Rink at Yerba Buena Gardens, San Francisco

DESIGN ARCHITECT: Santos Prescott and Associates

EXECUTIVE ARCHITECT: LDA Architects

STRUCTURAL ENGINEER: Johnson and Neilsen/SOH&A

The original project concept was to make the skating rink at Yerba Buena Gardens a great urban room, intimately connected with its civic surroundings. Early in design, as part of their strategy to achieve this goal, architects Santos Prescott and Associates envisioned the north wall of the rink constructed entirely of glass. This would maximize the rink's connection to the neighboring outside garden and capitalize on views of the city skyline beyond. Finding an economical and elegant solution to the design and construction of this curtain wall became one of the essential ingredients in realizing the project (see Figure A).

As the design developed, translucent glazing panels were proposed for the remaining sides of the rink enclosure to provide balanced natural illumination. As a consequence, Santos Prescott was now looking for a glazing system that could accommodate both 2¾-inch-thick translucent panels and as standard 1-inch clear glazing for the south wall. As the building continued to take shape, additional criteria were established:

- At 30 feet in height, the south wall was too tall for the aluminum curtain wall to span on its own. A structural steel framework would be required behind the curtain wall to provide support against wind and other loads.
- Once the decision was made to fully support the curtain wall, structural requirements for the mullion system itself became minimal. This freed the architect to choose a relatively simple mullion profile based primarily on its ability to accommodate the glazing materials of varying thicknesses.
- To maximize transparency and fit the curtain wall within the spacing of the roof supports, a mullion grid 11 feet 3 inches wide by 5 feet high was established. These dimensions were approaching the upper size limits for sealed insulating glass units. However, pushing for larger glass units also permitted the least number of curtain wall mullions and steel backup elements to be used.

Once a curtain wall system had been selected. work proceeded in close coordination with the structural engineer to develop the steel backup structure in greater detail (see Figure B). Still striving for transparency, the goal was to find the most slender support elements possible. Since deflection under wind load was the limiting factor in the curtain wall design, a tube section was selected for its efficient sectional properties, and final sizes were determined: $3 \times 8 \times \frac{1}{2}$ inches for the vertical tubes and $3 \times 3 \times \frac{3}{8}$ inches for the horizontal tubes.

In the construction documents, the aluminum curtain wall and steel backup assembly was detailed (see Figure C) and its various components were specified. In the case of the aluminum mullion system, the specification excepert included here illustrates how performance requirements were use to delegate final engineering of this proprietary part of the assembly to its supplier. Specified criteria included resistance to structural loads, accommodation of movement, allowance for dimensional tolerances within the supporting elements, and others not shown here.

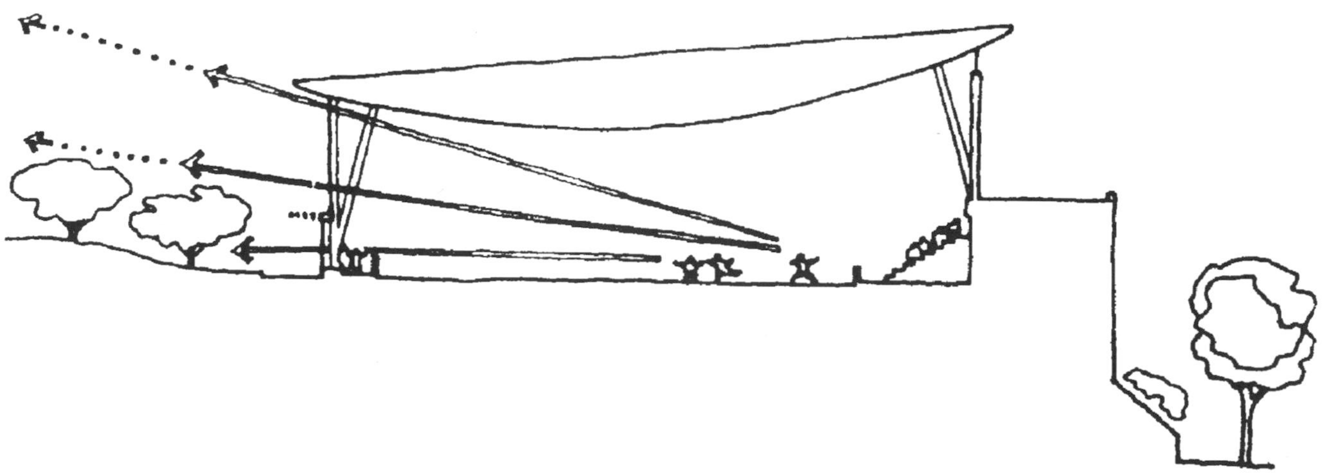

Figure A
Concept diagram.

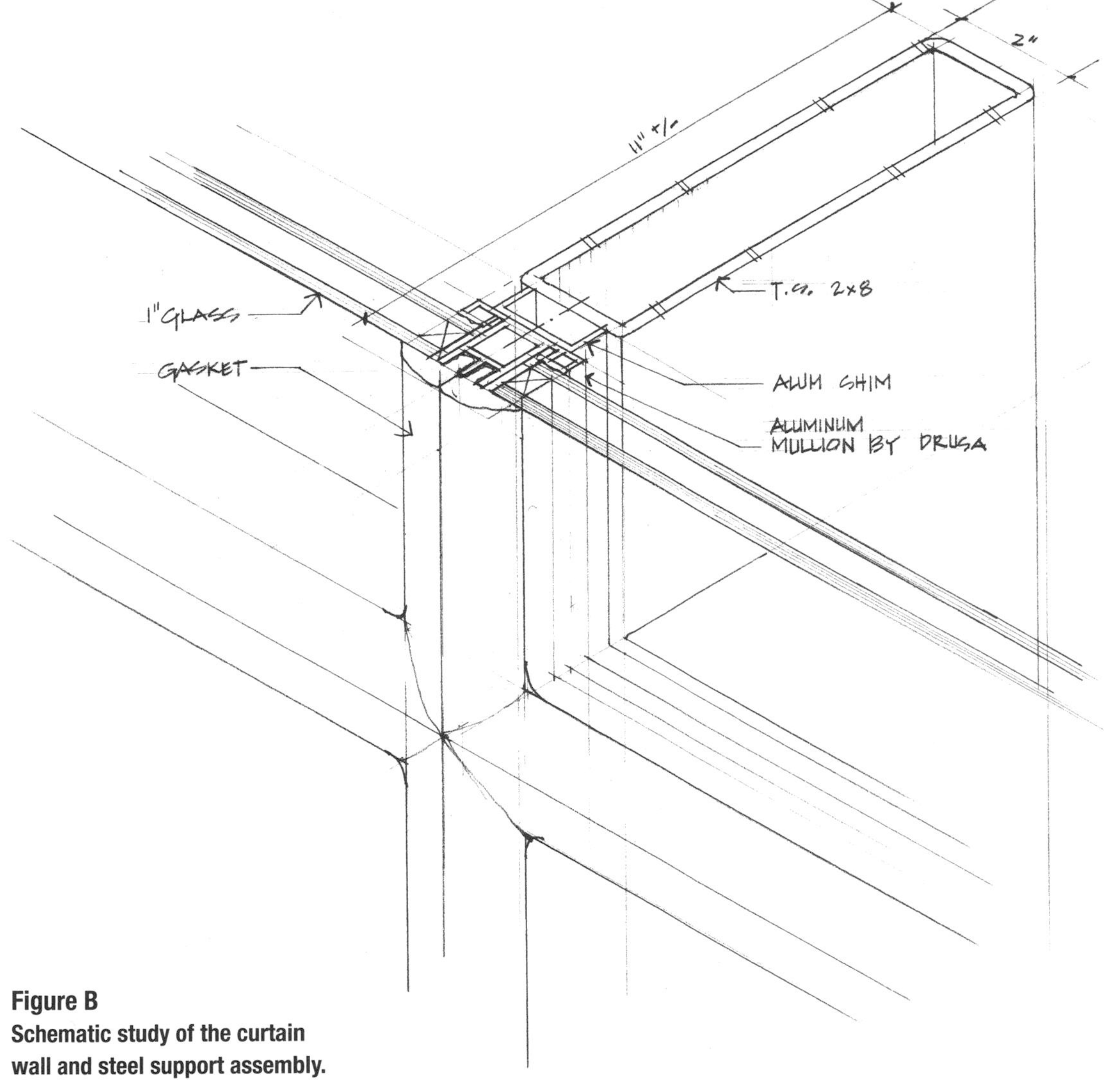

Figure B
Schematic study of the curtain wall and steel support assembly.

> **SECTION 08910**
>
> **GLAZED ALUMINUM CURTAIN WALLS**
>
> **PART 1— GENERAL**
>
> **1.03 PERFORMANCE REQUIREMENTS**
>
> **A. General: Provide glazed aluminum curtain wall systems, including anchorage, capable of withstanding, without failure, the effects of the following:**
>
> **1. Structural loads**
>
> **2. Movements of supporting structure indicated on drawings including, but not limited to, story drift, twist, column shortening, long-term creep, and deflection from uniformly distributed and concentrated live loads**
>
> **3. Dimensional tolerances of building frame and other adjacent construction**

Also included in the specification were requirements for the supplier to submit shop drawings and product information describing in detail the proposed curtain wall components and their attachment to the structural backup. Later in the project, these documents were reviewed by the general contractor, the architects, and the structural engineer. This review ensured that the proposed system indeed met the specified requirements and that the various parts of the curtain wall/glazing/structural system would come together as intended.

During the construction phase, a change was required. While preparing shop drawings, the curtain wall installer uncovered potential problems in attaching the curtain wall to the steel structure. When erection tolerances for the steel tubes were taken into consideration, it became apparent that it might not be possible to position the tubes with sufficient precision to meet the fastening requirements of the curtain wall. This problem was exacerbated by the ½-inch-thick side walls of the vertical steel tubes, which reduced the allowable area for fasteners on the face of the tubes to a narrow central strip. At this relatively late stage in the project, the decision was made to switch to a 4-inch-wide tube with a thinner side wall to provide a wider

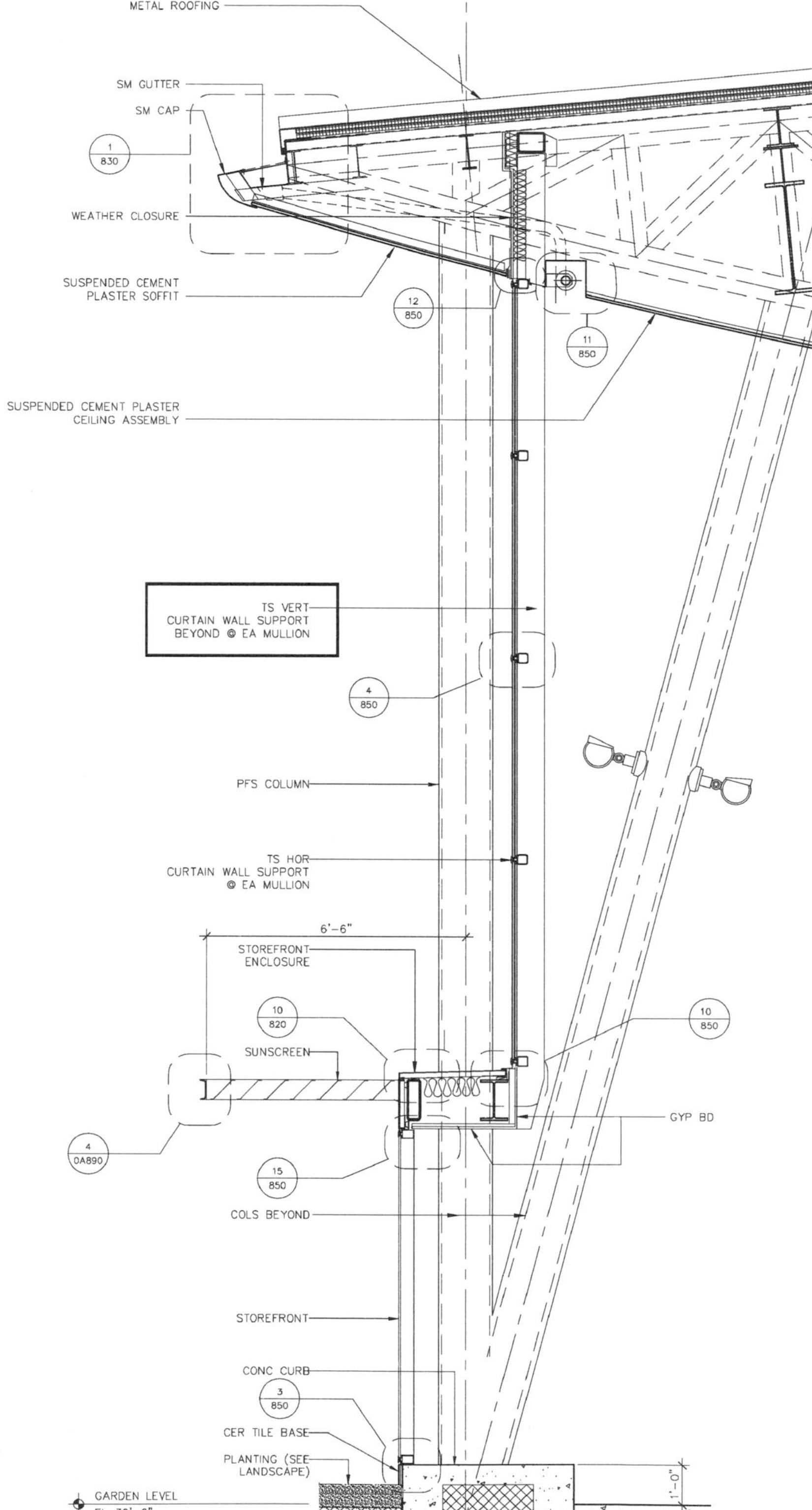

Figure C
Wall section from the construction drawing set.

target area for fasteners (see Figure D). In the end, the change had little if any noticeable visual impact. This example illustrates how a successful project outcome depends on good communication among the various team members throughout the design and construction of the project, and how even seemingly minor technical considerations can affect essential aspects of a project's design.

Special thanks to Santos Prescott and Associates, and Bruce Prescott, Principal, for assistance with the preparation of this case study.

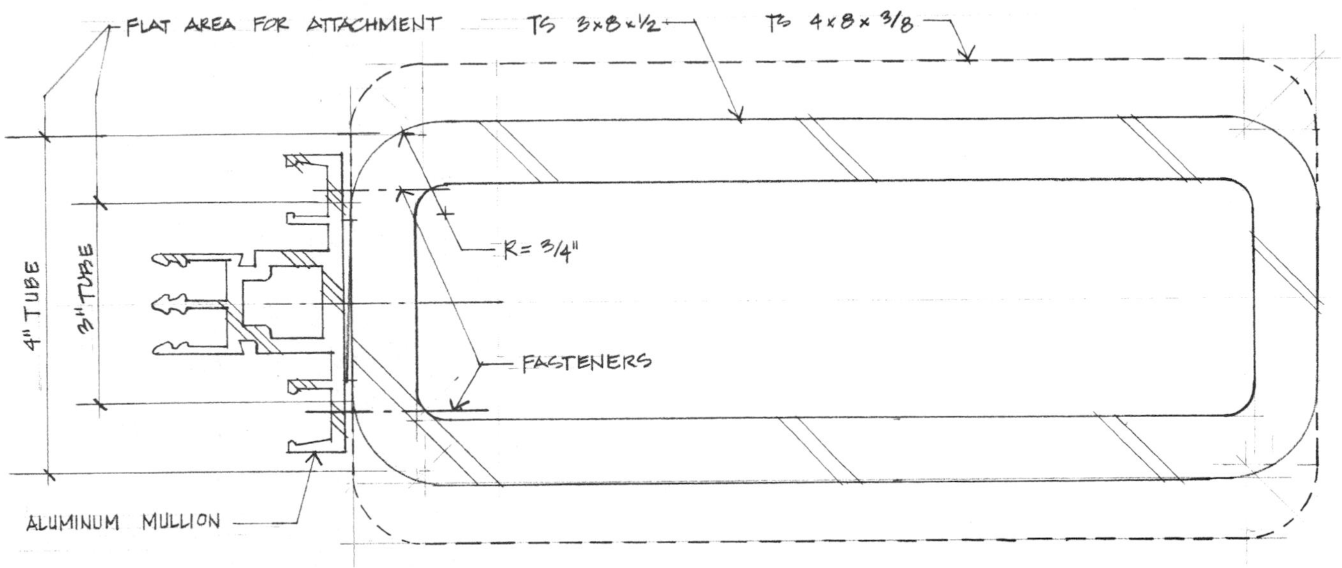

Figure D
Detail study of the curtain wall attachment to tube steel.

Figure E
Interior View.

18

WINDOWS AND DOORS

The mullion patterns of steel-framed windows are matched to the rusticated joints of the concrete cladding on the Arizona State University School of Architecture. *(Architect: Hillier Group. Photograph © Jeff Goldberg/Esto)*

Doors and windows are very special components of walls. Doors permit people, goods, and, in some cases, vehicles to pass through walls. Windows allow for simultaneous control of the passage of light, air, and sight through walls. Windows and doors, in addition to performing these important functions, play a large role in establishing the character and personality of a building, much as our eyes, nose, and mouth play a large role in our personal appearance. At the same time, windows and doors are the most complex, expensive, and potentially troublesome parts of a wall. To achieve satisfactory results, the experienced designer exercises great care and wisdom in the selection of doors and windows and in the specification of their installation.

Windows

The word *window* is thought to have originated in an old English expression that means "wind eye." The earliest windows in buildings were open holes through which smoke could escape and fresh air could enter. Devices were soon added to the holes to give greater control: hanging skins, mats, or fabric to regulate airflow; shutters for shading and to keep out burglars; translucent membranes of oiled paper or cloth, and eventually of glass, to admit light while preventing the passage of air, water, and snow. When a translucent membrane was eventually mounted in a moving sash, light and air could be controlled independently of each other. With the addition of woven insect screens, windows permitted air movement while keeping out mosquitoes and flies. Further improvements followed over the centuries. A typical window today is an intricate, sophisticated mechanism with many layers of control: curtains, shade or blind, sash, glazings, insulating airspace, low-emissivity and other coatings, insect screen, weatherstripping, and perhaps a storm sash or shutters.

Windows were formerly made on the construction site by highly skilled carpenters, but today nearly all of them are produced in factories. The reasons for factory production are higher production efficiency, lower cost, and, most important, better quality. Windows must be made to a very high standard of precision if they are to operate easily and maintain a high degree of weathertightness for many years. In cold climates especially, a loosely fitted window with single glass and a frame that is highly conductive of heat will significantly increase heating fuel consumption for a building, cause noticeable discomfort to the occupants, and condense large quantities of water that will stain and decay materials in and around the windows.

A *prime window* is one that is made to be installed permanently in a building. A *storm window* is a removable auxiliary unit that is added seasonally to a prime window to improve its thermal performance. A *combination window*, which is an alternative to a storm window, is an auxiliary unit that incorporates both glass and insect screening; a portion of the glass is mounted in a sash that can be opened in summer to allow ventilation through the screening. A combination window is normally left in place year round. Some windows are designed and manufactured specifically as *replacement windows* that install easily in the openings left by deteriorated windows removed from older buildings.

Types of Windows

Figure 18.1 illustrates in diagrammatic form the window types used most commonly in residential buildings, and Figure 18.2 shows additional types that are found largely in commercial and institutional buildings. *Fixed windows* are the least expensive and the least likely to leak air or water because they have no operable components. *Single-hung* and *double-hung windows* have one or two moving *sashes*, which are the frames in which the glass is mounted (Figure 18.3). The sashes slide up and down in tracks that are part of the window frame. In older windows, the sashes were held in position by cords and counterweights, but today's double-hung windows usually rely on a system of springs to counterbalance the weight of the sashes. A *sliding window* is essentially a single-hung window on its side, and shares with single-hung and double-hung windows the advantage that tracks in the frame hold the sashes securely along two opposite sides. This inherently stable construction allows single-hung, double-hung, and sliding windows to be designed in an almost unlimited range of sizes and proportions. It also allows the sashes to be more lightly built than those in *projected windows*, a category that includes principally *casement windows*, *awning windows*, *hopper windows*, *inswinging windows*, and *pivot windows*. All projected windows have sashes that rotate outward or inward from their frames and therefore must have enough structural stiffness to resist wind loads while being supported only at two corners.

With the exception of the rare *triple-hung window*, no window with sashes that slide can be opened to more than half of its total area. By contrast, many projected windows can be opened to virtually their full area. Casement windows assist in catching passing breezes and inducing ventilation through the building. They are generally narrow in width but can be joined to one another and to sashes

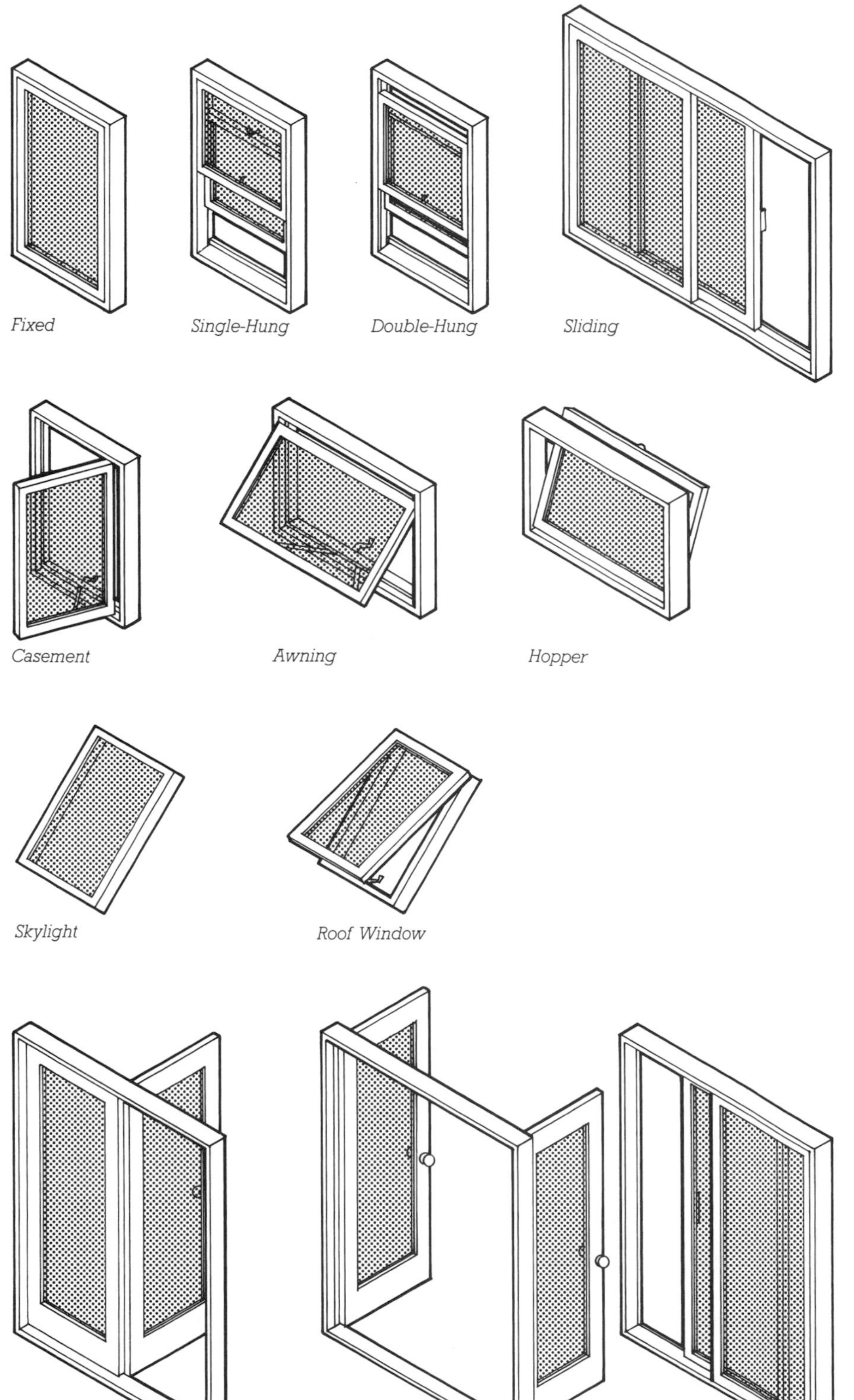

FIGURE 18.1
Basic window types.

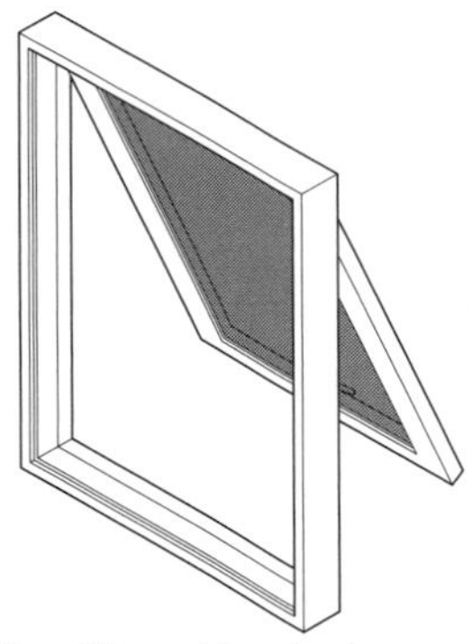

Top-Hinged Inswinging

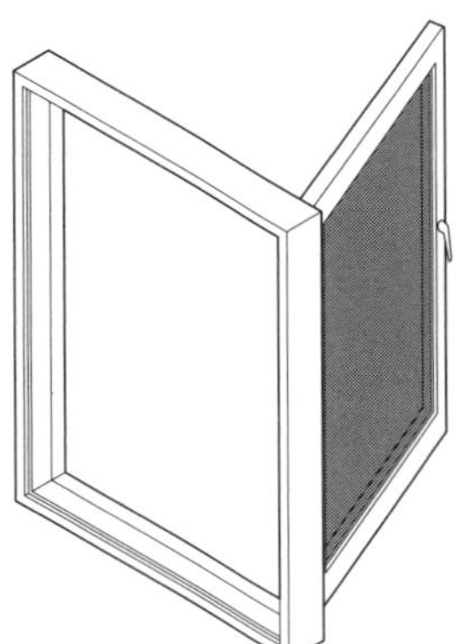

Side-Hinged Inswinging

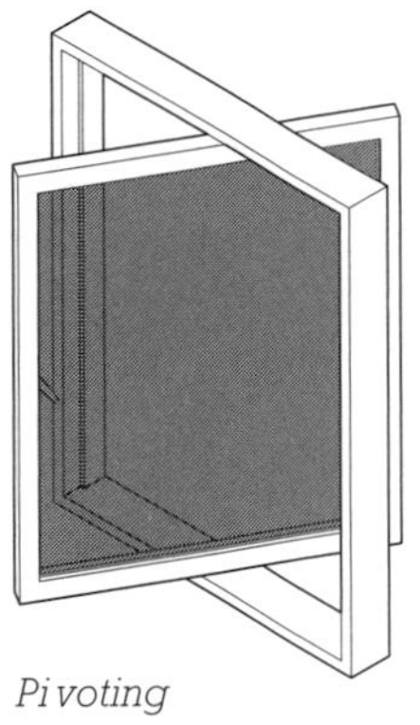

Pivoting

FIGURE 18.2
Additional window types that are used mainly in larger buildings.

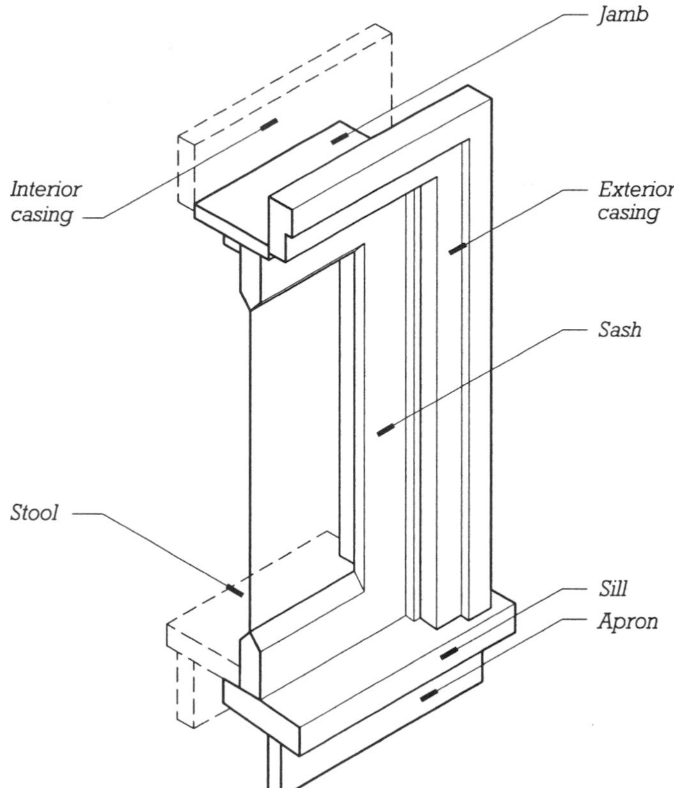

FIGURE 18.3
Basic window nomenclature follows a tradition that has developed over many centuries. The *jamb* consists of the *head jamb* across the top of the window and the *side jamb* to either side. In practice, the head jamb is usually referred to simply as the *head* and the side jambs as jambs. The *sill* frames the bottom of the opening on the exterior side, and the *stool* does the same on the interior. *Interior casings* and *exterior casings* cover the gaps between the jambs and the rough opening, and *aprons* do the same below the sill and stool.

of fixed glass to fill wider openings. Awning windows can be broad but are not usually very tall. They have the advantages of protecting an open window from water during a rainstorm and of lending themselves to a building-block approach to the design of window walls (Figure 18.4). Hopper windows are more common in commercial buildings than in residential ones. Like awning windows, they will admit little or no rainwater if left open during a rainstorm (since they are inswinging). *Tilt/turn windows* (not illustrated) are a type of projected window with clever but concealed hardware that allows each window to be operated either as an inswinging side-hinged window or a hopper.

A projected window is usually provided with pliable synthetic rubber weatherstripping that seals by compression around all the edges of the sash when it is closed. Single-hung, double-hung, and sliding windows generally must rely on brush-type weatherstripping because it does not exert as much friction against a sliding sash as rubber does. Brush-type materials do not seal as tightly as compression weatherstripping, and they are also subject to more wear than rubber weatherstripping over the life of the window. As a result, projected windows are generally somewhat more resistant to air leakage than windows that slide in their frames.

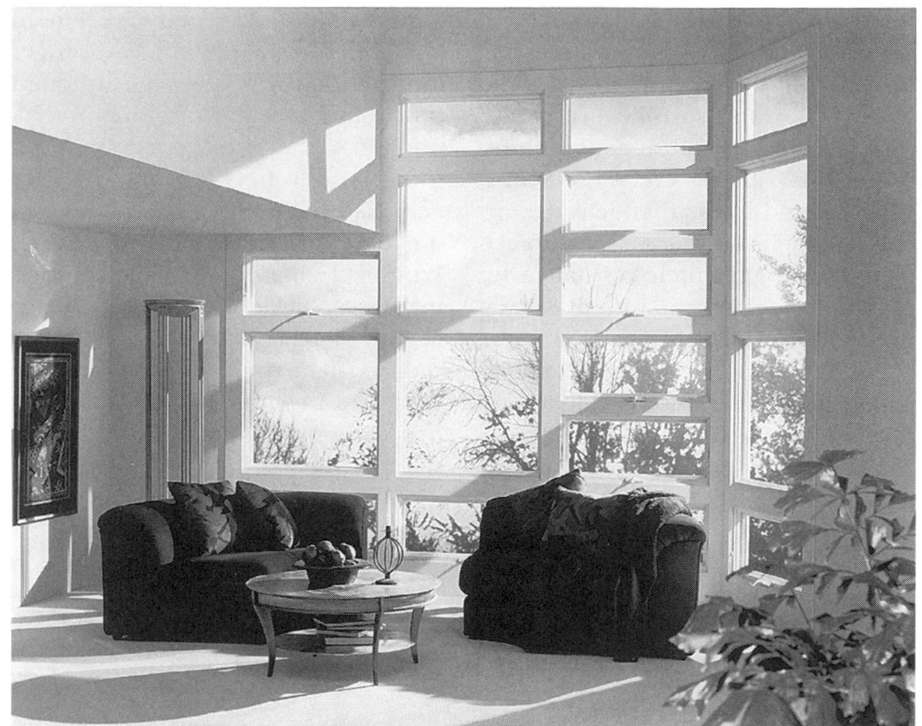

FIGURE 18.4
Awning and fixed windows in coordinated sizes offer the architect the possibility of creating patterned walls of glass. *(Photo courtesy of Marvin Windows and Doors)*

Glazed units for installation in roofs are specially constructed and flashed for watertightness. *Skylights* (also called *unit skylights*, to distinguish them from larger, framed-in-place skylights) may be either fixed or operable (venting). The term *roof window* is also sometimes applied to any venting skylight; at other times, it is applied only to operable windowlike units that include some kind of inward rotation capability to make outside glass surfaces accessible from the inside for easier cleaning.

Large glass doors (which are most often supplied by window manufacturers) may slide in tracks or swing open on hinges. The hinged *French door* opens fully and, with its arms flung wide, is a more welcoming type of door than the *sliding door*, but it cannot be used to regulate airflow through the room unless it is fitted with a doorstop that can hold it securely in an open position. The French door is prone to air leakage along its seven separate edges, which must be carefully fitted and weatherstripped. The *terrace door*, with only one operating door, minimizes this problem but, like the sliding door, can open to only half its area.

Window types that are used almost exclusively in commercial and institutional work include horizontally and vertically *pivoting windows* and *side-hinged inswinging* and *top-hinged inswinging windows* (Figure 18.2) These types all allow for inside washing of exterior glass surfaces. Because they project inward rather than outward, they are much less subject to the damage that could occur if, for example, casement windows were opened into the high winds that swirl around tall buildings. To further limit wind damage, windows in tall buildings are often fitted with devices that limit the extent to which they can be opened. In most tall buildings, if the windows operate at all, they are not meant to be used for ventilation unless the building's air conditioning system is inoperative.

Insect screens may be mounted only inside the sash in casement and awning windows (since the sash swing outward). Screens are usually positioned to the exterior side of other window types. Sliding patio doors and terrace doors have exterior sliding screens, and French doors require a pair of hinged screen doors on the exterior. Pivoting windows cannot be fitted with insect screens.

Glass must be washed at intervals if it is to remain transparent and attractive. Inside surfaces of glass are usually easy for window washers to reach. Outside surfaces are often hard to reach, requiring ladders, scaffolding, or window-washing platforms that hang from the top of the building on cables. Accordingly, most operable windows are designed to allow personnel to wash the outside surface of glass while standing inside the building. Casement and awning windows are usually hinged in such a way that there is sufficient space between the hinged edge of the sash and the frame when the window is open to allow one's arm to reach the outer surface of glass. Double-hung and sliding windows are often designed to allow sash to be rotated or tilted out of their tracks to allow easy cleaning of exterior glass from the building interior (Figure 18.13). See also the discussion of self-cleaning glass in the previous chapter.

Windows and glass doors may also be combined side-by-side or stacked vertically to create larger glazed areas with various fixed and operable components (Figures 18.4, 18.6, 18.7, 18.18).

Window Frames

Wood

The traditional frame material for windows is wood, but aluminum, steel, plastics, and combinations of these four materials have also come into widespread use. Wood is a fairly good thermal insulator, changes size relatively little with changes in temperature, and, if free of knots, is easily worked and consistently strong. But in service, wood shrinks and swells with changing moisture content and requires repainting every few years. When wetted by weather, leakage, or condensate, *wood windows* are subject to decay, though their resistance to decay can be improved with preservative treatments. Knot-free wood is becoming increasingly rare and expensive, so composite wood products are increasingly used for window frames. These include lumber made of short lengths of defect-free wood finger jointed and glued together, oriented strand lumber, and laminated veneer lumber. These materials, although functionally satisfactory, are not attractive, so they are normally covered with wood veneer on the interior and clad with plastic or aluminum on the exterior (Figures 18.5–18.7). Window frames made of solid wood may also be clad on the exterior for improved weather resistance and to reduce maintenance requirements. *Clad wood windows*, at the time of this writing, account for the largest share of the market for wood-framed windows.

FIGURE 18.5
Cutaway sample of an aluminum-clad wood-framed window. *(Photo courtesy of Marvin Windows and Doors)*

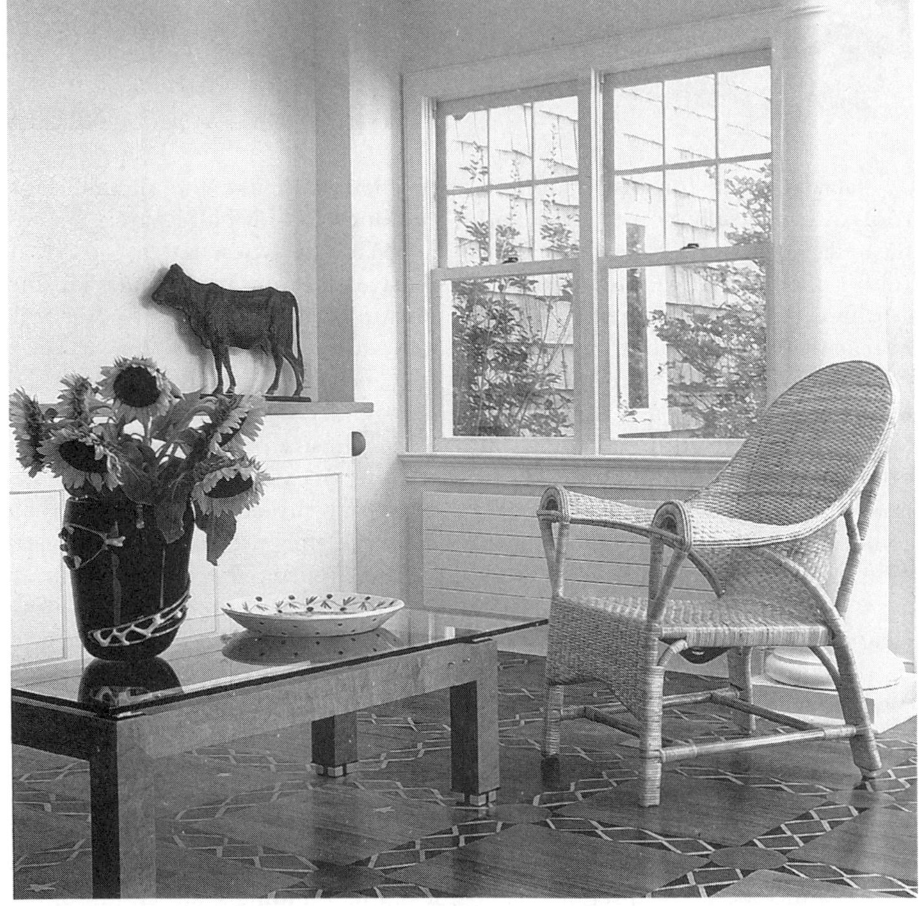

FIGURE 18.6
A pair of double-hung wood-framed windows in a dwelling. *(Photo courtesy of Marvin Windows and Doors)*

Aluminum

Aluminum used in window construction is strong, easy to form and join, and, in comparison to wood, much less vulnerable to moisture damage. The extrusion process by which aluminum sections are formed results in shapes with crisp, attractive profiles, and durable factory finishes eliminate the need for periodic repainting after installation. Aluminum conducts heat so rapidly, however, that unless an aluminum frame is constructed with a *thermal break* made of plastic or synthetic rubber components to interrupt the flow of heat through the metal, condensate and sometimes even frost will form on interior window frame surfaces during cold winter weather. *Aluminum windows* are also more costly than wood or plastic windows. The majority of commercial and institutional windows, as well as many residential windows, are framed with aluminum (Figures 18.8–18.10). Aluminum frames are usually anodized or permanently coated, as described in Chapter 21.

Plastics

Plastic window frames, though relatively new, now account for more than half of all windows sold in the U.S. residential market. Plastic windows never need painting, and they are fairly good thermal insulators. Many also cost less than wood or clad wood windows. The disadvantages of plastics as window frame materials are that they are not as stiff or strong as other window materials and they have very high coefficients of thermal expansion (Figure 18.11). The most common material for plastic window frames is polyvinyl chloride (PVC, vinyl), which is formulated with a high proportion of inert filler material to minimize thermal expansion and contraction.

FIGURE 18.7
Large double-hung wood windows and a triangular fixed window bring sunlight and views.
(Photo courtesy of Marvin Windows and Doors)

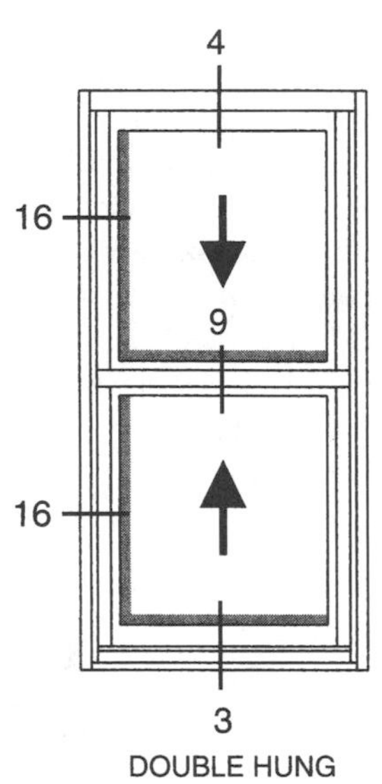

DOUBLE HUNG

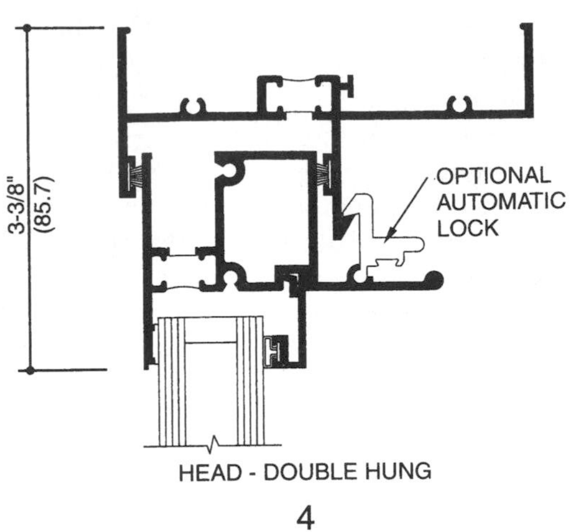

HEAD - DOUBLE HUNG

4

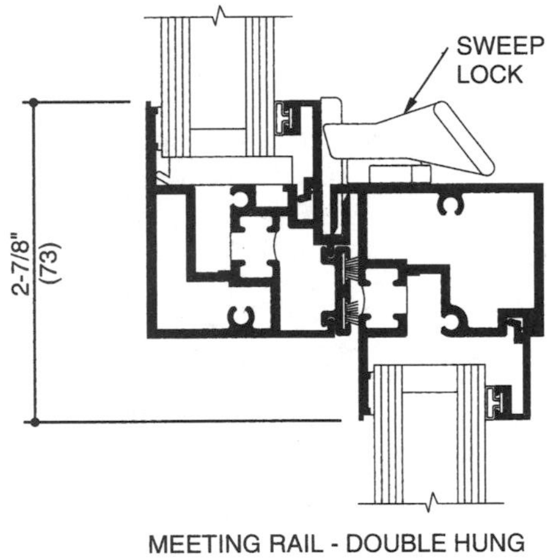

MEETING RAIL - DOUBLE HUNG

9

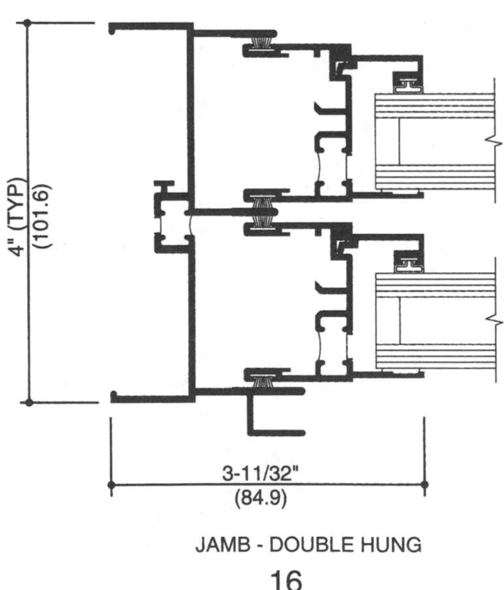

JAMB - DOUBLE HUNG

16

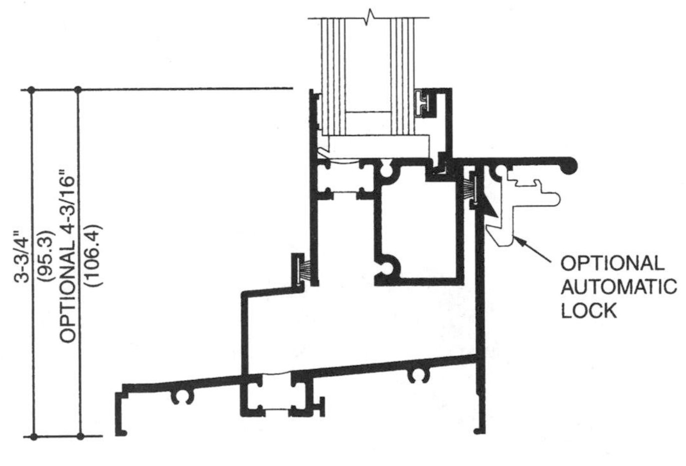

SILL
SINGLE / DOUBLE HUNG

3

FIGURE 18.8
The details of this commercial-grade double-hung aluminum window are keyed to the numbers on the small elevation view at the upper left. Cast and debridged thermal breaks, which are shown on the drawings as small white areas gripped by a "claw" configuration of aluminum on either side, separate the outdoor and indoor portions of all the sash and frame extrusions. Pile weatherstripping seals against air leaks at all the interfaces between sashes and frame. For help in understanding the complexities of aluminum extrusions, see Chapter 21. *(Courtesy of Kawneer Company, Inc.)*

(*a*)

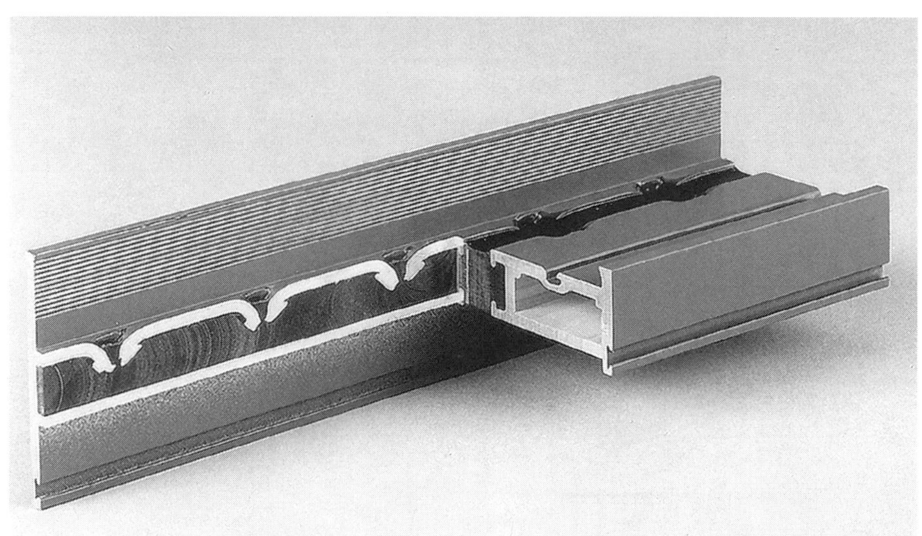

(*b*)

FIGURE 18.9
(a) **Two aluminum double-hung window units in back, with an aluminum sliding window in front. *(b)* A cutaway demonstration sample of a cast and debridged plastic thermal break in an aluminum window frame. In this proprietary design, the aluminum is "lanced" into the plastic at intervals to lock the two materials securely together.** *(Courtesy of Kawneer Company, Inc.)*

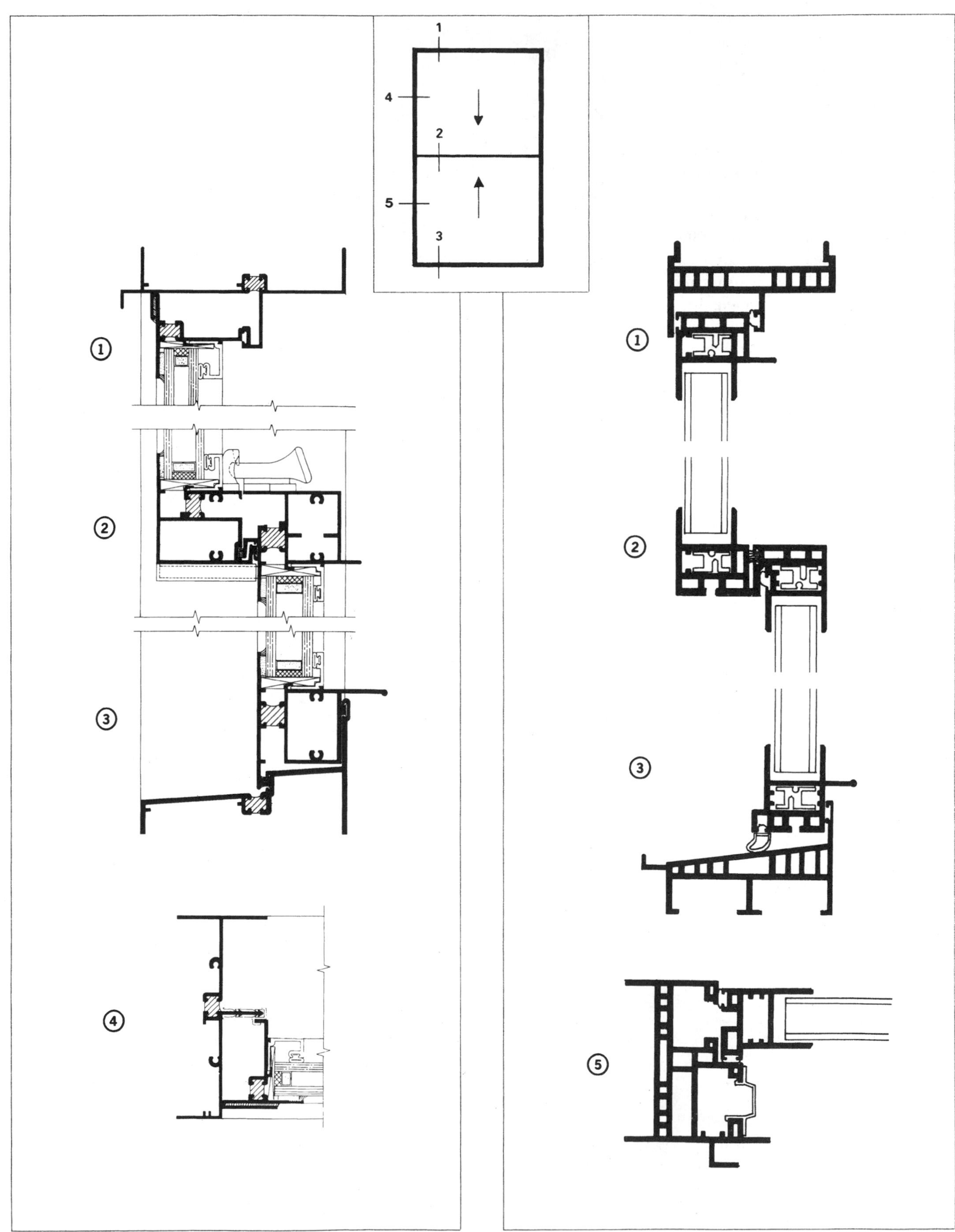

1
4
2
5
3
①
②
③
④
⑤

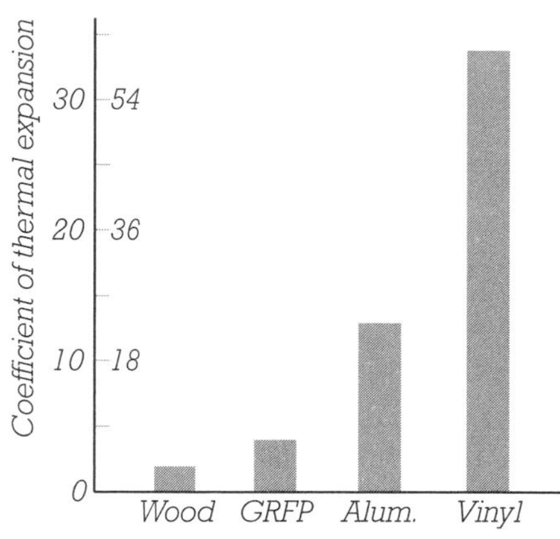

FIGURE 18.11
A comparison of the coefficients of thermal expansion of wood, glass-fiber-reinforced plastic (GFRP), aluminum, and vinyl. Vinyl expands 15 times as much as wood, 8 times as much as GFRP, and 3 times as much as aluminum. Units on the graph are in./in./°F x 10^{-6} on the left of the vertical axis and mm/mm/°C x 10^{-6} on the right.

FIGURE 18.10
Comparative details of a single-hung residential window with an aluminum frame (*left*) and a double-hung residential window with a PVC plastic frame (*right*). The small inset drawing at the top center of the illustration shows an elevation view of the window with numbers that are keyed to the detail sections below. The comparatively thick sections of plastic are indicative of the greater stiffness of aluminum as a material. The diagonally hatched areas of the aluminum details are plastic thermal breaks that interrupt the flow of heat through the highly conductive metal frame. The inherently low thermal conductivity of the PVC and the multichambered construction eliminate the need for thermal breaks in the plastic window. Compare the thickness of the aluminum in this residential window to that of the aluminum in the commercial window in Figure 18.8. *(Reprinted with permission from AAMA Aluminum Curtain Wall Design Guide Manual)*

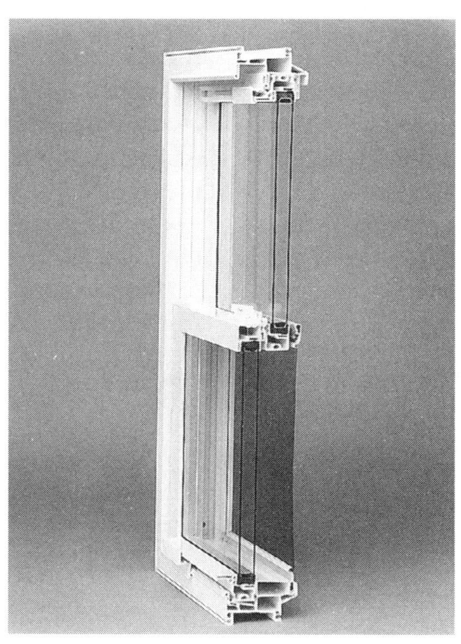

FIGURE 18.12
Cutaway sample of a plastic double-hung window with double glazing and an external half-screen. *(Courtesy of Vinyl Building Products, Inc.)*

FIGURE 18.13
For ease of washing the exterior surfaces of the glass, the sashes of this plastic window can be unlocked from the frame and tilted inward. *(Courtesy of Vinyl Building Products, Inc.)*

Some typical PVC window details are shown in Figures 18.10, 18.12, and 18.13.

Glass-fiber-reinforced plastic (GFRP) windows, frequently referred to as *fiberglass windows,* are the newest product in the window market. GFRP frame sections are produced by a process of *pultrusion*: Continuous lengths of glass fiber are pulled through a bath of plastic resin, usually polyester, and then through a shaped, heated die in which the resin hardens. The resulting sash pieces are strong, stiff, and relatively low in thermal expansion. Like PVC, they are fairly good thermal insulators. However, GFRP windows are more expensive than those made of wood or plastic.

The thermal performance of both vinyl and GRFP window frames can be enhanced with foam insulation injected into the hollow spaces within the frame sections.

Plastics in Building Construction

The first plastic was formulated more than a century ago. The major development of plastic materials has taken place since 1930, during which time plastics have come increasingly into use in buildings. Presently, the U.S. construction industry uses more than 10 billion pounds (5 billion kg) of plastics each year in hundreds of applications.

Plastics in this context may be loosely defined as synthetically produced giant molecules (polymers or copolymers) made up of large numbers of small, repetitive chemical units. Most plastics are based on carbon chemistry (Figure A), except for the silicones, which are based on silicon (Figure B). The various *synthetic rubber* compounds are usually considered a different class of materials from plastics, although chemically they are similar; they are often referred to as *elastomers.* Plastics and elastomers are manufactured largely from organic molecules obtained from oil, natural gas, and coal.

A *polymer* is composed of many identical chemical units or *monomers.* Polyvinyl Chloride (PVC), for example, is a polymer that is produced by polymerizing vinyl chloride monomers into a long chemical chain (Figure C). A *copolymer* consists of repeating patterns of two or more monomers. High-impact polystyrene is a copolymer made up of both polystyrene and polybutadiene (Figure D).

The molecular structure of a polymer or copolymer bears an important relationship to its physical properties. A high-density polyethylene molecule, for instance, is a long single chain containing up to 200,000 carbon atoms (Figure E). Low-density polyethylene has a branching molecular structure that does not pack together as readily as the single chain (Figure F).

There are two broad classes of plastics. *Thermoplastic* plastics consist generally of linear molecules and may be softened by reheating at any time after their manufacture. Upon cooling, they regain their original properties. *Thermosetting* plastics have a molecular structure that is strongly crosslinked in three dimensions. They cannot be remelted after manufacture. Thermosetting plastics are generally harder, stronger, and more chemically stable than thermoplastics.

Many modifiers are added to various polymers to change their properties or reduce their cost. *Plasticizers* are organic compounds that impart flexibility and softness to otherwise brittle plastics. *Stabilizers* are added to resist deterioration of polymers from the effects of sunlight, heat, oxygen, and electromagnetic radiation. *Fillers* are inexpensive nonreacting materials such as talc or marble dust that are added to reduce the cost or to improve toughness or resistance to high temperatures. *Extenders* are waxes or oils that add bulk to the plastic at low cost. *Reinforcing* fibers of glass, metal, carbon, or minerals can increase strength, impact resistance, stiffness, abrasion resistance, hardness, and other mechanical proper-

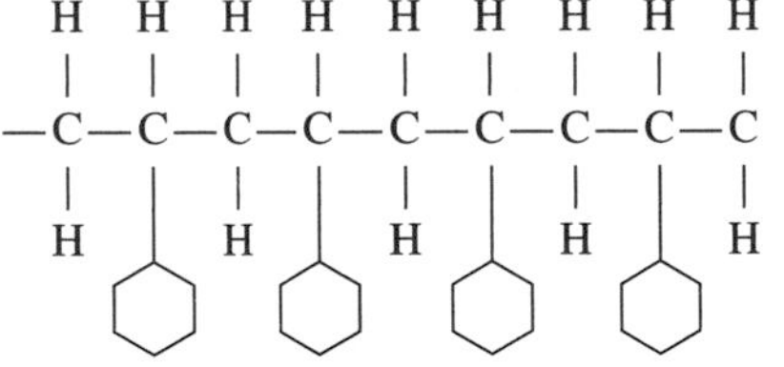

Figure A
Polystyrene.

Figure B
A silicone.

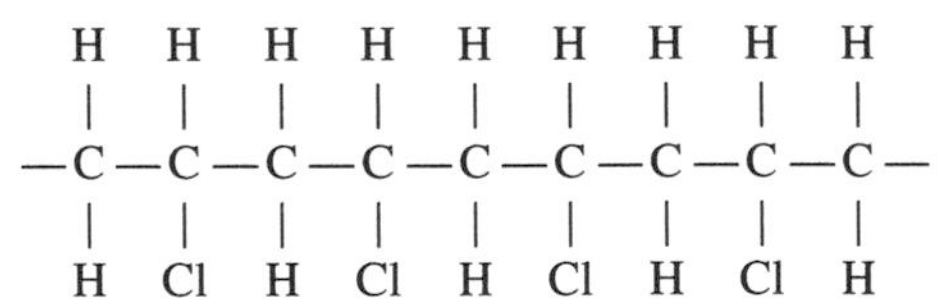

Figure C
Polyvinyl chloride (PVC, vinyl)

Figure D
High-impact polystyrene.

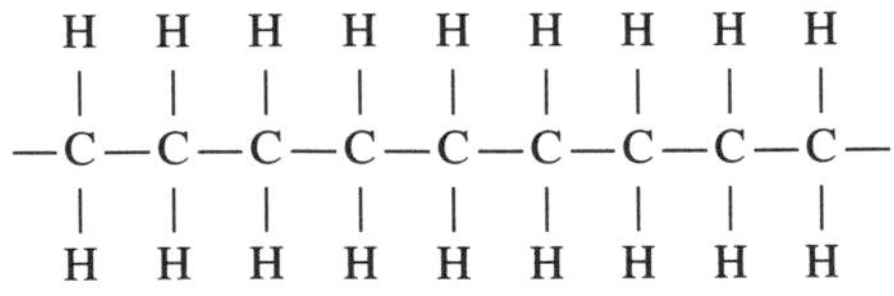

Figure E
High-density polyethylene.

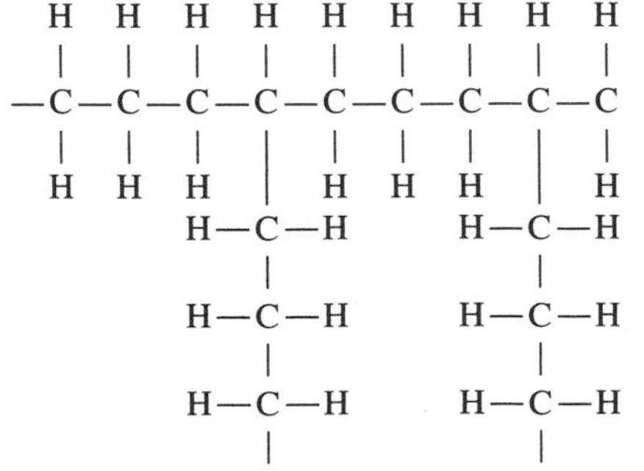

Figure F
Low-density polyethylene.

ties. *Flame retarders* are often introduced into plastics that are destined for interior use in buildings. Color can be added to plastics with dyes or pigments. Some examples of extremely useful modifiers are the carbon black that is often added to polyethylene as a stabilizer to improve its resistance to sunlight, the lead carbonate that is used to stabilize PVC products for use outdoors, the di-isooctyl plasticizer that converts brittle PVC to a stretchable, rubbery compound, and the chopped glass fibers that reinforce polyester to make it suitable for use in boat hulls and building components.

The plastics used in buildings range from dense solids such as those used in floor tiles to the lightweight cellular foams used for thermal insulation. They include the soft, pliable sheets used for roofing membranes and flashings and the hard, rigid plastics used for plumbing pipes. Glazing sheets are made from highly transparent plastics, whereas most plastics manufactured for other purposes are opaque.

People readily recognize the solid plastic materials used in construction, but most tend not to realize that liquid plastics are major ingredients of many protective paints and coatings. Plastics show up also in the form of *composites*, in which they are teamed with nonplastic materials: *laminates* consisting of paper and melamine formaldehyde, used for countertops and facings; *sandwiches* such as the foam-core plywood panels used as cladding for heavy timber frame buildings (see Chapter 3); and mixes of plastics with particulate materials, such as polyester concrete (stone aggregates cemented with a polyester binder rather than portland cement) and particleboard (wood chips and phenolic resins).

Plastics are given form through any of an almost endless list of processes. *Extrusion* manufactures long shaped sections by forcing the plastic through a shaped die. Pultrusion, used for certain fiber-reinforced products, is much the same as extrusion, except that the section is pulled through the die rather than pushed. A host of *molding* processes cast plastic into shaped cavities to give it form. Many foam plastics can be foamed directly into the mold, which they expand to fill. Polyethylene film is produced by *film blowing*, in which air is pumped into a small extruded tube of plastic to expand and stretch it into a very thin-walled tube many feet in diameter. The large tube is then slit lengthwise, folded, and rolled for distribution. Films and sheets can be made by casting plastic onto a chilled roller or a chilled moving belt. Some plastic sheet products and many plastic laminates are produced by *calendaring*, a process in which a material or a sandwich of materials is pressed first through hot rollers, then cold rollers.

Thermoplastic sheets may be further formed by heating them and pressing them against shaped dies. The pressing force may be furnished mechanically by a matching die or by air pressure. If compressed air pushes the plastic into the die, the process is called *blow forming*. In *vacuum forming*, a pump draws the air from between the heated plastic sheet and the die, and atmospheric pressure does the rest.

Many plastics are amenable to machining processes—sawing, drilling, milling, planing, turning, sanding—like those used to shape wood or metal. Nylon and acrylics are often given final form in this way. Various plastics can also be joined with adhesives, or with heat or solvents that

Some Synthetic Rubbers Used in Construction

Chlorinated Polyethylene, Chlorosulfonated Polyethylene (Hypalon)	Roof membranes
Ethylene-Propylene-Diene Monomer (EPDM)	Roof membranes, flashings
Isobutylene-Isoprene Copolymer (butyl rubber)	Flashings, waterproofing
Polychloroprene (Neoprene)	Gaskets, waterproofing
Polyisobutylene (PIB)	Roof membranes
Polysiloxane (silicone rubber)	Sealants, adhesives, coatings, roof membranes
Polyurethanes	Sealants, insulation
Sodium Polysulfide (polysulfide, thiokol)	Sealants

Plastics in Building Construction (continued)

Some Plastics Used in Construction

Chemical Name (common names and trade names in parentheses)	Thermoplastic or Thermosetting	Characteristics	Some Uses in Buildings
Acrylonitrile-butadiene-styrene terpolymer (ABS)	tp	Hard, dimensionally stable	Plumbing pipes, shower stalls
Alkyd	ts	Tough, transparent, weather resistant	Coatings
Epoxy	ts	Tough, hard, high adhesion	Adhesives, coatings, binders
Melamine formaldehyde	ts	Hard, very durable	Plastic laminates
Phenol formaldehyde (phenolic, Bakelite)	ts	Hard, very durable	Coatings, laminates, insulating foams, electrical boxes
Polyamides (Nylon)	tp	Tough, strong, elastic, machinable	Membranes for air-supported and tensile roofs, hardware for doors and cabinets
Polybutene	tp	Sticky, soft, weather resistant	Sealant tapes, sealants
Polycarbonate (Lexan)	tp	Hard, transparent, unbreakable	Glazing sheet, lighting fixtures, door sills
Polyesters (Mylar, Dacron)	ts	Durable, combustible	Fabrics for roofing membranes and geotextiles, matrix for glass-fiber reinforced plasticss
Polyethylene (polythene)	tp	Tough, flexible, impermeable	Vapor retarders, moisture barriers, piping, tarpaulins
Polyisocyanurate	ts	Heat resistant, fire retardant	Insulating foams
Polymethyl methacrylate (acrylic, Plexiglas, Lucite)	tp	Tough, transparent, machinable	Glazing sheet, dome skylights, lighting fixtures, illuminated signs, coatings, adhesives
Polypropylene	tp	Tough, hard, does not fatigue	Crack control fibers in concrete, one-piece molded hinges
Polystyrene (Styrofoam, bead foam)	tp	Foam is excellent thermal insulator, solid is transparent	Insulating foams, glazing sheet
Polytetrafluorethylene (PTFE, Teflon)	tp	Heat resistant, low friction	Sliding joint bearings, pipe thread tape
Polyurethanes	ts or tp	Vary widely with form of plastic, from low-density foams to hard, dense coatings to synthetic rubbers	Coatings, sealants, insulating foams, adhesives
Polyvinyl chloride (PVC, vinyl), polyvinyl acetate (PVA, vinyl)	tp	Vary widely with form of plastic, from flexible rubberlike sheet to hard, rigid solid	Pipes, conduits, ductwork, siding, gutters, window frames, flooring sheets and tiles, coatings, wallcoverings, geotextiles, roof membranes
Polyvinyl fluoride (PVF, Tedlar)	tp	Tough, inert	Exterior coatings
Polyvinylidene fluoride (PVF, Kynar, fluoropolymer)	tp	Crystalline, translucent	Exterior coatings
Silicones	ts	Weather resistant, low surface tension	Masonry water repellants, Sealants
Thermoplastic polyolefin (TPO)	tp	Weather resistant, resistant to UV radiation	Roof membranes
Urea formaldehyde (UF)	ts	Strong, rigid, stable	Coatings, plywood adhesives, binders for wood panel products, insulating foams

soften the two surfaces to be joined so that they can be pressed together to reharden as a single piece.

Plastics can be joined mechanically with screws or bolts. Many plastic products are designed with ingenious snap-together features so that they can be joined without fasteners.

As a group, plastics exhibit some common advantages and some common problems when used as building materials. Among the advantages, plastics in general are low in density, they are often cheaper than other materials that will do the same job, they have good surface and decorative qualities, and, because their molecules are made to order for each end use, they can often offer a better solution to a building problem than more traditional materials. Plastics generally are little affected by water or biological decay, and they do not corrode galvanically. They tend to have low thermal and electrical conductivity. Many have high strength-to-weight ratios. Most plastics are essentially impermeable to water and water vapor. Many are very tough and resistant to abrasion.

Their disadvantages are also numerous. All plastics can be destroyed by fire, and many give off toxic combustion products. Some burn very rapidly, but others are slow-burning, self-extinguishing, or do not ignite at all, so careful selection of polymers and modifiers can be crucial in building applications. Flame-spread ratings, smoke-developed ratings, and toxicity of combustion products should be checked for each use of plastics within a building.

Plastics have much higher coefficients of thermal expansion than other building materials and require close attention to the detailing of control joints, expansion joints, and other devices for accommodating volume change. Plastics tend not to be very stiff. They deflect considerably more under load than most conventional materials. Taken together with their combustibility, this severely limits their application as primary structural materials. Many also creep under prolonged loading, especially at elevated temperatures. In strength, plastics vary from fiber-reinforced composites that are as strong as many metals (but not nearly as stiff) to cellular foams that can be crushed easily between two fingers.

Plastics tend to degrade in the outdoor environment. They are especially susceptible to attack by the ultraviolet component of sunlight, oxygen, and ozone. In many plastic products (acrylic glazing sheets and some synthetic rubbers, for example), these problems have largely been solved by adjustments to the chemistry of the material that have resulted from a prolonged program of testing and research. However, it is usually a mistake for a designer to use an untried plastic material in an exposed location without consulting its manufacturer.

The accompanying table lists the plastics most commonly used in construction.

Selected References

1. Dietz, Albert G. H. *Plastics for Architects and Builders.* Cambridge, MA, MIT Press, 1969.

Despite its age, this deservedly famous little book is still the best introduction to the subject for building professionals.

2. Hornbostel, Caleb. Construction Materials: *Types, Uses and Applications* (2nd ed.). New York, John Wiley & Sons, Inc., 1991.

The section on plastics of this monumental book gives excellent summaries of the plastics most used in buildings.

Steel

The chief advantage of steel as a frame material for windows is its strength, which permits steel sash sections to be much more slender than those of wood and aluminum (Figures 18.14–18.18). Usually, *steel windows* are permanently coated to present a pleasing appearance and prevent corrosion. If they are not, they will need periodic repainting. Steel is less conductive of heat than aluminum, so it is unlikely to form condensation under most weather conditions. Thus, steel window frames are rarely provided with thermal breaks, although, where greater thermal performance is required, thermal break frames are also available from some manufacturers. They are, however, more conductive of heat than wood or plastic frames.

Muntins

In earlier times, because of the difficulty of manufacturing large lights of glass that were free of significant defects, windows were necessarily divided into small lights by *muntins*, thin wooden bars in which the glass was mounted within each sash. (The upper sashes in Figure 18.6 have muntins.) A typical double-hung window had its upper sash and lower sash each divided into six lights and was referred to as a *six over six*. Muntin arrangements changed with changing architectural styles and improvements in glass manufacture. Today's windows, glazed with large, virtually flawless lights of float glass, need no muntins at all, but many building owners and designers prefer the look of traditional muntined windows. This desire for muntins is greatly complicated by the necessity of using insulated (double or triple) glazing to meet building code energy conservation requirements. Some manufacturers offer the option of individual small lights of double glazing held in deep muntins. This is relatively expensive, and the muntins tend to look thick and heavy. The least expensive option utilizes grids of imitation muntin bars, made of wood or plastic, that are clipped into each sash against the interior surface of the glass. These are designed to be removed easily for washing the glass. Other alternatives are imitation muntin grids between the sheets of glass, which are not very convincing replicas of the real thing, and grids, either removable or permanently bonded to the glass, on both the outside and inside faces of the window. Another option is to use a prime window with authentic divided lights of single glazing and to increase its thermal performance with a storm sash. Of all the options, this one looks best from the inside, but reflections in the storm sash largely obscure the muntins from the outside.

Glazing

A number of glazing options are available for residential windows. Single

FIGURE 18.14
These samples of hot-rolled steel window frame sections demonstrate a range of permanent finishes. The nearest sample includes a snap-in aluminum bead for holding the glass in place. *(Steel windows by Hope's; photography by David Moog)*

FIGURE 18.15
Cutaway sample of a steel-framed window with aluminum glazing beads and a permanent finish. The details of this window are shown in Figure 18.17. *(Steel windows by Hope's; photography by David Moog)*

FIGURE 18.16
With fire-rated glazing such as wired glass, shown here, a steel window can be fire rated. This window features an awning sash below a fixed light. *(Steel windows by Hope's; photography by David Moog)*

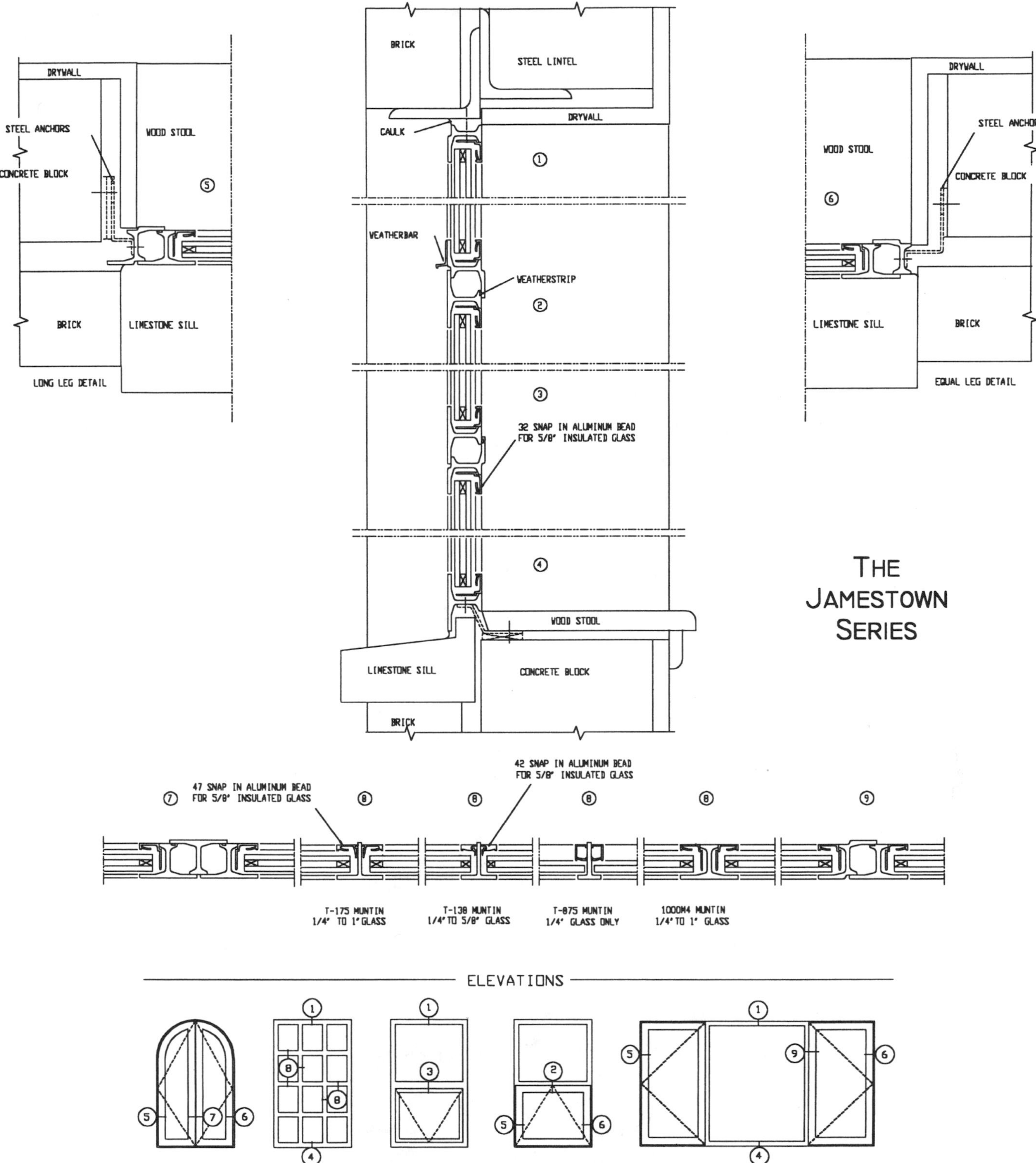

FIGURE 18.17
Manufacturer's details of a steel window in a masonry wall. The details are keyed by number to the elevation views at the bottom. Notice the bent steel anchors, shown with broken lines, that fasten the window unit to the masonry. *(Steel windows by Hope's)*

glazing is acceptable only in the mildest climates because of its low resistance to heat flow and the likelihood that moisture will condense on its interior surface in cool weather. Sealed double glazing or single glazing with storm windows is the minimum acceptable glazing under most building codes. Storm windows must be taken down periodically and cleaned, which is a nuisance that can be avoided by using sealed double or triple glazing. More than 90 percent of all residential windows sold today in North America have two or more layers of glass. Double glazing with a low-emissivity (low-e) coating on one glass surface performs at least as well as triple glazing. See the previous chapter for a more in-depth discussion of glazing types.

Figure 18.19 lists thermal transmittance properties for some example combinations of window frame and glazing options. The listed U-Factors are overall values for complete window assemblies, accounting for differences in the thermal properties of the center of glass, edge of glass, and frames. When selecting actual windows, whole-window U-Factors for the particular window are provided by the window manufacturer as determined by laboratory testing or computer simulation. In addition to thermal transmittance, solar heat gain coefficient (SHGC) and visible light transmittance (VT) are other important measures of a window system's performance. See the previous chapter for a discussion of these properties.

FIGURE 18.18
The narrow sight lines of steel windows and doors are evident in this photograph. *(Steel windows and doors by Hope's; photography by David Moog)*

Installing Windows

Some catalog pages for windows are reproduced in Figures 18.20–18.22 to give an idea of the information on window configurations that is available to the designer. Important dimensions given in catalogs are those of the *rough opening* and *masonry opening.* The rough opening height and width are the dimensions of the hole that must be left in a framed wall for installation of the window. They are slightly larger than the corresponding outside dimensions of the window unit itself to allow the installer to locate and level the unit accurately and to ensure that the window unit is isolated from structural stresses within the wall system. The masonry opening dimensions indicate the size of the hole that must be provided if the window is mounted in a masonry wall.

A rough opening or masonry opening should be flashed carefully before the window is installed in order to avoid later problems with leakage of water or air (for example, Figures 6.12 and 6.13). At a minimum, this flashing may be done with asphalt-saturated felt. A better result is obtained by using adhesive-backed window flashing materials or metal. Adhesive-backed flashings may be made of rubberized asphalt, similar in composition to the rubberized roof underlayment frequently used along the eaves of the roof and described in Chapter 7, reinforced plastic, or synthetic fibers designed for compatibility with proprietary building wrap products. Metal used for flashings must be corrosion resistant.

Most factory-made windows are extremely easy to install, often requiring only a few minutes per window. Windows that are framed or clad in aluminum or plastic are usually provided with a continuous flange around the perimeter of the window unit. When the unit is pushed into the rough opening from the outside, the flange bears against the sheathing along all four edges. After the unit has been located and plumbed (made level and square) in the opening, it is attached to the frame by means of nails driven through the flanges. Then all the edges of the flanges should be made airtight, as shown in Figure 6.12. The flanges are eventually concealed by the exterior cladding or trim.

Methods for anchoring window units into masonry walls vary widely, from nailing the unit to wood strips that have been fastened inside the masonry with bolts or powder-driven fasteners to attaching the unit to steel clips that have been laid into the mortar joints of the masonry. The window manufacturer's recommendations should be followed in each case.

	Overall U-Factor[a]		
Window Frame	**Single-Glazed**	**Double-Glazed, Clear**	**Double-Glazed, Low-e, Argon Gas**
Aluminum, without thermal break	1.2 *6.8*	0.76 *4.3*	0.60 *3.4*
Thermal break aluminum	1.0 *5.7*	0.63 *3.6*	0.48 *2.7*
Steel	0.92 *5.2*	0.55 *3.1*	0.41 *32.3*
Wood, clad wood, vinyl	0.84 *4.8*	0.49 *2.8*	0.35 *2.0*
GFRP	0.65 *3.7*	0.44 *2.5*	0.27 *1.5*

[a]U-factor: Btu/ft^2-hr-°F followed by *W/m^2-°K.*

Figure 18.19

Comparative U-Factors for some window frame and glazing combinations. Lower values correspond to better thermal performance. Note that the frame material has a significant impact on the overall performance of the window assembly. The figures listed here are examples only. The thermal transmittance of any particular window is influenced by the properties of the glazing and frame, the size of the window, whether the window is operable or fixed, and other factors.

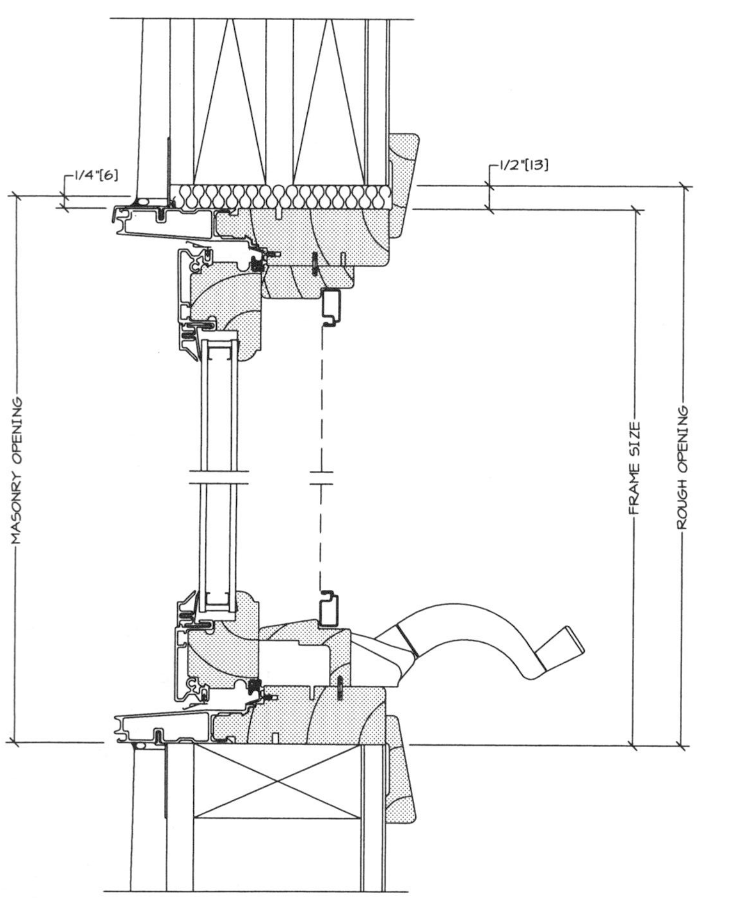

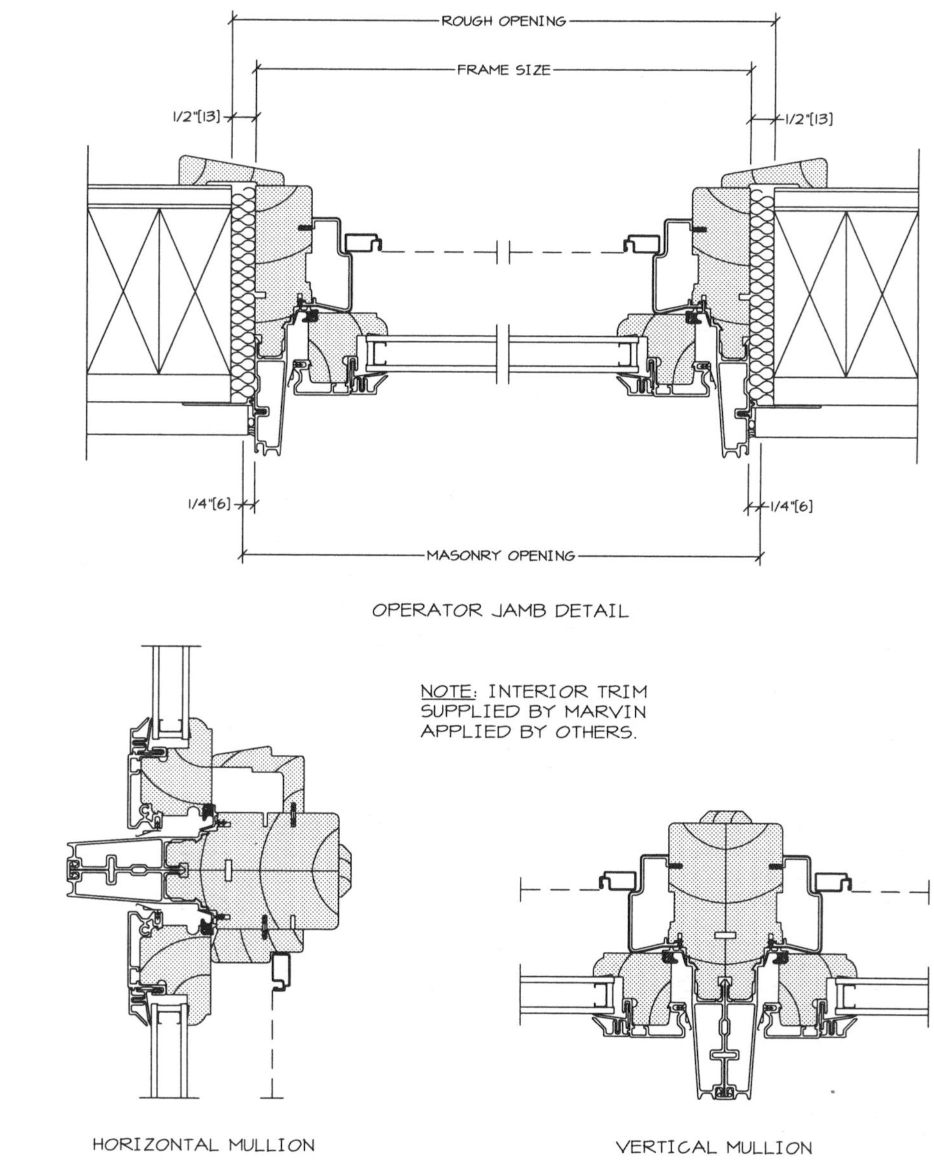

Figure 18.20
Manufacturer's catalog details for an aluminum-clad, wood-framed casement window with double glazing and an interior insect screen.
(Courtesy of Marvin Windows and Doors)

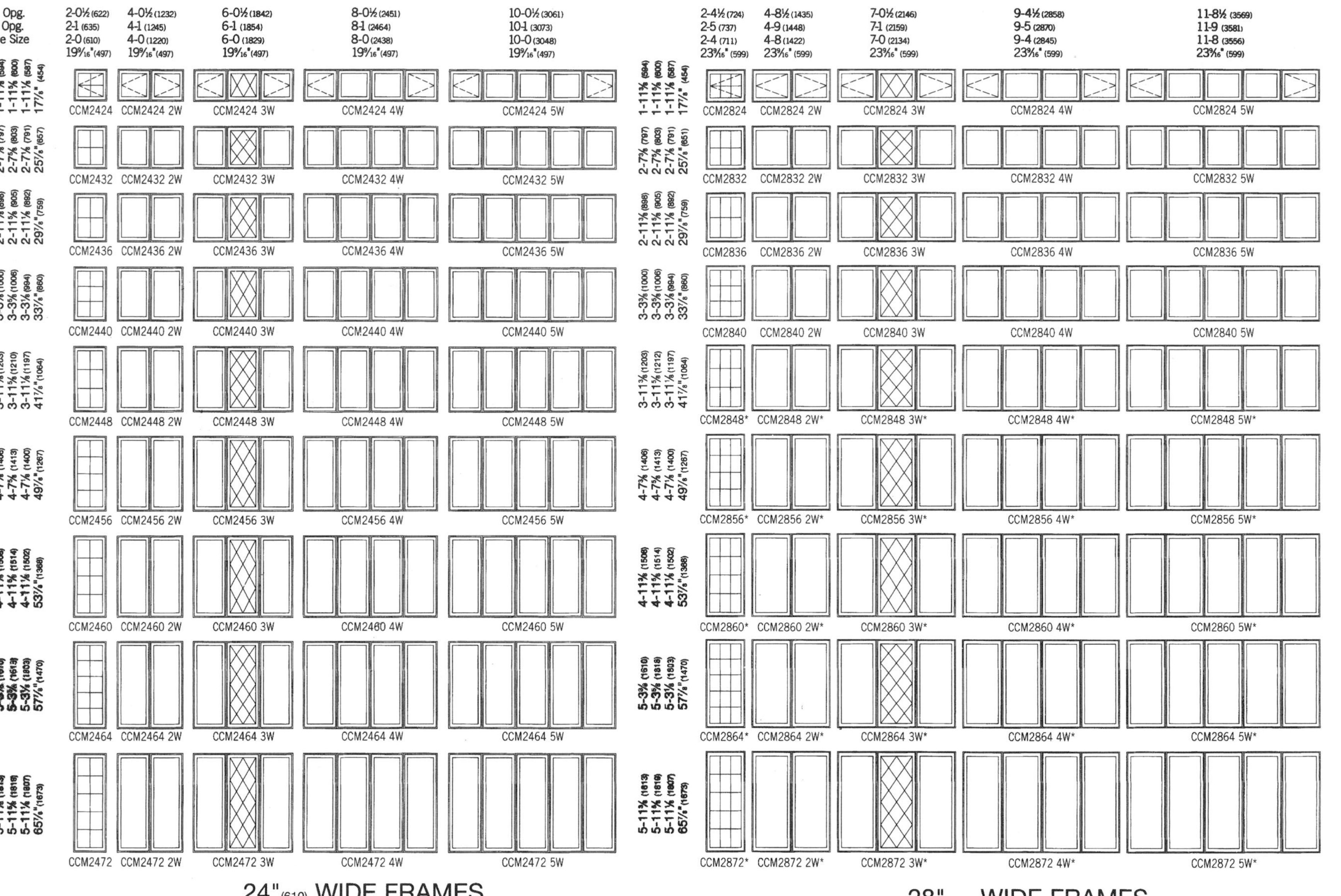

FIGURE 18.21
One of several pages in a manufacturer's catalog that show stock sizes and configurations of aluminum-clad wood casement windows. *(Courtesy of Marvin Windows and Doors)*

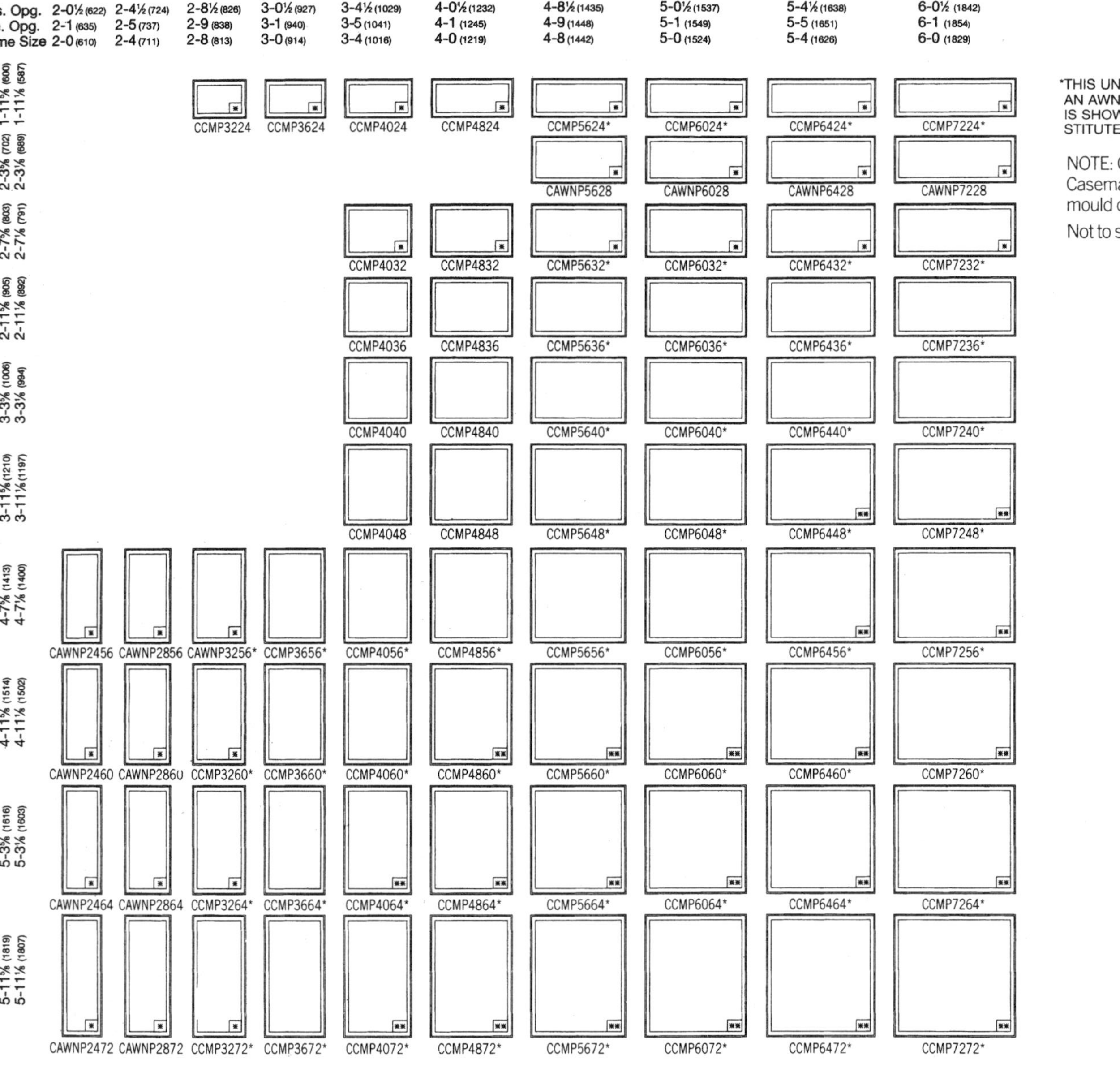

FIGURE 18.22
These fixed windows are sized to match stock sizes of casement windows by the same manufacturer, allowing the designer to mix and match. *(Courtesy of Marvin Windows and Doors)*

Considerations of Sustainability Relating to Windows and Doors

Chapter 17 covers issues of sustainability that pertain to glass and glazing. With regard to windows and doors specifically:

- According to the National Renewable Energy Laboratory, solar heat gain and wintertime heat losses through windows account for roughly 30 percent of U.S. building heating and cooling electrical loads.
- In addition to the thermal properties of the glass in a window, the thermal conductivity of the frame and the air leakage of the window or door unit have very significant effects on the amount of energy that will be required to heat and cool the building.
- Doors can leak significant quantities of heat by conduction through the material of the door. Foam-core doors have better thermal performance than other types. The performance of any exterior residential door can be improved substantially by adding a storm door during the cold season of the year. Air-lock vestibules can limit the amount of outdoor air that enters a building when the exterior door is open, as well as improve the comfort of building occupants in the vicinity of the vestibule. Revolving doors, which maintain an air seal regardless of their position, are suitable alternatives to vestibules. All doors should be tightly weatherstripped to limit loss of conditioned air.
- With respect to frame materials:
  - Issues of sustainability of wood production are covered in Chapter 3. When a building with wood windows is demolished, the windows are generally sent to landfills or incinerators and are not recycled.
  - Aluminum frames must be thermally broken for the sake of energy efficiency. They are often recycled during demolition and should be recycled in every case. Chapter 21 discusses sustainability issues relating to the material aluminum itself.
  - PVC window frames are thermally efficient. They can be recycled during demolition, and a significant percentage is being recycled at the present time.
  - Steel window and door frames are made from recycled steel and can be recycled again when a building is demolished. Their thermal performance is moderate and can be improved greatly by the insertion of thermal breaks.

Doors

Doors fall into two general categories, *exterior* and *interior.* Weather resistance is usually the most important functional factor in choosing exterior doors, whereas resistance to the passage of sound or fire and smoke are frequently important criteria in the selection of interior doors. Many different modes of door operation are possible (Figure 18.23).

There are numerous types of exterior doors: solid entrance doors, entrance doors that contain glass, storefront doors that are mostly or entirely made of glass, storm doors, screen doors, vehicular doors for residential garages and industrial use, revolving doors, and cellar doors, to name just a few. Interior doors come in dozens of additional types. To simplify our discussion, we will focus on swinging doors for both residential and commercial use.

Wood Doors

At one time, nearly all doors were made of wood. In simple buildings, primitive doors made of planks and *Z-bracing* were once common. In more finished buildings, *stile-and-rail doors* gave a more sophisticated appearance while avoiding the worst problems of moisture expansion and contraction to which plank doors are subject (Figures 7.23 and 7.24). The panels are not glued to the stiles and rails, but instead "float" in unglued grooves that allow them to move. The doors may be made of solid wood or of wood composite materials with veneered faces and edges. In either case, they are available in many different species of woods.

In recent decades, stile-and-rail doors have continued to be popular in higher-quality buildings. However, *flush doors* have captured the majority of the market, chiefly because they are easier to manufacture and therefore less costly. For exterior use in small buildings, and for both exterior and interior use in institutional and commercial buildings, flush doors are constructed with a *solid core* of wood blocks or wood composite material (Figure 7.24). Interior doors in residences often have a *hollow core.* These consist of two veneered wood faces that are bonded to a concealed grid of interior spacers made of paperboard or wood. The perimeters of the faces are glued to wood edge strips. Flush doors with wood faces are also available with a solid *mineral core* that qualifies them as fire doors.

Flush wood doors may be manufactured and specified according to any one of several industry standards for door appearance and durability, the most commonly used of which is the Window and Door Manufacturers *ANSI/WDMA I.S.1-A-04 Architectural Wood Flush Doors.* This standard includes three performance grades—Standard Duty, Heavy Duty, and Extra

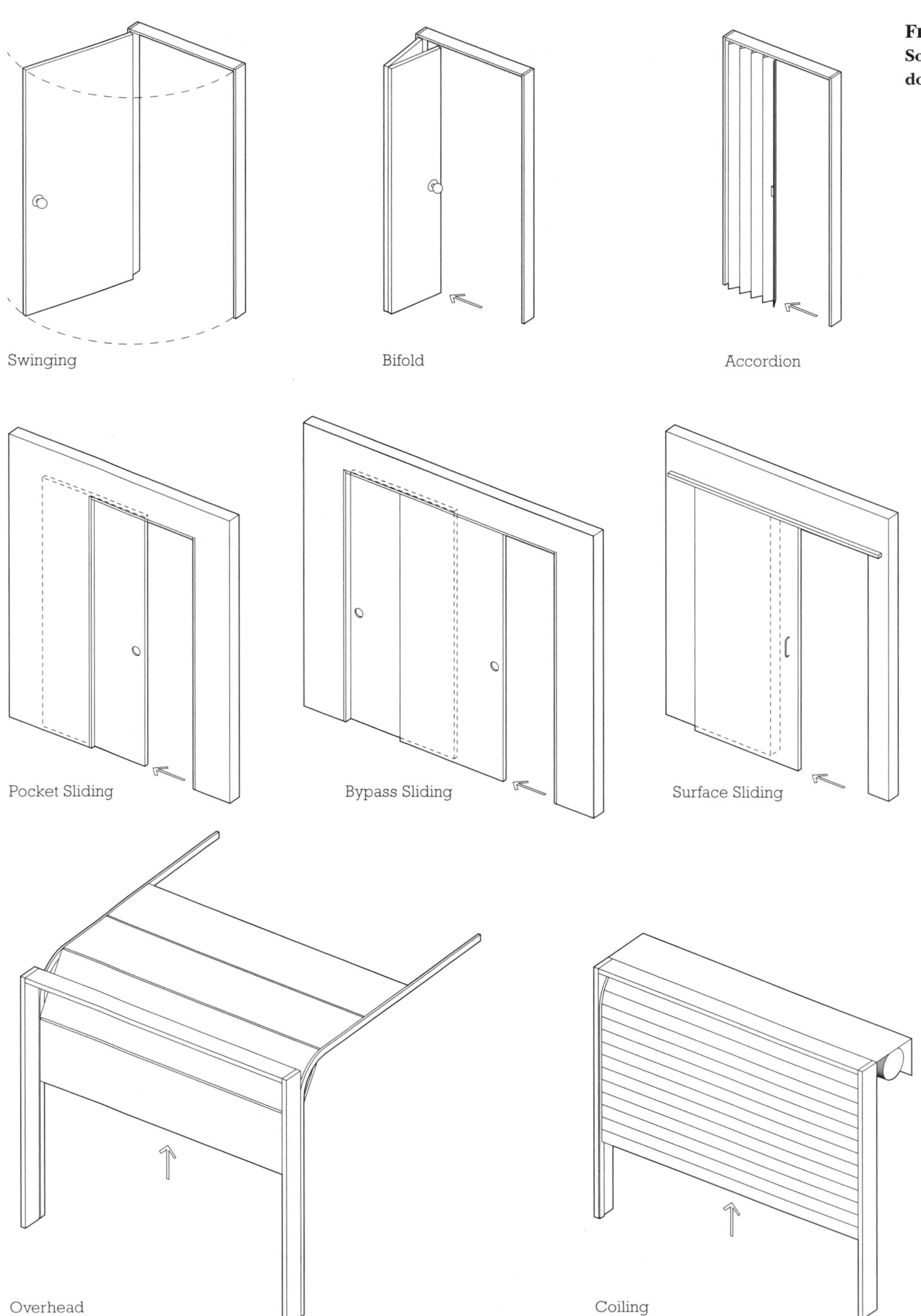

FIGURE 18.23 Some modes of door operation.

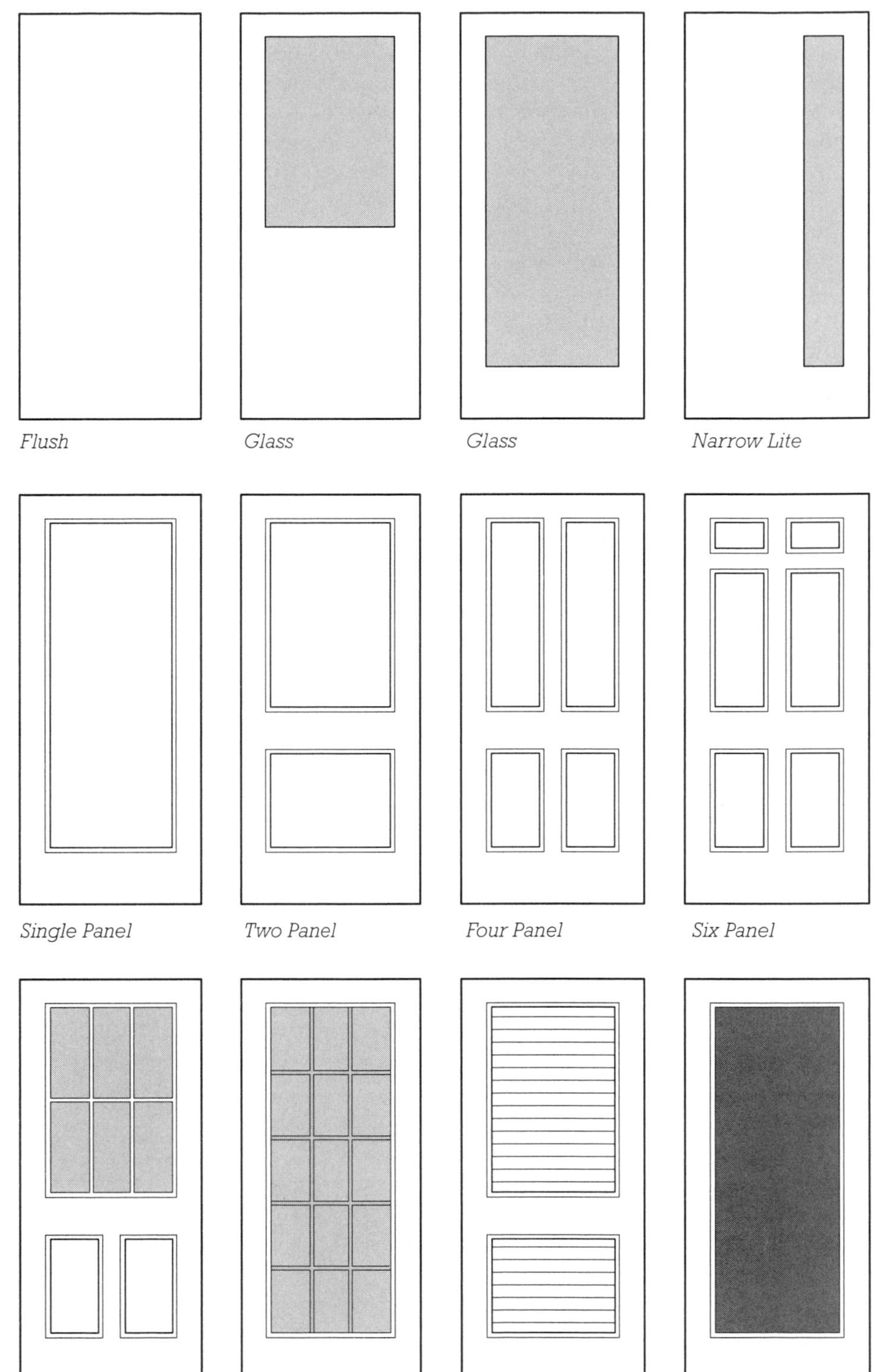

FIGURE 18.24
Some typical configurations for wood doors. The top row consists of flush doors. The middle row is made up of stile-and-rail doors.

Heavy Duty—intended for doors used in applications of increasingly heavy usage, as well as varying appearance grades that control the quality of face veneers.

A relatively recent development is a door made of wood fiber composite material that is pressed into the shape of a stile-and-rail door. Usually, the faces of the door may be given an artificial wood grain texture or faced with real wood veneer.

Entrance doors must be well constructed and tightly weatherstripped if they are not to leak air and water. Properly installed and finished wood panel or solid-core doors are excellent for exterior residential use (Figures 6.14 and 6.15). Pressed *sheet metal doors* and molded *GFRP doors,* usually embossed to resemble wood stile-and-rail doors, are popular alternatives to wood exterior residential doors. Their cores are filled with insulating plastic foam, making their thermal performance superior to that of wood doors. They do not suffer from moisture expansion and contraction, as wood doors do. They are often furnished *prehung,* meaning that they are already mounted on hinges in a surrounding frame, complete with weatherstripping, ready to install by merely nailing the frame into the wall. Wood doors can also be purchased prehung, although many are still hung and weatherstripped on the building site. The major disadvantage of metal and plastic exterior doors is that they do not have the satisfying appearance, feel, or sound of a wood door.

Residential entrance doors almost always swing inward and are mounted on the interior side of the door frame. This makes them less vulnerable to thieves who would remove hinge pins or use a thin blade to push back the latch to gain entrance. In cold climates, it also prevents snow that may accumulate against the door from preventing the door from opening. For improved wintertime thermal performance of the entrance, a *storm door* may be mounted on the outside of the same frame, swinging outward. The storm door usually includes at least one large panel of tempered glass.

In summer, a *screen door* may be substituted for the storm door. A *combination door,* which has easily interchangeable screen and storm panels, is more convenient than separate screen and storm doors.

Steel Flush Doors

Flush doors with faces of painted sheet steel are the most common type of door in nonresidential buildings (Figure 18.25). For economy, interior steel doors in many situations have hollow cores. Solid-core doors are required for exterior use and in situations that demand increased fire resistance, more rugged construction, or better acoustical privacy between rooms.

Metal doors and most nonresidential wood doors are usually hinged to *hollow steel door frames,* although wood and aluminum frames can also be used. Many different types of anchors are available for mounting frames to partitions of various materials (Figure 18.26). Where hollow metal door frames are installed within masonry walls, they may be filled with cementitious grout to improve sound deadening and to make the door frame more resistant to tampering or forced entry.

Steel doors and frames are commonly manufactured and specified

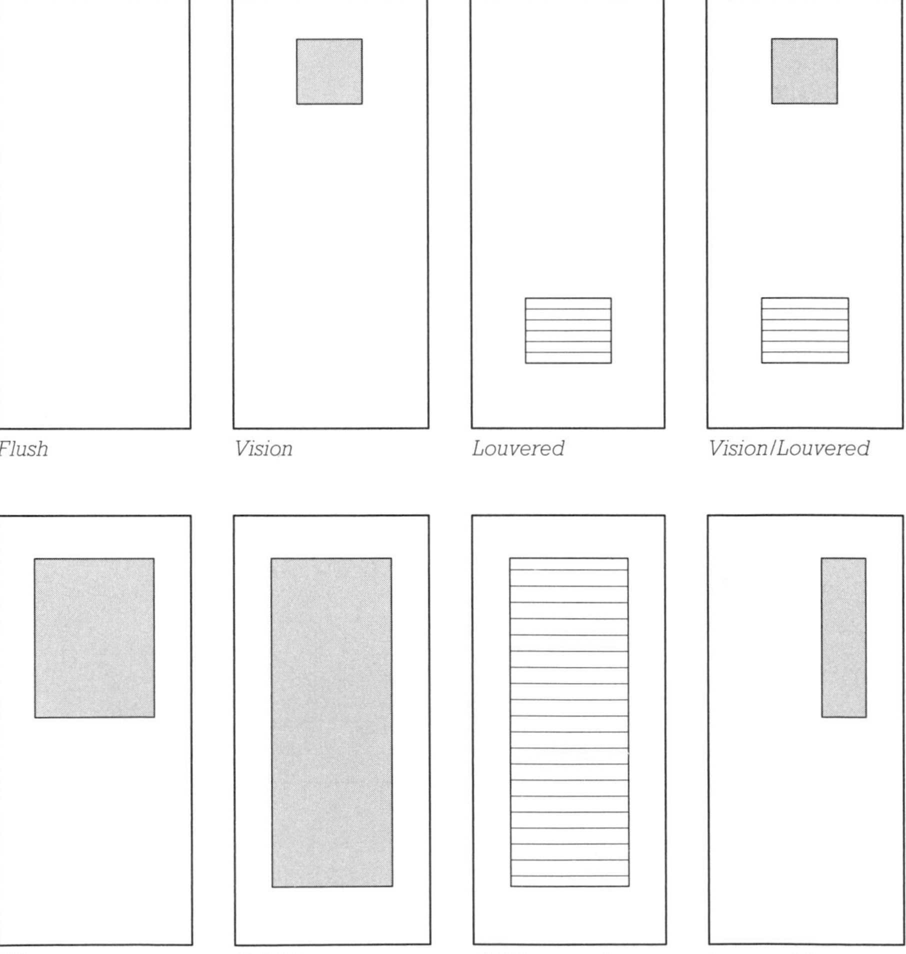

FIGURE 18.25
Some typical configurations for steel doors.

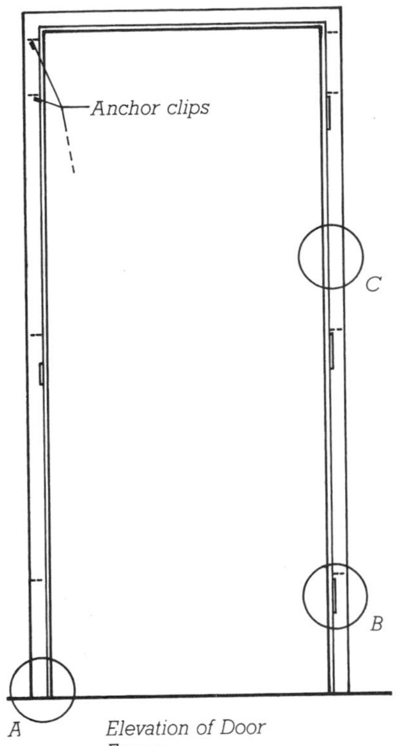

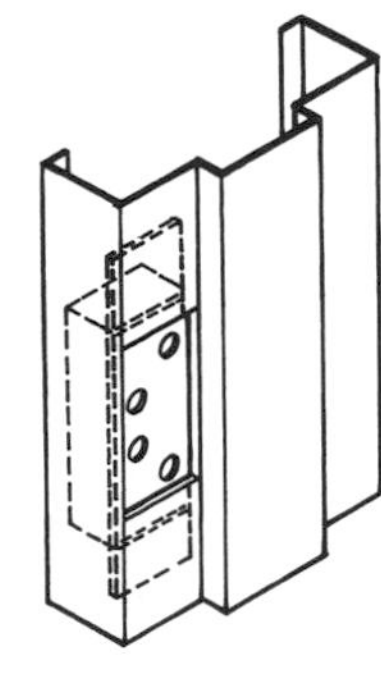

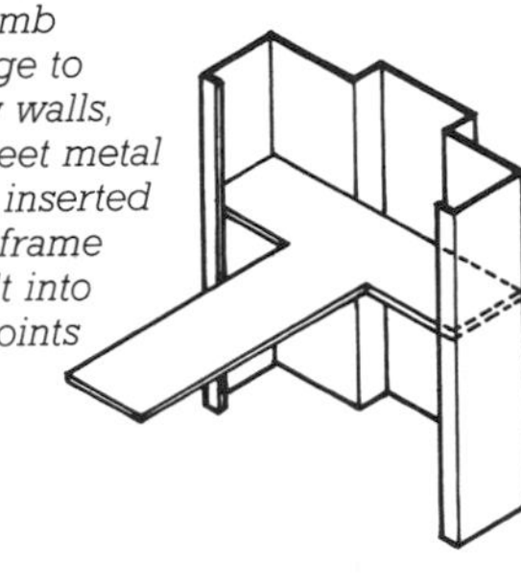

Elevation of Door Frame

B. Reinforcement of jamb at hinge attachments

C. For jamb anchorage to masonry walls, loose sheet metal tees are inserted into the frame and built into mortar joints

FIGURE 18.26

Details of hollow steel door frames. The lettered circles on the elevation at the upper left correspond to the details on the rest of the page.

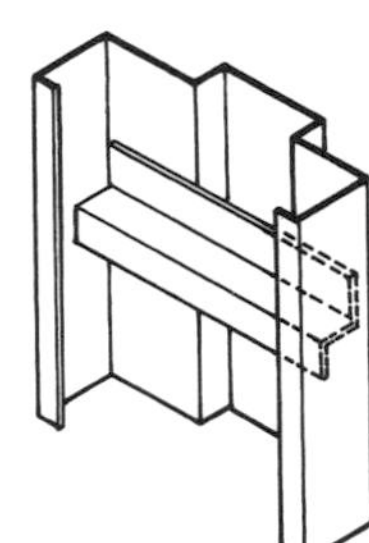

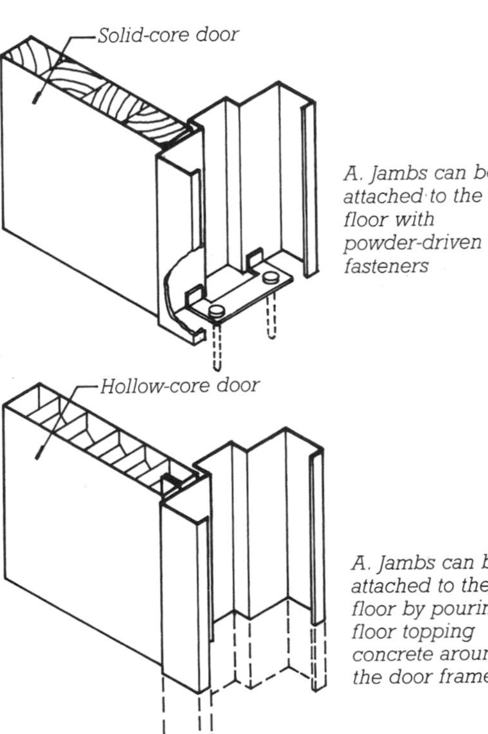

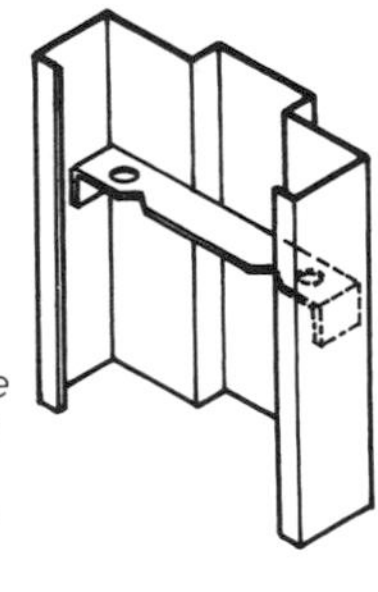

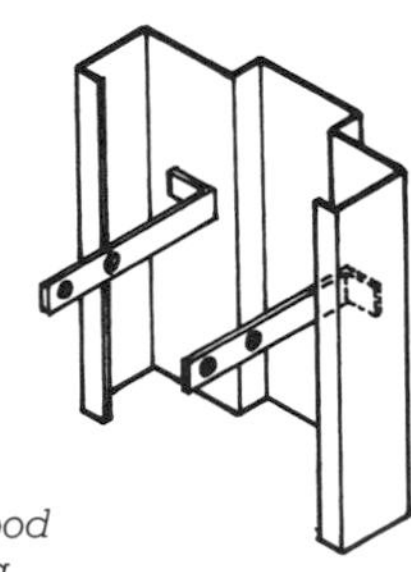

C. For jamb anchorage to steel studs, sheet metal zees are factory-welded to the jambs to receive screws driven through the studs

A. Jambs can be attached to the floor by pouring floor topping concrete around the door frame

C. For jamb anchorage to wire truss studs, notches key to the vertical members of the studs, and holes provide for tie wires

C. Jambs are anchored to wood studs by nailing through holes in the jamb inserts

according to one of two standards: the Steel Door Institute's *ANSI/SDI A250.8 Recommended Specifications for Standard Steel Doors and Frames* or the Hollow Metal Manufacturers Association's *ANSI/NAAMM-HMMA 861 Guide Specifications for Commercial Hollow Metal Doors and Frames.* The first of these is intended for *standard steel doors,* manufactured to a standardized set of sizes, configurations, and quality levels. The second standard is for *custom steel doors,* those generally manufactured to a high quality standard in custom sizes and configurations.

Fire Doors

Fire doors have a noncombustible mineral core and are rated according to the period of time for which they are able to resist specified time and temperature conditions, as defined by *NFPA 252 Standard Methods of Fire Tests of Door Assemblies,* or by several similar tests defined by Underwriters Laboratories. In general, doors within fire-resistance-rated walls must themselves also be fire rated. However, because doors constitute only a limited area of most walls, and because combustible furnishings or materials are not normally located directly in front of door openings, the required fire resistance rating for fire doors is often less than that required for the walls in which they are located. Figure 18.27 gives fire resistance ratings for fire doors as required by the International Building Code (IBC). For example, a door in a 2-hour rated exit stairway enclosure must be 1½-hour rated, a door in a 1-hour rated exit stairway enclosure must be 1-hour rated, and a door in a 1-hour rated exit corridor must be 20-minute (⅓-hour) rated. Doors in walls that are 2-, 3-, and 4-hour fire resistance rated, such as those separating uses within a building or separating buildings from one another, must be 1½- or 3-hour rated. A standardized label is permanently affixed to the edge of each fire door at the time of manufacture to designate its degree of fire resistance. (The building code requires that these labels not be painted over during construction so that the fire rating of the door can always be verified during subsequent building inspections.)

Glass used in fire doors must itself be fire rated (see the previous chapter) so that it will not break and fall out of the opening for a specified length of time when exposed to the heat of fire. The maximum size of glass may also be restricted, depending on the fire classification of the door and the properties of the particular type of glass used. Like glass in any door, glass in fire doors must also meet the requirements of safety glazing so that if broken, it does not create life-threatening shards.

Egress Doors and Accessible Doors

Many doorways act as components of a building's egress system, the path that occupants take when exiting a

TABLE 715.4
FIRE DOOR AND FIRE SHUTTER FIRE PROTECTION RATINGS

TYPE OF ASSEMBLY	REQUIRED ASSEMBLY RATING (hours)	MINIMUM FIRE DOOR AND FIRE SHUTTER ASSEMBLY RATING (hours)
Fire walls and fire barriers having a required fire-resistance rating greater than 1 hour	4 3 2 1 1/2	3 3[a] 1 1/2 1 1/2
Fire barriers having a required fire-resistance rating of 1 hour: Shaft, exit enclosure and exit passageway walls Other fire barriers	 1 1	 1 3/4
Fire partitions: Corridor walls Other fire partitions	1 0.5 1 0.5	1/3 [b] 1/3 [b] 3/4 1/3
Exterior walls	3 2 1	1 1/2 1 1/2 3/4
Smoke barriers	1	1/3 [b]

a. Two doors, each with a fire protection rating of 1 1/2 hours, installed on opposite sides of the same opening in a fire wall, shall be deemed equivalent in fire protection rating to one 3-hour fire door.

b. For testing requirements, see Section 715.4.3.

FIGURE 18.27
Required fire resistance ratings for doors according to the International Building Code.
(Portions of this publication reproduce tables from the 2006 International Building Code, *International Code Council, Inc., Washington, D.C. Reproduced with Permission. All rights reserved.)*

building during a fire or other emergency. Building codes require that such doorways be sufficiently wide to allow occupants to exit in a timely manner, with the width of any particular door dependent on the number of occupants served. For ease of operation, most egress doors must be side-hinged, they must not be too large, and when equipped with closers, they must not require too much force to swing open. The International Building Code requires egress doors serving 50 or more building occupants, as well as doors serving Hazardous Occupancy spaces, to swing in the direction of egress travel so that they do not become impediments to occupants attempting to exit quickly. Other building codes have similar requirements. Even when locked, egress doors must remain readily openable from the side from which occupants may approach the door when exiting. To ensure the simplest possible operation under emergency conditions, some egress doors are required to be fitted with *panic hardware*, horizontal bars or similar devices installed across the face of the door that unlock and unlatch the door whenever the bar is depressed.

Doorways along buildings routes that must be accessible to persons with disabilities must meet requirements for minimum width, ease of operation, maximum height of sill, and adequate clearance for approaching and opening the door.

There are many types of special-purpose doors. Among the most common are X-ray shielding doors, which contain a layer of lead foil; electric field shielding doors, with an internal layer of metal mesh that is electrically grounded through the hinges; heavily insulated cold storage doors; and bank vault doors.

Safety Considerations in Windows and Doors

In order to prevent accidental breakage and injuries, building codes require glass within doors, and large lights within windows that are near enough to the floor or to doors to be mistaken for open doorways, to be made of breakage-resistant material. Tempered glass is most often used for this purpose, but laminated glass and plastic glazing sheets are also permitted. For more information, see the discussion of safety glazing in the previous chapter.

In residences, buildings codes require at least one *emergency escape and rescue opening* in each bedroom, consisting of either a door to the exterior or a window that can be opened to an aperture large enough to permit occupants of the bedroom to escape through it and firefighters to enter through it. In the International Building Code, where a window is used for this purpose, it must have a net clear opening area of at least 5.7 square feet (0.53 m^2), a clear width of at least 20 inches (510 mm), a clear height of at least 24 inches (610 mm), and no sill higher than 44 inches (1.12 m) above the floor.

Where operable windows in apartments, residential dwelling units, and similar residential occupancies are more than 6 feet (1829 mm) above the exterior finished grade, the International Building Code requires that they be designed to minimize the risk of a child accidentally falling through them. Such windows must have sills not less than 24 inches (610 mm) above the interior finish floor. Alternatively, where glazing in such windows is located below this level, it must either be fixed, it must have openings sufficiently limited in size that a 4-inch (102-mm)-diameter sphere cannot pass through, or it must be protected with guards or other fall prevention devices.

Casement and awning windows should not be used adjacent to porches or walkways, where someone might be injured by running into the projecting sash. Similarly, inswinging windows should not be used in corridors unless they are above head level.

Fenestration Testing and Standards

The designer's task in selecting windows, doors, and skylights is facilitated by testing programs that allow objective comparisons of the structural, thermal, and other performance requirements of products of different types and from different manufacturers.

Structural Performance and Resistance to Wind and Rain

The first of these standards is the Standard/Specification for Windows, Doors, and Unit Skylights, jointly published by the American Architectural Manufacturers Association (AAMA), the Window and Door Manufacturers Association (WDMA), and the Canadian Standards Association (CSA), officially designated as AAMA/WDMA/CSA 101/I.S.2/A440. This specification establishes minimum requirements for air leakage, water penetration, structural strength, operating force, and forced-entry resistance of aluminum, plastic, and wood-framed windows, doors, and unit skylights.

The AAMA/WDMA/CSA Standard/Specification uses a letter

designation called *Performance Class* and a numeric designation called *Performance Grade* to indicate the minimum capabilities of fenestration products. Performance Classes, in order of increasing capability, are R, LC, C, HC, and AW. In previous editions of the standard, these letter designations were associated with the terms "residential," "light commercial," "commercial," "heavy commercial," and "architectural," respectively. Although these plain word descriptions have been removed from the newest version of the standard, knowledge of them is still helpful in recalling the intended ranking of the letter designations themselves. Each Performance Class sets criteria for resistance to wind loads, resistance to water penetration, and maximum air leakage.

Numeric Performance Grades correspond to maximum design wind pressures. For example, Grade 30 indicates a unit suitable for design wind pressures up to 30 psf (1440 Pa). Grades are specified starting at 15 psf (720 Pa) and increasing in increments of 5 psf (240 Pa). Each Class (letter designation) has a minimum acceptable Grade (design wind pressure), and higher than minimum Grades can be specified where resistance to higher wind forces is required.

An example of a manufacturer's complete tested product designation is C-R30 760 × 1520 (30 × 60), where C stands for casement window, R is the Performance Class, 30 is the Performance Grade, and the final pairs of numbers indicate the maximum size of the tested unit that meets these criteria, expressed as width by height, first in millimeters and then, in parentheses, in inches. In practice, the designer may choose a Performance Class based on the building type and general expectations for durability of the system. For example, a Class LC window may be specified for a low-rise multifamily building, a Class HC window for a hospital or school, or a Class AW window for a large institutional or high-rise building. The required Performance Grade should be determined based on the design wind pressures acting on the building, information that is usually provided by the structural engineer.

Thermal Performance

With regard to energy efficiency, the National Fenestration Rating Council (NFRC) sponsors a program of testing and labeling based on the standards Procedure for Determining Fenestration Product U-Factors, NFRC 100 and Procedure for Determining Fenestration Product Solar Heat Gain Coefficient and Visible Transmittance at Normal Incidence, NFRC 200. The two most important properties included in these standards are thermal transmittance (U-Factor) and solar heat gain coefficient, both of which directly impact building energy consumption and are regulated by energy codes. Importantly, U-Factors represent the *overall thermal transmittance*, or *whole product heat loss*, of complete window, door, and skylight products. That is, they account for the combined contributions to thermal transmittance of the center of glass, edge of glass, framing, and any other components. Visible light transmittance air leakage, and condensation resistance ratings may also be included in NFRC ratings. An example of a standard label that is affixed to each NFRC-rated window is shown in Figure 18.28.

Both AAMA/WDMA/CSA and NFRC standards are referenced by the International Building Code, making them the de facto standards for the selection of most building fenestration products sold in United States.

Impact Resistance

Buildings in hurricane-prone regions can be subject to extremely powerful and destructive winds, and glazed openings in such buildings are especially vulnerable. The pressure of high-speed winds can cause glass to break, or it can suck whole lights out of their sashes, whole sashes out of their surrounding frames, or whole frames out of their rough openings. Glass can also be broken by rocks, severed tree limbs, and other debris launched by the wind with missilelike force. Once openings in a building are breached, the force of the wind can, in extreme cases, literally blow the roof off a structure. Even where a building structure remains intact, failed openings can admit large amounts of rainwater that can severely damage the building and its contents.

In the International Building Code, glazed openings in *wind-borne debris regions* must meet special standards for resistance to high wind forces and debris impact. These regions include portions of the U.S. Atlantic Ocean and Gulf of Mexico coasts, the islands of Hawaii, and certain other U.S. territorial islands that are frequently subjected to hurricane-force winds. In these regions, most glazed openings must meet the requirements of two tests, ASTM E1996 and E1886, which subject assemblies to airborne "missiles" and cyclical air pressures to determine their ability to remain in place under hurricane-like conditions. The testing can be quite dramatic. For windows destined for installed locations not more than 30 feet (9.1 m) above grade, a 9-pound, roughly 8-foot-long (4.1 kg, roughly 2.4 m long) 2 × 4 is fired endwise toward the window from a special cannon at a speed of 34 mph (55 kph). While the glass is permitted to crack, it must survive in place, without being penetrated by the 2 × 4.

Such *impact-resistant openings* (also sometimes referred to as *hurricane-rated openings*) are fitted with laminated glass with a heavy interlayer of PVB or other similarly tough, viscous plastic. They also have stronger glazing (gasketing) systems to better hold the glass units in place, their frames are structurally reinforced, and they are fastened into their rough openings with extra-strong attachment hardware. As an alternative to providing impact-resistant openings in one- and two-story buildings, the code permits the use of precut plywood or OSB panels that can be fastened into place over the outside of such openings when needed to act as temporary storm shutters.

Blast Resistance

In buildings subject to special security requirements, windows, curtain walls, and other glazing systems can be designed for resistance to the force of explosive blasts. At this time, standards for *blast-resistant glazing systems* are published by several U.S. federal agencies, including the Department of Defense and the General Services Administration, as well AAMA, one of the copublishers of the *Standard/Specification for Windows, Doors, and Unit Skylights* discussed previously. Design for blast resistance involves defining the size of the blast and its distance from the glazing system, as well as the glazing system's response to the blast. Of particular concern is the extent to which glass remains intact in the system or is shattered and dispersed as hazardous fragments. Like impact-resistant glazing, blast-resistant glazing typically relies on laminated glass and reinforced framing and attachment systems.

World's Best
Window Co.

Millennium 2000+
Vinyl-Clad Wood Frame
Double Glazing • Argon Fill • Low E
Product Type: **Vertical Slider**

ENERGY PERFORMANCE RATINGS

U-Factor (U.S./I-P)	Solar Heat Gain Coefficient
0.34	**0.25**

ADDITIONAL PERFORMANCE RATINGS

Visible Transmittance	Air Leakage (U.S./I-P)
0.41	**0.2**

Manufacturer stipulates that these ratings conform to applicable NFRC procedures for determining whole product performance. NFRC ratings are determined for a fixed set of environmental conditions and a specific product size. Consult manufacturer's literature for other product performance information.
www.nfrc.org

FIGURE 18.28
A sample NFRC certification label that is affixed to a window unit so that buyers may compare energy efficiencies. The U-Factor is the reciprocal of the R-value, which means that the R-value of this window unit is about 3.

FIGURE 18.29
Custom wood doors on architect Steven Holl's The Chapel of St. Ignatius. *(Photo by Joseph Iano)*

FIGURE 18.30
The Blanchard Road Alliance Church in Wheaton, Illinois, silhouettes laminated wood trusses against a wall of vinyl-clad wood frame fixed windows. *(Architect: Walter C. Carlson Associates. Photo courtesy Andersen Windows, Inc. Andersen is a registered trademark of Andersen Corporation, copyright 1997. All rights reserved)*

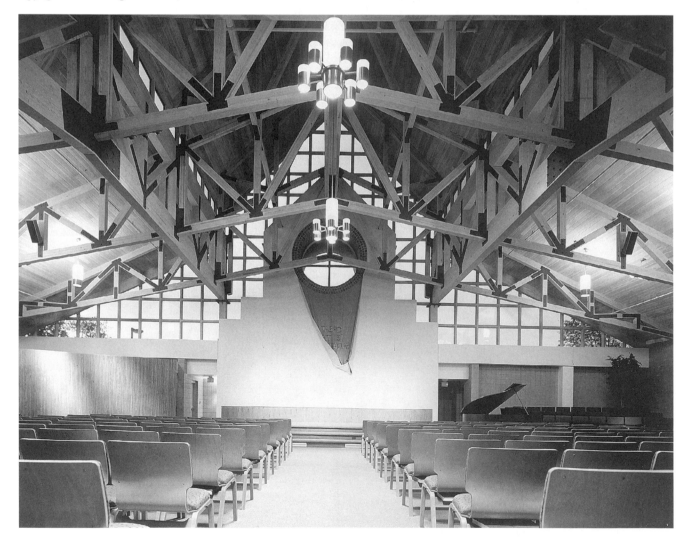

CSI/CSC

MasterFormat Sections for Windows and Doors

08 10 00	DOORS AND FRAMES
08 11 00	Metal Doors and Frames
	Hollow Metal Doors and Frames
	Metal Screen and Storm Doors and Frames
08 14 00	Wood Doors
	Flush Wood Doors
	Clad Wood Doors
	Stile and Rail Wood Doors
08 15 00	Plastic Doors
08 50 00	WINDOWS
08 51 00	Metal Windows
	Aluminum Windows
	Steel Windows
08 52 00	Wood Windows
08 53 00	Plastic Windows
08 54 00	Composite Windows
	Fiberglass Windows
08 60 00	ROOF WINDOWS AND SKYLIGHTS
08 61 00	Roof Windows
08 62 00	Unit Skylights

Selected References

1. Carmody, John, Stephen Selkowitz, Dariush Arasteh, and Lisa Heschong. *Residential Windows: A Guide to New Technologies and Energy Performance* (2nd ed.). New York, W. W. Norton & Company, 2000.

This book is a clearly written, well-illustrated introduction to considerations of energy efficiency in residential windows.

2. Selkowitz , Stephen, Eleanor S. Lee, Dariush Arasteh, Todd Willmert, John Carmody, and Eleanor Lee. *Window Systems for High-Performance Buildings.* New York, W. W. Norton & Company, 2003.

This book addresses the myriad performance requirements and selection criteria for commercial glazing and window systems.

3. Hollow Metal Manufacturers Association. *Hollow Metal Manual.* Chicago, various dates.

This binder includes standard ANSI/NAAMM-HMMA 861, discussed in this chapter, as well as numerous other standards and guides useful to the designer and specifier of steel doors and frames.

4. Window & Door Manufacturers Association. *Specifiers Guide to Windows and Doors.* Des Plaines, IL, various dates.

This is a compilation of documents useful to the designer and specifier of window and door systems, including two important standards discussed in this chapter, AAMA/WDMA/CSA 101/I.S.2/A440 and ANSI/WDMA I.S.1-A.

5. American Architectural Manufacturers Association. *Window Selection Guide (AAMA WSG.1-95).* Palatine, IL, 1995.

In this booklet, AAMA sets forth and explains the standards under which aluminum and plastic windows are manufactured and specified. In addition to this guide, AAMA publishes a variety of standards and guides related to the manufacturing, selection, and installation of window systems.

Web Sites

Windows and Doors

Author's supplementary web site: **www.ianosbackfill.com/18_windows_and_doors**
American Architectural Manufacturers Association: **www.aamanet.org**
Andersen Windows: **www.andersenwindows.com**
Ceco Steel Doors: **www.cecodoor.com**
Hollow Metal Manufacturers Association: **www.naamm.org/hmma**
Hope's Steel Windows & Doors: **www.hopeswindows.com**
Impact Grade Windows (EFCO Corporation): **www.impactgrade.com**
Marvin Windows: **www.marvin.com**
Morgan Wood Doors: **www.morgancorp.com**
National Fenestration Rating Council: **www.nfrc.org**
Steel Door Institute: **www.steeldoor.org**
Steel Window Institute: **www.steelwindows.com**
Whole Building Design Guide, Windows: **www.wbdg.org/design/env_fenestration_win.php**
Window & Door Manufacturers Association: **www.wdma.com**
Windows and Daylighting (Lawrence Berkeley National Laboratory): **windows.lbl.gov**
Window Systems for High-Performance Buildings (Center for Sustainable Building Research): **www.commercialwindows.umn.edu Efficient Windows Collaborative: www.efficientwindows.org**

Key Terms and Concepts

window
prime window
storm window
combination window
replacement window
fixed window
single-hung window
double-hung window
sash
sliding window
projected window
casement window
awning window
hopper window
inswinging window
pivot window
jamb
head jamb, head
side jamb
sill
stool
interior casing
exterior casing
apron
triple-hung window
tilt/turn window
skylight, unit skylight
roof window
French door
sliding door
terrace door
pivoting window
side-hinged inswinging window
top-hinged inswinging window
wood window
clad wood window
thermal break
aluminum window
plastic window
glass-fiber-reinforced plastic (GFRP) window, fiberglass window
pultrusion
plastic
synthetic rubber, elastomer
polymer
monomer
copolymer
thermoplastic
thermosetting
plasticizer
stabilizer
filler
extender
reinforcing (fibers)
flame retarder
composite
laminate
sandwich
extrusion
molding
film blowing
calendaring
blow forming
vacuum forming
steel window
muntin
six over six
rough opening
masonry opening
door
exterior door
interior door
Z-bracing
stile-and-rail door
flush door
solid core door
hollow-core door
mineral core door
sheet metal door
GFRP door
prehung door
storm door
screen door
combination door

hollow steel door frame
standard steel door
custom steel door
fire door
panic hardware
emergency escape and rescue opening
AAMA/WDMA/CSA 101/I.S.2/A440 Performance Class
AAMA/WDMA/CSA 101/I.S.2/A440 Performance Grade
overall thermal transmittance, whole product heat loss
wind-borne debris region
impact-resistant opening, hurricane-rated opening
blast-resistant glazing system

Review Questions

1. List in detail the primary functional requirements for windows in each of the following situations:

a. A residential bathroom in an urban apartment building

b. A jail cell

c. A display window in a department store

d. A teller's window in a drive-in banking facility

e. A bedroom in Nome, Alaska

f. A living room in Hilo, Hawaii

2. Select a type of window operation for each of the following situations:

a. A window that can be left open in the rain

b. A window that must induce the maximum possible ventilation from passing breezes

c. A window in a high-rise office building

d. A window to frame an expansive view of distant mountains

e. A window that can be operated as either a casement or a hopper

3. Compare the advantages and disadvantages of wood, plastic-clad wood, PVC, aluminum, and steel as window frame materials.

4. A well-insulated residential wall has a U-value of 0.05 and an R-value of 20 (in U.S. units). Compare the heat loss per square foot of the worst- and best-performing window and glazing combinations listed in Figure 18.19 with that of this wall.

5. Select a type of door for each of the following situations:

a. Your bedroom closet

b. A front door of a house

c. A front door of a department store

d. A door between the industrial arts shops and the cafeteria in a high school

e. A door on a warehouse loading dock

Exercises

1. Obtain a copy of the building code that applies to the area in which you currently live. What fire resistance ratings are required for doors in the following situations?

a. A door between a hotel room and a public corridor

b. A door in an egress stair enclosure

c. A door between an iron foundry and an office building

d. A door between a single-family residence and its attached garage

2. Obtain catalogs from several window manufacturers. From them, select a set of windows for a one-room wilderness cabin of your own design.

3. Examine closely the windows in the room in which you are now sitting. What type of glazing do they have? What type of frame? How do they operate? How are they weatherstripped? Do these windows make sense to you in terms of today's energy efficiency requirements and your own feelings about the room? How would you change them?

19

Designing Exterior Wall Systems

The beautifully detailed cladding of the Hypolux Bank includes both granite blocks and metal panels. *(Architect: Richard Meier. Photograph © Scott Frances/Esto. All rights reserved)*

The *exterior wall enclosure* (also called the *building envelope*) is the part of a building that must defend the interior spaces against invasion by water, wind, sunlight, heat and cold, and all the other forces of nature. Its design is an intricate process that merges art, science, and craft to solve a long list of difficult problems. The exterior wall also typifies a paradox of building: Those parts of a building that are exposed to our view are also those that are exposed to wear and weathering. The outermost layer of the exterior wall is the most visible part of a building, one to which architects devote a great deal of time to achieve the desired visual effect. It is also the part of the building that is most subject to attack by natural forces that can spoil its appearance.

Figure 19.1
A steel-framed Chicago office building during the installation of its aluminum, stainless steel, and glass curtain wall cladding. Notice the diagonal wind braces in the steel frame. *(Architects: Kohn Pedersen Fox/Perkins & Will. Photo by Architectural Camera. Permission of American Institute of Steel Construction)*

Design Requirements for the Exterior Wall

Primary Functions of the Exterior Wall

The major purpose of the exterior wall is to separate the indoor environment of a building from the outdoors in such a way that indoor environmental conditions can be maintained at levels suitable for the building's intended use. This translates into a number of separate and diverse functional requirements.

Keeping Water Out

The exterior wall must prevent the entry of rain, snow, and ice into a building. This requirement is complicated by the fact that water on the face of a building is often driven by wind at high velocities and high air pressures, not just in a downward direction but in every direction, even upward. Water problems are especially acute on tall buildings, which present a large profile to the wind at altitudes where wind velocities are much higher than at ground level. Enormous amounts of water must be drained from the windward face of a tall building during a heavy rainstorm, and the water, pushed by wind, tends to accumulate in crevices and against projecting mullions, where it will readily penetrate the smallest crack or hole and enter the building. We will devote a considerable portion of this chapter to methods for keeping water out.

Preventing Air Leakage

The exterior wall of a building must prevent the unintended passage of air between indoors and outdoors. At a gross scale, this is necessary to regulate air velocities within the building. Smaller air leaks are harmful because they waste conditioned (heated or cooled) air, carry water through the wall, allow water vapor to condense inside the wall, and allow noise to penetrate the building from outside. Building code requirements for airtightness of building enclosures are growing more stringent. Sealants, gaskets, weatherstrips, and air barrier membranes of various types are all used to prevent air leakage through the exterior wall.

Controlling Light

The exterior wall of a building must control the passage of light, especially sunlight. Sunlight is heat that may be welcome or unwelcome. Sunlight is visible light, useful for illumination but bothersome if it causes glare within a building. It includes destructive ultraviolet wavelengths that must be kept off human skin and away from interior materials that will fade or deteriorate. Windows should be placed and proportioned with these considerations in mind. Exterior wall systems sometimes include external shading devices to keep light and solar heat away from windows. The glass in windows is often selected to control light and heat, as discussed in Chapter 17. Interior shades, blinds, and curtains may be added for further control.

Controlling the Radiation of Heat

Beyond its role in regulating the flow of radiant heat from the sun, the exterior wall of a building should also present interior surfaces at temperatures that will not cause radiant discomfort. A very cold interior surface will make people feel chilly when they are near the wall even if the air in the building is warmed to a comfortable level. A hot interior surface or direct sunlight in summer can cause overheating of the body despite the coolness of the interior air. External sun shading devices, adequate thermal insulation and thermal breaks, and appropriate selection of glass are potential strategies in controlling heat radiation.

Controlling the Conduction of Heat

The exterior wall of a building must resist the conduction of heat into and out of the building. This requires not merely satisfactory overall resistance of the wall to the passage of heat, but also avoidance of *thermal bridges*, wall components such as metal framing members that are highly conductive of heat and therefore likely to cause localized condensation on interior surfaces. Thermal insulation, appropriate glazing, and thermal breaks are used to control heat conduction through the exterior wall, as we will

FIGURE 19.2
The curtain wall of Chicago's Reliance Building, built in 1894–1895, has spandrels constructed of white terracotta tiles. *(Architect: Charles Atwood, of Daniel H. Burnham and Company. Photo by Wm. T. Barnum. Courtesy of Chicago Historical Society ICHi-18294)*

observe in the two chapters that follow. Building codes specify minimum values of thermal resistance of wall components as a way of limiting the conduction of heat and also as a way of controlling the condensation of moisture on cold interior surfaces.

Controlling Sound

The exterior wall serves to isolate the inside of a building from noises outside and vice versa. Noise isolation is best achieved by walls that are airtight, massive, and resilient. The required degree of noise isolation varies from one building to another, depending on the noise levels and noise tolerances of the inside and outside environments. The exterior wall for a hospital near a major airport requires a high level of noise isolation. The exterior wall for a commercial office in a suburban office park need not perform to as high a standard.

Secondary Functions of the Exterior Wall

Fulfillment of the primary functional requirements of the exterior wall leads unavoidably to a secondary but equally important set of requirements.

Resisting Wind Forces

The exterior wall of a building must be adequately strong and stiff to sustain the pressures and suctions that will be placed upon it by wind. For low buildings, which are exposed to relatively predictable winds, this requirement is fairly easily met. The upper reaches of taller buildings are beset by much faster winds whose directions and velocities are often determined by aerodynamic effects from surrounding buildings. High suction forces can occur on some

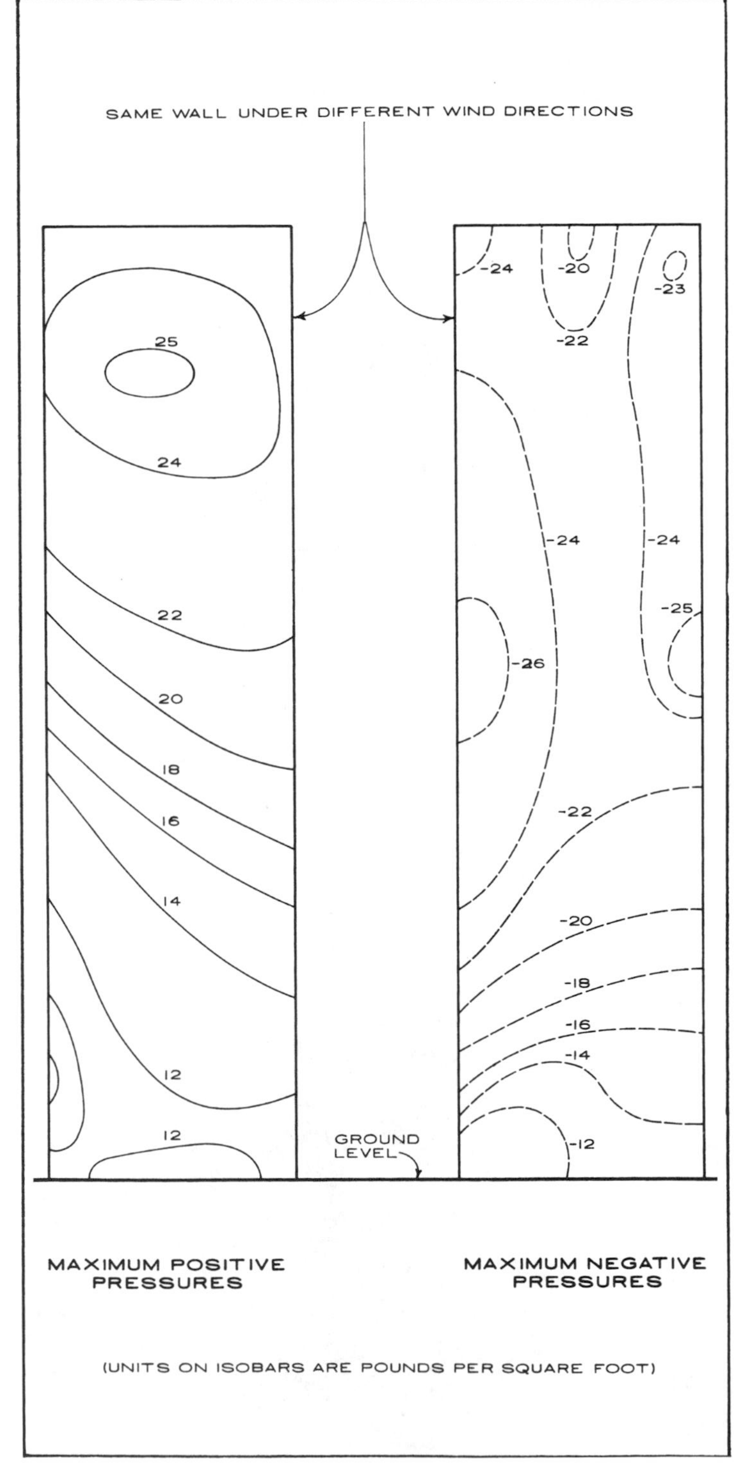

FIGURE 19.3
An example of expected positive and negative wind pressures on the cladding of a tall building, shown here in elevation, as predicted by wind tunnel testing. The building in this case is 64 stories tall and triangular in plan. Notice the high negative pressures (suctions) that can occur on the upper regions of the facade. The wind pressures on a building are dependent on many factors, including the shape of the building, its orientation, topography, wind direction, and surrounding buildings. Each building must be modeled and tested individually to determine the pressures it is expected to undergo. *(Reprinted with permission from AAMA Aluminum Curtain Wall Design Guide Manual)*

portions of the exterior wall, especially near corners of the building (Figure 19.3).

Controlling Water Vapor

The exterior wall of a building must retard the passage of water vapor. In the heat of summer or the cold of winter, vapor moving through a wall assembly may condense inside the assembly and cause staining, loss of insulating value, corrosion of metals, and decay of wood. The exterior wall must be constructed to resist the diffusion of water vapor and to restrict the leakage of moisture-laden air in order to prevent the transfer of water vapor to parts of the wall where it may condense.

Adjusting to Movement

Several different kinds of forces are always at work throughout a building, tugging and pushing both the frame and the exterior wall: thermal expansion and contraction, moisture expansion and contraction, and structural deflections. These forces must be anticipated and allowed for in designing a system of building enclosure.

Thermal Expansion and Contraction

The exterior wall of a building has to accommodate movements due to changes in temperature at several levels: Indoor/outdoor temperature differences can cause warping of cladding panels due to differential expansion and contraction of their inside and outside faces (Figure 19.4*a*). The exterior wall as a whole, exposed to outdoor temperature variations, expands and shrinks constantly with respect to the frame of the building, which is usually protected by the exterior wall from temperature extremes. And the building frame itself expands and contracts to some extent, especially between the time the exterior wall is installed and the time the building is first occupied and its indoor temperature is controlled.

Moisture Expansion and Contraction

Masonry and concrete exterior wall materials must accommodate their own expansion and contraction that is caused by varying moisture content. Bricks and building stone generally expand slightly after they are installed. Concrete blocks and precast concrete shrink slightly after installation in a building as their curing is completed and excess moisture is given off. These movements are small but can accumulate to significant and potentially troublesome quantities in long or tall panels of masonry or concrete. In smaller buildings, wood cladding components are the types of components most susceptible to moisture movement, as discussed in Chapter 3.

Structural Movements

The exterior wall must adjust to structural movements in the frame of the building. Building foundations may settle unevenly, causing distortions of the frame. Gravity forces shorten

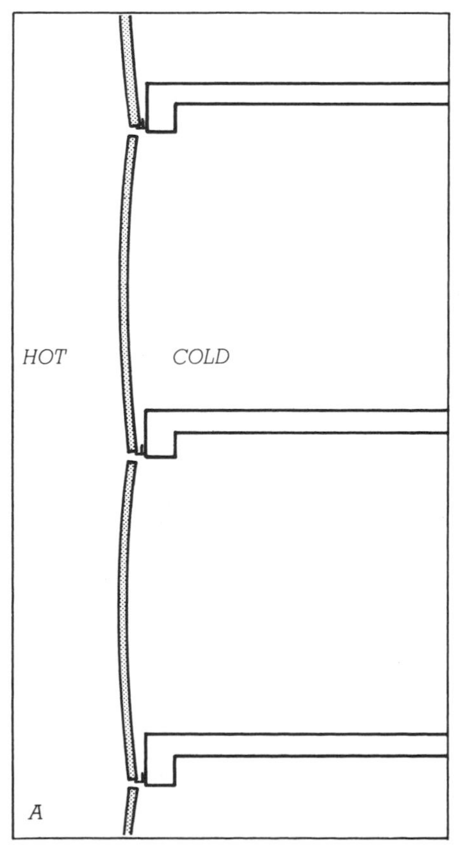

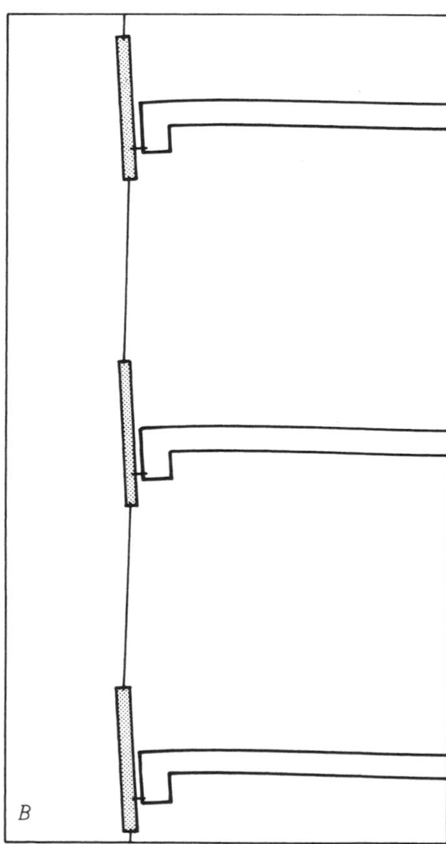

FIGURE 19.4
Distortions of curtain wall panels, illustrated in cross section. (*a*) Bowing caused in this case by greater thermal expansion of the outside skin of the panels than of the inside skin under hot summertime conditions. (*b*) Twisting of spandrel beams because of the weight of the curtain wall.

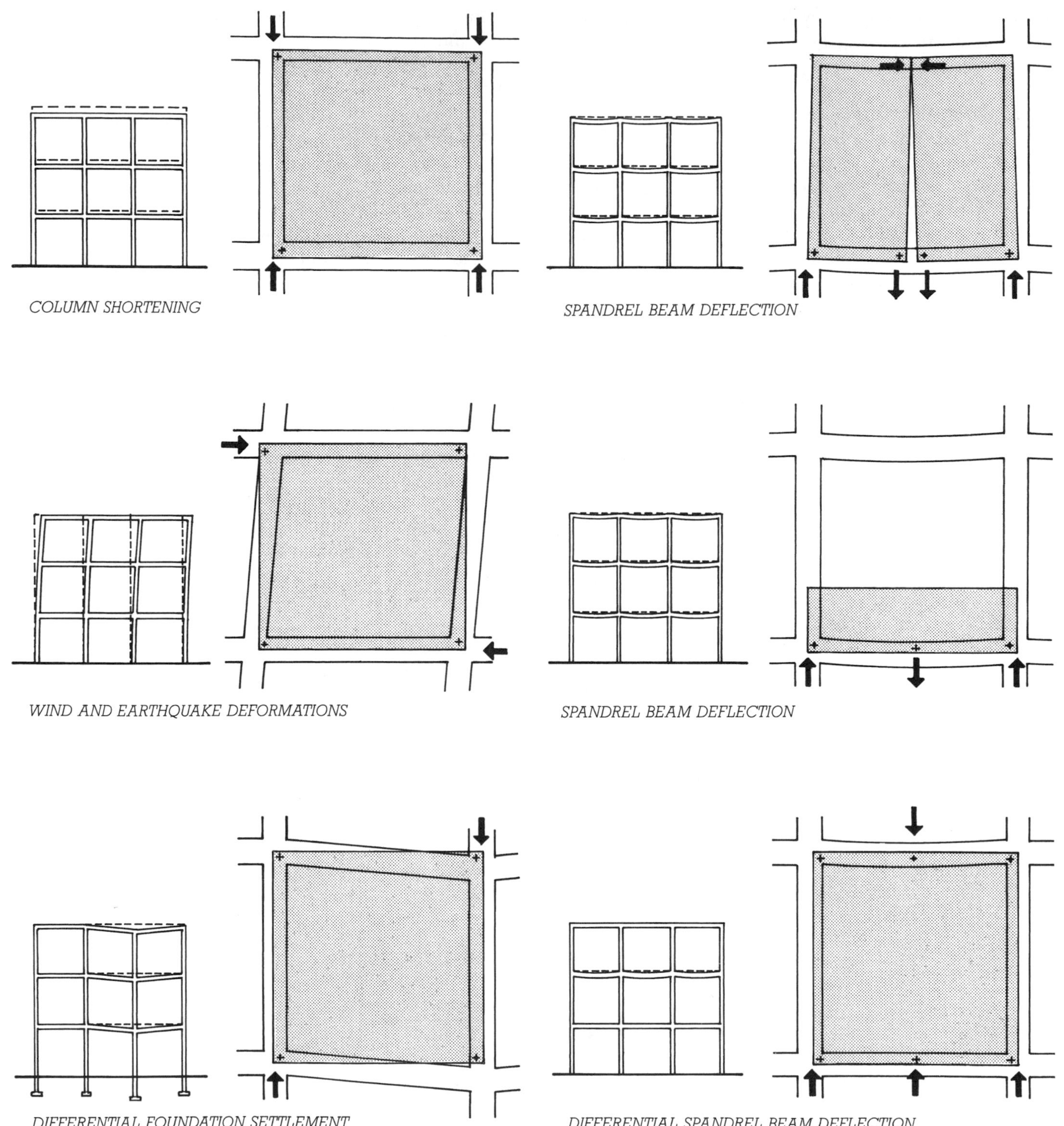

FIGURE 19.5
Forces on curtain wall panels caused by movements in the frame of the building, illustrated in elevation. In each of the six examples, the drawing to the left shows the movement in the overall frame of the building, and the larger-scale drawing to the right shows its consequences on the curtain wall panels (shaded in gray) covering one bay of the building. Points of attachment between the panels and the frame are shown as crosses. The black arrows indicate forces on the wall panels caused by the movement in the structure. The magnitude of the structural movements is exaggerated for clarity, and some inadvisable attachment schemes are shown to demonstrate their consequences. Forces such as these, if not taken into account in the design of the frame and cladding, can result in glass breakage, panel failures, and failure of the attachments between the panels and the frame.

Considerations of Sustainability in Exterior Wall Systems

For many if not most buildings, the design of the exterior wall has a greater effect on lifetime energy consumption than any other factor. A poorly designed all-glass box loses excessive amounts of heat in winter and gains excessive solar heat in summer. Its undifferentiated faces show no awareness on the part of the designer of the effects of orientation on energy transactions through the walls of a building.

- Glass should be used where it can supply daylighting and provide views. If it cannot be effectively shaded, it should be avoided where summertime overheating could otherwise occur or where occupants could be subject to excessive glare at times of the day when the sun is low in the sky.
- In many buildings, windows that can be opened and closed by the occupants can help reduce energy costs.
- Opaque areas of the exterior wall should be well insulated.
- Thermal bridges should be eliminated from the exterior wall.
- The entire building envelope should be detailed for airtightness. Fresh air should be provided by the building's ventilation system, not by air leakage through the exterior wall.
- Where appropriate, south-facing glass can be used to provide solar heat to the building in winter, but care must be taken to avoid glare, local overheating, and ultraviolet deterioration of interior surfaces and furnishings that are exposed to sunlight.
- As photovoltaic cells become more economical, consideration should be given to using south-facing surfaces of the exterior wall to generate electrical energy.

columns and cause beams and girders to which the exterior wall system is attached to sag slightly. Wind and earthquake forces push laterally on building frames and wrack panels attached to the faces. Long-term creep causes significant shortening of concrete columns and sagging of concrete beams and slabs during the first year or two of a building's life.

If building movements due to temperature differences, moisture differences, structural stresses, and creep are allowed to be transmitted between the frame and the exterior wall, unexpected things may happen. Wall system components may be subjected to forces for which they were not designed, which can result in broken glass, buckled cladding, sealant failures, and broken cladding attachments (Figure 19.5). In extreme cases, the building frame may end up supported by the exterior wall, rather than the reverse, or pieces of cladding may fall off the building. A number of provisions for dealing with movement from all these causes are evident in the details of exterior wall systems presented in the two chapters that follow.

Resisting Fire

The exterior wall of a building can interact in several ways with building fires. This has resulted in a number of building code provisions relating to the construction of building exterior wall systems, as summarized at the end of this chapter.

Weathering Gracefully

To maintain the visual quality of a building, its cladding must weather gracefully. The inevitable dirt and grime should accumulate evenly, without streaking or splotching. Functional provisions must be made for maintenance operations such as glass and sealant replacement and for periodic cleaning, including scaffolding supports and safety equipment attachment points for window washers. The cladding must resist oxidation, ultraviolet degradation, breakdown of organic materials, corrosion of metallic components, chemical attack from air pollutants, and freeze-thaw damage of stone, brick, concrete, concrete block, and tile.

Installation Requirements for the Exterior Wall

The exterior wall system should be easy to install. There should be secure places for the installers to stand, preferably on the floors of the building rather than on scaffolding outside. There must be built-in adjustment mechanisms in all the fastenings of components of the wall system to the frame to allow for the inaccuracies that are normally present in the structural frame of the building and the wall components themselves. Dimensional clearances must be provided to allow the wall components to be inserted without binding against adjacent components. And most important, there must be forgiving features that allow for a lifetime of trouble-free enclosure function despite all the lapses in workmanship that inevitably occur—features such as air barriers and drainage channels to get rid of moisture that has leaked through a faulty sealant joint or generous edge clearances that keep a sheet of glass from contacting the hard material of the frame even if the glass is installed slightly askew.

Conceptual Approaches to Watertightness in the Exterior Wall

In detailing the exterior wall for water resistance, we work from a secure theoretical base, which can be stated as follows:

In order for water to penetrate a wall, three conditions must be satisfied simultaneously:

1. There must be water present at the outer face of the wall.
2. There must be an opening through which the water can move.
3. There must be a force to move the water through the opening.

If any one of these conditions is not satisfied, the wall will not leak. This suggests three conceptual approaches to making a wall watertight:

1. We can try to keep water completely away from the wall. A very broad roof overhang can keep a one- or two-story wall dry under most conditions. When designing the exterior wall of a taller building, however, we must either shelter each opening with its own small roof—frequently not a realistic option—or else assume that the wall will get wet.

2. We can try to eliminate the openings in a wall. We can build very carefully, sealing every seam in the wall with membranes, sealants, or gaskets, attempting to eliminate every hole and crack.

This approach, which is called the "barrier wall approach," works fairly well if done well, but it has inherent problems. In a wall made up of sealant-jointed components, the joints are unlikely to be perfect. If a surface is a bit damp, dirty, or oily, sealant may not stick to it. If the worker applying the sealant is insufficiently skilled or has to reach a bit too far to finish a joint, he or she may fail to fill the joint completely. Even if the joints are all made perfectly, building movements can tear the sealant or pull it loose. Because, in this approach, the sealant is on the outside of the building, it is exposed to the full destructive forces of sun, wind, water, and ice and may fail prematurely from weathering. And whatever the cause of sealant failure, because the sealant joint is on the outside face of the wall, it is difficult to reach for inspection and repair. Thus, in practice, the barrier wall approach proves unreliable.

In response to these problems, exterior wall designers often employ a strategy of *internal drainage* or secondary defense, which accepts the uncertainties of external sealant joints by providing internal drainage channels within the wall to carry away any leakage or condensate and backup sealant joints to the inside of the drainage channels. The ordinary masonry cavity wall facing exemplifies this strategy: The cavity, flashings, and weep holes constitute an internal drainage system for any moisture that finds its way through the facing bricks. Internal drainage systems are an important component of every metal-and-glass curtain wall system on the market, as we will see in Chapter 21.

3. We can try to eliminate or neutralize all the forces that can move water through the wall. These forces are five in number: gravity, momentum, surface tension, capillary action, and air currents (Figure 19.6).

Gravity is a factor in pulling water through a wall only if the wall contains an inclined plane that slopes into, rather than out of, the building. It is usually a simple matter to detail the exterior wall system so that no such inclined planes exist, though sometimes a loose gasket or an errant bead of sealant can create one despite the best efforts of the designer.

The *momentum* of falling raindrops can drive water through a wall only if there is a suitably oriented slot or hole that goes completely through the wall. Momentum is easily neutralized by applying a cover to each joint in the wall or by designing each joint as a simple *labyrinth.*

The *surface tension* of water, which causes it to adhere to the underside of a cladding component, allows water to be drawn into the building. The provision of a simple *drip* on any underside surface to which water might adhere will eliminate the problem.

Capillary action is the surface tension effect that pulls water through any opening that can be bridged by a water drop. It is the primary force that transports water through the pores of a masonry wall. It can be eliminated as a factor in the entry of water through a wall by making each of the openings in a wall wider than a drop of water can bridge or, if this is not feasible or desirable, by providing a concealed *capillary break* somewhere inside the opening. In porous materials such as brick, capillary action can be counteracted by applying an invisible coating of silicone-based water repellant, which destroys the adhesive force between water and the walls of the pores in the brick.

The solutions described in the four preceding paragraphs are easy to implement. With relatively straightforward geometric manipulations of the joint, the possibility of leakage caused by four of the five forces that can move water through an opening in a wall can be eliminated. The fifth force, *wind currents,* is the force most difficult to deal with in designing a wall for watertightness. We can neutralize it by employing pressure-equalized wall design.

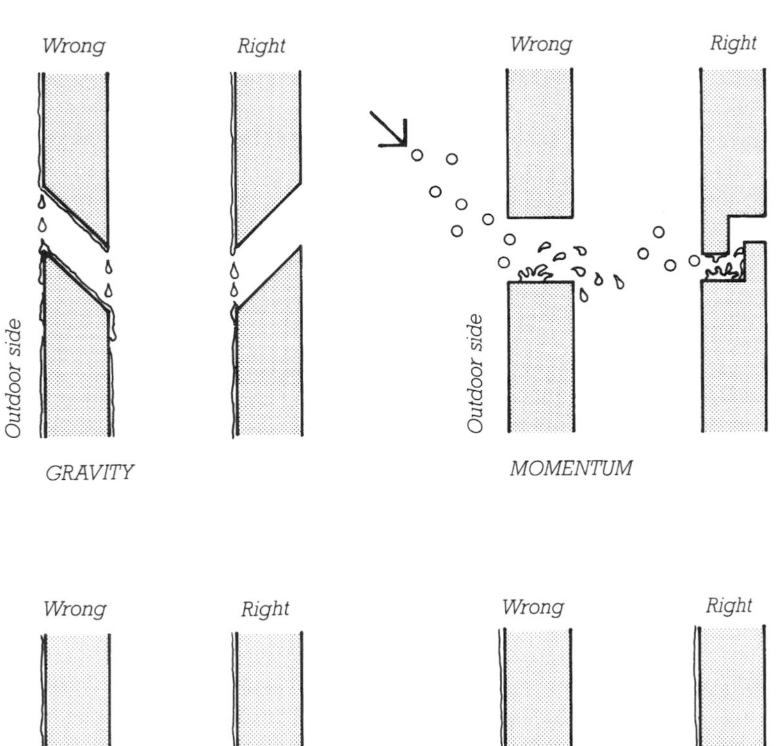

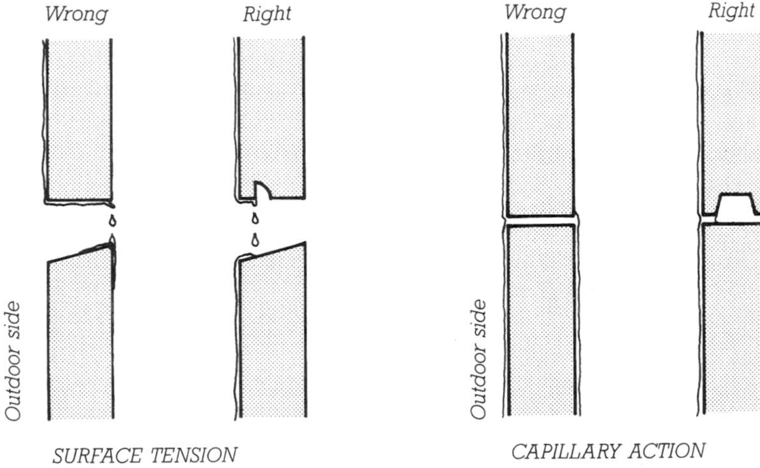

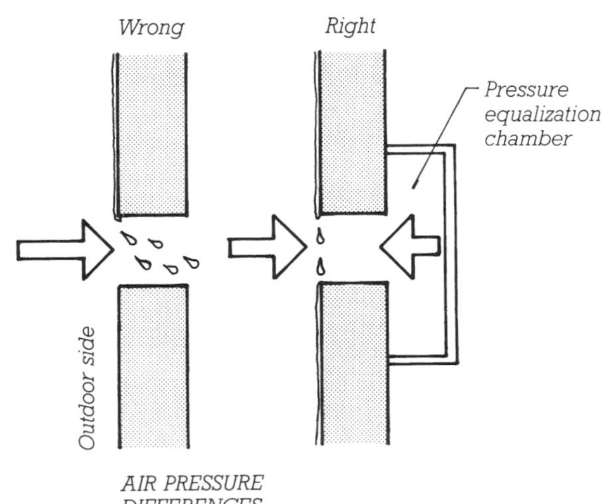

FIGURE 19.6

Five forces that can move water through an opening in a wall, illustrated in cross section with the outdoors to the left. Each pair of drawings shows first a horizontal joint between curtain wall panels in which a force is causing water leakage through the wall, then an alternative design for the joint that neutralizes this force. Leakage caused by gravity is avoided by sloping internal surfaces of joints toward the outside; the slope is called a "wash." Momentum leakage can be prevented with a simple labyrinth, as shown. A drip and a capillary break are shown here as means for stopping leakage from surface tension and capillary action, which are closely related forces. Air pressure differences between the outside and inside of the joint will result in air currents that can transport water through the joint. This is prevented by closing the area behind the joint with a pressure equalization chamber (PEC), as shown. When wind strikes the face of the building, a slight movement of air through the joint raises the pressure in the PEC until it is equal to the pressure outside the wall, after which all air movement ceases. Each joint in an exterior wall, window, or door must be designed to neutralize all five of these forces.

Rainscreen Cladding and Pressure-Equalized Wall Design

The generic solution to the wind current problem is to allow wind pressure differences between the outside and inside of the exterior wall to neutralize themselves through a concept known as *pressure-equalized wall design*. This involves the creation of an airtight plane, the *air barrier*, behind the outer face of the wall. The air barrier is protected from direct exposure to the outdoors by an unsealed, labyrinth-jointed layer known as the *rainscreen*. Between the rainscreen and the air barrier is a space known as the *pressure equalization chamber (PEC)*.

As wind pressures on the exterior wall build up and fluctuate, small currents of air pass back and forth through each unsealed joint in the rainscreen, just enough to equalize the pressure inside the PEC with the pressure immediately outside it (Figure 19.6). These currents are far too weak to carry water with them. A small flaw in the air barrier, such as a sealant bead that has pulled away from one side of the joint, is unlikely to cause a water leak because the volume of air that can pass through the flaw is still relatively small and is probably insufficient to carry water. By contrast, any flaw, no matter how small, in an external sealant joint without an air barrier behind it will cause a water leak, because the sealant joint itself is wetted (Figure 19.7).

Because wind pressures across the face of a building may vary considerably at any given moment between one area of the face and another, the PEC must be divided into airtight compartments small enough so that volumes of air cannot rush through the joints in higher-pressure areas of the face and flow across the air chamber to lower-pressure areas, carrying water with them as they go. The appropriate size of these chambers may vary considerably, depending on the design of the wall system and the wind forces to which it is exposed. Broadly speaking, PECs are normally no taller than one story or wider than one or two columns bays. In some applications they may be significantly smaller.

The term *rainscreen principal* originated with the concept of pressure-equalized wall design, and at one time it was used exclusively in reference to pressure-equalized cladding systems. More recently, the term *rainscreen cladding* has come to be applied more broadly to any cladding system with a system of internal drainage, regardless of the extent of compartmentalization of the drainage space and the degree of pressure equalization that can be achieved. In practice, varying degrees of pressure equalization are achievable, and the line between cladding systems best characterized as simple rainscreens or pressure-equalized walls is often indistinct.

A Pressure-Equalized Wall Design

Figure 19.8 depicts a cladding design that embodies the rainscreen and pressure-equalization principles in very simple form. No surface joint sealants or gaskets are used. The metal rainscreen panels do not touch one another, but are separated by generous gaps that preclude capillary movement of water, provide installation clearances, and allow for expansion and contraction. All four edges of each panel are shaped so as to create labyrinth joints. The forces of surface tension and gravity are counteracted by sloping surfaces and drips.

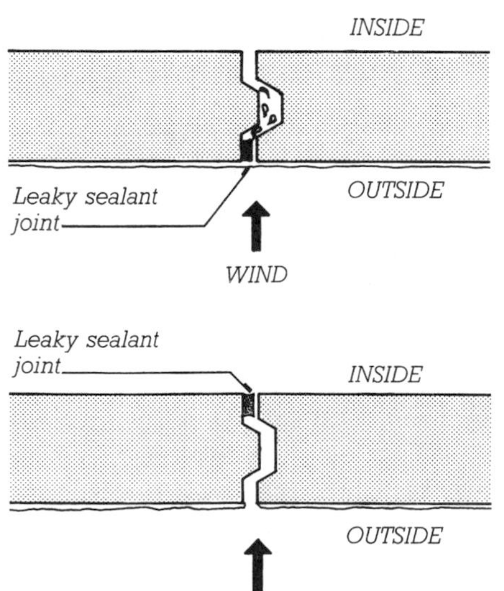

FIGURE 19.7
Leakage through a defective vertical sealant joint between curtain wall panels, shown in plan view. In the upper example, the sealant joint is at the outside face of the panels, where it is wetted during a storm. Even a small current of air passing through the defective joint carries water with it. In the lower example, with the defective sealant located on the inside of the panels where it remains dry, air leakage through the joint is insufficient to transport water through the joint, and no water penetrates.

FIGURE 19.8
The rainscreen in this exemplary cladding system is made up of metal panels, each formed from sheet metal. The design team included Wallace, Floyd Associates, Inc., Bechtel/Parsons Brinckerhoff, Stull and Lee, Inc., Gannett Fleming/URS/TAMS Consultants, and the Massachusetts Highway Department.

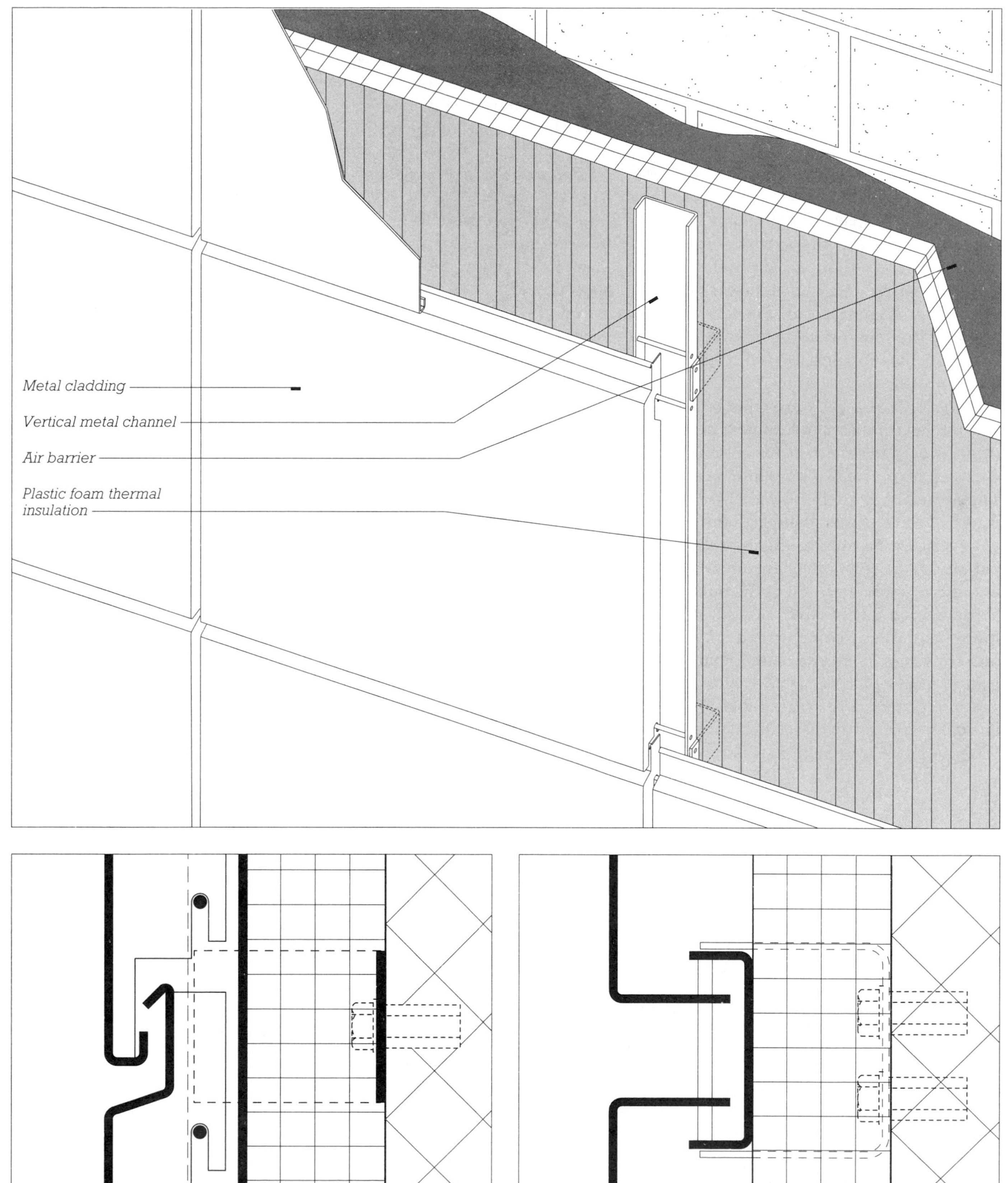

SECTION THROUGH HORIZONTAL JOINT

PLAN OF VERTICAL JOINT

Installation is simple and forgiving of minor lapses in workmanship: Metal U-shaped clips are bolted to the backup wall, which is coated with an airtight mastic to create an air barrier. Rigid insulation panels are adhered to the wall, allowing the clips to project through. Vertical metal channels are bolted to the clips. Finally, the metal panels that make up the rainscreen are simply hung on horizontal rods that are supported by the channels, much as pictures are hung on hooks on a wall. The space between the metal rainscreen panels and the insulation acts as an internal drainage space.

To achieve a pressure-equalized design, horizontal metal angles, not shown, are installed between the channels at one- or two-story intervals. The vertical channels further divide the PEC into narrow compartments. (The spaces between the edges of the panels and the channels are narrow enough to restrict airflow sufficiently to achieve a pressurized design. If more complete compartmentalization is needed, compressible foam rods or gaskets can be installed alongside the channels to create more airtight boundaries at these locations.) When wind drives rain against this wall, small quantities of air flow through the open joints in the rainscreen until the pressure in the PEC equals the pressure outside. These airflows are insufficient in volume or velocity to carry water with them.

Pressure Equalization at Smaller Scale

The principles of rainscreen design and pressure equalization may also be applied on a small scale to guide us in many aspects of exterior detailing of buildings. Figure 19.9 demonstrates how this practice is embodied in the placement of weatherstrip in a window sill detail. In the correct detail, the weatherstrip, whose function is to act as an air barrier, is placed to the inside of the lower rail of the sash. The open joint under the sash rail, which is provided with a capillary break, acts as the PEC. Unless the weatherstrip is grossly defective, water cannot be blown through the joint by air pressure differentials. Notice how the other forces that could transport water through the joint are counteracted: A slope on the sill (called by architects a *wash*) prevents gravity from pulling water in. The groove in the lower edge of the sash that acts as a capillary break also acts as a drip to counteract surface tension. The L-shaped joint between the sash and the sill acts as a labyrinth to prevent entry by momentum.

In the incorrect detail, the weatherstrip can be wetted by rain. Any minor flaw in the weatherstrip will allow water to be blown through the joint.

Relatively few buildings rely completely on the rainscreen principle and pressure-equalized wall design for watertightness. However, there are very few contemporary cladding systems that do not employ these principles as an important part of their defense against water penetration. Consider again the familiar example of the masonry cavity wall: These principles can be seen in the brick facing wythe acting as a rainscreen, the backup wall as an air barrier, and the cavity, partially pressurized through weep and vent holes, as the PEC. Nevertheless, the surface of the outer masonry veneer is also frequently sealed with compounds to reduce its absorbency (an application of the barrier wall approach), and the cavity is flashed and provided with weep holes so that water that does penetrate the veneer can be safely channeled back to the exterior (an application of internal drainage).

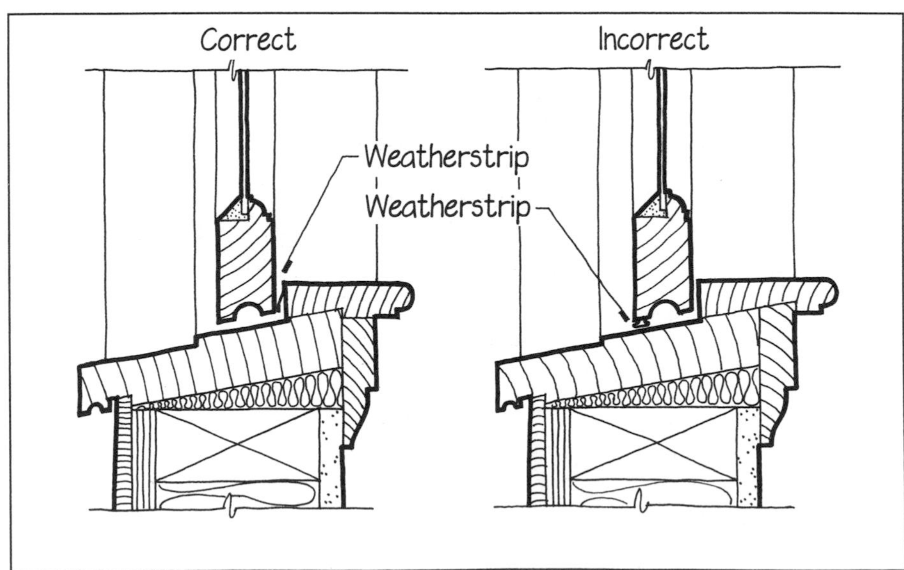

FIGURE 19.9
Applying the rainscreen principle to the detailing of the sill of a double-hung window. (*From Edward Allen,* Architectural Detailing: Function, Constructibility, Aesthetics, *New York, John Wiley & Sons, Inc., reproduced by permission of the publisher*)

Sealant Joints in the Exterior Wall

Most exterior wall systems require *sealant joints*, seams that are closed with rubberlike compounds. Systems that do not use sealants as water barriers in the face of the wall frequently use them to seal joints in the air barrier behind the face. The role of a sealant is to fill the joints between wall components, preventing the flow of air and/or water while still allowing reasonable dimensional tolerances for assembly and reasonable amounts of subsequent movement between the components. Sealant joint widths are usually ⅜ to ¾ inch (9–19 mm) but can be as small as ¼ inch (6 mm) and sometimes range up to 1 inch (25 mm) or more.

Sealants are often used to seal joints between panels of stone or precast concrete in a curtain wall (Figures 20.8 and 20.13), to seal the joint beneath the shelf angle in a brick curtain wall (Figure 20.3), and to seal joints between dissimilar materials, such as where a metal-and-glass cladding system ends against a masonry wall (Figure 21.12, details 6, 9, and 9A). Specially formulated sealants are used to seal between lights of glass and the frames that support them (Figure 17.17) and even to prevent the passage of sound around the edges of interior partitions (Figures 23.22, 23.23, 23.35, and 23.38).

Figure 19.10
Applying polysulfide, a high-range gunnable sealant, to a joint between exposed-aggregate precast concrete curtain wall panels, using a sealant gun. The operator moves the gun slowly so that a bulge of sealant is maintained just ahead of the nozzle. This exerts enough pressure on the sealant so that it fully penetrates the joint. Following application, the operator will return to smooth and compress the wet sealant into the joint with a convex tool, much as a mason tools the mortar joints between masonry units. (*Courtesy of Morton Thiokol, Inc., Morton Chemical Division*)

Sealant Materials

Gunnable Sealant Materials

Gunnable sealant materials are viscous, sticky liquids that are injected into the joints of a building with a *sealant gun* (Figure 19.10). They cure within the joint to become rubberlike materials that adhere to the surrounding surfaces and seal the joint against the passage of air and water. Gunnable sealants can be grouped conveniently in three categories according to the amount of change in joint size that each can withstand safely after curing:

- *Low-range sealants,* also called *caulks,* are materials with very limited *elongation* (stretching and squeezing) capabilities, up to plus or minus 5 percent of the width of the joint. They are used mainly for filling minor cracks or nonmoving joints, especially in preparation for painting. Most caulks cure by evaporation of water or an organic solvent and shrink substantially as they do so. None are used for sealing of joints in building exterior wall systems. (Although the term "caulk" is properly applied only to low-range sealants, in common usage it is frequently applied more broadly to mean any type of sealant, regardless of elongation capability.)
- *Medium-range sealants* are materials such as butyl rubber or acrylic that have safe elongations in the plus or minus 5 to 10 percent range. They are used in the building exterior wall for sealing nonworking joints (joints that are fastened together mechanically as well as being filled with sealant, as shown in Figure 19.11). Because these sealants cure by the evaporation of water or an organic solvent, they undergo some shrinkage during curing.
- *High-range sealants* can safely sustain elongations up to plus or minus 50 to 100 percent. They include various *polysulfides,* which are usually site mixed from two components to effect a chemical cure; *polyurethanes,* which may also cure by a two-component reaction or by reacting with moisture vapor from the air, depending on the formulation; and *silicones,* which all cure by reacting with moisture vapor from the air. None of these sealants shrink upon curing because none relies on the evaporation of water or a solvent to effect a cure. All adhere tenaciously to the sides of properly prepared joints. All are highly resilient rubberlike materials that return to their original size and shape after being stretched or compressed, and all are durable for 20 years or more if properly formulated and installed. Sealants for the working joints in exterior wall systems are selected from among this group. Polysulfide sealants have the longest history of use in such applications. However, improved formulations of polyurethanes and silicones now account for 90 percent or more of the high-range construction sealant market, with silicones generally considered the longest-lasting and highest-performing of the three.

Gunnable joint sealants are specified according to ASTM standard C920, which defines designations for sealant *Type, Grade, Class,* and *Use.* Type S sealants are *single-component* and require no jobsite mixing. Type M sealants are *multi-component* and must be mixed on the job-site before installation. Multicomponent sealants generally cure faster than single-component sealants. They also allow a greater variety in color choice, as dye packs can be added during mixing. Grade P sealants, also called *self-leveling,* are *pourable.* They are easily installed in horizontal paving joints. But for vertical wall joints, Type NS, *nonsag,* sealants must be used. Class defines the elongation capability of a sealant. A Class 25 sealant can tolerate 25 percent expansion and contraction under normal usage. A Class 100/50 sealant (the highest Class designation in the current standard) can tolerate 100 percent expansion and 50 percent contraction. A Use T, *traffic,* sealant can tolerate wear and physical abuse of pedestrian or vehicular traffic (most pourable sealants are also Use T); a Use NT, *nontraffic,* sealant is not suitable for traffic exposure and is normally intended for use in vertical wall joints; a Use I, *immersible,* sealant is suitable for sealing applications that will be submerged once the sealant has cured. Sealants may also be classified as Use M, G, A, or O, meaning that they have passed a series of tests demonstrating satisfactory adherence to mortar, glass, aluminum, or other materials, respectively. As an example, a multicomponent sealant intended for expansion joints between sections of aluminum curtain wall, which must be capable of 50 percent elongation, can be specified as Type M, Grade NS, Class 50, Uses NT and A.

Solid Sealant Materials

In addition to the gunnable sealants, several types of solid materials are used for sealing seams in the building exterior wall (Figure 19.11):

- *Gaskets* are strips of various fully cured elastomeric (rubberlike) materials manufactured in several different configurations and sizes for different purposes. They are either compressed into a joint to seal tightly against the surfaces on either side or inserted in the joint loose and then expanded with a *lockstrip* insert, as illustrated in Figures 17.17–17.19.
- *Preformed cellular tape sealant* is a strip of polyurethane sponge material that has been impregnated with a mastic sealant. It is delivered to the construction site in an airtight wrapper, compressed to one-fifth or one-sixth of its original volume. When a strip is unwrapped and inserted, it expands to fill the joint, and its sealant material cures with moisture from the air to form a watertight seam.
- *Preformed solid tape sealants* are used only in lap joints, as in mounting glass in a metal frame or overlapping two thin sheets of metal at a cladding seam. They are thick, sticky ribbons of polybutene or polyisobutylene that adhere to both sides of the joint to seal and cushion the junction. They are so sticky that they cannot be inserted into a joint, but must be applied to one side of the joint before it is assembled.

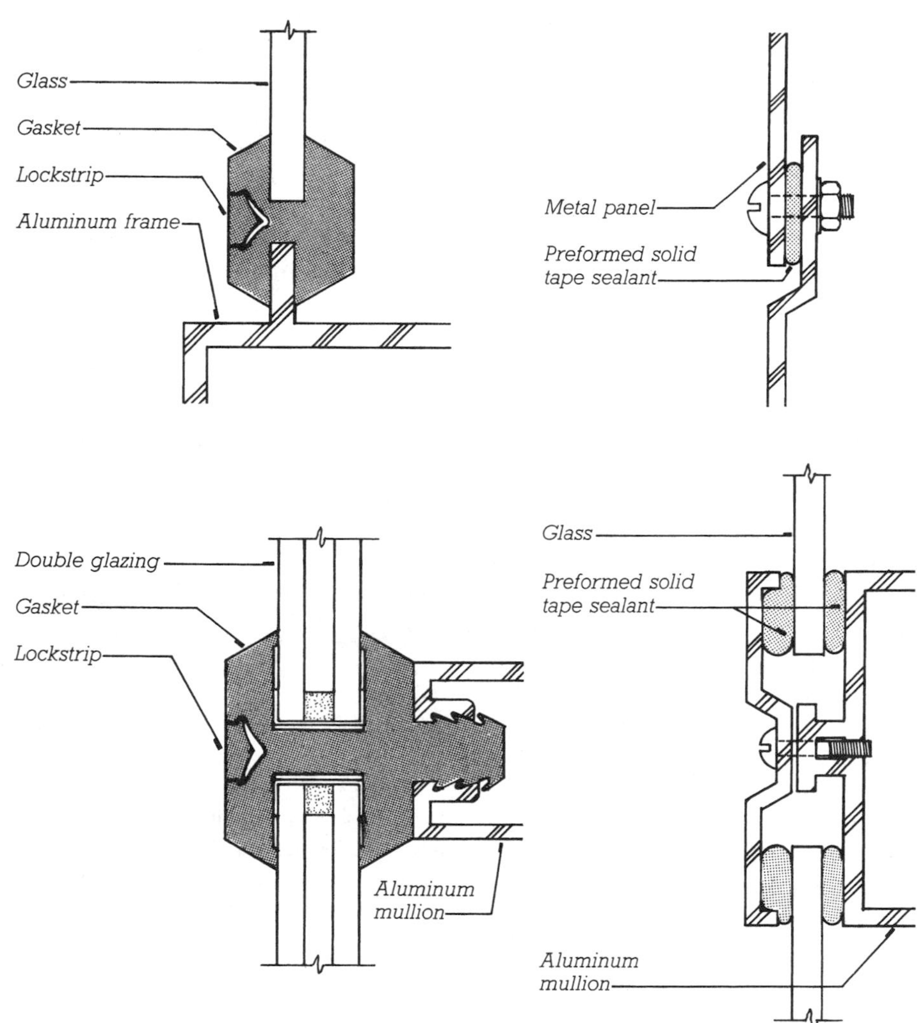

FIGURE 19.11
Some solid sealant materials. At the left, two examples of lockstrip gaskets. At the right, preformed solid tape sealants.

Sealant Joint Design

Figures 19.12–19.14 show the major principles that need to be kept in mind while designing a gunnable sealant joint. For a joint between materials with high coefficients of expansion, the time of year when the sealant is to be installed must be taken into account when specifying the size of the joint and the type of sealant. Sealant installed in cold weather will have to stretch less during its lifetime but will have to compress more in summer than the same sealant installed in hot weather, which will have to stretch more and compress less.

Installation procedures are also critical to the success of gunnable sealant joints in an exterior wall system. Each joint must be carefully cleaned of oil, dirt, oxide, moisture, or concrete form release compound. If it is necessary to improve adhesion between the sealant and the substrate, the edges of the joint are *primed* with a suitable coating. Then the *backer rod* or *backup rod* is inserted. This is a cylindrical strip of highly compressible, very flexible plastic foam material that is just a bit larger in diameter than the width of the joint. It is pushed into the joint, where it holds its place by friction, to limit the depth to which the sealant will penetrate in order to maintain the optimum proportions of the sealant bead and avoid waste of sealant material. Backer rod material is available in a large range of diameters to fit every joint.

The sealant is extruded into the joint from the nozzle of a sealant gun, filling completely the portion of the joint outside the backer rod. Lastly, the sealant is mechanically tooled, much as a masonry mortar joint is tooled, to compress the sealant material firmly against the sides of the joint and the backer rod. The tooling also gives the desired surface profile to the sealant. (The backer rod's role is now finished but, being inaccessible, the rod stays in the joint.)

Gasket sealants have generally proved to be less sensitive to installation problems than gunnable sealants. For this reason, they are widely used in proprietary cladding systems.

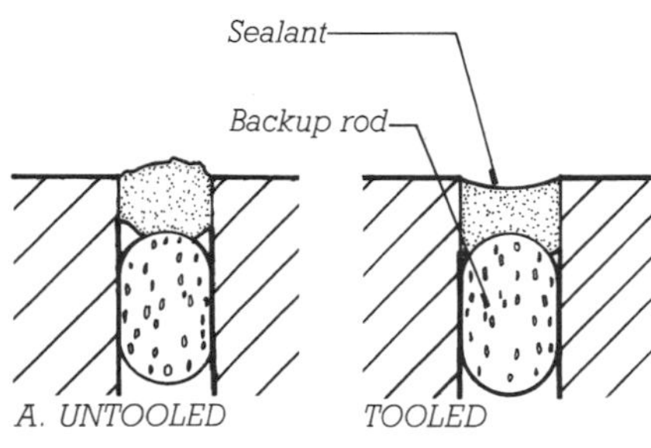

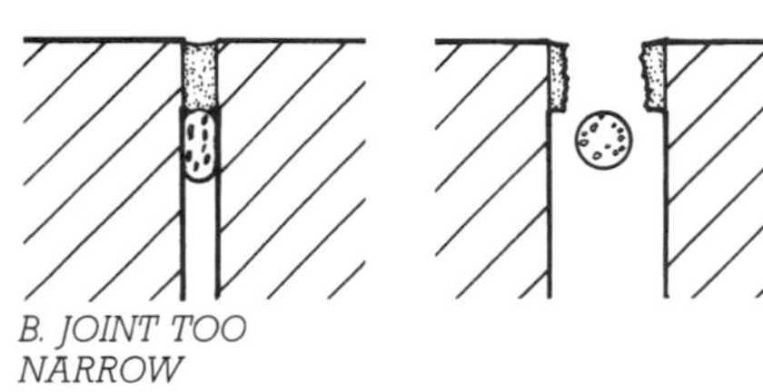

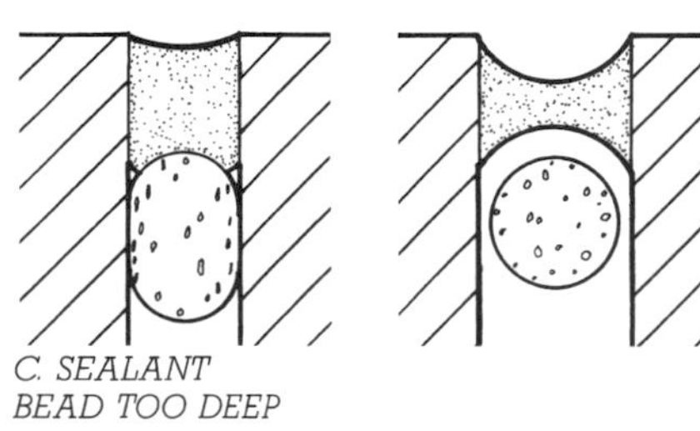

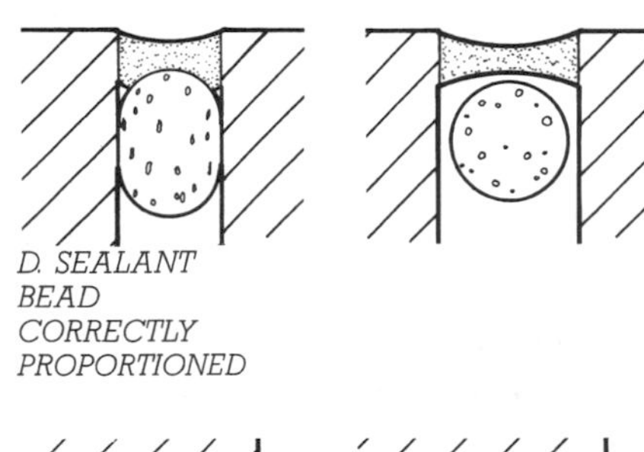

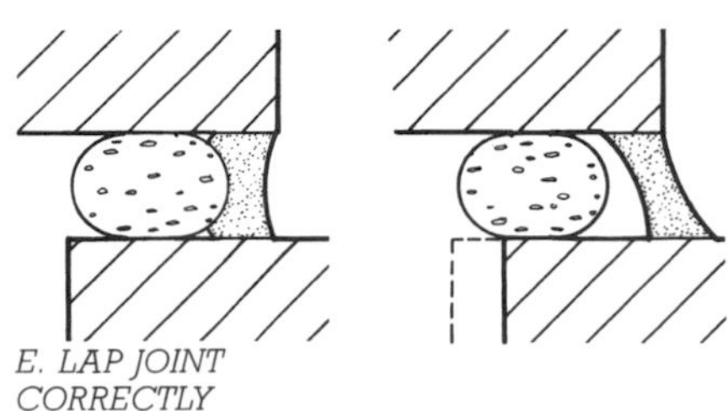

FIGURE 19.12
Good and bad examples of sealant joint design. *(a)* This properly proportioned joint is shown both untooled and tooled. The untooled sealant fails to penetrate completely around the backer rod and does not adhere fully to the sides of the joint. *(b)* A narrow joint may cause the sealant to elongate beyond its capacity when the panels on either side contract, as shown to the right. *(c)* If the sealant bead is too deep, sealant is wasted, and the four edges of the sealant bead are stressed excessively when the joint enlarges. *(d)* A correctly proportioned sealant bead. The backer rod, made of a spongy material that does not stick to the sealant, is inserted into the joint to maintain the desired depth. The width is calculated so that the expected elongation will not exceed the safe range of the sealant, and the depth is between ⅛ and ⅜ inch (3 and 9.5 mm). *(e)* A correctly proportioned lap joint. The width of the joint (the distance between the panels) should be twice the depth of the sealant bead and twice the expected movement in the joint.

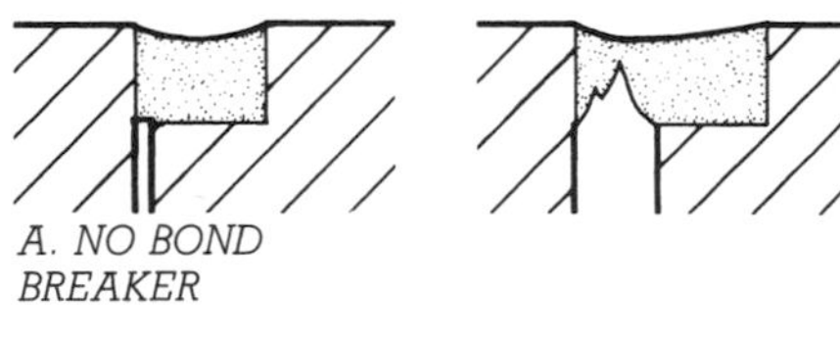

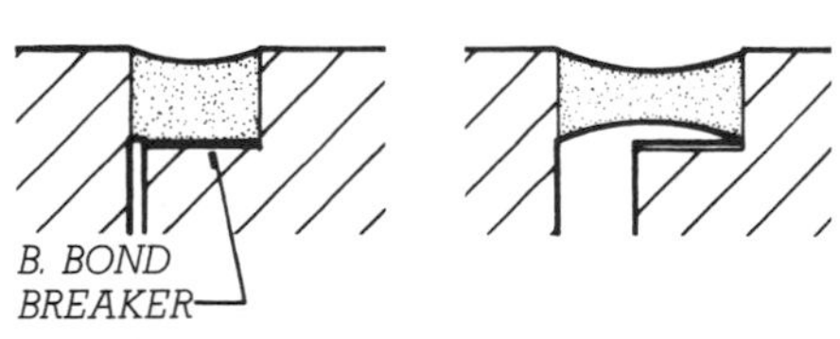

FIGURE 19.13
In three-sided joints, the sealant is likely to tear unless a nonadhering plastic bond breaker strip is placed in the joint before the sealant.

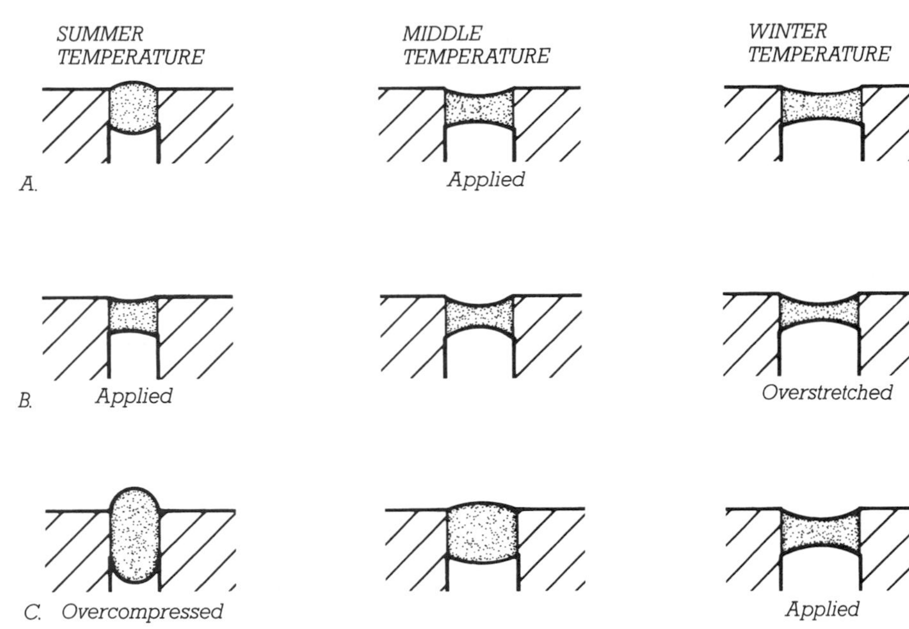

FIGURE 19.14
Sealants are best applied at temperatures that are neither excessively hot nor excessively cold. If cold- or hot-weather applications of sealants are anticipated, the joints should be proportioned to minimize overstretching or overcompression. Row *A* shows the behavior of a sealant applied at a medium temperature. Rows *B* and *C* show sealant applied at summer and winter temperatures, respectively.

Basic Concepts of Exterior Wall Systems

The Loadbearing Wall

Until the late 19th century, nearly all large buildings were built with loadbearing exterior walls. These walls supported a substantial portion of the floor and roof loads of the building, as well as separating the indoor environment from the outdoors. In noncombustible buildings, these walls were built of brick or stone masonry. Functionally, such walls had several inherent limitations. They were poor thermal insulators, and they were heavy, requiring large foundations and limiting their height to a few stories.

The *loadbearing wall* has been brought up to date with higher-strength masonry and concrete; components such as thermal insulating materials, cavities, flashings, air barriers, and vapor retarders have been added to make the wall more resistant to the passage of water, air, and heat; and the addition of steel reinforcing has allowed the wall to become thinner, lighter, and better able to resist wind and seismic loads. Loadbearing masonry and concrete exterior walls are often attractive and economical for low- and medium-rise buildings. High-rise residential towers with exterior loadbearing masonry walls also continue to be built, especially in Asia. These types of construction are illustrated and discussed in more detail in Chapters 8, 10, and 14.

The Curtain Wall

The first steel-framed skyscrapers, built in the late 19th century, introduced the concept of the *curtain wall*, an exterior wall supported at each story by the frame. The name "curtain wall" derives from the idea that the wall is thin and "hangs" like a curtain on the structural frame. (Most curtain wall panels do not actually hang in tension from the frame but are supported from the bottom at each floor level.) The earliest curtain walls were constructed of masonry (Figure 19.2). The principal advantage of the curtain wall is that, because it bears no vertical load, it can be thin and lightweight regardless of the height of the building, compared to a masonry loadbearing wall, which may become prohibitively thick and weighty at the base of a very tall building.

Curtain walls may be constructed of any noncombustible material that is suitable for exposure to the weather. They may be either constructed in place or prefabricated. In the next chapter, we will examine curtain walls that are made of masonry and concrete. In Chapter 21, we will look at curtain walls that are made of metal and glass. In both chapters, we will see that some types of walls are constructed in place and others are prefabricated, but all are supported by the frame of the building.

Building Enclosure Essentials: Air Barrier

Air Leakage

Air can move through a building assembly wherever air pressure differences exist between one side of the assembly and the other. Such pressure differences can be created by wind forces acting on the external surfaces of a building, by *stack effect* (the tendency of tall buildings to act somewhat like chimneys, drawing air in at either the top or bottom and expelling it at the other end), and by building mechanical equipment such as exhaust fans and air handling systems (Figure A).

When outside air infiltrates a building through exterior walls and roofs, it increases building energy consumption. A 2005 study by the U.S. National Institute of Standards and Technology estimated that air infiltration can account for as much as 40 percent of a building's heating and cooling costs. Outside air infiltration also introduces unfiltered air pollutants and unconditioned air into the building's interior, where it can compromise indoor air quality and reduce occupant comfort. Air leakage transports water vapor into insulated walls and roofs, increasing the risk of condensation and moisture damage to building components. When air leaks between spaces within a building, it can disrupt pressure differentials maintained by HVAC systems for the purpose of controlling the spread of odors or contaminants between separately zoned parts of a building. For example, unpleasant cooking odors can be drawn from one apartment building living unit to another, car exhaust or gas fumes from a parking garage can infiltrate adjacent occupied areas, dust particles can be carried into a laboratory clean room, or bacteria can be introduced into a hospital operating suite.

Air Barriers

Air barrier materials act to reduce air leakage through a building assembly. Examples of air barrier materials include building wrap, gypsum wallboard, polyethylene sheet plastic, rigid foam insulation, liquid-applied membranes of various formulations, caulking, sealants, gaskets, tapes, and more. To function as an air barrier, a material must be resistant to the passage of air; it must have sufficient strength and rigidity to withstand the air pressure differentials that act upon it; where it spans movement joints, it must be sufficiently resilient to accommodate movement without failure; and it must be durable enough to perform its function throughout the life of the building.

The greater a material's resistance to the passage of air, the lower its *air permeance* and the better its performance as an air barrier. Air permeance is measured according to ASTM E2178 and is expressed as cubic feet of air per minute per square foot of area at 1.57 pounds per square foot (or 0.3 inch of water) of air pressure, or, in metric units, as liters per second per square meter of area at 75 Pascals of air pressure. An air permeance of 1 cfm/sf@1.57 psf is equal to approximately 5 L/s-m^2@75 Pa. The most commonly cited standard for air barrier materials is an air permeance not greater than 0.004 cfm/sf@1.57 psf or 0.02 L/s-m^2@75 Pa (Figure B).

An air barrier material must be able to resist air pressure, acting either inward or outward across the building assembly, without damage or excessive deflection. Flexible sheet materials such as building wraps, plastic sheets, roofing membranes, and flexible flashings are especially vulnerable. If not properly supported or adequately fastened, these materials can tear, stretch, or pull loose and become

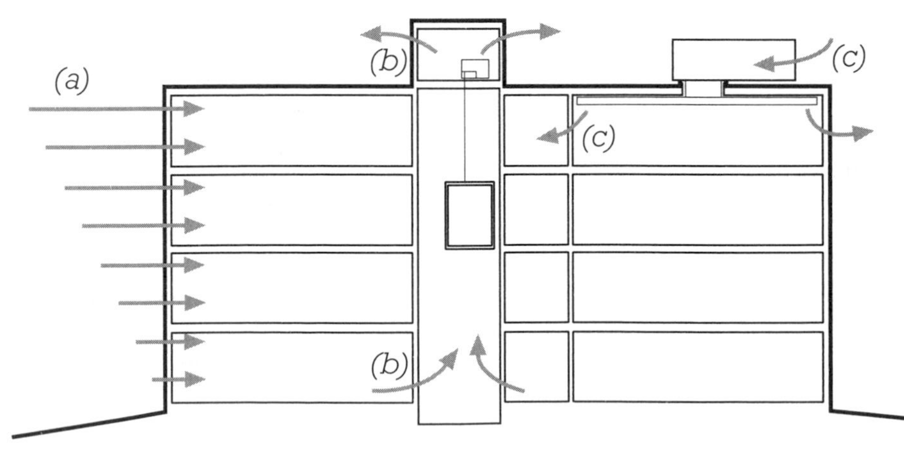

Figure A
Air pressure differences in a building can be caused by (*a*) wind, (*b*) the stack effect, such as within an elevator shaft, and (*c*) building mechanical systems.

Material	Air Permeance	
	cfm/sf@1.57 psf	L/s-m²@75 Pa
Polyethylene sheet, 6 mil (0.15 mm)	~0	~0
Aluminum foil, 1 mil (0.025 mm)	~0	~0
Self-adhered modified asphalt membranes	~0	~0
Plywood, ⅜ inch (9.5 mm)	~0	~0
Extruded polystyrene rigid foam insulation, 1½ inch (38 mm)	~0	~0
Most low-slope roofing membranes	~0–0.002	~0–0.01
Proprietary high-density spray polyurethane foam insulation, 1½ inches (38 mm)	0.0002	0.001
Proprietary fluid-applied, vapor-permeable air barrier membrane	<0.0004	<0.002
Proprietary nonperforated polyolefin building wrap		
Commercial grade	0.001	0.005
Residential grade	0.007	0.036
Low-density spray polyurethane foam insulation, 3 inches (75 mm)	0.002–0.32	0.01–1.6
Gypsum wallboard, ½ inches (12 mm)		
Exterior sheathing	0.002	0.0091
Interior wallboard, unpainted	0.004	0.0196
Oriented strand board, ⅜ inch (11 mm)	<0.004	<0.02
Roofing felt, #30	0.037	0.1873
Nonperforated asphalt felt, #15	0.078	0.3962
Asphalt-impregnated fiberboard	0.163	0.8285
Tongue-and-groove planks	3.7	19
Glass fiber batt insulation	7.3	37
Cellulose insulation, sprayed	17	87

FIGURE B
Air permeance of common building materials. Shaded rows indicate materials exceeding recommended air permeance for air barrier materials.

ineffective as air barriers. Damage caused by air pressures can also lead to the failure of materials to perform other important functions, such as keeping liquid water out of the building or resisting the diffusion of water vapor. If an air barrier material remains intact but deflects excessively under alternating cycles of positive and negative air pressure, it can pump air in and out of a building assembly, reducing the effectiveness of insulation and increasing the risk of water vapor transport into the assembly. Air barrier materials must also be able to accommodate the normal thermal and structural movements that occur within building systems without undue wear or failure.

Air Barrier Systems

To limit air leakage into and out of a building, its conditioned space must be completely surrounded by air barrier materials, creating an uninterrupted *air barrier system* of surfaces, membranes, manufactured components, gap fillers, and joint sealers that can effectively resist air pressure differentials acting across this boundary. Careful attention to detail during design and construction is required to achieve this goal. All potential discontinuities in the air barrier system—gaps between panels, laps in sheet materials, transitions between dissimilar substrates,

Building Enclosure Essentials: Air Barrier *(continued)*

fastener penetrations, movement joints, penetrations for structure or services, installation space around window and doors frames, junctions between foundation, wall, and roof assemblies, gaps between operable doors and windows and their frames, and so on—must be made airtight by the use of tapes, sealants, caulks, flashing, gaskets, and other materials that can themselves meet the air permeance limits and structural requirements of an air barrier material. Due to the significant air pressure differentials acting across the air barrier system, even small gaps can allow large volumes of air and water vapor to pass through the building enclosure and must therefore be minimized to the greatest extent possible.

Air permeance standards for the performance of assembled air barrier materials are less stringent than those for individual materials, reflecting the reality that flawless, continuous sealing between materials and components is never possible. Recommendations for the maximum air permeance of *air barrier assemblies*, that is, collections of materials responsible for the air barrier performance of a complete wall, roof, or floor system, are in the range of 0.01 to 0.04 cfm/sf@1.57 psf (0.05 to 0.2 L/s-m^2@75 Pa). Acceptable air leakage rates for whole building air barrier systems, reflecting the combined performance of a building's connected air barrier assemblies, fenestration, and other enclosing elements, can be expected to be even greater.

Air Barrier Location

Air barrier materials can be located anywhere in an assembly as long as they form an interconnected airtight system. At the inside surface of the building enclosure, the Airtight Drywall Approach and Simple Caulk and Seal are air barrier systems consisting of gypsum wallboard combined with caulks, sealants, and gaskets to seal leakage paths around wallboard penetrations and between underlying framing members (Figure 7.22). These systems are relatively easy and inexpensive to install, making them especially popular for residential construction. They are less favored for commercial building types where frequent changes to interior partitions, finishes, and wiring make it unlikely that the continuity of a system depending on the careful detailing of these elements will be maintained over the life of the building.

Plastic sheeting, frequently used as a vapor retarder behind gypsum wallboard, also has low air permeance and can act as an air barrier. However, difficulty in sealing plastic sheet seams and penetrations, as well as a tendency for the plastic to stretch and deflect between supporting framing, limit this material's suitability in air barrier systems, especially for taller buildings or wherever else high air pressure differentials are expected.

Toward the middle of a building enclosure assembly, foam insulation can be sprayed into the space between studs, joists, and rafters, acting as part of an air barrier system in combination with caulks or sealants to seal leakage paths around framing members. Air barrier materials located close to the interior side of the building enclosure or within framing cavities also benefit from being protected from the exterior elements.

Toward the outside of the building enclosure, air barrier materials are frequently installed over sheathing in framed construction or on the exterior face of masonry or concrete backup walls. Building wraps, plywood or gypsum board sheathing panels, and fluid-applied or fully adhered sheet membranes may all be used in combination with various sealing and taping materials. In this location, air barrier materials are easy to install, with a minimum of complex intersections. Where penetrations occur for the anchoring of cladding or sheathing, they are usually easily sealed to ensure airtightness (Figure 20.1*b*).

Curtain Wall Testing and Standards

Structural Performance and Resistance to Wind and Rain

For any new curtain wall design, it is advisable to build and test a full-scale section of wall to determine its resistance to infiltration of air and water and its structural performance under heavy wind loadings. There are several outdoor laboratories in North America that are equipped to conduct these tests. A full-scale specimen of the wall system, often two stories high and a bay wide, is constructed as the exterior wall of a chamber that can be pressurized or evacuated by a calibrated blower system.

Curtain wall testing is conducting according to the American Architectural Manufacturers Association standard *AAMA 501 Methods of Tests for Exterior Walls*, which itself references numerous other standards for specific aspects of the testing. The specimen is tested first for air infiltration, using ASTM E283, in which it is subjected to a static air pressure that corresponds to the pressure that will be created by the anticipated maximum wind velocity in the vicinity of the building. Air that leaks through the wall is carefully measured, and the rate of leakage is compared to specified standards.

Where air barriers fall to the exterior side of the building insulation, they can also protect against *wind washing*, in which exterior air currents within the assembly reduce insulation effectiveness. However, materials close to the exterior side of a building assembly must also be especially durable and able to perform satisfactorily while exposed to the effects of penetrating rainfall and large temperature fluctuations over the life of the building. In taller buildings, systems consisting of liquid-applied or sheet membranes that are fully adhered to rigid substrates are superior to flexible sheets such as building wrap. As noted earlier, loose sheet materials can be compromised by tearing or excessive deflection. Exterior-side air barrier materials also frequently play an important role in keeping liquid water out of the building enclosure, forming water-resistant drainage surfaces behind the outer cladding.

Unlike vapor retarders, there is no harm in installing multiple air barriers within one assembly. Multiple air barriers can provide the particular advantages of each type of system and can also provide redundancy, lessening the chance of a flaw in one material compromising a building's overall air leakage performance.

Air Barriers and Water Vapor Control

When air passes through a building assembly, the water vapor in the air is transported through the assembly as well. Where significant air pressure differentials exist between one side of an assembly and the other, the amount of water vapor transported through a building assembly by uncontrolled air leakage can be one or two orders of magnitude greater than that transported by water vapor diffusing directly through building materials. By controlling the flow of warm, moist air toward the cooler side of a building assembly, air barriers can play an important role in protecting against condensation within the assembly.

When designing air barrier systems, the water vapor permeability of the air barrier material must be considered. For example, in a heating-driven climate, an air barrier material located toward the outer, cooler side of an insulated building assembly must be vapor permeable to prevent the trapping of moisture within the assembly. Traditional building paper and breathable building wraps are good choices for use in this application; by contrast, a bituminous membrane, with low vapor permeability, would be a poor choice.

As a general rule, air barrier materials located on the cooler, lower-vapor-pressure side of a building assembly should always be vapor permeable. Conversely, air barriers located on the warmer, higher-vapor-pressure side of an assembly, may consist of materials with low vapor permeance and be designed to function as both an air barrier and a vapor retarder. For more information on vapor retarders and the control of water vapor diffusion in insulated building assemblies, see pages 658–661.

Building Code Air Barrier Requirements

The National Building Code of Canada (2005) sets quantifiable air permeance limits for air barrier materials and requires the control of air leakage with continuous air barrier systems in most buildings. The International Energy Conservation Code (2006) has more limited requirements for controlling air leakage, calling for the sealing of gaps through building envelope assemblies but without setting measurable criteria. At the time of this writing, several U.S. states have adopted more comprehensive air barrier system requirements based on Canada's model code, and it is likely that requirements for such systems will continue to spread in the United States over the coming years.

A static test for water penetration is next, using ASTM E331: The wall is subjected to a static air pressure while being wetted uniformly across its surface at a rate of 5 gallons per hour per square foot (3.4 L/m^2-min). Points of water leakage are noted, and leaking water is carefully collected and measured. A dynamic water penetration test may also be performed in accordance with AAMA 501.1, using an aircraft engine and propeller to drive water against the wall.

The structural performance of the wall is tested according to ASTM E330, in which a calibrated blower subjects the wall specimen to air pressures and suctions as high as 50 percent over the specified wind load, and the deflections of the structural members in the wall are measured. Optionally, tests for thermal performance, sound transmission, and the effects of thermal cycling, seismic loads, and movement of the structure to which the curtain wall is attached may also be performed.

While all these tests yield numerical results, it is also important that the behavior of the specimen be observed closely during each test so that specific problems with the design, materials, detailing, and installation can be identified and corrected. Most wall system specimens fail one or more of the tests for air and water leakage on the first attempt. By observing the sources of leakage during the test, it is usually possible to modify the flashings, sealants, weep holes, or other components of the design so that the modified specimen will pass the subsequent test. These modifications are then incorporated into the final details for the actual building.

After testing has been completed and final design adjustments have been made, production of the wall components begins, and deliveries to the site can commence as soon as the frame is ready to receive the system.

Curtain wall systems require careful inspection during installation to be sure that there are no defects in workmanship. Even seemingly small imperfections in assembly can lead to large, expensive problems later. As work progresses, installed portions of the curtain wall can be checked for water leakage according to AAMA 501.2. This involves directing water at the joints in the wall with a hose that has a specified nozzle and following specified procedures to isolate the causes of any leaks. Where deemed necessary, more elaborate instrumented field tests for water and air leakage can also be performed.

Thermal Performance and Other Properties

The thermal properties of curtain wall systems are most commonly tested according to AAMA 1503, for thermal transmittance, and AAMA 507 for solar heat gain coefficient, visible transmittance, and condensation resistance, although comparable NFRC standards may also sometimes be used. Curtain wall systems are adaptable to a great variety of glass types, frame sizes, and configurations. For this reason, determining precise U-factors and other properties for a particular system design usually requires more detailed analysis than, for example, when standard window configurations are specified.

Where impact or blast resistance is required, curtain wall systems can be tested to the same standards described in Chapter 18 for doors and windows.

The Exterior Wall and the Building Codes

The major impact of building codes on the design of the exterior wall is in the areas of structural strength, fire resistance, and energy efficiency. Strength requirements relate to the strength and stiffness of the wall system itself and to the adequacy of its attachments to the building frame, with special reference to wind and seismic loadings.

Fire requirements are concerned with the combustibility of the wall materials, the fire resistance ratings and vertical dimensions of parapets and spandrels, the fire resistance ratings of exterior walls facing other buildings that are near enough to raise questions of fire spread from one building to the other, and the closing off (*firestopping*) of any vertical passages in the wall that are more than

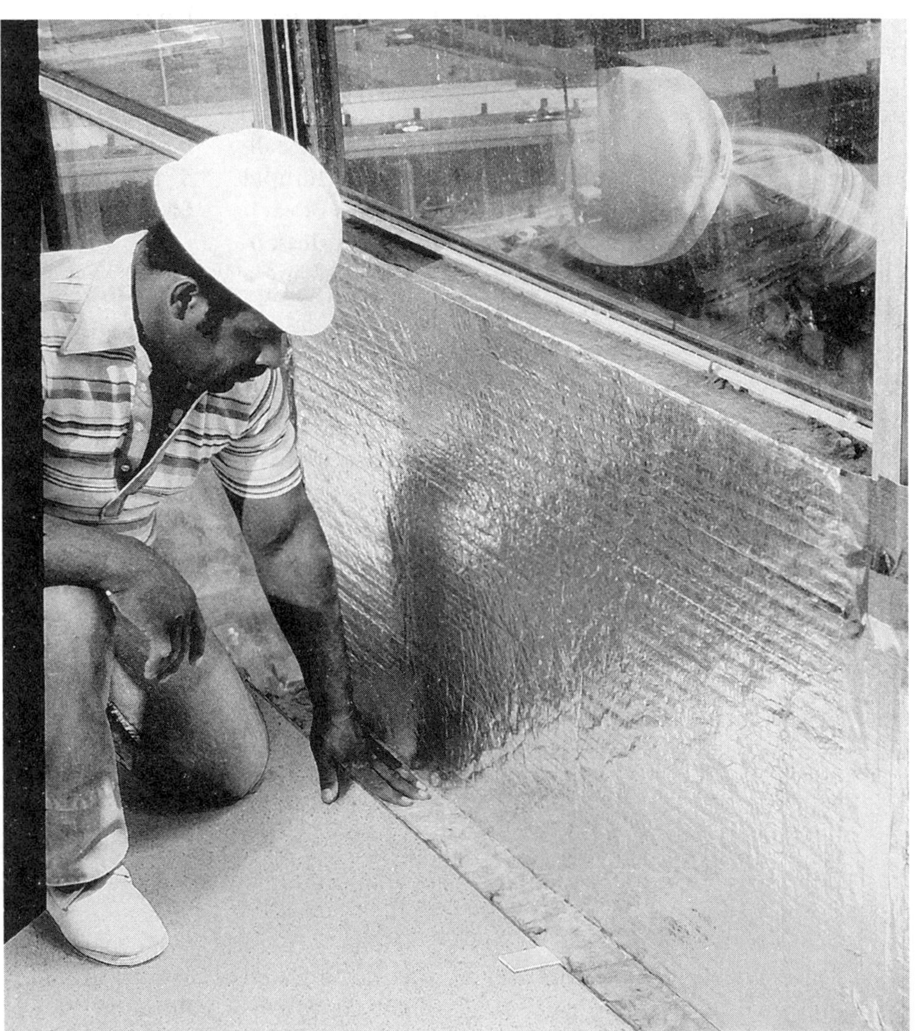

Figure 19.15
Safing is a high-temperature, highly fire-resistant mineral batt material that is inserted between a curtain wall panel and the edge of the floor slab to block the passage of fire from one floor to the next. It is seen here behind a metal-and-glass curtain wall with insulated spandrel panels. The safing is held in place by metal clips such as the one seen in the foreground. *(Courtesy of United States Gypsum Company)*

one story in height. At each floor, the space inside column covers and the space between the exterior wall system and the edges of floors must be firestopped, using a steel plate and grout, metal lath and plaster, mineral wool *safing*, or other material that can restrict the passage of smoke and fire through these gaps (Figure 19.15).

Energy conservation requirements are becoming more and more demanding. Most energy codes allow several alternative approaches to demonstrating compliance. In the prescriptive approach, minimum thermal resistances of panels, spandrels, and glass; vapor retarder performance; and maximum levels of air leakage are specified. For example, in the International Energy Conservation Code's prescriptive approach for commercial buildings, up to 40 percent of the above-grade walls may be glazed, with a maximum U-factor ranging from 1.20 to 0.35 (6.8-2.0 W/m^2-°K), depending on the climate zone in which the building is located.

Component trade-off and systems analysis approaches give the building designer more flexibility in selecting enclosure systems while demonstrating that the overall energy performance of the proposed design is equal or superior to that of the same building constructed to meet the prescriptive code requirements.

CSI/CSC
MasterFormat Sections for Designing Exterior Wall Systems

07 25 00	WEATHER BARRIERS
07 27 00	Air Barriers
07 90 00	JOINT PROTECTION
07 91 00	Preformed Joint Seals
	Compression Seals
	Joint Gaskets
	Backer Rods
	Joint Fillers
07 92 00	Joint Sealants
	Elastomeric Joint Sealants
07 80 00	FIRE AND SMOKE PROTECTION
07 84 00	Firestopping
	Fire Safing

Selected References

1. The Institute for Research in Construction of the National Research Council Canada has done pioneering work in theorizing about exterior wall design and performing tests and field observations to back up the theory. This work is summarized in a large library of reports on specific topics that may be viewed online at www.nrc.ca/irc/ircpubs. See, for example, the document entitled *Evolution of Wall Design for Controlling Rain Penetration.*

2. Brock, Linda. *Designing the Exterior Wall: An Architectural Guide to the Vertical Envelope.* Hoboken, NJ, John Wiley & Sons, Inc., 2005.

This book covers the building science underlying the performance of the exterior wall and provides examples of its application to the design of wall types ranging from light wood construction to metal curtain wall.

3. Brookes, Alan. *Cladding of Buildings* (3rd ed.). London, Spon Press, 1998.

This book provides a clear general introduction to cladding principles and material types.

4. Anderson, J. M., and J. R. Gill. *Rainscreen Cladding: A Guide to Design Principles and Practice.* London, Butterworth-Heinemann, 1988.

This clear, succinct summary of rainscreen cladding principles includes an extensive bibliography on the subject.

5. Amstock, Joseph S. *Handbook of Adhesives and Sealants in Construction.* New York, McGraw-Hill, 2000.

This book offers detailed information on all types of sealants, construction adhesives, joint and crack control in concrete, and firestopping, as well as the design, specification, and testing of sealant joints.

Web Sites

Designing Exterior Wall Systems

Author's supplementary web site: **www.ianosbackfill.com/19_designing_cladding_systems**

Dow-Corning sealants: **www.dowcorning.com**

GE sealants: **www.gesealants.com**

Institute for Research in Construction: **www.nrc.ca/irc/ircpubs**

Whole Building Design Guide—Wall Systems: **www.wbdg.org/design/env_wall.php**

Key Terms and Concepts

exterior wall enclosure
building envelope
thermal bridge
stack effect
air barrier
air permeance
air barrier system
air barrier assembly
wind washing
internal drainage
gravity
momentum
labyrinth
surface tension
drip
capillary action
capillary break
wind current
pressure-equalized wall design
air barrier
rainscreen
pressure-equalization chamber (PEC)
rainscreen principle
rainscreen cladding
wash
sealant joint
gunnable sealant
sealant gun
low-range sealant, caulk
elongation
medium-range sealant
high-range sealant
polysulfide sealant
polyurethane sealant
silicone sealant
sealant Type
sealant Grade
sealant Class
sealant Use
single-component sealant
multicomponent sealant
self-leveling sealant, pourable sealant
nonsag sealant
traffic sealant
nontraffic sealant
immersible sealant
gasket
lockstrip
preformed cellular tape sealant
preformed solid tape sealant
priming (of sealant joints)
backer rod, backup rod
loadbearing wall
curtain wall
firestopping
safing

Review Questions

1. Why is it so difficult to make cladding watertight?

2. List the functions that cladding performs and list one or two ways in which each of these functions is typically satisfied in a cladding design.

3. Using a series of simple sketches, explain the principles of sealant joint design. List several sealant materials suitable for use in the joints that you have shown.

4. What are the forces that can move water through a joint in an exterior wall? How can each of these forces be neutralized?

Exercises

1. Examine the cladding of a building with which you are familiar. Look especially for features that have to do with insulation, condensation, drainage, and movement. Sketch a detail of how this cladding is installed and how it works. You will probably have to guess at some of the hidden features, but try to produce a complete, plausible detail. Add explanatory notes to make everything clear.

2. Work out a way to add a rainscreen window with fixed double glazing to the cladding system shown in Figure 19.8.

3. Prepare a sample sealant joint, using a backer rod and silicone sealant obtained from a hardware store or building materials supplier. Apply the sealant to two parallel pieces of quarry tile or glass that are taped together with a spacer between them. After the sealant has had time to cure (a week or so), remove the tape and spacer and test the joint by pulling and twisting it to find out how elastic it is and how well the sealant has adhered to the substrate.

20

Cladding with Masonry and Concrete

Architects Thompson Ventulett Stainback & Associates created a bold pattern of precast concrete sunshades and spandrel panels for the facades of the United Parcel Service Headquarters Building in Atlanta, Georgia. *(Photo by Brian Gassel/TVS & Associates)*

Buildings framed with structural steel or concrete are often clad with brick masonry, stone masonry, cut stone panels, or precast concrete. These substantial materials, though in fact they are supported by the loadbearing frame of the building, impart a sense of solidity and permanence. Thin facings of masonry or concrete, however, do not behave the same way as solid loadbearing walls. When mounted on a frame, these brittle materials must adjust to the movements of the frame and maintain weathertightness despite being applied in a layer only a few inches thick. Careful detailing and good construction practices are required to make this possible.

Masonry Veneer Curtain Walls

Figure 20.1 shows in a series of steps how a brick *masonry veneer* (a single wythe of brick masonry separated by a cavity from a structural backup wall) may be applied to a reinforced concrete frame. The veneer may also be made of stone. The veneer wythe is erected brick by brick or stone by stone with conventional mortar, starting from a steel *shelf angle* that is attached

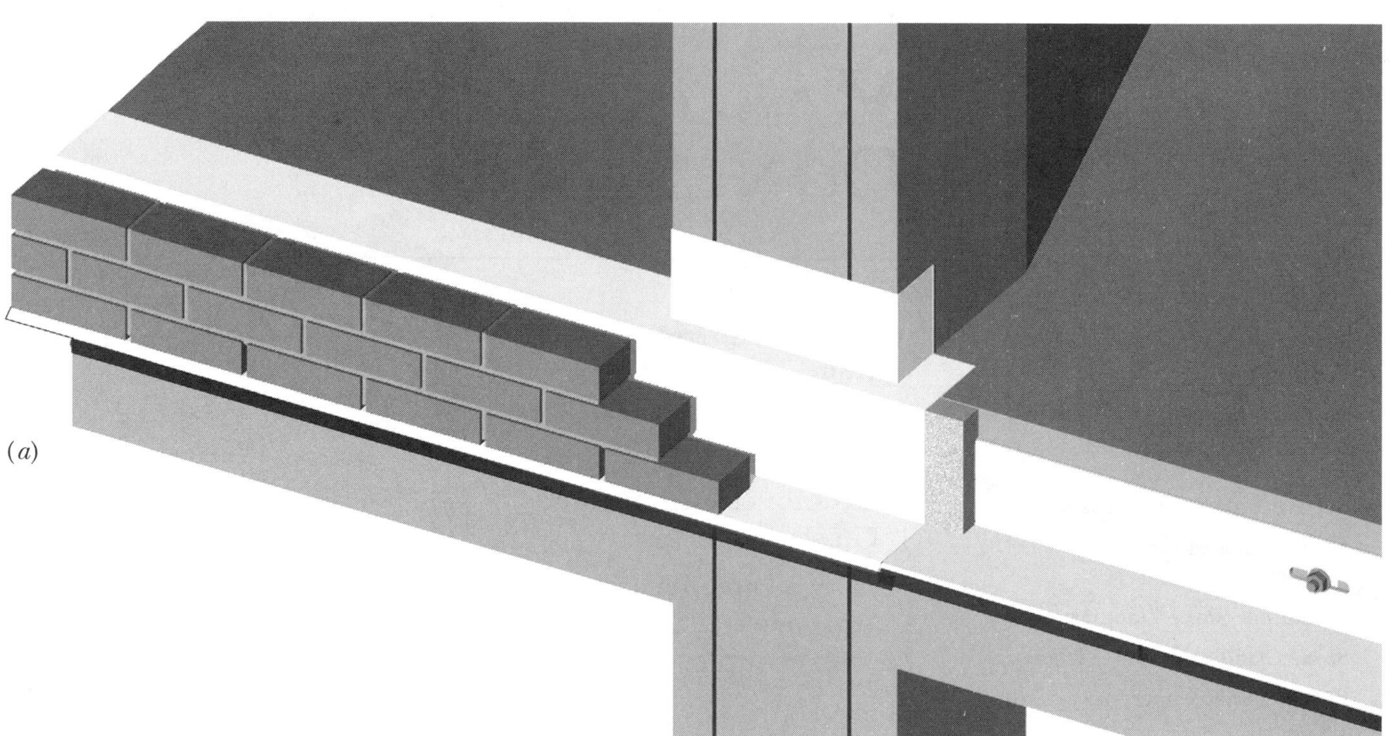
(*a*)

Figure 20.1
Construction sequence for a brick veneer curtain wall supported by a reinforced concrete frame. (*a*) Before the concrete frame of the building was cast, inserts were put into the formwork to form attachments for the brick veneer, including wedge anchors along the line of each shelf angle, two vertical dovetail slots in each column, short vertical dovetail slots in the spandrel beams, and horizontal reglets in the centers of the spandrel beams to accept the inner edge of a flashing over each window head (see pages 624 and 625 for pictures of these inserts). To begin installation of the brick veneer, a steel shelf angle is bolted to each spandrel beam, using malleable iron wedge inserts as shown in Figure 20.2. A slab of polystyrene foam thermal insulation (gray) is placed over the upright leg of the shelf angle, and a continuous flashing (white) is installed over the shelf angle, the foam, and the edge of the floor slab. This flashing also wraps around the front of the column. All the seams in the flashing are overlapped and made watertight with sealant. The first course of brickwork is laid directly on the shelf angle and flashing, without a bed joint of mortar. Every third head joint is left open in this first course to form a weep hole. Three courses of brickwork bring the veneer up to the level of the floor slab. (*b*) The first course of the concrete masonry backup wall is laid. Vertical reinforcing bars are grouted into the hollow cores of the backup wall at intervals specified by the structural engineer. An asphaltic coating is applied to the backup wall to act as an air and moisture barrier. Three more courses of brick veneer bring the top of the veneer up to the level of the top of the first course of concrete masonry. Polystyrene foam thermal insulation is placed against the concrete masonry. A combination joint reinforcing and masonry tie made of heavy steel wires is laid on top of the masonry, tying the brick veneer to the backup wall. Plastic clips are snapped onto the tie rods

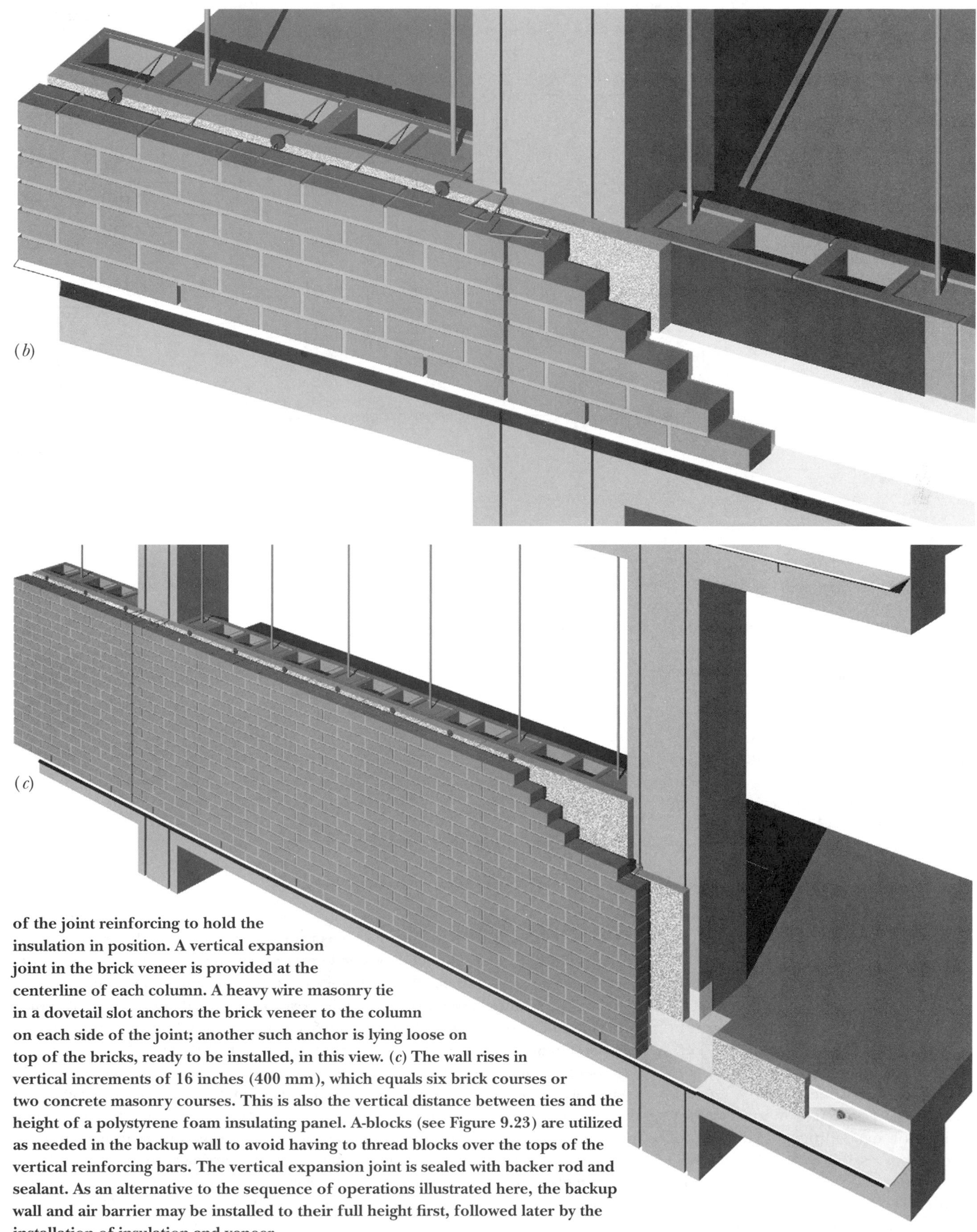

of the joint reinforcing to hold the insulation in position. A vertical expansion joint in the brick veneer is provided at the centerline of each column. A heavy wire masonry tie in a dovetail slot anchors the brick veneer to the column on each side of the joint; another such anchor is lying loose on top of the bricks, ready to be installed, in this view. (*c*) The wall rises in vertical increments of 16 inches (400 mm), which equals six brick courses or two concrete masonry courses. This is also the vertical distance between ties and the height of a polystyrene foam insulating panel. A-blocks (see Figure 9.23) are utilized as needed in the backup wall to avoid having to thread blocks over the tops of the vertical reinforcing bars. The vertical expansion joint is sealed with backer rod and sealant. As an alternative to the sequence of operations illustrated here, the backup wall and air barrier may be installed to their full height first, followed later by the installation of insulation and veneer.

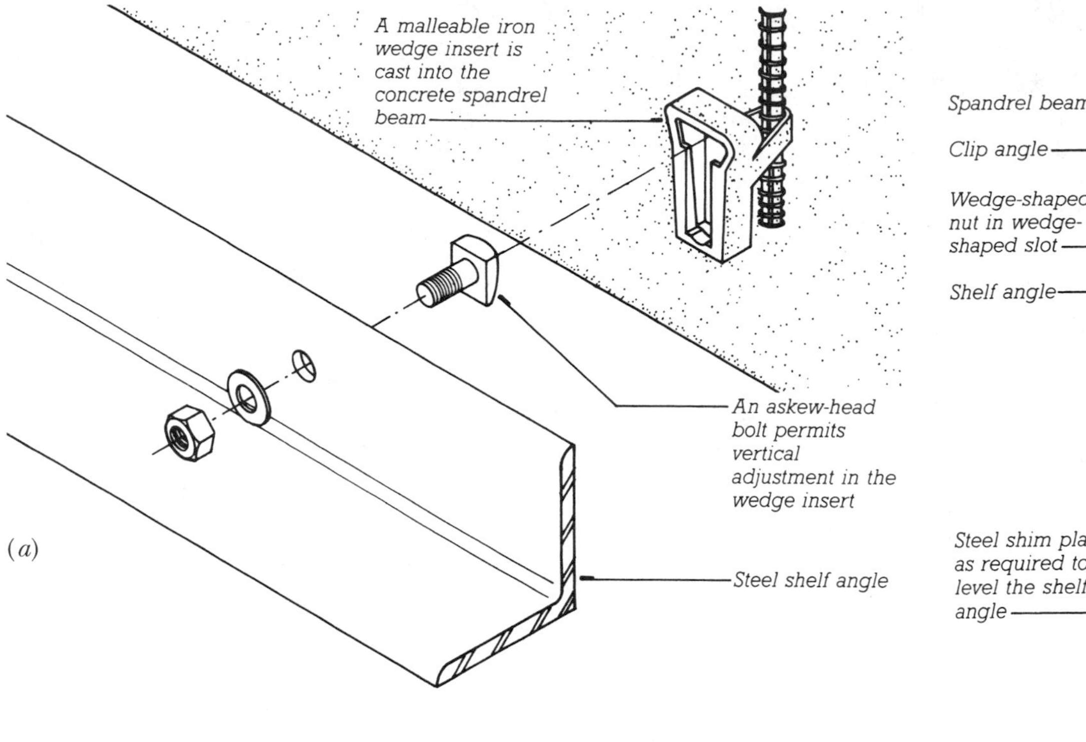

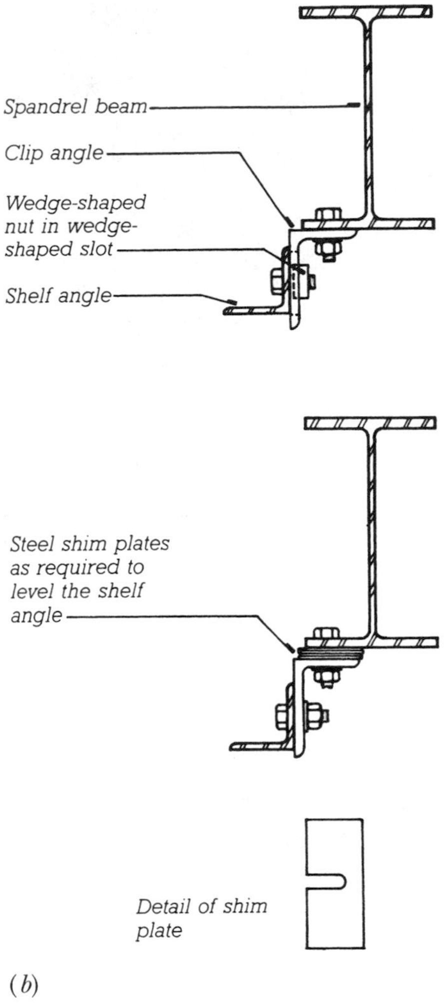

FIGURE 20.2
(*a*) An example of a cast-in-place anchoring system for attaching a steel shelf angle to a concrete spandrel beam. Steel shims are added as necessary between the shelf angle and the spandrel beam to place the angle exactly in the plane of the facing wythe.
(*b*) The traditional method for attaching shelf angles to steel spandrel beams uses steel clip angles with shim plates as needed to make up for dimensional inaccuracies in the components. In practice, providing anchoring systems with adequate adjustability to account for deviations in the structural frame is often a difficult challenge.

to the structural frame at each floor (Figure 20.2). The construction process and details are essentially the same as for a masonry cavity wall of a single-story building, but there are some crucial differences: To prevent normal movements in the frame of the building from stressing the masonry veneer, and to allow the veneer to expand and contract without distress, there must be a *soft joint* (*horizontal expansion joint*) beneath each shelf angle (Figure 20.3). This joint must be dimensioned to absorb the maximum anticipated sum of column creep, brick expansion, spandrel beam deflection, and a dimensional tolerance to allow for construction inaccuracies while not exceeding the maximum safe compressibility of the sealant. Masonry curtain walls also must be divided vertically by movement joints (*vertical expansion joints*) to allow the frame and the masonry cladding to expand and contract independently of one another (Figure 20.4).

A *backup wall* of light gauge steel studs covered with water-resistant sheathing panels of gypsum or cementitious materials is often considered to be interchangeable with a concrete masonry backup wall for a masonry facing. The stud wall even has certain advantages over masonry in its lighter weight, its greater ability to contain thermal insulation and electrical wiring, and its greater

FIGURE 20.3
A complete detail section of the brick veneer wall that was begun in Figure 20.1 shows how the top of the backup wall is fastened to the underside of the spandrel beam with a series of steel restraint clips that brace the top of the wall against wind loads but allow the spandrel beam to deflect under load. Two lines of backer rod and sealant along the columns and across the top of the backup wall make the backup wall airtight. A soft joint of sealant beneath the shelf angle permits the spandrel beam to deflect without applying force to the brick veneer. The brick ties nearest the underside of the shelf angle are anchored to dovetail slots in the spandrel beam. An additional plastic clip on each wire tie in the center of the cavity acts as a drip to prevent water from clinging to the tie and running toward the backup wall. The interior of the building is finished with gypsum veneer plaster mounted on steel furring channels, similar to the assembly shown in Figure 23.5. *(Drawing from Allen, Edward,* Architectural Detailing: Function, Constructibility, Aesthetics, *New York, John Wiley & Sons, Inc., 1993, reproduced by permission of the publisher)*

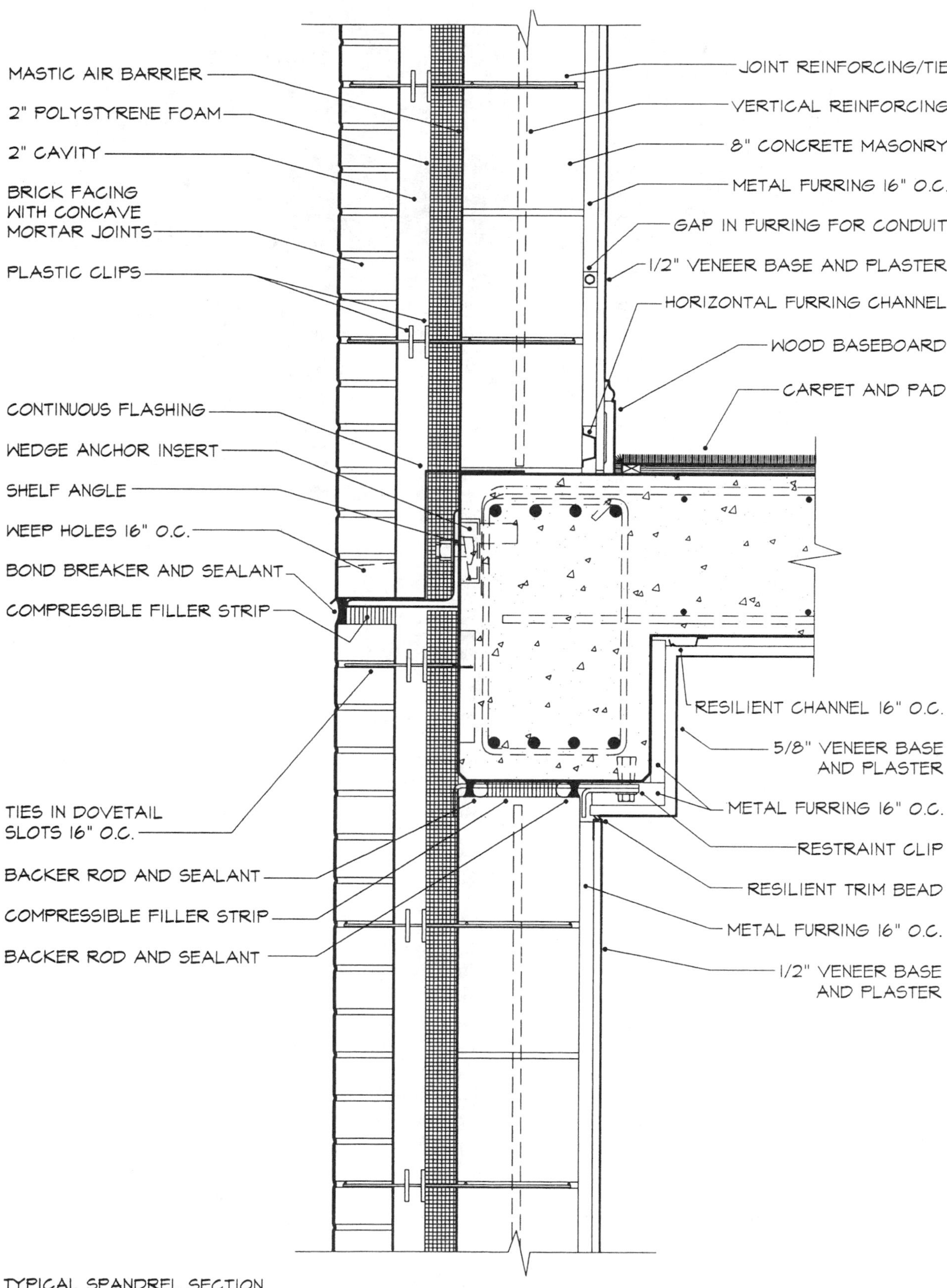
MASTIC AIR BARRIER
2" POLYSTYRENE FOAM
2" CAVITY
BRICK FACING WITH CONCAVE MORTAR JOINTS
PLASTIC CLIPS
CONTINUOUS FLASHING
WEDGE ANCHOR INSERT
SHELF ANGLE
WEEP HOLES 16" O.C.
BOND BREAKER AND SEALANT
COMPRESSIBLE FILLER STRIP
TIES IN DOVETAIL SLOTS 16" O.C.
BACKER ROD AND SEALANT
COMPRESSIBLE FILLER STRIP
BACKER ROD AND SEALANT
JOINT REINFORCING/TIE
VERTICAL REINFORCING
8" CONCRETE MASONRY
METAL FURRING 16" O.C.
GAP IN FURRING FOR CONDUIT
1/2" VENEER BASE AND PLASTER
HORIZONTAL FURRING CHANNEL
WOOD BASEBOARD
CARPET AND PAD
RESILIENT CHANNEL 16" O.C.
5/8" VENEER BASE AND PLASTER
METAL FURRING 16" O.C.
RESTRAINT CLIP
RESILIENT TRIM BEAD
METAL FURRING 16" O.C.
1/2" VENEER BASE AND PLASTER

TYPICAL SPANDREL SECTION

(*a*)

(*b*)

FIGURE 20.4
(*a*) A carefully detailed brick curtain wall by architects Kallman, McKinnell and Wood covers the steel frame of Hauser Hall at Harvard University. Notice the vertical expansion joint near the far-right corner. (*b*) At the base of Hauser Hall, the facing wythe is made of limestone blocks. The backup wall consists of steel studs and gypsum sheathing panels. A vertical expansion is visible in the far-left corner in this view. *(Photos © Steve Rosenthal)*

receptivity to a variety of interior finish materials. However, the steel studs and their fastenings are inherently more flexible than a concrete masonry wall and may deflect enough under maximum wind pressures to cause cracking of brittle masonry veneers. Such cracking often leads to water leakage. Furthermore, if there is leakage through the facing layer of the wall because of cracking, porous masonry, or poor workmanship, the steel studs and fasteners are susceptible to corrosion, and the gypsum sheathing panels are subject to water deterioration.

A concrete masonry backup wall is usually stiffer than the veneer that it supports, so the veneer is unlikely to crack under wind loadings. A concrete masonry backup wall can also, if necessary, maintain its structural integrity despite prolonged periods of wetting. For these reasons, a concrete masonry backup wall is generally preferable to a wall of steel studs. If a steel stud backup system is selected, the studs, masonry ties, and fasteners should be sized very conservatively so as to be stiff enough against wind loadings that the veneer material will not crack. The sheathing material and fasteners should be selected for their durability under damp conditions. Each metal tie that connects the masonry veneer to the studs should be attached directly to a stud with at least two corrosion-resistant screws. The wall must be detailed carefully to keep leaked water away from the backup components. Constant inspection is required during construction to be certain that all these details are faithfully executed and the cavity is kept clean so that it will drain freely.

The structural frame of a building is never absolutely flat or plumb. Thus, the attachment system for the shelf angles must allow for adjustments so that the masonry veneer may be constructed in a precisely vertical plane with level courses. Figure 20.2 shows how this is usually done for both concrete and steel

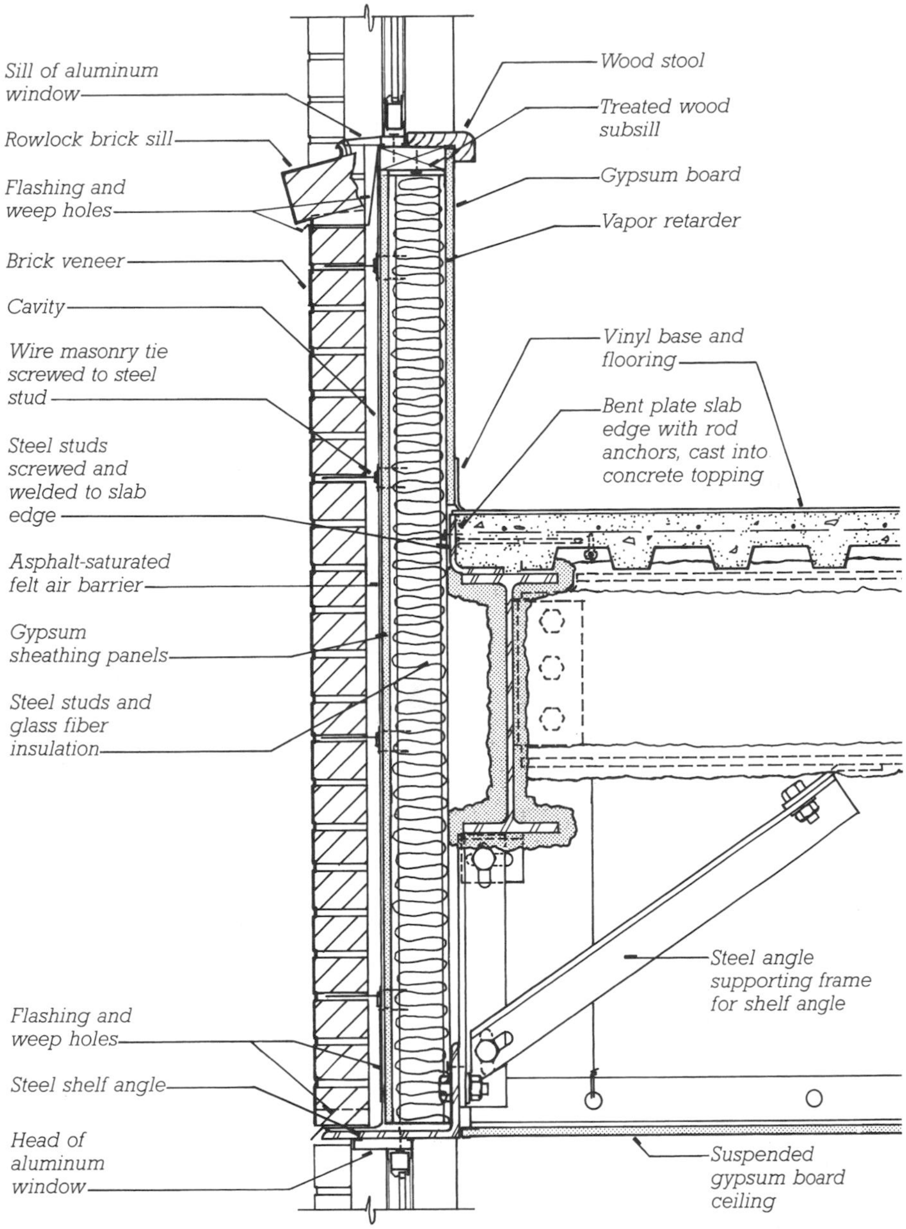

Sill of aluminum window
Rowlock brick sill
Flashing and weep holes
Brick veneer
Cavity
Wire masonry tie screwed to steel stud
Steel studs screwed and welded to slab edge
Asphalt-saturated felt air barrier
Gypsum sheathing panels
Steel studs and glass fiber insulation
Flashing and weep holes
Steel shelf angle
Head of aluminum window
Wood stool
Treated wood subsill
Gypsum board
Vapor retarder
Vinyl base and flooring
Bent plate slab edge with rod anchors, cast into concrete topping
Steel angle supporting frame for shelf angle
Suspended gypsum board ceiling

FIGURE 20.5
A detail section of a brick curtain wall that is supported below the level of the spandrel beam on a frame made of steel angles. The supporting frame becomes necessary when continuous horizontal bands of windows are to be installed between brick spandrels. All the connections in the supporting frame are made with bolts in slotted holes to allow for exact alignment of the shelf angle. After the frame has been aligned and before the masonry work begins, the connections are welded to prevent slippage. Shelf angle constructions for masonry curtain walls require careful engineering to accommodate expected loads and structural deflections.

FIGURE 20.6
The detail shown in Figure 20.5 allows the construction of brick spandrels between continuous horizontal bands of glass. *(Photo by Edward Allen)*

frames. The attachment system in Figure 20.5, which is designed to suspend a masonry veneer spandrel wall over a continuous band of windows, also provides for free adjustment of the shelf angle location.

The flashing above the shelf angle should project beyond the face of the masonry by 1 inch (25 mm) or so and should be bent downward at a 45-degree angle to form a drip. In this way, it is able to conduct water that has leaked into the cavity back to the outdoors and to drain it safely away from the wall. If a flexible plastic or composite flashing is used, it should be cemented to a strip of sheet metal flashing over the shelf angle, with the sheet metal forming the projecting drip.

Many architects, because they wish to maintain the fiction that a masonry veneer is actually a solid masonry wall, find the soft joint and projecting flashing objectionable. They use specially molded bricks with a face lip that hangs down over the shelf angle to conceal it from view, and they do not allow the flashing to project out of the wall. The color of the sealant that they use in the soft joint is matched as closely as possible to the color of the mortar. Unfortunately, the use of lipped bricks and recessed flashings is very risky. The recessed flashing allows water to accumulate around the toe of the angle, causing the angle to rust. Freeze-thaw action and the expansion of the steel as it rusts are likely to cause the lips to spall off of the bricks. Eventually, the deterioration along the line of the shelf angle becomes unsightly. Worse yet, the stability of the veneer may be endangered by failure of the

FIGURE 20.7
Fabrication and installation of a brick panel curtain wall. (*a*) Masons construct the panels in a factory, using conventional bricks and mortar. Both horizontal and vertical reinforcing are used, the vertical bars being grouted into the hollow cores of the bricks. (*b*) Brick panels are stored to await shipment, complete with thermal insulation. The welded metal brackets are for attachment to the building; the structural strength of the panel comes primarily from the reinforced masonry, not the brackets. (*c*) A crane lifts a parapet panel to its final position. (*d*) Corners can be constructed as single panels. *(Panelized masonry by Vet-O-Vitz Masonry Systems, Inc., Brunswick, Ohio)*

(*a*)

(*b*)

(*c*)

(*d*)

corroded shelf angle. A better strategy for the conscientious architect is to find a way to express visually the presence of the shelf angle, flashing, and soft joint, and to make them positive features of the building facade. A soldier course or cut stone sill above each shelf angle is a good start toward a frank expression of a constructional necessity.

Prefabricated Brick Panel Curtain Walls

Figure 20.7 shows the use of *prefabricated reinforced brick panels* for cladding. Masons construct the panels while working comfortably at ground level in a factory. Horizontal reinforcing may be laid into the mortar joints or grouted into channel-shaped bricks. Vertical reinforcing bars are placed in grouted cavities of hollow-core bricks. These panels are self-rigid; they need no structural backup and can be fastened to the building in much the same way as precast concrete panels. A steel stud backup wall is required to carry thermal insulation, electrical wiring, and an interior finish layer, but it has no structural role.

Stone Curtain Walls

Chapter 9 discusses types of stone and illustrates conventionally set stone facing systems that tie relatively small blocks of cut stone set in mortar to a concrete masonry backup wall. Slabs of stone that are larger in surface area may be fastened to framed buildings in several different ways.

Stone Panels Mounted on a Steel Subframe

Figure 20.8 shows a system for mounting stone panels on a steel subframe, called *grid-system-supported stone cladding*. The vertical members of the subframe are erected first. They are designed to transmit gravity and wind loads from the stone slabs to the frame of the building. The horizontal members are aluminum shapes that engage slots in the upper and

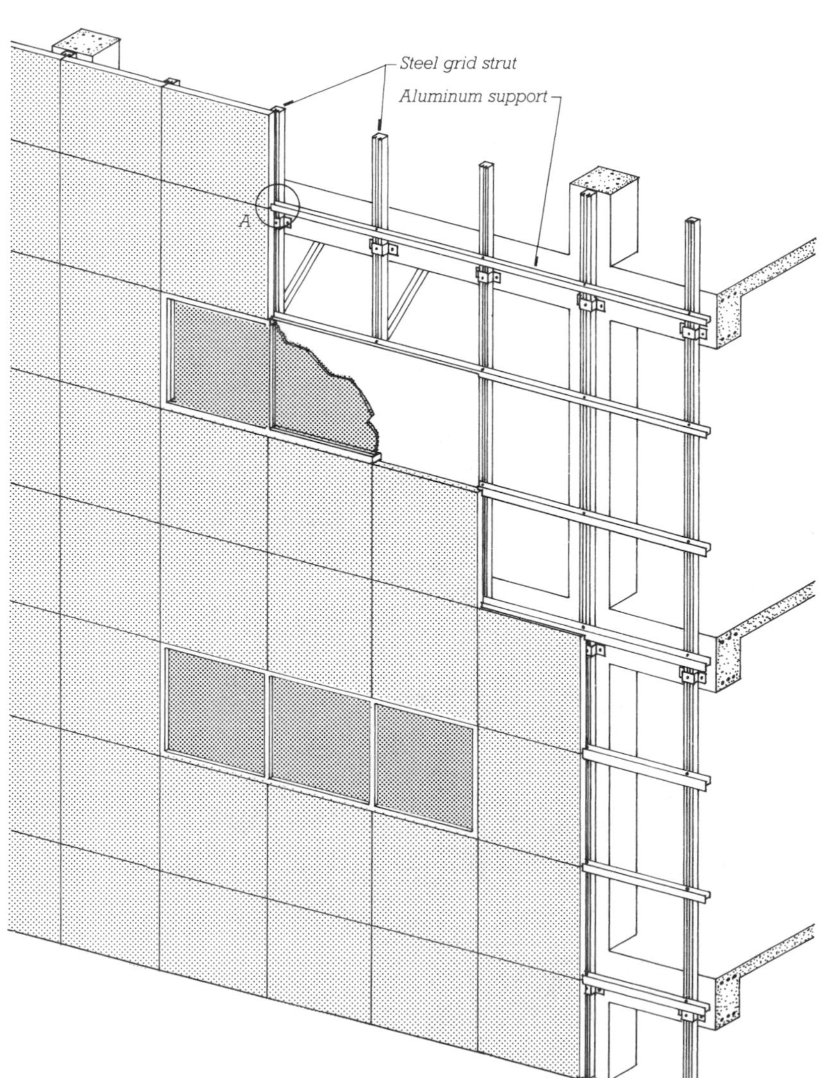

(*a*)

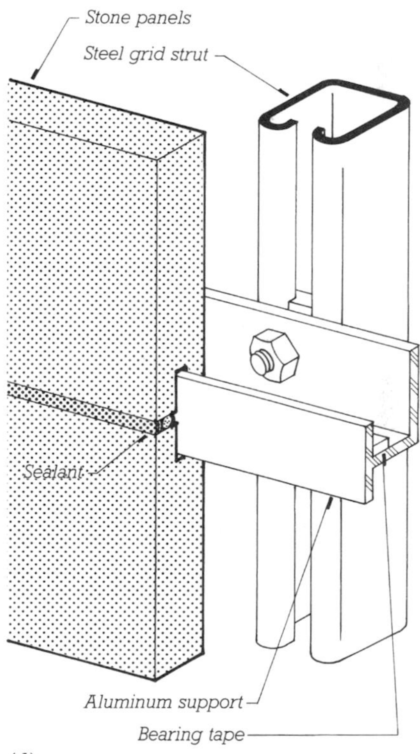

(*b*)

FIGURE 20.8
A subframe of vertical steel struts supports a facing of stone panels by means of horizontal metal clips that engage slots in the upper and lower edges of the panels. In order to avoid corrosion and staining problems, the steel struts should be galvanized, and the clips should be made of a nonferrous metal (aluminum or stainless steel) that is chemically compatible with the type of stone that is used.

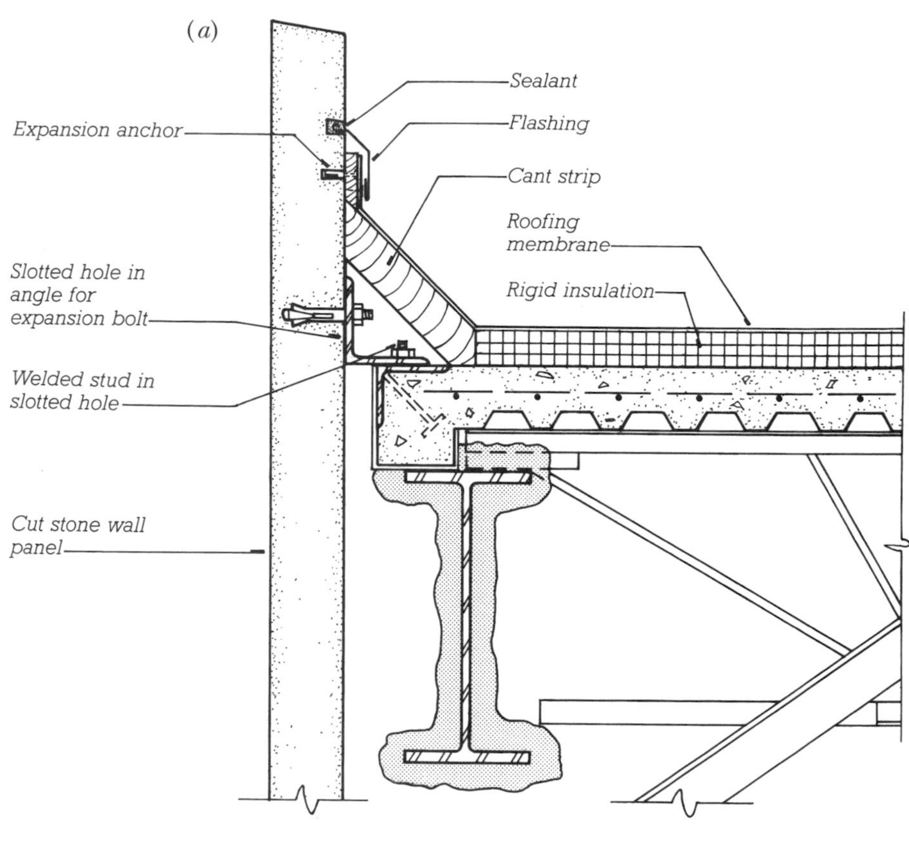

FIGURE 20.9
(*a*) Parapet and (*b*) spandrel details for a stone panel curtain wall made of limestone, marble, or granite. The broken lines indicate the outline of the interior finish and thermal insulation components, which are not shown. Each support plate holds edges of two adjacent wall panels, which are pocketed as shown to rest on the plate. The plate should be made of a noncorroding metal. The vertical joints between panels are closed with a backer rod and sealant.

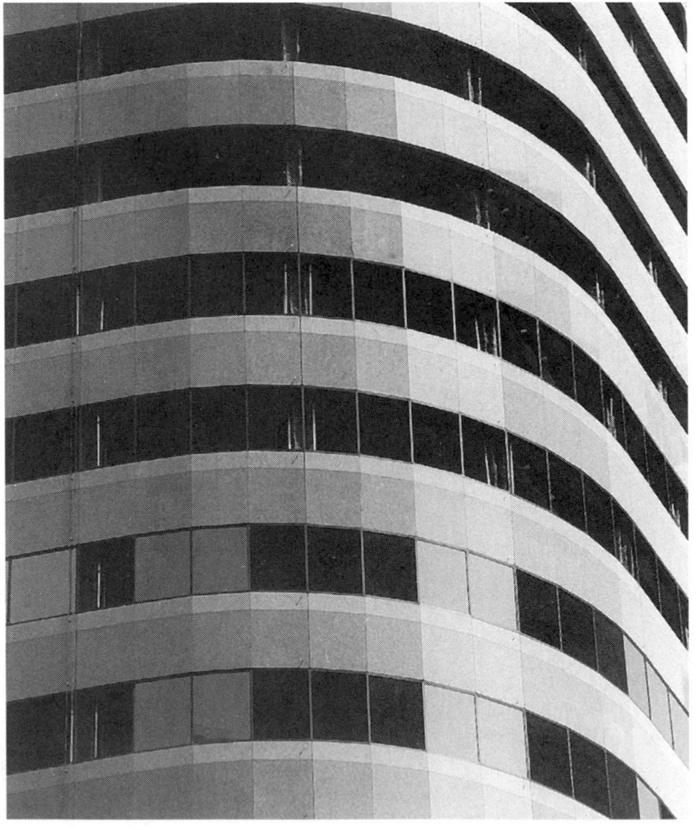

FIGURE 20.10
A granite panel curtain wall of the type illustrated in Figure 20.9 wraps around the corner of a Boston office building. The upper-floor windows have not yet been installed, but the window frames have been mounted in two of the middle floors, and the lower floors have been glazed. *(Architect: Hugh Stubbins and Associates. Photo by Edward Allen)*

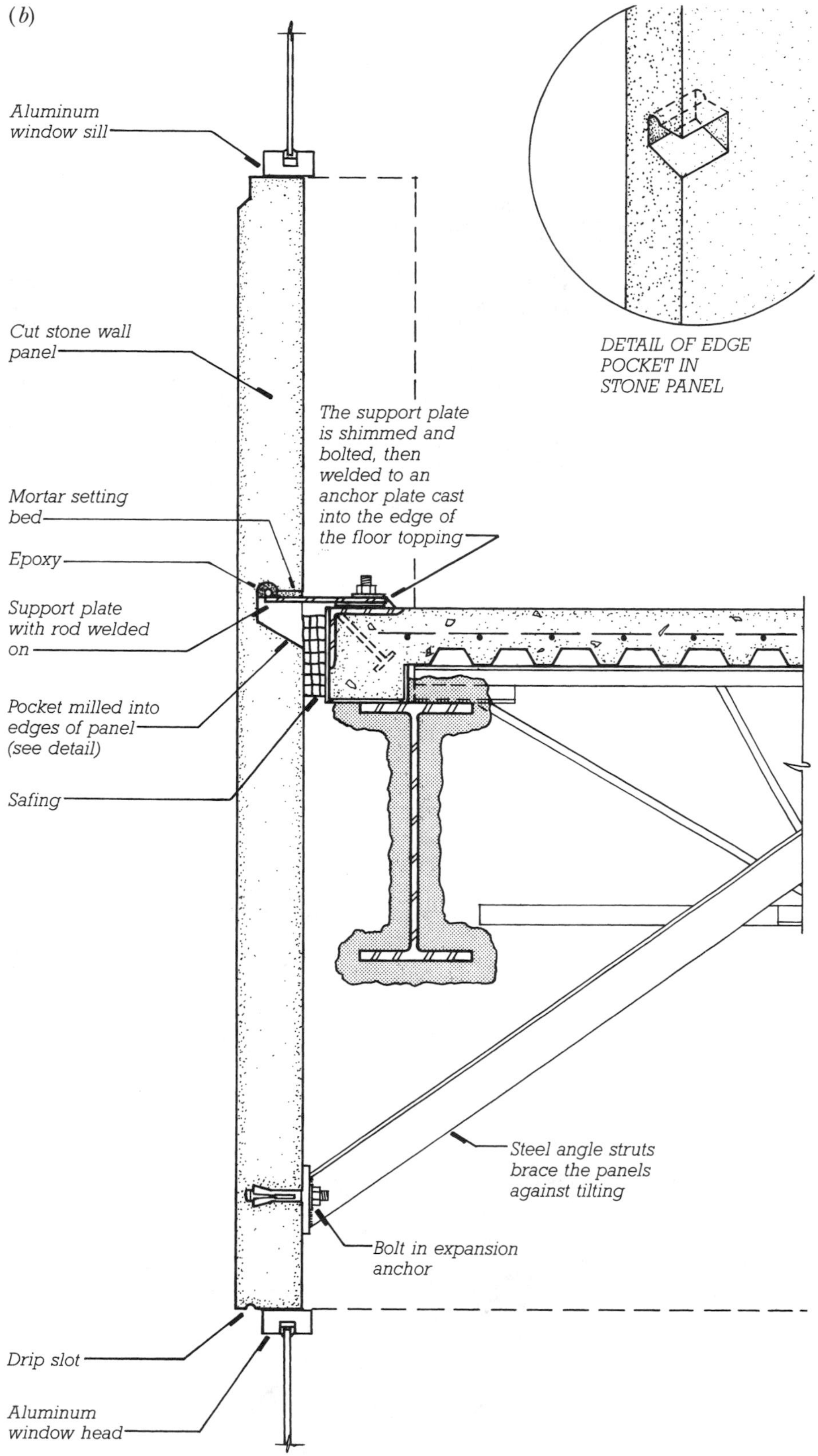

lower edges of each panel to attach them firmly to the building. They are added as the installation of the stone panels progresses. Backer rods and sealant fill the spaces between the panels, allowing for a considerable range of movement. A nonstructural backup wall, usually made of steel studs and gypsum sheathing panels, is constructed within the frame of the building but is not attached to the subframe. Its functions are to provide an air barrier, to house thermal insulation batts and electrical wiring, and to support the interior wall finish layer, which is usually plaster or gypsum board.

A weakness of this system is its dependence on the integrity of the sealant joints. If a sealant joint leaks, water may accumulate in the slots in the tops of the stone panels, and freeze-thaw deterioration may ensue.

Monolithic Stone Cladding Panels

Figures 20.9 and 20.10 illustrate the use of *monolithic stone cladding panels* that are fastened directly to the frame of the building. The weight of each panel is transferred to two steel support plates by means of edge pockets that are cut into both sides of each panel at the stone mill. Each panel is stabilized by a pair of steel angle struts that are bolted to the stone with expansion anchors in drilled holes. Joints are closed with backer rod and sealant, and a nonstructural backup wall is required.

FIGURE 20.11
A steel truss system for stone cladding. (*a*) Masons working in a fabrication yard attach thin sheets of stone to welded steel trusses. The vertical joints are closed with backer rods and sealant. (*b*) The fabricated spandrel panel is lifted onto a truck using a crane. The metal clips that are just visible along the top and bottom edges of the panel engage slots in the edges of the sheets of stone to hold the stone securely to the truss. The steel angle clips at the two upper corners of the truss will support the panel on brackets welded to the steel columns of the building frame. (*c*) The panel is installed. *(Courtesy of International Masonry Institute, Washington, DC)*

(*a*)

(*b*)

(*c*)

Stone Cladding on Steel Trusses

In *truss-supported stone cladding,* sheets of stone are combined into large prefabricated panels by mounting them on structural steel trusses (Figure 20.11). Each truss is designed to carry both wind loads and the dead load of the stone to steel connection brackets that transfer these loads to the frame of the building. Sealant joints and a nonstructural backup wall finish the installation.

Posttensioned Limestone Spandrel Panels

Thick blocks of limestone may be joined with adhesives into long spandrel panels and posttensioned with high-strength steel tendons so that the assembly is self-supporting between columns (Figure 20.12). Such *posttensioned limestone spandrel panels* are a relatively costly type of panel because of their use of comparatively large quantities of stone per unit area of cladding.

Very Thin Stone Facings

Extremely thin sheets of stone (as thin as ¼ inch, or 6.5 mm, for granite) may be stiffened with a structural backing such as a metal honeycomb and mounted as spandrel panels in an aluminum mullion system such as those described in Chapter 21.

Very thin sheets of stone may also be used as facings for precast concrete curtain wall panels. The stone sheets are laid face down in the forms. Stainless steel clips are inserted into holes drilled in the backs of the stone. A grid of steel reinforcing bars is added, and then the concrete is poured and cured to complete the panel. The clips anchor the stone to the concrete.

When specifying the thickness of stone for any exterior cladding application, the designer should work closely with the stone supplier and also consult the relevant standards of the building stone industry. Stone that has been sliced thinner than industry standards has caused a number of failures of cladding systems.

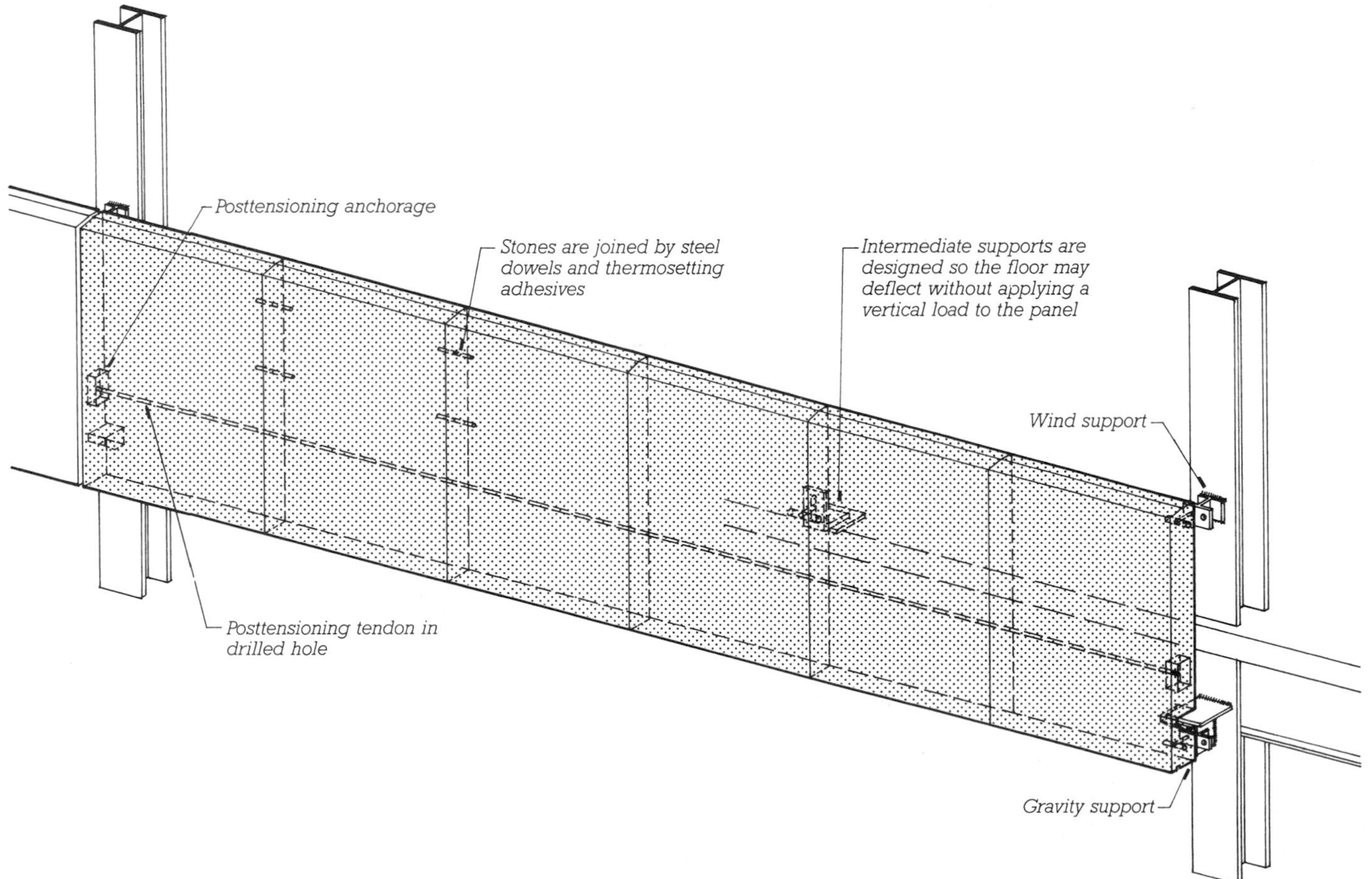
Posttensioning anchorage
Stones are joined by steel dowels and thermosetting adhesives
Intermediate supports are designed so the floor may deflect without applying a vertical load to the panel
Wind support
Posttensioning tendon in drilled hole
Gravity support

FIGURE 20.12
Thicker blocks of Indiana limestone may be posttensioned together to make spandrel panels that span from column to column but require little steel. The posttensioning tendon is threaded through matching holes that are drilled in the individual stones prior to assembly.

Precast Concrete Curtain Walls

Precast concrete cladding panels, both conventionally reinforced and prestressed (page 808 and Figures 20.13–20.18), are simple in concept but require close attention to matters of surface finish, mold design, thermal insulation, attachment to the building frame, and sufficient strength and rigidity in the building frame to support the weight of the panels.

The factory production of concrete cladding panels makes it possible to utilize very high-quality molds and a variety of surface finishes, from glassy smooth to rough, exposed aggregates. Ceramic tiles, thin bricks, or thin stone facings may be attached to precast concrete panels. In *precast concrete sandwich panels*, thermal insulation is incorporated as an inner layer of the panel (Figures 20.17 and 20.18). Alternatively, insulation may be affixed to the back of the panel or may be provided in a nonstructural backup wall that is constructed in place. Reinforcing or prestressing of the panel must be designed to resist wind, gravity, and seismic forces and to control cracking of the concrete. Attachments must transfer all these forces to the building frame while

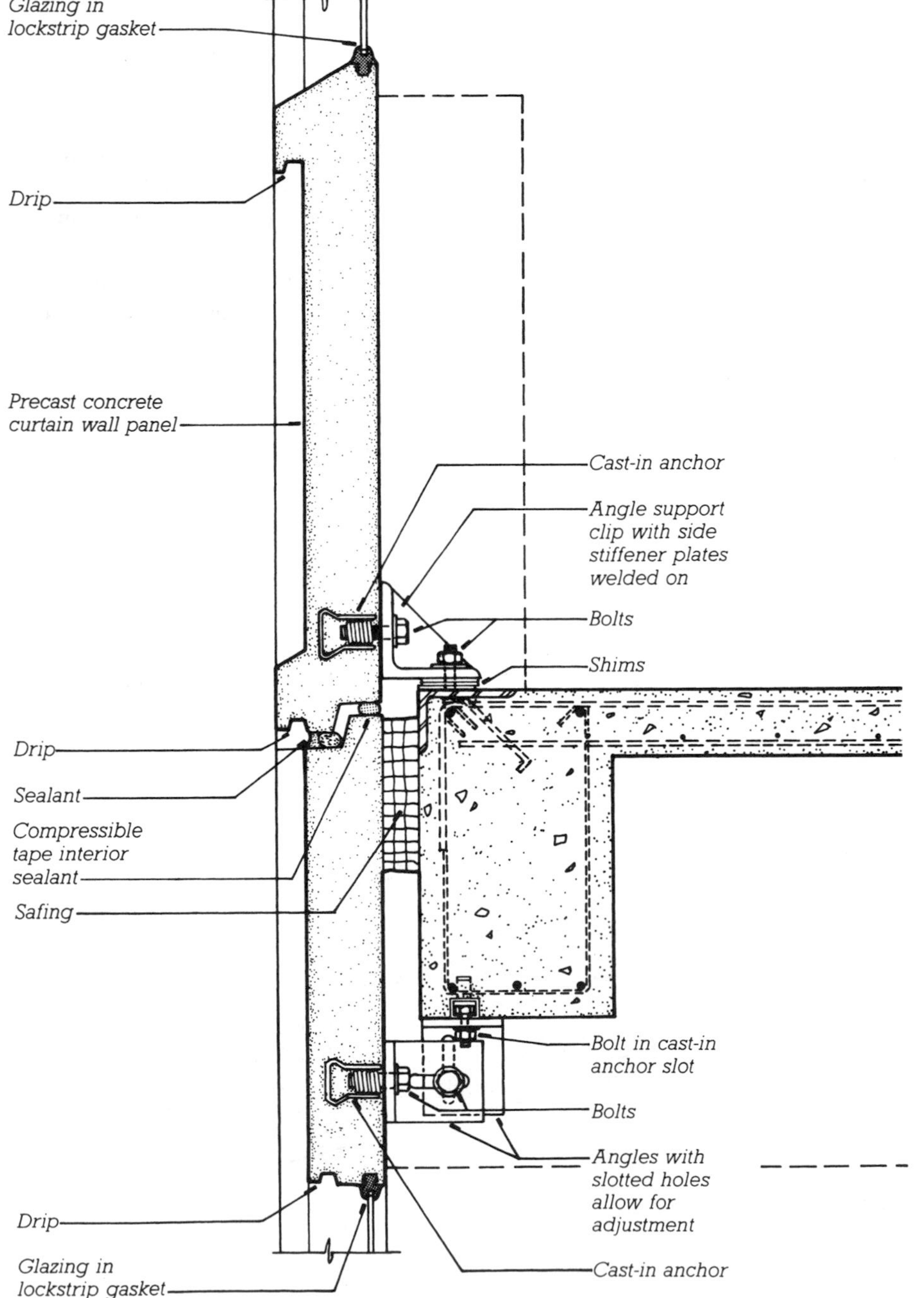

Figure 20.13
A typical detail of a precast concrete curtain wall on a sitecast concrete frame. Panels in this example are a full story high, each containing a fixed window. The reinforcing has been omitted from the panel for the sake of clarity, and the outline of the thermal insulation and interior finishes, which are not shown, is indicated by the broken lines.

FIGURE 20.14
Workers install a precast concrete curtain wall panel. *(Architects and engineers: Andersen-Nichols Company, Inc. Photo by Edward Allen)*

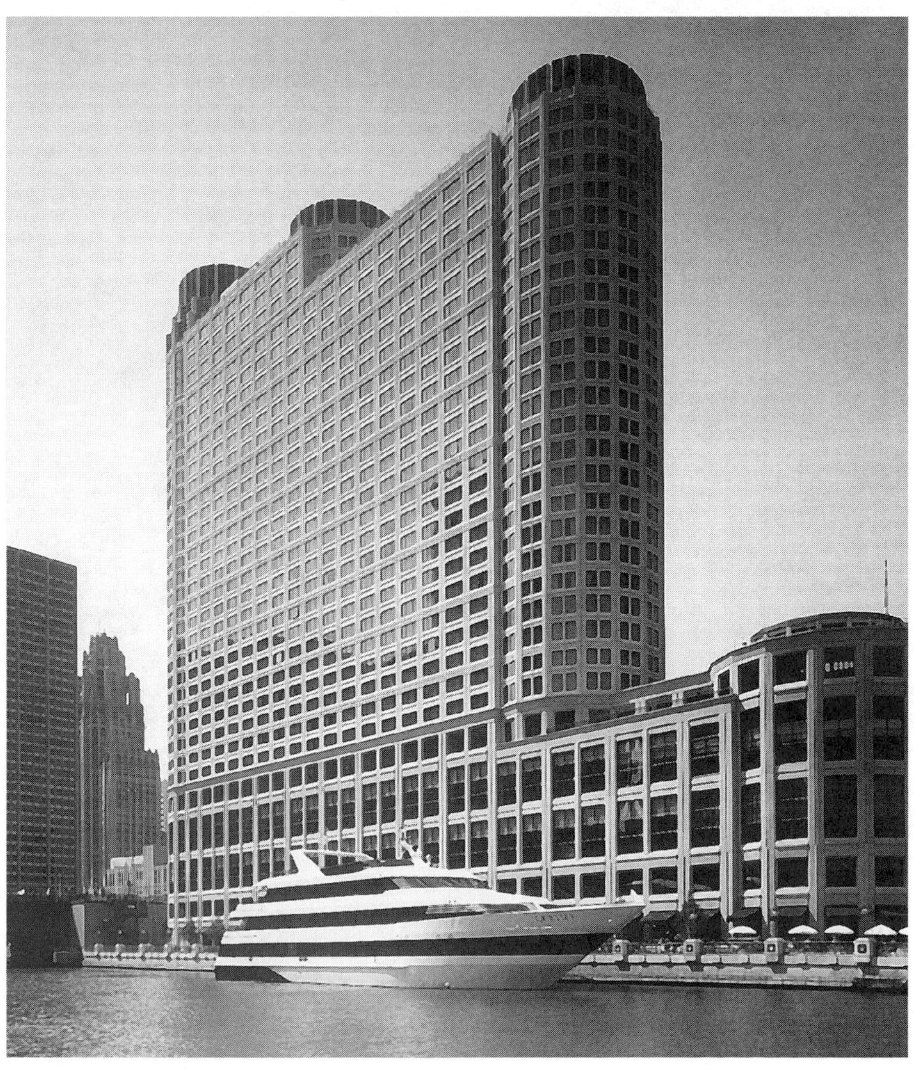

FIGURE 20.15
A Chicago hotel is clad in precast concrete panels.
(Architects: Solomon Cordwell Buenz Associates. Photo by Hedrich Blessing)

allowing for installation adjustment and for relative movements of the frame and the cladding.

More recently developed materials, such as carbon fiber reinforcing or ultra-high-performance concrete (see Chapter 15), allow the manufacture of panels that are thinner and lighter than those made of conventional materials.

Glass-Fiber-Reinforced Concrete Curtain Walls

Glass-fiber-reinforced concrete (GFRC) is a relatively new cladding material that has several advantages over conventional precast concrete panels. Its admixture of short glass fibers furnishes enough tensile strength that no steel reinforcing is required. Panel thicknesses and weights are about one-quarter of those for conventional precast concrete panels, which saves money on shipping, makes the panels easier to handle, and allows the use of lighter attachment hardware. The light weight of the cladding also allows the loadbearing frame of the building to be lighter and less expensive. GFRC can be molded into three-dimensional forms with intricate detail and an extensive range of colors and textures (Figures 20.19 and 20.20).

The fibers in GFRC must be manufactured from a special alkali-resistant type of glass to prevent their disintegration in the concrete. The panels may be self-stiffened with GFRC ribs, but the usual practice is to attach a welded frame made of light gauge steel studs to the back of each GFRC facing in the factory. The attachment is made by means of thin steel rod anchors that flex slightly as needed to permit small amounts of relative movement between the

FIGURE 20.16
Horizontal bands of smooth and textured precast concrete create the facade pattern of this suburban office building. *(Architect: ADD, Inc. Courtesy of Precast/Prestressed Concrete Institute)*

FIGURE 20.17
Manufacturing Corewall®, a proprietary foam-core precast concrete sandwich panel. Rollers apply a ribbed texture to the outside of a panel that includes a layer of foam plastic insulation sandwiched between layers of concrete. *(Courtesy of Butler Manufacturing Co.)*

FIGURE 20.18
A completed panel is lifted from the casting bed. *(Courtesy of Butler Manufacturing Co.)*

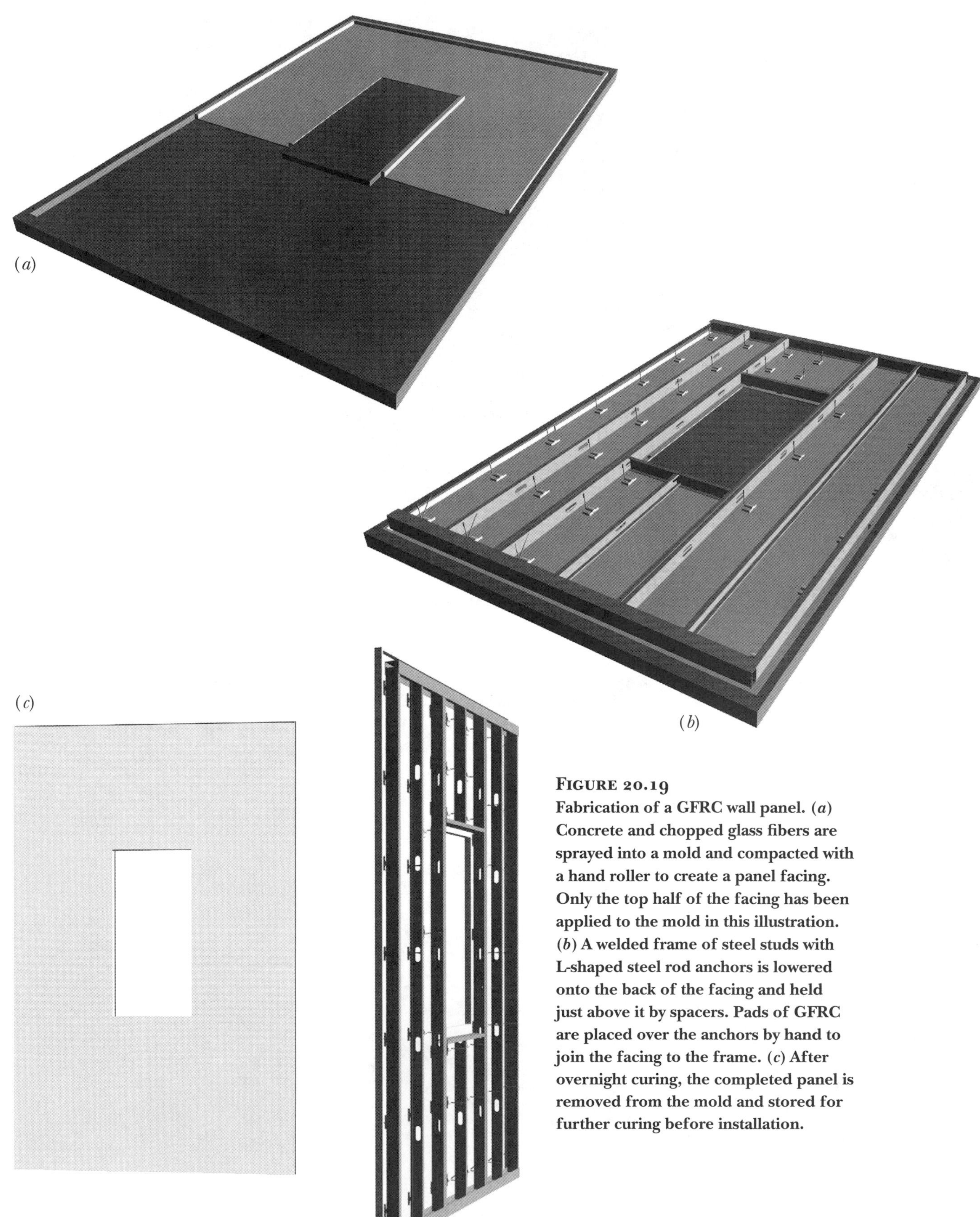

FIGURE 20.19
Fabrication of a GFRC wall panel. (*a*) Concrete and chopped glass fibers are sprayed into a mold and compacted with a hand roller to create a panel facing. Only the top half of the facing has been applied to the mold in this illustration. (*b*) A welded frame of steel studs with L-shaped steel rod anchors is lowered onto the back of the facing and held just above it by spacers. Pads of GFRC are placed over the anchors by hand to join the facing to the frame. (*c*) After overnight curing, the completed panel is removed from the mold and stored for further curing before installation.

(*a*)

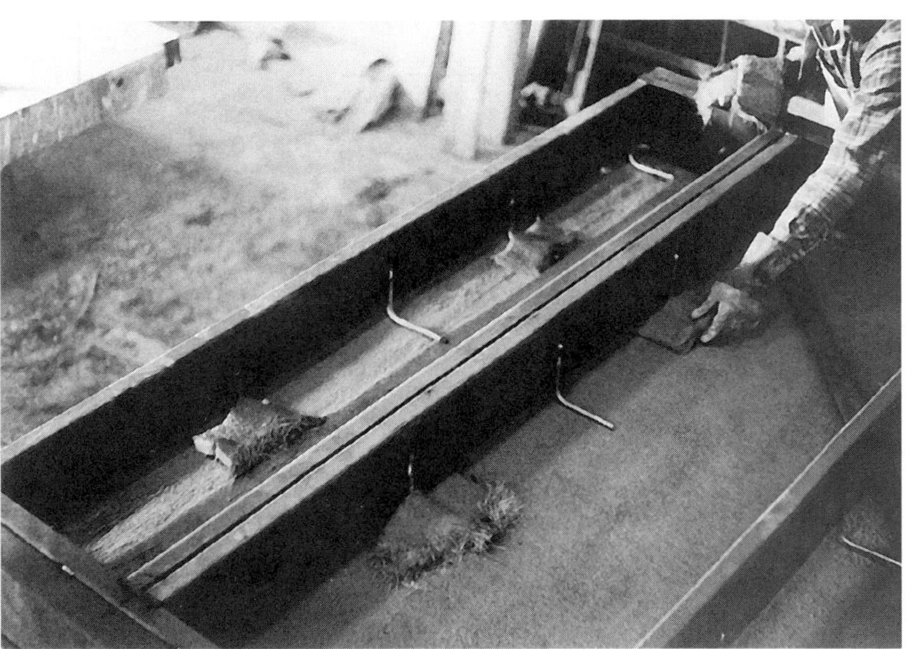
(*b*)

Figure 20.20
Fabrication of a GFRC curtain wall panel. (*a*) A special gun deposits a layer of sand–cement slurry simultaneously with 1.5-inch (38-mm) lengths of alkali-resistant glass fiber reinforcing. Three layers are usually required to make up the full thickness of the panel facing; each is compacted with a small hand roller before the next layer is applied. The overall thickness is usually ½ inch (13 mm). (*b*) After the GFRC facing layer has been completed, the steel frame is lowered over it and the operator hand-applies pads of wet GFRC over the rod anchors to bond the frame to the GFRC facing. *(Courtesy of Precast/Prestressed Concrete Institute)*

facing and the frame. Figure 20.21 shows typical ways of attaching metal-framed GFRC panels to the building. The edges of the GFRC facing, which is usually only about ½ inch (13 mm) thick, are flanged as shown in Figure 20.22 so that backer rods and sealant may be inserted between panels.

Exterior Insulation and Finish System

An *exterior insulation and finish system (EIFS)* consists of a layer of plastic foam insulation that is adhered or mechanically fastened to a backup wall, a reinforcing mesh that is applied to the outer surface of the foam by embedment in a base coat of a stuccolike material, and an exterior finish coat of a similar stuccolike material that is troweled over the reinforced base coat. In most cases, EIFS is constructed in place over a backup wall made either of concrete masonry or of steel studs and water-resistant sheathing (Figures 20.23 and 20.24), but the

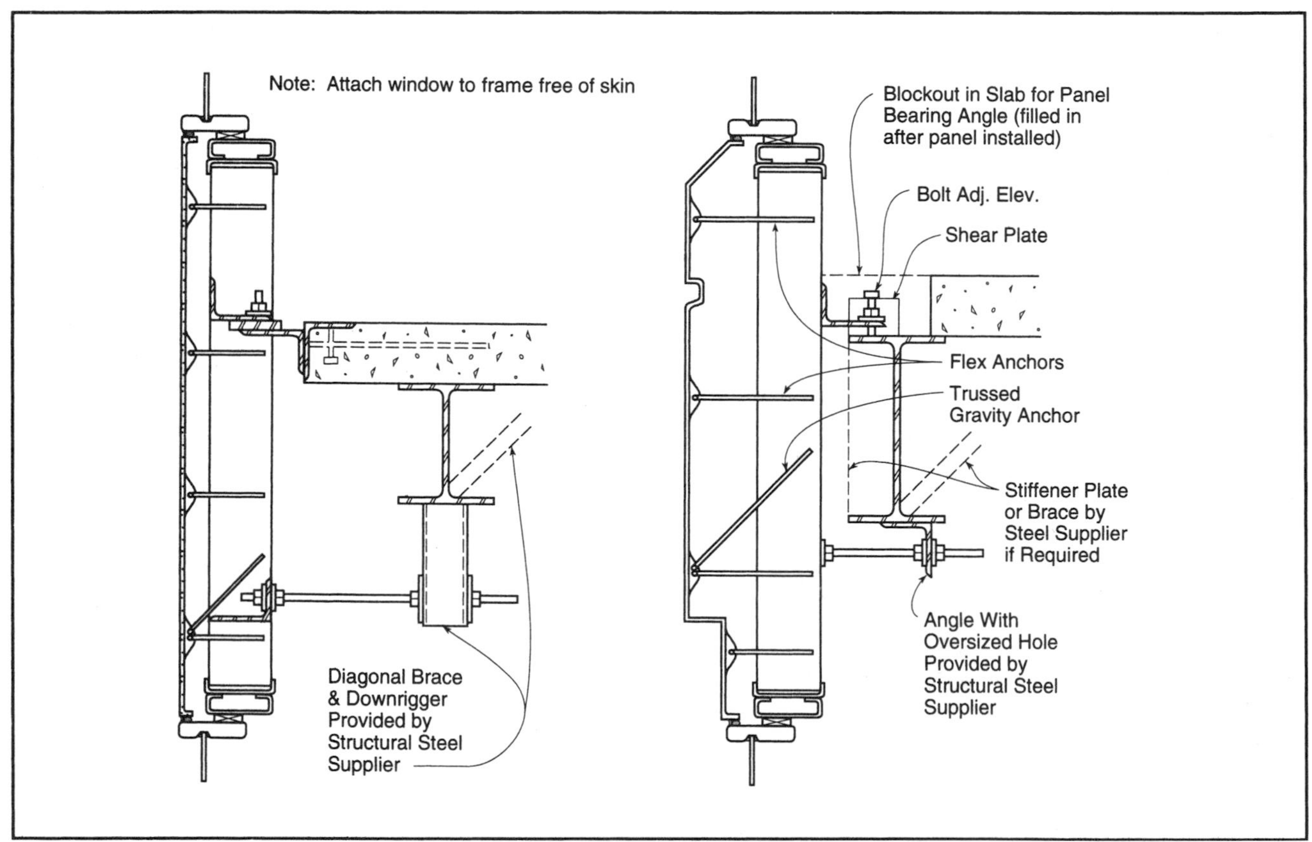

Figure 20.21
Typical connections of GFRC panels to a steel building frame. The lower connection in each case is a threaded rod that can flex as necessary as the height of the upper connection is adjusted with shims. *(Courtesy of Precast/Prestressed Concrete Institute)*

system also adapts readily to prefabrication (Figure 20.25). EIFS finds wide use over wood light framing as well, where it is used for small commercial and residential buildings.

There are two generic types of EIFS, polymer based and polymer modified. *Polymer-based EIFS* uses a very low density expanded polystyrene bead foam insulation, a glass fiber reinforcing mesh embedded in a base coat that is formulated primarily from either portland cement or acrylic polymer, and a finish coat that consists of texture granules in an acrylic polymer vehicle. The foam insulation is adhered to the backup wall. *Polymer-modified EIFS* uses a slightly higher density, extruded polystyrene foam insulation rather than expanded bead foam. The foam panels are mechanically attached to the backup wall with metal or plastic screws (plastic screws minimize thermal bridging through the insulation). A metal reinforcing mesh is embedded in a relatively thick portland cement base

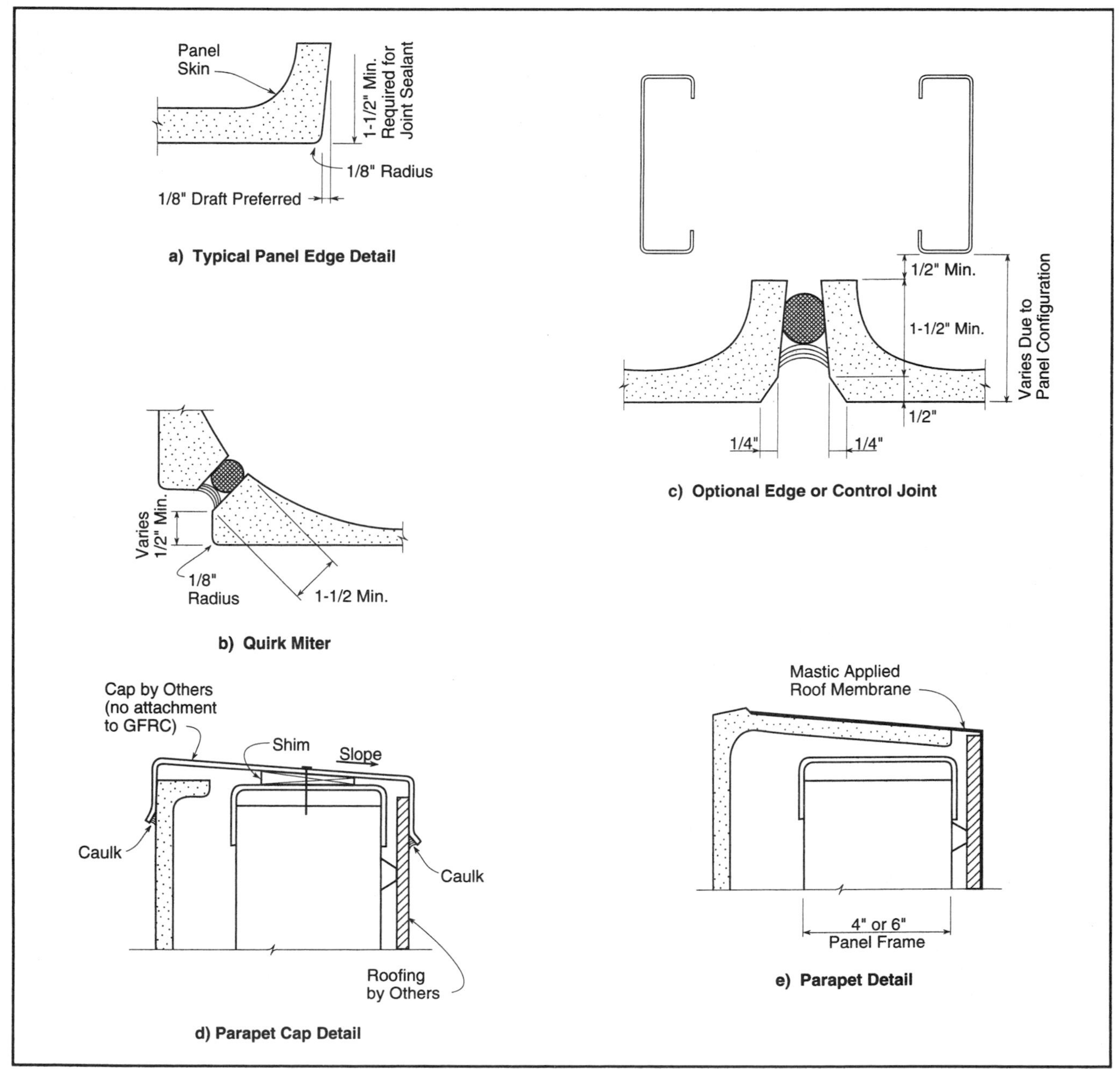

FIGURE 20.22
Typical edge details for GFRC cladding panels. *(Courtesy of Precast/Prestressed Concrete Institute)*

coat, and the finish coat is formulated of portland cement with acrylic modifiers. Polymer-modified systems are more durable (and more expensive) than polymer-based systems. They are more susceptible to shrinkage cracking during curing but are much less susceptible to denting or puncture. Polymer-based systems, on the other hand, have a very thin coating that is more elastic and less prone to cracking but relatively easy to dent or puncture when applied to areas of a building that are within reach of passersby or vehicles.

EIFS is an unusually versatile type of cladding, used for building types as diverse as single-family residences of wood or masonry construction as well as the largest buildings of noncombustible construction. It is used both for new construction and for refacing and insulating existing buildings. The insulating foam layer may be up to 4 inches (100 mm) thick, and there is little or no thermal bridging. The finish layer may be applied in a range of colors and textures. In appearance, at least from a distance, EIFS is virtually indistinguishable from conventional stucco.

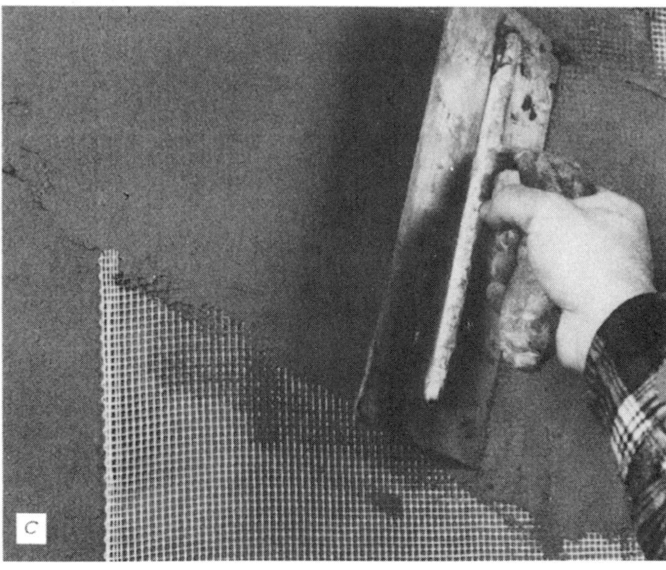

FIGURE 20.23
Four steps in installing an EIFS over a building with walls of masonry or solid sheathing. (*a*) A panel of foam is daubed with polymer-modified portland cement mortar. The foam may be as thick as required to achieve the desired thermal performance. (*b*) The foam panel is pressed into place, where it is held permanently by the daubs of mortar. (*c*) A thin base coat of polymer-modified stucco is applied to the surface of the foam panels, with an embedded mesh of glass fiber to act as reinforcing. (*d*) After the base coat has hardened, a finish coat in any desired color is troweled on. *(Used by permission of Dryvit® System, Inc.)*

A weakness of conventional EIFS of either type is that it is designed as a barrier system, without any means of internal drainage to prevent damage to the backup material if water leakage occurs at joints or through damaged areas. There have been numerous cases of extensive water damage in EIFS-faced buildings that have experienced leakage through poor detailing, faulty sealant joints, or failed coatings, especially around windows and doors. In response to this problem (and to the extensive litigation that arose out of it), EIFS producers now market *water-managed* or *water-drainage* EIFS. These systems utilize a layer of drainage matting behind the foam insulation that can capture water that does leak past the outer layer and conduct it to plastic flashings and weeps above wall openings and at the base of the wall, lessening the

FIGURE 20.24
A new bank building clad in EIFS. *(Architect: Paul Thoryk. Photo by John Bare. Used by permission of Dryvit® System, Inc.)*

FIGURE 20.25
EIFS cladding can be shop fabricated and erected in panel form. (*a*) Steel studs are welded together to make panel frames. (*b*) Rigid sheathing is screwed to the panel frames and finished with EIFS as shown in Figure 20.23. (*c*) The finished panels are bolted to the frame of the building. *(Used by permission of Dryvit® System, Inc.)*

risk of water penetrating deeper into the system, where it can cause greater damage (Figure 20.26).

Another weakness of polymer-based EIFS is the ease with which it is dented or punctured; this problem may be overcome by specifying a polymer-modified system or a specially reinforced polymer-based system in areas subject to damage. Damaged spots are easily and unobtrusively patched.

Because of these weaknesses and problems, the designer is advised to proceed with extreme care in detailing and specifying EIFS cladding and to avoid the use of barrier EIFS except in combination with backup systems, such as cast-in-place concrete, that are highly tolerant of moisture intrusion.

FIGURE 20.26
A mockup demonstrates the features of a proprietary EIFS system with internal drainage. From interior to exterior, the layers are an asphalt-saturated felt air and moisture barrier, a drainage mat composed of plastic fibers, plastic foam insulation, reinforcing mesh, a base coat, and a finish coat. A continuous plastic flashing under a gap at the bottom of the wall drains any leakage out to the face of the wall. *(Photograph of Senergy CD System courtesy of Senergy, Cranston, Rhode Island)*

FUTURE DIRECTIONS IN MASONRY AND STONE CLADDING

The alert reader will have noticed that most of the cladding systems shown in this chapter are detailed as barrier systems, not rainscreen systems (the notable exception being masonry veneer systems with drainage cavities and backup walls). This means that they are entirely dependent on good installation and careful maintenance if they are to remain watertight. If a sealant joint fails or a stone cracks in any of these systems, there is little to keep water from getting behind the cladding. There is also no well-organized system of secondary drainage in these systems: Although most have cavities behind their facings, the cavities are interrupted by framing and attachment components that are likely to splatter draining water onto the building frame and backup wall, where it can cause serious damage.

There have been many isolated efforts to design true rainscreen details for stone and concrete cladding systems, and a number of successful projects have been constructed. So far, however, no standard rainscreen details have emerged for these materials. This is an area to which trade associations and researchers should direct a great deal of effort, because reliably watertight details that are not heavily dependent on good workmanship and maintenance would save tens of millions of dollars each year in repair and reconstruction costs.

CSI/CSC

MasterFormat Sections for Masonry and Concrete Cladding

03 40 00	PRECAST CONCRETE
03 45 00	Precast Architectural Concrete Faced Architectural Precast Concrete
03 49 00	Glass-Fiber-Reinforced Concrete
04 20 00	UNIT MASONRY
04 21 00	Clay Unit Masonry Brick Veneer Masonry
04 25 00	Unit Masonry Panels Metal-Supported Unit Masonry Panels
04 40 00	STONE ASSEMBLIES
04 42 00	Exterior Stone Cladding Grid-System-Supported Stone Cladding Stone Panels for Curtain Walls
07 20 00	THERMAL PROTECTION
07 24 00	Exterior Insulation and Finish Systems Polymer-Based Exterior Insulation and Finish System Polymer-Modified Exterior Insulation and Finish System Water-Drainage Exterior Insulation and Finish System
07 40 00	ROOFING AND SIDING PANELS
07 42 00	Wall Panels Fabricated Wall Panel Assemblies

Selected References

1. Brick Industry Association. *Technical Notes on Brick Construction*, Nos. 18, 18A, 21, 21A, 21B, 21C, 27, 28B. Reston, VA, various dates.

These detailed pamphlets cover every aspect of brick veneer cladding systems.

2. All the references on stone and concrete masonry listed at the end of Chapter 9 are also relevant to this chapter.

3. Precast/Prestressed Concrete Institute. *Architectural Precast Concrete* (3rd ed.). Chicago, 2007.

This is a well-illustrated hardbound book that covers all aspects of the design, manufacture, and installation of precast concrete curtain walls. Also available from the same source is *Architectural Precast Concrete—Color and Texture Selection Guide* (2003), an extensive set of full-color plates of finishes for precast concrete panels.

4. Precast/Prestressed Concrete Institute. *GFRC: Recommended Practice for Glass Fiber Reinforced Concrete Panels* (4th ed.). Chicago, 2001.

This 104-page booklet is a clear, complete guide to the design and manufacture of GFRC cladding systems. (Address for ordering: See reference 3.)

5. Sands, Herman. *Wall Systems: Analysis by Detail*. New York, McGraw-Hill, 1986.

Though dated, this volume remains a valuable collection of case studies of cladding systems, illustrated comprehensively, with detailed drawings, many of them in three dimensions.

Web Sites

Cladding with Masonry and Concrete

Author's supplementary web site: **www.ianosbackfill.com/20_cladding_with_masonry_and_concrete**

Brick Industry Association: **www.bia.org**

Dry-Vit Systems: **www.dryvit.com**

EIFS Industry Members Association: **www.eima.com**

Precast/Prestressed Concrete Institute: **www.pci.org**

Whole Building Design Guide, Wall Systems: **www.wbdg.org/design/env_wall.php**

Key Terms and Concepts

masonry veneer
shelf angle
soft joint, horizontal expansion joint
vertical expansion joint
backup wall
prefabricated reinforced brick panels
grid-system-supported stone cladding
monolithic stone cladding panel
truss-supported stone cladding
posttensioned limestone spandrel panels
precast concrete cladding panel
precast concrete sandwich panel
glass-fiber-reinforced concrete (GFRC) cladding panel
exterior insulation and finish system (EIFS)
polymer-based EIFS
polymer-modified EIFS
water-managed EIFS, water-drainage EIFS

Review Questions

1. List all the common ways of attaching stone cladding to a building. Make a simple sketch to explain each system.

2. Working from memory, sketch all the details of a brick veneer wall over a concrete frame.

3. What are some options of surface finishes for precast concrete cladding panels?

4. Describe the process of producing GFRC panels, illustrating your account with simple sketches.

5. Name two types of EIFS. Describe two ways of applying EIFS to a building. Why should barrier wall EIFS be avoided?

Exercises

1. Design and detail a brick veneer cladding for a multistory building that you are designing. Rather than trying to conceal the flashings and soft joints, work out a way of expressing them boldly as part of the architecture of the building.

2. Visit one or more buildings under construction that are being clad with masonry, concrete, GFRC, or EIFS. Make sketches of how the materials are detailed, especially how they are anchored to the building. What will happen to any water that leaks through the cladding?

3. Adapt the brick veneer details in this chapter to installation on a building framed with structural steel.

PROJECT: Seattle University School of Law, Seattle, Washington

ARCHITECT: Olson Sundberg Kundig Allen Architects

ASSOCIATED ARCHITECT: Yost Grube Hall Architecture

STRUCTURAL ENGINEER: Putnam Collins Scott Associates

When Seattle University founded its new school of law, it commissioned Olson Sundberg Kundig Allen Architects to design the program's new home. The university wanted a building that would give the school prominence on the campus, one that would both fit within the context of existing masonry structures, and reflect the contemporary roots of this newly instituted program.

In response, the architects chose a building cladding system combining brick veneer and aluminum curtain wall—a balance of old and new materials. As the design took shape, they also strove to design and detail the veneer in such a way as to express the modern skinlike qualities of this cladding system. Brick panels were to be arranged in horizontal bands, visibly supported on structural steel channels. Integrated exterior steel sunshades would reinforce the ribbonlike character of the system.

An important decision came when the architects had to design the veneer's structural steel channel supports. One option was to support the brick with steel shelf angles concealed behind the veneer and then add a visible channel shape that would express the loadbearing function while not actually acting in a structural role. The second option was to design the steel channel as the true means of veneer support, allowing the channel to perform in both the architectural and structural roles.

There were pros and cons to both options. The first option relied on a more conventional method of construction, one that would be easier to communicate to the builder. In addition, since the channel would be nonstructural, it could be fabricated from materials that would be lighter and would accept a wider range of finish options, such as sheet metal formed to imitate the shape of a structural channel. On the other hand,

Figure A
Seattle University School of Law.

Figure B
Assembly sequence studies.

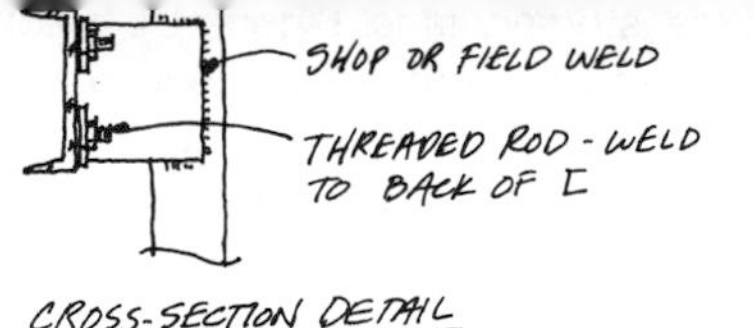

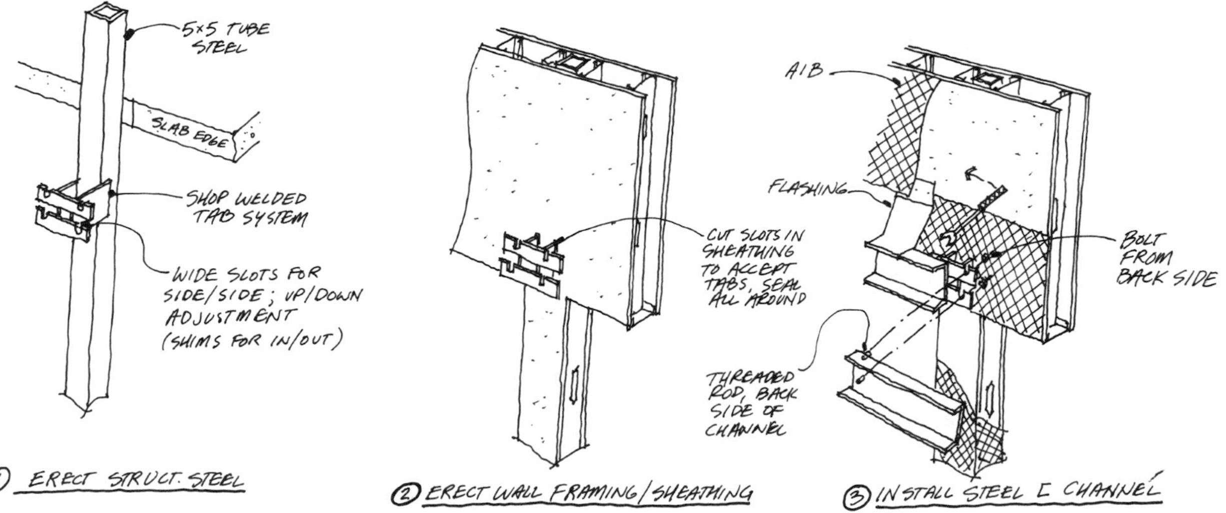

BRICK LEDGER CHANNEL CONSTRUCTION "A"

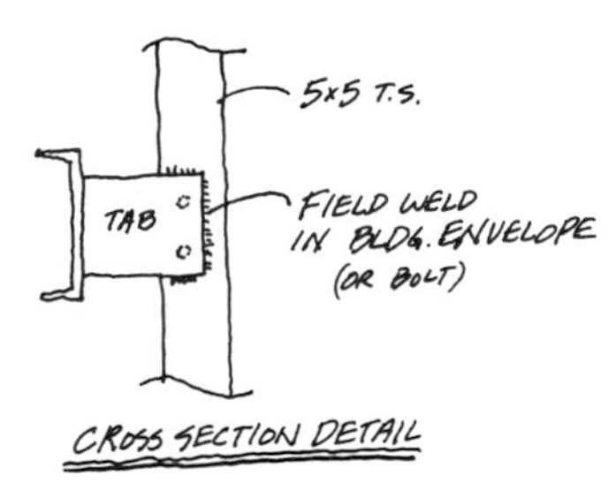

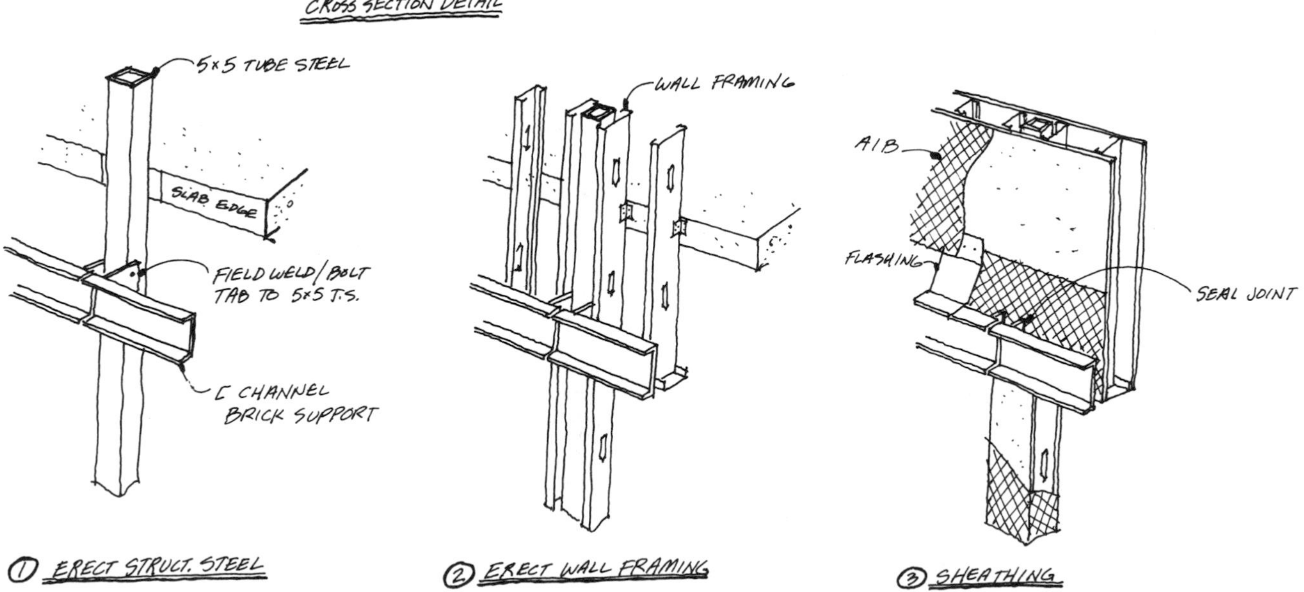

BRICK LEDGER CHANNEL CONSTRUCTION "B"

a structural channel would have a heavier and more genuine appearance. It would not be at risk of being seen as an unnecessary adornment that could be removed from the project as a budget-cutting measure. And to the architects, it seemed a more authentic way to express the weight of the veneer and its means of support.

If the second option was to be considered, the architect needed to validate this unconventional method of veneer support. Working with the structural engineer, design of the channel and options for its attachment to the building frame were studied (see Figure B). Sequences of assembly were explored to verify that the various building systems, including the structural frame, steel stud backup, sheathing, exterior enclosure materials, and veneer, could be erected in a logical way. These studies convinced the architect that the structural channel was practical, and the decision was made to proceed with this option. The final construction drawings showed a structural steel channel blind bolted to connection plates that were, in turn, welded to the structural steel frame (see Figure C).

Figure C
Construction drawing set detail.

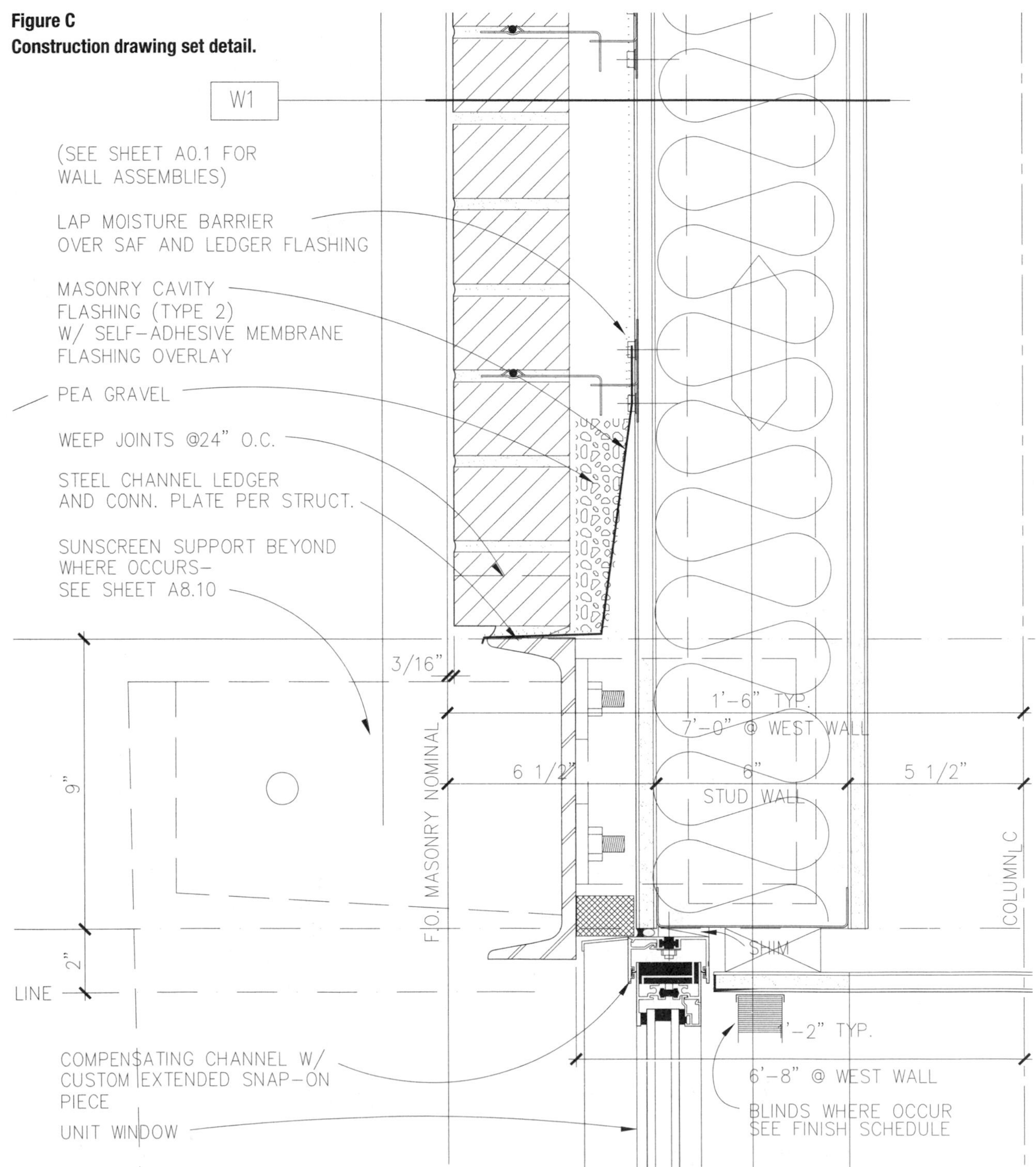

As expected, extra effort was required on the part of the architect to address questions that arose during construction. One issue concerned the sequence of assembly. After studying various options during the design phase, the architect had completed the design and detailing based on the assumption that the supporting channels would be erected along with the brickwork, not until after the backup steel studs and sheathing had been put in place. The contractors based their construction schedule on attachment of the structural channels much earlier in the process, at the same time as the erection of the structural steel frame. While both methods were feasible, adopting this sequence required the architect to make revisions to various other parts of the construction documents inorder to coordinate the effects of this change with other aspects of the exterior wall assembly.

A second difficulty was related to maintaining erection tolerances for the structural channels. The channel supports needed to be located with sufficient precision to achieve the dimensional accuracy required by the brick masons for their work. Recognizing that tolerances for the placement of the structural steel columns were significantly greater than could be allowed for the channels themselves, attachments between the columns and channels had been designed to allow for adjustability between these elements. Despite these preparations, column misalignments in the field in some cases exceeded the limits of adjustment that had been designed into the connections.

This problem was compounded by unanticipated irregularities in the channels themselves. The channels had been specified to be hot-dip galvanized, a process by which a heavy zinc coating is applied to the steel to protect against corrosion. Heating of the channels during the hot-dip process had caused them to warp subtly. Although not sufficient to make the channels unusable, this added distortion made it even more difficult to mount the channels on the columns with the required precision. To solve these problems, the connection design was revised to allow for welding as an alternative to the original bolted design, thereby permitting greater adjustability.

In the end, constructional difficulties were resolved, the brick veneer was successfully completed, and the final wall system achieved the functional and architectural goals of both the client and the architect. This project illustrates the important role that planning for constructability can play in the smooth and successful outcome of a project and the realization of the project's design goals.

Special thanks to Olson Sundberg Kundig Allen Architects, and John Kennedy, Associate, for assistance with the preparation of this case study.

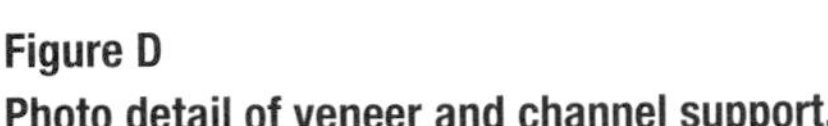

Figure D
Photo detail of veneer and channel support.

21

CLADDING WITH METAL AND GLASS

Kawneer 2800 Trusswall® aluminum mullions support the tall glass walls of the lobby of the Zentrum Building at the BMW Manufacturing Corporation in Spartanburg, South Carolina. *(Architects: Simons/AKA. Photo courtesy of Kawneer Company, Inc.)*

The contemporary metal-and-glass curtain wall is a descendant of the cast-iron-and-glass walls that were common features of commercial buildings in the 19th century. Today's walls, however, are vastly more sophisticated in every respect. They are carefully isolated from the frame of the building so that they support only their own weight and the forces of the wind. They are insulated and thermally broken to maximize comfort and minimize heating and cooling costs and moisture condensation. They utilize advanced glazing and spandrel materials that offer precise control of luminous and thermal properties. They are carefully gasketed and drained to discourage water leaks. Their intricate inner features are concealed behind smooth snap-on covers. They are designed for easy, forgiving installation and maintenance. And they are made of aluminum, light and strong, in the form of sleek extrusions that glisten with long-lasting anodized or organic finishes.

Aluminum Extrusions

Aluminum is the metal of choice for metal cladding systems for three primary reasons: It protects itself against corrosion. It accepts and holds a variety of attractive surface finishes. And it can be fabricated economically into elaborately detailed shapes by means of the process of extrusion.

The principle of *extrusion* is easily visualized: It is like squeezing toothpaste from a tube. In response to the squeezing pressure, the tube extrudes a column of toothpaste that is cylindrical in shape because the orifice of the tube is round. If the shape of the orifice were changed, one could produce many other shapes of toothpaste as well—square, triangular, flat, and so on.

To manufacture an aluminum extrusion, a large cylindrical *billet* of aluminum is heated to a temperature at which the metal flows under pressure but still retains its shape when the pressure is released. The heated billet is placed in the cylinder of a large press where a piston squeezes it under enormous pressure through a *die*, a steel plate with a shaped metal orifice. The orifice imparts its shape to a long extruded column of aluminum that is supported on rollers, cooled, straightened as necessary, and cut into convenient lengths (Figures 21.2 and 21.3).

Very intricate aluminum sections can be extruded for a variety of purposes, including not only curtain wall components but also door frames, window frames, entrances and storefronts, handrails, grillwork, and structural shapes such as wide flanges, channels, and angles. The high precision of the extrusion process permits it to be used for close-tolerance details such as snap-in glazing beads, snap-on mullion covers, *screw slots*, and *screw ports* (Figures 21.4–21.7). Hollow shapes may be extruded by mounting the portion of the die that forms the interior of the shape on a steel "spider" that is attached to the inside of the die. The metal flows around the legs of the spider before it passes through the orifice. The accompanying illustrations show some of the ways in which extruded aluminum details are utilized in building cladding. Extrusion dies are easily produced for custom-designed sections if there will be a long enough production run to amortize their expense.

FIGURE 21.1
The basic components of the Trusswall mullions shown on p. 838 are aluminum extrusions. Aluminum plate webs, welded to the tubular extrusions, increase the depth of the mullion and make it stiff enough to support very tall glass walls against wind loads. *(Photo courtesy of Kawneer Company, Inc.)*

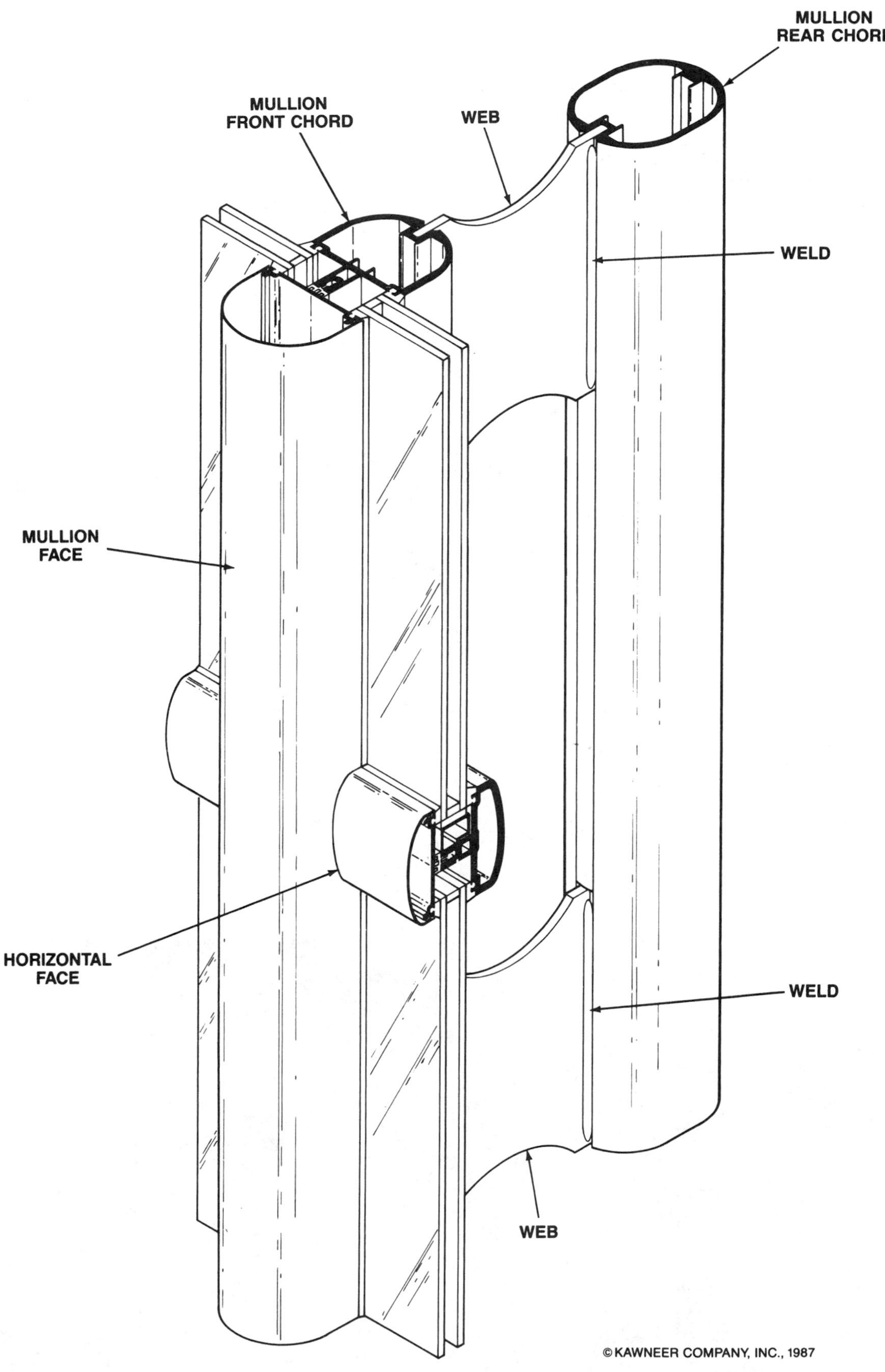
MULLION
REAR CHORD
MULLION
FRONT CHORD
WEB
WELD
MULLION
FACE
HORIZONTAL
FACE
WELD
WEB
©KAWNEER COMPANY, INC., 1987

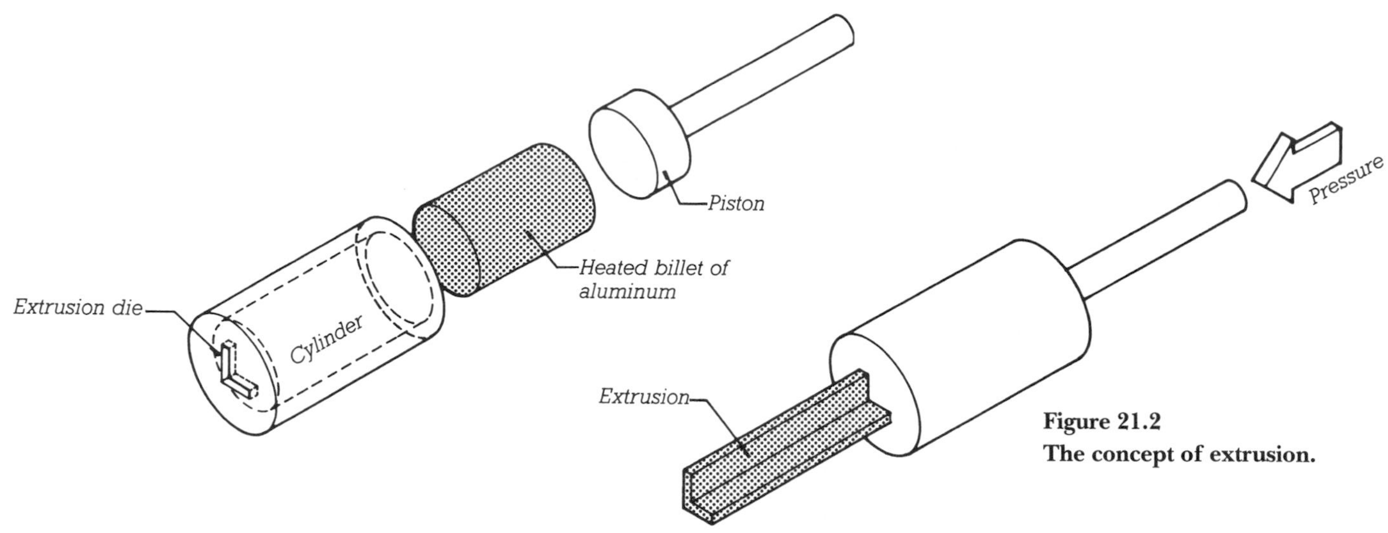

Figure 21.2
The concept of extrusion.

(*a*)

(*b*)

(*c*)

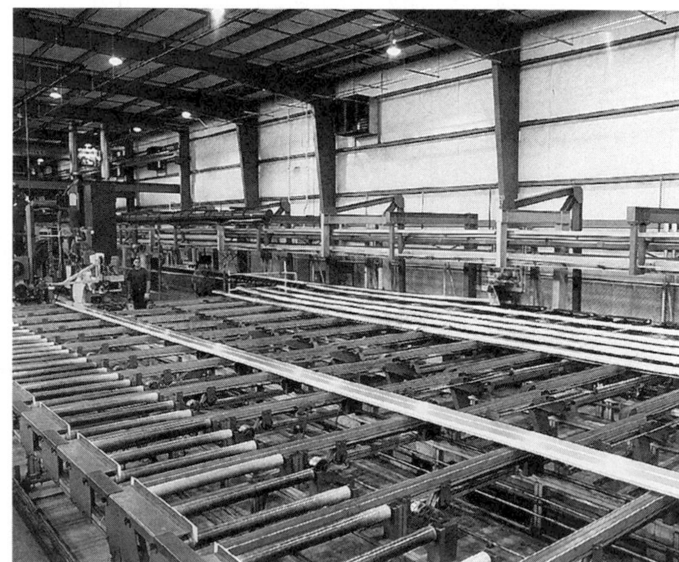

(*d*)

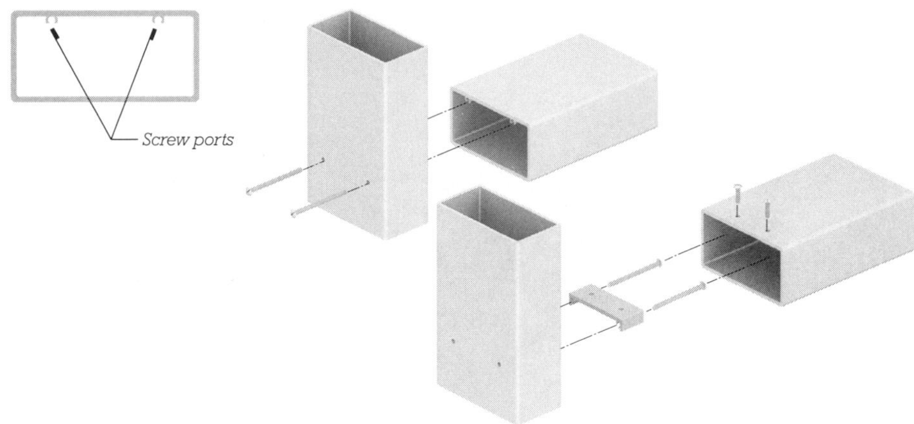

FIGURE 21.3
Making aluminum extrusions for curtain wall components. (*a*) Hundreds of extrusion dies for the many components of curtain walls are organized in racks. (*b*) A heated billet of aluminum is inserted into the cylinder of the extrusion press. (*c*) An extrusion emerges from the die. (*d*) Long aluminum extrusions cool on rollers, ready for straightening and cutting to length. *(Photos courtesy of Kawneer Company, Inc.)*

FIGURE 21.4
Screw ports are extruded cylindrical features that allow a screw to be driven parallel to the long axis of an extrusion. In the upper example, the screws pass through slightly oversized holes drilled through one box-shaped extrusion into screw ports in another box-shaped extrusion that are slightly smaller in diameter than the screws, thus engaging the screw threads tightly in the screw ports and making a snug joint. In the lower example, the screw ports are slightly larger than the outside diameter of the screws. The screws pass through the ports in a short piece of aluminum known as a "shear block" and engage slightly undersized holes in the vertical extrusion. Then the horizontal member is slipped over the shear block and screwed to it through drilled holes.

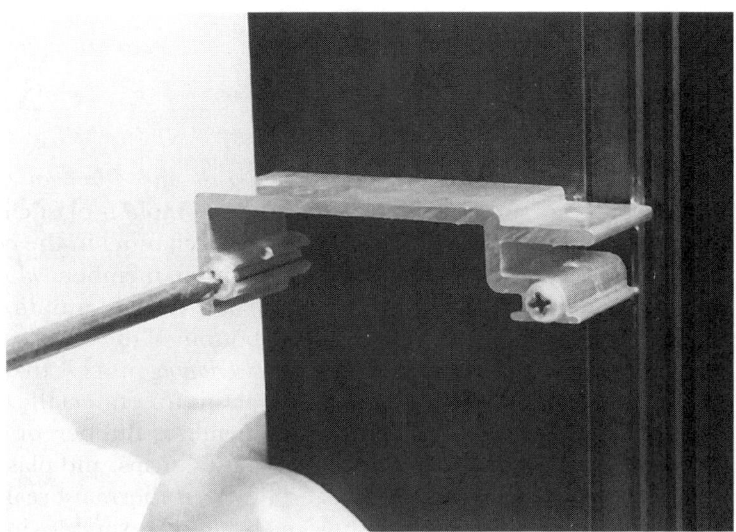

FIGURE 21.5
Self-tapping screws are driven through screw ports to fasten an extruded aluminum shear block to a vertical mullion. *(Courtesy of Kawneer Company, Inc.)*

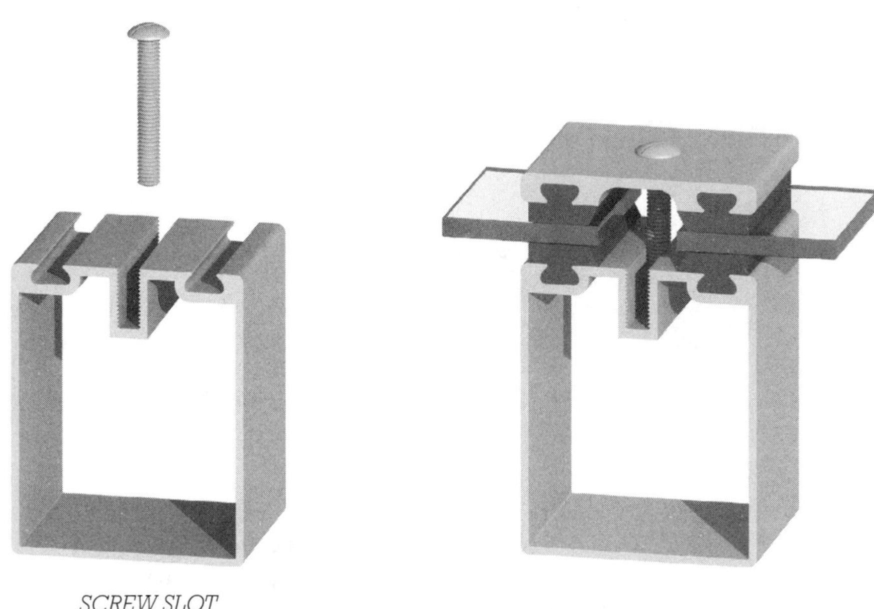

FIGURE 21.6
Screw slots allow screws to be driven perpendicular to the long axis of an extrusion. In this example, screws pass through oversized holes in an extruded aluminum pressure plate and pull the plate down toward the screw slot. Extruded gaskets of synthetic rubber are pressed into channels in both extrusions to seal tightly against the glass.

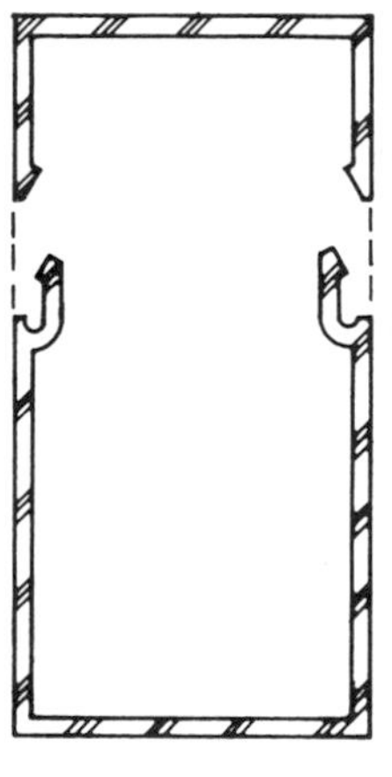

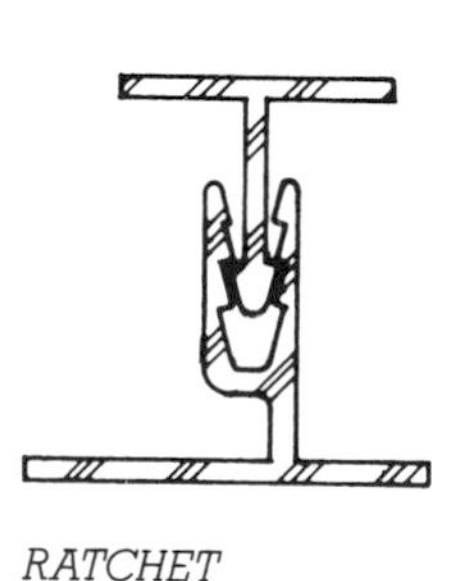

FIGURE 21.7
Snap-on and snap-together features are commonly used in extruded aluminum curtain wall components. Assembly is accomplished simply by aligning the components and tapping firmly with a rubber mallet or squeezing with a rubber-cushioned clamp.

Thermal Breaks

Aluminum conducts heat rapidly. In very cold weather, the indoor surfaces of an aluminum member that passes from the outside of the building to the inside, such as a window frame, would be so cold that moisture and possibly frost would condense on them. In very hot, humid weather, in an air-conditioned building, the outdoor surfaces of the same window frame might be cool enough to condense moisture from the air. This is why all but the simplest aluminum framing systems are manufactured with *thermal breaks,* which are internal components of insulating material that isolate the aluminum on the interior side of the component from aluminum on the exterior side. These dramatically reduce the flow of heat through the member (Figure 21.8).

There are several ways of creating thermal breaks; Figure 21.8 shows a *cast and debridged thermal break,* in which molten plastic is poured into a deep channel in the center of an aluminum member, where it hardens. Then the aluminum that forms the bottom of the channel is cut away in a *debridging* process that leaves only the plastic to connect the two halves of the member. Rubber or plastic gaskets, plastic strips, and plastic clips are also used as thermal breaks, as illustrated by several examples in this chapter.

Considerations of Sustainability in Aluminum Cladding

Manufacture

- The ore from which aluminum is refined, bauxite, is finite but relatively plentiful. The richest deposits are generally found in tropical areas, often where rain forests must be clearcut to facilitate mining operations.
- Aluminum is refined from bauxite by an electrolytic process that uses huge quantities of electricity. Aluminum smelters are often located near plentiful supplies of inexpensive hydroelectric power for this reason.
- The embodied energy in aluminum is roughly 100,000 BTU per pound (230 MJ/kg), seven times that of steel, making it one of the most energy-intensive materials used in construction.
- Large volumes of water are required for smelting. Wastewater from aluminum manufacture contains cyanide, antimony, nickel, fluorides, and other pollutants.
- Aluminum is recycled at a very high rate, due largely to industry efforts. Recycled aluminum is produced using only a fraction of the energy, approximately 5000 BTU per pound (12 MJ/kg), required to convert ore to aluminum.
- Aluminum extrusions are easy to produce and to form into cladding components. Their light weight saves transportation energy.
- Powder coatings for aluminum, which release no solvents into the atmosphere, are preferable environmentally to solvent-based coatings.

Construction

- Aluminum cladding is easy to erect because of its light weight and simple connections. Little waste or pollution is associated with the process. Scrap is readily recycled.

In Service

- Aluminum cladding seldom needs maintenance, lasts for a very long time, and can be recycled when a building is demolished.
- Because aluminum is highly conductive of heat, cladding components must be thermally broken.
- Aluminum foils used as vapor retarders, components of insulation systems, and radiant heat barriers save large amounts of heating and cooling energy. They are so thin that they consume little metal relative to the energy they can save over the lifetime of the building.

FIGURE 21.8
This thermally broken aluminum mullion extrusion is sitting on a block of dry ice in a humid room. Frost has formed on the portion of the mullion nearest the dry ice, but the cast and debridged thermal break keeps the rest of the mullion warm enough that moisture does not condense on it. (*Photo compliments of the H. B. Fuller Company, St. Paul, Minnesota*)

Surface Finishes for Aluminum

Aluminum, though it is a very active metal chemically, does not corrode away in service because it protects itself with a thin, tenacious oxide film that seals the surface of the metal and discourages further oxidation. While this film does an adequate job of protecting the aluminum, in the outdoor environment it develops a chalky or spotty appearance that looks rather shabby.

Anodizing is a manufacturing process that produces an integral oxide coating on aluminum that is thousands of times thicker and more durable than the natural oxide film that would otherwise form. The component to be anodized is immersed in an acid bath and becomes the anode in an electrolytic process that takes oxygen from the acid and combines it with the aluminum. Color can be added to the coating by means of dyes, pigments, special electrolytes, or special aluminum alloys. The colors most frequently used in buildings are the natural aluminum color, golds, bronzes, grays, and black, but other colors are possible. The advantages of anodized finishes are their extreme hardness and, in most colors, their extreme resistance to weather and fading.

Aluminum cladding components can also be finished with a variety of organic coatings. *Fluoropolymer coatings* are based on highly inert synthetic resins, such as *polyvinylidene fluoride* (*PVDF*), which are exceptionally colorfast and resistant to all forms of weathering, including ultraviolet deterioration. In a typical finishing process, the aluminum is first chemically cleaned and then primed. Next, one or two fluoropolymer finish coats, first a color coat and, optionally, a second clear coat, are spray applied. After the application of each of these finish coats, the aluminum piece is passed through an oven and baked at 450 degrees Fahrenheit (230°C), a process that causes the resin molecules to intertwine and fuse into a tightly bonded matrix. Fluoropolymer coatings are available in a broad spectrum of colors, including bright metallic finishes. They are the most expensive of the organic coatings and the longest lasting, the best of which can be expected to last 20 years or more under normal service conditions.

Powder coatings are manufactured with thermosetting powders that are composed of plastic resins, such as polyester, and pigments. The powder is electrically charged and then sprayed onto the aluminum component, which is grounded so that the powder adheres to it electrostatically. The component is then passed through an oven, where the powder fuses to produce a hard, resistant coating, usually in a single application. Among the advantages of powder coatings are their lower cost in comparison to fluoropolymers, their durability, the wide range of colors and finishes in which they are available, and their freedom from organic solvents that cause air pollution. In fact, the application of powder coatings generates no volatile organic compounds (VOCs) whatsoever.

Baked enamel coatings, consisting of spray-applied acrylic or polyester polymers, sometimes modified with silicone, are also used as aluminum coatings. They provide finishes with very high glosses in a wide selection of colors.

Aluminum finishes are specified according to standards published by the American Architectural Manufacturers Association (AAMA) that establish minimum requirements for coating thickness, colorfastness, gloss retention, corrosion resistance, resistance to chalking, abrasion resistance, and other characteristics related to the durability and performance of the finish. Anodized finishes are specified according to AAMA 611. Class I anodized finishes are thicker and longer-lasting than Class II anodized finishes. Class I finishes are suitable, for example, for curtain wall components on a tall building, whereas Class II anodized finishes are more commonly specified for residential aluminum products or low-rise aluminum storefront systems. Organic coatings are specified according to AAMA standards 2605, 2604, and 2603. Fluoropolymers with a high PVDF content and some powder coatings can meet the requirements of AAMA 2605, the most stringent of the three. Coatings meeting the requirements of either AAMA 2604 or 2603 are progressively less expensive and less long-lasting.

A wide range of surface effects may be applied to aluminum by mechanical and chemical processes. Mechanical finishes are produced by such means as wire brushing, wheel or belt polishing, buffing, grinding, burnishing, barrel tumbling, sandblasting, blasting with steel shot or glass beads, and abrasive blasting. Each produces a different surface texture. Chemical finishes include bright dipping, which produces mirrorlike surfaces; etching; and chemical conversion coatings such as oxides, phosphates, or chromates. Mechanical and chemical finishes may be done in preparation for the application of other types of finishes or, in some cases, may act as final finishes.

Other Curtain Wall Frame Metals

While the vast majority of contemporary curtain wall framing is made from aluminum extrusions, systems manufactured from other metals, such as extruded bronze or cold formed galvanized or stainless steel are also available.

Aluminum and Glass Framing Systems

Aluminum framing and glass are used to construct enclosure systems of various types, including windows, entrances, storefronts, and curtain walls. Aluminum framed windows are discussed in Chapter 18. *Entrances* are systems of aluminum framed doors, hardware, aluminum framing, and glass typically used for commercial buildings. They may also include framing for vestibules, sidelights, transoms, and the like.

Aluminum *storefronts*, though often similar in appearance to curtain wall systems, are based on simplified, lighter framing elements that are less expensive and quicker to assemble. Most storefront systems span vertically no more than 10 to 12 feet (3.0–3.7 m), although some, with heavier framing members or internal reinforcing, can span farther. Rather than being hung off the face of the building, as is common with curtain wall systems, storefront framing is typically installed between floor slabs or within wall openings (Figure 21.9).

FIGURE 21.9
As the name implies, storefront systems are intended primarily for ground-level commercial glazing applications. These simple, field-assembled systems lack the greater capacity to resist wind loads and the more sophisticated water management features of curtain wall systems.

Most importantly, the use of lighter framing elements and simplified internal construction makes storefront systems, in comparison to curtain wall systems, more limited in their ability to resist wind loads and water penetration. For these reasons, even improved storefront framing systems are not used more than three to four stories high on a building. Storefront framing is used not only for floor-to-ceiling glass enclosures but also sometimes as a substitute for conventional commercial aluminum window applications.

Aluminum *curtain wall* systems are the most high-performance, sophisticated, and costly of the aluminum and glass framing systems. Curtain wall systems can be readily applied to any number of stories. They use stronger, stiffer aluminum sections that can resist the higher wind pressures that act on taller buildings. Their methods of joining and attachment to the building structure can tolerate significant movements that arise from structural deflections as well as from thermal expansion and contraction of the aluminum itself. And they include more sophisticated internal construction and systems of sealing and water drainage that enable them to resist water leakage even under the much more severe conditions of wind-driven rain to which they are exposed in a tall building. It is to these types of systems that the remainder of this chapter is devoted.

Modes of Assembly

Metal curtain wall systems can be classified according to their degree or mode of assembly at the time of installation on the building (Figure 21.10). Many metal-and-glass curtain walls are furnished as *stick systems* whose principal components are metal mullions and rectangular panels of glass and spandrel material that are assembled in place on the building (Figure 21.10*a*). Stick systems have the advantages of low shipping bulk and high degree of adjustability to unforeseen site conditions, but they must be assembled on site, under highly variable conditions, rather than in a factory with its ideal tooling, controlled environmental conditions, and lower wage rates. Aluminum entrances and storefronts are also typically installed as stick systems.

The *unit system* of curtain wall installation takes full advantage of factory assembly and minimizes on-site labor, but the units require more space during shipping and more protection from damage than stick system components (Figure 21.10*b*). The *unit-and-mullion system* (Figure 21.10*c*), which is seldom used today, offers a middle ground between the stick and unit systems.

The *panel system* (Figure 21.10*d*) is made up of homogeneous units that are formed from metal sheet. Its advantages and disadvantages are similar to those of the unit system, but its production involves the higher tooling costs of a custom-made die or mold, which makes it advantageous only for a building that requires a large number of identical panels.

The *column-cover-and-spandrel system* (Figure 21.10*e*) emphasizes the structural module of the building rather than creating its own grid on the facade, as the previously described systems do. A custom design must be created for each project because there is no standard column or floor spacing for buildings. Special care is required in detailing the spandrel panel support to ensure that the panels do not deflect when loads are applied to the spandrel beams of the building frame; otherwise, the window strips could be subjected to loadings that would deform the mullions and crack the glass.

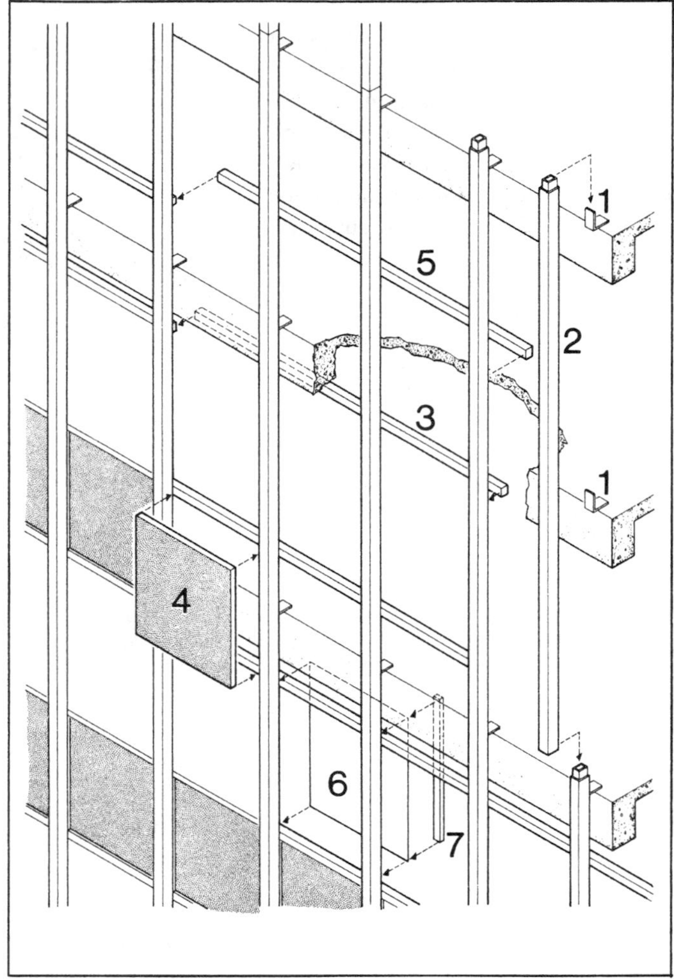

STICK SYSTEM—Schematic of typical version

1: Anchors. 2: Mullion. 3: Horizontal rail (gutter section at window head). 4: Spandrel panel (may be installed from inside building). 5: Horizontal rail (window sill section). 6: Vision glass (installed from inside building). 7: Interior mullion trim.

Other variations: Mullion and rail sections may be longer or shorter than shown. Vision glass may be set directly in recesses in framing members, may be set with applied stops, may be set in sub-frame, or may include operable sash.

(*a*)

Outside Glazing and Inside Glazing

A metal cladding system may be designed to be *outside glazed*, which means that glass must be installed or replaced by workers standing on scaffolding or staging outside the building. Alternatively, it may be designed to be *inside glazed* by workers who stand inside the building. Inside glazing is more convenient and is more economical for a tall building, but it requires a somewhat more elaborate set of extrusions. Outside glazing systems utilize a relatively simple set of shapes and are less expensive for a building that is only one to three stories tall. Some curtain wall systems are designed so that they may be glazed from either side.

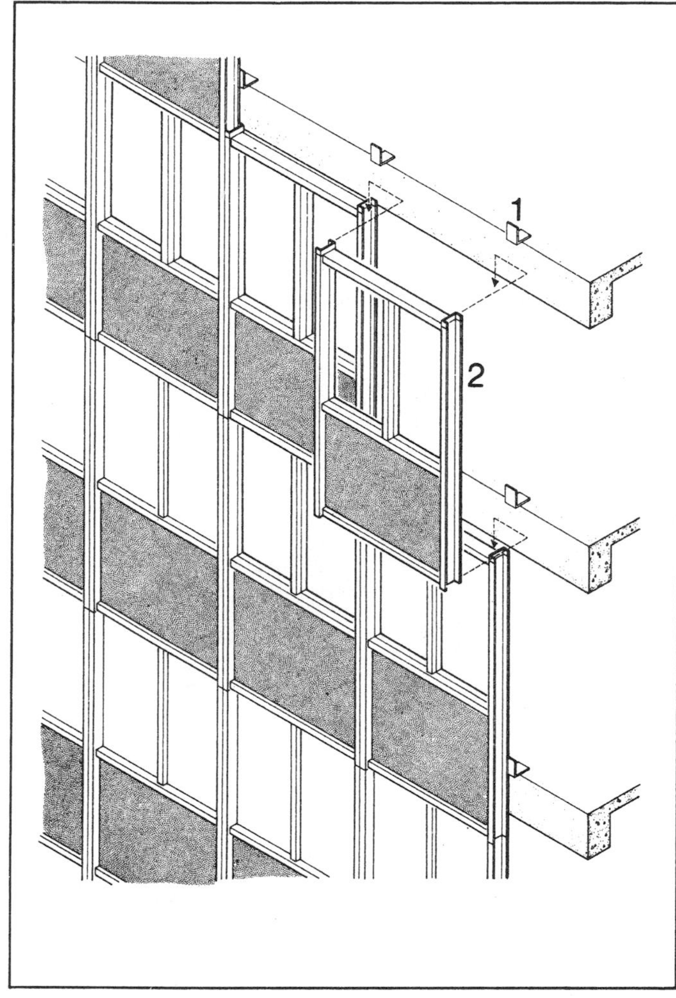

UNIT SYSTEM—Schematic of typical version

1: Anchor. 2: Pre-assembled framed unit.

Other variations: Mullion sections may be interlocking "split" type or may be channel shapes with applied inside and outside joint covers. Units may be unglazed when installed or may be pre-glazed. Spandrel panel may be either at top or bottom of unit.

(*b*)

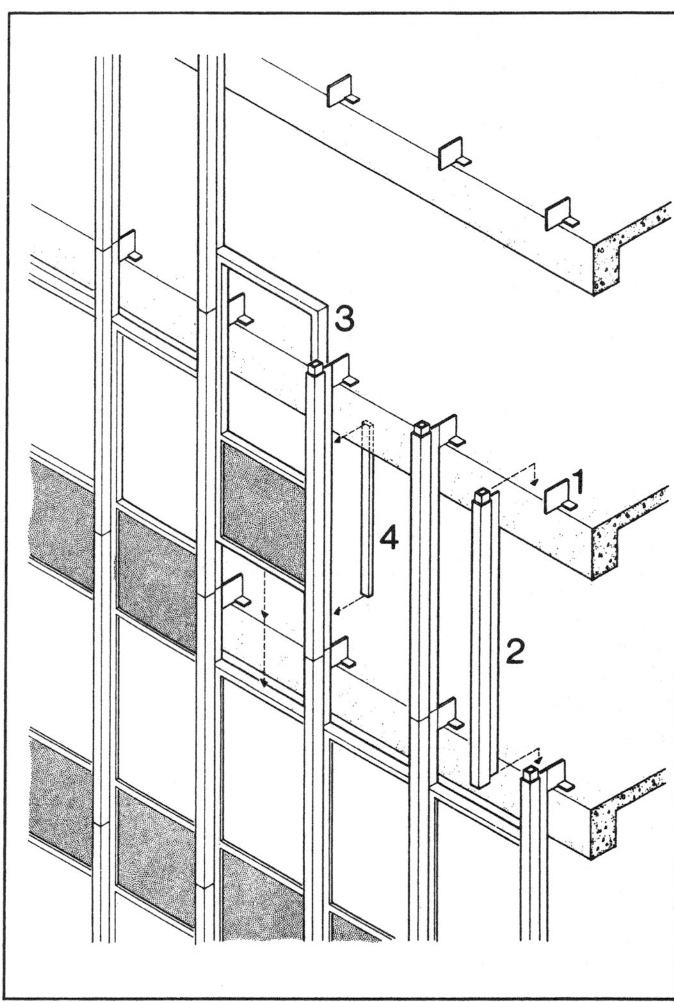

UNIT-AND-MULLION SYSTEM—Schematic of typical version

1: Anchors. 2: Mullion (either one- or two-story lengths). 3: Pre-assembled unit —lowered into place behind mullion from floor above. 4: Interior mullion trim.

Other variations: Framed units may be full-story height (as shown), either unglazed or pre-glazed, or may be separate spandrel cover units and vision glass units. Horizontal rail sections are sometimes used between units.

(*c*) (*continued*)

FIGURE 21.10

Modes of assembly for curtain walls. (*a*) Stick system. (*b*) Unit system. (*c*) Unit-and-mullion system.

An Outside Glazed Curtain Wall System

An off-the-shelf, externally glazed stick system for aluminum-and-glass curtain walls is illustrated in Figures 21.11–21.16. This system is suitable for a low building whose walls workers can easily reach from external scaffolding.

An Inside Glazed Curtain Wall System

An off-the-shelf, internally glazed stick system for aluminum-and-glass curtain walls is illustrated in Figures 21.17–21.23. This system is suitable for use on tall buildings, because it is installed entirely by workers who stand inside the building. Replacement of glass may also be done from within the building.

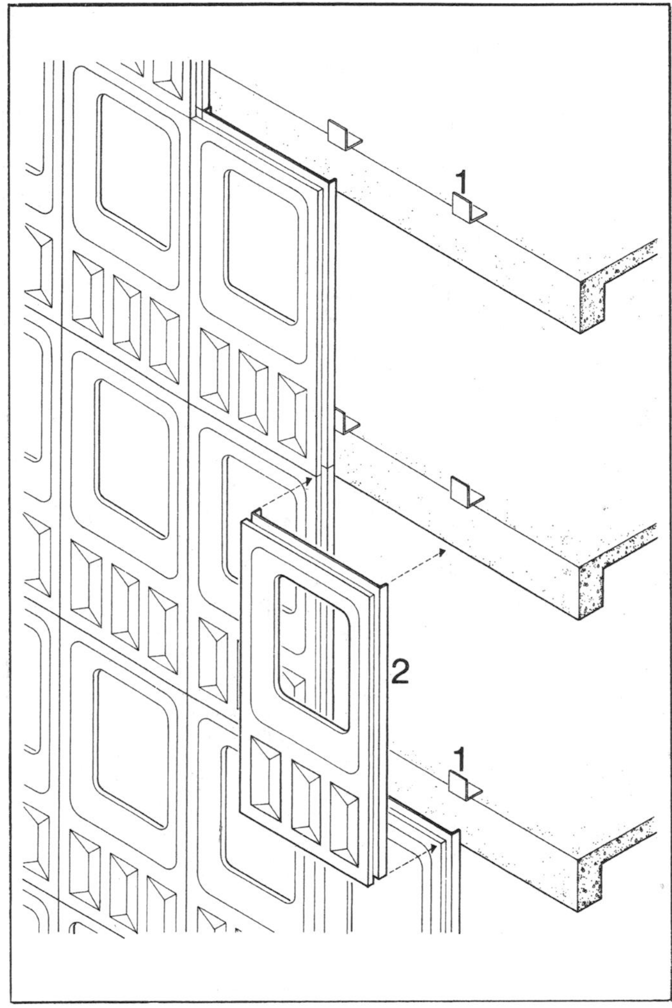

PANEL SYSTEM—Schematic of typical version

1: Anchor. 2: Panel.

Other variations: Panels may be formed sheet or castings, may be full story height (as shown) or smaller units, and may be either pre-glazed or glazed after installation.

(*d*)

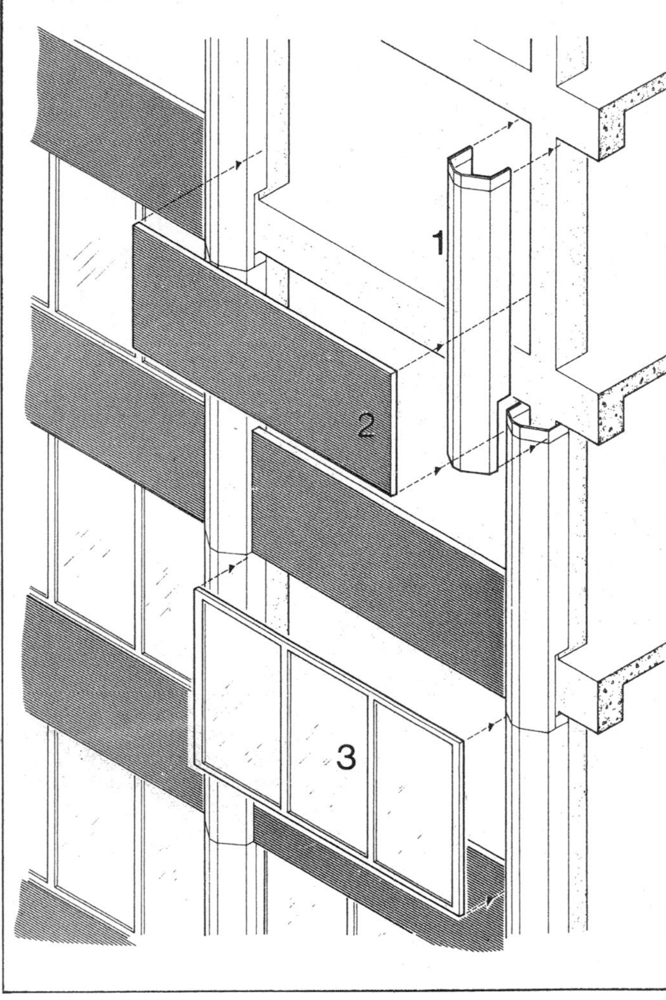

COLUMN COVER AND SPANDREL SYSTEM—Schematic of typical version

1: Column cover section. 2: Spandrel panel. 3: Glazing infill.

Other variations: Column covers may be one piece or an assembly, may be of any cross-sectional profile, and either one or two stories in height. Spandrel panel may be plain, textured or patterned. Glazing infill may be a pre-assembly, either glazed or unglazed, or be assembled in place.

(*e*)

FIGURE 21.10 (*continued*)

(*d*) Panel system. (*e*) Column-cover-and-spandrel system. *(Reprinted with permission from AAMA* Aluminum Curtain Wall Design Guide*)*

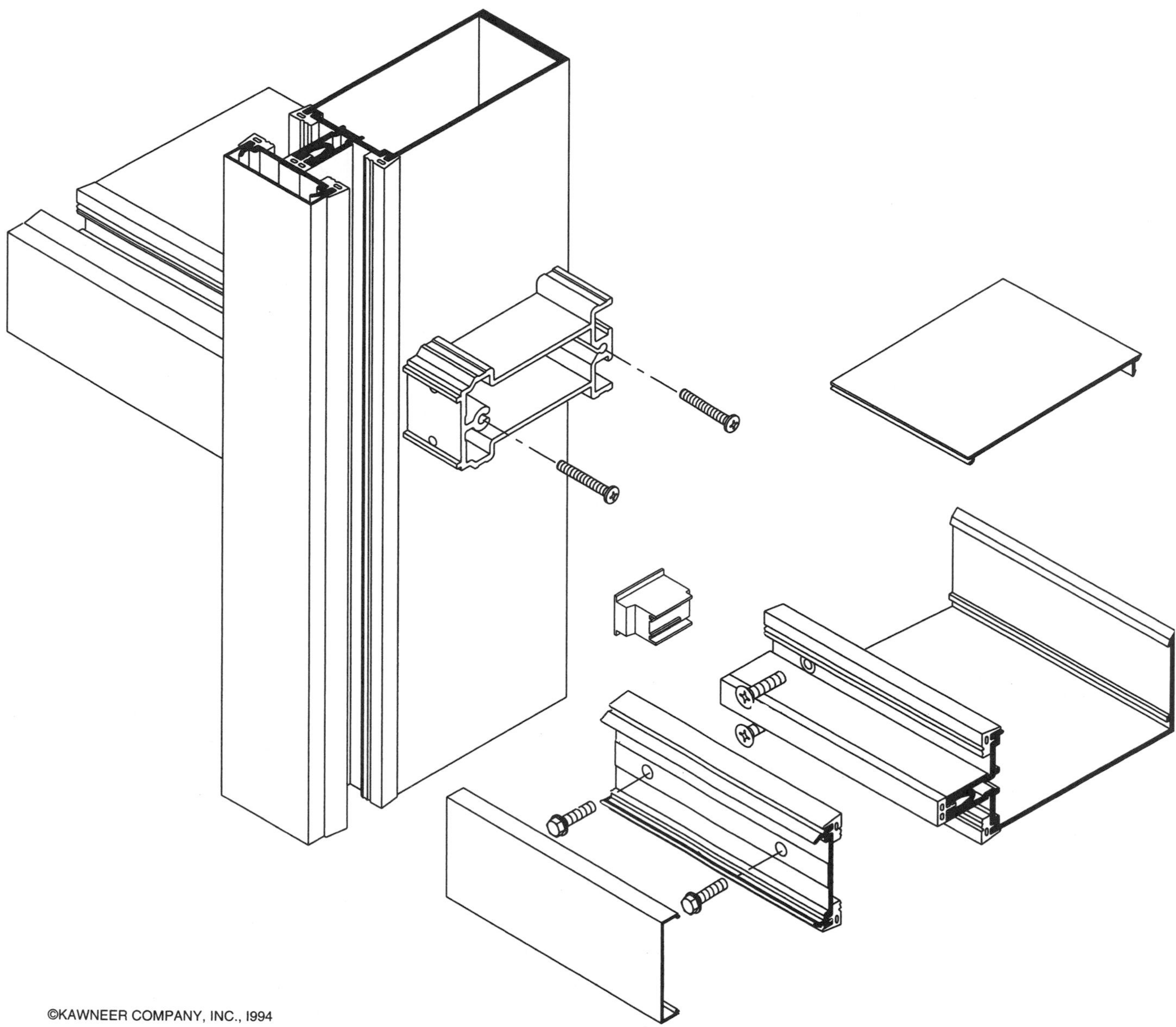

FIGURE 21.11

The Kawneer 1600 System1® is an outside glazed stick system. Vertical mullions that run continuously from floor to floor support discontinuous horizontal mullions that are connected to them by means of shear blocks and screws. Each light of glass sits on two rubber glazing blocks in the gutter of the horizontal mullion (not shown in this drawing). The inner surface of the glass rests against extruded rubber glazing gaskets pressed into small channels in both the vertical and horizontal mullions. Extruded aluminum pressure plates with rubber glazing gaskets are applied to both horizontal and vertical mullions to clamp the glass into place and create a weathertight seal. Each pressure plate is attached by means of screws that pass through drilled holes into an extruded screw slot. A thick rubber gasket in the screw slot acts as a thermal break. Snap-on covers conceal the screw heads and give a neat exterior appearance. A molded rubber plug at either end of each horizontal mullion contains any leakage or condensate within the horizontal mullion, from which it escapes via 5/16-inch-diameter (8-mm) weep holes (not shown here) drilled through the pressure plate and the bottom edge of the snap-on cover. Figures 21.11–21.15 further illustrate this system. *(All the drawings of this system in Figures 21.11–21.14 are courtesy of Kawneer Company, Inc.)*

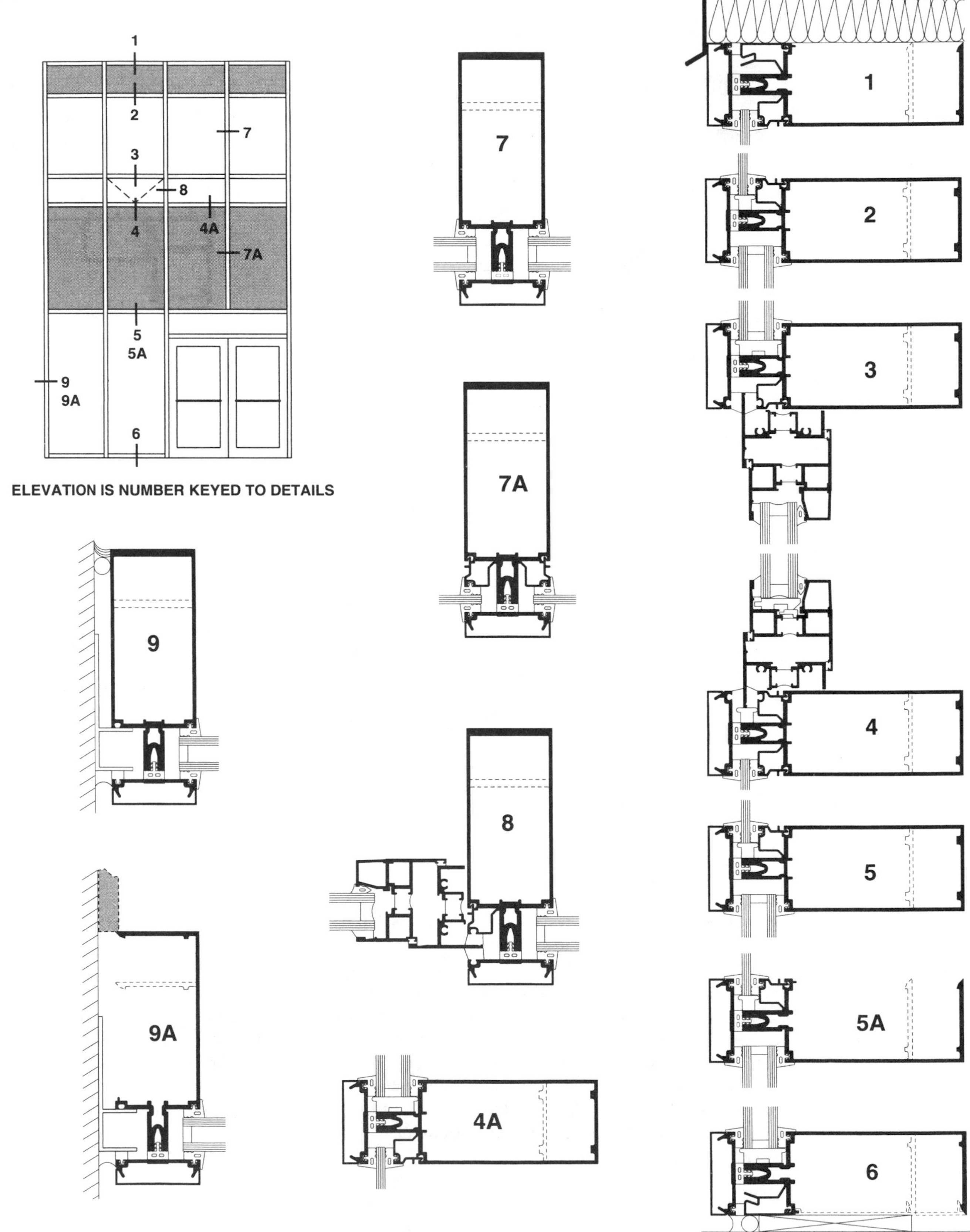
SCALE 3"= 1'– 0"
1
2
3
4
4A
5
5A
6
7
7A
8
9
9A
ELEVATION IS NUMBER KEYED TO DETAILS
©KAWNEER COMPANY, INC., 1994

FIGURE 21.12
The manufacturer's details for the Kawneer 1600 System1 curtain wall are all keyed to the small elevation view at the upper left corner of this page. The details are reproduced here at one-quarter of their actual size.

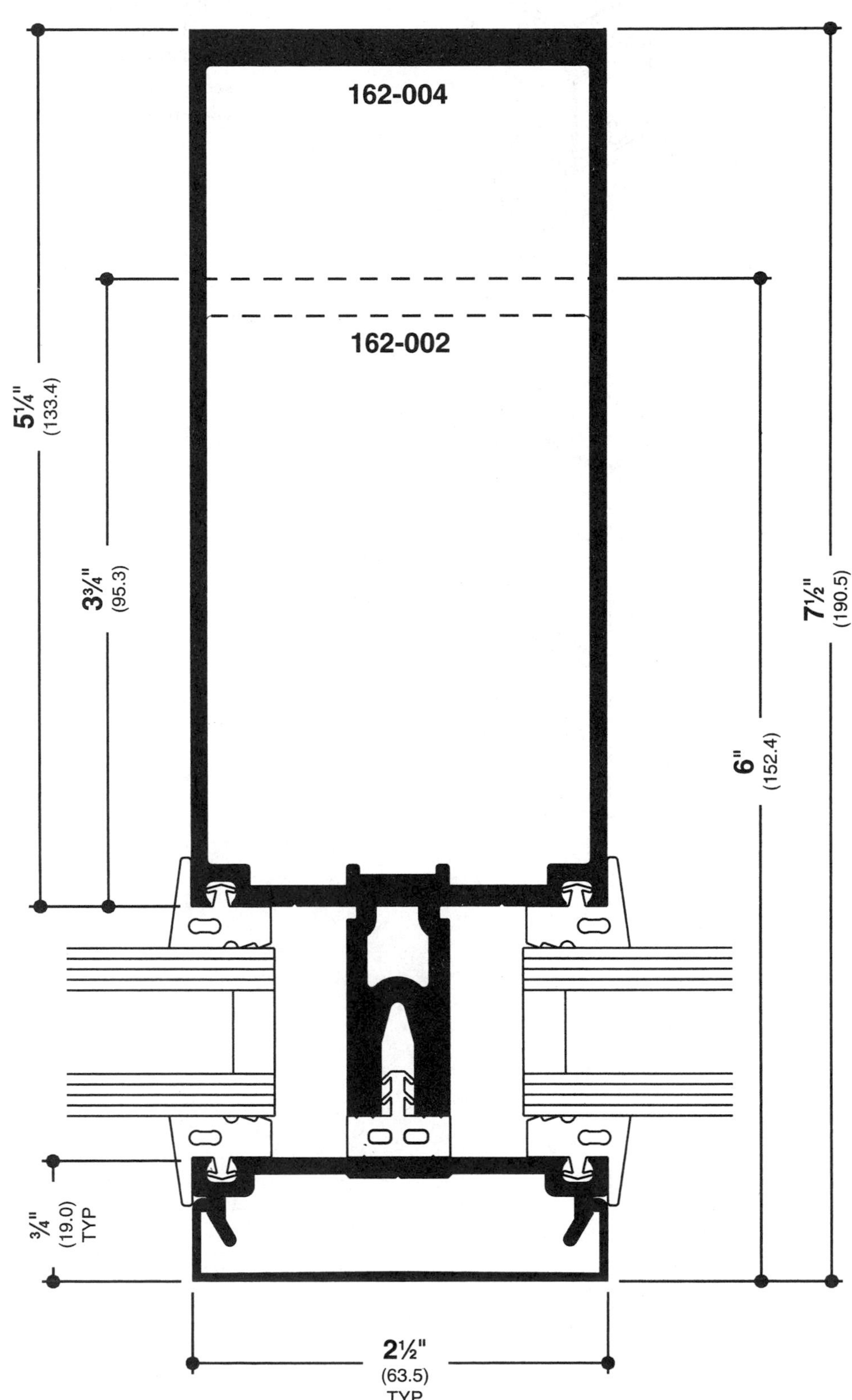

FIGURE 21.13
In this full-size detail of the vertical mullion for the Kawneer 1600 System1, we see that the basic extrusion is a rectangular box shape that is structurally stiff and presents a neat appearance inside the building. The broken lines indicate a smaller box mullion (162-002) that may be used for buildings with shorter floor-to-floor heights or smaller wind loads. The extruded plastic thermal break attaches to the mullion with a projecting "pine tree" spline that is pushed into the screw slot. Holes are drilled through the thermal break for the screws (not shown) that attach the pressure plate. The four rubber glazing gaskets attach to the aluminum pieces with projecting splines.

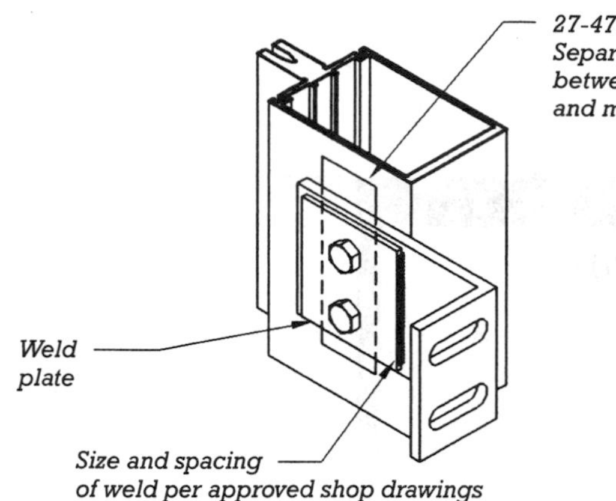

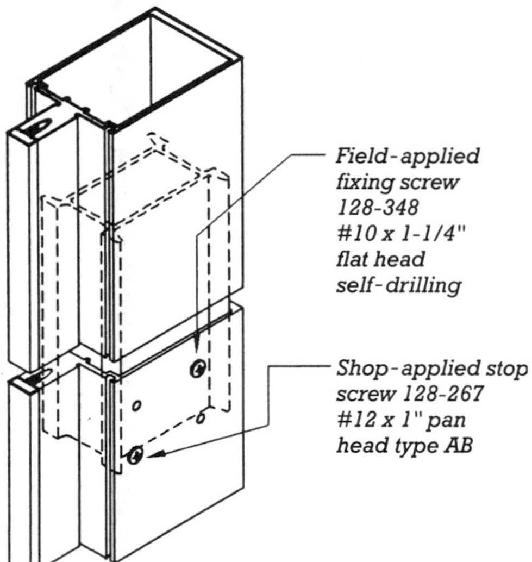

FIGURE 21.14
Vertical mullions of the Kawneer 1600 System1 are attached to the edges of the building's floors with the angle anchor shown in the top drawing. Sections of vertical mullion are spliced with the aluminum internal spline shown in the lower drawing. The spline is screwed to the lower section of mullion but the upper section is free to slide, which allows for thermal expansion and contraction.

FIGURE 21.15
Two different horizontal mullion extrusions are available for the Kawneer 1600 System1. The closed extrusion (*top*) is used in most situations. The open-back extrusion (*bottom*) allows access to the shear block on the vertical mullion for purposes of assembly; it is used for portions of the wall that butt against other materials, such as heads, sills, and end bays. A snap-on cover closes the open back of the mullion, leaving only a virtually invisible seam. Rubber setting blocks are shown beneath the edges of the glass in both mullions. In the upper-left corner of the upper detail, a small aluminum extrusion has been added to allow the mullion to hold a single layer of spandrel glass or vision glass rather than the standard 1-inch (25-mm) double-glazing assembly. Weep holes are not shown; they are drilled horizontally through the pressure plate just above the thermal break and vertically through the bottom edge of the outside snap-on cover.

162-001
162-003
6"
(152.4)
7½"
(190.5)

162-008
162-009

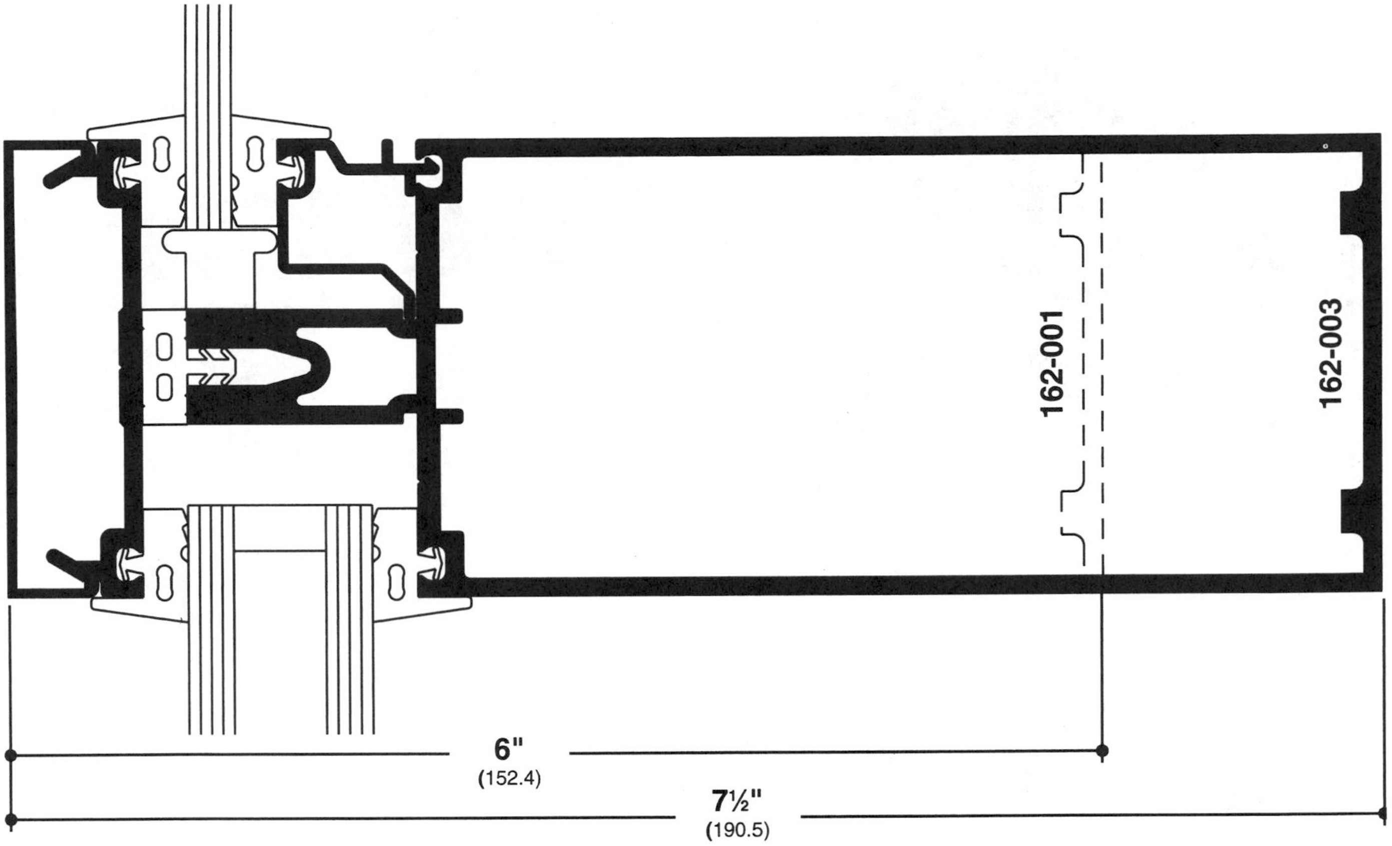

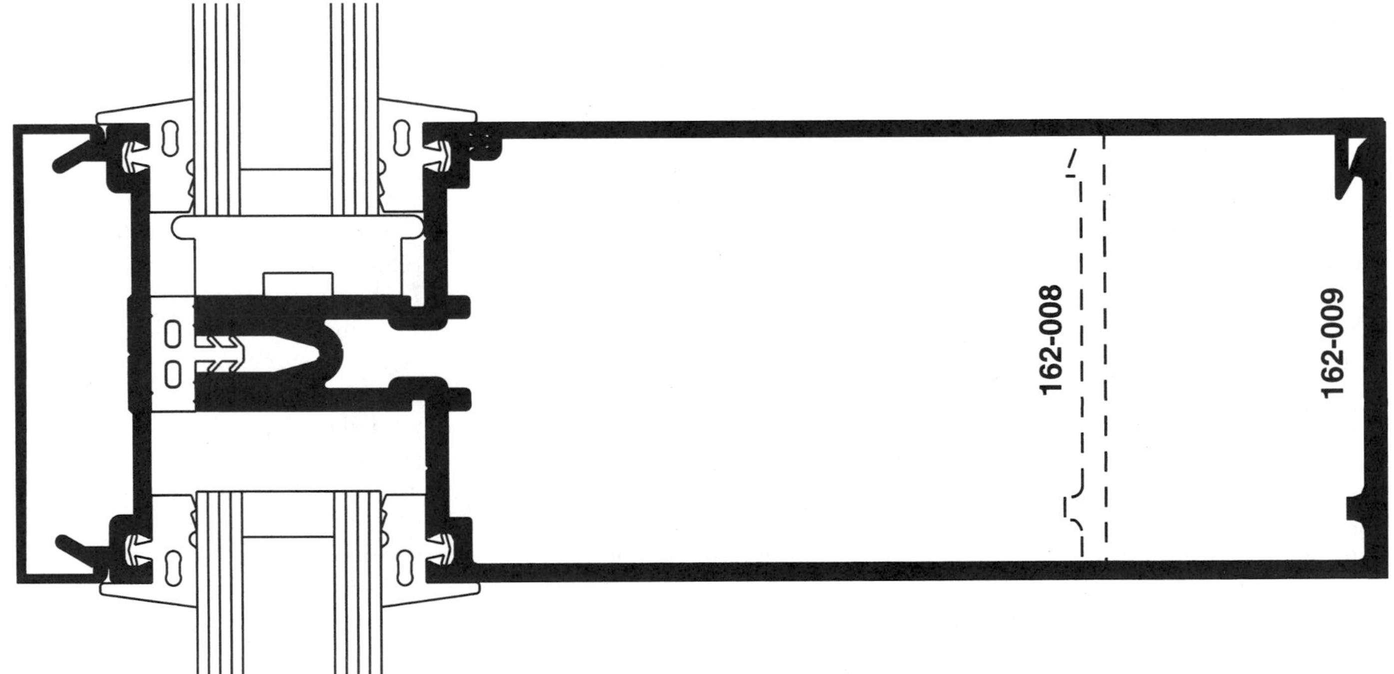

FIGURE 21.16
This building for the Harley-Davidson Motor Company in Wauwatosa, Wisconsin, features the Kawneer 1600 System1, whose details are shown in Figures 21.11–21.14.
(Architect: Flad & Associates, Madison, Wisconsin. Photo courtesy of Kawneer Company, Inc.)

The Rainscreen Principle in Metal-and-Glass Cladding

At first glance, neither of the curtain wall systems presented in this chapter might seem to be a rainscreen design, because both use rubber gaskets to seal around the glass on the outside of the wall as well as on the inside. However, consider what would happen in either of these systems if both the inner and outer gaskets around a light of glass were defective: During a wind-driven rain, gravity and capillary action would be likely to draw some water past the outer gasket into the spaces between the edges of the glass and the aluminum. However, as we discovered in Chapter 19, even a defective inner gasket is very unlikely to allow air currents strong enough to carry water farther than this toward the interior of the building. If water has leaked in along a vertical edge of the glass, it is contained within the vertical mullion and will fall by gravity to the bottom, where it drains out through weep holes. If water accumulates in the horizontal mullion, it is prevented from running out of the ends of the mullion by rubber end plugs, which are seen in Figure 21.11. Its only recourse is to drain to the outdoors through weep holes that are drilled horizontally through the pressure plate and vertically through the bottom edge of the external snap-on cover. Thus, the external gasket need serve only as a *deterrent seal*, essentially a rainscreen, to discourage water from entering without necessarily barring its entry altogether. The internal gasket serves as an air barrier, and the hollow spaces between the edges of the glass and the mullions act as pressure equalization chambers. The entire system functions as a rainscreen assembly. In practice, of course, the manufacturer and installer take every precaution to ensure that the external gaskets are properly installed and will act as positive barriers to the passage of water, but these curtain wall systems do not depend on a perfect seal to retain their watertightness.

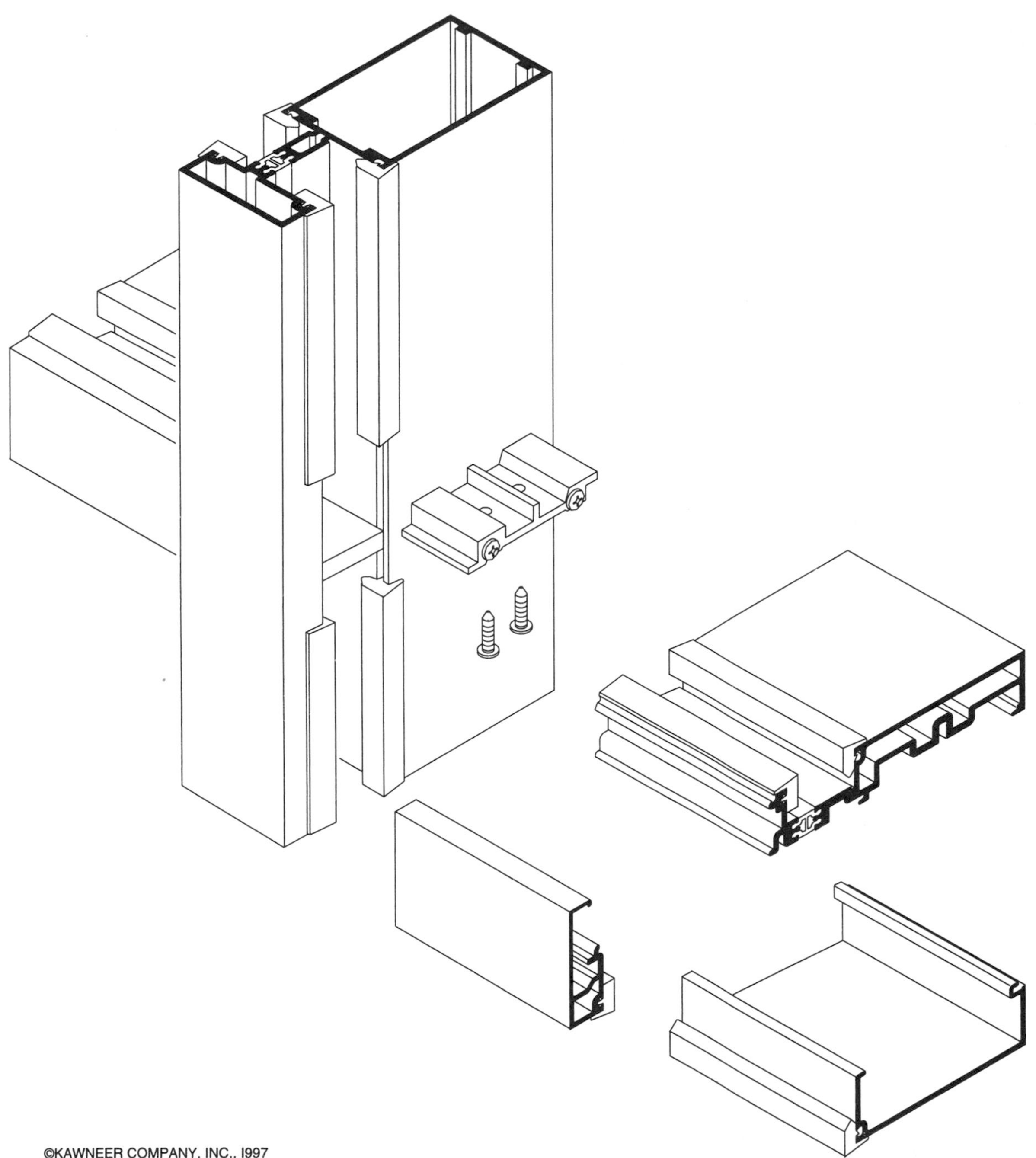

FIGURE 21.17
The Kawneer 1600 System4® may be glazed from either inside or outside the building. Inside glazing is usually preferred in taller buildings for its speed, safety, and convenience. *(All the drawings of this system in Figures 21.17–21.22 are courtesy of Kawneer Company, Inc.)*

SCALE 3" = 1' 0"

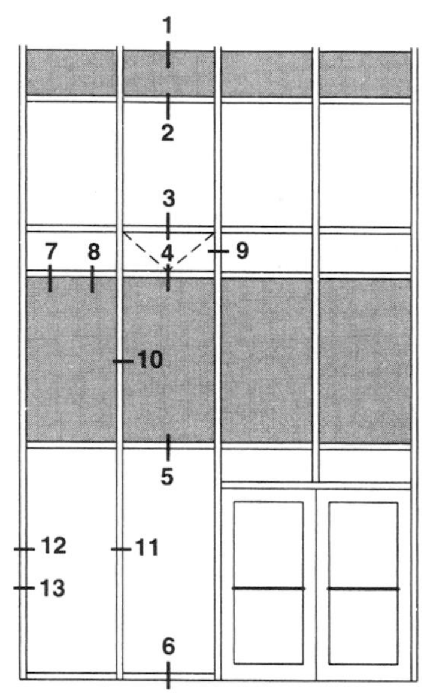

ELEVATION IS NUMBER KEYED TO DETAILS

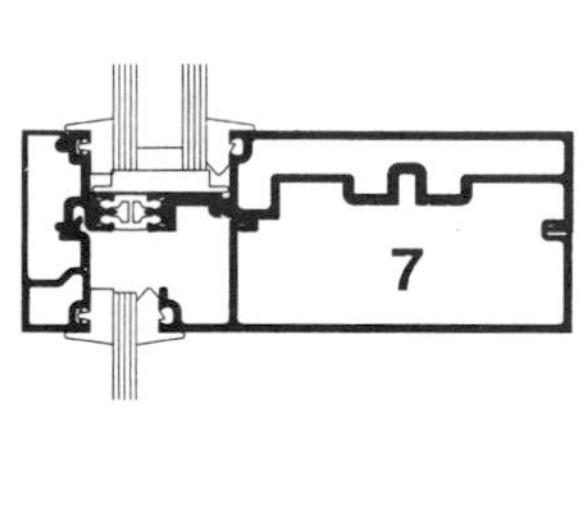

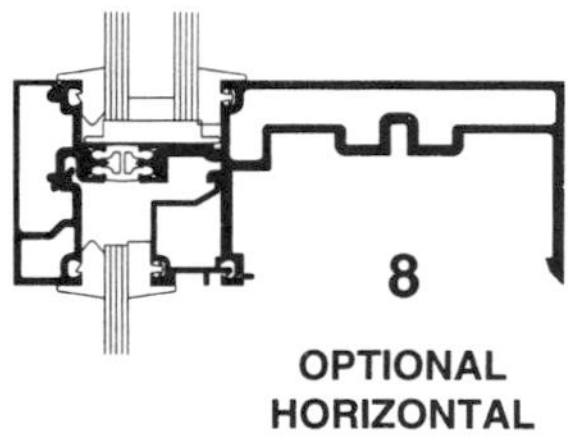

OPTIONAL HORIZONTAL

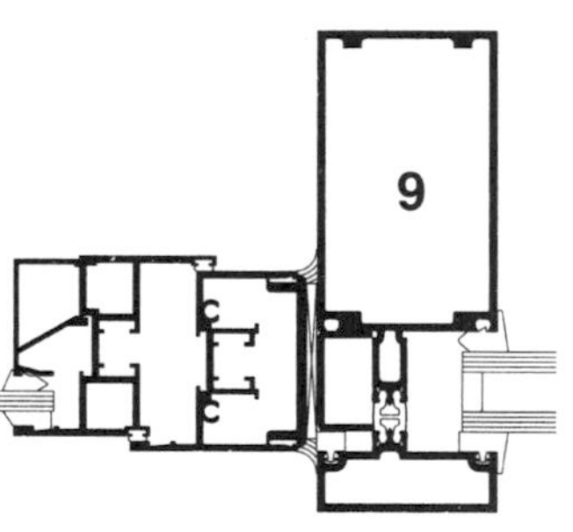

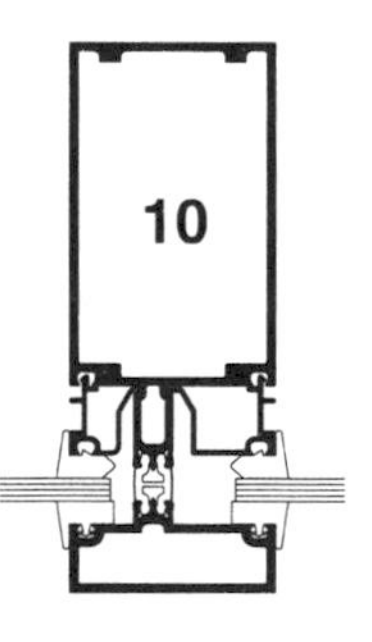

ALTERNATE JAMB DETAIL

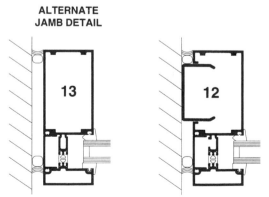

13

12

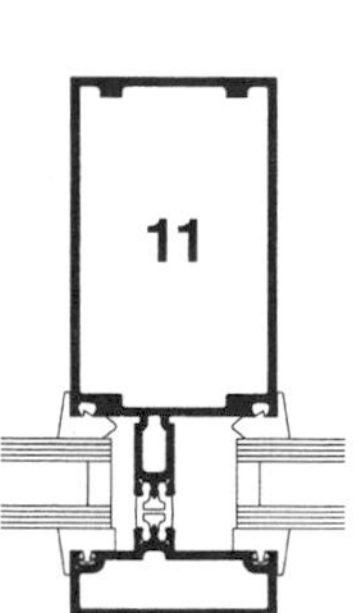

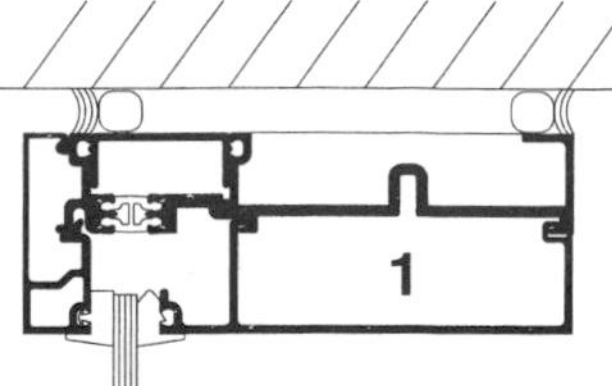

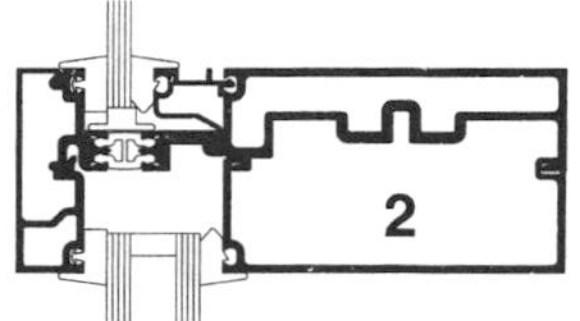

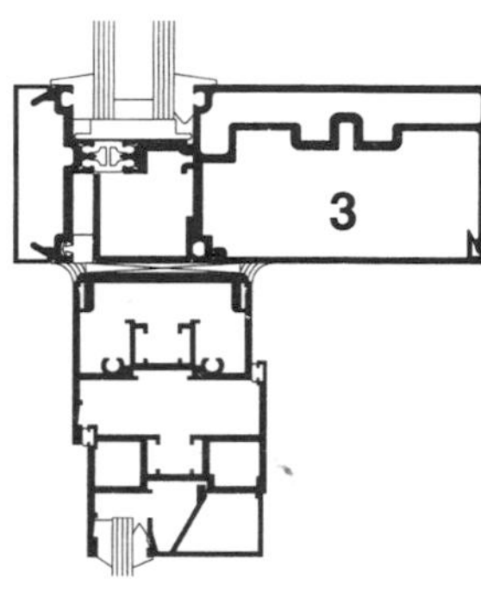

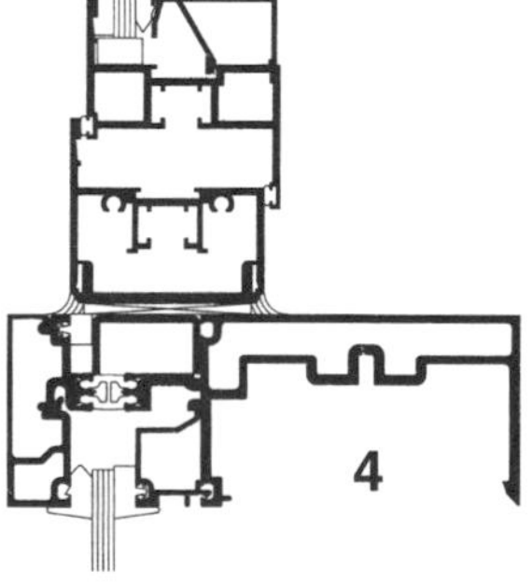

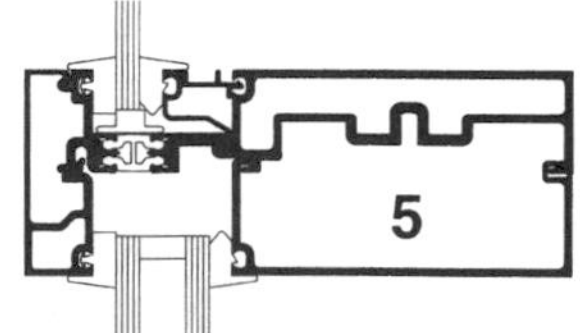

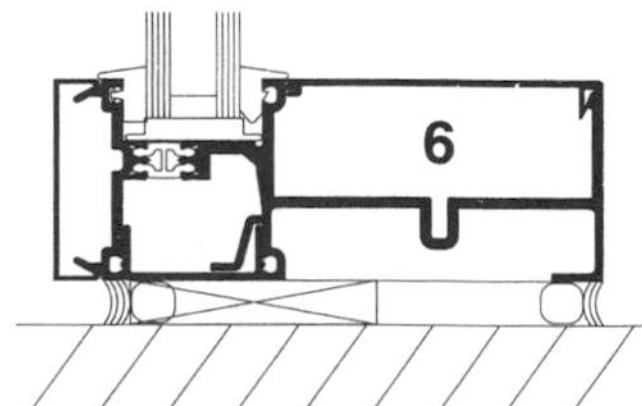

FIGURE 21.18
A complete set of details for the Kawneer 1600 System4, reproduced in the manufacturer's detail book at one-quarter of the full size.

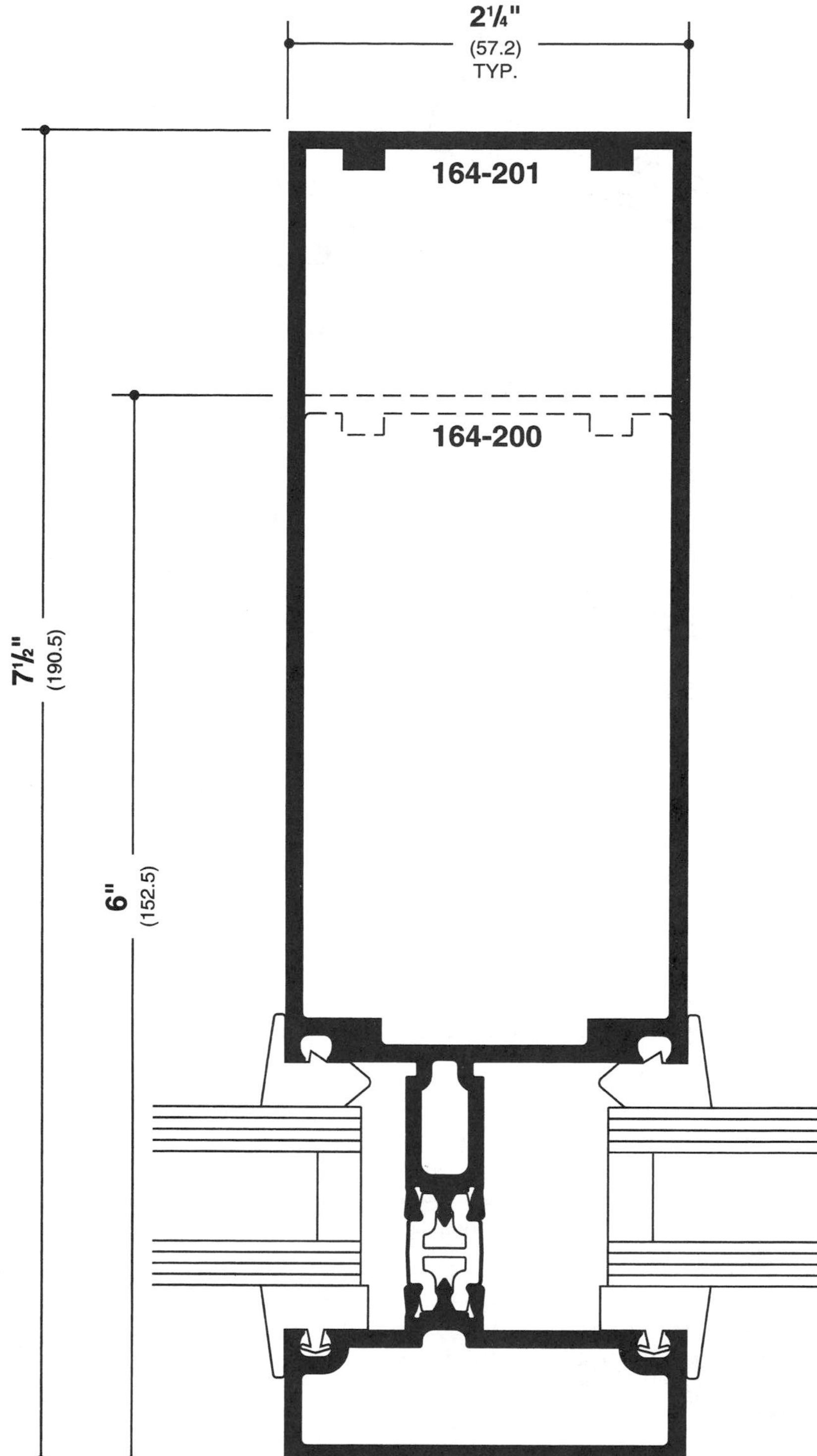

FIGURE 21.19
The vertical mullion for the Kawneer 1600 System4 is rigid, with no removable components except the glazing gaskets. The thermal break is an H-shaped extrusion of rigid plastic that is crimped securely into small grooves in the inner and outer portions of the mullion. Notice that the glazing pockets are of two different depths. This facilitates inside glazing, as explained in the next figure.

Vision

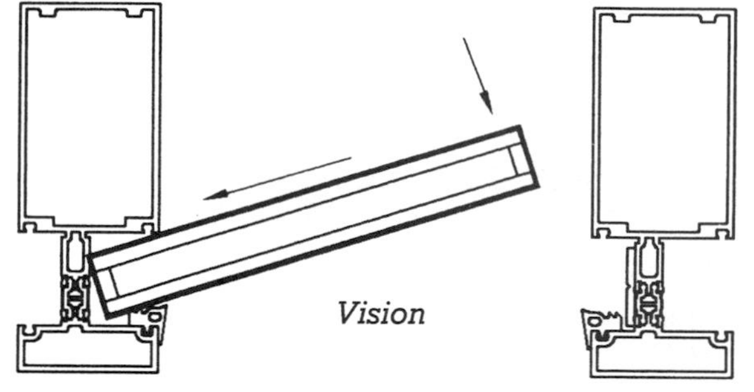

FIGURE 21.20
To glaze the Kawneer 1600 System4 from inside the building, rubber setting blocks are applied inside the shallow glazing pocket on the vertical mullion to cushion the edge of the glass. Rubber setting blocks are also installed in the horizontal mullion, and all the exterior-side glazing gaskets are inserted. Then the glass unit is pushed obliquely into the deep glazing pocket, squared away, and backed into the shallow pocket.

Anchors at spandrels

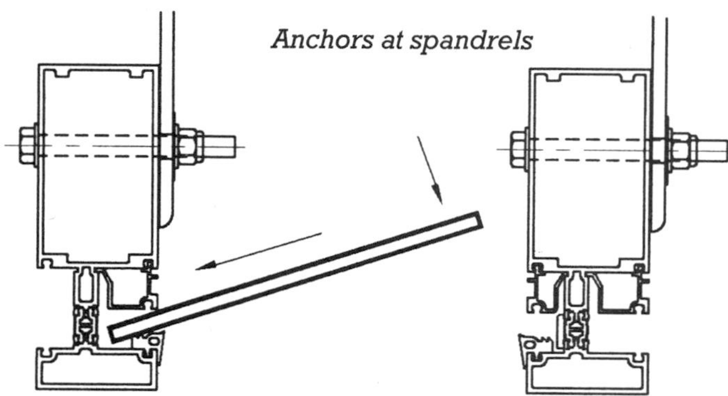

FIGURE 21.21
In order to make this mode of glass installation possible, the bottom half of the interior of each horizontal mullion is left off until the glass has been pushed into place. Then the bottom-half extrusion is added, and the interior glazing gaskets are installed. The interior gaskets are called "glazing wedges" because of their blunt wedge shape; this allows them to be installed simply by pressing them into place, wedging them between the glass and the mullion, after the glass has been installed. Notice how this shape is different from that of the exterior gaskets, which are installed before the glass.

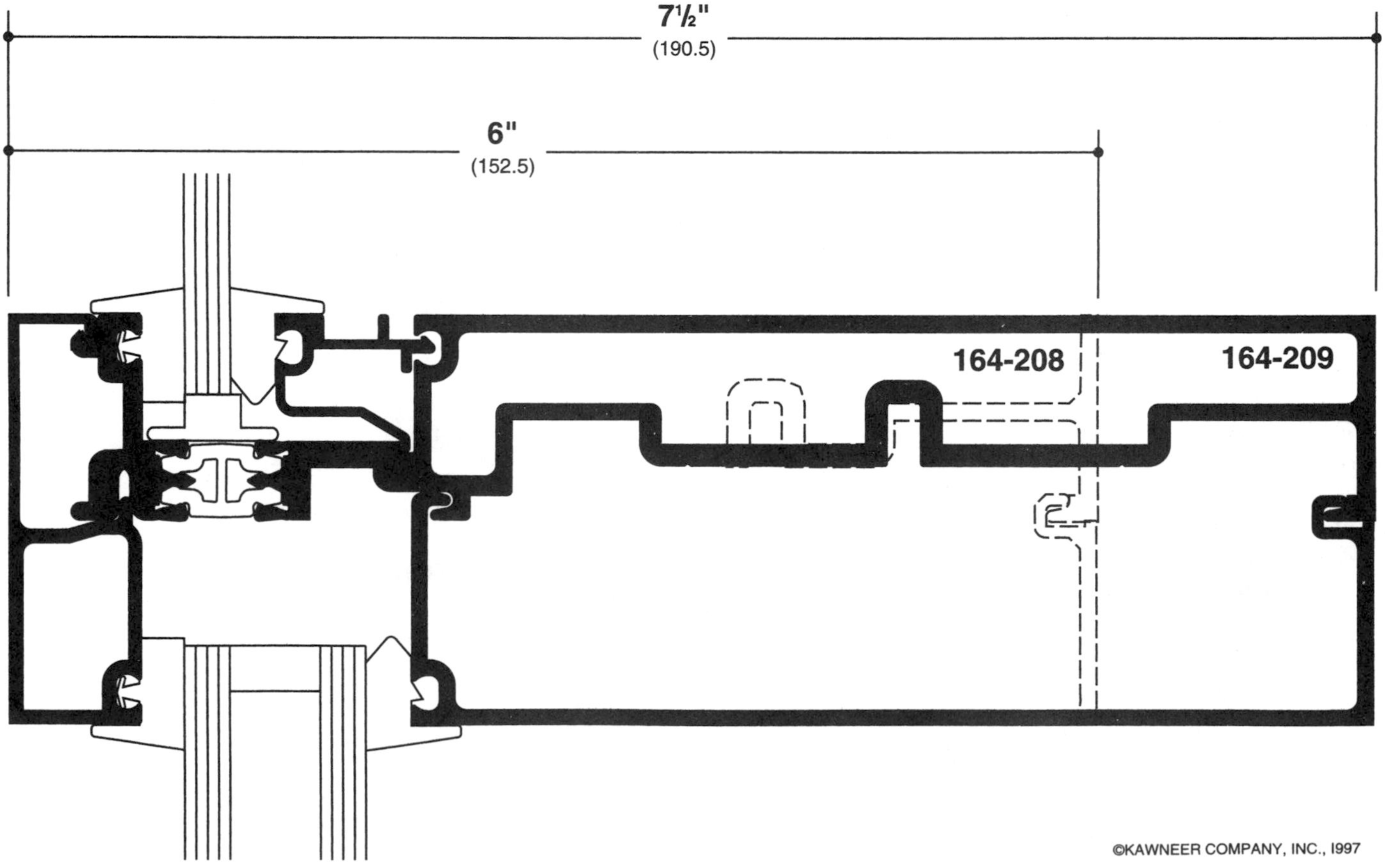

7½"
(190.5)
6"
(152.5)
164-208
164-209

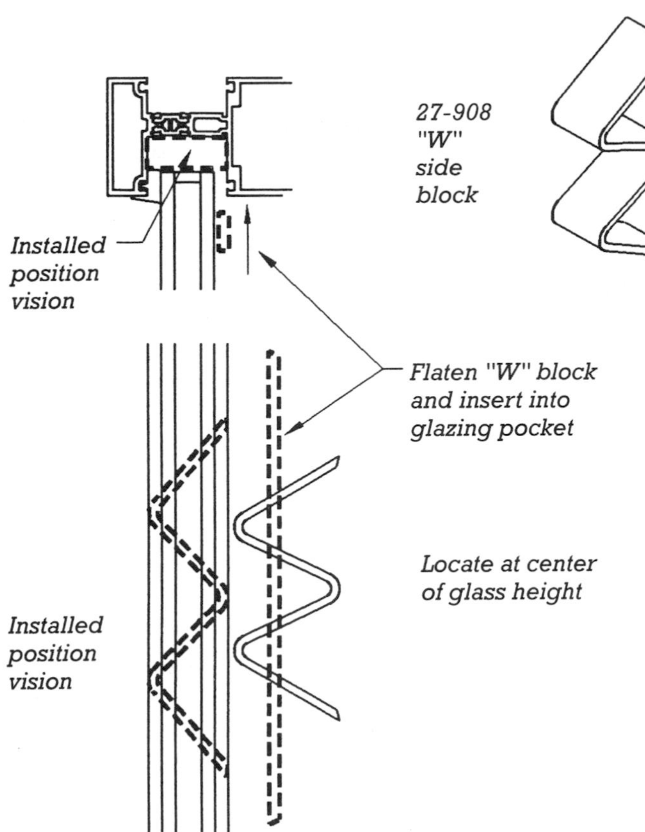

FIGURE 21.22
Before the glazing wedges are installed in the deep-pocket side of the vertical mullion, a W-shaped rubber block is flattened and slipped between the glass and the interior side of the mullion until it passes into the empty pocket. There it springs back into shape and prevents the glass from moving too deeply into the pocket.

FIGURE 21.23
This Las Vegas casino hotel features the Kawneer 1600 System4, whose details are shown in Figures 21.17–21.22. *(Photo courtesy of Kawneer Company, Inc.)*

Expansion Joints in Metal-and-Glass Walls

Aluminum has a relatively high coefficient of thermal expansion. The coefficient for glass is less than half as much. Because the cladding of a building is exposed to air temperature fluctuations as well as direct heating by the sun, it must be provided with expansion joints to allow thermal movement to occur without damaging the cladding or the frame of the building.

The differences in thermal movement between the glass and the aluminum are generally accommodated by very small sliding and flexing motions that occur between the glass and the gaskets in which it is mounted. Rubber blocks placed between the edge of the glass and the mullion on either side of each light prevent the glass from "walking" too far in either direction during repeated cycles of heating and cooling.

In the curtain wall systems illustrated in this chapter, vertical thermal movement in the aluminum is absorbed by telescoping joints that are provided at regular intervals in the vertical mullions (Figure 21.14). Horizontal thermal movement is accommodated by intentionally cutting horizontal aluminum components slightly short by a calculated fraction of an inch at each vertical mullion. Because the horizontal mullions are interrupted at each vertical mullion, there are many of these joints to work together in absorbing horizontal expansion and contraction.

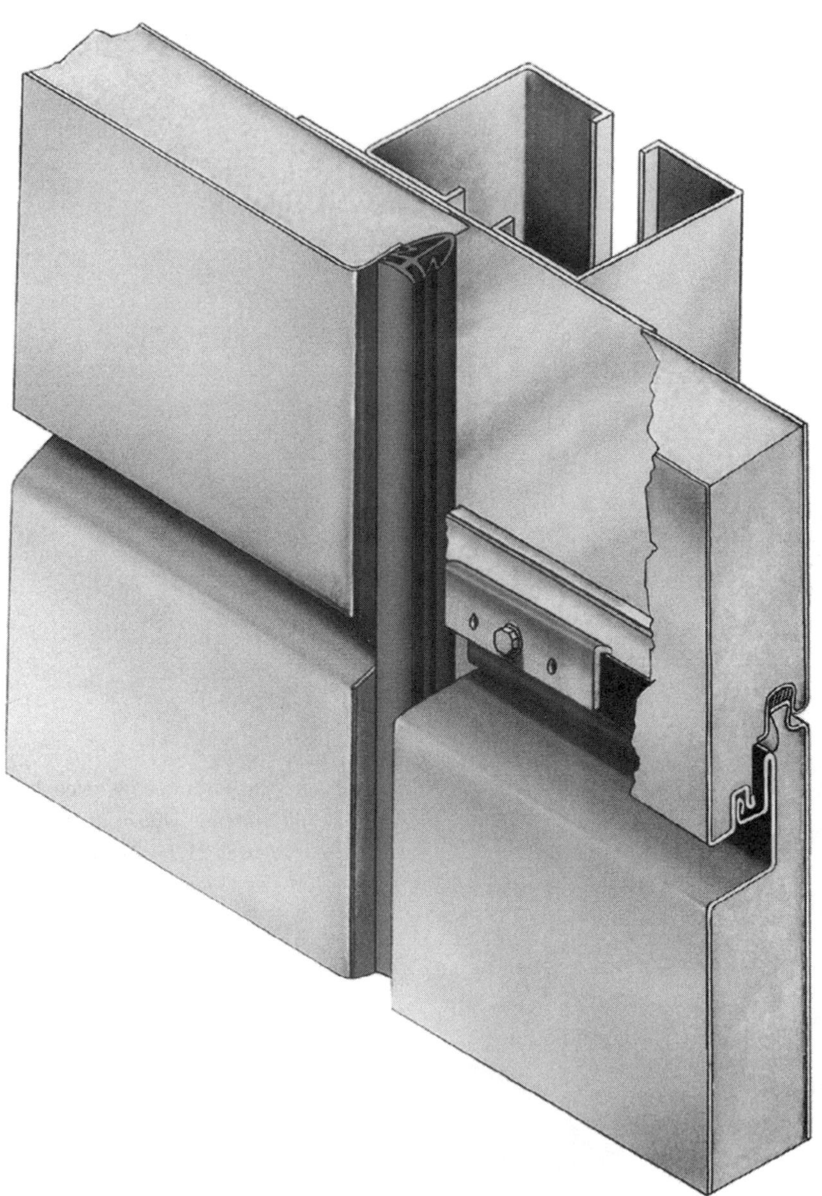

Figure 21.24
The CENTRIA Form-A-Wall® aluminum panel system for curtain walls is supported by vertical framing members behind the vertical joints between panels. Each panel is fastened to the frame with a screw at each top corner of the panel, seen here in the middle of the picture. The bottom edge of each panel is secured by an interlocking horizontal joint with the panel below. *(Courtesy of CENTRIA)*

FIGURE 21.25
CENTRIA Form-A-Wall panels present a neat appearance that features the pattern of the open joints. Notice that special panels are used for corners; there is no seam at the corner itself. *(Courtesy of CENTRIA)*

SLOPED GLAZING

Many buildings feature glass roofs over such amenities as lobbies, restaurants, cafes, swimming pools, and garden courtyards. A glass roof presents particular problems with respect to potential water leakage, because it is impossible to neutralize the force of gravity on a surface that is not vertical. Furthermore, moisture that condenses on the interior surfaces of the glass is likely to accumulate and drip onto the occupants of the space beneath. Therefore, every *sloped glazing* system is designed by its manufacturer to include an internal drainage system. This system collects any water that results from leakage or condensation and drains it to the outdoors. The glass surface is sloped rather than flat because the slope enables gravity to assist in keeping water from ponding on the roof, in causing condensate to run to the lower edge of each light of glass before dripping off, and in moving water through the drainage channels to the weep holes through which it is conducted back to the outdoors.

A proprietary sloped glazing system is illustrated in Figures 21.26 and 21.27. This system is designed to adapt to a range of slopes from 15 to 60 degrees. Water leakage is discouraged by a well-designed system of glazing

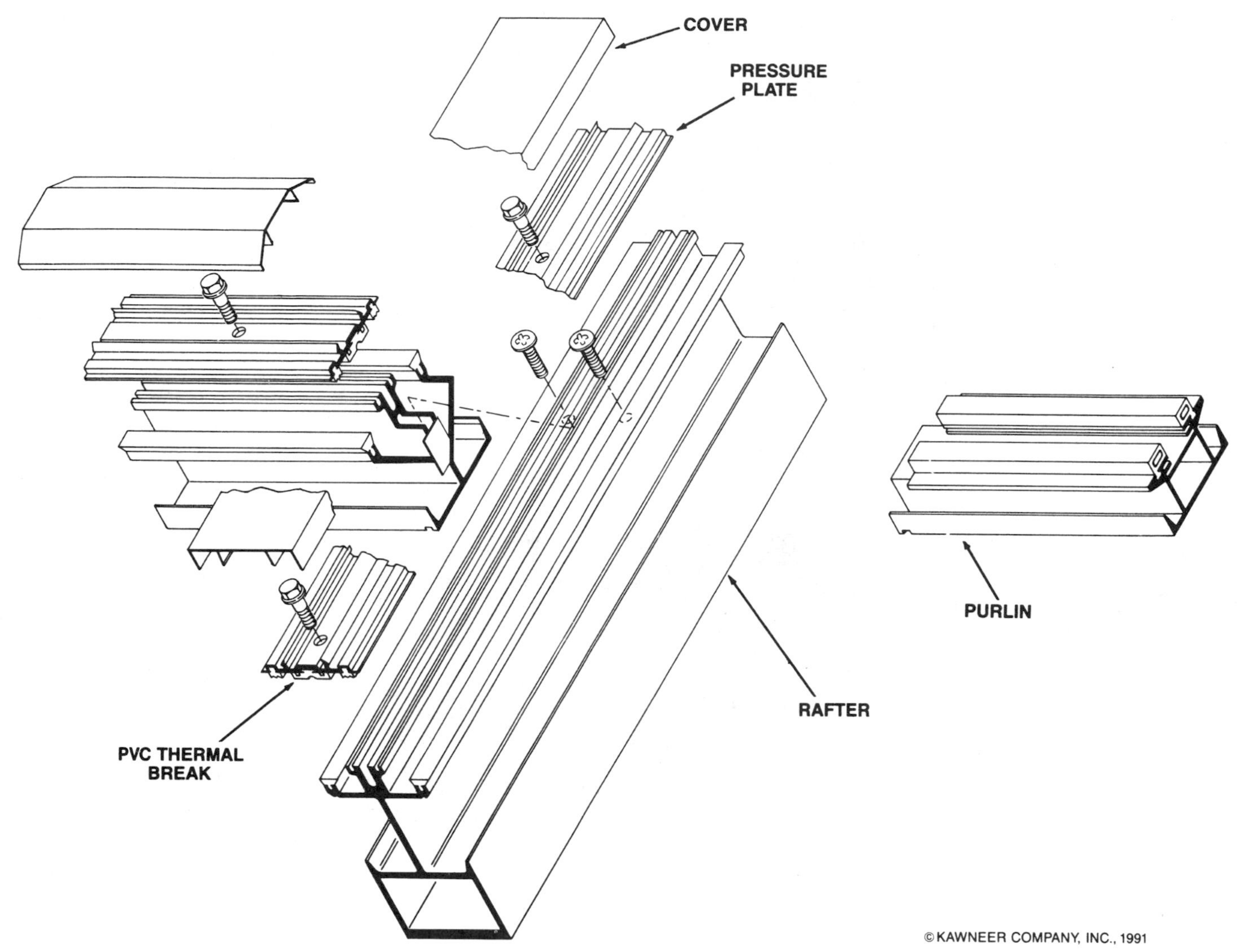

FIGURE 21.26
The Kawneer 1600 S.G.® sloped glazing system is framed with aluminum extrusions that incorporate internal gutters to drain away condensate and incidental leakage. *(Courtesy of Kawneer Company, Inc.)*

gaskets, but if leakage should occur because of gasket deterioration or faulty installation, the internal drainage system will catch it along any purlin or rafter and drain it away. Moisture condensation is minimized by double glazing and thermal breaks, but any condensate that may form under extreme conditions is also caught and drained by the same system of channels in the aluminum members.

The International Building Code places restrictions, intended to prevent falling glass from injuring occupants below, on glass sloped more than 15 degrees from vertical. The only glazing material types permitted without any limitations are laminated glass and plastic. Other glass types are permitted in some circumstances, depending on the height of the glass above the floor, the size of the lights, the type of occupancy, and other factors. Alternatively, a metal screen may be installed below the glass to catch the shards if a light breaks.

Dual-Layered Glass Cladding

In a *double-skin façade,* the wall system consists of two distinct glass cladding systems separated by an airspace that is wide enough in most cases to allow service personnel to pass between them. Such systems, also variously labeled *dual-wall facades* or *double-skin walls,* have long been popular in European building construction and

Figure 21.27
This hotel in Orlando, Florida, includes massive areas of Kawneer 1600 S.G. sloped glazing. Reflective glass is used to minimize solar overheating of the interior space.
(Photo courtesy of Kawneer Company, Inc.)

have more recently begun to appear in North American buildings as well. Various configurations of glazing and treatment of the space between the two skins are possible. Most often, one of the two skins is double-glazed, while the other relies on single glass. The interstitial airspace may be ventilated to the interior of the building (when the outer skin is double-glazed) or to the exterior (when the inner skin is double-glazed). Natural ventilation of the building interior through operable openings in both skins is also possible. Ventilation of the interstitial space may be passive, that is, powered by natural convection, or active, accomplished with the aid of fans, and in some cases may be coupled to the building's HVAC system. Shading devices, such as louvered blinds or roller shades, that may be manually controlled or driven by automatic timers or solar tracking devices, are also frequently integrated into the interstitial space.

Double-skin facades can improve the performance of the building enclosure in a variety of ways: They can reduce unwanted solar heat gain, increase daylighting potential, minimize thermal conduction and radiation through the wall system, provide a space for shading elements that is protected from the weather, permit natural ventilation designs in tall buildings, and create a quieter building interior than is possible with conventional glazing systems.

Against the potential benefits of a dual-skin facade system must be weighed its disadvantages: The airspace may subtract as much as 5 to 10 percent of the usable floor area of the building. Though the second skin of the facade may be constructed of less expensive framing and glass than the first skin, this still constitutes a significant increase in construction cost over a traditional curtain wall. And during the life of the building, twice as much glass surface must be washed on a regular basis.

Given these trade-offs, dual-skin facade systems are most appropriate for glass walls that must meet very stringent energy conservation targets, where the building owner's financial planning allows the higher cost of construction to be amortized over a relatively long time period in which it can be offset by increased energy savings, and where the creation of a high-performance building is an implicit goal of the building program.

Curtain Wall Design and Construction: The Process

Metal curtain wall design is not undertaken by an architect alone; it is far too complex and specialized a process. For many buildings, the architect simply adopts a proprietary system that meets the design requirements. This puts the primary responsibility on the manufacturer and installer of the system to ensure that the proper components are chosen and the wall is correctly installed.

When an architect sets out to design a new system of metal cladding, as is often done for large, important buildings, other professionals are brought into the process. An independent cladding consultant can bring a great deal of experience and expertise to this effort and minimize the risk to the architect. The structural engineer of the building is involved at least to the extent of understanding the weight and attachment requirements of the system. A curtain wall manufacturer is also brought into the design team early in the process.

The architect and building owner may elect either of two methods for selecting a manufacturer: One curtain wall manufacturer may be chosen on the basis of reputation and past experience. Alternatively, the architect and cladding consultant may prepare a rough design and performance specifications and submit these to several manufacturers for proposals. Each manufacturer will then submit a more detailed design and a financial proposal, and one will be selected on the basis of these proposals to proceed with the project.

The manufacturer understands better than any other member of the team the manufacturing, assembly, installation, and cost implications of a new curtain wall design. The manufacturer often does the installation as well as the manufacturing of the components, or it subcontracts the installation work to companies that it certifies to be well qualified and familiar with the manufacturer's products and standards. This installation experience is also invaluable during the design process.

From conceptual drawings prepared by the architect and the cladding consultant, the manufacturer prepares a more detailed set of design drawings as a basis for reaching preliminary agreement on the design. The manufacturer then prepares a very detailed set of shop drawings and installation drawings. These are checked carefully by the architect, engineer, and cladding consultant to ensure compliance with the design intentions and the structural capabilities of the building frame. Full-scale testing of the curtain wall, as described in Chapter 19, is usually carried out before the manufacture of a custom-designed curtain wall system is authorized to begin. The curtain wall manufacturer may also visit the construction site during the erection of the building frame to become familiar with the level of dimensional accuracy of the structural surfaces to which the curtain wall will be fastened.

CSI/CSC

MasterFormat Sections for Cladding with Metal and Glass

08 40 00	**ENTRANCES, STOREFRONTS, AND CURTAIN WALL**
08 41 00	**Entrances and Storefronts** **Aluminum-Framed Entrances and Storefronts** **Bronze-Framed Storefronts** **Stainless-Steel-Framed Entrances and Storefronts** **Steel-Framed Entrances and Storefronts**
08 44 00	**Curtain Wall and Glazed Assemblies** **Glazed Aluminum Curtain Walls** **Glazed Bronze Curtain Walls** **Glazed Stainless-Steel Curtain Walls** **Structural Glass Curtain Walls** **Sloped Glazing Assemblies**

Selected References

1. American Architectural Manufacturers Association. *Aluminum Curtain Wall Design Guide Manual.* Schaumburg, IL, 2005.

This publication covers its topic in exemplary fashion with clear text and beautifully prepared illustrations. The same organization also publishes a series of more specialized publications on various aspects of metal-and-glass curtain walls: rainscreen design, care and handling of aluminum components, design of wind loads and wind tunnel testing, fire safety, test methods, installation procedures, finishes, and so on.

Web Sites

Cladding with Metal and Glass

Author's supplementary web site: **www.ianosbackfill.com/21_cladding_with_metal_and_glass**

Aluminum Extruders Council: **www.aec.org**

American Architectural Manufacturers Association: **www.aamanet.org**

Centria curtain wall: **www.centria.com/CAS**

EFCO curtain wall: **www.efcocorp.com**

Formacore Curtain Walls: **www.portafab.com/curtain_walls.shtml**

Kawneer curtain wall and windows: **www.kawneer.com**

Whole Building Design Guide, Curtain Walls: **www.wbdg.org/design/env_fenestration_cw.php**

Key Terms and Concepts

extrusion
billet
die
screw slot
screw port
thermal break
cast and debridged thermal break
debridging
anodizing
fluoropolymer coating
polyvinylidene fluoride (PVDF)
powder coating
baked enamel coating
entrance
storefront
curtain wall
stick system
unit system
unit-and-mullion system
panel system
column-cover-and-spandrel system
outside glazed
inside glazed
deterrent seal
sloped glazing
double-skin facade, dual-wall facade, double-skin wall

Review Questions

1. For what reasons is aluminum the metal most frequently used in metal-and-glass cladding systems?

2. What are the relative advantages and disadvantages of a stick system and a panel system?

3. What are some characteristic features of aluminum extrusions, and what is the function of each?

4. What are the most common finishes applied to aluminum curtain wall components? What are the advantages of each?

5. How would you choose between an inside glazed system and an outside glazed one?

6. Explain the differences between aluminum framed entrances, storefronts, and curtain wall. What is an appropriate application for each?

7. In what way is a slope glazing system more difficult to make watertight than a wall cladding system? How is this reflected in the details of a slope glazing system?

Exercises

1. Make photocopies of full-scale details of the vertical and horizontal mullions of a metal curtain wall system from a manufacturer's catalog. Paste the details on a larger sheet of paper and add notes and arrows to explain every aspect of them—the features of the extrusions, the gaskets and sealants, the glazing materials and methods, drainage, insulation, thermal breaks, rainscreen features, and so on. From your examination of the details, list the order in which the components would be assembled. Is the system glazed from inside the building or outside? How can you tell?

2. Design a coffee table that is made up of aluminum extrusions and a glass top. Design and draw complete details of the extrusions and connections. Select a surface finish for the aluminum. What type of glass will you use?

22

Selecting Interior Finishes

Workers complete installation of an elaborate ceiling made of gypsum board. The corners of the steplike construction have been reinforced with metal corner bead, and the joints and nail heads have been filled and sanded, ready for painting. *(Courtesy of United States Gypsum Company)*

The installation of interior finish materials—ceilings, walls, partitions, floors, casework, finish carpentry—cannot proceed at full speed until the roof and exterior walls of a building are complete and mechanical and electrical services have been installed. The roof and walls are needed to shelter the moisture-sensitive finish materials. The mechanical and electrical services generally must be covered by the interior finish materials, and thus must precede them. The finish materials themselves must be selected to meet a bewildering range of functional parameters: durability, acoustical performance, fire safety, relationship to mechanical and electrical services, changeability over time, and fire resistance. They must also look good, presenting a neat appearance and meeting the architectural goals of the building.

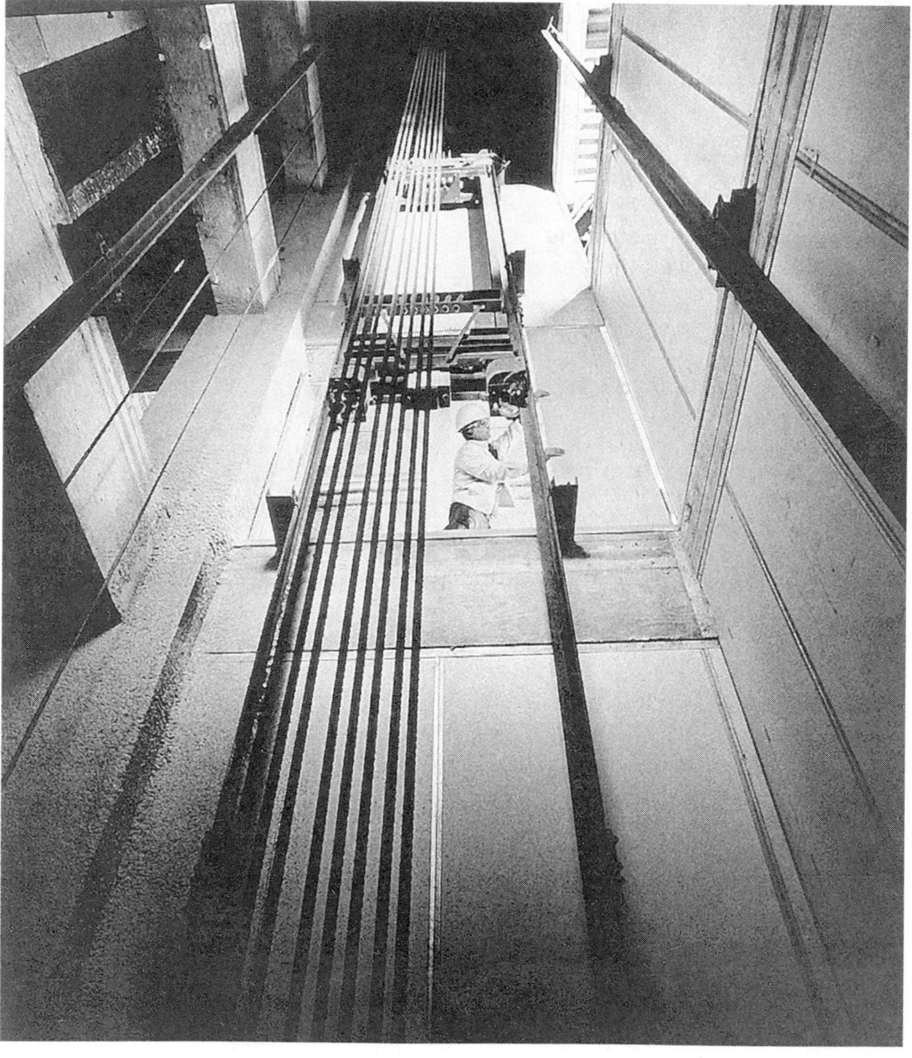

FIGURE 22.1
A worker constructs a fire-resistant wall around an elevator shaft, using gypsum panels and steel C–H studs. Chapter 23 contains more detailed information on shaft walls. *(Courtesy of United States Gypsum Company)*

INSTALLATION OF MECHANICAL AND ELECTRICAL SERVICES

When a building has been roofed and most of its exterior cladding has been installed, its interior is sufficiently protected from the weather that work can begin on the mechanical and electrical systems. The waste lines and water supply lines of the plumbing system are installed and, if specified, the pipes for an automatic sprinkler fire suppression system. The major part of the work for the heating, ventilating, and air conditioning system is carried out, including the installation of boilers, chillers, cooling towers, pumps, fans, piping, and ductwork. Electrical, communications, and control wiring are routed through the building. Elevators and escalators are installed in the structural openings provided for them.

The vertical runs of pipes, ducts, wires, and elevators through a multistory building are made through vertical *shafts* whose sizes and locations were determined at the time the building was designed. Before the building is finished, each shaft will be enclosed with fire-resistive walls to prevent the vertical spread of fire (Figure 22.1).

FIGURE 22.2
Three diagrammatic plans for an actual three-story suburban office building show the principal arrangements for plumbing, communications, electricity, heating, and cooling. Heating and cooling are accomplished by means of air ducted downward through two shafts from equipment mounted on the roof. The conditioned air from the vertical ducts is distributed around each floor by a system of horizontal ducts that run above a suspended ceiling, as shown on the plan of the intermediate floor. A row of doubled columns divides the building into two independent structures at the building separation joint to allow for differential foundation settlement and thermal expansion and contraction. *(Courtesy of ADD Incorporated, Architects)*

Air conditioning equipment
Elevator Overtravel
Skylights
Building separation joint
Air conditioning equipment
Mechanical penthouse
Elevator Overtravel
Plumbing vents
Air conditioning equipment

ROOF

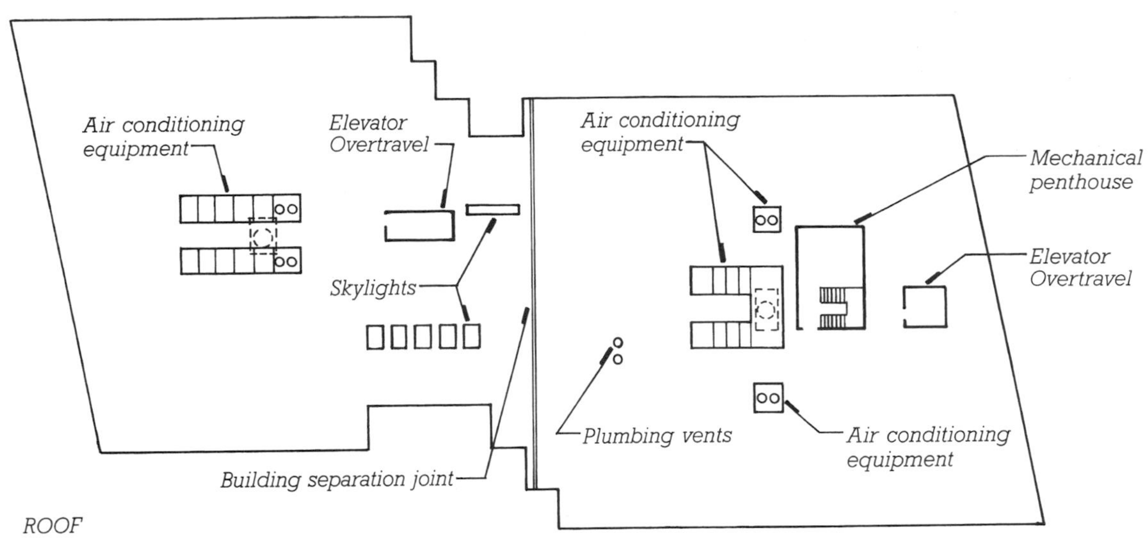

Air conditioning shaft
Elevators
Air conditioning shaft
Electric room with electrical riser
Electric room with electrical riser
Toilets
Freight elevator
Plumbing space
Typical branch ducts at ceiling
Main air conditioning supply duct at ceiling

INTERMEDIATE FLOOR

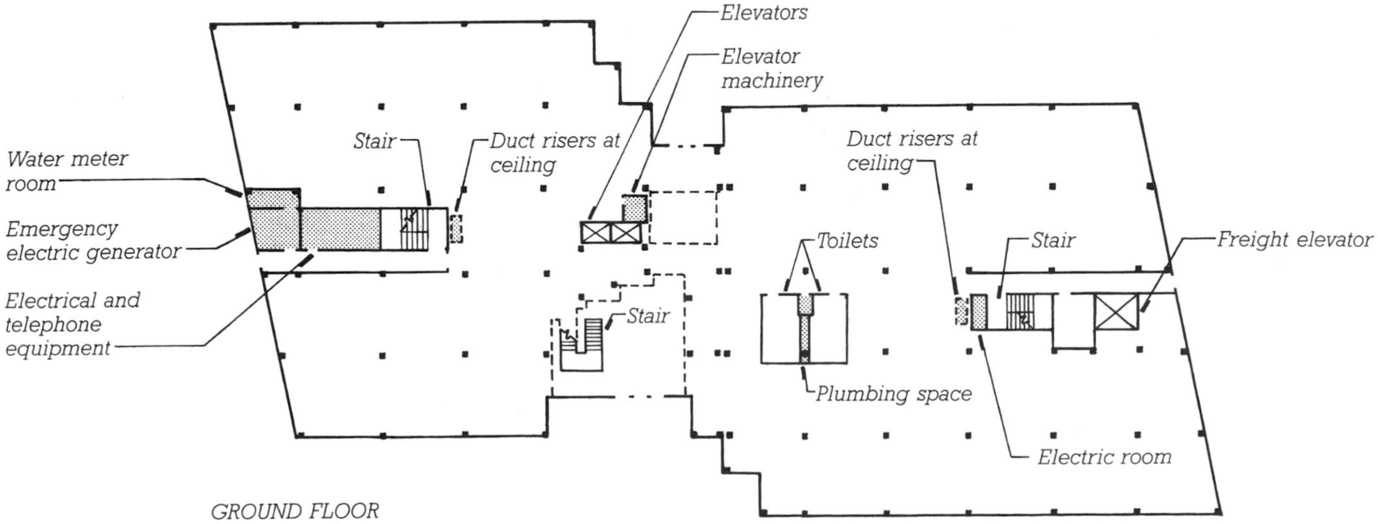

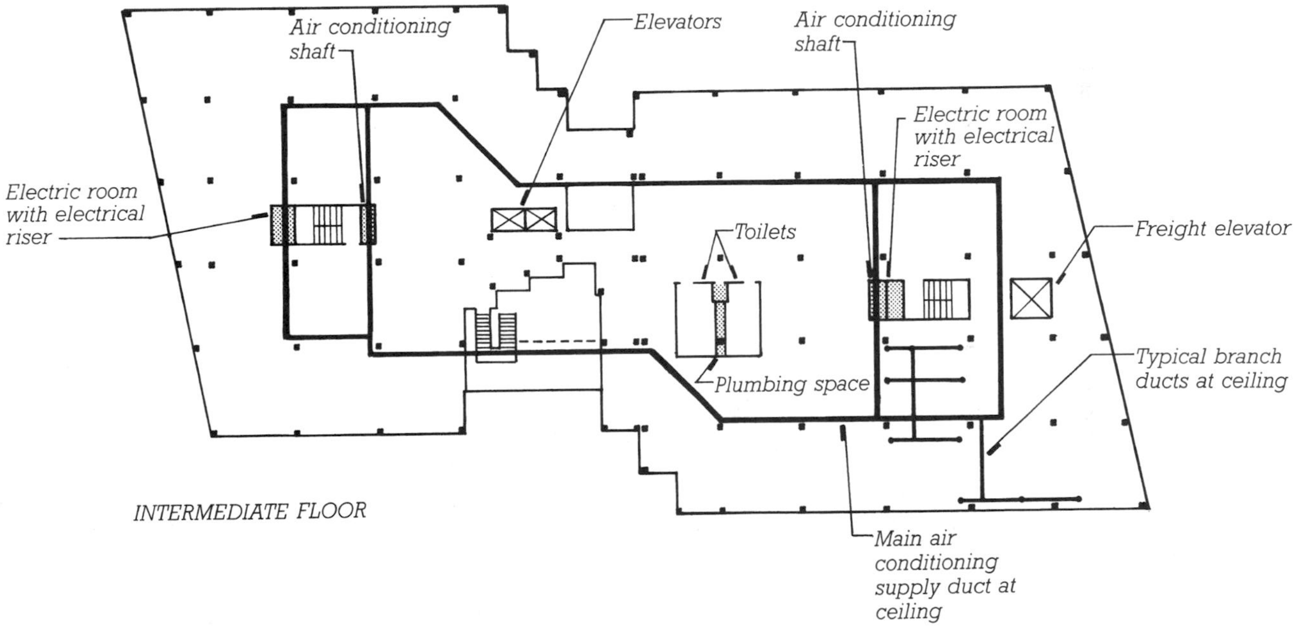

Elevators
Elevator machinery
Water meter room
Stair
Duct risers at ceiling
Duct risers at ceiling
Emergency electric generator
Toilets
Stair
Freight elevator
Electrical and telephone equipment
Stair
Plumbing space
Electric room

GROUND FLOOR

Horizontal runs of pipes, ducts, and wires are usually located just below each floor slab, above the ceiling of the floor below, to keep them up out of the way. These may be left exposed in the finished building or, as is more common, hidden above suspended ceilings. Sometimes these services, especially wiring, are concealed within a hollow floor structure such as cellular metal decking or cellular raceways. Sometimes services are run between the structural floor deck and a raised access flooring system. (For a more complete explanation of suspended ceilings, cellular floors, and access flooring, see Chapter 24.) To house the pipes where several plumbing fixtures are lined up along a wall, a plumbing space is created by constructing a double wall with space between.

Specific floor areas are reserved for mechanical and electrical functions in larger buildings (Figure 22.2). Distribution equipment for electrical and communications wiring and fiberoptic networks is housed in special rooms or closets. Fan rooms are often provided on each floor for air handling machinery. In a large multistory building, space is set aside, usually at a basement or subbasement level, for pumps, boilers, chillers, electrical transformers, and other heavy equipment. At the roof are penthouses for elevator machinery and such components of the mechanical systems as cooling towers and ventilating fans. In very tall buildings, one or two entire intermediate floors may be set aside for mechanical equipment, and the building is zoned vertically into groups of floors that can be reached by ducts and pipes that run up and down from each of the mechanical floors.

The Sequence of Interior Finishing Operations

Interior finishing operations follow a carefully ordered sequence that varies somewhat from one building to another, depending on the specific requirements of each project. The first finish items to be installed are usually hanger wires for suspended ceilings, and full-height partitions and enclosures, especially those around mechanical and electrical shafts, elevator shafts, mechanical equipment rooms, and stairways. *Firestopping* is inserted around pipes, conduits, and ducts where they penetrate floors and fire-rated walls (Figure 22.3). The full-height partitions and enclosures, firestopping, joint covers (Figure 22.4), and safing around the perimeters of the floors constitute a very important system for keeping fire from spreading through the building.

After the major horizontal electrical conduits and air ducts have been installed, the grid for the suspended ceiling is attached to the hanger wires so that the lights and ventilating louvers can be mounted to the grid. Then, typically, the ceilings are finished, and framing for the partitions that do not penetrate the finish ceiling is installed. Electrical

(*a*)

(*b*)

(*c*)

Figure 22.3

Applying firestopping materials to floor penetrations. (*a*) Within a plumbing wall, a layer of safing insulation is cut to fit and inserted by hand into a large slab opening around a cast iron waste pipe. Then a mastic firestopping compound is applied over the safing to make the opening airtight (*b*). (*c*) Applying firestopping compound around an electrical conduit at the base of a partition. *(Courtesy of United States Gypsum Company)*

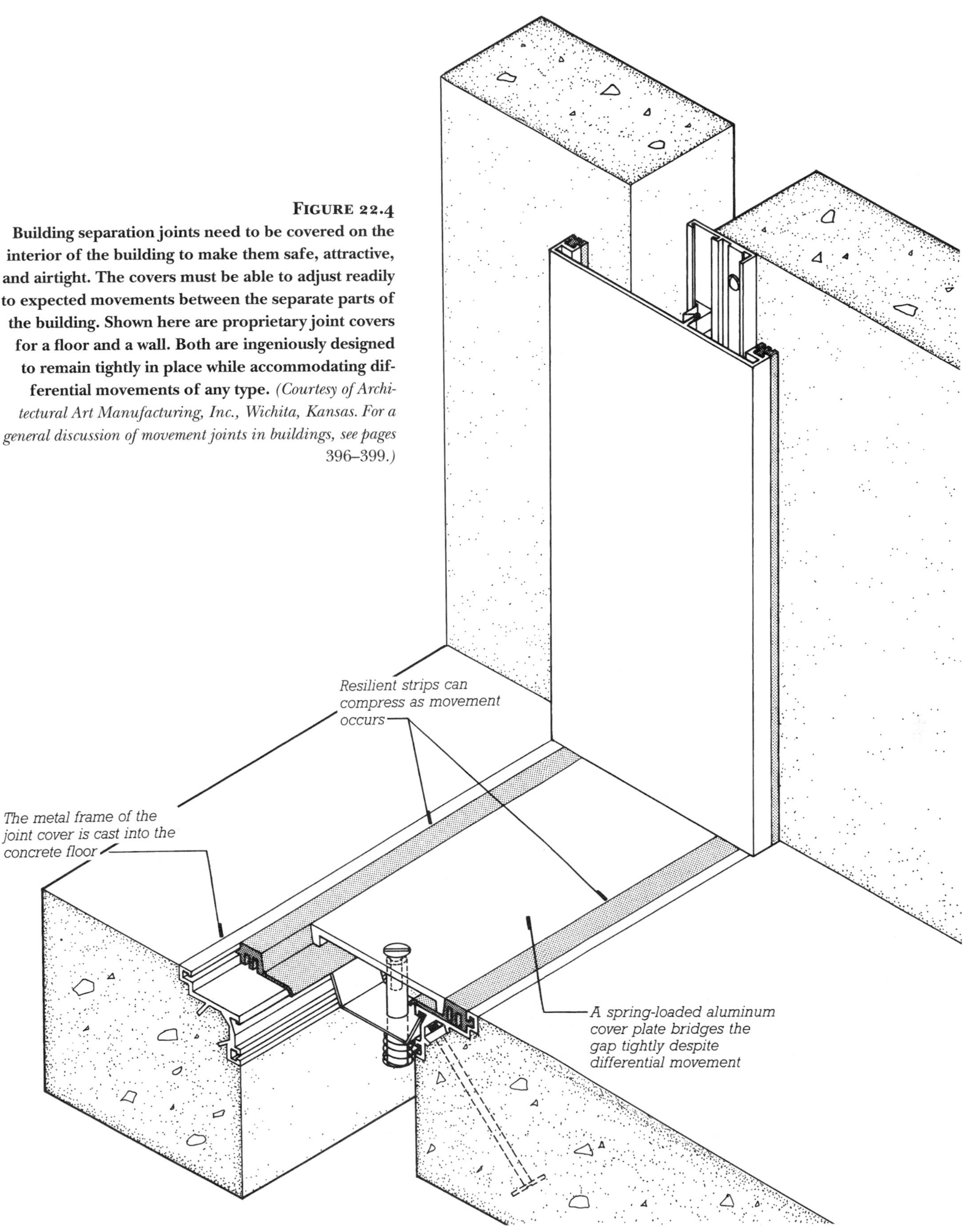
Resilient strips can compress as movement occurs
The metal frame of the joint cover is cast into the concrete floor
A spring-loaded aluminum cover plate bridges the gap tightly despite differential movement

FIGURE 22.4
Building separation joints need to be covered on the interior of the building to make them safe, attractive, and airtight. The covers must be able to adjust readily to expected movements between the separate parts of the building. Shown here are proprietary joint covers for a floor and a wall. Both are ingeniously designed to remain tightly in place while accommodating differential movements of any type. *(Courtesy of Architectural Art Manufacturing, Inc., Wichita, Kansas. For a general discussion of movement joints in buildings, see pages* 396–399.*)*

Considerations of Sustainability in Selecting Interior Finishes

In addition to sustainability considerations related to specific materials and systems discussed in the next two chapters, selection of interior finishes for sustainable buildings should be guided by the following general principals:

Materials

- Finish materials that have a high recycled content reduce the demand for virgin materials and make productive use of materials that would otherwise be treated as waste. The availability of finish materials with recycled content continues to increase in many finish product categories.
- Finish materials that can be recycled when they exceed their useful life also reduce waste. Some manufacturers, such as those of ceiling tile, carpet, and gypsum products, have established recycling or reclamation programs to divert such products from the waste stream. At present, however, recycling remains in its infancy, and most construction and demolition waste continues to end up in landfills.
- Interior finishes derived from rapidly renewable sources, such as bamboo flooring, or from certified woods reduce the depletion of raw materials of limited supply and protect forest ecosystems.
- Finish materials that are extracted, processed, and manufactured locally require less energy to transport, and their use helps to support local economies.

Indoor Air Quality

- Indoor finish materials and coatings present large surface areas to the interior environment of a building, making them potentially significant sources of emissions and indoor air quality problems. Potential emitters include glues and binders used in wood panels and other manufactured wood products, leveling compounds applied to subflooring, carpet fabrics and backings, carpet cushions, carpet adhesives, antimicrobial and mothproofing carpet treatments, wall covering adhesives, resilient flooring adhesives, vinyl in all its forms, gypsum board joint compounds, curtain and upholstery fabrics, paints, varnishes, stains, and more.
- Formaldehyde gas is an irritant to building occupants, causes nausea and headaches, and can exacerbate asthma. Potential sources include processed wood products, glues, adhesives, carpets, and permanent press fabrics.
- Volatile organic compounds are air pollutants, can act as irritants, and some are significant greenhouse gases. Common emitters include processed wood products, glues, adhesives, paints and other coatings, carpets, and plastic welding processes. The chemical 4-phenylcyclohexene, emitted by rubber binders used in some carpets and pads, is carcinogenic.
- Increasingly, manufacturers are publishing emissions data for their products, offering products with reduced emissions, and participating in rating systems meeting the low emission standards of LEED and other green building programs, making it easier for designers and specifiers to select green products. Examples of such programs include the Resilient Floor Covering Institute's *FloorScore* rating for resilient flooring, the Carpet and Rug Institute's *Green Label Plus* for carpet, and Greenguard Environmental Institute's *GreenGuard Indoor Air Quality Certification* for interior materials, finishes, and furnishings.
- Mold and mildew growth in carpets, wall coverings, gypsum board assemblies, and fabrics can cause acute respiratory distress in many people. Generally, this problem occurs only when these materials are repeatedly wetted by leakage or condensation. In response to this concern, many manufacturers now offer finish materials with improved resistance to moisture and mold growth.
- Construction dust, if not fully removed before the building is occupied, can be a source of irritating particulates after occupancy.

Design

- Floor plans that are flexible and easily adapted to new uses and partition systems that are easy to modify encourage building reuse.
- The strategic use of high ceilings, low partitions, transparency, reflective surfaces, and light colors can maximize daylighting potential and views to the exterior.
- Spaces designed with exposed structure and without suspended ceilings save materials.

and communications wiring is brought down from the conduits above the ceilings to serve outlets in the partitions. The walls are finished and painted. The last major finishing operation is the installation of the finish flooring materials. This is delayed as long as possible to let the other trades complete their work and get out of the building; otherwise, the floor materials could be damaged by dropped tools, spilled paint, heavy construction equipment, weld spatter, coffee stains, and construction debris ground underfoot.

Selecting Interior Finish Systems

Appearance

A major function of interior finish components is to make the interior of the building look neat and clean by covering the rougher and less organized portions of the framing, insulation, vapor retarder, electrical wiring, ductwork, and piping. Beyond this, the architect designs the finishes to carry out a particular concept of interior space, light, color, pattern, and texture. The form and height of the ceiling, changes in floor level, interpenetrations of space from one floor to another, and the configurations of the partitions are primary factors in determining the character of the interior space. Light originates from windows and electric lighting fixtures and is propagated by successive reflections off the interior surfaces of the building. Lighter-colored materials raise interior levels of illumination; darker colors and heavier textures result in a darker interior. Patterns and textures of interior finish materials are important in bringing the building down to a scale of interest that can be appreciated readily by the human eye and hand. No two buildings have the same requirements: Deep carpets and rich, polished marbles in muted tones may be chosen to give an air of affluence to a corporate lobby, brightly colored surfaces to create a happy atmosphere in a day care center, or slick plastic and highly reflective surfaces to provide a trendy ambience for the sale of designer clothing.

Durability and Maintenance

Expected levels of wear and tear must be considered carefully in selecting finishes for a building. Highly durable finishes generally cost more than shorter-lived ones and are not always required. In a courthouse, a transportation terminal, a recreation building, or a retail store, traffic is intense, and long-wearing materials are essential. In a private office or an apartment, more economical finishes are usually adequate. Water resistance is an important attribute of finish materials in kitchens, bathrooms, locker and shower rooms, entrance lobbies, and some industrial buildings. In hospitals, medical offices, kitchens, and laboratories, finish surfaces must not trap dirt and must be easily cleaned and disinfected. Maintenance procedures and costs should be considered in selecting finishes for any building: How often will each surface be cleaned, with what type of equipment, and how much will this procedure add to the cost of owning the building? How long will each surface last, and what will it cost to replace it?

Acoustic Criteria

Interior finish materials strongly affect noise levels, the quality of listening conditions, and levels of acoustic privacy inside a building. In noisy environments, interior surfaces that are highly absorptive of sound can decrease the noise intensity to a tolerable level. In lecture rooms, classrooms, meeting rooms, theaters, and concert halls, acoustically reflective and absorptive surfaces must be proportioned and placed so as to create optimum hearing conditions.

Between rooms, acoustic privacy is created by partitions that are both heavy and airtight. The acoustic isolation properties of lighter-weight partitions can be enhanced by partition details that damp the transmission of sound vibrations by means of resilient mountings on one of the partition surfaces and sound-absorbing batts of mineral wool in the interior cavity of the partition. Manufacturers test full-scale sample partitions of every type of material for their ability to reduce the passage of sound between rooms in a procedure outlined in ASTM E90. The results of this test are converted to *Sound Transmission Class* (*STC*) numbers that can be related to accepted standards of acoustic privacy. In an actual building, however, if the cracks around the edges of a partition are not completely sealed, or if a loosely fitted door or even an unsealed electric outlet is inserted into the partition, its airtightness is compromised and the published STC value is meaningless. Similarly, a partition with a high STC is worthless if the rooms on both sides are served by a common air duct that also acts incidentally as a conduit for sound, or if the partition reaches only to a lightweight, porous suspended ceiling that allows sound to pass over the top of the partition.

Transmission of impact noise from footsteps and machinery through floor–ceiling assemblies can be a major problem. Impact noise transmission is measured according to ASTM E492, in which a standard machine taps on a floor above while instruments in a chamber below record sound levels. The results are reported as *Impact Isolation Class* (*IIC*) ratings. Impact noise transmission can be reduced by floor details that rely on soft materials that do not transmit vibration readily, such as carpeting, soft underlayment boards, or resilient underlayment matting.

Fire Criteria

A building code devotes many pages to provisions that control the materials and details for interior finishes in buildings. These code requirements are aimed at several important characteristics of interior finishes with respect to fire.

Combustibility

The surface burning characteristics of interior wall and ceiling finish materials are tested in accordance with ASTM E84, also called the *Steiner Tunnel Test*. In this test, a sample of material 20 inches wide by 24 feet long (500 × 7300 mm) forms the ceiling of a rectangular furnace into which a controlled flame is introduced at one end. The time the flame takes to spread across the face of the material from one end of the furnace to the other is recorded, along with the density of smoke developed. The results of this test are given as a *flame-spread rating*, which indicates the rapidity with which fire can spread across a surface of a given material, and a *smoke-developed rating*, which classifies a material according to the amount of smoke it gives off when it burns.

Figure 22.5 defines allowable flame-spread and smoke-developed ratings for interior finish materials for various occupancy groups of buildings according to the International Building Code (IBC). It assigns each material to one of three classes: A, B, or C. Class A materials are those with flame-spread ratings between 0 and 25, Class B between 26 and 75, and Class C between 76 and 200. (The scale of flame-spread numbers is established arbitrarily by assigning a value of 0 to cement–asbestos board and 100 to a Red oak board.) For all three classes, the smoke-developed rating may not exceed 450. Materials with higher smoke-developed ratings are not permitted to be used inside buildings, because smoke, not heat or flame, is the primary killer in building fires. Interior trim materials, if their total surface area in a room does not exceed 10 percent of the total wall and ceiling area of the room, may be of Class A, B, or C in any type of building.

Some especially flammable wall and ceiling finish materials, such as textile or vinyl coverings, are also subject to testing according to NFPA 265, a test that measures their *room fire-growth contribution*, that is, their

TABLE 803.5
INTERIOR WALL AND CEILING FINISH REQUIREMENTS BY OCCUPANCY[k]

GROUP	SPRINKLERED[l]			NONSPRINKLERED		
	Exit enclosures and exit passageways[a,b]	Corridors	Rooms and enclosed spaces[c]	Exit enclosures and exit passageways[a,b]	Corridors	Rooms and enclosed spaces[c]
A-1 & A-2	B	B	C	A	A[d]	B[e]
A-3[f], A-4, A-5	B	B	C	A	A[d]	C
B, E, M, R-1, R-4	B	C	C	A	B	C
F	C	C	C	B	C	C
H	B	B	C[g]	A	A	B
I-1	B	C	C	A	B	B
I-2	B	B	B[h, i]	A	A	B
I-3	A	A[j]	C	A	A	B
I-4	B	B	B[h, i]	A	A	B
R-2	C	C	C	B	B	C
R-3	C	C	C	C	C	C
S	C	C	C	B	B	C
U	No restrictions			No restrictions		

For SI: 1 inch = 25.4 mm, 1 square foot = 0.0929 m^2.

a. Class C interior finish materials shall be permitted for wainscotting or paneling of not more than 1,000 square feet of applied surface area in the grade lobby where applied directly to a noncombustible base or over furring strips applied to a noncombustible base and fireblocked as required by Section 803.4.1.
b. In exit enclosures of buildings less than three stories in height of other than Group I-3, Class B interior finish for nonsprinklered buildings and Class C interior finish for sprinklered buildings shall be permitted.
c. Requirements for rooms and enclosed spaces shall be based upon spaces enclosed by partitions. Where a fire-resistance rating is required for structural elements, the enclosing partitions shall extend from the floor to the ceiling. Partitions that do not comply with this shall be considered enclosing spaces and the rooms or spaces on both sides shall be considered one. In determining the applicable requirements for rooms and enclosed spaces, the specific occupancy thereof shall be the governing factor regardless of the group classification of the building or structure.
d. Lobby areas in Group A-1, A-2 and A-3 occupancies shall not be less than Class B materials.
e. Class C interior finish materials shall be permitted in places of assembly with an occupant load of 300 persons or less.
f. For places of religious worship, wood used for ornamental purposes, trusses, paneling or chancel furnishing shall be permitted.
g. Class B material is required where the building exceeds two stories.
h. Class C interior finish materials shall be permitted in administrative spaces.
i. Class C interior finish materials shall be permitted in rooms with a capacity of four persons or less.
j. Class B materials shall be permitted as wainscotting extending not more than 48 inches above the finished floor in corridors.
k. Finish materials as provided for in other sections of this code.
l. Applies when the exit enclosures, exit passageways, corridors or rooms and enclosed spaces are protected by a sprinkler system installed in accordance with Section 903.3.1.1 or 903.3.1.2.

FIGURE 22.5
A table of surface-burning limits for interior finish materials, taken from the IBC. Distinctions between the three class ratings are explained in the accompanying text. *Portions of this publication reproduce tables from the* 2006 International Building Code, *International Code Council, Inc., Washington, D.C. Reproduced with Permission. All rights reserved.*

potential to add fuel to an incipient fire. The building code sets limits on flame spread, flashover, and smoke generated during this test. In some Occupancies, restrictions on combustible draperies, wall hangings, and other decorative materials also apply.

The combustibility of some flooring materials used in exits, corridors, and areas connected to these spaces must be tested according to NFPA 253 for *minimum critical radiant flux exposure*. The purpose of this test is to ensure that flooring materials in essential parts of the egress system cannot be easily ignited by the radiant heat of fire and hot gases in adjacent spaces. Materials must meet either Class I (most resistant to radiant heat) or Class II (moderately resistant) ratings, depending on the Occupancy Group of the spaces and whether or not the area is protected with an automatic sprinkler system. Some traditional flooring materials, including solid wood, resilient materials, and terrazzo, which have historically demonstrated satisfactory resistance to ignition during building fires, are not required to meet this test standard.

In other areas of the building, flooring materials are subject to the *pill test* (Consumer Product Safety Commission DOC FF-1), which evaluates the material's propensity for flame spread when exposed to a burning tablet intended to simulate a dropped lit cigarette, match, or similar hazard.

Fire Resistance

Fire resistance of a wall, ceiling, or floor assembly refers not to the assembly's own combustibility, but rather to its ability to resist the passage of fire from one side of the assembly to the other. The building code regulates the fire resistance of assemblies used to protect the structure of the building, to separate various parts of a building from one another, and to separate one building from another.

Figure 22.6 is a table from the International Building Code that specifies the required *fire resistance rating*, in hours of separation, between different Occupancies housed in the same building. (See the figure caption for comments regarding when these requirements apply.) Fire resistance rating requirements found elsewhere in the code for various types of separations, such as shaft walls, exit hallways, exit stairs, dwelling unit separations,

TABLE 508.3.3
REQUIRED SEPARATION OF OCCUPANCIES (HOURS)

	A[e], E		I		R[d]		F-2, S-2[c,d], U[d]		B[b], F-1, M[b], S-1		H-1		H-2		H-3, H-4, H-5	
OCCUPANCY	S	NS	S	NS	S	NS	S	NS	S	NS	S	NS	S	NS	S	NS
A[e], E[e]	N	N	1	2	1	2	N	1	1	2	NP	NP	3	4	2	3[a]
I	—	—	N	N	1	NP	1	2	1	2	NP	NP	3	NP	2	NP
R[d]	—	—	—	—	N	N	1	2	1	2	NP	NP	3	NP	2	NP
F-2, S-2[c,d], U[d]	—	—	—	—	—	—	N	N	1	2	NP	NP	3	4	2	3[a]
B[b], F-1, M[b], S-1	—	—	—	—	—	—	—	—	N	N	NP	NP	2	3	1	2[a]
H-1	—	—	—	—	—	—	—	—	—	—	N	NP	NP	NP	NP	NP
H-2	—	—	—	—	—	—	—	—	—	—	—	—	N	NP	1	NP
H-3, H-4, H-5	—	—	—	—	—	—	—	—	—	—	—	—	—	—	N	NP

For SI: 1 square foot = 0.0929 m^2.
S = Buildings equipped throughout with an automatic sprinkler system installed in accordance with Section 903.3.1.1.
NS = Buildings not equipped throughout with an automatic sprinkler system installed in accordance with Section 903.3.1.1.
N = No separation requirement.
NP = Not permitted.
a. For Group H-5 occupancies, see Section 903.2.4.2.
b. Occupancy separation need not be provided for storage areas within Groups B and M if the:
1. Area is less than 10 percent of the floor area;
2. Area is equipped with an automatic fire-extinguishing system and is less than 3,000 square feet; or
3. Area is less than 1,000 square feet.
c. Areas used only for private or pleasure vehicles shall be allowed to reduce separation by 1 hour.
d. See Section 406.1.4.
e. Commercial kitchens need not be separated from the restaurant seating areas that they serve.

FIGURE 22.6
IBC requirements for fire resistance ratings, in hours, for fire separation assemblies between Occupancy Groups. This code allows several alternative approaches to the design of buildings that contain more than one Occupancy Group. For larger buildings, parts of the building containing different occupancies must be separated by fire-resistance-rated walls and floor-ceilings as indicated in this table. For smaller buildings, an alternative approach allows occupancy areas to be nonseparated, that is, no fire-resistance-rated separations are required. Consult Chapter 5 of the IBC for more information. *Portions of this publication reproduce tables from the* 2006 International Building Code, *International Code Council, Inc., Washington, D.C. Reproduced with Permission.*

FIGURE 22.7
This table summarizes IBC fire resistance rating requirements for various types of assemblies not included in other code tables reproduced in this book. Fire areas, listed in the last row of the table, are portions of a building that are limited in area or occupant number for the purpose of determining sprinkler requirements.

Assembly Type	Required Fire Resistance Rating
Shaft enclosures	2 hours where connecting four or more stories; 1 hour where connecting fewer stories
Exit stairway and exit passageway enclosures	Same as shaft enclosures
Corridor and elevator lobby separations	1 hour in unsprinklered buildings; 0–1 hour in sprinklered buildings, depending on the number of occupants served and the Occupancy Group
Separations between dwelling or sleeping units in multiunit buildings	½–1 hour, depending on the Construction Type and the presence of automatic sprinklers
Mall tenant space separations	1 hour
Fire area separations	1–4 hours, depending on the Occupancy Group

and other nonbearing partitions, are summarized in Figure 22.7. Requirements for the protection of structure, for the fire resistance of exterior walls, and for fire walls that separate buildings are shown in Figures 1.3 and 1.7. Such requirements can be related to fire resistance information provided in manufacturers' literature similar to the examples shown in Figures 1.4–1.6.

Fire resistance ratings are determined by full-scale fire endurance tests conducted in accordance with ASTM E119, which applies not only to partitions and walls, but also to beams, girders, columns, and floor–ceiling assemblies. In this test, the assembly is constructed in a large laboratory furnace and subjected to the structural load (if any) for which it is designed. The furnace is then heated according to a standard time–temperature curve, reaching 1700 degrees Fahrenheit (925°C) at 1 hour and 2000 degrees Fahrenheit (1093°C) after 4 hours. To achieve a given fire resistance rating in hours, an assembly must safely carry its design structural load for the designated period, must not develop any openings that permit the passage of flame or hot gases, and must insulate sufficiently against the heat of the fire to maintain surface temperatures on the side away from the fire within specified maximum levels. Wall and partition assemblies must also pass a test, called the *hose stream test,* intended to assess their durability while exposed to fire conditions. A duplicate sample of the assembly is subjected to half of its rated fire exposure, then sprayed with water from a calibrated fire nozzle for a specified period at a specified pressure. To pass this test, the assembly must not allow passage of the water stream.

Openings in floors, ceilings, and partitions with required fire resistance ratings are restricted in size by most codes and must be protected against the passage of fire in various ways. Doors must be rated for fire resistance in accordance with a table such as that shown in Figure 18.27. Ducts that pass through rated assemblies must be equipped with sheet metal dampers (*fire dampers*) that close automatically if hot gases from a fire enter the duct. Penetrations for pipes and conduits must be closed tightly with fire-resistive material.

As an example of the use of these tables, consider a multistory vocational high school building of Type IIA construction that includes both a number of classrooms and a woodworking shop and that is fully sprinklered. For purposes of the table reproduced in Figure 22.6, the International Building Code places classrooms in Occupancy Group E (Educational) and the shop into Occupancy Group F-1 (Industrial, Moderate Hazard). Assuming that the building is large enough that the requirements of this table apply, the two uses must be separated from each other by walls (called "fire barriers") and, if applicable, floor–ceiling assemblies, of 1-hour construction. Doors through such walls must be rated at ¾ of an hour (Figure 18.27). Figure 22.5 indicates that, in the Occupancy E portion of the building, finish materials in exit stairway enclosures must have at least a Class B rating, while Class C finish materials are permitted throughout the remainder of the building. According to Figure 22.7, walls and floor–ceiling assemblies separating corridors from adjacent spaces must have a fire resistance rating between 0 and 1 hours, and exit stairways must be enclosed in construction rated between 1 and 2 hours (the final determination of these requirements depending also on other provisions of the code). Referring to Figure 1.3, we can see that the building's structural system must be protected with 1-hour rated assemblies or, as explained in the footnotes to this table and depending on other requirements of the code, it may be permissible to leave the structure unprotected due to the presence of a sprinkler system.

Relationship to Mechanical and Electrical Services

Interior finish materials join the mechanical and electrical services of a building at the points of delivery of the services—the electrical outlets, the lighting fixtures, the ventilating diffusers and grills, the convectors, the lavatories and water closets. Leading up to these points, the services may or may not be concealed by the finish materials. If the service lines are to be concealed, the finish systems must provide space for them, as well as for maintenance access points in the form of access doors, panels, hatches, cover plates, or ceiling or floor components that can be lifted out to expose the lines. If service lines are to be left exposed, the architect should organize them visually and specify a sufficiently high standard of workmanship in their installation so that their appearance will be satisfactory.

Changeability

How often are the use patterns of a building likely to change? In a concert hall, a chapel, or a hotel, major changes will be infrequent, so fixed, unchangeable interior partitions are appropriate. Appropriate finishes include many of the heavier, more expensive, more luxurious materials such as tile, marble, masonry, and plaster, which are considered highly desirable by many building owners. In a rental office building or a retail shopping mall, changes will be frequent; lighting and partitions should be easily and economically adjustable to new use patterns without serious delay or disruption. The likelihood of frequent change may lead the designer to select either relatively inexpensive, easily demolished construction such as gypsum wallboard partitions or relatively expensive but durable and reusable construction such as proprietary systems of modular, relocatable partitions. The functional and financial choices must be weighed for each building.

Cost

The cost of interior finish systems may be measured in two different ways. *First cost* is the installed cost. First cost is often of paramount importance when the construction budget is tight or the expected life of the building is short. *Life-cycle cost* is a cost figured by any of several formulas that take into account not only first cost, but also the expected lifetime of the finish system, maintenance costs and fuel costs (if any) over that lifetime, replacement cost, an assumed rate of economic inflation, and the time value of money. Life-cycle cost is important to building owners who expect to retain ownership for an extended period of time. Because of its higher maintenance and replacement costs, a material that is inexpensive to buy and install may be more costly over the lifetime of a building than a material that is initially more expensive.

Toxic Emissions from Interior Materials

A number of common construction materials give off substances that may be objectionable in interior environments. Many synthetics and wood panel products emit formaldehyde fumes for extended periods of time after the completion of construction. Solvents from paints, varnishes, and carpet adhesives often permeate the air of a new building. Airborne fibers of asbestos and glass can constitute health hazards. Some materials harbor molds and mildews whose airborne spores many people cannot tolerate. In isolated instances, stone and masonry products have proven to be sources of radon gas. Construction dust, even from chemically inert materials, can inflame respiratory passages. There is increasing pressure, both legal and societal, on building designers to select interior materials that do not create objectionable odors or endanger the health of building occupants. Compliance with these criteria is complicated by the fact that data on the toxicity of various indoor air pollutants are inconclusive. But data on emissions of pollutants from interior materials are becoming increasingly available to designers, and it is wise to select materials that give off the smallest possible quantities of irritating or unhealthful substances.

Trends in Interior Finish Systems

Interior finish systems have undergone a transformation over the past roughly 70 years. Formerly, the installation of finishes for a commercial office began with the construction of partitions of heavy clay tiles or gypsum blocks set in mortar. These were covered with two or three coats of plaster and joined to a three-coat plaster ceiling. The floor was commonly made of hardwood strips with a wood baseboard, or perhaps of poured terrazzo with an integral terrazzo base. Today, the same office might be framed in light metal studs and walled with gypsum board. The ceiling might be a separate assembly of lightweight, acoustically absorbent tiles, and the floor might be a thin layer of vinyl composition tile glued to a smooth concrete slab.

Several trends can be discerned in these changes. One trend is away from an integral, single-piece system of finishes toward a system made up of discrete components. In the old office, the walls, ceiling, and floor were all joined, and none could be changed without disrupting the others. In the new office, ceiling and floor finishes often extend uninterrupted from one side of the building to the other, so that partitions can be changed at will without affecting either the ceiling or the floor. The trend toward discrete components is epitomized by partitions made of modular, demountable, relocatable panels.

Another discernible trend is away from heavy finish materials to lighter ones. A partition of metal studs and gypsum board has a fraction of the weight of one of clay tiles and plaster, and a vinyl composition tile installation is many times lighter than a traditional terrazzo one of equal area. Lighter finishes dramatically reduce the dead load the structure of the building must carry. This enables the structure itself to be lighter and less expensive. Lighter finish materials reduce shipping, handling, and installation costs, and are easier to move or remove when changes are required.

"Wet" systems of interior finish, made of materials mixed with water on the building site, are steadily being replaced by "dry" ones, which are installed in rigid form. Plaster has been mostly replaced by gypsum board and ceiling tiles in most areas of new buildings, and tile and terrazzo floors by vinyl composition tiles and carpet. The installation of dry systems is fast and not heavily dependent on weather conditions. Dry systems require less skill on the part of the installer than wet systems by transferring the skilled work from the job site to the factory, where it is done by machines. All these differences tend to result in a lower installed cost.

More recently, growing interest in sustainable building has brought increased attention to the selection of finish materials whose use not only minimizes long-term impacts on the planet's resources but also contributes to more healthful living and working environments.

The traditional finishes, nevertheless, are far from obsolete. Gypsum board cannot rival three-coat plaster over metal lath for surface quality, durability, or design flexibility. Tile and terrazzo floorings are unsurpassed for wearing quality and appearance. In many situations, the life-cycle costs of traditional finishes compare favorably with those of lighter-weight finishes whose first cost is considerably less. And the aesthetic qualities of, for example, marble floors, wood wainscoting, and sculpted plaster ceilings, where such qualities are called for, cannot be imitated by any other material.

Selected References

1. International Code Council, Inc. *International Building Code.* Falls Church, VA, updated regularly.

The reader is referred to Chapters 7 and 8 of this model code, which deal with fire-resistive construction requirements for interior finishing systems.

2. Juracek, Judy A. *Surfaces: Visual Research for Artists, Architects, and Designers.* New York, W. W. Norton & Company, 1996.

This book is a wonderful catalog of the vast expressive potential of the architectural surface, containing over 1200 color photographs of finish surfaces of differing material types, patterns, textures, colors, and forms.

3. Allen, Edward, and Joseph Iano. *The Architect's Studio Companion* (4th ed.). Hoboken, NJ, John Wiley & Sons, Inc., 2006.

The fourth section of this book gives extensive information on providing space for mechanical and electrical equipment in buildings.

Web Sites

Selecting Interior Finishes

Author's supplementary web site: **www.ianosbackfill.com/22_selecting_interior_finishes**

Greenguard Environmental Institute: **www.greenguard.org**

Key Terms and Concepts

shaft
firestopping
Sound Transmission Class (STC)
Impact Isolation Class (IIC)
Steiner Tunnel Test
flame-spread rating
smoke-developed rating
room fire-growth contribution
minimum critical radiant flux exposure
pill test
fire resistance
fire resistance rating
hose stream test
fire damper
first cost
life-cycle cost

Review Questions

1. Draw a flow diagram of the approximate sequence in which finishing operations are carried out on a large building of Type IIA construction.

2. List the major considerations that an architect should keep in mind while selecting interior finish materials and systems.

3. What are the two major types of fire tests conducted on interior finish systems? What measures of performance are derived from each?

4. What is the difference between first cost and life-cycle cost?

Exercises

1. You are designing a 31-story apartment building in a large city. Assume that it will be fully sprinklered. What types of construction are you permitted to use under the International Building Code? What fire resistance rating will be required for the separation between the apartment floors and the retail stores on the ground floor, assuming the different occupancy areas must be separated? What classes of finish materials can you use in the exit stairway? In the corridors to those stairways? Within the individual apartments? If a Red oak board has a flame-spread rating of 100, can you panel an apartment in Red oak? What fire resistance ratings are required for partitions between apartments? Between an exit corridor and an apartment? What type of fire door is required between the apartment and the exit corridor? What is the required fire resistance rating for the walls around the elevator shafts? For the purposes of this exercise, when referring to Figure 22.7, assume the highest rating requirement where a range of requirements is provided.

23

Interior Walls and Partitions

A plasterer applies a scratch coat of gypsum plaster to expanded metal lath. The partition is framed with open-truss wire studs. Soft-annealed galvanized steel wire is used to make all the connections in this type of partition. *(Courtesy of United States Gypsum Company)*

There is more to interior walls and partitions than meets the eye. Behind their simple surfaces lie assemblies of materials carefully chosen and combined to meet specific performance requirements relating to structural strength, fire resistance, durability, and acoustical isolation. A partition may be framed with steel or wood studs and faced with plaster or gypsum board. Alternatively, masons may construct it of concrete blocks or structural clay tiles. For improved appearance or durability, a partition may be faced with ceramic tiles, masonry veneer, wood paneling, or any of a long list of other finish materials to tailor it to the requirements of a specific application.

Types of Interior Walls

Fire Walls

A *fire wall* is a wall that forms a required separation to restrict the spread of fire through a building and extends continuously from the foundation to or through the roof. A fire wall is used to divide a single building into smaller units, each of which may be considered as a separate building when calculating allowable heights and areas under the building code. A fire wall must either meet a noncombustible roof structure at the top or extend through and above the roof by a specified minimum distance, 30 inches (762 mm) in the case of the International Building Code (IBC). In this code, a fire wall must also extend horizontally at least 18 inches (457 mm) beyond the exterior walls of the building unless these exterior walls meet certain fire resistance and combustibility requirements. Except in buildings of Type V construction, a fire wall must be framed with noncombustible materials such as steel studs. A fire wall must also have sufficient structural stability during a building fire to allow collapse of the construction on either side without itself collapsing.

Openings in fire walls are restricted in size and aggregate area and must be closed with fire doors or fire-rated glass. The required fire resistance ratings for fire walls under the International Building Code are defined in Figure 1.7.

Shaft Walls

A *shaft wall* is used to enclose a multistory opening through a building, such as an elevator shaft or a shaft for ductwork, conduits, or pipes. In the International Building Code, a shaft wall connecting four or more floors must have a fire resistance rating of 2 hours or, if connecting fewer floors, a rating of 1 hour. Walls for elevator shafts must be able to withstand the air pressure and suction loads placed on them by the movements of the elevator cars within the shaft, and should be designed to prevent the noise of the elevator machinery from reaching other areas of the building.

Fire Barriers and Fire Partitions

Fire-rated walls are also used to restrict the spread of fire and smoke within a single building. Depending on the type of separation, the International Building Code requires such walls to be constructed as either *fire barriers* or *fire partitions*. Unlike fire walls, these wall types do not necessarily extend from foundation to roof. A fire barrier must extend vertically from the top of one floor slab to the underside of the next. Requirements for a fire partition are even less stringent. In some cases, a fire partition may terminate at the underside of a suspended ceiling. Fire barriers are used to protect exit stair enclosures, to separate different occupancies, and to limit the extent of *fire areas* (areas bounded by fire-resistant construction, the size and location within the building of which are related to automatic sprinkler requirements). Fire partitions are used to enclose corridors and to separate tenant spaces in mall buildings or dwelling units in hotels, dormitories, and other multidwelling unit buildings.

Openings in fire barriers and fire partitions are restricted in size and must be closed with fire doors or fire-rated glass. Required fire resistance ratings for the various types of fire barriers and partitions according to the International Building Code are listed in Figures 22.6 and 22.7.

The structural elements that support fire barriers and fire partitions must have a fire resistance rating at least as great as that of the wall being supported. For example, in a building of Type IIB construction, the structure is normally permitted to be unprotected. However, columns, bearing walls, and portions of the floor structure that support a 1-hour rated corridor wall must themselves also be protected with at least a 1-hour fire resistance rating.

Smoke Barriers and Smoke Partitions

In certain institutional occupancies such as hospitals and prisons, where occupants are unable to leave the building in case of fire, special partitions called *smoke barriers* are required. This type of wall divides floors of buildings in such a way that occupants may take refuge in case of fire by moving to the side of the smoke barrier that is away from the fire without having to exit the building. A smoke barrier is a 1-hour rated partition that is continuous from one side of the building to the other and from the top of a floor slab to the bottom of the slab on the floor above. It must be sealed at all edges. Penetrations for air ducts must be protected with dampers that close automatically if smoke is detected in the air. Other penetrations, such as those for pipes and conduits, must be sealed airtight. Doors through smoke barriers are necessary to allow movement of people in case of fire. They must be close fitting, without grilles or louvers, and must close automatically.

A *smoke partition* is a wall constructed—like a smoke barrier—to resist the passage of smoke, but without any fire resistance rating. For example, when walls for corridors and elevator lobbies need not be fire-rated, they are constructed as smoke partitions.

Other Nonbearing Partitions

Many of the partitions in a building neither bear a structural load nor are required as fire separation walls. These may be made of any material that meets the combustibility provisions of the building code for the selected type of construction, as explained in the next section.

Framed Partition Systems

Partition Framing

Partitions that will be finished in plaster or gypsum board are usually framed with wood or metal studs (Figures 23.1–23.3). Framing with wood studs is permitted by the building code only in buildings of certain combustible construction types, including Types III and V. Partitions in buildings of Type I or Type II (noncombustible) construction must be framed with metal studs. Partitions in Type IV, Heavy Timber construction, must either be framed with metal studs or constructed with wood members assembled into solid, laminated partitions as specified in the code. With certain limitations, fire-retardant-treated wood is also permitted for partition framing in Types I, II, and IV construction.

Metal partition framing is directly analogous to wood light framing, but constructed of *light gauge steel studs* and *runner channels* made of galvanized steel sheet metal 0.018 to 0.054 inch (0.45-1.37 mm) thick (Figure 12.2). Light gauge steel members and framing methods are detailed in Chapter 12. Because of its noncombustibility, metal partition framing is permitted in all building code construction types.

If plaster or gypsum board surfaces are to be applied over a masonry wall, they may be spaced away from the wall with either wood or metal *furring strips* (Figures 23.4–23.6). Furring allows for the installation of a flat wall finish over an irregular masonry surface, and provides a concealed space between the finish and the masonry for installing plumbing, wiring, and thermal insulation.

Fireblocking of Combustible Concealed Spaces

Building codes require that concealed hollow spaces within combustible partition or wall assemblies (that is, partitions framed with wood studs, but not those framed with light gauge steel) be internally partitioned into spaces sufficiently small in size to limit the ability of fire to travel undetected within such spaces. Materials used for this purpose are called *fireblocking* and may consist of solid lumber, plywood, OSB, particleboard, gypsum board, cement fiberboard, or even glass fiber insulation batts when securely installed. The International Building Code requires fireblocking in combustible concealed spaces at all ceiling and floor levels, as well as at horizontal intervals not exceeding 10 feet (3 m), such as may occur behind furred finish systems or within double stud walls constructed to accommodate plumbing piping or to improve acoustical separation.

Plaster

Plaster is a generic term that refers to any of a number of cementlike substances that are applied to a surface in paste form and then harden into a solid material. Plaster may be applied directly to a masonry surface or, more generally, to any of a group of plaster bases known collectively as *lath* (rhymes with "math"). Plastering began in prehistoric times with the smearing of mud over masonry walls or over a mesh of woven sticks and vines to create a construction known as *wattle and daub*, the wattle being the mesh and the daub the mud. The early Egyptians and Mesopotamians developed finer, more durable plasters based on gypsum and lime. Portland cement plasters evolved in the 19th century. It is from these latter three materials—gypsum, lime, and portland cement—that the plasters used in buildings today are prepared.

Gypsum Plaster

Gypsum is an abundant mineral in nature, a crystalline hydrous calcium sulfate. It is quarried, crushed, dried, ground to a fine powder, and heated to 350 degrees Fahrenheit (175°C) in a process known as *calcining* to drive off about three-quarters of its water of hydration. The *calcined gypsum*, ground to a fine white powder, is known as *plaster of Paris*. When plaster of Paris is mixed with water, it rehydrates and recrystallizes rapidly to return to its original solid state. As it hardens, it gives off heat and expands slightly.

Gypsum is a major component of interior finish materials in most buildings. It has but one major disadvantage—its solubility in water. Among its advantages are that it is durable and light in weight compared to many other materials. It resists the passage of sound better than most materials. It has a very fine grain, is easily worked in either its wet or dry state, and can be fashioned into surfaces that range from smooth to heavily textured. But above all it is

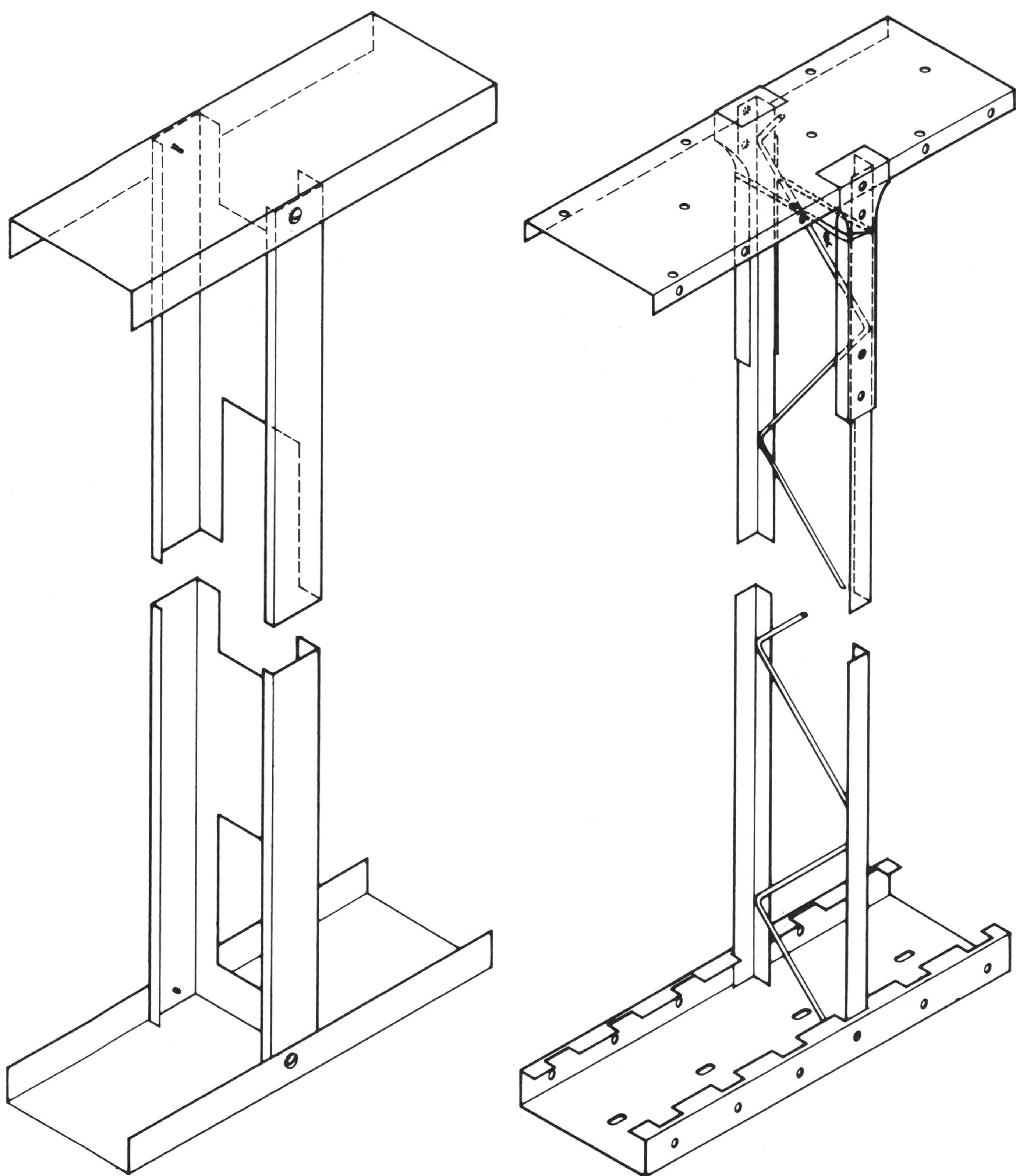

FIGURE 23.1
Partitions of light gauge steel studs, at the left, and open-truss wire studs on the right. The light gauge steel studs are assembled with self-drilling, self-tapping screws and may be used with any type of lath or panel. Open-wire truss studs are no longer used in new construction but may still be found in older buildings undergoing renovation. They were made specifically to minimize the need for screw fasteners. Studs were secured at their bottom ends into notches in the bottom track, and at their tops, with metal shoes tied with wire. Metal lath was attached with wire, and gypsum lath with special metal clips.

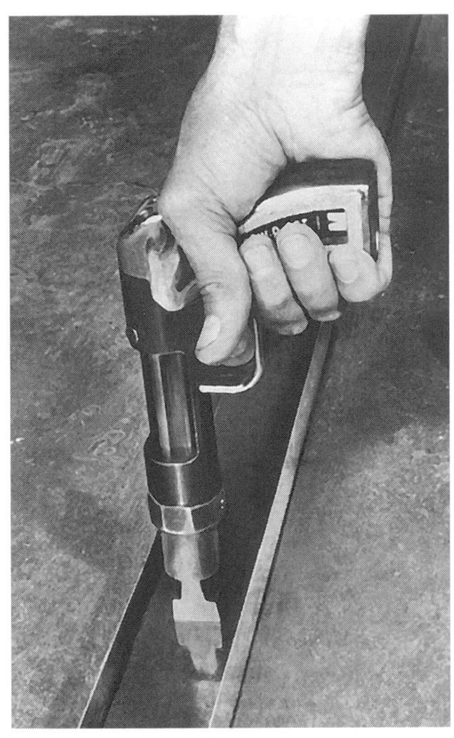

FIGURE 23.2
Attaching a runner to a concrete floor using powder-driven fasteners. The gun explodes a small charge of gunpowder to drive a steel pin through the metal and into the concrete to make a secure connection. *(Courtesy of United States Gypsum Company)*

FIGURE 23.3
Inserting studs into the runners to frame a partition of light gauge steel studs. The cutouts in the webs of the studs provide a passage for electrical conduits. On the right side of the photograph, a stack of gypsum board awaits installation. *(Courtesy of United States Gypsum Company)*

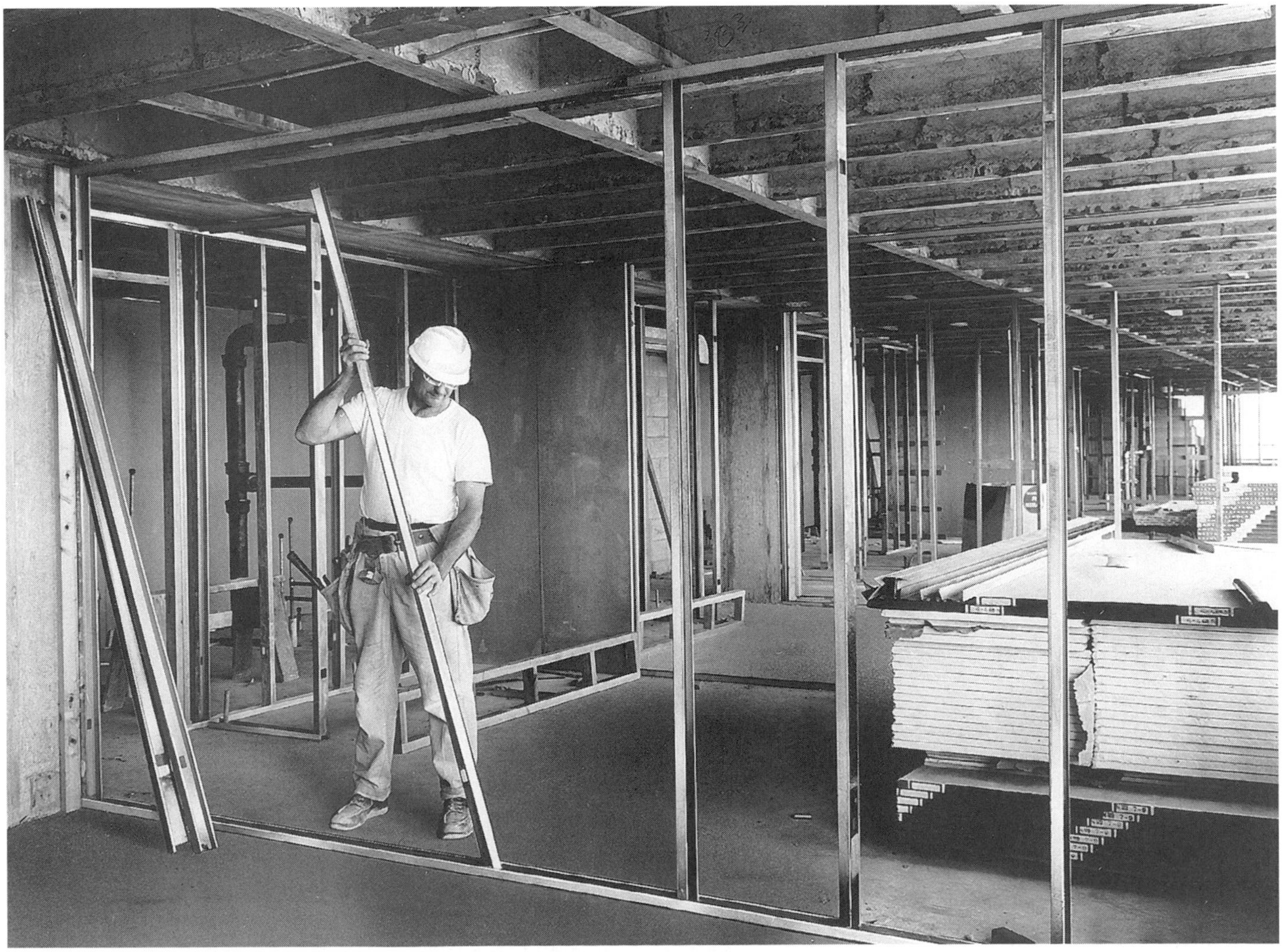

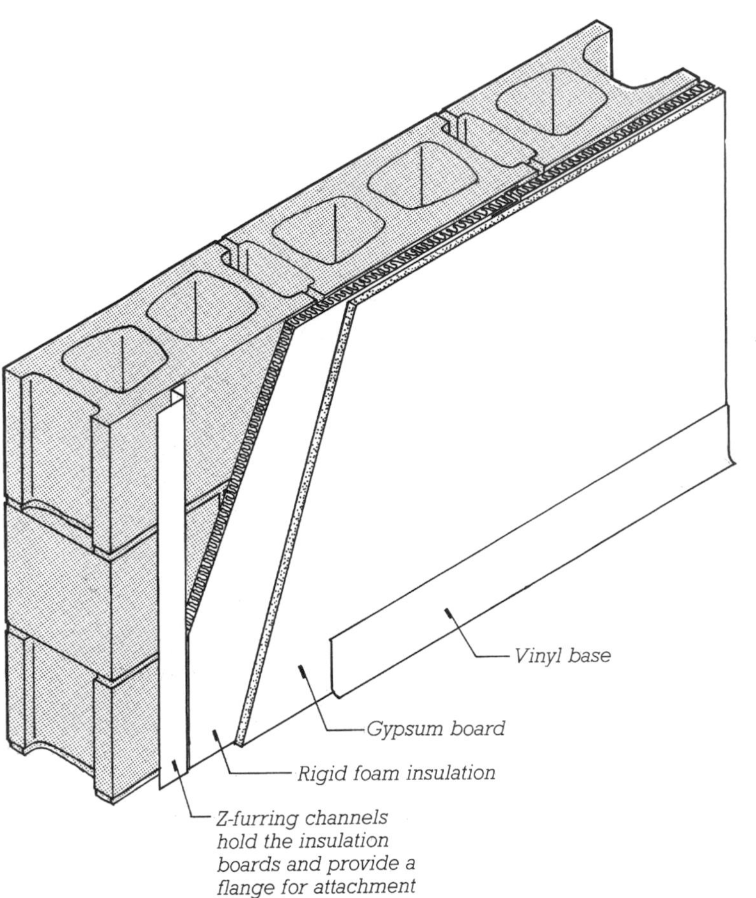
Vinyl base
Gypsum board
Rigid foam insulation
Z-furring channels hold the insulation boards and provide a flange for attachment of the gypsum board

FIGURE 23.4
A furred gypsum board finish over a concrete block wall. The Z-furring channels are attached to the masonry with powder-driven fasteners. The plastic foam insulation is tucked in behind the flange of the channel, and the gypsum board is screwed to the face of the flange. Long slots punched from the web of the channel (not visible in this drawing) help to reduce the thermal bridging effect of the Z-furring channel.

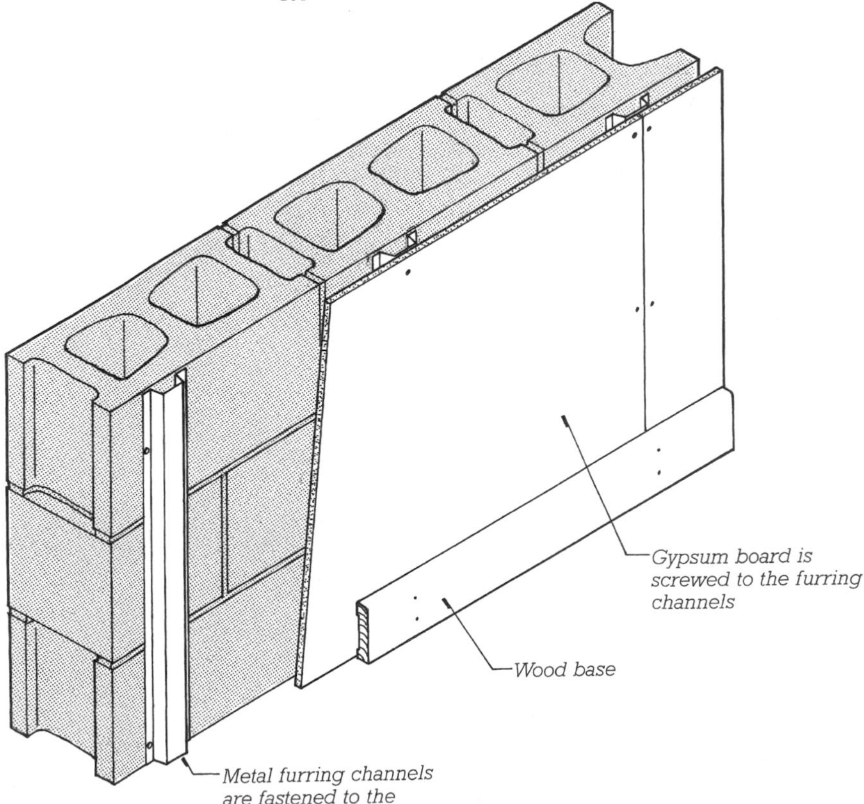
Gypsum board is screwed to the furring channels
Wood base
Metal furring channels are fastened to the masonry with powder-driven fasteners

FIGURE 23.5
A furred gypsum board finish using a standard hat-shaped metal furring channel *(hat channel)*.

inexpensive, and it is highly resistant to the passage of fire.

When a gypsum building component is subjected to the intense heat of a fire, a thin surface layer is calcined and gradually disintegrates. In the process, it absorbs considerable heat and gives off steam, both of which cool the fire (Figure 23.7). Layer by layer, the fire works its way through the gypsum, but the process is slow. The uncalcined gypsum never reaches a temperature more than a few degrees above the boiling point of water, so areas behind the gypsum component are well protected from the fire's heat. Any required degree of fire resistance can be created by increasing the thickness of the gypsum as necessary. The fire resistance of gypsum can also be increased by adding lightweight aggregates to reduce its thermal conductivity and by adding reinforcing fibers to retain the calcined gypsum in place as a fire barrier.

For use in construction, calcined gypsum is carefully formulated with various admixtures to control its setting time and other properties. Gypsum plaster is made by mixing the appropriate dry plaster formulation with water and an aggregate, either fine sand or a lightweight aggregate such as perlite or vermiculite. Because of its expansion during setting, gypsum plaster is very resistant to cracking.

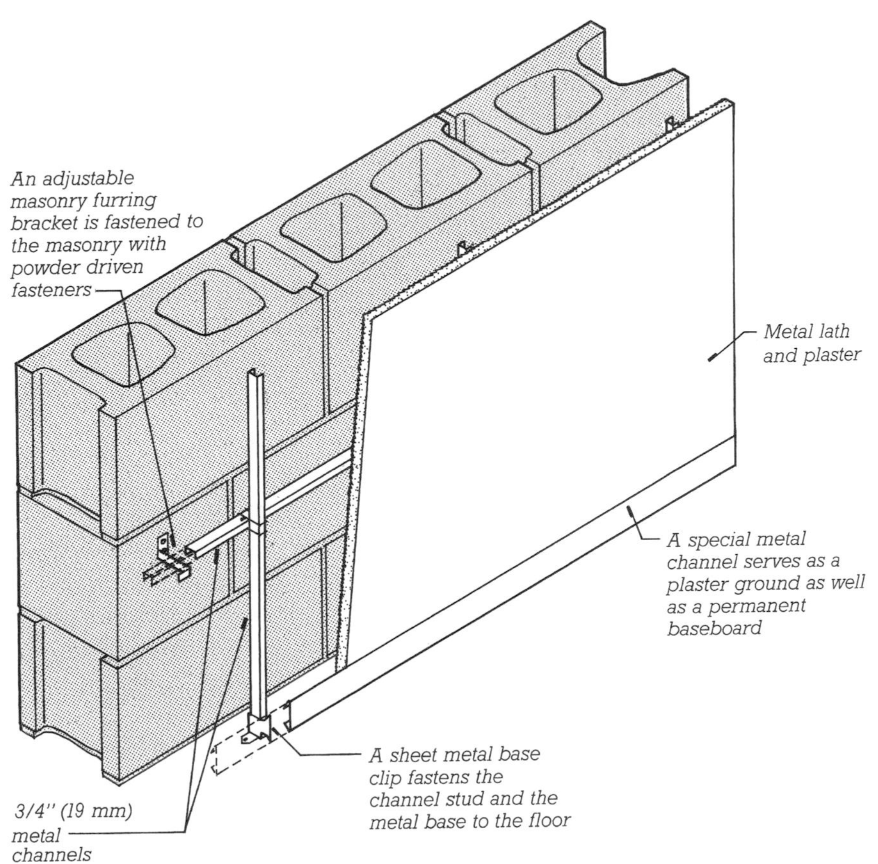
An adjustable masonry furring bracket is fastened to the masonry with powder driven fasteners
Metal lath and plaster
A special metal channel serves as a plaster ground as well as a permanent baseboard
A sheet metal base clip fastens the channel stud and the metal base to the floor
3/4" (19 mm) metal channels

FIGURE 23.6
A furred plaster finish using adjustable furring brackets. Each bracket has a series of teeth along its upper edge so that a metal channel can be wired securely to it in any of a number of positions, allowing the lather to produce a flat wall regardless of the surface quality of the masonry.

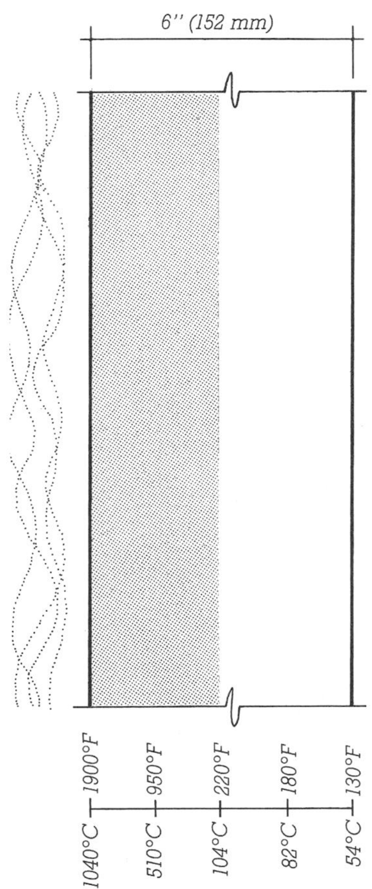
6" (152 mm)
1040°C 1900°F
510°C 950°F
104°C 220°F
82°C 180°F
54°C 130°F

FIGURE 23.7
The effect of fire on gypsum, based on data from Underwriters Laboratories, Inc. After a 2-hour exposure to heat following the ASTM E119 time–temperature curve, less than half of the gypsum on the side toward the fire, shown here by shading, has calcined. The portions of the gypsum to the right of the line of calcination remain at temperatures below the boiling point of water.

Considerations of Sustainability in Gypsum Products

Sources of Gypsum

• Naturally occurring gypsum is not renewable, but it is plentiful and widely distributed geographically.

• The majority of newly extracted gypsum is quarried in surface mines, with attendant risks of loss of wildlife habitat, surface erosion, and water pollution, as well as the problem of disposing of overburden and mine tailings.

• There is increasing use of *synthetic gypsum*, material recovered from power plant flue gases that would otherwise be sent to landfills, in the manufacture of gypsum construction materials. According to the Gypsum Association, approximately 1.5 million tons (1.4 million metric tons) of synthetic gypsum is used annually to produce about 7 percent of the U.S. construction industry's calcined gypsum. Some synthetic gypsums, however, contain toxic byproducts from the manufacturing processes in which they are produced and cannot be safely recycled into new construction materials.

Gypsum Products Manufacturing

• The calcining of gypsum involves temperatures that are not much higher than the boiling point of water, which means that the embodied energy of gypsum is relatively low, about 1200 BTU per pound (2.8 MJ/kg) for plaster and 2600 BTU per pound (6.0 MJ/kg) for gypsum board.

• The calcining process emits particulates of calcium sulfate, an inert, benign chemical, as dust.

• The paper faces of gypsum board are composed primarily of recycled newspapers.

• Some manufacturers produce gypsum board products made with as much as 95 percent recycled materials, including synthetic gypsum and recycled postconsumer waste paper.

Gypsum Products on the Building Site

• Approximately 15 million tons (14 million metric tons) of gypsum board are manufactured annually in the United States. On a typical construction site, about 10 to 12 percent of this material becomes waste.

• Gypsum board waste generated during construction can be minimized by sizing walls and ceilings to make efficient use of whole boards or by ordering custom-sized boards for nonstandard-size surfaces.

• Gypsum board scrap can be permanently stored in the hollow cavities of finished walls, eliminating disposal and transportation costs and reducing the amount of material destined for landfills (though care must be taken not to create interference with the pulling of electrical wires at a later date).

• Some dust is generated by the cutting and sanding of gypsum board and plaster. This dust has not been tied to any specific illnesses, but it is a nuisance and a source of discomfort until the work is done and all the dust has been swept up and removed from the building. Remodeling and demolition also create large quantities of gypsum dust.

• Most installed gypsum products have extremely low emissions. Some joint compounds, however, may also be sources of emissions.

• Additives used in the manufacture of moisture-resistant and fire-resistant gypsum board are potential sources of volatile organic compound (VOC) emissions.

• Paints, wallcovering adhesives, and other products used to finish gypsum surfaces can be significant emitters of VOCs, and thus require care in selection and specification.

Gypsum Disposal and Recycling

• Gypsum board waste can be recycled back into the manufacture of new gypsum board products. Current efforts limit recycled content to no more than 15 or 20 percent, due to the amount of paper waste that can be safely introduced into the new gypsum without impairing its fire resistance.

• Gypsum board waste from the demolition of older buildings may be contaminated with nails, drywall tape, joint compound, and paint. Gypsum board demolished from buildings constructed prior to 1978 may be coated with lead-based paint. These foreign materials must be removed from the waste; their presence may limit the material's recycling potential.

• Gypsum board waste can be used as a soil amendment and plant nutrient. With the recent advent of mobile grinders, construction site recycling of gypsum board waste for use as a soil amendment on the same building site is now feasible.

• Gypsum is an ingredient in many manufacturing and industrial processes. Studies and small-scale tests currently underway to identify potential uses of gypsum board waste in such processes are likely to lead to additional recycling opportunities in the future.

Gypsum plasters are manufactured in accordance with ASTM C28 and fall into two general categories: *base-coat plasters*, used for the underlying preparatory coats of a plaster application (as explained below), and *finish-coat plasters.* Base coat plasters are provided either *mill-mixed,* also called *ready-mixed,* with aggregate added at the manufacturing plant, or *neat,* for use with aggregate added at the job site. The more common base-coat gypsum plasters are:

- Ordinary *gypsum plaster,* in various formulations suitable for either hand or machine application
- *Wood-fibered gypsum plaster,* gypsum plaster blended with chopped wood fibers for lighter weight, and greater strength and fire resistance
- *Lightweight gypsum plaster* with perlite or vermiculite aggregate, for lighter weight and greater fire resistance
- *High-strength base-coat plaster,* for use under high-strength finish coats

Finish-coat plasters are typically a blend of gypsum plaster and lime. The *lime* provides superior workability and finishing qualities, while the gypsum provides greater hardness and strength and prevents shrinkage cracking. (For more information about the manufacture of lime and its use in plasters, see Chapter 8.) Common finish-coat plasters include:

- *Ready-mixed finish plaster,* with factory-blended lime and other ingredients
- *Gauging plaster,* gypsum plaster for job-site mixing with *hydrated* (prewetted on the job site) *finishing lime* (also called *lime putty*)
- *High-strength gauging plaster,* formulated to produce a finish plaster with higher compressive strength
- *Keenes cement,* a proprietary gauging plaster that produces an exceptionally dense, crack-resistant, low-absorbency finish
- *Molding plaster,* a fast-setting, fine-textured material for molding plaster ornament and running cornices (see the sidebar on pages 902 and 903)

Retarders and accelerators can also be added to plaster mixes on the job site to adjust the setting time to job-site temperature and humidity conditions.

Portland Cement Plasters

Portland cement–lime plaster, also known as *stucco,* is similar to masonry mortar. It is used where the plaster is likely to be subjected to moisture, as on exterior wall surfaces or in commercial kitchens, industrial plants, and shower rooms. Because freshly mixed stucco is not as buttery and smooth as gypsum and lime plasters, it is not as easy to apply and finish. It shrinks slightly during curing, so it should be installed with frequent control joints to regulate cracking.

Plastering

Plaster can be applied either by machine or by hand. Machine application is essentially a spraying process (Figure 23.8). Hand application is done with two very simple tools: a *hawk* in one hand to hold a small quantity of plaster ready for use and a *trowel* in the other hand to lift the plaster from

FIGURE 23.8
Spray applying a scratch coat of plaster to gypsum lath.
(Courtesy of United States Gypsum Company)

the hawk, apply it to the surface, and smooth it into place (Figures 23.9, 23.17, and 23.21). Plaster is transferred from the hawk to the trowel with a quick, practiced motion of both hands, and the trowel is moved up the wall or across the ceiling to spread the plaster, much as one uses a table knife to spread soft butter. After a surface is covered with plaster, it is leveled by drawing a straightedge called a *darby* across it, after which the trowel is used again to smooth the surface.

Lathing

Until a few decades ago, the most common form of lath consisted of thin strips of wood nailed to wood framing with small spaces left between the strips to allow *keying* of the plaster. Most lath today is made of either expanded metal or preformed gypsum boards. The skilled tradesperson who applies lath and trim accessories is known as a *lather.*

Expanded metal lath is made from thin sheets of steel that are slit and stretched in such a way as to produce a mesh of diamond-shaped openings (Figure 23.10). It is applied to light gauge steel studs with *self-drilling, self-tapping screws* or to wood studs with large-headed *lathing nails.* Lath used with portland cement stucco is galvanized to deter corrosion.

Gypsum lath is made in gypsum board sheets (see below) 16 by 48 inches (406 × 1220 mm) and ⅜ inch (9.5 mm) thick. It consists of sheets of hardened gypsum plaster faced with outer layers of a special absorbent paper to which fresh plaster readily adheres and inner layers of water-resistant paper to protect the gypsum core. Gypsum lath is attached to steel or wood studs with screws (Figure 23.11). Gypsum lath cannot be used as a base for gauged gypsum-lime finish plasters or portland cement stucco, as these materials will not bond adequately to the paper facing.

Veneer plaster base (*gypsum veneer base*) is a paper-faced gypsum board that comes in sheets 4 feet (1220 mm) wide, 8 to 14 feet (2440–4270 mm) long, and ½ or ⅝ inch (13–16 mm) thick. It is screwed to wood or steel

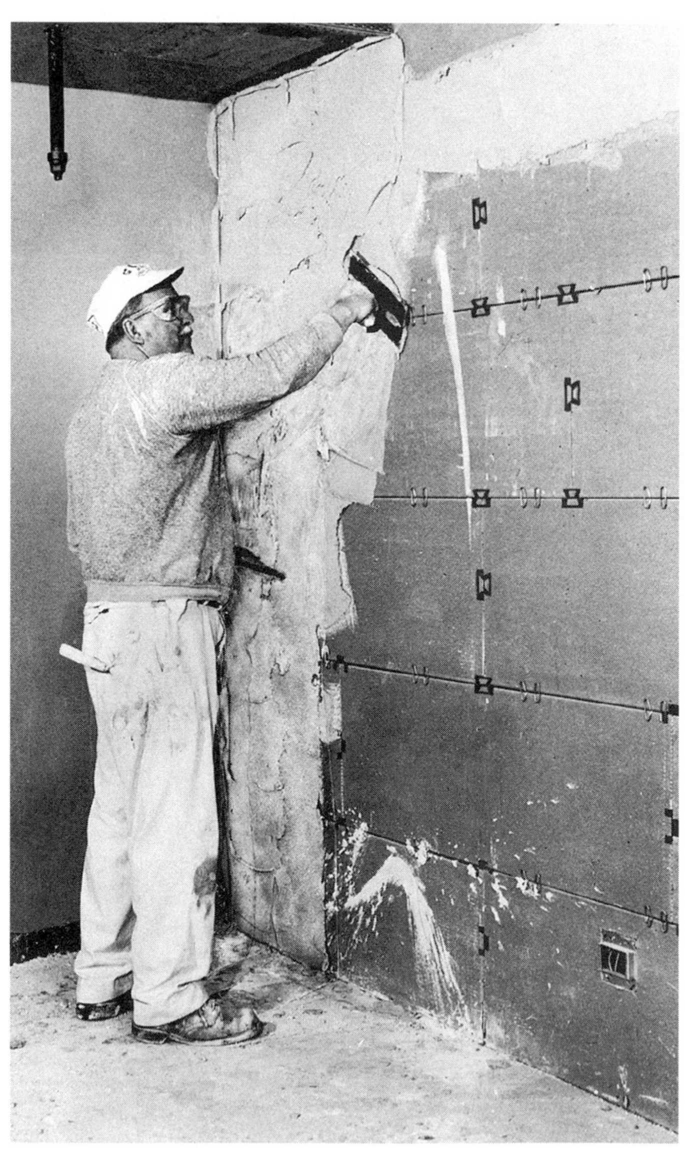

FIGURE 23.9
Applying a scratch coat over gypsum lath with a hawk (a corner of which is visible behind the plasterer's stomach) and trowel. Notice the wire clips that hold the lath to the open-truss studs and the sheet metal clips that strengthen the end joints between panels. The end joints do not occur over studs. (*Courtesy of United States Gypsum Company*)

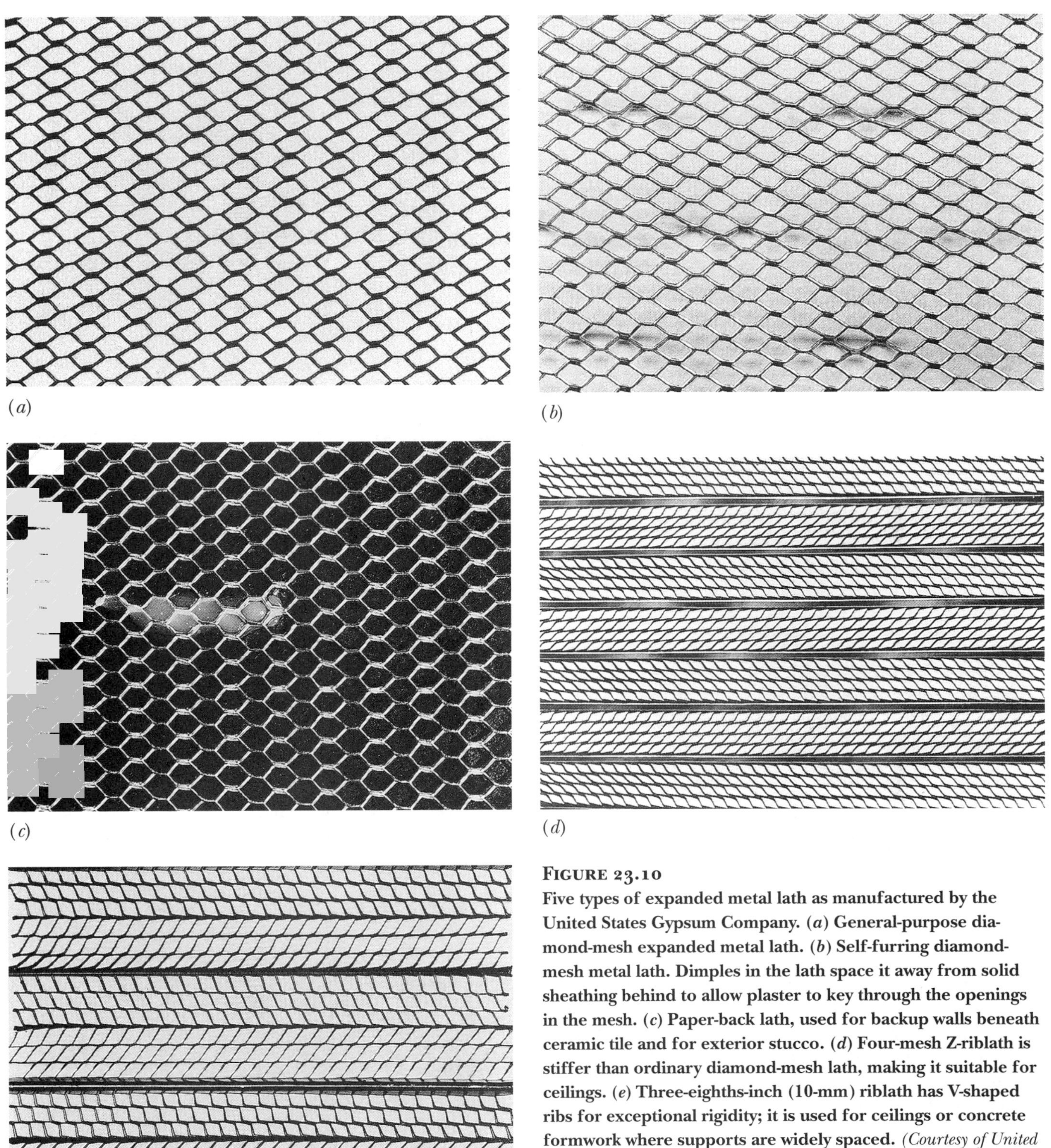
(a)
(b)
(c)
(d)
(e)

FIGURE 23.10
Five types of expanded metal lath as manufactured by the United States Gypsum Company. (*a*) General-purpose diamond-mesh expanded metal lath. (*b*) Self-furring diamond-mesh metal lath. Dimples in the lath space it away from solid sheathing behind to allow plaster to key through the openings in the mesh. (*c*) Paper-back lath, used for backup walls beneath ceramic tile and for exterior stucco. (*d*) Four-mesh Z-riblath is stiffer than ordinary diamond-mesh lath, making it suitable for ceilings. (*e*) Three-eighths-inch (10-mm) riblath has V-shaped ribs for exceptional rigidity; it is used for ceilings or concrete formwork where supports are widely spaced. *(Courtesy of United States Gypsum Company)*

studs, or nailed to wood studs, and used as a base specifically for the application of gypsum veneer plaster (discussed below).

Various lathing *trim accessories,* most frequently made of galvanized steel, are used at the edges of a plaster surface to make a neat, durable edge or corner (Figure 23.12). These are installed by the lather at the same time as the lath. In very long or tall plaster surfaces, metal *control joint* accessories are mounted over seams in the lath at predetermined intervals to control cracking. Trim accessories are also designed to act as lines that gauge the proper thickness and plane of plaster surfaces. A straightedge may be run across them to level the wet plaster. In this role, the trim accessories are known collectively as *grounds.* Trim accessories are made in several different thicknesses to match the required plaster thicknesses over the different types of lath.

In lathing I was pleased to be able to send home each nail with a single blow of the hammer, and it was my ambition to transfer the plaster from the board to the wall neatly and rapidly. . . . I admired anew the economy and convenience of plastering, which so effectually shuts out the cold and takes a handsome finish. . . . I had the previous winter made a small quantity of lime by burning the shells of the *Unio fluviatilis*, which our river affords. . . .

Henry David Thoreau, *Walden*, 1854.

Trim accessories are also produced as extrusions of plastic or aluminum. The aluminum accessories and some of the plastic ones are designed for improved precision and appearance when used in innovative details for bases, edges, and shadow lines in plaster walls.

Plaster Systems

Plaster Over Expanded Metal Lath Plaster is applied over expanded metal lath in three coats (Figure 23.13). The first, called the *scratch coat,* is troweled on rather roughly and cannot be made completely flat because the uncoated lath moves in and out considerably under the pressure of the trowel. This first coat is scratched while still wet, using a notched darby, a broom, or a special rake, to create a rough surface to which the second coat can bond mechanically (Figure 23.14).

FIGURE 23.11
Installing gypsum lath over light gauge steel studs with self-drilling, self-tapping screws. The electric screw gun disengages automatically from the screw head when the screw has reached the proper depth. *(Courtesy of United States Gypsum Company)*

FIGURE 23.12
Trim accessories for lath and plaster construction, as manufactured by the United States Gypsum Company. *(Courtesy of United States Gypsum Company)*

USG Corner Beads, Trim, Control Joints

description

USG Corner Beads and Trim, made from top-quality galvanized steel, enjoy the industry's top acceptance because of their dependability and continual improvement in design. Corner beads are available in 8 and 10-ft. lengths, metal trim in 7 and 10-ft. lengths, casing beads in 7, 8 and 10-ft. lengths.

1-A Expanded Corner Bead has 2⅞" wide expanded flanges that are easily flexed. Preferred for irregular corners. Provides increased reinforcement close to nose of bead. Made from galvanized steel or zinc alloy for exterior applications.

X-2 Corner Bead has full 3¼" flanges easily adjusted for plaster depth on columns. Ideal for finishing corners of structural tile and rough masonry. Has perforated stiffening ribs along expanded flange.

4-A Flexible Corner Bead is an economical general purpose bead. By snipping flanges, this bead may be bent to any curved design (for archways, telephone niches, etc.).

800 Corner Bead gives 1/16" grounds needed for one-coat veneer finishes. Approx. 90 keys per lin. in. provide superior bonding and strong, secure corners. The 1¼" fine-mesh flange eliminates shadowing, is easily nailed or stapled.

900 Corner Bead is used with two-coat veneer systems, gives 3/32" grounds. Its 1¼" fine-mesh flange can be either stapled or nailed. Provides superior plaster key and eliminates shadowing.

Cornerite and Striplath are strips of painted Diamond Mesh Lath used as reinforcement. **Cornerite** is bent lengthwise in the center to form a 100° angle. It should be used in all interior angles where metal lath is not lapped or carried around, over non-ferrous lath anchored to the lath, and over internal angles of masonry constructions to reduce plaster cracking. **Sizes:** 2" x 2" x 96" and 3" x 3" x 96". **Striplath** is a similar flat strip, used as a plaster reinforcement over joints of non-metallic lathing bases and where dissimilar bases join; also to span pipe chases. **Size:** 4" x 96".

USG Metal Trim comes in two styles and two grounds to provide neat edge protection for veneer finishing at cased openings and ceiling or wall intersections. All have fine-mesh expanded flanges to strengthen plaster bond and eliminate shadowing. **No. 701-A,** channel-type, and **No. 701-B,** angle edge trim, provide 3/32" grounds for two-coat systems. **No. 801-A,** channel-type, and **No. 801-B,** angle edge trim, provide 1/16" grounds for one-coat systems. **Sizes:** for ½" and ⅝" IMPERIAL Gypsum Base.

USG P-1 Vinyl Trim is a channel-shaped rigid trim with flexible vinyl fins which compress on installation to provide a positive acoustical seal comparable in performance to one bead of acoustical sealant. For veneer finish partition perimeters. **Lengths:** 8, 9 and 10 ft. **Sizes:** for ½" and ⅝" gypsum base.

USG P-2 Vinyl Trim is a channel-shaped vinyl trim with a pressure-sensitive adhesive backing for attachment to the wall at wall-ceiling intersections. Provides positive perimeter relief in radiant heat and veneer finish systems. Allow ⅛" to ¼" clear space for insertion. **Length:** 10 ft.

USG Control Joint relieves stresses of expansion and contraction in large plastered areas. Made from roll-formed zinc, it is resistant to corrosion in both interior and exterior uses with gypsum or portland cement plaster. An open slot, ¼" wide and ½" deep, is protected with plastic tape which is removed after plastering is completed. The perforated short flanges are wire-tied to metal lath, screwed or stapled to gypsum lath. Thus the plaster is key-locked to the control joint, which not only provides plastering grounds but can also be used to create decorative panel designs. **Limitations:** Where sound and/or fire ratings are prime considerations, adequate protection must be provided behind the control joint. USG Control Joints should not be used with magnesium oxychloride cement stuccos or stuccos containing calcium chloride additives. **Sizes and grounds: No. 50,** ½"; **No. 75,** ¾"; **No. 100,** 1" (for exterior stucco curtain walls)—10-ft. lengths.

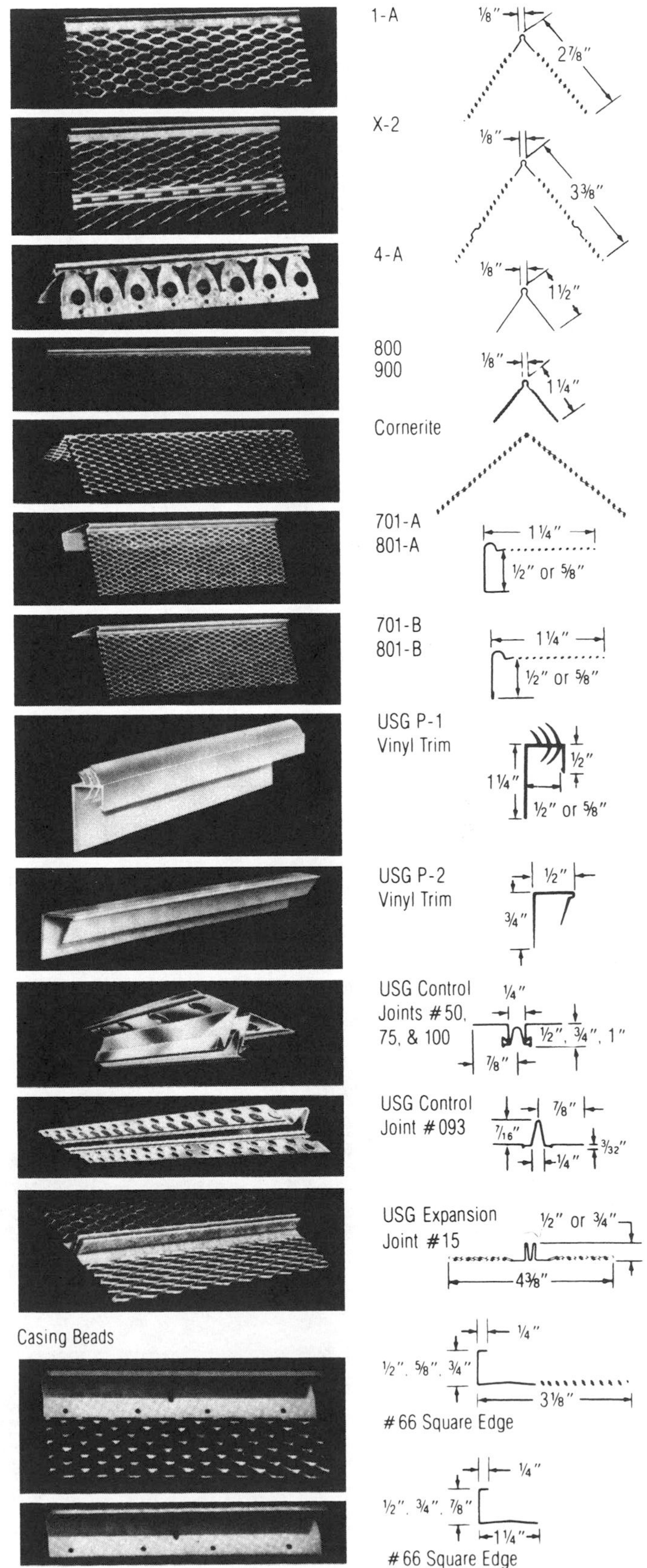

After the scratch coat has hardened, it works together with the lath as a rigid base for the second application of plaster, which is called the *brown coat.* The purpose of the brown coat is to build strength and thickness and to present a level surface for the application of the third or finish coat. The level surface is produced by drawing a long straightedge across the surfaces of the grounds (the *edge beads*, *corner beads*, and control joints) to strike off the wet plaster. On large, uninterrupted plaster surfaces, *plaster screeds*, intermittent spots or strips of plaster, are leveled up to the grounds in advance of brown coat plastering to serve as intermediate reference points for setting the thickness of the plaster during the striking-off operation. Base-coat plasters are used for scratch coats and brown coats.

The *finish coat* is very thin application of finish-coat plaster, about 1⁄16 inch (1.5 mm) thick. It may be troweled smooth or worked into any

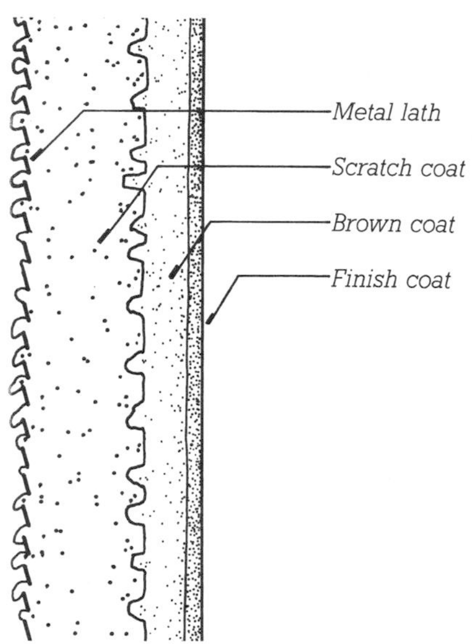

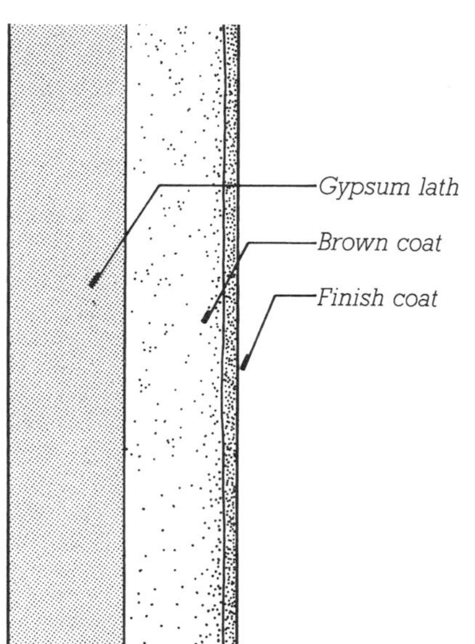

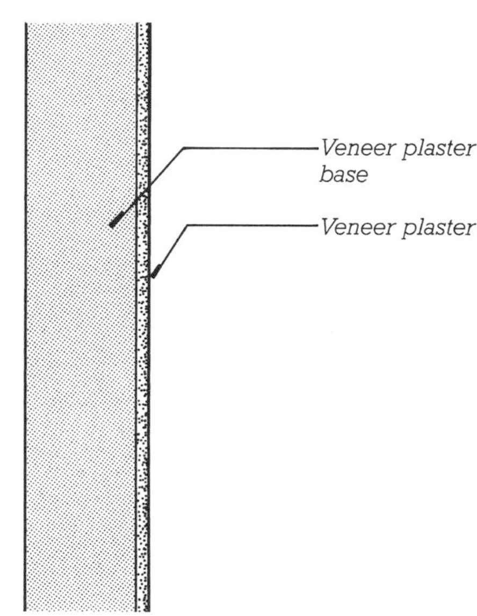

FIGURE 23.13
Sections through the three common lath-and-plaster systems, reproduced at full scale. Metal lath (*left*) requires three coats of plaster; the surface of the first coat is scratched for a better bond to the second coat. Gypsum lath (*middle*) may be finished with three coats or with two, as shown. Veneer plaster (*right*) usually consists of only a single thin finish coat, although two thin coats may be applied over rougher or more uneven substrates, such as concrete masonry or sitecast concrete.

FIGURE 23.14
Scratching the scratch coat while it is still soft to create a better bond to the brown coat. (*Courtesy of United States Gypsum Company*)

desired texture (Figures 23.15 and 23.16). The total thickness of the plaster that results from this three-coat process, as measured from the face of the lath, is about 5/8 inch (16 mm). Three-coat work over metal lath is the premium-quality plaster system, extremely strong and resistant to fire. The only disadvantage of three-coat plaster work is its cost, which can be attributed largely to the labor involved in applying the lath and the three separate coats of plaster.

Plaster Over Gypsum Lath The best plaster work over gypsum lath is applied in three coats, but gypsum lath is sufficiently rigid that if it is firmly mounted to the studs, only a brown coat and a finish coat need be applied. The elimination of the scratch coat has obvious economic advantages. Even with three coats of plaster, gypsum lath is often less expensive than metal lath because the gypsum in the lath replaces much of the plaster that would otherwise have to be mixed and applied by hand in the scratch coat. The total thickness of plaster applied over gypsum lath is 1/2 inch (13 mm).

Plaster Applied to Masonry Where plaster is applied directly over brick, concrete masonry, or poured concrete walls, the walls should be dampened thoroughly in advance of plastering to prevent premature dehydration of the plaster. A bonding agent may have to be applied

FIGURE 23.15
A sponge-faced float can be used to create various rough surface textures on plaster. *(Reprinted with permission of the Portland Cement Association from* Design and Control of Concrete Mixtures, *12th edition; Photos: Portland Cement Association, Skokie, IL)*

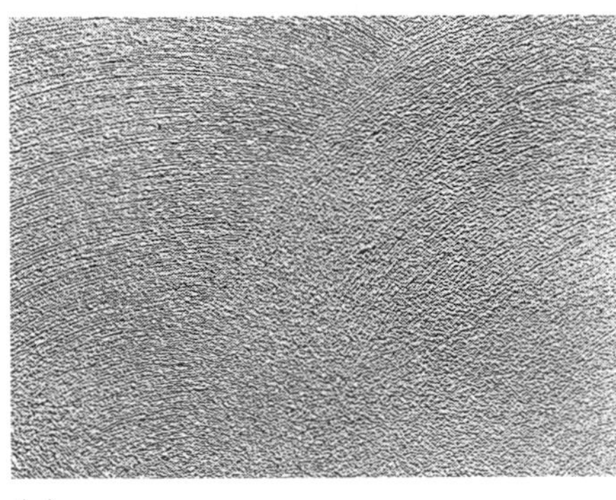

(*a*)

(*b*)

(*c*)

FIGURE 23.16
Three different plaster surface textures from among many. (*a*) Float finish. (*b*) Spray finish. (*c*) Texture finish. (*Courtesy of United States Gypsum Company*)

to some types of smooth masonry surfaces to ensure good adhesion of the plaster. The number of coats of plaster required to cover a wall is determined by the degree of unevenness of the masonry surface. For the best work, three coats totaling ⅝ inch (16 mm) should be applied, but for many walls two coats will suffice (Figure 23.17).

Veneer Plaster *Veneer plaster* is the least expensive of the gypsum plaster systems and is competitive in price with gypsum board finishes in many regions. The veneer base and accessories create a very flat surface that can be finished with a layer of a specially formulated dense gypsum plaster (manufactured to a separate standard, ASTM C587) that is applied in one or, occasionally, two coats usually no more than 1⁄16 to ⅛ inch (2–3 mm) in total thickness (Figures 23.18–23.21). A typical single-coat application is applied in a "double-back" process in which a thin coat is followed immediately by a second "skim" coat that is finish troweled to the desired texture. The plaster veneer hardens and dries so rapidly that it may be painted the following day. A two-coat application of veneer plaster can also be directly applied to surface of sitecast concrete or concrete masonry walls.

Stucco Stucco is applied over galvanized metal lath, using accessories of galvanized steel, or in wet areas or exterior applications, of solid zinc or plastic, which are less prone to corrosion than galvanized steel. Whereas gypsum plaster expands during hardening and is therefore highly resistant to cracking, portland cement stucco shrinks and is prone to cracking. Stucco walls should be provided with control joints at frequent intervals to channel the shrinkage into predetermined lines rather than allowing it to cause random cracks. The curing reaction in stucco is the same as that of concrete and is very slow relative to that of gypsum plaster. Stucco must be kept moist for at least a week before it is allowed to dry in order to attain maximum hardness and strength through full hydration of its portland cement binder.

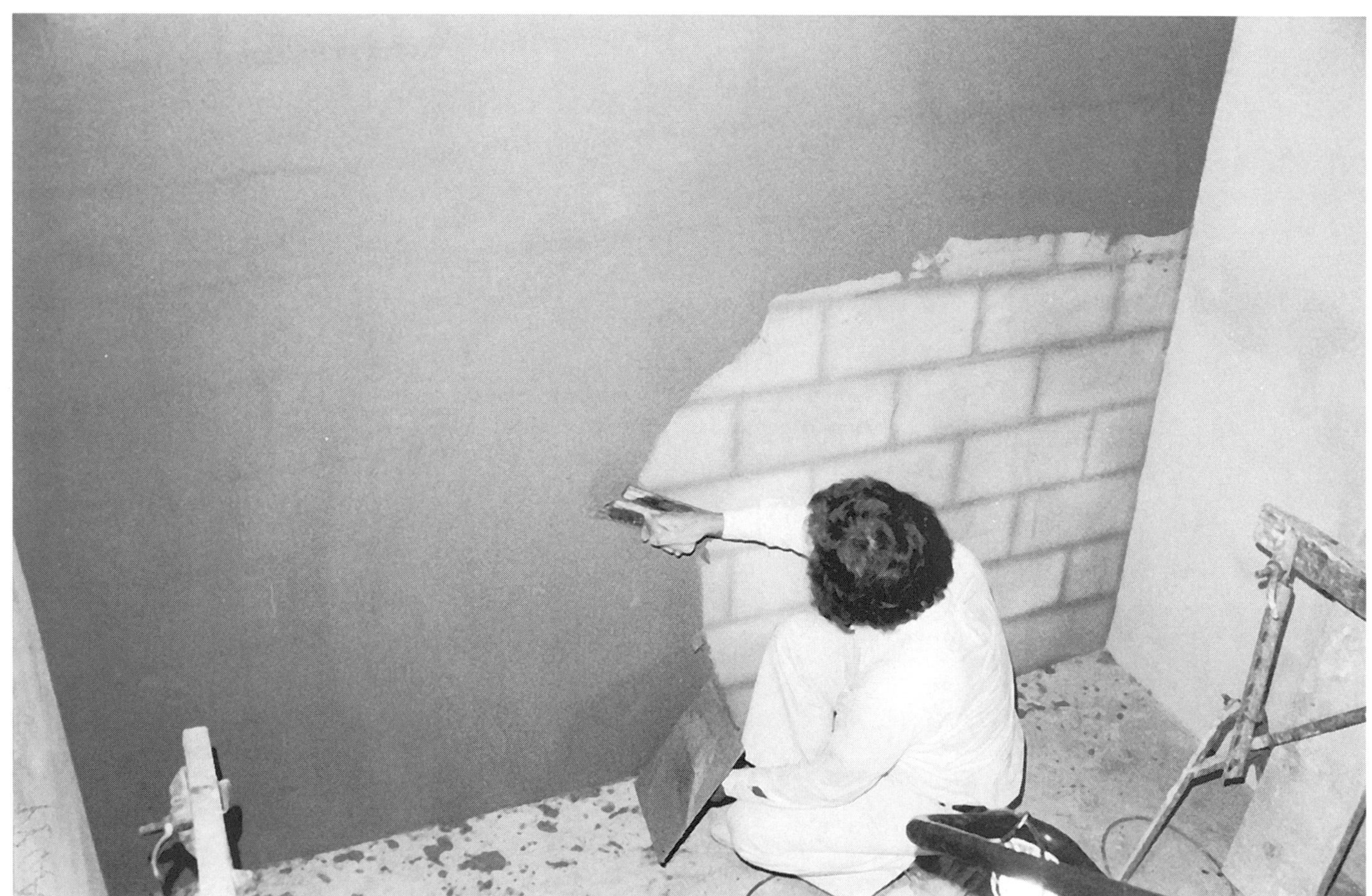

FIGURE 23.17
Applying a finish coat of portland cement plaster over a concrete masonry partition. The block joints are visible in the base coat of plaster because of a difference in the rate of water absorption between the blocks and the mortar joints. (*Reprinted with permission of the Portland Cement Association from* Design and Control of Concrete Mixtures, *12th edition; Photos: Portland Cement Association, Skokie, IL*)

In exterior applications over metal or wood studs, stucco may be applied over sheathing or without sheathing. Over sheathing, one or preferably two layers of asphalt-saturated building felt or building paper are first applied as an air and moisture barrier. Then a *self-furring metal lath*, which is formed with "dimples" that hold the lath away from the surface of the wall a fraction of an inch to allow the stucco to key to the lath, is attached with nails or screws (Figure 23.10*b*). If no sheathing is used, the wall is laced tightly with strands of *line wire* a few inches

FIGURE 23.18
Installing veneer plaster base with a screw gun. *(Courtesy of United States Gypsum Company)*

FIGURE 23.19
Stapling a corner bead to veneer plaster base to create a straight, durable corner. *(Courtesy of United States Gypsum Company)*

FIGURE 23.20
Reinforcing the panel joints of veneer plaster base with a self-adhesive glass fiber mesh tape. A panel opening for access to mechanical equipment is visible behind the installer. *(Courtesy of United States Gypsum Company)*

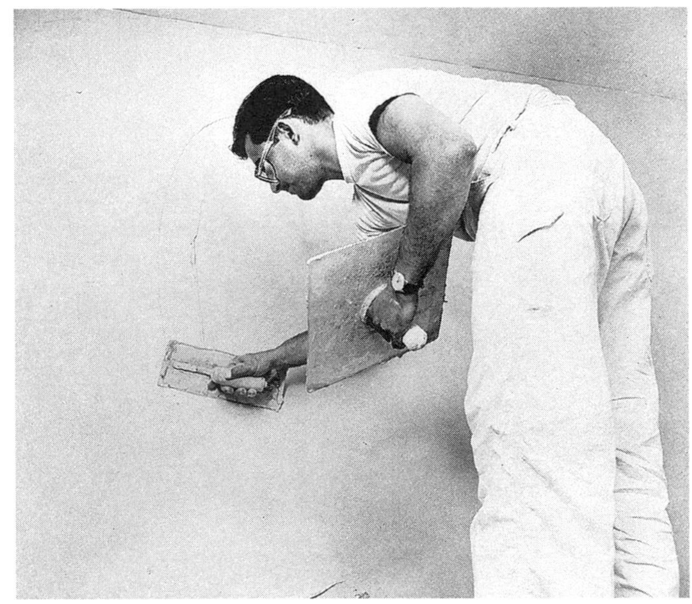

FIGURE 23.21
Applying veneer plaster with a hawk and trowel. *(Courtesy of United States Gypsum Company)*

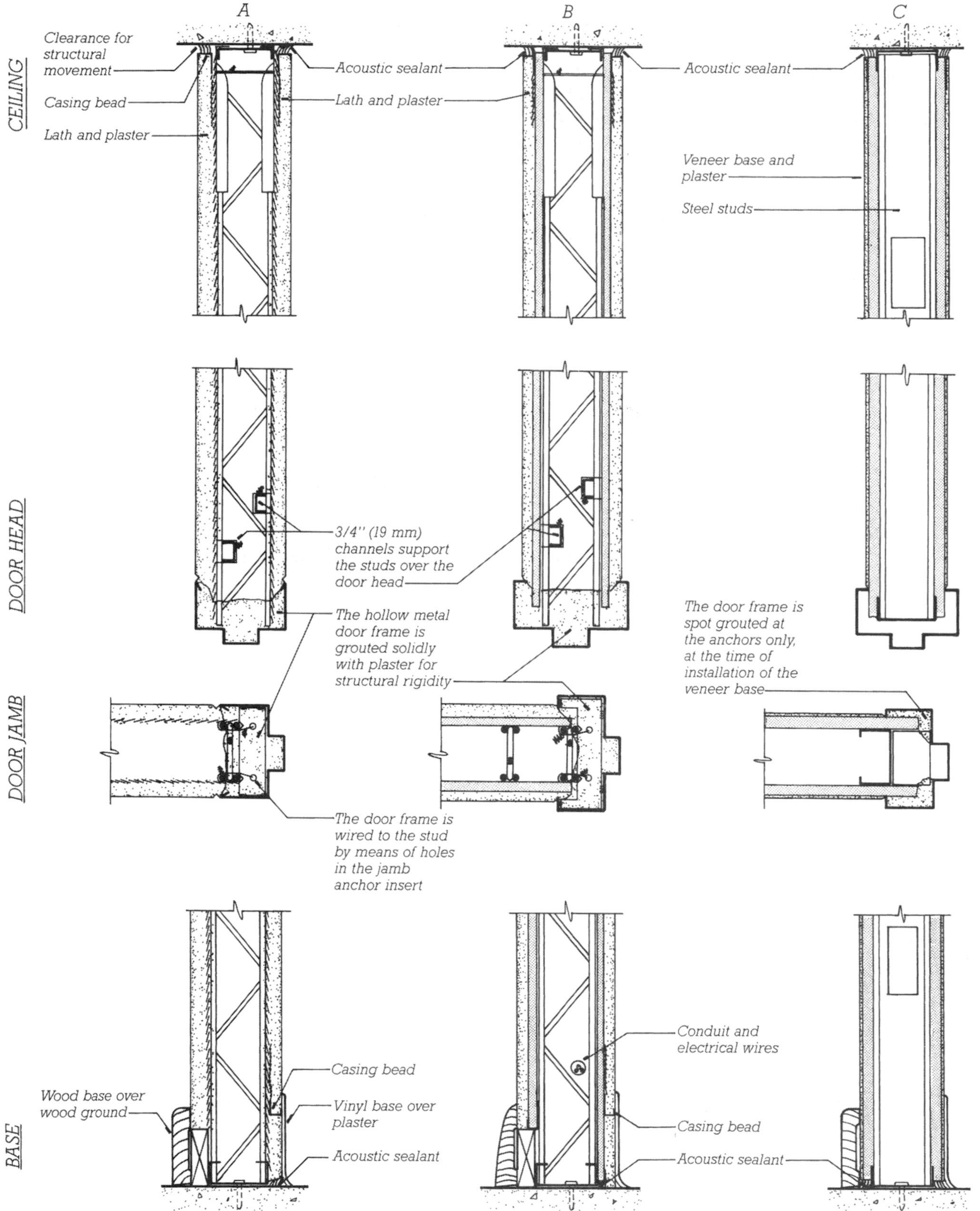

FIGURE 23.22

Three traditional plaster partition systems. (*a*) Three coats of plaster on metal lath and open-truss wire studs, rated at 1 hour of fire resistance and STC 39. (*b*) Two coats of plaster over gypsum lath and open-truss wire studs, rated at 1 hour and STC 41. (*c*) Veneer plaster on light gauge steel studs, rated at 1 hour and STC 40. Notice especially the provisions for airtightness and structural movement at the top and bottom of each partition and the methods of attachment used for hollow metal door frames. Doors weighing more than 50 pounds (23 kg) require special reinforcing details around the frames. In contemporary construction, conventional C-shaped light gauge steel studs are substituted for the open-truss wire studs shown in (*a*) and (*b*).

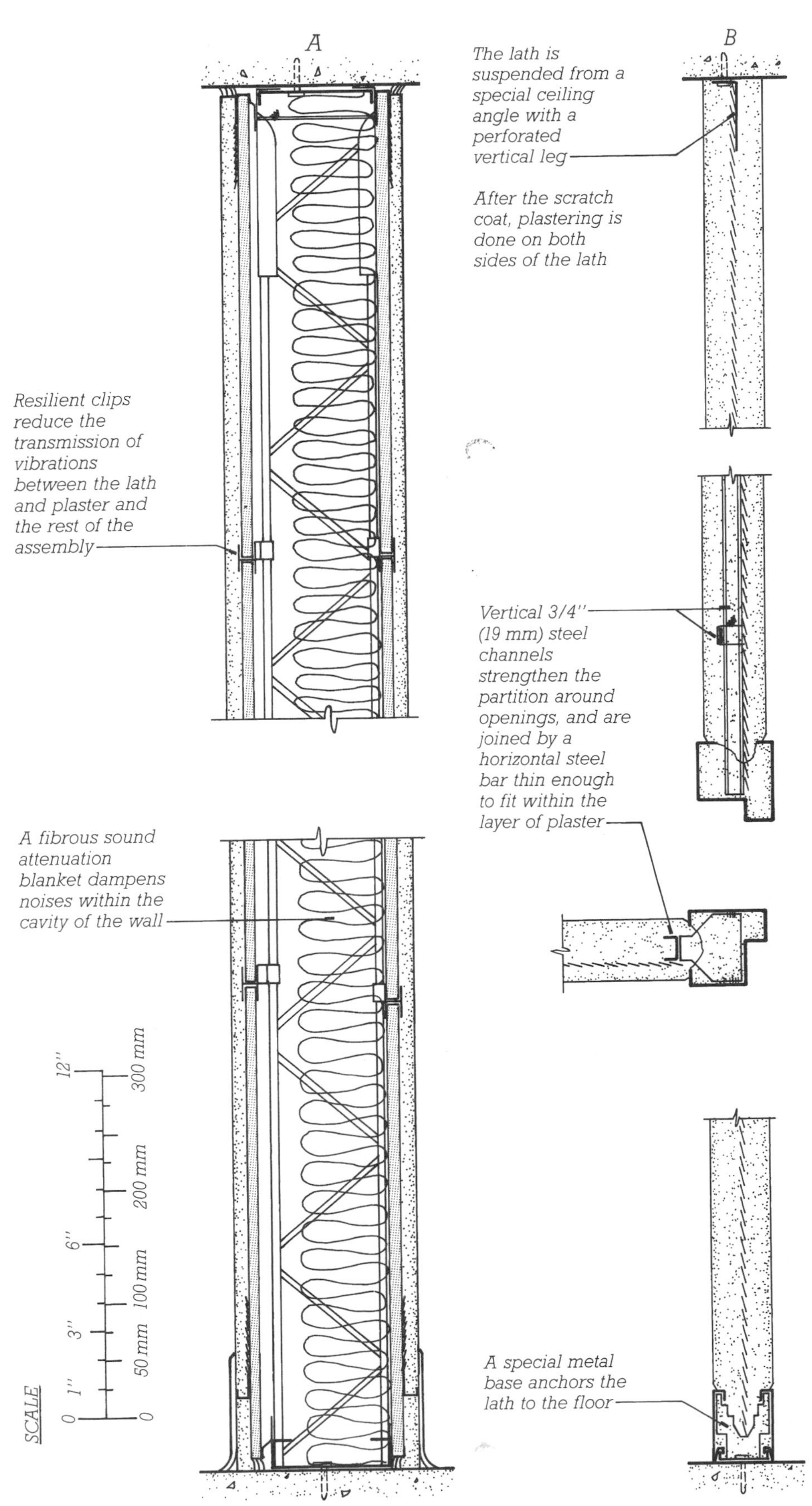
A
B
The lath is suspended from a special ceiling angle with a perforated vertical leg
After the scratch coat, plastering is done on both sides of the lath
Resilient clips reduce the transmission of vibrations between the lath and plaster and the rest of the assembly
Vertical 3/4" (19 mm) steel channels strengthen the partition around openings, and are joined by a horizontal steel bar thin enough to fit within the layer of plaster
A fibrous sound attenuation blanket dampens noises within the cavity of the wall
A special metal base anchors the lath to the floor
SCALE
0 1" 3" 6" 12"
0 50 mm 100 mm 200 mm 300 mm

apart and paperbacked metal lath is attached to the line wire, after which stucco is applied to encase the building in a thin layer of what amounts to reinforced concrete.

In exterior applications, where stucco is exposed to wind-driven rain, it may be applied over a thin, tangled filament matting or vertical furring strips to create a space behind the stucco that improves drainage and reduces the risk of water penetration further into the wall assembly.

Stucco is usually applied in three coats over metal lath, with a total thickness of ⅞ inch (22 mm), or in two coats when applied directly to the surface of concrete or concrete masonry, with a thickness of ⅜ to ½ inch (10–13 mm). It is applied either with a hawk and trowel or by spraying (Figure 6.31). In exterior work, pigments or dyes are often added to stucco to give an integral color, and rough textures are frequently used.

Plaster Partition Systems

Several types of plaster partitions are detailed in Figures 23.22 and 23.23. These diagrams show some of the ways in which the various trim accessories are used and the precautions taken to isolate the partitions from structural or thermal movements in the loadbearing frame of the building.

FIGURE 23.23
Two traditional plaster partition systems. (*a*) This partition increases its STC to 51 by mounting the gypsum lath on one side with resilient metal clips and filling the hollow space in the partition with a sound attenuation blanket. (*b*) The solid plaster partition has an STC of only 38 but is used in situations where floor space must be conserved. Both of these partitions are rated at 1 hour. In contemporary construction, conventional C-shaped light gauge steel studs are substituted for the open-truss wire studs shown in (*a*).

Plaster Ornament

Gypsum plaster, with its fine grain and even texture, has more sculptural potential than any other material used in architecture. While wet, it is easily formed with trowels and spatulas, molds, or templates. When dry, it is readily worked by sawing, sanding, machining, and carving. Plaster building ornament has been created for many centuries by two economical but powerful techniques, *casting* and *running*, and continues to be used in buildings of every size and every historic style.

Cast plaster ornament is made by pouring soupy plaster into molds. The plaster hardens in a few minutes, allowing the mold to be stripped and reused. Both rigid molds and soft rubber molds are used. The rubber molds are very flexible, so even undercut shapes can be cast without encountering difficulties in mold removal. Traditional rubber molds are created by first carving a plaster original, then brushing layers of latex over the original to build up the required wall thickness. More recently, two-component synthetic rubber compounds have replaced latex in most applications; their advantage is that they can be spread over the original in a single application rather than in layers.

Cast ornament is adhered to the brown coat of plaster in walls and ceilings with gobs of wet plaster or a mixture of plaster and glue. Once the ornament has been securely fastened in place, the finish coat of plaster is applied around it, and the plaster surfaces and adjacent pieces of ornament are merged by skillful trowel work and sanding to create a single-piece finish.

Running is used to make linear ornaments (*run plaster ornament*) such as classic cornice moldings. A rigid *blade* made of sheet metal or sheet plastic is cut to the profile of the molding. The blade is attached to a sliding wooden frame to create a *template*. The template is pushed back and forth along a guide strip mounted temporarily on the wall or ceiling while a mixture of lime putty and gauging plaster is inserted in front of the blade, which strikes it off to the desired profile. Repeated passes of the template are required to finish the molding smoothly and perfectly. These passes must be completed before the plaster begins to harden, or the setting expansion of the gypsum will cause the template to bind and spoil the plaster surface. The template may also be attached to a radius guide to produce circular moldings.

Casting and running are often used to reproduce plaster ornament during restoration of historic buildings. Rubber molds for casting can be made directly from existing ornaments, and templates for running duplicates of existing profiles are easily and cheaply produced.

New designs for plaster ornament are readily translated from the architect's paper sketches into carved plaster originals, from which rubber molds are made and

Figure A
Removing the flexible rubber mold from a cast plaster ornament. (*Courtesy of Dovetail, Inc.*)

Figure B
Running a plaster cornic[e] molding in place. (*Court[esy] of Dovetail, Inc.*)

duplicates cast. New profiles for moldings are quickly converted into template blades. The possibilities are almost limitless, yet few contemporary architects have chosen to explore them. This is surprising because plaster ornament is inexpensive compared to ornament carved from wood or stone, and there are few technical constraints on what can be accomplished.

Cast plaster ornament can be reinforced with short fibers of alkali-resistant glass. These greatly increase its strength and toughness and allow it to be produced in much thinner sections and much larger pieces than unreinforced plaster. This recent development has dramatically changed the economics and methods of ornamentation in plaster that is based on stock designs. A number of manufacturers issue catalogs of stock designs for ornaments made by this process. Much of the on-site assembly work for elaborate ornaments can be eliminated by combining what were formerly a number of small, thick, brittle castings into a single, larger, thinner casting that is light in weight and highly resistant to breakage. On the construction site, the lightweight castings are glued in place over gypsum board or veneer plaster base, using an ordinary mastic adhesive. The edges of the ornaments are feathered into the wall or ceiling surfaces with joint compound or veneer plaster, and the joints between pieces of ornament are smoothed over and sanded.

FIGURE C
Sculptor David Flaharty runs a circular plaster medallion on a bench in his shop.
(© Brian McNeill)

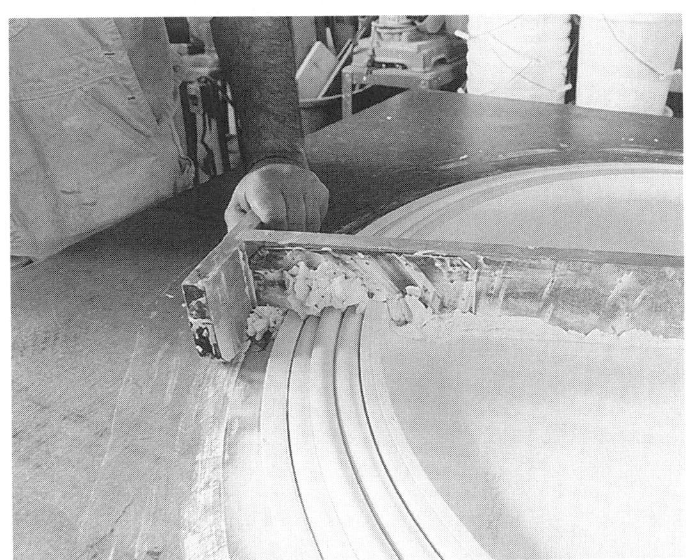

FIGURE D
A close-up of Flaharty's blade and template. The template rides on the sledlike runner of the portion under his hand, which is called the "slipper." The long portion of the template, called the "stock," is a radius guide that is fastened to a pin at the end to create the circular form of the medallion. *(© Brian McNeill)*

FIGURE E
The medallion having been removed from the bench and glued in place on the ceiling, Flaharty adds cast components to complete the ornament.
(© Brian McNeill)

Gypsum Board

Gypsum board is a prefabricated plaster sheet material that is manufactured in widths of 4 feet (1220 mm) and lengths of 8 to 14 feet (2440–4270 mm). It is also known as *gypsum wallboard*, *plasterboard*, and *drywall*. (The term "sheetrock" is a registered trademark of one manufacturer of gypsum board and should not be used in a generic sense.)

Gypsum board is the least expensive of all interior finishing materials for walls and ceilings. For this reason alone, it has found wide acceptance throughout North America as a substitute for plaster in buildings of every type. It retains the fire-resistive characteristics of gypsum plaster, but it is installed with less labor by less skilled workers than lathers and plasterers. And because it is installed largely in the form of dry materials, it eliminates some of the construction delay that may be associated with the curing and drying of plaster.

The core of gypsum board is formulated as a slurry of calcined gypsum, starch, water, pregenerated foam to reduce the density of the mixture, and various additives. This slurry is sandwiched between special paper faces and passed between sets of rollers to reduce it to the desired thickness. Within 2 or 3 minutes, the core material has hardened and bonded to the paper faces. The board is cut to length and heated to drive off residual moisture, then bundled for shipping (Figure 23.24).

Types of Gypsum Board

Gypsum board is produced in a variety of types suitable to a wide range of requirements:

• Regular gypsum board is used wherever the unique properties of other special board types are not required.

FIGURE 23.24
Sheets of gypsum board roll off the manufacturing line, trimmed and ready for packaging. *(Courtesy of United States Gypsum Company)*

• For most types of fire-rated assemblies, *Type X gypsum board* is required. The core material of Type X board is reinforced with short glass fibers. In a severe fire, the fibers hold the calcined gypsum in place to continue to act as a barrier to fire rather than permitting it to erode or fall out.

• *Type C gypsum board* is a proprietary formulation that is more fire resistant than Type X. Often a thinner board of Type C may generally be substituted for a thicker one of Type X.

• In locations exposed to moderate amounts of moisture, *water-resistant gypsum backing board*, with facings of water-repellant paper or glass matt, and a moisture-resistant core formulation is used.

• *Abuse-resistant* or *impact-resistant gypsum board* provides greater resistance to indentation and penetration and is intended for use in buildings that are exposed to rough usage. It may be manufactured with heavier facing paper and backing sheets, may have its core reinforced with cellulose fibers, may be faced with glass fiber mesh, or may be backed with polycarbonate film.

• *Mold-resistant gypsum board* is manufactured to resist moisture and mold growth. The common paper facing on gypsum board can, when wetted, provide conditions suitable to the growth of mold spores. Mold-resistant boards combine moisture-resistant cores with chemically treated paper facings or glass-matt facings that are not conducive to mold growth. At least one panel product is manufactured with a gypsum core blended with cellulose fibers that greatly strengthen the board and eliminate the need for facing materials of any kind.

• *Coreboard* is a 1-inch (25.4-mm)-thick panel that is used for shaft walls and solid gypsum board partitions. To facilitate handling, it is fabricated in sheets 24 inches (610 mm) wide.

• Sag-resistant *ceiling gypsum board*, ½ inch (13 mm) thick, is substantially lighter than, but just as resistant to sagging when used in ceiling assemblies as ⅝-inch (15-mm) panels.

• *Foil-backed gypsum board* can be used to eliminate the need for a separate vapor retarder in outside wall assemblies. If the back of the board faces a dead airspace at least ¾ inch (19 mm) thick, the bright foil also acts as a thermal insulator.

• *Predecorated gypsum board* is board that has been covered with paint, printed paper, or decorative plastic film. If handled carefully and installed with small nails, it needs no further finishing. This product is used in a number of demountable office partitioning systems.

• Weather-resistant *exterior gypsum soffit board* is used for exterior soffits, carport ceilings, undersides of exterior canopies, and other sheltered exterior applications.

• Gypsum lath and gypsum veneer base, as previously discussed.

Most gypsum board product types are manufactured in conformance with ASTM standard C1396. (Some related products, such as fiber cement backing board or glass matt–faced gypsum board, are manufactured to separate standards.)

Gypsum board is manufactured with a variety of edge profiles along its longer sides. The most common by far is the *tapered edge*, which permits sheets to be joined with a flush, invisible seam by means of subsequent joint finishing operations. Rounded and beveled edges are useful in predecorated panels, and tongue-and-groove edges serve to join coreboard panels in concealed locations. The edges of

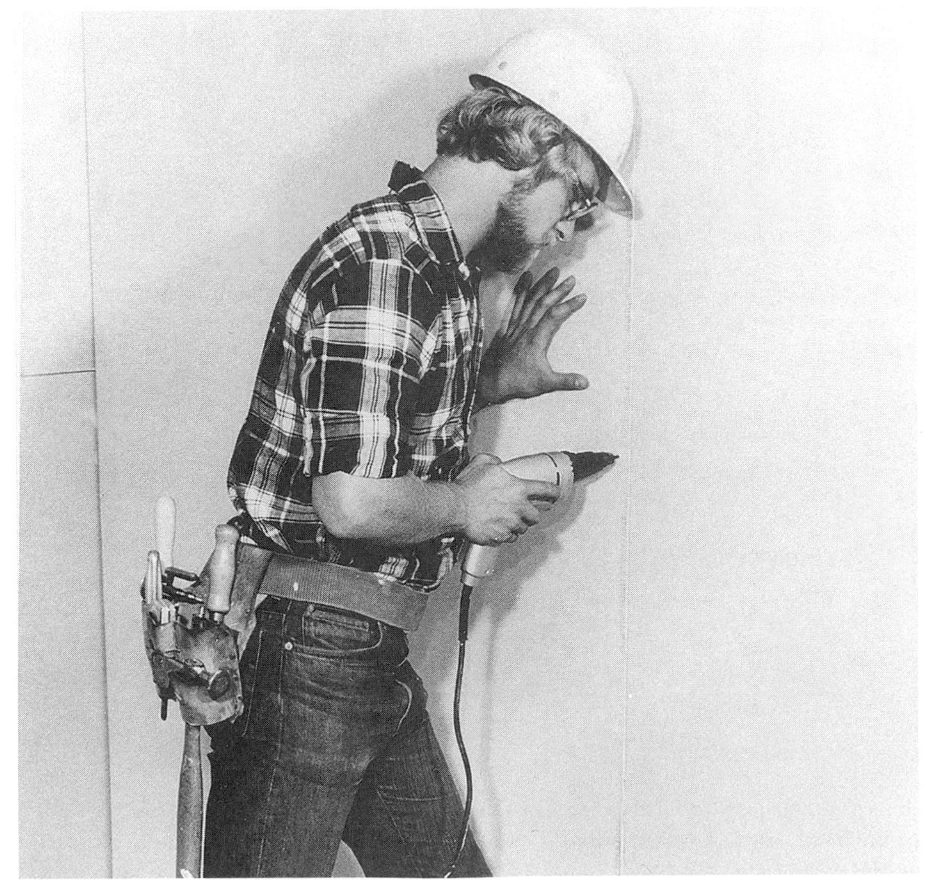

FIGURE 23.25
Attaching gypsum board to studs with a screw gun. *(Courtesy of United States Gypsum Company)*

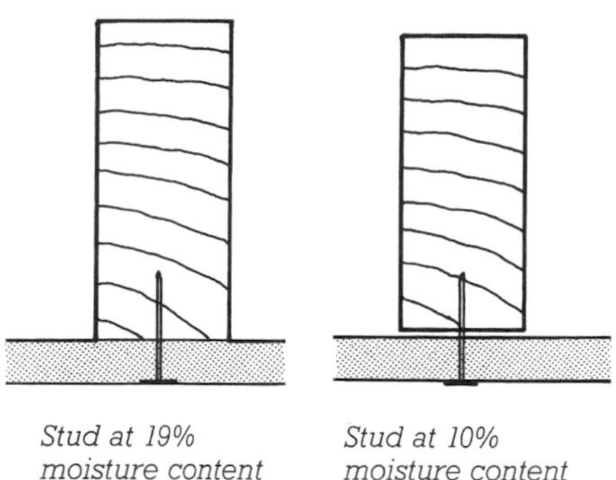

FIGURE 23.26
When wood studs dry and shrink during a building's first heating season, nail heads may pop through the surface of gypsum board walls.

the shorter ends of gypsum board panels are not tapered or shaped, since the individual panels are cut from longer continuous sheets during manufacturing.

Many different thicknesses of board are produced:

- ¼-inch (6.4-mm) board is used as a *gypsum backing board* in certain sound control applications. A special ¼-inch board is also produced by some manufacturers to be used for tight-radius bends (Figure 23.29).
- 5⁄16-inch (8-mm) board is made for manufactured housing, where weight reduction to facilitate shipping is an important consideration.
- ⅜-inch (9.5-mm) board is largely used in double-layer wall finishes. Though less durable than thicker panels, it may also sometimes be applied as a single-layer finish over ceiling joists or wall studs spaced no more than 16 inches (400 mm) apart.
- ½-inch (12.7-mm) board is the most common thickness. It is used for stud and joist spacings up to 24 inches (610 mm).
- ⅝-inch (16-mm) board is also limited to stud spacings of not more than 24 inches (610 mm), but it is often used where additional fire resistance, structural stiffness, durability, or sound deadening is required.

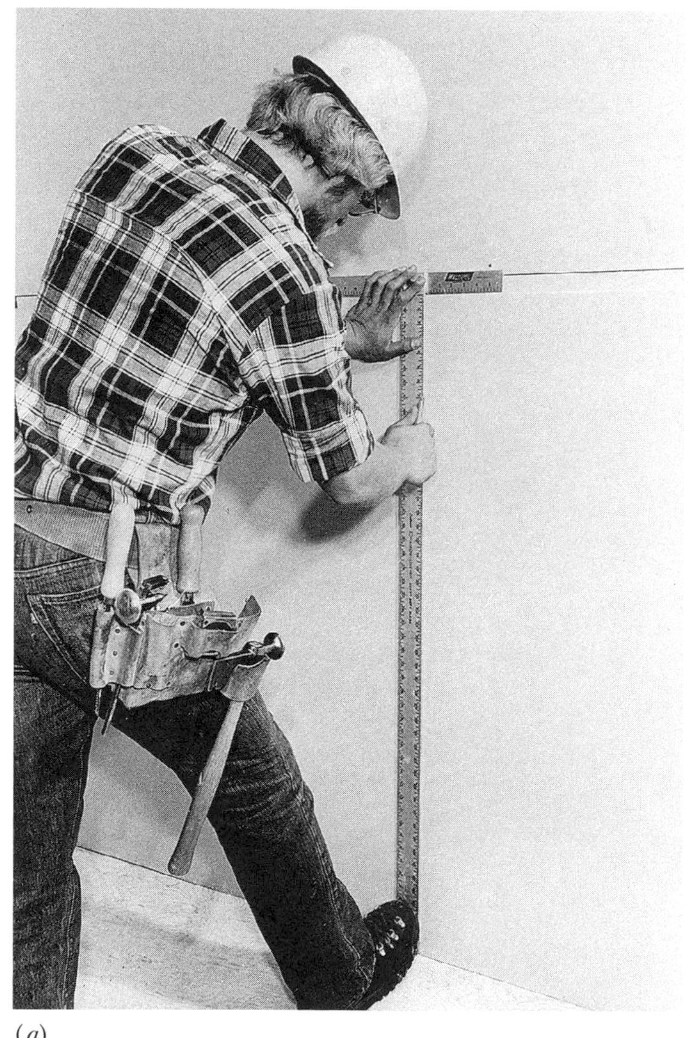
(*a*)

(*b*)

FIGURE 23.27
Cutting gypsum board. (*a*) A sharp knife and metal T-square are used to score a straight line through one paper face of the panel. (*b*) The scored board is easily "snapped," and the knife is used a second time to slit the second paper face. *(Courtesy of United States Gypsum Company)*

• ¾-inch (19-mm) Type X board is produced by some manufacturers. It is used to create 2-hour partitions with a single layer of gypsum board on each face and 4-hour partitions that include two layers on each face.

• 1-inch (25.4-mm)-thick board is made only as coreboard for use, for example, in shaft wall construction (Figure 23.38).

Installing Gypsum Board

Hanging the Board Gypsum board may be installed over either wood or light gauge steel studs, using self-drilling, self-tapping screws to fasten to steel and either screws or nails to fasten to wood (Figure 23.25). Wood studs can be troublesome with gypsum board because they usually shrink somewhat after the board is installed, which can cause nails to loosen slightly and "pop" through the finished surface of the board (Figure 23.26). *Nail popping* can be minimized by using only fully dried framing lumber, by using ring-shank nails that have extra gripping power in the wood, and by using the shortest nail that will do the job. Screws have less tendency to pop than nails. When screws or nails are driven into gypsum board, their heads are driven to a level slightly below the surface of the board but not enough to tear the paper surface.

To minimize the length of joints that must be finished and to create the stiffest wall possible, gypsum board is usually installed with the long dimension of the boards horizontal. The longest possible boards are used to eliminate or at least minimize end joints between boards, which are difficult to finish because the ends of boards are not tapered. Gypsum board is cut rapidly and easily by scoring one paper face with a sharp knife, snapping the brittle core along the score line with a blow from the heel of the hand, and cutting the other paper face along the fold created by the snapped core (Figure 23.27). A special metal T-square is used to make straight cuts that are perpendicular to the edges of the board. Notches, irregular cuts, and holes for electric boxes are made with a small saw or small electric router.

When two or more layers of gypsum board are installed on a surface, the joints between layers are staggered to create a stiffer wall, and a mastic adhesive is often used to join the layers to one another. Adhesive is also sometimes used between the studs and the gypsum board in single-layer installations to make a stronger joint.

Gypsum board can be curved when a design requires it. For gentle curves, the board can be bent into place dry (Figure 23.28). For somewhat sharper curves, the paper faces are moistened to decrease the stiffness of the board before it is installed. When the paper dries, the board is

FIGURE 23.28
Gypsum board can be curved to a large radius by simply bending it around a curving line of studs. *(Courtesy of United States Gypsum Company)*

as stiff as before. Special high-flex ¼-inch (6.5-mm) board is available that can be bent dry to a relatively small radius and bent wet to an even smaller radius (Figure 23.29).

Metal trim accessories are required at exposed edges and external corners to protect the brittle board and present a neat edge (Figure 23.30). These are similar to lathing accessories for plaster.

Finishing the Joints and Fastener Holes Joints and holes in gypsum board are finished to create the appearance of a monolithic surface almost indistinguishable from plaster. The finishing process is based on the use of a *joint compound* that resembles a smooth, sticky plaster. For most purposes, a *drying-type joint compound* is used; this is a mixture of marble dust, binder, and admixtures, furnished either as a dry powder to be mixed with water or as a premixed paste. In some high-production commercial work, *setting compounds* that cure rapidly by chemical reaction are used to minimize the waiting time between applications. Joint compounds of different weights and strengths may be used for different stages of the joint finishing process, or a single all-purpose compound may be used for all steps.

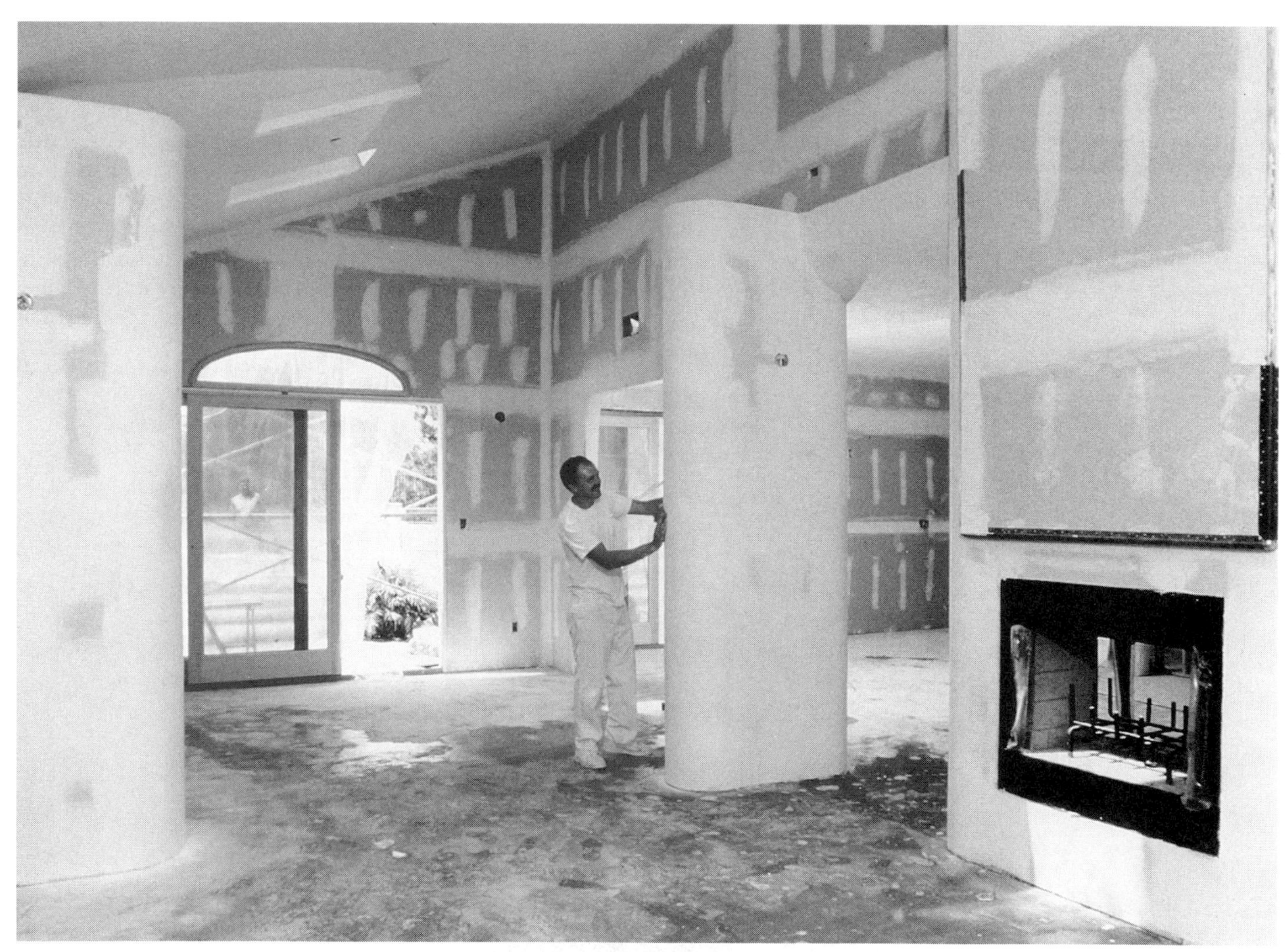

FIGURE 23.29
Tighter radii of curvature can be achieved by using ¼-inch (6.5-mm) high-flex gypsum board. *(Photo courtesy of National Gypsum Company)*

The finishing of a joint between panels of gypsum board begins with the troweling of a layer of joint compound into the tapered edge joint and the bedding of a paper or glass fiber reinforcing tape in the compound (Figures 23.31–23.34). Compound is also troweled over the nail or screw holes. After drying (usually overnight), a second layer of compound is applied to the joint to bring it level with the face of the board and to fill the space left by the slight drying shrinkage of the joint compound. When this second coat is dry, the joints are lightly sanded before a very thin final coat is applied to fill any remaining voids. The final coat is featured out (tapered down to zero thickness) to create an invisible edge. Before painting, the wall is again sanded lightly to remove any roughness or ridges. If the finishing is done properly, the painted or papered wall will show no signs at all that it is made of discrete panels of material.

Gypsum board has a smooth surface finish, but a number of spray-on textures and textured paints can be applied to give a rougher surface. Most gypsum board contractors prefer to texture ceilings; the texture

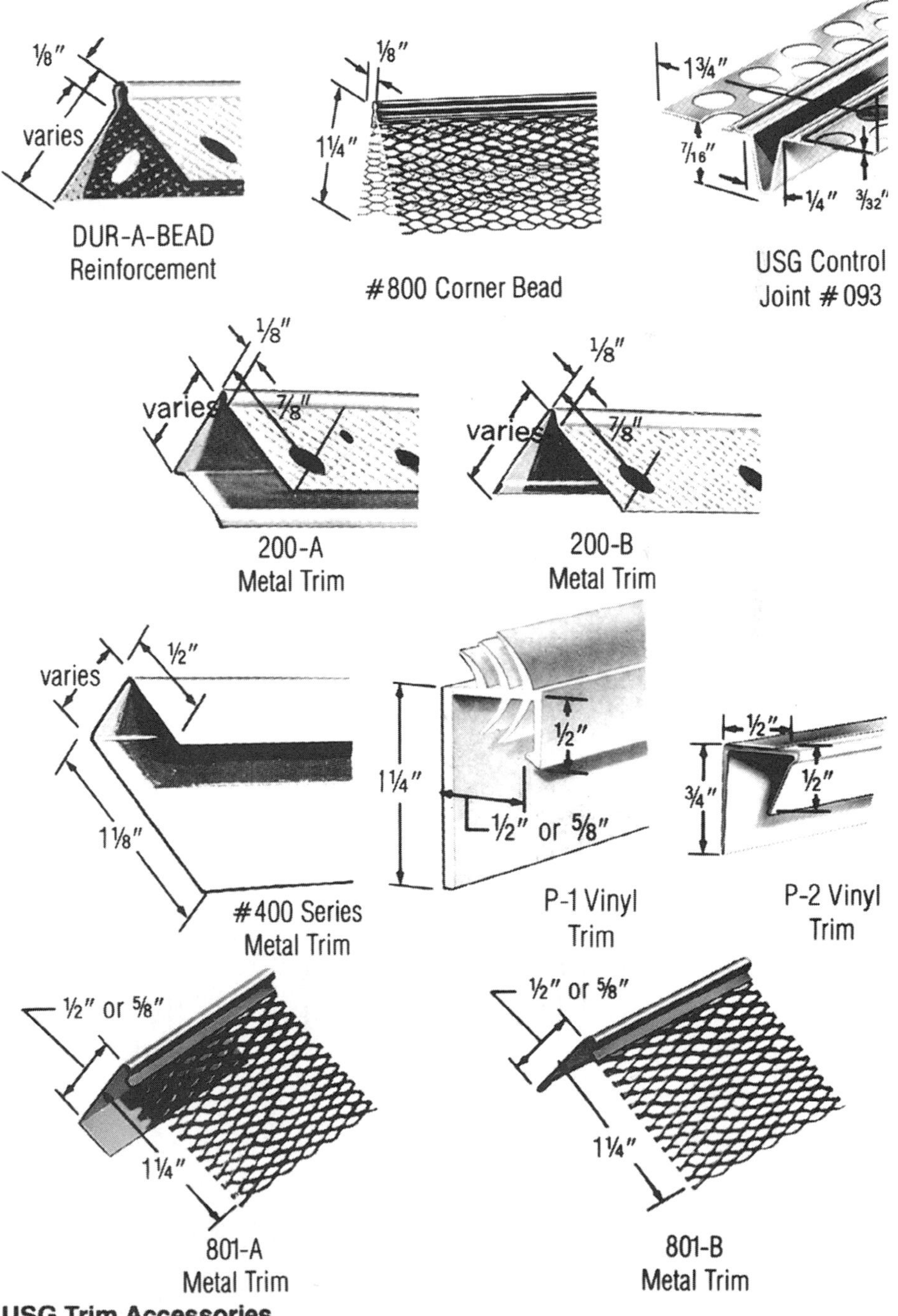

FIGURE 23.30
Accessories for gypsum board construction, as manufactured by the United States Gypsum Company. *(Courtesy of United States Gypsum Company)*

conceals the minor irregularities in workmanship that are likely to occur because of the difficulty of working in the overhead position.

Standardized gypsum board *finish levels* have been developed by the Gypsum Association and are also specified in ASTM C840. These enable the designer to specify quickly and simply the minimum level of finish that is acceptable for any project or portion of a project.

- Level 0, the minimum, consists of just the boards, without taping, finishing, or accessories. It is usually used only for temporary construction or where finishing is postponed until a later date.
- Level 1 requires only that joints be covered with tape set in joint compound. Its primary use is in areas of the building that are not open to view, such as above ceilings, in attics, and in service corridors. Level 1 is also the minimum finish level for fire-resistance-rated gypsum board assemblies, in which applications it may also be referred to as *fire-taping*.
- Level 2 adds to a Level 1 finish a coat of joint compound over the accessories and fasteners. After joint tape is set in compound, these joints are also immediately wiped with a joint knife a second time to add a thin coat of compound over the tape. A Level 2 finish is appropriate in garages, warehouses, and storage areas, and for boards used as a backer for ceramic tile.
- Level 3 adds a full second coat of compound over tape, accessories, and fasteners after the first coat has dried. It is intended for surfaces that will be textured or covered with heavy wallcoverings.

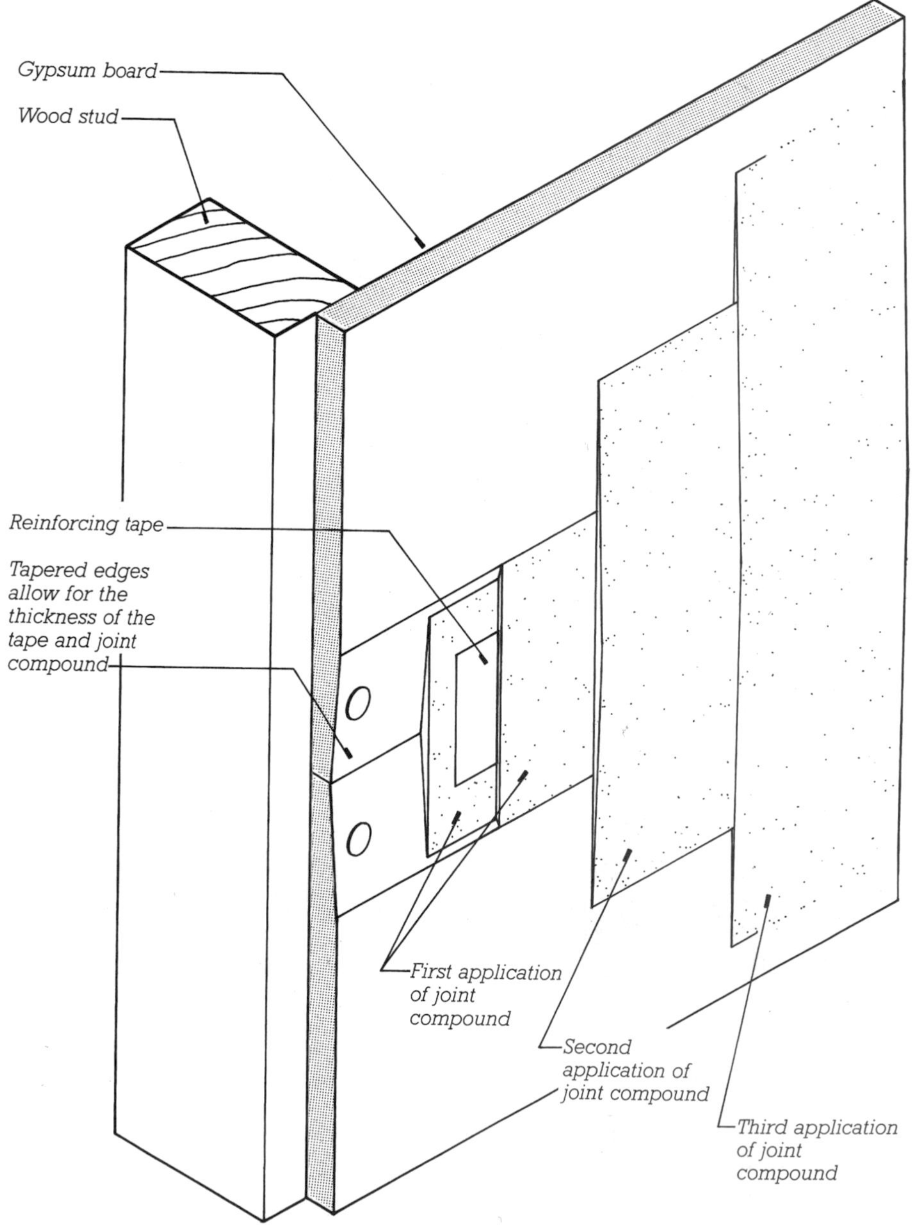

FIGURE 23.31
Finishing a joint between panels of gypsum board.

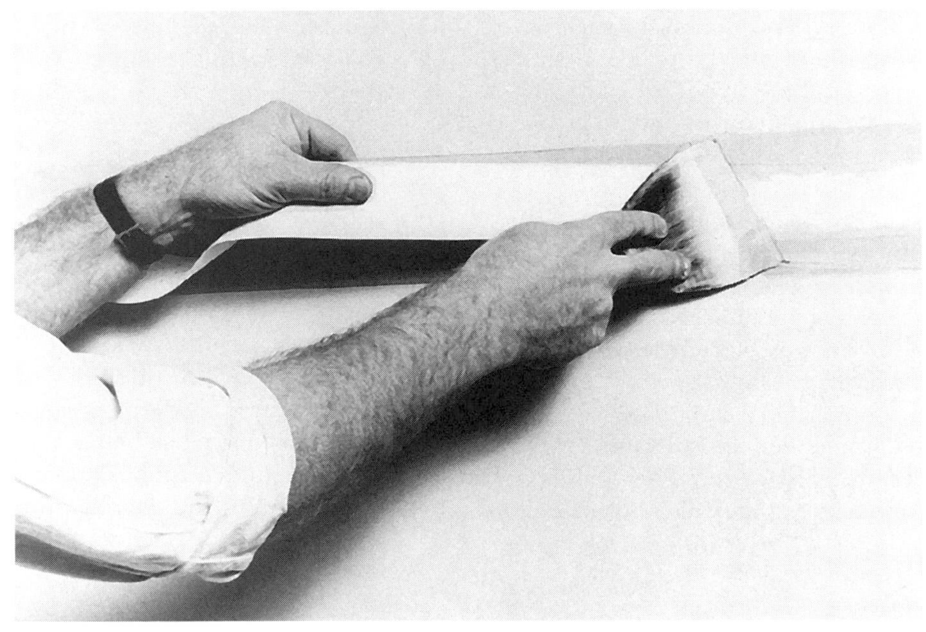

FIGURE 23.32
Applying paper joint reinforcing tape to gypsum board. *(Courtesy of United States Gypsum Company)*

FIGURE 23.33
An automatic taper simultaneously applies tape and joint compound to gypsum board joints. *(Courtesy of United States Gypsum Company)*

- Level 4 is designed for surfaces to be finished with flat paints, light textures, or thin wallcoverings. It adds a third distinct coat of joint compound over taped seams, fasteners, and accessories.
- Level 5, the highest, adds a very thin *skim coat* of joint compound over the entire surface of the board. The skim coat has no measurable thickness, because its purpose is only to fill pores and low spots in the wall to produce a very smooth surface. It is recommended for surfaces that will receive gloss or semigloss paints and for surfaces that will be lit in such a way as to cast shadows that can highlight even slight imperfections.

Gypsum Board Partition Systems

Gypsum board partition assemblies have been designed and tested with fire resistance ratings of up to 4 hours and sound transmission performance of up to at least STC 69. A selection of such partitions is shown in Figure 23.35. Notice in these details how provisions are made to prevent the gypsum panels from being subjected to structural loadings caused by movements in the loadbearing frame of the building—structural deflections, concrete creep, moisture expansion, and temperature expansion and contraction. Notice also the use of sealant to eliminate sound transmission around the edges of the partitions.

Demountable partition systems of gypsum board have also been developed, using concealed mechanical fasteners that can be disassembled and reassembled easily without damage to the panels (Figure 23.37). These systems are used in buildings whose partitions must be rearranged at frequent intervals.

FIGURE 23.34
Plasterers and drywall finishers often work on stilts to avoid having to erect and move scaffolding. The stilts, which are strapped to the legs of the workers, are very sophisticated, stable devices that keep the worker's foot fully supported and parallel to the floor. *(Photo by Rob Thallon)*

FIGURE 23.35
Four gypsum board partition systems. (*a*) A 1-hour partition, STC 40, using Type X gypsum board over light gauge steel studs. (*b*) This 1-hour partition on wood studs achieves an STC of 60 to 64 through heavy laminations of gypsum board, a sound attenuation blanket, and resilient channel mounting for one face of the partition. (*c*) A 2-hour partition with an STC of 48. (*d*) A 4-hour partition, STC 58. These are representative of a large number of gypsum board partition systems in a range of fire ratings and sound transmission classes.

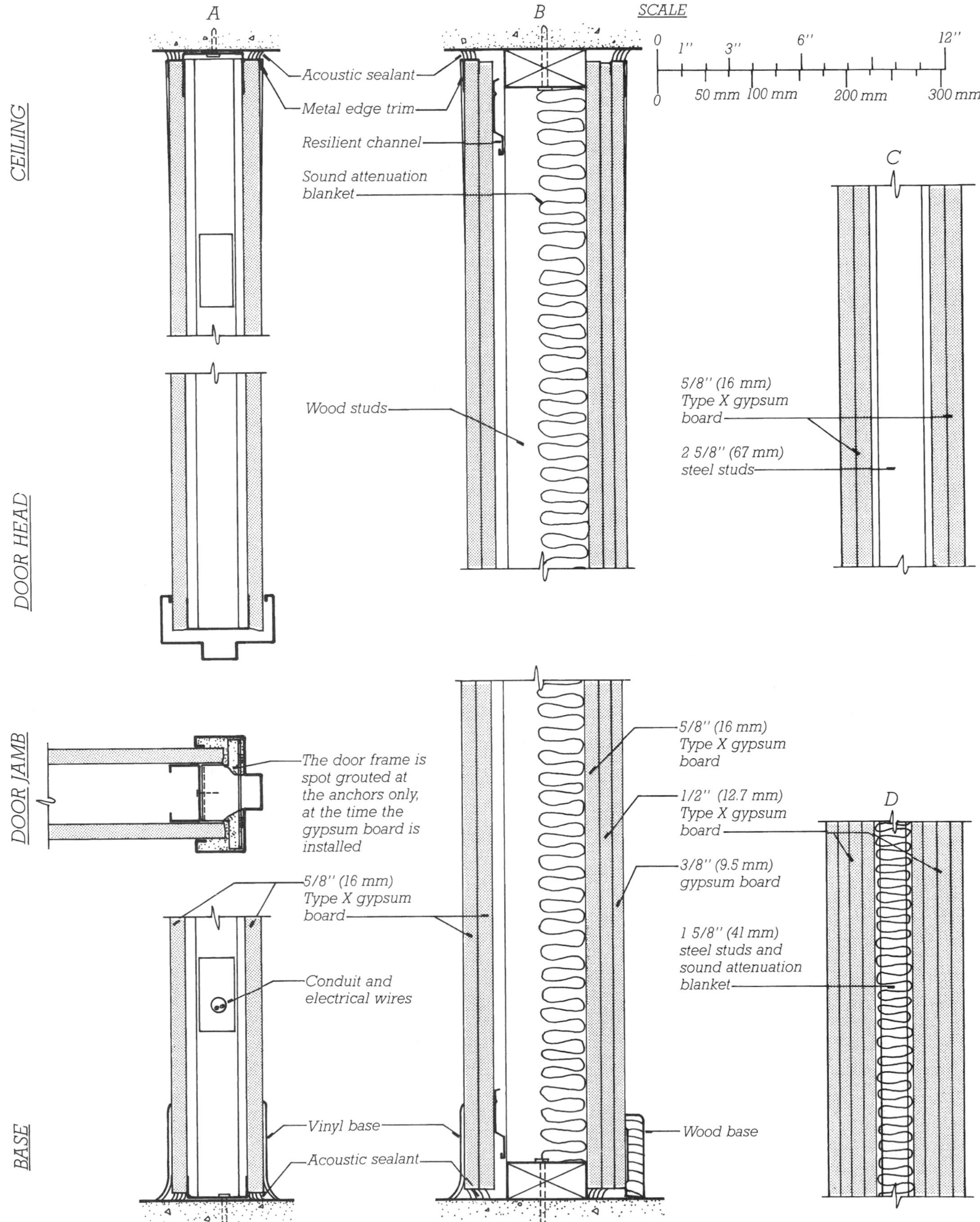
A
B
SCALE
0
1''
3''
6''
12''
0
50 mm
100 mm
200 mm
300 mm
CEILING
Acoustic sealant
Metal edge trim
Resilient channel
Sound attenuation blanket
C
Wood studs
5/8'' (16 mm) Type X gypsum board
2 5/8'' (67 mm) steel studs
DOOR HEAD
DOOR JAMB
The door frame is spot grouted at the anchors only, at the time the gypsum board is installed
5/8'' (16 mm) Type X gypsum board
1/2'' (12.7 mm) Type X gypsum board
D
3/8'' (9.5 mm) gypsum board
5/8'' (16 mm) Type X gypsum board
1 5/8'' (41 mm) steel studs and sound attenuation blanket
Conduit and electrical wires
BASE
Vinyl base
Wood base
Acoustic sealant

FIGURE 23.36
Installing a sound attenuation blanket. *(Courtesy of United States Gypsum Company)*

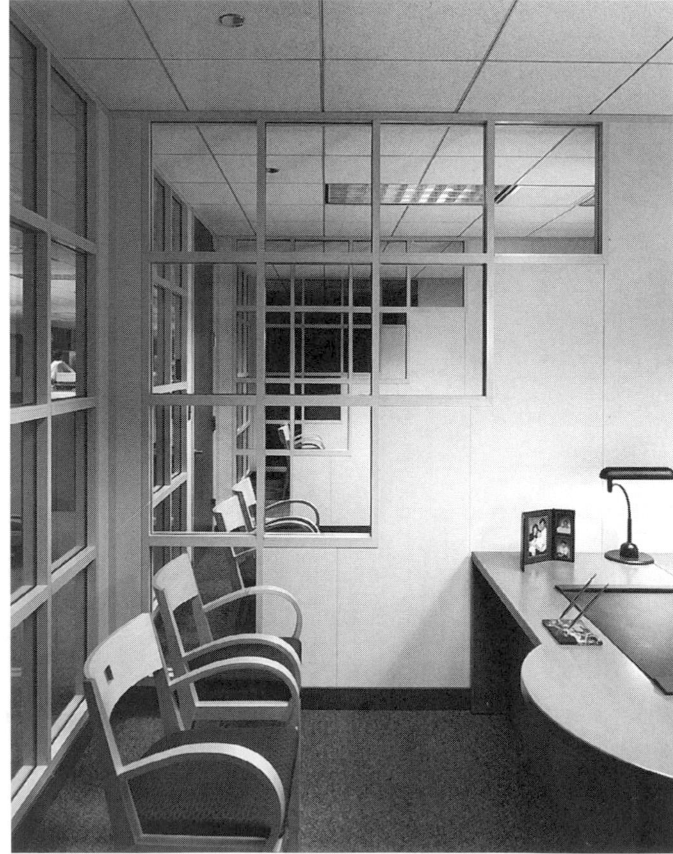

FIGURE 23.37
Two photographs of relocatable (demountable) partition systems. *(Courtesy of United States Gypsum Company)*

Gypsum Shaft Wall Systems

Walls around elevator shafts, stairways, and mechanical chases can be made of any masonry, lath and plaster, or gypsum panel assembly that meets fire resistance and structural requirements. Gypsum shaft wall systems offer several advantages over the alternatives: They are lighter in weight, they are installed dry, and they are built entirely from the floor outside the shaft, with no need to erect scaffolding inside. Depending on the requirements for fire resistance rating, air pressure resistance, STC, and floor-to-ceiling height, any of several designs may be used. Figure 23.38 shows representative shaft wall details, and Figure 22.1 shows a shaft wall being installed.

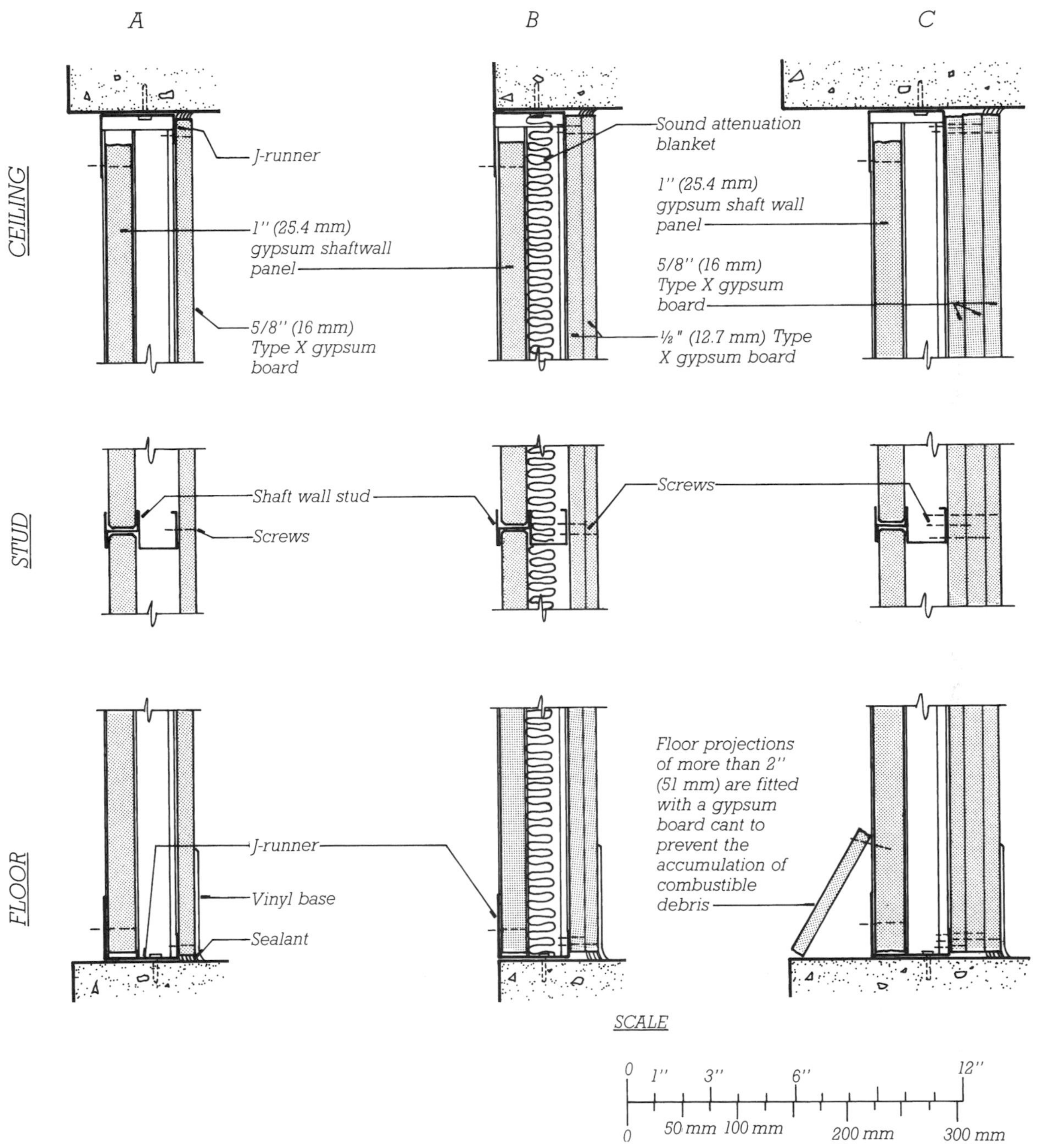

FIGURE 23.38
Three gypsum shaft wall systems, all framed with steel C–H studs. The H portion of the stud holds a 1-inch (25-mm) shaft wall coreboard panel, while the C portion accepts the screws used to attach finish layers of gypsum board. *(a)* A 1-hour system. *(b)* A 2-hour system, STC 47. *(c)* A 3-hour system.

Masonry Partition Systems

A century ago, interior partitions were often made of common brick masonry plastered on both sides. These had excellent STC and fire resistance ratings but were labor intensive and heavy. Partition systems of hollow clay tile and hollow gypsum tile (Figure 23.39) were developed to meet these objections and continued to be used extensively until the 1950s. Both have now become obsolete in North America, replaced by plaster, gypsum board, and concrete masonry, although they are still frequently encountered in the restoration of older buildings.

Concrete masonry partitions may be plastered or faced with gypsum board but are more often left exposed, either painted or unpainted. Several types of lightweight aggregate may be used to reduce the dead weight of the partition. Decorative concrete masonry units (architectural concrete unit masonry), as described in Chapter 9, may also be used. Electrical wiring is relatively difficult to conceal in concrete block partitions; the electrician and the mason must coordinate their work closely, or the wiring must be mounted on the surface of the wall after the mason has finished.

Glazed structural clay tiles make very durable partitions, especially in areas with heavy wear, moisture problems, or strict sanitation requirements (Figure 23.40). The ceramic glazes are nonfading and virtually indestructible.

Wall and Partition Facings

The vast majority of gypsum board partitions are finished with several coats of paint. For more information about paints and coatings, see pages 234–237. *Ceramic tile* facings are often added to walls for reasons of appearance, durability, sanitation, or moisture resistance. In a *thickset* or *mortar bed* application, tile is applied to a base of portland cement mortar (Figure 23.41).

Lower-cost tile wall facings eliminate the mortar base and are *thin-set* onto *tile backing boards*, also called *backer boards*, most frequently made of fiber-reinforced lightweight cement or glass-mat faced water-resistant gypsum board, similar to the floor tile assembly illustrated in Figure 24.29. Cement backer board is the more water resistant, but is more difficult to cut and handle than lighter-weight gypsum backer board. Water-resistant paper-faced gypsum backer board, used in the past for low-cost thin-set tile setting applications, is no longer considered sufficiently durable for tile backing applications, particularly in wet areas such as shower surrounds.

Tiles are bonded to the backer board with a variety of compounds, the most common of which are dry-set mortar, latex/polymer modified

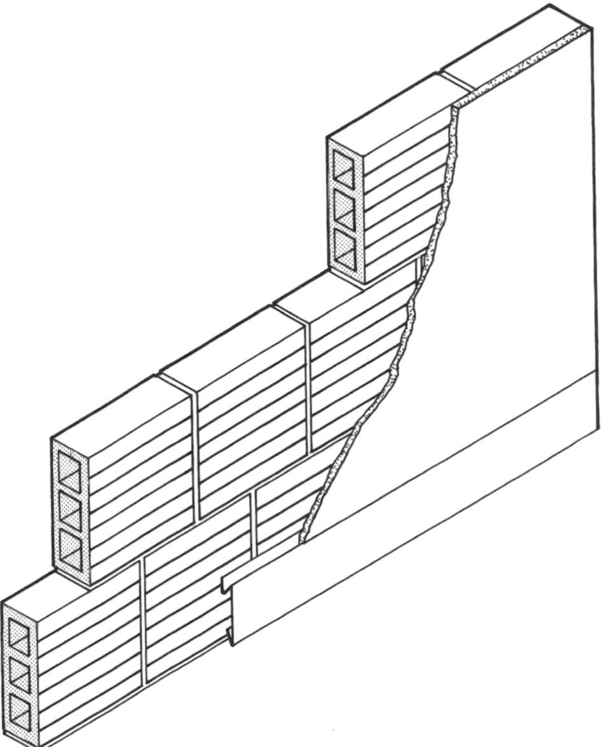

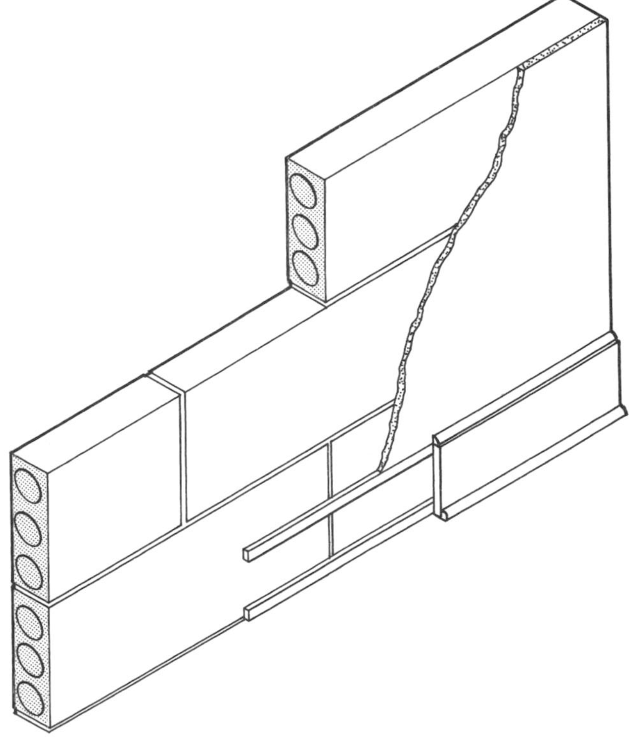

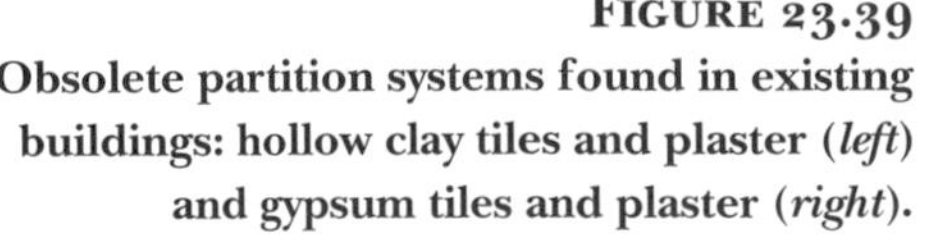

Figure 23.39
Obsolete partition systems found in existing buildings: hollow clay tiles and plaster (*left*) and gypsum tiles and plaster (*right*).

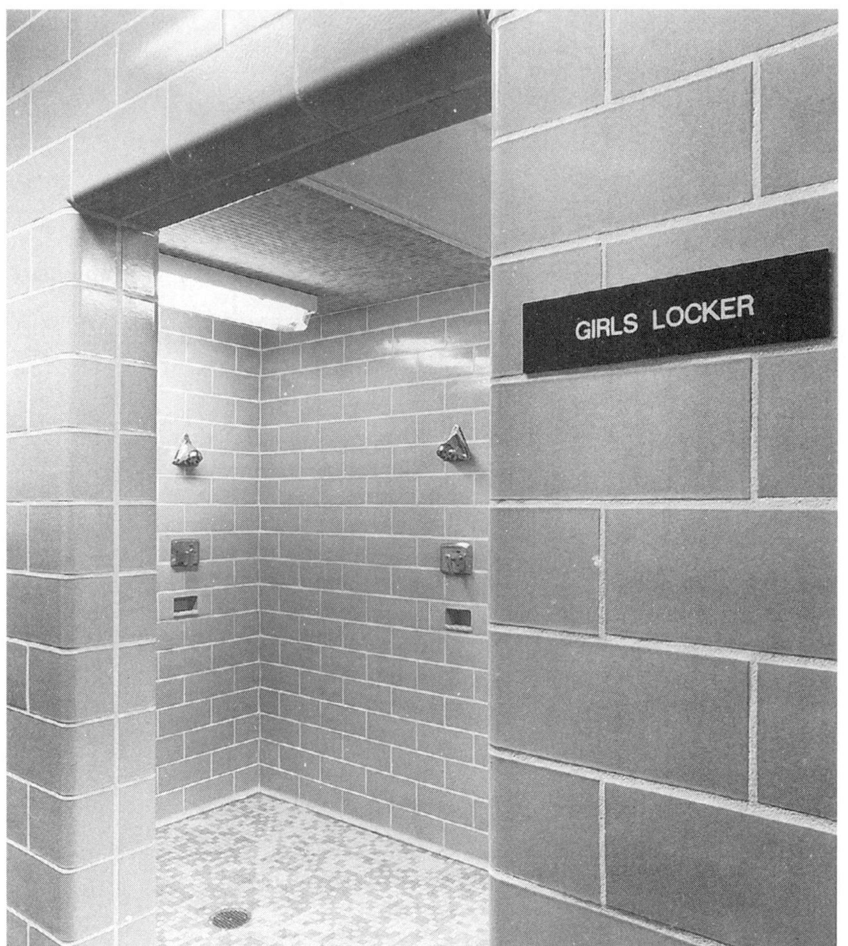

FIGURE 23.40
A glazed structural clay tile partition installation. The floor is finished with glazed ceramic tiles. *(Courtesy of Stark Ceramics, Inc.)*

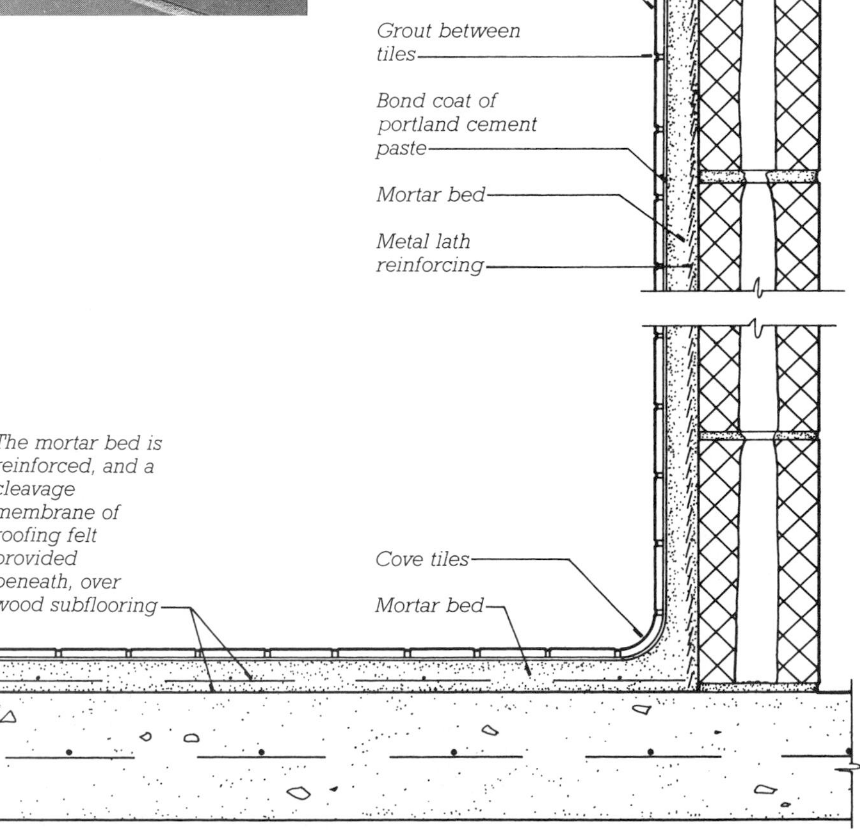

FIGURE 23.41
Example details of a thickset ceramic tile installation. Thickset applications are used where the face of the partition or the surface of the floor is cracked, coated, rough, unstable, or so uneven as to make it unsuitable for direct bonding of tile using thin-set methods. Depending on the quality of the substrate, metal reinforcing of the portland cement mortar base may or may not be required, and the mortar base may or may not be isolated from the substrate with a layer of felt paper acting as a slip sheet. Thickset mortar beds for wall tiling are typically ¾ to 1 inch (19–25 mm) thick, while those for floors are typically 1¼ to 2 inches (32–50 mm) thick.

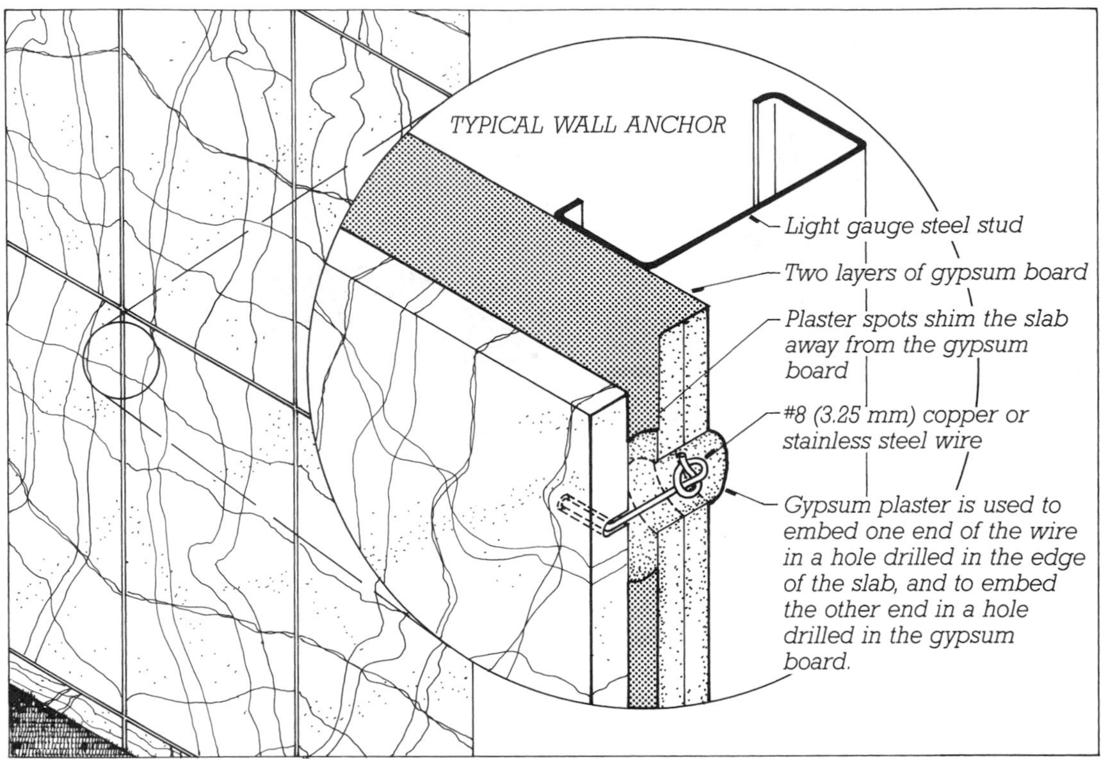
TYPICAL WALL ANCHOR
Light gauge steel stud
Two layers of gypsum board
Plaster spots shim the slab away from the gypsum board
#8 (3.25 mm) copper or stainless steel wire
Gypsum plaster is used to embed one end of the wire in a hole drilled in the edge of the slab, and to embed the other end in a hole drilled in the gypsum board.

FIGURE 23.42
Attaching a stone facing over a backup of gypsum boards and steel studs.

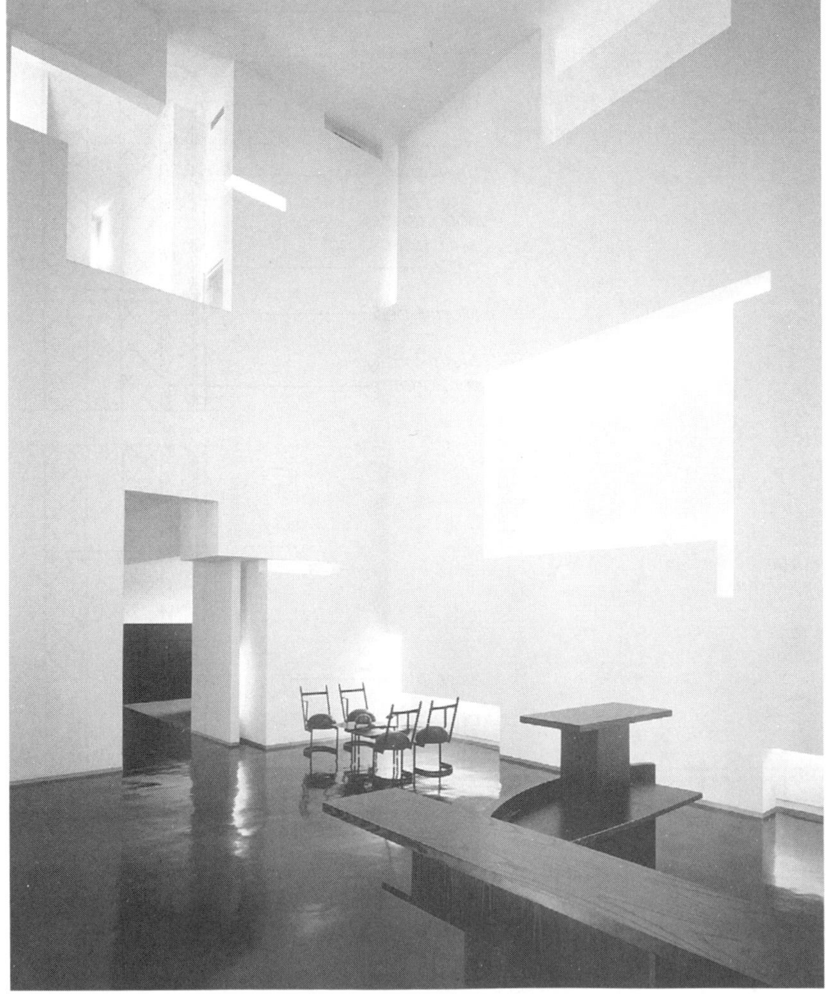

FIGURE 23.43
Imaginative use of gypsum products is coordinated with concealed sources of light to create a singularly dramatic space for the D. E. Shaw & Company office and trading area in New York.
(Stephen Holl Architects. Photo by Paul Warchol)

portland cement mortar, and organic adhesive. *Dry-set mortar* is a mix of cement, fine sand, and water retention compounds that allow the thin mortar layer to cure properly. *Latex/polymer modified portland cement mortar* is similar to dry-set mortar, but with additives that improve the cured mortar's freeze-thaw resistance, flexibility, and adhesion. *Organic adhesives,* various proprietary synthetic polymer adhesives, are used for light-duty tile applications. After the tiles have become fully adhered, a cementitious *grout* of any desired color is wiped into the tile joints with a rubber-faced trowel. Thin-set compounds and grouts formulated with epoxies or furan resins (colorless, highly volatile solvents distilled from wood) may also be used for tiling applications where greater strength, impact resistance, or chemical resistance is required.

In showers, steam rooms, and other wet locations, a *waterproofing membrane* should be added to the tile assembly to prevent water from seeping through the tile and into the wall behind. Either liquid-applied or flexible sheet membranes may be used, positioned either behind the base material or between the base and the finish tile.

Facings of granite, limestone, marble, or slate are sometimes used in public areas of major buildings. A common method of mounting is shown in Figure 23.42.

Wood wainscoting and paneling may be used in limited quantities in fire-resistant buildings, as discussed on pages 877–878. They are mounted over a backing of plaster or gypsum board to retain the fire-resistive qualities of the partition.

CSI/CSC

MasterFormat Sections for Interior Walls and Partitions

04 20 00	UNIT MASONRY
04 21 00	Clay Unit Masonry
	Glazed Structural Clay Tile Masonry
04 22 00	Concrete Unit Masonry
	Architectural Concrete Unit Masonry
06 40 00	ARCHITECTURAL WOODWORK
06 42 00	Wood Paneling
09 20 00	PLASTER AND GYPSUM BOARD
09 22 00	Supports for Plaster and Gypsum Board
	Metal Furring
	Non-Structural Metal Framing
	Gypsum Lath
	Metal Lath
	Veneer Plaster Base
09 23 00	Gypsum Plastering
09 24 00	Portland Cement Plastering
09 26 00	Veneer Plastering
09 28 00	Backing Boards and Underlayments
	Cementitious Backing Boards
	Glass-Mat Faced Gypsum Backing Boards
09 29 00	Gypsum Board
09 30 00	TILING
09 31 00	Thin-Set Tiling
09 32 00	Mortar-Bed Tiling
09 70 00	WALL FINISHES
09 75 00	Stone Facing
09 90 00	PAINTING AND COATING
09 91 00	Painting
	Interior Painting
10 20 00	INTERIOR SPECIALTIES
10 22 00	Partitions
	Demountable Partitions

Selected References

1. Gypsum Association. *GA-600, Fire Resistance Design Manual.* Washington, DC, updated regularly.

Fire resistance ratings and STCs are given in this booklet for a large number of wall and ceiling assemblies that use either gypsum plaster or gypsum board. This and many other useful resources are also available for free download from the Gypsum Association's web site.

2. USG Company. *Gypsum Construction Handbook.* Chicago, revised regularly.

This manual represents manufacturers' literature at its best—close to 600 well-illustrated pages crammed with every important fact about gypsum wallboard, gypsum plaster, and associated products. Available for purchase or as a free download from the USG web site.

3. USG Company. *SA100 – Fire-Resistance Assemblies Brochure.* Chicago, revised regularly.

Though limited to this company's products, this brochure provides an extensive and easy to use listing of fire-rated and acoustically rated wall and ceilings assemblies. Along with many other useful technical documents, it is available for free download from the USG web site.

4. Portland Cement Association. *Portland Cement Plaster (Stucco) Manual.* Skokie, IL, updated regularly.

A complete, illustrated guide to stucco.

5. Pegg, Brian, and W. Stagg. *Plastering: An Encyclopedia* (4th ed). Oxford, Blackwell Publishing Limited; New York, Crown Publishers, 2007.

Techniques of ornamental plastering are covered in complete detail in this 276-page reference work.

6. Tile Council of North America, Inc. *TCA Handbook for Ceramic Tile Installation.* Anderson, SC, updated regularly.

This is the definitive standard for ceramic tile installation materials and methods. Over 100 methods of installation for floors and walls, both interior and exterior, are illustrated and specified. Guidelines for selecting appropriate installation methods based on project requirements are also included.

7. Marble Institute of America. *Dimension Stone Design Manual.* Cleveland, OH, updated regularly.

This binder provides comprehensive guidance on the design, specification, and installation of all types of stone wall facing.

Web Sites

Interior Walls and Partitions

Author's supplementary web site: **www.ianosbackfill.com/23_interior_walls_and_partitions**

Dietrich Metal Framing: **www.dietrichindustries.com**

Drywallrecyling.org: **www.drywallrecycling.org**

Georgia-Pacific: **www.gp.com/build**

Gypsum Association: **www.gypsum.org**

Marble Institute of America: **www.marble-institute.com**

National Gypsum: **www.national-gypsum.com**

Portland Cement Association: **www.pca.org**

Temple-Inland: **www.temple.com/gypsum**

Tile Council of North America (TCNA): **www.tileusa.com**

USG: **www.usg.com**

Key Terms and Concepts

fire wall
shaft wall
fire barrier
fire partition
fire area
smoke barrier
smoke partition
light gauge steel stud
runner channel
furring strip
hat channel
fireblocking
synthetic gypsum
plaster
lath
wattle and daub
gypsum
calcining
calcined gypsum
plaster of Paris
base-coat plaster
finish-coat plaster
mill-mixed base-coat plaster, ready-mixed base-coat plaster
neat base-coat plaster
gypsum plaster
wood-fibered gypsum plaster
lightweight gypsum plaster
high-strength base-coat plaster
lime
ready-mixed finish plaster
gauging plaster
hydrated finishing lime, lime putty
high-strength gauging plaster
Keenes cement
molding plaster
portland cement–lime plaster, stucco
hawk
trowel
darby
keying
lather
expanded metal lath
self-drilling, self-tapping screw
lathing nail
gypsum lath
veneer plaster base, gypsum veneer base
trim accessories
control joint
ground
scratch coat
brown coat
edge bead
corner bead
plaster screed
finish coat
veneer plaster
self-furring metal lath
line wire
casting
running
cast plaster ornament
run plaster ornament
blade
template
gypsum board, gypsum wallboard, plasterboard, drywall
Type X gypsum board
Type C gypsum board
water-resistant gypsum backing board
abuse-resistant gypsum board, impact-resistant gypsum board
mold-resistant gypsum board
coreboard
ceiling gypsum board
foil-backed gypsum board
predocorated gypsum board
exterior gypsum soffit board
tapered edge
gypsum backing board
nail popping
joint compound
drying-type joint compound
setting compound
finish level
fire-taping
skim coat
demountable partition system
glazed structural clay tile
ceramic tile
thickset tile, mortar bed tile
thin-set tile
tile backing board, tile backer board
dry-set mortar
latex/polymer modified portland cement mortar
organic adhesive
grout
waterproofing membrane

Review Questions

1. What are the major types of interior walls and partitions in a larger building, such as a hospital, classroom building, apartment building, or office building? How do these types differ from one another?

2. Why is gypsum used so much in interior finishes?

3. Name the coats of plaster used over expanded metal lath and explain the role of each.

4. Under what circumstances would you specify the use of portland cement plaster? Keenes cement plaster? Veneer plaster?

5. Describe step by step how the joints between sheets of gypsum board are made invisible.

Exercises

1. Determine the construction of a number of partitions in the places where you live and work. What materials are used? What accessories? Why were these chosen for their particular situations? Sketch a detail of each partition.

2. Sketch typical details showing how the various metal lath trim accessories are used in a plaster wall.

3. Repeat Exercise 2 for gypsum board trim accessories.

4. What type of gypsum wall finish system would you specify for a major art museum? For a low-cost rental office building? Outline a complete specification of wall and partition construction for a building you are presently working on.

USG Interiors, Inc.
USG Interiors, Inc.
Acoustone
417
W
USG Interiors, Inc.
Acoustone
417
W

24

Finish Ceilings and Floors

A worker lays acoustical panels in a recessed grid to create a suspended ceiling. *(Courtesy of United States Gypsum Company)*

As the ceilings and finish floors are being installed, the construction of a building is drawing to a close. The components of the mechanical and electrical systems that remain exposed are either finished or concealed, intersections of interior surfaces are neatly trimmed, and painters work their magic to reveal for the first time the interior character of the building. The architect, engineers, and municipal building officials make their last inspections, and, following last-minute corrections of minor defects, the contractor turns the building over to its owner.

Finish Ceilings

Functions of Finish Ceilings

The ceiling surface is an important functional component of a room. It helps control the diffusion of light and sound about the room. It may play a role in preventing the passage of sound vertically between the rooms above and below, and horizontally between rooms on either side of a partition. It is often designed to resist the passage of fire and must itself be appropriately noncombustible. Frequently, it is called upon to assist in the distribution of conditioned air, artificial light, and electrical energy. In many buildings, it must accommodate sprinkler heads for fire suppression and loudspeakers for intercommunication systems. And its color, texture, pattern, and shape are prominent in the overall visual impression of the room. A ceiling can be a simple, level plane, a series of sloping planes that give a sense of the roof above, a luminous surface, a richly coffered ornamental ceiling, or even a frescoed plaster vault such as Michelangelo's famous ceiling in the Sistine Chapel in Rome; the possibilities are endless.

Types of Ceilings

Exposed Structural and Mechanical Components

In many buildings, it makes sense to omit finished ceiling surfaces altogether and simply expose the structural and mechanical components of the floor or roof above (Figure 24.1). In industrial and agricultural buildings, where appearance is not of prime importance, this approach offers the advantages of economy and ease of access for maintenance. Many types of floor and roof structures are inherently attractive if left exposed, such as heavy timber beams and decking, concrete waffle slabs, and steel trusses.

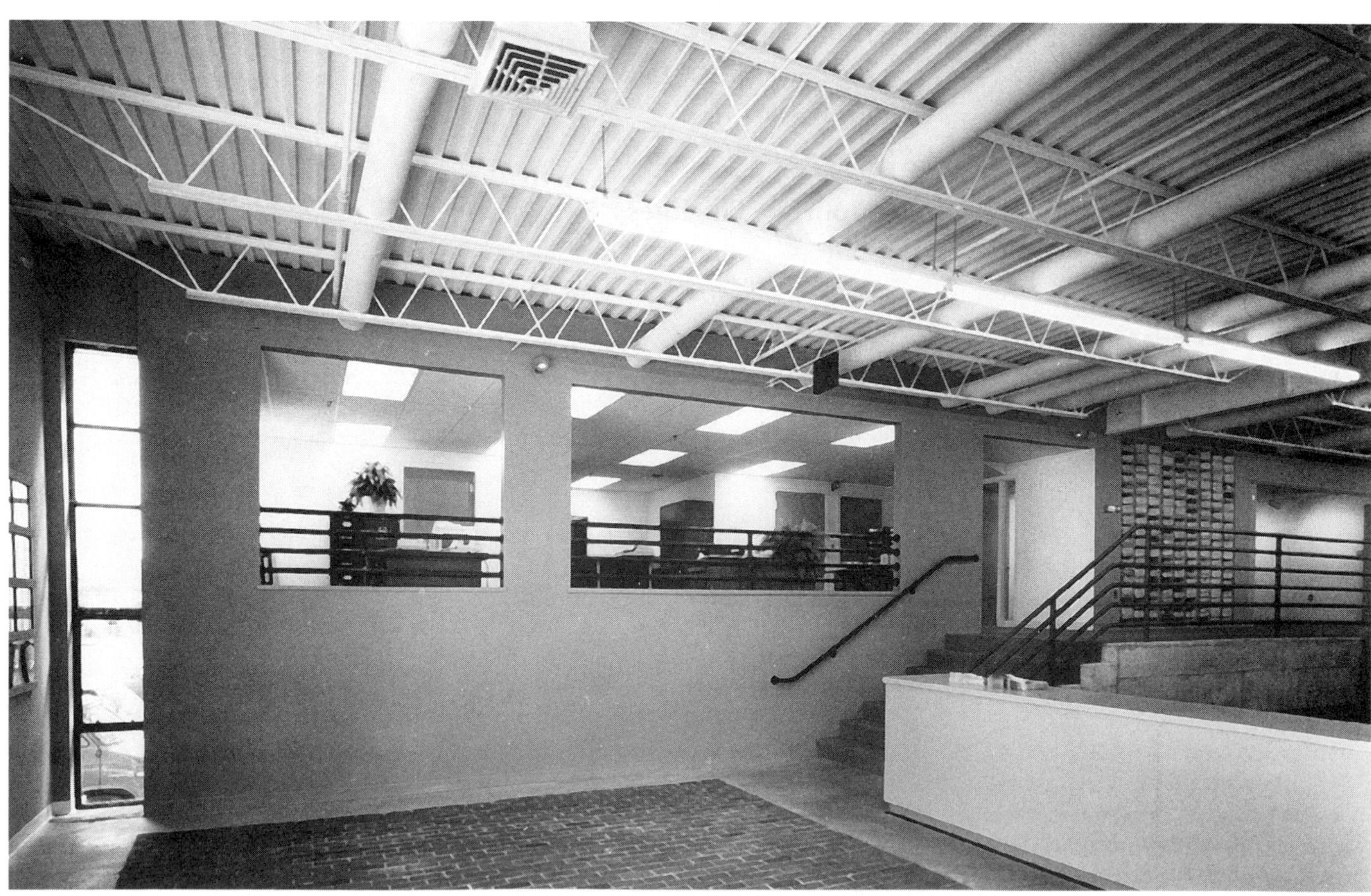

Other types of structures, such as concrete flat plates and precast concrete planks, have little visual interest, but their undersurfaces can be painted and left exposed as finished ceilings in apartment buildings and hotels, which have little need for mechanical services at the ceiling. This saves money and reduces the overall height of the building. In some buildings, the structural and mechanical elements at the ceiling, if carefully designed, installed, and painted, can create a powerful aesthetic of their own.

Exposing structural and mechanical components rather than covering them with a finished ceiling does not always save money. Mechanical and structural work is not normally done in a precise, attractive fashion because it is not usually expected to be visible, and it is less expensive for workers to take only as much care in installation as is required for satisfactory functional performance. To achieve perfectly straight, neatly sealed ductwork that is free of dents, steel decks without rust and weld spatter, and square, well-organized runs of electrical conduit and plumbing, the drawings and specifications for the project must tell exactly the results that are expected, and a higher labor cost must be anticipated.

Tightly Attached Ceilings

Ceilings of any material may be attached tightly to wood joists, wood rafters, steel joists, or concrete slabs (Figure 24.2). Special finishing arrangements must be worked out for any beams and girders that protrude through the plane of the ceiling, and for ducts, conduits, pipes, and sprinkler heads that fall below the ceiling.

Suspended Ceilings

A ceiling that is suspended on wires some distance below the floor or roof structure can hang level and flat despite varying sizes of girders, beams, joists, and slabs above, and even under, a roof structure that slopes toward roof drains. Ducts, pipes, and conduits can run freely in the *plenum* space between the ceiling and the structure above. Lighting fixtures, sprinkler heads, loudspeakers, and fire detection devices may be recessed into the ceiling. Such a ceiling can also, at additional cost, serve as *membrane fire protection* for the floor or roof structure above, eliminating the need for fussy individual fireproofing of steel joists, or imparting a higher fire resistance rating to wood or precast concrete structures. For these reasons, *suspended ceilings* are a popular and economical feature in many types of buildings, especially office and retail structures.

Suspended ceilings can be made of almost any material; the ones most widely used are gypsum board, plaster, and various proprietary panels and tiles composed of incombustible fibers. Each of these materials is supported on its own special system of small steel framing members that hang from the structure on heavy steel wires.

FIGURE 24.1
Sprinkler pipes, air conditioning ductwork, electrical conduits, and lighting fixtures are exposed in a ceiling structure of painted open-web steel joists and corrugated steel decking. The office space behind has a suspended ceiling of lay-in acoustical panels with recessed lighting fixtures. The floor is of brick with a border of concrete. *(Woo & Williams, Architects. Photographer: Richard Bonarrigo)*

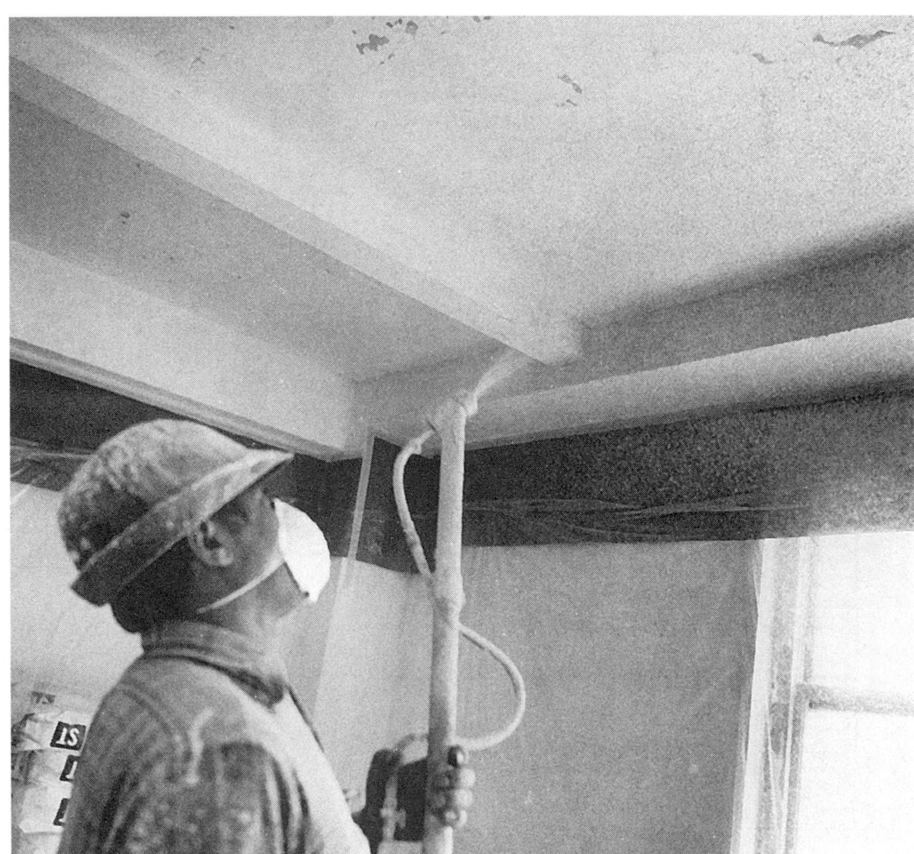

FIGURE 24.2
Spraying a textured finish onto the underside of a concrete slab in a residential building, where there are no pipes, ducts, or wires to be concealed below the plane of the floor structure. *(Courtesy of United States Gypsum Company)*

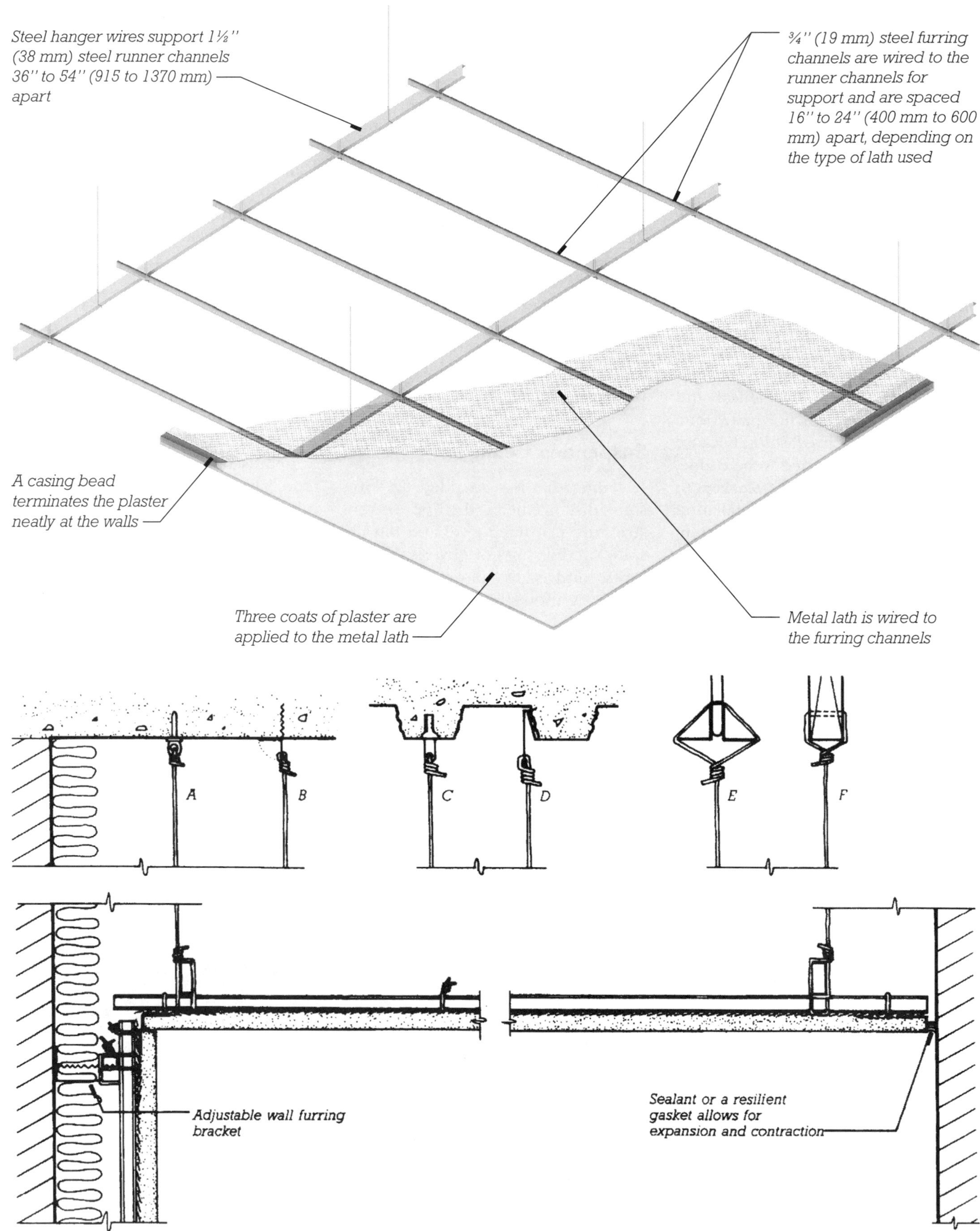

Steel hanger wires support 1½" (38 mm) steel runner channels 36" to 54" (915 to 1370 mm) apart
¾" (19 mm) steel furring channels are wired to the runner channels for support and are spaced 16" to 24" (400 mm to 600 mm) apart, depending on the type of lath used
A casing bead terminates the plaster neatly at the walls
Three coats of plaster are applied to the metal lath
Metal lath is wired to the furring channels
A
B
C
D
E
F
Adjustable wall furring bracket
Sealant or a resilient gasket allows for expansion and contraction

FIGURE 24.3
A suspended plaster ceiling on metal lath. At the top of the page is a cutaway isometric drawing, as viewed from below, of the essential components of the ceiling. Across the center of the page are details of six ways of supporting the hanger wires: (*a*) A pin is powder driven into a concrete structure. (*b*) A corrugated sheet metal tab with a hole punched in it is nailed to the formwork before the concrete is poured. When the formwork is stripped, the tab bends down and the hanger wire is threaded through the hole. (*c*) A sharp, daggerlike tab of sheet metal is driven through the corrugated metal decking before the concrete topping is poured. (*d*) A sheet metal hook is hung onto the lap joints in the metal decking. (*e*) The hanger wire is wrapped around the lower chord of an open-web steel joist. (*f*) The hanger wire is passed through a hole drilled near the bottom of a wood joist. At the bottom of the page is a section through a furred, insulated plaster wall and a suspended plaster ceiling.

With automated fabrication of custom elements using computerized modeling techniques (see Chapter 1), an expanded variety of panelized ceilings systems are becoming available to designers. These systems can economically produce panels made of sheet metal and, optionally, thin wood veneers, which are unique in appearance and shape, lightweight, noncombustible, easy to install, and highly effective at reducing ambient noise levels.

Suspended Gypsum Board and Plaster Ceilings

Gypsum board suspended ceilings may be screwed to ordinary light gauge steel cee channels that are suspended on wires. Special framing components have been developed that make it easy to suspend more complex shapes of gypsum board ceilings, such as cylindrical vaults, undulating surfaces, and deep coffers.

Suspended plaster ceilings have been in use for many decades; some typical details are shown in Figure 24.3. Although most suspended plaster ceilings are flat, lathers are capable of constructing ceilings that are richly sculpted, ranging from configurations resembling highly ornamented Greek or Roman coffered ceilings to nearly any form that the contemporary designer can draw. This capability is especially useful in auditoriums, theaters, lobbies of public buildings, and other uniquely shaped rooms.

Suspended Acoustical Ceilings

Ceilings made from fibrous materials in the form of lightweight tiles or panels are customarily referred to as *acoustical ceilings* because most of them are highly absorptive of sound energy, unlike plaster and gypsum board, which are highly reflective of sound. They are also often less costly than either plaster or even gypsum board ceilings. The sound absorption performance of a ceiling material is measured and published in the trade literature as its *Noise Reduction Coefficient* (*NRC*). NRC is a number between 0 and 1, with higher values representing higher levels of sound absorption at four specific frequencies ranging from 250 Hz to 2000 Hz. An NRC of 0.85 indicates that the material absorbs 85 percent of the sound that reaches it and reflects only 15 percent back into the room. NRCs for most acoustical ceiling materials range from 0.50 to 0.90, compared to values below 0.10 for plaster and gypsum board ceilings. This makes acoustical ceilings valuable for reducing noise levels in lobbies, office spaces, restaurants, retail stores, recreational spaces, and noisy industrial environments.

The lightweight, porous materials that produce high NRC ratings allow most sound energy to pass through. In other words, a ceiling made of porous materials will not furnish very good acoustic privacy between adjacent rooms unless a suitable full-height wall separates the rooms and blocks the ceiling plenum. The ability of a ceiling system to reduce sound transmission from one room to another through a shared plenum is measured by its *Ceiling Attenuation Class* (*CAC*). CAC is measured in decibels, with higher values representing greater reductions in sound transmission. For closed offices with shared ceiling plenums, a ceiling system with a CAC of not less than of 35 to 40 is recommended. Dense, nonporous ceiling materials tend to have higher CACs than lighter, more porous materials.

A third measure of ceiling acoustical performance is *Articulation Class* (*AC*). Like NRC, AC is a measure of sound reflection and absorption. However, AC is intended specifically to measure a ceiling system's contribution to speech clarity and privacy in a typical open office environment. It measures a ceiling's absorption and reflection of sound over a 60-inch (1500-mm)-high partition at frequencies ranging from 500 to 4000 Hz, those particularly critical to normal speech. Higher AC values represent greater acoustical clarity and privacy, with minimum recommended values falling in the range of 170 to 200.

Where both noise reduction within a space and sound attenuation

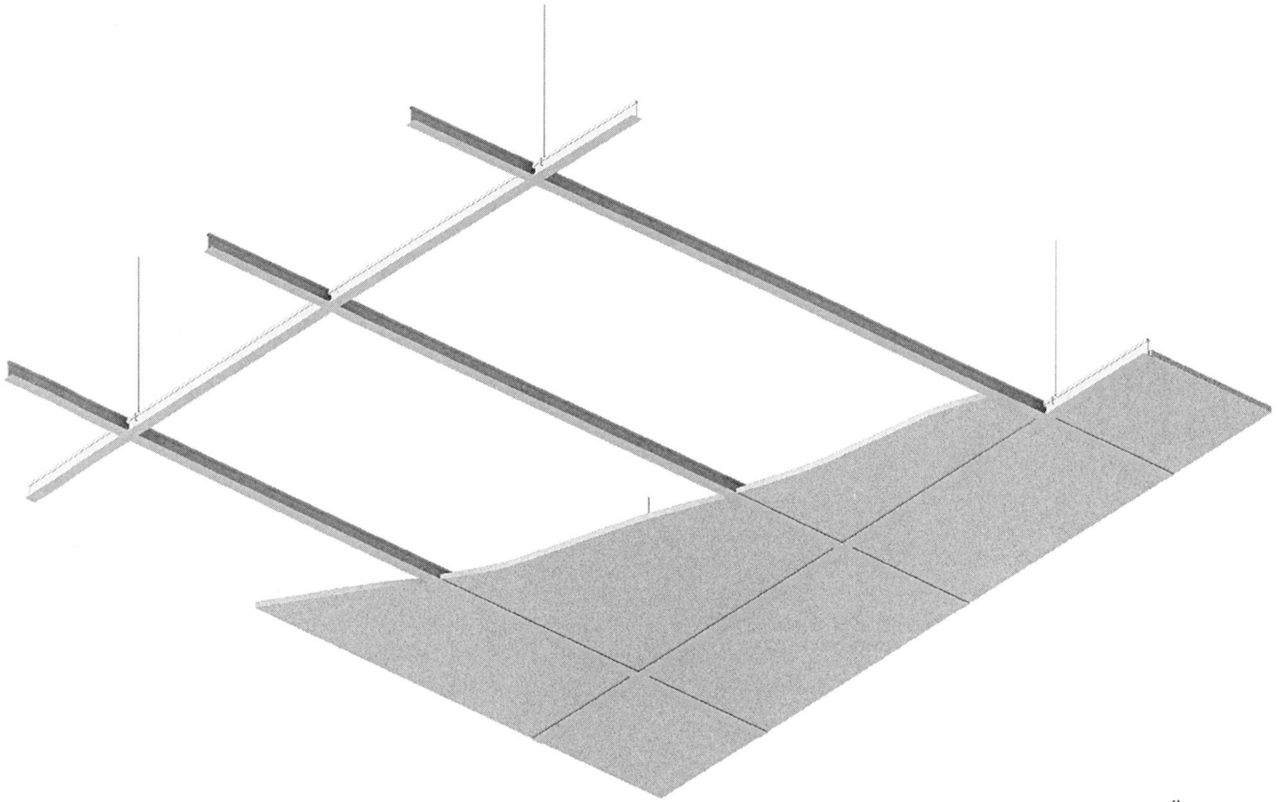

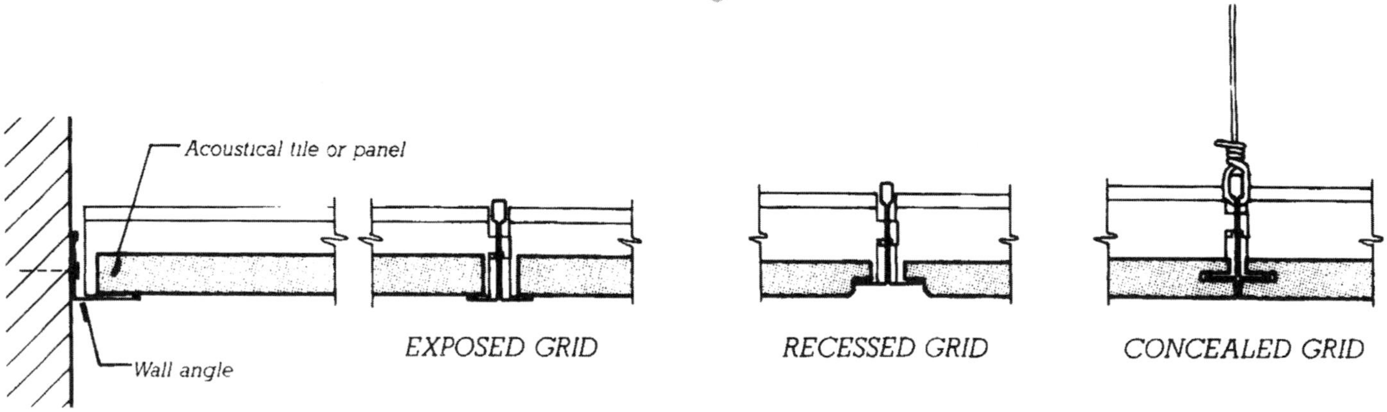

Acoustical tile or panel

Wall angle

EXPOSED GRID

RECESSED GRID

CONCEALED GRID

FIGURE 24.4
Acoustical ceilings are supported on suspended grids of tees formed from sheet metal. At the top of the figure is a cutaway view looking up at an acoustical ceiling of lay-in panels. Below are sections illustrating how the grid may be exposed, recessed, or concealed for different visual appearances.

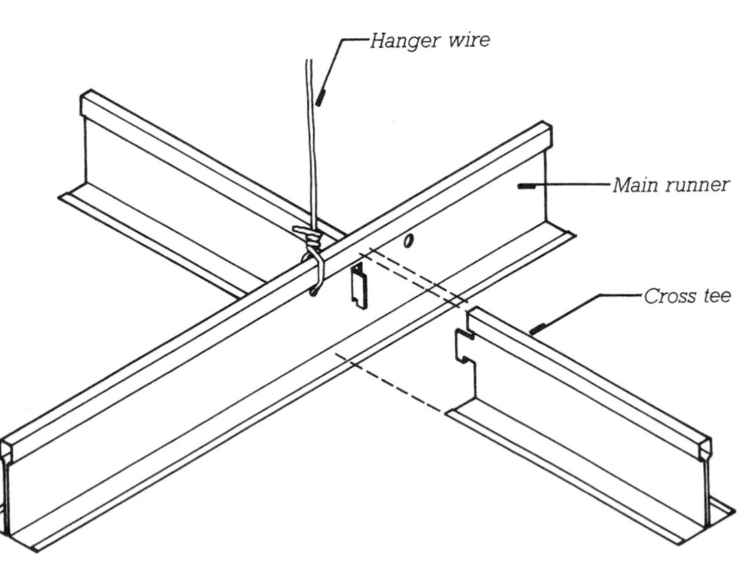

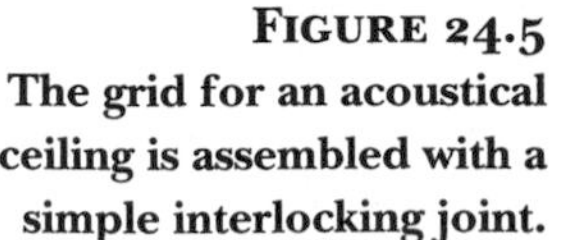
FIGURE 24.5
The grid for an acoustical ceiling is assembled with a simple interlocking joint.

between spaces are required simultaneously, composite ceiling panels with a highly absorbent material laminated to a dense substrate may be used; these have high values for both noise reduction (NRC or AC) and sound attenuation (CAC). The same result can be achieved by mounting acoustically absorbent tiles on a suspended ceiling of plaster or gypsum board.

The most economical acoustical ceiling systems consist of *lay-in panels* that are supported by an *exposed grid* (Figures 24.4 and 24.5). Any panel in the ceiling can be lifted and removed for access to services in the plenum space. For a smoother appearance, a *concealed grid* system may be used instead. Concealed grid systems require special panels for plenum access. Suspended acoustical ceilings are available in hundreds of different designs. Figures 24.6–24.11 show some typical examples.

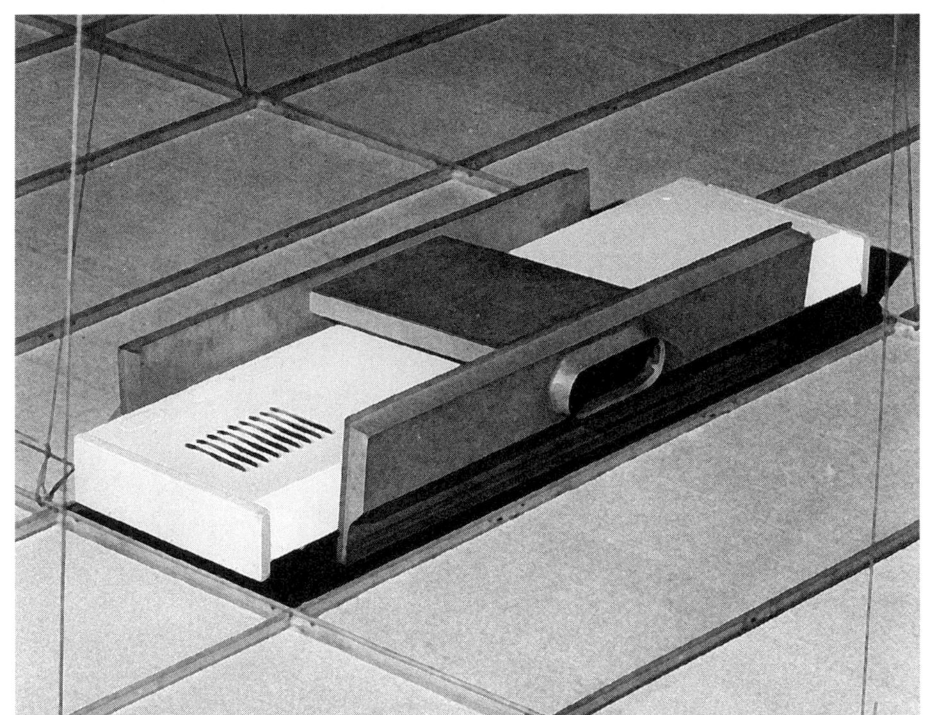

FIGURE 24.6
Many acoustical ceilings are manufactured as *integrated ceiling systems* that incorporate the lighting fixtures and air conditioning outlets into the module of the grid. In this integrated ceiling, viewed from above, the hanger wires, grid, acoustical panels, fluorescent lighting fixture, and distribution boot for conditioned air have been installed. The boot will be connected to the main ductwork with a flexible oval duct. *(Photo courtesy Armstrong World Industries)*

FIGURE 24.7
As viewed from below, the integrated ceiling shown in Figure 24.6 has a slot around the lighting fixture through which air is distributed from the boot above. The roughly textured acoustical panels used in this example are patterned with two cross grooves that work with the recessed grid to create the look of a ceiling composed of smaller square panels. *(Photo courtesy Armstrong World Industries)*

FIGURE 24.8
Air distribution in this integrated acoustical ceiling is through slots that occur between panels. Lighting fixtures, loudspeakers, and smoke detectors are incorporated simply and unobtrusively. *(Courtesy of United States Gypsum Company)*

FIGURE 24.9
Coffered acoustical ceilings that act as light diffusers are designed and marketed as integrated systems. *(Photo courtesy Armstrong World Industries)*

Economy has worked so great a change in our dwellings, that their ceilings are, of late years, little more than miserable naked surfaces of plaster. [A discussion of ceiling design will] possess little interest in the eye of speculating builders of the wretched houses erected about the suburbs of the metropolis, and let to unsuspecting tenants at rents usually about three times their actual value. To the student it is more important, inasmuch as a well-designed ceiling is one of the most pleasing features of a room.

Joseph Gwilt, *The Encyclopedia of Architecture*, London, 1842

FIGURES 24.10 AND 24.11
A variety of patterns and textures are available in acoustical ceilings. *(Photos courtesy Armstrong World Industries)*

Suspended Linear Metal Ceilings

Figure 24.12 illustrates a suspended *linear metal ceiling* made of long elements that are formed from sheet aluminum attached to a special type of concealed grid.

Suspended Fire Resistance Rated Ceilings

Suspended ceilings that are part of a fire resistance rated floor–ceiling or roof–ceiling assembly may be made of gypsum board, plaster, or lay-in panel and grid systems that are specially designed to have the necessary resistance to the passage of fire. Penetrations in such membrane ceilings must be detailed so as to maintain the required degree of fire resistance throughout the ceiling: Lighting fixtures must be backed up with fire-resistive material, air conditioning grills must be isolated from the ducts that feed them by means of automatic fire dampers, and any access panels provided for maintenance of above-ceiling services must themselves meet requirements for fire resistance.

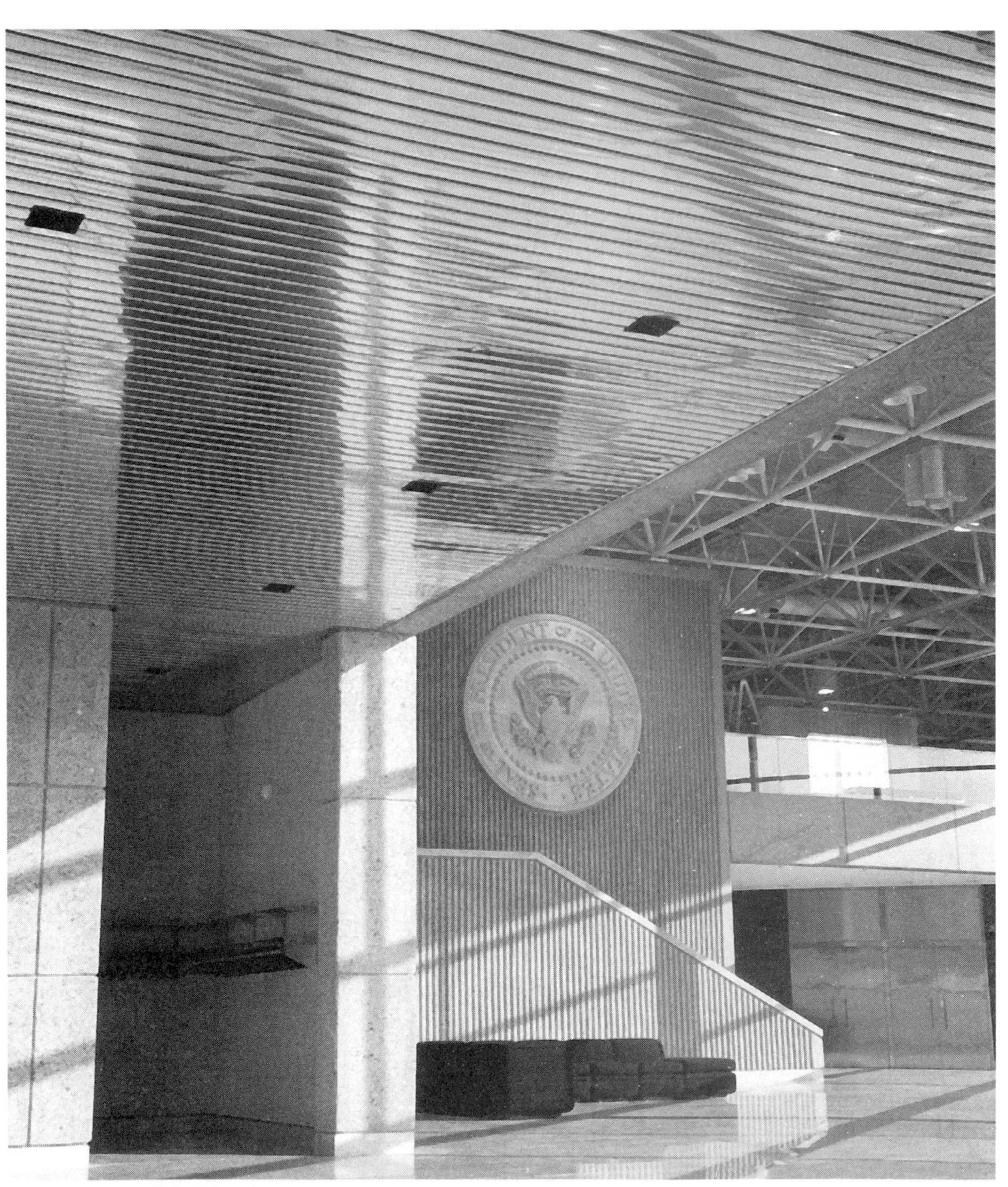

FIGURE 24.12
A mirror finish linear metal suspended ceiling, with an exposed steel space truss structure beyond. *(Architect: DeWinter and Associates. Photo courtesy of Alcan Building Products, a division of Alcan Aluminum Corporation)*

Interstitial Ceilings

Many hospital and laboratory buildings have extremely elaborate mechanical and electrical systems, including not just the usual air conditioning ducts, water and waste piping, and electrical and communications wiring, but also such services as fume hood ducting, fuel gas lines, compressed air lines, oxygen piping, chilled water piping, vacuum piping, and chemical waste piping. These ducts and tubes occupy a considerable volume of space in the building, often in an amount that virtually equals the inhabited volume. Furthermore, all these systems require continual maintenance and are subject to frequent change. As a consequence, many laboratory and hospital buildings are designed with *interstitial ceilings*. An interstitial ceiling is suspended at a level that allows workers to travel freely in the plenum space, usually while walking erect, and is structured strongly enough to support safely the weight of the workers and their tools. In effect, the plenum space becomes another floor of the building, slipped in between the other floors, and the overall height of the building must be increased accordingly. Its advantage is that maintenance and updating work on the mechanical and electrical systems of the building can be carried on without interrupting the activities below. Interstitial ceilings are made of gypsum or lightweight concrete and combine the construction details of poured gypsum roof decks and suspended plaster ceilings. Figure 24.13 shows the installation of an interstitial ceiling.

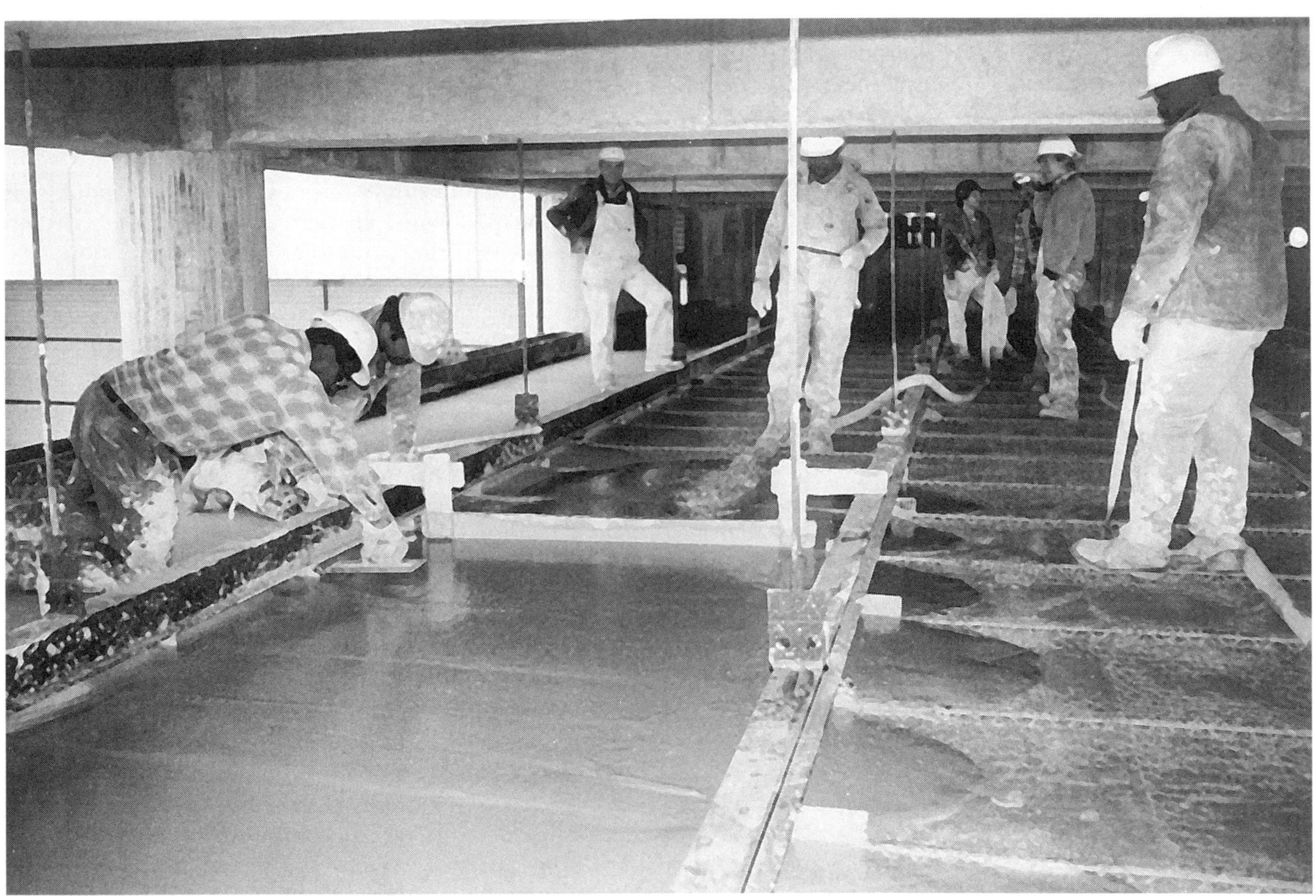

FIGURE 24.13
The final steps in constructing an interstitial ceiling: The ceiling plane consists of gypsum reinforced with hexagonal steel mesh. It is framed with steel truss tee subpurlins, the tops of which are visible at the right of the picture, supported by steel wide-flange beams suspended on rods from the sitecast concrete framing above. The final layer of gypsum is being pumped onto the ceiling from the hose near the center of the picture. The wet gypsum is struck off level with the wooden straightedge seen here hanging on the beams and troweled to a smooth walking surface. When the gypsum has hardened, installation of the ductwork, piping, and wiring in the interstitial plenum space can begin, with workers using the gypsum ceiling as a walking surface. *(Photo compliments Keystone Steel & Wire Co.)*

Considerations of Sustainability in Finish Ceilings and Floors

Acoustical Ceilings

- Acoustical ceiling tile can be a source of volatile organic compound (VOC) emissions as well as a reservoir for emissions from other sources.
- Low-emitting acoustic tiles are available, as are tiles with high recycled materials content.
- When the space above suspended ceilings is used as a return air plenum for the building's HVAC system, contaminants and emissions from the acoustical tile can be introduced into the system's air stream and redistributed to other parts of the building.

Hard Flooring Materials

- Concrete, stone, masonry, ceramic tile, and cementitious mortars and grouts are chemically inert and generally free of emissions.
- Organic adhesives used in tiling and resins used in thin-set terrazzo may be sources of emissions.
- Sealers applied to hard flooring materials to provide water repellency and protection from staining are potential sources of emissions. Solvent-based sealers generally have higher emissions than water-based products.

Wood and Bamboo Flooring

- Considerations of sustainability in wood products are discussed in Chapter 3 of this volume.
- Bamboo is harvested in a 4- to 6-year cycle and is considered a rapidly renewable material. Some bamboo flooring products are produced with urea-formaldehyde glues, a potential source of emissions. Others are manufactured with alternative low-emitting adhesives.
- Finishes for wood and bamboo flooring are potential sources of emissions. Water-based finishes and waxes are generally lower emitting than solvent- based finishes, but they may not be as durable or as easy to apply.

Resilient Flooring

- Self-leveling cements used to prepare subfloors for resilient flooring coverings are potential sources of emissions.
- Vinyl (polyvinyl chloride) is a component of many resilient floor coverings and other interior finish products. It has an embodied energy of approximately 30,000 BTU per pound (70 MJ/kg). At the time of this writing, it remains a controversial material from a sustainability standpoint:
  - Vinyl manufacture releases significant amounts of toxic air pollutants. Plasticizers in vinyl products tend to outgas over time, and some are toxic. The more flexible a vinyl product is, the more plasticizers it contains. When vinyl burns, it can release hydrochloric acid and dioxins. Although vinyl is a thermoplastic, and therefore readily recyclable, today most vinyl building products are disposed of in landfills at the end of their normal lifespan.
  - Vinyl materials are lightweight, strong, and relatively low in embodied energy in comparison to many available alternative materials. The vinyl manufacturing

Finish Flooring

Functions of Finish Flooring

Floors have a lot to do with our visual and tactile appreciation of a building. We sense their colors, patterns, and textures, their "feel" underfoot, and the noises they make in response to footsteps. Floors affect the acoustics of a room, contributing to a noisy quality or a hushed quality, depending on whether a hard or soft flooring material is used. Floors also interact in various ways with light: Some floor materials give mirrorlike reflections; others give diffuse reflections or none at all. Dark flooring materials absorb most of the light incident upon them and contribute to the creation of a darker room, whereas light materials reflect most incident light and help create a brighter room.

Floors are also a major functional component of a building. They are its primary wearing surfaces, subject to water, grit, dust, and the abrasive and penetrating actions of feet and furniture. They require more cleaning and maintenance effort than any other component of a building. They must be designed to deal with problems of skid resistance, sanitation, noise reduction between floors of a building, and even electrical conductivity in occupancies such as computer rooms and hospital operating rooms, where static electricity would pose a threat. And like other interior finish components, floors must be selected with an eye to combustibility, fire resistance ratings, and the structural loads that they will place on the frame of the building.

Underfloor Services

Floor structures are frequently used for the distribution of electrical and communications wiring, especially in floor areas that are broad and have

industry continues to improve manufacturing processes and to reduce vinyl-related health risks. Studies of the full life-cycle impacts of vinyl building materials compared to available alternatives, undertaken by the U.S. Green Building Council and The Natural Step (a United Kingdom sustainability organization), have recommended against the elimination of vinyl materials from building construction at the present time.

- Some vinyl tiles are manufactured from recycled vinyl.

- The embodied energy of synthetic rubber is about 50,000 BTU per pound (120 MJ/kg), and that of natural latex rubber is about 30,000 BTU per pound (70 MJ/kg).
- Some rubber flooring is manufactured from recycled car tires. Natural rubber is a renewable resource extracted from tropical rubber plants without harming the plant.
- Cork is a renewable material harvested in a 9-year cycle.
- VOC emissions from the stripping and rewaxing of resilient flooring over its lifetime can be many times greater than those originating from the flooring material itself.
- Resilient flooring meeting the Resilient Floor Institute's *FloorScore* requirements for low VOC emissions satisfy current LEED requirements for low-emitting floor materials.

Carpet

- The embodied energy of nylon carpet is roughly 65,000 BTU per pound (150 MJ/kg), that of polypropylene is 40,000 BTU per pound (93 MJ/kg), and that of wool is 45,000 BTU per pound (105 MJ/kg). In general, natural fibers tend to be lower in embodied energy than synthetic fibers but also less durable.
- Many carpets and cushions are made with at least some recycled material.
- Carpet, cushions, and adhesives meeting the requirements of the Carpet and Rug Institute's *Green Label Plus* program satisfy current LEED requirements for low-emitting carpet systems.
- Factory-applied adhesives tend to have lower VOC emissions than adhesives applied on the construction site.
- Stretch-in installation of sheet carpet and free-lay installation of carpet tile (methods of installation explained later in this chapter) eliminate the need for carpet adhesive.
- Carpets can become reservoirs for VOCs emitted by other materials, as well as for bacteria, microbes, dust mites, and other contaminants.
- Carpet tile, which allows easy spot replacement, lessens the need for full carpet replacement when a small area becomes worn or damaged, thus extending the life of the carpet installation and reducing waste.
- Recycling of used carpet is complicated by the fact that many different fibers, which must be recycled separately, are used. Some manufacturers are introducing "close-the-loop" reclamation programs in which they will recycle or repurpose their own products to prevent them from being disposed of in landfills.

few fixed partitions. If the needs for electrical and telephone service are minimal and predictable, the most economical horizontal distribution system for wiring in a floor consists of conventional conduits of metal tubing that are embedded in the floor slab or concrete topping. In most commercial buildings, however, greater flexibility is required to accommodate wiring changes that will need to occur from time to time during the life of the building. There are several alternative systems for creating this flexibility. In buildings with concrete structural systems, *cellular raceways* may be cast into the floor slabs (Figure 24.14). These are sheet metal ducts that can carry many wires. Working through access boxes that reach from the top of the raceway to the surface of the slab, electricians can add or remove wiring at any time. Electrical and communications outlets can be installed in any of the access boxes.

In steel-framed buildings, *cellular steel decking* provides the same functional advantages as cellular raceways (Figures 24.15–24.17). *Poke-through fittings* (Figure 24.18) allow wiring flexibility over time without the need for raceways or cellular decking. Poke-through systems, however, require the electrician to work from the floor below the one on which changes are being made, which can be an inconvenience for the tenants of the lower floor.

Raised access flooring is advantageous in buildings where wiring changes are frequent and unpredictable, such as computer rooms and offices with a large number of electronic machines (Figures 24.19 and 24.20). Raised access flooring has a virtually unlimited capability to meet future wiring needs, and changes in wiring are extremely easy to make. If the access floor is raised high enough, ductwork for air distribution can be run beneath it, possibly

FIGURE 24.14
A worker levels a cellular raceway over steel dome formwork for a concrete waffle slab floor structure. The concrete will be poured so that its surface is level with the tops of the access boxes, completely embedding the raceways and allowing each box to be opened merely by removing its metal cover. Electrical and communications outlets can be installed in any access box. The covers of unused access boxes will be concealed beneath vinyl composition floor tiles or carpeting. *(Courtesy of American Electric—Construction Materials Group)*

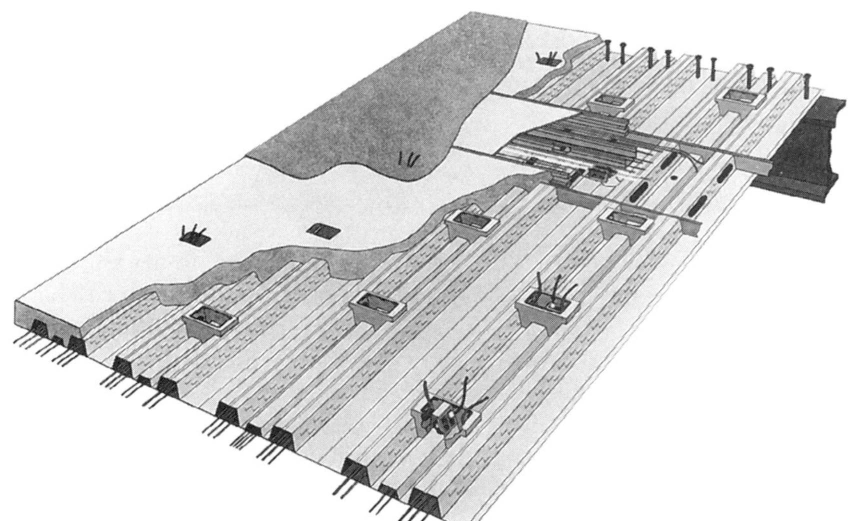

FIGURE 24.15
Cellular steel decking is often used for underfloor electrical and communications wiring. A transverse feeder trench, near the top of the picture, brings the wiring across the floor from the electrical risers to the cells in the deck. Boxes cast into the topping give access to the cells for the installation of electrical outlets. Notice the shear studs on the steel beam at the top of the picture. *(Photo courtesy of H. H. Robertson Company)*

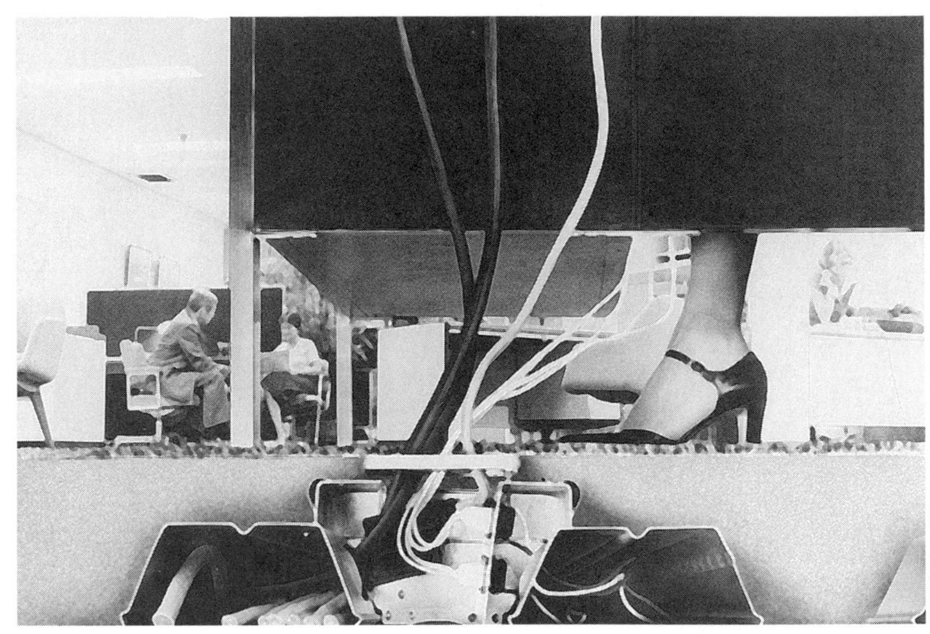

FIGURE 24.16
A sectional view of cellular steel decking, showing telephone and other low-voltage communications wiring in the cell to the left, electrical power wiring in the cell to the right, and access to these services through the box in the middle. *(Photo courtesy of H. H. Robertson Company)*

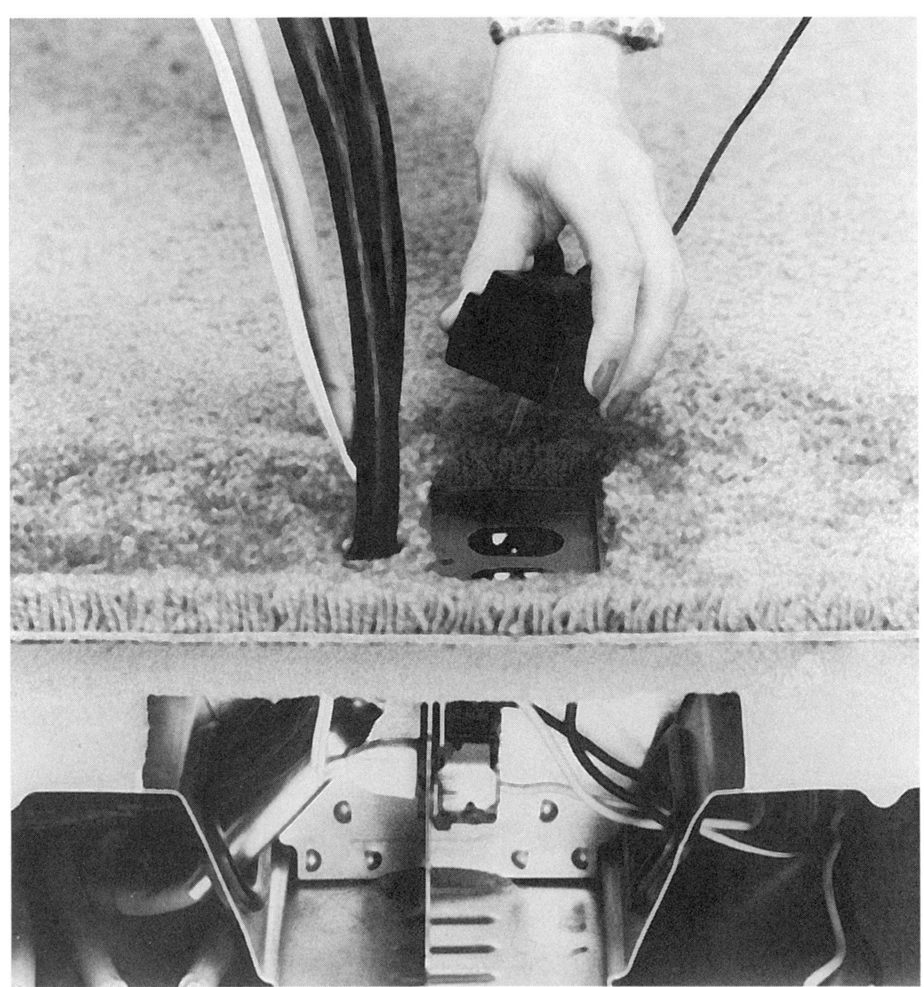

FIGURE 24.17
Plugging into a floor-mounted electric receptacle served by wires from the cells of the floor decking. *(Photo courtesy of H. H. Robertson Company)*

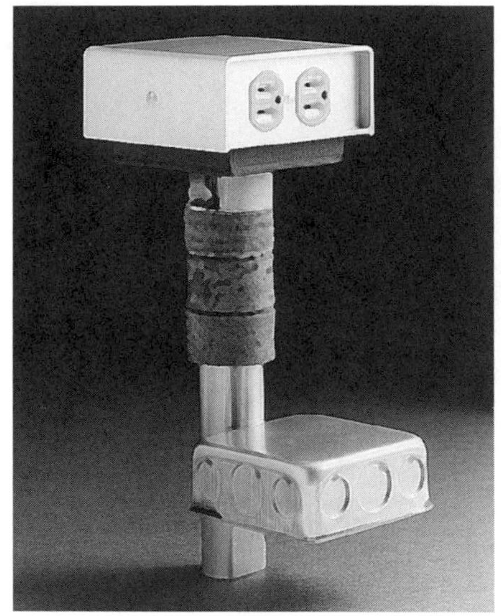
(*a*)

(*b*)

eliminating the need for a suspended ceiling. This can be useful in buildings where the structural system is meant to be left exposed as a finish ceiling, or in older buildings with beautiful plaster or wood ceilings. Raised access flooring also works well in older buildings because its pedestal heights can be adjusted to compensate for uneven floor surfaces. Several systems for providing individualized control of air conditioning in large office buildings utilize the space below a raised access floor as a large distribution chamber for conditioned air; each workstation is provided with a small outlet diffuser in the floor surface.

Undercarpet wiring systems that use flat conductors rather than conventional round wires are appropriate in many buildings for both electrical power and communications wiring (Figures 24.21 and 24.22). These are applicable to both new and retrofit projects.

FIGURE 24.18 (above)
(*a*) A poke-through fitting is designed to be installed in a hole drilled through a concrete floor slab. The junction box at the bottom connects to a wiring conduit above the suspended ceiling of the floor below. Wires pass from the junction box through the vertical tube to the outlet above. (*b*) Fire-retardant gaskets seal the hole through the slab so that fire cannot pass, thus maintaining the fire resistance rating of the floor structure. If a poke-through fitting is removed in a later renovation, a special fire-resistant abandonment plug is installed to seal the hole. *(Courtesy of American Electric—Construction Materials Group)*

FIGURE 24.19 (left)
Raised access flooring provides unlimited capacity for wiring, piping, and ductwork. The space below the flooring can serve as a plenum for air distribution. Changes in any of the underfloor systems are easily made, and wiring outlets can be installed at any point in the floor. *(Courtesy of Tate Architectural Products, Inc.)*

FIGURE 24.20
Conditioned air is supplied to this computer room through the space below the raised access flooring and is fed upward through perforated floor panels. Air is returned through slots in the suspended ceiling. *(Photo courtesy Armstrong World Industries)*

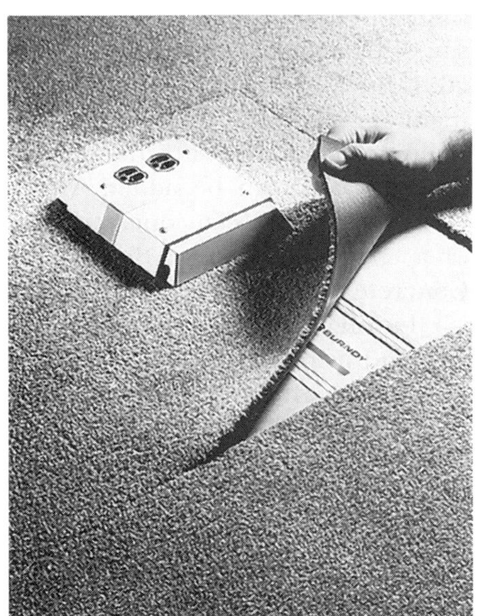

FIGURE 24.21
The flat conductors for an undercarpet wiring system lie unseen beneath the carpet and are accessed through projecting boxes. *(Photo: Courtesy of Burndy Corporation)*

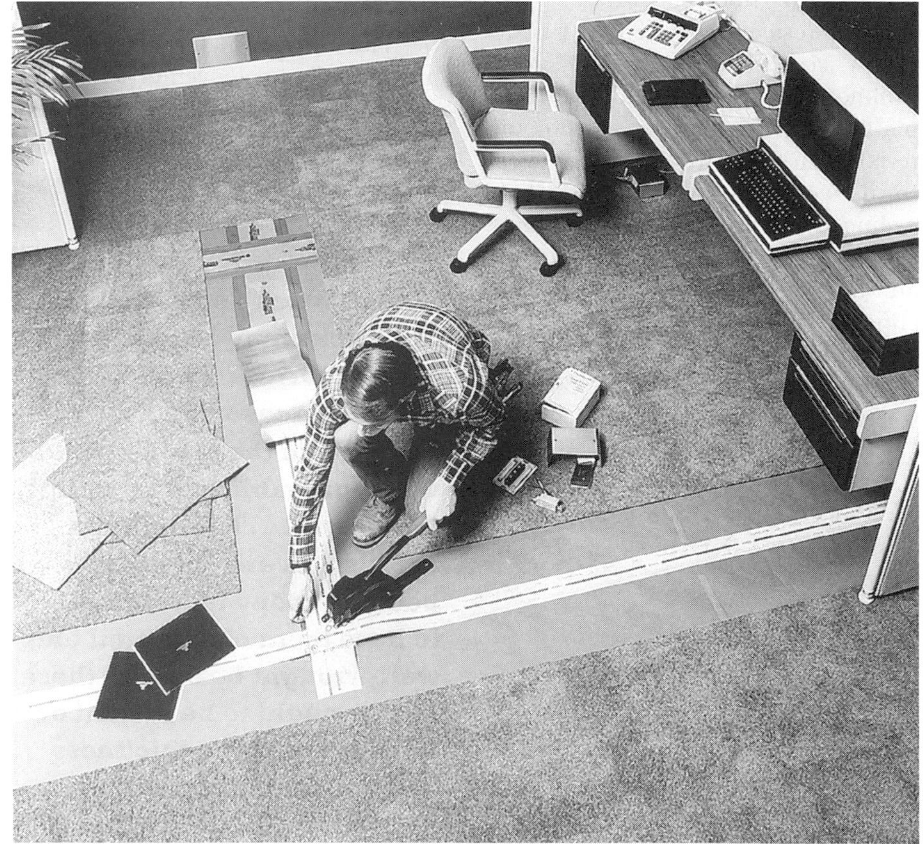

FIGURE 24.22
The flat conductors are ribbons of copper laminated between insulating layers of plastic sheet. These conductors are connected as necessary with the splicing tool shown in this photograph and covered with a grounded metallic shield before being taped to the floor. *(Photo: Courtesy of Burndy Corporation)*

Reducing Noise Transmission Through Floors

In multistory buildings, it is sometimes necessary to take precautions to reduce the amount of impact noise transmitted through a floor to the room below. This is particularly true of hotels and apartment buildings, where people are sleeping in rooms below the rooms of others who may be awake and moving about. Impact noise is generated by footsteps or machinery and is transmitted as *structure-borne vibration* through the material of the floor to become airborne noise in the room below.

There are several strategies for dealing with impact noise; these may be employed individually or in various combinations. One is to use padded carpeting or cushioned resilient flooring to reduce the amount of impact noise that is generated. A second is to underlay the flooring material with a layer of resilient material that is not highly conductive of impact noise. Cellulose fiber panels and nonwoven plastic filament matting are two materials marketed for this purpose. A third mechanism is to make an airtight ceiling below of a heavy, dense material such as plaster or gypsum board, and to mount this ceiling on resilient clips or on *hanger wires* with springs. The springs or clips absorb most of the sound energy that would otherwise travel through the structure.

Many floor–ceiling assemblies have been tested for sound transmission and rated for both Sound Transmission Class (STC), which is concerned with transmission of airborne sound, and Impact Insulation Class (IIC). (The ratings are explained in more detail in Chapter 22.) Product ratings will be found in trade literature concerned with various types of floor construction, and offer a ready comparison of acoustical performance.

Skid Resistance and Fire Resistance of Flooring Materials

The skid resistance of a flooring material is measured by its *Static Coefficient of Friction* (*SCOF*). Particular care must be taken when specifying a very smooth, polished material for a floor, especially in entrances and lobby areas where people may have wet feet. An SCOF of 0.50 or more is desirable to minimize accidents caused by slipping. SCOFs are published by manufacturers for all floor materials.

Floor finish materials must also meet building code requirements for resistance to ignition by radiant heat (NFPA 253) and flame spread (DOC FF-1), as explained in Chapter 22.

The brick floors, because the bricks may be made of diverse forms and of diverse colors by reason of the diversity of the chalks, will be very agreeable and beautiful to the eye. . . . The ceilings are also diversely made, because many take delight to have them of beautiful and well-wrought beams . . . these beams ought to be distant one from another one thickness and a half of the beam, because the ceilings appear thus very beautiful to the eye.

Andrea Palladio, *The Four Books of Architecture*, 1570.

Types of Finish Flooring Materials

Hard Flooring Materials

Hard finish flooring materials (concrete, stone, brick, tile, and terrazzo) are often chosen for their resistance to wear and moisture. Being rigid and unyielding, they are not comfortable to stand on for extended periods of time, and they contribute to a live, noisy acoustic environment. Many of these materials, however, are so beautiful in their colors and patterns, and so durable, that they are considered among the most desirable types of flooring by designers and building owners alike.

Concrete

With a lightly textured wood float finish for traction, concrete makes an excellent finish floor for parking garages and many types of agricultural and industrial buildings. With a smooth, hard, steel trowel finish, concrete finds its way into a vast assortment of commercial and institutional buildings, and even into homes and offices. Color can be added with a colorant admixture, a concrete stain, or a couple of coats of floor paint. Concrete's chief advantages as a finish flooring material are its low initial cost and its durability. On the minus side, extremely good workmanship is required to make an acceptable floor finish, and, unless applied as a finish topping very close to the end of construction, even the best concrete surface is likely to sustain some damage and staining during construction.

Stone

Many types of building stone are used as flooring materials, in surface textures ranging from mirror-polished marble and granite to split-face slate and sandstone (Figures 24.23 and 24.24). Installation is a relatively simple but highly skilled procedure of bedding the stone pieces in mortar and filling the joints with grout (Figure 24.25). Most stone floorings are coated with multiple applications of a clear sealer coating, and are waxed periodically throughout the life of the building to bring out the color and figure of the stone.

Bricks and Brick Pavers

Both bricks and half-thickness bricks called *pavers* are used for finish flooring, with pavers often preferred because they add less thickness and dead weight to the floor (Figures 9.41 and 24.26). Bricks may be laid with their largest surface horizontal or on edge. As with stone and tile flooring, decorative joint patterns can be designed especially for each installation.

Quarry Tiles

Quarry tiles are not, as might be guessed from their name, made of stone. They are simply large, fired clay tiles, usually square but sometimes rectangular, hexagonal, octagonal, or other shapes (Figures 24.27 and 24.28). Sizes range from about 4 inches (100 mm) to 12 inches (300 mm) square, with thicknesses ranging from ⅜ inch (9 mm) to a full inch (25 mm) for some handmade tiles. Quarry tiles are available in myriad earth colors, as well as in certain kiln-applied colorations. They are usually set in a reinforced mortar bed, although in residential work they may be thin-set directly to a subfloor of wood panels or tile backer boards (Figure 24.29). (Tile-setting methods are discussed in more detail in the previous chapter.) It is important that any subfloor to which tiles are glued should be exceedingly stiff; otherwise, flexing of the subfloor under changing loads will pop the tiles loose. Additional subfloor thickness and/or a stiff underlayment are advisable.

FIGURE 24.23
A slate floor in an automobile showroom. *(Photo by Bill Engdahl, Hedrich Blessing. Courtesy of Buckingham-Virginia Slate Company)*

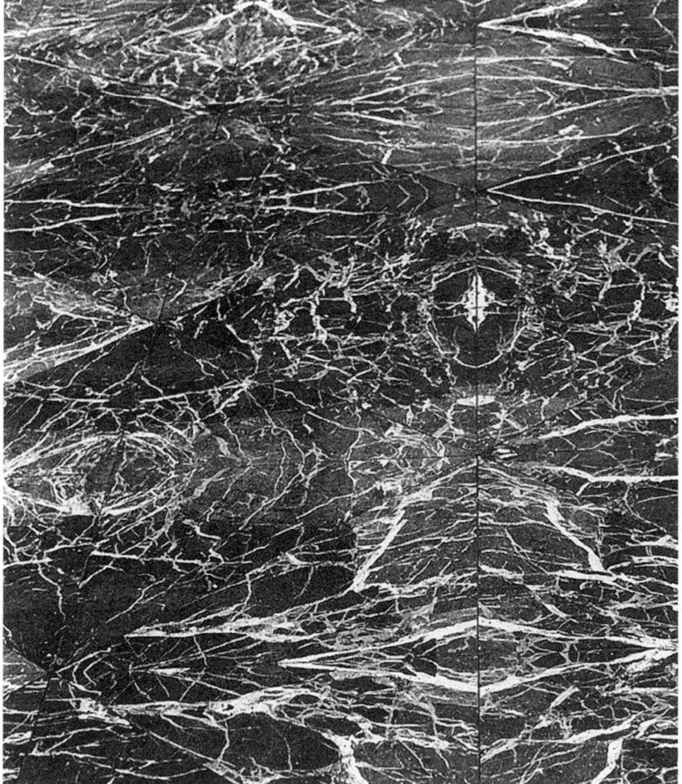

FIGURE 24.24
Flooring of matched, polished triangles of white-veined red marble gives a kaleidoscopic effect. *(Architect: The Architects Collaborative. Photo by Edward Allen)*

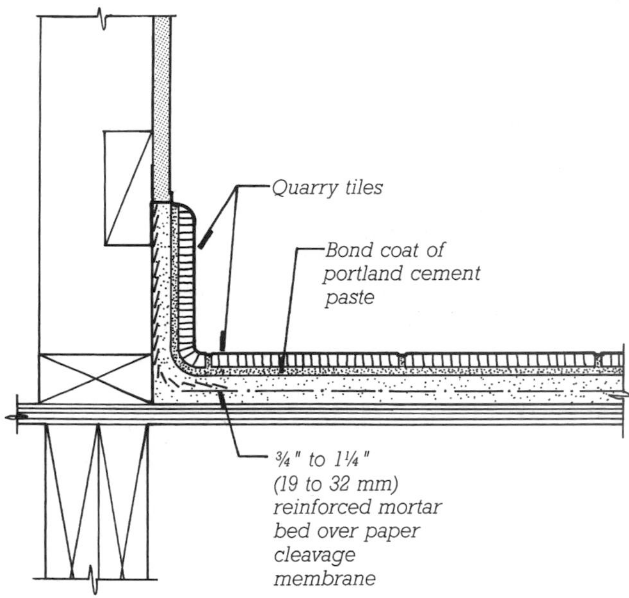

QUARRY TILE ON WOOD SUBFLOOR

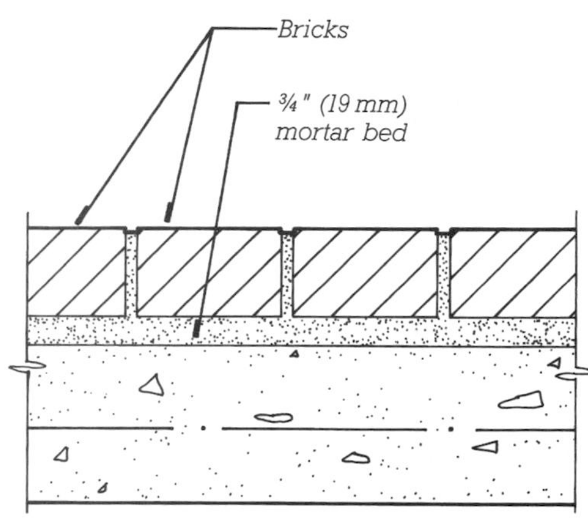

BRICKS ON CONCRETE SLAB

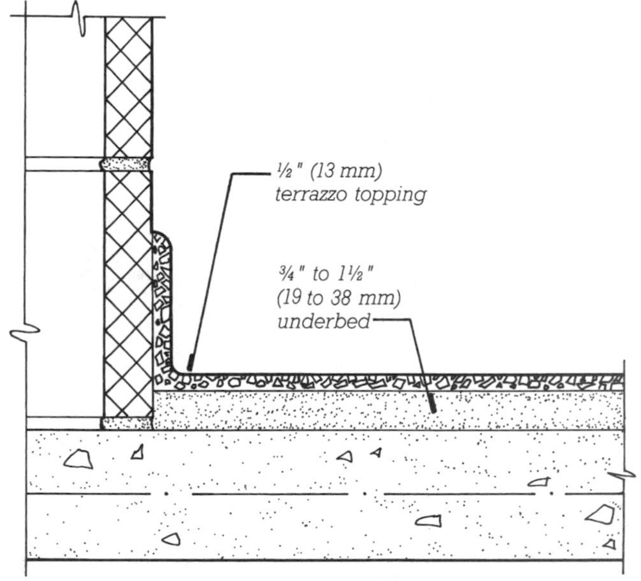

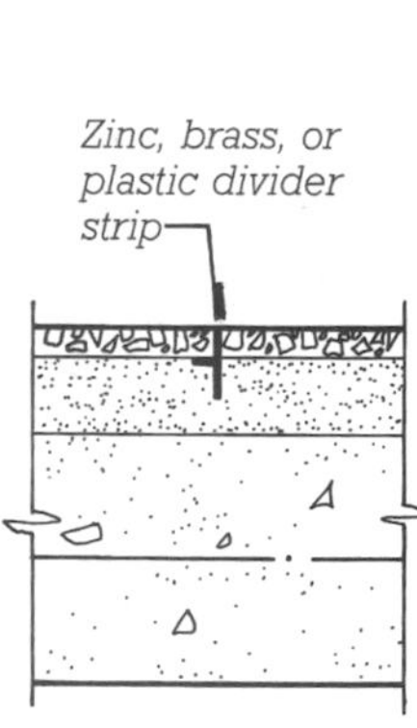

BONDED TERRAZZO

½" (13 mm) terrazzo topping

Divider strip in saw-cut control joint

MONOLITHIC TERRAZZO

FIGURE 24.25
Typical details of tile, brick, stone, and terrazzo flooring. Traditional sand cushion terrazzo flooring is not illustrated. The terrazzo systems shown here are thinner and lighter than sand cushion terrazzo, but they perform as well on a properly engineered floor structure. Thin-set terrazzo, the lightest of the systems, uses very small stone chips in a mortar that is strengthened with epoxy, polyester, or polyacrylate.

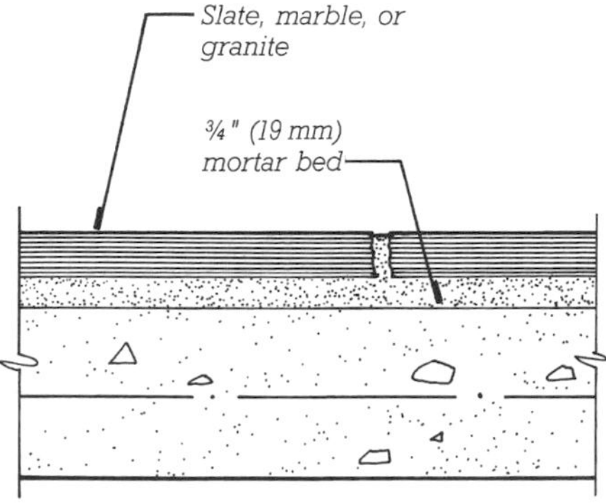

STONE ON CONCRETE SLAB

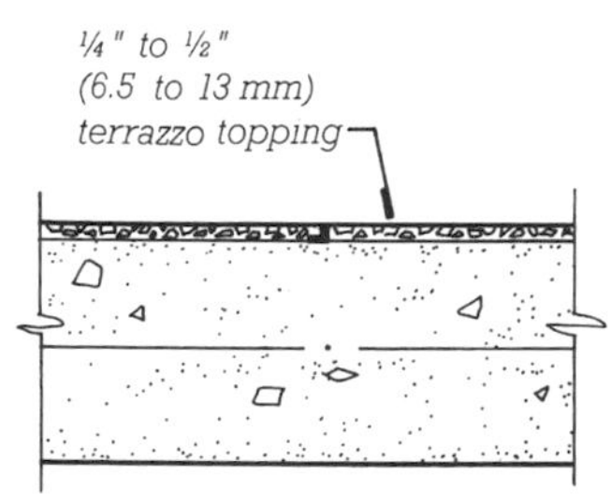

THIN SET TERRAZZO

FIGURE 24.26
A floor of glazed brick pavers meets a planting bed constructed of brick masonry.
(Courtesy of Stark Ceramics, Inc.)

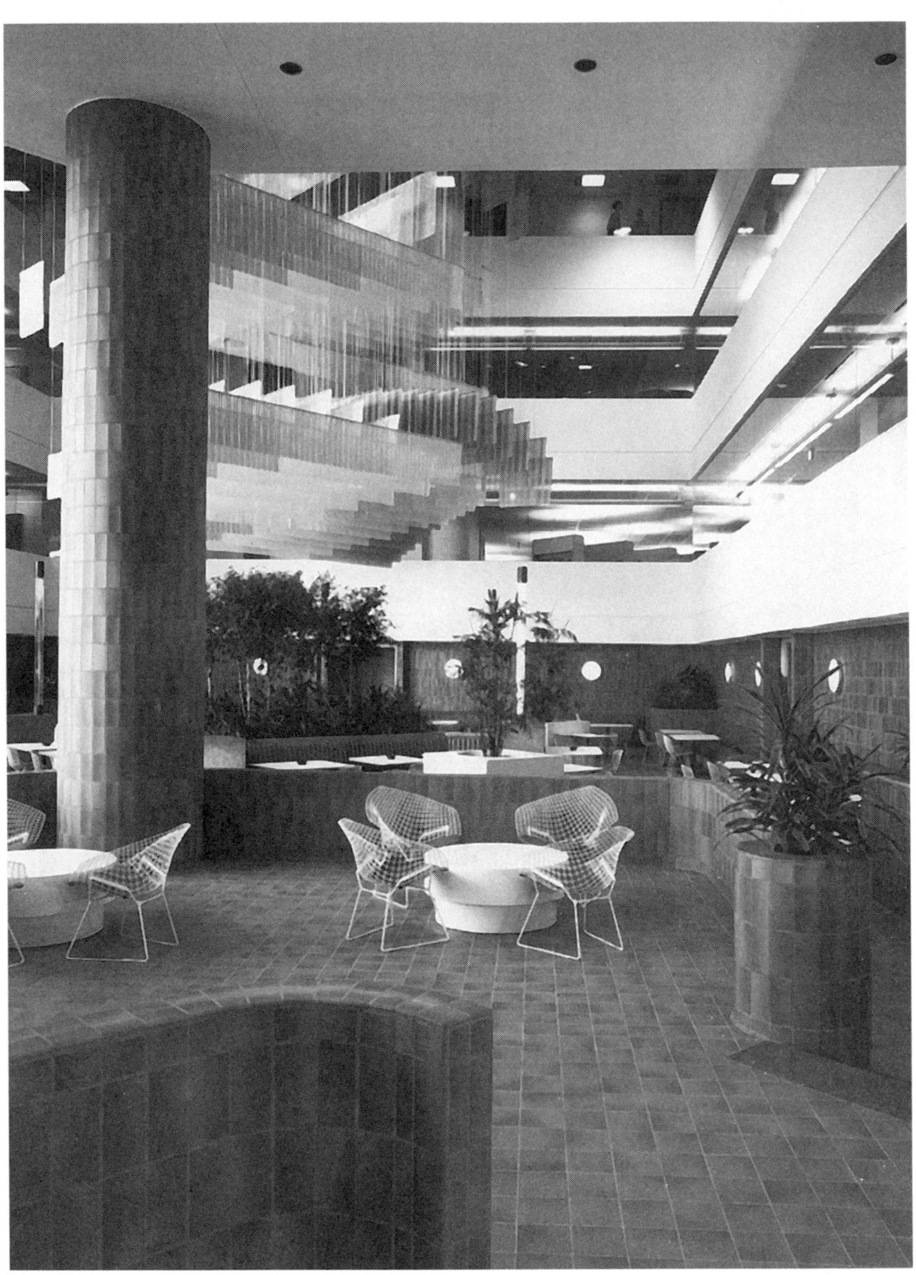

FIGURE 24.27
Unglazed quarry tiles used as flooring and as a facing for columns and railings. *(Architect: Skidmore, Owings & Merrill. Interior designer: Duffy, Inc. Courtesy of American Olean Tile Company)*

Ceramic Tiles

Fired clay tiles that are smaller than quarry tiles are referred to collectively as *ceramic tiles.* Ceramic tiles are usually glazed. The most common shape is square, but rectangles, hexagons, circles, and more elaborate shapes are also available (Figure 24.30). Sizes range from ½ inch (13 mm) to 4 inches (100 mm) and more. Smaller-sized tiles are shipped from the factory with their faces adhered to large backing sheets of plastic mesh or perforated paper. The tilesetter lays whole sheets of 100 or more tiles together in a single step rather than as individual units (Figure 7.39). The backing sheet is easily removed by wetting it after the tile adhesive has cured.

Grout color has a strong influence on the appearance of tile surfaces, just as it does for brick and stone. Many different premixed colors are available or the tilesetter may color a grout with pigments.

FIGURE 24.28
Square and rectangular quarry tiles in contrasting colors create a pattern on the floor of a retail mall. *(Architect: Edward J. DeBartolo Corp. Courtesy of American Olean Tile Company)*

Methods of installing ceramic tile on interior wall surfaces, discussed in the previous chapter, apply to floor tiling as well, and an example of ceramic tile installation on floors and walls is shown in Figure 23.41. As in wall tiling applications, waterproofing membranes may be integrated into floor tile assemblies in wet-use locations. Where tile are thickset over substrates that are cracked or prone to excessive deflection, a *slip sheet* or *cleavage membrane,* usually consisting of ordinary building felt, may be inserted between the mortar base and the substrate to isolate the tile assembly from the substrate and reduce the chance of tile cracking. Where tile are thin-set over problematic substrates, *crack isolation membranes* or *uncoupling membranes* that preserve the necessary bond between the thin-set compound and the substrate, but limit the transfer of stresses into the tile assembly, may be used.

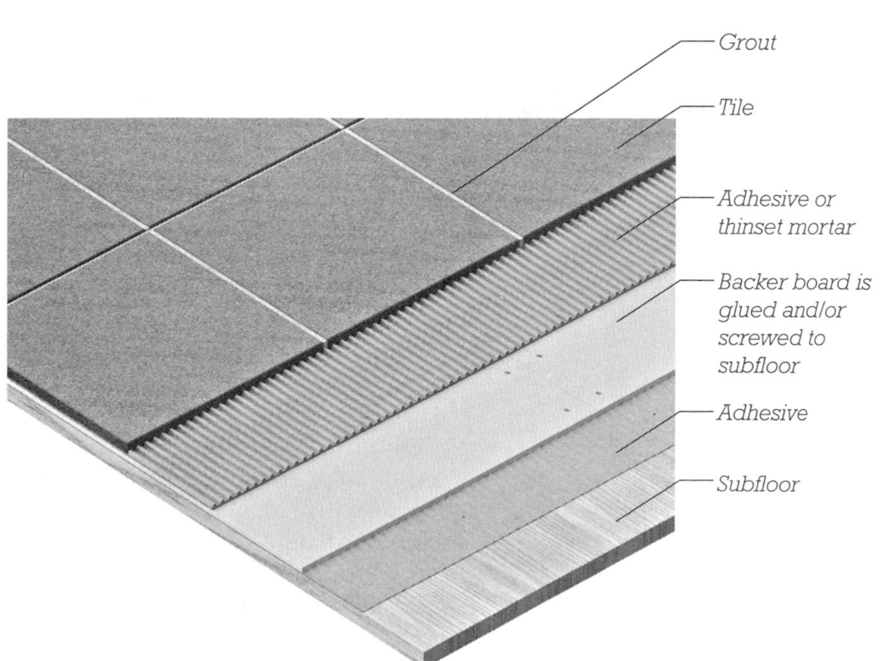

FIGURE 24.29
Quarry tiles may be adhered to a cementitious backer board that is applied as an underlayment to a wood panel subfloor.

Terrazzo

Terrazzo is an exceptionally durable flooring. It is made by grinding and polishing a concrete that consists of marble or granite chips selected for size and color in a matrix of colored portland cement or another binding agent. The polishing brings out the pattern and color of the stone chips. A sealer is usually applied to further enhance the appearance of the floor (Figure 24.31). Terrazzo may be formed in place or installed as factory-made tiles. For stair treads, window sills, and other large components, terrazzo is often precast. Because of its endless variety of colors and textures, terrazzo is often used in decorative flooring patterns. The colors are separated from one another by *divider strips* of metal, plastic, or marble. The divider strips are installed in the underbed prior to placing the terrazzo, and are ground and polished flush in the same operation as the terrazzo itself.

Traditionally, terrazzo is installed over a thin bed of sand that isolates it from the structural floor slab, thus protecting it to some extent from movements in the building frame. This *sand cushion terrazzo*, due to its thickness, usually 2½ inches (64 mm), is heavy. For greater economy and reduced thickness, the sand bed may be eliminated to produce *bonded terrazzo*, or both the sand bed and underbed may be eliminated with *monolithic terrazzo*. *Thin-set terrazzo*, made from epoxy resins, polyester resins, or polymer-modified cements, is the thinnest of all terrazzo installation methods (Figure 24.25). In any of these systems, a terrazzo baseboard can be formed and finished as an integral part of the floor, thus eliminating a dirt-catching seam where the floor meets the wall.

FIGURE 24.30
Ceramic tile wainscoting and flooring in a bar. *(Architect: Daughn/Salisbury, Inc. Designer: Morris Nathanson Design. Courtesy of American Olean Tile Company)*

Wood and Bamboo Flooring

Wood Flooring

Wood is used in several different forms as a finish flooring material, the most common of which is solid wood tongue-and-groove *strip flooring*, typically ¾ inch (19 mm) thick and 1½ to 2¼ inches (38–57 mm) wide. Strip flooring can be made from many hardwood and softwood species, some of the more commonly used being White oak, Red oak, Pecan, and Maple (Figures 24.32–24.34). The wood strips are held tightly together and *blind nailed* by driving nails diagonally through the upper interior corners of the tongues, where they are concealed from view once the next strip is installed. The entire floor is then sanded smooth, stained if desired, and finished with a varnish or other clear coating. When its surface becomes worn, the flooring can be restored to a new appearance by sanding and refinishing.

Solid wood flooring is also available in widths ranging from 3 to 8 inches (75–200 mm), in which case it

FIGURE 24.31
A terrazzo floor in a residential entry uses divider strips and contrasting colors to create a custom floor pattern. *(Courtesy of National Terrazzo and Mosaic Association, Inc.)*

is referred to as *plank flooring*. Because the wider planks are more prone to distortion with changes in moisture content, they are often fastened through the face with countersunk and plugged screws in addition to or instead of being blind nailed along their edges.

For greater economy, factory-made wood flooring consisting of finish wood veneers laminated to a plywoodlike core, called *engineered wood flooring*, is available. Typically ⅜ or ½ inch (10 or 13 mm) thick, it is usually glued to the subfloor with a mastic adhesive rather than nailed. The laminated construction of engineered wood flooring makes it less sensitive to the effects of moisture and more dimensionally stable than solid wood flooring. For this reason, it is considered better suited for use on basement slabs on grade or in other

FIGURE 24.32
Details of hardwood strip flooring installation. At the left, the flooring is applied to a wood joist floor, and at the right, to wood sleepers over a concrete slab. The blind nailing of the flooring is shown only for the first several strips of flooring at the left. The baseboard makes a neat junction between the floor and the wall, covering up the rough edges of the wall material and the flooring. The three-piece baseboard shown at the left does this job somewhat better than the one-piece baseboard, but it is more expensive and elaborate. For a more stable, higher-quality installation, an additional sheet of plywood underlayment or sheathing may be added under the wood flooring in each of these examples.

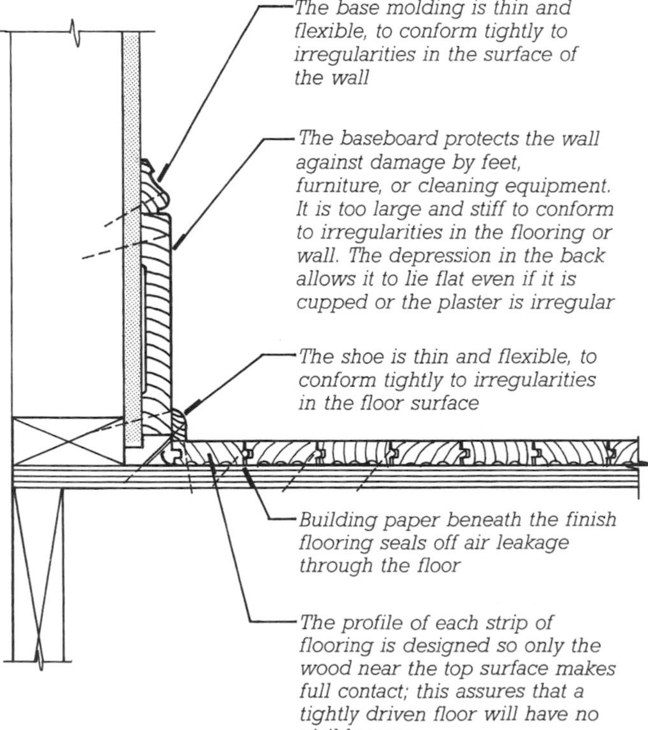
The base molding is thin and flexible, to conform tightly to irregularities in the surface of the wall
The baseboard protects the wall against damage by feet, furniture, or cleaning equipment. It is too large and stiff to conform to irregularities in the flooring or wall. The depression in the back allows it to lie flat even if it is cupped or the plaster is irregular
The shoe is thin and flexible, to conform tightly to irregularities in the floor surface
Building paper beneath the finish flooring seals off air leakage through the floor
The profile of each strip of flooring is designed so only the wood near the top surface makes full contact; this assures that a tightly driven floor will have no visible gaps

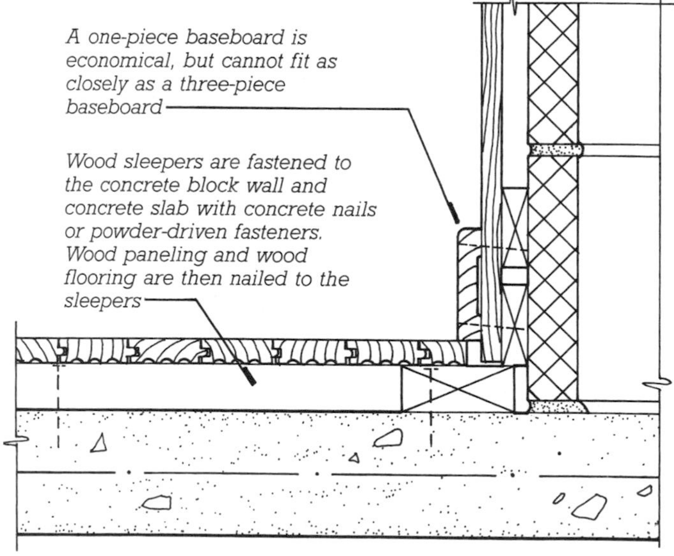
A one-piece baseboard is economical, but cannot fit as closely as a three-piece baseboard
Wood sleepers are fastened to the concrete block wall and concrete slab with concrete nails or powder-driven fasteners. Wood paneling and wood flooring are then nailed to the sleepers
If the concrete slab lies directly on grade, a sheet of polyethylene is laid beneath the sleepers to prevent moisture from entering the building

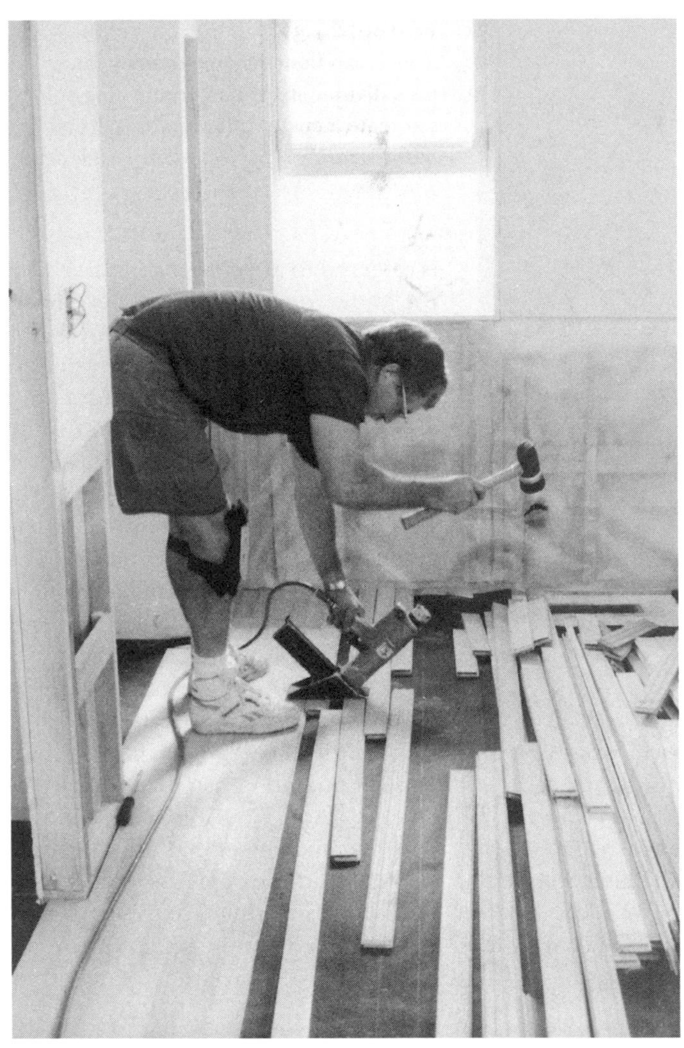

locations exposed to high humidity levels. Because their finish surface is a thin veneer, most engineered wood floorings are not able to withstand subsequent sanding and refinishing.

Parquet is wood flooring of varying hues arranged in patterns. It may be made of solid wood strips assembled in the field, of factory-preassembled blocks made from solid wood strips, or of engineered wood.

Some proprietary wood floorings are not nailed or glued to the subfloor, but instead "float" above it on a thin pad of resilient foam. These so-called *floating floors* are made by connecting the individual pieces of flooring together at the edges to make one continuous piece as large as the room in which the flooring is laid. Edge gluing is the most common way of making this connection, but systems are available that use metal clips or interlocking edge details. A gap is left at the edges to permit expansion and contraction

FIGURE 24.33
An installer of hardwood strip flooring uses a pneumatic nail gun to drive the diagonal blind nails that fasten the flooring to the subfloor. The nail gun is a special type that is activated by a blow from a mallet that also serves to drive the flooring pieces tightly together. Asphalt-saturated felt paper cushions the flooring and helps to prevent squeaking. *(Photo by Rob Thallon)*

FIGURE 24.34
Oak strip flooring in a hair salon. Notice the use of exposed ducts and a lighting track at the ceiling. *(Architects: Michael Rubin and Henry Smith-Miller in association with Kenneth Cohen. Courtesy of Oak Flooring Institute)*

of the floor. This is later covered with baseboard trim. Most floating floors are made of engineered wood flooring, which is more dimensionally stable than solid wood flooring.

Many types of wood flooring are available with factory-applied finishes. When the installation of conventional unfinished flooring begins on a project, it may require the complete cessation of other construction activities for a number of days while the flooring is installed and sanded, and several coats of finish are applied and allowed to dry. With factory-finished flooring, the time required for a complete installation and the impact on the other activities are minimized. Prefinished flooring, however, cannot be sanded after installation, so these products are typically supplied with eased edges (a slight beveling of the corners along the finish face edges) that hide any minor differences in level between adjacent pieces of flooring after installation. Factory finishing is especially common with engineered wood flooring. Because the finish veneer is thin, and can never be sanded and refinished, an especially hard, wear-resistant acrylic resin finish is applied in the factory that helps to prolong the life of the floor.

The fastest-growing segment of the flooring market is *plastic laminate flooring*, which is almost always laid as a floating floor. It is composed of planks or large tiles that have a wood composite core and a wearing layer of high-density plastic laminate much like that used on countertops. The laminate is usually patterned to resemble wood, but other patterns are also available. Most laminate floorings may be used in kitchens or baths if a sealant is applied around the perimeter of the floor to keep water from getting beneath it.

Exceptionally long-wearing industrial *wood block flooring* is made of small blocks of wood set in adhesive with their grain oriented vertically. Although this type of floor is relatively high in first cost, it is economical for heavily used floors and is sometimes chosen for use in public spaces because of the beauty of its pattern and grain.

Bamboo Flooring

Bamboo, a rapidly growing grass, can be manufactured into flooring products very much like those made of wood. The manufacturing process entails slicing the hollow bamboo shoots into strips, processing the strips to remove starch, laminating and gluing the strips under pressure, and then machining the laminated stock into the final flooring profile. Laminations may be oriented either vertically or horizontally within the strip, creating a surface appearance analogous to either flat grain or edge grain in solid woods.

Bamboo flooring is harder and more dimensionally stable than flooring made of conventional wood. Its natural color is light, akin to that of maple. A darker amber hue can also be achieved by pressure steaming, a process called "carbonization." Like wood, bamboo flooring can be provided either as solid strip, made entirely of bamboo laminations, or as an engineered product consisting of a roughly 1/8-inch (3-mm)-thick bamboo finish layer adhered to a laminated base of conventional wood. It may be provided unfinished or factory finished.

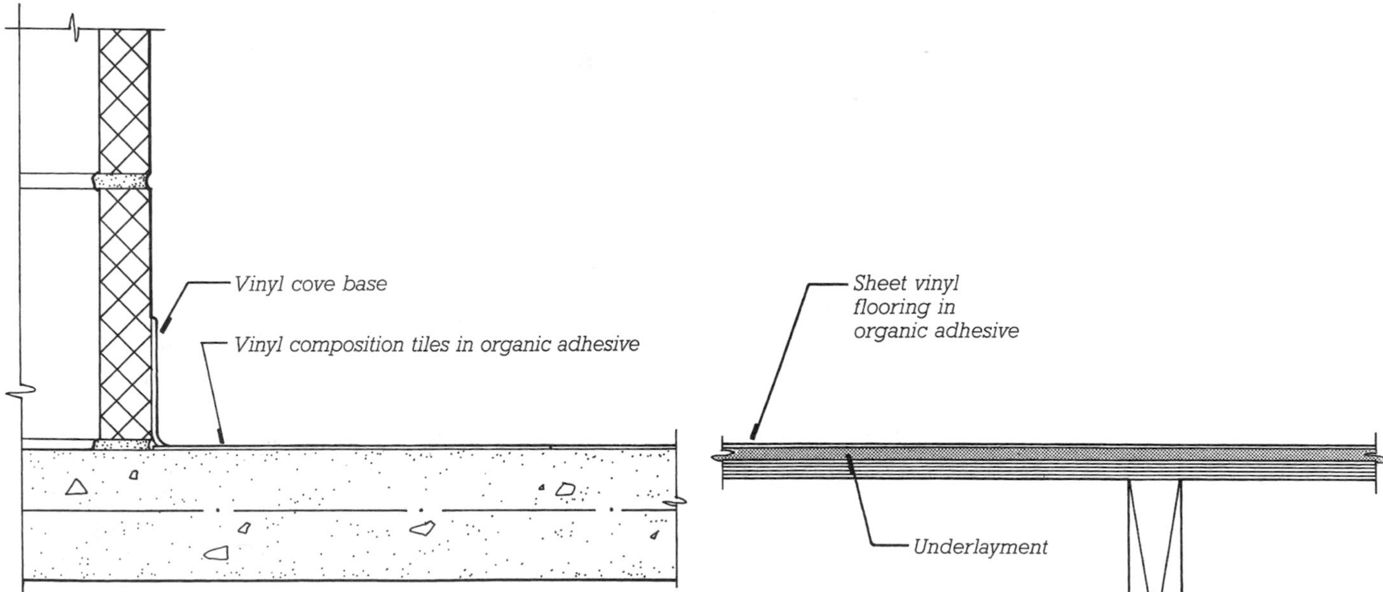

FIGURE 24.35
Typical installation details for resilient flooring. At the left, vinyl composition tiles applied directly to a steel-trowel-finish concrete slab, with a vinyl cove base adhered to a concrete masonry partition. To the right, sheet vinyl flooring on underlayment and a wood joist floor structure.

Resilient Flooring

The oldest *resilient flooring* material is *linoleum*, a sheet material made of ground cork in a linseed oil binder over a burlap backing. *Asphalt tiles* were later developed as an alternative to linoleum, but the majority of today's resilient sheet floorings and tiles are made of compounds of vinyl or rubber. The primary advantages of resilient floorings are the wide range of available colors and patterns, moderately high durability, and low initial cost.

Vinyl composition tile (*VCT*), made of one or more vinyl resins in combination with binders, pigments, and fillers (VCT may consist of as much as 85 percent limestone filler), has the lowest installed cost of any flooring material except concrete and is used in vast quantities on the floors of residences, offices, classrooms, and retail spaces. Other common resilient tile flooring materials include *solid vinyl tile* (*SVT*), with higher vinyl content and greater durability than VCT, and *rubber floor tile*, made from vulcanized natural or synthetic rubber compounds and various additives. Floor tile thickness is typically 1/8 inch (3 mm) or slightly less. The most common tile size is 12 inches (305 mm) square, although other sizes, up to 36 inches (914 mm) square, are also available.

The most common *resilient sheet flooring* materials are solid vinyl and rubber. Resilient floorings of linoleum, cork, and other materials are also available. Each offers particular characteristics of durability and appearance. Thicknesses of sheet flooring are also on the order of 1/8 inch (3 mm), slightly thinner for lighter-duty floorings, and slightly thicker if a cushioned back is added to the product. Sheet floorings are furnished in rolls 4.9 to 12 feet (1.5–3.6 m) wide. If they are skillfully installed, the seams between sheet strips are virtually invisible.

Most resilient flooring materials are glued to the concrete or wood of the structural floor (Figures 24.35–24.37). The resilient materials themselves are so thin and deformable that they show even the slightest irregularities in the floor deck beneath. Concrete surfaces to which resilient materials will be applied must first be scraped clean of construction debris and spatters. Wood panel decks are covered with a layer of smooth *underlayment panels*, usually of hardboard, particleboard, or sanded plywood, to provide a smoother substrate for the resilient flooring materials. Joints between underlayment panels are offset from joints in the subfloor to eliminate soft spots at joints. The thickness of the underlayment is chosen to make the surface of the resilient flooring level with the surfaces of flooring

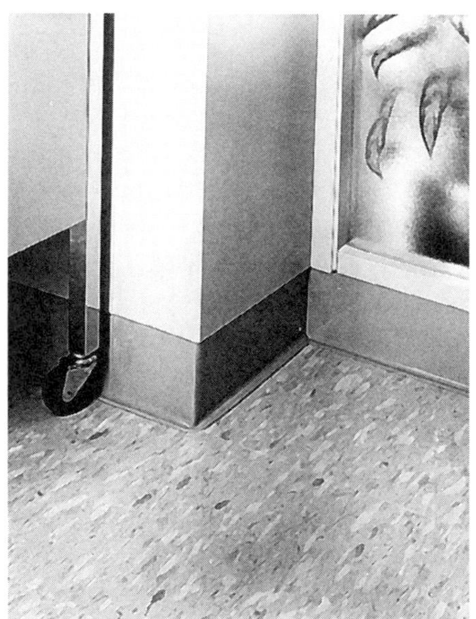

FIGURE 24.36
Vinyl composition tile and a vinyl cove base. Notice how the base is simply folded around the outside corner of the partition. *(Photo courtesy Armstrong World Industries)*

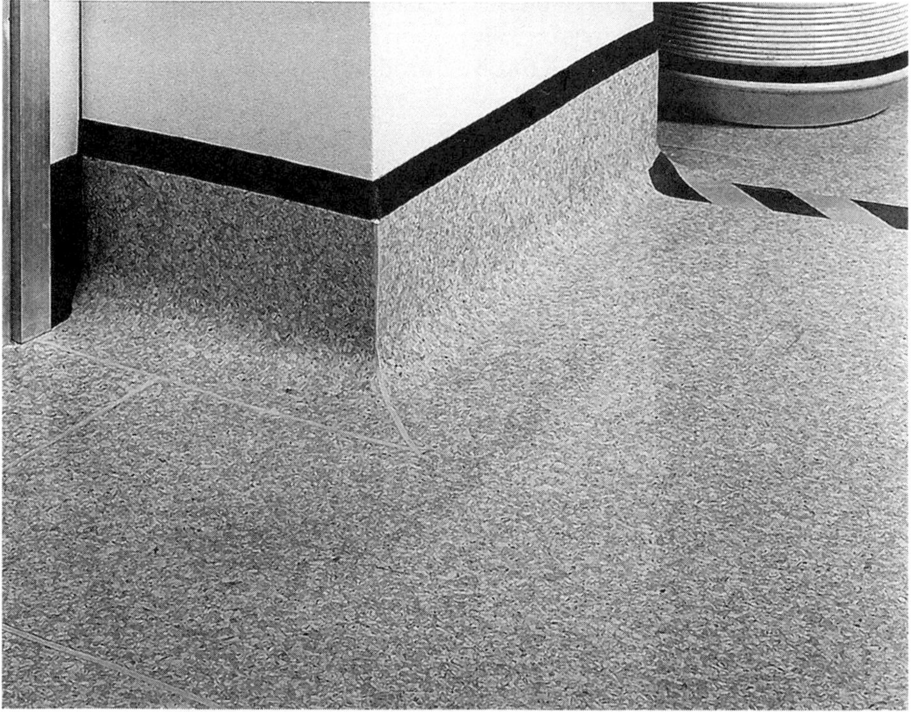

FIGURE 24.37
Sheet vinyl flooring can be flash coved to create an integral base that is easily cleaned, for use in health care facilities, kitchens, and bathrooms. The seams are welded to eliminate dirt-catching cracks. *(Photo courtesy Armstrong World Industries)*

materials such as hardwood and ceramic tile that are used in surrounding areas of the building. Leveling compounds, gypsum or cement-based plasterlike materials that can be feathered to a very thin edge, are also used to fill in minor low spots in the floor surface prior to installing the resilient materials.

Various flooring accessories such as bases, stair treads, and stair nosings are also made of vinyl and rubber compounds. *Cove base* is the most common base used with resilient flooring (Figure 24.35). *Straight base,* also called *flat* or *toeless base,* has no cove or toe and is most commonly used with carpet flooring. *Fit-to-floor base,* or *butt-to base,* has a square-edged toe the same thickness as the floor covering; it butts tightly to the finish flooring, creating a flush transition between the two.

Carpet

Carpet is manufactured in fibers, styles, and patterns to meet almost any flooring requirement, indoors or out, except for rooms that need thorough sanitation, such as hospital rooms, food processing facilities, and toilet rooms. Some carpets are tough enough to wear for years in public corridors (Figure 24.38); others are soft enough for intimate residential interiors. The costs of carpeting are often competitive with those of other flooring materials of similar quality, whether they are measured on an installed-cost or life-cycle-cost basis. Slightly more than one-half of all carpet sold in North America is made with nylon fibers; more than an additional one-third is made with polypropylene; the remainder is made with other synthetics and natural fibers.

There are four ways to install carpet: Most commonly, they are either glued directly to the floor deck (*direct glue-down installation*) or stretched over a *carpet pad* or *cushion* and attached around the perimeter of the room by means of a *tackless strip* (*stretch-in installation*). The tackless strip (also called a *tackstrip*) is a continuous length of wood, fastened to the floor, that has protruding spikes along the top to catch the backing of the carpet and hold it taut as it is stretched into place by the installers. Alternatively, the carpet pad can be glued to the floor deck and then the carpet can be glued to the pad (*double glue-down installation*), or a carpet with a factory-attached pad can be glued to the floor deck (*attached cushion installation*).

If carpet or carpet tile is laid directly over a wood panel subfloor such as plywood, the panel joints perpendicular to the floor joists should be blocked beneath to prevent movement between sheets. Tongue-and-groove plywood subflooring accomplishes the same result without blocking. Alternatively, a layer of underlayment panels may be nailed over the subfloor with its joints offset from those in the subfloor.

Carpets, carpet pads, and adhesives are all potentially significant sources of VOC emissions. Products meeting the requirements of the Carpet & Rug Institute's *Green Label Plus* program are certified as meeting low-emission standards complying with current requirements of LEED NC and other green building programs. A newer standard is the NSF International's *NSF/ANSI Standard 140 Sustainable Carpet Assessment Standard.* NSF 140 provides a more comprehensive life-cycle rating system for carpet that addresses not only indoor air quality, but also other public health and environmental concerns, energy efficiency, recycled or biobased materials content, end-of-life reclamation, and other considerations.

Carpet Tile

Carpet materials are also manufactured in tile form, with typical sizes ranging from 18 to 36 inches (457–914 mm) square. In comparison to sheet carpet, carpet tiles are easier to deliver, store, handle, and install; they allow easier spot replacement; they permit ready underfloor access; and they are compatible with raised-access flooring systems. Carpet tile installation methods include *glue-down,* in which every tile is adhered to the subfloor; *partial glue-down,* in which only periodically spaced tiles are adhered to the subfloor; and *free-lay,* in which interlocking tiles are laid without any adhesive.

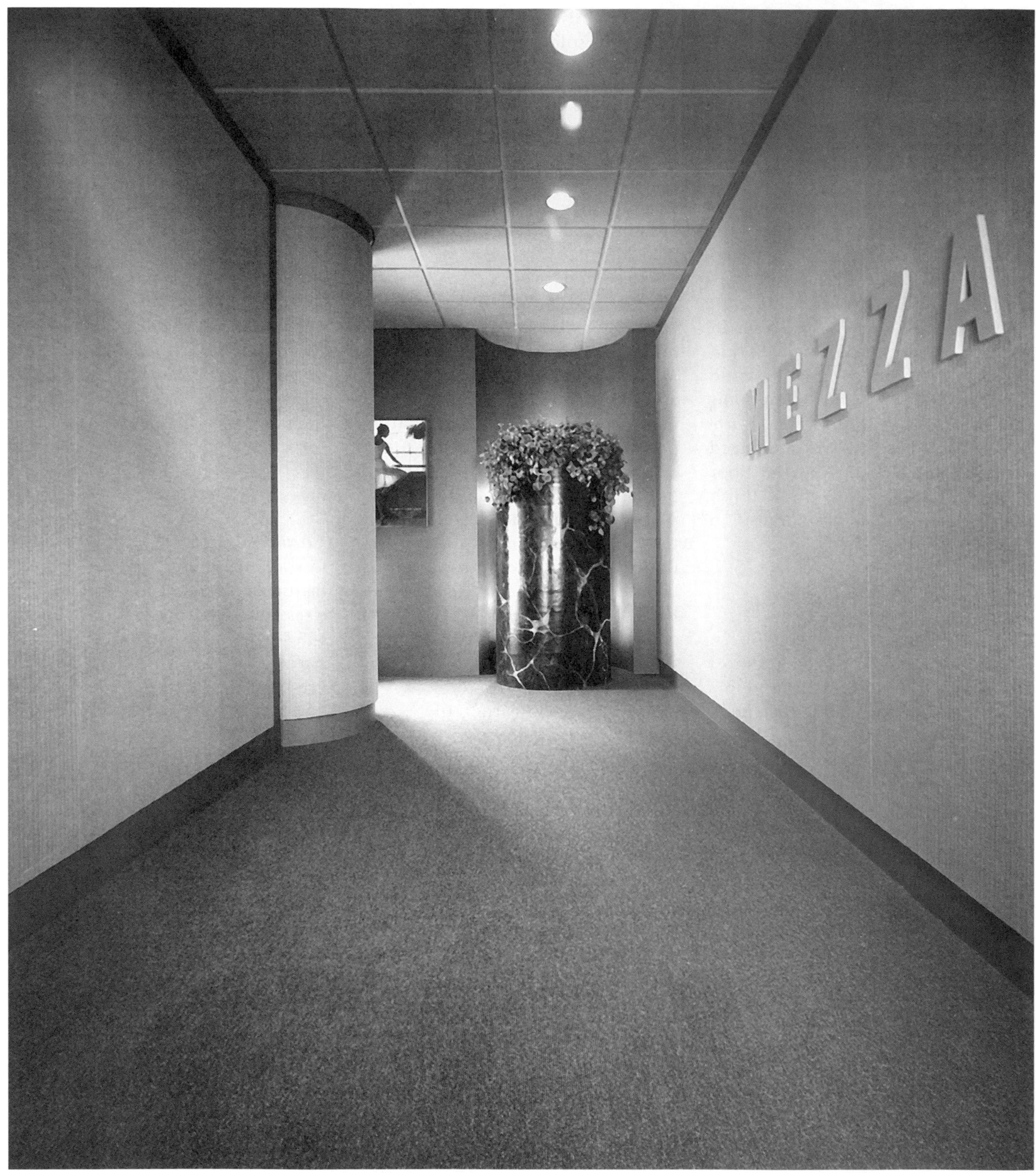

FIGURE 24.38
Wall-to-wall carpeting in a commercial installation. *(Photo courtesy Armstrong World Industries)*

Flooring Thickness

Thicknesses of floor finishes vary from 1/8 inch (3 mm) or less for resilient flooring to 3 inches (76 mm) or more for brick flooring. Frequently, several different types of flooring are used on different areas of the same floor level of a building. If the differences in thickness of the flooring materials are not great, they can be resolved by using tapered edgings or thresholds at changes of material. Greater differences may be resolved with variations in the thickness of underlayment panels. Alternatively, gypsum or cementitious *self-leveling toppings* up to several inches thick can be poured over some portions of the subfloor to raise the finish floor level.

If none of these solutions can satisfactorily resolve the differences, the level of the top of the floor deck must be adjusted from one part of the building to the next to bring the finish floor surfaces to the same elevation. The architect should work out the necessary level changes in advance and indicate them clearly on the construction drawings. In many cases, special structural details must be drawn to indicate how the level changes should be made. In wood framing, they can usually be made either by notching the ends of the floor joists to lower the subfloor in parts of the building with thicker floor materials or by adding sheets of underlayment material of the proper thickness to areas of thinner flooring. In steel and concrete buildings, slab or topping thicknesses can change, or whole areas of the structure can be raised or lowered by the necessary amount.

CSI/CSC

MasterFormat Sections for Finish Ceilings and Floors

09 20 00	PLASTER AND GYPSUM BOARD
09 22 00	Supports for Plaster and Gypsum Board
	Suspension Systems
09 23 00	Gypsum Plastering
09 24 00	Portland Cement Plastering
09 26 00	Veneer Plastering
09 28 00	Backing Boards and Underlayments
09 29 00	Gypsum Board
09 30 00	TILING
09 31 00	Thin-Set Tiling
09 32 00	Mortar-Bed Tiling
09 50 00	CEILINGS
09 51 00	Acoustical Ceilings
09 53 00	Acoustical Ceiling Suspension Assemblies
09 54 00	Specialty Ceilings
	Linear Metal Ceilings
09 60 00	FLOORING
09 62 00	Specialty Flooring
	Bamboo Flooring
	Cork Flooring
09 63 00	Masonry Flooring
	Brick Flooring
	Stone Flooring
09 64 00	Wood Flooring
	Wood Block Flooring
	Wood Parquet Flooring
	Wood Strip and Plank Flooring
	Laminated Wood Flooring
09 65 00	Resilient Flooring
	Resilient Base and Accessories
	Resilient Sheet Flooring
	Resilient Tile Flooring
09 66 00	Terrazzo Flooring
	Portland Cement Terrazzo Flooring
	Sand Cushion Terrazzo Flooring
	Monolithic Terrazzo Flooring
	Bonded Terrazzo Flooring
	Resinous Matrix Terrazzo Flooring
09 68 00	Carpeting
	Tile Carpeting
	Sheet Carpeting
09 69 00	Access Flooring

Selected References

Because so many ceiling and flooring materials and systems are proprietary, much of the best information on ceiling and floor finishes is to be found in manufacturers' literature. Certain generic products are, however, well documented in trade association literature:

1. Ceilings and Interior Systems Construction Association. *Ceiling Systems Handbook.* St. Charles, IL, updated frequently.

Aimed at construction workers, this manual covers everything pertaining to the installation of ceilings.

2. Marble Institute of America. *Dimension Stone Design Manual.* Cleveland, OH, updated regularly.

This binder provides comprehensive guidance on the design, specification, and installation of all types of stone flooring.

3. Tile Council of North America, Inc. *TCA Handbook for Ceramic Tile Installation.* Anderson, SC, updated regularly.

This is the definitive standard for ceramic tile installation materials and methods. Over 100 methods of installation for floors and walls, both interior and exterior, are illustrated and specified. Guidelines for selecting appropriate installation methods based on project requirements are also included.

4. National Wood Flooring Association. *Hardwood Flooring Installation Guidelines.* Chesterfield, MO, updated regularly.

This manual provides industry standard recommendations for the installation of all kinds of wood flooring.

5. National Oak Flooring Manufacturers Association. *Installing Hardwood Flooring* and *Finishing Hardwood Flooring.* Memphis, TN, updated frequently.

These two brochures, available as free downloads, offer well-illustrated and thorough guidelines for the installation and finishing of wood flooring of all common types. Many other useful technical references are also available for download from this association's web site (listed below).

5. Carpet and Rug Institute. *Commercial Carpet Installation Standard* and *Residential Carpet Installation Standard.* Dalton, GA, updated regularly.

Industry standard recommendations for the installation of carpet of all types are included in these two guides, available from this association's web site as free downloads.

Web Sites

Finish Ceilings and Floors

Author's supplementary web site: **www.ianosbackfill.com/24_finish_ceilings_and_floors**

Armstrong Floors and Ceilings: **www.armstrong.com**

Carpet and Rug Institute (CRI): **www.carpet-rug.com**

Ceilings and Interior Systems Construction Association: **www.cisca.org**

Forbo Floor Coverings: **www.themarmoleumstore.com**

H. H. Robertson: **www.hhrobertson.com**

Mannington Floors: **www.mannington.com**

Maple Flooring Manufacturers Association (MFMA): **www.maplefloor.org**

Marble Institute of America (MIA): **www.marble-institute.com**

National Oak Flooring Manufacturers Association (NOFMA): **www.nofma.org**

National Terrazzo & Mosaic Association (NTMA): **www.ntma.com**

National Wood Flooring Association (NWFA): **www.nwfa.org**

NSF International (for NSF 140 Sustainable Carpet Assessment Standard): **www.nsf.org**

Resilient Floor Covering Institute (RFI): **www.rfci.com**

Tile Council of North America (TCNA): **www.tileusa.com**

Key Terms and Concepts

plenum
membrane fire protection
suspended ceiling
acoustical ceiling
Noise Reduction Coefficient (NRC)
Ceiling Attenuation Class (CAC)
Articulation Class (AC)
lay-in panel
exposed grid
concealed grid
integrated ceiling system
linear metal ceiling
interstitial ceiling
cellular raceway
cellular steel deck
poke-through fitting
raised access flooring
undercarpet wiring system
structureborne vibration
hanger wire
Static Coefficient of Friction (SCOF)
paver
quarry tile
ceramic tile
slip sheet, cleavage membrane
crack isolation membrane, uncoupling membrane
terrazzo
divider strip
sand cushion terrazzo
bonded terrazzo
monolithic terrazzo
thin-set terrazzo
strip flooring
blind nail
plank flooring
engineered wood flooring
parquet
floating floor
plastic laminate flooring
wood block flooring
resilient flooring
linoleum
asphalt tile
vinyl composition tile (VCT)
solid vinyl tile (SVT)
rubber floor tile
resilient sheet flooring
underlayment panel
cove base
straight base, flat base, toeless base
fit-to-floor base, butt-to base
carpet
direct glue-down installation
carpet pad, cushion
tackless strip, tackstrip
stretch-in installation
double glue-down installation
attached cushion installation
glue-down tile
partial glue-down tile
free-lay tile
self-leveling topping

Review Questions

1. List the potential range of functions of a finish ceiling and of a finish floor.

2. What are the advantages and disadvantages of a suspended ceiling compared to those of a tightly attached ceiling?

3. When designing a building with its structure and mechanical equipment left exposed at the ceiling, what precautions should you take to ensure a satisfactory appearance?

4. What does underlayment do? How should it be laid?

5. List several different approaches to the problem of running electrical and communications wiring beneath a floor of a building framed with steel or concrete.

Exercises

1. Visit some rooms that you happen to like in nearby buildings: classrooms, auditoriums, theaters, restaurants, bars, museums, shopping centers—both new buildings and old. For each room, list the ceiling and floor finishes used. Why was each material chosen? Sketch details of critical junctions between materials. What does each room sound like—noisy, hushed, "live"? How does this acoustical quality relate to the floor and ceiling materials? What is the quality of the illumination in the room, and what role do ceiling and floor materials play in creating this quality?

2. Consult current architectural magazines for interior photographs of buildings that appeal to you. What ceiling and floor materials are used in each? Why?

Densities and Coefficients of Thermal Expansion of Common Building Materials

Material	Density		Coefficient of Thermal Expansion		
	lb/ft³	kg/m³	in./in.-°F × 10^{-6}	mm/mm-°C × 10^{-6}	
Wood (seasoned)					
Douglas fir	32	510	2.1	3.8	parallel to grain
			32	58	perpendicular to grain
Pine	26	415	3.0	5.4	parallel to grain
			19	34	perpendicular to grain
Oak, red or white	41–46	655–735	2.7	4.9	parallel to grain
			30	54	perpendicular to grain
Masonry					
Limestone	130–170	2080–2720	4.4	7.9	
Granite	165–170	2640–2720	4.7	8.5	
Marble	165–170	2640–2720	7.3	13.1	
Brick	100–140	1600–2240	3.6	6.5	
Concrete masonry units	75–145	1200–2320	5.2	9.4	
Concrete					
Normalweight concrete	145	2320	5.5	9.9	
Metals					
Steel	490	7850	6.5	11.7	
Stainless steel, Type 304	490	7850	9.9	17.8	
Aluminum	165	2640	12.8	23.1	
Copper	556	8900	9.3	16.8	
Finish Materials					
Gypsum board	43–50	690–800	9.0	16.2	
Gypsum plaster, sand	105	1680	7.0	12.6	
Glass	156	2500	5.0	9.0	
Acrylic glazing sheet	72	1150	41.0	74.2	
Polycarbonate glazing sheet	75	1200	44.0	79.6	
Polyethylene	57–61	910–970	85.0	153	
Polyvinyl chloride (PVC)	75–106	1200–1700	40.0	72.0	

APPENDIX

U.S. Customary/Metric Conversions

U.S. Customary	Metric	Metric	U.S. Customary
1 in.	25.4 mm	1 mm	0.394 in.
1 ft	304.8 mm	1 m	39.37 in.
1 ft	0.3048 m	1 m	3.2808 ft
1 lb	0.4536 kg	1 kg	2.205 lb
1 ton	0.9072 metric ton (tonne)	1 metric ton (tonne)	2205 lb
1 ft^2	0.0920 m^2	1 m^2	10.76 ft^2
1 ft^3	0.02832 m^3	1 m^3	35.31 ft^3
1 ft^3	28.32 L	1 L	0.03531 ft^3
1 psi	6.894 kPa	1 kPa	0.1451 psi
1 psi	0.006894 MPa	1 MPa	145.1 psi
1 lb/ft^2	4.882 kg/m^2	1 kg/m^2	0.2048 lb/ft^2
1 lb/ft^2	0.04788 kPa	1 kPa	20.89 lb/ft^2
1 lb/ft^3	16.02 kg/m^3	1 kg/m^3	0.06243 lb/ft^3
1 lb/ft^3	0.01602 g/cm^3	1 g/cm^3	62.43 lb/ft^3
1 ft/min	0.00508 m/s	1 m/s	197 ft/min
1 cfm (cubic feet per minute)	0.0004719 m^3/s	1 m^3/s	2119 cfm (cubic feet per minute)
1 cfm	0.4739 L/s	1 L/s	2.119 cfm
1 BTU	0.2522 kcal	1 kcal	3.966 BTU
1 BTU	1.055 kJ	1 kJ	0.9478 BTU
1 BTUH	0.2931 W	1 W	3.412 BTUH
Thermal resistance and conductance			
R 1 (ft^2-hr-°F/BTU)	RSI 0.1761 (m^2-°K/W)[a]	RSI 1 (m^2-°K/W)[a]	R 5.678 (ft^2-hr-°F/BTU)
U 1 (BTU/ft^2-hr-°F)	U 5.678 (W/m^2-°K)[a]	U 1 (W/m^2-°K)[a]	U 0.1761 (BTU/ft^2-hr-°F)
Vapor permeance			
1 perm (grains/hr-ft^2-in. Hg)	57.21 ng/s-m^2-Pa	1 ng/s-m^2-Pa	0.01748 perms (grains/hr-ft^2-in. Hg)
Air permeance			
1 cfm/sf@1.57 psf	5.080 L/s-m^2@75 Pa	1 L/s-m^2@75 Pa	0.1968 cfm/sf@1.57 psf

1 mil = 0.0010 in.
1 in. = 1000 mil
1 Pa = 0.1020 kg/m^2
1 kg/m^2 = 9.807 Pa

[a]Metric thermal conductance and resistance units may be expressed with either "°K" or "°C." In either case, the conversion factors are the same, because a 1-degree difference in either the Celsius or Kelvin temperature scale is the same.

Note: Within the text, units are converted to a degree of precision consistent with the number being converted.

GLOSSARY

A

AAMA American Architectural Manufacturers Association, a trade organization that develops standards for windows, doors, skylights, storefront, and curtain-wall systems.

Abutment joint A *surface divider joint* designed to allow free movement between new and existing construction or between differing materials.

AC *See* **Articulation class**.

ACC *See* **Autoclaved aerated concrete.**

Accelerating admixture An *admixture* that causes concrete or mortar to cure more rapidly.

Access flooring A raised finish floor surface consisting of small, individually removable panels beneath which wiring, ductwork, and other building services may be installed.

Access standard A set of regulations ensuring that buildings are accessible and usable by physically handicapped members of the population.

Acoustical ceiling A ceiling of fibrous tiles that are highly absorbent of sound energy.

ACQ *See* **Alkaline copper quat**.

Acrylic A transparent plastic material widely used in sheet form for glazing windows and skylights.

Active metal A metal relatively high on a galvanic series, tending to act as an *anode* in galvanic couples.

Actual dimension The true dimension of a material, as distinct from its *nominal dimension.*

ADA *See* **Airtight drywall approach**.

Admixture In a concrete or mortar, a substance other than cementitious material, water, and aggregates included in the mixture for the purpose of altering one or more properties of the mixed material, either in its plastic working state or after it has hardened.

Advanced framing techniques A wood light framing system that minimizes redundant framing members, reducing the amount of lumber required and increasing the thermal efficiency of the insulated frame.

Aerogel A silicon-based foam used to fill the interstitial space in insulating glazing units.

AESS *See* **Architecturally exposed structural steel**.

Aggregate Inert particles, such as sand, gravel, crushed stone, or expanded minerals, in a concrete, mortar, or plaster.

Air barrier A material that reduces air leakage through a building assembly.

Air barrier assembly An interconnected assemblage of materials that can effectively resist air pressure differentials acting across the boundary of a wall, roof, or floor assembly.

Air barrier system An interconnected collection of air barrier assemblies responsible for the overall air leakage performance of a completed building.

Air-entraining admixture An admixture that causes a controlled quantity of stable microscopic air bubbles to form in concrete or mortar during mixing, usually for the purposes of increasing workability and resistance to freeze-thaw conditions.

Air permeance A measure of a material's permeability to airflow. A low air permeance is a desirable characteristic of an air barrier material.

Air-supported structure A structure, usually long-span, with a fabric roof supported by an increase in air pressure inside the structure.

Airtight drywall approach (ADA) An air barrier system relying on gypsum board interior finish and the sealing of joints between framing members of a light frame building, serving to reduce the flow of air through the exterior walls and roof.

Air-to-air heat exchanger A device that exhausts air from a building while recovering much of the heat from the exhausted air and transferring it to the incoming air.

AISC American Institute of Steel Construction.

AISC Type I construction *See* **Fully restrained moment connection**.

AISC Type II construction *See* **Simple connection**.

AISC Type III construction *See* **Partially restrained moment connection**.

Albedo *See* **Solar reflectance**.

Alkaline copper quat (ACQ) A chemical used to preserve wood against attack by decay and insects.

Alloy A substance composed of two or more metals or of a metal and a nonmetallic constituent.

Aluminum A silver-colored, nonferrous metal that naturally forms a self-protecting oxide layer.

Anchorage The device that fastens the end of a posttensioning tendon to the end of a concrete slab or beam.

Anchor bolt A bolt embedded in concrete for the purpose of fastening a building frame to a concrete or masonry foundation.

Angle A structural section of steel or aluminum whose profile resembles the letter *L*.

Angle of repose The steepest angle at which an excavation may be sloped so that the soil will not slide back into the hole.

Annealed Cooled under controlled conditions to minimize internal stresses.

Anode The metal in a *galvanic couple* that experiences accelerated corrosion.

Anodizing An electrolytic process that forms a permanent protective oxide coating on aluminum, with or without added color.

ANSI American National Standards Institute, an organization that fosters the establishment of voluntary industrial standards.

Anticlastic Saddle-shaped or having curvature in two opposing directions.

APP *See* **Atactic polypropylene**.

Appearance grading The grading of wood for its appearance properties, as distinct from its structural properties; not to be confused with *visual grading*.

Apron The finish piece that covers the joint between a window stool and the wall finish below.

Arch A structural device that supports a vertical load by translating it into axial inclined forces at its supports.

Architectural concrete Concrete intended as a finish surface and produced to a higher-quality standard.

Architectural sheet metal roofing A roof covering made up of sheets of metal in a traditional shop- or site-fabricated pattern such as standing seam, flat seam, or batten seam.

Architecturally exposed structural steel (AESS) Structural steel intended to be left exposed in the finished building and fabricated and installed to a higher-quality standard.

Architectural Woodwork Institute (AWI) A trade organization that develops standards for custom millwork.

Arc welding A process of joining two pieces of metal by melting them together at their interface with a continuous electric spark and adding a controlled additional amount of molten metal from a metallic electrode.

Area divider A curb used to partition a large roof membrane into smaller areas to allow for expansion and contraction in the deck and membrane.

Articulation class (AC) A measure of a finish ceiling's absorption and reflection of sound, particularly with regard to speech clarity in an open office environment.

Ash dump A door in the underfire of a fireplace that permits ashes from the fire to be swept into a chamber beneath, from which they may be removed at a later time.

Ashlar Squared stonework.

Asphalt A tarry brown or black mixture of hydrocarbons; one type of *bitumen*.

Asphalt roll roofing A continuous sheet of the same roofing material used in asphalt shingles. *See* **Asphalt shingle.**

Asphalt-saturated felt A water-resistive sheet material, available in several different thicknesses, usually consisting of matted cellulose fibers that have been impregnated with asphalt; used to provide a protective layer in an exterior wall or roof assembly; also called *building felt*.

Asphalt shingle A roofing unit composed of a heavy organic or inorganic felt saturated with asphalt and faced with mineral granules.

ASTM American Society for Testing and Materials, an organization that promulgates standards for testing, materials, and methods of building construction.

Atactic polypropylene (APP) An amorphous form of polypropylene used as a modifier in modified bitumen roofing.

Auger A helical tool for creating cylindrical holes.

Autoclaved aerated concrete (AAC) Concrete formulated so as to contain a large percentage of gas bubbles as a result of a chemical reaction that takes place in an atmosphere of steam.

AWI *See* **Architectural Woodwork Institute**.

Awning window A window that pivots on a horizontal axis at the top edge of the sash and projects toward the outdoors.

Axial In a direction parallel to the long axis of a structural member.

B

Backer board A fiber-reinforced cement board or glass-mat-faced gypsum board used as a base for *thin-set tile* applications.

Backer rod A flexible, compressible strip of plastic foam inserted into a joint to limit the depth to which sealant can penetrate.

Backfill Earth or earthen material used to fill the excavation around a foundation; the act of filling around a foundation.

Backup, backup wall A vertical plane of masonry, concrete, or wood framing used to support a thin facing such as a single wythe of brickwork.

Backup bar A small rectangular strip of steel applied beneath a joint to provide a solid base for beginning a weld between two steel structural members.

Ballast A heavy material installed over a roof membrane to prevent wind uplift and shield the membrane from sunlight.

Balloon frame A wooden building frame composed of closely spaced members nominally 2 inches (51 mm) thick, in which the wall members are single pieces that run from the sill to the top plates at the eave.

Baluster A small vertical member that serves to fill the opening between a handrail and a stair or floor.

Band joist A wooden joist running perpendicular to the primary direction of the joists in a floor and closing off the floor platform at the outside face of the building.

Bar A small rolled steel shape, usually round or rectangular in cross section; a rolled steel shape used for reinforcing concrete.

Barrel shell A scalloped roof structure of reinforced concrete that spans in one direction as a barrel vault and in the other as a folded plate.

Barrel vault A segment of a cylindrical surface that spans as an arch.

Barrier wall An exterior wall of a building whose watertightness depends on its freedom from passages through the wall.

Basalt A very dense and durable igneous rock, usually dark gray in color; classified by ASTM C119 in the Granite group.

Baseboard A strip of finish material placed at the junction of a floor and a wall to create a neat intersection and to protect the wall against damage from feet, furniture, and floor-cleaning equipment.

Base-coat plaster One or more preparatory plaster coats that provide a flat, solid surface suitable for the application of the final finish coat plaster. *See also* **Scratch coat** and **Brown coat**.

Base flashing The flashing at the edges of a low-slope roof membrane that turns up against the adjacent face of a parapet or wall, and frequently overlapped by a *counterflashing*.

Base isolator A device at foundation level that diminishes the transmission of seismic motions to a building.

Baseplate A steel plate inserted between a column and a foundation to spread the concentrated load of the column across a larger area of the foundation.

Basic oxygen process A steel-making process in which a stream of pure oxygen is introduced into a batch of molten iron so as to remove excess carbon and other impurities.

Batten A strip of wood or metal used to cover the crack between two adjoining boards or panels.

Batten-seam A seam in a sheet metal roof that encloses a wood batten.

Batter board Boards mounted on stakes outside the excavation area of a building, used to preserve locations for string lines marking the corners of the building foundation.

Bay A rectangular area of a building defined by four adjacent columns; a portion of a building that projects from a facade.

Bead A narrow line of weld metal or sealant; a strip of metal or wood used to hold a sheet of glass in place; a narrow, convex molding profile, a metal edge or corner accessory for plaster.

Beam A straight structural member that acts primarily to resist nonaxial loads.

Beam blank *See* **Bloom**.

Bearing A point at which one building element rests upon another.

Bearing block A piece of wood fastened to a column to provide support for a beam or girder.

Bearing pad A block of plastic or synthetic rubber used to cushion the point at which one precast concrete element rests upon another.

Bearing wall A wall that supports floors or roofs.

Bed *See* **Casting bed.**

Bed joint The horizontal layer of mortar beneath a masonry unit.

Bedrock A solid stratum of rock.

Bending moment The combination of tension and compression forces that cause a beam or other structural member to bend. *See also* **Moment.**

Bending stress A compressive or tensile stress resulting from the application of a nonaxial force to a structural member.

Bent A plane of framing consisting of beams and columns joined together, often with rigid joints.

Bentonite clay An absorptive colloidal clay that swells to several times its dry volume when saturated with water; the primary ingredient in bentonite waterproofing.

Bessemer process An early method of steel manufacturing in which air was blown into a vessel of molten iron to burn out impurities.

Bevel An end or edge that is cut at an angle other than a right angle.

Bevel siding Wood cladding boards that taper in cross section.

Billet A large cylinder or rectangular solid of material.

BIM *See* **Building information modeling**.

Birdsmouth cut An angled notch cut into a rafter to allow the rafter to seat securely on the top plate of a wall.

Bite The depth to which the edge of a piece of glass is held by its frame.

Bitumen A tarry mixture of hydrocarbons, such as asphalt or coal tar.

Bituminous roof membrane A *low-slope roof* membrane made from bituminous materials, either a *built-up roof* membrane or *a modified bitumen roof membrane.*

Blast furnace slag A hydraulic cementitious material formed as a byproduct of iron manufacture, used in mortar and concrete mixtures; also called slag cement.

Blast-resistant glazing Window, storefront, or curtainwall systems designed for resistance to the force of explosive blasts.

Bleed water In freshly placed concrete, water that rises to the top surface of the concrete as the solid cement and aggregate particles settle.

Blended hydraulic cement Hydraulic cement made from a mixture of cementitious materials such as portland cement, other hydraulic cements, and pozzolans for the purpose of altering one or more properties of the cement or reducing the energy required in the cement manufacturing process.

Blind nailing Attaching boards to a frame, sheathing, or subflooring with toe nails driven through the edge of each piece so as to be completely concealed by the adjoining piece.

Blind-side waterproofing An impervious layer or coating on the outside of a foundation wall that, for reasons of inaccessibility, was installed before the wall was constructed.

Blocking Pieces of wood inserted tightly between joists, studs, or rafters in a building frame to stabilize the structure, inhibit the passage of fire, provide a nailing surface for finish materials, or retain insulation.

Bloom A rectangular solid of steel formed from an ingot as an intermediate step in creating rolled steel structural shapes.

Blooming mill A set of rollers used to transform an ingot into a bloom.

Bluestone A sandstone that is gray to blue-gray in color and splits readily into thin slabs; classified by ASTM C119 in the Quartz-Based Stone group.

Board foot A unit of lumber volume, a rectangular solid nominally 12 square inches in cross-sectional area and 1 foot long.

Board siding Wood cladding made up of boards, as differentiated from shingles or manufactured wood panels.

Bolster A long chair used to support reinforcing bars in a concrete slab.

Bolt A fastener consisting of a cylindrical metal body with a head at one end and a helical thread at the other, intended to be inserted through holes in adjoining pieces of material and closed with a threaded nut.

Bond In masonry, the adhesive force between mortar and masonry units, or the pattern in which masonry units are laid to tie two or more wythes together into a structural unit. In reinforced concrete, the adhesion between the surface of a reinforcing bar and the surrounding concrete.

Bond breaker A strip of material to which sealant does not adhere.

Bonded posttensioning A system of prestressing in which the tendons are grouted after stressing so as to bond them to the surrounding concrete.

Bonded terrazzo Terrazzo flooring whose underbed is poured directly upon the structural floor.

Bottom bar A reinforcing bar that lies close to the bottom of a beam or slab.

Bottom plate *See* **Sole plate**.

Bound water In wood, the water held within the cellulose of the cell walls. *See also* **Free water.**

Box beam A bending member of metal or plywood whose cross section resembles a closed rectangular box.

Box girder A major spanning member of concrete or steel whose cross section is a hollow rectangle or trapezoid.

Box nail A nail with a more slender shank than a *common nail*, used for fastening framing members in wood light frame construction.

Braced frame A structural building frame strengthened against lateral forces with diagonal members.

Bracing Diagonal members, either temporary or permanent, installed to stabilize a structure against lateral loads.

Brad A small *finish nail*.

Brake A machine used to form lengths of sheet metal into bent shapes.

Brazing A process that uses molten, nonferrous metal to join two pieces of metal. The brazing metal is melted at a temperature below that of the metals being joined, so that, unlike in welding, the joined metals remain in a solid state throughout the process.

Breather mat A wiry plastic matting placed within a roof or wall assembly to create a space for drainage and ventilation.

Bridging Bracing or blocking installed between steel or wood joists at midspan to stabilize them against buckling and, in some cases, to permit adjacent joists to share loads.

British thermal unit (BTU) The quantity of heat required to raise the temperature of 1 pound of water 1 degree Fahrenheit.

Broom finish A skid-resistant texture imparted to an uncured concrete surface by dragging a stiff-bristled broom across it.

Brown coat The second of two basecoat plaster applications in a three-coat plaster.

Brownstone A brownish or reddish sandstone; classified by ASTM C119 in the Quartz-Based Stone group.

BTU *See* **British thermal unit**.

Buckling Structure failure by gross lateral deflection of a slender element under compressive stress, such as the sideward buckling of a long, slender column or the upper edge of a long, thin floor joist.

Building brick Brick used for concealed masonry work where appearance is not a concern.

Building code A set of regulations intended to ensure a minimum standard of health and safety in buildings.

Building enclosure The parts of the building, principally its walls, roofs, and fenestration, that separate the interior of the building from the exterior, and that must effectively control the flow of heat, air, and moisture; also called the *thermal envelope* or the *building envelope*.

Building envelope *See* **Building enclosure**.

Building felt *See* **Asphalt-saturated felt**.

Building information modeling (BIM) The computerized three-dimensional modeling of building systems, with the linking of model components to a database of properties and relationships.

Building paper A water-resistive, asphalt-saturated paper used similarly to asphalt-saturated felt, to provide a protective layer in an exterior wall assembly.

Building separation joint A plane along which a building is divided into separate structures that may move independently of one another.

Built-up roof (BUR) A multi-ply roof membrane, made from layers of asphalt-saturated felt or other fabric, bonded together with bitumen.

Bull float A long-handled tool used for the initial floating of a freshly poured concrete slab. *See also* **Darby.**

Buoyant uplift The force of water or liqufied soil that tends to raise a building foundation out of the ground.

BUR *See* **Built-up roof**.

Butt The thicker end, such as the lower edge of a wood shingle or the lower end of a tree trunk; a joint between square-edged pieces; a weld between square-edged pieces of metal that lie in the same plane; a type of door hinge that attaches to the edge of the door.

Butt-joint glazing A type of glass installation in which the vertical joints between lights of glass do not meet at a mullion, but are made weathertight with a sealant.

Button head A smooth, convex bolt head with no provision for engaging a wrench.

Buttress A structural device of masonry or concrete that resists the diagonal forces from an arch or vault.

Butyl rubber A synthetic rubber compound.

C

CA *See* **Copper boron azole.**

CAC *See* **Ceiling Attenuation Class.**

CAD *See* **Computer-aided design.**

Caisson A cylindrical sitecast concrete foundation unit that penetrates through unsatisfactory soil to rest upon an underlying stratum of rock or satisfactory soil; an enclosure that permits excavation work to be carried out underwater. Also called a drilled pier.

Calcined gypsum. Gypsum which has been ground to a fine powder and heated to drive off most of its water of hydration; used in the manufacture of gypsum board and as the principal ingredient in gypsum plasters; a nonhydraulic cementitious material; also called *plaster of Paris.*

Calcining The driving off of the water of hydration from gypsum by the application of heat.

Camber A slight, intentional initial curvature in a beam or slab.

Cambium The thin layer beneath the bark of a tree that manufactures cells of wood and bark.

Cantilever A beam, truss, or slab that extends beyond its last point of support.

Cant strip A strip of material with a sloping face used to ease the transition from a horizontal to a vertical surface at the edge of a membrane roof.

Capillary action The pulling of water through a small orifice or fibrous material by the adhesive force between the water and the material.

Capillary break A slot or groove intended to create an opening too large to be bridged by a drop of water and, thereby, to eliminate the passage of water by capillary action; the coarse aggregate layer under a concrete slab on grade which reduces the migration of water from the ground below into the concrete slab above.

Carbide-tipped tools Drill bits, saws, and other tools with cutting edges made of an extremely hard alloy.

Carbonation The process by which lime mortar reacts with atmospheric carbon dioxide to cure.

Carbon fiber reinforcing In precast concrete, an open grid fabric of carbon fibers bonded with epoxy resin, used as a substitute for *welded-wire reinforcing.*

Carbon steel Low-carbon or mild steel.

Carpenter One who makes things of wood.

Casement window A window that pivots on an axis at or near a vertical edge of the sash.

Casing The wood finish pieces surrounding the frame of a window or door; a cylindrical steel tube used to line a drilled or driven hole in foundation work.

Castellated beam A steel wide-flange section whose web has been cut along a zigzag path and reassembled by welding in such a way as to create a deeper section.

Casting Pouring a liquid material or slurry into a mold whose form it will take as it solidifies.

Casting bed A permanent, fixed form in which precast concrete elements are produced.

Cast in place Concrete that is poured in its final location; sitecast.

Cast iron Iron with too high a carbon content to be classified as steel.

Cathode The metal in a *galvanic couple* that experiences a decreased rate of corrosion.

Caulk A low-range *sealant.*

Cavity drainage material A material placed in the air space of a cavity wall to catch mortar droppings and prevent clogging of weep holes at the bottom of the cavity.

Cavity wall A masonry wall that includes a continuous airspace between its outermost wythe and the remainder of the wall.

CBA *See* **Copper boron azole.**

CCA *See* **Chromated copper arsenate.**

Cee A metal framing member whose cross-sectional shape resembles the letter *C.*

Ceiling Attenuation Class (CAC) An index of the ability of a ceiling construction to obstruct the passage of sound between rooms through the plenum.

Ceiling joist *See* **Joist.**

Cellular decking Panels made of steel sheets corrugated and welded together in such a way that hollow longitudinal cells are created within the panels.

Cellular raceway A rectangular tube cast into a concrete floor slab for the purpose of housing electrical and communications wiring.

Cellulose A complex polymeric carbohydrate of which the structural fibers in wood are composed.

Celsius A temperature scale on which the freezing point of water is established as 0 and the boiling point as 100 degrees.

Cement Generally, any substance used to adhere material together. In concrete, masonry, and plastering work, any of a number of inorganic materials that have cementing properties when combined with the water. *See also* **Cementitious materials.**

Cementitious materials In concrete, masonry, and plastering, inorganic materials that, when mixed with water, produce hardened products with adhesive and cohesive (cementing) properties; frequently used to refer exclusively to hydraulic cements (such as portland cement), to the exclusion of nonhydraulic (lime and gypsum) cements.

Cement-lime mortar Mortar made from portland cement, hydrated lime, aggregate, and water, the most traditional formulation of modern masonry mortars. *See also* **Masonry cement**, **Mortar cement**.

Centering Temporary formwork for an arch, dome, or vault.

Centering shims Small blocks of synthetic rubber or plastic used to hold a sheet of glass in the center of its frame.

Ceramic tile Small, flat, thin clay tiles intended for use as wall and floor facings.

Chair A device used to support reinforcing bars.

Chamfer A flattening of a longitudinal edge of a solid member on a plane that lies at an angle of 45 degrees to the adjoining planes.

Channel A steel or aluminum section shaped like a rectangular box with one side missing.

Chemically strengthened glass Glass strengthened by immersion in a molten salt bath, causing an ion exchange at the surfaces of the glass that creates a prestress in much the same manner as heat-treated glass.

Chlorinated polyethylene (CPE) A plastic material used in roof membranes.

Chlorosulfonated polyethylene (CSPE) A plastic material used in roof membranes.

Chord A top or bottom member of a truss.

Chromated copper arsenate (CCA) A chemical used to protect wood against attack by decay and insects. Due to toxicity concerns, this chemical has been phased out of most treated wood used in residential and commercial buliding construction.

Chromogenic glass Glass that can change its optical properties, such as *thermochromic, photochromic,* or *electrochromic glass.*

C–H stud A steel wall framing member whose profile resembles a combination of the letters *C* and *H,* used to support gypsum panels in shaft walls.

Chuck A device for holding a steel wire, rod, or cable securely in place by means of steel wedges in a tapering cylinder.

Churn drill A steel tool used with an up-and-down motion to cut through rock at the bottom of a steel pipe caisson.

Cladding A material used to cover the exterior of a building.

Clamp A tool for holding two pieces of material together temporarily; unfired bricks piled in such a way that they can be fired without using a kiln.

Class A, B, C roofing Roof covering materials classified according to their resistance to fire when tested in accordance with ASTM E108. Class A is the most resistant, and Class C is the least.

Clay A fine-grained soil with plate-shaped particles less than 0.0002 inch (0.005 mm) in size, which properties are significantly influenced by the structural arrangements of the particles and the electrostatic forces acting between them.

Cleanout hole An opening at the base of a masonry wall through which mortar droppings and other debris can be removed prior to grouting the interior cavity of the wall.

Clear dimension, clear opening The dimension between opposing inside faces of an opening.

Cleavage membrane A resilient sheet placed underneath a finish tile assembly to prevent movement stresses in the underlying substrate from telegraphing into the finish assembly.

Climbing crane A heavy-duty lifting machine that raises itself as the building rises.

Clinker A fused, pebblelike mass that is an intermediate product of cement manufacture; a brick that is overburned.

Closer The last masonry unit laid in a course; a partial masonry unit used at the corner of a header course to adjust the joint spacing; a mechanical device for regulating the closing action of a door.

CLSM *See* **Controlled low-strength material**.

CMU *See* **Concrete masonry unit**.

Coarse aggregate Gravel or crushed stone in a concrete mix.

Coarse-grained soil Soil with particles ranging in size from roughly 0.003 to 3 inches (0.075–75 mm); sands and gravels.

Code *See* **Building code**.

Cohesionless soil *See* **Frictional soil**.

Cohesive soil A soil such as clay whose particles are able to adhere to one another by means of cohesive and adhesive forces.

Cold-formed steel *See* **Cold-worked steel.**

Cold-rolled steel Steel rolled to its final form at a temperature at which it is no longer plastic.

Cold-worked steel Steel formed at a temperature at which it is no longer plastic, as by rolling or forging at room temperature.

Collar joint The vertical mortar joint between wythes of masonry.

Collar tie A piece of wood nailed across two opposing rafters near the ridge to resist wind uplift.

Collated nails Nails glued together in a strip for insertion into a nail gun.

Column An upright structural member acting primarily in compression.

Column bar *See* **Vertical bar**.

Column cage An assembly of vertical reinforcing bars and ties for a concrete column.

Column-cover-and-spandrel system A system of cladding in which panels of material cover the columns and spandrels, with horizontal strips of windows filling the remaining portion of the wall.

Column spiral A continuous coil of steel reinforcing used to tie a concrete column.

Column tie A single loop of steel bar, usually bent into a rectangular configuration, used to tie a concrete column.

Combination door A door with interchangeable inserts of glass and insect screening, usually used as a second, exterior door and mounted in the same opening with a conventional door.

Combination window A sash that holds both insect screening and a retractable sheet of glass, mounted in the same frame with a window and used to increase its thermal resistance.

Commercial wrap A synthetic sheet material, heavier than *housewrap*, with water-resistive and air-resistive properties used to provide a protective layer in an exterior wall assembly.

Common bolt An ordinary carbon steel bolt.

Common bond Brickwork laid with five courses of stretchers followed by one course of headers.

Common nail A standard-sized nail used for the fastening of framing members in wood light frame construction.

Common rafter A roof *rafter* that runs parallel to the main slope of the roof. *See also* **Hip rafter.**

Composite A material or assembly made up of two or more materials bonded together to act as a single structural unit.

Composite column An upright structural member, acting primarily in compression, that is composed of concrete and a steel structural shape, usually a wide flange or a tube.

Composite construction Any element in which concrete and steel, other than reinforcing bars, work as a single structural unit.

Composite metal decking Corrugated steel decking manufactured in such a way that it bonds securely to the concrete floor fill to form a reinforced concrete deck.

Composite wall A masonry wall that incorporates two or more different types of masonry units, such as clay bricks and concrete blocks.

Composition shingle See *Asphalt shingle.*

Compression A squeezing force.

Compression gasket A synthetic rubber strip that seals around a sheet of glass or a wall panel by being squeezed tightly against it.

Compressive strength The ability of a structural material to withstand squeezing forces.

Computer-aided design (CAD) The digital two-dimensional representation of building systems.

Concave joint A mortar joint tooled into a curved, indented profile.

Concealed flashing *See* **Internal flashing.**

Concealed grid A suspended ceiling framework that is completely hidden by the tiles or panels it supports.

Concrete A structural material produced by mixing predetermined amounts of cement, aggregates, and water and allowing this mixture to cure under controlled conditions.

Concrete block A concrete masonry unit, usually hollow, that is larger than a brick.

Concrete masonry unit (CMU) A block of hardened concrete, with or without hollow cores, designed to be laid in the same manner as a brick or stone; a concrete block.

Condensate Water formed as a result of *condensation.*

Condensation The process of changing from a gaseous to a liquid state, especially as applied to water.

Conduit A steel or plastic tube through which electrical wiring is run.

Consolidate In freshly poured concrete, eliminate trapped air and cause the concrete to fill completely around the reinforcing bars and into all the corners of the formwork, usually by vibrating the concrete.

Construction documents The graphic *construction drawings* and written *specifications* to which a building is constructed.

Construction Drawings The graphic instructions from an architect or an engineer concerning the construction of a building.

Construction manager An entity that assists the owner in the procurement of construction services.

Construction Type In the International Building Code, any of five major systems of building construction that are differentiated by their relative resistance to fire.

Continuous ridge vent *See* **Ridge vent**.

Contraction joint *See* **Control joint.**

Contractor A person or organization that undertakes a legal obligation to do construction work.

Control joint An intentional, linear discontinuity in a structure or component designed to form a plane of weakness where cracking can occur in response to various forces so as to minimize or eliminate cracking elsewhere in the structure. Also called a *contraction joint.*

Controlled low-strength material (CLSM) A concrete that is purposely formulated to have a very low but known strength, used primarily as a backfill material.

Convector A heat exchange device that uses the heat in steam, hot water, or an electric resistance element to warm the air in a room; often called, inaccurately, a *radiator.*

Cool color A coating applied to a roofing material that is nonwhite, yet reflects a relatively high percentage of the sun's thermal energy.

Cool roof A roof covering that reflects a substantial portion of the sun's thermal energy.

Cope The removal of a flange at the end of a steel beam in order to facilitate connection to another member.

Coped connection A joint in which the end of one member is cut to match the profile of the other member.

Coping A protective cap on the top of a masonry wall.

Coping saw A handsaw with a thin, narrow blade, used for cutting detailed shapes in the ends of wood moldings and trim.

Copolymer A large molecule composed of repeating patterns of two or more chemical units.

Copper A soft, nonferrous metal, orange-red in color, that oxidizes to a color ranging from blue-green to black.

Copper boron azole (CBA, CA) A chemical used to preserve wood against attack by decay and insects.

Corbel A spanning device in which masonry units in successive courses are cantilevered slightly over one another; a projecting bracket of masonry or concrete.

Coreboard A thick gypsum panel used primarily in shaft walls.

Corner bead A metal or plastic strip used to form a neat, durable edge at an outside corner of two walls of plaster or gypsum board.

Cornice The exterior detail at the meeting of a wall and a roof overhang; a decorative molding at the intersection of a wall and a ceiling.

Corrosion Oxidation, such as rust.

Corrosion inhibitor A concrete admixture intended to prevent oxidation of reinforcing bars.

Corrugated Formed into a fluted or ribbed profile.

Counterflashing A flashing turned down from above to overlap another flashing turned up from below so as to shed water.

Course A horizontal layer of masonry units one unit high; a horizontal line of shingles or siding.

Coursed In masonry, laid in courses with straight bed joints.

Cove base A flexible strip of plastic or synthetic rubber used to finish the junction between resilient flooring and a wall.

Cover In concrete, a specified thickness of concrete surrounding steel reinforcing bars to provide full embedment for the bars and protect them against fire and corrosion.

CPE *See* **Chlorinated polyethylene**.

Crawlspace A space that is not tall enough to stand in, located beneath the bottom of a building.

Creep A permanent inelastic deformation in a material due to changes in the material caused by the prolonged application of a structural stress, common in wood and concrete.

Cripple stud A wood wall framing member that is shorter than full-length studs because it is interrupted by a header or sill.

Critical path The sequence of tasks that determines the least amount of time in which a construction project can be completed.

Cross-grain wood Wood incorporated into a structure in such a way that the direction of its grain is perpendicular to the direction of the principal loads on the structure.

Crosslot bracing Horizontal compression members running from one side of an excavation to the other, used to support sheeting.

Crown glass Glass sheet formed by spinning an opened hollow globe of heated glass.

Cruck A framing member cut from a bent tree so as to form one-half of a rigid frame.

CSPE *See* **Chlorosulfonated polyethylene**.

Cup A curl in the cross section of a board or timber caused by unequal shrinkage or expansion between one side of the board and the other.

Curing The hardening of concrete, plaster, gunnable sealant, or other wet materials. Curing can occur through evaporation of water or a solvent, hydration, polymerization, or chemical reactions of various types, depending on the formulation of the material.

Curing compound A liquid that, when sprayed on the surface of newly placed concrete, forms a water-resistant layer to prevent premature dehydration of the concrete.

Curtain wall An exterior building wall that is supported entirely by the frame of the building, rather than being self-supporting or loadbearing.

Cylinder glass Glass sheet produced by blowing a large, elongated glass cylinder, cutting off its ends, slitting it lengthwise, and opening it into a flat rectangle.

D

d *See* **Penny**.

Damper A flap to control or obstruct the flow of gases; specifically, a metal control flap in the throat of a fireplace or in an air duct.

Dampproofing A coating intended to resist the passage of water, commonly applied to the outside face of basement walls or to the inner face of a cavity in a masonry cavity wall.

Dap A notch at the end of a piece of material.

Darby A stiff straightedge of wood or metal used to level the surface of wet plaster or concrete.

Daylighting Illuminating the interior of a building by natural means.

Dead load Permanent loads on a building, including the weight of the building itself and any permanently attached equipment.

Deadman A large and/or heavy object buried in the ground as an anchor.

Decking A material used to span across beams or joists to create a floor or roof surface.

Deep foundation A building *foundation* that extends through upper strata of incompetent soil to reach deeper strata with greater bearing capacity.

Deformation A change in the shape of a structure or structural element caused by a load or force acting on the structure.

Deformed reinforcing bar Steel reinforcing bars with surface ribs for better bonding to concrete.

Depressed strand A pretensioning tendon that is pulled to the bottom of the beam at the center of the span to follow more closely the path of tensile forces in the member.

Derrick Any of a number of devices for hoisting building materials on the end of a rope or cable.

Design/bid/build project delivery A method of providing design and construction services in which the design and construction phases of the project are provided by different entities, usually used in combination with *sequential construction.*

Design/build project delivery A method of providing design and construction services in which the design and construction phases of the project are provided by a single entity, frequently used in combination with *fast track construction.*

Dew point The temperature at which water will begin to condense from a mass of air with a given moisture content.

Dewatering The extraction of water from an excavation or its surrounding soil.

Diagonal bracing *See* **Bracing.**

Diamond saw A tool with a moving chain, belt, wire, straight blade, or circular blade whose cutting action is carried out by diamonds.

Diaphragm action A bracing action that derives from the stiffness of a thin plane of material when it is loaded in a direction parallel to the plane. Diaphragms in buildings are typically floor, wall, or roof surfaces of plywood, reinforced masonry, steel decking, or reinforced concrete.

Die An industrial tool for giving identical form to repeatedly produced or continuously generated units, such as a shaped orifice for giving form to a column of clay, a steel wire, or an aluminum extrusion; a shaped punch for making cutouts of sheet metal or paper; or a mold for casting plastic or metal.

Die-cut Manufactured by punching from a sheet material.

Differential settlement Subsidence of the various foundation elements of a building at differing rates.

Diffuser A louver shaped so as to distribute air about a room.

Dimension lumber Lengths of wood, rectangular in cross section, sawed directly from the log.

Dimension stone Building stone cut to a rectangular shape.

Direct tension indicator washer *See* **Load indicator washer**.

Distribution rib A transverse beam at the midspan of a one-way concrete joist structure, used to allow the joists to share concentrated loads.

Divider strip A strip of metal or plastic embedded in terrazzo to form control joints and decorative patterns.

Dome An arch rotated about its vertical axis to produce a structure shaped like an inverted bowl; a form used to make one of the cavities in a concrete waffle slab.

Dormer A structure protruding through the plane of a sloping roof, usually containing a window and having its own smaller roof.

Double glazing Two parallel sheets of glass with an airspace between.

Double-hung window A window with two overlapping sashes that slide vertically in tracks.

Double shear Acting to resist shear forces at two locations, such as a bolt that passes through a steel supporting angle, a beam web, and another supporting angle.

Double-skin facade An exterior wall system consisting of two separate glass skins separated by an interstitial space; also called dual-wall facade or double-skin wall.

Double-skin wall *See* **Double-skin facade**.

Double-strength glass Glass that is approximately 1/8 inch (3 mm) thick.

Double tee A precast concrete slab element that resembles the letters *TT* in cross section.

Dovetail slot anchor A system for fastening to a concrete structure that uses metal tabs inserted into a slot that is small at the face of the concrete and larger behind.

Dowel A short cylindrical rod of wood or steel; a steel reinforcing bar that projects from a foundation to tie it to a column or wall, or from one section of a concrete slab or wall to another.

Downspout A vertical pipe for conducting water from a roof to a lower level, also called a leader.

Drainage Removal of water.

Drainage backfill Crushed stone or gravel backfill materials with good drainage characteristics, placed around a foundation to facilitate drainage.

Draped tendon A posttensioning strand that is placed along a curving profile that approximates the path of the tensile forces in a beam.

Drawing Shaping a material by pulling it through an orifice, as in the drawing of steel wire or the drawing of a sheet of glass.

Drawings *See* **Construction Drawings**.

Drawn glass Glass sheet pulled directly from a container of molten glass.

Drift Lateral deflection of a building caused by wind or earthquake loads.

Drift pin A tapered steel rod used to align bolt holes in steel connections during erection.

Drilled pier *See* **Caisson**.

Drip A discontinuity formed into the underside of a window sill or wall component to force adhering drops of water to fall free of the face of the building rather than to move farther toward the interior.

Dropchute A flexible hoselike tube for placing concrete; used to break the fall of the concrete and prevent *segregation.*

Drop panel A thickening of a two-way concrete structure at the head of a column.

Drying shrinkage Shrinkage of concrete, mortar, or plaster, that occurs as excess water evaporates from the material.

Dry-pack grout A low-slump cementitious mixture tamped into the space in a connection between precast concrete members.

Dry-press process A method of molding slightly damp clays and shales into bricks by forcing them into molds under high pressure.

Dry-set mortar A tile setting mortar formulated with portland cement, sand, and water retention compounds, used in thin-set tile applications.

"Dry" systems Systems of construction that use little or no water during construction, as differentiated from systems such as masonry, plastering, and ceramic tile work.

Drywall *See* **Gypsum board**.

Dry well An underground pit filled with broken stone or other porous material from which rainwater from a roof drainage system can seep into the surrounding soil.

Dual-wall facade *See* **Double-skin facade**.

Duct A hollow conduit, commonly of sheet metal, through which air can be circulated; a tube used to establish the position of a posttensioning tendon in a concrete structure.

DWV Pipe Drain–waste–vent pipe; the part of the plumbing system of a building that removes liquid wastes and conducts them to the sewer or sewage disposal system.

E

Earlywood *See* **Springwood**.

Earth material Rock or soil.

Eave The horizontal edge at the low side of a sloping roof.

Eco-roof *See* **Green roof**.

Edge bead A strip of metal or plastic used to make a neat, durable edge where plaster or gypsum board abuts another material.

Edge-grain lumber *See* **Vertical gain lumber.**

Edge spacer The material used to separate lights of glass in an insulating glass unit, also called a spline.

Efflorescence A powdery deposit on the face of a structure of masonry or concrete, caused by the leaching of chemical salts by water migrating from within the structure to the surface.

EIFS *See* **Exterior insulation and finish system.**

Elastic Able to return to its original size and shape after removal of stress.

Elastomer A rubber or synthetic rubber.

Elastomeric Rubberlike. In low-slope roofing, a *thermosetting* membrane material.

Electrochromic glass Glass that changes its optical properties in response to the application of electric current.

Electrode A consumable steel wire or rod used to maintain an arc and furnish additional weld metal in electric arc welding.

Electrogalvanizing A method of *galvanizing*, in which an electric current is used to deposit zinc from a liquid bath onto steel.

Elevation A drawing that views a building from any of its sides; a vertical height above a reference point such as sea level.

Elongation Stretching under load; growing longer because of temperature expansion.

Embodied energy The total life cycle energy expended in extraction of raw materials, processing, fabrication, and transportation of a material or product to its point of use in a building; in some calculations, may also include energy required to dispose of or recycle the material.

EMC *See* **Equilibrium moisture content**.

Enamel A glossy or semigloss paint.

End dam The turned-up end of a flashing that prevents water from running out of the end; a block inserted into the space within a horizontal aluminum mullion for the same purpose.

End nail A nail driven through the side of one piece of lumber and into the end of another.

Engineered fill Earth compacted into place in such a way that it has predictable physical properties, based on laboratory tests and specified, supervised installation procedures.

Engineered lumber See **Structural composite lumber**.

English bond Brickwork laid with alternating courses, each consisting entirely of headers or stretchers.

EPDM Ethylene propylene diene monomer, a synthetic rubber thermosetting material used in low-slope roofing membranes.

Equilibrium moisture content (EMC) The moisture content at which wood stabilizes after a period of time in its destination environment.

Erector The subcontractor who raises, connects, and plumbs up a building frame from fabricated steel or precast concrete components.

Ethylene propylene diene monomer *See* **EPDM**.

Expanded metal lath A thin sheet of metal that has been slit and stretched to transform it into a mesh, used as a base for the installation of plaster.

Expanded shale aggregate A *structural lightweight aggregate* made from ground shale particles that have been heated to the point that moisture within the particles vaporizes, causing the particles to expand.

Expansion joint A *surface divider joint* that provides space for the surface to expand. In common usage, a building separation joint.

Exposed aggregate finish A concrete surface in which the coarse aggregate is revealed.

Exposed grid A framework for an acoustical ceiling that is visible from below after the ceiling is completed.

Exposure durability classification A system for rating the expected resistance of a wood panel product to weathering.

Extended-life admixture A substance that retards the onset of the curing reaction in mortar so that the mortar may be used over a protracted period of time after mixing.

Extended set-control admixture A substance that retards the onset of the curing reaction in concrete or mortar so that the material may be used over a protracted period of time after mixing.

Extensive green roof A *green roof* with a relatively shallow soil, planted with low-maintenance, drought-tolerant plant materials.

Exterior insulation and finish system (EIFS) A cladding system that consists of a thin layer of reinforced stucco applied directly to the surface of an insulating plastic foam board.

External flashing In masonry, a flashing that is not concealed within the wall, usually at the roof level or top of the wall.

External gutter A *gutter* hung off the edge of a roof, external to the roof construction itself.

Extrados The convex surface of an arch.

Extrusion The process of squeezing a material through a shaped orifice to produce a linear element with the desired cross section; an element produced by this process.

F

Fabricator The company that prepares structural steel members for erection; any entity that assembles building components prior to arrival on the construction site.

Facade An exterior face of a building.

Face brick A brick selected on the basis of appearance and durability for use in the exposed surface of a wall.

Face nail A nail driven through the side of one wood member into the side of another.

Face shell The portion of a hollow concrete masonry unit that forms the face of the wall.

Fahrenheit A temperature scale on which the boiling point of water is fixed at 212 degrees and the freezing point at 32 degrees.

Fanlight A semicircular or semielliptical window above an entrance door, often with radiating muntins that resemble a fan.

Fascia The exposed vertical face of an *eave.*

Fast track construction A method of providing design and construction services in which design and construction overlap in time; also called *phased construction.*

Faying surface The contacting surfaces of steel members joined with a *slip-critical connection.*

Felt A thin, flexible sheet material made of soft fibers pressed and bonded together. In building practice, a thick paper or a sheet of glass or plastic fibers.

Ferrous metal Any iron-based metal.

Ferrous steel In common usage, steel unprotected from corrosion by either galvanizing or alloying.

Fibrous reinforcing Short fibers of glass, steel, or polypropylene mixed into concrete to act as either *microfiber reinforcing* or *macrofiber reinforcing.*

Fieldstone Rough building stone gathered from river beds and fields.

Figure The surface pattern of the grain of a piece of smoothly finished wood or stone.

Filigree precast concrete A hybrid concrete system in which *precast concrete* sections are used as *permanent formwork* for *cast in place* concrete.

Fillet A rounded inside intersection between two surfaces that meet at an angle.

Fillet weld A weld at the inside intersection of two metal surfaces that meet at right angles.

Fine aggregate Sand used in concrete, mortar, or plaster mixes.

Fine-grained soil Soil with particles 0.003 inch (0.075 mm) or less in size; silts and clays.

Finger joint A glued end connection between two pieces of wood, using an interlocking pattern of deeply cut "fingers." A finger joint creates a large surface for the glue bond, to allow it to develop the full tensile strength of the wood it connects.

Finial A slender ornament at the top of a roof or spire.

Finish Exposed to view; material that is exposed to view.

Finish carpenter One who does finish carpentry.

Finish carpentry The wood components exposed to view on the interior of a building, such as window and door casings, baseboards, bookshelves, and the like; may also refer to exterior finish carpentry, such as exterior trim, deck railings, and similar items.

Finish coat The final coat of a paint or other finishing system.

Finish-coat plaster The final coat of plaster, applied over gypsum base or one or more applications of base-coat plaster.

Finish floor The floor material exposed to view, as differentiated from the *subfloor,* which is the loadbearing floor surface beneath.

Finish lime A fine grade of *quicklime* used in finish-coat gypsum plasters and in ornamental plaster work; also called *lime putty.*

Finish nail A relatively thin nail with a very small head, used for fastening trim and other finish woodwork items.

Fire area In the International Building Code, an area within a building bounded by fire-resistant construction. Fire area size, occupant load, and location within the building are used to determine automatic sprinkler requirements.

Fire barrier In the International Building Code, a fire-resistant wall intended to deter the spread of fire, used to separate exit stair enclosures, differing occupancies, and fire areas.

Fireblocking Wood or other material used to partition concealed spaces within combustible framing, intended to restrict the spread of fire within such spaces.

Firebox The part of a fireplace, stove, or furnace in which fuel is combusted.

Firebrick A brick made to withstand very high temperatures, as in a fireplace, furnace, or industrial chimney.

Firecut A sloping end cut on a wood beam or joist where it enters a masonry wall. The purpose of the firecut is to allow the wood member to rotate out of the wall without prying the wall apart if the floor or roof structure burns through in a fire.

Fire door A fire-resistant door, used in fire-resistance rated partitions and walls.

Fire partition In the International Building Code, a fire-resistant wall intended to deter the spread of fire, used to separate tenant spaces, dwelling units, and corridors from surrounding areas of a building.

Fireproofing Material used around a steel (or concrete) structural element to insulate it against excessive temperatures in case of fire.

Fire protective glazing *Fire-rated glass* for use in fire doors, fire windows, and other protected openings that does not meet all of the requirements for use as a fire-resistance rated wall assembly.

Fire-rated glass Glass that is capable of retaining its integrity in an opening after being exposed to fire. *See also* **Fire protective glazing, glass fire wall.**

Fire resistance rating The time, in hours or fractions of an hour, that a material or assembly will resist fire exposure as determined by ASTM E119.

Fire resistant Noncombustible; slow to be damaged by fire; forming a barrier to the passage of fire.

Fire-resistive glazing *See* **Glass fire wall.**

Firestopping A component or mastic installed in an opening through a floor or around the edge of a floor to retard the passage of fire; frequently used interchangeably with *fireblocking.*

Fire wall A wall extending from foundation to roof, required under a building code to separate buildings, or parts of buildings, as a deterrent to the spread of fire.

Fire zone A legally designated area of a city in which construction must meet established standards of fire resistance and/or combustibility.

Firing The process of converting dry clay into a ceramic material through the application of intense heat.

First cost The cost of construction, not including operational costs.

Fixed window Glass that is immovably mounted in a wall.

Flagstone Flat stones used for paving or flooring.

Flame-spread rating A measure of the rapidity with which fire will spread across the surface of a finish material as determined by ASTM standard E84.

Flange A projecting crosspiece of a wide-flange or channel profile; a projecting fin.

Flash cove A detail in which a sheet of resilient flooring is turned up at the edge and finished against the wall to create an integral baseboard.

Flashing A thin, continuous sheet of metal, plastic, rubber, or waterproof paper used to prevent the passage of water through a joint in a wall, roof, or chimney.

Flat-grain lumber Dimension lumber sawed in such a way that annual rings are oriented close to parallel with the face. *See also* **Vertical grain lumber.**

Flat seam A sheet metal roofing seam that is formed flat against the surface of the roof.

Flemish bond Brickwork laid with each course consisting of alternating headers and stretchers.

Flitch-sliced veneer A thin sheet of wood cut by passing a block of wood vertically against a long, sharp knife.

Float A small platform suspended on ropes from a steel building frame to permit ironworkers to work on a connection; a trowel with a slightly rough surface used in an intermediate stage of finishing a concrete slab; as a verb, to use a float for finishing concrete.

Float glass Glass sheet manufactured by cooling a layer of molten glass on a bath of molten tin.

Floating floor Wood or laminate flooring that is not fastened or adhered to the subfloor.

Floating foundation A foundation placed at depth such that the weight of the soil removed is close to the weight of the building being supported.

Flocculated Having a "fluffy" microstructure such as that of clay particles in which the platelets are randomly oriented.

Flood test The submersion of a horizontal waterproofing system, usually for an extended period of time, to check for leaks.

Floor joist *See* **Joist**.

Flue A passage for smoke and combustion products from a furnace, stove, water heater, or fireplace.

Fluid-applied roof membrane A roof membrane applied in one or more coats of a liquid that cure to form an impervious sheet.

Fluoropolymer A highly stable organic compound used as a finish coating for building cladding.

Flush Smooth, lying in a single plane.

Flush door A door with smooth planar faces.

Flush glazing *See* **Structural silicone flush glazing**.

Flux A material added to react chemically with impurities and remove them from molten metal. Fluxes are used both in steelmaking and in welding. Welding fluxes serve the additional purpose of shielding the molten weld metal from the air to reduce oxidation and other undesirable effects.

Fly ash Dust collected in the stacks of coal-fired power plants, used as a *supplementary cementitious material* in concrete and mortar.

Flying formwork Large sections of slab formwork that are moved by crane.

Fly rafter A rafter in a rake overhang.

F-number An index number expressing the statistical flatness or levelness of a concrete slab.

Foil-backed gypsum board Gypsum board with aluminum foil laminated to its back surface to act as a vapor retarder and thermal insulator.

Folded plate A roof structure whose strength and stiffness derive from a pleated or folded geometry.

Footing The part of a foundation that spreads a load from the building across a broader area of soil.

Forced-air system A furnace and/or cooling coil and ductwork that heat and/or cool air and deliver it, driven by a fan, to the rooms of a building.

Form deck Thin, corrugated steel sheets that serve as *permanent formwork* for a reinforced concrete deck.

Form release compound A substance applied to concrete formwork to prevent concrete from adhering.

Form tie A steel or plastic rod with fasteners on either end, used to hold together the two surfaces of formwork for a concrete wall.

Form tie hole A depression, typically conical in shape, in a cast-in-place concrete wall that remains after the protruding portions of a form tie are removed.

Formwork Structures, usually temporary, that serve to give shape to poured concrete and to support it and keep it moist as it cures.

Foundation The portion of a building that transmits structural loads from the building into the earth.

Framed connection A *shear connection* between steel members made by means of steel angles or plates connecting to the web of the beam or girder.

Framing plan A diagram showing the arrangement and sizes of the structural members in a floor or roof.

Framing square An L-shaped measuring tool used by carpenters to lay out right-angle cuts as well as more complicated cuts, such as those required for stairs and sloping roof rafters.

Freestone Fine-grained sedimentary rock that has no planes of cleavage or sedimentation along which it is likely to split.

Free water In wood, water held within the cavities of the cells. *See also* **Bound water.**

Freeze protection admixture A concrete or mortar additive, used to allow curing under conditions of low ambient temperature.

French door A symmetrical pair of glazed doors hinged to the jambs of a single frame and meeting at the center of the opening.

Frictional soil A soil such as sand that has little or no attraction between its particles and derives its strength from geometric interlocking of the particles; also called a *cohesionless soil.*

Friction connection *See* **Slip-critical connection**.

Frit Ground-up colored glass that is heat-fused to lights of glass to form functional or decorative patterns.

Frost line The depth in the earth to which the soil can be expected to freeze during a severe winter.

Fully restrained moment connection A steel frame *moment connection* sufficiently rigid such that the geometric angles between connected pieces remain unchanged during normal loading; previously referred to as an "AISC Type 1" connection.

Furring channel A formed sheet metal *furring strip.*

Furring strip A length of wood or metal attached to a masonry or concrete wall to permit the attachment of finish materials to the wall using screws or nails; any linear material used to create a spacial separation between a finish material and an underlying substrate.

G

Gable The triangular wall beneath the end of a *gable roof.*

Gable roof A roof consisting of two oppositely sloping planes that intersect at a level ridge.

Gable vent A screened, louvered opening in a gable, used for exhausting excess heat and humidity from an attic.

Gage *See* **Gauge**.

Galling Chafing or tearing of one material against another under extreme pressure.

Galvanic couple A pair of metals with differing electrochemical potential, between which electrical current will flow if the metals are placed in a conducting medium.

Galvanic series A list of metals in order of their relative electrochemical potential when immersed in a given conducting medium.

Galvanized steel Steel with a zinc coating for the purpose of providing protection from corrosion.

Galvanizing The application of a zinc coating to steel.

Gambrel A roof shape consisting of two superimposed levels of gable roofs with the lower level at a steeper pitch than the upper.

Gantt chart A graphic representation of a construction schedule, using a series of horizontal bars representing the duration of various tasks or groups of tasks that make up the project.

Gasket A dry, resilient material used to seal a joint between two rigid assemblies by being compressed between them.

Gauge A measure of the thickness of sheet material. Lower gauge numbers signify thicker sheets; also spelled "gage."

Gauged brick A brick that has been rubbed on an abrasive stone to reduce it to a trapezoidal shape for use in an arch.

Gauging plaster A gypsum plaster formulated for use in combination with finishing lime in finish coat plaster.

General contractor A construction entity with responsibility for the overall conduct of a construction project.

Geotextile A synthetic cloth used beneath the surface of the ground to stabilize soil or promote drainage.

GFRC *See* **Glass-fiber-reinforced concrete.**

Girder A horizontal beam that supports other beams; a very large beam, especially one that is built up from smaller elements.

Girt A horizontal beam that supports wall cladding between columns.

Glass block A hollow masonry unit made of glass.

Glass fiber batt A thick, fluffy, nonwoven insulating blanket of filaments spun from glass.

Glass-fiber-reinforced concrete (GFRC) Concrete with a strengthening admixture of short, alkali-resistant glass fibers.

Glass fire wall *Fire-rated glass* meeting all of the requirements for use as a fire-resistance rated wall assembly; also called *fire-resistive glazing.*

Glass mullion system A method of constructing a large glazed area by stiffening the sheets of glass with perpendicular glass ribs.

Glaze A glassy finish on a brick or tile; as a verb, to install glass.

Glazed structural clay facing tile A hollow clay block with glazed faces, usually used for constructing interior partitions.

Glazier One who installs glass.

Glazier's points Small pieces of metal driven into a wood sash to hold glass in place.

Glazing The act of installing glass; the transparent material (most often glass) in a glazed opening; as an adjective, referring to materials used in installing glass, for example "glazing tape."

Glazing compound Any of several types of mastic used to bed small lights of glass in a frame.

Glue-laminated wood A wood member made up of a large number of small strips of wood glued together.

Glulam A short expression for glue-laminated wood.

Grade A classification of size or quality for an intended purpose; to classify as to size or quality.

Grade The surface of the ground; to move earth for the purpose of bringing the surface of the ground to an intended level or profile.

Grade beam A reinforced concrete beam that transmits the load from a bearing wall into spaced foundations such as pile caps or caissons.

Grain In wood, the direction of the longitudinal axes of the wood fibers or the figure formed by the fibers. In stone, *see* **Quarry bed.**

Granite Igneous rock with visible crystals of quartz and feldspar.

Green building Sustainable building; energy-efficient building; *See also* **Sustainability.**

Green roof A roof covered with soil and plant materials; *eco-roof; vegetated roof.*

Groove weld A weld made in a groove, created by beveling or milling the edges of the mating pieces of metal.

Ground A strip attached to a wall or ceiling to establish the level to which plaster should be applied.

Grout A high-slump mixture of portland cement, aggregates, and water, which can be poured or pumped into cavities in concrete or masonry for the purpose of embedding reinforcing bars and/or increasing the amount of loadbearing material in a wall; a specially formulated mortarlike material for filling under steel baseplates and around connections in precast concrete framing; a mortar used to fill joints between ceramic tiles or quarry tiles.

Gunnable sealant A sealant material that is extruded in liquid or mastic form from a *sealant gun.*

Gusset plate A flat steel plate used to connect the members of a truss; a stiffener plate.

Gutter A channel to collect rainwater and snowmelt at the eave of a roof.

GWB Gypsum wallboard; *see* **Gypsum board.**

Gypsum An abundant mineral; chemically, hydrous calcium sulfate.

Gypsum backing board A lower-cost gypsum panel intended for use as an interior layer in multilayer constructions of gypsum board.

Gypsum board An interior facing panel consisting of a gypsum core sandwiched between paper faces. Also called *drywall, plasterboard.*

Gypsum lath Sheets of gypsum board manufactured specifically for use as a plaster base.

Gypsum plaster Plaster whose cementing substance is calcined gypsum; used almost exclusively for interior finish plaster work.

Gypsum sheathing panel A water-resistant, gypsum-based sheet material used for exterior sheathing.

Gypsum wallboard (GWB) *See* **Gypsum board.**

H

Hammerhead boom crane A heavy-duty lifting device that uses a tower-mounted horizontal boom that may rotate only in a horizontal plane.

Hardboard A very dense panel product, usually with at least one smooth face, made of highly compressed wood fibers.

Hardwood Wood from deciduous (broadleaf) trees.

Hawk A square piece of sheet metal with a perpendicular handle beneath, used by a plasterer to hold a small quantity of wet plaster and transfer it to a trowel for application to a wall or ceiling.

HDO *See* **High-density overlay**.

Head The horizontal top portion of a window or door.

Header In framed construction, a member that carries other perpendicular framing members, such as a beam above an opening in a wall or a joist supporting other joists where they are interrupted by a floor opening. In steel construction, a beam that spans between girders. In masonry construction, a brick or other masonry unit that is laid across two wythes with its end exposed in the face of the wall.

Head jamb *See* **Head**.

Head joint The vertical layer of mortar between ends of masonry units.

Hearth The noncombustible floor area outside a fireplace opening.

Heartwood The dead wood cells in the center region of a tree trunk.

Heat-fuse To join by softening or melting the edges with heat and pressing them together.

Heat of hydration The thermal energy given off by concrete or gypsum as it cures.

Heat-strengthened glass *Heat-treated glass* that is not as strong as tempered glass, and that may not be used as *safety glazing.*

Heat-treated glass Glass that is strengthened by a heat treatment process; either *heat-strengthened glass* or *tempered glass.*

Heaving The forcing upward of ground or buildings by the action of frost or pile driving.

Heavy timber construction A type of wood construction made from large wood members and solid timber decking in a post and beam configuration; in the International Building Code, buildings of *Type IV HT construction*, consisting of heavy timber interior construction and noncombustible exterior walls, are considered to have moderate fire-resistive properties.

High-density overlay (HDO) A heavy weight, resin-treated overlay applied to plywood panels to achieve a smoother, more durable face.

High-lift grouting A method of constructing a reinforced masonry wall in which the reinforcing bars are embedded in grout in story-high increments.

High-range sealant A *sealant* that is capable of a high degree of elongation without rupture.

High-range water-reducing admixture *See* **Superplasticizer**.

High-reactivity kaolin *See* **Metakaolin**.

High-strength bolt A bolt designed to connect steel members by clamping them together with sufficient force that the load is transferred between them by friction.

High-volume fly ash concrete (HVFA concrete) A concrete in which a high percentage of cementing substance is fly ash rather than portland cement.

Hip The diagonal intersection of planes in a hip roof.

Hip rafter A roof *rafter* at the intersection of two sloping roof planes. *See also* **Common rafter.**

Hip roof A roof consisting of four sloping planes that intersect to form a pyramidal or elongated pyramid shape.

Hollow brick Clay brick with up to 60 percent void area.

Hollow concrete masonry Concrete masonry units that are manufactured with open cores, such as ordinary concrete blocks.

Hollow-core door A door consisting of two face veneers separated by an airspace, with solid wood spacers around the four edges. The face veneers are usually connected by a grid of thin spacers within the airspace.

Hollow-core slab A precast concrete slab element that has internal longitudinal cavities to reduce its self-weight.

Hollow structural section (HSS). Hollow steel cylindrical or rectangular shapes made to be used as structural members; also called *structural tubing.*

Hook A semicircular bend in the end of a reinforcing bar, made for the purpose of anchoring the end of the bar securely into the surrounding concrete.

Hopper window A window whose sash pivots on an axis along or near the sill and that opens by tilting toward the interior of the building.

Horizontal force A force whose direction of action is horizontal or nearly horizontal. *See also* **Lateral force.**

Horizontal reinforcing Steel reinforcing that runs horizontally in a masonry wall in the form of either welded grids of small diameter metal rods or larger conventional reinforcing bars.

Hose stream test A standard laboratory test to determine the relative ability of a building assembly to stand up to water from a fire hose after a specified period of fire testing.

Hot-dip galvanizing A method of *galvanizing* in which a steel member or assembly is dipped into a bath of molten zinc.

Hot-rolled steel Steel formed into its final shape by passing it between rollers while it is very hot.

Housewrap A synthetic sheet material with water-resistive and air-resistive properties used as a substitute for asphalt-saturated felt or building paper to provide a protective layer in an exterior wall assembly.

HSS *See* **Hollow structural section.**

HVFA concrete *See* **High-volume fly ash concrete**.

Hydrated lime Quicklime mixed with water, either in the factory or on the job site; an ingredient in masonry mortars, portland cement plaster, and gypsum plasters, to which materials it imparts properties such as workability, bulk, and smoothness; chemically, calcium hydroxide; also called *slaked lime.*

Hydration The process by which cements combine chemically with water to harden.

Hydraulic cements Cementitious materials, such as portland cement or blast furnace slag, that harden by reacting with water and whose hardened products are not water soluble. Nonhydraulic cements, such as lime, can also be mixed with pozzolans to create cements with hydraulic properties.

Hydronic heating system A system that circulates warm water through convectors to heat a building.

Hydrostatic pressure Pressure exerted by standing water.

Hygroscopic Readily absorbing and retaining moisture.

Hyperbolic paraboloid shell A concrete roof structure with a saddle shape.

I

IBC *See* **International Building Code** and **International Residential Code.**

I-beam (obsolete term) An American Standard section of hot-rolled steel.

Ice and water shield *See* **Rubberized underlayment**.

Ice barrier A sheet material, usually *rubberized underlayment* or sheet metal, applied to the lower portions of sloped roofs in cold climates to protect against *ice dams*.

Ice dam An obstruction along the eave of a roof, caused by the refreezing of water emanating from melting snow on the roof surface above.

ICF *See* **Insulating concrete form**.

Igneous rock Rock formed by the solidification of magma.

IIC *See* **Impact Isolation Class.**

I-joist A manufactured wood framing member whose cross-sectional shape resembles the letter *I*.

Impact Isolation Class (IIC) An index of the extent to which a floor assembly transmits *impact noise* from a room above to the room below.

Impact noise Noise generated by footsteps or other impacts on a floor.

Impact wrench A device for tightening bolts and nuts by means of rapidly repeated torque impulses produced by electrical or pneumatic energy.

Incising Short, repetitive cuts made in the surface of a wood member to increase its absorption of treatment chemicals.

Ingot A large block of cast metal.

Insulating concrete form (ICF) A system of lightweight components, most commonly made of rigid polystyrene insulating foam, used as *permanent formwork* for the casting of concrete walls.

Insulating glass A glazing unit made up of two or more sheets of glass with an airspace in between.

Insulating glass unit (IGU) *See* **Insulating glass**.

Intensive green roof A *green roof* with relatively deep soil capable of supporting a broad variety of plants and shrubs.

Internal drainage Providing a curtain wall with hidden channels and weep holes to remove any water that may penetrate the exterior layers of the wall.

Internal flashing In masonry, a flashing concealed with the masonry; also called a *concealed* or *through-wall flashing*.

Internal gutter A *gutter* built into a roof assembly.

International Building Code (IBC) and **International Residential Code (IRC)** The predominant U.S. *model building codes*.

Interstitial ceiling A suspended ceiling with sufficient structural strength to support workers safely as they install and maintain mechanical and electrical installations above the ceiling.

Intrados The concave surface of an arch.

Intumescent coating A paint or mastic that expands to form a stable, insulating char when exposed to fire.

Inverted roof A membrane roof assembly in which the thermal insulation lies above the membrane.

IRC *See* **International Building Code** and **International Residential Code.**

Iron In pure form, a metallic element. In common usage, ferrous alloys other than steels, including cast iron and wrought iron.

Iron dog A heavy U-shaped staple used to tie the ends of heavy timbers together.

Ironworker A skilled laborer who erects steel building frames or places reinforcing bars in concrete construction.

Isocyanurate foam A thermosetting plastic foam with thermal insulating properties.

Isolation joint A type of *structure/enclosure joint* used with concrete slabs on grade to allow differential movement where they abut adjacent walls and columns.

J

Jack A device for exerting a large force over a short distance, usually by means of screw action or hydraulic pressure.

Jack rafter A shortened rafter that joins a hip or valley rafter.

Jack stud A shortened stud that carriers a header above a wall opening; also called a *trimmer stud*.

Jamb The vertical side of a door or window.

Jet burner A torch that burns fuel oil and compressed air, used in quarrying granite.

Joist One of a parallel array of light, closely spaced beams used to support a floor deck (*floor joist*) or low-slope roof (*ceiling joist*).

Joist band A broad, shallow concrete beam that supports one-way concrete joists whose depths are identical to its own.

Joist girder A light steel truss used to support open-web steel joists.

Joist hanger A sheet metal device used to create a structural connection where a joist is framed into a *header* or a *ledger*.

K

Keenes cement A proprietary, dense, crack-resistant gypsum plaster formulation.

Key A slot formed into a concrete surface for the purpose of interlocking with a subsequent pour of concrete; a slot at the edge of a precast member into which grout will be poured to lock it to an adjacent member; a mechanical interlocking of plaster with lath.

Kiln A furnace for firing clay or glass products; a heated chamber for seasoning wood; a furnace for manufacturing quicklime, gypsum hemihydrate, or portland cement.

King stud A full-length stud nailed alongside a *jack stud*.

Knee wall A short wall under the slope of a roof.

Knot A growth characteristic in wood, occurring where a branch joins the trunk of the tree from which the wood has been sawed.

kPa Kilopascal, a unit of pressure equal to 1 kilonewton per square meter.

L

Labyrinth A cladding joint design in which a series of interlocking baffles prevents drops of water from penetrating the joint by momentum.

Lacquer A coating that dries extremely quickly through evaporation of a volatile solvent.

Lagging Planks placed between soldier beams to retain earth around an excavation.

Lag screw A large-diameter wood screw with a square or hexagonal head.

Laminate As a verb, to bond together in layers; as a noun, a material produced by bonding together layers of material.

Laminated glass A glazing material consisting of outer layers of glass laminated to an inner layer of transparent plastic.

Laminated strand lumber (LSL) Wood members made up of long shreds of wood fiber joined with a binder.

Laminated veneer lumber (LVL) Structural composite lumber made up of thin wood veneers joined with glue.

Laminated wood *See* **Glue-laminated wood**.

Landing A platform in or at either end of a stair.

Lap joint A connection in which one piece of material is placed partially over another piece before the two are fastened together.

Lateral force A force acting generally in a horizontal direction, such as wind, earthquake, or soil pressure against a foundation wall.

Lateral thrust The horizontal component of the force produced by an arch, dome, vault, or rigid frame.

Latewood *See* **Summerwood**.

Latex caulk A low-range *sealant* based on a synthetic latex.

Latex/polymer modified portland cement mortar A tile setting mortar similar to *dry-set mortar*, but with additives that improve the cured mortar's freeze-thaw resistance, flexibility, and adhesion; used for *thin-set tile* applications.

Lath (rhymes with "math") A base material to which plaster is applied.

Lathe (rhymes with "bathe") A machine in which a piece of material is rotated against a sharp cutting tool to produce a shape, all of whose cross sections are circles; a machine in which a log is rotated against a long knife to peel a continuous sheet of veneer.

Lather (rhymes with "rather") One who applies lath.

Lay-in panel A finish ceiling panel that is installed merely by lowering it onto the top of the metal grid components of the ceiling.

Lead (rhymes with "bed") A soft, easily formed nonferrous metal, dull gray in color.

Lead (rhymes with "bead") In masonry work, a corner or wall end accurately constructed with the aid of a spirit level to serve as a guide for placing the bricks in the remainder of the wall.

Leader (rhymes with "feeder") *See* **Downspout**.

Leaf The moving portion of a door.

Lean construction Methods of construction and its management that emphasize efficiency, elimination of waste, and continuous improvement in quality.

Ledger A horizontal wood member fastened to a wall or beam to which the ends of joists may be connected.

Lehr A chamber in which glass is annealed.

Let-in bracing Diagonal bracing that is nailed into notches cut in the face of the studs so as not to increase the thickness of the wall.

Level cut A saw cut that produces a level surface in a sloping rafter when the rafter is in its final position. *See also* **Plumb cut.**

Leveling plate A steel plate placed in grout on top of a concrete foundation to create a level bearing surface for the lower end of a steel column.

Lewis A device for lifting a block of stone by means of friction exerted against the sides of a hole drilled in the top of the block.

Life-cycle cost A cost that takes into account both the first cost and the costs of maintenance, replacement, fuel consumed, monetary inflation, and interest over the life of the object being evaluated.

Lift-slab construction A method of building multistory sitecast concrete buildings by casting all the slabs in a stack on the ground, then lifting them up the columns with jacks and welding them in place.

Light A sheet of glass, also spelled "lite."

Light gauge steel stud A length of thin sheet metal formed into a stiff shape and used as a wall framing member.

Light to solar gain (LSG) ratio The visible light transmittance of a glazing unit divided by the solar heat gain coefficient, a measure of the energy-conserving potential of the unit.

Lightweight aggregate Low-density aggregate used to make lightweight concrete, mortar, and plaster; in concrete, aggregate with a density of less than 70 lb/ft^3 (1120 kg/m^3).

Lignin The natural cementing substance that binds together the cellulose in wood.

Lime A nonhydraulic cementitious material, used as an ingredient in mortars and plasters. *See also* **Hydrated lime**, **Quicklime**.

Lime mortar Masonry mortar made from a mix of lime, sand, and water; used principally in the restoration of historic structures.

Lime putty *See* **Finish lime**.

Limestone A sedimentary rock consisting of calcium carbonate, magnesium carbonate, or both.

Linear metal ceiling A finish ceiling whose exposed face is made up of long, parallel elements of sheet metal.

Liner A piece of marble doweled and cemented to the back of another sheet of marble.

Line wire Wire stretched across wall studs as a base for the application of metal mesh and stucco.

Linoleum A resilient floor covering material composed primarily of ground cork and linseed oil on a burlap or canvas backing.

Lintel A beam that carries the load of a wall across a window or door opening.

Liquid sealant Gunnable *sealant.*

Lite *See* **Light**.

Live load Nonpermanent loads on a building caused by the weights of people, furnishings, machines, vehicles, and goods in or on the building.

Load A weight or force acting on a structure.

Loadbearing Supporting a superimposed weight or force.

Loadbearing wall *See* **Bearing wall.**

Load indicator washer A disk placed under the head or nut of a high-strength bolt to indicate sufficient tensioning of the bolt by means of the deformation of ridges on the surface of the disk; also called a *direct tension indicator washer.*

Lockpin and collar fastener A boltlike device that is passed through holes in structural steel components, held in very high tension, and closed with a steel ring that is squeezed onto its protruding shank.

Lockstrip gasket A synthetic rubber strip compressed around the edge of a piece of glass or a wall panel by inserting a spline (lockstrip) into a groove in the strip.

Longitudinal shrinkage In wood, shrinkage along the length of the log.

Lookout A short rafter, running perpendicular to the other rafters in the roof, which supports a rake overhang.

Louver A construction of numerous sloping, closely spaced slats used to diffuse air or to prevent the entry of rainwater into a ventilating opening.

Low-e coating *See* **Low-emissivity coating**.

Low-emissivity coating A surface coating for glass that selectively reflects solar radiation of different wavelengths so as to permit high visible light transmittance while reflecting some or all types of infrared (heat) radiation.

Low-lift grouting A method of constructing a reinforced masonry wall in which the reinforcing bars are embedded in grout in increments not higher than 4 feet (1200 mm).

Low-range sealant A *sealant* that is capable of only a slight degree of elongation prior to rupture; a *caulk*.

Low-slope roof A roof that is pitched so near to horizontal that it must be made waterproof with a continuous membrane rather than shingles; commonly and inaccurately referred to as a "flat roof." In the International Building Code, a roof with a slope of less than 2:12 (17 percent).

LSG *See* **Light to solar gain ratio**.

LSL *See* **Laminated strand lumber.**

Luffing-boom crane A heavy-duty lifting device that uses a tower-mounted boom that may rotate in any vertical plane as well as in a horizontal plane.

LVL *See* **Laminated veneer lumber**.

M

Machine grading The grading of wood for its structural properties, performed by automated machinery, as distinct from *visual grading*.

Macrofiber reinforcing In concrete, fibrous reinforcement capable of providing resistance to drying shrinkage and thermal stresses, and in some specialized concretes, also capable of acting as primary reinforcing. *See also* **Microfiber reinforcing.**

Mandrel A stiff steel core placed inside the thin steel shell of a sitecast concrete pile to prevent it from collapsing during driving.

Mansard A roof shape consisting of two superimposed levels of hip roofs with the lower level at a steeper pitch than the upper.

Manufactured home. A transportable house that is entirely factory built on a steel underframe supported by wheels; euphemistically referred to as a *mobile home*.

Marble A metamorphic rock formed from limestone by heat and pressure.

Mason One who builds with bricks, stones, or concrete masonry units; one who works with concrete.

Masonry Brickwork, concrete blockwork, and stonework.

Masonry cement A hydraulic cement made from a blend of portland cement, lime and other dry admixtures designed to increase the workability of the mortar. *See also* **Cement-lime mortar.**

Masonry opening The clear dimension required in a masonry wall for the installation of a specific window or door unit.

Masonry unit A brick, stone, concrete block, glass block, or hollow clay tile intended to be laid in mortar.

Masonry veneer A single wythe of masonry used as a facing over a frame of wood or metal.

MasterFormat The copyrighted title of a uniform indexing system for construction specifications, as created by the Construction Specifications Institute and Construction Specifications Canada.

Mastic A viscous, doughlike, adhesive substance; can be any of a large number of formulations for different purposes such as sealants, adhesives, glazing compounds, or roofing cements.

Mat foundation A single concrete footing that is essentially equal in area to the area of ground covered by the building.

Medium-density fiberboard (MDF) A fine-grained wood fiber and resin panel product.

Medium-density overlay (MDO) A medium-weight, resin-treated overlay applied to plywood panels to achieve a smoother, more durable face.

Medium-range sealant A *sealant* material that is capable of a moderate degree of elongation before rupture.

Meeting rail The wood or metal bar along which one sash of a double-hung, single-hung, or sliding window seals against the other.

Member An element of a structure such as a beam, girder, column, joist, piece of decking, stud, or component of a truss.

Membrane A sheet material that is impervious to water or water vapor.

Membrane fire protection A ceiling used to provide fire protection to the structural members above.

Metakaolin A white-colored natural *pozzolan* that enhances appearance, workability, and hardened properties of concrete; also called *high reactivity metakaolin*.

Metal decking Corrugated metal sheets used as the structural base for floors ("floor decking") and roofs ("roof decking") in steel frame construction. *See also* **Celluar decking** and **Composite metal decking**.

Metal lath A steel mesh used primarily as a base for the application of plaster.

Metallic-coated steel Steel sheet coated with zinc or zinc-aluminum for improved corrosion resistance.

Metamorphic rock A rock created by the action of heat or pressure on a sedimentary rock or soil.

Microfiber reinforcing In concrete, fibrous reinforcement against plastic shrinkage cracking. *See also* **Macrofiber reinforcing.**

Microsilica *See* **Silica fume**.

Middle strip The half-span-wide zone of a two-way concrete slab that lies midway between columns.

Mild steel Ordinary structural steel, containing less than three-tenths of 1 percent carbon.

Mill construction The traditional name for a construction type consisting of exterior masonry bearing walls and an interior framework of heavy timbers and solid timber decking; also called *slow-burn construction. See also* **Heavy timber construction.**

Milling Shaping or planing by using a rotating cutting tool.

Millwork Wood interior finish components of a building, including moldings, windows, doors, cabinets, stairs, mantels, and the like.

Minimum critical radiant flux exposure A measure of a material's resistance to ignition by the radiant heat of fire and hot gases in adjacent spaces, usually applied to flooring materials.

Miter A diagonal cut at the end of a piece; the joint produced by joining two diagonally cut pieces at right angles.

Mobile home *See* **Manufactured home.**

Model building code A code that is offered by a recognized national organization as worthy of adoption by state or local governments.

Modified bitumen A natural *bitumen* with admixtures of synthetic compounds to enhance such properties as flexibility, plasticity, and durability.

Modified bitumen roof membrane A multi-ply *bituminous roof membrane* made from plies of factory-manufactured *modified bitumen* sheets.

Modular Conforming to a multiple of a fixed dimension.

Modular green roof A *green roof* system in which all components are provided in self-contained, easily transported and installed trays or modules.

Modular home A house assembled on the site from boxlike factory-built sections.

Modulus of elasticity An index of the stiffness of a material, derived by measuring the elastic deformation of the material as it is placed under stress and then dividing the stress by the deformation.

Moisture barrier A membrane used to resist the migration of liquid water through a floor, wall, or roof.

Molding A strip of wood, plastic, or plaster with an ornamental profile.

Molding plaster A fast-setting gypsum plaster used for the manufacture of cast ornament.

Moment A force acting at a distance from a point in a structure so as to cause a tendency of the structure to rotate about that point. *See also* **Bending moment, Moment connection.**

Moment connection A connection between two structural members that is highly resistant to rotation between the members and therefore capable of transmitting *bending moments* between the connected members, as differentiated from a *shear connection,* which allows (slight) rotation. *See also* **Fully restrained moment connection, Partially restrained moment connection**, and **Simple connection**.

Moment-resisting frame A structural building frame, strengthened to resist lateral forces with moment connections between beams and columns.

Momentum The tendency of a moving body to continue to move in the same direction unless acted on by an outside force.

Monolithic Of a single massive piece.

Monolithic terrazzo A thin terrazzo topping applied to a concrete slab without an underbed.

Mortar A substance used to join masonry units, consisting of cementitious materials, fine aggregate, and water. *See also* **Cement-lime mortar, Lime mortar.**

Mortar bed tile *See* **Thickset tile**.

Mortar cement In masonry, a blend of portland cement, lime, and other additives, that produces mortar comparable in its bond strength properties to cement-lime mortar. *See also* **Cement-lime mortar.**

Mortise and tenon A joint in which a tonguelike protrusion (tenon) on the end of one piece is tightly fitted into a rectangular slot (mortise) in the side of the other piece.

Movement joint A line or plane along which movement is allowed to take place in a building or a surface of a building in response to such forces as moisture expansion and contraction, thermal expansion and contraction, foundation settling, and seismic forces.

MPa Megapascal, a unit of pressure equal to 1 meganewton per square meter.

Mud set tile *See* **Thickset tile**.

Mud slab A slab of weak concrete placed directly on the ground to provide a usually temporary working surface that is hard, level, and dry.

Mullion A vertical or horizontal bar between adjacent window or door units. A framing member in a metal-and-glass curtain wall.

Muntin A small vertical or horizontal bar between small lights of glass in a sash.

Muriatic acid Hydrochloric acid.

Mushroom capital A flaring conical head on a concrete column.

N

Nail A sharp-pointed metal pin used for fastening of wood.

Nail-base sheathing A sheathing material, such as wood boards or plywood, to which siding can be attached by nailing, as differentiated from one such as cane fiber board or plastic foam board that is too soft to hold nails.

Nail popping The loosening of nails holding gypsum board to a wall, caused by drying shrinkage of the studs.

Nail set A hardened steel punch used to drive the head of a nail to a level flush with or below the surface of the wood.

National Building Code of Canada (NBCC) The predominant Canadian *model building code.*

NBCC *See* **National Building Code of Canada.**

Near-infrared (NIR) radiation An invisible portion of the solar spectrum that accounts for over half of the total heat energy in solar radiation.

Needle beam A steel or wood beam threaded through a hole in a bearing wall and used to support the wall and its superimposed loads during underpinning of its foundation.

Needling The use of *needle beams.*

Negative-side waterproofing Waterproofing applied to the inner side of a wall, acting to resist water passage from the opposite side.

Neoprene Polychloroprene, a synthetic rubber.

NIR radiation *See* **Near-infrared radiation**.

Noble metal *See* **Passive metal**.

Noise Reduction Coefficient (NRC) An index of the proportion of incident sound that is absorbed by a surface, expressed as a decimal fraction of 1.

Nominal dimension An approximate dimension assigned to a piece of material as a convenience in referring to the piece.

Nonaxial In a direction not parallel to the long axis of a structural member.

Nonbearing Not carrying a load.

Nonhydraulic cements *Cementitious materials,* such as gypsum and lime, that remain water soluble after curing. *See also* **Hydraulic cements.**

Nonmovement joint A connection between materials or elements that is not designed to allow for movement.

Nonworking joint A joint that is not subjected to significant deformations.

Nosing The projecting forward edge of a stair tread.

NRC *See* **Noise Reduction Coefficient**.

Nut A steel fastener with internal helical threads, used to close a bolt.

O

o.c. Abbreviation for "on center," meaning that the spacing of framing members is measured from the center of one member to the center of the next, rather than the clear spacing between members.

Occupancy group In the International Building Code, a definition of the types of activities that occur within the building or a part of the building, relating to considerations of life safety.

Ogee An S-shaped curve.

Oil-based caulk A low-range *sealant* made with linseed oil.

One-way action The structural action of a slab that spans between two parallel beams or bearing walls.

One-way concrete joist system A reinforced concrete framing system in which closely spaced concrete joists span between parallel beams or bearing walls.

One-way solid slab A reinforced concrete floor or roof slab that spans between parallel beams or bearing walls.

Open-truss wire stud A wall framing member in the form of a small steel truss.

Open-web steel joist A lightweight, prefabricated, welded steel truss used at closely spaced intervals to support floor or roof decking.

Optical-quality ceramic A transparent, glasslike material that is used as a fire-rated glazing sheet.

Optimum value engineering *See* **Advanced framing techniques**.

Ordinary construction A traditional building type with exterior masonry bearing walls and an interior structure of balloon framing

Organic soil Soil containing decayed vegetable and/or animal matter; topsoil.

Oriented strand board (OSB) A building panel composed of long shreds of wood fiber oriented in specific directions and bonded together under pressure.

Oriented strand lumber (OSL) *Structural composite lumber* made from shredded wood strands, coated with adhesive, and pressed into a rectangular cross section.

OSB *See* **Oriented strand board**.

OSL *See* **Oriented strand lumber**.

Oxidation Corrosion; rusting; rust; chemically, the combining with oxygen.

P

Package fireplace A factory-built fireplace that is installed as a unit.

Paint A heavily pigmented coating applied to a surface for decorative and/or protective purposes.

Pan A form used to produce the cavity between joists in a one-way concrete joist system.

Panel A broad, thin piece of wood; a sheet of building material such as plywood or particleboard; a prefabricated building component that is broad and thin, such as a curtain wall panel; a rectangular area within a truss bounded by two vertical interior members.

Panel door A wood door in which one or more thin panels are held by stiles and rails.

Panelized construction A method of prefabricated wood light frame construction, in which whole sections of walls or floors are framed and sheathed in the factory and then transported to the construction site for erection.

Panic hardware A mechanical device that opens a door automatically if pressure is exerted against the device from the interior of the building.

Parallel strand lumber (PSL) *Structural composite lumber* made of wood shreds oriented parallel to the long axis of each piece and bonded together with adhesive.

Parapet The region of an exterior wall that projects above the level of the roof.

Parging Portland cement plaster applied over masonry to make it less permeable to water.

Partially restrained moment connection A steel frame *moment connection* that is less rigid than a *fully restrained moment connection* but that still possesses a usable degree of resistance to rotation; previously referred to as an "AISC Type 3" connection.

Particleboard A building panel composed of small particles of wood bonded together under pressure.

Parting compound *See* **Form release compound**.

Partition An interior nonloadbearing wall.

Passive metal A metal relatively low on a galvanic series, tending to act as an *cathode* in galvanic couples; also called a *noble metal.*

Patterned glass Glass into which a texture has been rolled during manufacture.

Pattern rafter A wood rafter cut to size and shape and then used to trace cuts onto additional wood members so as to assure consistent dimensions among all rafters.

Paver A half-thickness brick used as finish flooring.

PEC *See* **Pressure equalization chamber**.

Pediment The gable end of a roof in classical architecture.

Penetrometer A device for testing the resistance of a material to penetration, usually used to make a quick, approximate determination of its compressive strength.

Penny (d) A designation of nail size.

Performance grade A rating used to indicate the relative weather resistance of a window.

Periodic kiln A kiln that is loaded and fired in discrete batches, as differentiated from a tunnel kiln, which is operated continuously.

Perlite Expanded volcanic glass, used as a *lightweight aggregate* in concrete and plaster and as an insulating fill.

Perm A unit of vapor permeance, a measure of a material's permeability to the diffusion of water vapor.

Permanent formwork Concrete formwork that remains permanently in place after concrete is poured and cured and becomes part of the finished construction.

Phased construction *See* **Fast track construction**.

Photochromic glass Glass that changes it optical properties in response to light intensity.

Photovoltaic Capable of converting light into electricity.

Pier A caisson foundation unit.

Pilaster A vertical, integral stiffening rib in a masonry or concrete wall.

Pile A long, slender piece of material driven into the ground to act as an element of a foundation.

Pile cap A thick slab of reinforced concrete poured across the top of a pile cluster to cause the cluster to act as a unit in supporting a column or grade beam.

Piledriver A machine for driving piles.

Pill test A test of a flooring material's propensity for flame spread when exposed to a burning tablet intended to simulate a dropped lit cigarette, match, or similar hazard.

Pintle A metal device used to transmit compressive forces between superimposed columns in Mill construction.

Pitch The slope of a roof or other plane, often expressed as inches of rise per foot of run; a dark, viscous hydrocarbon distilled from coal tar; a viscous resin found in wood.

Pitched roof A sloping roof.

Pivoting window A window that opens by rotating around its vertical centerline.

Plainsawing Sawing a log into dimension lumber without regard to the direction of the annual rings.

Plain slicing Cutting a log into veneers without regard to the direction of the annual rings.

Plan An architectural drawing, representing the layout of walls and floor areas as seen from above ("floor plan") or ceilings as seen from below ("ceiling plan").

Planing Smoothing the surface of a piece of wood, stone, or steel with a cutting blade.

Plank flooring Solid wood finish flooring members 3 inches (75 mm) or more in width.

Plaster A cementitious material, usually based on gypsum or portland cement, applied to lath or masonry in paste form to harden into a finish surface.

Plasterboard *See* **Gypsum board**.

Plaster of Paris *See* **Calcined gypsum**.

Plaster screeds Intermittent spots or strips of plaster used to establish the level to which a large plaster surface will be finished.

Plastic A synthetically produced giant molecule, mostly based on carbon chemistry.

Plasticity The ability to retain a shape attained by pressure deformation.

Plastic laminate flooring A finish material for floors that consists of a thin decorative and wearing layer of melamine laminate glued to a wood composite substrate.

Plastic lumber Lumberlike products with a plastic content of 50 percent or more. *See also* **Structural-grade plastic lumber.**

Plastic shrinkage cracking Cracking in freshly mixed concrete, most commonly in slabs, that occurs when the surface of the concrete dries too rapidly.

Plate A broad sheet of rolled metal 1/4 inch (6.35 mm) or more thick; a two-way concrete slab; a horizontal top or bottom member in a platform frame wall structure.

Plate girder A large beam made up of steel plates, sometimes in combination with steel angles, that are welded, bolted, or riveted together.

Plate glass Glass of high optical quality, produced by grinding and polishing both faces of a glass sheet.

Platform frame A wooden building frame composed of closely spaced members nominally 2 inches (51 mm) thick in which the wall members do not run past the floor framing members.

Plenum The space between the ceiling of a room and the structural floor above, used as a passage for ductwork, piping, and wiring.

Plumb Vertical.

Plumb cut A saw cut that produces a vertical (plumb) surface in a sloping rafter after the rafter is in its final position. *See also* **Level cut**.

Plumbing up The process of making a steel building frame vertical and square.

Ply A layer, such as a layer of felt in a built-up roof membrane or a layer of veneer in plywood.

Plywood A wood panel composed of an odd number of layers of wood veneer bonded together under pressure.

PMR *See* **Protected membrane roof.**

Pneumatically placed concrete *See* **Shotcrete.**

Pointing The process of applying mortar to the surface of a mortar joint after the masonry has been laid, either as a means of finishing the joint or to repair a defective joint. *Raked mortar joints* can also be pointed with elastomeric sealant.

Pointing mortar Mortar used for the pointing of masonry joints, generally of relatively low strength and with good workability and adhesion characteristics.

Poke-through fitting An electrical outlet that is installed by drilling a hole through a floor, inserting the outlet from above, and bringing in the wiring from the plenum below.

Polybutene tape A sticky, masticlike tape used to seal nonworking joints, especially between glass and mullions.

Polycarbonate An extremely tough, strong, usually transparent plastic used for window and skylight glazing, light fixture globes, door sills, and other applications.

Polyethylene A thermoplastic widely used in sheet form for vapor retarders, moisture barriers, and temporary construction coverings.

Polyisocyanurate foam *see* **Isocyanurate foam.**

Polymer A large molecule composed of many identical chemical units.

Polypropylene A plastic formed by the polymerization of propylene.

Polystyrene foam A thermoplastic foam with thermal insulating properties.

Polysulfide A high-range gunnable *sealant.*

Polyurethane Any of a large group of resins and synthetic rubber compounds used in sealants, varnishes, insulating foams, and roof membranes.

Polyurethane foam A thermosetting foam with thermal insulating properties.

Polyvinyl butyral (PVB) interlayer A transparent plastic used in the fabrication of *laminated glass.*

Polyvinyl chloride (PVC) A thermoplastic material widely used in construction products, including plumbing pipes, floor tiles, wall coverings, and roof membranes. Called "vinyl" for short.

Ponding The accumulation of standing water on a low-slope roof due to inadequate drainage.

Poorly graded soil *See* **Well-sorted soil**.

Poorly sorted soil See **Well-graded soil.**

Portal frame A rigid frame; two columns and a beam attached to one another with moment connections.

Portland cement A gray or white powder, composed principally of calcium silicates, which, when combined with water, hydrates to form the binder in concrete, mortar, and stucco.

Posttensioning Compressing the concrete in a structural member by tensioning high-strength steel tendons against it after the concrete has cured.

Pour To cast concrete; an increment of concrete casting carried out without interruption.

Powder coating A coating produced by applying a powder consisting of thermosetting resins and pigments, adhered to the substrate by electrostatic attraction, and fused into a continuous film in an oven.

Powder driven Inserted by a gunlike tool using energy provided by an exploding charge of gunpowder.

Pozzolan A *supplementary cementitious material*, such as fly ash, silica fume, and some naturally occurring shales and clays, that has few or no inherent cementitious properties but that, in the presence of moisture, can react with calcium hydroxide released by other cementitious materials to create a hydraulic cement product. The Romans mixed natural pozzolans with lime to make the first hydraulic cement.

Precast concrete Concrete cast and cured in a location other than its final position in the structure.

Predecorated gypsum board Gypsum board finished at the factory with a decorative layer of paint, paper, or plastic.

Prefabrication Construction that takes place in a factory or shop, rather than on the building site.

Preformed cellular tape sealant A *sealant* inserted into a joint in the form of a compressed sponge impregnated with compounds that cure to form a watertight seal.

Preformed joint filler A strip of rubbery or spongelike material designed to fit snugly into a gap between two materials.

Preformed solid tape sealant A *sealant* inserted into a joint in the form of a flexible strip of solid material.

Prehung door A door that is hinged to its frame in a factory or shop.

Prescriptive building code A set of legal regulations that mandate specific construction details and practices rather than establish performance standards.

Preservative-treated wood Wood that has been impregnated with preservative chemicals to increase its resistance to decay and biological attack; also commonly called *pressure-treated wood.*

Pressure equalization chamber (PEC) The wind-pressurized cavity in a rainscreen wall.

Pressure-equalized wall design Curtain wall design that relies on neutralization of wind pressures on both sides of a the exterior cladding to control water entry into the wall system. *See also* **Rainscreen principle.**

Pressure-treated wood Wood that has been impregnated with chemicals under pressure for the purpose of retarding decay or reducing combustibility.

Prestressed concrete Concrete that has been pretensioned or posttensioned.

Prestressing Applying an initial compressive stress to a concrete structural member, either by *pretensioning* or *posttensioning.*

Pretensioning Compressing the concrete in a structural member by pouring the concrete for the member around stretched high-strength steel strands, curing the concrete, and releasing the external tensioning force on the strands.

Prime window A window unit that is made to be installed permanently in a building.

Priming Covering a surface with a coating that prepares it to accept another coating or a sealant.

Protected membrane roof (PMR) A membrane roof assembly in which the thermal insulation lies above the membrane.

Protection board Semirigid sheet material used to cushion the outside of a foundation wall, particularly its waterproofing layer, from damage caused by rocks in the backfill material.

PSL *See* **Parallel strand lumber.**

Pultrusion The process of producing a shaped linear element by pulling glass fibers through a bath of uncured plastic, then through a heated, shaped die in which the plastic hardens.

Punty A metal rod used in working with hot glass.

Purlin A beam that spans across the slope of a steep roof to support the roof decking.

Putty A simple glazing compound used to seal around a small light.

PVB interlayer *See* **Polyvinyl butyral interlayer**.

PVC *See* **Polyvinyl chloride**.

Pyrolitic coating A coating applied at a very high temperature.

Q

Quarry An excavation from which building stone is obtained; the act of taking stone from the ground.

Quarry bed A plane in a building stone that was horizontal before the stone was cut from the quarry; also called *grain.*

Quarry sap Excess water found in rock at the time of its quarrying.

Quarry tile A large clay floor tile, usually unglazed.

Quartersawn Lumber sawn in such a way that the annual rings run roughly perpendicular to the face of each piece.

Quartersliced Veneer sliced in such a way that the annual rings run roughly perpendicular to the face of each veneer.

Quenching The rapid cooling of metal so as to alter its physical properties; a form of heat treatment.

Quicklime Produced by burning calcium carbonate found in limestone or sea shells; once hydrated, used as an ingredient in mortars and plasters; chemically, calcium oxide. *See also* **Hydrated lime.**

Quoin (pronounced "coin") A corner reinforcing of cut stone or bricks in a masonry wall, usually done for decorative effect.

R

Radial shrinkage In wood, shrinkage perpendicular to the growth rings.

Rabbet A longitudinal groove cut at the edge of a member to receive another member; also called a *rebate.*

Radiant barrier A reflective foil or metal placed adjacent to an airspace in roof or wall assemblies as a deterrent to the passage of infrared energy.

Radiant heating system Providing heat to spaces and their inhabitants by heating one or more surfaces of each room. The heated surface is usually either the floor or the ceiling. Heat is usually provided either by electric resistance coils or by hot-water tubing.

Radiator *See* **Convector**.

Raft A mat footing.

Rafter A framing member that runs up and down the slope of a steep roof.

Rail A horizontal framing piece in a panel door; a handrail.

Rainscreen principle A theory by which wall cladding is made watertight by providing wind-pressurized air chambers behind joints to eliminate air pressure differentials between the outside and the inside that might transport water through the joints. *See also* **Pressure-equalized wall design.**

Rainscreen siding Siding systems that include a system of internal drainage, but not necessarily meeting all the criteria of a *pressure-equalized wall design.*

Raised access flooring *See* **Access flooring.**

Rake The sloping edge of a steep roof.

Raked mortar joint A mortar joint in which mortar has been removed from the portion of the joint closest to the surface of the masonry.

Raker A sloping brace for supporting sheeting around an excavation.

Ram A hydraulic piston device used for bending steel, tensioning steel strands in prestressed concrete, or lifting heavy loads.

Random stone masonry *See* **Uncoursed stone masonry.**

Ratchet A mechanical device with sloping teeth that allows one piece to be advanced against another in small increments, but not to move in the reverse direction.

Ray A tubular cell that runs radially in a tree trunk.

RBM *See* **Reinforced brick masonry**.

Rebar *See* **Steel reinforcing bars**.

Rebate *See* **Rabbet.**

Reflective coated glass Glass onto which a thin layer of metal or metal oxide has been deposited to reflect light and/or heat.

Reglet A slot, usually horizontal, and inclined in cross section, into which a flashing or roof membrane may be inserted in a concrete or masonry surface.

Reinforced brick masonry (RBM) Brickwork into which steel bars have been embedded to impart tensile strength to the construction.

Reinforced concrete Concrete work into which steel bars have been embedded to impart tensile strength to the construction.

Relative humidity A percentage representing the ratio of the amount of water vapor contained in a mass of air to the maximum amount of water it could contain under the existing conditions of temperature and pressure.

Relieved back A longitudinal groove or series of grooves cut from the back of a flat wood molding or flooring strip to minimize cupping forces and make the piece easier to fit to a flat surface.

Removable glazing panel A framed sheet of glass that can be attached to a window sash to increase its thermal insulating properties.

Replacement window A window unit that is designed to be installed easily in an opening left in a wall by a deteriorated window unit that has been removed.

Repointing The process of removing deteriorated mortar from the zone near the surface of a brick wall and inserting fresh mortar. *See also* **Tuckpointing.**

Reshoring Inserting temporary supports under concrete beams and slabs after the formwork has been removed to prevent overloading prior to full curing of the concrete.

Resilient clip A springy mounting device for plaster or gypsum board that helps reduce the transmission of sound vibrations through a wall or ceiling.

Resilient flooring A manufactured sheet or tile flooring made of asphalt, polyvinyl chloride, linoleum, rubber, or other elastic material.

Resin A natural or synthetic, solid or semisolid organic material of high molecular weight, used in the manufacture of paints, varnishes, and plastics.

Restraightening A step in the finishing of concrete slabs for the purpose of removing minor undulations produced during floating or troweling.

Retaining wall A wall that resists horizontal soil pressures at an abrupt change in ground elevation.

Retarding admixture An admixture used to slow the curing of concrete, mortar, or plaster.

Ridge beam A structural beam supporting the upper ends of rafters in a sloped roof, required where the rafters are not tied at their lower ends.

Ridge board A nonstructural framing member against which the upper ends of rafters are fastened.

Ridge vent A screened, water-shielded ventilation opening that runs continuously along the ridge of a gable roof.

Rigid connection A *fully restrained moment connection.*

Rigid frame Two columns and a beam or beams attached to one another with moment connections; a *moment-resisting frame.*

Rim joist *See* **Band joist**.

Rise A difference in elevation, such as the rise of a stair from one floor to the next or the rise per foot of run in a sloping roof.

Riser A single vertical increment of a stair; the vertical face between two treads in a stair; a vertical run of plumbing, wiring, or ductwork.

Rivet A structural fastener on which a second head is formed after the fastener is in place.

Rock anchor A posttensioned rod or cable inserted into a rock formation for the purpose of tying it together.

Rock wool An insulating material manufactured by forming fibers from molten rock.

Roofer One who installs roof coverings.

Roofing The material used to make a roof watertight, such as shingles, slate, tiles, sheet metal, or a roof membrane; the act of applying roofing.

Roof membrane A waterproof sheet or multiply assembly that protects a low-slope roof from water penetration.

Roof window Either an openable glazed unit installed in the sloping surface of a roof or, more specifically, a glazed roof unit with the inward sash operation to allow easy cleaning.

Rotary-sliced veneer A thin sheet of wood produced by rotating a log against a long, sharp knife blade in a lathe.

Rough arch An arch made from masonry units that are rectangular rather than wedge- shaped.

Rough carpentry Framing carpentry, as distinguished from *finish carpentry.*

Roughing in The installation of mechanical, electrical, and plumbing components that will not be exposed to view in the finished building.

Rough opening The clear dimensions of the opening that must be provided in a wall frame to accept a given door or window unit.

Rowlock A brick laid on its long edge, with its end exposed in the face of the wall.

RSI-value The metric equivalent of *R-value.*

Rubberized underlayment An adhered bituminous sheet material that self-heals around nails, applied to roof sheathing to prevent the entry of water; also called *ice and water shield.*

Rubble Unsquared stones.

Run Horizontal dimension in a stair or sloping roof.

Runner channel A steel member from which furring channels and lath are supported in a suspended plaster ceiling.

Running bond Brickwork consisting entirely of stretchers.

Runoff bar One of a pair of small rectangular steel bars attached temporarily at the end of a prepared groove for the purpose of permitting the groove to be filled to its very end with weld metal.

Run plaster ornament A linear molding produced by passing a profiled sheet metal or plastic template back and forth across a mass of wet plaster.

R-value A numerical measure of resistance to the flow of heat; the reciprocal of *U-factor*.

S

Safety glazing Glass or plastic glazing material that, when broken, does not create hazardous shards, permitted for use in locations in buildings at risk of occupant impact; typically tempered glass, laminated glass, or plastic.

Safing Fire-resistant material inserted into a space between a curtain wall and a spandrel beam or column to retard the passage of fire through the space.

Sand cushion terrazzo Terrazzo with an underbed that is separated from the structural floor deck by a layer of sand.

Sand-mold brick, sand-struck brick A brick made in a mold that was wetted and then dusted with sand before the clay was placed in it.

Sandstone A sedimentary rock formed from sand; classified by ASTM C119 in the Quartz-Based Dimension Stone group.

Sandwich panel A panel consisting of two outer faces of wood, metal, gypsum, or concrete bonded to a core of insulating foam.

Sapwood The living wood in the outer region of a tree trunk or branch.

Sash A frame that holds glass.

SBS *See* **Styrene-butadiene-styrene**.

SBX *See* **Sodium borate**.

Scab A piece of framing lumber nailed to the face of another piece of lumber.

Scarf joint A glued end connection between two pieces of wood, using a sloping cut to create a large surface for the glue bond, to allow it to develop the full tensile strength of the wood that it connects.

SCC *See* **Self-consolidating concrete**.

SCOF *See* **Static Coefficient of Friction**.

Scratch coat The first of two base-coat plaster applications in a three-coat plaster.

Screed A strip of wood, metal, or plaster that establishes the level to which concrete or plaster will be placed.

Screw port A three-quarter circular profile in an aluminum extrusion, made to accept a screw driven parallel to the long axis of the extrusion.

Screw slot A serrated slot profile in an aluminum extrusion, made to accept screws driven at right angles to the long axis of the extrusion.

Scupper An opening through a parapet through which water can drain over the edge of a flat roof.

Sealant A rubberlike, adhesive material, usually applied in liquid or tape form, used to seal a joint, gap, or crack against the passage of air and moisture.

Sealant gun A tool for injecting sealant into a joint.

Sealer A coating used to close the pores in a surface, usually in preparation for the application of a finish coating.

Seasoning The drying of wood, to bring its moisture content into equilibrium with ambient conditions.

Seated connection A connection in which a steel beam rests on top of a steel angle or tee that is fastened to a column or girder.

Section An architectural drawing representing a vertically cut plane through a whole building, part of a building, or detail.

Security glass A glazing sheet with multiple laminations of glass and plastic, designed to stop bullets.

Sedimentary rock Rock formed from materials deposited as sediments, such as sand or sea shells, which form sandstone and limestone, respectively.

Segregation Separation of the constituents of wet concrete caused by excessive handling or vibration.

Seismic Relating to earthquakes.

Seismic load A force on a structure caused by movement of the earth relative to the structure during an earthquake.

Seismic separation joint A building separation joint that allows adjacent building masses to oscillate independently during an earthquake.

Self-adhered flashing A flexible, self-sticking flashing material, usually made of polymer-modified asphalt laminated to a plastic backing, with preapplied adhesive on one side.

Self-consolidating concrete (SCC) Concrete formulated so that it is highly flowable and fills formwork completely without needing *consolidation*.

Self-consolidating grout Grout formulated so that it is highly flowable.

Self-drilling Drills its own hole.

Self-furring metal lath Metal lath with dimples that space the lath away from the sheathing behind to allow plaster to penetrate the lath and key to it.

Self-tapping Creates its own screw threads on the inside of the hole.

Self-weight The weight of a beam or slab.

Semirigid connection A *partially restrained moment connection*.

Sequential construction A method of providing design and construction services in which each major phase of design and construction is completed before the next phase is begun.

Set To cure; to install; to recess the heads of nails; a punch for recessing the heads of nails.

Setting block A small block of synthetic rubber or lead used to support the weight of a sheet of glass at its lower edge.

Settlement joint A *building separation joint* that allows the foundations of adjacent building masses to settle at different rates.

SFRM *See* **Spray-applied fire-resistive material**.

SGPL *See* **Structural-grade plastic lumber**.

Shading coefficient The ratio of total solar heat passing through a given sheet of glass to that passing through a sheet of clear double-strength glass; mostly replaced in contemporary energy calculations by *solar heat gain coefficient*.

Shaft An unbroken vertical passage through a multistory building, used for elevators, wiring, plumbing, ductwork, and so on.

Shaft wall A wall surrounding a *shaft*.

Shake A shingle split from a block of wood.

Shake-on hardener A dry powder that is dusted onto the surface of a concrete slab before troweling to react with the concrete and produce a hard wearing surface for industrial use.

Shale A rock formed from the consolidation of clay or silt.

Shallow foundation A building *foundation* located at the base of a wall or a column, bearing on soil relatively close to the ground surface.

Shear A deformation in which planes of material slide with respect to one another.

Shear connection A connection designed to resist only the tendency of one member to slide past the other, and not, as in a *moment connection*, to resist any tendency of the members to rotate with respect to one another; in steel frame construction, a *simple connection*.

Shear panel A wall, floor, or roof surface that acts as a deep beam to help stabilize a building against deformation by lateral forces.

Shear stud A piece of steel welded to the top of a steel beam or girder so as to become embedded in the concrete fill over the beam and cause the beam and the concrete to act as a single structural unit.

Shear wall A stiff wall that imparts lateral force resistance to a building frame.

Sheathing The rough covering applied to the outside of the roof, wall, or floor framing of a structure.

Shed A building or dormer with a single-pitched roof.

Sheeting A stiff material used to retain the soil around an excavation; a material such as polyethylene in the form of very thin, flexible sheets.

Sheet metal Flat rolled metal generally less than $\frac{1}{4}$ inch (6.35 mm) thick.

Shelf angle A horizontal steel angle attached to the wall or spandrel of a building to support a masonry facing.

SHGC *See* **Solar heat gain coefficient.**

Shim A thin piece of material placed between two components of a building to adjust their relative positions as they are assembled; to insert shims.

Shingle A small unit of water-resistant material nailed in overlapping fashion with many other such units to render a wall or sloping roof watertight; to apply shingles.

Shiplap A board with edges rabbeted so as to overlap flush from one board to the next.

Shop drawings Detailed drawings prepared by a fabricator to guide the shop production of such building components as cut stonework, steel or precast concrete framing, curtain wall panels, and cabinetwork.

Shoring Temporary vertical or sloping supports of steel or timber.

Shotcrete A low-slump concrete mixture that is deposited by being blown from a nozzle at high speed with a stream of compressed air; *pneumatically placed concrete.*

Shrinkage-compensating cement Specially formulated cement, used to counteract the drying shrinkage that normally occurs during curing.

Shrinkage-reducing admixture A concrete additive that reduces drying shrinkage and the cracking that results.

Shrinkage–temperature steel Reinforcing bars laid at right angles to the principal bars in a one-way slab for the purpose of preventing excessive cracking caused by drying shrinkage or temperature stresses in the concrete.

Side-hinged inswinging window A window that opens by pivoting inward on hinges at or near a vertical edge of the sash.

Side jamb *See* **Jamb**.

Sidelight A tall, narrow window alongside a door.

Siding The exterior wall finish material applied to a light frame structure.

Siding nail A nail with a small head, used to fasten siding to a building.

Silica fume Very finely divided silicon dioxide, a *pozzolan*, used as an admixture in the formulation of high-strength, low-permeability concrete; also called *microsilica.*

Silicone A polymer used for high-range sealants, roof membranes, and masonry water repellents.

Sill The horizontal bottom portion of a window or door; the exterior surface, usually sloped to shed water, below the bottom of a window or door.

Sill plate The strip of wood that lies immediately on top of a concrete or masonry foundation in wood frame construction.

Sill sealer A compressible material placed between a foundation and a sill to reduce air infiltration between the outdoors and indoors.

Simple caulk and seal A method similar to the *airtight drywall approach* for constructing a light frame building enclosure that is resistant to the free flow of air, but requiring less coordination between framing and sealing operations than the airtight drywall approach.

Simple connection A steel frame *shear connection* with no useable resistance to rotation; previously referred to as an "AISC Type 2" connection.

Single-hung window A window with two overlapping sashes, the lower of which can slide vertically in tracks and the upper of which is fixed.

Single-ply roof membrane A sheet of plastic or synthetic rubber used as a membrane for a low-slope roof.

Single-strength glass Glass approximately $\frac{3}{32}$ inch (2.5 mm) thick.

Single tee A precast slab element whose profile resembles the letter *T*.

SIP *See* **Structural insulated panel**.

Sitecast Concrete that is poured and cured in its final position in a building; *cast in place.*

Skip-joist system *See* **Wide-module concrete joist system.**

Skylight A glazed unit installed in a roof; also referred to as a *unit skylight.*

Slab band A very broad, shallow beam used with a one-way solid slab.

Slab on grade A concrete surface lying upon, and supported directly by, the ground beneath.

Slag The mineral waste that rises to the top of molten iron or steel or to the top of a weld.

Slag cement See **Blast furnace slag.**

Slaked lime *See* **Hydrated lime**.

Slate A metamorphic form of clay, easily split into thin sheets.

Sliding window A window with one fixed sash and another that moves horizontally in tracks.

Slip-critical connection A structural steel connection in which the members are clamped together by high-strength bolts with sufficient force that the loads on the members are transmitted between them by friction along their mating (*faying*) surfaces; also called a *friction connection.*

Slip forming Building multistory sitecast concrete walls with forms that rise up the wall as construction progresses.

Slip sheet A thin sheet of paper, plastic, or felt, placed between two materials to eliminate friction or bonding of the materials.

Slurry A watery mixture of insoluble materials with a high concentration of suspended solids.

Sloped glazing A system of metal and glass components used to make an inclined, transparent roof; in the International Building Code, glass sloped more than 15 degrees from vertical.

Slump test A test in which wet concrete or plaster is placed in a cone-shaped metal mold of specified dimensions and allowed to sag under its own weight after the cone is removed. The vertical distance between the height of the mold and the height of the slumped mixture is an index of its working consistency.

Slurry A watery mixture of insoluble materials with a high concentration of suspended solids.

Smoke-developed rating An index of the toxic fumes generated by a material as it burns, as determined by ASTM standard E84.

Smoke shelf The horizontal area behind the damper of a fireplace.

Slow-burn construction *See* **Mill construction**.

Sodium borate (SBX) A chemical used to preserve wood against attack by decay and insects.

Soffit The undersurface of a horizontal element of a building, especially the underside of a stair or a roof overhang.

Soffit vent An opening under the eave of a roof used to allow air to flow into the attic or the space below the roof sheathing.

Soft mud process Making bricks by pressing wet clay into molds.

Softwood Wood from coniferous (evergreen) trees.

Soil Any particulate earth material, excluding rock.

Solar heat gain coefficient (SHGC) The ratio of solar heat admitted through a particular glass to the total heat energy striking the glass.

Solar reflectance A unitless index, ranging from 0 to 1, expressing a material's tendency to absorb or reflect solar radiation; also called *albedo.*

Soldering A low-temperature form of *brazing.*

Soldier A brick laid on its end, with its narrow face toward the outside of the wall.

Sole plate The horizontal piece of dimension lumber at the bottom of the studs in a wall in a light frame building; also called a *bottom plate.*

Solid-core door A flush door with no internal cavities.

Solid masonry Masonry walls without cavities; historically, thick, monolithic masonry walls that rely primarily on their mass for their strength, durability, and tempering of the flow of heat and moisture from inside to outside.

Solid slab A concrete slab, without ribs or voids, that spans between beams or bearing walls.

Solid tape sealant *See* **Preformed solid tape sealant**.

Solvent A liquid that dissolves another material.

Sound Transmission Class (STC) An index of the resistance of a wall or partition to the passage of sound.

Space truss, space frame A truss that spans with two-way action.

Spalling The cracking or flaking of the surface of concrete or masonry units, caused, for example, by freeze-thaw action, corroding reinforcing, or pointing mortars that are harder and stronger than the mortar deeper in the masonry joint.

Span The distance between supports for a beam, girder, truss, vault, arch, or other horizontal structural device; to carry a load between supports.

Spandrel The wall area between the head of a window on one story and the sill of a window on the floor above; the area of a wall between adjacent arches.

Spandrel beam A beam that runs along the outside edge of a floor or roof.

Spandrel glass Opaque glass manufactured especially for use in spandrel panels.

Spandrel panel A curtain wall panel used in a spandrel.

Span rating The number stamped on a sheet of plywood or other wood building panel to indicate how far in inches it may span between supports.

Specifications The written instructions from an architect or engineer concerning the quality of materials and execution required for a building.

Spirit level A tool in which a bubble in an upwardly curving cylindrical glass vial indicates whether a building element is level or not level, plumb or not plumb.

Splash block A small precast block of concrete or plastic used to divert water at the bottom of a downspout.

Spline A thin strip inserted into grooves in two mating pieces of material to hold them in alignment; a ridge or strip of material intended to lock to a mating groove. In glazing, the *edge spacer* in an insulating glass unit.

Split jamb A door frame fabricated in two interlocking halves, to be installed from the opposite sides of an opening.

Spray-applied fire-resistive material (SFRM) Fibrous or cementitious insulation applied to steel or concrete with a sprayer to provide protection against the heat of fire.

Springwood In wood, the portion of the growth ring comprised of relatively larger, less dense cells; also called earlywood.

SSP *See* **Stressed-skin panel**.

Staggered truss system A steel framing system in which story-high trusses, staggered one-half bay from one story to the next, support floor decks on both their top and bottom chords.

Stain A coating intended primarily to change the color of wood or concrete without forming an impervious film.

Stainless steel A steel alloy, silvery in color, with superior corrosion resistance due principally to high chromium and nickel content.

Standing and running trim Door and window casings and baseboards.

Standing seam A sheet metal roofing seam that projects at right angles to the plane of the roof.

Static Coefficient of Friction (SCOF) A measure of the slip resistance of a flooring material.

Stay A sloping cable used to stabilize a structure.

STC *See* **Sound Transmission Class.**

Steam curing Aiding and accelerating the setting reaction of concrete by the application of steam.

Steel Iron with a controlled amount of carbon, generally less than 2 percent.

Steel reinforcing bars Hot-rolled, deformed steel bars used to impart tensile strength and ductility to concrete structures; *rebar.*

Steel trowel A metal bladed tool used in the final stages of finishing of a concrete slab.

Steep roof A roof with sufficient slope to be made waterproof with shingles. In the

International Building Code, a roof with a slope of 2:12 (17 percent) or greater.

Sticking The cementing together of defects in marble slabs.

Stick system A metal curtain wall system that is largely assembled in place.

Stiffener plate A steel plate attached to a structural member to support it against heavy localized loading or stresses.

Stiff mud process A method of molding bricks in which a column of damp clay is extruded from a rectangular die and cut into bricks by fine wires.

Stile A vertical framing member in a panel door.

Stirrup A vertical loop of steel bar used to reinforce a concrete beam against diagonal tension forces.

Stirrup-tie A stirrup that forms a complete loop, as differentiated from a U-stirrup, which has an open top.

Stool The interior horizontal plane at the sill of a window.

Storm window A sash added to the outside of a window in winter to increase its thermal resistance and decrease air infiltration.

Story pole A strip of wood marked with the exact course heights of masonry for a particular building, used to make sure that all the leads are identical in height and coursing.

Straightedge To strike off the surface of a concrete slab using screeds and a straight piece of lumber or metal; as a noun, a long, straight item, used to perform straightedging, test the flatness of a surface, or trace a straight line.

Strain Deformation under stress; expressed as a ratio of the change in length over the original length.

Stress Force per unit area.

Stressed-skin panel (SSP) A panel consisting of two face sheets of wood, metal, or concrete bonded to perpendicular spacer ribs or framing members such that the panel can act as a composite structural panel.

Stretcher A brick or masonry unit laid in its most usual position, with the broadest surface of the unit horizontal and the length of the unit parallel to the surface of the wall.

Striated Textured with parallel scratches or grooves.

Stringer The sloping wood or steel member that supports the treads of a stair.

Strip flooring Solid wood finish flooring members less than 3 inches (75 mm) in width, usually in the form of tongue-and-groove boards.

Stripping Removing formwork from concrete; sealing around a roof flashing with layers of felt and bitumen.

Structural bond The interlocking pattern of masonry units used to tie two or more wythes together in a wall.

Structural composite lumber Substitutes for solid lumber made from wood veneers or wood fiber strands and glue; also called *engineered lumber.*

Structural glazed facing tile A hollow clay masonry unit with glazed faces.

Structural-grade plastic lumber (SGPL) Lumberlike plastic members, reinforced with glass fibers, and formulated to be roughly as strong as conventional solid wood.

Structural insulated panel (SIP) A panel consisting of two face sheets of wood panel bonded together by plastic foam core.

Structural lightweight aggregate Lightweight aggregate with sufficient density and strength for use in structural concrete.

Structural mill The portion of a steel mill that rolls structural shapes.

Structural silicone flush glazing Glass secured to the face of a building with strong, highly adhesive silicone sealant to eliminate the need for any metal to appear on the exterior of the building.

Structural standing-seam metal roofing Sheets of folded metal that serve both as decking and as the waterproof layer of a roof.

Structural terra cotta Molded components, often highly ornamental, made of fired clay, designed to be used in the facades of buildings.

Structural tubing *See* **Hollow structural section.**

Structure/enclosure joint A connection designed to allow the structure of a building and its cladding or partitions to move independently.

Stucco Plaster made from a mixture of portland cement, lime, sand, and water; commonly used as an exterior finish material.

Stud One of an array of small, closely spaced, parallel wall framing members; a heavy steel pin.

Styrene-butadiene-styrene (SBS) A copolymer of butadiene and styrene used as a modifier in polymer-modified bitumen roofing.

Subcontractor A contractor who specializes in one area of construction activity and who works under a general contractor.

Subfloor The loadbearing surface beneath a finish floor.

Subpurlin A very small roof framing member that spans between joists or purlins.

Substrate The base to which a coating, veneer, or finish material is applied.

Substructure The occupied below-ground portion of a building.

Summerwood In wood, the portion of the growth ring comprised of relatively smaller, denser cells; also called latewood.

Sump A pit designed to collect water for removal from an excavation or basement.

Superflat floor A concrete slab finished to a high degree of flatness and levelness according to a recognized system of measurement.

Superplasticizer An admixture that makes wet concrete or grout extremely fluid without additional water.

Superstructure The above-ground portion of a building.

Supplementary cementitious material Hydraulic cementitious material or pozzolan mixed with portland cement to modify the cement product's properties or lower the energy required to manufacture the cement.

Supply pipe A pipe that brings clean water to a plumbing fixture.

Supporting stud A wall framing member that extends from the sole plate to the underside of a header and supports the header.

Surface-bonded masonry Concrete block laid without mortar and then plastered on both sides with a fiber-reinforced cement plaster so as to make a structurally sound masonry wall.

Surfacing Smoothing the surface of a material, usually by *planing.*

Surface divider joint A line along which a surface may expand and/or contract without damage.

Surface number In glazing assemblies, the distinct faces of glazing, counting from the outermost to the innermost of a glazing unit, including each face of each glazing material.

Suspended ceiling A finish ceiling that is hung on wires from the structure above.

Suspended glazing Large sheets of glass hung from clamps at their top edges to eliminate the need for metal mullions.

Sustainability Providing for the needs of the current generation without compromising the ability of future generations to provide for their needs. Providing healthy, resource- conserving, energy-efficient buildings. "Green" building.

Swedge bolt *See lockpin and collar fastener.*

Synthetic gypsum Chemically manufactured gypsum made from the byproducts of various industrial processes, such as the desulfurization of power plant flue gasses.

T

Tackstrip *See* **Tackless strip**.

Tackless strip A wood strip with projecting points used to fasten a carpet around the edge of a room; also called a *tackstrip.*

Tagline A rope attached to a building component to help guide it as it is lifted by a crane or derrick.

Tap To cut internal threads, such as in a hole or nut.

Tangential shrinkage In wood, shrinkage along the circumference of the log.

Tapered edge The longitudinal edge of a sheet of gypsum board, which is recessed to allow room for reinforcing tape and joint compound.

Tee A metal or precast concrete member with a cross section resembling the letter *T.*

Tempered glass Heat-treated glass that is stronger than heat-strengthened glass and is suitable for use as *safety glazing.*

Tempering Controlled heating and cooling of a material to alter its mechanical properties; a form of heat-treatment.

Tendon A steel strand used for prestressing a concrete member.

Tensile strength The ability of a structural material to withstand stretching forces.

Tensile stress A stress caused by stretching of a material.

Tension A stretching force; to stretch.

Tension control bolt A bolt tightened by means of a splined end that breaks off when the bolt shank has reached the required tension.

Termite shield A metal flashing placed on top of a concrete foundation to prevent termites from traveling undetected from the ground into the superstructure.

Terne An alloy of lead and tin, used to coat sheets of carbon steel or stainless steel, used in the past for metal roofing sheet.

Terrace door A double glass door, one leaf of which is fixed and the other hinged to the fixed leaf at the centerline of the door.

Terrazzo A finish floor material consisting of concrete with an aggregate of marble chips selected for size and color, which is ground and polished smooth after curing.

Thatch A thick roof covering of reeds, straw, grasses, or leaves.

Thermal break A section of material with low thermal conductivity installed between metal components to retard the passage of heat through a wall or window assembly.

Thermal bridge A component of higher thermal conductivity that conducts heat more rapidly through an insulated building assembly, such as a steel stud in an insulated stud wall.

Thermal conductivity The rate at which a material conducts heat.

Thermal emittance A unitless index, from 0 to 1, expressing a material's tendency to radiate thermal energy as its temperature rises in relation to surrounding surfaces.

Thermal envelope *See* **Building enclosure**.

Thermal insulation A material that greatly retards the passage of heat.

Thermal resistance The resistance of a material or assembly to the conduction of heat.

Thermochromic glass Glass that changes its optical properties in response to changes in temperature.

Thermoplastic In plastics, having the property of softening when heated and rehardening when cooled; weldable by heat or solvents.

Thermoplastic polyolefin (TPO) A thermoplastic single-ply roof membrane material, made from blends of polyethylene, polypropylene, and ethylene-propylene rubber polymers.

Thermosetting In plastics, not having the property of softening when heated; not heat-fusible.

Thickset tile Ceramic tile installed on a thick bed of portland cement mortar; also called *mortar bed* or *mud set tile.*

Thin-set tile Ceramic tile bonded to a solid base with a thin application of portland cement mortar or organic adhesive.

Through-wall flashing *See* **Internal flashing**.

Thrust A lateral or inclined force resulting from the structural action of an arch, vault, dome, suspension structure, or rigid frame.

Thrust block A wooden block running perpendicular to the stringers at the bottom of a stair, whose function is to hold the stringers in place.

Tie A device for holding two parts of a construction together; a structural device that acts in tension.

Tieback A tie, one end of which is anchored in the ground, with the other end used to support sheeting around an excavation.

Tie beam A reinforced concrete beam cast as part of a masonry wall, whose primary purpose is to hold the wall together, especially against seismic loads, or cast between a number of isolated foundation elements to maintain their relative positions.

Tier The portion of a multistory steel building frame supported by one set of fabricated column pieces, commonly two stories in height.

Tie rod A steel rod that acts in tension.

Tile A fired clay product that is thinner in cross section than a brick, either a thin, flat element (ceramic tile or quarry tile), a thin, curved element (roofing tile), or a hollow element with thin walls (flue tile, tile pipe, structural clay tile); also a thin, flat element of another material, such as an acoustical ceiling unit or a resilient floor unit.

Tilt/turn window A window that opens either by rotating its sash about its vertical centerline or as a hopper.

Tilt-up construction A method of constructing concrete walls in which panels are cast and cured flat on a floor slab, then tilted up into their final positions.

Timber Standing trees; a large piece of dimension lumber.

Tinted glass Glass that is colored with pigments, dyes, or other admixtures.

Titanium A strong, corrosion-resistant, nonferrous metal, silver gray in color.

Toe nailing Fastening with nails driven at an angle.

Tongue and groove An interlocking edge detail for joining planks or panels.

Tooling The finishing of a mortar joint or sealant joint by pressing and compacting it to create a particular profile.

Toothed plate A multipronged fastener made from stamped sheet metal, used to join members of a lightwood wood truss.

Top-hinged inswinging window A window that opens inward on hinges on or near its head.

Topping A thin layer of concrete cast over the top of a floor deck.

Topping-out Placing the last member in a building frame.

Top plate The horizontal member at the top of the studs in a wall in a light frame building.

Topside vent A water-protected opening through a roof membrane to relieve pressure from water vapor that may accumulate beneath the membrane.

Torque Twisting action; moment.

Torsional stress Stress resulting from the twisting of a structural member.

Touch sanded In plywood, lightly sanded to produce a smoother, flatter surface.

TPO *see* **Thermoplastic polyolefin.**

Tracheids The longitudinal cells in a softwood.

Traffic deck A walking surface placed on top of a roof membrane.

Transit-mixed concrete Concrete mixed in a drum on the back of a truck as it is transported to the building site.

Travertine A richly patterned, marble-like form of limestone; classified by ASTM C119 in the Other Stone group.

Tread One of the horizontal planes that make up a stair.

Tremie A large funnel with a tube attached, used to deposit concrete in deep forms or beneath water or slurry.

Trim accessories Casing beads, corner beads, expansion joints, and other devices used to finish edges and corners of a plaster wall or ceiling.

Trimmer joist A joist that supports a header around an opening in a floor or roof frame.

Trimmer stud. *See* **Jack stud**.

Trowel A thin, flat steel tool, either pointed or rectangular, provided with a handle and held in the hand, used to manipulate mastic, mortar, plaster, or concrete. Also, a machine whose rotating steel blades are used to finish concrete slabs; to use a trowel.

Truss A triangulated arrangement of structural members that reduces nonaxial external forces to a set of axial forces in its members. *See also* **Vierendeel truss.**

Tuckpointing Traditionally, a method of finishing masonry joints using mortars of different colors to artificially create the appearance of a more refined joint; in contemporary usage, may be used interchangeably with *repointing*.

Tunnel kiln A kiln through which clay products are passed on railroad cars.

Turn-of-nut method A method of achieving the correct tightness in a *high-strength bolt* by first tightening the nut snugly, then turning it a specified additional fraction of a turn.

Two-way action Bending of a slab or deck in which bending stresses are approximately equal in the two principal directions of the structure.

Two-way concrete joist system A reinforced concrete framing system in which columns directly support an orthogonal grid of intersecting joists.

Two-way flat plate A reinforced concrete framing system in which columns directly support a two-way slab that is planar on both of its surfaces.

Two-way flat slab A reinforced concrete framing system in which columns with mushroom capitals and/or drop panels directly support a two-way slab that is planar on both of its surfaces.

Type IV HT construction *See* **Heavy timber construction.**

Type X gypsum board A fiber-reinforced gypsum board used where greater fire resistance is required.

U

U-Factor A measure of the thermal conductance of a material or assembly; the mathematical reciprocal of *R-value.*

Unbonded construction Posttensioned concrete construction in which the tendons are not grouted to the surrounding concrete.

Uncoursed stone masonry Stone masonry laid without continuous horizontal joints; random.

Undercarpet wiring system Flat, insulated electrical conductors that run under carpeting, and their associated outlet boxes and fixtures.

Undercourse A course of shingles laid beneath an exposed course of shingles at the lower edge of a wall or roof in order to provide a waterproof layer behind the joints in the exposed course.

Underfire The floor of the firebox in a fireplace.

Underlayment A panel laid over a subfloor to create a smooth, stiff surface for the application of finish flooring. Or, a water-resistant material applied under shingled roofing.

Underpinning The process of placing new foundations beneath an existing structure.

Unfinished bolt An ordinary carbon steel bolt.

Uniformly graded soil A special instance of a *well-sorted soil* in which the soil particles are mostly of one size.

Uniform settlement Subsidence of the various foundation elements of a building at the same rate, resulting in no distress to the structure of the building.

Unit-and-mullion system A curtain wall system consisting of prefabricated panel units secured with site-applied mullions.

Unit skylight *See* **Skylight**.

Unit system A curtain wall system consisting entirely of prefabricated panel units.

Unreinforced Constructed without steel reinforcing bars or welded wire fabric.

Up–down construction A sequence of construction activity in which construction proceeds downward on the sublevels of a building at the same time as it proceeds upward on the superstructure.

Upside-down roof A membrane roof assembly in which the thermal insulation lies above the membrane.

U-stirrup An open-top, U-shaped loop of steel bar used as reinforcing against diagonal tension in a concrete beam.

V

Valley A trough formed by the intersection of two roof slopes.

Valley rafter A diagonal rafter that supports a valley.

Vapor barrier *See* **Vapor retarder**.

Vapor permeability Vapor permeance per unit of thickness.

Vapor permeance A measure of the ease with which water vapor can diffuse through a material.

Vapor pressure A measure of the pressure exerted by water molecules in a gaseous state, generally higher with higher relative humidity and higher air temperature.

Vapor retarder A layer of material intended to resist the diffusion of water vapor through a building assembly. Also called, less accurately, *vapor barrier.*

Varnish A slow-drying transparent coating.

Vault An arched surface. A strongly built room for such purposes as housing large electrical equipment or safeguarding money.

Vee joint A joint whose profile resembles the letter V.

Vegetated roof *See* **Green roof.**

Veneer A thin layer, sheet, or facing.

Veneer plaster A wall finish system in which a thin finish layer of gypsum plaster is applied over a special gypsum board base.

Veneer plaster base The special gypsum board over which veneer plaster is applied.

Vent spacer A device used to maintain a free air passage above the thermal insulation in an attic or roof.

Vermiculite Expanded mica, used as an insulating fill or lightweight aggregate.

Vertical bar An upright reinforcing bar in a concrete column; also called a *column bar.*

Vertical grain lumber Dimension lumber sawed in such a way that the annual rings run mostly perpendicular to the faces of each piece; also called *edge-grain lumber. See also* **Flat-grain lumber.**

Vierendeel truss A truss with rectangular panels and rigid joints but no diagonal members. The members of a Vierendeel truss are subjected to strong nonaxial forces.

Vinyl *See* **Polyvinyl chloride**.

Visible light transmittance (VT) The ratio of visible light that passes through a sheet of glass or a glazing unit to the amount of light striking the glass or unit.

Visual grading The grading of wood for its structural properties, based on visual inspection, as distinct from *machine grading*; not to be confused with *appearance grading.*

Vitrification The process of transforming a material into a glassy substance by means of heat.

Volatile organic compound (VOC) Organic (carbon-based) chemical compound that evaporates readily, is a significant air pollutant, a potential irritant to building occupants, and, in some cases, a greenhouse gas.

Volume change joint A building separation joint that allows for expansion and contraction of adjacent portions of a building without distress.

Voussoir A wedge-shaped element of an arch or vault.

VT *See* **Visible light transmittance**.

W

Waferboard A building panel made by bonding together large, flat flakes of wood.

Waffle slab A two-way concrete joist system.

Wainscoting A wall facing, usually of wood, cut stone, or ceramic tile, that is carried only partway up a wall.

Waler A horizontal beam used to support sheeting or concrete formwork.

Wane An irregular rounding of a long edge of a piece of dimension lumber caused by cutting the lumber from too near the outside surface of the log.

Warm edge spacer A glazing *edge spacer* with improved thermal resistance.

Washer A steel disk with a hole in the middle, used to spread the load from a bolt, screw, or nail across a wider area of material.

Water–cement ratio An expression of the relative proportions, by weight, of water and cement in a concrete mixture.

Waterproofing Material acting as a barrier to the flow of water and capable of withstanding hydrostatic pressure.

Water-reducing admixture Concrete admixture that allows a reduction in the amount of mixing water while retaining the same workability, resulting in higher-strength concrete.

Water-resistant gypsum board A gypsum board designed for use in locations where it may be exposed to occasional dampness.

Water smoking The process of applying heat to evaporate the last water from clay products before they are fired.

Waterstop A metal, synthetic rubber, bentonite clay, or sealant strip used to seal joints in concrete foundation walls.

Water-struck brick A brick made in a mold that was wetted before the clay was placed on it.

Water table The level at which the pressure of water in the soil is equal to the atmospheric pressure; effectively, the level to which groundwater will fill an excavation; a wood molding or shaped brick used to make a transition between a thicker foundation and the wall above.

Water vapor Water in its gaseous phase.

Wattle and daub Mud plaster (daub) applied to a primitive lath of woven twigs or reeds (wattle).

Waxing Filling of voids in marble slabs.

Weathered joint A mortar joint finished in a sloping, planar profile that tends to shed water to the outside of the wall.

Weathering steel A steel alloy that forms a tenacious, self-protecting rust layer when exposed to the atmosphere.

Weather-resistive barrier A membrane used to resist the passage of liquid water or air through the exterior enclosure of a building.

Weatherstrip A ribbon of resilient, brushlike, or springy material used to reduce air infiltration through the crack around a sash or door.

Web A cross-connecting piece, such as the portion of a wide-flange shape that is perpendicular to the flanges or the portion of a concrete masonry unit that is perpendicular to the face shells.

Web stiffener A metal rib used to support the web of a light gauge steel joist or a structural steel girder against buckling.

Weep hole A small opening whose purpose is to permit drainage of water that accumulates inside a building component or assembly.

Weld A joint between two pieces of metal formed by fusing the pieces together by the application of intense heat, usually with the aid of additional metal melted from a rod or electrode.

Welded wire fabric (WWF) *See* **Welded wire reinforcing**.

Welded wire reinforcing (WWR) A welded grid of steel reinforcing wires or bars, used most commonly for reinforcing of slabs; also called *welded wire fabric (WWF)*.

Welding The process of making a weld.

Weld plate A steel plate anchored into the surface of concrete, to which another steel element can be welded.

Well-graded soil Coarse-grained soil with a full range of particle sizes; also called *poorly sorted soil*.

Well-sorted soil Soil with less than a full range of particle sizes; also called *poorly graded soil*.

"Wet" systems Construction systems that utilize considerable quantities of water on the construction site, such as masonry, plaster, sitecast concrete, and terrazzo.

White portland cement A portland cement that is white in color; used for architectural concrete where greater color control is required.

Wide-flange shape Any of a wide range of structural steel components rolled in the shape of the letter *I* or *H*.

Wide-module concrete joist system A one-way concrete framing system with joists that are spaced more widely than those in a conventional one-way concrete joist system.

Wind brace A diagonal structural member whose function is to stabilize a frame against lateral forces.

Winder (rhymes with "reminder") A stair tread that is wider at one end than at the other.

Wind load A force on a building caused by wind pressure and/or suction.

Wind uplift Upward forces on a structure caused by negative aerodynamic pressures that result from certain wind conditions.

Wired glass Glass in which a wire mesh is embedded during manufacture, principally for fire resistance.

Wood-plastic composites (WPC) Wood-like products made from wood fibers, plastics of various types, and other additives, with a plastic content not exceeding 50 percent

Wood–polymer composite planks Linear strips intended for outdoor decking and other outdoor uses that are made of wood fiber and a plastic binder.

Workability agent Admixture for concrete that improves the plasticity of wet material to make it easier to place in forms and to finish.

Working construction joint A connection that is designed to allow for small amounts of relative movement between two pieces of a building assembly.

WPC *See* **Wood-plastic composites**.

Wracking Forcing out of *plumb*.

Wrought iron A form of iron that is soft, tough, and fibrous in structure, containing about 0.1 percent carbon and 1–2 percent slag.

WWF *See* **Welded wire fabric**.

WWR *See* **Welded wire reinforcing**.

Wythe (rhymes with "scythe" and "tithe") A vertical layer of masonry that is one masonry unit thick.

Y

Yield strength The stress at which a material ceases to deform in a fully elastic manner and begins to deform irreversibly.

Z

Z-brace door A door made of vertical planks held together and braced on the back by three pieces of wood whose configuration resembles the letter *Z*.

Zero-slump concrete A concrete mixed with so little water that it does not sag when piled vertically.

Zinc A relatively weak and brittle nonferrous metal used, most notably, as a protective galvanic coating for steel.

Zoning ordinance A law that specifies in detail how land within a municipality may be used.

Index